Duckweed: Research Meets Applications

Duckweed: Research Meets Applications

Editors

Viktor Oláh
Klaus-Jürgen Appenroth
K. Sowjanya Sree

MDPI

Basel • Beijing • Wuhan • Barcelona • Belgrade • Novi Sad • Cluj • Manchester

Editors

Viktor Oláh
Department of Botany
University of Debrecen
Debrecen
Hungary

Klaus-Jürgen Appenroth
Mathias Schleiden
Institute—Plant Physiology
Friedrich Schiller University
Jena
Germany

K. Sowjanya Sree
Dept. of Environmental Science
Central University of Kerala
Periye
India

Editorial Office
MDPI
St. Alban-Anlage 66
4052 Basel, Switzerland

This is a reprint of articles from the Special Issue published online in the open access journal *Plants* (ISSN 2223-7747) (available at: www.mdpi.com/journal/plants/special_issues/duckweed).

For citation purposes, cite each article independently as indicated on the article page online and as indicated below:

Lastname, A.A.; Lastname, B.B. Article Title. *Journal Name* **Year**, *Volume Number*, Page Range.

ISBN 978-3-0365-9069-1 (Hbk)
ISBN 978-3-0365-9068-4 (PDF)
doi.org/10.3390/books978-3-0365-9068-4

Cover image courtesy of K. Sowjanya Sree
Central University of Kerala, India
A pond inhabited by *Spirodela polyrhiza* (Great duckweed).

Contents

About the Editors

Viktor Oláh

Viktor Oláh is an Associate Professor at the Department of Botany, University of Debrecen, Debrecen, Hungary. He received his Ph.D. in 2010 at the University of Debrecen in the fields of plant ecophysiology and ecotoxicology by studying the physiological responses of duckweeds (Lemnaceae) to heavy metals. His main research area is plant ecophysiology; his work focuses on traits, environmental responses, and potential utilization of duckweeds. In addition, he has also been long involved in long-term ecological monitoring (LTER) and projects focusing on forest ecology and tree ecophysiology. He has been a Section Board Member of "Plant Response to Abiotic Stress and Climate Change" for the journal *Plants* of MDPI since 2020.

Klaus-Jürgen Appenroth

Klaus-Juergen Appenroth is a Senior Guest Scientist at the Matthias Schleiden Institute, Plant Physiology, Friedrich Schiller University, Jena, Germany. With a study of Chemistry, Dr. rer. nat. (equivalent to PhD) in Photochemistry (1978), he earned his Habilitation (equivalent to Doctor of Sciences) in Plant Physiology (1991). He was a recipient of the prestigious Alexander von Humboldt fellowship in 1991/1992. Dr. Appenroth's research focus for the past 40 years has been on the plant family Lemnaceae, commonly known as duckweeds. Current focal activities include the genomics of duckweeds and the nutritive properties of duckweeds. He is also the member of the International Steering Committee on Duckweed Research and Applications. Out of the many awards he has received, the most recent and significant are the Simon Schwendener Medal award (1995) and Honorary member of the German Botanical Society (2022). He also serves as a Guest Scientist in India.

K. Sowjanya Sree

Dr. K. Sowjanya Sree serves as an Assistant Professor at the Central University of Kerala in the Department of Environmental Science, Periye, India. With a Masters in Plant Science from the University of Hyderabad (2004) and a PhD in Botany from Andhra University (2009), she has been working for over a decade with duckweeds (Lemnaceae). Dr. Sree had chaired the 4th International Conference on Duckweed Research and Applications held in Kerala, India, in 2017. She is the recipient of the prestigious Max Planck Germany-India grant and also currently serves as a member of the International Steering Committee on Duckweed Research and Applications.

plants

MDPI

Editorial

Duckweed: Research Meets Applications

Viktor Oláh [1,*], Klaus-Juergen Appenroth [2,*] and K. Sowjanya Sree [3,*]

1 Department of Botany, Institute of Biology and Ecology, Faculty of Science and Technology, University of Debrecen, 4032 Debrecen, Hungary
2 Matthias Schleiden Institute–Plant Physiology, Friedrich Schiller University Jena, 07743 Jena, Germany
3 Department of Environmental Science, Central University of Kerala, Periye 671320, India
* Correspondence: olahviktor@unideb.hu (V.O.); klaus.appenroth@uni-jena.de (K.-J.A.); ksowsree@gmail.com or ksowsree9@cukerala.ac.in (K.S.S.)

1. Introduction

The Special Issue "Duckweed: Research Meets Applications" of the journal *Plants* (ISSN 2223-7747) presents a comprehensive update of the current progress in the field. It includes a total of 38 articles, 29 original research papers, 5 reviews, 2 conference reports and 2 communications, encompassing almost all areas of research and applications related to the aquatic monocotyledonous plants duckweeds. The content of this Special Issue reflects the diversity of the duckweed community well in terms of the focal areas of research (Figure 1) as well as the international linkages (Figure 2). The authors are affiliated to a total of 18 countries: Belgium, Canada, China, Czechia, Germany, Hungary, India, Ireland, Israel, Italy, Poland, Russia, Slovenia, Spain, Thailand, Ukraine, the USA and Vietnam (in alphabetic order), and exactly half of the papers (19 out of 38) were an outcome of international collaborations. The original deadline of article submission to this Special Issue was extended in order to provide an opportunity to the participants of the 6th International Conference on Duckweed Research and Applications (ICDRA), which was held in Gatersleben, Germany, from 29 May to 1 June 2022. A report on this conference, "Sixth International Conference on Duckweed Research and Applications Presents Lemnaceae as a Model Plant System in the Genomics and Postgenomics Era" [1], presents the enormous progress made in duckweed research and applications since the first ICDRA in 2011 [2]. Interestingly, also in 2022, a workshop was held at the University of Jena, comparing the stress responses of duckweeds (aquatic plants) and terrestrial plants, a report of which is presented in this Special Issue as well [3]. In a similar direction, Ziegler et al. [4] reviewed the present knowledge of "Survival Strategies of Duckweeds". The survival strategies in duckweeds represent the natural potential of these plants to withstand several unfavourable conditions.

Together with a recently published review article [5], the articles in this Special Issue offer a fast overview of the present state of the art in duckweed research. So far, the papers published in this Special Issue have been accessed >75,000 times through the journal's homepage alone and received 161 citations in total as of 4 August 2023.

Citation: Oláh, V.; Appenroth, K.-J.; Sree, K.S. Duckweed: Research Meets Applications. *Plants* **2023**, *12*, 3307. https://doi.org/10.3390/plants12183307

Received: 11 August 2023
Accepted: 14 September 2023
Published: 19 September 2023

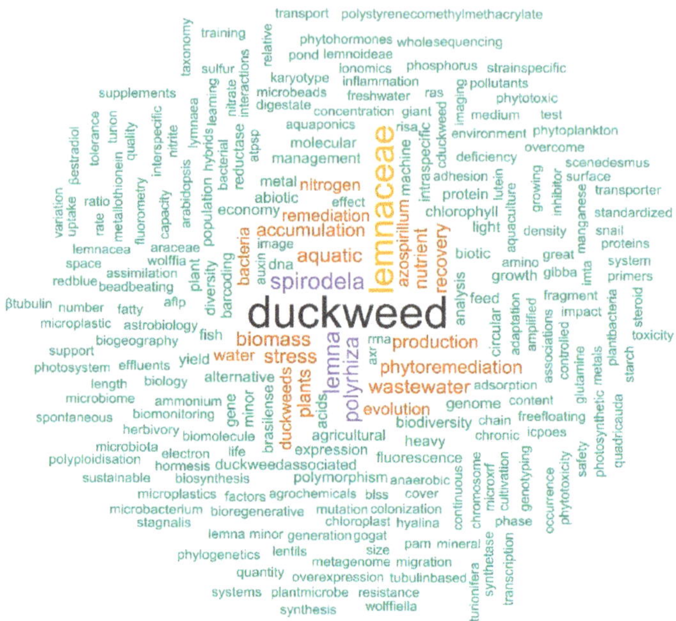

Figure 1. A word cloud of the keywords used in the studies published in the Special Issue "Duckweed: Research meets applications". The font size of the words indicates their occurrence frequency throughout the Special Issue.

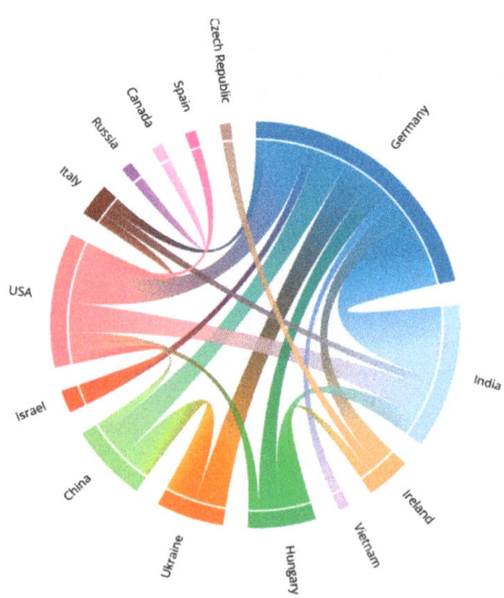

Figure 2. International linkages between different countries cooperating on diverse fields of duckweed research and applications, based on the publications in this Special Issue.

2. Molecular Characterization and Taxonomy

The family Lemnaceae was circumscribed by Ivan Martinov (1771–1833) as early as 1820. Therefore, the valid name of the family is Lemnaceae Martinov [6]. With the beginning of the era of molecular taxonomy, the closeness of this family to Araceae became evident [7] and the Angiosperm Phylogeny Group (APG) integrated this plant family as the subfamily Lemnoideae into the expanded family Araceae. Tippery et al. [8] demonstrated that this is not a vital step following the taxonomic rules. They suggested instead to restore the Araceae subfamily Orontioideae as the family Orontiaceae, which makes it possible to keep the family rank of the long-recognized family Lemnaceae. This results in three distinct lineages as families: Araceae s.s., Lemnaceae and Orontiaceae [8]. These authors also showed that the change of family Lemnaceae into a subfamily of Araceae was not well accepted by the scientific community, and this holds true for all papers published in this Special Issue. More than 10 years after the suggestion to treat duckweeds as an Araceae subfamily, it might be time to revise this on the basis of de facto use and as further advocated for by Tippery et al. [8]. A landmark in duckweed taxonomy was presented by the group of Laura Morello [9], describing the discovery of interspecific hybrids in the *Lemna* genus (*L. minor*, *L. gibba*, *L. turionifera*) using the method of tubulin-gene-based polymorphism (TBP; cf. also [10]). This provides a new insight into the evolution of duckweeds. Besides the identification of duckweed species, the identification of clones of the same species is also important, especially for several practical applications, including patenting. Bog et al. [11] used five orthogonal molecular methods, NB-ARC-related genes, TBP, simple sequence repeats (SSRs), multiplexed inter-simple sequence repeat genotyping by sequencing (MIG-seq) and genotyping by sequencing (GBS), for this purpose. Whereas TBP could distinguish only 7 clones out of 23 of *Spirodela polyrhiza*, the other four methods could distinguish 20 to 22 genotypes. *Spirodela polyrhiza* was selected for these test methods because it is known that it has especially low intraspecific variation [12]. This also became evident in a project where samples of *S. polyrhiza* and *L. minor*, collected around small ponds, were investigated using amplified fragment length polymorphism (AFLP). Whereas several distinct clones were identified within the populations of each pond in the case of *L. minor*, *S. polyrhiza* clones only showed genetic differences between the ponds [13]. Using plastid barcoding (*atp*F-*atp*H and *psb*K-*psb*I), Chen et al. [14] identified six different species from Ukraine and six different species from Eastern China. Interestingly, *Lemna aequinoctialis* did not form a uniform taxon, which might be a hint for the existence of hybrids. The same plastidic markers for barcoding were used by Yosef et al. [15], identifying six different species from Israel. For several investigations, the knowledge of the number of chromosomes in the different duckweed species and clones is important, and Hoang et al. [16] summarized it for all 36 duckweed species. However, only half of them were investigated by this group, whereas the other half were taken from results published mainly by Urbanska [17] and Geber [18]. These two authors reported unusually high variations in chromosome numbers within the same species, which need to be reviewed with care. A spontaneous mutant of *L. gibba* G3 was shown to be tetraploid [19]. This mutant is significantly larger in several parameters at morphological and anatomical levels. Also, at the physiological level, differences were found in the flower induction patterns, wherein the tetraploid plants flowered under conditions that were non-inductive for the diploid plants. Moreover, the transcript levels of nuclear genes of the photosynthetic apparatus were expressed at a higher level in the tetraploid plants when compared to the diploid ones.

3. Phytoremediation: Wastewater

Water pollution and meeting the ever-increasing clean water demands are interconnected problems that our modern society must tackle. Duckweeds have long been considered as potent candidates for wastewater management. As reviewed by Zhou et al. [20], these fragile plants were true giants in reclaiming contaminated waters while providing valuable biomass at the end of the process.

Six case studies examined the performance and pitfalls of duckweed-based wastewater remediation systems. Using multi-tiered indoor bioreactors, Coughlan et al. [21] studied how to design the water depth and flow rate of such systems to minimize the physical disturbance of plants while maintaining efficient mixture of the medium. Devlamynck et al. [22] and Lambert et al. [23] studied flow-through and recirculating systems for treating swine manure effluents with duckweeds and discussed the risks and possible solutions for avoiding or managing the accumulation of various chemical elements. Walsh et al. [24] and O'Mahoney et al. [25] focused on the possible valorisation of waste streams from dairy processing by studying whether this kind of medium could support the cultivation of duckweeds and how plant density affected the nutrient removal efficiency of the system. Paolacci et al. [26] addressed another applied field by analysing the actual contribution of duckweeds, phytoplankton and biofilm to the nutrient-removing performance of a duckweed-based aquaculture wastewater remediation system.

A series of studies addressed basic physiological processes that make duckweeds efficient in water remediation and waste valorisation. Zhou et al. [27], by studying species from four genera, analysed whether the preference for either NO_3^- or NH_4^+ as inorganic nitrogen sources was general or rather species-specific amongst Lemnaceae. In addition, they provided insights to the complex regulation of nitrogen assimilation in these plants by reporting the molecular structure and differential expression of several key enzymes in response to different inorganic nitrogen sources. Nitrogen availability is not only crucial in the plants' nutritional status, but may also modulate responses to other stressors, such as the presence of heavy metals. In their study, Kishchenko et al. [28] focused on the mechanisms by which NH_4^+ could mitigate manganese toxicity in *S. polyrhiza*, including the interactions between Mn availability and the transcription of ammonium transporters. Besides remediation, metal accumulation may also gain significance when future duckweed-based feed and food production is considered. Accordingly, Pakdee et al. [29] identified metallothionein-like genes and analysed dynamics in their transcript abundances under Cu and Cd stress, while, in another study, Oláh et al. [30] compared changes in the biomass ionomic composition of different duckweed species in response to acute Ni and Cr(VI) stress and the localization of heavy metal accumulation in the fronds.

4. Applications: Accumulation of Protein or Starch

Application of duckweed on an industrial or commercial scale requires the production of large amounts of biomass. This can be attained either by using very large water surface areas or by using a large number of smaller facilities in a modular arrangement. Wastewater cleaning with typically very high volumes of liquid waste might be preferentially carried out outdoors in large ponds or waterways [31]—although indoor treatment might be possible. Modular arrangement might be especially useful when specific environmental conditions should be applied in order to produce biomass with a specific quality. This is the domain of indoor cultivation. Petersen et al. [32] investigated the growth rates and protein contents of *L. minor* and *Wolffiella hyalina* in an indoor experiment under the influence of different nitrate-to-ammonium ratios. The best results were obtained in a 50% diluted N medium with a nitrate-N to ammonium-N ratio of 3:1. In an additional set of experiments, the influence of light intensity and light source spectrum was investigated in a "small-scale, re-circulating indoor vertical farm" [33] providing the pre-condition for further upscaling of the platform (cf. [34]). The same group used the produced biomass successfully for feeding broiler chickens [35]. Romano et al. [36] discussed a somewhat uncommon realm of indoor cultivation in their review paper, that is, the potentials of *Wolffia* species in space applications. As the authors pointed out, the world's smallest plants are promising candidates to be incorporated into bioregenerative life support systems in long-term space missions, and they are able to recycle water and oxygen for astronauts while also providing them fresh vegetable biomass.

A group from Italy substituted different amounts of the standard feed of rainbow trout on the protein basis and reported that 20% substitution did not have any negative effects on the fish, but 28% substitution did show effects such as reduced fish body weight and a few other parameters [37]. Demmig-Adams et al. [38] stressed that duckweeds (*Lemna*) have very high growth rates combined with unusually high levels of zeaxanthin, which is important for human nutrition. Moreover, *Lemna* plants can respond to elevated CO_2

concentrations with increasing growth rates. It has been known for a long time that under stress conditions, the protein content of duckweeds decreases as strongly as the starch content increases. The accumulation of starch in a large number of duckweed species under nutrient-limited cultivation conditions (phosphate, nitrate, sulphate) has been shown [39]. Starch contents of 50% per dry weight can be reached, making the biomass suitable for energy production, e.g., through saccharification and fermentation to bioalcohols. As a link between carbon, nitrogen and sulphur homeostasis, overexpression of the phosphoserine phosphatase-encoding gene was found to promote starch accumulation in *Lemna turionifera* under sulphur limitation in the study of Wang et al. [40].

5. Interaction with Microorganisms

The interaction of duckweeds with microorganisms, especially plant-growth-promoting bacteria, can result in enhanced plant growth [41]. The identification of bacteria associated with the root–frond interface on several duckweed species is important in this field. Acosta et al. [42] described the development of a generic fingerprinting assay that included genomics-based bacterial-strain-specific primers, making it possible to distinguish strains from the same genus. Quantification was possible by using plant reference genes. Many of these bacterial strains produced indole-related compounds like auxins. Gilbert et al. [43] identified suitable bacterial strains and investigated morphological responses of duckweeds as consequences of this interaction. Communities of duckweed-associated bacteria depend on external factors, e.g., those connected with stress induced by nutrient deficiency. Bunyoo et al. [44] were able to show that the relative abundance of bacteria, e.g., from the genus *Rhodobacter*, changes after transfer from conditions in nature to nutrient-deficient conditions. Some herbivores feed on duckweeds, e.g., on *S. polyrhiza*. Schaefer and Xu [45] quantified the influence of duckweed-associated bacteria on the fitness of plants evaluated using the extent of herbivory by the great pond snail. The six genotypes of *S. polyrhiza* tested did not differ in their resistance toward the herbivore. However, after outdoor inoculation with microbiota associated with the same plant species, altered herbivory tolerance was observed in a genotype-specific manner.

6. Physiology and Phytomonitoring

Duckweeds are ideal for studying hormonal effects in plants, as they can be cultured axenically and take up substances directly into the shoot. As a vivid example of this traditional role, Chmur and Bajguz [46] analysed growth responses to brassinosteroids in parallel with other biochemical parameters in *Wolffia arrhiza* treated with brassinolide and brassinazole, a synthetic brassinosteroid inhibitor. Kozlova and Levin [47], on the other hand, studied plant responses by *L. minor* to a fish steroid, 17β-Estradiol, that has been released at a large scale by intensive fish farms, and they found growth-promoting effects.

Besides plant physiology research, bioindication of pollutants is another classical field of duckweed applications. Microplastics are emerging contaminants and, as such, their potential risks need to be explored urgently. Two studies aimed to fill this knowledge gap: Rozman and Kalčíková [48] tested if duckweeds could be used in monitoring microplastic contamination of freshwaters by adsorbing these particles on their surface. Ceschin et al. [49] focused on the morphological and biochemical responses of these plants to microplastics, comparing acute effects to the chronic ones over an extended exposure period. Two further studies addressed methodological aspects of working with duckweeds: Oláh et al. [50] reviewed the vast diversity of various chlorophyll fluorescence-derived phytotoxicity endpoints that were reported in the literature on the application of the chlorophyll fluorescence imaging method on duckweeds and compared the responsivity of some widely used parameters to different toxicants. As another approach, Romano et al. [51] developed a workflow for automated frond surface area quantification by using digital images and machine learning. This method promises to significantly reduce the need for human input in applications that rely on duckweed growth.

7. Future Outlook

The present Special Issue gives an update on the state of the art of duckweed research and applications and, together with the two previous Special Issues on this topic [52,53], demonstrates the enormous progress made in this field during the last decade. Nevertheless, there are serious challenges both in research and applications. Duckweeds propagate dominantly by vegetative means but, being angiosperms, they are capable of flowering and seed production, which has been shown in several species. Artificial cross-fertilization and production of seeds has been demonstrated as a viable technique [54]; however, it is not yet a routine procedure. Additionally, genetic modification has been demonstrated by several research groups using various duckweed species but is also not yet a routine method. In most cases, the efficiency is still too low and needs more research. A bottleneck for the transition from research to applications is still the production of large amounts of biomass for any kind of industrial use, be it for food, feed or as energy source. Important steps were already made by the contributions of Petersen et al. [33,34] and by the group of Marcel Jansen (e.g., [25]). Several steps ahead in research and applications were mentioned in the conference report of the 6th ICDRA in Gatersleben, Germany, in 2022 [1]. The 7th International Duckweed Conference will be held in November 2024 in Thailand and we can expect interesting new results in the field.

Author Contributions: All authors contributed equally to this article. All authors have read and agreed to the published version of the manuscript.

Funding: This research received no external funding.

Conflicts of Interest: The authors declare no conflict of interest.

References

1. Oláh, V.; Appenroth, K.-J.; Lam, E.; Sree, K.S. Sixth International Conference on Duckweed Research and Applications Presents Lemnaceae as a Model Plant System in the Genomics and Postgenomics Era. *Plants* **2023**, *12*, 2134. [CrossRef]
2. Zhao, H.; Appenroth, K.; Landesman, L.; Salmeán, A.A.; Lam, E. Duckweed Rising at Chengdu: Summary of the 1st International Conference on Duckweed Application and Research. *Plant Mol. Biol.* **2012**, *78*, 627–632. [CrossRef]
3. Sree, K.S.; Appenroth, K.J.; Oelmüller, R. Sustainable Stress Management: Aquatic Plants vs. Terrestrial Plants. *Plants* **2023**, *12*, 2208. [CrossRef]
4. Ziegler, P.; Appenroth, K.J.; Sree, K.S. Survival Strategies of Duckweeds, the World's Smallest Angiosperms. *Plants* **2023**, *12*, 2215. [CrossRef]
5. Acosta, K.; Appenroth, K.J.; Borisjuk, L.; Edelman, M.; Heinig, U.; Jansen, M.A.K.; Oyama, T.; Pasaribu, B.; Schubert, I.; Sorrels, S.; et al. Return of the Lemnaceae: Duckweed as a Model Plant System in the Genomics and Postgenomics Era. *Plant Cell* **2021**, *33*, 3207–3234. [CrossRef]
6. Martinov, I. *Techno-Botanical Dictionary (Техно-Ботанический Словарь)*; Imperial Russian Academy: St. Petersburg, Russia, 1820.
7. Cabrera, L.I.; Salazar, G.A.; Chase, M.W.; Mayo, S.J.; Bogner, J.; Dávila, P. Phylogenetic Relationships of Aroids and Duckweeds (Araceae) Inferred from Coding and Noncoding Plastid DNA. *Am. J. Bot.* **2008**, *95*, 1153–1165. [CrossRef]
8. Tippery, N.P.; Les, D.H.; Appenroth, K.J.; Sree, K.S.; Crawford, D.J.; Bog, M. Lemnaceae and Orontiaceae Are Phylogenetically and Morphologically Distinct from Araceae. *Plants* **2021**, *10*, 2639. [CrossRef]
9. Braglia, L.; Breviario, D.; Gianì, S.; Gavazzi, F.; De Gregori, J.; Morello, L. New Insights into Interspecific Hybridization in *Lemna* L. Sect. *Lemna (Lemnaceae Martinov)*. *Plants* **2021**, *10*, 2767. [CrossRef]
10. Braglia, L.; Lauria, M.; Appenroth, K.J.; Bog, M.; Breviario, D.; Grasso, A.; Gavazzi, F.; Morello, L. Duckweed Species Genotyping and Interspecific Hybrid Discovery by Tubulin-Based Polymorphism Fingerprinting. *Front. Plant Sci.* **2021**, *12*, 625670. [CrossRef]
11. Bog, M.; Braglia, L.; Morello, L.; Noboa Melo, K.I.; Schubert, I.; Shchepin, O.N.; Sree, K.S.; Xu, S.; Lam, E.; Appenroth, K.J. Strategies for Intraspecific Genotyping of Duckweed: Comparison of Five Orthogonal Methods Applied to the Giant Duckweed *Spirodela polyrhiza*. *Plants* **2022**, *11*, 3033. [CrossRef]
12. Xu, S.; Stapley, J.; Gablenz, S.; Boyer, J.; Appenroth, K.J.; Sree, K.S.; Gershenzon, J.; Widmer, A.; Huber, M. Low Genetic Variation Is Associated with Low Mutation Rate in the Giant Duckweed. *Nat. Commun.* **2019**, *10*, 1243. [CrossRef]
13. Bog, M.; Appenroth, K.-J.; Schneider, P.; Sree, K.S. Intraspecific Diversity in Aquatic Ecosystems: Comparison between *Spirodela polyrhiza* and *Lemna minor* in Natural Populations of Duckweed. *Plants* **2022**, *11*, 968. [CrossRef]
14. Chen, G.; Stepanenko, A.; Lakhneko, O.; Zhou, Y.; Kishchenko, O.; Peterson, A.; Cui, D.; Zhu, H.; Xu, J.; Morgun, B.; et al. Biodiversity of Duckweed (Lemnaceae) in Water Reservoirs of Ukraine and China Assessed by Chloroplast DNA Barcoding. *Plants* **2022**, *11*, 1468. [CrossRef]

15. Yosef, A.; Ghazaryan, L.; Klamann, L.; Kaufman, K.S.; Baubin, C.; Poodiack, B.; Ran, N.; Gabay, T.; Didi-Cohen, S.; Bog, M.; et al. Diversity and Differentiation of Duckweed Species from Israel. *Plants* **2022**, *11*, 3326. [CrossRef] [PubMed]
16. Hoang, P.T.N.; Fuchs, J.; Schubert, V.; Tran, T.B.N.; Schubert, I. Chromosome Numbers and Genome Sizes of All 36 Duckweed Species (Lemnaceae). *Plants* **2022**, *11*, 2674. [CrossRef]
17. Urbanska-Worytkiewitz, K. Cytological Variation within the Family of Lemnaceae. In *Veröffentlichungen des Geobotanischen Institutes der Eidg. Tech. Hochschule, Siftung Rübel, in Zurich*; ETH Library: Zurich, Switzerland, 1980; pp. 30–101.
18. Geber, G. Zur Karyosystematik der Lemnaceae. Ph.D. Thesis, University of Vienna, Vienna, Austria, 1989.
19. Pasricha Sarin, L.; Sree, K.S.; Bóka, K.; Keresztes, Á.; Fuchs, J.; Tyagi, A.K.; Khurana, J.P.; Appenroth, K.-J. Characterisation of a Spontaneous Mutant of *Lemna gibba* G3 (Lemnaceae). *Plants* **2023**, *12*, 2525. [CrossRef] [PubMed]
20. Zhou, Y.; Stepanenko, A.; Kishchenko, O.; Xu, J.; Borisjuk, N. Duckweeds for Phytoremediation of Polluted Water. *Plants* **2023**, *12*, 589. [CrossRef] [PubMed]
21. Coughlan, N.E.; Walsh, É.; Ahern, R.; Burnell, G.; O'Mahoney, R.; Kuehnhold, H.; Jansen, M.A.K. Flow Rate and Water Depth Alters Biomass Production and Phytoremediation Capacity of *Lemna minor*. *Plants* **2022**, *11*, 2170. [CrossRef] [PubMed]
22. Devlamynck, R.; de Souza, M.F.; Leenknegt, J.; Jacxsens, L.; Eeckhout, M.; Meers, E. *Lemna minor* Cultivation for Treating Swine Manure and Providing Micronutrients for Animal Feed. *Plants* **2021**, *10*, 1124. [CrossRef]
23. Lambert, M.; Devlamynck, R.; Fernandes de Souza, M.; Leenknegt, J.; Raes, K.; Eeckhout, M.; Meers, E. The Impact of Salt Accumulation on the Growth of Duckweed in a Continuous System for Pig Manure Treatment. *Plants* **2022**, *11*, 3189. [CrossRef]
24. Walsh, É.; Coughlan, N.E.; O'Brien, S.; Jansen, M.A.K.; Kuehnhold, H. Density Dependence Influences the Efficacy of Wastewater Remediation by *Lemna minor*. *Plants* **2021**, *10*, 1366. [CrossRef] [PubMed]
25. O'Mahoney, R.; Coughlan, N.E.; Walsh, É.; Jansen, M.A.K. Cultivation of *Lemna minor* on Industry-Derived, Anaerobically Digested, Dairy Processing Wastewater. *Plants* **2022**, *11*, 3027. [CrossRef] [PubMed]
26. Paolacci, S.; Stejskal, V.; Toner, D.; Jansen, M.A.K. Integrated Multitrophic Aquaculture; Analysing Contributions of Different Biological Compartments to Nutrient Removal in a Duckweed-Based Water Remediation System. *Plants* **2022**, *11*, 3103. [CrossRef] [PubMed]
27. Zhou, Y.; Kishchenko, O.; Stepanenko, A.; Chen, G.; Wang, W.; Zhou, J.; Pan, C.; Borisjuk, N. The Dynamics of NO_3^- and NH_4^+ Uptake in Duckweed Are Coordinated with the Expression of Major Nitrogen Assimilation Genes. *Plants* **2022**, *11*, 11. [CrossRef] [PubMed]
28. Kishchenko, O.; Stepanenko, A.; Straub, T.; Zhou, Y.; Neuhäuser, B.; Borisjuk, N. Ammonium Uptake, Mediated by Ammonium Transporters, Mitigates Manganese Toxicity in Duckweed, *Spirodela polyrhiza*. *Plants* **2023**, *12*, 208. [CrossRef] [PubMed]
29. Pakdee, O.; Tshering, S.; Pokethitiyook, P.; Meetam, M. Examination of the Metallothionein Gene Family in Greater Duckweed *Spirodela polyrhiza*. *Plants* **2023**, *12*, 125. [CrossRef]
30. Oláh, V.; Irfan, M.; Szabó, Z.B.; Sajtos, Z.; Ragyák, Á.Z.; Döncző, B.; Jansen, M.A.K.; Szabó, S.; Mészáros, I. Species- and Metal-Specific Responses of the Ionome of Three Duckweed Species under Chromate and Nickel Treatments. *Plants* **2023**, *12*, 180. [CrossRef]
31. Jansen, M.A.K.; Paolacci, S.; Stejskal, V.; Walsh, É.; Kühnhold, H.; Coughlan, N.E. Duckweed Research and Applications on the Emerald Isle. *Duckweed Forum* **2023**, *11*, 2–7.
32. Petersen, F.; Demann, J.; Restemeyer, D.; Ulbrich, A.; Olfs, H.-W.; Westendarp, H.; Appenroth, K.-J. Influence of the Nitrate-N to Ammonium-N Ratio on Relative Growth Rate and Crude Protein Content in the Duckweeds *Lemna minor* and *Wolffiella hyalina*. *Plants* **2021**, *10*, 1741. [CrossRef]
33. Petersen, F.; Demann, J.; Restemeyer, D.; Olfs, H.-W.; Westendarp, H.; Appenroth, K.-J.; Ulbrich, A. Influence of Light Intensity and Spectrum on Duckweed Growth and Proteins in a Small-Scale, Re-Circulating Indoor Vertical Farm. *Plants* **2022**, *11*, 1010. [CrossRef]
34. Petersen, F.; Demann, J.; Von Salzen, J.; Olfs, H.-W.; Westendarp, H.; Wolf, P.; Appenroth, K.-J.; Ulbrich, A. Re-Circulating Indoor Vertical Farm: Technicalities of an Automated Duckweed Biomass Production System and Protein Feed Product Quality Evaluation. *J. Clean. Prod.* **2022**, *380*, 134894. [CrossRef]
35. Demann, J.; Petersen, F.; Dusel, G.; Bog, M.; Devlamynck, R.; Ulbrich, A.; Olfs, H.-W.; Westendarp, H. Nutritional Value of Duckweed as Protein Feed for Broiler Chickens—Digestibility of Crude Protein, Amino Acids and Phosphorus. *Animals* **2022**, *13*, 130. [CrossRef] [PubMed]
36. Romano, L.E.; Aronne, G. The World Smallest Plants (*Wolffia* sp.) as Potential Species for Bioregenerative Life Support Systems in Space. *Plants* **2021**, *10*, 1896. [CrossRef] [PubMed]
37. Fiordelmondo, E.; Ceschin, S.; Magi, G.E.; Mariotti, F.; Iaffaldano, N.; Galosi, L.; Roncarati, A. Effects of Partial Substitution of Conventional Protein Sources with Duckweed (*Lemna minor*) Meal in the Feeding of Rainbow Trout (*Oncorhynchus Mykiss*) on Growth Performances and the Quality Product. *Plants* **2022**, *11*, 1220. [CrossRef]
38. Demmig-Adams, B.; López-Pozo, M.; Polutchko, S.K.; Fourounjian, P.; Stewart, J.J.; Zenir, M.C.; Adams, W.W. Growth and Nutritional Quality of Lemnaceae Viewed Comparatively in an Ecological and Evolutionary Context. *Plants* **2022**, *11*, 145. [CrossRef]
39. Sree, K.S.; Appenroth, K.-J. Starch Accumulation in Duckweeds (Lemnaceae) Induced by Nutrient Deficiency. *Emir. J. Food Agric.* **2022**, *34*, 204–212. [CrossRef]

40. Wang, L.; Kuang, Y.; Zheng, S.; Tong, Y.; Zhu, Y.; Wang, Y. Overexpression of the Phosphoserine Phosphatase-Encoding Gene (AtPSP1) Promotes Starch Accumulation in *Lemna turionifera* 5511 under Sulfur Deficiency. *Plants* **2023**, *12*, 1012. [CrossRef]
41. Yamaga, F.; Washio, K.; Morikawa, M. Sustainable Biodegradation of Phenol by *Acinetobacter Calcoaceticus* P23 Isolated from the Rhizosphere of Duckweed *Lemna aoukikusa*. *Environ. Sci. Technol.* **2010**, *44*, 6470–6474. [CrossRef]
42. Acosta, K.; Sorrels, S.; Chrisler, W.; Huang, W.; Gilbert, S.; Brinkman, T.; Michael, T.P.; Lebeis, S.L.; Lam, E. Optimization of Molecular Methods for Detecting Duckweed-Associated Bacteria. *Plants* **2023**, *12*, 872. [CrossRef]
43. Gilbert, S.; Poulev, A.; Chrisler, W.; Acosta, K.; Orr, G.; Lebeis, S.; Lam, E. Auxin-Producing Bacteria from Duckweeds Have Different Colonization Patterns and Effects on Plant Morphology. *Plants* **2022**, *11*, 721. [CrossRef]
44. Bunyoo, C.; Roongsattham, P.; Khumwan, S.; Phonmakham, J.; Wonnapinij, P.; Thamchaipenet, A. Dynamic Alteration of Microbial Communities of Duckweeds from Nature to Nutrient-Deficient Condition. *Plants* **2022**, *11*, 2915. [CrossRef] [PubMed]
45. Schaefer, M.; Xu, S. The Effects of Microbiota on the Herbivory Resistance of the Giant Duckweed Are Plant Genotype-Dependent. *Plants* **2022**, *11*, 3317. [CrossRef] [PubMed]
46. Chmur, M.; Bajguz, A. Brassinolide Enhances the Level of Brassinosteroids, Protein, Pigments, and Monosaccharides in *Wolffia arrhiza* Treated with Brassinazole. *Plants* **2021**, *10*, 1311. [CrossRef] [PubMed]
47. Kozlova, T.A.; Levin, D.B. Effect of 17β-Estradiol on Growth and Biosynthesis of Microalgae Scenedesmus Quadricauda (CPCC-158) and Duckweed *Lemna minor* (CPCC-490) Grown in Three Different Media. *Plants* **2022**, *11*, 1669. [CrossRef]
48. Rozman, U.; Kalčíková, G. The Response of Duckweed *Lemna minor* to Microplastics and Its Potential Use as a Bioindicator of Microplastic Pollution. *Plants* **2022**, *11*, 2953. [CrossRef] [PubMed]
49. Ceschin, S.; Mariani, F.; Di Lernia, D.; Venditti, I.; Pelella, E.; Iannelli, M.A. Effects of Microplastic Contamination on the Aquatic Plant *Lemna minuta* (Least Duckweed). *Plants* **2023**, *12*, 207. [CrossRef]
50. Oláh, V.; Hepp, A.; Irfan, M.; Mészáros, I. Chlorophyll Fluorescence Imaging-Based Duckweed Phenotyping to Assess Acute Phytotoxic Effects. *Plants* **2021**, *10*, 2763. [CrossRef]
51. Romano, L.E.; Iovane, M.; Izzo, L.G.; Aronne, G. A Machine-Learning Method to Assess Growth Patterns in Plants of the Family Lemnaceae. *Plants* **2022**, *11*, 1910. [CrossRef]
52. Appenroth, K.-J.; Crawford, D.J.; Les, D.H. After the Genome Sequencing of Duckweed—How to Proceed with Research on the Fastest Growing Angiosperm? *Plant Biol.* **2015**, *17*, 1–4. [CrossRef]
53. Edelman, M.; Appenroth, K.J.; Sree, K.S. (Eds.) *Duckweed: Biological Chemistry and Applications;* Frontiers Research Topics; Frontiers Media SA: Lausanne, Switzerland, 2021; ISBN 978-2-88966-429-0.
54. Fu, L.; Huang, M.; Han, B.; Sun, X.; Sree, K.S.; Appenroth, K.-J.; Zhang, J. Flower Induction, Microscope-Aided Cross-Pollination, and Seed Production in the Duckweed *Lemna gibba* with Discovery of a Male-Sterile Clone. *Sci. Rep.* **2017**, *7*, 3047. [CrossRef]

plants

MDPI

Conference Report

Sixth International Conference on Duckweed Research and Applications Presents Lemnaceae as a Model Plant System in the Genomics and Postgenomics Era

Viktor Oláh [1], Klaus-Juergen Appenroth [2], Eric Lam [3,*] and K. Sowjanya Sree [4,*]

[1] Department of Botany, Faculty of Science and Technology, University of Debrecen, 4032 Debrecen, Hungary; olahviktor@unideb.hu
[2] Matthias Schleiden Institute—Plant Physiology, University of Jena, 07743 Jena, Germany; klaus.appenroth@uni-jena.de
[3] Department Plant Biology, Rutgers State University of New Jersey, New Brunswick, NJ 08901, USA
[4] Department of Environmental Science, Central University of Kerala, Periye 671320, India
* Correspondence: ericl89@hotmail.com (E.L.); ksowsree@gmail.com or ksowsree9@cukerala.ac.in (K.S.S.)

Abstract: The 6th International Conference on Duckweed Research and Applications (6th ICDRA) was organized at the Institute of Plant Genetics and Crop Plant Research (IPK) located in Gatersleben, Germany, from 29 May to 1 June 2022. The growing community of duckweed research and application specialists was noted with participants from 21 different countries including an increased share of newly integrated young researchers. The four-day conference focused on diverse aspects of basic and applied research together with practical applications of these tiny aquatic plants that could have an enormous potential for biomass production.

Keywords: biomass production; duckweed; Lemnaceae; whole genome sequencing

check for updates

Citation: Oláh, V.; Appenroth, K.-J.; Lam, E.; Sree, K.S. Sixth International Conference on Duckweed Research and Applications Presents Lemnaceae as a Model Plant System in the Genomics and Postgenomics Era. *Plants* **2023**, *12*, 2134. https://doi.org/10.3390/plants12112134

Academic Editor: Vagner A. Benedito

Received: 31 March 2023
Revised: 8 May 2023
Accepted: 24 May 2023
Published: 28 May 2023

1. Introduction

Duckweeds are aquatic monocots that exhibit a high degree of miniaturization and simplification of their plant body (Figure 1). By efforts from the community of researchers and application specialists, these plants have been re-established as model plants in the current genomics and postgenomics era [1]. Over this course, the first International Conference on Duckweed Research and Applications (ICDRA) was organized in 2011 at the Chinese Academy of Science in Chengdu, China [2]. A crucial step in the evolution of this community was the establishment of the International Steering Committee on Duckweed Research and Applications (ISCDRA) in 2013 at the second ICDRA organized at the Rutgers University, New Brunswick, NJ, USA [3]. Steered by the ISCDRA, the biennial meetings continued to be organized at the Kyoto University, Japan (2015), the Central University of Kerala, India (2017), and the Weizmann Institute of Science at Rehovot, Israel, (2019), discussing and deliberating on the advances made in the field over each two-year interval. Most recently, the 6th ICDRA was organized on behalf of the ISCDRA (Figure 2) at the Institute of Plant Genetics and Crop Plant Research (IPK) in Gatersleben, Germany, from 29 May to 1 June 2022, which has a one-year delay because of the COVID-19 pandemic. Ingo Schubert from the IPK served as the Chair, and Manuela Nagel, also of the IPK, and Klaus-J. Appenroth, University of Jena, Germany, were the Co-Chairs (Figure 3). A special issue "Duckweeds: Research meets applications" was organized by three guest editors for the journal *Plants* (Basel) (Figure 4).

The four-day conference included forty-two oral presentations and thirty-one posters, most of them presented by young researchers or Ph.D. students, with a total of ninety-four participants from twenty-one different countries. The increasing number of research groups entering the duckweed research field is an important development for our community as this brings in expertise from varied fields to work on duckweeds.

Figure 1. Colonies of *Spirodela polyrhiza* showing the frond architecture.

Figure 2. International Steering Committee on Duckweed Research and Applications (ISCDRA). From left to right: Klaus-J. Appenroth (University of Jena, Germany), K. Sowjanya Sree (Central University of Kerala, India), Tsipi Shoham (GreenOnyx, Tel Aviv, Israel), Marcel Jansen (Cork University, Ireland), Eric Lam (Rutgers University, New Brunswick, NJ, USA; Chair of the committee). (Photo courtesy: IPK Archives).

Apart from the lecture and poster sessions, a public lecture was also delivered by Klaus-J. Appenroth on the topic "A new crop plant with great potential for nutrition, water treatment and energy", in view of the potential use of duckweeds in various practical applications. This session was intended for and attended by the public in and around the IPK as outreach to the society at large. As customary for the final day of the conference, a General Assembly of the participants voted and decided for the next conference to be held in 2024 at Bangkok, Thailand. This will be organized by Arinthip Thamchaipenet from Kasetsart University, Thailand, and Metha Meetam of Mahidol University, Thailand.

Figure 3. Scientific organization committee of the 6th ICDRA. From left to right: Ingo Schubert (IPK Gatersleben, Germany; Chair of the conference), Manuela Nagel (IPK Gatersleben, Germany) and Klaus-J. Appenroth (University of Jena, Germany), Co-Chairs. (Photo courtesy: IPK Archives).

Figure 4. Guest editors of the Special Issue "Duckweeds: Research meets applications" of the journal *Plants*. From left to right: Klaus-J. Appenroth (University of Jena, Germany), K. Sowjanya Sree (Central University of Kerala, India), Viktor Oláh (University of Debrecen, Hungary). (Photo courtesy: IPK Archives).

The following sections on (1) genomics, evolution and genome organization; (2) differentiation and diversity; (3) physiology, metabolism and microbiome; (4) stress, toxicology, cryopreservation; (5) wastewater remediation; (6) large-scale cultivation; and (7) feed and food, will highlight recent advances in the diverse areas of duckweed research and applications.

2. Genomics, Evolution and Genome Organization

The report about "Genomic and epigenomic consequences of clonal growth habit in the Lemnaceae" (Rob Martienssen, Cold Spring Harbor, NY, USA), the first "Invited talk" of the conference, provided evidence that the genomes of Lemnaceae selectively lost genes required for RNA-directed DNA methylation that are involved in transposon silencing. Further, triploid hybrids that have arose from *Lemna minor* and *Lemna turionifera* are commonly found. He also reported that they presumably form through hybridization with unreduced gametes, and that divergent "ZMM class" mismatch repair genes could support polyploid meiosis. Several intraspecific hybrids were analyzed by the method of **T**ubulin-**B**ased **P**olymorphism (TBP). Species described by the late Elias Landolt as *Lemna japonica* turned out to be hybrids between *L. minor* and *L. turionifera*. In some cases, also hybrids between *Lemna gibba* and *L. minor* were uncovered ("TBP fingerprinting unveiled interspecific hybridization in the genus *Lemna*", Laura Morello, Milano, Italy). Transcriptomic data revealed the absence and non-expression of key components including the RNA-directed DNA methylation (RdDM) pathway methyltransferase (DRM2), as well as the lack of siRNA associated with transposable elements ("Epigenetic regulation of transposable elements in duckweeds", by Rodolphe Dombey, Gregor Mendel Institute, Austria). Some preliminary results from ongoing long-term experimental evolution studies in mesocosms that aim to investigate plant evolution in multitrophic communities were reported by Shuqing Xu (University of Mainz, Germany). In the report via virtual zoom link on "Duckweed genome architecture" it was reported that whole genome sequence data of at least one species from all genera of Lemnaceae are now available, very recently supplemented by that of *Landoltia punctata*, *Wolffia australiana* and *Wolffiella neotropica* (Todd P. Michael, Salk Institute for Biological Studies, USA). Whereas *Spirodela intermedia* and *Spirodela polyrhiza* share a similar genome architecture [4], *Wolffia* and *Lemna* revealed dynamic evolutionary trajectories, whole genome duplication events and polyploidization. For the first time, a survey on genome size measurement for all 36 duckweed species, on chromosome counts for 31 species and on their evolutionary impact was performed using flow cytometry and cytogenetic approaches (Phuong Thi Nhu Hoang, Dalat University, Vietnam). While chromosome numbers for all 36 species has been published [5], only half of them are based on recent, more reliable measurements. Soon progress in this field should be expected from additional advances in high-throughput single molecule long-read sequencing. Due to the high copy number of ribosomal DNA in higher plants, conserved coding sequence and more rapidly evolving spacer sequences, the rDNA has become a favorite subject for studies related to plant systematics, evolution and biodiversity. Investigating rDNA in a large number of duckweed species, the Borisjuk group (Nikolai Borisjuk, Huaiyin Normal University, China) concluded that these put duckweeds in the spotlight for research on the molecular evolution of the rDNA.

3. Differentiation and Diversity

The series of oral presentations started with a highlight as a plenary talk about "Duckweed hibernation: unravelling the molecular basis of the turion induction switch in *Spirodela polyrhiza*" (Eric Lam, Rutgers University, NJ, USA). "Hibernation" means the formation of hibernacles, better known as turions. Using resources with a newly assembled genome for *S. polyrhiza* 9512, which has a high turion yield coupled with a fast turion formation rate, the transcriptome of mature turions (survival organs) were compared with that of normal vegetative fronds. This was enabled by a new method developed for isolation of high quality RNA from turions despite their high content of starch and tannins. Identification and informatics analysis of global transcripts with changes of more than eight-fold between turions and fronds, it was found that genes in the pathways for stress responses, dormancy and in several biosynthetic pathways to increase starch and lipid synthesis are likely to play a special role in turion biology [6]. Especially convincing was the comparison of the results of *S. polyrhiza* to those of *S. intermedia*, a species with very similar properties [4,7] but unable to produce turions, where many of the turion-specific genes are no longer induced by the low phosphate trigger in spite of the presence of their orthologs in the genome. Finally, the first evidence for epigenetic changes in the transition between frond and turion tissues were also presented from a global analysis of cytosine methylation patterns in these two states.

Applying light and confocal microscopy with high levels of spatial resolution, the process of proliferating daughter fronds in *Wolffia globosa*, *W. australiana* and *L. gibba* was investigated (Ljudmilla Borisjuk, Leibnitz-Institute Gatersleben, Germany). This work showed that budding of new meristems started from very early stage onwards during formation of the nascent daughter fronds. Metabolome analysis (LC/MS) and spatially resolved infrared imaging (FTIR) in five genotypes of *Wolffia* with distinct growth characteristics were carried out. The integrative analysis of structural organization, meristematic activity and metabolism is expected to provide a better understanding of growth dynamics in duckweed. The life cycle of *S. polyrhiza* has been subjected to numerous studies that have revealed several key processes involved in turion formation and function. As reviewed and concluded by Paul Ziegler (University of Bayreuth, Germany), these processes might be a model for other turion-producing aquaphytes [8]. The developmental switch from frond propagation to turion formation is not primarily due to the decreasing day lengths and temperatures, but rather to nutrient depletion (e.g., phosphate) of the water habitat. The newly formed turions are dormant and require a prolonged exposure to cold water to be able to germinate, after which the plant's early development (germination and growth of the newly formed fronds) will be fueled by two distinct carbohydrate reserves, sugars and starch. Development of roots in duckweeds was investigated in face of the fact that species in two genera (*Wolffiella*, *Wolffia*) do not have roots at all. The authors (Alexander Ware, University of Nottingham, UK) could show by careful comparative cytological studies that roots in duckweeds are vestigial as suggested already on a morphological basis by Gorham in 1941 [9]. Different clones of *Wolffiella hyalina* have different photoperiodic requirements for flowering induction and respond differently to exogenous application of Salicylic acid or Benzoic acid (Minako Isoda, Kyoto University, Japan). These results suggest natural variation of floral inductive pathways between accessions in *W. hyalina*. Moreover, flowering of non-flowering *W. hyalina* was induced when co-cultured with flowering *W. hyalina* even in the absence of SA or BA in the medium. This result suggests the existence of plant-to-plant communication for floral induction. Duckweed sexual reproduction is still one of the most unknown aspects in this plant family. For the first time, transcriptomic analysis was conducted on duckweed pistils, anthers, seeds and fronds from the short-day plant *Lemna aequinoctialis* ("Unravelling the genetic mechanisms of plant sexual reproduction in duckweed", Cristian Mateo-Elizalde, Cold Spring Harbor, USA). These results should improve our understanding of the molecular pathways involved in vegetative and generative propagation in duckweeds.

A basic requirement to investigate biodiversity in the field of duckweeds is the reliable identification of species and clones. A significant progress in the genus *Lemna* was made possible by the application of the method of tubulin-based polymorphism [10]. Except for the species *S. polyrhiza* [11], it is not yet known how this method will perform in other genera (Laura Morello, IBBA CNR, Milano, Italy). Anton Stepanenko and co-worker (Huaiyin Normal University, China) investigated the species diversity in some regions of Ukraine and Eastern China. Only by using molecular methods, i.e., a two-barcode protocol with the chloroplast atpH–atpF and psbK–psbI spacer sequences as described previously [12], six species from Ukraine and six species from China could be identified reliably. *Lemna aequinoctialis* does not form a uniform taxon and therefore, the phylogenetic status of this species requires further investigations.

4. Physiology, Metabolism and Microbiome

Duckweeds have long earned their position as models in plant physiology research, and a series of presentations gave diverse examples of the current state-of-the-art research topics that were addressed with these tiny plants. In his invited talk, Tokitaka Oyama

(Kyoto University, Japan) presented a study on the circadian regulation in duckweeds using transient reporter-gene expression-based bioluminescence monitoring. The evolutionary and physiological significance of such circadian traits and regulation has been less explored so far, but potentially has high relevance in plant adaptive mechanisms. The presented system allows analyses at frond- and cell-specific levels to track synchronization or uncoupling between cells in the circadian regulation amongst and within duckweed fronds [13]. They found the stability of circadian cycle to be genotype-dependent: *Lemna* species performed more stable rhythms, while those in the *Wolffiella* genus showed arrhythmia under constant irradiation and high temperature [14]. Similarly, intraspecific populations of *L. aequinoctialis* collected in diverse latitudes showed differences in critical daylengths and circadian periods, thus suggesting microevolutionary adaptation of circadian traits to local environmental factors [15].

Wisuwat Songnuan (Mahidol University, Thailand) focused on characterizing the potential of two duckweed species from Thailand to sequester CO_2 on the basis of their high growth rates. Moshe T. Halpern (ARO, Volcani Institute, Israel) depicted another exciting field of application for duckweed models. One of the current concerns in agriculture is understanding the physiological background of declining protein content in crops under elevated ambient CO_2 levels [16]. This can be caused by both decreasing nitrogen uptake and restricted nitrate reduction, and duckweeds, especially species of the *Wolffia* genus, may facilitate resolving this urgent problem. *Wolffia* offers an ideal system, in which the lack of roots excludes any possible influence of the root-to-shoot transport and leaves the shoot as the only organ for N-uptake and NO_3^- assimilation. In addition, duckweeds may not only tolerate but prefer NH_4^+ as a nitrogen source and can also be cultured under aseptic conditions, thus limiting interference by such factors as nitrification normally taking place in soil-based systems. Speaking of nitrogen metabolism, revealing mechanisms of uptake and assimilation of nitrogen by duckweeds is of high practical significance on its own. Yet, despite its importance for wastewater remediation and biomass production, this field needs to be developed further. Olena Kishchenko (Huaiyin Normal University, China) presented their latest results on the topic with six duckweed species, shedding light on the complex regulation of nitrogen assimilation genes and confirming that these plants, though can easily utilize both nitrogen sources, prefer NH_4^+ to NO_3^- [17].

The aquatic lifestyle, simple anatomy and ability to thrive under axenic conditions also makes duckweeds well suited to study plant–microbe interactions [18]. Three presentations addressed this topic from various aspects. An invited talk by Asaph Aharoni introduced ongoing research at Weizmann Institute of Science (Israel) with *L. minor* to identify metabolic pathways and key metabolites that regulate the colonization and composition of microbiome community through plant exudates. What does a microbiome look like, however, when there are no roots? Osnat Gillor's presentation (Ben Gurion University of the Negev, Israel) focused on this question by studying the rootless *W. globosa*. It was hypothesized that the microbiome of this unique plant may be a composite of functionally distinct communities that normally inhabit separate parts of the plant body. The metagenomic analyses revealed a bacteria-dominated microbiome with a characteristic endophyte community. This community was not only different from that of the surrounding medium but also from that of epiphytes, and besides aiding phosphorus and iron uptake, it was likely capable of fixing CO_2 and atmospheric nitrogen, as well as synthesizing vitamin B12. Besides friendly microbes, however, there are always hostile ones surrounding the plants. Dynamics and interactions of two duckweed-associated bacteria strains (*Bacillus* sp. MRB10 and *Chryseobacterium* sp. 27AL) with the cyanobacterium (also known as blue-green algae) *Microcystis aeruginosa* and duckweed *L. gibba* suggest that proper growth of host plants requires an appropriate population mixture and size, as it was described by Masaaki Morikawa (Hokkaido University, Japan). They found that even though duckweed-associated bacteria could support growth of plants while modulating that of *M. aeruginosa*, they could also become deleterious to the host plants when they were suspended instead of being attached to the plant surface.

5. Stress, Toxicology, Cryopreservation

As it was mentioned earlier, many presentations in the conference discussed turions of *S. polyrhiza* from various aspects. In her invited talk, K. Sowjanya Sree (Central University of Kerala, India) reported turion formation in another duckweed species for the first time. The enigmatic species *Wolffia microscopica* has recently been re-discovered [19] and, so far, was the only *Wolffia* species that had no known turion formation in the genus. Besides turion formation, K. Sowjanya Sree discussed another typical duckweed response to adverse conditions, which is starch accumulation. She presented that, while phosphate and nitrogen starvation can induce rapid carbohydrate accumulation in all genera of the Lemnaceae family, sulphate limitation had only marginal effects in that regard [20].

Besides stressors of natural origin, anthropogenic factors can also threaten duckweeds. Philippe Juneau (University of Quebec in Montreal, Canada) reported that even if aminomethylphosphonic acid (AMPA), which is one of the primary degradation products of the herbicide glyphosate, did not disturb growth and photosynthesis of duckweeds at environmentally-relevant concentrations, it nevertheless decreased the chlorophyll content of plants by interfering with chlorophyll biosynthesis. Another talk on the physiological effects of anthropogenic freshwater pollutants on duckweeds was presented by Darlielva do Rosario Freitas (Federal University of Viçosa, Brazil). She discussed the significance of the chemical form of iron supply in alleviating cadmium toxicity to *Lemna valdiviana*. Iron modulated the antioxidant system of duckweed plants depending on whether it was applied as zerovalent nanoparticles or in ionic form and also affected plant responses to cadmium treatments. With regard to duckweed stress responses, Viktor Olah (University of Debrecen, Hungary) approached this topic from a different aspect by analyzing spatial patterns in photosynthetic responses, starch and anthocyanin content at the level of individual fronds. His results indicated that there were considerable differences in duckweed responses depending on frond ontogeny, species and stressor applied.

By what mechanisms can clonal plants improve their stress resistance? One possible way to achieve higher diversity can be polyploidization [5]. Quinten Bafort (Ghent University, Belgium) gave insights in his invited talk to the phenotypic responses of tetraploid *S. polyrhiza* strains to gradients of environmental factors and their implications on fitness of the neopolyploid populations [21]. One other way to cope with stressors in monoclonal populations can be the inheritance of non-genetic phenotypic traits, such as through epigenetic mechanisms or incorporation of a microbiota [22]. Another invited talk by Alexandra Chávez (University of Muenster, Germany) addressed this topic by analyzing responses of many *S. polyrhiza* genotypes under biotic and abiotic stress for several frond generations. It was concluded that both formation of non-genetic based phenotypes as well as selection of these phenotypic variants improved stress resistance of clonal *S. polyrhiza* populations, and this effect can last for multiple generations even after stress release.

Whether focusing on basic research topics or practical utilization, many duckweed applications rely on the diversity of Lemnaceae. Thus, maintaining and further expanding duckweed stock collections with naturally evolved or artificially created clones is a necessity of the job [23]. Maintenance of large duckweed collections may, however, become very demanding in terms of infrastructure and human resource. Two presentations by Shogo Ito (Kyoto University, Japan) and by Manuela Nagel (Leibniz Institute of Plant Genetics and Crop Plant Research, Germany) discussed the potentials and pitfalls of long-term cryopreservation of duckweed. This method, besides its promise of a more cost-efficient way to maintain large-scale duckweed stock collections, can also decrease the risk of mixing stocks, or inadvertent infections of the cultures with microbes. Both groups tested a wide range of cryopreservation protocols and parameter settings to achieve successful preservation and regeneration of duckweed plantlets after their storage under cryogenic conditions. The presented results suggest that long-term cryopreservation of duckweed strains is achievable, but species-specific protocols may be needed, while some clones from different geographical latitudes showed different success rates in regeneration after cold storage.

6. Wastewater Remediation

Duckweeds are the fastest growing flowering plants [24], and as aquatic plants, they are especially suitable for wastewater treatment. The importance of these applications was emphasized in a contribution via zoom link about the "New Circular Economy" (Paul Skillicorn, Skillicorn Technologies, Austin, TX, USA) that described application of duckweeds in the universe of volatile wastes produced by humans and farm animals, industrial volatile wastes and farm-crop residuals, as well as the waters that accompany them. The plants' potential roles in cleaning up the global environment and helping to remove carbon from the atmosphere, solving global malnutrition, providing rural jobs and bestowing wealth on smallholder farmers, were explored. A research group from Cork, Ireland, also developed an "Integrated MultiTrophic Aquaculture"-system (IMTA) and applied it for dairy wastewater remediation. A new cascading system for valorization of dairy wastewater couples microbial-based technologies of anaerobic digestion and/or aerobic dynamic feeding with duckweed cultivation. As part of the indoor system, duckweed is grown in a stacked flow-through system to minimize the spatial footprint. This system is not only useful for cleaning large volumes of dairy wastewater, it produces also large amount of biomass all year round (Marcel Jansen, University College Cork, Ireland; see [25]). In another report, duckweed-based IMTA was used to test the removal of total nitrogen and phosphorus from wastewater. This system united large-scale wastewater purification with production of biomass that was used for fish feeding (rainbow trout and European perch) (Simona Paolacci, Bantry Marine Research station, Ireland, see [26]). Both reports from Ireland represent a milestone in the practical application of duckweeds.

7. Large Scale Cultivation

The required up-scaling from lab-scale cultivation of duckweeds in the range of some grams of fresh weight, to production in the range of tonnes of biomass being generated, is presently a bottleneck for duckweed applications. Some successful developments were already described in the previous section (Paul Skillicorn, USA; Marcel Jansen, Ireland; Simona Paolacci, Ireland). Under the brand name "LemnaCore 1.0", a group from Germany presented a mobile duckweed cultivation system, which is based on recycling and utilization of industrial process and wastewater. A vertical aquafarming technology was devised that cultivates biomass on connected levels to reduce the footprint required. The goal is to create an environment for duckweed in which it can grow well, and the production is reproducible and automated (Marko Dietz, Carbon Clouds, Germany). In another project, the development of a large-scale, vertically-integrated duckweed production system was started with optimizing abiotic factors (nutrient medium, light conditions) for the cultivation of *L. minor* 9441 and *W. hyalina* 9525 from a small culture. Goals of this optimization were not only high growth rates and biomass with high protein content but also prevention of competing growth of algae [27,28]. On the basis of these results, a recirculating, large indoor vertical farm for duckweed biomass production was developed [29] possessing nine levels for cultivation. The water is re-circulating within the system, while nutrients, temperature and light settings are controlled and regulated automatically. Per day more than 900 g of fresh biomass is harvested. Further up-scaling and improved levels of regulation are in preparation (Finn Petersen, University for Applied Sciences Osnabrück, Germany).

8. Feed and Food

Another highlight of the 6th ICDRA was the report that duckweed for human nutrition could be cultivated under sterile conditions and in large scale, using lighted cabinets that are sealed off from the outside environment. The company GreenOnyx, Israel, introduced a breakthrough farming technology to grow and deliver duckweed vegetables, called Wanna greens®, to the global market. Whereas any duckweed species can theoretically be used for this technology, the focus at the moment is on two species from the genus *Wolffia*, *W. globosa* and *W. arrhiza*. Based on an array of compact-modular growing systems that are sealed and sterile, a fully automated supply chain was introduced (Tsipi Shoham,

GreenOnyx, Israel). It should be mentioned that this company obtained the permission from the European Commission to use and sell fresh plants of *W. globosa* and *W. arrhiza* as "traditional food", and not considered as "novel food" [30]. Investigations to obtain the same permission for *Lemna* species (*L. gibba*, *L. minor*) are on the way (Ingrid van der Meer, Wageningen University, Netherlands; see [31]). Studies on the nutritional value and digestibility of duckweed protein were carried out and provided promising results (Jurriaan Mes, Wageningen Food and Biobased Research, Netherlands; see [32]).

Known for a long time, duckweeds can also be used for animal feeding. The nutritional value of duckweed as feed has already been confirmed in several studies with broiler chickens. However, contrasting effects of using duckweed in complete diets on growth and especially feed intake have been found. The digestibility of crude protein can differ strongly along with the quality of duckweed batches, ranging from 68 to 90% for methionine. The differences are most probably caused by higher contents of tannins and fiber. Different cultivation conditions for duckweeds are being tested further to evaluate the factors more critically (Johannes Demann, University for Applied Sciences, Osnabrück, Germany).

9. Duckweed Futures: Visions of Things to Come

It is fitting in closing this meeting summary to reflect on the advances in duckweed research and applications that have been made over the past eleven years since the first ICDRA in Chengdu, China [2]. At that time, the first draft genome of *S. polyrhiza* had just been completed to the sequencing phase by the JGI (Joint Genome Institute, USA) using a combination of 454 and Illumina (San Diego, CA, USA) sequencing technologies. In those days, BAC-end sequencing of large genomic fragments was being carried out to use as scaffolding tools for genome assembly [33]. Today, we have in hand reference quality genomes for all five genera of duckweeds and for species such as *S. polyrhiza*, multiple large-scale population genomic datasets with hundreds of accessions are available to all. This fast-forward genomics approach would likely accelerate even more over the next decade due to rapid improvements in long-read, single-molecule sequencing technologies, notably Oxford Nanopore and PacBio being orthogonal approaches that are increasing their fidelity while decreasing time and costs. In addition, further diversification of these technologies to rapidly quantifying epigenetic marks on DNA can be used to broaden their impact in biology as well. The high quality reference genome collection will likely be completed for all 36 known species within the next decade, if not sooner. Their availability will enable researchers to begin a quantitative understanding of how this family has evolved over the millions of years since its divergence from the early monocot ancestor. Furthermore, their comparative analysis to other terrestrial plants should open new windows into alternative strategies for these miniature plants to adapt to vastly diverse environments. A case in point is the discovery of the severe loss of most of the immune-related NLR genes in *W. australiana* while this species has elaborated more genes in the PRR-dependent basal immunity pathways [18]. The advance in genomics will likely drive new discoveries in the role of epigenetic pathways in duckweed biology. As the comparison of turion and frond DNA methylation in *S. polyrhiza* has shown [6], there are clearly significant differences in the way DNA are being methylated during this developmental transition, even in a duckweed species that has the lowest overall cytosine methylation in plants studied so far [34]. Future examination of this behavior in other species of duckweeds and response pathways will likely open many new avenues for epigenetic studies in duckweed, as well as providing new understanding of how things can be performed differently for similar outcomes.

The areas of duckweed farming and product development will likely forge close relationships with many types of basic research interests in ecology, evolution and chemical genomics in the coming years. One approach that has gained popularity in the past few years is the interest in modular indoor growth system for duckweed, in contrast to the more traditional usage of raceway ponds and lagoons that have been deployed in earlier academic research and by companies such as Parabel USA Inc. In this ICDRA, there are at least five groups who have presented talks or posters involving vertically integrated duckweed growth systems for scalable indoor production, which is a sign that this approach is being adopted and developed by our community at large. While the indoor hydroponic growth system for plant production is not novel per se, the parameters and precise constraints posed by an aquatic plant such as duckweed are new and unique. The automation and plant maintenance knowledge for optimized duckweed growth in this scenario would need to be established for these new experimental systems, and their availability down the line as user-friendly growth systems will not only open new opportunities for cropping duckweed, but also will provide useful platforms for basic research under controlled environmental conditions. For example, this could help accelerate phenotyping approaches to identify important traits and test for effects by chemicals or biological agents such as beneficial microbes on duckweed health and productivity.

Finally, there are two bottlenecks for duckweed research and applications that we see as potential breakthroughs in the coming years that would help to further transform the community. One is the functional understanding of flowering control in duckweed that will enable effective and routine genetic manipulation through crossing and selection for varieties of interest. This capability could be facilitated by a molecular understanding of the basis for key flowering inducing genes and compounds. Its success would usher in an era of duckweed breeding via molecular genetics. The second hurdle that we would like to see overcame is the optimization and deployment of a robust and rapid in planta transformation technology for duckweeds. This ability will enable rapid molecular studies for genes of interest that have been defined from a rapidly growing database of transcriptomics data being generated by our community. A reliable and facile method that can overcome the often-observed strain and species dependence of transformation protocols [35] to investigate gene functions and pathways will truly open the door to the fascinating world of duckweeds in the coming decade.

Author Contributions: All authors contributed to preparation of the draft, review and revision of the publication. All authors have read and agreed to the published version of the manuscript.

Funding: No funding was received for writing this report.

Data Availability Statement: All the data related to this publication are mentioned in the text.

Conflicts of Interest: The authors declare no conflict of interest.

References

1. Acosta, K.; Appenroth, K.J.; Borisjuk, L.; Edelman, M.; Heinig, U.; Jansen, M.A.K.; Oyama, T.; Pasaribu, B.; Schubert, I.; Sorrels, S.; et al. Return of the Lemnaceae: Duckweed as a model plant system in the genomics and postgenomics era. *Plant Cell* **2021**, *33*, 3207–3234. [CrossRef] [PubMed]
2. Zhao, H.; Appenroth, K.; Landesman, L.; Salmeán, A.A.; Lam, E. Duckweed rising at Chengdu: Summary of the 1st International Conference on Duckweed Application and Research. *Plant Mol. Biol.* **2012**, *78*, 627–632. [CrossRef]
3. Lam, E.; Appenroth, K.J.; Michael, T.; Mori, K.; Fakhoorian, T. Duckweed in bloom: The 2nd International Conference on Duckweed Research and Applications heralds the return of a plant model for plant biology. *Plant Mol. Biol.* **2014**, *84*, 737–742. [CrossRef] [PubMed]
4. Hoang, P.T.N.; Schubert, I. Reconstruction of chromosome rearrangements between the two most ancestral duckweed species Spirodela polyrhiza and S. intermedia. *Chromosoma* **2017**, *126*, 729–739. [CrossRef] [PubMed]
5. Hoang, P.T.N.; Fuchs, J.; Schubert, V.; Tran, T.B.N.; Schubert, I. Chromosome Numbers and Genome Sizes of All 36 Duckweed Species (*Lemnaceae*). *Plants* **2022**, *11*, 2674. [CrossRef]
6. Pasaribu, B.; Acosta, K.; Aylward, A.; Liang, Y.; Abramson, B.W.; Colt, K.; Hartwick, N.T.; Shanklin, J.; Michael, T.P.; Lam, E. Genomics of turions from the Greater Duckweed reveal its pathways for dormancy and re-emergence strategy. *New Phytol.* **2023**. [CrossRef]
7. Hoang, P.N.T.; Michael, T.P.; Gilbert, S.; Chu, P.; Motley, S.T.; Appenroth, K.J.; Schubert, I.; Lam, E. Generating a high-confidence reference genome map of the Greater Duckweed by integration of cytogenomic, optical mapping, and Oxford Nanopore technologies. *Plant J.* **2018**, *96*, 670–684. [CrossRef]
8. Adamec, L. Ecophysiological characteristics of turions of aquatic plants: A review. *Aquat. Bot.* **2018**, *148*, 64–77. [CrossRef]
9. Braglia, L.; Breviario, D.; Giani, S.; Gavazzi, F.; De Gregori, J.; Morello, L. New insights into interspecific hybridization in Lemna L. Sect. Lemna (Lemnaceae Martinov). *Plants* **2021**, *10*, 2767. [CrossRef]

10. Gorham, P.R. Measurement of the response of Lemna to growth promoting substances. *Am. J. Bot.* **1941**, *28*, 98–101. [CrossRef]
11. Bog, M.; Braglia, L.; Morello, L.; Noboa Melo, K.I.; Schubert, I.; Shchepin, O.N.; Sree, K.S.; Xu, S.; Lam, E.; Appenroth, K.J. Strategies for Intraspecific Genotyping of Duckweed: Comparison of Five Orthogonal Methods Applied to the Giant Duckweed *Spirodela polyrhiza*. *Plants* **2022**, *11*, 3033. [CrossRef]
12. Borisjuk, N.; Chu, P.; Gutierrez, R.; Zhang, H.; Acosta, K.; Friesen, N.; Sree, K.S.; Garcia, C.; Appenroth, K.J.; Lam, E. Assessment, validation and deployment strategy of a two-barcode protocol for facile genotyping of duckweed species. *Plant Biol.* **2015**, *17* (Suppl. S1), 42–49. [CrossRef] [PubMed]
13. Ueno, K.; Ito, S.; Oyama, T. An endogenous basis for synchronisation characteristics of the circadian rhythm in proliferating Lemna minor plants. *New Phytol.* **2022**, *233*, 2203–2215. [CrossRef] [PubMed]
14. Isoda, M.; Ito, S.; Oyama, T. Interspecific divergence of circadian properties in duckweed plants. *Plant Cell Environ.* **2022**, *45*, 1942–1953. [CrossRef]
15. Muranaka, T.; Ito, S.; Kudoh, H.; Oyama, T. Circadian-period variation underlies the local adaptation of photoperiodism in the short-day plant *Lemna aequinoctialis*. *iScience* **2022**, *25*, 104634. [CrossRef]
16. Taub, D.R.; Miller, B.; Allen, H. Effects of elevated CO2 on the protein concentration of food crops: A meta-analysis. *Glob. Chang. Biol.* **2008**, *14*, 565–575. [CrossRef]
17. Zhou, Y.; Kishchenko, O.; Stepanenko, A.; Chen, G.; Wang, W.; Zhou, J.; Pan, C.; Borisjuk, N. The dynamics of NO3− and NH4+ uptake in duckweed are coordinated with the expression of major nitrogen assimilation genes. *Plants* **2022**, *11*, 11. [CrossRef] [PubMed]
18. Acosta, K.; Xu, J.; Gilbert, S.; Denison, E.; Brinkman, T.; Lebeis, S.; Lam, E. Duckweed hosts a taxonomically similar bacterial assemblage as the terrestrial leaf microbiome. *PLoS ONE* **2020**, *15*, e0228560. [CrossRef] [PubMed]
19. Sree, K.S.; Maheshwari, S.C.; Boka, K.; Khurana, J.P.; Keresztes, Á.; Appenroth, K.-J. The duckweed Wolffia microscopica: A unique aquatic monocot. *Flora* **2015**, *210*, 31–39. [CrossRef]
20. Sree, K.S.; Appenroth, K.-J. Starch accumulation in duckweeds (Lemnaceae) induced by nutrient deficiency. *Emir. J. Food Agric.* **2022**, *34*, 204–212.
21. Bafort, Q.; Wu, T.; Natran, A.; De Clerck, O.; Van de Peer, Y. The immediate effects of polyploidization of *Spirodela polyrhiza* change in a strain-specific way along environmental gradients. *Evol. Lett.* **2023**, *7*, 37–47. [CrossRef]
22. Huber, M.; Gablenz, S.; Hoefer, M. Transgenerational non-genetic inheritance has fitness costs and benefits under recurring stress in the clonal duckweed *Spirodela polyrhiza*. *Proc. Royal Soc. B* **2021**, *288*, 20211269. [CrossRef]
23. Sree, K.S.; Appenroth, K.-J. Worldwide genetic resources of duckweed: Stock collections. In *The Duckweed Genomes*; Cao, X.H., Fourounjian, P., Wang, W., Eds.; Compendium of Plant Genomes; Springer International Publishing: Cham, Switzerland, 2020; pp. 39–46. ISBN 978-3-030-11045-1.
24. Ziegler, P.; Adelmann, K.; Zimmer, S.; Schmidt, C.; Appenroth, K.J. Relative in vitro growth rates of duckweeds (Lemnaceae)—The most rapidly growing higher plants. *Plant Biol.* **2015**, *17*, 33–41. [CrossRef]
25. Paolacci, S.; Stejskal, V.; Toner, D.; Jansen, M.A.K. Integrated Multitrophic Aquaculture; Analysing Contributions of Different Biological Compartments to Nutrient Removal in a Duckweed-Based Water Remediation System. *Plants* **2022**, *11*, 3103. [CrossRef]
26. Stejskal, V.; Paolacci, S.; Toner, D.; Jansen, M.A.K. A novel multitrophic concept for the cultivation of fish and duck-weed: A technical note. *J. Clean. Prod.* **2022**, *366*, 132881. [CrossRef]
27. Petersen, F.; Demann, J.; Restemeyer, D.; Ulbrich, A.; Olfs, H.-W.; Westendarp, H.; Appenroth, K.-J. Influence of the Nitrate-N to Ammonium-N Ratio on Relative Growth Rate and Crude Protein Content in the Duckweeds *Lemna minor* and *Wolffiella hyalina*. *Plants* **2021**, *10*, 1741. [CrossRef] [PubMed]
28. Petersen, F.; Demann, J.; Restemeyer, D.; Olfs, H.-W.; Westendarp, H.; Appenroth, K.-J.; Ulbrich, A. Influence of Light Intensity and Spectrum on Duckweed Growth and Proteins in a Small-Scale, Re-Circulating Indoor Vertical Farm. *Plants* **2022**, *11*, 1010. [CrossRef] [PubMed]
29. Petersen, F.; Demann, J.; von Salzen, J.; Olfs, H.-W.; Westendarp, H.; Wolf, P.; Appenroth, K.-J.; Ulbrich, A. Re-circulating indoor vertical farm: Technicalities of an automated duckweed biomass production system and protein feed product quality evaluation. *J. Clean. Prod.* **2022**, *380*, 134894. [CrossRef]
30. Shoham, T. Current state of regulatory approval for duckweeds as human food. *Duckweed Forum* **2022**, *10*, 155–163. Available online: http://www.ruduckweed.org/ (accessed on 26 May 2023).
31. Mes, J.J.; Esser, D.; Oosterink, E.; van den Dool, R.T.M.; Engel, J.; de Jong, G.A.H.; Wehrens, R.; van der Meer, I.M. A controlled human intervention trial to study protein quality by amino acid uptake kinetics with the novel Lemna protein concentrate as case study. *Int. J. Food Sci. Nutr.* **2022**, *73*, 251–262. [CrossRef]
32. Mes, J.J.; Esser, D.; Somhorst, D.; Oosterink, E.; van der Haar, S.; Ummels, M.; Siebelink, E.; van der Meer, I.M. Daily intake of Lemna minor or spinach as vegetable does not show significant difference on health parameters and taste preference. *Plant Foods Hum. Nutr.* **2022**, *77*, 121–127. [CrossRef]
33. Wang, W.; Haberer, G.; Gundlach, H.; Glaesser, C.; Nussbaumer, T.; Luo, M.C.; Lomsadze, A.; Borodovsky, M.; Ker-stetter, R.A.; Shanklin, J.; et al. The *Spirodela polyrhiza* genome reveals insights into its neotenous reduction fast growth and aquatic lifestyle. *Nat. Commun.* **2014**, *5*, 3311. [CrossRef] [PubMed]

34. Michael, T.P.; Bryant, D.; Gutierrez, R.; Borisjuk, N.; Chu, P.; Zhang, H.; Xia, J.; Zhou, J.; Peng, H.; El Baidouri, M.; et al. Comprehensive definition of genome features in *Spirodela polyrhiza* by high-depth physical mapping and short-read DNA sequencing strategies. *Plant J.* **2017**, *89*, 617–635. [CrossRef] [PubMed]
35. Liu, Y.; Wang, Y.; Xu, S.; Tang, X.; Zhao, J.; Yu, C.; He, G.; Xu, H.; Wang, S.; Tang, Y.; et al. Efficient genetic transformation and CRISPR/Cas9-mediated genome editing in Lemna aequinoctialis. *Plant Biotechnol. J.* **2019**, *17*, 2143–2152. [CrossRef] [PubMed]

plants

Conference Report

Sustainable Stress Management: Aquatic Plants vs. Terrestrial Plants †

K. Sowjanya Sree [1,*], Klaus J. Appenroth [2,*] and Ralf Oelmüller [2,*]

1 Department of Environmental Science, Central University of Kerala, Periye 671320, India
2 Matthias Schleiden Institute—Plant Physiology, Friedrich Schiller University of Jena, 07743 Jena, Germany
* Correspondence: ksowsree@gmail.com or ksowsree9@cukerala.ac.in (K.S.S.); klaus.appenroth@uni-jena.de (K.J.A.); ralf.oelmueller@uni-jena.de (R.O.)

† Workshop of the Indo-German Science and Technology Centre at the University of Jena, Germany.

Abstract: The Indo-German Science and Technology Centre (IGSTC) funded an Indo-German Workshop on *Sustainable Stress Management: Aquatic plants* vs. *Terrestrial plants* (*IGW-SSMAT*) which was jointly organized at the Friedrich Schiller University of Jena, Germany from 25 to 27 July 2022 by Prof. Dr. Ralf Oelmüller, Friedrich Schiller University of Jena, Germany as the German coordinator and Dr. K. Sowjanya Sree, Central University of Kerala, India as the Indian Coordinator. The workshop constituted researchers working in this field from both India and Germany and brought together these experts in the field of sustainable stress management for scientific discussions, brainstorming and networking.

Keywords: abiotic stress management; biotic stress management; duckweed

check for updates

Citation: Sree, K.S.; Appenroth, K.J.; Oelmüller, R. Sustainable Stress Management: Aquatic Plants vs. Terrestrial Plants. *Plants* **2023**, *12*, 2208. https://doi.org/10.3390/plants12112208

Academic Editor: Karl H. Mühling

Received: 17 April 2023
Revised: 27 May 2023
Accepted: 30 May 2023
Published: 3 June 2023

1. Introduction

Climate change and the rapid growth of the human population are major global challenges. Evolving sustainable strategies to counter their effects will be the tasks for the future. We need to produce more food, which includes higher crop yields per area, with plants that are better adapted to the upcoming threats, while nature and climate need to be protected simultaneously (cf. United Nations Sustainable Development Goals). Besides crop production for food and feed, the growing human population and modern technology require plant biomass also for energy production, in the chemical and pharmaceutical industries, in housing and in the textile industry. Further to the promotion of our knowledge on resistance mechanisms in utilized crop species, higher crop yields and biomass production require novel approaches such as the use of new non-conventional crop species and the exploration of new agriculturally suitable areas. Finally, the scientific achievements obtained with conventional crop plants must be tested and eventually transferred to newly introduced crop species, if possible, or new strategies need to be developed for new crop species and new agricultural systems.

In this context, the Indo-German Workshop on *Sustainable Stress Management: Aquatic plants* vs. *Terrestrial plants* (*IGW-SSMAT*) was organized at the Friedrich Schiller University of Jena, Germany from 25 to 27 July 2022. This workshop was funded by the Indo-German Science and Technology Centre (IGSTC) with Prof. Dr. Ralf Oelmüller, Friedrich Schiller University of Jena, Germany, as the German coordinator and Chair of the workshop and Dr. K. Sowjanya Sree, Central University of Kerala, India, as the Indian Coordinator and Co-chair of the workshop. The scientific organization of the workshop was supported by Dr. Klaus J. Appenroth, Friedrich Schiller University of Jena, Germany. Interestingly, the workshop was held in a seminar room of the Department of Indo-German Studies, University of Jena, Germany, which was an apt location for the exchange of scientific ideas between the researchers of the two countries (Figure 1). The workshop included 25 oral presentations, 4 e-poster presentations, and a total of 38 invited and additional participants (Figure 2), which was an ideal group for one-to-one scientific discussions, group discussions, brainstorming sessions and for networking between the scientists from the two countries. The efforts of the IGSTC (Figure 3) in bringing together leading scientists in this field from both countries had synergistic effects on concept development, project coordination, grant applications, and contacts to industry.

(**A**)

(**B**)

Figure 1. Inauguration of the Indo-German Workshop—SSMAT. (**A**) K. Sowjanya Sree (Indian coordinator) felicitating Ralf Oelmüller (German coordinator). (**B**) K. Sowjanya Sree felicitating Klaus-J. Appenroth (Right) in the presence of Ralf Oelmüller (Center).

Figure 2. Group photo of the IGW-SSMAT participants.

Figure 3. Scientific discussion between the IGW-SSMAT team and IGSTC team. From left to right: Saquib Shaikh, Deputy Scientific Officer, IGSTC; R. Madhan, Director, IGSTC; K. Sowjanya Sree, Central University of Kerala, India; P. V. Lalitha, Senior Scientific Officer, IGSTC; Klaus-J. Appenroth, University of Jena, Germany.

The introduction of aquatic plants such as duckweeds into agricultural concepts for the future has opened novel research and application fields. Aquatic plants take up the nutrients from water for their growth and utilize space that otherwise is not used for agriculture. Therefore, comparative analyses of these crops with conventional land crops may contribute to a faster introduction of aquatic plants into agricultural applications. Besides knowledge transfer, novel strategies for large scale applications were also discussed. For instance, space limitation for the growth of land plants has promoted strategies which act on individuals (more biomass and better resistance), whereas strategies that allow the faster propagation of the plant population under different environmental conditions might be more important for aquatic plants.

Duckweeds (Lemnaceae) are on their way to becoming an important new crop plant. This is the result of more than a decade of intensive research, which revealed interesting properties of this family of aquatic plants [1]. Under optimal growth conditions, they contain high amounts of protein and an amino acid spectrum that fits very well to the nutritional requirements of humans, and also possess a high-quality fatty acid spectrum (although low in content), rich mineral composition, antioxidants and phytosterols [2–4]. Very recently, the European Union Commission permitted the sale of fresh duckweed plants (*Wolffia arrhiza* and *Wolffia globosa*) for human nutrition under the law of novel food [5]. Therefore, in the present, their use for human nutrition, and perhaps for the feeding of selected animals like pets, seems to be the most promising application of duckweeds. These qualitative advantages meet one important physiological property: duckweeds represent the fastest growing angiosperms [6,7]. Beside their possible use in nutrition, there is an alternative application: under certain stress conditions, their protein content decreases dramatically but their starch content increases by up to 50% of their dry weight or even higher [8,9]. Starch can be degraded to low-molecular-weight carbohydrates and thereafter fermented to bioalcohols, i.e., ethanol or even higher alcohols like butanol [10].

With the intention to cultivate duckweeds under optimal conditions for nutrition, i.e., with a high protein content, one has to deal with stress factors in the environment. To make large scale production of duckweeds possible, as required for a novel nutrition crop, farmers have to learn how to deal with these stressors. These include both abiotic factors, like high temperature, salinity, heavy metals, xenobiotics [11] and biotic factors [12] similar to terrestrial crop plants.

In the current era of global climate change, it is extremely important to foster strategies in line with the principles of sustainability in agriculture. One of the major fields is stress management from both abiotic and biotic stressors. The objective of this workshop was to bring together scientists from these fields and to discuss strategies of how terrestrial and aquatic plants can be integrated in a concept of agriculture for the future. In the following four sections—Abiotic stress management in terrestrial plants, beneficial interaction of fungi with plants, biotic stress management in terrestrial plants, and stress in duckweeds and other aquatic plants—we have reported the findings presented by different speakers at the workshop as indicated by their name and place in braces.

2. Abiotic Stress Management in Terrestrial Plants

Abiotic stresses reduce the crop yield significantly, partially because of the effects of global climate change. Rice (*Oryza sativa*) is an important staple food for over half the world's population, and also for India. It has been discovered that several stress factors cause the accumulation of a cytotoxic metabolite, methylglyoxal. The concentration of this metabolite was reduced by engineering the glyoxalase pathway, and stress tolerance was enhanced in rice [13]. Additionally, it was demonstrated that Na^+/H^+ antiporters help in the sequestration of salt and enhance salt tolerance (Sneh Lata Singla-Pareek, New Delhi, India). In fact, almost two-and-a-half decades back whilst elaborating the role of Na^+/H^+ antiporters in the salt tolerance of halophytes, Glenn et al. [14] were already reporting on the potential application of halophytes, non-conventional crops, in the remediation of saline-struck soils and wastewater. With the present-day global challenges, this application could contribute to the development of sustainable strategies. Under environmental conditions, often several stress factors affect the plants, either in a combined or sequential manner. Besides drought and high temperature, it is expected that flooding, connected with the rivers Ganga and Brahmaputra, will create stress on crop plants. The use of modern methods of plant breeding like gene modification tools and mutation breeding is expected to provide a solution to manage stress in rice (Ashwani Pareek, Mohali, India; [15]). Drought stress responses in resistant and sensitive rice were investigated using high-throughput multi-omics (transcriptome, proteome, and metabolome) technologies. An integrated analysis revealed the importance of carbohydrate metabolism through glycolysis, the pentose phosphate pathway, and L-phenylalanine biosynthesis in the resistance response. Moreover, the dominant role of epigenetic regulations became evident (Mukesh Jain, New Delhi, India; [16]). Deepwater rice is naturally resistant to flooding and even seed germination under submerged conditions is possible. During the investigation of the involved mechanisms, it was discovered that the phytoglobin-NO cycle plays an important role. As a practical measure it could be shown that the addition of nitrite stimulates anaerobic germination and reduces ROS production, indicating the alleviation of stress (Kapuganti Jagadis Gupta, New Delhi, India; [17]). The cryopreservation of genetic resources is an ex situ conservation technique, commonly used for vegetatively propagated plants. The meristematic tissues that are used in the cryopreservation process have to survive through several stressors such as cold temperature, osmosis and wounding. Studies on the effect of these stressors on the success of the cryopreservation process are very crucial for the long-term storage of plants genetic resources. In order to develop successful protocols for the cryopreservation of plants, identifying the signalling cascades involved in cryopreservation could be significant. Whereas good progress has been obtained with *Arabidopsis* using shoot tips, progress is urgently awaited for duckweeds (Lemnaceae), which rarely flower, and the production of seeds is not yet established as a routine method (Manuela Nagel, Gatersleben, Germany). Brassinosteroids play an important role in the regulation of plant development and in their interaction with the environment. The dynamics of the brassinosteroid response pathway at the plasma membrane and the initiation of cellular responses related to cell elongation were identified using recurrently combined computational modelling together with quantitative cell physiology. The model shows the hyperpolarization of the plasma membrane that is induced by brassinosteroids and the swelling of the cell wall subsequent to the apoplast acidification (Klaus Harter, Tübingen, Germany; [18]). "Chlorophyll is a Janus-faced molecule of life and death". Electrons from chlorophyll can be lethal to the cells, acting via the reactive oxygen species. Insect pests on plants have developed mechanisms to defend themselves from the phototoxicity of the ingested chlorophyll. It was found that two special proteins are involved in this process. Insects with knock-outs of these proteins could not survive in the presence of light after being fed on green leaves, unlike in the dark. This deepens the understanding of insect adaptation to herbivory (David G. Heckel, Jena, Germany; "Naked chlorophyll stresses insects as well as plants"). In intensive discussions, biological systems were proposed in which the relevance of these new findings could be tested for future agricultural applications.

3. Beneficial Interaction of Fungi with Plants

The association of endophytic microbes with crop plants is beneficial in building a sustainable agriculture system, owing to the ability of these microbes to quickly adapt to changing environmental conditions in comparison to their host plants. This association stimulates the physiological traits of the crop plants which results in significant improvement in the crop's stress tolerance. Endophyte-enrichment technology can be a booster to plant productivity (Karaba N Nataraja, Bengaluru, India; [19]). Microbial interactions of the endophytic fungus *Piriformospora indica* with plants may alter or destroy the structure of the plant cell wall, connecting cell wall integrity maintenance to immune responses. Plant cell wall breakdown generates short-chain cellooligomers which induce Ca^{2+}-dependent responses. These responses require the malectin domain-containing CELLOOLIGOMER-

RECEPTOR KINASE 1 (CORK1] in *Arabidopsis*. Two conserved phenylalanine residues in the malectin domain are crucial for CORK1 function (Ralf Oelmüller, Jena, Germany; [20]). *Piriformospora indica* colonizes the roots of a broad host range and, additional to a higher biomass production, the symbiosis also confers resistance to abiotic and biotic stress conditions like high salinity or infection by pathogenic fungi. Furthermore, *P. indica* provides the host not only with water and phosphate, but also with other important nutrients like nitrogen. By use of an N-15-isotope-labelled fungus, it was demonstrated that the fungus directly transfers nitrogen from the hyphae to the host plant (Sandra Scholz, Jena, Germany). *Cyanodermella asteris*, is a fungal endophyte isolated from *Aster tataricus*, a plant known in Chinese traditional medicine that produces astin as an active molecule. The interaction of non-natural hosts *A. thaliana*, Chinese cabbage, rapeseed, tomato, maize, or sunflower with *C. asteris* results in phenotypes like a decrease in the main root length, extensive lateral root growth, increased leaf and root biomass, and enhanced anthocyanin levels—properties of high agricultural interest [21]. It was found that auxin secreted by *C. asteris* was involved in the phenotype of the root whereas signalling pathways for salicylic acid, jasmonic acid or ethylene were not found to be involved in the plant-microbe interaction (Jutta Ludwig-Müller, Dresden, Germany). Association of plants with Arbuscular mycorrhizal fungi help the plants with sourcing nutrients and also with increasing their resistance and tolerance to varied stresses. However, the plant–mycorrhizal interactions are not always positive. The current research is directed towards the performance of arbuscular mycorrhiza under the changing and challenging conditions of agricultural and horticultural plant production systems (Philipp Franken, Erfurt and Jena, Germany; [22]). Adverse effects of salinity on crops are mainly osmotic toxicities and the uptake of ions in concentrations that are not beneficial for the plants. Soil bacteria can help plants to better adapt to saline soils with high osmotic stress, poor physical conditions, nutritional disorders, and toxicities. A newly isolated and characterized Rhizobium from the common bean nodule from Kenya showed high symbiotic efficiency to relieve its host plant from acute salt stress (Clabe Wekesa, Jena, Germany; [23]). Non-brassinoid steroids were investigated in several terrestrial plants as well as in species of all five genera of duckweeds. It was shown that the infection of *A. thaliana* with *Alternaria brassicicola* (a phytopathogenic Ascomycota) specifically increases the amounts of dehydroepiandrosterone (an early androgen) in infected leaves, while the amounts of other steroids did not change. This steroid is a strong inhibitor of fungal growth (Jan Klein, Jena, Germany). This workshop session provided multiple strategies and novel chemical compounds relevant for agricultural applications. We discussed a common field experiment. The harvested material should be analysed by members of the workshop using their specialized knowledge.

4. Biotic Stress Management in Terrestrial Plants

Autotrophs are constantly challenged by abiotic as well as biotic stressors. Insect pests of plants are one of the major biotic stressors. Plants have evolved several strategies to protect themselves from insect infestation. However, the signal transduction pathways induced early on that connect the process of wounding as a result of insect herbivory to the plant defence responses are yet to be well comprehended. Perceived by specific chemicals (herbivore-associated molecular patterns or damage-associated molecular patterns), plants respond with local and systemic signalling processes including $[Ca^{2+}]_{cyt}$ changes and phytohormones like jasmonates. Moreover, volatile-mediated plant–plant communication play an important role in the biotic stress response (Axel Mithoefer, Jena, Germany; [24,25]). Plant- and microbe-based biologicals can boost crop productivity as well as mitigate the effect of stressors on plants. Prime Verdant (produced by *BioPrime AgriSolutions*, India) improved the productivity of tomato crops both qualitatively and quantitatively. Prime-Verdant-applied plants performed well also under stress conditions (Rahul Jog, Pune, India). Furthermore, vegetable crop plants can become heavily infected by different plant viruses which results in damage to the crops. It is difficult to mitigate RNA and DNA viral diseases in vegetable crops. *Piriformospora indica*-colonized tomato plants developed increased resistance to *Tomato yellow leaf curl virus*, and the colonization of fruit plants like banana and papaya with *P. indica* effectively reduced the disease symptoms induced by RNA viruses like *Banana bract mosaic virus* and *Papaya ring spot virus*, respectively. Additionally, vegetable crops (yard long beans, tomato and lady's finger) in association with *P. indica* developed more resistance against RNA virus (*Cowpea aphid borne mosaic virus, Blackeye cowpea mosaic virus*), and DNA virus (*Tomato leaf curl virus* and *Bhindi yellow vein mosaic virus*) infections. Thus, *P. indica* colonization protects plants also against viral diseases (Joy Michal Johnson, Kollam, India). To understand the molecular basis of this *P. indica*-induced resistance, colonized and uncolonized crops after viral infections should be analysed with the molecular and biochemical tools available in the consortium.

5. Stress in Duckweeds and Other Aquatic Plants

Under suboptimal cultivation conditions, several duckweed species and clones accumulate starch, useful finally in biotechnology to produce bioalcohols via degradation and fermentation. Suboptimal cultivation conditions mean that the plants are under stress. Several stress factors are effective but in each case it is necessary to optimize their duration and strength for the different species and clones or ecotypes of duckweeds. The most obvious stress response in duckweeds is the inhibition of vegetative propagation (growth), i.e., by increasing salinity, a lack of nutrients or their exposure to heavy metals [26]. Evidently, in many cases, growth is more sensitive to stressors than photosynthesis is. As a consequence, not all photosynthetic products are required to support growth and the surplus is used to synthesize starch. The induction of starch-synthesizing enzymes like ADP-glucose pyrophosphorylase has been shown by transcriptomics and immunoblots. This inhibition of growth can be used also for the biomonitoring of toxic compounds in the aquatic environment as shown recently by *Lemna* root re-growth test (Klaus-J. Appenroth, Jena, Germany; [27]). The management of stressors can be executed by the integration of strategies like genetic engineering, e.g., of Na^+/H^+ antiporters in these unconventional crop plants that might increase their stress (salt stress)-tolerance capacity. The application of such management approaches is important, especially in the scenario where the ability of these plants to produce high amounts of biomass at a fast pace and the increasing practical applicability of this biomass in a circular economy are transforming duckweeds into a sustainable cropping system. The anatomical and molecular basis for the fast growth rate of these tiny aquatic plants was also revealed (K. Sowjanya Sree, Periye, India; [6,28,29]). With the intention to make duckweed biomass available on a large scale, an Indoor Vertical Farm was constructed, which produces duckweed biomass in a continuous manner and manages stress by nutrients and light (Finn Petersen, Osnabrück, Germany; [30]). Flowering in angiosperms is a robust process. Under stress conditions, when several of the vegetative parts of the plant show marks of stress on their morphology, flower development is intact. Each angiosperm species has a specific floral structure. The extensive diversity of flowers has evolved over "variation of the floral theme". The ABCDE and Floral Quartet models were developed for model plants. However, these standard model themes have been modified during the evolution of some land plants, such as orchids, and also in some aquatic plants (Günter Theißen, Jena, Germany; [31]). In angiosperms, the evolution of aquatic plants had occurred several times independently during the transition of these plants from a terrestrial to an aquatic habit. The authors of this presentation made use of the enormous amount of available genome data for water plants, like different duckweed species, *Nelumbo nucifera*, and so on in order to annotate the MADS-box genes. The transcription factors encoded by these genes are involved in diverse plant developmental processes. The molecular bases for the morphological adaptations of water plants to the aquatic habit have been identified by the first ever genome-wide analysis of MADS-box genes in these aquatic plants (Lydia Gramzow, Jena, Germany; [31]). In a stand, the land plant communities show strong gradients of light intensity and light quality. This happens because of their competitive light absorption strategy. Plants under the stress of such light gradients, e.g., the exposure to sub-optimal light intensity that effects the plant's photosynthetic efficiency, have evolved mechanisms to acclimatize and mitigate the diverse effects of these stressors. A light system that preferentially excites either photosystem I or photosystem II was established so as to study the acclimation responses of these stressed plants. This light system can induce short-term and long-term acclimation responses that include structural changes in the thylakoid membrane, changes in protein phosphorylation, photosystem II supercomplex formation, the structure of light-harvesting complex II and so on [32]. This understanding from the land plants, from the point of view of the knowledge and the experimentation abilities gathered has been systematically applied to carry out studies on aquatic plants like duckweeds. The effect of the aquatic habit and the two-dimensional canopy on the photosynthetic acclimation responses in duckweeds was investigated (Thomas Pfannschmidt, Hannover, Germany). Furthermore, the remarkable difference in the flowering behaviour of Lemnaceae and land plants offers various molecular approaches to investigate how investments into flower development or vegetative growth impact crop yield.

6. Conclusions

Knowledge flow between the researchers from both India and Germany on sustainable stress management in terrestrial and aquatic plants has helped to widen the understanding of the mechanism of stress management in general and also allowed inputs and learning into each field in order to work towards developing sustainable aquatic and terrestrial cropping systems that can deal with stress in a more efficient manner. The development of well-organized sustainable agriculture systems, which will enable us to create innovative technologies that can be integrated into circular economy, is the future. Our efforts to understand the stress-management strategies in both terrestrial plants and aquatic plants facilitated a holistic understanding of the requirements and strategies for building sustainable cropping systems.

Author Contributions: All authors have contributed to writing, reviewing, editing and proofreading of the manuscript. All authors have read and agreed to the published version of the manuscript.

Funding: The indo-German workshop (IGW-SSMAT) was funded by the Indo German Science and Technology Centre (IGSTC).

Data Availability Statement: All the required data are presented in the manuscript.

Conflicts of Interest: The authors declare no conflict of interest.

References

1. Acosta, K.; Appenroth, K.J.; Borisjuk, L.; Edelman, M.; Heinig, U.; Jansen, M.A.K.; Oyama, T.; Pasaribu, B.; Schubert, I.; Sorrels, S.; et al. Return of the *Lemnaceae*: Duckweed as a model plant system in the genomics and postgenomics era. *Plant Cell* **2021**, *33*, 3207–3234. [CrossRef]
2. Appenroth, K.J.; Sree, K.S.; Boehm, V.; Hammann, S.; Vetter, W.; Leiterer, M.; Jahreis, G. Nutritional value of duckweeds (*Lemnaceae*) as human food. *Food Chem.* **2017**, *217*, 266–273. [CrossRef]
3. Appenroth, K.J.; Sree, K.S.; Bog, M.; Ecker, J.; Seeliger, C.; Boehm, V.; Lorkowski, S.; Sommer, K.; Vetter, W.; Tolzin-Banasch, K.; et al. Nutritional value of the duckweed species of the genus *Wolffia* (*Lemnaceae*) as human food. *Front. Chem.* **2018**, *6*, 483. [CrossRef]
4. Xu, J.; Shen, Y.T.; Zheng, Y.; Smith, G.; Sun, X.Z.S.S.; Wang, D.H.; Zhao, Y.; Zhang, W.; Li, Y.H. Duckweed (*Lemnaceae*) for potentially nutritious human food: A review. *Food Rev. Int.* **2021**. [CrossRef]
5. Shoham, T. Current state of regulatory approval for duckweeds as human food. *Duckweed Forum* **2022**, *10*, 154–163.
6. Sree, K.S.; Sudakaran, S.; Appenroth, K.J. How fast can angiosperms grow? Species and clonal diversity of growth rates in the genus *Wolffia* (*Lemnaceae*). *Acta Physiol. Plant.* **2015**, *37*, 204. [CrossRef]
7. Ziegler, P.; Adelmann, K.; Zimmer, S.; Schmidt, C.; Appenroth, K.J. Relative in vitro growth rates of duckweeds (*Lemnaceae*)- the most rapidly growing higher plants. *Plant Biol.* **2015**, *17*, 33–41. [CrossRef] [PubMed]
8. Sree, K.S.; Appenroth, K.J. Starch accumulation in duckweeds (*Lemnaceae*) induced by nutrient deficiency. *Emir. J. Food Agric.* **2022**, *34*, 204–212.
9. Fang, Y.; Guo, L.; Wang, S.; Xiao, Y.; Ding, Y.; Jin, Y.; Tian, X.; Du, A.; Liao, Z.; He, K.; et al. Artificially cultivated duckweed: A high-efficiency starch producer. *bioRxiv* **2023**. [CrossRef]
10. Chen, G.; Zhao, K.; Li, W.; Yan, B.; Yu, Y.; Li, J.; Zhang, Y.; Xia, S.; Cheng, Z.; Lin, F.; et al. A review on bioenergy production from duckweed. *Biomass Bioenergy* **2022**, *161*, 106468. [CrossRef]

11. Ziegler, P.; Sree, K.S.; Appenroth, K.J. Duckweeds for water remediation and toxicity testing. *Toxicol. Environ. Chem.* **2016**, *98*, 1127–1154. [CrossRef]
12. Appenroth, K.J.; Ziegler, P.; Sree, K.S. Duckweed as a model organism for investigating plant-microbe interactions in an aquatic environment and its applications. *Endocytobiosis Cell Res.* **2016**, *27*, 94–106.
13. Gupta, B.K.; Sahoo, K.K.; Ghosh, A.; Tripathi, A.K.; Anwar, K.; Das, P.; Singh, A.K.; Pareek, A.; Sopory, S.K.; Singla-Pareek, S.L. Manipulation of glyoxalase pathway confers tolerance to multiple stresses in rice. *Plant Cell Environ.* **2018**, *41*, 1186–1200. [CrossRef] [PubMed]
14. Glenn, E.P.; Brown, J.J.; Blumwald, E. Salt tolerance and crop potential of halophytes. *Crit. Rev. Plant Sci.* **1999**, *18*, 227–255. [CrossRef]
15. Chen, T.X.; Shabala, S.; Niu, Y.A.; Chen, Z.H.; Shabala, L.; Meinke, H.; Venkataraman, G.; Pareek, A.; Xu, J.L.; Zhou, M.X. Molecular mechanisms of salinity tolerance in rice. *Crop J.* **2021**, *9*, 506–520. [CrossRef]
16. Shankar, R.; Dwivedi, A.K.; Singh, V.; Jain, M. Genome-wide discovery of genetic variations between rice cultivars with contrasting drought stress response and their potential functional relevance. *Physiol. Plant.* **2023**, *175*, e13879. [CrossRef] [PubMed]
17. Kumari, A.; Singh, P.; Kaladhar, V.C.; Manbir; Paul, D.; Pathak, P.K.; Gupta, K.J. Phytoglobin-NO cycle and AOX pathway play a role in anaerobic germination and growth of deepwater rice. *Plant Cell Environ.* **2022**, *45*, 178–190. [CrossRef]
18. Grosseholz, R.; Wanke, F.; Rohr, L.; Glockner, N.; Rausch, L.; Scholl, S.; Scacchi, E.; Spazierer, A.J.; Shabala, L.; Schumacher, K.; et al. Computational modeling and quantitative physiology reveal central parameters for brassinosteroid-regulated early cell physiological processes linked to elongation growth of the Arabidopsis root. *eLife* **2022**, *11*, 73031. [CrossRef]
19. Byregowda, R.; Prasad, S.R.; Oelmuller, R.; Nataraja, K.N.; Kumar, M.K.P. Is endophytic colonization of host plants a method of alleviating drought stress? Conceptualizing the hidden world of endophytes. *Int. J. Mol. Sci.* **2022**, *23*, 9194. [CrossRef]
20. Tseng, Y.H.; Scholz, S.S.; Fliegmann, J.; Krueger, T.; Gandhi, A.; Furch, A.C.U.; Kniemeyer, O.; Brakhage, A.A.; Oelmueller, R. CORK1, a LRR-Malectin Receptor Kinase, is required for cellooligomer-induced responses in *Arabidopsis thaliana*. *Cells* **2022**, *11*, 2960. [CrossRef]
21. Jahn, L.; Storm-Johannsen, L.; Seidler, D.; Noack, J.; Gao, W.; Schafhauser, T.; Wohlleben, W.; van Berkel, W.J.H.; Jacques, P.; Kar, T.; et al. The Endophytic Fungus *Cyanodermella asteris* Influences Growth of the Nonnatural Host Plant *Arabidopsis thaliana*. *Mol. Plant-Microbe Interact.* **2022**, *35*, 49–63. [CrossRef] [PubMed]
22. Bitterlich, M.; Mercy, L.; Arato, M.; Franken, P. Arbuscular mycorrhizal fungi as biostimulants for sustainable crop production. In *Biostimulants for Sustainable Crop Production*; Rouphael, Y., Ed.; Burleigh Dodds Science Publishing: Cambridge, UK, 2020; pp. 227–272.
23. Wekesa, C.; Asudi, G.O.; Okoth, P.; Reichelt, M.; Muoma, J.O.; Furch, A.C.U.; Oelmuller, R. Rhizobia contribute to salinity tolerance in Common Beans (*Phaseolus vulgaris* L.). *Cells* **2022**, *11*, 3628. [CrossRef]
24. Malabarba, J.; Meents, A.; Reichelt, M.; Scholz, S.; Peiter, E.; Rachowka, J.; Konopka-Postupolska, D.; Wilkins, K.A.; Davies, J.M.; Oelmueller, R.; et al. ANNEXIN1 mediates calcium-dependent systemic defense in *Arabidopsis* plants upon herbivory and wounding. *New Phytol.* **2021**, *231*, 243–254. [CrossRef]
25. Heyer, M.; Scholz, S.S.; Reichelt, M.; Kunert, G.; Oelmuller, R.; Mithofer, A. The Ca^{2+} sensor proteins CML37 and CML42 antagonistically regulate plant stress responses by altering phytohormone signals. *Plant Mol. Biol.* **2022**, *109*, 611–625. [CrossRef]
26. Appenroth, K.J.; Ziegler, P.; Sree, K.S. Accumulation of starch in duckweeds (*Lemnaceae*), potential energy plants. *Physiol. Mol. Biol. Plants* **2021**, *27*, 2621–2633. [CrossRef] [PubMed]
27. Park, J.; Yoo, E.J.; Shin, K.; Depuydt, S.; Li, W.; Appenroth, K.J.; Lillicrap, A.D.; Xie, L.; Lee, H.; Kim, G.; et al. Interlaboratory validation of toxicity testing using the duckweed *Lemna minor* root-regrowth test. *Biology* **2022**, *11*, 37. [CrossRef]
28. Sree, K.S.; Maheshwari, S.C.; Boka, K.; Khurana, J.P.; Keresztes, Á.; Appenroth, K.-J. The duckweed *Wolffia microscopica*: A unique aquatic monocot. *Flora* **2015**, *210*, 31–39. [CrossRef]
29. Michael, T.P.; Ernst, E.; Hartwick, N.; Chu, P.; Bryant, D.; Gilbert, S.; Ortleb, S.; Baggs, E.L.; Sree, K.S.; Appenroth, K.J.; et al. Genome and time-of-day transcriptome of *Wolffia australiana* link morphological minimization with gene loss and less growth control. *Genome Res.* **2021**, *31*, 225–238. [CrossRef]
30. Petersen, F.; Demann, J.; Restemeyer, D.; Olfs, H.-W.; Westendarp, H.; Appenroth, K.J.; Ulbrich, A. Influence of light intensity and spectrum on duckweed growth and proteins in a small-scale, re-circulating indoor vertical farm. *Plants* **2022**, *11*, 1010. [CrossRef]
31. Gramzow, L.; Theissen, G. Stranger than fiction: Loss of MADS-box genes during evolutionary miniaturization of the duckweed body plan. Loss of MADS-box genes in duckweeds. In *The Duckweed Genomes, Compendium of Plant Genomes*; Cao, X.H., Fourounjian, P., Wang, W., Eds.; Springer Nature: Cham, Switzerland, 2020.
32. Hommel, E.; Liebers, M.; Offermann, S.; Pfannschmidt, T. Effectiveness of light-quality and dark-white growth light shifts in short-term light acclimation of photosynthesis in *Arabidopsis*. *Front. Plant Sci.* **2022**, *12*, 615253. [CrossRef]

plants

MDPI

Review

Survival Strategies of Duckweeds, the World's Smallest Angiosperms

Paul Ziegler [1], Klaus J. Appenroth [2] and K. Sowjanya Sree [3,*]

[1] Department of Plant Physiology, University of Bayreuth, 95440 Bayreuth, Germany; paul.ziegler@uni-bayreuth.de
[2] Matthias Schleiden Institute—Plant Physiology, University of Jena, 07743 Jena, Germany; klaus.appenroth@uni-jena.de
[3] Department of Environmental Science, Central University of Kerala, Periye 671320, India
* Correspondence: ksowsree9@cukerala.ac.in

Abstract: Duckweeds (Lemnaceae) are small, simply constructed aquatic higher plants that grow on or just below the surface of quiet waters. They consist primarily of leaf-like assimilatory organs, or fronds, that reproduce mainly by vegetative replication. Despite their diminutive size and inornate habit, duckweeds have been able to colonize and maintain themselves in almost all of the world's climate zones. They are thereby subject to multiple adverse influences during the growing season, such as high temperatures, extremes of light intensity and pH, nutrient shortage, damage by microorganisms and herbivores, the presence of harmful substances in the water, and competition from other aquatic plants, and they must also be able to withstand winter cold and drought that can be lethal to the fronds. This review discusses the means by which duckweeds come to grips with these adverse influences to ensure their survival. Important duckweed attributes in this regard are a pronounced potential for rapid growth and frond replication, a juvenile developmental status facilitating adventitious organ formation, and clonal diversity. Duckweeds have specific features at their disposal for coping with particular environmental difficulties and can also cooperate with other organisms of their surroundings to improve their survival chances.

Keywords: abiotic stress; biotic stress; duckweed; Lemnaceae; turion

check for updates

Citation: Ziegler, P.; Appenroth, K.J.; Sree, K.S. Survival Strategies of Duckweeds, the World's Smallest Angiosperms. *Plants* **2023**, *12*, 2215. https://doi.org/10.3390/plants12112215

Academic Editor: Bernhard Huchzermeyer

Received: 18 April 2023
Revised: 26 May 2023
Accepted: 31 May 2023
Published: 3 June 2023

1. Introduction

Duckweeds are small, simply constructed aquatic higher plants or macrophytes that represent an extreme and highly successful adaptation to life on or just below the surface of quiet fresh water. Their integration into the realm of vascular aquatic plants [1] and their anatomical, morphological, physiological, ecological, and distributional features have long been described [2,3] and recently revisited [4]. Particular anatomical and physiological features enable them to grow and maintain themselves on ponds, ditches, slowly flowing streams, and other small bodies of water worldwide in all climate zones. The success of duckweeds in colonizing and persisting on quiet water surfaces is based on extensive reduction of the well-developed root and shoot systems that are characteristic of most higher plants for taking up nutrients and for exposing assimilation and reproductive organs to light and the airspace. Duckweed individuals consist primarily of leaf-like assimilatory organs, or fronds. The duckweed frond is a thallus-like structure of less than 1 to 15 mm in diameter or length and only a few cells in thickness that represents a fusion of leaves and stem and, thus, the extreme reduction of an entire vascular plant. The fronds consist largely of spongy mesophyll with large air spaces that make them buoyant, and they are either rootless or bear one to several simple hairless adventitious roots on the underside. Duckweeds are thought by some groups of researchers to represent a subfamily (Lemnoideae) of the Araceae (see [5] for a new publication), and this has also been suggested by the Angiosperm Phylogeny III decision. However, our research indicates that duckweeds better constitute a family (Lemnaceae) in its own right and that this is also in agreement with basic taxonomic rules [6]. Although the Lemnaceae have been considered until recently to consist of 37 species (e.g., [7]), the number of species has recently been revised to 36 [8]). These species are distributed among five genera (*Spirodela*: abbreviation *S.*, *Landoltia*: *La.*, *Lemna*: *Le.*, *Wolffiella*: *Wa.*, *Wolffia*: *Wo.*), which differ in the size and complexity of the fronds and in the number of roots they bear [2,4,7,9–13]. The differences reflect an evolutionary progression from *Spirodela* to *Wolffia* in terms of morphological reduction and genome augmentation [4,14].

Despite their small size, simple structure, and inconspicuous appearance, duckweeds are widespread on Earth, inhabiting all climate zones except the very cold polar regions and extremely dry deserts. Some species are quite cosmopolitan, such as *S. polyrhiza* and *Le. aequinoctialis*, which inhabit all continents, whereas others are confined to certain continents (e.g., *Wo. brasiliensis* in both North and South America), to a single continent (e.g., *Le. perpusilla* in North America), or to much more restricted regions (e.g., *Le. tenera* in southeast Asia, *Wa. denticulata* in southeast Africa, and *Wo. microscopica* in the northern part of the Indian subcontinent) [2,7,12]. Multiple duckweed species can inhabit particular regions: six different species have been identified in each of China, the Ukraine and Israel [15,16]. Whatever the regions inhabited by the various species, the ability of duckweeds to successfully colonize compatible water bodies and to persist in these habitats is due in large part to pronounced growth potential, juvenile organization, and clonal vegetative propagation. These attributes, together with small size and floating habit, provide duckweeds with a unique means to productively respond to environmental challenges.

1.1. Growth and Vegetative Propagation

The restriction of duckweeds to small floating assimilatory organs facilitates rapid growth. Duckweed fronds consist mainly of photosynthetic tissue, and the channelling of produced photosynthate into the production of new, simply constructed photosynthetic tissue constitutes streamlined utilization resulting in rapid augmentation of frond biomass. Indeed, duckweeds have been shown to be the most rapidly growing higher plants in laboratory experiments [17,18] and produce large amounts of biomass under natural conditions and in agricultural/industrial contexts that can be utilized for, e.g., bio-energy production [3,4,9–11,19]. This strong growth potential is coupled with vegetative propagation to result in rapid frond production. Although duckweeds can, in principle, flower, and some indeed do so regularly, the main means of propagation of all duckweed species is the budding of daughter fronds from one or two pouches in the mother fronds while remaining attached for a time via stipes to form colonies of 2 to 50 connected fronds [2,4,10,11,20]. The growth of duckweeds is, therefore, often quoted as an increase in frond number, as well as an increase in frond weight. Rapid growth of duckweeds thus manifests itself in the production of numerous colonies of interconnected fronds that spread out over the water surface. Frond colonies tend to distribute themselves equidistantly over free water surfaces, probably by exuding surface-active repellent substances into the surrounding water, thus ensuring optimal access to water nutrient substances [9]. The potential for rapid growth and the vegetative reproduction of fronds and frond collectives (colonies) enables duckweeds to successfully colonize stretches of quiet open water without having to resort to time- and internal resource-consuming sexual propagation.

1.2. Juvenile Organization and Adventitious Development

The vegetative propagation of duckweeds can be understood in the context of restriction of development to a juvenile stage and adventive organ formation. The lack of differentiation of the duckweed assimilatory axis or "shoot", i.e., the frond, into the distinct classical shoot and leaf systems is reminiscent of embryonal or seedling organization, and the organs developing from this juvenile shoot (quite generally daughter shoots or fronds, but also roots, flowers, and bracts, when present) can be regarded as irregularly formed or adventive organs [9]. Along with the small size of the duckweed frond, the simplified juvenile structure to be reproduced is an important factor in enabling the rapid propagation of duckweed fronds. In addition, the adventitious propagation of juvenile assimilatory shoots can be readily modified upon the impact of appropriate signalling to enable the development of frond derivatives that can help a duckweed cope with highly unfavourable climatic conditions. This is evident in the formation of resting fronds for overwintering and of flowers for the production of seeds, to be discussed in the following.

1.3. Clonal Diversity

The vegetative propagation of duckweed fronds gives rise to clones of the mother fronds, i.e., all the progenies of a particular frond have the same genetic makeup as the mother frond. However, clonal diversity is a characteristic of duckweeds, which becomes evident when certain attributes are compared among isolates of a single species collected from different geographical regions. Not only species themselves but also different clones of individual species can show considerable variation in growth potential [17,18], salt tolerance [21], the ability to accumulate starch under nutrient deficiency [22], and the ability to grow on agricultural wastewater [23]. Even the genotypes of 22 of 23 investigated clones of *S. polyrhiza* could be differentiated by several orthogonal genotyping methods [24]. Clonal differences in the specific turion yields of *S. polyrhiza* have been found to be stable after years of in vitro cultivation and are assumed to be genetically determined [25]. However, intraspecific genetic diversity in *S. polyrhiza* is extremely low, in association with a low mutation rate [26,27]. Clonal diversity thus represents a largely asexual adaptation to different surroundings and may be an example of epigenetic acclimation as an alternative to adaptation through natural selection [28]. Stress-induced DNA methylation can be an important factor in the epigenetic background of clonal diversity [29], which may be enhanced by spontaneous polyploidization that can create a fitness increase for some already existent strains in some stressful environments [30].

2. Duckweed Survival

As small, free-floating aquatic plants, duckweeds can easily be displaced or removed from their habitat by the action of moving water and wind, foraging by water animals, and gathering by man. They are also susceptible to incapacitation or destruction of their habitat by the impact of unfavourable environmental conditions such as excessive cold, water pollution, or competition for the water surface (see [2]). A quite fundamental factor in ensuring duckweed survival in general is thus the ability to establish themselves in new surroundings. This requires the ability to reach these new surroundings and then proliferate in them.

Although duckweeds have the potential to grow and propagate themselves rapidly, they can only do so under propitious, non-limiting conditions. These include favourable temperatures, adequate lighting conditions, a sufficient supply of mineral salts, and a lack of serious competition for the water surface. When these requirements are met in nature—as they are to at least some extent during the growing seasons of the various climate zones—duckweeds can successfully colonize their surroundings. However, numerous factors can encroach upon these favourable constellations to impede or even stop growth and propagation or to damage or even kill the fronds. Insufficient mineral salt nutrition and low temperatures can severely curtail growth and metabolism; frost and desiccation can be lethal to the fronds. The surface of the water body can be overgrown by other macrophytes and by duckweeds themselves; the fronds can be subject to microbial attack and exposed to toxic water-contaminating substances.

In the context of their adaptation to life on the water surface in many different climate zones, duckweeds have evolved structural and physiological features and developmental patterns that go beyond the mere potential for rapid growth and serve to cope with the manifold influences that can compromise growth and propagation. These attributes are discussed in Section 4 in terms of how duckweeds can maintain their distribution status on the water surface during the growing season in the face of adverse influences. These include coping with the prevailing temperature and light regimes, ensuring sufficient nutrition, resisting microbial attack and cooperating productively with aquatic microorganisms, and coping with overcrowding and water pollution. On the other hand, the means by which duckweeds can withstand conditions that effectively preclude any growth at all and can be lethal to the organisms are discussed in Sections 5 and 6. The normal growing season fronds of duckweeds faced with critical conditions best exemplified by winter cold can produce quiescent "resting" fronds that can tolerate and "wait out" the unfavourable conditions and resume "normal" vegetative growth and propagation when conditions improve. Duckweeds can withstand drought—along with other unfavourable conditions—by flowering and forming resilient seeds that can later germinate to form a new, sexually recombinant generation of fronds under appropriate conditions.

3. Colonization of New Habitats

The colonization of new habitats by duckweeds depends on the ability to disperse from already-occupied habitats, proliferate in newly reached habitats, and compete with already-established species there [31].

Duckweed fronds growing at a particular location on a water body can be transported to another part of the water body or to another water body by water currents, flooding, wave action, and being blown across the water by wind. However, the main means of duckweed relocation is via adherence to animals that live in or near water, such as muskrats, and especially birds [2,32]. This dispersal is facilitated by the small size of duckweed fronds, but especially transport out of the water may be limited by the inability of the fronds to survive severe desiccation [2,31–33]. Nevertheless, *Le. minuta* fronds were found to retain moisture and viability for a prolonged period between duck feathers, supporting the idea of epizoochorous transport by birds [34]. Transport by birds can also occur by endozoochory, as fronds of *Wo. columbiana*, *Le. minor*, and *Le. gibba* were found to survive passage through the guts of waterfowl [35,36]. In time, repeated short-range relocation events can result in far-reaching dispersal of the fronds [2,37], and long-distance dispersal by birds may also occur [7,38]. Duckweed seeds can, in principle, be transported in the same ways as fronds, but the tendency of the seeds to sink to the bottom of the water body and the predominantly vegetative propagation of duckweed fronds indicate only a minor role for seed relocation and dispersal in the colonization of new habitats.

When duckweed fronds have arrived at a new location, they must be able to propagate rapidly to successfully establish themselves in the new habitat. This can be achieved based on the pronounced growth potential and clonal vegetative reproduction characteristic of duckweeds under conditions of sufficient mineral salt nutrition and light, favourable temperature, sufficient water space on the surface, lack of toxic water substances, and lack of serious competition. Specific growth potential may determine success in colonizing new water bodies when conditions are otherwise comparable: the higher growth rate of *Le. minor* in comparison with *Le. trisulca* was regarded to be decisive in the far greater frequency of the former in regions where both species were distributed [31]. However, the degree to which superior growth potential can be realized upon interaction with the environment has been only anecdotally investigated. Nutrient availability is a primary factor in enabling a duckweed to establish itself on the surface of a water body, with especially nitrogen driving the initial phases of clonal expansion of *Le. minor* [39,40]. Light availability strongly interacts with nutrient availability in determining *Le. minor* dominance of the water surface [41]. Duckweeds must be able to survive in regions characterized by seasons of particularly harsh conditions. An example is the ability of frost-sensitive *S. polyrhiza* fronds to survive freezing winter temperatures by developing frond derivatives (turions) that can withstand the cold season in contrast to the closely related *S. intermedia* with equally frost-sensitive fronds that do not develop turions [38]. Success in colonization is also dependent on the absence of potentially lethal seasonal developments such as summer increases in water pH to values above 8 [31].

The consolidation of the colonization of a new habitat by a duckweed, i.e., persistence on the newly occupied water body, depends on the ability of the duckweed to ensure the growth and propagation required to maintain the duckweed population during the growing season. This includes coping with high temperatures, low and high light intensities and nutritional deficiency, competing successfully with other aquatic plants, resisting attack by microorganisms, productive cooperation with aquatic microbiota, and withstanding the presence of harmful substances in the water (see Section 4). Duckweeds must also be able to cope with low temperatures and drought that prevent growth and may be life-threatening if they are features of the inhabited region. This includes the production of resting fronds to withstand the cold season (see Section 5) and flowering and the production of seeds to avoid severe drought (Section 6). Examples of successful colonization by duckweeds are provided by alien invasive species such as *Le. minuta*, which has spread across much of western and central Europe in the past 6–7 decades [42], or *Le. aequinoctialis*, which has recently migrated into the Ukraine [43]. Substantial genetic diversity exhibited by *Le. minuta* having colonized Ireland is thought to reflect repeated invasions across an extensive open-water barrier from continental Europe [44]. Studies of *Le. minuta* have illustrated some ways in which this alien species asserts itself against the resident duckweed *Le. minor* upon its arrival at new locations. *Le. minuta* was found to make better use of high light intensities than *L. minor* [45] and to be more tolerant of drought and the presence of metals in the water [46]. The latter study indicated, however, that the relative performances of an alien and a native species depend on multi-faceted differences between the species and on the nature of the stressors that are involved.

4. Coping with the Growing Season

If a duckweed has established itself in a given environment, its growth and propagation must be compatible with the prevailing temperatures, light regime, nutrient supply, and pH value of the water. In addition, the duckweed must be able to fend off a microbial attack or at least contain it to an acceptable extent, tolerate the presence of harmful substances in the water, and assert itself in the face of intra- and interspecific competition for light, nutrients, and space. The ability of the various species of duckweed to adapt to widely differing regimes of temperature, light, nutrient availability, and medium pH is basically rooted in the specific attributes of the species in question that have developed in the course of the evolution of that particular species. The extent of the ability of a particular species to tolerate changes in its environmental parameters is also basically delineated by the characteristic attributes of the species. The adaptive and tolerative potential of a duckweed species is further diversified by the clonal diversity that the species can exhibit. The ability of a species to resist a microbial attack can similarly be enhanced by clonal diversity. The ability of duckweeds to adapt to an environment and tolerate its potentially harmful influences can include active metabolic reactions. This can take the form of measures to combat the ill effects of excessive light, the presence of harmful substances in the water, and competition. Duckweeds also actively contribute to the interaction with aquatic microorganisms that can be beneficial to both organisms. The sum of the species-specific adaptive attributes diversified by clonal variation and inducible reactions to cope with harmful influences and to promote mutualism are vital for duckweed survival.

4.1. Temperature, Light, and pH

4.1.1. Temperature

Duckweed growth results from many interacting temperature-dependent chemical processes, including nutrient uptake, nutrient assimilation and transport, photosynthesis, and respiration, as well as many other processes incorporating enzymatic activities [3]. The survival of a duckweed under a particular temperature regime depends on the genetically determined intrinsic ability of the organism to grow well and propagate at the temperatures in question that have evolved with the formation of the species and its clonal derivatives. The optimum temperatures for the growth of numerous duckweed species and clones were found to vary between 20 °C and 30 °C; minimum temperatures that just enable a very slow permanent growth rate were found to range between <8 °C and 16–20 °C, and long-term maximum temperatures at which growth could still proceed slowly ranged between <30 °C and >34 °C [3,47,48]. The success of a species in a particular climate can depend on adaptation to either higher or lower average temperatures: duckweeds having a high optimum temperature (e.g., *S. polyrhiza*, *Le. aequinoctialis*, *Wo. globosa*) are better suited to warm climates, whereas those showing a lower optimum temperature (e.g., *La. punctata*, *Le. trisulca*, *Le. minuscula*, now *Le. minuta*) fare better in cooler and oceanic climates. Duckweeds exhibiting low minimum temperatures for growth (e.g., *Le. minor*, *Le. gibba*, *Le. trisulca*) have a better chance of survival in cooler climates, and those having a high maximum temperature for growth (e.g., *S. polyrhiza*, *Le. aequinoctialis*) will do well in more tropical surroundings [3]. The ability to tolerate very high temperatures over a relatively long period (e.g., 24 h at 50 °C and one week at 45 °C for *S. polyrhiza* [32]) is particularly advantageous for success in hot climates. Duckweeds must be able to react constructively to the heat stress that results when temperatures become dangerously high. Transcriptome analysis of the reaction of *S. polyrhiza* upon exposure to 45 °C demonstrated alterations in the expression of numerous genes, as well as increased superoxide dismutase activity parallel to malondialdehyde accumulation at the physiological level [49]. Exposure of *Le. minor* to 30 °C, which is a high temperature for this organism, resulted in DNA methylation that persisted over numerous frond generations and represents a long-term, transgenerational stress memory not observed in sexually reproducing plant species [29].

Very low temperatures that can threaten duckweed survival generally do not occur during the growing season. Some duckweeds can cope with the advent of such temperatures in the autumn by the formation of resting fronds, as is discussed in Section 5. The distribution of, e.g., *S. polyrhiza* in almost all climate zones [2,12] is a function of the high maximum temperatures for growth and the formation of turions upon the advent of cold winters exhibited by this species.

4.1.2. Light

Duckweed growth and propagation are driven by the photosynthetic utilization of light, which is dependent on the temperature and nutrient and CO_2 supply [47]. A duckweed requires sufficient light for suitable growth, whereby the highest rates of photosynthesis and growth possible for a particular species or clone take place at high light intensities. This is advantageous for the growth and propagation on open, unshaded water often observed for duckweeds. However, the maximum growth rates that can be achieved and the light intensities at which they are reached are strongly dependent on the temperature [47], and they vary considerably, depending more on the clone than on the species [48]. Duckweed success in growth and proliferation on a particular water body is thus not species-specific as much as requiring the presence of a clone well suited to the light intensity and temperature regimes at hand.

Very high light intensities can inhibit duckweed growth and damage the organisms, especially in terms of photoinhibition and oxidative damage [50]. The ability of duckweeds to grow rapidly at high light intensities depends on protective physiological features, such as the ability to convert much of the xanthophyll cycle pool to zeaxanthin and to dissipate much of the absorbed light non-photochemically, as shown by *Le. gibba* [51]. Growth can also be problematic at low light intensities, as in shading, in which case the light intensity required for light saturation and the compensation point of photosynthesis are important for growth. In a comparison of growth rates at different light intensities, clones of *Le. aequinoctialis*, *Le. valdiviana*, and *Le. minuscula* (now *Le. minuta*) showed the lowest optimum light intensity [47]. These clones would be expected to be the most shade tolerant, and the fact that they were collected from shady places illustrates that certain duckweeds can assert themselves well under limited light conditions. *Spirodela polyrhiza* responds to shading by increasing its frond surface area to optimize light capture, while *Le. minor* increases its chlorophyll content [52], and *Le. gibba* and *Le. minor* tolerate deep shade on the basis of large light-harvesting complexes and high photochemical efficiency [51]. The ability to grow better in shady conditions has the advantages of less exposure to high temperatures, better access to organic nutrient material (see following chapter), and usually quieter water conditions [2]. The advantage of a low compensation point for especially duckweed species that live below the water surface is illustrated by the occurrence of *Le. trisulca* at a depth of 12–14 m [53]. *Le. gibba* and *Le. minor* are exceptional in that their pronounced growth potential combined with pigment and photochemical characteristics of both shade and sun plants enables them to thrive under a wide range of high light intensities and ensures their success in dynamic light environments [51].

Duckweeds possess a differentiated cuticle to interface both the atmosphere on the adaxial side of the fronds and the water surface on the abaxial side. The biochemical composition of the cuticular waxes of *S. polyrhiza* is unique and may be of particular importance for the protection of the duckweed fronds under high light intensities, as it consists of up to 60% phytosterols that are important in the absorption of UV radiation and the scavenging of UV-generated radicals [54].

4.1.3. pH Value

Many duckweeds are able to grow well at pH values of between 5 and 8 [3], although duckweeds have been found in natural waters with pH values between 3.5 and 10.4 [2]. Species found in nature at pH < 5 include *Le. minor*, *Le. aequinoctialis*, and *Wo. globosa*, and those observed at pH > 9 include *S. polyrhiza*, *Le. minuscula* (now *Le. minuta*), and *Wo. brasiliensis* [2]. Three species (*La. punctata*, *Le. minor*, and *Wo. arrhiza*) have been shown to tolerate pH values of up to 10 in laboratory experiments [55]. The lower pH value limits for the growth of essentially all duckweed species range between 3 and 4. A few species, such as *La. punctata*, *Le. turionifera*, and *Le. perpusilla*, can grow at pH 3.2–3.5, whereas others, including *S. polyrhiza*, *Le. trisulca*, and *Wa. hyalina*, cannot tolerate pH values of less than 4 [2]. The success of duckweeds in growing and proliferating on waters with especially extreme pH values can accordingly be dependent on the ability to tolerate these values. As an example, pH values above 8 have been reported to preclude both *Le. minor* and *Le. trisulca* growth [56] and thus cause local and temporal extinctions in the populations of these two species that are otherwise widely distributed in southern Ontario lake waters [31].

High temperatures, light intensities, and pH values can all disrupt duckweed growth and propagation and can, as such, be seen as stress factors that can induce flowering to ensure survival by the setting of viable seeds (see Section 6).

4.2. Ensuring Sufficient Nutrition

4.2.1. Ensuring Mineral Salt Uptake and Storage

As facultative photoautotrophic organisms, duckweeds must have access to sufficient mineral ions, especially those of nitrogen, phosphorus, and sulphur. The mineral requirements of duckweeds have been summarized by Landolt and Kandeler [3]. Since phosphate (Pi) was the limiting mineral factor for floating aquatic plants under natural conditions in the pre-anthropogenic era, duckweeds, along with other macrophytes, have evolved to be particularly proficient in assimilating and storing this ion [9]. The priority of Pi uptake for duckweeds is illustrated by the uncoupling of this uptake from growth, i.e., the maintenance of Pi uptake by *Le. minor*/*japonica* at temperatures too low for growth [57].

Pi, which is the form of phosphorus usually taken up and assimilated by duckweeds [3], is made available to the plants by the action of phosphatases, which release Pi from organic material. Pi deficiency in the medium has long been known to inhibit the growth of and have other far-reaching effects on *S. oligorrhiza* (now *La. punctata*) [58], including strong enhancement of phosphatase activity [59]. Phosphatases and ribonucleases induced by Pi deficiency were observed in membrane-bound form at the water–plant interface and as exuded soluble enzymes [60,61]. The major phosphatase induced by low Pi supply in *S. oligorrhiza* (now *La. punctata*: [62]) was shown to be a glycosylphosphatidylinositol-anchored membrane protein [63] that was purified [64] and characterized as a purple acid phosphatase (PAP: [65]). The activity of this alkaline phosphatase may complement the induction of a high-affinity Pi transporter in the plasma membrane of *La. punctata* [66] in effecting the highly enhanced Pi uptake activity shown by this species under phosphate deficiency. The synthesis of PAPs and high-affinity Pi transporters are features of the PSR for optimizing external Pi acquisition.

La. punctata can store assimilated Pi in the vacuole as a reserve for growth upon the onset of Pi deficiency in the medium [67]. Linear oligophosphates and cyclic metaphosphates can function as short-term Pi reserves in *Le. minor* [68] and phytin as a long-term reserve in *Le. gibba* [69].

Plants can acclimatize to extended periods of Pi deprivation by eliciting a complex array of morphological, physiological, and biochemical/metabolic adaptations collectively known as the Pi-starvation response (PSR). The PSR arises in part from the coordinated induction of Pi-starvation-inducible genes encoding enzymes that reprioritize internal Pi use and maximize external Pi acquisition [70]; it may be stimulated in *S. polyrhiza* by SPX genes that are expressed in response to Pi (and nitrate) deficiency stress [71]. Interestingly, in this regard, starch accumulation—which is an expression of limited interior Pi availability—is strongly induced in duckweeds by deficiency of the mineral nutrient elements nitrogen and phosphorus [22], as well as sulphur [72], in the medium. Starch accumulation due to mineral nutrient deficiency is important in the formation of resting fronds and especially turions for overwintering (see Section 5). The accumulation, which may represent a depot of carbohydrate skeletons for use when mineral salts become more available again, reflects a reprioritization of available interior Pi. The accumulation of starch by Pi-deficient plant cells may largely arise from the release of ADP-glucose pyrophosphorylase, the gateway enzyme of starch synthesis, from allosteric inhibition by Pi, owing to the large reductions in cytoplasmic Pi pools that accompany long-term Pi deprivation [70]. Indeed, high starch accumulation in *La. punctata* has been shown to be a function of high-efficiency Pi recycling [73]. In addition, Pi and nitrogen deficiency were shown to increase the expression of starch-synthesizing enzymes [74] in addition to Pi transporters and phosphatases [73].

4.2.2. Diet Supplementation with Organic Material

Although duckweeds generally grow photoautotrophically, using light and mineral salts for photosynthesis, they can also grow mixotrophically in light with sugars and even heterotrophically in the dark if sufficient sugars, amino acids, and vitamins are available in the medium [2,3,47,75]. The ability to transition between different trophic conditions was shown to endow *S. polyrhiza* with great metabolic flexibility [76]. Duckweed mixotrophy and heterotrophy are of commercial interest in the context of the production of starch-rich biomass [72], and especially mixotrophy is thought to be much more widespread in nature than previously thought [77]. Lake waters have been shown to contain sugars and other organic substances, and especially duckweeds living in shaded habitats such as *Le. trisulca* can supplement their photoautotrophic nutrition by the uptake of such substances [2]. Large amounts of organic substances can emanate from aging and dying water organisms, including duckweeds themselves when these form thick mats covering the water surface [9].

Mixotrophic nutrition requires the possession of the necessary systems for the uptake of organic substances, and the ability to compete effectively with ubiquitous aquatic microorganisms in assimilating organic substances from the medium. *Le. gibba* was shown to possess a constitutive active hexose uptake system [78], *Le. aequinoctialis* fronds have been shown to dispose of multiple carriers for taking up a large variety of small organic molecules against concentration gradients [79,80], and high-affinity transport systems for neutral/acidic and basic amino acids were described for *S. polyrhiza* [81]. Organic substances in the vicinity of the duckweeds are conserved by the release of phenolic substances. As shown for *La. punctata*, a number of flavonoid substances leach out into the medium from aging and dying fronds that exhibit antibacterial activity. Intact fronds also exude phenolic substances, as shown for *S. polyrhiza* and indicated for some other species [9].

4.3. Protection against Microbial and Insect Damage

Duckweeds have adapted to be able to thrive in aqueous environments rich in organic materials, as illustrated by their ability to grow on organic wastewaters (e.g., [23]) and their value in the remediation of such waters [10,11]. These environments can accordingly have a high microbial load, and since plants have bacterial virulence factors in common with animals, duckweeds are susceptible to microbial attack. This has been illustrated in the development of *Le. minor* as a model plant system for studying human microbial pathogenesis, with which *Staphylococcus aureus, Pseudomonas aeruginosa,* and several other bacteria known to be pathogenic to humans were shown to be severely detrimental to duckweed growth and viability [82]. However, the ability of duckweeds to tolerate highly microbial surroundings indicates that they may have particularly effective disease resistance function [4].

It is not clear how duckweed plants persist in a wide range of environments in the light of their susceptibility to bacterial phytopathogens in an experimental context. However, genetic analysis has shown that duckweed defence responses against pathogens differ from those of most plants [4]. *S. polyrhiza* and especially *Wo. australiana* contain significantly fewer of the nucleotide-binding leucine-rich repeat domain genes (NLRs) that encode many disease-resistant proteins than do other plant model organisms, which indicates that they do not require a large variety of NLRs for pathogen immunity and survival. Nevertheless, NLR genes are more important for the pathogen response of *S. polyrhiza* than of *Wo. australiana*, in which pattern-recognition receptors (PRRs) may play a more dominant role. Genes encoding the antimicrobial proteins (AMPs), lipid transfer proteins (LTP), defensins, and snakins were indicated to be vital for the pathogen resistance of the duckweeds. These findings were complemented by the determination that duckweeds lack the enhanced disease susceptibility gene ESD1 responsible for inducing anti-pathogen defence in most plants and that they feature the upregulation of AMPs absent from the model plant *Arabidopsis thaliana* upon pathogen attack [83].

Aquatic plants can be exposed to saprophytic and parasitic bacteria and fungi in the water that exude enzymes capable of degrading certain components of the plant cell walls. Duckweeds are protected against such microbial attack to an extent in that the composition of their cell wall substances differs considerably from that of most plants and is characterized by high contents of apiose and xylose [9]. Duckweeds (*Le. minor* and *Wo. arrhiza*) have long been known to be rich in apiose [84], which was found to be a component of the cell wall in *Le. gibba* and *Le. minor* [85]. The cell wall polysaccharide apiogalacturonan has been detected only in duckweeds and seagrasses [86,87]. In *Le. minor*, it has been found to contain about 25% apiose with some xylose [88,89], and the apio- and xylogalacturonans of the cell walls of *S. polyrhiza, Le. gibba,* and *Wo. australiana* constitute 48–57% of the cell wall mass of these species [90]. A substantial fraction of the apiogalacturonan fraction of *Le. minor* cell walls with a high apiose content was resistant to pectinase degradation, illustrating how apiose may protect pectic substances from the cell wall polysaccharide-degrading action of infecting pathogens [91]. Xylose and possibly also arabinose may have a function in the cell walls of duckweeds complementary to that of apiose. This is shown by the finding that the apiogalacturonan content in *Wolffiella* and *Wolffia* cell walls is far lower than that in *Spirodela, Landoltia,* and *Lemna* cell walls, but xyloglacturonan is far more abundant than apiogalacturonan in *Wolffia* cell walls, and *Wolffiella* cell walls have a high arabinose content [92].

Chemical defence strategies may also be involved in the response of duckweeds to pathogens. Cell extracts of *Le. minor* have been shown to have antibacterial and antifungal properties in that they inhibited the growth of strains of several bacterial and fungal species isolated from human patients, foods, or fish that can be pathogenic to humans or animals [93–95]. However, it is not clear if the extracted compounds that were detrimental to the microbes in biotests are actually involved in the resistance of intact duckweeds to pathogens. Flavonoids are well known to contribute to pathogen resistance in plants [96], and duckweeds contain large numbers of these compounds [97]. The effect of flavonoids on the duckweed weevil provides evidence that these substances can indeed be important in protecting duckweeds from biotic attack. *Le. minor* contains appreciable amounts of the flavones isoorientin, vitexin, and isovitexin that significantly decrease the survival rate of the larvae of the herbivore insect *Tanysphyrus lemnae* that feeds on the duckweed [98].

4.4. Cooperation with Microorganisms

Aquatic microorganisms do not only pose a threat to aquatic plants: they also engage in mutually advantageous cooperation with the macrophytes. Plants quite generally host structured communities of microorganisms, or microbiomes, that confer fitness advantages, including growth enhancement, nutrient uptake, stress tolerance, and pathogen resistance to the host [99]. Duckweeds have long been known to bear epiphytic bacteria on their fronds and roots [2], and more recent studies have revealed that their microbiome can stimulate growth, improve the removal of nutrients, heavy metals and xenobiotics from waters, and inhibit gas release from aquatic communities [91]. This has stimulated great interest in the duckweed microbiome in terms of optimizing duckweed biomass yields for the production of biofuel and improving duckweed-mediated water remediation. In conjunction with the advantages provided by its small size, rapid growth, ease of cultivation and analysis, and increasing genomic resources, duckweed has become a promising model organism for investigating plant–microbe interactions in aquatic environments [100–102].

A total of 24 genera of bacteria of the phylum Proteobacteria (now Pseudomonadota) constitute a highly consistent core microbiome over the four duckweed genera *Spirodela*, *Landoltia*, *Lemna*, and *Wolffia* [103]. An important point of inquiry is how such a microbial community is assembled. Microbiomes of *S. polyrhiza* and *Le. minor* collected at different locations were determined, and their similar compositional profiles—including the predominant Proteobacteria—were established even when surface-sterilized fronds were exposed to wastewaters quite different to the waters of their original habitats. In addition, these profiles were quite similar to those of the leaves of terrestrial plants [102]. This indicates that duckweeds actively assemble and maintain their microbiomes in a manner conserved among all plant leaves. Further investigation of microbiome assembly can be carried out with duckweed-based synthetic microorganism communities.

The association of bacteria with the duckweed frond is an important factor in the ability of the duckweed to survive or thrive in a given aqueous environment. If a duckweed associates with bacteria that increase its innate growth potential, it will have an enhanced ability to colonize open water and compete with other surface macrophytes for space, light, and mineral resources. The first plant growth-promoting bacterium (PGPB) identified was a strain closely resembling *Acinetobacter calcoaceticus* isolated from *Le. aoukikusa* (now *Le. aequinoctialis*) that was able to enhance the growth rate of the host duckweed while degrading phenol present in the medium [104]. Subsequently, numerous studies have been carried out for the improvement of duckweed yield by the application of PGPBs such as strains of *Sinorhizobium*, *Exiguobacterium* [100], *Pseudomonas* [105], and *Acidobacter* [106] in addition to *Acinetobacter*. Duckweed/bacteria associations can give rise to mutualistic growth promotion. The association of *Le. gibba* and an *Acinetobacter* strain resulted in the promotion of the growth of both the bacterium the duckweed [107]. This was also the case with the association of the nitrogen-fixing bacterium *Azotobacter vinelandii* and *Le. minor*. The bacterium provided growth promotion factors and fixed nitrogen for the duckweed, which enhanced the nitrogen-fixing activity and the cell number of the bacterium [108].

The probability of establishing a productive PGPB/duckweed association depends on the ability of the bacteria to attach to and remain adhered to the macrophyte. A strain of the PGPB *Aquitalea magnusonii* isolated from *Le. minor* proved to be very successful in colonizing the duckweed even in the presence of much higher titres of growth-inhibiting bacteria that also associate with the duckweed [109]. However, the growth-improving effect of the addition of a PGPB was—as has often been observed—only short-lived, due to the strong resilience of the natural duckweed microbial community [110]. If PGPBs play a role in duckweeds under natural conditions, they may be water constituents that temporarily attach and adhere to the duckweed or remained attached as components of the natural microbiome. Several bacteria in pond water attached to axenic *Le. minor* and were able to promote the growth of the duckweed [105].

The microbiome of duckweeds can help the macrophytes to improve the quality of their medium. The bacteria of the microbiome can assist in the removal of excess nutrients, heavy metals, and organic xenobiotics from the aqueous surrounding of the duckweed [100]. Recent examples are the synergistic action of *Le. gibba* and an *Acinetobacter* strain in removing ammonium nitrogen from aquaculture water [107], the identification of six bacterial strains adhered to *Le. minor* that could all efficiently remove phenol from the medium as well stimulate the growth of the duckweed [111], and the improvement of tolerance of *S. polyrhiza* to cadmium by the action of rhizobacteria native to the duckweed roots [112].

The microbiome can also respond constructively to changes in the environment. The relative abundance of many of the bacteria constituting the core microbiome of *Spirodela*, *Landoltia*, *Lemna*, and *Wolffia* species underwent marked changes upon the onset of nutrient deficiency in the medium, corresponding to indications of increased motility, biofilm formation, nitrogen metabolism, and biodegradative ability of the microbiome [94]. The presence of the PGPB *A. magnusonii* mitigated the inhibitory effect of copper and zinc on the growth of *Le. minor* and enhanced the duckweed's ability to accumulate and tolerate these heavy metals [113]. Although this may not reflect processes occurring in nature, it illustrates how the duckweed microbiome interacts in a clonally dependent manner with environmental factors [114].

The microbiome can also influence the resistance of a duckweed to herbivory. Three of six different genotypes of *S. polyrhiza* inoculated with microbiota associated with the duckweed growing outdoors exhibited increased resistance by up to 41% to feeding by the pond snail *Lymnaea stagnalis*. However, three other genotypes showed *decreased* resistance to the herbivore attack [115], illustrating how clonal differences complicate the interpretation of duckweed cause/effect relationships, and that a beneficial effect on one clone may not be mirrored in another clone of the same species.

4.5. Coping with Water Pollution

Duckweeds grow on quiet or only slowly flowing waters, which are susceptible to contamination by numerous organic and inorganic substances from municipal, agricultural, and industrial wastewaters and run-off from fertilized fields. Many of the contaminating substances are toxic to duckweeds, and indeed duckweeds—especially *Le. minor* and *Le. gibba*—have long been used as test organisms in established test protocols for testing toxicity to aquatic higher plants [10,116]. The effects of water contaminants on duckweeds are illustrated by the biomarkers of effect that result from exposure to these substances [117,118]. These biomarkers can, on the one hand, show the harmful effects of water contaminants on a duckweed and can also, on the other hand, illustrate how the duckweed reacts constructively to the harmful influence of the contaminant to improve its survival chances in the presence of the contaminant.

Water contaminants can be classified into three groups: excess nutrients, metals, and organic xenobiotics. Nutrient water contaminants encompass the plant macronutrient ions NH_4^+, NO_3^- PO_4^{3-}, and SO_4^{2-} that can accumulate in surface waters from fertilizer washout and microbial action on organic wastewater. Contaminating metals comprise mainly heavy metals in dissolved ionic form or suspended as nanoparticles, as well as the metalloids As and Se, from industrial wastewaters and, to some extent, geological and solid waste leaching. A multitude of organic xenobiotic substances, including industrial chemicals, natural toxins, pesticides, pharmaceuticals, and personal care products, can also contaminate water. These myriad water pollutants detrimentally affect duckweeds on developmental, morphological, anatomical, physiological, biochemical, and molecular levels. Duckweeds respond to excessive nutrient supply with exaggerated growth leading to eutrophication, whereas other contaminants usually result in growth inhibition. Exposure to some metals can lead to frond disintegration, chloroplast damage, and frond starch accumulation. Oxidative damage due to the production of reactive oxygen species is very widespread, especially in conjunction with inhibition of photosynthetic activity and damage to the photosynthetic apparatus. These and numerous further observations of biochemical and molecular effects due to water contaminants are documented in Ziegler et al. [118]. In addition to the determination of specific biomarkers of effect, comprehensive transcriptomic analyses have illustrated the far-reaching differential gene expression and metabolic alterations occasioned by the deleterious effects of NH_4^+ [119], Cd^{2+} [120], and streptomycin [121] on *Le. minor*, *La. punctata*, and *Le. aequinoctialis*, respectively.

Duckweeds can react to alleviate damages caused by water contaminants. Several responses to deleterious impingement of water pollutants include increased activities of antioxidant and detoxification enzymes and enhancement of thiol protectant, flavonoid, phytochelatin, and heat shock protein synthesis (numerous examples in [118]). Such responses and the widespread physiological and molecular reactions to water contaminants revealed in the transcriptome studies mentioned above cannot be regarded as being duckweed-specific but are rather representative of remediative measures common to higher plants deleteriously affected by toxic substances. The formation of pectinous cell wall thickenings in *Le. trisulca* mesophyll cells that sequester lead taken up by the duckweed is an example of a widespread strategy in many plants to compartmentalize accumulated heavy metals away from sensitive sites in the protoplast [122]. Nevertheless, they illustrate that duckweeds can cope with water pollution as well as other plants to the extent that it does not prove to be too debilitating. However, a physiological and transcriptomic analysis of salt stress in *S. polyrhiza* revealed some mechanisms with respect to particularly hormone-related responses to salinity that appear to be different from those operative in other plants [123]. This may signify that duckweeds do have some unique means of coping with water pollution.

The coexistence of different duckweed species can be of mutual advantage to the involved organisms in coping with heavy metal stress. Both *S. polyrhiza* and *Le. aequinoctialis*, which frequently occur together in nature, grown together grew more rapidly when exposed to various concentrations of a mixture of copper, cadmium, and zinc than when grown separately. This was accompanied by an increase in the activities of antioxidant enzyme activities in both species, which increases tolerance to the metals. Metal uptake was thereby not limited so much as differentially accumulated: *Le. aequinoctialis* accumulated Cd and Zn preferentially, whereas *S. polyrhiza* accumulated mainly Cu and Cd [124]. In another study with the same two duckweed species, the presence of *S. polyrhiza* improved the growth of *Le. aequinoctialis* at high copper concentrations and decreased the environmental load of the heavy metal by increasing sequestration of Cu in the cell walls of *Le. aequinoctialis* [125].

The ability of duckweeds to withstand the deleterious effects of metals can be improved by the presence of growth-promoting bacteria that associate with the duckweed (see Section 4.4). An example is the alleviation of the harmful effect of chromium (Cr(VI)) on *Le. minor* in the presence of the rhizobacterium *Exiguobacterium* sp. MH3 by enhancing the growth of the duckweed and preventing the duckweed from taking up excessive amounts of the metal [126]. The presence of the PGPB *A. magnusonii* mitigated the inhibitory effect of copper and zinc on the growth of *Le. minor* and enhanced the duckweed's ability to accumulate and tolerate these heavy metals [39]. The alleviation of the multiple heavy metal toxicity by the coexistence of *S. polyrhiza* and *Le. aequinoctialis* described above was accompanied by increased duckweed-associated microbial activity compared with that exhibited by the duckweed by itself and is indicative of regulation of the activities of the bacterial communities associated with the individual species [127].

Duckweeds may protect themselves from the harmful effects of water contaminants in water by degrading the toxic substances to non-toxic forms with the aid of bacteria in their microbiome. This has been illustrated by the colonization of sterilized *Le. aoukikusa* (now *Le. aequinoctialis*) roots by a phenol-degrading *Acinetobacter* strain P23 that was isolated from the rhizosphere of the duckweed. A long-term continuous degradation of phenol in the medium was attributed to the beneficial symbiotic interaction between the duckweed and the bacterium [104].

Duckweed communities may experience pulse—in contrast to long-term—exposure to harmful water contaminants, following which surviving members of the community may recover to regain their original vitality and distribution. Both *Le. minor* and *Le. gibba* suffered significant inhibition of growth rate and biomass production upon exposure to >100 mg/L diuron for 7 days, after which they recovered completely when transferred to non-contaminated medium. This suggested that duckweed can withstand short-term exposure to environmentally relevant concentrations of herbicides at significant risk levels [128].

There is evidence that duckweeds may actually be able to develop resistance to herbicides such as diquat, which is used to control *Le. minor* and *Wo. columbiana* spreading in an unwanted manner [129]. *La. punctata* was found to be very susceptible to diquat if it had not previously been exposed to the herbicide but quite resistant if it had a prior history of exposure to diquat [130]. This also illustrates the ability of a duckweed to overcome anthropogenic management efforts to suppress it and thus increase its chances of survival.

A truly duckweed-specific means of coping with the presence of a heavy metal water contaminant is the production of turions by *S. polyrhiza* upon exposure to cadmium at a concentration inhibiting the growth of the fronds ([131]; see also Section 5.2.1). In this way, fronds threatened by Cd^{2+} produced robust, dormant derivatives that can avoid the deleterious effects of the metal. It would be interesting to determine if this is a Cd-specific effect or if it reflects a general defensive response to exposure to heavy metals.

4.6. Competition

Duckweeds often occur together with other floating water plants (see [1,2]). If they are then to sustain themselves, they must be able to assert themselves in the face of competition from these other macrophytes, as well as from algae and cyanobacteria, for space, light, and nutrients. Their most basic "trump card" in this respect is their ability to grow and propagate themselves rapidly. This enables them to quickly cover any open-water space available to them and consolidate their areas of dispersion by forming multi-layered mats. Their rapid, surface-covering growth can deprive other photosynthetic aquatic organisms of space, light, and nutrients, thus diminishing their competitive ability (see [132]). This is illustrated by the designation of *La. punctata*, *Le. minor*, and *Wo. columbiana* as problematic weeds that overgrow waterways [129,130] and the prevention of weed growth in rice fields by the introduction of *S. polyrhiza* and *La. punctata* [133].

Excessive rapid growth can, however, also lead to intraspecific competition in duckweeds and a decline in vitality. When a duckweed proliferates rapidly for a long time in a confined area, the fronds will bunch together to form mats of various thicknesses after having initially covered the water surface. This overcrowding leads to growth inhibition and the production of smaller and more uniform but morphologically modified fronds in *Le. minor* [39] and *S. polyrhiza* [134]. Contact between previously separated fronds has also been observed to result in a burst of ethylene release in *S. polyrhiza*, *Le. gibba*, and *Le. aequinoctialis* [135]. The ethylene formation may cause crowding-associated growth retardation, as well as the promotion of aerenchym formation in *S. polyrhiza* and especially in *Le. gibba* providing the fronds with greater buoyancy to help them surface in crowded surroundings [9]. When overcrowding persists and growth stagnates, flowering/seed set and turion formation can provide possibilities for escape and renewed growth at more opportune times. Crowding has been found to enhance turion formation in *S. polyrhiza* (see Section 5.2) when this has been initiated [136]. It also inhibits the turion germination when it is still in effect when the turions have lost their dormancy [137], thus precluding a precocious return to the growth mode.

In some cases, the success of a duckweed in the face of a potential competitor is dependent upon the extent to which the environmental conditions are conducive to the growth of each species. Free-floating *Le. gibba* and the submerged, rootless hornwort *Ceratophyllum demersum* are both common in temperate eutrophic waters but are mutually exclusive. Sufficient mineral nutrient availability and a neutral water pH value favoured the success of the duckweed over the hornwort, whereas a low inorganic nitrogen supply and a high water pH value led to takeover by *C. demersum* [138]. The relative success of competing duckweed and non-duckweed species is not merely a matter of growth, however. In monitoring the presence, abundance, and growth rates of *Le. minor*, *Le. minuta*, and the water fern *Azolla filiculoides*, it was concluded that the distribution of the macrophytes did not associate with nutrient or light levels. Although *A. filiculoides* had the highest growth rate, it occurred least frequently, in contrast to *Le. minor*, which grew the most slowly but had the widest distribution. The ability to persist under winter conditions and to disperse after disturbances appeared to be the major determinant of competitive success [139].

Specific morphological and physiological characteristics can enable certain duckweed species to survive in regions not supportive of other Lemnaceae. An example is the ability of frost-sensitive *S. polyrhiza* fronds to survive freezing winter temperatures by developing frond derivatives (turions: see Section 5.2) that can withstand the cold season in comparison with equally frost-sensitive fronds of the otherwise very similar *S. intermedia*, which do not develop turions [38].

An important factor in the competition between duckweeds and other aquatic plants that is not based on growth success is allelopathy, or the ability of an organism to influence other organisms sharing the same habitat by means of exuding chemical substances. This has particular significance when the duckweed and its competitor have a similar ability to grow rapidly and have similar requirements for light and nutrients. In some cases, duckweeds appear to have a competitive disadvantage in cohabitation with non-duckweeds due to allelopathy. The ability of the water soldier *Stratiotes aloides* to compete successfully with *S. polyrhiza* was concluded to result from an allelopathic influence of *S. aloides*, resulting in an inhibition of frond production and concomitant induction of turion formation (see Section 5.2) in the duckweed [140]. The ability of the green alga *Cladophora glomerata* to dominate *Le. minor* was concluded to be due to the production of phenolic compounds acting in an allelopathic manner [132]. Nevertheless, the cessation of growth under the production of turions represents a means of coping with a competitive disadvantage, and *Le. minor* was also observed to form potentially allelopathic phenols in competition with *C. glomerata*. Indeed, another report has also indicated that duckweeds may have allelopathic potential in that extracts of *Le. minor* fronds show inhibitory activity on the root and shoot growth of several terrestrial plant species [141]. These authors also identified (3R)-(-)-hydroxy-β-ionone as the active ingredient of a *Le. minor* extract that inhibited the growth of cress [142]. However, these findings are no proof of the actual allelopathic activity of duckweeds.

Cyanobacteria compete with aquatic plants not only in terms of the removal of nutrients from the water due to their capacity for rapid growth but also because of the toxic substances, especially microcystins, that they excrete [143,144]. *Microcystis aeruginosa* is a widely distributed cyanobacterium that can have harmful allelopathic effects on duckweeds. Microcystins have been observed to inhibit the growth of *Le. minor* [145–147], *Le. gibba* [148], *La. punctata* [149], and *Wo. arrhiza* [146]. However, microcystin has not always been observed to detrimentally affect *Le. gibba* [150], and susceptibility to microcystin toxicity has been shown to be clone-specific in *Le. minor* [151].

Besides developing microcystin-resistant clones, duckweeds have some means of counteracting the competitive disadvantage resulting from microcystin action. As illustrated with *Le. gibba*, duckweeds can take up and detoxify the cyanobacterial toxin [143]. Although *Le. japonica* growth was inhibited by co-cultivation with *M. aeruginosa*, the presence of the duckweed also inhibited the growth of cyanobacterium, presumably by excreting allelopathic chemicals of its own [152]. The possibility of allelopathic duckweed competition against cyanobacteria is lent plausibility by the detrimental effect extracts of *La. punctata* on *M. aeruginosa* [153].

In contrast to cases of dominance or exclusion, two (or more) species may stably coexist with one another even though they have similar requirements for growth and would be expected to compete openly for dominance. An analysis of the widespread common presence of *S. polyrhiza* and *Le. minor* indicates that while this coexistence requires fluctuating environmental conditions, it is not primarily dependent on interspecific differences in such characters as thermal reaction norms or dormancy behaviour. Rather, it requires subtle niche differences causing negative frequency-dependent growth that acts consistently across environmental gradients [154].

5. Coping with Winter Cold: The Formation of Resting Fronds

Temperatures ranging from below 8 °C to about 17 °C are sufficiently low to completely prevent frond growth of various groups of duckweeds that inhabit regions exhibiting temperate to very warm growing seasons, and although most fronds can tolerate temperatures down to the freezing point or somewhat lower for at least short periods, they usually cannot withstand prolonged or severe frost [2,3]. Fronds of *Le. minor* and *S. polyrhiza* have been observed to survive even when encased in ice for a prolonged period [3,155], but duckweeds usually respond to the onset of winter cold by forming resting fronds.

Resting fronds are generally smaller and more robust than the fronds characteristic of the growing season and have fewer air spaces, as well as higher starch contents [2–4,9]. Their extremely reduced or completely arrested growth and propagation is key to the survival of duckweeds under extended periods of winter cold. The very low metabolic activity of the resting state enables the quiescent fronds to endure long periods of conditions inimical to growth and propagation. The formation of resting fronds and their subsequent "reactivation", i.e., germination and sprouting to give rise to new, growing fronds when conditions improve at some later point, constitute a scheme of survival in a purely vegetative mode. The survival that these fronds convey under cold conditions is based on avoidance of severe freezing temperatures and tolerance of temperatures not significantly below the freezing point. Two principal types of resting fronds can develop.

5.1. *Resting Fronds Still Capable of Growth*

Some resting fronds basically resemble the "normal" fronds of the growing season, although they are generally thicker and fleshier in appearance than the latter. Despite their restricted metabolism, they can still grow and even reproduce slowly when the adverse conditions are not too severe [1]. They can resume normal growth and propagation when conditions improve.

La. punctata, Le. perpusilla, Le. gibba, Le. minor, most strains of *Le. aequinoctialis,* and some strains of *Le. japonica* form resting fronds capable of growth that remain on the water surface. This surface location generally renders them suitable for survival only in winters not characterized by freezing temperatures. They may indeed avoid the effects of such temperatures when these do occur, however, by being pressed beneath ice forming on the water surface or by remaining attached via stipes to the pouches of mother fronds that have died and sunk to the bottom of the water body [2].

Le. trisulca, Wa. gladiata, and *Wo. arrhiza* form resting fronds capable of growth that sink to the bottom of the water body on account of their density due to reduced air spaces and high starch content. In their submerged surroundings, they avoid severe frost temperatures that may be in effect at the water surface since the water temperatures on the bottom hardly go below the freezing point [2]. They thus provide for survival even in very cold winters.

It has recently been described that 90% of *Le. minor* fronds—which are generally thought to overwinter on the water surface—growing on a pond in Quebec, Canada, survived very cold winters beneath massive ice layers [156]. Since neither the anatomy nor the actual location of the fronds beneath the ice were investigated, it is unclear to which category of resting fronds this remarkable rate of survival can be attributed.

Little is known of the mechanisms involved in the formation of the resting fronds still capable of growth or about their resumption of "normal" growth when conditions improve. The developmental cycle of resting fronds has been thoroughly investigated only on the example of the turions of *S. polyrhiza*. Since the resting fronds still capable of growth resemble turions functionally [3], the principles elucidated with regard to *S. polyrhiza* turions may also be relevant for the formation and activation of these fronds.

5.2. Turions

Duckweed turions are resting fronds that emerge from meristematic pockets in the "normal" mother fronds giving rise to them. They separate from the mother fronds and sink to the bottom of the body of water on which the "normal" fronds grow on account of their density. They are particular examples of detachable, truly dormant modified green shoots that are widespread in aquatic plants [2,157]. According to Landolt [2], they are found in *S. polyrhiza*, *Le. turionifera*, some clones of *Le. aequinoctialis*, and many species of *Wolffia* (*Wo. brasiliensis*, *Wo. borealis*, *Wo. angusta*, *Wo. australiana*, *Wo. arrhiza*, *Wo. columbiana*, *Wo. globosa*). They also occur in *Wo. microscopica* (our unpublished observation). Duckweed turions are morphologically different from the "normal" fronds that give rise to them. The turions of *S. polyrhiza* and *Le. turionifera* are flat and rounded, while those of *Wolffia* are very small and spherical [2]. As is typical for turion-bearing duckweeds, the turions of *S. polyrhiza* have smaller air spaces, smaller vacuoles, thicker cell walls, and much more starch than the "normal" fronds giving rise to them [158–160].

Turions of *S. polyrhiza* (Figure 1) are more tolerant of low temperatures than are the "normal" growing season fronds of this species. However, this is not true of all duckweeds: turions of *Wo. arrhiza* are as sensitive to cold as are the "normal" fronds of this species [161]. Although turions cannot tolerate severe frost, they can withstand long periods of intense cold at the bottom of the water body where the water temperatures fall scarcely below the freezing point, as in the case of the submerged resting fronds still capable of growth. Turions are truly, or innately, dormant upon their formation in that they do not and cannot grow, although they do exhibit some respiration and are capable of photosynthesis [162]. Duckweed turions become capable of resuming growth once more after a prolonged period of exposure to low but not freezing temperatures. This "after-ripening" (turion formation can be termed "ripening") breaks the dormancy and allows the turion to germinate and sprout to form new "normal" fronds when conditions again become conducive to growth and propagation [163].

Figure 1. Fronds *of Spirodela polyrhiza* growing under non-limiting conditions (left-hand photos (**A**)) and under nutrient stress (right-hand photos (**B**)). The upper photos show colonies made up of several interconnected fronds, and the lower photos show single fronds that also exist alongside the multi-frond colonies. The fronds under nutrient stress produce dark turions are indicated by the white arrows.

The formation and overwintering of turions, as well as the subsequent germination and sprouting of these propagules to resume "normal" frond growth and vegetative propagation in the spring, has been thoroughly investigated only with *S. polyrhiza* (see [3]). The knowledge that has been amassed with this species is summarized below and provides a suitable picture of how turion formation enables a particular duckweed to survive cold winters.

5.2.1. Turion Formation

Turion formation is a consequence of a switch in the developmental program of frond primordia from the formation of new fronds characteristic of the growing season to the production of resting turions [164]. In *S. polyrhiza*, a shortage of phosphate in the water is the prime environmental factor bringing about this switch, and low temperatures have the same effect when phosphate concentrations are higher [165,166]. The formation of turions is thus initiated in nature by the exhaustion of water resources at the end of a season of profuse aquatic plant growth and the approach of cold weather in the autumn. High light intensities and CO_2 concentrations, as well as the presence of carbohydrates, can enhance turion formation in *S. polyrhiza* once this has been induced. This is due, however, to an increment in turion-producing biomass rather than representing a switch in the developmental program of the frond primordia [165,166] and is irrelevant for turion formation under natural conditions.

Turions are not formed exclusively in the context of overwintering. They can be produced upon phosphate deficiency at any time and may also be formed upon exposure to the heavy metal cadmium (see Section 4.5), as well as upon overcrowding and allelopathic influence (see Section 4.6). They can even be formed during the summer under conditions of high temperature and light intensity [167]. Turions can thus be seen as vegetative propagules formed in answer to various types of stress that must all act in a common or similar manner to re-program duckweed shoot primordia development. Abscisic acid is thought to be involved in this re-programming of *S. polyrhiza* turion formation [164,168,169].

The photomorphogenic effects of light (mediated by the photoreceptor phytochrome) can enhance or modulate turion formation in *S. polyrhiza* [170], but no critical day length, and thus no inductive effect of photoperiod, has been observed with this species [171]. It is remarkable that short days, which also herald the onset of the winter season, do not induce turion formation. Decreasing mineral nutrient availability in conjunction with decreasing temperatures thus gives rise to *S. polyrhiza* turion formation in nature in place of the low temperatures and short photoperiods usually responsible for turion formation upon the approach of winter in other hydrophytes [157,172–174].

Turion formation in *S. polyrhiza* shows great clonal variation when expressed as the specific turion yield (SY), i.e., the number of turions formed per frond under inductive conditions [25,175]. The SY is important in an ecological context as an indicator of the number of turions available to support the survival of the duckweed under adverse conditions, i.e., in winter [173]. Variability in SY represents adaptations to local climatic conditions and is presumably genetically determined [166]. The mean annual temperature of a site inhabited by an *S. polyrhiza* clone has an important influence on the SY of that clone. Low temperatures result in increased SY to offset the reduced survival rate of the turions under these conditions [25]. Clonal differences in turion formation as SY are independent of the specific signals that induce turion formation and are located in the transduction chain leading to the developmental switch from "normal" frond replication to turion production [173].

5.2.2. Turion Dormancy

The innate dormancy that characterizes newly formed turions of *S. polyrhiza* is the key to the survival of duckweed in cold winters. Innately dormant turions in nature become able to germinate and resume normal vegetative growth after prolonged exposure to cold but not freezing temperatures ("chilling"). This "after-ripening" is a gradual response, the length of which depends on the conditions the turions are subjected to [3,32,176]. For *S. polyrhiza*, after-ripening must proceed for at least two weeks at water temperatures of 0–5 °C to remove the dormancy, as has been demonstrated by quantitative measuring the influence of the duration of after-ripening on the germination response [163]. This requirement for prolonged chilling ensures that the turion will not germinate or sprout precociously before the cold season has passed and conditions again become once more suitable for growth. *S. polyrhiza* turions may be formed in the late summer or early autumn in response to nutrient deficiency while temperatures are still warm and ample light is available. Without dormancy and the requirement of a protracted cold period to break it (i.e., resting on the bottom of the water body throughout the winter), the turions could germinate immediately after their formation with no prospect of appreciable growth and renewed turion formation before the onset of fatal winter water surface conditions.

The dormancy of newly formed turions represents a metabolic block, or state of "self-arrest" [177], that prevents the response of the turions to growth-promoting signals. It is not due to a lack of nutrient reserves to fuel metabolism, as the turions contain up to over 70% starch in terms of weight (e.g., [178]). However, this high carbohydrate reserve may initially not be accessible for turion metabolism. The prolonged dormancy of freshly formed turions may be related to a gradual breakdown of the highly polymeric starch molecules to soluble carbohydrates required for later germination metabolism. Freshly harvested *S. polyrhiza* turions indeed germinate to a certain extent, even without after-ripening in the presence of an external sugar supply [179]. Accordingly, newly formed turions may not normally contain levels of soluble, readily metabolizable carbohydrates sufficient to permit germination to take place. A gradual breakdown of the starch stored in newly formed turions has been observed to take place upon extensive storage of the turions under cold aqueous conditions [179]. Quantification of soluble sugars during turion after-ripening showed that this starch degradation resulted in the accumulation of soluble, readily metabolizable carbohydrates [180].

When after-ripened turions have lost their dormancy, they are in principle able to germinate in the presence of appropriate conditions of temperature and light. However, they will not germinate until these conditions actually apply. In their absence, the after-ripened turions remain quiescent in "imposed" dormancy (able to germinate but prevented from doing this by environmental constraints). This imposed dormancy persists after completion of after-ripening on the bottom of the water body until the water temperature has increased sufficiently to permit germination and ensure a successful resumption of growth.

5.2.3. Turion Germination and Sprouting: The Resumption of Growth

Turion Rising: Bubble Formation

Turions that have waited out the cold of winter on the bottom of water bodies must surface in the spring to germinate and resume "normal" growth on the water surface to be able to re-establish themselves and propagate in their aquatic environment. How they do this is not clear, but submerged turions of *S. polyrhiza* have been observed to expel a small bubble of gas upon light incidence when the water temperature had increased to >15 °C. This bubble adheres to the junction between the pocket sheath and the upper surface of the turion and provides the turion with the buoyancy necessary to rise [32,181].

Germination

The actual resumption of growth commences with germination. "Germination" is the onset of developmental processes in quiescent turions as observed in terms of the reflection of leaves or scales and a slight elongation of the internodes [157]. The first indication of this in after-ripened *S. polyrhiza* turions is a slight swelling, after which 2 to 5 roots push through the root shield. When the first new shoot then pushes aside the pocket sheath as it emerges from the pocket, the turion is considered to have germinated. Germination normally begins shortly after the turions have reached the surface of the water and is dependent on temperatures of about 15 °C or higher and light [2].

Light has long been known to trigger turion germination [2,32], and the germination response of surfaced *S. polyrhiza* turions to light is mediated by phytochrome [182]. A single pulse of red light ("Rp") induces germination: it can be reversed by a subsequent pulse of far-red light ("FRp" [182]) and is a low fluence-type, "classical" phytochrome response [183]. Germination can also be induced to a similar extent by repeated red light pulses or continuous red light ("cR": [178,180]), which indicates a special low-fluence response that requires the presence of newly formed phytochrome in its far-red light absorbing, physiologically active form over an extended period [178].

Under natural conditions, germination is closely followed by sprouting, and the breakdown of the considerable reserves of starch stored in the turions (see [160]) would appear to be predestined to provide energy and carbon skeletons for the course of both developmental processes. However, germination can be induced by a red light pulse without starch breakdown and is, in this case, fuelled by soluble sugars having accumulated within the turion from the slow breakdown of storage starch during dormancy and after-ripening ([179,180]; see also Section 5.2.2).

Sprouting

Once turions have germinated, they "sprout" to resume vegetative growth, i.e., the production of new "normal" fronds. "Sprouting" commences with the distinct elongation of the still very short internodes of the germinated turions to enable better access to light, gas, and solute exchange for the emerging tissues, followed by the formation of new "normal" frond structures in the apical meristems (see [157]). Water temperatures favourable for germination (i.e., $\geq$15 °C) and light are key ecological requirements for turion sprouting.

Freshly germinated turions in *S. polyrhiza* are already equipped with effective photosynthetic and respiratory machinery [162], but the assimilative potential of the newly sprouted fronds is limited. Although a single red light pulse results in suitable germination of cold after-ripened *S. polyrhiza* turions, it leads to only very limited growth of the emergent sprouts. The weight of turions germinated in response to an Rp only doubled in the two weeks following the irradiation, whereas the growth of the newly emerging shoots progressed much more rapidly under cR irradiation [180]. This rapid sprouting is enabled by the breakdown of the reserve starch of the turions that is initiated by the cR treatment. The effect of cR in triggering *S. polyrhiza* turion starch breakdown lags only about 12 h behind germination, with the starch reserves of the turion being exhausted within a week [174]. Of course, sunlight in nature ensures both germination and starch degradation with its cR component.

The rapid mobilization of turion storage starch in nature occasioned by the cR component of sunlight thus provides young fronds emerging from turions upon germination with a supply of readily metabolizable carbohydrates sufficient to support the rapid frond growth and development of sprouting. This, together with the early surfacing and germination of after-ripened turions, is propitious for enabling the newly formed fronds to occupy the water surface before other plants in the spring.

5.2.4. The Molecular Biology of the S. polyrhiza Turion Developmental Cycle

A very recent publication describes the results of an RNA-seq analysis carried out on mature turions and actively growing fronds from *S. polyrhiza* [184]. Differentially expressed transcripts between the mature turion and frond tissues revealed how the re-programming of frond meristems for turion formation involved the mobilization of major pathways related to the development of turion dormancy and to the starch and lipid metabolism that builds up nutrient reserves in the developing turions and remobilizes them again during turion germination and sprouting. It was also shown that dormant turions store numerous mRNA transcripts for use in mobilizing metabolic pathways required during the resumption of growth. DNA methylation appeared to represent an epigenetic component of turion tissue formation, and it was indicated that regulatory elements known to be involved in seed setting and germination have been reworked for analogous function in turions. This study provides a comprehensive conception of the molecular background of the turion-based overwintering strategy of *S. polyrhiza*.

5.2.5. *Spirodela polyrhiza* as a Model for Turion-Based Duckweed Overwintering?

Experimental findings as to the developmental cycle of *S. polyrhiza* turions provide detailed insight into how one species of duckweed can survive winter cold by means of vegetative propagules that are formed under climatic conditions heralding the approach of winter, bridge long periods of low temperatures in a dormant state, and resume growth upon the onset of favourable conditions. The comprehensive picture might be regarded as a model for the overwintering of all duckweeds that form turions or functionally equivalent resting fronds. However, a model organism should be truly representative of a given set of organisms in a particular biological context, and very little information from other resting frond-bearing duckweeds is available for comparison with the extensive information pertaining to *S. polyrhiza*. Much further information along the lines of that presented here for *S. polyrhiza* must be gathered from these other species to evaluate how representative *S. polyrhiza* turions are for turion- or resting frond-based duckweed overwintering.

Of particular interest in this regard is how widespread the primary induction of turion formation by nutrient deficiency and low temperature evidenced in *S. polyrhiza* is. It is notable that Adamec [157] limited his comprehensive discussion of macrophyte turion physiology to mainly non-duckweed species on the grounds that turion formation in *S. polyrhiza*—as a representative of the duckweeds—was based on nutrient deficiency rather than the short photoperiods that are otherwise responsible for turion formation. If the formation of all duckweed turions is induced by mineral deficiency and low temperatures, this will represent a signature turion-based survival "strategy".

6. Flowering and Seed Setting

Duckweeds do undergo sexual reproduction despite their more visible and widespread asexual vegetative propagation. Flowering and the production of viable seeds can always be a means for duckweeds to deal with situations inimical to growth and even life itself and is the only possibility of survival and reproduction when the duckweed habitat dries out completely or becomes too salty [9,185]. Flowering in duckweeds has long been of interest to researchers due to the fact that it is a question of the smallest flowering plants on Earth that are only rarely seen to flower when cultured under laboratory conditions. There have nevertheless been numerous observations of duckweeds flowering in the field, including all species except *Le. obscura*, *Wo. elongata*, and *Wo. australiana*. Some species flower relatively often (e.g., *Le. gibba*, *Le. perpusilla*, *Wa. lingulata*), whereas others do so only occasionally (e.g., *La. punctata*, *Le. minor*, *Wo. brasiliensis*) or very infrequently (e.g., *S. polyrhiza*, *Wo. borealis*) [2]. The actual frequency of flowering in nature may be higher than that observed, keeping in mind that a very small flower on a very small plant may be quite inconspicuous. Why particular duckweeds flower more frequently may not be easy to understand because many environmental factors can be involved in the induction of the flowering. These include crowding, light intensity and light duration, temperature, and the chemical composition of the water. Landolt [2] has tabulated the influence of these factors on the flowering of a number of duckweed species. In some locations, several duckweed species have been observed to flower at the same time, which indicates that environmental requirements may be similar for different species, but in other cases, flowering appears to be species-, season-, and location-specific [2].

Duckweeds have male and female floral organs, and two whorls of a typical flower—the sepals and petals—are missing. In the species belonging to the genera *Spirodela*, *Landoltia*, and *Lemna*, the floral organs develop into one of the two lateral pouches, normally giving rise to the vegetative buds that produce new fronds, whereupon the budding of daughter fronds from that pouch pauses. However, the daughter fronds continue to bud from the second lateral pouch present in these species. In the species belonging to the genera *Wolffiella* and *Wolffia*, the floral organs develop in a specialized cavity that opens to the dorsal side of the frond; the budding of daughter fronds thereby continues from the single vegetative pouch present in these species [2,18].

The sporadic occurrence of flowering and the ease of investigating duckweeds led to the use of duckweeds as a model organism for the investigation of flowering and have provoked numerous studies of flowering physiology and of the influence of environmental factors such as light, temperature, and the chemical makeup of the water in inducing the flowering [2–4,177]. Kandeler [186] rightly pointed out in 1984 that "Lemnaceae are one of the pilot systems to investigate the physiological basis of flowering". The use of duckweeds in this respect features ease of maintenance and growth of gnotobiotic cultures in an aqueous medium, the uptake of investigatory chemicals directly from the aqueous culture medium, and the expeditious observation of effects on successive generations due to the rapid vegetative propagation of fronds [187].

The geobotanical occurrence of the different species of duckweeds correlates with flowering behaviour as well as with growth. The differing photoperiodic and temperature requirements for the flowering of the various species of duckweeds are in coherence with the occurrence of the species in a widespread or a specific climate zone, e.g., day-neutral species exhibiting a cosmopolitan distribution or long-day species being distributed in the temperate regions [2]. Exposure to low temperatures (22 °C) induced flowering in *W. microscopica* [188,189] and has been able to induce flowering in this species even under continuous white light illumination (Sree and Appenroth, unpublished; Figure 2).

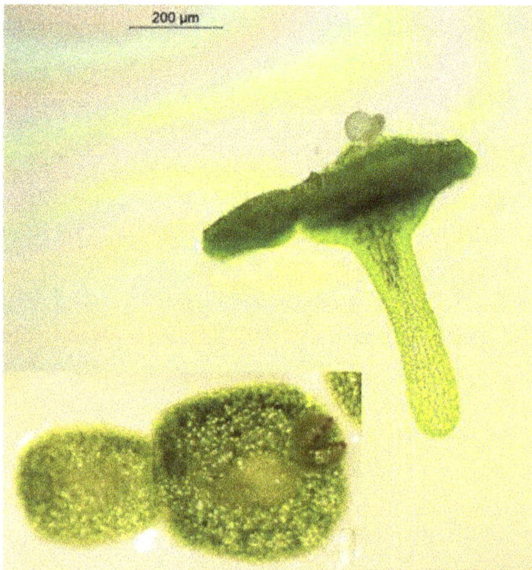

Figure 2. Flowering fronds of *Wolffia microscopica* with stigma and anther lobes seen on the exterior (lateral view). Inset: top view.

Several chemical compounds have been successfully used for initiating flowering in different duckweed species. The effects of molecules such as phytohormones and metabolites on flower initiation were investigated during the 1960s to 1980s [190,191]. Two compounds warrant special attention in this regard. Ethylenediamine-di-o-hydroxyphenylacetic acid (EDDHA) was shown to be a floral inducer in *S. polyrhiza* in 1966 [192], and 8 years later, the effect of salicylic acid (SA) on inducing flowering in duckweeds was established [193]. EDDHA and SA were thereafter successfully used for floral induction in several duckweed species under even non-inductive laboratory conditions [3]. EDDHA was able to induce flowering in plants sensitive to different photoperiods [192,194,195]. It was initially hypothesized that EDDHA acts by chelating metal ions that might be required for floral induction in duckweeds; however, it was subsequently suggested that the breakdown of EDDHA releases an SA-like active molecule that induces flowering [185,196]. With the current understanding of the role of SA in plant defence, Pieterse [185] suggested that flowering could be a stress response and that endogenously produced SA induces flowering upon exposure of the plant to stress conditions. Interestingly, crowding of plants, which is also a stress factor, has been suggested to induce flowering [2]. Crowding has been shown to increase ethylene production in *S. polyrhiza* [197], but Pieterse [198] found that ethylene did not induce flowering in *Le. gibba*. The bioassays originally planned for investigating florigen in duckweeds led to the identification of the floral-inducing capacity of SA [193,195]. Almost three decades later, the photoperiod-dependent flowering mechanism had been unfolded to a certain extent. The mobile florigen signal that is transported from the leaf to the shoot apical meristem has been identified as the FLOWERING LOCUS T protein (FT protein) that migrates through the phloem [199,200]. Two functional FT genes have been identified in *Le. aequinoctialis* that promote or suppress flowering [201], and the induction timing of an FT gene was shown to be important in connecting the phase of the circadian clock to photoperiodism at the molecular level in the same species [202]. These are important steps in understanding how flowering is initiated in duckweeds. The availability of ever more whole genome sequences of different clones and species of duckweeds has enabled the detection of the loss of five clades of MADS-box genes in duckweeds. This categorizes duckweeds as the angiosperms possessing the smallest number of clades of MIKC-type MADS-box genes [203]. Of the five, three of them, *AGL9*, *AGL12*, and *OsMADS32*, have been specifically lost in duckweeds. The authors have correlated this high number of missing clades to the simple architecture of the duckweed body and have suggested that the loss of *AGL9*-like genes may be responsible for the rarity of flowering in the world's smallest angiosperms and, thus, for limited use duckweeds make of flowering and seed production to cope with untenable situations.

Flowering and the production of seeds is a strategy of duckweeds for the survival of drought or dry seasons. The seeds are able to tolerate desiccation on account of their anatomical structure, and they germinate upon the return of favourable conditions to develop into seedlings and establish a fresh duckweed culture in their environment [2]. It must be kept in mind, however, that induction of flowering does not ensure the production of viable seeds. Flowers may be aborted, or the floral organs may be sterile [204]. Pollination, which in nature can occur with assistance from wind, water, or small animals or by direct flower contact, must be successful, and self-pollination can result in sterility [1]. This is of present concern in the quest for the breeding of elite duckweed varieties for commercial applications and is being addressed by artificial cross-pollination [204].

7. Conclusions

Although they are very small, simply constructed, and apparently fragile aquatic higher plants lacking attachment to any substrate, duckweeds have proved to be very successful in colonizing new habitats and persisting on them in almost every part of the world. To do this, they cope with environmental conditions that are often less than optimal for growth and proliferation and may even prevent growth and be life-threatening. The means by which they do this reflects, in many instances, general patterns of plant response to environmental challenges. This is evident in how duckweeds adjust themselves to varied regimes of temperature, light, and pH value, ensure sufficient mineral and organic nutrient uptake, resist microbial and herbivore attacks, cooperate with microorganisms, and cope with water contaminants and competition for living space and nutrients during the growing season. The response of duckweeds to winter cold by forming resting fronds and turions is common to many macrophytes, and the flowering of duckweeds to survive life-threatening conditions is common to most higher plants. This indicates that duckweed responses to environmental stresses in the main reflect conserved survival strategies rather than unique mechanisms. Duckweeds do exhibit some highly specific survival characteristics, however, in terms of defence gene expression and cell wall composition in the face of microbial attack and the induction of turion formation by a mineral salt deficiency in place of photoperiodic effect otherwise evidenced by aquatic plants. However, the state of our knowledge about duckweed survival means is fragmentary: many relevant investigations have been carried out on only one or very few of the 36 duckweed species and have often not been carried out in depth. Conclusions about the extent to which responses of duckweeds as a plant family to environmental challenges have a unique status among plants are, therefore, premature.

What can be considered unique about the "survival strategies" of duckweeds is that much of the widespread success of these macrophytes can be recognized in an exceptional growth potential coupled with a primarily vegetative mode of frond propagation that gives rise to pronounced, epigenetically driven clonal diversity. The juvenile developmental status of the fronds that underlies the vegetative expansion also enables flowering and the development of overwintering propagules for surviving conditions that prevent growth and are potentially lethal.

In order to better understand the "survival strategies" of the family of duckweeds, future research must incorporate a more comprehensive selection of duckweeds into investigations of how these macrophytes react to environmental challenges. This may include comparing the findings respective of numerous duckweed species with those of selected "model" duckweed species to determine common duckweed traits and with those of established plant models such as *Arabidopsis* to assess how unique duckweed survival responses really are. This goal can be profitably approached by employing modern transcriptomic, proteomic, and metabolomic methods in the investigations wherever possible. In addition to increasing the understanding of duckweed responses to particular environmental stresses, the molecular information obtained with these techniques can identify via informatics how representative these responses are of general plant mechanisms.

Author Contributions: P.Z and K.S.S. developed the concept. P.Z. prepared the original draft, which was reviewed and edited by K.J.A. Chapter 6 was written by K.S.S. All authors have read and agreed to the published version of the manuscript.

Funding: This research received no external funding.

Institutional Review Board Statement: Not applicable.

Informed Consent Statement: Not applicable.

Data Availability Statement: Not appliable.

Conflicts of Interest: The authors declare no conflict of interest.

References

1. Sculthorpe, C.D. *The Biology of Aquatic Vascular Plants;* Edward Arnold: London, UK, 1967; 610p.
2. Landolt, E. *The Family of Lemnaceae—A Monographic Study, Vol. 1. Biosystematic Investigations in the Family of Duckweeds (Lemnaceae);* Des Geobotanischen Institutes der Eidgenössischen Technischen Hochschule, Stiftung Rubel: Zürich, Switzerland, 1986.
3. Landolt, E.; Kandeler, R. *The Family of Lemnaceae—A Monographic Study, Vol. 2. Biosystematic Investigations in the Family of Duckweeds (Lemnaceae);* Des Geobotanischen Institutes der Eidgenössischen Technischen Hochschule, Stiftung Rubel: Zürich, Switzerland, 1987.
4. Acosta, K.; Appenroth, K.J.; Borisjuk, L.; Edelman, M.; Heinig, U.; Jansen, M.A.K.; Oyama, T.; Pasaribu, B.; Schubert, I.; Sorrels, S.; et al. Return of the Lemnaceae: Duckweed as a model plant system in the genomics and postgenomics era. *Plant Cell* **2021**, *33*, 3207–3234. [CrossRef]
5. Haigh, A.L.; Gibernau, M.; Maurin, O.; Bailey, P.; Carlsen, M.M.; Hay, A.; Leempoel, K.; McGinnie, C.; Mayo, S.; Morris, S.; et al. Target sequence data shed new light on the infrafamilial classification of Araceae. *Am. J. Bot.* **2023**, *110*, 16117. [CrossRef] [PubMed]
6. Tippery, N.P.; Les, D.H.; Appenroth, K.J.; Sree, K.S.; Crawford, D.J.; Bog, M. Lemnaceae and Orontiaceae are phylogenetically and morphologically distinct from Araceae. *Plants* **2021**, *10*, 2639. [CrossRef]
7. Tippery, N.P.; Les, D.H. Tiny Plants with Enormous Potential: Phylogeny and Evolution of Duckweeds. In *The Duckweed Genomes. Compendium of Plant Genomes;* Cao, X., Fourounjian, P., Wang, W., Eds.; Springer: Cham, Switzerland, 2020. [CrossRef]
8. Bog, M.; Sree, K.S.; Fuchs, J.; Hoang, P.T.N.; Schubert, I.; Kuever, J.; Rabenstein, A.; Paolacci, S.; Jansen, M.A.K.; Appenroth, K.-J. A taxonomic revision of *Lemna* sect. *Uninerves* (Lemnaceae). *Taxon* **2020**, *69*, 56–66. [CrossRef]
9. Kandeler, R. Überlebensstrategien bei Wasserlinsen. *Biol. Rundsch.* **1988**, *26*, 347–354.
10. Ziegler, P.; Sree, K.S.; Appenroth, K.-J. Duckweeds for water remediation and toxicity testing. *Toxicol. Environ. Chem.* **2016**, *98*, 1127–1154. [CrossRef]

11. Ziegler, P.; Sree, K.S.; Appenroth, K.-J. The uses of duckweed in relation to water remediation. *Desalination Water Treat.* **2017**, *63*, 327–342. [CrossRef]
12. Bog, M.; Appenroth, K.-J.; Sree, K.S. Key to the determination of taxa of Lemnaceae: An update. *Nordic J. Bot.* **2020**, *38*, e02658. [CrossRef]
13. Bog, M.; Appenroth, K.-J.; Sree, K.S. Duckweed (Lemnaceae): Its molecular taxonomy. *Front. Sust. Food Syst.* **2019**, *3*, 117. [CrossRef]
14. Hoang, P.T.N.; Fuchs, J.; Schubert, V.; Tran, T.B.N.; Schubert, I. Chromosome numbers and genome sizes of all 36 duckweed species (*Lemnaceae*). *Plants* **2022**, *11*, 2674. [CrossRef]
15. Chen, C.; Stepanenko, A.; Lakhneko, O.; Zhou, Y.; Kischenko, O.; Peterson, A.; Cui, D.; Zhu, H.; Xu, J.; Morgun, B.; et al. Biodiversity of duckweed (Lemnaceae) in water reservoirs of Ukraine and China assessed by chloroplast DNA barcoding. *Plants* **2022**, *11*, 1468. [CrossRef]
16. Yosef, A.V.; Ghazaryan, L.; Klamann, L.; Kaufman, K.S.; Baubin, B.; Poodiack, B.; Ran, N.; Gabay, T.; Didi-Cohen, S.; Bog, M.; et al. Diversity and differentiation of duckweed species from Israel. *Plants* **2022**, *11*, 3326. [CrossRef] [PubMed]
17. Ziegler, P.; Adelmann, K.; Zimmer, S.; Schmidt, C.; Appenroth, K.-J. Relative in vitro growth rates of duckweeds (Lemnaceae)—The most rapidly growing higher plants. *Plant Biol.* **2015**, *17* (Suppl. S1), 33–41. [CrossRef]
18. Sree, K.S.; Sudakaran, S.; Appenroth, K.-J. How fast can duckweeds grow? Species and clonal diversity of growth rates in the genus *Wolffia* (Lemnaceae). *Acta Physiol. Plant.* **2015**, *37*, 204. [CrossRef]
19. Cheng, J.J.; Stomp, A.-M. Growing duckweed to recover nutrients from wastewaters and for production of fuel ethanol and animal feed. *Clean* **2009**, *37*, 17–26. [CrossRef]
20. Topp, C.; Henke, R.; Keresztes, A.; Fischer, W.; Eberius, M.; Appenroth, K.-J. A novel mechanism of abscission in fronds of *Lemna minor* L. and the effect of silver ions. *Plant Biol.* **2011**, *13*, 517–523. [CrossRef]
21. Sree, K.S.; Adelmann, K.; Garcia, C.; Lam, E.; Appenroth, K.-J. Natural variance in salt tolerance and induction of starch accumulation in duckweeds. *Planta* **2015**, *241*, 1395–1404. [CrossRef] [PubMed]
22. Sree, K.S.; Appenroth, K.-J. Starch accumulation in duckweeds (Lemnaceae) induced by nutrient deficiency. *Emir. J Food Agric.* **2022**, *34*, 24–212. [CrossRef]
23. Bergmann, B.A.; Cheng, J.; Classen, J.; Stomp, A.-M. In vitro selection of duckweed geographical isolates for potential use in swine lagoon effluent renovation. *Bioresour. Technol.* **2000**, *73*, 13–30. [CrossRef]
24. Bog, M.; Braglia, L.; Morello, L.; Melo, K.I.N.; Schubert, I.; Shchepin, O.N.; Sree, K.S.; Xu, S.; Lam, E.; Appenroth, K.-J. Strategies for intraspecific genotyping of duckweed: Comparison of five orthogonal methods applied to the Giant Duckweed *Spirodela polyrhiza*. *Plants* **2022**, *11*, 3033. [CrossRef]
25. Kuehdorf, K.; Jetschke, G.; Ballani, L.; Appenroth, K.-J. The clonal dependence of turion formation in the duckweed *Spirodela polyrhiza*—An ecogeographical approach. *Physiol. Plant.* **2014**, *150*, 46–54. [CrossRef] [PubMed]
26. Ho, E.K.H.; Bartkowska, M.; Wright, S.J.; Agrawal, A.F. Population genomics of the facultatively asexual duckweed *Spirodela polyrhiza*. *New Phytol.* **2019**, *224*, 1361–1371. [CrossRef]
27. Xu, S.; Stapely, J.; Gablenz, S.; Boyer, J.; Appenroth, K.J.; Sree, K.S.; Gershenzon, J.; Widmer, A.; Huber, M. Low genetic variation is associated with low mutation rate in the giant duckweed. *Nat. Commun.* **2019**, *10*, 1857. [CrossRef] [PubMed]
28. Douhovnikoff, V.; Dodd, R.S. Epigenetics: A potential mechanism for clonal plant success. *Plant Ecol.* **2015**, *216*, 227–237. [CrossRef]
29. Van Antro, M.; Prelovsek, S.; Ivanovic, S.; Gawehns, F.; Wagemaker, N.C.A.M.; Mysara, M.; Horemans, N.; Vergeer, P.; Verhoeven, K.J.F. DNA methylation in clonal duckweed (*Lemna minor* L.) lineages reflects current and historical environmental exposures. *Mol. Ecol.* **2023**, *32*, 428–443. [CrossRef]
30. Bafort, Q.; Wu, T.; Natran, A.; De Clerk, O.; Van de Peer, Y. The immediate effects of polyploidization of *Spirodela polrhiza* change in a strain-specific way along environmental gradients. *Evol. Lett.* **2023**, *7*, 37–47. [CrossRef]
31. Keddy, P.A. Lakes as islands: The distributional ecology of two aquatic plants, *Lemna minor* L., and *L. trisulca* L. *Ecology* **1976**, *57*, 353–359. [CrossRef]
32. Jacobs, D.L. An ecological life-history of *Spirodela polyrhiza* (Greater Duckweed) with emphasis on the turion phase. *Ecol. Monogr.* **1947**, *17*, 437–469. [CrossRef]
33. Landolt, E. How do the Lemnaceae (duckweed family) survive dry conditions? *Bull. Geobot. Inst. ETH* **1997**, *63*, 25–31. [CrossRef]
34. Coughlan, N.E.; Kelly, T.C.; Jansen, M.A.K. Mallard duck (*Anas platyrhynchos*)-mediated dispersal of Lemnaceae: A contribution factor to the spread of invasive *Lemna minuta*? *Plant Biol.* **2015**, *17* (Suppl. S1), 108–114. [CrossRef]
35. Silva, G.G.; Green, A.J.; Weber, V.; Hoffmann, P.; Lovas-Kiss, A.; Stenert, C.; Maltchik, L. Whole angiosperms *Wolffia columbiana* disperse by gut passage through wildfowl in South America. *Biol. Lett.* **2018**, *14*, 20180703. [CrossRef]
36. Paolacci, S.; Jansen, M.A.K.; Stejskal, V.; Kelly, T.C.; Coughlan, N.E. Metabolically active angiosperms survive passage through the digestive tract of a large-bodied waterbird. *R. Soc. Open Sci.* **2023**, *10*, 230090. [CrossRef]
37. Coughlan, N.E.; Kelly, T.C.; Jansen, M.A.K. "Step by step": High frequency short-distance epizoochorous dispersal of aquatic macrophytes. *Biol. Invasions* **2017**, *19*, 625–634. [CrossRef]
38. Les, D.H.; Crawford, D.J.; Kimball, R.T.; Moody, M.L.; Landolt, E. Biogeography of discontinuously distributed hydrophytes: A molecular appraisal of intercontinental disjunctions. *Int. J. Plant Sci.* **2003**, *164*, 917–932. [CrossRef]

39. Kufel, L.; Strzalek, M.; Konieczna, A.; Izdebska, K. The effect of *Stratiotes aloides* L. and nutrients on the growth rate of *Lemna minor* L. *Aquat. Bot.* **2010**, *92*, 168–172. [CrossRef]
40. Smith, S.D.P. The roles of nitrogen and phosphorous in regulating the dominance of floating and submerged aquatic plants in a field mesocosm experiment. *Aquat. Bot.* **2014**, *112*, 1–9. [CrossRef]
41. Smith, S.D.P. The influence of light and nutrient availability on floating plant dominance in forested temporary and semipermanent wetlands. *Hydrobiologia* **2022**, *849*, 2595–2608. [CrossRef]
42. Ceschin, S.; Abati, S.; Thomas, N.; Ellwood, W.; Zuccarello, V. Riding invasion waves: Spatial and temporal patterns of the invasive *Lemna minuta* from its arrival to its spread across Europe. *Aquat. Bot.* **2018**, *150*, 1–8. [CrossRef]
43. Fedoniuk, T.; Bog, M.; Orlov, O.; Appenroth, K.J. *Lemna aequinoctialis* migrates further into temperate continental Europe—A new alien aquatic plant for the Ukraine. *Feddes Repert.* **2022**, *133*, 305–312. [CrossRef]
44. Paolacci, S.; Bog, M.; Lautenschlager, U.; Bonfield, R.; Appenroth, K.-J.; Oberprieler, C.; Jansen, M.A.K. Clonal diversity amongst island populations of alien, invasive *Lemna minuta* Kunth. *Biol. Invasions* **2021**, *23*, 2649–2660. [CrossRef]
45. Paolacci, S.; Harrison, S.; Jansen, M.A.K. The invasive duckweed *Lemna minuta* Kunth displays a different light utilization strategy than native *Lemna minor* Linnaeus. *Aquat. Bot.* **2018**, *146*, 8–14. [CrossRef]
46. Paolacci, S.; Harrison, S.; Jansen, M.A.K. Are alien species necessarily stress-sensitive? A case study on *Lemna minuta* and *Lemna minor*. *Flora* **2018**, *249*, 31–39. [CrossRef]
47. Landolt, E. Physiologische und ökologische Untersuchungen an Lemnaceen. *Ber. Schweiz. Bot. Ges.* **1957**, *67*, 271–410.
48. Docauer, D.M. A Nutrient Basis for the Distribution of the Lemnaceae. Ph.D. Thesis, University of Michigan, Ann Arbor, MI, USA, 1983; 223p.
49. Shang, S.; Zhang, Z.; Li, L.; Chen, J.; Zhang, Y.; Li, X.; Wang, J.; Tang, X. Transcriptome analysis reveals gene expression pattern of *Spirodela polyrhiza* response to heat stress. *Int. J. Biol. Macromol.* **2023**, *225*, 767–775. [CrossRef] [PubMed]
50. Shi, Y.; Ke, X.; Yang, X.; Liu, Y.; Hou, X. Plants response to light stress. *J. Genet. Genom.* **2022**, *49*, 735–747. [CrossRef] [PubMed]
51. Stewart, J.J.; Adams, W.W., III; Lopez-Pozo, M.; Garcia, N.D.; McNamara, M.; Escobar, C.M.; Demmig-Adams, B. Features of the duckweed *Lemna* that support rapid growth under extremes of light intensity. *Cells* **2021**, *10*, 1481. [CrossRef]
52. Strzalek, M.; Kufel, L. Light intensity drives different growth strategies in two duckweed species: *Lemna minor* L. and *Spirodela polyrhiza* (L.) Schleiden. *PeerJ* **2021**, *9*, e21698. [CrossRef]
53. Pip, E.; Simmons, K. Aquatic angiosperms at unusual depths in Shoal Lake. *Canad. Field-Nat.* **1986**, *100*, 354–358.
54. Borisjuk, N.; Peterson, A.A.; Lv, J.; Qu, G.; Luo, Q.; Shi, L.; Chen, G.; Kishchenko, O.; Zhou, Y.; Shi, J. Structural and biochemical properties of duckweed surface cuticle. *Front. Chem.* **2018**, *6*, 317. [CrossRef]
55. McLay, C.L. The effect of pH on the population growth of three species of duckweed: *Spirodela oligorrhiza, Lemna minor* and *Wolffia arrhiza*. *Freshw. Biol.* **1976**, *6*, 125–136. [CrossRef]
56. Hicks, L.E. Ranges of pH-tolerance of the Lemnaceae. *Ohio J. Sci.* **1932**, *32*, 227–234.
57. Paterson, J.B.; Camargo-Valero, M.A.; Baker, A. Uncoupling growth from phosphorus uptake in *Lemna*: Implications for use of duckweed in wastewater remediation and P recovery in temperate climates. *Food Energy Secur.* **2020**, *9*, e124. [CrossRef] [PubMed]
58. Reid, M.S.; Bieleski, R.L. Changes in phosphatase activity in phosphorus-deficient *Spirodela*. *Planta* **1970**, *94*, 273–281. [CrossRef] [PubMed]
59. Reid, M.S.; Bieleski, R.L. Response of *Spirodela oligorrhiza* to phosphorus deficiency. *Plant Physiol.* **1970**, *46*, 609–613. [CrossRef] [PubMed]
60. Bieleski, R.L.; Johnson, P.N. The external location of phosphatase activity in phosphorus-deficient *Spirodela oligorrhiza*. *Aust. J. Biol. Sci.* **1972**, *25*, 707–720. [CrossRef]
61. Knypl, J.S. Molecular forms of phosphatase and ribonuclease in phosphate deficient and N,N-dimethylmorpholinium chloride treated *Spirodela olgorrhiza* (Lemnaceae). *Acta Soc. Bot. Pol.* **1979**, *48*, 65–85. [CrossRef]
62. Les, D.H.; Crawford, D.J. *Landoltia* (Lemnaceae), a new genus of duckweed. *Novon* **1999**, *9*, 530–533. [CrossRef]
63. Morita, N.; Nakazato, H.; Okuyama, H.; Kim, Y.; Thompson, G.A. Evidence for a glycosylinositolphospholipid-anchored alkaline phosphatase in the aquatic plant *Spirodela oligorrhiza*. *Biochim. Biophys. Acta (BBA)-Gen. Subj.* **1996**, *1290*, 53–62. [CrossRef]
64. Nakazato, H.; Okamoto, T.; Ishikawa, K.; Okuyama, H. Purification and characterization of phosphatase inducibly synthesized in *Spirodela oligorrhiza* grown under phosphate-deficient conditions. *Plant Physiol. Biochem.* **1997**, *35*, 437–446.
65. Nakazato, H.; Okamoto, T.; Nishikoori, M.; Washio, K.; Moriza, N.; Haraguchi, K.; Thompson, G.A.; Okuyama, H. The glycosylphosphatidylinositol-anchored phosphatase from *Spirodela oligorrhiza* is a purple acid phosphatase. *Plant Physiol.* **1998**, *118*, 1015–1020. [CrossRef]
66. Hase, A.; Nishikoori, M.; Okuyama, H. Induction of high affinity phosphate transporter in the duckweed *Spirodela oligorrhiza*. *Physiol. Plant.* **2004**, *120*, 271–279. [CrossRef]
67. Bieleski, R. L Effect of phosphorus deficiency on levels of phosphorus compounds in Spirodela. *Plant Physiol.* **1968**, *43*, 1309–1316. [CrossRef] [PubMed]
68. Inhuelsen, D.; Niemeyer, R. Kondensierte Phosphate in *Lemna minor* L. und ihre Beziehungen zu den Nucleinsäuren. *Planta* **1975**, *124*, 159–167. [CrossRef]
69. Scheiner, O.; Pittner, F.; Bollmann, O.; Kandeler, R. Effect of nitrogen deficiency and other factors on phytic acid accumulation in *Lemna gibba* G1. *Z. Pflanzenphysiol.* **1978**, *88*, 295–303. [CrossRef]

70. Plaxton, W.C.; Tran, H.T. Metabolic adaptations of phosphate-starved plants. *Plant Physiol.* **2011**, *156*, 1006–1015. [CrossRef] [PubMed]
71. Yang, J.; Zhao, X.; Chen, Y.; Li, G.; Li, X.; Xia, M.; Sun, Z.; Chen, Y.; Li, Y.; Yao, L.; et al. Identification, structural, and expression analysis of *SPX* genes in Giant Duckweed (*Spirodela polyrhiza*) reveal its role in response to low phosphorus and nitrogen stresses. *Cells* **2022**, *11*, 1167. [CrossRef]
72. Sun, Z.; Guo, W.; Zhao, X.; Chen, Y.; Yang, J.; Xu, S.; Hou, H. Sulfur limitation boosts more starch accumulation than nitrogen or phosphorus in duckweed (*Spirodela polyrhiza*). *Ind. Crops Prod.* **2022**, *185*, 115098. [CrossRef]
73. Li, J.-M.; Du, A.-P.; Liu, P.-H.; Tia, X.-P.; Jin, Y.-L.; Yi, Z.-L.; He, K.; Fang, Y.; Zhao, H. High starch accumulation mechanism and phosphorus utilization efficiency of duckweed (*Landoltia punctata*) under phosphate starvation. *Ind. Crops Prod.* **2021**, *167*, 113529. [CrossRef]
74. Zhao, Z.; Shi, H.-J.; Wang, M.-L.; Cui, L.; Zhao, H.; Zhao, Y. Effect of nitrogen and phosphorus deficiency on transcriptional regulation of genes encoding key enzymes of starch metabolism in duckweed (*Landoltia punctata*). *Plant Physiol. Biochem.* **2015**, *86*, 72–81. [CrossRef]
75. Frick, H. Heterotrophy in the Lemnaceae. *J. Plant Physiol.* **1994**, *144*, 189–193. [CrossRef]
76. Sun, Z.; Zhao, X.; Li, G.; Yang, J.; Chen, Y.; Xia, M.; Hwang, I.; Hou, H. Metabolic flexibility during a trophic transition reveals the phenotypic plasticity of greater duckweed (*Spirodela polyrhiza* 7498). *New Phytol.* **2023**, *238*, 1386–1402. [CrossRef]
77. Firmin, A.; Selosse, M.-A.; Dunand, C.; Elger, A. Mixotrophy in aquatic plants, an overlooked ability. *Trends Plant Sci.* **2022**, *27*, 147–157. [CrossRef]
78. Ullrich-Eberius, C.I.; Novacky, A.; Luettge, U. Active hexose uptake in *Lemna gibba* G1. *Planta* **1978**, *139*, 149–153. [CrossRef] [PubMed]
79. Datko, A.H.; Mudd, S.H. Uptake of amino acids and other organic compounds by *Lemna paucicostata* Hegelm. 6746. *Plant Physiol.* **1985**, *77*, 770–778. [CrossRef]
80. Datko, A.H.; Mudd, S.H. Uptake of choline and ethanolamine by *Lemna paucicostata* Hegelm. 6746. *Plant Physiol.* **1986**, *81*, 285–288. [CrossRef] [PubMed]
81. Borstlap, A.C.; Meenks, J.L.D.; van Eck, W.F.; Bicker, J.T.E. Kinetics and specificity of amino acid uptake by the duckweed *Spirodela polyrhiza* (L.) Schleiden. *J. Exp. Bot.* **1986**, *37*, 1020–1033. [CrossRef]
82. Zhang, Y.; Hu, Y.; Yang, B.; Ma, F.; Lu, P.; Li, L.; Wan, C.; Rayner, S.; Chen, S. Duckweed (*Lemna minor*) as a model plant system for the study of human microbial pathogenesis. *PLoS ONE* **2010**, *5*, e13527. [CrossRef]
83. Baggs, E.L.; Tiersma, M.B.; Abramson, B.W.; Michael, T.P.; Krasileva, K.V. Characterization of defense responses against bacterial pathogens in duckweeds lacking EDS1. *New Phytol.* **2022**, *236*, 1838–1855. [CrossRef]
84. Duff, R.B. The occurrence of apiose in *Lemna* (duckweed) and other angiosperms. *Biochem. J.* **1965**, *94*, 768–772. [CrossRef]
85. Beck, E.; Kandler, O. Apiose als Bestandteil der Zellwand höherer Pflanzen. *Z. Naturforsch. B* **1965**, *20*, 62–67. [CrossRef]
86. Picmanova, M.; Moller, B.L. Apiose: One of nature's witty games. *Glycobiology* **2016**, *26*, 430–442. [CrossRef]
87. Smith, J.; Yang, Y.; Levy, S.; Adelus, O.O.; Hahn, M.G.; O'Neill, M.A.; Bar-Peled, M. Functional characterization of UDP-apiose synthases from bryophytes and green algae provides insight into the appearance of apiose-containing glycans during plant evolution. *J. Biol. Chem.* **2016**, *291*, 21434–21447. [CrossRef]
88. Beck, E. Isolierung und Charakterisierung eines Apigalacturonans aus der Zellwand von *Lemna minor*. *Z. Für Pflanzenphysiol.* **1967**, *57*, 444–461.
89. Ovodova, R.G.; Golovchenko, V.V.; Shashkov, A.S.; Popov, S.V.; Ovodov, Y.S. Structural studies and physiological activity of lemnan, a pectin from *Lemna minor* L. *Russ. J. Bioorganic Chem.* **2000**, *26*, 743–751. [CrossRef]
90. Sowinski, E.E.; Gilbert, S.; Lam, E.; Carpita, N.C. Linkage structure of cell-wall polysaccharides from three duckweed species. *Carbohydr. Polym.* **2019**, *223*, 115119. [CrossRef] [PubMed]
91. Hart, D.A.; Kindel, P.K. Isolation and partial characterization of apigalacturonans from the cell wall of *Lemna minor*. *Biochem. J.* **1970**, *116*, 569–579. [CrossRef]
92. Avci, U.; Pena, M.J.; O'Neill, M.A. Changes in the abundance of cell wall apigalacturonen and xylogalacturonan and conservation of rhamnogalacturonan II structure during the diversification of the Lemnoideae. *Planta* **2018**, *247*, 953–973. [CrossRef]
93. Gülcin, I.; Kirecci, E.; Akkemik, E.; Topal, F.; Hisar, O. Antioxidaant, antibacterial, and anticandidal activities of an aquatic plant: Duckweed (*Lemna minor* L. Lemnaceae). *Turk. J. Biol.* **2010**, *34*, 175–188. [CrossRef]
94. Tan, L.P.; Hamdan, R.H.; Mohamed, M.; Choong, S.S.; Chan, Y.Y.; Lee, S.H. Antibacterial activity and toxicity of duckweed, *Lemna minor* L. (Arales: Lemnaceae) from Malaysia. *Malays. J. Microbiol.* **2018**, *14*, 387–392. [CrossRef]
95. Gonzalez-Renteria, M.; Monroy-Dosta, M.d.C.; Guzman-Garcia, X.; Hernandez-Calderas, I.; Ramos-Lopez, M.A. Antibacterial activity of *Lemna minor* extracts against *Pseudomonas fluorescens* and safety evaluation in a zebrafish model. *Saudi J. Biol. Sci.* **2022**, *27*, 3465–3473. [CrossRef]
96. Treutter, D. Significance of flavonoids in plant resistance and enhancement of their biosynthesis. *Plant Biol.* **2005**, *7*, 581–591. [CrossRef] [PubMed]
97. Pagliuso, D.; Jara, C.E.P.; Grandis, A.; Lam, E.; Ferreira, M.J.P.; Buckeridge, M.S. Flavonoids from duckweeds: Potential applications in the human diet. *RSC Adv.* **2020**, *10*, 44981–44988. [CrossRef]
98. Lee, G.; Choi, H.; Joo, Y.; Kim, S.-G. Flavone-associated resistance of two *Lemna* species to duckweed weevil attack. *Ecol. Evol.* **2022**, *12*, e9459. [CrossRef]

99. Trivedi, P.; Leach, J.E.; Tringe, S.G.; Sa, T.; Singh, B.K. Plant-microbiome interactions: From community assembly to plant health. *Nat. Rev. Microbiol.* **2020**, *18*, 607–621. [CrossRef] [PubMed]
100. Appenroth, K.-J.; Ziegler, P.; Sree, K.S. Duckweed as a model organism for investigating plant-microbe interactions in an aquatic environment and its applications. *Endocytobiosis Cell Res.* **2016**, *27*, 94–106.
101. Ishizawa, H.; Tada, M.; Kuroda, M.; Inoue, D.; Futama, H.; Ike, M. Synthetic bacterial community of duckweed: A simple and stable system to study plant-microbe interactions. *Microbes Environ.* **2020**, *35*, ME20112. [CrossRef] [PubMed]
102. Acosta, K.; Xu, J.; Gilbert, S.; Denison, E.; Brinkman, T.; Lebeis, S.; Lam, E. Duckweed hosts a taxonomically similar bacterial assemblage as the terrestrial leaf microbiome. *PLoS ONE* **2020**, *15*, e228560. [CrossRef]
103. Bunyoo, C.; Roongsattham, P.; Khumwan, S.; Phonakham, J.; Wonnapinij, P.; Thamchaipenet, A. Dynamic alteration of microbial communities of duckweeds from nature to nutrient-deficient conditions. *Plants* **2022**, *11*, 2915. [CrossRef]
104. Yamaga, F.; Wasio, K.; Morikawa, M. Sustainable biodegradation of phenol by *Acinetobacter calcoaceticus* P23 isolated from the rhizosphere of duckweed *Lemna aoukikusa*. *Environ. Sci. Technol.* **2010**, *44*, 6470–6474. [CrossRef]
105. Yamakawa, Y.; Jog, R.; Morikawa, M. Effects of co-inoculation of two different plant growth-promoting bacteria on duckweed. *Plant Growth Regul.* **2018**, *86*, 287–296. [CrossRef]
106. Yoneda, Y.; Yamamoto, K.; Mkino, A.; Taanak, Y.; Meng, X.-Y.; Hashimoto, J.; Shin-ya, K.; Satoh, N.; Fujie, M.; Toyama, T.; et al. Novel plant-associated *Acidobacteria* promotes growth of common floating aquatic plants, duckweeds. *Microorganisms* **2021**, *9*, 1133. [CrossRef]
107. Shen, M.; Yin, Z.; Xia, D.; Zhao, Q.; Kang, Y. Combination of heterotrophic nitrifying bacterium and duckweed (*Lemna gibba* L.) enhances ammonium nitrogen removal efficiency in aquaculture water via mutual growth promotion. *J. Gen. Appl. Microbiol.* **2019**, *65*, 151–160. [CrossRef] [PubMed]
108. Shuvro, S.K.; Jog, R.; Morikawa, M. Diazotropic bacterium *Azotobacter vinelandii* as a mutualistic growth promoter of an aquatic plant: *Lemna minor*. *Plant Growth Regul.* **2023**, *100*, 171–180. [CrossRef]
109. Ishizawa, H.; Kuroda, M.; Inoue, K.; Inoue, D.; Morikawa, M.; Ike, M. Colonization and competition dynamics of plant growth-promoting/inhibiting bacteria in the phytosphere of the duckweed *Lemna minor*. *Microb. Ecol.* **2019**, *77*, 440–450. [CrossRef] [PubMed]
110. Ishizawa, H.; Kuroda, M.; Inoue, D.; Morikawa, M.; Ike, M. Community dynamics of duckweed-associated bacteria upon inoculation of plant growth-promoting bacteria. *FEMS Microbiol. Ecol.* **2020**, *96*, fiaa101. [CrossRef]
111. Radulovic, O.; Stankovic, S.; Uzelac, B.; Tadic, V.; Trifunovic-Momcilov, M.; Lozo, J.; Markovic, M. Phenol removal capacity of the common duckweed (*Lemna minor* L.) and six phenol-resistant bacterial trains from its rhizosphere: In vitro evaluation at high phenol concentrations. *Plants* **2020**, *9*, 599. [CrossRef]
112. Mu, D.; Lin, W.; Luo, J. Non-negligible effect of native rhizobacteria on co-operation with plant growth regulators improve tolerance to cadmium: A case study using duckweed *Spirodela polyrhiza* as indicating plant. *J. Plant Growth Regul.* **2023**. [CrossRef]
113. Ishizawa, H.; Tad, M.; Kuroda, M.; Inoue, D.; Ike, M. Performance of plant growth-promoting bacterium of duckweed under different kinds of abiotic stress factors. *Biocatal. Agric. Biotechnol.* **2019**, *19*, 101146. [CrossRef]
114. O'Brien, A.M.; Laurich, J.; Lash, E.; Frederikson, M.E. Mutualistic outcomes across plant populations, microbes, and environments in the duckweed *Lemna minor*. *Microb. Ecol.* **2020**, *80*, 384–397. [CrossRef]
115. Schaefer, M.; Xu, S. The effects of microbiota on the herbivory resistance of the giant duckweed are plant genotype-dependent. *Plants* **2022**, *11*, 3317. [CrossRef]
116. Mkandawire, M.; Teixeira da Silva, J.A.; Dudel, G. The *Lemna* bioassay: Contemporary issues as the most standardized plant bioassay for aquatic ecotoxicology. *Crit. Rev. Environ. Sci. Technol.* **2014**, *44*, 154–191. [CrossRef]
117. Brain, R.A.; Cedergreen, N. Biomarkers in aquatic plants: Selection and utility. *Rev. Environ. Contam. Toxicol.* **2008**, *198*, 49–109. [CrossRef]
118. Ziegler, P.; Appenroth, K.J.; Sree, K.S. Duckweed biomarkers for identifying toxic water contaminants? *Environ. Sci. Pollut. Res.* **2019**, *26*, 14797–14822. [CrossRef] [PubMed]
119. Wang, W.; Li, R.; Zhu, Q.; Tang, X.; Zho, Q. Transcriptomic and physiological analysis of common duckweed *Lemna minor* responses to NH_4^+ toxicity. *BMC Plant Biol.* **2016**, *16*, 92. [CrossRef]
120. Xu, H.; Yu, C.; Xia, X.; Li, M.; Li, H.; Wang, Y.; Wang, S.; Wang, C.; Ma, Y.; Zhou, G. Comparative transcriptome analysis of duckweed (*Landoltia punctata*) in response to cadmium provides insights into molecular mechanisms underlying hyperaccumulation. *Chemosphere* **2018**, *190*, 150–165. [CrossRef]
121. Huang, W.; Kong, R.; Chen, L.; An, Y. Physiological responses and antibiotic-degradation capacity of duckweed (*Lemna aequinoctialis*) exposed to streptomycin. *Front. Plant Sci.* **2022**, *13*, 1065199. [CrossRef]
122. Krzeslowska, M.; Rabeda, I.; Basinska, A.; Lewandowski, M.; Mellerowicz, E.J.; Napieralska, A.; Samardakiewicz, W.A. Pectinous cell wall thickenings formation—A common defense strategy to cope with Pb. *Environ. Pollut.* **2016**, *214*, 354–361. [CrossRef] [PubMed]
123. Fu, L.; Ding, Z.; Sun, X.; Zhang, H. Physiological and transcriptomic analysis reveals distorted ion homeostasis and responses in the freshwater plant *Spirodela polyrhiza* L. under salt stress. *Genes* **2019**, *10*, 743. [CrossRef]
124. Zhao, Z.; Shi, H.; Kang, X.; Liu, C.; Chen, L.; Liang, X.; Jin, L. Inter- and intra-specific competition of duckweed under multiple heavy metal contaminated water. *Aquat. Toxicol.* **2017**, *192*, 216–223. [CrossRef]

125. Shi, H.; Duan, M.; Li, C.; Zhang, Q.; Liu, C.; Liang, S.; Guan, Y.; Kang, X.; Zhao, Z.; Xiao, G. The change of accumulation of heavy metal drive interspecific facilitation und copper and cold stress. *Aquat. Toxicol.* **2020**, *225*, 105550. [CrossRef] [PubMed]
126. Tang, J.; Zhang, Y.; Cui, Y.; Ma, J. Effects of a rhizobacterium on the growth and chromium remediation by *Lemna minor*. *Environ. Sci. Pollut. Res.* **2015**, *22*, 9686–9693. [CrossRef]
127. Zhao, Z.; Shi, H.; Liu, C.; Kang, X.; Chen, L.; Liang, X.; Jin, L. Duckweed diversity decreases heavy metal toxicity by altering the metabolic function of associated microbial communities. *Chemosphere* **2018**, *203*, 76–82. [CrossRef] [PubMed]
128. Burns, M.; Hanson, M.L.; Prosser, R.S.; Crossan, A.N.; Kennedy, I.R. Growth recovery of *Lemna gibba* and *Lemna minor* following a 7-day exposure to the herbicide diuron. *Bull. Environ. Contam. Toxicol.* **2015**, *95*, 150–156. [CrossRef]
129. Wersal, R.M.; Turnage, G. Using contact herbicides for control of duckweed and watermeal with implications for management. *J. Aquat. Plant Manag.* **2021**, *59*, 40–45.
130. Koschnick, T.J.; Haller, W.T.; Glasgow, L. Documentation of Landoltia (*Landoltia punctata*) resistance to diquat. *Weed Sci.* **2006**, *54*, 615–619. [CrossRef]
131. Olah, V.; Hepp, A.; Lakatos, G.; Meszaros, I. Cadmium-induced turion formation of *Spirodela polyrhiza* (L.) Schleiden. *Acta Biol. Szeged.* **2014**, *58*, 103–108.
132. Gostynska, J.; Pankiewicz, R.; Romanowska-Duda, Z.; Messyasz, B. Overview of allelopathic potential of *Lemna minor* L. obtained from a shallow eutrophic lake. *Molecules* **2022**, *27*, 3428. [CrossRef]
133. Wang, F.; Wang, S.; Xu, S.; Shen, J.; Cao, L.; Sha, Z.; Chu, Q. A non-chemical weed control strategy, introducing duckweed into the paddy field. *Pest Manag. Sci.* **2022**, *78*, 3654–3663. [CrossRef] [PubMed]
134. Zhang, L.-M.; Jin, Y.; Yao, S.-M.; Lei, N.-F.; Cen, J.-S.; Zhang, Q.; Yu, F.-H. Growth and morphological responses of duckweed to clonal fragmentation, nutrient availability, and population density. *Front. Plant Sci.* **2020**, *11*, 618. [CrossRef]
135. Faerber, E.; Koenigshofer, H.; Kandeler, R. Ethylene production and overcrowding in Lemnaceae. *J. Plant Physiol.* **1986**, *124*, 379–384. [CrossRef]
136. Czopek, M. Studies on the external factors inducing the formation of turions in *Spirodela polyrrhiza* (L.) Schleiden. *Acta Soc. Bot. Pol.* **1963**, *32*, 199–206. [CrossRef]
137. Appenroth, K.-J.; Hertel, W.; Augsten, H. Photophysiology of turion germination in *Spirodela polyrhiza* (L.) Schleiden. The cause of germination inhibition by overcrowding. *Biol. Plant.* **1990**, *32*, 420–428. [CrossRef]
138. Szabo, S.; Koleszar, G.; Zavanyi, G.; Nagy, P.T.; Braun, M.; Hilt, S. The mechanisms sustaining a stable state of submerged macrophyte dominance against free-floating competitors. *Front. Plant Sci.* **2022**, *13*, 963579. [CrossRef]
139. Paolacci, S.; Jansen, M.A.K.; Harrison, S. Competition between *Lemna minuta*, *Lemna minor*, and *Azolla filiculoides*. Growing fast or being steadfast? *Front. Chem.* **2018**, *6*, 207. [CrossRef] [PubMed]
140. Strzalek, M.; Kufel, L.; Wysokinska, U. How does *Stratiotes aloides* L. affect the growth and turion formation of *Spirodela polyrhiza* (L.) Schleiden? *Aquat. Bot.* **2019**, *154*, 45–52. [CrossRef]
141. Bich, T.T.N.; Kato-Noguchi, H. Allelopathic potential of two aquatic plants, duckweed (*Lemna minor* L.) and water lettuce (*Pistia stratiotes* L.) on terrestrial plant species. *Aquat. Bot.* **2012**, *103*, 30–36. [CrossRef]
142. Bich, T.T.N.; Ohno, O.; Suenaga, K.; Kato-Noguchi, H. Isolation and identification of an allelopathic substance from duckweed (*Lemna minor* L.). *Allelopath. J.* **2013**, *32*, 213–222.
143. Bittencourt-Oliveira, M.C.; Hereman, T.C.; Cordeiro-Araujo, M.K.; Macedo-Silva, I.; Dias, C.T.; Sasaki, F.F.C.; Mura, A.N. Phytotoxicity associated to microcystins: A review. *Braz. J. Biol.* **2014**, *74*, 753–760. [CrossRef] [PubMed]
144. Pham, T.-L.; Utsumi, M. An overview of the accumulation of microcystins in aquatic ecosystems. *J. Environ. Manag.* **2018**, *213*, 520–529. [CrossRef]
145. Weiß, J.; Liebert, H.-P.; Braune, W. Influence of microcystin-RR on growth and photosynthetic capacity of the duckweed *Lemna minor* L. *J. Appl. Bot.* **2000**, *74*, 100–105.
146. Mitrovic, S.M.; Allis, O.; Furey, A.; James, K.J. Bioaccumulation and harmful effects of microcystin-LR in the aquatic plants *Lemna minor* and *Wolffia arrhiza* and the filamentous alga *Cladophora fracta*. *Ecotoxicol. Environ. Saf.* **2005**, *61*, 345–352. [CrossRef]
147. Wan, X.; Steinman, A.D.; Shu, X.; Cao, Q.; Yao, L.; Xie, L. Combined toxic effects of microcystin-LR and phenanthrene on growth and antioxidant system of duckweed (*Lemna gibba* L.). *Ecotoxicol. Environ. Saf.* **2019**, *185*, 109668. [CrossRef]
148. Sagrane, S.; El ghazali, I.; Ouahid, Y.; El Hassnib, M.; El Hadami, I.; Bouarab, L.; del Campo, F.F.; Oudra, B.; Vasconcelos, V. Phytotoxic effects of cyanobacteria extract on the aquatic plant Lemna gibba: Microcystin accumulation, detoxification and oxidative stress induction. *Aquat. Toxicol.* **2007**, *85*, 284–294. [CrossRef] [PubMed]
149. Yang, G.-L.; Huang, M.-J.; Tan, A.-J.; Ly, S.-M. Joint effects of naphthalene and microcystin-LR on physiological responses and toxin bioaccumulation of *Landoltia punctata*. *Aquat. Toxicol.* **2021**, *231*, 105710. [CrossRef] [PubMed]
150. LeBlanc, S.; Pick, F.R.; Aranda-Rodriguez, R. Allelopathic effects of the toxic cyanobacterium *Microcystis aeruginosa* on duckweed, *Lemna gibba* L. *Environ. Toxicol.* **2005**, *20*, 67–73. [CrossRef] [PubMed]
151. Kovats, N.; Acs, A.; Paulovits, G.; Vasas, G. Response of *Lemna minor* clones to *Microcystis* toxicity. *Appl. Ecol. Environ. Res.* **2011**, *9*, 17–26. [CrossRef]
152. Jang, M.-H.; Ha, K.; Takamura, N. Reciprocal allelopathic responses between toxic cyanobacteria (*Microcystis aeruginosa*) and duckweed (*Lemna japonica*). *Toxicon* **2007**, *49*, 727–739. [CrossRef]
153. Dan, L.; Peng, L.; Zhiquiang, Y.; Na, L.; Lunguang, Y.; Lingling, C. Allelopathic inhibition of the extracts of *Landoltia punctata* on *Microcystis aeruginosa*. *Plant Signal. Behav.* **2022**, *17*, e2058256. [CrossRef]

154. Armitage, D.W.; Jones, S.E. Negative frequency-dependent growth underlies the stable coexistence of two cosmopolitan aquatic plants. *Ecology* **2019**, *100*, e02657. [CrossRef]
155. Appenroth, K.-J.; Schubert, I.; Sree, K.S. Das Comeback der Wasserlinsen. *Nat. Rundsch.* **2022**, *78*, 70–75.
156. Jewell, M.D.; Bell, G. Overwintering and re-emergence in *Lemna minor*. *Aquat. Bot.* **2023**, *186*, 103633. [CrossRef]
157. Adamec, L. Ecophysiological characteristics of turions of aquatic plants: A review. *Aquat. Bot.* **2018**, *148*, 64–77. [CrossRef]
158. Smart, C.C.; Trewavas, A.J. Abscisic-acid-induced turion formation in *Spirodela polyrrhiza* L. II. Ultrastructure of the turion. *Plant Cell Environ.* **1983**, *6*, 515–522. [CrossRef]
159. Appenroth, K.-J.; Bergfeld, R. Photophysiology of turion germination in *Spirodela polyrhiza* (L.) Schleiden. XI. Structural changes during red light induced responses. *J. Plant Physiol.* **1993**, *141*, 583–588. [CrossRef]
160. Appenroth, K.-J.; Keresztes, A.; Krzysztofowicz, E.; Gabrys, H. Light-induced degradation of starch granules in turions of *Spirodela polyrhiza* studied by electron microscopy. *Plant Cell Physiol.* **2011**, *52*, 384–391. [CrossRef]
161. Godziemba-Czyz, J. Characteristic of vegetative and resting forms in *Wolffia arrhiza* (L.) Wimm. *Acta Soc. Bot. Pol.* **1970**, *39*, 421–443. [CrossRef]
162. Czopek, M. Photosynthesis and respiration of turions and vegetative fronds of *Spirodela polyrrhiza*. *Acta Soc. Bot. Pol.* **1967**, *36*, 87–96. [CrossRef]
163. Appenroth, K.-J.; Opfermann, J.; Hertel, W.; Augsten, H. Photophysiology of turion germination in *Spirodela polyrhiza* (L.) Schleiden. II. Influence of after-ripening on germination kinetics. *J. Plant Physiol.* **1989**, *135*, 274–279. [CrossRef]
164. Smart, C.C.; Trewavas, A.J. Abscisic-acid-induced turion formation in *Spirodela polyrrhiza* L. I. Production and development of the turion. *Plant Cell Environ.* **1983**, *6*, 507–514. [CrossRef]
165. Appenroth, K.-J. Co-action of temperature and phosphate in inducing turion formation in *Spirodela polyrhiza* (Great Duckweed). *Plant Cell Environ.* **2002**, *25*, 1079–1085. [CrossRef]
166. Appenroth, K.-J.; Nickel, G. Turion formation in *Spirodela polyrhiza*: The environmental signals that induce the developmental process in nature. *Physiol. Plant.* **2010**, *138*, 312–320. [CrossRef] [PubMed]
167. Das, R.R.; Gopal, B. Vegetative propagation in *Spirodela polyrhiza*. *Trop. Ecol.* **1969**, *10*, 270–277.
168. Smart, C.C.; Fleming, A.J.; Chaloupkova, K.; Hanke, D.E. The physiological role of abscisic acid in eliciting turion morphogenesis. *Plant Physiol.* **1995**, *108*, 623–632. [CrossRef] [PubMed]
169. Wang, W.; Wu, Y.R.; Messing, J. RNA-Seq transcriptome analysis in *Spirodela* dormancy without reproduction. *BMC Genom.* **2014**, *15*, 60. [CrossRef]
170. Appenroth, K.-J.; Hertel, W.; Augsten, H. Phytochrome control of turion formation in *Spirodela polyrhiza* L. Schleiden. *Ann. Bot.* **1990**, *66*, 163–168. [CrossRef]
171. Appenroth, K.-J. No photoperiodic control of the formation of turions in eight clones of *Spirodela polyrhiza*. *J. Plant Physiol.* **2003**, *160*, 1329–1334. [CrossRef]
172. Bartley, M.R.; Spence, D.H.N. Dormancy and propagation in helophytes and hydrophytes. *Arch. Hydrobiol. Beih. Ergeb. Limnol.* **1987**, *27*, 139–155.
173. Appenroth, K.-J. Clonal differences in the formation of turions are independent of the specific turion-inducing signal in *Spirodela polyrhiza* (Great Duckweed). *Plant Biol.* **2002**, *4*, 688–693. [CrossRef]
174. Appenroth, K.-J.; Gabrys, H. Ion antagonism between calcium and magnesium in phytochrome-mediated degradation of storage starch in *Spirodela polyrhiza*. *Plant Sci.* **2003**, *165*, 1261–1265. [CrossRef]
175. Appenroth, K.-J.; Adamec, L. Specific turion yields of different clones of *Spirodela polyrhiza* depend on external phosphate thresholds. *Plant Biol.* **2015**, *17* (Suppl. S1), 125–129. [CrossRef]
176. Henssen, A. Die Dauerorgane von *Spirodela polyrrhiza* (L.) SCHLEID. in physiologischer Betrachtung. *Flora* **1954**, *141*, 525–566.
177. van der Schoot, C.; Rinne, P.L.H. Dormancy cycling at the shoot apical meristem: Transitioning between self-organisation and self-arrest. *Plant Sci.* **2011**, *180*, 120–131. [CrossRef] [PubMed]
178. Appenroth, K.-J.; Gabrys, H. Light-induced starch degradation in non-dormant turions of *Spirodela polyrhiza*. *Photochem. Photobiol.* **2001**, *73*, 77–82. [CrossRef] [PubMed]
179. Ley, S.; Dolger, K.; Appenroth, K.-J. Carbohydrate metabolism as a possible physiological modulator of dormancy in turions of *Spirodela polyrhiza* (L.) Schleiden. *Plant Sci.* **1997**, *129*, 1–7. [CrossRef]
180. Appenroth, K.-J.; Palharini, L.; Ziegler, P. Low-molecular weight carbohydrates modulate dormancy and are required for post-germination growth in turions of *Spirodela polyrhiza*. *Plant Biol.* **2013**, *15*, 284–291. [CrossRef] [PubMed]
181. Newton, R.J.; Shelton, D.R.; Disharoon, S.; Duffey, J.E. Turion formation and germination in *Spirodela polyrhiza*. *Am. J. Bot.* **1978**, *65*, 421–428. [CrossRef]
182. Augsten, H.; Kunz, E.; Appenroth, K.-J. Photophysiology of turion germination in *Spirodela polyrrhiza* (L.) SCHLEIDEN. I. Phytochrome-mediated responses of light- and dark-grown turions. *J. Plant Physiol.* **1988**, *132*, 90–93. [CrossRef]
183. Appenroth, K.-J.; Augsten, H. Photophysiology of turion germination in *Spirodela polyrhiza* (L.) Schleiden–V. Demonstration of a calcium-requiring phase during phytochrome-mediated germination. *Photochem. Photobiol.* **1990**, *52*, 61–65. [CrossRef]
184. Pasaribu, B.; Acosta, K.; Aylward, A.; Liang, L.; Abramson, B.W.; Colt, K.; Hartwick, N.T.; Shanklin, J.; Michael, T.P.; Lam, E. Genomics of turions from the Greater Duckweed reveal its pathways for dormancy and re-emergence strategy. *New Phytol.* **2023**, *239*, 116–131. [CrossRef]
185. Pieterse, A.H. Is flowering in Lemnaceae stress-induced? A review. *Aquat. Bot.* **2013**, *104*, 1–4. [CrossRef]

186. Kandeler, R. Flowering in the *Lemna* system. *Phyton* **1984**, *24*, 113–124.
187. Kandeler, R. Lemnaceae. In *Handbook of Flowering*; Halevy, A.H., Ed.; CRC Press: Boca Raton, FL, USA, 1985.
188. Rimon, D.; Galun, E. Morphogenesis of *Wolffia microscopica*: Frond and flower development. *Phytomorphology* **1968**, *18*, 364–372.
189. Sree, K.S.; Maheshwari, S.C.; Boka, K.; Khurana, J.P.; Keresztes, A. The duckweed *Wolffia microscopica*: A unique aquatic monocot. *Flora* **2015**, *210*, 31–39. [CrossRef]
190. Appenroth, K.-J. Historical account: Riklef Kandeler (1927–2015). *Duckweed Forum* **2021**, *9*, 5–8.
191. Appenroth, K.-J.; Sharma, R.P.; Sree, K.S. Obituary: Jitendra, P. Khurana (30. 10. 1954–72. 10. 2021). *Duckweed Forum* **2022**, *10*, 16–20.
192. Maheshwari, S.C.; Seth, P.N. Induction of flowering in *Wolffia microscopica* by the iron salt of ethylenediamine-di-o-hydroxyphenylacetic acid (Fe-EDDHA). *Z. Pflanzenphysiol.* **1966**, *55*, 89–91.
193. Cleland, C.F.; Ajami, A. Isolation and identification of the flower-inducing factor from aphid honeydew as being salicylic acid. *Plant Physiol.* **1974**, *54*, 904–906. [CrossRef]
194. Pieterse, A.H.; Mueller, L.J. Induction of flowering in *Lemna gibba* G3 under short-day conditions. *Plant Cell Physiol.* **1977**, *18*, 45–53. [CrossRef]
195. Khurana, J.P.; Maheshwari, S.C. A comparison of the effects of chelates, salicylic acid and benzoic acid on growth and flowering of *Spirodela polyrrhiza*. *Plant Cell Physiol.* **1986**, *27*, 919–924. [CrossRef]
196. Tanaka, O.; Cleland, C.F.; Hillman, W.S. Inhibition of flowering in the long-day plant *Lemna gibba* G3 by Hutner's medium and its reversal by salicylic acid. *Plant Cell Physiol.* **1979**, *20*, 839–846. [CrossRef]
197. Faerber, E.; Kandeler, R. Significance of calcium ions in the overcrowding effect in *Spirodela polyrrhiza* P 143. *J. Plant Physiol.* **1989**, *135*, 94–98. [CrossRef]
198. Pieterse, A.H. Specific interactions in the physiology of flowering and gibbosity of *Lemna gibba* G3. *Plant Cell Physiol.* **1976**, *17*, 713–720. [CrossRef]
199. Cleland, C.F. Isolation of flower-inducing and flower-inhibitory factors from aphid honeydew. *Plant Physiol.* **1974**, *54*, 899–903. [CrossRef] [PubMed]
200. Kobayashi, Y.; Weigel, D. Move on up, it's time for change—Mobile signals controlling photoperiod-dependent flowering. *Genes Dev.* **2007**, *21*, 2371–2384. [CrossRef] [PubMed]
201. Yoshida, A.; Taoka, K.-I.; Hoska, A.; Tanaka, K.; Kobayashi, H.; Muranaka, T.; Toyooka, K.; Oyama, T.; Tsuji, H. Characterization of frond and flower development and identification of FT and FD genes from duckweed *Lemna aequinoctialis* Nd. *Front. Plant Sci.* **2021**, *12*, 697206. [CrossRef] [PubMed]
202. Muranaka, T.; Ito, S.; Kudoh, H.; Oyama, T. Circadian-period variation underlies the local adaptation of photoperiodism in the short-day plant *Lemna aequinoctialis*. *iScience* **2022**, *25*, 104634. [CrossRef]
203. Gramzow, L.; Theissen, G. Stranger than fiction: Loss of MADS-box genes during evolutionary miniaturization of the duckweed body plan. In *The Duckweed Genomes, Compendium of Plant Genomes*; Cao, X.H., Fourounjian, P., Wang, W., Eds.; Springer Nature: Cham, Germany, 2020.
204. Fu, L.; Huang, M.; Han, B.; Sun, X.; Sree, K.S.; Appenroth, K.-J.; Zhang, J. Flower induction, microscope-aided cross-pollination, and seed production in the duckweed *Lemna gibba* with discovery of a male-sterile clone. *Sci. Rep.* **2017**, *7*, 3047. [CrossRef]

plants

MDPI

Review

Lemnaceae and Orontiaceae Are Phylogenetically and Morphologically Distinct from Araceae

Nicholas P. Tippery [1], Donald H. Les [2], Klaus J. Appenroth [3], K. Sowjanya Sree [4], Daniel J. Crawford [5,6] and Manuela Bog [7,*]

1 Department of Biological Sciences, University of Wisconsin-Whitewater, Whitewater, WI 53190, USA; tipperyn@uww.edu
2 Department of Ecology and Evolutionary Biology, University of Connecticut, Storrs, CT 06269, USA; don.les@uconn.edu
3 Matthias Schleiden Institute—Plant Physiology, University of Jena, D-07743 Jena, Germany; klaus.appenroth@uni-jena.de
4 Department of Environmental Science, Central University of Kerala, Periye 671320, India; ksowsree@gmail.com or ksowsree9@cukerala.ac.in
5 Department of Ecology and Evolutionary Biology, University of Kansas, Lawrence, KS 66045, USA; dcrawfor@ku.edu
6 Natural History Museum and Biodiversity Research Center, University of Kansas, Lawrence, KS 66045, USA
7 Institute of Botany and Landscape Ecology, University of Greifswald, D-17489 Greifswald, Germany
* Correspondence: manuela.bog@uni-greifswald.de; Tel.: +49-38344204146

Abstract: Duckweeds comprise a distinctive clade of pleustophytic monocots that traditionally has been classified as the family Lemnaceae. However, molecular evidence has called into question their phylogenetic independence, with some authors asserting instead that duckweeds should be reclassified as subfamily Lemnoideae of an expanded family Araceae. Although a close phylogenetic relationship of duckweeds with traditional Araceae has been supported by multiple studies, the taxonomic disposition of duckweeds must be evaluated more critically to promote nomenclatural stability and utility. Subsuming duckweeds as a morphologically incongruent lineage of Araceae effectively eliminates the family category of Lemnaceae that has been widely used for many years. Instead, we suggest that Araceae subfamily Orontioideae should be restored to family status as Orontiaceae, which thereby would enable the recognition of three morphologically and phylogenetically distinct lineages: Araceae, Lemnaceae, and Orontiaceae.

Keywords: aquatic plants; Araceae; duckweeds; Lemnoideae; molecular phylogenetics; taxonomy

check for updates

Citation: Tippery, N.P.; Les, D.H.; Appenroth, K.J.; Sree, K.S.; Crawford, D.J.; Bog, M. Lemnaceae and Orontiaceae Are Phylogenetically and Morphologically Distinct from Araceae. *Plants* **2021**, *10*, 2639. https://doi.org/10.3390/plants10122639

Academic Editor: Pablo García Murillo

Received: 14 November 2021
Accepted: 26 November 2021
Published: 30 November 2021

Publisher's Note: MDPI stays neutral with regard to jurisdictional claims in published maps and institutional affiliations.

1. Introduction

1.1. Taxonomic History of Araceae

Araceae Juss. (aroids) are one of the larger angiosperm families, comprising around 5000 species that are distributed primarily in tropical latitudes [1,2]. These plants have a variety of identifying characteristics, including calcium oxalate crystals and tiny flowers that are borne on a distinctive spadix inflorescence [1] (Figure 1). The application of molecular data to angiosperm phylogenetic analyses has sparked greater confidence in angiosperm classification, and Araceae are no exception. Molecular data have validated several monophyletic subfamilies and enabled a richer interpretation of their morphological evolution [3–6]. Molecular data also produced a somewhat unexpected result, namely that duckweeds, long classified as the separate family Lemnaceae Martinov, nom. cons. [7,8], were descended from the same common ancestor as Araceae [3]. Because a number of molecular phylogenetic analyses have grouped duckweeds with Araceae, modern taxonomic treatments have begun to assign duckweeds to an aroid subfamily (Lemnoideae Engler) in order to preserve Araceae as monophyletic [3–6,9,10]. However, many of the same studies have shown Lemnaceae to be phylogenetically and morphologically distinct [5,11,12], with the duckweed lineage diverging around 104 Ma (Figure 2) [6]. Although the expansion of Araceae to include duckweeds is one solution to reconcile the phylogenetic observations, there also are other options that would allow the primary taxonomic categories to remain consistent with phylogenetic lineages; yet, these seem to have been given little consideration.

Figure 1. Species of Arales, divided among the family-level categories that are described herein: Orontiaceae (**A–D**), Lemnaceae (**E**,**F**), and Araceae (**G–K**). (**A**) Vegetative growth and inflorescence (inset) of *Orontium aquaticum* (photo credit: Wolfgang Pomper), (**B**) *Symplocarpus foetidus* (photo credit: Claire O'Neill), (**C**) Inflorescence of *Gymnostachys anceps* (photo credit: Leith Woodall), (**D**) *Lysichiton americanus* with inflorescence (photo credit: Ryan Kurtz), (**E**) Fronds of *Spirodela polyrhiza* surrounded by *Wolffia globosa*, (**F**) Light micrograph of *Wolffia microscopica* bearing the floral organs on the dorsal side (anther lobes seen on the top), (**G**) *Anthurium andraeanum* with inflorescence, (**H**) *Monstera deliciosa* with inflorescence (photo credit: Wolfgang Pomper), (**I**) *Calla palustris* with inflorescence, (**J**) *Amorphophallus konjac* with inflorescence (photo credit: Wolfgang Pomper), (**K**) Inflorescence of *Pistia stratiotes* (photo credit: Bo-Fu Sun).

An alternative option that would preserve both Araceae and Lemnaceae as monophyletic would be to restore the araceous lineages of Gymnostachydoideae Bogner and Nicolson and Orontioideae Mayo, Bogner and Boyce collectively to the family level as Orontiaceae Bartl. These plants, often referred to as 'proto-Araceae', are the phylogenetic sibling lineage of duckweeds plus the remaining Araceae, having diverged around 122 Ma (Figure 2) [3–6]. For clarity, we will refer to the clade that includes Gymnostachydoideae, Lemnoideae, and Orontioideae as Araceae s.l. (*sensu lato*) (as suggested by APG [10]), and the clade of Araceae lacking these subfamilies (sometimes referred to as the 'true Araceae' [1,4,5,13] or the 'core Araceae' [14–17]) as Araceae s.s. (*sensu stricto*) (i.e., our recommended taxonomic disposition). We propose that taxa in Araceae s.l. should be divided among the established families Araceae s.s., Lemnaceae, and Orontiaceae. Besides being separated by ancient and long evolutionary branches, these families are morphologically divergent and well suited to a classification scheme that highlights their distinctness while similarly preserving more traditional morphological concepts for both groups.

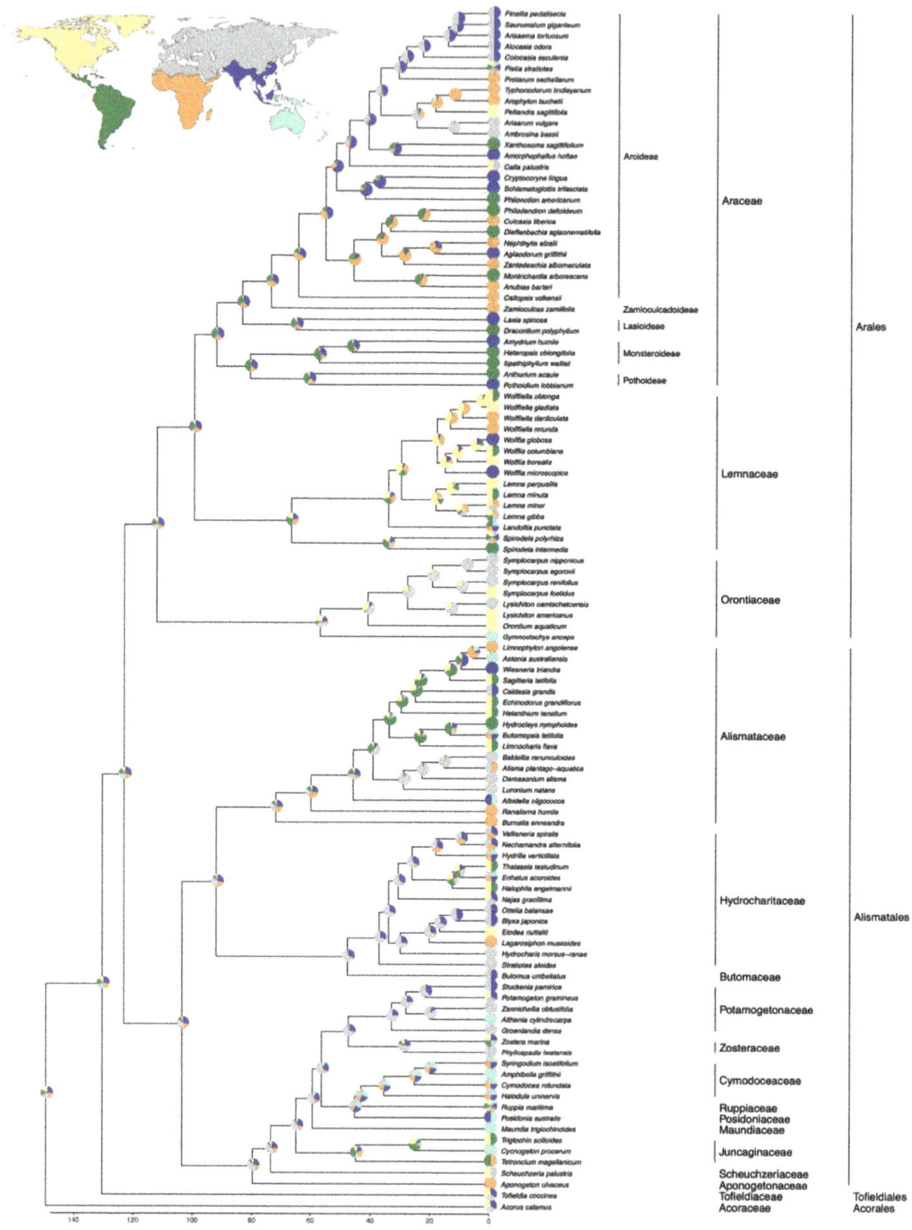

Figure 2. Phylogeny of Arales, constructed using combined DNA sequence data from five plastid regions (*matK*, *ndhF*, *rbcL*, *rps16*, and *trnL-F*). The tree represents a maximum likelihood tree, constructed in IQ-TREE [18] using default settings. Taxa were trimmed to include only one representative species for most genera. Branch lengths were adjusted to be ultrametric using the *chronopl* function in the R package *ape* [19], and the x-axis was scaled to approximate node ages that were reported previously [20]. Ancestral biogeography was reconstructed using the *ace* function in the *ape* package, to estimate the likelihood of an ancestor occupying one or more of the biogeographical realms indicated by the colored regions in the inset map. The ancestral likelihood values are shown as pie charts at the nodes of the phylogeny.

A great deal of diversity is contained in Araceae s.l., comprising morphological diversity in the extreme growth forms of duckweeds and terrestrial aroids, as well as evolutionary diversity in the large number of species and the ancient origin of the lineage. Some of the more prominent angiosperm and monocot phylogenetic studies have advocated for the Araceae s.l. circumscription, but a universal criterion does not exist for defining the boundaries of a plant family. We believe that the ultimate taxonomic disposition of aroids and duckweeds should integrate data from multiple disciplines and perspectives, and not be based simply on the opinions of either the broad-scale angiosperm phylogenetic community or the group of scientists who have devoted their careers to studying the traditional Araceae (i.e., Araceae s.s. plus Orontiaceae). Although it has been nearly 30 years since molecular data first suggested a close relationship between aroids and duckweeds [3], the usage of 'Lemnaceae' has remained quite prevalent in the literature (Figure 3). Thus, it remains necessary to provide an objective, equitable, and stable solution to the taxonomic disposition of these exciting and diverse angiosperms.

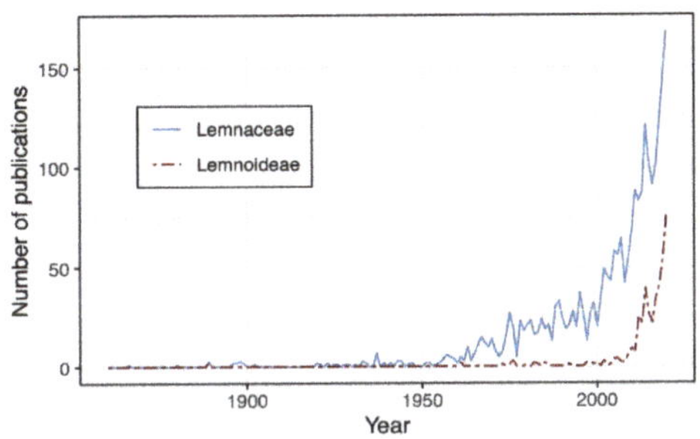

Figure 3. Usage of 'Lemnaceae' and 'Lemnoideae' in publications over time. Important events that may have influenced usage include the initial plastid data suggesting that duckweeds belong to the Araceae clade in 1995 [3] and the APG recommendation to include duckweeds in Araceae s.l., in 1998 [10]. Data were obtained from the Dimensions website (https://app.dimensions.ai/; accessed on 20 August 2021) by searching for each keyword anywhere in an article.

1.2. *Orontiaceae*

In several recent phylogenetic studies, Gymnostachydoideae (*Gymnostachys* R. Br.) and Orontioideae (*Lysichiton*, *Orontium* L., *Symplocarpus* Salisb. ex W. P. C. Barton) are considered as subfamilies of Araceae s.l. [4–6]. Araceae originally included *Orontium* [21], and eventually, the genera *Gymnostachys* [22], *Lysichiton* [23], and *Symplocarpus* [24] also were described within the aroids. Infrafamilial classifications of the aroids have varied considerably, but most have acknowledged the distinctness of the 'proto-Araceae' genera [25–28]. The category Orontiaceae, originally erected to accommodate several exceptional aroid genera [29], would be appropriate for encompassing the taxa currently designated as the subfamilies Gymnostachydoideae and Orontioideae.

Prior to the advent of molecular systematics, the classification by Engler [26] closely approximated the current phylogenetic hypothesis by placing *Lysichiton*, *Orontium*, and *Symplocarpus* together (in Calloideae: Symplocarpeae), yet this classification also grouped *Gymnostachys* in a separate subfamily (Pothoideae: Acoreae) with *Acorus* L. Further analysis of morphological data led to the segregation of Acoraceae Martinov, while retaining *Gymnostachys* within Araceae [30]. Molecular data clearly have advanced our understanding of character evolution since that time, and the consistent phylogenetic placement of subfamilies Gymnostachydoideae and Orontioideae enables a more confident evaluation of their shared morphological characters [3–6,31].

Whereas the overall inflorescence morphology in *Lysichiton* and *Symplocarpus* resembles the spathe + spadix model of Araceae s.s., the inflorescences of *Gymnostachys* are branched, and those of *Gymnostachys* and *Orontium* lack a subtending spathe [26,32]. It should be noted also that the spadix inflorescence is not uniquely synapomorphic to aroids, with plants in the order Acorales Martinov producing a similar inflorescence [30,33]. Several features, such as stomata type and flavonoid profile, unite Orontioideae and distinguish this group from most other Araceae s.l. [28,34,35]. Shared features of Gymnostachydoideae and Orontioideae include seedlings with cataphylls [36], rhizomatous growth [28], inflorescences with uniformly hermaphroditic flowers [5], dimerous flowers (trimerous in *Orontium* [37]), pollen walls with ektexine [38,39], unilocular and uniovulate gynoecia (bilocular with 1–2 ovules per locule in *Lysichiton* [28]), orthotropous ovules (hemianatropous in *Orontium* [40]), and apical placentation (basal in *Orontium* [28,37]. Orontiodeae plants lack vessels entirely [41], and this feature has not been investigated in Gymnostachydoideae. Although some of the morphological features found in Gymnostachydoideae and Orontioideae also occur in various Araceae s.s. taxa, the combination of these features unites the 'proto-Araceae' (i.e., Orontiaceae according to our suggestion) reasonably well.

Just nine species are classified within the four Orontiaceae genera, and their distinctness extends beyond morphology and anatomy to include biogeography and habitat [31]. *Lysichiton*, *Orontium*, and *Symplocarpus* are denizens of temperate wetland or aquatic habitats in the northern hemisphere, which contrast with the tropical distributions and terrestrial habitats of most Araceae s.s. [42,43]. *Gymnostachys* grows in moist forest habitats in eastern Australia, another location where relatively few other Araceae s.s. are found [1]. Consequently, we believe that the phylogenetic, morphological, and ecological divergence of these genera is more than sufficient to justify their recognition as an independent family. Also, the taxonomic reassignment of ten 'proto-Araceae' species to Orontiaceae would be relatively minor, in sharp contrast with the migration of 36 duckweed species [44] from Lemnaceae into Araceae s.l., which became necessary upon acceptance of the APG classifications [10].

1.3. Lemnaceae

Duckweeds (Lemnaceae) are an aquatic monocot lineage that had been difficult to classify using morphological data, owing to their extreme reduction in size and complexity [11,45]. Morphological classifications have long supposed a close relationship between duckweeds and the araceous pleustophyte genus *Pistia* L. [26,46], yet the similar aquatic habits in these groups are convergent rather than synapomorphic [3]. Duckweeds are not particularly close relatives of Araceae s.s., with their common ancestor coalescing over 100 Ma, but their phylogenetic position as descendants from the shared ancestor of Araceae s.l. has caused them to lose their independent family designation [10]. Faced with molecular evidence that the two families were not reciprocally monophyletic [3], botanists soon developed a consensus that Araceae and Lemnaceae should be merged [10]. Although it makes evolutionary sense to infer the extreme morphological reduction of duckweeds from an araceous ancestor, nevertheless, the morphology of extant duckweeds cannot easily be equated with many features that are shared between Orontiaceae and Araceae s.s.

Duckweeds are diverse and ecologically unique, occupying the surface or subsurface layer of water bodies [11], reproducing extremely effectively through vegetative means [47–49], and even dispersing as whole-plant units by adhering to the bodies of aquatic fauna [50]. The uniqueness of the duckweed growth strategy has caused them to be regarded as a separate plant family for nearly the entire time since their discovery [51–53]. There are abundant features that distinguish Lemnaceae from both Araceae s.s. and Orontiaceae beyond the obvious reduction in morphological size and complexity. Duckweeds have unique ulcerate, spinose pollen [5,54], and there are numerous taxonomically informative characters that differentiate duckweed genera and species [11,55,56]. The category of Lemnaceae has been in use for two centuries [7,51,57], and importantly, there is extensive literature written by researchers who specialize on duckweeds.

Morphological and molecular analyses of Lemnaceae continue to validate this lineage as distinct and worthy of independent taxonomic recognition [12,15,44,58,59]. Whereas an initial molecular phylogenetic study showed duckweeds to be deeply nested within Araceae s.l. [3], subsequent and more thorough studies have consistently placed them as rather distantly related to Araceae s.s., with only Orontiaceae preventing them from being reciprocally monophyletic with the traditional Araceae (i.e., Araceae s.s. plus Orontiaceae) [4–6]. In more recent years, genome sequencing of several duckweed species has provided additional molecular and cytological data, with authors largely continuing to promote the taxonomic independence of Lemnaceae [60–64]. In the 20 years since the APG revision [10], the taxonomic term 'Lemnaceae' has been used consistently more than the term 'Lemnoideae' (Figure 3), and studies focusing specifically on duckweeds predominantly refer to them as an independent family [12,15,44,47,48,55,56,58,59]. It is reasonable to suggest that a taxonomic solution that preserves the nomenclature currently used by duckweed biologists would be preferable.

1.4. Related Lineages and Ordinal Classification

According to widely accepted phylogenetic analyses of plastid data, the sister lineage to Araceae s.l. is diverse and comprises several wholly aquatic lineages such as Hydrocharitaceae and Potamogetonaceae [20,65]. More distantly related to these, the family Tofieldiaceae resolves as the sister lineage of the clade containing Araceae, Hydrocharitaceae, and Potamogetonaceae.

The most recent APG ordinal classification scheme [66] recognizes a single order, Alismatales Dumort. that includes the diverse families Araceae, Hydrocharitaceae, Potamogetonaceae, and Tofieldiaceae. Opinions differ regarding the appropriate ordinal classification of these lineages, however, and other authors prefer to divide the alismatid monocots among two or more orders. Several recent publications [17,67,68] recognize Arales Dumort. as separate from Alismatales and Tofieldiales Reveal & Zomlefer [69], and other authors [17,65,70] additionally consider Potamogetonales Dumort. (=Zosterales Nakai) to be distinct from Alismatales. To avoid confusion, we will refer to the expanded order comprising Alismataceae and Araceae as Alismatales s.l., and the more narrow-sense order, limited to include Alismataceae, Hydrocharitaceae, and related lineages (but not Araceae s.l.) as Alismatales s.s.

As advocated by the APG [66], the order Alismatales s.l. is equivalent to subclass Alismatidae Takht. [71] plus Araceae s.l. and Tofieldiaceae [10,20]. Prior to molecular phylogenetic studies, Araceae and Lemnaceae (along with Acoraceae) comprised a separate order, Arales [72]. With the expansion of Araceae, the ordinal category of Arales became synonymous with Araceae s.l. and fell out of favor [10]. Although the diverse families of Araceae s.s., Alismataceae, Lemnaceae, Orontiaceae, Tofieldiaceae, and others are indeed monophyletic [4,58,65,73], their common ancestor extends back to ca. 130 Ma, not long after the ca. 139 Ma crown age for all extant angiosperms [20]. (Note that estimates of ancient diversification events are bounded by considerable uncertainty. We will use the time scale established in Ref. [20], but a range of other estimates exist for alismatid monocots [14,68]. Regardless of the dating method used, the relative ages of phylogenetic nodes are consistent across studies that use plastid sequence data). The immense morphological and ecological diversity contained in this lineage is rather difficult to conceptualize as a single order. Instead, we propose that it would be simpler and clearer to differentiate four categories at the ordinal rank: Alismatales (crown age 100 Ma), Potamogetonales (103 Ma), Tofieldiales (100 Ma), and Arales (122 Ma) [6,20,74]. The crown ages of these lineages would then be closer to the average range of values for other angiosperm orders.

The practice of using the Alismatales category to contain such diverse lineages as Araceae, Potamogetonaceae, and Tofieldiaceae [10,66] makes it more difficult to refer specifically to distinct evolutionary units within monocots. A broader diversity of taxonomic categories would enable a more nuanced discussion of the diversity of alismatid monocots. To facilitate discussion, in this paper we will consider Arales to be synonymous with Araceae s.l. (as described above), and we will limit the Alismatales s.s. category to the clade containing Alismataceae, Hydrocharitaceae, and related lineages (Figure 2). We also will use the categories Acorales, Potamogetonales, and Tofieldiales, as appropriate. The ordinal classification scheme used in this paper is aligned with ordinal categories that were widely accepted prior to the undeniably influential APG publications [10,66].

1.5. Objectives

Aroids and duckweeds have been categorized inconsistently in the time since the first molecular phylogenetic analyses were conducted, with some authors preferring to sink duckweeds within Araceae s.l. and others preserving the more traditional Lemnaceae category. The debate about taxonomic categories has not necessarily considered all evidence in an objective approach that promotes nomenclatural stability, universal criteria for taxonomic boundaries, and morphological diagnostics. Therefore, we endeavored to evaluate the available morphological, phylogenetic, and other evidence in order to determine the most appropriate classification scheme for these diverse and economically important plants.

2. Molecular Phylogenetic Evidence

The phylogenetic relationships among Orontiaceae, Lemnaceae, and Araceae s.s. are widely accepted, but the comprehensive phylogenetic analyses to date have focused primarily on plastid data. After some initial uncertainty about the phylogenetic position of Lemnaceae, they now consistently resolve as a strongly supported clade that is sister to Araceae s.s., with Orontiaceae being sister to the clade of Araceae s.s. plus Lemnaceae. All three families are separated by substantial branch lengths and receive high statistical support. The phylogenetic relationship of Lemnaceae with Araceae and Orontiaceae has provided the foundation for classification schemes that prefer to lump all three families into one large Araceae s.l.

2.1. Plastid Molecular Data

Araceae s.s. comprise a large number of species with a most recent common ancestor that diversified roughly in the last 100 Ma [6]. Recent studies have used a combination of plastid sequence data from the more variable spacer and intron regions (e.g., *trnK* introns, *trnL-trnF* spacer) and more conserved protein-coding sequences (e.g., *matK*, *ndhF*, *rbcL*, and *rps16* genes) to investigate relationships among species [4–6]. These plastid regions have become the backbone of evidence for phylogenetic relationships in Araceae s.l., and the taxonomic sampling from these regions has been extensive [4,5]. Recent years have seen a rapid expansion of studies that compare whole plastid genomes (e.g., [75–77]), and while these provide a wealth of informative molecular data, the sampling remains limited in many cases.

We conducted updated phylogenetic analyses using plastid, mitochondrial, and nuclear DNA sequence data. The majority of sequence data used in this study were published previously, with some sequences newly generated to augment the taxon sampling for groups of interest (Supplementary Table S1). Phylogenetic trees were obtained by conducting maximum likelihood analyses using IQ-TREE version 1.6.12 [18,78] with integrated model selection. Five plastid regions were included: *matK*, *ndhF*, *rbcL*, *rps16*, and *trnL-trnF* [5,58,79,80], and the plastid data matrix was trimmed to include a maximum of four species per genus. Where possible, sequence data were obtained from complete plastid genome sequences, which are becoming increasingly available [75,76,81–87].

The plastid phylogeny corroborates prior evidence for the monophyly of orders Alismatales s.s., Arales, and Tofieldiales, as well as the families Orontiaceae, Lemnaceae, and Araceae s.s. (Figure 2). Araceae subfamilies also are monophyletic on the tree, and a more thoroughly sampled tree shows the tribal categories to be monophyletic as well (Figure 4). These relationships have been shown previously and are generally accepted [4–6]. Figure 2 represents an ultrametric tree, where branch lengths are approximately equal to time. Branch lengths on the maximum likelihood tree were transformed using the penalized likelihood method [88], employed in R version 4.1.1 [89] using the *chronopl* function in the *ape* package version 5.0 [19]. Besides the sequence data that support the phylogenetic relationships of Araceae s.s., Lemnaceae, and Orontiaceae, there are molecular patterns that also support the independence of these lineages. For example, a comparison of Arales genomes has identified duplications of the *rps15* and *ycf1* genes that are synapomorphic to Lemnaceae [75,85,87]. Additional patterns of this sort may be identified as more plastid genomes are sequenced across Arales.

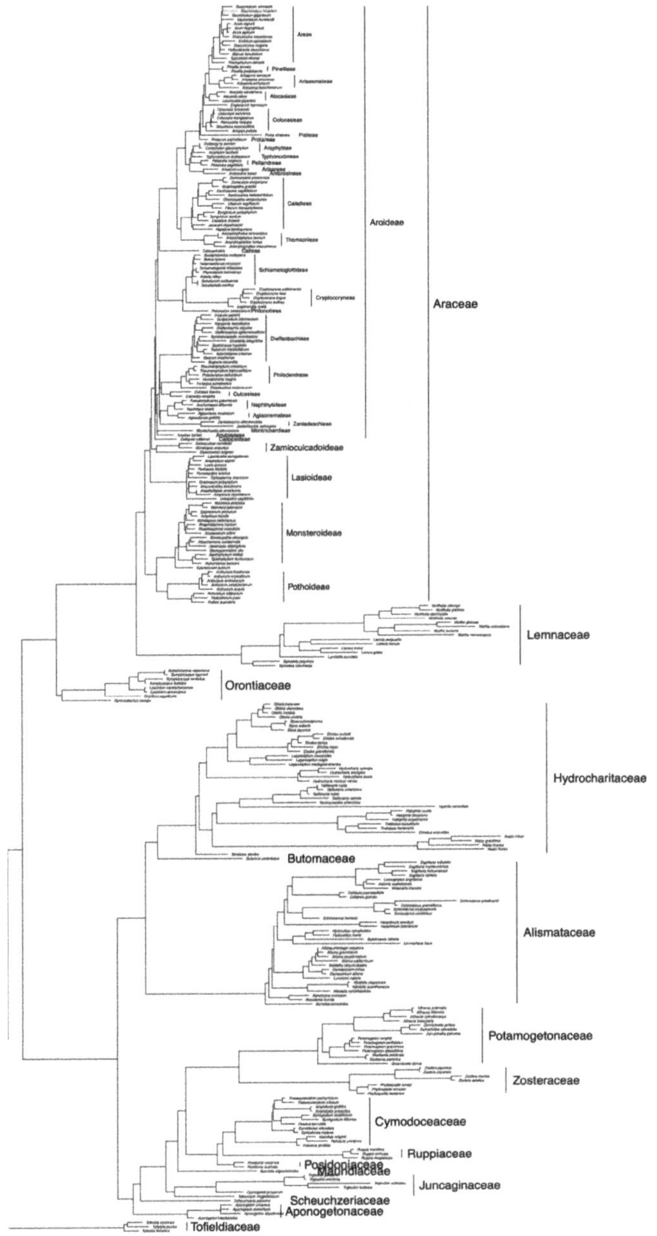

Figure 4. Phylogeny of Arales, constructed using combined DNA sequence data from five plastid regions (*matK*, *ndhF*, *rbcL*, *rps16*, and *trnL-F*), showing more complete taxon sampling from Arales genera. The tree represents a maximum likelihood tree, constructed in IQ-TREE [18] using default settings.

2.2. Nuclear Molecular Data

In contrast to the relative abundance of plastid sequence data, far fewer studies in Arales have included nuclear DNA sequence data. The nuclear ribosomal internal transcribed spacer (nrITS) region commonly is used to reconstruct relationships among species, because the sequences evolve fairly quickly and can be sequenced with relative ease [90]. The nrITS region is situated between the large and evolutionarily conserved 18S and 26S ribosomal RNA genes, which themselves can be useful for determining larger-scale evolutionary patterns such as the relationships among families and orders [91,92]. Potential reasons to avoid nuclear sequence data include the phylogenetic uncertainty of reconstructing trees using biparentally inherited markers that may show evidence of hybridization, introgression, incomplete lineage sorting, or other potential challenges [93]. Another challenge that may hinder the usefulness of the nrITS region in Arales is that the sequences are difficult to obtain in some taxa, most notably the Lemnaceae lineage that apparently has an exceptionally firm secondary structure that resists molecular data acquisition methods [59].

We conducted an updated phylogenetic investigation using published and newly generated sequence data for the 18S ribosomal RNA gene and the nrITS region. Novel data were obtained using published methods [94–97], and GenBank accession numbers are provided in Supplementary Table S1. Prior to our study, there were very few 18S sequences available for Arales, with the exception of a complete sampling of Lemnaceae species [59]. We were able to add nuclear DNA data for 15 Arales species, including heretofore unavailable sequences for Araceae subfamily Zamioculcadoideae and other species spanning the breadth of diversity in Araceae s.s. The phylogenetic relationships that were determined using nuclear sequence data (Figure 5) recapitulate the same major relationships that are depicted on the plastid phylogeny (Figures 2 and 4). Orontiaceae resolve as the sister to the clade containing Araceae s.s. and Lemnaceae, and the latter two families are reciprocally monophyletic.

Although they are useful for inferring nuclear DNA evolution, the nuclear ribosomal genes and internal transcribed spacer regions reflect the extremes of conserved and variable sequences, respectively. Efforts are ongoing to expand the number of nuclear gene regions that can be used for phylogenetic reconstruction, including the ambitious One Thousand Plant Transcriptomes Initiative [98]. However, the taxonomic sampling from Araceae s.l. remains limited for such an analysis.

Targeted efforts are underway to increase the available nuclear sequence data for Lemnaceae [45], and it would be valuable to enact a parallel approach to studying other Arales and Alismatales taxa. Sequencing additional nuclear genes for taxa in Araceae s.s., Lemnaceae, and Orontiaceae may even further corroborate the phylogenetic distinctness of these lineages and potentially provide insights into their genome evolution.

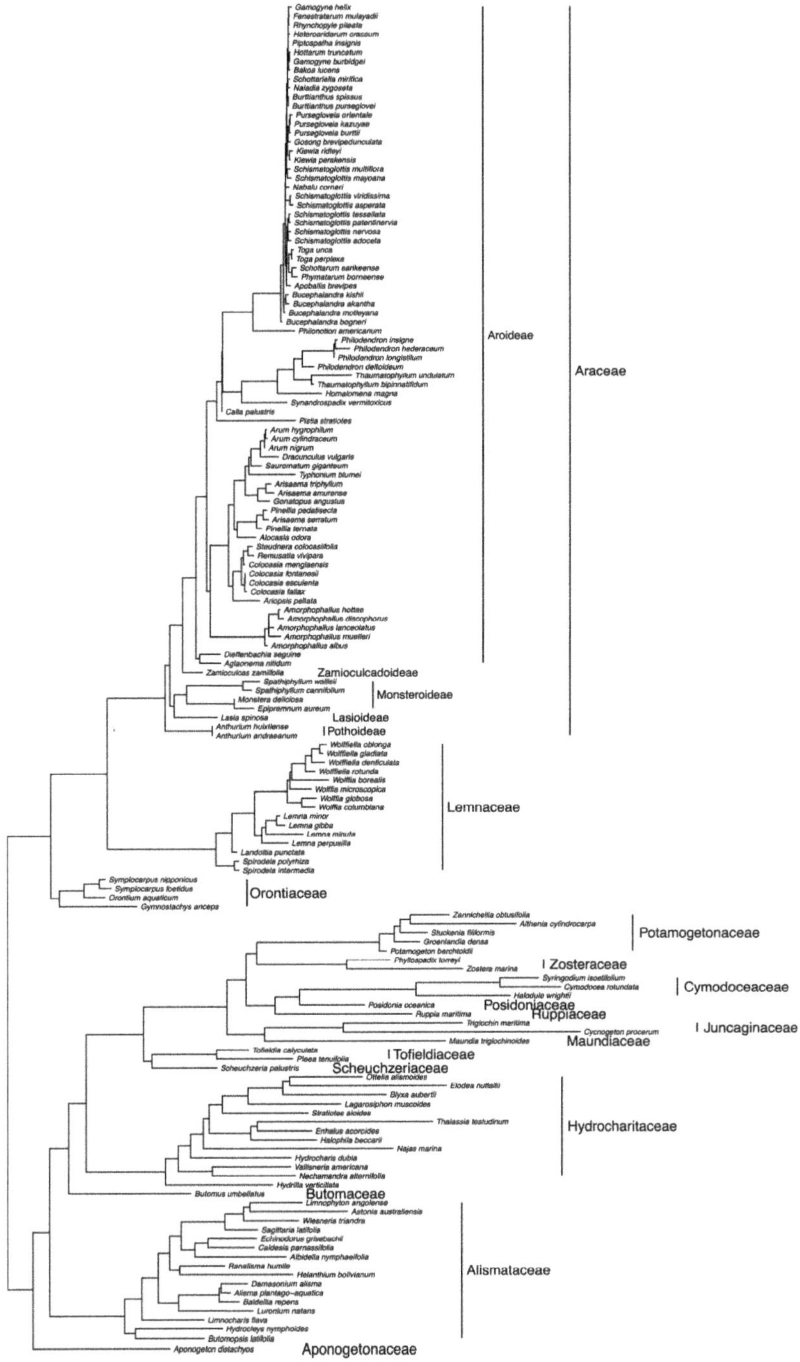

Figure 5. Phylogeny of Arales, constructed using combined DNA sequence data from nuclear ribosomal genes and spacers (18S and 5.8S rRNA, and the ITS-1 and ITS-2 spacers). The tree represents a maximum likelihood tree, constructed in IQ-TREE [18] using default settings.

2.3. Mitochondrial Molecular Data

Mitochondrial gene data are rarely used in angiosperm phylogenetic studies, but they potentially represent an independent source of phylogenetic data that are predominantly inherited uniparentally like in the plastid [99]. Some of the earliest attempts at reconstructing plant phylogenies with mitochondrial data used the *cox1* gene (cytochrome c oxidase subunit 1), a homolog of the most widely used phylogenetic marker in animals (where it is commonly known as COI) [100]. Unfortunately, the angiosperm *cox1* sequences were determined to have relatively few nucleotide polymorphisms in the coding region and a variable intron, the presence or absence of which is not phylogenetically informative in many taxa [101,102]. Because of this and the contrasting high utility of the plastid sequence data, mitochondrial genes were largely abandoned as phylogenetic markers. Although *cox1* turned out to be minimally useful, other mitochondrial genes have shown promise for recapitulating the broad topology of flowering plants [103,104]. Protein-coding and intron regions of four genes (*atp1*, *matR*, *nad5*, and *rps3*) are potentially effective for building an independent phylogenetic hypothesis about the evolution of angiosperms.

Preliminary phylogenetic analyses using mitochondrial genes [103,104] resolved a monophyletic Arales and also a larger clade that also included Alismataceae, Juncaginaceae, and Potamogetonaceae (i.e., Alismatales s.s.), as well as Tofieldiaceae (Tofieldiales). The Alismatales clade is characterized by curiously large branch lengths relative to comparable lineages (e.g., using the plastid phylogeny for reference), and considerable sequence data have been generated for this group to help illuminate the interesting evolutionary history of their mitochondrial genomes [105–107].

Our updated phylogenetic analysis using expanded sampling of Arales s.l. taxa (Figure 6) produced a phylogeny that supported many of the same relationships as the plastid tree (Figures 2 and 4). Thus, the mitochondrial data may become a useful complement to the organellar sequence data contained in the plastid. Mitochondrial genomes can be obtained using the same next-generation sequencing techniques that are enabling so many plastid genomes to be published [108]. A recently published mitochondrial genome for *Spirodela polyrhiza* (Lemnaceae) [109] may represent the beginning of a surge in similar data from other Arales species.

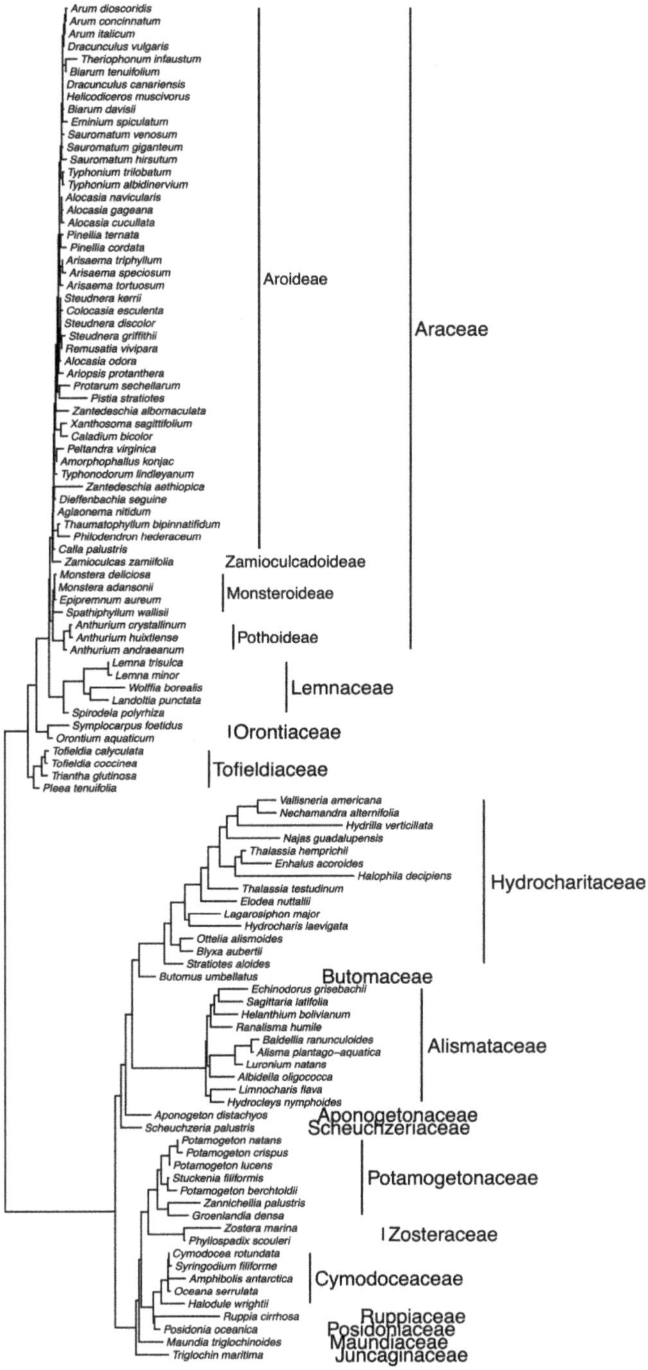

Figure 6. Phylogeny of Arales, constructed using combined DNA sequence data from four mitochondrial regions (*atp1*, *matR*, *rps3*, and *nad5*). The tree represents a maximum likelihood tree, constructed in IQ-TREE [18] using default settings.

3. Morphological Data

The morphological distinctness of the aroids is undeniable. They have cells with calcium oxalate crystals, and a distinctive inflorescence type consisting of a typically showy bract (spathe) subtending a thick spike of tiny flowers (spadix) [1]. The inflorescence similarity between Araceae s.s. and Orontiaceae enabled them to be classified together, however, there are some exceptions to the idea that all aroids can be identified by the spathe + spadix inflorescence type. Firstly, two Orontiaceae genera, *Orontium* and *Gymnostachys*, lack spathes (Figure 1A,C), and the latter genus even exhibits a branched inflorescence (Figure 1C) unlike any known in Araceae s.s. Additionally, a distinctive inflorescence architecture, with continuation shoots produced in the penultimate leaf axil, characterizes nearly all the Araceae s.s. taxa, whereas this trait is absent from Orontiaceae [5]. The evolutionary origins of superficially similar organs are important to consider. In a phylogenetic context, the inflorescence spathe is reconstructed to be absent from the common ancestor of Orontiaceae, and thus its evolution at the root of Araceae s.s. is independent of the origin in *Lysichiton* and *Symplocarpus* [5]. Secondly, several more distantly related angiosperm groups have a spadix-like inflorescence, such as Acoraceae (Acorales), Cyclanthaceae Poit. ex A.Rich. (Pandanales R.Br. ex Bercht. and J.Presl), and even Piperaceae Giseke (Piperales Bercht. & J.Presl) [110]. Moreover, Hydrocharitaceae inflorescences are subtended by one or two bracts that also are termed 'spathes' [111]. Therefore, neither spathe nor spadix is unique to Araceae s.l.

The extreme morphological reduction that characterizes Lemnaceae has always made them an awkward fit for Araceae s.l., and their highly reduced morphology undermines the morphological characters that otherwise might unify plants in this group. If duckweeds are considered to belong to Araceae s.l., then the family must be characterized as having a distinctive inflorescence and vegetative features, except for the duckweed lineage that has no such features. If Lemnaceae and Orontiaceae instead are retained as separate from Araceae s.s., then each family can be identified readily by its distinctive features. Araceae s.s. and Lemnaceae each have a wealth of characteristics that unify their species, leaving only Orontiaceae as relatively difficult to classify.

The Orontiaceae clade includes two genera (*Lysichiton* and *Symplocarpus*) with inflorescences that are more like those found in Araceae s.s. (compare Figure 1B,D to Figure 1G–J), and two other genera (*Gymnostachys* and *Orontium*) that do not quite conform to the spathe+spadix morphology. Looking beyond the more obvious inflorescence features, there are in fact morphological traits that unite Orontiaceae and can be used to diagnose its constituent species from Araceae s.s. taxa. A morphological phylogenetic analysis of extant Arales [5] identified several features that are diagnostic or nearly so for Orontiaceae, including a collenchyma type [112] that is found only in *Lysichiton* and *Symplocarpus*, coincidentally the same two Orontiaceae genera whose inflorescence spathes otherwise make them appear superficially more similar to Araceae s.s. (petiole collenchyma is absent in *Gymnocarpus* and *Orontium*). Perhaps most noteworthy, there are leaf shape and venation patterns that can enable confident identification of Orontiaceae species, even in fossil material [113,114]. Thus, under more careful examination, there are more than a few characters that contradict the apparent similarity between Araceae s.s. and Orontiaceae.

4. Chromosome Number Evolution

Another superficial similarity that has been noted between Orontiaceae and Araceae s.s. is their range of chromosome numbers. Chromosome numbers have been reported for a large number of Arales taxa, including all genera of Lemnaceae and Orontiaceae. The reported base chromosome numbers for Orontiaceae ($x = 12$, 13, 14, or 15) have all been observed in Araceae s.s. genera, whereas Lemnaceae were reported to have a base $x = 10$ chromosome number that is nearly unique among other Arales [5]. However, the simplified $x = 10$ value fails to account for the wide variety of chromosome numbers that have been reported for Lemnaceae [115], many of which are not divisible by 10. An approach aimed specifically at reconstructing chromosome number evolution in Arales produced different numbers for the most recent common ancestors of Orontiaceae (n = 17), Lemnaceae (n = 22), and Araceae s.s. (n = 15) [13]. It should be noted, however, that these numbers were selected as the most likely among several competing values that also were highly probable.

A more precise reconstruction of chromosome evolution requires evaluating the synteny of homologous chromosome regions. Cao et al. [62] developed a method to visualize syntenic chromosome regions across Lemnaceae species and inferred that seven 'ancestral chromosome blocks' later became duplicated and distributed across n = 20 chromosomes in Lemnaceae [62,115,116]. In Araceae s.s., the full genome sequence has been obtained for *Colocasia esculenta*, and a synteny analysis indicates n = 14 linkage groups [117]. Synteny analyses will be important for developing appropriate chromosome comparisons among Arales lineages and for reconstructing evolutionary changes. Fortunately, additional genome sequences are forthcoming, and these will enable a thorough comparison among species at every taxonomic level.

5. Biogeography

One final evolutionary aspect of comparison is that of biogeography or the ancient dispersal processes that are manifest in the geographic distributions of extant species. A thorough evaluation of ancestral biography has been conducted for Lemnaceae and determined that the ancestor of the family likely diversified in the Americas [15]. We applied a similar approach to the phylogeny for Arales, Alismatales, and Tofieldiales, using the 'realm' division of the terrestrial ecoregions of the world [118]. Ancestral distribution ranges were reconstructed for species that are represented on the plastid phylogeny, which has the most extensive taxon sampling in our study and also represents the phylogenetic topology that has been used in most other studies of Arales evolution [4,5]. Native ranges for species were obtained from the Kew Plants of the World Online database [119]. Ecological realm boundaries were approximated onto the geopolitical boundaries identified by the Taxonomic Databases Working Group [120], and membership in one or more realms was determined according to the inset map in Figure 2.

The biogeography analysis illustrates the temperate northern distribution of Orontiaceae taxa and reconstructs the ancestor of the family in North America or Eurasia (Figure 2). The temperate northern distribution of Orontiaceae and the generally northern distribution of Lemnaceae contrast with the decidedly tropical distribution of most Araceae s.s. species. Our analysis estimates that the common ancestor of Araceae s.s. diversified in South America or southeastern Eurasia, where many of the extant species are found today. A small number of Araceae s.s. genera have northern temperate species (e.g., *Arisaema* Mart., *Calla* L., *Peltandra* Raf.), but the vast majority are tropical. The dispersals to northern temperate habitats were independent, and they are scattered across the phylogeny (Figure 2). Thus, even geographic distributions can be useful for distinguishing Araceae s.s. from Orontiaceae, as the latter are almost entirely temperate and the former are relatively rare at temperate latitudes.

6. Discussion

6.1. Nomenclatural Stability and Utility

Modern botanical taxonomy aims to circumscribe natural groups that descended from common ancestors and, if possible, groups that are clearly differentiated using morphological or anatomical characters. Molecular phylogenetic data have been invaluable for identifying monophyletic groups, but these data are unable to prescribe the appropriate size of the clade that should constitute an order, family, or genus. To address this question, it is instructive to compare taxa at the same rank, while considering their respective ages and degrees of morphological divergence. The 122 Ma crown age of Araceae s.l. is among the oldest of any angiosperm family [14,16,121] and even falls at the older range of crown ages for angiosperm orders (27.1–128.9 Ma, $\overline{x}$ = 88.2 Ma) [20]. In contrast, the crown ages of Lemnaceae (73 Ma), Orontiaceae (96 Ma), and Araceae s.s. (97 Ma) (i.e., the family-level categories that we propose in this paper) would be more consistent with the ages of other angiosperm families (0.0–139.4 Ma, $\bar{x} = 45.1$ Ma) [6,121]. As previously demonstrated, the morphological divergence of Lemnaceae is unmistakable [12,58], and the independence of Orontiaceae is supported by numerous morphological characteristics [31].

In the era when relationships inferred from molecular phylogenetic studies frequently differ from traditional taxonomy, in the interest of stability it can be valuable to make the fewest reassignments from traditional categories. Lemnaceae are a firmly established group that molecular data have validated as monophyletic. Even amateur botanists are familiar with duckweeds and their distinctive growth form, and a diverse assemblage of duckweed genera all fall cleanly within the Lemnaceae category. In addition to maintaining nomenclatural stability, augmenting the number of family-level categories enables more effective discussions about the defining traits of each family. In contrast, amalgamating a large amount of taxonomic and morphological diversity into one large Araceae s.l. family obscures the features that unify Araceae s.s. and prevents facile discussion of Lemnaceae and Orontiaceae.

6.2. *The Nature of a Plant Family*

The advent of molecular phylogenetics has caused many traditional taxonomic categories to be reorganized, as categories are widely expected to reflect monophyletic evolutionary lineages. Adherents to prior categories have understandably resisted some of the recent taxonomic changes, but in general, the categories are trending toward greater stability and an enriched evolutionary perspective. In some cases, a small number of taxa can be reassigned to maintain monophyletic categories, but for other groups, a large number of reassignments are required. Among the more noteworthy examples are the monocot order Asparagales Link, which decreased the number of included families by half in the time between the initial [10] and the most recent APG publication [66], and the eudicot family Scrophulariaceae Juss., where the majority of major lineages (at the rank of tribe) ended up being assigned to other families [122]. Changes in the latter group became necessary because the morphological similarity of some species was not consistent with their evolutionary relationships as supported by molecular evidence. Throughout the taxonomic upheaval that resulted from the age of molecular systematics, taxonomists have strived largely to retain categories that are informative (in terms of morphology), equivalent to other categories at the same rank, and consistent as much as possible with traditional taxonomic categories [66].

There is not a strictly defined set of criteria for determining the boundaries of a plant family that would take into account, for example, evolutionary age, degree of morphological or molecular divergence, or the number of subordinate taxa. The most general guideline seems to be that plant families should be roughly equivalent to one another in these aspects, so that one might develop a general sense of what constitutes a typical plant family. In many respects, the family-level categories advocated herein (i.e., Araceae s.s., Lemnaceae, and Orontiaceae) are more in line with the 'typical' plant family, when compared against the alternative Araceae s.l. classification scheme. When compared to Alismatales s.s. families and families across the angiosperms, our proposed family categories for Arales are more similar in evolutionary age, morphological divergence, and diagnosability. Moreover, the recommended categories require fewer reassignments of genera or species than the Araceae s.l. alternative. Therefore, we maintain that the categories of Araceae s.s., Lemnaceae, and Orontiaceae are more stable and more useful.

7. Conclusions

Classification schemes are necessarily subjective, but ideally, they strive to achieve consistency, clarity, and utility for botanists. The subsumption of Lemnaceae into Araceae s.l. effectively has removed a useful taxonomic category that duckweed biologists have used for many years, and a category that was among the most clearly defined of any angiosperm family. In contrast, we argue that there is less of a need to preserve the four Orontiaceae genera within Araceae s.l., as these plants already have been recognized as distinct in the 'proto-Araceae' category and surely merit their own taxonomic category at the family level. Classifications that relied heavily upon plastid molecular data are now bolstered by data from the nuclear and mitochondrial genomes, which support the same major evolutionary relationships. The categories of Araceae s.s., Lemnaceae, and Orontiaceae as proposed herein are informative and stable, and their usage will promote a better understanding of each respective group by professional and amateur botanists alike.

Supplementary Materials: The following are available online at https://www.mdpi.com/article/10.3390/plants10122639/s1. Table S1.xlsx (Accessions used for tree reconstruction), Arales-cpDNA-alignment.nex (Sequence alignment of plastid DNA for Arales), Arales-mtDNA-alignment.nex (Sequence alignment of mitochondrial DNA for Arales), Arales-nrDNA-alignment.nex (Sequence alignment of nuclear DNA for Arales).

Author Contributions: Conceptualization, N.P.T.; writing—original draft preparation, N.P.T.; writing—review and editing, N.P.T., D.H.L., K.J.A., K.S.S., D.J.C. and M.B. All authors have read and agreed to the published version of the manuscript.

Funding: We acknowledge support for the Article Processing Charge from the DFG (German Research Foundation, 393148499) and the Open Access Publication Fund of the University of Greifswald.

Data Availability Statement: All sequences used for tree reconstruction are available from GenBank (https://www.ncbi.nlm.nih.gov/; accessed on 12 November 2021). The respective accession numbers can be found in Supplementary Table S1.

Conflicts of Interest: The authors declare no conflict of interest.

References

1. Mayo, S.J.; Bogner, J.; Boyce, P.C. *The Genera of Araceae*; Royal Botanic Gardens: Kew, UK, 1997.
2. Boyce, P.C.; Croat, T.B. The Überlist of Araceae, Totals for Published and Estimated Number of Species in Aroid Genera. 2018. Available online: http://www.aroid.org/genera/180211uberlist.pdf (accessed on 14 April 2020).
3. French, J.C.; Chung, M.G.; Hur, Y.K. Chloroplast DNA phylogeny of the Ariflorae. In *Monocotyledons: Systematics and Evolution*; Rudall, P.J., Cribb, P.J., Cutler, D.F., Humphries, C.J., Eds.; Royal Botanic Gardens: Kew, UK, 1995; Volume 1, pp. 255–275.
4. Cabrera, L.I.; Salazar, G.A.; Chase, M.W.; Mayo, S.J.; Bogner, J.; Dávila, P. Phylogenetic relationships of aroids and duckweeds (Araceae) inferred from coding and noncoding plastid DNA. *Am. J. Bot.* **2008**, *95*, 1153–1165. [CrossRef]
5. Cusimano, N.; Bogner, J.; Mayo, S.J.; Boyce, P.C.; Wong, S.Y.; Hesse, M.; Hetterscheid, W.L.; Keating, R.C.; French, J.C. Relationships within the Araceae: Comparison of morphological patterns with molecular phylogenies. *Am. J. Bot.* **2011**, *98*, 654–668. [CrossRef] [PubMed]
6. Nauheimer, L.; Metzler, D.; Renner, S.S. Global history of the ancient monocot family Araceae inferred with models accounting for past continental positions and previous ranges based on fossils. *New Phytol.* **2012**, *195*, 938–950. [CrossRef]
7. Martinov, I. *Techno-Botanical Dictionary (Техно-БотаническийСловарь)*; Pechashano v Imperatorskoĭ Tipografii: Saint Petersburg, Russia, 1820.
8. *International Code of Nomenclature for Algae, Fungi, and Plants (Shenzhen Code)*; Wiersema, J.H.; Turland, N.J.; Barrie, F.R.; Greuter, W.; Hawksworth, D.L.; Herendeen, P.S.; Knapp, S.; Kusber, W.-H.; Li, D.-Z.; Marhold, K.; et al. (Eds.) International Association for Plant Taxonomy: Bratislava, Slovakia, 2018; Appendices I–VII.
9. Bogner, J. The free-floating Aroids (*Araceae*)–living and fossil. *Zitteliana A* **2009**, *48*, 113–128. [CrossRef]
10. The Angiosperm Phylogeny Group (APG). An ordinal classification for the families of flowering plants. *Ann. Missouri Bot. Gard.* **1998**, *85*, 531–553. [CrossRef]
11. Landolt, E. The family of Lemnaceae—A monographic study. *Ver. Geobot. Inst. ETH Stift. Rübel* **1986**, *71*, 1–563.
12. Les, D.H.; Landolt, E.; Crawford, D.J. Systematics of the Lemnaceae (duckweeds): Inferences from micromolecular and morphological data. *Plant Syst. Evol.* **1997**, *204*, 161–177. [CrossRef]
13. Cusimano, N.; Sousa, A.; Renner, S.S. Maximum likelihood inference implies a high, not a low, ancestral haploid chromosome number in Araceae, with a critique of the bias introduced by 'x'. *Ann. Bot.* **2012**, *109*, 681–692. [CrossRef] [PubMed]
14. Iles, W.J.D. The Phylogeny and Evolution of Two Ancient Lineages of Aquatic Plants. Ph.D. Thesis, University of British Columbia, Vancouver, BC, Canada, 2013.
15. Tippery, N.P.; Les, D.H. Tiny plants with enormous potential: Phylogeny and evolution of duckweeds. In *The Duckweed Genomes. Compendium of Plant Genomes*; Cao, X., Fourounjian, P., Wang, W., Eds.; Springer Nature Switzerland: Cham, Switzerland, 2020; pp. 19–38.
16. Stevens, P.F. Angiosperm Phylogeny Website, 2001; Version 14. July 2017, Last Updated 24 February 2020. Available online: http://www.mobot.org/MOBOT/research/APweb (accessed on 8 April 2020).
17. Les, D.H. *Aquatic Monocotyledons of North America: Ecology, Life History, and Systematics*; CRC Press: Boca Raton, FL, USA, 2020.

18. Nguyen, L.T.; Schmidt, H.A.; Haeseler, A.; Minh, B.Q. IQ-TREE: A fast and effective stochastic algorithm for estimating maximum likelihood phylogenies. *Mol. Biol. Evol.* **2015**, *32*, 268–274. [CrossRef]
19. Paradis, E.; Schliep, K. ape 5.0: An environment for modern phylogenetics and evolutionary analyses in R. *Bioinformatics* **2019**, *35*, 526–528. [CrossRef]
20. Magallón, S.; Gómez-Acevedo, S.; Sánchez-Reyes, L.L.; Hernández-Hernández, T. A metacalibrated time-tree documents the early rise of flowering plant phylogenetic diversity. *New Phytol.* **2015**, *207*, 437–453. [CrossRef]
21. Jussieu, A.L. *Genera Plantarum Secundum Ordines Naturales Disposita juxta Methodum in Horto Regio Parisiensi Exaratam, Anno 1774*; Herissant: Paris, France, 1789.
22. Brown, R. *Prodromus Florae Novae Hollandiae*; Johnson & Soc.: London, UK, 1810; Volume 1.
23. Schott, H.W. Aroideae. *Österr. Bot. Wochenbl.* **1857**, *7*, 61–62. [CrossRef]
24. Barton, W.P.C. *Vegetable Materia Medica of the United States*; H. C. Carey & I. Lea: Philadelphia, PA, USA, 1817.
25. Schott, H.W. *Genera Aroidearum*; Carl Ueberreuter: Vienna, Austria, 1858.
26. Engler, A. *Das Pflanzenreich. Araceae–Pars Generalis et Index Familiae Generalis*; Wilhelm Engelmann: Leipzig, Germany, 1920; Volume IV.23A.
27. Hutchinson, J. *The Families of Flowering Plants*; Clarendon Press: London, UK, 1973.
28. Grayum, M.H. Evolution and phylogeny of the Araceae. *Ann. Missouri Bot. Gard.* **1990**, *77*, 628–697. [CrossRef]
29. Bartling, F.G. *Ordines Naturales Plantarum*; Sumptibus Dieterichianis: Göttingen, Germany, 1830.
30. Grayum, M.H. A summary of evidence and arguments supporting the removal of *Acorus* from the Araceae. *Taxon* **1987**, *36*, 723–729. [CrossRef]
31. Lee, J.S.; Kim, S.-H.; Lee, S.; Maki, M.; Otsuka, K.; Kozhevnikov, A.E.; Kozhevnikova, Z.V.; Wen, J.; Kim, S.-C. New insights into the phylogeny and biogeography of subfamily Orontioideae (Araceae). *J. Syst. Evol.* **2019**, *57*, 616–632. [CrossRef]
32. Ray, T.S. Survey of shoot organization in the Araceae. *Am. J. Bot.* **1988**, *75*, 56–84. [CrossRef]
33. Buzgo, M.; Endress, P.K. Floral structure and development of Acoraceae and its systematic relationships with basal angiosperms. *Int. J. Plant Sci.* **2000**, *161*, 23–41. [CrossRef]
34. Williams, C.A.; Harborne, J.B.; Mayo, S.J. Anthocyanin pigments and leaf flavonoids in the family Araceae. *Phytochemistry* **1981**, *20*, 217–234. [CrossRef]
35. Iwashina, T. Sebuah tinjauan: Senyawa-senyawa flavonoid pada suku Araceae. *Bul. Kebun Raya* **2020**, *23*, 1–24. [CrossRef]
36. Tillich, H.-J. Seedling diversity in Araceae and its systematic implications. *Feddes Repert.* **2003**, *114*, 454–487. [CrossRef]
37. Buzgo, M. Flower structure and development of Araceae compared with alismatids and Acoraceae. *Bot. J. Linn. Soc.* **2001**, *136*, 393–425. [CrossRef]
38. Weber, M.; Halbritter, H.; Hesse, M. The basic pollen wall types in Araceae. *Int. J. Plant Sci.* **1999**, *160*, 415–423. [CrossRef]
39. Hesse, M. Pollen wall ultrastructure of Araceae and Lemnaceae in relation to molecular classifications. *Aliso* **2006**, *22*, 204–208. [CrossRef]
40. Grayum, M.H. Systematic embryology of the Araceae. *Bot. Rev.* **1991**, *57*, 167–203. [CrossRef]
41. Carlquist, S.; Schneider, E.L. Origins and nature of vessels in Monocotyledons. 14. Vessellessness in Orontioideae (Araceae): Adaptation or relictualism? *Nord. J. Bot.* **2014**, *32*, 493–502. [CrossRef]
42. Mayo, S.J.; Bogner, J.; Boyce, P.C. Araceae. In *The Families and Genera of Vascular Plants*; Kubitzki, K., Ed.; Springer: Berlin/Heidelberg, Germany, 1998; Volume 4, pp. 26–74.
43. Nie, Z.L.; Sun, H.; Li, H.; Wen, J. Intercontinental biogeography of subfamily Orontioideae (*Symplocarpus*, *Lysichiton*, and *Orontium*) of Araceae in eastern Asia and North America. *Mol. Phyl. Evol.* **2006**, *40*, 155–165. [CrossRef]
44. Bog, M.; Appenroth, K.J.; Sree, K.S. Duckweed (Lemnaceae): Its molecular taxonomy. *Front. Sustain. Food Syst.* **2019**, *3*, 117. [CrossRef]
45. Acosta, K.; Appenroth, K.J.; Borisjuk, L.; Edelman, M.; Heinig, U.; Jansen, M.A.K.; Oyama, T.; Pasaribu, B.; Schubert, I.; Sorrels, S.; et al. Return of the Lemnaceae: Duckweed as a model plant system in the genomics and postgenomics era. *Plant Cell* **2021**, *33*, 3207–3234. [CrossRef]
46. Maheshwari, S.C. *Spirodela polyrrhiza*: The link between the aroids and the duckweeds. *Nature* **1958**, *181*, 1745–1746. [CrossRef]
47. Sree, K.S.; Maheshwari, S.C.; Boka, K.; Khurana, J.P.; Keresztes, Á.; Appenroth, K.J. The duckweed *Wolffia microscopica*: A unique aquatic monocot. *Flora* **2015**, *210*, 31–39. [CrossRef]
48. Sree, K.S.; Sudakaran, S.; Appenroth, K.J. How fast can angiosperms grow? Species and clonal diversity of growth rates in the genus *Wolffia* (Lemnaceae). *Acta Physiol. Plant.* **2015**, *37*, 204. [CrossRef]
49. Ziegler, P.; Adelmann, K.; Zimmer, S.; Schmidt, C.; Appenroth, K.J. Relative in vitro growth rates of duckweeds (*Lemnaceae*), the most rapidly growing higher plants. *Plant Biol.* **2015**, *17* (Suppl. S1), 33–41. [CrossRef] [PubMed]
50. Coughlan, N.E.; Kelly, T.C.; Jansen, M.A. "Step by step": High frequency short-distance epizoochorous dispersal of aquatic macrophytes. *Biol. Invasions* **2017**, *19*, 625–634. [CrossRef]
51. Dumortier, B.C.J. *Florula Belgica*; J. Casterman: Tournai, Belgium, 1827.
52. Schleiden, M.J. Prodromous monographiae Lemnacearum. *Linnaea* **1839**, *13*, 385–392.
53. Hegelmaier, F. *Die Lemnaceen. Eine Monographische Untersuchung*; Wilhelm Engelmann: Leipzig, Germany, 1868.
54. Grayum, M.H. Comparative external pollen ultrastructure of the Araceae and putatively related taxa. *Monogr. Syst. Bot. Missouri Bot. Gard.* **1992**, *43*, 1–167.

55. Bog, M.; Appenroth, K.J.; Sree, K.S. Key to the determination of taxa within the family of Lemnaceae: An update. *Nord. J. Bot.* **2020**, *38*, e02658. [CrossRef]
56. Bog, M.; Sree, K.S.; Fuchs, J.; Hoang, P.T.N.; Schubert, I.; Kuever, J.; Rabenstein, A.; Paolacci, S.; Jansen, M.A.K.; Appenroth, K.J. A taxonomic revision of *Lemna* sect. *Uninerves (Lemnaceae)*. *Taxon* **2020**, *69*, 56–66. [CrossRef]
57. Gray, S.F. *A Natural Arrangement of British Plants*; Baldwin, Cradock, and Joy: London, UK, 1821; Volume 2.
58. Les, D.H.; Crawford, D.J.; Landolt, E.; Gabel, J.D.; Kimball, R.T. Phylogeny and systematics of Lemnaceae, the duckweed family. *Syst. Bot.* **2002**, *27*, 221–240. [CrossRef]
59. Tippery, N.P.; Les, D.H.; Crawford, D.J. Evaluation of phylogenetic relationships in Lemnaceae using nuclear ribosomal data. *Plant Biol.* **2015**, *17* (Suppl. S1), 50–58. [CrossRef] [PubMed]
60. Wang, W.; Haberer, G.; Gundlach, H.; Gläßer, C.; Nussbaumer, T.; Luo, M.-C.; Lomsadze, A.; Borodovsky, M.; Kerstetter, R.A.; Shanklin, J.; et al. The genome of the primordial monocotyledonous *Spirodela polyrhiza*: Neotenous reduction, fast growth, and aquatic lifestyle. *Nat. Commun.* **2014**, *5*, 3311. [CrossRef]
61. Van Hoeck, A.; Horemans, N.; Monsieurs, P.; Cao, H.X.; Vandenhove, H.; Blust, R. The first draft genome of the aquatic model plant *Lemna minor* opens the route for future stress physiology research and biotechnological applications. *Biotechnol. Biofuels* **2015**, *8*, 188. [CrossRef]
62. Cao, H.X.; Vu, G.T.; Wang, W.; Appenroth, K.J.; Messing, J.; Schubert, I. The map-based genome sequence of *Spirodela polyrhiza* aligned with its chromosomes, a reference for karyotype evolution. *New Phytol.* **2016**, *209*, 354–363. [CrossRef] [PubMed]
63. An, D.; Li, C.S.; Zhou, Y.; Wu, Y.R.; Wang, W.Q. Genomes and transcriptomes of duckweeds. *Front. Chem.* **2018**, *6*, 230. [CrossRef]
64. Hoang, P.T.N.; Michael, T.P.; Gilbert, S.; Chu, P.; Motley, S.T.; Appenroth, K.J.; Schubert, I.; Lam, E. Generating a high-confidence reference genome map of the Greater Duckweed by integration of cytogenomic, optical mapping, and Oxford Nanopore technologies. *Plant J.* **2018**, *96*, 670–684. [CrossRef]
65. Les, D.H.; Tippery, N.P. In time and with water ... the systematics of alismatid monocotyledons. In *Early Events in Monocot Evolution*; Wilkin, P., Ed.; Cambridge University Press: Cambridge, UK, 2013; pp. 118–164.
66. Chase, M.W.; Christenhusz, M.J.M.; Fay, M.F.; Byng, J.W.; Judd, W.S.; Soltis, D.E.; Mabberly, D.J.; Sennikov, A.N.; Soltis, P.S.; Stevens, P.F.; et al. An update of the Angiosperm Phylogeny Group classification for the orders and families of flowering plants: APG IV. *Bot. J. Linn. Soc.* **2016**, *181*, 1–20. [CrossRef]
67. Kubitzki, K.; Rudall, P.J.; Chase, M.C. Systematics and Evolution. In *Flowering Plants Monocotyledons. The Families and Genera of Vascular Plants*; Kubitzki, K., Ed.; Springer: Berlin/Heidelberg, Germany, 1998; Volume 3. [CrossRef]
68. Eguchi, S.; Tamura, M.N. Evolutionary timescale of monocots determined by the fossilized birth-death model using a large number of fossil records. *Evolution* **2016**, *70*, 1136–1144. [CrossRef]
69. Reveal, J.L.; Zomlefer, W.B. Two New Orders for Monocotyledonous Plants. *Novon* **1998**, *8*, 176–177. [CrossRef]
70. Haynes, R.R.; Les, D.H. Alismatales (Water Plantains). In *eLS*; Wiley: Hoboken, NJ, USA, 2005. [CrossRef]
71. Takhtajan, A. *Diversity and Classification of Flowering Plants*; Columbia University Press: New York, NY, USA, 1997.
72. Cronquist, A. *An Integrated System of Classification of Flowering Plants*; Columbia University Press: New York, NY, USA, 1981.
73. Tamura, M.N.; Yamashita, J.; Fuse, S.; Haraguchi, M. Molecular phylogeny of monocotyledons inferred from combined analysis of plastid matK and rbcL gene sequences. *J. Plant. Res.* **2004**, *117*, 109–120. [CrossRef]
74. Janssen, T.; Bremer, K. The age of major monocot groups inferred from 800+rbcL sequences. *Bot. J. Linn. Soc.* **2004**, *146*, 385–398. [CrossRef]
75. Wang, W.; Messing, J. High-Throughput Sequencing of Three Lemnoideae (Duckweeds) Chloroplast Genomes from Total DNA. *PLoS ONE* **2011**, *6*, e24670. [CrossRef] [PubMed]
76. Henriquez, C.L.; Arias, T.; Pires, J.C.; Croat, T.B.; Schaal, B.A. Phylogenomics of the plant family Araceae. *Mol. Phylogenet. Evol.* **2014**, *75*, 91–102. [CrossRef] [PubMed]
77. Ruhfel, B.R.; Gitzendanner, M.A.; Soltis, P.S.; Soltis, D.E.; Burleigh, J.G. From algae to angiosperms–inferring the phylogeny of green plants (Viridiplantae) from 360 plastid genomes. *BMC Evol. Biol.* **2014**, *14*, 23. [CrossRef]
78. Trifinopoulos, J.; Nguyen, L.T.; von Haeseler, A.; Minh, B.Q. W-IQ-TREE: A fast online phylogenetic tool for maximum likelihood analysis. *Nucleic Acids Res.* **2016**, *44*, W232–W235. [CrossRef] [PubMed]
79. Li, H.T.; Yi, T.S.; Gao, L.M.; Ma, P.F.; Zhang, T.; Yang, J.B.; Gitzendanner, M.A.; Fritsch, P.W.; Cai, J.; Luo, Y.; et al. Origin of angiosperms and the puzzle of the Jurassic gap. *Nat. Plants* **2019**, *5*, 461–470. [CrossRef]
80. Li, H.T.; Yi, T.S.; Gao, L.M.; Ma, P.F.; Zhang, T.; Yang, J.B.; Gitzendanner, M.A.; Fritsch, P.W.; Cai, J.; Luo, Y.; et al. Data from: Origin of angiosperms and the puzzle of the Jurassic gap. *Dryad Dataset* **2019**, *5*, 461–470. [CrossRef]
81. Ross, T.G.; Barrett, C.F.; Soto Gomez, M.; Lam, V.K.; Henriquez, C.L.; Les, D.H.; Davis, J.I.; Cuenca, A.; Petersen, G.; Seberg, O.; et al. Plastid phylogenomics and molecular evolution of Alismatales. *Cladistics* **2016**, *32*, 160–178. [CrossRef]
82. Choi, K.S.; Park, K.T.; Park, S. The Chloroplast Genome of *Symplocarpus renifolius*: A Comparison of Chloroplast Genome Structure in Araceae. *Genes* **2017**, *8*, 324. [CrossRef]
83. Kim, S.-H.; Yang, J.; Park, J.; Yamada, T.; Maki, M.; Kim, S.-C. Comparison of Whole Plastome Sequences between Thermogenic Skunk Cabbage *Symplocarpus renifolius* and Nonthermogenic *S. nipponicus* (Orontioideae; Araceae) in East Asia. *Int. J. Mol. Sci.* **2019**, *20*, 4678. [CrossRef]
84. Henriquez, C.L.; Ahmed, I.; Carlsen, M.M.; Zuluaga, A.; Croat, T.B.; McKain, M.R. Evolutionary dynamics of chloroplast genomes in subfamily Aroideae (Araceae). *Genomics* **2020**, *112*, 2349–2360. [CrossRef] [PubMed]

85. Henriquez, C.L.; Mehmood, F.; Shahzadi, I.; Ali, Z.; Waheed, M.T.; Croat, T.B.; Poczai, P.; Ahmed, I. Comparison of Chloroplast Genomes among Species of Unisexual and Bisexual Clades of the Monocot Family Araceae. *Plants* **2020**, *9*, 737. [CrossRef]
86. Henriquez, C.L.; Mehmood, F.; Carlsen, M.M.; Islam, M.; Waheed, M.T.; Poczai, P.; Croat, T.B.; Ahmed, I. Complete Chloroplast Genomes of *Anthurium huixtlense* and *Pothos scandens* (Pothoideae, Araceae): Unique Inverted Repeat Expansion and Contraction Affect Rate of Evolution. *J. Mol. Evol.* **2020**, *88*, 562–574. [CrossRef]
87. Henriquez, C.L.; Croat, T.B.; Poczai, P.; Ahmed, I. Mutational Dynamics of Aroid Chloroplast Genomes II. *Front. Genet.* **2021**, *11*, 610838. [CrossRef]
88. Sanderson, M.J. Estimating Absolute Rates of Molecular Evolution and Divergence Times: A Penalized Likelihood Approach. *Mol. Biol. Evol.* **2002**, *19*, 101–109. [CrossRef] [PubMed]
89. R Core Team. *R: A Language and Environment for Statistical Computing*; R Foundation for Statistical Computing: Vienna, Austria, 2021.
90. Baldwin, B.G.; Sanderson, M.J.; Porter, J.M.; Wojciechowski, M.F.; Campbell, C.S.; Donoghue, M.J. The ITS Region of Nuclear Ribosomal DNA: A Valuable Source of Evidence on Angiosperm Phylogeny. *Ann. Mo. Bot. Gard.* **1995**, *82*, 247–277. [CrossRef]
91. Soltis, D.E.; Soltis, P.S.; Nickrent, D.L.; Johnson, L.A.; Hahn, W.J.; Hoot, S.B.; Sweere, J.A.; Kuzoff, R.K.; Kron, K.A.; Chase, M.W.; et al. Angiosperm Phylogeny Inferred from 18S Ribosomal DNA Sequences. *Ann. Mo. Bot. Gard.* **1997**, *84*, 1–49. [CrossRef]
92. Maia, V.H.; Gitzendanner, M.A.; Soltis, P.S.; Wong, G.K.S.; Soltis, D.E. Angiosperm Phylogeny Based on 18S/26S rDNA Sequence Data: Constructing a Large Data Set Using Next-Generation Sequence Data. *Int. J. Plant Sci.* **2014**, *175*, 613–650. [CrossRef]
93. Álvarez, I.; Wendel, J.F. Ribosomal ITS sequences and plant phylogenetic inference. *Mol. Phylogenet. Evol.* **2003**, *29*, 417–434. [CrossRef]
94. Medlin, L.; Elwood, H.J.; Stickel, S.; Sogin, M.L. The characterization of enzymatically amplified eukaryotic 16S-like rRNA-coding regions. *Gene* **1988**, *71*, 491–499. [CrossRef]
95. Blattner, F.R. Direct Amplification of the Entire ITS Region from Poorly Preserved Plant Material Using Recombinant PCR. *BioTechniques* **1999**, *27*, 1180–1186. [CrossRef]
96. Shoup, S.; Lewis, L.A. Polyphyletic origin of parallel basal bodies in swimming cells of chlorophycean green algae (Chlorophyta). *J. Phycol.* **2003**, *39*, 789–796. [CrossRef]
97. Cheng, T.; Xu, C.; Lei, L.; Li, C.; Zhang, Y.; Zhou, S. Barcoding the kingdom Plantae: New PCR primers for ITS regions of plants with improved universality and specificity. *Mol. Ecol. Resour.* **2016**, *16*, 138–149. [CrossRef]
98. One Thousand Plant Transcriptomes Initiative. One thousand plant transcriptomes and the phylogenomics of green plants. *Nature* **2019**, *574*, 679–685. [CrossRef] [PubMed]
99. Barr, C.M.; Neiman, M.; Taylor, D.R. Inheritance and recombination of mitochondrial genomes in plants, fungi and animals. *New Phytol.* **2005**, *168*, 39–50. [CrossRef]
100. Pentinsaari, M.; Salmela, H.; Mutanen, M.; Roslin, T. Molecular evolution of a widely-adopted taxonomic marker (COI) across the animal tree of life. *Sci. Rep.* **2016**, *6*, 35275. [CrossRef] [PubMed]
101. Cho, Y.; Palmer, J.D. Multiple acquisitions via horizontal transfer of a group I intron in the mitochondrial cox1 gene during evolution of the *Araceae* family. *Mol. Biol. Evol.* **1999**, *16*, 1155–1165. [CrossRef]
102. Cusimano, N.; Zhang, L.B.; Renner, S.S. Reevaluation of the cox1 Group I Intron in Araceae and Angiosperms Indicates a History Dominated by Loss rather than Horizontal Transfer. *Mol. Biol. Evol.* **2008**, *25*, 265–276. [CrossRef] [PubMed]
103. Qiu, Y.; Li, L.; Hendry, T.A.; Li, R.; Taylor, D.W.; Issa, M.J.; Ronen, A.J.; Vekaria, M.L.; White, A.M. Reconstructing the basal angiosperm phylogeny: Evaluating information content of mitochondrial genes. *Taxon* **2006**, *55*, 837–856. [CrossRef]
104. Qiu, Y.L.; Li, L.; Wang, B.; Xue, J.Y.; Hendry, T.A.; Li, R.Q.; Brown, J.W.; Liu, Y.; Hudson, G.T.; Chen, Z.D. Angiosperm phylogeny inferred from sequences of four mitochondrial genes. *J. Syst. Evol.* **2010**, *48*, 391–425. [CrossRef]
105. Petersen, G.; Seberg, O.; Davis, J.I.; Stevenson, D.W. RNA editing and phylogenetic reconstruction in two monocot mitochondrial genes. *Taxon* **2006**, *55*, 871–886. [CrossRef]
106. Cuenca, A.; Petersen, G.; Seberg, O.; Davis, J.I.; Stevenson, D.W. Are substitution rates and RNA editing correlated? *BMC Evol. Biol.* **2010**, *10*, 349. [CrossRef]
107. Petersen, G.; Cuenca, A.; Zervas, A.; Ross, G.T.; Graham, S.W.; Barrett, C.F.; Davis, J.I.; Seberg, O. Mitochondrial genome evolution in Alismatales: Size reduction and extensive loss of ribosomal protein genes. *PLoS ONE* **2017**, *12*, e0177606. [CrossRef]
108. Mower, J.P. Variation in protein gene and intron content among land plant mitogenomes. *Mitochondrion* **2020**, *53*, 203–213. [CrossRef] [PubMed]
109. Wang, W.; Wu, Y.; Messing, J. The Mitochondrial Genome of an Aquatic Plant, *Spirodela polyrhiza*. *PLoS ONE* **2012**, *7*, e46747. [CrossRef]
110. Thompson, S.A. Araceae. In *Flora of North America Editorial Committee. Flora of North America, North of Mexico. Magnoliophyta: Alismatidae, Arecidae, Commelinidae (in Part), and Zingiberidae*; Oxford University Press: New York, NY, USA, 2000; Volume 22.
111. Haynes, R.R. Hydrocharitaceae. In *Flora of North America Editorial Committee. Flora of North America, North of Mexico. Magnoliophyta: Alismatidae, Arecidae, Commelinidae (in Part), and Zingiberidae*; Oxford University Press: New York, NY, USA, 2000; Volume 22.
112. Keating, R.C. Acoraceae and Araceae. In *Anatomy of the Monocotyledons*; Gregory, M., Cutler, D.F., Eds.; Oxford University Press: Oxford, UK, 2002; Volume 9.

113. Sender, L.M.; Doyle, J.A.; Upchurch, G.R., Jr.; Villanueva-Amadoz, U.; Diez, J.B. Leaf and inflorescence evidence for near-basal Araceae and an unexpected diversity of other monocots from the late Early Cretaceous of Spain. *J. Syst. Palaeontol.* **2019**, *17*, 1313–1346. [CrossRef]
114. Stockey, R.A.; Hoffman, G.L.; Rothwell, G.W. Fossil evidence for Paleocene diversification of Araceae: *Bognerospadix* gen. nov. and *Orontiophyllum grandifolium* comb. nov. *Am. J. Bot.* **2021**, *108*, 1417–1440. [CrossRef]
115. Hoang, P.T.N.; Schubert, V.; Meister, A.; Fuchs, J.; Schubert, I. Variation in genome size, cell and nucleus volume, chromosome number and rDNA loci among duckweeds. *Sci. Rep.* **2019**, *9*, 3234. [CrossRef]
116. Hoang, P.T.N.; Schubert, I. Reconstruction of chromosome rearrangements between the two most ancestral duckweed species *Spirodela polyrhiza* and *S. intermedia*. *Chromosoma* **2017**, *126*, 729–739. [CrossRef] [PubMed]
117. Yin, J.; Jiang, L.; Wang, L.; Han, X.; Guo, W.; Li, C.; Zhou, Y.; Denton, M.; Zhang, P. A high-quality genome of taro (*Colocasia esculenta* (L.) Schott), one of the world's oldest crops. *Mol. Ecol. Resour.* **2021**, *21*, 68–77. [CrossRef]
118. Olson, D.M.; Dinerstein, E.; Wikramanayake, E.D.; Burgess, N.D.; Powell, G.V.N.; Underwood, E.C.; D'amico, J.A.; Itoua, I.; Strand, H.E.; Morrison, J.C.; et al. Terrestrial Ecoregions of the World: A New Map of Life on Earth: A new global map of terrestrial ecoregions provides an innovative tool for conserving biodiversity. *BioScience* **2001**, *51*, 933–938. [CrossRef]
119. POWO. Plants of the World Online. Facilitated by the Royal Botanic Gardens, Kew. Available online: http://www.plantsoftheworldonline.org (accessed on 15 October 2021).
120. Brummitt, R.K. *World Geographical Scheme for Recording Plant Distributions;* For International Working Group on Taxonomic Databases for Plant Sciences (TDWG) 153; Hunt Institute for Botanical Documentation; Carnegie Mellon University: Pittsburgh, PA, USA, 2001.
121. Smith, S.A.; Brown, J.W. Constructing a broadly inclusive seed plant phylogeny. *Am. J. Bot.* **2018**, *105*, 302–314. [CrossRef] [PubMed]
122. Olmstead, R.G.; de Pamphilis, C.W.; Wolfe, A.D.; Young, N.D.; Elisons, W.J.; Reeves, P.A. Disintegration of the *Scrophulariaceae*. *Am. J. Bot.* **2001**, *88*, 348–361. [CrossRef] [PubMed]

plants

MDPI

Article

New Insights into Interspecific Hybridization in *Lemna* L. Sect. *Lemna* (Lemnaceae Martinov)

Luca Braglia, Diego Breviario, Silvia Gianì, Floriana Gavazzi, Jacopo De Gregori and Laura Morello *

Istituto Biologia e Biotecnologia Agraria, Via Alfonso Corti 12, 20131 Milano, Italy; braglia@ibba.cnr.it (L.B.); breviario@ibba.cnr.it (D.B.); giani@ibba.cnr.it (S.G.); gavazzi@ibba.cnr.it (F.G.); degregori@ibba.cnr.it (J.D.G.)
* Correspondence: morello@ibba.cnr.it; Tel.: +39-02-2369-9686

Abstract: Duckweeds have been increasingly studied in recent years, both as model plants and in view of their potential applications as a new crop in a circular bioeconomy perspective. In order to select species and clones with the desired attributes, the correct identification of the species is fundamental. Molecular methods have recently provided a more solid base for taxonomy and yielded a consensus phylogenetic tree, although some points remain to be elucidated. The duckweed genus *Lemna* L. comprises twelve species, grouped in four sections, which include very similar sister species. The least taxonomically resolved is sect. *Lemna*, presenting difficulties in species delimitation using morphological and even barcoding molecular markers. Ambiguous species boundaries between *Lemna minor* L. and *Lemna japonica* Landolt have been clarified by Tubulin Based Polymorphism (TBP), with the discovery of interspecific hybrids. In the present work, we extended TBP profiling to a larger number of clones in sect. *Lemna*, previously classified using only morphological features, in order to test that classification, and to investigate the possible existence of other hybrids in this section. The analysis revealed several misidentifications of clones, in particular among the species *L. minor*, *L. japonica* and *Lemna gibba* L., and identified six putative '*L. gibba*' clones as interspecific hybrids between *L. minor* and *L. gibba*.

Keywords: duckweeds; Lemnaceae; interspecific hybrids; tubulin-based polymorphism; β-tubulin

Citation: Braglia, L.; Breviario, D.; Gianì, S.; Gavazzi, F.; De Gregori, J.; Morello, L. New Insights into Interspecific Hybridization in *Lemna* L. Sect. *Lemna* (Lemnaceae Martinov). *Plants* **2021**, *10*, 2767. https://doi.org/10.3390/plants10122767

Academic Editors: Viktor Oláh, Klaus-Jürgen Appenroth and K. Sowjanya Sree

Received: 15 November 2021
Accepted: 9 December 2021
Published: 15 December 2021

1. Introduction

The genus *Lemna* L. (Lemnaceae, Martinov) [1] is thought to have originated around 41.7 MYA (crown age) from a common ancestor which separated from the Wolffioideae Engl. (genera *Wolffia* Horkel ex Schleiden and *Wolffiella* Hegelm.) branch around 54.4 MYA [2]. Its most probable origin has been established as around 16.4–41.7 MYA in North America. According to the latest taxonomic revision, uniting *Lemna minuta* Kuntz with *Lemna valdiviana* Phil., the genus *Lemna* comprises 12 species [3], grouped in four monophyletic sections: *Alatae* Hegelm., *Uninerves* Hegelm., *Biformes* Landolt, and *Lemna* [4]: this categorisation into sections is also supported by GBS data [5]. Sect. *Lemna* includes seven species, among which we can find the most cosmopolitan *Lemna gibba* L. and *Lemna minor* L., as well as geographically restricted species such as *Lemna disperma* Hegelm. (Oceania) and *Lemna obscura* (Austin) Daubs (South-East coast of North America and Ecuador). This is the most problematic section within the taxonomically complex genus *Lemna*, in which boundaries between some species seem to blur, due to extremely similar morphology. Species can be distinguished from each other based on few, recently updated, key features [3], some of which refer to flowers or fruits, rarely observed in many species. Although sufficient in most cases, key features may vary within the same species among clones or under different growth conditions, particularly upon in vitro cultivation, making identification difficult.

For example, *L. gibba* is usually easily recognized for the inflated (gibbous) form of the frond caused by enlarged air spaces in the aerenchyma tissue. However, morphological variability in *L. minor* and *L. gibba* has long been known, and discrimination between the two becomes particularly difficult when air spaces of *L. gibba* are reduced, making fronds flat [6–8]. This has led to grouping the two species in the *L. minor–L. gibba* group, or complex, which identifies a continuum between the two [8,9]. Both species are long-day plants which sometimes share the same habitat. Variations in frond morphology in *L. gibba* may be seasonal. but the occurrence of mixed populations of gibbous and permanently 'flat forms' of *L. gibba* (*L. minor*-like) has been repeatedly reported in The Netherlands, together with supposed transition forms between the two species, which could not be assigned with certainty to either of the two [9,10].

Species with intermediate traits between *L. minor* and *L. gibba*, but bearing some distinctive features have been described in the past, such as *Lemna parodiana* Giardelli in Argentina [11] and *Lemna symmeter* Giuga, in Southern Italy [12]. In the absence of further evidence, such species have been considered conspecific with *L. gibba* [13].

Intraspecific variability in *L. minor* is represented by reported differences in chromosome numbers, from 40 to 50, and by different genome size reported among clones [14,15], although data are not always consistent because of the small chromosome size and variability in genome size estimation by different methods of measurements. It is therefore difficult to say how much the wide morphological intraspecific diversity observed could be due to phenotypic plasticity or to genetic diversity.

Lemna disperma shows a combination of characters of *L. gibba* and *L. minor* [16], but it is restricted to Australia and New Zealand. Similarly, *L. obscura*, was previously identified either as *L. minor* or *L. gibba* but is limited to temperate regions of North America [16]. In both cases, morphological classification is often supported by geographical distribution. A further species in this section, former *Lemna ecuadoriensis* Landolt, is now considered conspecific with *L. obscura* [17].

Moreover, *Lemna turionifera* Landolt, of Northern Asia and America, can be occasionally confused with *L. minor* and *L. gibba* by frond morphology, but fruits and seed characters, together with turion-forming ability, provide distinctive traits. *Lemna japonica* was described as a new species in 1980 as a biotype of *L. minor* with a limited geographical distribution and posited as a possible interspecific hybrid between *L. minor* and *L. turionifera* [18]. This hypothesis was supported by intermediate allozyme pattern shown by *L. japonica* clones with respect to other similar specimens collected in Japan and likely corresponding to *L. minor* and *L. turionifera* respectively [19]. This evidence was recently supported by genetic proofs based on intron length polymorphism in the β-tubulin genes (also known as Tubulin Based Polymorphism, TBP) and AFLP. Accordingly, *L. japonica* is hereafter indicated as *Lemna* ×*japonica* to indicate the hybrid status of this taxon, assessed as *L. minor*-SubCluster II (*L. minor sensu lato*) in our previous work [20]. Its similarity with *L. minor* is therefore evident and is the cause of frequent misclassifications.

The species most recently included in the section is *Lemna trisulca*, formerly separated in the single-species sect. *Hydrophylla* Dumort. This species has a unique morphology with submerged, narrowly ovate fronds connected to a green stalk, often forming branched chains. Despite this peculiarity, the combined data cladogram (morphological, flavonoid, allozyme and DNA sequence data) first obtained by Les [4] clearly placed *L. trisulca* within sect. *Lemna*, as later confirmed by nuclear and plastid molecular markers [21]. In some cases, even the more distantly related species like *L. minuta*, native to America but invasive in Europe [22], can be distinguished from the European native *L. minor* only by quantitative morphometry [23]. The advent of molecular taxonomy has greatly facilitated species delimitation among duckweeds by AFLP fingerprinting [24,25] and plastid barcoding sequences as *psbK-psbI* and *atpF-atpH*, which are now commonly used for accurate identification of clones [26,27]. Nuclear and plastid molecular markers have also bolstered phenetics in improving phylogenetic studies [4,21] on duckweeds. However, despite this progress, phylogenetic uncertainty still persists among some lineages and some nodes resolved incongruently by using plastid and nuclear ribosomal sequences [21]. This has been repeatedly attributed to potential interbreeding or incomplete divergence, although neither has ever been demonstrated. The application of high-throughput methods as genotyping-by-sequencing (GBS) has helped to resolve those problematic species boundaries in the genus where plastid sequences alone were inadequate as in the case of *L. minor* and *L. ×japonica* [5]. This observation is in accordance with the maternal inheritance of plastids from *L. minor*, in the light of the finding that *L. ×japonica* is of hybrid origin [20]. On the one hand, some interpretive problems may have arisen because some accessions were not identified accurately [4] while, on the other, when many clones of the same species were compared, molecular marker analysis enabled uncovering possible misclassifications of clones [20,24,26], particularly in sect. *Lemna*.

To this end, TBP fingerprinting has been particularly suitable for species delimitation in the genus [20]. This genetic profiling method had been successfully applied to species and subspecies level discrimination in different plant families [28–30]. The Landolt Duckweed Collection (LDC, Rotterdam, The Netherlands, http://www.duckweed.ch (accessed on 8 June 2021)) is perhaps the most important historical collection worldwide, with over 500 clones left of the more than 1000 collected and morphologically classified over 70 years by Professor Elias Landolt in Zürich, CH. Many replicated clones are also present in the collections of the Rutgers Duckweed Stock Cooperative (http://www.ruduckweed.org/ (accessed on 14 December 2021)) and at the University of Jena and are being investigated in many laboratories worldwide. Although many clones have been investigated by molecular markers, a large part of the collection remains genetically unexplored. As TBP profiling provides a simple way for duckweed species discrimination without sequencing, we planned to investigate all LDC's clones, under an agreement with Mr. W. Lämmler, the manager of the LDC.

We started with a large selection of clones, about 100 in the problematic *Lemna* sect. *Lemna*, with the dual aim of verifying the morphological classification of each clone and of finding evidence for the existence of other interspecific hybrids in this section. Given the importance of the LDC as a fundamental resource for scientists working in the field, our data provide useful information for further investigations and for a critical revision of the literature. Interspecific relationships within sect. *Lemna* are also investigated by length and sequence similarity in β-tubulin introns. Moreover, leveraging the high genetic variability of such regions, introns are also used as a suitable source of SNPs for the evaluation of intraspecific variability.

2. Results

2.1. TBP Profiling of Duckweed Clones in Lemna Sect. Lemna

TBP was demonstrated to be a reliable tool for clustering *Lemna* clones according to the respective species, as validated by plastid markers [20]. Distinctive amplification profiles are obtained for each species, with some intraspecific allelic variations. Ninety-eight duckweed clones belonging to sect. *Lemna* were analyzed by TBP profiling of the first and second β-tubulin introns (Table 1 and Supplementary Table S1). Fifty-seven clones belonging to the same section and analyzed in the previously mentioned work [20] were added to the cluster analysis, with *Landoltia punctata* clone 9354 used as an outgroup. The scoring of the Capillary Electrophoresis TBP (CE-TBP) peaks revealed 139 polymorphic markers across the seven *Lemna* species (87 and 52 from the 1st and the 2nd intron region, respectively). The derived dendrogram is shown in Figure 1.

The separation of the seven *Lemna* species forming the section, according to their genetic similarity by TBP, allowed the unequivocal reclassification of those clones which do not correspond to their morphological description. The hybrid status of *L. ×japonica* is confirmed here by this larger dataset. In fact, all the clones in this cluster, which includes the *L. ×japonica* holotype 7182, showed hybrid TBP profiles between *L. minor* and *L. turionifera*, reflecting the duplicate set of six β-tubulin genes, whereas all clones in the *L. minor* cluster have similar pattern among each other, with just six main peaks. Despite the low bootstrap values (<50%) of this branch, the tree topology is clearly due to the fact that *L. ×japonica* shares alleles with both putative parental species *L. minor* and *L. turionifera*. Interestingly, through the whole tree, high probability support was given to some sub-clusters indicating intraspecific allelic variance among populations (in Figure 1). Separation was in agreement with the geographical origin of clones, as in the case of the American clusters of *L. ×japonica*, *L. turionifera* (the only American clone) and *L. gibba*, the East Asian clusters of *L. ×japonica* and *L. turionifera* and a Mediterranean group of *L. gibba* (Figure 1).

Table 1. List of the plant material and reclassification of clones by TBP analysis.

Clone ID	Collection	Continent/Region	Country	Classification System	
				Morphological Characters	TBP Analysis
0050	Landolt Collection	Asia	China	*L. japonica*	*L. ×japonica*
0078	Landolt Collection	Asia	China	*L. japonica*	*L. ×japonica*
0150	Landolt Collection	Asia	China	*L. japonica*	*L. ×japonica*
0190	Landolt Collection	North America	USA	*L. japonica*	*L. gibba* *
0198	Landolt Collection	Asia	China	*L. japonica*	*L. ×japonica*
6580	Landolt Collection	North America	USA	*L. minor*	*L. ×japonica* *
6591	Landolt Collection	North America	USA	*L. minor*	*L. minor*
6619	Jena University	North America	USA	*L. turionifera*	*L. turionifera*
6728	Jena University	North America	USA	*L. turionifera*	*L. ×japonica* *
6742	Landolt Collection	North America	USA	*L. japonica*	*L. ×japonica*
6745	Landolt Collection	North America	USA	*L. gibba*	*L. gibba*
6853	Jena University	North America	Canada	*L. turionifera*	*L. turionifera*
6861	Landolt Collection	Europe	Italy	*L. gibba*	*L. gibba × L. Minor* *
7018	Landolt Collection	Asia	Turkey	*L. minor*	*L. minor*
7021	Landolt Collection	Europe	Spain	*L. gibba*	*L. ×japonica* *
7123	Landolt Collection	North America	Canada	*L. minor*	*L. ×japonica* *
7182	Landolt Collection	East Asia	Japan	*L. japonica*	*L. ×japonica*
7295	Landolt Collection	Africa	Libya	*L. minor*	*L. minor*
7320	Landolt Collection	Africa	Egypt	*L. gibba*	*L. gibba × L. Minor* *
7427	Landolt Collection	East Asia	Japan	*L. turionifera*	*L. turionifera*
7432	Landolt Collection	East Asia	Japan	*L. japonica*	*L. turionifera* *
7537	Landolt Collection	Europe	Spain	*L. gibba*	*L. ×japonica* *
7641	Landolt Collection	Asia	Israel	*L. gibba*	*L. gibba × L. Minor* *
7683	Landolt Collection	Asia	South Korea	*L. turionifera*	*L. turionifera*
7705	Landolt Collection	India	India	*L. gibba*	*L. gibba*
7767	Landolt Collection	Oceania	Australia	*L. disperma*	*L. disperma*
7777	Landolt Collection	Oceania	Australia	*L. disperma*	*L. disperma*
7798	Landolt Collection	South America	Peru	*L. gibba*	*L. gibba*
7816	Landolt Collection	Oceania	Australia	*L. disperma*	*L. disperma*
7856	Landolt Collection	North America	USA	*L. obscura*	*L. obscura*
7868	Jena University	Europe	Ireland	*L. japonica*	*L. ×japonica*
7922	Landolt Collection	South America	Argentina	*L. gibba*	*L. gibba*
7951	Landolt Collection	Asia	China	*L. turionifera*	*L. turionifera*
8227	Landolt Collection	North America	USA	*L. obscura*	*L. obscura*
8428	Landolt Collection	Europe	Switzerland	*L. gibba*	*L. gibba*
8434	Landolt Collection	North America	Canada	*L. minor*	*L. ×japonica* *
8653	Landolt Collection	Asia	China	*L. japonica*	*L. ×japonica*
8697	Landolt Collection	East Asia	Japan	*L. japonica*	*L. ×japonica*
8717	Landolt Collection	Oceania	Australia	*L. L. disperma*	*L. disperma*
8760	Landolt Collection	Europe	Czech Republic	*L. turionifera*	*L. turionifera*
8892	Landolt Collection	North America	USA	*L. obscura*	*L. obscura*
9016	Landolt Collection	East Asia	Japan	*L. japonica*	*L. ×japonica*
9109	Jena University	Europe	Poland	*L. turionifera*	*L. turionifera*
9223	Landolt Collection	Europe	United Kingdom	*L. minor*	*L. minor*
9240	Landolt Collection	Europe, Asia	Russia	*L. minor*	*L. minor*
9248	Landolt Collection	Europe	Italy	*L. gibba*	*L. gibba × L. Minor* *
9250	Landolt Collection	Europe	Finland	*L. japonica*	*L. ×japonica*
9253	Landolt Collection	Europe	Finland	*L. minor*	*L. minor*
9254	Landolt Collection	Europe	Finland	*L. turionifera*	*L. turionifera*
9285	Landolt Collection	Asia	China	*L. japonica*	*L. ×japonica*
9330	Landolt Collection	Asia	China	*L. japonica*	*L. ×japonica*

Table 1. *Cont.*

Clone ID	Collection	Continent/Region	Country	Classification System	
				Morphological Characters	TBP Analysis
9345	Landolt Collection	Europe	Switzerland	*L. minor*	*L. minor*
9352	Landolt Collection	Europe	Albania	*L. gibba*	*L. gibba*
9421	Landolt Collection	North America	USA	*L. japonica*	*L. ×japonica*
9424	Landolt Collection	Europe	Germany	*L. minor*	*L. minor*
9429	Jena University	Europe, Asia	Russia	*L. turionifera*	*L. ×japonica* *
9435	Landolt Collection	Europe	Albania	*L. gibba*	*L. gibba*
9438	Landolt Collection	Europe	Czech Republic	*L. minor*	*L. minor*
9439	Landolt Collection	Europe	Germany	*L. minor*	*L. ×japonica* *
9470	Landolt Collection	Europe	United Kingdom	*L. turionifera*	*L. turionifera*
9471	Jena University	Europe	United Kingdom	*L. turionifera*	*L. turionifera*
9478	Jena University	Europe	Poland	*L. turionifera*	*L. turionifera*
9480	Landolt Collection	Europe, Asia	Russia	*L. turionifera*	*L. turionifera*
9482	Landolt Collection	Europe	Italy	*L. minor*	*L. minor*
9483	Landolt Collection	Europe	Albania	*L. minor*	*L. ×japonica* *
9485	Landolt Collection	Europe	Ireland	*L. minor*	*L. minor*
9532	Landolt Collection	Europe	Macedonia	*L. minor*	*L. ×japonica* *
9534	Landolt Collection	Europe	Germany	*L. minor*	*L. minor*
9542	Landolt Collection	Europe	Italy	*L. minor*	*L. ×japonica* *
9561	Landolt Collection	Europe	Sweden	*L. minor*	*L. minor*
9562	Jena University	Europe	Germany	*L. gibba*	*L. gibba × L. Minor* *
9574	Landolt Collection	Oceania	New Zealand	*L. minor*	*L. minor*
9577	Landolt Collection	Europe	Italy	*L. gibba*	*L. gibba*
9591	Landolt Collection	Europe	Hungary	*L. gibba*	*L. ×japonica* *
9598	Landolt Collection	Europe	Germany	*L. gibba*	*L. gibba*
9660	Landolt Collection	Asia	China	*L. japonica*	*L. ×japonica*
9942	Landolt Collection	Europe	Norway	*L. minor*	*L. minor*
9951	Landolt Collection	Europe	France	*L. gibba*	*L. ×japonica* *
9952	Landolt Collection	Europe	France	*L. minor*	*L. minor*
9961	Landolt Collection	Europe	Germany	*L. minor*	*L. minor*
9965	Landolt Collection	Europe	Switzerland	*L. gibba*	*L. ×japonica* *
9967	Landolt Collection	Europe	Switzerland	*L. minor*	*L. minor*
9969	Landolt Collection	Europe	Switzerland	*L. minor*	*L. ×japonica* *
9973	Landolt Collection	Europe	Germany	*L. minor*	*L. minor*
9977	Landolt Collection	Europe	Germany	*L. minor*	*L. minor*
9978	Landolt Collection	Europe	Switzerland	*L. minor*	*L. ×japonica* *
9979	Landolt Collection	Europe	Germany	*L. minor*	*L. minor*
9980	Landolt Collection	Europe	Germany	*L. minor*	*L. ×japonica* *
9982	Landolt Collection	North America	USA	*L. japonica*	*L. ×japonica*
9983	Landolt Collection	Europe	Switzerland	*L. japonica*	*L. ×japonica*
9986	Landolt Collection	North America	USA	*L. minor*	*L. ×japonica* *
9991	Landolt Collection	North America	USA	*L. japonica*	*L. ×japonica*
8784b	Landolt Collection	Europe	Sweden	*L. japonica*	*L. ×japonica*
9425a	Landolt Collection	Europe	Italy	*L. gibba*	*L. gibba × L. Minor* *
BOG0024	Greifswald University	Europe	Germany	*L. turionifera*	*L. turionifera*
BOG0071	Greifswald University	Europe	Germany	*L. turionifera*	*L. turionifera*
BOG0072	Greifswald University	Europe	Germany	*L. turionifera*	*L. turionifera*
KJA007	Jena University	Europe, Asia	Russia	*L. turionifera*	*L. turionifera*

* reclassified according to TBP analysis.

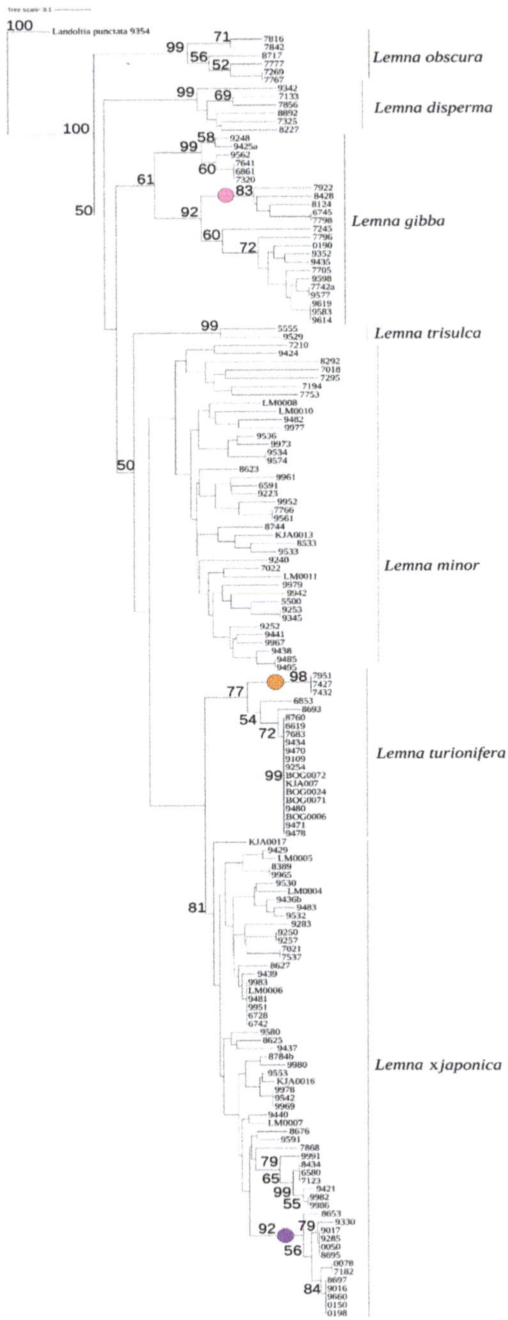

Figure 1. Neighbor joining similarity tree of 156 duckweed clones belonging to *Lemna* sect. *Lemna*, inferred through TBP fragment analysis (1st and 2nd intron regions). *Landoltia punctata* 9354 was used as outgroup to root the tree. The estimated bootstrap values (1000 replicates, >50%) are reported at the branch node. Colored dots highlight sub-cluster grouping clones with shared geographic origin: pink, *L. gibba* from America; orange, *L. turionifera* from East Asia; violet, *L.* ×*japonica* from East Asia).

The TBP data were also used to infer a principal component analysis (PCA) to further describe the genetic diversity among clones belonging to the four most represented species, *L. minor*, *L.* ×*japonica*, *L. gibba* and *L. turionifera* (Figure 2). The cumulative contribution of the first three principal components explains 63% of the total variation, providing a clear description of the relationships among species, concurrently revealing hybrid entries, which grouped separately. This was the case of *L.* ×*japonica* clones, including those originally classified as *L. minor*, which were clearly separated from the putative parental species *L. minor* and *L. turionifera*, along the plot axes.

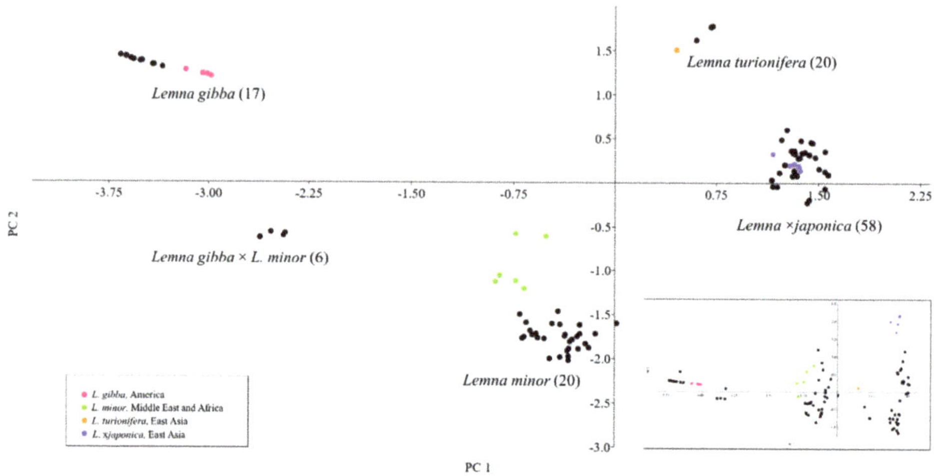

Figure 2. Principal components analysis (PCA) plot based on genetic distances between clones of *L. minor*, *L.* ×*japonica*, *L. gibba* and *L. turionifera*, inferred from TBP analysis (PC1 and PC2). In the insert, the plot of PC1 and PC3 shows the distribution of *L.* ×*japonica* clones. Clone IDs are omitted and the total number of analyzed clones per species is reported in brackets. Colored dots highlight only clones of each species forming the subclusters with shared geographic origin shown in Figure 1. The color code of dots is in accordance.

Similarly, an additional group, with respect to the four recognized species, was formed by six clones, mentioned above as the Mediterranean group, classified as *L. gibba* by morphology and representing one of two sister clades of the *L. gibba* cluster observed in the dendrogram of Figure 1. The bi-dimensional plot placed this group of clones in an intermediate position between *L. minor* and *L. gibba*, suggesting shared alleles with both species. This prompted us to further investigate if the aforementioned group of six clones could be considered a separate taxon, possibly a hybrid. In the plot and thereafter we then refer to this group as *Lemna gibba* × *Lemna minor* (see below).

In addition, within each species, isolated subgroups can be recognized as spread apart from the main clusters by the first three components of the PCA (Figure 2). In accordance with the dendrogram in Figure 1, the subgrouping distribution shown by the PCA is congruent with the geographical origin of clones (colored dots in Figure 2). Notably, in *L. minor* a group of clones from the Middle East and Africa was significantly spread apart from the respective main group.

2.2. Reclassification of Clones by TBP

The correspondence between the original morphological characterization of the analyzed clones and TBP results is summarized in Table 2, which reports misidentifications for each species and the kind of error involved. *L. obscura*, *L. trisulca* and *L. disperma* were the most easily identified species with a 100% correct assignment, although a reduced number of clones was available, and results are not included in the Table. The overall misidentification rate by morphology was 28.6% considering five species (treating *L. gibba* × *L. minor* hybrid as a separate taxon). Classification by morphology failed to correctly identify 48% of *L.* ×*japonica* clones, considering them as *L. minor* (30%), *L. gibba* (14%), and *L. turionifera* (5%). Fifty-five percent of supposed '*L. gibba*' turned out to be either *L.* ×*japonica* (25%) or the newly described hybrid *L. minor* × *L. gibba* (30%). In just two cases each, clones of *L. turionifera* and *L. gibba* were exchanged for *L.* ×*japonica*. It is therefore clear that most incorrect identifications involved hybrids, which were often classified as one of the parental species. The new classification of all clones according to TBP is given in Table 1.

Table 2. Misidentification rate: correlation between morphological classification and TBP analysis.

		TBP						
	Species	***L. minor***	***L. turionifera***	***L.* ×*japonica***	***L. gibba***	***L. gibba* × *L. minor***	**Tot. n**	**% Incorrect**
morphology	***L. minor***	21		11			32	34.3
	L. turionifera		16	2			18	11.1
	L.* ×*japonica		1	19	1		21	9.5
	L. gibba			5	9	6	20	55.0
	L. gibba* × *L. minor					0	0	0.0
	tot. n	21	17	37	10	6	91	28.6
	% incorrect	0.0	5.9	48.6	10.0	100.0	28.6	

The high rate of misidentification of *L.* ×*japonica* has at least two direct implications:

- The abundance of *L.* ×*japonica* populations was highly underestimated, as well as its geographical distribution, which is not limited to Japan, Korea and the east coast of China, as reported by Landolt for *L. japonica* [16]. The actual distribution, deduced from investigated clones and shown in Figure 3, covers all the temperate regions from Eastern Asia to Central Asia, Europe and North America, although their invasive origin in the different regions remains to be elucidated. One clone was even found in South Africa (Figure 3).
- At least part of the huge variability observed in *L. minor*, e.g., in genome size, ploidy or physiological parameters etc. could be due to erroneous classification of clones.

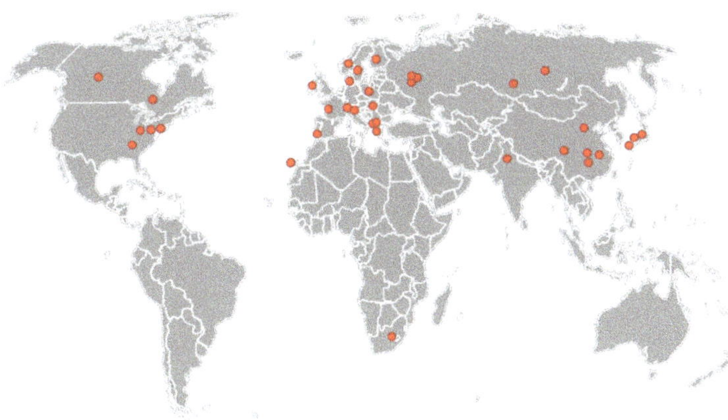

Figure 3. Geographical distribution of the analyzed clones of *L.* ×*japonica*.

2.3. *Lemna gibba* × *Lemna Minor* Hybrids

Comparison of electrophoretic TBP profiles of the six clones in the *L. gibba* subcluster with those of other species revealed additional peaks attributable to *L. minor*, definitively confirming the hypothesis of interspecific hybridization between the two species. As shown in Figure 4, the additivity of the peak profiles in the hybrid clones witness the presence of two subgenomes. However, karyotyping will be needed in order to determine if hybrids are allotetraploids, as usually expected for interspecific hybrids, or homoploids, as it can be the case for asexually reproducing plants. Each TBP amplicon was assigned to the correspondent β-tubulin locus, based on the sequences retrieved from Whole-genome sequencing (WGS) data of *L. minor* 5500 [31] (https://genomevolution.org/r/ik6h (accessed on 14 November 2021), ID 27408), *L. minor* 8627, here reclassified as *L. ×japonica*, and *L. gibba* 7742a (https://www.lemna.org/ (accessed on 14 November 2021)). The two sets of six parologous *TUBB* loci in *L. minor* and *L. gibba* were arbitrarily numbered *TUBB*1-*TUBB*6, in the absence of rules for tubulin gene numbering in plants. Corresponding positions on chromosomes or contigs of WGS data are given in Supplementary Table S2. An additional β-tubulin sequence, likely a pseudogene, was retrieved only from the *L. gibba* WGS data. In fact, its sequence lacks the canonical two introns and has a short deletion in the second exon, leading to its interpretation as a retrotransposed copy of *TUBB*2 by sequence similarity, and therefore named Ψ*TUBB*2. This sequence is nevertheless amplified by TBP primers, producing a short fragment of 250 base pairs distinctive of *L. gibba* (Figure 4).

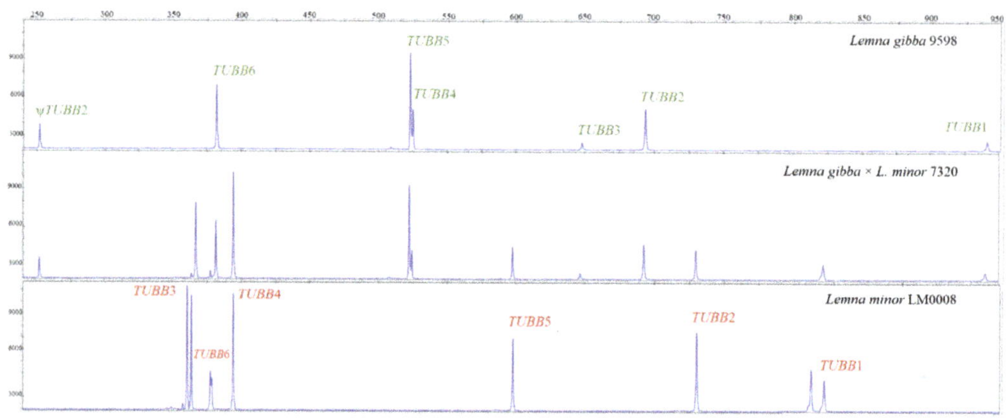

Figure 4. TBP profiles of representative clones of *L. gibba* × *L. minor* hybrids and the two putative parental species *L. minor* and *L. gibba*. Peak size is expressed in base pairs and peak height in Relative Fluorescence units. *TUBB* loci corresponding to each peak, as deduced from expected amplicon size, are indicated. Doublets indicate length variant heterozygosity at *TUBB*3 and *TUBB*1 loci in LM0008.

The possibility of an artefact originating from the analysis of cross-contaminated clones was excluded by sub-cloning twelve single fronds of each clone and performing the TBP analysis on each clonal population after one week's cultivation. Profiles were identical to those of the original clones (not shown). On this base, we therefore concluded that we have identified a new interspecific hybrid in the genus, between *L. minor* and *L. gibba*, to add to *L. ×japonica*.

Plastid intergenic spacers *psb*K-*psb*I and *atp*F-*atp*H were used to identify the maternal parent of each putative hybrid clone. Interestingly, both reciprocal crosses were observed: clones 9425a and 9248 have maternal inheritance of *L. gibba*, whereas plastid markers of clones 7641, 7320, 6861, 9562 matched the *L. minor* sequences (Supplementary Table S3). This seems different from what was found in *L. ×japonica*, where all clones so far investigated by plastid barcoding sequencing have *L. minor* as the maternal parent. Interestingly, the two reciprocal crosses were separated by cluster analysis.

2.4. Frond Morphology and Flowering

All six *L. gibba* × *L. minor* clones showed a flat morphology of the frond under standard culture conditions. Frond shape was quite heterogeneous among clones, from almost round to ovate, generally asymmetrical. An overview is given in Figure 5. They were also different in size, with the two *L. minor* × *L. gibba* (9425a and 9248) showing larger fronds than the reverse crosses (Average length 4.45 vs. 3.88 mm, average width 3.01 vs. 2.62 mm). We will present elsewhere a more detailed, formal description of the new hybrid species *L. gibba* × *L. minor*.

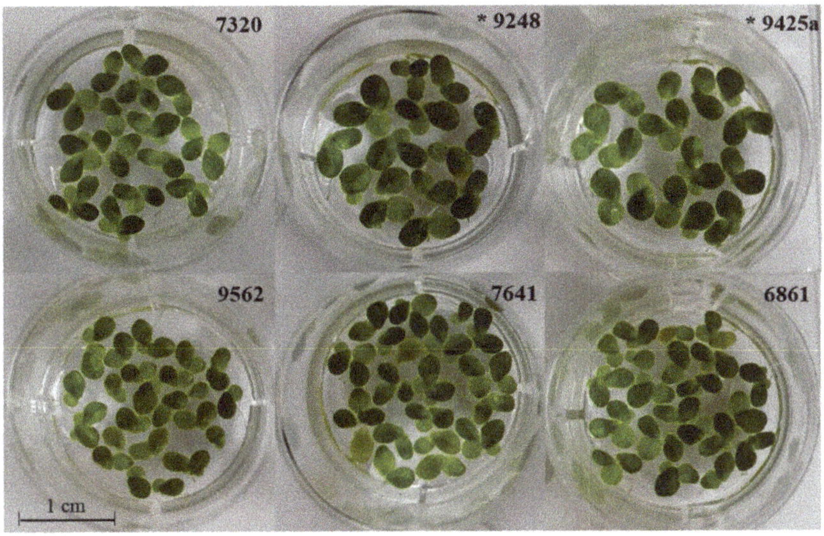

Figure 5. Cultures of each of the six hybrid clones *L. gibba* × *L. minor:* 9562, 7641, 6861, 7320, 9425a and 9248. Asterisks (*) indicate clones with *L. gibba* as the female maternal parental.

Descriptions of supposed intermediate forms between *L. gibba* and *L. minor* are reported in the literature. In particular, a new species from Southern Italy was described in 1973 as a possible hybrid between the two species and was named *L. symmeter* Giuga, species nova [12] is now considered as a synonym of *L. gibba* [32]. The name comes from the most distinctive trait reported for the new species, which is the symmetric (simultaneous) development of stamens during flower development. This is different from what described by Kandeler [7] and by Giuga himself about flowering in *L. gibba*, characterized by the appearance of the first anther together or soon after the stigma, later followed by the second anther. The new species was described as sterile, as fruit formation and seed setting was never observed.

We tried to verify if the hybrid clones that we have identified could correspond to putative '*L. symmeter*' on these criteria, by inducing flowering through treatment with salicylic acid (SA) as reported by others [33,34]. After four weeks of cultivation in 20 μM SA, we were able to induce flowering in two out of three *L. gibba* clones (7742a and 9598, not 8124), but neither the six hybrid clones, nor two *L. minor* clones (9977 and 9942) showed reproductive organs after six weeks. Therefore, we have not yet been able to provide proof of the identity of the hybrid clones with the previously described species *L. symmeter.*

2.5. Infrasectional Structure of Lemna Sect. Lemna

The infrageneric structure of the genus *Lemna* is not unequivocally defined because of conflicting results obtained by the nuclear ribosomal coding and noncoding sequences, and plastid markers [21] and/or possible misclassification of some clones used for these studies. Nevertheless, a phylogenetic tree was obtained by merging a large number of sequence data [4,21]. The finding of hybrids calls for an update to that tree. Fast evolving intron regions are suitable for phylogenetic analysis of closely related taxa [35], particularly if a low recombination rate is expected, as in the case of predominant clonal propagation in duckweeds. We chose two β-tubulin introns showing little or no intraspecific length variability by TBP profiling and designed gene-specific primers for cross-species amplification, at their exon-intron borders. After optimization of primers and PCR conditions, the first intron of *TUBB2* and the second intron of *TUBB1* were amplified in all species of the section with the same two primer pairs. PCR amplification is shown in Figure 6, which also highlights the hybrid origin of *L.* ×*japonica* and *L. gibba* × *L. minor* by the presence of markers of the same length of those shown by the two parental sub-genomes. Intron size was quite conserved among the different species, with the exception of the second *TUBB1* intron of *L. gibba* and *L. obscura*, showing large deletions. No amplification was obtained on clones of other *Lemna* sections with the same primers (not shown).

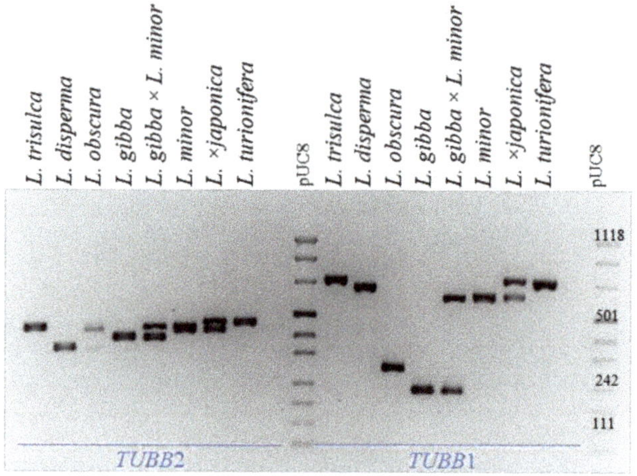

Figure 6. Cross-species PCR amplification of *TUBB2* and *TUBB1* in each species and hybrids of *Lemna* sect. *Lemna*.

Gene trees were obtained by concatenation of the two regions (sequences are provided in Supplementary Table S4). Intron sequences of both hybrid species, almost identical (>99% identity) to both parental species, were not included in the alignment. The *TUBB2* sequence of *Spirodela polyrhiza*, was used to root the tree. Equivalent tree topologies were obtained by both Maximum Parsimony and Maximum Likelihood clustering methods (not shown). The gene tree is congruent with that proposed by Tippery 2015 [21] showing *Lemna* sect. *Lemna* as split in two subclusters, one comprising *L. gibba*, *L. disperma* and *L. minor* and the second *L. turionifera*, *L. trisulca* and *L. obscura*. In that tree, *L.* ×*japonica* was the closest relative of *L. minor*, for the obvious influence of the *L. minor* plastid sequences. A phylogenetic reconstruction of the section based on the *TUBB* gene tree and showing the origin of the two interspecific hybrids is shown in Figure 7.

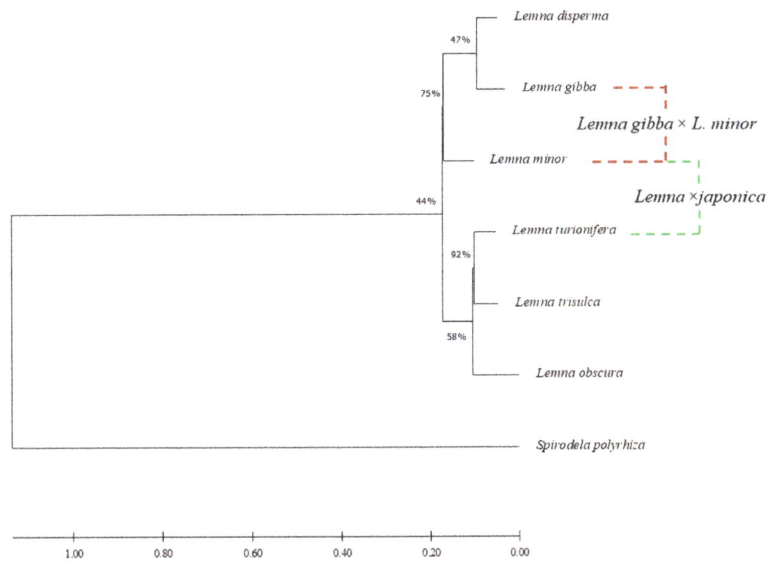

Figure 7. Maximum likelihood tree inferred from the alignment of concatenated loci *TUBB2* and *TUBB1* for all species of *Lemna* sect. *Lemna*. Bootstrap values are shown on the specific branches. Interspecific hybrids were manually inserted on the tree; relative branches are not to scale.

2.6. Intraspecific TUBB2 Polymorphism in Lemna gibba and Lemna minor

As clones from different geographical origin clustered separately by TBP, we searched for intraspecific polymorphisms in *TUBB2* first intron sequences by investigating 18 clones of *L. minor* and 13 clones of *L. gibba* (Supplementary Table S4). PCR amplicons were directly sequenced. All *L. minor* and most *L. gibba* (supposedly diploids), were homozygous at the *TUBB2* locus, generating unique sequence profiles. Sequences of six *L. gibba* clones could not be obtained due to the superimposition of two separate sequence chromatograms within the same frame, likely arising from length variant heterozygosis. We did not further investigate these clones. Fourteen polymorphic sites out of 410 positions investigated in *TUBB2* were identified among *L. minor* clones, producing four different sequence variants, here referred to as haplotypes (M1–M4). In *L. gibba*, eight polymorphic sites were found over 380 base pairs, defining three haplotypes, named G1–G3 (Table 3).

Table 3. *TUBB2* polymorphic sites and corresponding haplotypes.

	Haplotype	Polymorphic Sites														Alignment Length (bp)
		35	45	86	175	177	184	197	206	249	286	316	330	337	366	407
L. minor	M1	T	A	T	G	C	C	T	T	A	C	G	G	G	A	
	M2	C	G	G	T	G	T	A	G	T	T	A	A	A	T	
	M3	T	A	T	T	C	C	T	T	A	C	G	G	G	A	
	M4	T	A	T	T	C	C	T	T	A	C	A	G	G	A	
		40	68	109	224	250	255	302	329							380
L. gibba	G1	T	T	C	G	C	T	A	G							
	G2	–	T	C	G	T	T	A	G							
	G3	–	A	T	–	T	C	C	T							

Position numbers are given from the beginning of the aligned region.

The *TUBB2* sequences of clones *L. minor* 5500 and *L. gibba* 7742a, retrieved from WGS data, were used as references (M1 and G1) for the calculation of the number of polymorphisms. Association of the different haplotypes to each clone is reported in Table 4. In *L. minor*, the separation of an African genetic pool was evident from the sharing of twelve conserved SNPs defining haplotype M2 among the three investigated clones, evident, whereas the Euro-Asiatic clones showed by far larger sequence identity, with 1–2 SNPs.

Table 4. Association of the *TUBB* haplotypes to the analyzed clones.

	Clone	Origin	Haplotype	SNPs		Clone	Origin	Haplotype	SNPs
L. minor	5500	Ireland	M1	0	*L. gibba*	7742a	Italy	G1	0
	9961	Germany	M1	0		9598	Italy	G1	0
	9424	Germany	M1	0		6745	California—USA	G3	6
	7194	Uganda	M2	12		8124	Arizona—USA	G3	6
	7753	Ethiopia	M2	12		7705	India	G2	2
	7210	S. Africa	M2	12		190	USA	G2	2
	9495	Norway	M3	1		7796	Italy	G2	2
	LM0011	Russia	M3	1		9583	Poland	n.d	–
	7766	New Zealand	M3	1		7922	Argentina	n.d	–
	9536	Germany	M4	2		9614	Poland	n.d	–
	7022	Spain	M4	2		9619	Albany	n.d	–
	9482	Italy	M4	2		7245	S. Africa	n.d	–
	9942	Norway	M4	2		8428	Switzerland	n.d	–
	9533	Macedonia	M4	2	*L. gibba* × *L. minor*	7320	Egypt	G2 M4	
	8744	Albania	M4	2		6861	Italy	G2 M4	
	LM0010	Italy	M4	2		9562	Italy	G2 M4	
	LM0008	Russia	M4	2		9425a	Germany	G2 M4	
	9252	Finland	M4	2		7641	Israel	G2 M4	
	8292	Iran	M4	2		9248	Italy	G2 M1	

The number of substitutions is reported with respect to *L. minor* 5500 and *L. gibba* 7742a taken as references. n.d, sequence with overlapping peaks in the chromatogram.

In *L. gibba*, two clones from the American lineage, genetically distinct from others clones (Figures 1 and 2), showed the same holotype G3, characterized by the presence of six SNPs (Single Nucleotide Polymorphism).

We also looked for parental genome signatures in the six clones *L. gibba* × *L. minor* by sequence analysis of the *TUBB2* homoeologous loci, after amplification with selective primers for the two subgenomes G and M. Interestingly, five out of six clones showed the same haplotype at both loci suggesting their common origin and excluding contributions from the American *L. gibba* and African *L. minor* genomic pools. The sixth clone differs for its M haplotype. The high degree of sequence identity between hybrids and the parental species suggests a quite recent origin, with respect to the differentiation of *Lemna* species.

3. Discussion

Species delimitation by morphology can be highly challenging in duckweeds. Within the genus *Lemna* sect. *Lemna* is the most problematic, as it includes closely related species which show blurred boundaries due to high intraspecific and low interspecific variability of morphological traits, as *L. japonica, L. minor and L. gibba*. Even plastid barcoding markers, easily discriminating *L. minor* from *L. gibba*, fails to separate *L. japonica* from *L. minor*. In this work, we extended a previous molecular survey by TBP to 98 additional duckweed clones, mostly from the Landolt Duckweed Collection, classified by morphology as belonging to species of sect. *Lemna*. The analysis provided further support to the identification of *L.* ×*japonica* as a hybrid and led to the identification of a new interspecific hybrid within this section. This is clearly shown in the PCA plot obtained from the TBP dataset, with clear separation of two clusters of hybrid clones sharing *L. minor* as the donor of one subgenome and having *L. turionifera* and *L. gibba*, respectively, as the other parental species.

3.1. *Lemna ×japonica*

Natural interspecific hybrids between *L. minor* and *L. turionifera* were identified in our previous work [20] by TBP cluster analysis as a subcluster merging accessions classified as *L. japonica* and *L. minor* by morphology and nested within the *L. turionifera* branch. This larger analysis, including many more *L. turionifera, L. minor* and *L. japonica* clones, further supported previous evidence that all clones in this cluster are definitely genetically separated from true *L. minor*, despite their similar morphology. All of them have a duplicate set of six β-tubulin genes and highly similar TBP profiles (Supplementary Table S1). Therefore, we suggest referring the whole taxonomic unit to *L. ×japonica*, to account for its hybrid ancestry. Moreover, TBP profiling is a suitable method to distinguish *L. ×japonica* from *L. minor* on a molecular base, in addition to morphology. Accordingly, thirty-six percent of '*L. minor*' clones was identified as *L. ×japonica* by TBP analysis. Misidentifications of *L. ×japonica* with *L. minor* clones could not be easily recognized before, as plastid markers are not able to resolve the two species [27] and GBS, while detecting some genetic distance between the two species, is not manageable for single clone identification [5].

Although more powerful genetic approaches, including sequencing and karyotype analysis, must be used to address this point, a certain degree of genetic variability within *L. ×japonica* is already evident from TBP analysis, separating East Asian and American subclusters (Figure 1) distinct in some private alleles. The species *L. japonica* originally described by Landolt 1980 [18] as the *L. minor* biotype originally present in Japan and Eastern China could possibly coincide with the East-Asian subcluster. Accordingly, all *L. ×japonica* clones from China and Japan were correctly classified by morphology, suggesting that either they have more canonical traits or that the geographical origin was a determinant for classification. As a further consequence of misclassification, *L. ×japonica* has a broader geographical distribution than previously reported for *L. japonica* by Landolt, as it is present in all temperate regions of Eurasia and North America.

Misidentification of some *L. ×japonica* clones as *L. minor* could also partially explain the large variability in genome size (ranging from 356 to 604 Mb) and the variable chromosome numbers, 40–50, attributed to *L. minor* [14,15,36]. For example, at least four *L. minor* clones having higher DNA content according to Wang [14] (550–600 Mb; clone 9016, 9436b, 9439 and 9440) are classified as *L. ×japonica* by TBP (the present work), in line with the genome size of the *L. japonica* holotype clone 7182, assessed as 600 Mb [14]. Conversely, other clones also identified as *L. ×japonica* by TBP (7210, 7123 and 8434) showed a lower DNA content, around 400 Mbp. However, these clones have different geographical origin, coming either from the American (7123 and 8434) or the African continent (7210).

One possible explanation of these results is that hybridization between *L. minor* and *L. turionifera* has occurred more than once, thus originating independent lineages of *L. ×japonica* with different ploidy or chromosome rearrangements as reported for other plant genera, e.g., *Senecio* spp. [37,38], or *Tragopogon* spp. [39].

However, large differences in genome size can also be produced by different proliferation of Transposable Elements (TEs) in different lineages with the same ploidy level. This is widely documented for sunflower (*Helianthus annuus*) where three different homoploid hybrid species, independently originated from the same two parents, featured genome sizes larger then parental species (approximately 5.3–5.6 Gb with respect to 3.3 and 3.5 Gb), mainly represented by TEs [40]. Further work is ongoing in order to understand if *L. ×japonica* is monophyletic or may have originated from multiple hybridization events.

3.2. *Lemna gibba × Lemna minor*

The second interspecific cluster that was separated by the PCA plot consisted of six clones, all recorded as *L. gibba* by morphology, with overlapping TBP profiles between *L. minor* and *L. gibba*. While waiting for a formal description including morphological traits (manuscript in preparation), we will refer to it as *L. gibba × L. minor* even knowing that the six hybrid clones have originated from both reciprocal crosses of the two parental species.

The finding of this new hybrid was not completely unexpected as intermediate phenotypes between *L. minor* and *L. gibba* were already described in the past [7–9,11]. The difficulty in distinguishing these two species, when not flowering, has been widely reported in the literature, particularly when individuals of *L. gibba* do not show the typical gibbosity. In facts, 'flat forms' of *L. gibba* often occur under culture conditions or seasonally, at some sites in the wild. Confusion may also arise between *L. gibba* and *L.* ×*japonica*, as shown in our survey.

We believe that the *L. gibba* × *L. minor* hybrid could correspond to a *Lemna* species described as '*L. symmeter*' [12]. The description of this new species was reported in a monographic publication, written in Italian with only the abstract in English, entitled "Vita segreta di Lemnacee" (The secret life of Lemnaceae) [12]. According to the description, the new *Lemna* species was first observed by the author in 1968, along the coast of Campania (Southern Italy) and later at different sites in Southern Italy where, over the many years, was always found in mixed populations with *L. gibba* (30–90%), rarely associated with *L. minor* or *L. trisulca*. In these mixed populations, *L. gibba* could easily be distinguished by the pronounced gibbosity, typical of most Italian genotypes. Since then, the existence of '*L. symmeter*' has never been confirmed and the original clones went missing. Since the description was not supported by a formal diagnosis in Latin, the putative new species is not valid and '*L. symmeter*' is currently reported as a synonym of *L. gibba*. According to Giuga's description, flower development becomes the most distinctive trait in the absence of gibbosity. Flowering of '*L. symmeter*' was reported to differ from that of *L. gibba* for the simultaneous growth of the two anthers, occurring a few days after the stigma. Moreover, anthers were reported as indehiscent and seed setting, frequent in *L. gibba*, has never been observed.

Unfortunately, we have so far been unable to induce flowering in any of the hybrid clones to confirm the most important diagnostic trait described by Giuga. However, the present work has demonstrated, by molecular markers and by nuclear sequence analysis that a hybrid species compatible with the description of '*L. symmeter*' indeed exists, and is unlikely to be rare in the wild as six clones are present in the Landolt collection. All clones but one came from the Mediterranean basin, three of them (9248, 9562 and 6861) from different Italian regions, from North to South, supporting their correspondence with putative '*L. symmeter*'. This is also supported by a recent survey in Central Italy, in which 13 *Lemna* spp. samples, out of 56 collected at different stations, could not be unequivocally identified by morphology and were described as 'non-gibbous form of *L. gibba* or *L. minor*'. Plastid marker sequencing assigned all of them to *L. minor* [41]. One clone from that study (9562), originally sent by the authors to Prof. K.J. Appenroth at the University of Jena was then delivered to the Landolt Duckweed Collection, is one of the six clones recognized as hybrid in our work. It is therefore likely that such hybrids are often identified as 'flat forms' of *L. gibba* and are widely distributed in Italy. Hybrid discovery in Egypt (clone 7320) and Israel (7641) suggests they are adapted to the Mediterranean climate, while the clone collected in Hamburg Germany, (9425a), could be explained either as a recent invasion or the result of local hybridization occurred in that area. Mixed populations of *L. gibba* and *L. minor* have been repeatedly reported by De Lange in The Netherlands [9]. Despite their high morphological similarity to the parental species, *L. gibba* × *L. minor* must have some physiological and ecological features that confer increased fitness in competition with *L. gibba* in certain conditions. Therefore, physiological and ecological properties should be investigated, as well as the true geographical distribution of the hybrid must be recorded. To this end, we believe that TBP and/or the simpler PCR amplification with the gene-specific primer pairs described in this work provide a simple tool for fast identification of the hybrid.

In conclusion, hybridization could have been one of the driving mechanism of duckweed evolution as it has been observed in other plant families, amounting to an average of 25% of plant species that are known to hybridize naturally with at least one other species [42]. What was more unexpected is that such hybridization occurs in plants which are reported to reproduce clonally and rarely flower in nature. However, extensive interlineage hybridization has been recently reported also for the predominantly clonal, aquatic plant *Hydrilla verticillata* [43].

3.3. *Lemna sect. Lemna*

Phylogenetic relationships in duckweeds have been established initially by combinations of morphological, flavonoid, allozyme and plastid DNA sequences [4] and subsequently a larger dataset including 73 accessions, have been produced using both plastid and nuclear ribosomal [21]. Despite some node resolved incongruently using the two kind of molecular markers, a solid consensus tree was obtained by the combination of both. Major conflicts in the genus *Lemna* were actually found in the section *Lemna*.

The existence of hybrid species, particularly *L.* ×*japonica*, might explain conflicting results obtained with the use of plastid and nuclear loci and shed new light on the phylogenetic history of the genus. In this work, the phylogenetic tree of sect. *Lemna* was inferred from the alignment of two concatenated β-tubulin intron sequences, which deployed high interspecific variability, and further completed with the positioning of the two interspecific hybrid species. The tree topology was congruent with what reported by Tippery, 2015 [21], that place *L. minor, L. turionifera and L. trisulca* in a separate branch with respect to *L. gibba*, *L disperma* and *L. obscura*. Interbreeding between species from the two branches of the tree demonstrates absent or incomplete reproductive isolation mechanisms between species of sect. *Lemna*. If extended to other duckweed species, this suggests that other hybrids could have originated where different species occur, or have occurred in the past, in sympatry, possibly explaining incomplete species separation by molecular markers also in species of other allied genera like *Wolffia* [27].

3.4. *Infraspecific Variation*

Despite large phenotypic and physiological infraspecific variability observed in duckweeds [44], genetic diversity is barely detectable by plastid markers, while nuclear ribosomal sequences, such as ETS and ITS, are not easily amplified in all species [21]. Although genome-wide approaches such as genotyping by sequencing or whole genome comparison are now economically affordable, such techniques require managing of huge datasets. Our study revealed that infraspecific variation within β-tubulin intron sequences identifies haplotypes associated with a particular continent. This suggests that these sequences can detect greater genetic diversity than could be estimated using plastid markers. Further investigations based on β-tubulin introns may help revealing biogeographical patterns due to intraspecific variation in other species, e.g., *L.* ×*japonica*.

Conversely, *TUBB* intron sequences obtained for some hybrid clones revealed very high similarity to those of the parent species, suggesting either recent hybridization, or very low mutation rates, as it was shown for species in the allied genus *Spirodela* Schleiden [45].

Accurate identification and understanding of the genetic structure are important elements for application purposes or when managing populations of invasive species such as the native American species *L. minuta* Kunth, in Europe. Tubulin genes, combining highly conserved gene structure and exon sequences with highly variable intron regions, provided an effective toolkit for studying genetic relationships at both specific and subspecific level in duckweeds, providing new directions for further investigations by whole genome approaches.

4. Materials and Methods

4.1. *Plant Material and Cultivation*

Plant material, including 98 duckweed clones, mostly from the Landolt Duckweed Collection, and named with the four-digit code as defined by Landolt, is summarized in Table 1. Information about additional clones included in the cluster analyses of the present paper can be found in [20]. *Lemna* clones were aseptically cultivated in 90 mm Petri dishes on agarized Schenk and Hildebrandt (SH) medium (plant agar 8 g/L, Duchefa

Biochemia, Haarlem, NL) with 2 g/L sucrose, at pH 5.1 [46]. Fronds were maintained at 25 °C under a 16-h photoperiod with light flux of 41–45 μmol m^{-2} s^{-1} provided by white florescent lights.

4.2. DNA Extraction and TBP Profiling

We provide here a brief description of the protocol; the whole procedure is described in details in [20].

One hundred milligrams of fresh duckweed fronds were disrupted in 2 ml tubes with three 3-mm stainless steel beads, using a TissueLyser II apparatus at a frequency of 30 Hz for 1 min. The standard protocol of the DNeasy Plant Mini Kit (Qiagen, Valencia, CA, USA) was adopted for the extraction of total DNA, then the quality and amount were determined by UV absorbance with the Nanodrop 2000 C (Thermo Fisher Scientific, Inc., Waltham, MA, USA), and DNA was stored at −20°C until used.

Thirty nanograms of gDNA was used as template for the TBP profiling (1st and 2nd intron). PCR primers, amplification protocols, as well all subsequent steps including Capillary Electrophoresis (CE) and the data collection were performed as reported [20]. Amplicon lengths (expressed in base pairs) of the CE-TBP profiles were used to analyze clones and the sorting of the numerical data were performed according to them. Both TBP 1st and 2nd intron markers (peaks size) were scored in a binary matrix (1/0, presence and absence respectively). The genetic similarity among genotypes was estimated according to the Jaccard's index for binary data, using the open source software package Past v.4.07b [47]. The multivariate analysis including the cluster distribution and the principal component analysis (PCA) were also performed through the same software. The dendrogram was computed by the neighbor joining (NJ) algorithm, and bootstrap confidence values were obtained applying 1000 replications. Minor graphical editing was performed using the online tool Interactive Tree Of Life (iTOL) v.6.4 [48].

4.3. DNA Barcoding Analysis

The plastid markers *atp*F-*atp*H and *psb*K-*psb*I were amplified with a modified version of the primers according to [27]. PCR reactions were run in a total volume of 20 μL, with 1 Unit of Platinum Taq DNA Polymerase and 0.5 μM of each primer, using 20 ng of template gDNA. Reactions were carried out by incubation at 95 °C for 3 min followed by 30 cycles of 95 °C for 40 s, 52 °C for 50 s, 72 °C for 1 min and a final extension of 72 °C for 3 min. PCR products were checked on 2.0% agarose, 0.5× Tris Borate EDTA gels containing 1× Atlas ClearSight DNA Stain (Bioatlas, Tartu, EE, Estonia). PCR products were purified using the microCLEAN PCR/DNA Cleanup reagent, according to the manufacturer's instructions (Labgene Scientific SA, Châtel-Saint-Denis, CH, Switzerland). The amplified products were forward and reverse sequenced (Microsynth, Balgach, Switzerland) and the obtained consensus sequence (contig) were considered for the alignment. The NCBI BLASTn analysis was performed for clone identification by best match analysis (http://www.ncbi.nlm.nih.gov (accessed on 14 November 2021)).

4.4. β-. Tubulin Intron Amplification and Sequencing

Gene-specific, cross-species primer pairs were designed on the sequence alignment of the chosen β-tubulin genes (*TUBB*1 and *TUBB*2) of *L. minor* clones 5500 and 8623 with the corresponding orthologs of *L. gibba* 7742, and tested for the simultaneous amplification of all the seven species belonging to *Lemna* sect. *Lemna*. More conserved regions encompassing intron borders were chosen for the primer design and degenerated nucleotides were introduced at polymorphic sites. *TUBB*2 cross-species primers (forward: I-FwTUBB2 5′-CCT CCA GGG TAT GCG ATC-3′ and reverse: I-Rv_11-23_1 5′- GGA ATC CTG CAM KTA AAT GAY G-3) targeted the first intron, whereas *TUBB*1 primers (forward: II-FwTUBB1 5′-CAC YCC AAG CTG TAA GWT CC-3′ and reverse: II-Rv-TUBB15′- GAT CGC CGA CTA YAA GAA ATC-3′) amplified the second intron.

Furthermore, species-specific primers for the selective amplification of the *L. gibba* (forward I-FwTUBB2 5′-CCT CCA GGG TAT GCG ATC-3′ and reverse I-RvTUBB2g 5′-AAC TTG GAA TCC TGC AAG CA-3′) and *L. minor* (forward 1_Fw_11-23_15′-TTC AGG GTA TGC GAT CTA TTC-3′ and reverse 1_Rv_11-23_1 5′- GGA ATC CTG CAM KTA AAT GAY G-3′) *TUBB2* orthologous genes in interspecific hybrids were designed to target more specific regions of the same sequence alignment.

PCR was performed according to [20] after optimization of the annealing temperature; amplification products were purified using microCLEAN PCR/DNA Cleanup and directly sequenced on both strands. In the case of *TUBB1* amplification on *L. gibba* × *L. minor* clones, the two amplified bands were cut from gel and purified with the MinElute Reaction Cleanup (Qiagen, Valencia, CA, USA). Forward and reverse sequences were inspected, manually edited where necessary, and combined in single consensus sequences.

4.5. Bioinformatic Sequence Analysis

Sequences were subjected to multiple sequence alignment using MUSCLE [49] implemented in MEGA X v.10.1.8 [50]. Alignments were manually edited and the evolutionary analysis was inferred by using the Maximum Likelihood method and General Time Reversible models [51]. Initial tree(s) for the heuristic search were obtained by applying the Neighbor-Joining method to a matrix of pairwise distances estimated using the Maximum Composite Likelihood (MCL) approach. The percentage of trees in which the associated taxa clustered together were estimated by bootstrap test (1000 replicates) and shown next to the tree branches. A discrete Gamma distribution was used to model evolutionary rate differences among sites (+G, 2 categories) and the rate variation model (+I) allowed for some sites to be evolutionarily invariable. The tree with the highest log likelihood was drawn to scale, with branch lengths measured in the number of substitutions per site.

4.6. Flower Induction

Fifty fronds for each of the 11 selected clones of *L. minor* (9977, 9942), *L. gibba* (9598, 8124) and *L. gibba* × *L. minor* (9425a, 6861, 7641, 9562, 9248, 7320) were aseptically cultured in Sterivent Low Containers 107 mm × 94 mm × 65 mm (Duchefa Biochemia, Haarlem, The Netherlands), in liquid SH medium supplemented with 2 g/L sucrose and 20 µM salicylic acid (SA). The trial was performed in the laboratory from mid-July to August for 40 days. Plastic boxes were positioned indoor in front of the window, under natural daylight (approx. 15 h day length) and avoiding direct sunlight, at room temperature (23–28 °C). Three replicates were set up for each clone, with an additional sample without SA as negative control. Flowers were observed from the 4th week.

Supplementary Materials: The following are available online at https://www.mdpi.com/article/10.3390/plants10122767/s1. Table S1: Alignment of the numerical output derived from the scoring of TBP fragment by capillary electrophoresis separation. Table S2: *TUBB* gene locations on chromosomes or contigs as from WGS data. Table S3: Plastid marker sequences from the analyzed clones. Table S4: *TUBB* sequences from the analyzed clones.

Author Contributions: Conceptualization, L.B. and L.M.; methodology, F.G. and S.G.; formal analysis, L.B. and J.D.G.; investigation, F.G., S.G. and J.D.G.; resources, D.B.; writing—original draft preparation, L.B.; writing—review and editing, L.M. and D.B.; visualization, F.G.; supervision, D.B.; funding acquisition, D.B. All authors have read and agreed to the published version of the manuscript.

Funding: This research received no specific funding.

Acknowledgments: We wish to thank W. Lämmler for providing most of the clones to be analyzed and useful information. We also thank KJ Appenroth and M. Bog for providing some additional clones. We are indebted to David Mabberley for valuable suggestions about the taxonomic treatment of the hybrid species and other useful comments.

Conflicts of Interest: The authors declare no conflict of interest.

References

1. Martinov, I. *Techno-Botanical Dictionary (Техно-ботаническийСловарь)*; Pechashano v Imperatorskoї Tipografii: Saint Petersburg, Russia, 1820.
2. Tippery, N.P.; Les, D.H. Tiny Plants with Enormous Potential: Phylogeny and Evolution of Duckweeds. In *The Duckweed Genomes*; Cao, X.H., Fourounjian, P., Wang, W., Eds.; Springer: Berlin, Germany, 2020; pp. 19–38. [CrossRef]
3. Bog, M.; Appenroth, K.; Sree, K. Key to the determination of taxa of Lemnaceae: An update. *Nord. J. Bot.* **2020**, *38*, e02658. [CrossRef]
4. Les, D.; Crawford, D.; Landolt, E.; Gabel, J.; Kemball, R. Phylogeny and systematics of *Lemnaceae*, the duckweed family. *Syst. Bot.* **2002**, *27*, 221–240.
5. Bog, M. Genotyping-By-Sequencing for species delimitation in *Lemna* section Uninerves Hegelm (*Lemnaceae*). In *The Duckweed Genomes*; Cao, X.H., Fourounjian, P., Wang, W., Eds.; Springer: Berlin/Heidelberg, Germany, 2020; pp. 115–123. [CrossRef]
6. De Lange, L. Gibbosity in the complex *Lemna gibba/Lemna minor*: Literature survey and ecological aspects. *Aquat. Bot.* **1975**, *1*, 1327–1332. [CrossRef]
7. Kandeler, R. Species delimitation in the genus *Lemna*. *Aquat. Bot.* **1975**, *1*, 365–376. [CrossRef]
8. Landolt, E. Morphological differentiation and geographical distribution of the *Lemna gibba-Lemna minor* group. *Aquat. Bot.* **1975**, *1*, 345–363. [CrossRef]
9. de Lange, L.; Pieterse, A.H.; van Baarsen-Beckers, I. The occurrence of mixed populations of different genotypes of the *Lemna gibba—Lemna minor* complex. *Acta Bot. Neerl.* **1981**, *30*, 191–197. [CrossRef]
10. De Lange, L.; Pieterse, A.H. A comparative study of the morphology of *Lemna gibba* L. and *Lemna minor* L. *Acta Bot. Neerl.* **1973**, *22*, 510–517. [CrossRef]
11. Giardelli, M.L. Una nueva especia de Lemnacea de la Flora Argentina. *Notas Del Mus. De La Plata* **1937**, *2*, 97–100.
12. Giuga, G. Vita segreta di Lemnacee. In *Lemna Symmeter G. Giuga-Species Nova*; Blasio: Napoli, Italy, 1973; p. 19.
13. Bog, M.; Appenroth, K.; Sree, K. Duckweed (*Lemnaceae*): Its Molecular Taxonomy. *Front. Sustain. Food Syst.* **2019**, *3*, 117. [CrossRef]
14. Wang, W.Q.; Kerstetter, R.; Michael, T. Evolution of genome size in duckweeds (*Lemnaceae*). *J. Bot.* **2011**, *2011*, 570319. [CrossRef]
15. Hoang, P.; Schubert, V.; Meister, A.; Fuchs, J.; Schubert, I. Variation in genome size, cell and nucleus volume, chromosome number and rDNA loci among duckweeds. *Sci. Rep.* **2019**, *9*, 3234. [CrossRef] [PubMed]
16. Landolt, E. The family of Lemnaceae—A Monographic Study, Volume 1, Biosystematic investigations in the family of duckweeds (Lemnaceae). *Ver. Geobot: Inst. ETH Shift. Rübel* **1986**, *71*, 1–563.
17. Landolt, E. Contribution on the *Lemnaceae* of Ecuador. *Fragm. Flor. Geobot.* **2000**, *45*, 221–237.
18. Landolt, E. Description of six new species of *Lemnaceae*. *E-Period.* **1980**, *70*, 13–21.
19. Hirahaya, M.; Kadono, Y. Biosystematic study of *Lemna minor* L. sensu lato (*Lemnaceae*) in Japan with special reference to allozyme variation. *Acta Phytotaxon. Geobot.* **1995**, *46*, 117–129.
20. Braglia, L.; Lauria, M.; Appenroth, K.; Bog, M.; Breviario, D.; Grasso, A.; Gavazzi, F.; Morello, L. Duckweed Species Genotyping and Interspecific Hybrid Discovery by Tubulin-Based Polymorphism Fingerprinting. *Front. Plant Sci.* **2021**, *12*. [CrossRef] [PubMed]
21. Tippery, N.P.; Les, D.H.; Crawford, D.J. Evaluation of phylogenetic relationships in *Lemnaceae* using nuclear ribosomal data. *Plant Biol.* **2015**, *17*, 50–58. [CrossRef]
22. Ceschin, S.; Abati, S.; Leacche, I.; Zuccarello, V. Ecological comparison between duckweeds in central Italy: The invasive *Lemna minuta* vs the native *L. minor*. *Plant Biosyst.* **2018**, *152*, 674–683. [CrossRef]
23. Ceschin, S.; Leacche, I.; Pascucci, S.; Abati, S. Morphological study of *Lemna minuta* Kunth, an alien species often mistaken for the native *L. minor* L. (Araceae). *Aquat. Bot.* **2016**, *131*, 51–56. [CrossRef]
24. Bog, M.; Baumbach, H.; Schween, U.; Hellwig, F.; Landolt, E.; Appenroth, K. Genetic structure of the genus *Lemna* L. (*Lemnaceae*) as revealed by amplified fragment length polymorphism. *Planta* **2010**, *232*, 609–619. [CrossRef]
25. Bog, M.; Landrock, M.; Drefahl, D.; Sree, K.; Appenroth, K. Fingerprinting by amplified fragment length polymorphism (AFLP) and barcoding by three plastidic markers in the genus Wolffiella Hegelm. *Plant Syst. Evol.* **2018**, *304*, 373–386. [CrossRef]
26. Wang, W.; Wu, Y.; Yan, Y.; Ermakova, M.; Kerstetter, R.; Messing, J. DNA barcoding of the *Lemnaceae*, a family of aquatic monocots. *BMC Plant Biol.* **2010**, *10*, 205. [CrossRef] [PubMed]
27. Borisjuk, N.; Chu, P.; Gutierrez, R.; Zhang, H.; Acosta, K.; Friesen, N.; Sree, K.; Garcia, C.; Appenroth, K.; Lam, E. Assessment, validation and deployment strategy of a two-barcode protocol for facile genotyping of duckweed species. *Plant Biol.* **2015**, *17*, 42–49. [CrossRef] [PubMed]
28. Bardini, M.; Lee, D.; Donini, P.; Mariani, A.; Gianì, S.; Toschi, M.; Lowe, C.; Breviario, D. Tubulin-based polymorphism (TBP): A new tool, based on functionally relevant sequences, to assess genetic diversity in plant species. *Genome* **2004**, *47*, 281–291. [CrossRef] [PubMed]
29. Braglia, L.; Gavazzi, F.; Morello, L.; Gianì, S.; Nick, P.; Breviario, D. On the applicability of the Tubulin-Based Polymorphism (TBP) genotyping method: A comprehensive guide illustrated through the application on different genetic resources in the legume family. *Plant Methods* **2020**, *16*, 86. [CrossRef] [PubMed]
30. Gavazzi, F.; Braglia, L.; Mastromauro, F.; Giani, S.; Morello, L.; Breviario, D. The Tubulin-Based-Polymorphism Method Provides a Simple and Effective Alternative to the Genomic Profiling of Grape. *PLoS ONE* **2016**, *11*, e0163335. [CrossRef] [PubMed]

31. Van Hoeck, A.; Horemans, N.; Monsieurs, P.; Cao, H.; Vandenhove, H.; Blust, R. The first draft genome of the aquatic model plant *Lemna minor* opens the route for future stress physiology research and biotechnological applications. *Biotechnol. Biofuels* **2015**, *8*, 188. [CrossRef]
32. Sree, K.; Bog, M.; Appenroth, K. Taxonomy of duckweeds (*Lemnaceae*), potential new crop plants. *Emir. J. Food Agric.* **2016**, *28*, 291–302. [CrossRef]
33. Fu, L.; Huang, M.; Han, B.; Sun, X.; Sree, K.; Appenroth, K.; Zhang, J. Flower induction, microscope-aided cross-pollination, and seed production in the duckweed *Lemna gibba* with discovery of a male-sterile clone. *Sci. Rep.* **2017**, *7*, 3047. [CrossRef]
34. Fourounjian, P.; Slovin, J.; Messing, J. Flowering and Seed Production across the *Lemnaceae*. *Int. J. Mol. Sci.* **2021**, *22*, 2733. [CrossRef]
35. Creer, S. Choosing and using introns in molecular phylogenetics. *Evol. Bioinform. Online* **2007**, *3*, 99–108. [CrossRef]
36. Acosta, K.; Appenroth, K.J.; Borisjuk, L.; Edelman, M.; Heinig, U.; Jansen, M.A.K.; Oyama, T.; Pasaribu, B.; Schubert, I.; Sorrels, S.; et al. Return of the *Lemnaceae*: Duckweed as a model plant system in the genomics and postgenomics era. *Plant Cell* **2021**, *33*, 3207–3234. [CrossRef]
37. Abbott, R.; Albach, D.; Ansell, S.; Arntzen, J.; Baird, S.; Bierne, N.; Boughman, J.; Brelsford, A.; Buerkle, C.; Buggs, R.; et al. Hybridization and speciation. *J. Evol. Biol.* **2013**, *26*, 229–246. [CrossRef] [PubMed]
38. Lowe, A.; Abbott, R. Hybrid swarms: Catalysts for multiple evolutionary events in Senecio in the British Isles. *Plant Ecol. Divers.* **2015**, *8*, 449–463. [CrossRef]
39. Novak, S.; Soltis, D.; Soltis, P. Ownbey Tragopogons—40 Years Later. *Am. J. Bot.* **1991**, *78*, 1586–1600. [CrossRef]
40. Renaut, S.; Rowe, H.C.; Ungerer, M.C.; Rieseberg, L.H. Genomics of homoploid hybrid speciation: Diversity and transcriptional activity of long terminal repeat retrotransposons in hybrid sunflowers. *Philos. Trans. R. Soc. B Biol. Sci.* **2014**, *369*, 20130345. [CrossRef]
41. Marconi, G.; Landucci, F.; Rosellini, D.; Venanzoni, R.; Albertini, E. DNA barcoding as a tool for early warning and monitoring alien duckweeds (*Lemna* sp.pl.): The case of Central Italy. *Plant Biosyst.* **2019**, *153*, 660–668. [CrossRef]
42. Mallet, J. Hybrid speciation. *Nature* **2007**, *446*, 279–283. [CrossRef] [PubMed]
43. Benoit, L.K.; Les, D.H.; King, U.M.; Na, H.R.; Chen, L.; Tippery, N.P. Extensive interlineage hybridization in the predominantly clonal *Hydrilla verticillata*. *Am. J. Bot.* **2019**, *106*, 1622–1637. [CrossRef] [PubMed]
44. Sree, K.; Sudakaran, S.; Appenroth, K. How fast can angiosperms grow? Species and clonal diversity of growth rates in the genus Wolffia (*Lemnaceae*). *Acta Physiol. Plant.* **2015**, *37*, 204. [CrossRef]
45. Xu, S.; Stapley, J.; Gablenz, S.; Boyer, J.; Appenroth, K.; Sree, K.; Gershenzon, J.; Widmer, A.; Huber, M. Low genetic variation is associated with low mutation rate in the giant duckweed. *Nat. Commun.* **2019**, *10*, 1243. [CrossRef] [PubMed]
46. Schenk, R.; Hildebrandt, A. Medium and techniques for induction and growth of monocotyledonous and dicotyledonous Plant-Cell Cultures. *Can. J. Bot.* **1972**, *50*, 199–204. [CrossRef]
47. Hammer, Ø.; Harper, D.; Ryan, P. PAST: Paleontological statistics software package for education and data analysis. *Palaeontol. Electron.* **2001**, *4*, 9.
48. Letunic, I.; Bork, P. Interactive Tree Of Life (iTOL) v5: An online tool for phylogenetic tree display and annotation. *Nucleic Acids Res.* **2021**, *49*, W293–W296. [CrossRef] [PubMed]
49. Edgar, R. MUSCLE: Multiple sequence alignment with high accuracy and high throughput. *Nucleic Acids Res.* **2004**, *32*, 1792–1797. [CrossRef]
50. Kumar, S.; Stecher, G.; Li, M.; Knyaz, C.; Tamura, K. MEGA X: Molecular Evolutionary Genetics Analysis across Computing Platforms. *Mol. Biol. Evol.* **2018**, *35*, 1547–1549. [CrossRef]
51. Nei, M.; Kumar, S. *Molecular Evolution and Phylogenetics*; Oxford University Press: New York, NY, USA, 2000.

plants

MDPI

Article

Strategies for Intraspecific Genotyping of Duckweed: Comparison of Five Orthogonal Methods Applied to the Giant Duckweed *Spirodela polyrhiza*

Manuela Bog [1,*], Luca Braglia [2], Laura Morello [2], Karen I. Noboa Melo [1], Ingo Schubert [3], Oleg N. Shchepin [1], K. Sowjanya Sree [4], Shuqing Xu [5], Eric Lam [6] and Klaus J. Appenroth [7]

1 Institute of Botany and Landscape Ecology, University of Greifswald, 17489 Greifswald, Germany
2 Istituto Biologia e Biotecnologia Agraria, Via Bassini 15, 20131 Milano, Italy
3 Leibniz Institute of Plant Genetics and Crop Plant Research (IPK), Gatersleben, 06466 Stadt Seeland, Germany
4 Department of Environmental Science, Central University of Kerala, Periye 671320, India
5 Institute of Organismic and Molecular Evolution, Johannes Gutenberg University Mainz, 55128 Mainz, Germany
6 Department of Plant Biology, Rutgers the State University of New Jersey, New Brunswick, NJ 08901, USA
7 Matthias Schleiden Institute—Plant Physiology, University of Jena, 07743 Jena, Germany
* Correspondence: manuela.bog@uni-greifswald.de; Tel.: +49-38344204146

Abstract: The predominantly vegetative propagating duckweeds are of growing commercial interest. Since clonal accessions within a respective species can vary considerably with respect to their physiological as well as biochemical traits, it is critical to be able to track the clones of species of interest after their characterization. Here, we compared the efficacy of five different genotyping methods for *Spirodela polyrhiza*, a species with very low intraspecific sequence variations, including polymorphic NB-ARC-related loci, tubulin-gene-based polymorphism (TBP), simple sequence repeat variations (SSR), multiplexed ISSR genotyping by sequencing (MIG-seq), and low-coverage, reduced-representation genome sequencing (GBS). Four of the five approaches could distinguish 20 to 22 genotypes out of the 23 investigated clones, while TBP resolved just seven genotypes. The choice for a particular method for intraspecific genotyping can depend on the research question and the project budget, while the combination of orthogonal methods may increase the confidence and resolution for the results obtained.

Keywords: *Spirodela polyrhiza*; Lemnaceae; duckweed; genotyping; intraspecific variation

Citation: Bog, M.; Braglia, L.; Morello, L.; Noboa Melo, K.I.; Schubert, I.; Shchepin, O.N.; Sree, K.S.; Xu, S.; Lam, E.; Appenroth, K.J. Strategies for Intraspecific Genotyping of Duckweed: Comparison of Five Orthogonal Methods Applied to the Giant Duckweed *Spirodela polyrhiza*. *Plants* **2022**, *11*, 3033. https://doi.org/10.3390/plants11223033

Academic Editor: Andreas W. Ebert

Received: 13 October 2022
Accepted: 6 November 2022
Published: 9 November 2022

1. Introduction

In the last two decades, the evolution of molecular methods has revolutionized phytotaxonomy. These advances have also been applied to the duckweed family, Lemnaceae Martinov [1,2]. The taxonomic investigation of duckweeds using molecular methods started with the genotyping work presented in [3] and was extended with the application of several methods, such as amplified fragment length polymorphism (AFLP) and plastidic and nuclear barcoding (reviewed in [4]). Classification based on morphological markers was the sole option of botanists for centuries (for duckweeds, see [5,6]) and was, to a great extent, confirmed and extended by molecular taxonomic investigations, uncovering phylogenetic relationships. It should be emphasized that there is little in common between these two approaches, with the latter being a more quantitative method.

Presently, 30 out of 36 species of duckweed can be reliably identified by molecular taxonomy [7,8]. Thus, further progress in these methods is still required. Several areas of duckweed research require not only the identification of species but also that of specific clones from the same species, since some physiological properties of duckweeds are defined at the level of clones rather than the level of species and are sometimes linked with the respective ploidy level [9]. Such features include, e.g., the growth rate [10,11] and starch accumulation under the condition of the deficiency of essential nutrients, such as nitrate, phosphate, and sulphate [12], and other stresses [13,14]. Furthermore, in view of the growing commercial perspective of duckweeds, the patenting of specific clones [15] or their monitoring in commercial products would also require the ability to distinguish clones of interest from others of the same species. However, identification at the level of clones for duckweeds is extremely difficult using solely morphological markers due to their simple structure and abbreviated anatomical features. Therefore, suitable molecular methods for intraspecific genotyping must be developed.

Among the duckweed family, *Spirodela polyrhiza* (L.) Schleid (commonly called greater duckweed, Figure 1) displays unusually low intraspecific genetic variation. This was first suggested by the authors of [16], who found that there is no significant difference in genome size between the 38 clones of this species (in contrast to several other duckweed species; see [17], this Special Issue), and the investigation of three plastidic regions similarly detected hardly any variations. Subsequently, clonal variations between two reference genome quality assemblies of *S. polyrhiza* clones were found to be very low in terms of both intraspecific sequence variations as well as heterozygosity [18], with the single-nucleotide polymorphism positions (SNPs) being approximately six times fewer than those found between the ecotypes of *Arabidopsis* [19,20]. Using a population genomics approach with low-coverage sequencing reads, refs. [21,22] further extended the generality of these characteristics by finding very low genetic diversity between *S. polyrhiza* clones from a large number of locations across the globe. Moreover, the genome-wide spontaneous mutation rate in this species was estimated to be seven times lower than those of other multicellular eukaryotes characterized to date [21]. Using a novel approach to systematically identify and rank the polymorphic loci in the nuclear genome that may enable effective intraspecific genotyping, the authors of [23] used the genome sequences (40X or more coverage) from 10 clones of greater duckweed as a training set to identify loci among the NB-ARC-related gene family that could be used to discriminate between the clones of this species. This gene family, which is known to be involved in plant defence and immunity, was chosen because it displays the highest intraspecific polymorphism among the plant genomes that have been studied [19,24]. Validated primer sets were then used to uncover the intraspecific variations with an additional 13 clones of *S. polyrhiza*, bringing the total number to 23. From this work, 20 genotypes of these 23 clones could be distinguished. All 23 clones were selected from the list of 36 that were studied in [25] for their specific turion yield trait, based on their availability in the RDSC at the time. Three of the tested clones could not be resolved from each other using this genotyping technique but were distinct from each other in terms of the specific turion yield [25]. Nevertheless, the application of NB-ARC-related gene polymorphism represented a leap forward for the identification of intraspecific variations in Lemnaceae.

Figure 1. A colony of fronds of *Spirodela polyrhiza*, clone 7498: (**A**) dorsal view; (**B**) ventral view.

Here, we used the same clones as those studied by [23] for comparative studies using four additional orthogonal methods: fragment length polymorphisms (FLPs) and/or single-nucleotide polymorphisms (SNPs), tubulin-gene-based polymorphism (TBP; [26]), simple sequence repeat variants (SSR; [27]), multiplexed ISSR genotyping by sequencing (MIG-seq; [28]), and genotyping by genome-wide sequencing (GBS; [7]). In all five cases, we calculated dissimilarity trees to evaluate the different methods in terms of their genotyping efficacy and considered the advantages and limitations of each.

2. Results

2.1. NB-ARC-Related Genes (NB-ARC)

The NB-ARC approach is based on FLPs and SNPs of the DNA samples after their amplification with different primer sets, the sequences for which were generated and ranked by an informatic pipeline targeting genomic loci containing annotated NB-ARC genes [23]. Overall, 40 polymorphic characters were observed, with 17 based on FLPs and 23 based on SNPs (Table 1). The ambiguous, i.e., non-homozygous differences amounted to 26%. Figure 2 shows a single-linkage cluster and a heat map representation based on the uncorrected p-distance for the 23 *S. polyrhiza* clones used. All the numeric data, including the absolute and relative number of differences between clones, are given in the Supplementary Data S1. Three clones, 7379, 9503, and 9506, form a single data point and could not be distinguished from each other, but they are distinct from the other 20 clones. Thus, 21 genotypes were resolved from the 23 clones using this method. It should be noted that no error calculation is available for this approach, but based on the experience of one of the authors (E.L.), the repetition of the investigation of the same clone always produces identical results. One advantage of this method, as well as the TBP method, is that the number of fragments per sample tested is relatively small for the FLP and SNP analysis. Thus, the problem that usually affects other generic approaches in terms of the ambiguity in the band assignment becomes less of an issue. In any case, the results of this work revealed a geographic pattern associated with the dendrogram, i.e., clones from one continent were more likely to be similar to each other than clones from different continents (Figure 2).

Table 1. Comparison of the five tested methods for the genotyping of *Spirodela polyrhiza*, concerning the number of characters obtained and the proportion of ambiguous (non-homozygous) characters.

	NB-ARC	TBP	SSR	MIG-Seq	GBS
Marker Type [1]	FLP + SNP	FLP	FLP	SNP	SNP
Number of Characters	40 (17 + 23)	13	95	1292	6170
Percentage of Ambiguous Characters	26	0	0	29	14
Average Number of Ambiguous Characters per Sample ± Standard Deviation	11 ± 5	0 ± 0	0 ± 0	380 ± 69	894 ± 286
Error Rate % (Number of Replicates) [2]	n.d. (0)	0.00 (1)	0.26 (8)	0.00 (3)	0.17 (3)

[1] FLP—fragment length polymorphism; SNP—single-nucleotide polymorphism; [2] n.d.—not determined.

2.2. Tubulin-Gene-Based Polymorphism (TBP)

The TBP method uses FLPs from two introns of the conserved, multigene β-tubulin family. Since only a relatively small number of genomic loci were queried using this method, the number of variants available for the differentiation of the clones was also lowest when using this method. A total of 13 characters (fragment lengths), including four monomorphic clones, were detectable, with all of them originating from the first intron and with no polymorphism scored for the second intron (Table 1, Supplementary Data S1). Since the fragments were scored as either present or absent, no ambiguous character was detected. Due to the low number of the polymorphic scores, the resolution of this method for the 23 clones studied was very low, and only seven distinct genotypes could be resolved (Figure 2, Supplementary Data S1). The one sample that was run in replicates in this work yielded an identical result. The same was observed by two of the authors (L.B. and L.M.) for up to four replicates per sample of other *S. polyrhiza* clones not investigated in this study. In spite of the relatively low resolution of this method, a geographic pattern became obvious as well (Figure 2).

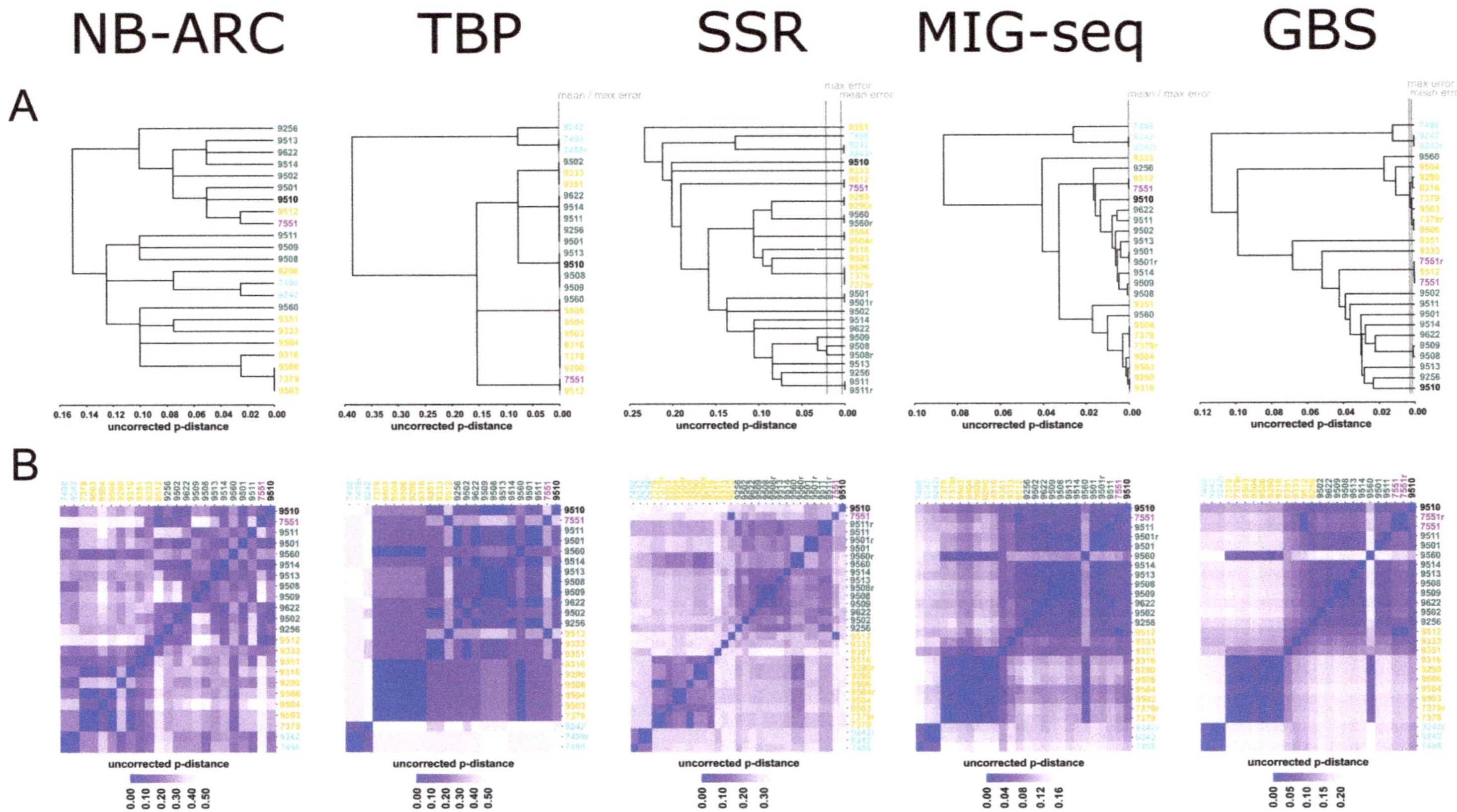

Figure 2. Comparison of five molecular methods for the genotyping of 23 clones of *S. polyrhiza*, including replicates (denoted by r). Colours denote the continent of origin of the clones: black—Africa, light blue—North + South America, yellow—Asia, violet—Australia, blue green—Europe. Upper panel (**A**) single-linkage clustering based on the uncorrected p-distances. The borders of resolution are given by the mean and max errors, calculated from the replicated clones. Lower panel (**B**) heat map representation of the uncorrected p-distances.

2.3. Simple Sequence Repeats (SSR)

The SSR method, like the TBP method, uses FLPs for the differentiation of different genotypes. In total, 95 amplification fragments resolved by their mobility on gels were scored as present or absent, which led, again, to no ambiguous characters (Supplementary Data S1). The error rate for the SSR was the highest of all the tested methods, but at 0.26%, it was still considered low (Table 1). Of the 23 investigated clones, 21 different genotypes could be detected. It was not possible to discriminate clone 7379 from clone 9506 and clone 7551 from clone 9512. A geographical pattern was, again, observable (Figure 2).

2.4. Multiplexed ISSR Genotyping by Sequencing (MIG-Seq)

The MIG-seq method is an SNP-based method. The final dataset (alignment can be found as Supplementary Data S2), chosen from the 20 tested parameter combinations, consisted of 1292 characters, of which 29% or, on average, 380 $\pm$ 69 characters per sample were ambiguous due to their heterozygosity and gaps (Supplementary Data S3). Surprisingly, there were no differences in homozygous position between the three samples that were run in replicates, which displayed an error rate of 0% (Table 1). The method revealed 20 genotypes. The clones 7379 and 9506, 7551 and 9512, and 9290 and 9316 could not be distinguished from one another. The geographical pattern was characterised by shorter branch lengths within the geographical subgroups rather than between the groups, in addition to the division between the continents of origin (Figure 2).

2.5. Genotyping-by-Sequencing (GBS)

The SNP-based GBS method yielded 6170 SNPs in total, which is the highest number of characters among all the tested methods (for the alignment, see Supplementary Data S4). The error rate (0.17%) and proportion of ambiguous characters (14%) are intermediate compared to the other methods (Table 1). Depending on the strictness of the error rate treatment, GBS could distinguish 22 genotypes (mean error rate as the threshold) or 17 genotypes (maximum error rate as the threshold) (Supplementary Data S1). Using the mean error rate as a threshold, only the clones 9506 and 9316 could not be distinguished from one another, while using the maximum error rate, the following clone pairs could not be distinguished from each other: 7551/9512, 7379/9506, 9290/9316, 9503/9506, 9506/9316, and 9509/9508. The clustering methods show a clear separation of the American clones from all the other clones (Figure 2). Additionally, the clones from Europe can be found in a separate cluster, although clone 9560 from Hungary clusters with the Asian clones, and this unusual pattern was found using all the investigated methods, including that of TBP. The Asian clones show a paraphyletic clustering, with clones 9333 (Hubei-China), 9351 (Hanoi-Vietnam), and 9512 (Irkutsk-Russia) being more similar to the clones from Europe and the one from Australia. The other monophyletic Asian clones are from India.

3. Discussion

In order to test the efficacy of the orthogonal molecular methods in distinguishing between the clones of *S. polyrhiza*, we compared the NB-ARC, TBP, SSR, MIG-seq, and GBS approaches. At least four of these methods are known to have a high capacity to distinguish genotypes. We added TBP in this work, because this method is experimentally easy to carry out, as just two PCR reactions, followed by capillary electrophoresis, are required. Moreover, the capacity of TBP for resolving certain clones of the same species has been successfully tested on different *Lemna* species [29].

In most cases, the tested methods detected 20 or 21 genotypes among the 23 investigated clones. By GBS, with a lower stringency error rate treatment, even 22 genotypes may be distinguishable. We were unable to discriminate between all 23 clones using any of these methods alone. Four methods failed to distinguish clones 7379 and 9506 from each other, while with GBS, at a higher stringency, these two clones also became difficult to resolve. The situation was almost the same for the pair of clones 9506 and 9503, where SSR was the only method able to resolve these two genotypes. All three of these clones with unusually high genome sequence identities, 7379, 9506, and 9503, are from India, but they exhibited a more than 3-fold range difference in their specific turion yields of 1.86, 0.97, and 0.51, respectively. We therefore suspect that relatively specific genomic sequence (or epigenome) variations might account for these trait differences, consistent with previous speculations [23]. Our data were analysed quite strictly, and only the homozygous sites were considered for the method comparison. This led to a weakening of the results, because even heterozygous differences could represent genotype-distinguishing features. Nevertheless, we decided to opt for this strict approach, which does not require any complicating assumptions about phasing.

As expected, the methods that yield a high number of total characters and, therefore, differences, i.e., MIG-seq and GBS, showed the best resolution, especially in the phylogeographical studies. The geographical pattern closely fitted the results obtained by whole-genome resequencing in [21], which also separated the Indian *S. polyrhiza* clones from the Southeast Asian ones and clustered clone 9560 from Hungary together with the Asian clones. A great advantage of these two methods is their easy applicability for a wide range of organisms, although both methods require Illumina sequencing and the related equipment [7] (see Table 2 for a general comparison of the applied methods). The costs, especially if a company is asked to perform the analyses, are relatively high compared to the other methods. Furthermore, MIG-seq has an advantage over GBS, because in the case of MIG-seq, degraded DNA can be used as a template, since the amplified fragments between the SSR-based primers are relatively short. In contrast, GBS requires very good-quality DNA, since the first step is a fragmentation using restriction enzymes, and the clustering success of the single short Illumina-sequenced fragments highly depends on the quality of the template.

Table 2. Pros and cons of the five different methods used for genotyping of 23 *Spirodela polyrhiza* clones with a worldwide distribution.

	NB-ARC	TBP	SSR	MIG-Seq	GBS
DNA requirements	degraded works as well	degraded works as well	degraded works as well	degraded works as well	high quality necessary
Establishment	elaborate, genome sequence information needed	easy	elaborate, sequence information needed	easy	easy
Universality *	no, primers can be species-specific	yes	(no), low cross-amplification	yes	yes
Resolution capacity	high	low	high	high	high
Costs per sample	low to moderate, depends on whether genome information is available	low	low	moderate	high

* indicates whether the established method can be easily transferred between species. The sequence-guided approach and pipeline created for NB-ARC is indeed universal, but specific primers need to be established for each species. The same applies for SSR.

The NB-ARC polymorphism-based genotyping approach has shown a very good performance in genotyping *S. polyrhiza* [23], as observed here, with many of the original key observations confirmed and extended by the present work. A drawback of this method is the genome assembly resource required at the outset to establish this marker system for the species of interest. A well-assembled reference genome for the species and the nine other sequenced accessions were used as a training set for *S. polyrhiza*, a species known to have unusually low sequence variations, in order to informatically generate a ranked list of suitable primers to amplify the most polymorphic regions among the NB-ARC genes in the genome [23]. To improve the resolving power of this method, one could include additional genome sequence data on challenging clones, such as 7379 and 9503, in our training set for the pipeline, since clone 9506 was already part of the original training set. One apparent limitation of this approach could affect genomes that have a limited number of NB-ARC genes, such as the recently sequenced *Wolffia australiana* genome that has only three to four copies of these genes remaining [30]. In these cases, other large-plant gene families with members containing highly conserved exons could be used in the same informatic pipeline as that established in [23] to generate the requisite primers. For example, we note that in *W. australiana*, there is an amplification of the genes encoding the LRR-RK type of pattern recognition receptors (PRRs) to about 90 copies [1]. This gene family may present an alternative to the NB-ARC genes in *Wolffia australiana* for the generation of intraspecific genotyping primer sets.

SSR is likewise powerful, but it requires at least partial genomic information as well as the selection of useful sequence repeats and the design of primers in the flanking regions, as conducted in [27]. Furthermore, the use of cross-amplification to find useful polymorphic sites in related species is rather inefficient. The authors of [27] demonstrated this for *Lemna perpusilla*, and tests on a clone of *S. intermedia* (the sister species of *S. polyrhiza*) were unsuccessful in the lab of one of the authors (M.B.). The screening of already available whole-genome sequences or shotgun genome information appears to be a straightforward method for developing SSR markers using tools such as GMATA and others ([31] and references therein). Once the proper primers are available, the experimental procedure is not highly demanding financially or with respect to the required equipment or the quality of the template DNA.

The TBP method yielded, in our analysis, only seven genotypes based on the 13 scored characters for the 23 *S. polyrhiza* clones studied here. The same method, however, provided a useful resolution for the presumed *Lemna minor* clones, with 34 characters scored for the first intron and 36 genotypes identified among the 40 analysed clones [32]. Successful intraspecific genotyping was previously reported in crop species such as grape and olive [33,34]. However, because *S. polyrhiza* generally shows very little intraspecific variation [21,22], this species is a tough match for the TBP method and could be an exception rather than the common rule. Thus, for a duckweed species that displays greater intraspecific genome variations, this simple approach could still be a useful genotyping tool.

In summary, we presented five methods, four of which are well-suited for genotyping *S. polyrhiza* and likely other species, even those beyond the duckweed family, with a low intraspecific variability. The decision regarding which of these methods should be applied depends on the question to be solved (and on the available budget and genome resources). In addition, while the data generated using the four most effective approaches are consistent, to a large extent, and thus provide strong support for their validity, they also suggest the potential advantage of combining two of these orthogonal methods to extend the power of the analysis. For example, using GBS analysis, under a lower stringency of error, only one pair of clones, 9506 and 9316, could not be resolved. These two clones, however, can be distinguished using any of the other three methods. Thus, if data based on GBS and one of these methods can be integrated together in a genotype matrix, the resolution of all 23 clones tested here could become feasible.

4. Materials and Methods

4.1. Plant Material and Cultivation

All 23 clones of *Spirodela polyrhiza* were acquired from the duckweed collection of the Matthias Schleiden Institute—Plant Physiology, University of Jena, Germany, and are available as living materials from the duckweed collections in Jena and at Rutgers State University of New Jersey (New Brunswick, NJ, U.S.A.) under the international four-digit code [35] given in Table 3. They were selected by the authors of [23] according to their geographical diversity and the wide range of their specific turion yield [25]. The clones were additionally characterised previously by measurements of their relative growth rate [11] and genome size [16], or these parameters were measured in the present project (Table 3). The species' identity was confirmed by barcoding using the plastidic markers *rpl16*, *rps16*, and *atpF-atpH* [16]. In total, 20 out of the 23 clones were whole-genome sequenced [21,23,36]. All the plants were cultivated under axenic conditions in N medium under standardised conditions for 10 days [37]: 8 mM KNO_3, 0.15 mM KH_2PO_4, 1 mM

$MgSO_4$, 1 mM $Ca(NO_3)_2$, 5 μM H_3BO_3, 0.4 μM Na_2MoO_4, 13 μM $MnCl_2$, and 25 μM Fe(III)NaEDTA. The plant material was stored at −80 °C for further use.

Table 3. Origin of investigated *Spirodela polyrhiza* clones and their specific properties.

Clone ID	Country	Area/State	Specific Turion Yield [1]	Specific Growth Rate (h^{-1}) [2]	Genome Size (Mbp/1C) [3]	Genome Sequencing [4]
7379	India	Tamil Nadu	1.86 ± 0.26	0.315 ± 0.008	158 ± 3	n.d.
7498	USA	North Carolina	2.37 ± 0.27	0.401 ± 0.016	157 ± 2	+
7551	Australia	Northern Territory	n.d.	0.376 ± 0.009	n.d.	+
9242	Ecuador	Guayas	1.37 ± 0.12	0.456 ± 0.003	n.d.	+
9256	Finland	Uusimaa	3.73 ± 0.25	0.296 ± 0.011	160 ± 6	+
9290	India	Delhi	1.40 ± 0.13	0.299 ± 0.010	n.d.	+
9316	India	Rajasthan	1.22 ± 0.13	0.281 ± 0.009	159 ± 2	+
9333	China	Hubei	n.d.	0.311 ± 0.033	n.d.	+
9351	Vietnam	Hanoi	n.d.	0.367 ± 0.024	n.d.	n.d.
9501	Albania	Fieri	5.93 ± 0.03	0.357 ± 0.005	164 ± 4	+
9502	Ireland	Leinster	1.64 ± 0.14	0.337 ± 0.008	164 ± 5	+
9503	India	Rajasthan	0.51 ± 0.03	0.312 ± 0.007	170 ± 5	+
9504	India	Rajasthan	0.34 ± 0.03	0.284 ± 0.012	168 ± 6	+
9506	India	Telangana	0.97 ± 0.07	0.313 ± 0.005	161 ± 7	+
9508	Poland	Cracow	0.66 ± 0.04	0.360 ± 0.007	160 ± 4	+
9509	Germany	Thuringia	0.51 ± 0.07	0.323 ± 0.006	157 ± 2	+
9510	Mozambique	Maputo	0.98 ± 0.10	0.289 ± 0.008	n.d.	+
9511	Russia	Moscow	2.25 ± 0.11	0.381 ± 0.009	161 ± 5	+
9512	Russia	Irkutsk	2.92 ± 0.49	0.386 ± 0.005	156 ± 3	+
9513	Czech	Jindřichův Hradec	1.16 ± 0.04	0.324 ± 0.006	161 ± 5	+
9514	Austria	Viena	1.17 ± 0.16	0.333 ± 0.012	159 ± 5	+
9560	Hungary	Bekes	n.d.	0.367 ± 0.008	160 ± 6	+
9622	Germany	Baden-Württemberg	n.d.	0.310 ± 0.011	159 ± 5	+

[1] [25], [2] [11] or measured in the present project, [3] [16], [4] [21,23,36], n.d.—not determined.

4.2. DNA Isolation and Downstream Lab Work

DNA was isolated from 100–200 mg fresh weight of duckweed plants using the CTAB method [38] and quantified using a NanoVue spectrophotometer (GE Healthcare Europe GmbH, Freiburg, Germany). For GBS, the DNA was extracted using silica columns, according to [39].

The wet lab methods used for the analysis by NB-ARC are described in [23].

TBP amplification was performed by targeting both the 1st and 2nd introns of the multigene β-tubulin family, using two universal primer pairs, Fex-Rex and Fin-Rin, respectively, and PCR conditions described in [29]. FAM-labelled amplified fragments were separated by capillary electrophoresis (CE) using a AB3500 Genetic Analyzer (Thermo Fisher Scientific Inc., Waltham, MA, USA). The numerical output of the CE was exported to Excel format, and the fragments were aligned by size across the samples, considering a one-nucleotide (± 0.5) approximation. The amplicon size was considered as a marker, and its presence/absence was scored in a binary matrix (1/0, respectively), which was used for the subsequent elaboration of the data. Both the TBP 1st and 2nd introns were scored. Further technical details about the data acquisition and analysis are reported in [26]. The sample 7498 was run in duplicate using the same DNA extract for the error estimation.

For the SSR investigations, a total of 12 SSR markers multiplexed in three sets of four primer pairs each were used, according to [27]. The forward primers were labelled with a fluorescence dye (set1: SP12[FAM], SP14[HEX], SP43[CY3], SP51[ROX]; set2: SP6[HEX], SP25[ROX], SP29[FAM], SP42[CY3]; set3: SP36[HEX], SP45[ROX], SP47[CY3], SP53[FAM]). PCR amplification was performed using a thermal cycler (SensoQuest GmbH, Göttingen, Germany) in a total volume of 10 μL containing 45 ng DNA template, 2.5 μL of 2 X KAPA 2G (Kapa Biosystems Pty Ltd., South Africa), and the final concentration of each primer (SP6[0.1 μM], SP12[0.2 μM], SP14[0.2 μM], SP25[0.2 μM], SP29[0.2 μM], SP36[0.1 μM], SP42[0.4 μM], SP43[0.2 μM], SP45[0.4 μM], SP47[0.4 μM], SP51[0.2 μM], SP53[0.1 μM]). The samples were amplified under the following conditions: 3 min at 95 °C, 35 cycles of 15 s at 95 °C, 35 s at 54 °C, and 30 s at 72 °C, and 5 min of the final extension at 72 °C. Subsequently, 2.5 μL of the diluted PCR products (set 1: undiluted, set 2: 1:10, set 3: 1:25) were mixed with 0.16 μL of GeneScan 500 LIZ size standard and 10.5 μL HiDi formamide (both Thermo Fisher Scientific). Finally, the fragment analysis was carried out using a 3130xl Genetic

Analyzer (Thermo Fisher Scientific). Eight samples (7379, 9242, 9290, 9501, 9504, 9508, 9511, and 9560) from the same DNA extracts were run in duplicate for the error estimation.

For MIG-seq, 25 µL of each DNA extract at a concentration of 30 ng/µL was sent to LGC Genomics GmbH (Berlin, Germany) for analysis. Sequencing by synthesis was performed with a 150 bp paired-end read chemistry using an Illumina NextSeq500/550 device (Illumina Inc., San Diego, CA, USA). Three samples (7379, 9242, 9501) from the same DNA extracts were run in duplicate for the error estimation. The sequence reads were uploaded to the sequence read archive (SRA) [40], PRJNA888369, accession numbers: SRR21845253-SRR21845278.

For the GBS method, DNA was used for the construction of barcoded libraries, essentially performed as described in [41]. Sequencing by synthesis (single read, 1×10^7 cycles, index read 8 cycles) using the Illumina HiSeq 2500 device was performed according to the protocols provided by the manufacturer (Illumina Inc.). Three samples (7379, 7551, and 9242) from the same DNA extracts were run in duplicate for the error estimation. The sequence reads were uploaded to the sequence read archive (SRA) [40], PRJNA888369, accession numbers: SRR21845363-SRR21845387 and SRR21845392.

4.3. Data Analysis

The dataset for the analysis by NB-ARC was taken from [23], and the scoring of the TBP fragments was performed as described in [26]. The fragment lengths based on the SSR analysis were scored using GeneMapper v5 (Thermo Fisher Scientific) and subsequently rounded using Tandem v1.09 [42]. Finally, the fragment length data were converted into a 0/1 matrix. For MIG-seq, demultiplexing and trimming of sequencing adapters and primers from the raw sequence reads were done by the sequencing facility. Around 8 million pre-processed paired-end reads (mean: 307.5 thousand reads per sample, SD: 45.4 thousand reads) were de novo assembled and analysed using the ipyrad v0.9.84 pipeline [43] run on the HPC Brain Cluster of the University of Greifswald (Supplementary Data S5). We optimized the core assembly parameters 'clustering threshold for de novo assembly' (ct) and 'minimum depth for statistical base calling' (md) by running the analysis with 10 different values of ct (0.81–0.99, with a step of 0.2) and two values of md (6 and 10). Following the method of [44], the datasets obtained with the parameters ct = 0.91 and md = 10 were determined as optimal for the downstream analyses (Supplementary Data S5). The SNP identification of the approximately 77 mio pre-processed single-end reads (mean: 3 mio reads per sample, SD: 600 thousand reads) for the GBS data was performed as described in [7], except for the fact that the *S. polyrhiza* genome available in [45] was used for the SNP mapping.

We attempted to make the data analysis for the comparison as similar as possible, although the different methods produced different data types, including FLP (scored as 0/1) and SNP data. Therefore, 0/1 data was recoded as c/a, and each of the five datasets was saved as a FASTA file, which was further processed using R v4.1.0 [46], with the additional libraries ape v5.6-2 [47], stringdist v0.9.8 [48], phangorn v2.10.0 [49], reshape2 v1.4.4 [50], and ggplot2 v3.3.6 [51]. In a first step, the uncorrected p distance and character differences were calculated using the 'dist.hamming' function. Ambiguous characters were not considered. Then, a single-linkage clustering was applied using the 'hclust' function. The resulting trees were saved in newick format and formatted in FigTree v1.4.4 [52]. Heat maps were plotted using the 'ggplot' function after the conversion of the distance matrices into a linear form using the 'melt' function. Ambiguous sites were counted using a custom python script.

Supplementary Materials: The following are available online at https://www.mdpi.com/article/10.3390/plants11223033/s1, Supplementary Data S1.xlsx (primer/adapter sequences and input data of the five investigated methods and calculated p-distance matrices with absolute number of differences and proportion of differences); Supplementary Data S2.fas (aligned SNP data from the MIG-seq analysis); Supplementary Data S3.xlsx (general statistics on MIG-seq results); Supplementary Data S4.fas (aligned SNP data from the GBS analysis); Supplementary Data S5.pdf (MIG-seq analysis script and results of ipyrad parameter optimisation).

Author Contributions: Conceptualization, K.J.A., K.S.S. and M.B.; methods and analysis, E.L. (NB-ARC), L.B. and L.M. (TBP), K.I.N.M. and M.B. (SSR), O.N.S. and M.B. (MIG-seq), S.X. (GBS); meta-analysis, M.B.; data curation, M.B.; writing—original draft preparation, K.J.A., K.S.S. and M.B.; editing, E.L., I.S., K.I.N.M., K.J.A., K.S.S., L.B., L.M., M.B., O.N.S. and S.X. All authors have read and agreed to the published version of the manuscript.

Funding: This research received no external funding.

Data Availability Statement: All the Illumina sequence data based on MIG-seq and GBS are available from the Sequence Read Archive (https://www.ncbi.nlm.nih.gov/, accessed on 12 October 2022). The respective project numbers and accession numbers can be found in the main text. The raw data and distance matrices for the other methods are available as Supplementary Data within this article.

Conflicts of Interest: The authors declare no conflict of interest.

References

1. Acosta, K.; Appenroth, K.J.; Borisjuk, L.; Edelman, M.; Heinig, U.; Jansen, M.A.K.; Oyama, T.; Pasaribu, B.; Schubert, I.; Sorrels, S.; et al. Return of the Lemnaceae: Duckweed as a model plant system in the genomics and postgenomics era. *Plant Cell* **2021**, *33*, 3207–3234. [CrossRef] [PubMed]
2. Tippery, N.P.; Les, D.H.; Appenroth, K.J.; Sree, K.S.; Crawford, D.J.; Bog, M. Lemnaceae and Orontiaceae are phylogenetically and morphologically distinct from Araceae. *Plants* **2021**, *10*, 2639. [CrossRef] [PubMed]
3. Les, D.H.; Crawford, D.J.; Landolt, E.; Gabel, J.D.; Kimball, R.T. Phylogeny and systematics of Lemnaceae, the duckweed family. *Syst. Bot.* **2002**, *27*, 221–240. [CrossRef]
4. Bog, M.; Appenroth, K.J.; Sree, K.S. Duckweed (Lemnaceae): Its molecular taxonomy. *Front. Sustain. Food Syst.* **2019**, *3*, 117. [CrossRef]
5. Landolt, E. *The Family of Lemnaceae—A Monographic Study. Vol. 1. Biosystematic Investigations in the Family of Duckweeds (Lemnaceae)*; Veröffentlichungen des Geobotanischen Institutes der ETH, Stiftung Ruebel: Zurich, Switzerland, 1986.
6. Bog, M.; Appenroth, K.J.; Sree, K.S. Key to the determination of taxa of Lemnaceae: An update. *Nordic J. Bot.* **2020**, *8*, e02658. [CrossRef]
7. Bog, M.; Xu, S.; Himmelbach, A.; Brandt, R.; Wagner, F.; Appenroth, K.J.; Sree, K.S. Genotyping-By-Sequencing for species delimitation in the section *Uninerves* Hegelm. (Genus *Lemna* L.). In *Compendium of Plant Genomes: The Duckweed Genomes*, 1st ed.; Cao, X.H., Fourounjian, P., Wang, W., Eds.; Springer Nature: Cham, Switzerland, 2020; pp. 115–123. [CrossRef]
8. Bog, M.; Sree, K.S.; Fuchs, J.; Hoang, P.T.N.; Schubert, I.; Kuever, J.; Rabenstein, A.; Paolacci, S.; Jansen, M.A.K.; Appenroth, K.J. A taxonomic revision of *Lemna* sect. *Uninerves (Lemnaceae)*. *Taxon* **2020**, *69*, 56–66. [CrossRef]
9. Vunsh, R.; Heinig, U.; Malitsky, S.; Aharoni, A.; Avidov, A.; Lerner, A.; Edelman, M. Manipulating duckweed through genome duplication. *Plant Biol.* **2015**, *17*, 115–119. [CrossRef]
10. Sree, K.S.; Sudakaran, S.; Appenroth, K.J. How fast can angiosperms grow? Species and clonal diversity of growth rates in the genus *Wolffia* (Lemnaceae). *Acta Physiol. Plant.* **2015**, *37*, 204. [CrossRef]
11. Ziegler, P.; Adelmann, K.; Zimmer, S.; Schmidt, C.; Appenroth, K.J. Relative in vitro growth rates of duckweeds (Lemnaceae), the most rapidly growing higher plants. *Plant Biol.* **2015**, *17*, 33–41. [CrossRef]
12. Sree, K.S.; Appenroth, K.J. Starch accumulation in duckweeds (Lemnaceae) induced by nutrient deficiency. *Emir. J. Food Agric.* **2022**, *34*, 204–212. [CrossRef]
13. Sree, K.S.; Adelmann, K.; Garcia, C.; Lam, E.; Appenroth, K.J. Natural variance in salt tolerance and induction of starch accumulation in duckweeds. *Planta* **2015**, *241*, 1395–1404. [CrossRef]
14. Appenroth, K.J.; Ziegler, P.; Sree, K.S. Accumulation of starch in duckweeds (Lemnaceae), potential energy plants. *Physiol. Mol. Biol. Plants.* **2021**, *27*, 2621–2633. [CrossRef]
15. Kaplan, A.; Zelicha, H.; Tsaban, G.; Yaskolka, M.A.; Rinott, E.; Kovsan, J.; Novack, L.; Thiery, J.; Ceglarek, U.; Burkhardt, R.; et al. Protein bioavailability of *Wolffia globosa* duckweed, a novel aquatic plant—A randomized controlled trial. *Clin. Nutr.* **2019**, *38*, 2576–2582. [CrossRef] [PubMed]
16. Bog, M.; Lautenschlager, U.; Landrock, M.F.; Landolt, E.; Fuchs, J.; Sree, K.S.; Oberprieler, C.; Appenroth, K.J. Genetic characterization and barcoding of taxa in the genera *Landoltia* and *Spirodela* (Lemnaceae) by three plastidic markers and amplified fragment length polymorphism (AFLP). *Hydrobiologia* **2015**, *749*, 169–182. [CrossRef]
17. Hoang, P.T.N.; Fuchs, J.; Schubert, V.; Tran, T.B.N.; Schubert, I. Chromosome Numbers and Genome Sizes of All 36 Duckweed Species (Lemnaceae). *Plants* **2022**, *11*, 2674. [CrossRef] [PubMed]
18. Michael, T.P.; Bryant, D.; Gutierrez, R.; Borisjuk, N.; Chu, P.; Zhang, H.; Xia, J.; Zhou, J.; Peng, H.; El Baidouri, M.; et al. Comprehensive definition of genome features in *Spirodela polyrhiza* by high-depth physical mapping and short-read DNA sequencing strategies. *Plant J.* **2017**, *89*, 617–635. [CrossRef]
19. Gan, X.; Stegle, O.; Behr, J.; Steffen, J.; Drewe, P.; Hildebrand, K.; Lyngsoe, R.; Schultheiss, S.J.; Osborne, E.J.; Sreedharan, K.L.; et al. Multiple reference genomes and transcriptomes for *Arabidopsis thaliana*. *Nature* **2011**, *477*, 419–423. [CrossRef]

20. Lu, P.; Han, X.; Qi, J.; Yang, J.; Wijeratne, A.J.; Li, T.; Ma, H. Analysis of *Arabidopsis* genome-wide variations before and after meiosis and meiotic recombination by resequencing Landsberg *erecta* and all four products of a single meiosis. *Genome Res.* **2012**, *22*, 508–518. [CrossRef]
21. Xu, S.; Stapley, J.; Gablenz, S.; Boyer, J.; Appenroth, K.J.; Sree, K.S.; Gershenzon, J.; Widmer, A.; Huber, M. Low genetic variation is associated with low mutation rate in the giant duckweed. *Nat. Commun.* **2019**, *10*, 1243. [CrossRef]
22. Ho, E.K.H.; Bartkowska, M.; Wright, S.I.; Agrawal, A.F. Population genomics of the facultatively asexual duckweed *Spirodela polyrhiza*. *New Phytol.* **2019**, *224*, 1361–1371. [CrossRef]
23. Chu, P.; Wilson, G.M.; Michael, T.P.; Vaiciunas, J.; Honig, J.; Lam, E. Sequence-guided approach to genotyping plant clones and species using polymorphic NB-ARC-related genes. *Plant Mol. Biol.* **2018**, *98*, 219–231. [CrossRef]
24. Clark, R.M.; Schweikert, G.; Toomajian, C.; Ossowski, S.; Zeller, G.; Shinn, P.; Warthmann, N.; Gu, T.T.; Fu, G.; Hinds, D.A.; et al. Common sequence polymorphisms shaping genetic diversity in *Arabidopsis thaliana*. *Science* **2007**, *317*, 338–342. [CrossRef] [PubMed]
25. Kuehdorf, K.; Jetschke, G.; Ballani, L.; Appenroth, K.J. The clonal dependence of turion formation in the duckweed *Spirodela polyrhiza*—An ecogeographical approach. *Physiol. Plant.* **2014**, *150*, 46–54. [CrossRef] [PubMed]
26. Braglia, L.; Gavazzi, F.; Morello, L.; Gianì, S.; Nick, P.; Breviario, D. On the applicability of the Tubulin-Based Polymorphism (TBP) genotyping method: A comprehensive guide illustrated through the application on different genetic resources in the legume family. *Plant Methods* **2020**, *16*, 86. [CrossRef] [PubMed]
27. Xu, N.; Hu, F.; Zhang, W.; Wang, M.; Zhu, M.; Ke, J. Characterization of 19 polymorphic SSR markers in *Spirodela polyrhiza* (Lemnaceae) and cross amplification in *Lemna perpusilla*. *Appl. Plant Sci.* **2018**, *6*, 1–5. [CrossRef]
28. Suyama, Y.; Matsuki, Y. MIG-seq: An effective PCR-based method for genome-wide single-nucleotide polymorphism genotyping using the next-generation sequencing platform. *Sci. Rep.* **2015**, *5*, 16963. [CrossRef] [PubMed]
29. Braglia, L.; Lauria, M.; Appenroth, K.J.; Bog, M.; Breviario, D.; Grasso, A.; Gavazzi, F.; Morello, L. Duckweed species genotyping and interspecific hybrid discovery by tubulin-based polymorphism fingerprinting. *Front. Plant Sci.* **2021**, *12*, 625670. [CrossRef]
30. Michael, T.P.; Ernst, E.; Hartwick, N.; Chu, P.; Bryant, D.; Gilbert, S.; Ortleb, S.; Baggs, E.L.; Sree, K.S.; Appenroth, K.J.; et al. Genome and time-of-day transcriptome of *Wolffia australiana* link morphological minimization with gene loss and less growth control. *Genome Res* **2021**, *31*, 225–238. [CrossRef]
31. Wang, X.; Wang, L. GMATA: An Integrated Software Package for Genome-Scale SSR Mining, Marker Development and Viewing. *Front. Plant Sci.* **2016**, *7*, 1350. [CrossRef]
32. Braglia, L.; Breviario, D.; Gianì, S.; Gavazzi, F.; De Gregori, J.; Morello, L. New Insights into Interspecific Hybridization in *Lemna* L. Sect. *Lemna (Lemnaceae Martinov)*. *Plants* **2021**, *10*, 2767. [CrossRef] [PubMed]
33. Gavazzi, F.; Braglia, L.; Mastromauro, F.; Gianì, S.; Morello, L.; Breviario, D. The Tubulin-Based-Polymorphism Method Provides a Simple and Effective Alternative to the Genomic Profiling of Grape. *PLoS ONE* **2016**, *11*, e0163335. [CrossRef] [PubMed]
34. Braglia, L.; Manca, A.; Gianì, S.; Hatzopoulos, P.; Breviario, D. A Simplified Approach for Olive (*Olea europaea* L.) Genotyping and Cultivars Traceability. *Am. J. Plant Sci.* **2017**, *8*, 3475–3489. [CrossRef]
35. Zhao, H.; Appenroth, K.J.; Landesman, L.; Salmean, A.A.; Lam, E. Duckweed rising at Chengdu: Summary of the 1st International Conference on Duckweed Application and Research. *Plant Mol. Biol.* **2012**, *78*, 627–632. [CrossRef] [PubMed]
36. Hoang, P.N.T.; Michael, T.P.; Gilbert, S.; Chu, P.; Motley, S.T.; Appenroth, K.J.; Schubert, I.; Lam, E. Generating a high-confidence reference genome map of the greater duckweed by integration of cytogenomic, optical mapping, and Oxford Nanopore technologies. *Plant J.* **2018**, *96*, 670–684. [CrossRef] [PubMed]
37. Appenroth, K.J.; Teller, S.; Horn, M. Photophysiology of turion formation and germination in *Spirodela polyrhiza*. *Biol. Plant.* **1996**, *38*, 95–106. [CrossRef]
38. Doyle, J.J.; Doyle, J.L. A rapid DNA isolation procedure for small quantities of fresh leaf tissue. *Phytochem. Bull.* **1987**, *19*, 11–15.
39. Valledor, L.; Escandón, M.; Meijón, M.; Nukarinen, E.; Canal, M.J.; Weckwerth, W. A universal protocol for the combined isolation of metabolites, DNA, long RNAs, small RNAs, and proteins from plants and microorganisms. *Plant J.* **2014**, *79*, 173–180. [CrossRef]
40. National Center for Biotechnolgy Information (NCBI). Available online: https://www.ncbi.nlm.nih.gov/ (accessed on 10 October 2022).
41. Wendler, N.; Mascher, M.; Nöh, C.; Himmelbach, A.; Scholz, U.; Ruge-Wehling, B.; Stein, N. Unlocking the secondary genepool of barley with next-generation sequencing. *Plant Biotechnol. J.* **2014**, *12*, 1122–1131. [CrossRef]
42. Matschiner, M.; Salzburger, W. TANDEM: Integrating automated allele binning into genetics and genomics workflows. *Bioinformatics* **2009**, *25*, 1982–1983. [CrossRef]
43. Eaton, D.A.R.; Overcast, I. ipyrad: Interactive assembly and analysis of RADseq datasets. *Bioinformatics* **2020**, *36*, 2592–2594. [CrossRef]
44. McCartney-Melstad, E.; Gidis, M.; Shaffer, H.B. An empirical pipeline for choosing the optimal clustering threshold in RADseq studies. *Mol. Ecol. Resour.* **2019**, *19*, 1195–1204. [CrossRef] [PubMed]
45. Spirodela polyrhiza Genome on Joint Genome Institute (JGI). Available online: https://data.jgi.doe.gov/refine-download/phytozome?organism=Spolyrhiza&expanded=290 (accessed on 13 October 2022).
46. R Core Team R: A Language and Environment for Statistical Computing. R Foundation for Statistical Computing, Vienna, Austria. 2021. Available online: https://www.R-project.org/ (accessed on 6 October 2022).

47. Paradis, E.; Schliep, K. ape 5.0: An environment for modern phylogenetics and evolutionary analyses in R. *Bioinformatics* **2019**, *35*, 526–528. [CrossRef] [PubMed]
48. van der Loo, M.P.J. The stringdist package for approximate string matching. *R J.* **2014**, *6*, 111–122.
49. Schliep, K.P. phangorn: Phylogenetic analysis in R. *Bioinformatics* **2011**, *27*, 592–593. [CrossRef]
50. Wickham, H. Reshaping Data with the reshape Package. *J. Stat. Soft* **2007**, *21*, 1–20. [CrossRef]
51. Wickham, H. *ggplot2: Elegant Graphics for Data Analysis*; Springer: New York, NY, USA, 2016.
52. FigTree—A Graphical Viewer of Phylogenetic Trees and as a Program for Producing Publication-Ready Figures. Available online: http://tree.bio.ed.ac.uk/software/figtree/ (accessed on 6 October 2022).

plants

MDPI

Article

Intraspecific Diversity in Aquatic Ecosystems: Comparison between *Spirodela polyrhiza* and *Lemna minor* in Natural Populations of Duckweed

Manuela Bog [1], Klaus-Juergen Appenroth [2,*], Philipp Schneider [2] and K. Sowjanya Sree [3,*]

[1] Institute of Botany and Landscape Ecology, University of Greifswald, D-17489 Greifswald, Germany; manuela.bog@uni-greifswald.de

[2] Matthias Schleiden Institute-Plant Physiology, University of Jena, D-07743 Jena, Germany; pilles@gmx.de

[3] Department of Environmental Science, Central University of Kerala, Periye 671320, India

[*] Correspondence: klaus.appenroth@uni-jena.de (K.-J.A.); ksowsree9@cukerala.ac.in or ksowsree@gmail.com (K.S.S.); Tel.: +49-3641-949233 (K.-J.A.); +91-9999-672921 (K.S.S.)

Abstract: Samples of two duckweed species, *Spirodela polyrhiza* and *Lemna minor*, were collected around small ponds and investigated concerning the question of whether natural populations of duckweeds constitute a single clone, or whether clonal diversity exists. Amplified fragment length polymorphism was used as a molecular method to distinguish clones of the same species. Possible intraspecific diversity was evaluated by average-linkage clustering. The main criterion to distinguish one clone from another was the 95% significance level of the Jaccard dissimilarity index for replicated samples. Within natural populations of *L. minor*, significant intraspecific genetic differences were detected. In each of the three small ponds harbouring populations of *L. minor*, based on twelve samples, between four and nine distinct clones were detected. Natural populations of *L. minor* consist of a mixture of several clones representing intraspecific biodiversity in an aquatic ecosystem. Moreover, identical distinct clones were discovered in more than one pond, located at a distance of 1 km and 2.4 km from each other. Evidently, fronds of *L. minor* were transported between these different ponds. The genetic differences for *S. polyrhiza*, however, were below the error-threshold of the method within a pond to detect distinct clones, but were pronounced between samples of two different ponds.

Keywords: amplified fragment length polymorphism (AFLP); biodiversity; duckweed; intraspecific diversity; *Lemna minor*; Lemnaceae; population analysis; *Spirodela polyrhiza*

check for updates

Citation: Bog, M.; Appenroth, K.-J.; Schneider, P.; Sree, K.S. Intraspecific Diversity in Aquatic Ecosystems: Comparison between *Spirodela polyrhiza* and *Lemna minor* in Natural Populations of Duckweed. *Plants* **2022**, *11*, 968. https://doi.org/10.3390/plants11070968

Academic Editor: Ornella Calderini

Received: 14 February 2022
Accepted: 30 March 2022
Published: 1 April 2022

Publisher's Note: MDPI stays neutral with regard to jurisdictional claims in published maps and institutional affiliations.

1. Introduction

Duckweeds belong to the monocotyledonous aquatic plant family Lemnaceae [1], comprising 36 species [2–4]. Although duckweeds can bear flowers, fruits and seeds, the most common mode of their reproduction is vegetative propagation. Daughter fronds arise from a mother frond by budding out of pockets [5]. The daughter fronds that have a common ancestral mother frond, together with their progeny, form a clone. This vegetative propagation proceeds with the fastest growth rates known in Angiosperms [6,7]. The two species, investigated in the present study, *Spirodela polyrhiza* (L.) Schleid. and *Lemna minor* L., have clone-specific doubling times of approximately 2.3 and 1.7 days, respectively. In laboratory experiments, both *S. polyrhiza* and *L. minor* were reported to flower [8,9]. In the present study, however, we have never observed the two species flowering, neither in the laboratory nor outdoors. Thus, clonal propagation is by far the dominating mechanism of propagation. However, in *S. polyrhiza*, sexual reproduction is assumed to be not too infrequent at the population level because of the very high number of individuals in natural populations. This was concluded by Ho et al. [10] as effects of recombination in their study.

During the past few years, it became clear that some of the physiological properties like growth rate [6,7], turion formation capacity [11], protein content [12], and starch accumulation capacities [13–15] vary between clones that belong to the same species. This raises the question whether a natural population of duckweed constitutes only a single clone, or whether several clones coexist. It is a pertinent ecological issue bearing consequences on both the basic science and the applications of duckweeds [16,17], wherein proper collection, isolation and maintenance of a clone becomes a prerequisite for reproducible results. Hence, this question led us to investigate the probable existence of clonal diversity or intraspecific diversity of the predominantly vegetative propagating species *S. polyrhiza* and *L. minor* in a pond ecosystem.

Because of the reduction in morphological and anatomical structures, differentiating the species of Lemnaceae on a morphological basis [3] is by itself a difficult task, even for highly specialized experts. Over the last fifteen years, notable progress has been made in the field of molecular taxonomy of duckweeds by employing different techniques including the use of nuclear and plastid markers [18–25] (for a review, see Bog et al. [26]). However, the characterization of intraspecific genetic variations, i.e., distinct identification of clones within a given species is still at its infant stage [27]. Most recent efforts in this direction are genotyping clones of *S. polyrhiza* by sequencing NB-ARC-related genes [28] and application of genotyping-by-sequencing [29], or even SSR markers [30,31]. Nevertheless, cross-species amplification for NB-ARC-related genes and Simple sequence repeat (SSR) markers is low within duckweeds [28,31].

The traditional definition of a duckweed clone refers to fronds that have originated from a common mother frond (ancestor) by vegetative propagation. The samples used in the present study were collected from natural populations, in which case, it was not possible to evaluate on a morphological basis whether the two samples had originated from a common mother frond or not; in other words, whether the samples belong to the same clone. Therefore, in order to carry out the present study on natural populations of the same species, it was necessary to employ molecular methods and, consequently, a molecular definition of the term clone had to be derived. In the present study, we suggest the molecular criterion that, in order to be considered as a clone, a sample should be distinguishable from another sample by molecular methods, which should be confirmed by statistical validation to be significantly above the experimental error.

In an initial effort to characterize intraspecific differences, Bog et al. [18–25] used the method of amplified fragment length polymorphism (AFLP) and detected, in many cases, clear intraspecific differences. Moreover, it was shown that this method is clearly superior to plastidic barcoding [26]. Therefore, we chose AFLP as an inexpensive method to analyse the intraspecific or clonal diversity of duckweed populations. This makes it possible to address the question of whether populations of a single species existing in small ponds of only a few hundred square meters, where duckweeds can be found frequently, are homogenous concerning the intraspecific diversity, i.e., comprising only a single clone, or if several clones of a species coexist in the same pond. Cole and Voshkuil [32] reported, for the first time, the co-existence of several clones of duckweed (*L. minor*) in natural populations. These authors investigated 285 fronds in total, from 11 pond populations (i.e., on average 26 fronds per pond) by allozyme variations resolving 16 putative loci. In the present project, we used AFLP because this method has much higher resolving power than allozyme variation (see Table 1). To the best of our knowledge, this is the first report of characterization of intraspecific variations within a natural population of duckweeds directly at the level of DNA.

Table 1. Band statistics of AFLP analysis for the investigated ponds. lo = Lotschen, mo = Moscow, gro = Groeben, sch = Schloeben, S = *Spirodela polyrhiza*, L = *Lemna minor*, SD = standard deviation.

	S. polyrhiza		*L. minor*		
	moS	loS	loL	Gro	sch
mean band presence per sample (mean ± SD)	53 ± 5	72 ± 7	92 ± 17	107 ± 15	94 ± 14
number polymorphic bands	7	3	70	56	61
number fixed bands	52	73	53	68	67
number private bands	5	22	4	0	4
number fixed private bands	2	19	0	0	0
Shannon's index	2.2	0.9	28.8	16.2	28.9
Phi_{ST}	0.992			0.241	

2. Results

2.1. Evaluation of AFLP Fingerprinting

AFLP fingerprinting of 2 × 12 clones of *S. polyrhiza* yielded a total number of 81 reproducible fragments for the four primer combinations: (i) 26 loci (118–564 bp), (ii) 21 loci (100–498 bp), (iii) 20 loci (102–420 bp), and (iv) 14 loci (117–460 bp); that of 3 × 12 clones of *L. minor* yielded a total number of 133 reproducible fragments: (i) 48 loci (111–546 bp), (ii) 40 loci (100–498 bp), (iii) 18 loci (113–384 bp), and (iv) 27 loci (115–411 bp). The Euclidean error rate, calculated from five parallel doublet runs of samples for each species, resulted in 1.5% for the data set of *S. polyrhiza*, and 1.7% for *L. minor*. Table 1 shows characteristic band statistics and diversity indices for the AFLP data sets. The high proportion of polymorphic bands suggests a higher genetic variability of samples from the ponds of *L. minor* than ponds of *S. polyrhiza*. This assumption is strengthened by the Shannon index, which is very low for the respective *S. polyrhiza* ponds (0.9–2.2) in comparison to *L. minor* ponds (16.2–28.9). Additionally, AMOVA shows that there is 99.2% of genetic variability between the two *S. polyrhiza* populations while it is 0.8% within the populations, which leads to a Phi_{ST} value of 0.992. For *L. minor*, there is 24.1% of genetic variation among the three populations and 75.9% within the populations, leading to a Phi_{ST} value of 0.241.

2.2. Average-Linkage Cluster Analysis

Results of AFLP analysis were presented by dendrograms. For the statistical evaluation of the results, a methodological error-threshold was calculated based on the investigated replicates. The grey bars in the dendrograms (Figure 1) indicate the mean dissimilarity and its 95% confidence interval of the replicates. Therefore, only the branches to the left side of the vertical grey bar represent separate clones. The other branches were caused, e.g., by erroneous AFLP bands on the gel and were below the threshold level. For *S. polyrhiza* (Figure 1a), the upper confidence interval at approximately 0.08 Jaccard dissimilarity units cuts the dendrogram in such a way that only the separation into samples from two different places of origin were accepted as distinct clones (Spirodela1, Spirodela2; see Figure 1a and Table 2). No genetic differentiation was detected between samples of the same origin. These data are based on the analysis of 2 × 12 collected samples out of several millions of fronds in the ponds.

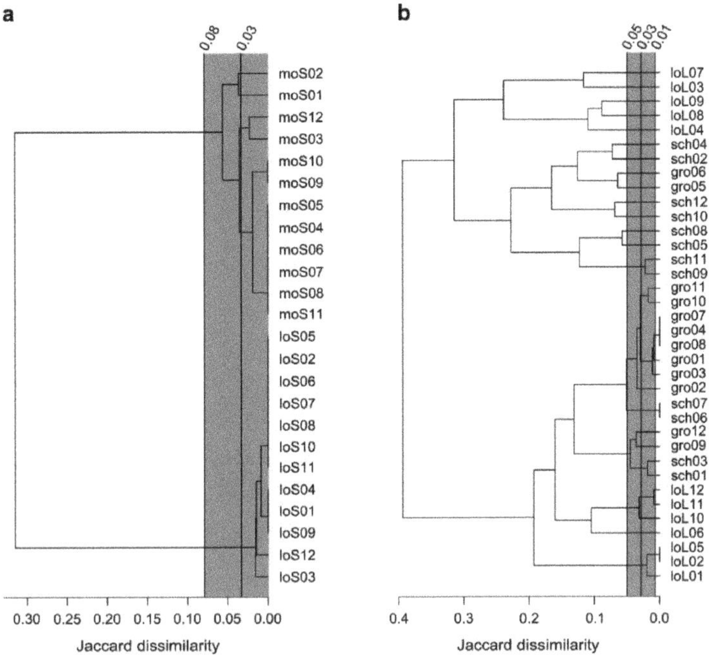

Figure 1. Average-linkage cluster analysis based on Jaccard dissimilarities for (**a**) *Spirodela polyrhiza* and (**b**) *Lemna minor*. Grey bar indicates mean Jaccard dissimilarity and its 95% confidence interval of the replicates. All samples separating to the left of the grey bar can be considered as different clones. The absolute Jaccard values are given at the top. lo = Lotschen, mo = Moscow, gro = Groeben, sch = Schloeben. S = *Spirodela polyrhiza*, L = *Lemna minor*. All samples from gro and sch are *Lemna minor*.

The total 36 samples of *L. minor* (out of several hundred million fronds in the ponds during the peak time) were separated into a larger number of groups out of the confidence interval (Figure 1b), and finally into 20 groups. Two main branches separated all 36 investigated samples into two large groups of equal numbers. At the bottom of the dendrogram there were three samples, i.e., loL01, 02 and 05. These three samples have to be considered as one single, distinct clone, termed as Lemna01 (see Figure 1b and Table 2). It follows sample loL06 in the tree (Figure 1b), forming a distinct clone, Lemna02. The next group is represented by the samples loL10, 11, and 12, forming the distinct clone, Lemna03. All clones mentioned until now are from the pond in Lotschen. The remaining part of the lower branch of the dendrogram consisted of 14 samples, all of them were separated into three distinct clones, i.e., clone Lemna04: schl01, 03, gro09, 12; clone Lemna05: sch06, 07; clone Lemna06: gro01–04, 07, 08, 10, 11. It should be stressed that clone Lemna04 has representatives in the pond at Groeben, as well as in the pond at Schloeben. The upper branch of the dendrogram starts with ten samples, subdivided by this analysis (Figure 1b) into nine different, distinct clones (Lemna07–Lemna15). Only samples sch09 and sch11 formed a common distinct clone, i.e., Lemna07. All other samples formed single distinct clones themselves, originating mainly from Schloeben (Lemna08, Lemna09, Lemna10, Lemna11, Lemna14, Lemna15), and two of them also from Groeben (Lemna12, Lemna13) (Figure 1b). The upper part of the dendrogram contained five samples that had originated from Lotschen, forming the distinct clones Lemna16, Lemna17, Lemna18, Lemna19 and Lemna20. The 20 clones of *Lemna* were consequently classified as distinct clones (Table 2).

Table 2. Distinct clones of the species *Spirodela polyrhiza* and *Lemna minor* and their origin by collected samples from ponds in Moscow, Lotschen, Groeben and Schloeben. For further explanations, cf. Table 1.

Distinct Clones	Moscow (mo)	Lotschen (lo)	Groeben (gro)	Schloeben (sch)
Spirodela1	moS01–12			
Spirodela2		loS01–12		
Lemna01		loL01, 02, 05		
Lemna02		loL06		
Lemna03		loL10–12		
Lemna04			gro09, 12	sch01, 03
Lemna05				sch06–07
Lemna06			gro01–04, 07–08, 10–11	
Lemna07				sch09, 11
Lemna08				sch05
Lemna09				sch08
Lemna10				sch10
Lemna11				sch12
Lemna12			gro05	
Lemna13			gro06	
Lemna14				sch02
Lemna15				sch04
Lemna16		loL04		
Lemna17		loL08		
Lemna18		loL09		
Lemna19		loL03		
Lemna20		loL07		

3. Discussion

3.1. No Clonal Diversity Could Be Detected within Natural Populations of Spirodela polyrhiza

The populations of *S. polyrhiza* in the two ponds in Moscow and Lotschen could be easily distinguished from one another, indicating the existence of two distinct clones of this species. This is in agreement with the results of Xu et al. [33], who reported whole genome sequencing of 68 world-wide collected clones of *S. polyrhiza*, including clone 9509 from Lotschen (see also [34]) and 9511 from Moscow, collected from the same ponds as the AFLP samples analysed here. These authors reported high similarity of both clones (belonging to the "European population"), but both clones from Lotschen and Moscow were distinguished. The first detailed comparison between the genomes of two *S. polyrhiza* clones (7498 from USA and 9509 from Lotschen, Germany [34,35]) demonstrated also very small differences. Similar conclusions about low genetic variation within *S. polyrhiza* were drawn by Ho et al. [10] on the basis of 38 clones of *S. polyrhiza*, almost all of them from Northern America, as well as from Bog et al. [20] and Feng et al. [25], where more than 40 defined *S. polyrhiza* clones, selected from world-wide stock collections, were used in each study. In the present context, it is important that the two investigated populations of *S. polyrhiza* were probably formed by only one genotype each. This was concluded on the basis of the low fraction of investigated plants with 2×12 samples from the two ponds using AFLP analysis. We evaluated the number of fronds of the total population in the pond in Lotschen to ca. 10^7. Such a large population can hardly be investigated in detail. The genome-wide mutation rate in *S. polyrhiza* was found to be within the range of mutation rates reported for unicellular eukaryotes and Eubacteria, but was more than seven times lower than the reported rates for multicellular eukaryotes [33]. Further, Xu et al. [33] found a much slower decay of the linkage disequilibrium with physical distance between linked loci for the "European population" compared to the "Southeast Asian population", and concluded that sexual reproduction might therefore be less frequent in the first mentioned population contrary to the latter one. It therefore could be concluded that low mutation rates and predominantly asexual reproduction might have led to the lack of genotype diversity in the two investigated populations from ponds in Lotschen and Moscow, both belonging to the "European population". As this holds true for samples from all over the world [33], environment might not have an essential influence on the mutation rate—in contrast to the flower frequency.

3.2. Natural Population of Lemna minor Comprises of Several Distinct Clones

The situation was quite different in three populations of *L. minor*, with conclusions based on the same number of investigated random samples per pond. It is impressive to note that in all the investigated natural populations existing in small ponds of few hundred square meter surface area, several distinct clones of *L. minor* co-existed. As in the case of *S. polyrhiza*, populations of *L. minor* could possess very high surface coverage rates. Despite investigating only a small fraction of this huge population, AFLP results demonstrate that the genetic variations within the species *L. minor* must be higher than in *S. polyrhiza*. It should be mentioned here that the genome of *L. minor* is approximately three to four times larger than that of *S. polyrhiza*, and that it shows much higher genomic variation as it is reflected, e.g., in its genome sizes between clones of this species compared to those of *S. polyrhiza* [20,36], which could be a possible explanation of the existence of a larger number of distinct clones. On the other hand, the higher mutation rate of *L. minor* in comparison to *S. polyrhiza* may be proposed; the reason for this is not known at the moment. Unfortunately, it is not yet possible to compare the genomic differences between clones of *L. minor* directly as presently only one high-quality draft genome is available (Rob Martienssen, www.lemna.org (accessed on 12 February 2022); for a review see [37]). A higher rate of sexual reproduction could be another possible explanation, but no quantitative data are available to compare the two species in this point under natural conditions. However, very recently, the existence of hybrids between *L. minor* and *L. turionifera* [38], and between *L. minor* and *L. gibba* [39], was demonstrated by molecular analysis (tubulin-based polymorphism), suggesting sexual propagation. Such hybrids are not known in *S. polyrhiza*, which might be a hint that sexual propagation in this species is rarer.

Tang et al. [40] investigated *S. polyrhiza*, *Landoltia punctata* (G. Mey.) Les & D.J.Crawford, and three *Lemna* species from Lake Tai (2250 km^2 large) using cpDNA markers. The authors could easily identify the duckweed species using this method, but could not detect any intraspecific differences. El-Kholy et al. [41] compared the genetic diversity between populations and collected samples from the Nile delta, Egypt, using the fingerprinting method of inter-simple sequence repeats (ISSRs). They detected intraspecific differences within *L. minor* and *L. gibba* L.; however, in both cases, the large size of the water bodies questions the existence of only a single population of duckweed in each of these study areas. Paolacci et al. [42] collected clones of *L. minor* from a restricted geographic area, i.e., the southern part of Ireland, and detected clearly intraspecific genetic variability by AFLP—as for the invasive species *Lemna minuta*. However, in this case also we cannot consider the samples as belonging to a single population. Hence, these results are rather similar to those of clones from stock collections from all over the world, implying geographical influence on different populations (for recent reviews, see [26]).

3.3. Natural Transport of Clones between Ponds

One distinct clone of *L. minor*, Lemna04, was detected in both lakes situated in Groeben and Schloeben, substantiating that there should have been a transfer of fronds either from one pond to the other, or an identical clone from a distant pond to the two ponds in Groeben and Schloeben, which is less probable. The distance between the two ponds is only 1 km. Thus, epizoochorous transport by water birds, but also by amphibians or rodents, seems to be possible. Beside mutation and sexual reproduction, both very rare processes in duckweeds, epizoochorous transport from other ponds seems to be a good candidate to explain the intraspecific variability in *L. minor*. The direction of the putative transfer is unknown as we had detected two samples of this clone in both the ponds. It has already been reported by Coughlan et al. [43–45] that duckweeds can be transported over certain distances by Mallard ducks, especially frequently over small distances [44]. The present results demonstrate the possible outcome of such natural transport between ponds, i.e., an increase in the intraspecific diversity of populations. However, this does not explain the different genetic variability between *S. polyrhiza* and *L. minor*. M. A. K. Jansen, University College Cork, Ireland, wrote: "The issue is that dispersal (of duckweeds) relies on entanglement between feathers, sticking to the bird, survival of drought stress during flight, and release upon arrival in a new water body. On balance, considering all these parameters, there are no clear differences between species" (personal communication to K.J.A.).

4. Materials and Methods

4.1. Taxon Sampling and Cultivation of Clones

Duckweed samples were collected from five populations in four ponds (Table 3). The circumference of each pond was divided in 12 equidistant sampling sites, and 12 samples (colonies) were collected from each population at these sites. Three of the ponds were in villages in Thuringia, Germany (Lotschen, Groeben, Schloeben; Figure 2) and one distant pond in Moscow, Russia (Table 3). The distance from Lotschen ("Ruttgersdorf-Lotschen") to Groeben and Schloeben is approximately 2.4 km and 1.9 km, respectively, from Groeben to Schloeben is approximately 1.0 km, and from Lotschen to Moscow is approximately 2000 km.

Table 3. Details of the location of ponds selected for investigation, duckweed species and samples collected. For further explanations, see Table 1.

Species	Location	Samples	Pond Size
Lemna minor	Groeben, Thuringia, Germany 50°53′06″ N 11°40′53″ E	gro01–gro12	900 m^2
Lemna minor	Schloeben, Thuringia, Germany 50°53′32″ N 11°41′26″ E	sch01–sch12	400 m^2
Lemna minor	Lotschen, Thuringia, Germany 50°53′07″ N 11°42′56″ E	loL01–loL12	700 m^2
Spirodela polyrhiza	Lotschen, 50°53′07″ N 11°42′56″ E	loS01–loS12	700 m^2
Spirodela polyrhiza	Botanical Garden Moscow, Russia 55°50′36″ N 37°35′23″ E	moS01–moS12	100 m^2

We extrapolated the density of a *S. polyrhiza* population by counting a square of 10×10 cm^2 in the pond of Lotschen to be ca. 6×10^4 m^{-2}, resulting in a total population of ca. 10^7 in the pond during the peak of the summer season. In the case of *L. minor*, Hicks [46] reported 10^6 fronds m^{-2}.

The late Elias Landolt, ETH Zurich, Switzerland, confirmed the species identities of the 3×12 *L. minor* samples and 2×12 *S. polyrhiza* samples using morphological markers. Thus, from each pond and each population, 12 samples were collected, and single colonies were sterilized [47]. All samples were tested in nutrient medium supplemented with 25 mM glucose for the absence of microbial contamination. Single colonies have different numbers of fronds connected by stolons, and all these fronds originate from a single mother frond. Offspring of a colony are therefore clonally-related fronds, and the progeny of a single colony each were used for analysis—called a "sample" in the present paper. The samples were cultivated under axenic conditions in continuous white light (100 μmol m^{-2} s^{-1}) at 25 ± 1 °C in N-medium [48]: 8 mM KNO_3, 0.15 mM KH_2PO_4 (increased in comparison to the original protocol), 1 mM $MgSO_4$, 1 mM $Ca(NO_3)_2$, 5 μM H_3BO_3, 0.4 μM Na_2MoO_4, 13 μM $MnCl_2$, and 25 μM Fe(III)NaEDTA, in order to get sufficient plant material from the collected clonal samples. In most cases, plants were harvested after 14 days of cultivation, frozen in liquid nitrogen, and stored at −80 °C for further use. One sample from each of the two *S. polyrhiza* populations were used for whole genome sequencing, as clone 9509 from Lotschen and 9511 from Moscow [33,34].

Figure 2. Map of the three closely located ponds in villages in Thuringia, Germany (Lotschen, Groeben, Schloeben), represented by red dots.

4.2. DNA Isolation and AFLP Analysis

Total DNA was isolated immediately after grinding in liquid nitrogen using the Cetyl Trimethyl Ammonium Bromide (CTAB) method, following the protocol of Doyle and Doyle [49]. DNA was quantified by a NanoVue spectrophotometer (GE Healthcare Europe GmbH, Freiburg, Germany) at 260 nm. The complete AFLP procedure, mainly according to the protocol of Vos et al. [50], as described by Bog et al. [19], was followed. Infra-red dye (IRD)-labelled primers were used for the selective PCR amplification that consequently labelled the electrophoretic bands. After testing a large number of primers, the following four primer combinations were selected for AFLP analysis: (i) *Eco*RI-ATT/*Mse*I-CAC, (ii) *Eco*RI-ATT/*Mse*I-CAT, (iii) *Eco*RI-ATT/*Mse*I-CCA, (iv) *Eco*RI-ATT/*Mse*I-CTA [15,16]. An automated DNA sequencer (model 4000 L; Li-Cor Biosciences, Bad Homburg, Germany) was used for electrophoretic separation and detection of generated fragments [51]. AFLP patterns were manually compiled into a 0/1-matrix ("1" for presence, "0" for absence of a band), assuming that bands of equal fragment size are homologous and represent independent loci. This matrix for the investigated clones has been made available as a supplementary material of the present paper (Supplementary Table S1). As a measure for reproducibility, the Euclidean error rate was assessed by making two parallel, independent preparations for five of the samples for each of the two investigated species. These replicates were not included in the final evaluation.

4.3. Data Analyses

We used "Cluster Analysis" as the method, and the confidence interval as a statistical measure. To avoid confusion, we designated the clones that were collected, propagated and maintained from a single colony of the natural population, as "sample" (we investigated $5 \times 12 = 60$ such clonal samples) and based on the molecular method, the genetically diverse samples of the same species were categorized and characterized as "distinct clones".

For AFLP, we calculated measures for band statistics, like mean band presence per sample, number of polymorphic bands, number of fixed bands (=band presence in all samples of a pond), number of private bands (=band presence in at least some clones of a pond), and number of fixed private bands (=exclusive band presence in all samples of a pond). Additionally, we calculated Jaccard dissimilarity matrices for the two data sets, *S. polyrhiza* and *L. minor*, that were used in subsequent analysis. An analysis of molecular variance (AMOVA), Phi_{ST} (analogue value of F_{ST}) and Shannon's index were calculated to evaluate the molecular variation and diversity within our data sets. All aforementioned calculations and analyses were done using the program FAMD v1.31 [52], which can also handle missing data. To check the clones that could be characterized as independent AFLP phenotypes ("distinct clones"), we performed an average-linkage cluster analysis based on the Jaccard dissimilarity matrices in R v. 3.2.3 [53]. Additionally, the mean Jaccard dissimilarity based on the replicates and its 95% confidence intervals were calculated, which were used as threshold values for characterization of distinct clones.

5. Conclusions

1. A remarkable intraspecific biodiversity exists in natural populations of *Lemna minor*, but not in populations of *Spirodela polyrhiza*, suggesting either a higher mutation rate or a higher rate of sexual reproduction in *Lemna minor*.
2. The intraspecific biodiversity in the ponds was further enhanced by the putative transfer (most probably epizoochorous by birds) of plants between closely spaced ponds.
3. The high intraspecific diversity of *Lemna minor* may have a role in the adaptation of the natural duckweed populations to the changing environmental conditions.

Supplementary Materials: The following are available online at https://www.mdpi.com/article/10.3390/plants11070968/s1, Table S1: 0/1-matrix of AFLP investigations of clones of *Spirodela polyrhiza* and *Lemna minor*. AFLP patterns were compiled into a 0/1-matrix ("1" for presence, "0" for absence of a band).

Author Contributions: The research plan was developed by K.-J.A., M.B. and K.S.S.; isolation of DNA was carried out by P.S.; AFLP investigations were performed by M.B.; interpretation was developed by K.-J.A., M.B. and K.S.S.; the manuscript was written part by part by all authors. All authors have read and agreed to the published version of the manuscript.

Funding: This research received no external funding.

Informed Consent Statement: Not applicable.

Data Availability Statement: All data are available in the manuscript and Supplementary Materials.

Conflicts of Interest: The authors declare that there is no conflict of interest.

References

1. Tippery, N.P.; Les, D.H.; Appenroth, K.J.; Sree, K.S.; Crawford, D.J.; Bog, M. Lemnaceae and Orontiaceae are phylogenetically and morphologically distinct from Araceae. *Plants* **2021**, *10*, 2639. [CrossRef] [PubMed]
2. Sree, K.S.; Bog, M.; Appenroth, K.J. Taxonomy of duckweed (Lemnaceae), potential crop plants. *Emir. J. Food Agric.* **2016**, *28*, 291–303. [CrossRef]
3. Bog, M.; Appenroth, K.J.; Sree, K.S. Key to the determination of taxa of Lemnaceae: An update. *Nord. J. Bot.* **2020**, *38*, e02658. [CrossRef]
4. Bog, M.; Sree, K.S.; Fuchs, J.; Phuong, T.N.H.; Schubert, I.; Kuever, J.; Rabenstein, A.; Paolacci, S.; Jansen, M.A.K.; Appenroth, K.J. A taxonomic revision of *Lemna* sect. *Uninerves* (Lemnaceae). *Taxon* **2020**, *69*, 56–66. [CrossRef]
5. Acosta, K.; Appenroth, K.J.; Borisjuk, L.; Edelman, M.; Heinig, U.; Jansen, M.A.K.; Oyama, T.; Pasaribu, B.; Schubert, I.; Sorrels, S.; et al. Return of the Lemnaceae: Duckweed as a model plant system in the genomics and post-genomics era (Review). *Plant Cell* **2021**, *33*, 3207–3234. [CrossRef]

6. Sree, K.S.; Sudakaran, S.; Appenroth, K.J. How fast can angiosperms grow? Species and clonal diversity of growth rates in the genus *Wolffia* (Lemnaceae). *Acta Physiol. Plant.* **2015**, *37*, 204. [CrossRef]
7. Ziegler, P.; Adelmann, K.; Zimmer, S.; Schmidt, C.; Appenroth, K.J. Relative in vitro growth rates of duckweeds (Lemnaceae)—The most rapidly growing higher plants. *Plant Biol.* **2015**, *17* (Suppl. S1), 33–41. [CrossRef]
8. Krajncic, B.; Devide, Z. Report on photoperiodic responses in Lemnaceae from Slovenia. *Ber. Geobot. Inst. ETH Zur.* **1980**, *47*, 75–86.
9. Pieterse, A.H. Is flowering in Lemnaceae stress-induced? A review. *Aquat. Bot.* **2013**, *104*, 1–4. [CrossRef]
10. Ho, E.K.H.; Bartkowska, M.; Wright, S.I.; Agrawal, A.F. Population genomics of the facultatively asexual duckweed *Spirodela polyrhiza*. *New Phytol.* **2019**, *224*, 1361–1371. [CrossRef]
11. Kuehdorf, K.; Jetschke, G.; Ballani, L.; Appenroth, K.J. The clonal dependence of turion formation in the duckweed *Spirodela polyrhiza*—An ecogeographical approach. *Physiol. Plant.* **2014**, *150*, 46–54. [CrossRef] [PubMed]
12. Appenroth, K.J.; Sree, K.S.; Bog, M.; Ecker, J.; Seeliger, C.; Boehm, V.; Lorkowski, S.; Sommer, K.; Vetter, W.; Tolzin-Basch, K.; et al. Nutritional value of the duckweed species of the genus *Wolffia* (Lemnaceae) as human food. *Front. Chem.* **2018**, *6*, 483. [CrossRef] [PubMed]
13. Sree, K.S.; Adelmann, K.; Garcia, C.; Lam, E.; Appenroth, K.-J. Natural variance in salt tolerance and induction of starch accumulation in duckweeds. *Planta* **2015**, *241*, 1395–1404. [CrossRef] [PubMed]
14. Ma, Y.B.; Zhu, M.; Yu, C.J.; Wang, Y.; Liu, Y.; Li, M.L.; Sun, Y.D.; Zhao, J.S.; Zhou, G.K. Large-scale screening and characterisation of *Lemna aequinoctialis* and *Spirodela polyrhiza* strains for starch production. *Plant Biol.* **2018**, *20*, 357–364. [CrossRef]
15. Appenroth, K.J.; Ziegler, P.; Sree, K.S. Accumulation of starch in duckweeds (Lemnaceae), potential energy plants. *Physiol. Mol. Biol. Plants* **2021**, *27*, 2621–2633. [CrossRef]
16. Appenroth, K.J.; Sree, K.S.; Fakhoorian, T.; Lam, E. Resurgence of duckweed research and applications: Report from the 3rd International Duckweed Conference. *Plant Mol. Biol.* **2015**, *89*, 647–654. [CrossRef]
17. Sree, K.S.; Khurana, J.P. (Eds.) Duckweed: Biological Chemistry and Applications. In *Frontiers Abstract Book, Proceedings of the Fourth International Conference on Duckweed Research and Applications, Kerala, India, 23–26 October 2017*; Frontiers Media SA: Lausanne, Switzerland, 2018.
18. Bog, M.; Baumbach, H.; Schween, U.; Hellwig, F.; Landolt, E.; Appenroth, K.-J. Genetic structure of the genus *Lemna* L. (Lemnaceae) as revealed by amplified fragment length polymorphism. *Planta* **2010**, *232*, 609–619. [CrossRef]
19. Bog, M.; Schneider, P.; Hellwig, F.; Sachse, S.; Kochieva, E.Z.; Martyrosian, E.; Landolt, E.; Appenroth, K.J. Genetic characterization and barcoding of taxa in the genus *Wolffia* Horkel ex Schleid. (Lemnaceae) as revealed by two plastidic markers and amplified fragment length polymorphism (AFLP). *Planta* **2013**, *237*, 1–13. [CrossRef]
20. Bog, M.; Lautenschlager, U.; Landrock, M.F.; Landolt, E.; Fuchs, J.; Sree, K.S.; Oberprieler, C.; Appenroth, K.-J. Genetic characterization and barcoding of taxa in the genera *Landoltia* and *Spirodela* (Lemnaceae) by three plastidic markers and amplified fragment length polymorphism (AFLP). *Hydrobiologia* **2015**, *749*, 169–182. [CrossRef]
21. Bog, M.; Landrock, F.M.; Drefahl, D.; Sree, K.S.; Appenroth, K.J. Fingerprinting by Amplified Fragment Length Polymorphism (AFLP) and barcoding by three plastidic markers in the genus *Wolffiella* Hegelm. *Plant Syst. Evol.* **2018**, *304*, 373–386. [CrossRef]
22. Les, D.H.; Crawford, D.J.; Landolt, E.; Gabel, J.D.; Kimball, R.T. Phylogeny and systematics of Lemnaceae, the duckweed family. *Syst. Bot.* **2002**, *27*, 221–240.
23. Tippery, N.P.; Les, D.H.; Crawford, D.J. Evaluation of phylogenetic relationships in Lemnaceae using nuclear ribosomal data. *Plant Biol.* **2015**, *17* (Suppl. S1), 50–58. [CrossRef] [PubMed]
24. Xu, Y.; Ma, S.; Huang, M.; Peng, M.; Bog, M.; Sree, K.S.; Appenroth, K.J.; Zhang, J. Species distribution, genetic diversity and barcoding in the duckweed family (Lemnaceae). *Hydrobiologia* **2015**, *743*, 75–87. [CrossRef]
25. Feng, B.; Fang, Y.; Xu, Z.; Xiang, C.; Thou, C.; Jiang, F.; Wang, T.; Zhao, H. Development of a new marker system for identification of *Spirodela polyrhiza* and *Landoltia punctata*. *Int. J. Genomics* **2017**, *2017*, 5196763. [CrossRef]
26. Bog, M.; Appenroth, K.J.; Sree, K.S. Duckweed (Lemnaceae): Its molecular taxonomy. A Review. *Front. Sustain. Food Syst.* **2019**, *3*, 117. [CrossRef]
27. Borisjuk, N.; Chu, P.; Gutierrez, R.; Zhang, H.; Acosta, K.; Friesen, N.; Sree, K.S.; Garcia, C.; Appenroth, K.J.; Lam, E. Assessment, validation and deployment strategy of a two barcode protocol for facile genotyping of duckweed species. *Plant Biol.* **2015**, *17*, 42–49. [CrossRef]
28. Chu, P.; Wilson, G.M.; Michael, T.P.; Vaiciunas, J.; Honig, J.; Lam, E. Sequence-guided approach to genotyping plant clones and species using polymorphic NB-ARC-related genes. *Plant Mol. Biol.* **2018**, *98*, 219–231. [CrossRef]
29. Bog, M.; Xu, S.; Himmelbach, A.; Brandt, R.; Wagner, F.; Appenroth, K.J.; Sree, K.S. Genotyping-by-Sequencing for species delimitation in the section *Uninerves* Hegelm. (genus *Lemna* L.). In *Compendium of Plant Genomes: The Duckweed Genomes*, 1st ed.; Cao, X.H., Fourounjian, P., Wang, W., Eds.; Springer: Berlin/Heidelberg, Germany, 2020; ISBN 455 978-3-030-11045-1.
30. Xu, N.; Hu, F.; Wu, J.; Zhang, W.; Wang, M.; Zhu, M.; Ke, J. Characterization of 19 polymorphic SSR markers in *Spirodela polyrhiza* (Lemnaceae) and cross-amplification in *Lemna perpusilla*. *Appl. Plant Sci.* **2018**, *6*, e1153. [CrossRef]
31. Fu, L.; Ding, Z.; Kumpeangkeaw, A.; Tan, D.; Han, B.; Sun, X.; Zhang, J. De novo assembly, transcriptome characterization, and simple sequence repeat marker development in duckweed *Lemna gibba*. *Physiol. Mol. Biol. Plants* **2020**, *26*, 133–142. [CrossRef]
32. Cole, C.T.; Voskuil, M.I. Population genetic structure in duckweed (*Lemna minor*, Lemnaceae). *Can. J. Bot.* **1996**, *74*, 220–230. [CrossRef]

33. Xu, S.; Stapley, J.; Gablenz, S.; Boyer, J.; Appenroth, K.J.; Sree, K.S.; Gershenzon, J.; Widmer, A.; Huber, M. Low mutation rate determines low genetic variation in the greater duckweed. *Nat. Commun.* **2019**, *10*, 1857. [CrossRef] [PubMed]
34. Hoang, P.N.T.; Michael, T.P.; Gilbert, S.; Chu, P.; Motley, T.S.; Appenroth, K.J.; Schubert, I.; Lam, E. Generating a high-confidence reference genome map of the Greater Duckweed by integration of cytogenomic, optical mapping and Oxford Nanopore technologies. *Plant J.* **2018**, *96*, 670–684. [CrossRef] [PubMed]
35. Michael, T.P.; Bryant, D.; Gutierrez, R.; Borisjuk, N.; Chu, P.; Zhang, H.; Xia, J.; Zhou, J.; Peng, H.; El Baidouri, M.; et al. Comprehensive definition of genome features in *Spirodela polyrhiza* by high-depth physical mapping and short-read DNA sequencing strategies. *Plant J.* **2017**, *89*, 617–635. [CrossRef] [PubMed]
36. Wang, W.; Kerstetter, R.A.; Michael, T.P. Evolution of Genome Size in Duckweeds (Lemnaceae). *J. Bot.* **2011**, *2011*, 570319. [CrossRef]
37. An, D.; Li, C.S.; Zhou, Y.; Wu, Y.R.; Wang, W.Q. Genomes and transcriptomes of duckweeds. *Front. Chem.* **2018**, *6*, 230. [CrossRef] [PubMed]
38. Braglia, L.; Lauria, M.; Appenroth, K.J.; Bog, M.; Breviario, D.; Grasso, A.; Gavazzi, F.; Morello, L. Duckweed species genotyping and interspecific hybrid discovery by tubulin-based polymorphism fingerprinting. *Front. Plant Sci.* **2021**, *12*, 625670. [CrossRef]
39. Braglia, L.; Breviario, D.; Giani, S.; Gavazzi, F.; De Gregori, J.; Morello, L. New insights into interspecific hybridization in *Lemna* L. Sect. *Lemna* (Lemnaceae Martinov). *Plants* **2021**, *12*, 2767. [CrossRef]
40. Tang, J.; Zhang, F.; Cui, W.; Ma, J. Genetic structure of duckweed population of *Spirodela*, *Landoltia* and *Lemna* from Lake Tai, China. *Planta* **2014**, *239*, 1299–1307. [CrossRef]
41. El-Kholy, A.S.; Youssef, M.S.; Eid, E.M. Genetic diversity of *Lemna gibba* L. and *Lemna minor* L. populations in Nile delta based in biochemical and ISSR markers. *Egypt. J. Exp. Biol. Bot.* **2015**, *11*, 11–19.
42. Paolaci, S.; Bog, M.; Lautenschlager, U.; Bonfield, R.; Appenroth, K.J.; Oberprieler, C.; Jansen, M.A.K. Clonal diversity amongst island populations of alien, invasive *Lemna minuta* Kunth. *Biol. Invasions* **2021**, *23*, 2649–2660. [CrossRef]
43. Coughlan, N.E.; Kelly, T.C.; Jansen, M.A.K. Mallard duck (*Anas platyrhynchos*)-mediated dispersal of Lemnaceae: A contributing factor in the spread of invasive *Lemna minuta*? *Plant Biol.* **2015**, *17* (Suppl. S1), 108–114. [CrossRef] [PubMed]
44. Coughlan, N.E.; Kelly, T.C.; Jansen, M.A.K. "Step by step": High frequency short-distance epizoochorous dispersal of aquatic macrophytes. *Biol. Invasions* **2017**, *19*, 625–634. [CrossRef]
45. Coughlan, N.E.; Kelly, T.C.; Davenport, J.; Jansen, M.A.K. Humid microclimates within the plumage of mallard ducks (*Anas platyrhynchos*) can potentially facilitate long distance dispersal of propagules. *Acta Oecol.* **2015**, *65–66*, 17–23. [CrossRef]
46. Hicks, L.E. The Lemnaceae of Indiana. *Am. Midl. Nat.* **1937**, *18*, 774–789. [CrossRef]
47. Appenroth, K.J. Sterilization of Duckweed. *Duckweed Forum* **2015**, *3*, 90–91. Available online: www.ruduckweed.org/ (accessed on 12 February 2022).
48. Appenroth, K.-J.; Teller, S.; Horn, M. Photophysiology of turion formation and germination in *Spirodela polyrhiza*. *Biol. Plant.* **1996**, *38*, 95–106. [CrossRef]
49. Doyle, J.J.; Doyle, J.L. A rapid DNA isolation procedure for small quantities of fresh leaf tissue. *Phytochem. Bull.* **1987**, *19*, 11–15.
50. Vos, P.; Hogers, R.; Bleeker, M.; Reijans, M.; Van de Lee, T.; Hornes, M.; Frijters, A.; Pot, J.; Peleman, J.; Kuiper, M.; et al. AFLP: A new technique for DNA fingerprinting. *Nucleic Acids Res.* **1995**, *23*, 4407–4414. [CrossRef]
51. Baumbach, H.; Hellwig, F.H. Genetic differentiation of metallicolous and non-metallicolous *Armeria maritima* (MILL.) WILLD. Taxa (Plumbaginaceae) in Central Europe. *Plant Syst. Evol.* **2007**, *269*, 245–258. [CrossRef]
52. Schlüter, P.M.; Harris, S.A. Analysis of multilocus fingerprinting data sets containing missing data. *Mol. Ecol. Notes* **2006**, *6*, 569–572. [CrossRef]
53. R Core Team. R: A Language and Environment for Statistical Computing. 2015. Available online: https://www.R-project.org (accessed on 12 February 2022).

plants

Article

Biodiversity of Duckweed (Lemnaceae) in Water Reservoirs of Ukraine and China Assessed by Chloroplast DNA Barcoding

Guimin Chen [1,†], Anton Stepanenko [1,2,†], Olha Lakhneko [2], Yuzhen Zhou [1], Olena Kishchenko [1,2,‡], Anton Peterson [1,2,‡], Dandan Cui [1,§], Haotian Zhu [1,‖], Jianming Xu [1], Bogdan Morgun [2], Dmitri Gudkov [3], Nikolai Friesen [4] and Mykola Borysyuk [1,*]

[1] Jiangsu Key Laboratory for Eco-Agricultural Biotechnology around Hongze Lake, School of Life Sciences, Huaiyin Normal University, Huai'an 223300, China; cgm@hytc.edu.cn (G.C.); stepanenko@hytc.edu.cn (A.S.); zyz@hytc.edu.cn (Y.Z.); o_kishchenko@hotmail.com (O.K.); peterson@hytc.edu.cn (A.P.); cuidandan.meizi@foxmail.com (D.C.); haotianzhu567@163.com (H.Z.); xjm@hytc.edu.cn (J.X.)

[2] Institute of Cell Biology and Genetic Engineering, National Academy of Sciences of Ukraine, 03143 Kyiv, Ukraine; olakhneko@icbge.org.ua (O.L.); bmorgun@icbge.org.ua (B.M.)

[3] Institute of Hydrobiology, National Academy of Sciences of Ukraine, 04210 Kyiv, Ukraine; digudkov@gmail.com

[4] Botanical Garden of the University of Osnabrück, 49074 Osnabrück, Germany; nfriesen@uni-osnabrueck.de

* Correspondence: nborisjuk@hytc.edu.cn; Tel.: +86-150-5264-7255

† These authors contributed equally to this work.

‡ Current Address: Leibniz Institute of Plant Genetics and Crop Plant Research (IPK), 06466 Gatersleben, Germany.

§ Current Address: Key Laboratory of Agricultural Genetically Modified Organisms Traceability of the Ministry of Agriculture and Rural Affairs, Oil Crops Research Institute, Chinese Academy of Agricultural Sciences, Wuhan 430062, China.

‖ Current Address: Jiangsu Provincial Key Laboratory of Crop Genetics and Physiology, Yangzhou University, Yangzhou 225009, China.

Citation: Chen, G.; Stepanenko, A.; Lakhneko, O.; Zhou, Y.; Kishchenko, O.; Peterson, A.; Cui, D.; Zhu, H.; Xu, J.; Morgun, B.; et al. Biodiversity of Duckweed (Lemnaceae) in Water Reservoirs of Ukraine and China Assessed by Chloroplast DNA Barcoding. *Plants* **2022**, *11*, 1468. https://doi.org/10.3390/plants11111468

Academic Editors: Viktor Oláh, Klaus-Jürgen Appenroth and K. Sowjanya Sree

Received: 5 April 2022
Accepted: 26 May 2022
Published: 30 May 2022

Abstract: Monitoring and characterizing species biodiversity is essential for germplasm preservation, academic studies, and various practical applications. Duckweeds represent a group of tiny aquatic plants that include 36 species divided into 5 genera within the Lemnaceae family. They are an important part of aquatic ecosystems worldwide, often covering large portions of the water reservoirs they inhabit, and have many potential applications, including in bioremediation, biofuels, and biomanufacturing. Here, we evaluated the biodiversity of duckweeds in Ukraine and Eastern China by characterizing specimens using the two-barcode protocol with the chloroplast *atpH–atpF* and *psbK–psbI* spacer sequences. In total, 69 Chinese and Ukrainian duckweed specimens were sequenced. The sequences were compared against sequences in the NCBI database using BLAST. We identified six species from China (*Spirodela polyrhiza*, *Landoltia punctata*, *Lemna aequinoctialis*, *Lemna minor*, *Lemna turionifera*, and *Wolffia globosa*) and six from Ukraine (*S. polyrhiza*, *Lemna gibba*, *Lemna minor*, *Lemna trisulca*, *Lemna turionifera*, and *Wolffia arrhiza*). The most common duckweed species in the samples from Ukraine were *Le. minor* and *S. polyrhiza*, accounting for 17 and 15 out of 40 specimens, respectively. The most common duckweed species in the samples from China was *S. polyrhiza*, accounting for 15 out of 29 specimens. *La. punctata* and *Le. aequinoctialis* were also common in China, accounting for five and four specimens, respectively. According to both *atpH–atpF* and *psbK–psbI* barcode analyses, the species identified as *Le. aequinoctialis* does not form a uniform taxon similar to other duckweed species, and therefore the phylogenetic status of this species requires further clarification. By monitoring duckweeds using chloroplast DNA sequencing, we not only precisely identified local species and ecotypes, but also provided background for further exploration of native varieties with diverse genetic backgrounds. These data could be useful for future conservation, breeding, and biotechnological applications.

Keywords: aquatic plants; duckweed; biodiversity; barcoding; chloroplast DNA; molecular evolution

1. Introduction

Monitoring and characterizing species biodiversity is essential for germplasm preservation, academic studies, and various practical applications [1]. Duckweed is an important element in aquatic ecosystems worldwide, often covering large portions of the still water surface they inhabit. This group of tiny aquatic plants is composed of 36 species divided into five genera in the Lemnaceae family [2,3], an early diverging family of monocotyledonous plants [4].

Duckweeds are a diverse group and provide many opportunities for genetic, physiological, biochemical, and practical research [5,6]. After being important model plants in the 1950s–1970s, duckweeds became popular again in the 2010s, primarily due to their potential as a biofuel feedstock because of their high biomass growth rate, low lignin content, and high starch content [7,8]. In addition to starch, duckweed biomass is rich in proteins, carbohydrates, crude fiber, minerals, and lipids. This biomass composition makes duckweed a potential food source for animals, fish, and humans [9]. Duckweeds have also been studied for their use in wastewater treatment [10,11], biosensing [12,13], and phytoremediation of water reservoirs contaminated with various toxic chemicals [14,15]. Several duckweed species have been genetically engineered with the eventual aim of producing pharmaceutical proteins such as antigens, peptide hormones, and antibodies [16–18].

The multiple potential applications of duckweed have led to an increasing interest in duckweed genetics, molecular evolution, and diversity. The genome size of duckweeds varies by 14-fold, from 160 Mb in great duckweed (*Spirodela polyrhiza* (L.) Schleid) to $\approx$2.2 Gb in *Wolffia arrhiza* (L.) Horkel ex Wimm [19]. Researchers have sequenced the whole genomes of two representative ecotypes of *Spirodela polyrhiza* [20,21] and *Spirodela intermedia* W. Koch [22], as well as genomes of *Lemna minor* L. [23] and *Wolffia australiana* (Benth.) Hartog and Plas [24]. Moreover, there are ongoing whole-genome sequencing projects for *Landoltia punctata* (G.Mey.) Les and D.J.Crawford and *Lemna gibba* L. [25]. Four biannual international conferences specifically on duckweed have taken place since 2012, and there has been a tremendous surge in diverse academic and applied studies of various aspects of duckweed biology [5,26–29].

Duckweeds include the smallest known flowering plants and often have reduced morphology, making some species difficult to identify using traditional botanical approaches, not even mentioning ecotypes [30]. Recently, molecular methods have been developed to aid in identifying duckweed species and distinguishing ecotypes [31]. The Consortium for the Barcode of Life (CBOL) [32] recommends seven chloroplast DNA (cpDNA) barcodes to identify land plants simply and reliably [33]. The recommended barcodes have been adapted for identification of duckweeds supported by the constantly growing number of reference sequences deposited in DNA sequence databases. Most of these sequences came from studies of the samples deposited at the world's largest live duckweed depository at the Rutgers University's Duckweed Stock Cooperative (RDSC) in New Brunswick, NJ, USA (www.ruduckweed.org, accessed on 31 March 2022), with the rest coming from smaller collections or random samplings. Additionally, 12 chloroplast genome sequences, representing 7 duckweed species [34–37], have been sequenced and deposited in the NCBI database.

In many parts of the world, including Ukraine and China, farmers practicing intensive agriculture use substantial amounts of fertilizers. Fertilizer that is not fully utilized by crops eventually ends up in water reservoirs surrounding agricultural fields. Due to its ability to quickly assimilate nitrogen, phosphorous, and other nutrients, duckweed can rapidly grow, producing on average of 13–38 dry tons of biomass/hectare/year [38], converting agricultural and municipal wastewater into clean water and a high-value biomass ideal for animal/fish feed and numerous other applications [5]. In both Ukraine and China, duckweed is the dominant vegetation in ponds and lakes. In contrast to China, where different aspects of duckweed research are relatively well developed (for example, the RDSC collection hosts more than 200 duckweed ecotypes originated from China), there is rather scarce information on duckweed in Ukraine and Eastern Europe in general.

In this work, we evaluated the biodiversity of duckweed in different regions of Eastern China and Ukraine on the basis of the two-barcode protocol for sequencing the chloroplast *atpH–atpF* (*ATP*) and *psbK–psbI* (*PSB*) spacers in the collected duckweed specimens. With this approach, we precisely identified local species and ecotypes. Our results provide a foundation for further exploring native varieties with diverse genetic backgrounds and for duckweed breeding and biotechnological applications.

2. Results

2.1. Genotyping of the Duckweed Specimens

We collected 69 duckweed specimens from across Ukraine and southeastern China. Several locations contained more than one species, as illustrated in Figure 1. From these specimens, plus RDSC clone 8656, we obtained 140 representative sequences, which we deposited in GenBank (Table S1). Because the PCR primers for the *atpH–atpF* spacer are located further into the corresponding *atpH* and *atpF* gene sequences compared to *psbK–psbI* spacer, the *ATP* barcodes contain longer portions of the coding sequences compared to *PSB*. The high reliability of the represented barcodes is based on sequences generated using both forward and reverse primers following careful nucleotide validation.

Figure 1. Location of duckweed sampling sites in Ukraine and China.Dot colors correspond to different species: red, *S. polyrhiza*; green, *La. punctata*; light blue, *Le. minor*; dark blue, *Le. aequinoctialis* Welw.; pink, *Le. gibba*; black, *Le. trisulca*; grey, *Le. turionifera*; yellow, *W. arrhiza*; and brown, *W. globosa*. The image at the bottom was taken at a pond in Huai'an, China. It illustrates a community of three different species, *S. polyrhiza* (S), *Le. aequinoctialis* (L), and *W. globosa* (W), growing together. The exact GPS coordinates of the sites are listed in Table S2. Geographic maps were taken from the websites located at https://www.d-maps.com/m/asia/china/chine/chine58.gif (accessed on 11 February 2022) and https://www.d-maps.com/m/europa/ukraine/ukraine50.gif (accessed on 11 February 2022).

2.1.1. Great Duckweed, *Spirodela polyrhiza*

Barcoding showed that 15 of the 39 specimens collected in Ukraine and 15 of the 30 specimens collected in China were *S. polyrhiza*. The *S. polyrhiza ATP* sequences from our study and the reference sequences from the whole chloroplast genome of U.S. *S. polyrhiza* ecotype 7498 [35,37] had high sequence conservation. The main detected sequence variations were T↔C transitions at defined positions along the sequence and a couple of T↔A transversions, with no biases related to the specimen's geographic origins (Figure S1A). The *PSB* sequences showed similar low sequence diversity but with different sequence polymorphisms, including single-nucleotide polymorphisms (SNPs), insertion/deletions (InDels), and more random nucleotide transitions/transversions (Figure S1B) compared to the *ATP* sequences. Our *PSB* sequences also contained single-nucleotide insertions of additional A nucleotides at positions 25 and 354 and an additional T at position 402, compared to the reference sequence of *S. polyrhiza* ecotype 7498.

2.1.2. Dotted Duckweed, *Landoltia punctata*

By chloroplast DNA barcoding, we identified five *La. punctata* ecotypes. Two ecotypes were collected near the Hongze lake (Jiangsu province, China) and kept in our *in vitro* collections (NB0014 and NB0022). Ecotype Ya3 was collected from Yanling and Gu1 from Guilin; the ecotype RDSC EL019 collected earlier in Kuhming was obtained from the RDSC (New Brunswick, USA). As *La. punctata* inhabits tropical areas [29], we did not find it in Ukraine. Sequence alignments showed a high stability of both the *ATP* and *PSB* sequences in *La. punctata*. The *ATP* sequences of the six ecotypes only shared two A↔G transitions, both in Gu1, and a single nucleotide deletion (Figure S2A); the six *PSB* sequences differed by a single G→T transversion in Ya3 (Figure S2B).

2.1.3. Common Duckweed, *Lemna minor*

Lemna minor was the most represented duckweed species in the Ukraine specimens. We identified 17 of the 39 specimens collected in Ukraine as *Le. minor*. We also identified one specimen from China, Ya2 collected in Yangling, as *Le. minor*. The *ATP* and *PSB* sequences of the *Le. minor* ecotypes had very low sequence divergence, with near 100% similarity to the corresponding 29 and 31 GenBank *ATP* and *PSB* sequences representing *Le. minor* ecotypes, respectively. We compared the sequences of our specimens with the corresponding sequences of a Russian ecotype for which the chloroplast genome was sequenced [34] and found only three nucleotide substitutions in the *PSB* sequences (Figure S3A) and five G→A transition and a single T→G transversion in the *ATP* sequences (Figure S3B).

2.1.4. Star Duckweed, *Lemna trisulca*

We identified four duckweed specimens from Ukraine as *Le. trisulca*. They had a 100% similarity (Figure S4A) to the *ATP* sequences previously reported for ecotypes from the USA and Canada [39]. However, alignment of *PSB* sequences clearly distinguished Ukrainian ecotypes from the North American ones on the basis of the duplication of an AT-rich 23-bp long DNA sequence in the North American ecotypes (Figure S4B). Moreover, alignment of a few *Le. trisulca ATP* and *PSB* sequences [39] revealed distinct variants in strain UTCC 399 of unknown origin, characterized by short 4–6-nucleotide insertions/deletions as compared to the Ukrainian and North American ecotypes.

2.1.5. Turion Duckweed, *Lemna turionifera*

We identified one specimen from Ukraine (from the southeast) and two collected from China (from near Hongze lake) as *Le. turionifera*. The *ATP* and *PSB* sequences of these specimens showed no sequence variation when aligned with *Le. turionifera* sequences from Canada, the Czech Republic, and Lake Tai in China [40]. The only variation we found was a single nucleotide deletion in the *ATP* sequence of the accession from the Czech Republic (Figure S5).

2.1.6. Swollen or Fat Duckweed, *Lemna gibba*

We identified one specimen from Ukraine (DW102) as *Le. gibba*. The *PSB* sequence showed homology with the corresponding sequences of four *Le. gibba* strains from the USA, Italy, Ethiopia, and Japan [39], as well as strain RDSC 5504, which originated from Shanghai, China. The *ATP* sequence of DW102 differed in two positions, a single insertion of A (which was also in the sequence of the Shanghai strain) and a unique C→T transition (Figure S6).

2.1.7. Lesser Duckweed, *Lemna aequinoctialis*

Lemna aequinoctialis had the highest variation in *ATP* and *PSB* sequences among the species analyzed in this study. We collected four *Le. aequinoctialis* strains: two from Huai'an city (I2 and NB0017), one from Shanghai (NB0007), and one from Fuzhou (Fu94). We divided the strains into two groups on the basis of their barcode sequences (Figure S7). Strains NB0017 and NB0007 differed from I2 and Fu94 by two tandem duplications of 21 and 5 bp, three specific SNPs in their *PSB* sequences (positions 71, 117, and 162), and five SNPs (positions 12, 189, 359, 364, and 395) in their *ATP* sequences. These two groups aligned with three American *Le. aequinoctialis* strains [39], with NB0017 and NB0007 having similar sequences to those of strains 6612 (Centerville, CA, USA) and 8656 (Argentina), whereas I2 and Fu94 aligned with strains 6746 (Plainsburg, CA, USA) and 7126 (Texas, USA) (Figure S7). All four Chinese strains had three specific SNPs in their *PSB* sequences (positions 357, 401, 444) compared to the three American strains.

To examine polymorphism in *Le. aequinoctialis* barcodes in more detail, we analyzed the phylogeny of the 21 *ATP* sequences available in NCBI GenBank together with our five sequences. We included two *La. punctata* accessions as an outgroup. In total, there were 513 characters, of which 453 were constant. Of the variable characters, 18 were parsimony uninformative and 42 were parsimony informative. Parsimony and Bayesian analyses yielded the same topology but with lower bootstrap percentages than posterior probabilities. A heuristic search found most-parsimonious trees that were 70 steps long (consistency index 0.9143, retention index 0.9259). The resultant dendrogram from this analysis is shown in Figure 2. All sequences were divided into four subclades: two with very strong support and two with little support. Our accessions are subordinate to the two strongly supported clades.

2.1.8. Least Duckweed, *Wolffia arrhiza*, and Watermeal Duckweed, *Wolffia globosa*

We identified two Ukrainian specimens as *W. arrhiza* and two Chinese specimens as *W. globosa*. Both specimens from Ukraine, DW32 and DW35, had high *ATP* sequence similarity with the homologous sequence of African and Italian specimens, but a high level of nucleotide mismatches with the sequence from a *W. arrhiza* specimens from Brazil (Figure S8A). There was 100% similarity between the *PSB* sequences of the two Ukrainian specimens, with the sequence blasting revealing a single hit in GenBank (Figure S8B).

There were more hits for *W. globosa* compared to *W. arrhiza*; 31 for *ATP* and 14 for *PSB*. The Chinese strains C2 and NB0015 (characterized in this study), together with strains DW2101-4 (Acc. KJ630544.1; Hainan) and LC49 (Lake Chao) [41], were more closely grouped with a specimen from the USA [39] than with those from other Asian countries India, Japan, and Thailand (Figure S9). This grouping was based on nucleotide substitutions at positions 248, 253, and 383 in the *ATP* sequence (Figure S9A) and, even more profoundly, by multiple SNPs and three deletions/insertions of short nucleotide sequences in the *PSB* sequence (Figure S9B).

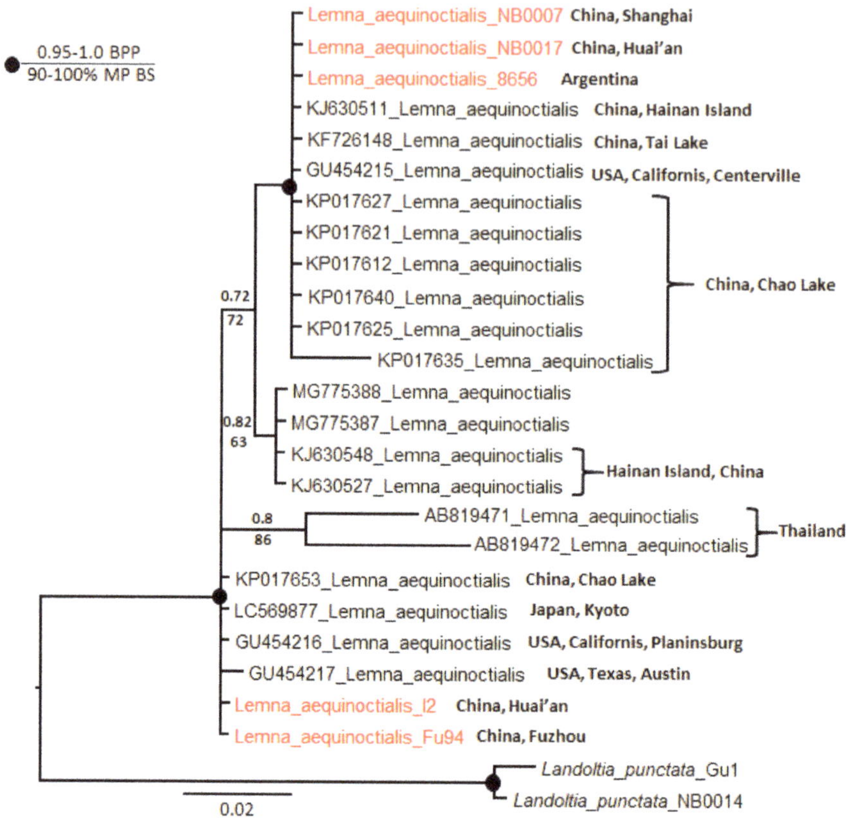

Figure 2. Bayesian consensus tree based on analysis of *atpH–atpF* intergenic spacer sequences of *Lemna aequinoctialis* and *Landoltia punctata* as an outgroup. Bayesian posterior probabilities (BPP) and maximum parsimony bootstrap values (MP BS) are shown above and below the branches, respectively. Strongly supported clades (MP BS > 90% and BPP > 0.95) are indicated with black circles at the branchpoints. For the origin of specimens, see Table S2.

2.2. Phylogenetic Analysis

Phylogenetic analysis of our 69 duckweed specimens using *ATP* and *PSB* sequences separately showed no conclusive results. Therefore, we performed a combined *ATP* and *PSB* analysis of 70 taxa, including *Pistia stratiotes* [42] as an outgroup species. The combined data matrix included 1197 characters divided in two partitions: 1–560 for *ATP* and 561–1197 for *PSB*, of which 751 were constant, 103 were parsimony uninformative, and 343 were parsimony informative.

Parsimony and Bayesian analyses yielded the same topology but with lower bootstrap percentages than posterior probabilities. The heuristic search found most-parsimonious trees that were 635 steps long (consistency index 0.8551, retention index 0.9691). The resultant dendrogram from this analysis is shown in Figure 3. All species studied built monophyletic and mostly not polymorphic clades, with few exceptions. The *S. polyrhiza* clade had several small subclades with weak support, and the clade with *L. aequinoctialis* accessions was divided into two subclades with strong support. Overall, this phylogeny of Lemnaceae is congruent with previous studies [3,43–45].

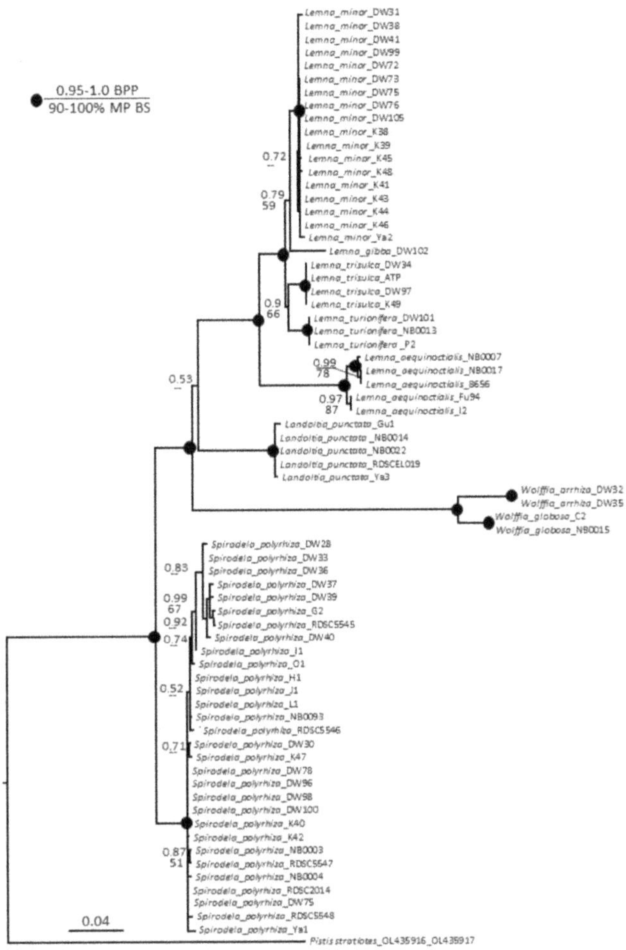

Figure 3. Bayesian consensus tree based on analysis of the combined chloroplast DNA dataset (*atpH–atpF* and *psbK–psbI* intergenic spacers) of Lemnaceae taxa and *Pistia stratiotes* as an outgroup. Bayesian posterior probabilities (BPP, top value) and maximum parsimony bootstrap values (MP BS, bottom value) are shown at the branches. Strongly supported clades (MP BS > 90% and BPP > 0.95) are indicated with black circles at the branchpoints.

3. Discussion

As the smallest known flowering plants, duckweeds have a reduced morphology, which makes them difficult to identify using traditional botanical approaches [30]. Therefore, molecular approaches [2] offer valuable alternatives for species monitoring of this ancient group of plants. Here, to provide additional data supporting molecular approaches for duckweed identification, as well as to examine duckweed diversity, we collected 39 duckweed specimens from Ukraine and 30 from China. Using DNA barcoding, we identified six duckweed species among the Ukrainian specimens (*S. polyrhiza, Le. gibba, Le. minor, Le. trisulca, Le. turionifera,* and *W. arrhiza*) and six species from China (*S. polyrhiza, La. punctata, Le. aequinoctialis, Le. minor, Le. turionifera,* and *W. globosa*). These species represent four out of the five genera in the Lemnaceae family (Figure 4). The only genus not represented was *Wolffiella*, which only occurs in the Americas and Africa [29].

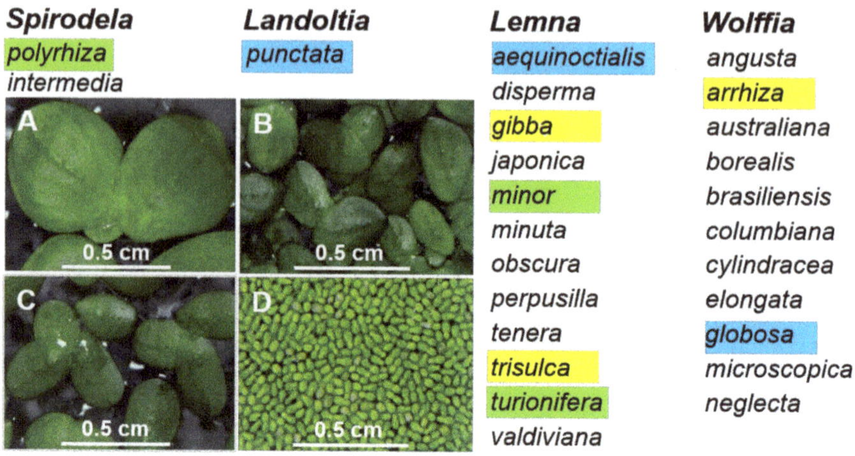

Figure 4. Four genera of the Lemnaceae plant family were represented by duckweed species in Ukraine and China. Species that were found in Ukraine and China are highlighted green, those found only in Ukraine are highlighted yellow, and those found only in China are highlighted blue. A, B, C, and D are representative images of *Spirodela polyrhiza, Landoltia punctata, Lemna minor,* and *Wolffia globose,* respectively, from the in vitro collection of Huaiyin Normal University.

The distribution of duckweed species in this study generally matched previously identified duckweed distributions, with *S. polyrhiza*, *Le. Minor*, and *W. arrhiza* being the most common species in Europe and *S. polyrhiza*, *La. Punctate*, *Le. aequinoctialis*, and *W. globoza* the usual species in China [29,46]. However, there is little information on duckweed biodiversity in Eastern Europe in general and in Ukraine in particular [47,48]. To the best of our knowledge, this study is the first chloroplast-barcoding-based record of duckweed biodiversity in an Eastern European country. Compared to Ukraine, duckweed biodiversity in China is relatively well investigated [31,40,41], and there are numerous ecotypes from China deposited in the RDSC world collection in the USA (New Brunswick, NJ) and in different institutions in China [49,50].

Our molecular identification of randomly sampled specimens agrees that the great duckweed, *S. polyrhiza*, is the most cosmopolitan of the 36 duckweed species recognized worldwide [29]. It was the most dominant species in East China and in Ukraine. Although the phylogenetic analysis demonstrated a certain degree of clustering of the *S. polyrhiza* specimens on the basis of the limited barcode sequence variations (Figure 3), it did not show any clear links to the geographic origin of the specimens. Generally, the *ATP* and *PSB* sequences of our specimens had almost 100% similarity with the corresponding sequences of strain 7498 [37], with a low sequence variability between the specimens. Similarly, *S. polyrhiza* nuclear genomes sequenced from 63 specimens collected worldwide had high sequence conservation [51].

La. punctata is the only representative of the genus *Landoltia*. It is considered to be closely related to *Spirodela* [36] but is not as widely distributed as *S. polyrhiza*. It mostly inhabits tropical and subtropical areas [29]. Therefore, it is not surprising that we collected *La. punctata* specimens in China but none in Ukraine. Genetic analysis of the six Chinese *La. punctata* specimens revealed few nucleotide substitutions. This stability of ATP and PSB sequences was also observed among specimens from different geographic origins, including India, Africa, America, and Australia (Figure S2).

We identified five *Lemna* species among the specimens collected from Ukraine and China. The common duckweed, *Le. minor*, was the most predominant duckweed species in Ukraine, represented by 17 out of 40 specimens, closely followed by *S. polyrhiza* (15 specimens) (Figure 1). However, we only identified one specimen from China as *Le. minor*, a specimen that was collected in north-central China. *Lemna turionifera* was a minor species in both China and Ukraine [40,45]. We identified *Le. gibba* and *Le. trisulca* only in Ukraine, and *Le. aequinoctialis* only in China (Figure 1).

Three *Lemna* species, *Le. minor, Le. gibba*, and *Le. turionifera*, have stable species-specific *ATP* and *PSB* sequences as reflected in the phylogenetic dendrogram (Figure 3), with very few variations compared to their counterparts from other parts of the world available in GenBank. The three *Le. trisulca* specimens from Ukraine had perfect *ATP* sequence similarity; however, there were clear differences in *PSB* sequences compared (Figure S4) with specimens from the USA and Canada [39]. The differences were due to a 23 bp duplication.

The most intriguing results from this study were on the phylogeny of *Le. aequinoctialis*. On the basis of the *ATP* and *PSB* sequences, we grouped the four *Le. aequinoctialis* specimens (collected in Fuzhou, Shanghai, and Huai'an) into two clades (Figure S7). Sequences of these clades were aligned with the *ATP* sequences of *Le. aequinoctialis* strain DW0101-3 (Hainan, China; Acc. KJ630511.1) and strain LC42 (Lake Chao, China) [41]. We constructed a phylogeny based on all *Le. aequinoctialis ATP* sequences in GenBank, and the resulting tree was complex (Figure 2). On the basis of these results, we suggest that the status of this species needs further careful examination.

We identified two *Wolffia* species in our study: *W. arrhiza*, a common species in Europe and Africa, and *W. globosa*, which inhabits Southeast Asia [29], were identified in Ukraine and China, respectively. The *W. arrhiza* sequences showed high sequence similarity with other *W. arrhiza* sequences of specimens from Europe and Africa (Figure S8). The *W. globosa* sequences had several characteristic SNPs, both in the ATP and PSB sequences, compared with sequences of specimens from India, Japan, and Thailand (Figure S9); however, they had a higher similarity to sequences of specimens from the USA [39]. The USA specimen is likely a recent invasion in addition to the native *Wolffia columbiana* [29].

4. Materials and Methods

4.1. Plant Material

Duckweed specimens were collected from various still water reservoirs, lakes, and ponds across eastern China and Ukraine during 2016–2019. Prior to genotyping, most specimens from China were sterilized and kept under aseptic conditions on agar medium according to previously described methods [43]. These specimens were kept at the duckweed *in vitro* collection recently organized at Huaiyin Normal University, Hui'an, China [47], barcoded and used in our previous studies [11,18,52,53]. The Ukrainian duckweed specimens were collected from water reservoirs, sorted according to their morphological characteristics, and directly subjected to DNA extraction for further chloroplast DNA barcoding. All analyzed duckweed specimens and locations are listed in Table S2.

4.2. DNA Extraction, Fragment Amplification, Sequencing, and Alignment

Total DNA was extracted from plant tissue using a modified CTAB method [54]. The PCR amplifications were carried out as recommended by the CBOL Plant Working Group [32], described in [39], using primers 5′-TTAGCATTTGTTTGGCAAG and 5′-AAAGTTTGAGAGTAAGCAT for the *psbK–psbI* intergenic spacer and 5′-ACTCGCACACA CTCCCTTTCC and 5′-GCTTTTATGGAAGCTTTAACAAT for the *atpH–atpF* intergenic spacer. Following amplification, the DNA fragments were sent to Sangon Biotech (Shanghai, China) for purification and sequencing. The raw sequences were preliminarily optimized using the 'Online Analysis Tools' package (http://molbiol-tools.ca, accessed on 25 March 2022), in particular, the program MAFFT, version 7 (https://mafft.cbrc.jp/alignment/server/ accessed on 25 March 2022). Multiple DNA sequence alignments were generated with ClustalW software [55], and the alignments were subsequently corrected manually in MEGA 5 [56].

For BLAST alignment analyses, a duckweed reference barcode set was compiled from *ATP* and *PSB* sequences that were generated in this study and those available from the

NCBI database as of January 2022. Queried sequences were trimmed to include only intergenic regions and used in BLASTN (version 2.2.26+) searches to identify homologies to other barcode sequences in the set. The number of top hits for each query are presented in Table S1.

4.3. Phylogenetic Analysis

Phylogenetic analysis was carried out on individual and combined *ATP* and *PSB* sequences using parsimony and Bayesian methods. *Pistia stratiotes*, from the Araceae family, was used as an outgroup for the Lemnaceae family. Parsimony analysis was performed with PAUP* 4.0b10 [57] using heuristic searches with tree bisection–reconnection and 100 random additional sequence replicates. Bootstrap support (BS) [58] was estimated with 100 bootstrap replicates, each with 100 random addition sequence searches. Bayesian analyses were implemented with MrBayes 3.1.23 [59]. Sequence evolution models were evaluated using the Akaike information criterion (AIC) with the aid of jModelTest2 v2.1.6 [60]. Two independent runs each of eight chains and 10 million generations, sampling every 1000 generations, were executed, and 25% of initial trees were discarded as burn-in. The remaining 15,000 trees were combined into a 50% majority-rule consensus tree.

5. Conclusions

Our survey of the duckweed species in Ukraine and China makes a solid contribution to monitoring the biodiversity of aquatic flora in these countries. In addition to precisely identifying six major species and their geographic distribution in each these countries by double chloroplast DNA barcoding, our data highlighted the need to re-examine the phylogenetic status of one of those species, *Lemna aequinoctialis*. The study added 138 new chloroplast *ATP* and *PSB* barcodes to the 1754 corresponding barcodes for these species available in the NCBI database as of March 2022 (Table S1). These new resources might fuel future research on plant molecular evolution, biodiversity conservation, breeding, and various biotechnological applications.

Supplementary Materials: The following supporting information can be downloaded at: https://www.mdpi.com/article/10.3390/plants11111468/s1, Figure S1: Nucleotide alignment of *atpH–atpF* and *psbK–psbI* spacers sequences of *Spirodela polyrhiza* collected in Ukraine and China (for ecotype details see supplemental Table S2). Matching residues are shown as dots. A. Nucleotide alignment of *atpH–atpF* barcodes including representative sequences of *S. polyrhiza* specimens available in GenBank: MN419335, USA; GU454201, Hong Kong; GU454202, India; GU454203, Malaysia; GU454204, USA; GU454205, Mexico; GU454206, Canada; GU454208, Europe. B. Nucleotide alignment of *psbK-psbI* barcodes including representative sequences of *S. polyrhiza* specimens available in GenBank: MN419335, USA; GU454297, Hong Kong; GU454298, India; GU454299, Malaysia; GU454300, USA; GU454301, Mexico; GU454302, Canada; GU454304, Europe. Figure S2: Nucleotide alignment of *atpH-atpF* and *psbK-psbI* spacers of *Landoltia punctata* (for ecotypes details see supplemental Table S2). Matching residues are shown as dots. A. Nucleotide alignment of *atpH-atpF* barcodes including the representative sequences of *La. punctata* from different parts of the world available in the GenBank: GU454209, South Africa; GU454211, India; GU454212, USA; GU454213, Australia. B. Nucleotide alignment of *psbK-psbI* barcodes including the representative sequences of *La. punctata* from different parts of the world available in the GenBank: GU454305, South Africa; GU454307, India; GU454308, USA; GU454309, Australia. Figure S3: Nucleotide alignment of *atpH-atpF* and *psbK-psbI* spacers of *Lemna minor* (for ecotypes details see supplemental Table S2). Matching residues are shown as dots. A. Nucleotide alignment of *atpH-atpF* barcodes including the representative sequences of *Le. minor* from different parts of the world available in the GenBank: DQ400350, Russia; GU454226, Turkey; GU454227, USA; GU454228, South Africa; GU454229, Japan; GU454230, Finland; GU454231, Germany. B. Nucleotide alignment of *psbK-psbI* barcodes including the representative sequences of *Le. minor* from different parts of the world available in the GenBank: DQ400350, Russia; GU454322, Turkey; GU454323, USA; GU454324, South Africa; GU454325, Japan; GU454326, Finland; GU454327, Germany. Figure S4: Nucleotide alignment of *atpH-atpF* and *psbK-psbI* spacers of *Lemna trisulca* (for ecotypes details see supplemental Table S2). Matching residues are shown as dots. A. Nucleotide alignment of *atpH-atpF* barcodes including the representative sequences of *Le. trisulca* from different parts of the world available in the GenBank: GU454236, Canada; GU454237, USA; GU454237, UTCC 399. B. Nucleotide alignment of *psbK-psbI* barcodes including the representative sequences of L. trisulca from different parts of the world available in the GenBank: GU454332, Canada; GU454333, USA; GU454334, UTCC 399. Figure S5: Nucleotide alignment of *atpH-atpF* and *psbK-psbI* spacers of *Lemna turionifera* (for ecotypes details see supplemental Table S2). Matching residues are shown as dots. A. Nucleotide alignment of *atpH-atpF* barcodes including the representative sequences of *Le. turionifera* from different parts of the world available in the GenBank: KF726146, China; MG000422, Canada; GU454240, Czech. B. Nucleotide alignment of *psbK-psbI* barcodes including the representative sequences of *Le. turionifera* from different parts of the world available in the GenBank: MG000496, Canada; GU454335, China; GU454336, Czech. Figure S6: Nucleotide alignment of *atpH-atpF* and *psbK-psbI* spacers of *Lemna gibba*, ecotype DW102 (Supplemental Table S2). Matching residues are shown as dots. A. Nucleotide alignment of the DW102 *atpH-atpF* barcode with the representative sequences of *Le. gibba* from different parts of the world available in the GenBank: KX212889, China; GU454219, USA; GU454220, Italy; GU454221, Ethiopia; GU454222, Japan. B. Nucleotide alignment of the DW102 *psbK-psbI* barcode with the representative sequences of *Le. gibba* from different parts of the world available in the GenBank: GU454315, USA; GU454316, Italy; GU454317, Ethiopia; GU454318, Japan. Figure S7: Nucleotide alignment of *atpH-atpF* and *psbK-psbI* spacers of *Lemna aequinoctialis* (for ecotypes details see supplemental Table S2). Matching residues are shown as dots. A. Nucleotide alignment of *atpH-atpF* barcodes including the representative sequences of *Le. aequinoctialis* from different parts of the world available in the GenBank: GU454215, USA, California, Centerville; GU454216, USA, California, Plainsburg; GU454217, USA, Texas, Austin. B. Nucleotide alignment of *psbK-psbI* barcodes including the representative sequences of *Le. aequinoctialis* from different parts of the world available in the GenBank: GU454311, USA, California, Centerville; GU454312, USA, California, Plainsburg; GU454313, USA, Texas, Austin. Figure S8: Nucleotide alignment of *atpH-atpF* and *psbK-psbI* spacers of *Wolffia arrhiza* (for ecotypes details see supplemental Table S2). Matching residues are shown as dots. A. Nucleotide alignment of *atpH-atpF* barcodes including the representative sequences of *W. arrhiza* from different parts of the world available in the GenBank: MG812317, Kenya; MG775412, Italy; MG775413, Brazil. B. Nucleotide alignment of *psbK-psbI* barcodes including the representative sequence of *W. arrhiza* available in the GenBank: GU454364, Hungary. Figure S9: Nucleotide alignment of *atpH-atpF* and *psbK-psbI* spacers of *Wolffia globosa* (for ecotypes details see supplemental Table S2). Matching residues are shown as dots. A. Nucleotide alignment of *atpH-atpF* barcodes including the representative sequences of *W. globosa* from different parts of the world available in the GenBank: GU454281, USA; KJ630544, China; KP017660, China; MG812325, India; MN881100, Japan. B. Nucleotide alignment of *psbK-psbI* barcodes including the representative sequences of *W. globosa* from different parts of the world available in the GenBank: GU454378, USA; GU454379, Thailand; MG812330, India; MN881100, Japan. References [61–71] are cited in the supplementary materials.

Author Contributions: Conceptualization, B.M., Y.Z. and M.B.; methodology, G.C., A.S., N.F. and M.B.; software and validation, A.S., N.F. and M.B.; formal analysis, G.C., A.S., N.F. and M.B.; investigation, G.C., A.S., O.L., O.K., A.P., D.C. and H.Z.; resources, O.K., A.P., Y.Z., B.M., D.G. and J.X.; data curation, A.S., N.F. and M.B.; writing—original draft preparation, A.S., N.F. and M.B.; writing—review and editing, M.B.; visualization, G.C., A.S., A.P. and M.B.; supervision, Y.Z., B.M. and M.B.; project administration, J.X., B.M., D.G. and M.B.; funding acquisition, M.B. All authors have read and agreed to the published version of the manuscript.

Funding: This work was supported by an individual grant provided by Huaiyin Normal University (Huai'an, China) to MB.

Institutional Review Board Statement: Not applicable.

Informed Consent Statement: Not applicable.

Data Availability Statement: GenBank accession numbers for the *atpF–atpH* (*ATP*) and *psbK–psbL* (*PSB*) barcodes are in the Table S1.

Acknowledgments: We thank the following people for their help collecting duckweed samples: Roman Volkov and Irina Panchuk (Yuriy Fedkovych University of Chernivtsi), Volodymyr Borysyuk (Kyiv, Ukraine), Raisa Stepanenko (Vysochne, Ukraine), Antonina Melnyk (Pasiky, Ukraine),

Oleksandr and Laura Chekanov (Sumy, Ukraine), and Vasyl Stepanenko (Korsun-Shevchenkivskyi, Ukraine), as well as many students from the Huaiyin Normal University and owners of fish and crab farms around Huai'an city (China).

Conflicts of Interest: The authors declare no conflict of interest.

References

1. Langridge, P.; Waugh, R. Harnessing the Potential of Germplasm Collections. *Nat. Genet.* **2019**, *51*, 200–201. [CrossRef] [PubMed]
2. Appenroth, K.J.; Borisjuk, N.; Lam, E. Telling Duckweed Apart: Genotyping Technologies for the Lemnaceae. *Chin. J. Appl. Environ. Biol.* **2013**, *19*, 1–10. [CrossRef]
3. Acosta, K.; Appenroth, K.J.; Borisjuk, L.; Edelman, M.; Heinig, U.; Jansen, M.A.K.; Oyama, T.; Pasaribu, B.; Schubert, I.; Sorrels, S.; et al. Return of the Lemnaceae: Duckweed as a Model Plant System in the Genomics and Postgenomics Era. *Plant Cell* **2021**, *33*, 3207–3234. [CrossRef] [PubMed]
4. Nauheimer, L.; Metzler, D.; Renner, S.S. Global History of the Ancient Monocot Family Araceae Inferred with Models Accounting for Past Continental Positions and Previous Ranges Based on Fossils. *New Phytol.* **2012**, *195*, 938–950. [CrossRef]
5. Zhou, Y.; Borisjuk, N. Small Aquatic Duckweed Plants with Big Potential for Production of Valuable Biomass and Wastewater Remediation. *Int. J. Environ. Sci. Nat. Resour.* **2019**, *16*, 555942. [CrossRef]
6. Baek, G.; Saeed, M.; Choi, H.-K. Duckweeds: Their Utilization, Metabolites and Cultivation. *Appl. Biol. Chem.* **2021**, *64*, 73. [CrossRef]
7. Xu, J.; Cui, W.; Cheng, J.J.; Stomp, A.-M. Production of High-Starch Duckweed and Its Conversion to Bioethanol. *Biosyst. Eng.* **2011**, *110*, 67–72. [CrossRef]
8. Cui, W.; Cheng, J.J. Growing Duckweed for Biofuel Production: A Review. *Plant Biol.* **2015**, *17* (Suppl. 1), 16–23. [CrossRef]
9. Appenroth, K.-J.; Sree, K.S.; Böhm, V.; Hammann, S.; Vetter, W.; Leiterer, M.; Jahreis, G. Nutritional Value of Duckweeds (Lemnaceae) as Human Food. *Food Chem.* **2017**, *217*, 266–273. [CrossRef]
10. Ge, X.; Zhang, N.; Phillips, G.C.; Xu, J. Growing Lemna Minor in Agricultural Wastewater and Converting the Duckweed Biomass to Ethanol. *Bioresour. Technol.* **2012**, *124*, 485–488. [CrossRef]
11. Zhou, Y.; Chen, G.; Peterson, A.; Zha, X.; Cheng, J.; Li, S.; Cui, D.; Zhu, H.; Kishchenko, O.; Borisjuk, N. Biodiversity of Duckweeds in Eastern China and Their Potential for Bioremediation of Municipal and Industrial Wastewater. *J. Geosci. Environ. Protect.* **2018**, *6*, 108–116. [CrossRef]
12. Cayuela, M.L.; Millner, P.; Slovin, J.; Roig, A. Duckweed (*Lemna gibba*) Growth Inhibition Bioassay for Evaluating the Toxicity of Olive Mill Wastes before and during Composting. *Chemosphere* **2007**, *68*, 1985–1991. [CrossRef] [PubMed]
13. Ziegler, P.D.; Sree, K.S.; Appenroth, K.J. Duckweeds for Water Remediation and Toxicity Testing. *Toxicol. Environ. Chem.* **2016**, *98*, 1127–1154. [CrossRef]
14. Bokhari, S.H.; Ahmad, I.; Mahmood-Ul-Hassan, M.; Mohammad, A. Phytoremediation Potential of *Lemna minor* L. for Heavy Metals. *Int. J. Phytoremediat.* **2016**, *18*, 25–32. [CrossRef] [PubMed]
15. Ziegler, P.; Sree, K.S.; Appenroth, K.-J. Duckweed Biomarkers for Identifying Toxic Water Contaminants? *Environ. Sci. Pollut. Res. Int.* **2019**, *26*, 14797–14822. [CrossRef] [PubMed]
16. Stomp, A.-M. The Duckweeds: A Valuable Plant for Biomanufacturing. *Biotechnol. Annu. Rev.* **2005**, *11*, 69–99. [CrossRef] [PubMed]
17. Rival, S.; Wisniewski, J.-P.; Langlais, A.; Kaplan, H.; Freyssinet, G.; Vancanneyt, G.; Vunsh, R.; Perl, A.; Edelman, M. *Spirodela* (Duckweed) as an Alternative Production System for Pharmaceuticals: A Case Study, Aprotinin. *Transgenic Res.* **2008**, *17*, 503–513. [CrossRef]
18. Peterson, A.; Kishchenko, O.; Zhou, Y.; Vasylenko, M.; Giritch, A.; Sun, J.; Borisjuk, N.; Kuchuk, M. Robust *Agrobacterium*-Mediated Transient Expression in Two Duckweed Species (Lemnaceae) Directed by Non-Replicating, Replicating, and Cell-to-Cell Spreading Vectors. *Front. Bioeng. Biotechnol.* **2021**, *9*, 5. [CrossRef]
19. Hoang, P.T.N.; Schubert, V.; Meister, A.; Fuchs, J.; Schubert, I. Variation in Genome Size, Cell and Nucleus Volume, Chromosome Number and RDNA Loci among Duckweeds. *Sci. Rep.* **2019**, *9*, 3234. [CrossRef]
20. Wang, W.; Haberer, G.; Gundlach, H.; Gläßer, C.; Nussbaumer, T.; Luo, M.C.; Lomsadze, A.; Borodovsky, M.; Kerstetter, R.A.; Shanklin, J.; et al. The *Spirodela polyrhiza* Genome Reveals Insights into Its Neotenous Reduction Fast Growth and Aquatic Lifestyle. *Nat. Commun.* **2014**, *5*, 3311. [CrossRef]
21. Michael, T.P.; Bryant, D.; Gutierrez, R.; Borisjuk, N.; Chu, P.; Zhang, H.; Xia, J.; Zhou, J.; Peng, H.; El Baidouri, M.; et al. Comprehensive Definition of Genome Features in *Spirodela polyrhiza* by High-Depth Physical Mapping and Short-Read DNA Sequencing Strategies. *Plant J.* **2017**, *89*, 617–635. [CrossRef] [PubMed]
22. Hoang, P.T.N.; Fiebig, A.; Novák, P.; Macas, J.; Cao, H.X.; Stepanenko, A.; Chen, G.; Borisjuk, N.; Scholz, U.; Schubert, I. Chromosome-Scale Genome Assembly for the Duckweed *Spirodela intermedia*, Integrating Cytogenetic Maps, PacBio and Oxford Nanopore Libraries. *Sci. Rep.* **2020**, *10*, 19230. [CrossRef] [PubMed]
23. Van Hoeck, A.; Horemans, N.; Monsieurs, P.; Cao, H.X.; Vandenhove, H.; Blust, R. The First Draft Genome of the Aquatic Model Plant *Lemna minor* Opens the Route for Future Stress Physiology Research and Biotechnological Applications. *Biotechnol. Biofuels* **2015**, *8*, 188. [CrossRef]

24. Michael, T.P.; Ernst, E.; Hartwick, N.; Chu, P.; Bryant, D.; Gilbert, S.; Ortleb, S.; Baggs, E.L.; Sree, K.S.; Appenroth, K.J.; et al. Genome and Time-of-Day Transcriptome of *Wolffia australiana* Link Morphological Minimization with Gene Loss and Less Growth Control. *Genome Res.* **2021**, *31*, 225–238. [CrossRef] [PubMed]
25. Vu, G.T.H.; Fourounjian, P.; Wang, W.; Cao, X.H. Future Prospects of Duckweed Research and Applications. In *The Duckweed Genomes*; Cao, X.H., Fourounjian, P., Wang, W., Eds.; Springer International Publishing: Cham, Switzerland, 2020; pp. 179–185.
26. Appenroth, K.-J.; Sree, K.S.; Fakhoorian, T.; Lam, E. Resurgence of Duckweed Research and Applications: Report from the 3rd International Duckweed Conference. *Plant Mol. Biol.* **2015**, *89*, 647–654. [CrossRef]
27. Cao, H.X.; Fourounjian, P.; Wang, W. The Importance and Potential of Duckweeds as a Model and Crop Plant for Biomass-Based Applications and Beyond. In *Handbook of Environmental Materials Management*; Hussain, C.M., Ed.; Springer International Publishing: Cham, Switzerland, 2018; pp. 1–16.
28. An, D.; Zhou, Y.; Li, C.; Xiao, Q.; Wang, T.; Zhang, Y.; Wu, Y.; Li, Y.; Chao, D.-Y.; Messing, J.; et al. Plant Evolution and Environmental Adaptation Unveiled by Long-Read Whole-Genome Sequencing of *Spirodela*. *Proc. Natl. Acad. Sci. USA* **2019**, *116*, 18893–18899. [CrossRef]
29. Tippery, N.P.; Les, D.H. Tiny Plants with Enormous Potential: Phylogeny and Evolution of Duckweeds. In *The Duckweed Genomes*; Cao, X.H., Fourounjian, P., Wang, W., Eds.; Springer International Publishing: Cham, Switzerland, 2020; pp. 19–38.
30. Les, D.H.; Crawford, D.J.; Landolt, E.; Gabel, J.D.; Kimball, R.T. Phylogeny and Systematics of Lemnaceae, the Duckweed Family. *Syst. Bot.* **2002**, *27*, 221–240. [CrossRef]
31. Zhang, J.; Azizullah, A. Genetic Diversity and DNA Barcoding in the Duckweed Family. In *The Duckweed Genomes*; Cao, X.H., Fourounjian, P., Wang, W., Eds.; Springer International Publishing: Cham, Switzerland, 2020; pp. 59–65.
32. CBOL Plant Working Group. A DNA Barcode for Land Plants. *Proc. Natl. Acad. Sci. USA* **2009**, *106*, 12794–12797. [CrossRef]
33. Hollingsworth, P.M. Refining the DNA Barcode for Land Plants. *Proc. Natl. Acad. Sci. USA* **2011**, *108*, 19451–19452. [CrossRef]
34. Mardanov, A.V.; Ravin, N.V.; Kuznetsov, B.B.; Samigullin, T.H.; Antonov, A.S.; Kolganova, T.V.; Skyabin, K.G. Complete Sequence of the Duckweed (*Lemna minor*) Chloroplast Genome: Structural Organization and Phylogenetic Relationships to Other Angiosperms. *J. Mol. Evol.* **2008**, *66*, 555–564. [CrossRef]
35. Wang, W.; Messing, J. High-Throughput Sequencing of Three Lemnoideae (Duckweeds) Chloroplast Genomes from Total DNA. *PLoS ONE* **2011**, *6*, e24670. [CrossRef] [PubMed]
36. Ding, Y.; Fang, Y.; Guo, L.; Li, Z.; He, K.; Zhao, Y.; Zhao, H. Phylogenic Study of Lemnoideae (Duckweeds) through Complete Chloroplast Genomes for Eight Accessions. *PeerJ* **2017**, *5*, e4186. [CrossRef] [PubMed]
37. Zhang, Y.; An, D.; Li, C.; Zhao, Z.; Wang, W. The Complete Chloroplast Genome of Greater Duckweed (*Spirodela polyrhiza* 7498) Using PacBio Long Reads: Insights into the Chloroplast Evolution and Transcription Regulation. *BMC Genom.* **2020**, *21*, 76. [CrossRef] [PubMed]
38. Fourounjian, P.; Fakhoorian, T.; Cao, X.H. Importance of Duckweeds in Basic Research and Their Industrial Applications. In *The Duckweed Genomes*; Cao, X.H., Fourounjian, P., Wang, W., Eds.; Springer International Publishing: Cham, Switzerland, 2020; pp. 1–17.
39. Wang, W.; Wu, Y.; Yan, Y.; Ermakova, M.; Kerstetter, R.; Messing, J. DNA Barcoding of the Lemnaceae, a Family of Aquatic Monocots. *BMC Plant Biol.* **2010**, *10*, 205. [CrossRef] [PubMed]
40. Tang, J.; Zhang, F.; Cui, W.; Ma, J. Genetic Structure of Duckweed Population of *Spirodela*, *Landoltia* and *Lemna* from Lake Tai, China. *Planta* **2014**, *239*, 1299–1307. [CrossRef]
41. Tang, J.; Li, Y.; Ma, J.; Cheng, J.J. Survey of Duckweed Diversity in Lake Chao and Total Fatty Acid, Triacylglycerol, Profiles of Representative Strains. *Plant Biol.* **2015**, *17*, 1066–1072. [CrossRef]
42. Stepanenko, A.; Chen, G.; Hoang, P.T.N.; Fuchs, J.; Schubert, I.; Borisjuk, N. The Ribosomal DNA Loci of the Ancient Monocot *Pistia Stratiotes* L. (Araceae) Contain Different Variants of the 35S and 5S Ribosomal RNA Gene Units. *Front. Plant Sci.* **2022**, *13*, 819750. [CrossRef]
43. Borisjuk, N.; Chu, P.; Gutierrez, R.; Zhang, H.; Acosta, K.; Friesen, N.; Sree, K.S.; Garcia, C.; Appenroth, K.J.; Lam, E. Assessment, Validation and Deployment Strategy of a Two-Barcode Protocol for Facile Genotyping of Duckweed Species. *Plant Biol.* **2015**, *17* (Suppl. 1), 42–49. [CrossRef]
44. Bog, M.; Baumbach, H.; Schween, U.; Hellwig, F.; Landolt, E.; Appenroth, K.-J. Genetic Structure of the Genus *Lemna* L. (Lemnaceae) as Revealed by Amplified Fragment Length Polymorphism. *Planta* **2010**, *232*, 609–619. [CrossRef]
45. Bog, M.; Schneider, P.; Hellwig, F.; Sachse, S.; Kochieva, E.Z.; Martyrosian, E.; Landolt, E.; Appenroth, K.-J. Genetic Characterization and Barcoding of Taxa in the Genus *Wolffia* Horkel Ex Schleid. (Lemnaceae) as Revealed by Two Plastidic Markers and Amplified Fragment Length Polymorphism (AFLP). *Planta* **2013**, *237*, 1–13. [CrossRef]
46. Landolt, E. The family of Lemnaceae—A monographic study. *Veröff. Geobot. Inst. Rübel* **1986**, *71*, 1–563.
47. Orlov, O.O.; Iakushenko, D.M. *Lemna turionifera* Landolt (Araceae) a new species for the flora of Ukraine. *Ukr. Bot. J.* **2013**, *70*, 224–231. [CrossRef]
48. Sabliy, L.; Konontsev, S.; Grokhovska, J.; Widomski, M.K.; Łagód, G. Nitrogen Removal from Fish Farms Water by *Lemna minor* and *Wolffia arrhiza*. *Proc. ECOpole* **2016**, *10*, 499–504. [CrossRef]
49. Lam, E.; Appenroth, K.J.; Ma, Y.; Shoham, T.; Sree, K.S. Registration of duckweed clones/strains-future approach. *Duckweed Forum* **2020**, *8*, 35–37. Available online: http://www.ruduckweed.org (accessed on 31 March 2022).

50. Sree, K.S.; Appenroth, K.-J. Worldwide Genetic Resources of Duckweed: Stock Collections. In *The Duckweed Genomes*; Cao, X.H., Fourounjian, P., Wang, W., Eds.; Springer International Publishing: Cham, Switzerland, 2020; pp. 39–46.
51. Xu, S.; Stapley, J.; Gablenz, S.; Boyer, J.; Appenroth, K.J.; Sree, K.S.; Gershenzon, J.; Widmer, A.; Huber, M. Low Genetic Variation Is Associated with Low Mutation Rate in the Giant Duckweed. *Nat. Commun.* **2019**, *10*, 1243. [CrossRef] [PubMed]
52. Borisjuk, N.; Peterson, A.; Lv, J.; Qu, G.; Luo, Q.; Shi, L.; Chen, Q.; Kishchenko, O.; Zhou, Y.; Shi, J. Structural and biochemical properties of duckweed surface cuticle. *Front. Chem.* **2018**, *6*, 317. [CrossRef] [PubMed]
53. Chen, G.; Stepanenko, A.; Borisjuk, N. Mosaic Arrangement of the 5S rDNA in the Aquatic Plant *Landoltia punctata* (Lemnaceae). *Front. Plant Sci.* **2021**, *12*, 678689. [CrossRef]
54. Murray, M.G.; Thompson, W.F. Rapid Isolation of High Molecular Weight Plant DNA. *Nucleic Acids Res.* **1980**, *8*, 4321–4325. [CrossRef]
55. Thompson, J.D.; Gibson, T.J.; Plewniak, F.; Jeanmougin, F.; Higgins, D.G. The CLUSTAL_X Windows Interface: Flexible Strategies for Multiple Sequence Alignment Aided by Quality Analysis Tools. *Nucleic Acids Res.* **1997**, *25*, 4876–4882. [CrossRef]
56. Tamura, K.; Peterson, D.; Peterson, N.; Stecher, G.; Nei, M.; Kumar, S. MEGA5: Molecular Evolutionary Genetics Analysis Using Maximum Likelihood, Evolutionary Distance, and Maximum Parsimony Methods. *Mol. Biol. Evol.* **2011**, *28*, 2731–2739. [CrossRef]
57. Wilgenbusch, J.C.; Swofford, D. Inferring evolutionary trees with PAUP*. *Curr. Protoc. Bioinform.* **2003**, *6*, 4. [CrossRef] [PubMed]
58. Felsenstein, J. Confidence Limits on Phylogenies: An Approach Using the Bootstrap. *Evolution* **1985**, *39*, 783–791. [CrossRef] [PubMed]
59. Ronquist, F.; Huelsenbeck, J.P. MrBayes 3: Bayesian Phylogenetic Inference under Mixed Models. *Bioinformatics* **2003**, *19*, 1572–1574. [CrossRef] [PubMed]
60. Darriba, D.; Taboada, G.L.; Doallo, R.; Posada, D. JModelTest 2: More Models, New Heuristics and Parallel Computing. *Nat. Methods* **2012**, *9*, 772. [CrossRef] [PubMed]
61. Xu, Y.; Ma, S.; Huang, M.; Peng, M.; Bog, M.; Sree, K.S.; Appenroth, K.-J.; Zhang, J. Species Distribution, Genetic Diversity and Barcoding in the Duckweed Family (Lemnaceae). *Hydrobiologia* **2015**, *743*, 75–87. [CrossRef]
62. Zhou, Y.; Kishchenko, O.; Stepanenko, A.; Chen, G.; Wang, W.; Zhou, J.; Pan, C.; Borisjuk, N. The Dynamics of NO^{3-} and NH^{4+} Uptake in Duckweed Are Coordinated with the Expression of Major Nitrogen Assimilation Genes. *Plants* **2021**, *11*, 11. [CrossRef]
63. Bog, M.; Lautenschlager, U.; Landrock, M.; Landolt, E.; Fuchs, J.; Sree, K.S.; Oberprieler, C.; Appenroth, K.-J. Genetic Characterization and Barcoding of Taxa in the Genera *Landoltia* and *Spirodela* (Lemnaceae) by Three Plastidic Markers and Amplified Fragment Length Polymorphism (AFLP). *Hydrobiologia* **2015**, *749*, 169–182. [CrossRef]
64. Kittiwongwattana, C.; Thawai, C. Rhizobium Lemnae Sp. Nov., a Bacterial Endophyte of *Lemna aequinoctialis*. *Int. J. Syst. Evol. Microbiol.* **2014**, *64*, 2455–2460. [CrossRef]
65. Lee, Y.; Choi, H.J.; Shiga, T. Taxonomic Identity of *Landoltia punctata* (Araceae, Lemnoideae) in Korea. *J. Asia-Pac. Biodivers.* **2020**, *13*, 494–498. [CrossRef]
66. Barks, P.; Dempsey, Z.; Burg, T.; Laird, R. Among-Strain Consistency in the Pace and Shape of Senescence in Duckweed. *J. Ecol.* **2018**, *106*, 2132–2145. [CrossRef]
67. Fu, L.; Huang, M.; Han, B.; Sun, X.; Sree, K.S.; Appenroth, K.-J.; Zhang, J. Flower Induction, Microscope-Aided Cross-Pollination, and Seed Production in the Duckweed *Lemna gibba* with Discovery of a Male-Sterile Clone. *Sci. Rep.* **2017**, *7*, 3047. [CrossRef] [PubMed]
68. Burgess, K.; Fazekas, A.; Kesanakurti, P.; Graham, S.; Husband, B.; Newmaster, S.; Percy, D.; Hajibabaei, M.; Barrett, S. Discriminating Plant Species in a Local Temperate Flora Using the RbcL+matK DNA Barcode. *Methods Ecol. Evol.* **2011**, *2*, 333–340. [CrossRef]
69. Bog, M.; Appenroth, K.-J.; Sree, K.S. Duckweed (Lemnaceae): Its Molecular Taxonomy. *Front. Sustain. Food Syst.* **2019**, *3*, 117. [CrossRef]
70. Appenroth, K.-J.; Sree, K.S.; Bog, M.; Ecker, J.; Seeliger, C.; Böhm, V.; Lorkowski, S.; Sommer, K.; Vetter, W.; Tolzin-Banasch, K.; et al. Nutritional Value of the Duckweed Species of the Genus *Wolffia* (Lemnaceae) as Human Food. *Front. Chem.* **2018**, *6*, 483. [CrossRef]
71. Park, H.; Park, J.H.; Jeon, H.H.; Woo, D.U.; Lee, Y.; Kang, Y.J. Characterization of the Complete Chloroplast Genome Sequence of *Wolffia globosa* (Lemnoideae) and Its Phylogenetic Relationships to Other Araceae Family. *Mitochondrial DNA Part B* **2020**, *5*, 1905–1907. [CrossRef]

plants

Article

Diversity and Differentiation of Duckweed Species from Israel

Avital Friedjung Yosef [1,*], Lusine Ghazaryan [1], Linda Klamann [1], Katherine Sarah Kaufman [1], Capucine Baubin [1,2], Ben Poodiack [1], Noya Ran [1], Talia Gabay [1,3], Shoshana Didi-Cohen [4], Manuela Bog [5], Inna Khozin-Goldberg [4] and Osnat Gillor [1,*]

1 Zuckerberg Institute for Water Research, J. Blaustein Institutes for Desert Research, Ben Gurion University, Midreshet Ben-Gurion 8499000, Israel
2 Department of Ecology and Evolutionary Biology, University of Colorado Boulder, Boulder, CO 80309, USA
3 Department of Life Sciences, Ben-Gurion University of the Negev, Be'er Sheva 8410501, Israel
4 French Associates Institute for Agriculture and Biotechnology of Drylands, The Jacob Blaustein Institutes for Desert Research, Ben-Gurion University of the Negev, Midreshet Ben-Gurion 8499000, Israel
5 Institute of Botany and Landscape Ecology, University of Greifswald, 17489 Greifswald, Germany
* Correspondence: avitush.f.y@gmail.com (A.F.Y.); gilloro@bgu.ac.il (O.G.)

Abstract: Duckweeds (Lemnaceae) are tiny plants that float on aquatic surfaces and are typically isolated from temperate and equatorial regions. Yet, duckweed diversity in Mediterranean and arid regions has been seldom explored. To address this gap in knowledge, we surveyed duckweed diversity in Israel, an ecological junction between Mediterranean and arid climates. We searched for duckweeds in the north and center of Israel on the surface of streams, ponds and waterholes. We collected and isolated 27 duckweeds and characterized their morphology, molecular barcodes (*atpF-atpH* and *psbK-psbI*) and biochemical features (protein content and fatty acids composition). Six species were identified—*Lemna minor*, *L. gibba* and *Wolffia arrhiza* dominated the duckweed populations, and together with past sightings, are suggested to be native to Israel. The fatty acid profiles and protein content further suggest that diverged functions have attributed to different haplotypes among the identified species. *Spirodela polyrhiza*, *W. globosa* and *L. minuta* were also identified but were rarer. *S. polyrhiza* was previously reported in our region, thus, its current low abundance should be revisited. However, *L. minuta* and *W. globosa* are native to America and Far East Asia, respectively, and are invasive in Europe. We hypothesize that they may be invasive species to our region as well, carried by migratory birds that disperse them through their migration routes. This study indicates that the duckweed population in Israel's aquatic environments consists of both native and transient species.

Keywords: duckweed; fatty acids; DNA barcoding; diversity; biogeography; nitrogen content; protein concentration; migration

Citation: Friedjung Yosef, A.; Ghazaryan, L.; Klamann, L.; Kaufman, K.S.; Baubin, C.; Poodiack, B.; Ran, N.; Gabay, T.; Didi-Cohen, S.; Bog, M.; et al. Diversity and Differentiation of Duckweed Species from Israel. *Plants* **2022**, *11*, 3326. https://doi.org/10.3390/plants11233326

Academic Editors: Viktor Oláh, Klaus-Jürgen Appenroth and K. Sowjanya Sree

Received: 11 October 2022
Accepted: 25 November 2022
Published: 1 December 2022

1. Introduction

The Lemnaceae (duckweeds) family comprises the world's smallest and fastest growing seed plants [1]. The duckweeds are miniscule plants that float on or below the surface of freshwater bodies. The duckweed family was found throughout the globe, except for polar regions, and was classified into five genera with 36 species [2]. Representatives of three genera contain one or few tiny roots emerging from the fronds (*Spirodela*, *Landoltia* and *Lemna*), while the two remaining genera are rootless and smaller (*Wolffiella* and *Wolffia*) [1]. These diminutive plants have gone through an extreme reduction in body size, with some species less than 0.5 mm in size, thereby minimizing the need for non-photosynthetic organs and selecting for rapid multiplication through budding [3]. As a consequence of their fast growth rate, biomass production is high, providing practical applications to the duckweeds in food [4,5], feed [6], water treatment [7,8] and biotechnology [9,10].

The first attempts to classify duckweed were based on their morphology [1]. However, due to the duckweed's diminutive size and organ reduction, morphological and anatomical classifications are challenging. Therefore, over the years, there was an attempt to classify duckweeds based on the chemical composition of the flavonoids [11], isoforms of enzymes (allozymes) [12] and fatty acids [13]. With the advancement of molecular taxonomy, molecular methods of identification have been developed, including molecular fingerprinting and sequencing [14]. The DNA-based molecular identification is based on the polymorphisms of target non-coding intron and gene spacer regions, mainly within the chloroplast genome [15,16]. DNA markers are considered the most reliable method for species classification and have been demonstrated to be capable of detecting polymorphisms among haplotypes of the same species [2]. The molecularly detected polymorphisms are supported by different physiological properties, including growth rate [3], protein and starch content [17], metabolite abundance [13] and turion formation [18]. Therefore, haplotype identification requires a combination of molecular and physiological-based methods [19].

Reliable identification methods are vital to establish the biogeography of duckweeds. Early studies have reported species-dependent biodiversity, ranging from regional to global distribution, with some species showing a broad distribution, while others were restricted to certain regions [16]. Yet, a species' global dispersal could result from the generalization of unique haplotypes within a species. In fact, it was reported that dispersal was not directly linked to taxonomy as closely related species—even haplotypes of the same species—were detected oceans apart [16,20].

Duckweed growth is mostly vegetative, whereas flowering and seed generation, i.e., sexual reproduction, is rarer [21]. The survival of duckweeds during winter in temperate regions is not only dependent on seeds; this is further evidenced by various duckweeds reportedly sinking to the bottom of water bodies, consequently morphing into turions that can survive freezing [18]. Moreover, dispersal was also attributed to biotic vectors, for example via migrating waterfowl that can carry the small duckweeds over great distances in their gastrointestinal tract or attached to their body [22,23]. When introduced to new aquatic habitats, the duckweed's fast vegetative growth facilitates their propagation and establishment [23].

The study of duckweed diversity in the context of biogeographical distribution is relevant to Israel because it is a meeting point between three continents: Africa, Asia and Europe, thereby forming a transitional region between arid and Mediterranean climates [24]. Moreover, the Jordan valley in the east of Israel is extended from the African Rift and serves as an important hub for migratory birds that winter in Africa and summer in Europe [25]. It is a part of the Afro-Palaearctic bird migration system, the largest land bird migration system in the world [26]. In spite of its small size, Israel's location between the Mediterranean Sea in the west and the Arabian Deserts in the east forms an ecological corridor and a bottleneck in the birds' flight path, making it an essential stop-over site during migration. The bird's stopover sites provide an opportunity for hitchhiking plants, such as duckweeds, to establish in new environments [27]. Nevertheless, the diversity of Israel's duckweeds was never systematically investigated, though sightings of duckweed have been documented. The following Lemnaceae species have been reported in Israel: in the genus *Lemna L. trisulca*, *L. gibba*, *L. aequinoctialis* and *L. minor*; in the genus *Spirodela*: *S. polyrhiza*; and in the genus *Wolffia*: *W. arrhiza* and *W. globosa* were reported (https://flora.org.il/plants/systematics/lemnaceae/ (accessed on 5 September 2022)). The following species are listed as endangered: *W. arrhiza* and *S. polyrhiza* (https://redlist.parks.org.il/plants/list/ (accessed on 5 September 2022)). The species *W. globosa* was reported but considered an invador. However, the difficulty of identifying duckweeds based solely on their morphology, questions the reliability of these observations.

In this study we conducted a systematic survey of duckweeds in northern and central Israel by following past sightings of duckweeds. This involved sampling the aquatic plants, then isolating them in the lab and identifying them through morphology, molecular methods, as well as biochemical features including fatty acid composition and nitrogen content. We hypothesized that the duckweed diversity in these sites would reflect the species reported in Africa, Europe and Central Asia, following birds' migration routes.

2. Materials and Methods

2.1. Survey of Duckweed Strains

During June 2021, duckweed species were collected from ponds, springs, streams and waterholes in northern and central Israel (Galilee, Hula Valley, Golan Heights and Sharon). The survey locations were selected based on previous observations from the last century taken from the Israel Nature and Parks Authority database (https://redlist.parks.org.il/plants/list/ (accessed on 11 September 2022)).

Duckweed plants were detected in 24 of the 67 reported locations detailed in the database. The verified duckweed locations are depicted in Figure 1 and detailed in Table S1.

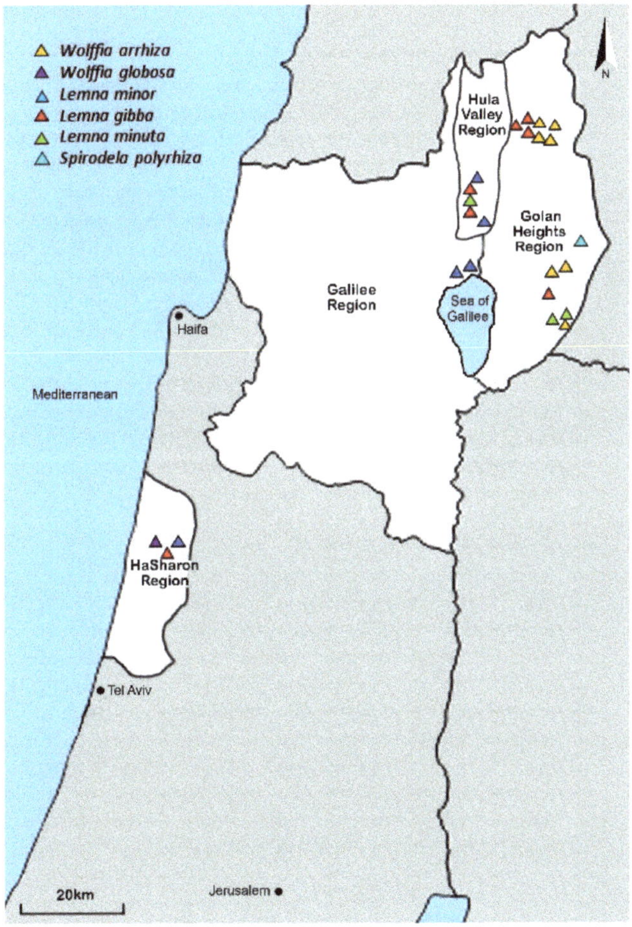

Figure 1. Confirmed duckweed sightings in northern and central Israel. White paint indicates the areas where the survey was conducted. Scale bar is 1:20,000.

2.2. Duckweed Collection

Plant samples were collected in duplicates in 100 mL plastic containers. The containers were stored in a refrigerated cooler (~10 °C) and transported to the lab up to 48 h after collection. For each confirmed collection site, the pH value of the water was measured by litmus paper and the results are detailed in Table S2. In the laboratory, electric conductivity (indicating water salinity) was measured using a conductivity meter (Cole-Parmer EW-19820-10, Vernon Hills, IL, USA) and the results are listed in Table S2.

2.3. Cultivation

In the laboratory, the collected duckweeds were sorted according to their morphology and each isolate was sterilized by rinsing the separated fronds with a 2% sodium hypochlorite solution (NaClO) for 2 min. Single fronds were picked and cultured in 0.5 × Schenk and Hildebrandt (SH) basal salt mixture (Sigma–Aldrich, St. Louis, MI, USA) supplemented by 1% sucrose at pH 5.8. The plants were grown in a controlled climate chamber under 25 °C, 16 h light/8 h dark cycles, and 200–250 μmol m^{-2} s^{-1} light intensity. Sterilization was repeated until a single unique isolate was detected. From each unique isolate, a single sterile frond was retrieved and cultured in 0.5 × SH agar with 0.5% sucrose and supplemented with 100 mg L^{-1} cefotaxime (Sigma) to avoid fungal contamination.

2.4. Morphological Identification

The morphology of the isolates was assessed according to Landolt 1986 [1] as well as Les et al. [11]. An M205 FCA fluorescence stereo microscope (Leica, Wetzlar, Germany) and Axio Imager 2 light microscope (Zeiss, Jena, Germany) was used.

2.5. DNA Extraction, Fragment Amplification and Sequencing

The isolates were cultivated as described above for 10–14 days then total DNA was extracted using DNeasy Plant Pro kit (Qiagen, Hilden, Germany) following the manufacturer's instructions. The extracted DNA was used as a template to amplify two plastid barcode loci of noncoding intergenic spacers: *atpF-atpH* (5′-ACTCGCACACACTCCCTTTCC-3′ and 5′-GCTTTTATGGAAGCTTTAACAAT-3′) and *psbK-psbI* (5′-TTAGCATTTGTT TGGCA AG-3′ and 5′-AAGTTTGAGAGTAAGCAT-3′). The amplification was performed as follows: 95 °C pre-denatured for 3 min, followed by 35 cycles of 95 °C for 45 s, 55 °C for 45 s, 72 °C for 45 s, and a further extension at 72 °C for 10 s. Purification of the resulting amplicons was carried out using the AccuPrep® PCR/Gel Purification Kit (Bioneer, Daejeon, S. Korea), according to the manufacturer's instructions. The purified PCR fragments were sequenced at McLab (San Francisco, CA, USA).

2.6. DNA Barcoding Analysis

DNA sequence alignment was generated using Geneious Prime version 2022.1.1 (http://www.geneious.com/prime/ (accessed on 15 September 2022)). Blast analysis was performed using NCBI database (https://www.ncbi.nlm.nih.gov/ (accessed on 17 September 2022)) and Rutgers database (http://epigenome.rutgers.edu/cgi-bin/duckweed/blast.cgi (accessed on 17 September 2022)). Duckweed species reference sequences of the two loci *atpF-atpH* and *psbK-psbI* were taken from the NCBI database and added to the tree analysis. Multiple alignments of both loci were performed using MUSCLE Alignment (https://www.ebi.ac.uk/Tools/msa/muscle/ (accessed on 20 September 2022)). A phylogenetic tree was constructed using Geneious Tree Builder using the Neighbor-Joining method with Tamura-Nei as the genetic distance model. Support values were calculated using bootstrapping with 1000 reiterations.

2.7. Fatty Acids Analysis

Plant fatty acid composition and content were analyzed using a direct transmethylation procedure. Plants were grown as described above for 14–21 days. After harvesting, the cultures were placed in 20 °C for 12 h, then dried for 48 h in a lyophilizer (VirTis, Gardiner, NY, USA). The dry material was ground (ULTRA-TURRAX, IKA, Merck) for 2 min at 6000 rpm. A total of ~10 mg of freeze-dried biomass was used in duplicate for analysis. Cellular fatty acids were converted into methyl esters (FAMEs) by incubation in 2 mL of 2% H_2SO_4 in dry methanol (*v*/*v*) for 1.5 h at 90 °C with continuous stirring under Argon gas atmosphere. Myristic acid ($C_{17:0}$) was used as an internal standard for FAME quantification. The reaction was terminated by the addition of 1 mL of water. A total of 1 mL of Hexane (Sigma) was then added for phase separation and extraction of FAMEs. Hexane fractions were evaporated under N_2 gas flow and resuspended in 400 μL of hexane.

FAMEs were analyzed by gas chromatography coupled with flame ionization detection (GC-FID) on a TRACE Ultra Gas Chromatograph (Thermo Electron, Milan, Italy) equipped with a programmed temperature vaporizing injector, a flame-ionization detector (FID) and a SUPELCO WAX 10 capillary column (L × I.D. 30 m × 0.25 mm, df 0.25 μm, Sigma), using a temperature gradient as follows: 1 min at 130 °C, 8 min of linear gradient to 220 °C and 10 min at 220 °C. Helium was used as a carrier gas. Retention times of FAMEs were compared with those of available commercial FAMEs standards (in-house library), and the literature data [28].

2.8. Nitrogen Content Analysis

For the nitrogen content analysis, plants were grown and dried as described above. For each sample, 2–3 mg of dry material was taken in triplicate. A Flash Smart elemental analyzer (OEA 2000, Thermo Fisher Scientific, Waltham, MA, USA) was used for the analysis. The nitrogen concentration results were adjusted to 1 g of dry weight. The estimation of protein was carried out as is commonly achieved by multiplying total nitrogen by the numeric factor 6.25 but with the addition of a correction factor specific to plants [29]. To validate our results, we measured the nitrogen concentrations in the same samples with FT-MIR (Fourier Transform Midinfrared Spectroscopy) described in the Supplementary Data (File S1).

3. Results

3.1. Distribution of Duckweed Species in Israel

In total, 24 locations in the north and center of Israel were found to have duckweed (Figure 1 and Table S1). These included 14 locations in the Golan Heights, five in the Hula Valley, two in the Galilee, and three in the Sharon. Plant samples were identified morphologically under a microscope (Table 1). Three species of *Lemna* have been identified: *L. gibba* at highest occurrence (seven observations), *L. minor* (five observations) and *L. minuta* (three observations). Two species of *Wolffia* have been identified: *W. arrhiza* at high occurrence (seven observations) and *W. globosa* at low occurrence (one observation). In addition, the species *S. polyrhiza* was observed once.

Table 1. Identification characteristics of duckweed species found in this study [1,2]. Images were taken using an epi-fluorescent microscope.

Species	Location	No of Strains	Morphology	General Occurrence	Micrograph
Wolffia arrhiza	Golan Heights	9	0.5–1.5 mm long, 0.4–1.2 mm wide; ellipsoid to spherical; upper surface is convex, opaque, bright green, with its greatest width slightly below the water surface; no veins; 30–100 stomata; no roots.	Widely distributed in temperate regions; native to Europe, South Africa; invasive in Brazil, Japan, and North America.	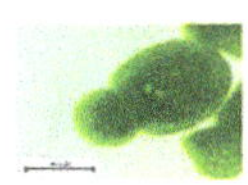
Wolffia globosa	HaSharon	2	0. 4–0.9 mm long, 0.3–0.6 mm wide; ellipsoid; upper surface convex, translucent pale green, with its greatest width well below the water surface; no veins; 8–25 stomata; no roots.	Tropical, subtropical, and warm temperate regions; native to eastern and southeast Asia and Africa; invasive in North America.	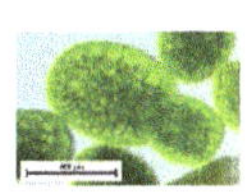
Lemna gibba	Golan Heights, Hula Valley, HaSharon	9	1–8 mm long, ~3.5 mm wide; lower surface of the fronds is usually gibbous; 4–5 veins extending from the nodes; >100 stomata; 1 root; difficult to identify due to high polymorphism.	Worldwide except Australia	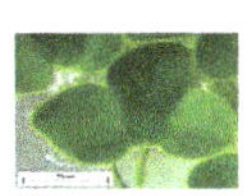

Table 1. *Cont.*

Species	Location	No of Strains	Morphology	General Occurrence	Micrograph
Lemna minor	Galilee, Hula Valley, HaSharon	5	1–10 mm long, 6–7 mm wide; upper surface shiny green, occasionally reddish; usually 3 veins, rarely 4–5; >100 stomata; 1 root.	Cooler oceanic regions; native to North America, Europe, Africa, and Western Asia.	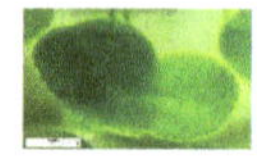
Lemna minuta	Golan Heights, Hula Valley	3	0.8–4 mm long, 0.5–2.5 mm wide; forming colonies of 2–4 fronds; circular with a slightly asymmetrical base; one vein, not very distinct; ~30 stomata; 1 root.	Temperate and subtropical regions, dry to moderately humid climate; native to America; invasive in Japan and Europe	
Spirodela polyrhiza	Golan Heights	1	Largest duckweed: 1.5–10 mm long, 1.5–8 mm wide; usually thin fronds, rarely gibbous; maximum 16 veins; >100 stomata; 7–21 roots	Worldwide	

3.2. DNA Barcoding Analysis

The sequences derived from the two markers *psbK-psbI* and *atpF-atpH* were used for identifying the species listed in Table 2. Once the sequences had been cleaned, corrected and aligned, they were blast-matched with sequences in the NCBI and Rutgers University databases. The results for species identification, along with their accession numbers, are presented in Table 2. All of the haplotypes were clearly identified as being of the same species, although there was not always a match between the barcodes *atpF-atpH* and *psbK-psbI*, as seen in the *S. polyrhiza* 19, *W. arrhiza* 56 and all *L. gibba* haplotypes.

Table 2. DNA identification of the duckweed isolates using *atpF-atpH* and *psbK-psbI* barcodes. Identification was conducted using the NCBI database.

Lab ID	Species	Region	Coordinates	Seasonality of the Water Source	*atpF-atpH* Strain Identification	Accession No	Identity (%)	*psbK-psbI* Strain Identification	Accession No	Identity (%)
32b	*L. gibba*	Golan Hights	33.137779, 35.725382	seasonal	RDSC 5504	KX212889.1	100	DW102	OM569589.1	100
58	*L. gibba*	HaSharon	32.333731, 34.876507	perennial	RDSC 5504	KX212889.1	100	DW102	OM569589.1	100
56	*L. gibba*	Golan Hights	33.09779, 35.81729	seasonal	RDSC 5504	KX212889.1	100	DW102	OM569589.1	100
7	*L. gibba*	Golan Hights	32.867917, 35.770194	seasonal	RDSC 5504	KX212889.1	100	DW102	OM569589.1	100
43b	*L. gibba*	Hulla Vally	33.060001, 35.615137	seasonal	RDSC 5504	KX212889.1	100	DW102	OM569589.1	100
43a	*L. gibba*	Hulla Vally	33.060002, 35.615138	seasonal	RDSC 5504	KX212889.1	100	DW102	OM569589.1	100
31a	*L. gibba*	Golan Heights	33.139682, 35.733806	perennial	RDSC 5504	KX212889.1	100	DW102	OM569589.1	100
31b	*L. gibba*	Golan Heights	33.139682, 35.733806	perennial	RDSC 5504	KX212889.1	100	DW102	OM569589.1	100
30	*L. gibba*	Golan Heights	33.138450, 35.734019	seasonal	RDSC 5504	KX212889.1	100	DW102	OM569589.1	100
45a	*L. minor*	Galille	32.912915, 35.569178	perennial	K46	OM569601.1	100	K46	OM569540.1	100
42	*L. minor*	Hulla Vally	33.015434, 35.629857	seasonal	K46	OM569601.1	100	K46	OM569540.1	100
44	*L. minor*	Hulla Vally	33.064187, 35.610817	seasonal	K46	OM569601.1	100	K46	OM569540.1	100
57	*L. minor*	HaSharon	32.363913, 34.958369	perennial	K46	OM569601.1	100	K46	OM569540.1	100
45b	*L. minor*	Galille	32.912915, 35.569178	perennial	K46	OM569601.1	100	K46	OM569540.1	100
55a	*L. minuta*	Golan Hights	32.801403, 35.783032	seasonal	5573	MK516255.1	100	5573	MK516236.1	100
55b	*L. minuta*	Golan Hights	32.801403, 35.783032	seasonal	5573	MK516255.0	100	5573	MK516236.2	100

Table 2. *Cont.*

Lab ID	Species	Region	Coordinates	Seasonality of the Water Source	*atpF-atpH* Strain Identification	Accession No	Identity (%)	*psbK-psbI* Strain Identification	Accession No	Identity (%)
43s	*L. minuta*	Hulla Vally	33.060002, 35.615138	seasonal	5573	MK516255.1	100	5573	MK516236.3	100
19	*S. polyrhiza*	Golan Hights	32.969559, 35.820036	seasonal	7498	MN419335.1	100	RDSC 2014	OM569580.1	100
58	*W. globosa*	HaSharon	32.333731, 34.876507	perennial	DW2101-4	KJ630544.1	100	5514	MG812327.1	100
11b	*W. arrhiza*	Golan Heights	32.894030, 35.775695	seasonal	DW35	OM569550.1	100	DW35	OM569611.1	99.01
30a	*W. arrhiza*	Golan Heights	33.138450, 35.734019	seasonal	DW35	OM569550.1	100	DW35	OM569611.1	99.01
31b	*W. arrhiza*	Golan Heights	33.139682, 35.733806	perennial	DW35	OM569550.1	100	DW35	OM569611.1	99.01
55	*W. arrhiza*	Golan Heights	32.801403, 35.783032	seasonal	DW35	OM569550.1	100	DW32	OM569610.1	99.29
32b	*W. arrhiza*	Golan Heights	33.137779, 35.725382	seasonal	DW35	OM569550.1	100	DW35	OM569611.1	99.31
30b	*W. arrhiza*	Golan Heights	33.138451, 35.734018	seasonal	DW35	OM569550.1	100			
11a	*W. arrhiza*	Golan Heights	32.895829, 35.776775	seasonal	DW35	OM569550.1	100	DW35	OM569611.1	99.31

Phylogenetic Tree

A phylogenetic tree based on *psbK-psbI* and *atpF-atpH* sequences was constructed after multiple alignment of the sequences (Figure 2) using a total of the 27 haplotypes collected in this study, as well as *W. globosa* "Mankai" (a domesticated duckweed haplotype that was isolated from the Golan Hights, Israel [5]) and additional reference sequences (*S. polyrhiza* 7498, *W. arrhiza* DW35, *W. globosa* 8789, *L. minuta* 5573, *L. gibba* 5504, *L. minor* 7123) taken from the NCBI database (one for each species).

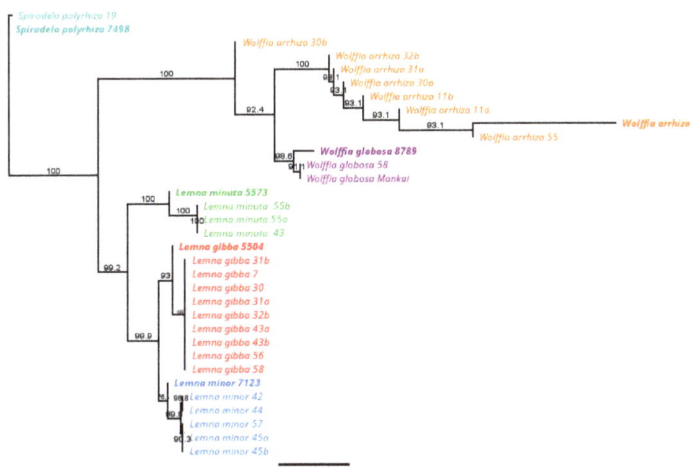

Figure 2. Neighbor-Joining tree using the concatenated *psbK-psbI* and *atpF-atpH* sequences and the Tamura-Nei distance. Based on 1000 reiterations. The numbers on the nodes represent the percentage of bootstrap values. Horizontal bars indicate genetic distances. Reference sequences were retrieved from the NCBI database and highlighted in bold.

3.3. Fatty Acid Analysis

Total fatty acid content measured by GC-FID varied between 2.83 and 6.34% of dry weight across the isolates. There were three major fatty acids that accounted for 80–90% of the total fatty acids: palmitic acid (16:0), linoleic acid (LA, 18:2*n*6) and α-linolenic acid (ALA, 18:3*n*3). The other fatty acids present at lower proportions are listed in Table 3 and Table S3.

Table 3. Fatty acid composition and content of duckweed plants collected in the current study. The strain *W. arrhiza* 9528 was added to the analysis as a reference. An isomer of 16:1 and 18:1*n7* are not shown. The full table is shown in the Supplementary Data (Table S3).

Haplotype	Fatty Acid (% of Total Fatty Acids)													ω3/ω6 18:3/18:2	TFA (%DW)
	16:0	16:1	16:2	16:3	18:0	18:1n9	18:2n6	18:3n6 GLA	18:3n3 ALA	18:4n3 SDA	20:0	22:0	24:0		
L. gibba 58	20.56 ± 0.43	3.91 ± 0.42	2.2 ± 0.05	0.52 ± 0.13	0.92 ± 0.04	1.13 ± 0.13	11.1 ± 0.33	0.61 ± 0.05	52.88 ± 0.65	2.62 ± 0.08	0.37 ± 0.01	0.44 ± 0.04	1.13 ± 0.03	4.77	5.25
L. gibba 30	19.72 ± 0.46	4.88 ± 0.03	3.27 ± 0.05	0.75 ± 0.04	1.01 ± 0.16	1.07 ± 0.09	10.46 ± 0.17	0.89 ± 0.05	50.32 ± 0.36	3.81 ± 0	0.45 ± 0.01	0.39 ± 0.02	1.15 ± 0.04	4.81	4.39
L. gibba 31a	21.32 ± 0.05	4.28 ± 0.04	2.06 ± 0.01	0.17 ± 0.02	1.25 ± 0.01	0.78 ± 0	11.27 ± 0.18	0.72 ± 0	50.26 ± 0	4.37 ± 0.03	0.53 ± 0	0.57 ± 0.01	1.05 ± 0.01	4.46	3.63
L. gibba 31b	21.61 ± 0.17	5.46 ± 0.31	2.18 ± 0.02	0.23 ± 0.07	1.07 ± 0.04	1.11 ± 0.01	13.93 ± 0.05	0.39 ± 0	48.98 ± 0.01	1.96 ± 0.03	0.39 ± 0	0.41 ± 0.01	0.84 ± 0.02	3.52	3.88
L. gibba 32b	20.01 ± 0.42	4.65 ± 1.09	2.11 ± 0.03	0.35 ± 0.21	1.36 ± 0.05	0.94 ± 0.01	15.37 ± 0.09	0.8 ± 0.01	48.09 ± 0.21	3.21 ± 0.01	0.55 ± 0.02	0.51 ± 0.05	0.93 ± 0.02	3.13	4.10
L. gibba 43a	20.96 ± 0.23	3.69 ± 0.42	2.07 ± 0.06	0.65 ± 0.08	1.01 ± 0.08	1.47 ± 0.24	13.83 ± 0.02	0.61 ± 0.07	49.87 ± 0.91	2.06 ± 0.14	0.48 ± 0.02	0.51 ± 0.01	1.1 ± 0.19	3.60	4.36
L. gibba 43b	21.4 ± 0.31	4.56 ± 0.77	1.96 ± 0.03	0.43 ± 0.13	0.97 ± 0.01	1.3 ± 0.02	13.69 ± 0.07	0.47 ± 0	49.92 ± 0.24	2.02 ± 0	0.39 ± 0.01	0.41 ± 0.02	0.76 ± 0.03	3.65	4.12
L. gibba 56	21.75 ± 2.27	5.39 ± 1.1	2.62 ± 0.02	0.23 ± 0.11	1.08 ± 0.31	1.08 ± 0.19	13.82 ± 1.37	0.52 ± 0.11	47.77 ± 3.37	2.35 ± 0.34	0.42 ± 0.14	0.45 ± 0.15	0.71 ± 0.2	3.49	4.09
L. gibba 7	21.49 ± 0.04	4.23 ± 0.22	2.24 ± 0.06	0.4 ± 0.03	1.24 ± 0.03	0.94 ± 0.01	15.3 ± 0.06	0.44 ± 0.05	47.78 ± 0.51	2.1 ± 0.03	0.53 ± 0.01	0.5 ± 0.01	1.4 ± 0.13	3.12	2.83
L. minor 42	22.73 ± 0.12	5.67 ± 0.37	1.62 ± 0.09	0.35 ± 0.06	1.8 ± 0.12	1.13 ± 0.02	16.01 ± 0.08	0.46 ± 0.01	45.8 ± 0.03	1.51 ± 0.14	0.55 ± 0	0.31 ± 0	1.01 ± 0.01	2.86	3.35
L. minor 44	19.85 ± 0.06	5.5 ± 0.01	2.01 ± 0.01	0.2 ± 0.01	1.28 ± 0.02	1.45 ± 0	18.4 ± 0.1	0.5 ± 0	46.75 ± 0.13	1.75 ± 0	0.31 ± 0.01	0.29 ± 0.01	0.82 ± 0.04	2.54	6.34
L. minor 45a	19.97 ± 0.33	2.89 ± 0.05	2.03 ± 0.02	0.75 ± 0.02	1.4 ± 0.04	1.69 ± 0.01	18.88 ± 0.05	0.6 ± 0.01	47.37 ± 0.39	1.88 ± 0.08	0.47 ± 0.01	0.36 ± 0.15	0.72 ± 0.78	2.51	3.77
L.minor 45b	22.49 ± 0.9	4.1 ± 0.79	2.54 ± 0.43	0.68 ± 0.37	1.4 ± 0.16	1.57 ± 0.03	17.64 ± 0.15	0.71 ± 0.05	42.74 ± 0.63	2.68 ± 0.21	0.45 ± 0.01	0.34 ± 0.14	1.16 ± 0.15	2.42	3.52
L. minor 57	18.28 ± 0.16	5.8 ± 0.18	2.77 ± 0.02	0.22 ± 0.04	0.77 ± 0.02	1.33 ± 0.01	16.1 ± 0.11	1.27 ± 0.01	48.36 ± 0.19	3.04 ± 0.01	0.17 ± 0.01	0.24 ± 0	0.71 ± 0.03	3.00	4.19
L. minuta 43	18.48 ± 0.45	4.64 ± 0.03	2.43 ± 0.05	0.52 ± 0.05	1.24 ± 0.39	1.27 ± 0.02	14.07 ± 0.21	0.23 ± 0.21	54.43 ± 1.13	0.09 ± 0.02	0.25 ± 0.08	0.29 ± 0.05	1.13 ± 0.1	3.87	4.78
L. minuta 55a	19.92 ± 0.2	4.97 ± 0.52	1.62 ± 0	0.31 ± 0.13	0.93 ± 0.03	1.08 ± 0.02	15.87 ± 0.14	0 ± 0	53.35 ± 0.23	0.12 ± 0.11	0.15 ± 0.02	0.19 ± 0.03	0.79 ± 0.06	3.36	3.95
L. minuta 55b	20.07 ± 0.12	3.44 ± 0.07	1.6 ± 0	0.66 ± 0.04	1.03 ± 0.02	1.35 ± 0.07	15.91 ± 0.07	0 ± 0	53.51 ± 0.09	0.13 ± 0.18	0.25 ± 0	0.15 ± 0	1.17 ± 0.02	3.36	4.19
S. polyrhiza 19	23.15 ± 0.24	5.43 ± 0.37	6.11 ± 0.06	0.35 ± 0.04	2.35 ± 0.07	1.17 ± 0.04	5.23 ± 0	0.04 ± 0.05	50.9 ± 0.48	0 ± 0	0.58 ± 0.04	0.62 ± 0.01	2.61 ± 0.05	9.73	4.27
W. arrhiza 11a	23.22 ± 1.1	5.64 ± 0.58	3.64 ± 0.23	0.55 ± 0.03	1.62 ± 0	2.17 ± 0.08	25.24 ± 0.32	0.28 ± 0.01	34.87 ± 0.64	0.1 ± 0.01	0.77 ± 0.01	0.64 ± 0.04	0.4 ± 0.03	1.38	4.38
W. arrhiza 11b	21.63 ± 0.26	5.36 ± 0.53	3.45 ± 0.12	0.59 ± 0.11	1.48 ± 0.07	1.87 ± 0.03	25.7 ± 0.02	0.15 ± 0.01	37.32 ± 0.56	0.1 ± 0.01	0.73 ± 0.01	0.53 ± 0.01	0.37 ± 0.08	1.45	4.55
W. arrhiza 30a	20.4 ± 0.01	5.32 ± 0.2	3.46 ± 0.02	0.48 ± 0	1.17 ± 0.02	1.73 ± 0.01	22.75 ± 0.02	0.14 ± 0	42.29 ± 0.14	0.08 ± 0	0.58 ± 0	0.47 ± 0.01	0.31 ± 0.02	1.86	4.70
W. arrhiza 31b	24.28 ± 0.14	3.11 ± 0.22	3.19 ± 0.14	1.03 ± 0.14	1.35 ± 0.04	1.64 ± 0.04	23.78 ± 0.03	0.11 ± 0.1	38.74 ± 0.33	0.14 ± 0.02	0.69 ± 0.02	0.4 ± 0	0.52 ± 0.03	1.63	3.18
W. arrhiza 55	21.85 ± 0.43	3.65 ± 0.25	3.33 ± 0.04	0.75 ± 0.04	1.37 ± 0.11	1.73 ± 0.04	25.7 ± 0.26	0.03 ± 0.01	38.97 ± 0.44	0.1 ± 0.01	0.6 ± 0.02	0.51 ± 0.04	0.34 ± 0.08	1.52	5.13
W. arrhiza 9528	19.05 ± 0.13	3.72 ± 0.46	2.48 ± 0.03	0.24 ± 0.02	1.81 ± 0.01	1.79 ± 0.01	24.29 ± 0.12	0.14 ± 0.01	44.41 ± 0.1	0.07 ± 0	0.86 ± 0.02	0.66 ± 0.05	0.29 ± 0.02	1.83	4.44
W. globosa 58	22.71 ± 0.05	4.04 ± 0.58	1.5 ± 0.05	0.35 ± 0.08	2.25 ± 0	2.41 ± 0.01	25.02 ± 0.01	0.07 ± 0.39	39.51 ± 0.04	0.02 ± 0.02	0.53 ± 0.01	0.39 ± 0.02	0.19 ± 0.01	1.58	4.38
W. globosa Mankai	21.73 ± 0.16	4.39 ± 0.64	2.34 ± 0.02	0.74 ± 0.21	2.14 ± 0.02	1.9 ± 0.01	19.88 ± 0.27	0.14 ± 0	44.64 ± 0.21	0.07 ± 0	0.41 ± 0.02	0.19 ± 0.01	0.33 ± 0.02	2.25	5.36

ALA (18:3n3) represented approximately 50% of the total fatty acids in most *Lemna* isolates, whereas *Wolffia* species had lower concentrations of this major n3 PUFA with a concurrent increase in 18:2n6. Accordingly, the ratio 18:3n3/18:2n6 attained higher values in *Lemna* species. *Spirodela* differed from other duckweed species by the lowest percentage of 18:2n6, resulting in the maximal 18:3n3/18:2n6 ratio. A variety of differences in fatty acid composition between duckweed genera, including *Lemna*, *Spirodela* and *Wolffia*, have already been reported [4,29]. Another apparent difference in this study was the presence of stearidonic acid (SDA,18:4n3), which is the product of a delta-6 (Δ6) desaturation on 18:3n3. In this study, all *L. minor* and *L. gibba* isolates featured the presence of SDA at ~2% to above 4% of total FA, as well as the detectable levels of another Δ6 C18 PUFA, γ-linolenic acid (GLA, 18:3n6). These data are in line with the presence of the Δ6 desaturase gene in *Lemna* [29] and *Wolffia* [30] species, enabling the biosynthesis of Δ6 C18 PUFA.

3.4. Nitrogen Analysis

Nitrogen content varied widely among strains, ranging from 2.2 to 5.4% of dry biomass (Figure 3). There was no clear pattern with respect to the different duckweed species: Nitrogen content appeared to vary depending on the species and even the strain. *W. globosa* "Mankai" had the highest nitrogen content of 5.82%; it translates to a high protein concentration of 25.52%–36.25%. The isolated *W. globosa* 58 showed a lower nitrogen concentration. *W. arrhiza* produced high N concentrations in all five samples: 4.12–5.42% nitrogen, translating to 18.28%–33.87% protein. The species *L. minuta* also produced high results in two samples: between 4.01–4.54% nitrogen, translating to 17.64–28.38% protein. Similar to *W. globosa*, nitrogen values were obtained in a wide range of values in the species *L. minor* and *L. gibba* as well. *S. polyrhiza* yielded an average nitrogen content of 2.95%, which translates into 12.98–18.44% protein (Figure 3). Because of the high variability, the analysis was validated using FT-MIR spectroscopy that yielded similar results (Figure S1).

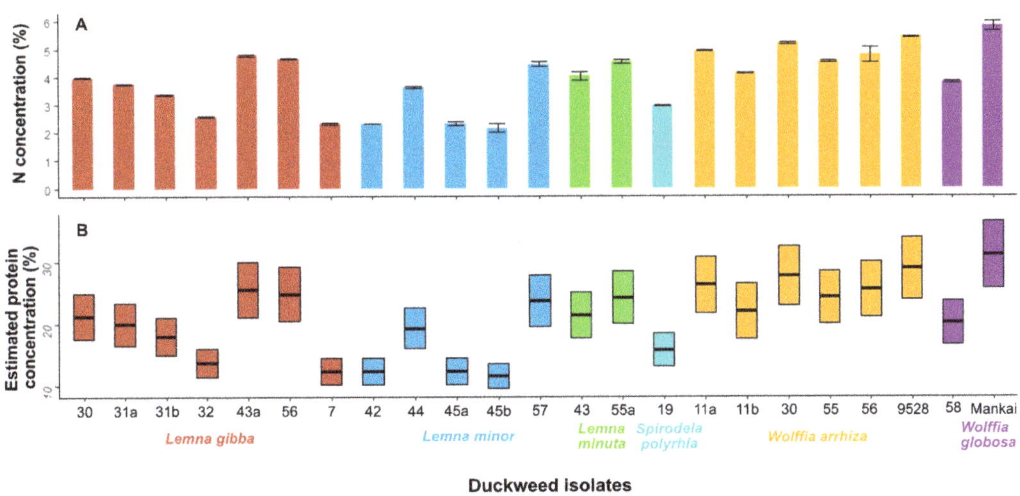

Figure 3. Nitrogen concentration (**A**) and estimated protein concentration (**B**) adjusted to duckweed dry biomass, which is calculated by multiplying nitrogen concentration by 4.4–6.25. The strain *W. arrhiza* 9528 was added to the analysis as a reference.

4. Discussion

In a campaign to explore the diversity of duckweed in Israel, we have followed almost 100 years of reports since 1926 (https://redlist.parks.org.il/plants/list/ (accessed on 22 September 2022)). We found duckweeds in ~40% of the seasonal and perennial water bodies explored (confirming 24 of 67 sights). In the confirmed sites, we collected the species, then we transferred them to the lab, there isolating and identifying the duckweeds. The identification was based on combined approaches that included morphology, molecular analyses and the biochemical features of the isolates (Tables 1 and 3). As a result, six species were confirmed (Figures 1 and 2), with five of the identified duckweed species previously reported in Israel. One of the identified species, *L. minuta*, was collected in Israel for the first time. In this approach, the two barcode regions previously proposed [31], *atpF-atpH* and *psbK-psbI*, have proven to well complement the morphological identification. Interestingly, this geographically limited study shows a pattern that was reported in geographically broader studies. Here, the two identified *Wolffia* species show stronger intraspecific differences between haplotypes than the other species [31], as was shown for the *W. globosa* haplotypes [14] identified by the two chloroplast markers in *Wolffia*. However, we note that the chloroplast markers do not allow for the identification of hybrids, which may well occur between the two species of *L. minor* and *L. gibba*, as recently described [32].

Three of the isolated species dominated the community: *L. gibba*, *L. minor* and *W. arrhiza* (accounting for 31, 17 and 31% of the isolates, respectively) and were previously sighted in Israel (https://biogis.huji.ac.il/heb/home.html (accessed on 25 September 2022)). We suggest that these species form the stable and established population of duckweeds in Israel. Their continuous presence in perennial and seasonal aquatic sites across Israel over the last century supports our hypothesis. Moreover, populations of *L. gibba*, *L. minor* and *W. arrhiza* were previously reported in North Africa, the Middle East and Europe (https://europlusmed.org/ (accessed on 25 September 2022)).

Three of the remaining species identified in this study, *L. minuta*, *W. globosa* and *S. polyrhiza*, were detected at a lower abundance (Figure 1 and Table 1). *S. polyrhiza* is native to the Middle East and Europe [33] and was previously sighted in Israel (https://biogis.huji.ac.il/heb/home.html (accessed on 26 September 2022)). Yet, *S. polyrhiza*'s low abundance detected in our survey (Figure 1 and Table 1) may reflect the species' growth inhibition by anthropogenic contamination, as was previously described [34,35]. The other two species identified in Israel, *L. minuta* and *W. globosa*, are not native to our region and are considered invasive species in Europe (https://europlusmed.org/ (accessed on 28 September 2022), [36]). *L. minuta* is native to America and is also an invasive species to Europe where it was first sighted in the 1990′s, probably through an accidental introduction [36]. *L. minuta*'s fast vegetative reproduction, high fitness and aggressive competition damages European lake ecosystems by inhibiting local duckweed populations such as *L. minor* [36]. *L. minuta* was spotted in the Hula Valley, a popular overnight break to many migrating waterfowl [25], hence, we hypothesize that the species was inadvertently carried by the migrating birds on their way from Europe to Africa and established there.

The species *W. globosa* is native to the Far East (Thailand, Cambodia and Laos [1]), China [37], and the Indian subcontinent [3]). Like *L. minor*, it was inadvertently introduced to Europe. Although *W. globosa* is the fastest propagating Angiosperm, with a doubling time as low as 72 h [3], it does not aggressively compete with native species but instead resides alongside them in water bodies across Europe [38]. In Israel, *W. globosa* was occasionally sited [5] but its presence in water bodies was transient (we were unable to find the species in the reported sites). The transient presence of *W. globosa* in Israel may suggest that it cannot survive the long Mediterranean summer drought and is reintroduced to various water bodies by birds migrating from Europe.

The molecular phylogeny of the duckweed species converged some isolates, suggesting a common haplotype (Figure 2), yet the biochemical features of species, including fatty acid profiles and nitrogen content, suggested diversification among the haplotypes (Table 3 and Figure 3). The nitrogen concentration results were validated by two independent methods (Elemental Analyzer (Figure 3) and FT-MIR (Figure S1)). The nitrogen contents of the haplotypes were diverse, but not always taxonomically associated. Diverging nitrogen concentrations among duckweeds were likewise reported for various haplotypes of the genera *Wolffia* [4,30] and *Lemna* [39]. There is a possibility that even if laboratory growth conditions are controlled, the optimal growth conditions for each haplotype and species can differ, which may have an effect on the biochemical composition, and specifically, on the nitrogen and protein content [4]. Fatty acid composition was consistent with previous duckweed surveys [15], showing the predominance of ALA (18:3*n*3) and the presence of a considerable amount of SDA (18:4*n*3) mostly in *Lemna* species. Some *Wolffia* species, yet not the ones isolated here, were shown to contain ALA and SDA (*W. australiana* and *W. microscopica* [30]). The presence of SDA, whose biosynthesis requires Δ6-desaturation, is restricted to only a few terrestrial plant families. However, this n-3 PUFA (SDA) widely occurs in cyanobacteria and algae, as well in some duckweeds [4,13,29], indicating the evolutionary radiation of the Δ6-desaturation, and this fatty acid's possible importance to duckweeds. Concomitant to the diverse nitrogen contents, some haplotypes of the same species showed variability in their fatty acid content, for instance SDA content in *L. gibba* (Table 3).

Our study suggests that both stable and transient duckweed communities are present side-by-side in Israel's water bodies. Almost 100 years of duckweed sightings suggest that *L. gibba*, *L. minor* and *W. arrhiza* inhabit the perennial water bodies but also survive the summer drought of seasonal ponds. Resilience of duckweed under drought gained little attention in contrast to survival during water freezing [40,41] and should be further explored, especially considering the currently changing climate. Three additional species were identified: *S. polyrhiza*, *L. minuta* and *W. globosa*. While the first was sighted in this region, the two latter are native to America and the Far East, respectively, and considered invasive species in Europe. We hypothesize that these species are transient in Israel, carried by migrating waterfowl on their way from Europe to Africa and established in water bodies as was previously proposed [16]. However, considering the aggressive nature of some invading species (like *L. minuta* [36]), their potential to endanger the fragile community of the endogenous duckweeds in Israel should be considered. However, to validate the community composition of duckweeds in Israel, additional surveys should be conducted covering wider spatial and temporal scales. These surveys should be accompanied by careful isolation of the species followed by a combination of methods to identify both their taxonomy and traits.

5. Conclusions

Here, we described the first survey of duckweed ever performed in Israel—a known junction between Mediterranean and arid environments. We isolated 27 duckweed haplotypes and used morphological and molecular approaches to identify them, resulting in six confirmed species. However, independent of the taxonomy, the haplotypes differed in their fatty acid profiles and protein contents. Three of the species were abundant among sites and confirmed by past sightings, thus, were proposed as being native to Israel. The other three species were rarer with two suspected invaders to our region. Thus, future surveys should be conducted to establish the identity and traits of the native duckweed communities in Mediterranean and arid regions.

Supplementary Materials: The following supporting information can be downloaded at: https://www.mdpi.com/article/10.3390/plants11233326/s1, Table S1. Duckweed plants collected in the present study—detailed location; Table S2. Water samples analysis: estimated pH and electric conductivity (EC); Table S3. Full table of fatty acid composition and content of duckweed plants isolates collected in the current study. The strain W. arrhiza 9528 was added to the analysis as a reference data continuance; Figure S1. Nitrogen concentration measured by N content FT-MIT; File S1. FTIR spectra collection.

Author Contributions: Conceptualization, A.F.Y. and O.G.; Data collection A.F.Y., L.G., L.K., K.S.K., C.B., B.P., N.R. and T.G.; methodology, A.F.Y., L.G., L.K. and M.B.; investigation, A.F.Y., S.D.-C. and I.K.-G.; resources, O.G. and I.K.-G.; data curation, A.F.Y.; writing—original draft preparation, A.F.Y., O.G. and I.K.-G.; writing—review and editing, A.F.Y., L.G., L.K., K.S.K., C.B., B.P., N.R., T.G., S.D.-C., M.B., I.K.-G. and O.G.; visualization, A.F.Y., B.P. and M.B.; supervision, O.G.; funding acquisition O.G. and I.K.-G. All authors have read and agreed to the published version of the manuscript.

Funding: This research was funded by Israel Ministry of Agriculture grant number 16-38-0038.

Institutional Review Board Statement: Not Applicable.

Informed Consent Statement: Not Applicable.

Data Availability Statement: Not Applicable.

Acknowledgments: This research was supported by the Israel Ministry of Agriculture Grants No 16-38-0038 to O.G. We would like to acknowledge the Negev Scholarship for the generous support to A.F.Y. We are grateful to Appenroth for his advice and for providing duckweed strains. We greatly appreciate Hernandez-Allica, Barrero-Sicilla, Mashanova, Espinosa-Montiel and Blachez for the FT-MIR analyses of the nitrogen concentration. We also greatly acknowledge the help and support from Hinoman, L.T.D. and especially Lapidot. We are grateful for the assistance given by the Israel Nature and Park Authority and for the information they provided (Data—Scientific Data Department, Israel Nature and Parks Authority).

Conflicts of Interest: The authors declare no conflict of interest.

References

1. Landolt, E. *Biosystematic Investigations in the Family of Duckweed (Lemnaceae)*; Geobotanischen Inst ETH: Zurich, Switzerland, 1986.
2. Bog, M.; Appenroth, K.-J.; Sree, K.S. Duckweed (Lemnaceae): Its Molecular Taxonomy. *Front. Sustain. Food Syst.* **2019**, *3*, 117. [CrossRef]
3. Sree, K.S.; Sudakaran, S.; Appenroth, K.-J. How Fast Can Angiosperms Grow? Species and Clonal Diversity of Growth Rates in the Genus Wolffia (Lemnaceae). *Acta Physiol. Plant* **2015**, *37*, 204. [CrossRef]
4. Appenroth, K.-J.; Sree, K.S.; Böhm, V.; Hammann, S.; Vetter, W.; Leiterer, M.; Jahreis, G. Nutritional Value of Duckweeds (Lemnaceae) as Human Food. *Food Chem.* **2017**, *217*, 266–273. [CrossRef] [PubMed]
5. Sela, I.; Yaskolka Meir, A.; Brandis, A.; Krajmalnik-Brown, R.; Zeibich, L.; Chang, D.; Dirks, B.; Tsaban, G.; Kaplan, A.; Rinott, E.; et al. Wolffia Globosa–Mankai Plant-Based Protein Contains Bioactive Vitamin B12 and Is Well Absorbed in Humans. *Nutrients* **2020**, *12*, 3067. [CrossRef] [PubMed]
6. Sońta, M.; Rekiel, A.; Batorska, M. Use of Duckweed (*Lemna* L.) in Sustainable Livestock Production and Aquaculture—A Review. *Ann. Anim. Sci.* **2019**, *19*, 257–271. [CrossRef]
7. Frédéric, M.; Samir, L.; Louise, M.; Abdelkrim, A. Comprehensive Modeling of Mat Density Effect on Duckweed (Lemna Minor) Growth under Controlled Eutrophication. *Water Res.* **2006**, *40*, 2901–2910. [CrossRef]
8. Iqbal, J.; Javed, A.; Baig, M.A. Growth and Nutrient Removal Efficiency of Duckweed (Lemna Minor) from Synthetic and Dumpsite Leachate under Artificial and Natural Conditions. *PLoS ONE* **2019**, *14*, e0221755. [CrossRef]
9. Xu, J.; Zhao, H.; Stomp, A.-M.; Cheng, J.J. The Production of Duckweed as a Source of Biofuels. *Biofuels* **2012**, *3*, 589–601. [CrossRef]
10. Ren, H.; Jiang, N.; Wang, T.; Mubashar Omar, M.; Ruan, W.; Ghafoor, A. Enhanced Biogas Production in the Duckweed Anaerobic Digestion Process. *J. Energy Resour. Technol.* **2018**, *140*, 041805. [CrossRef]
11. Les, D.H.; Landolt, E.; Crawford, D.J. Systematics of TheLemnaceae (Duckweeds): Inferences from Micromolecular and Morphological Data. *Plant Syst. Evol.* **1997**, *204*, 161–177. [CrossRef]
12. Crawford, D.J.; Landolt, E.; Les, D.H.; Kimball, R.T. Allozyme Studies in *Lemnaceae*: Variation and Relationships in *Lemna* Sections *Alatae* and *Biformes*. *Taxon* **2001**, *50*, 987–999. [CrossRef]
13. Tang, J.; Li, Y.; Ma, J.; Cheng, J.J. Survey of Duckweed Diversity in Lake Chao and Total Fatty Acid, Triacylglycerol, Profiles of Representative Strains. *Plant Biol. J.* **2015**, *17*, 1066–1072. [CrossRef] [PubMed]
14. Bog, M.; Schneider, P.; Hellwig, F.; Sachse, S.; Kochieva, E.Z.; Martyrosian, E.; Landolt, E.; Appenroth, K.-J. Genetic Characterization and Barcoding of Taxa in the Genus Wolffia Horkel Ex Schleid. (Lemnaceae) as Revealed by Two Plastidic Markers and Amplified Fragment Length Polymorphism (AFLP). *Planta* **2013**, *237*, 1–13. [CrossRef] [PubMed]
15. Wang, W.; Wu, Y.; Yan, Y.; Ermakova, M.; Kerstetter, R.; Messing, J. DNA Barcoding of the Lemnaceae, a Family of Aquatic Monocots. *BMC Plant Biol.* **2010**, *10*, 205. [CrossRef]
16. Tippery, N.P.; Les, D.H. Tiny Plants with Enormous Potential: Phylogeny and Evolution of Duckweeds. In *The Duckweed Genomes*; Cao, X.H., Fourounjian, P., Wang, W., Eds.; Compendium of Plant Genomes; Springer International Publishing: Cham, Switzerland, 2020; pp. 19–38. ISBN 978-3-030-11044-4.
17. Sree, K.S.; Adelmann, K.; Garcia, C.; Lam, E.; Appenroth, K.-J. Natural Variance in Salt Tolerance and Induction of Starch Accumulation in Duckweeds. *Planta* **2015**, *241*, 1395–1404. [CrossRef]
18. Kuehdorf, K.; Jetschke, G.; Ballani, L.; Appenroth, K.-J. The Clonal Dependence of Turion Formation in the Duckweed *Spirodela Polyrhiza*—An Ecogeographical Approach. *Physiol. Plantarum.* **2014**, *150*, 46–54. [CrossRef]
19. Sree, K.; Bog, M.; Appenroth, K. Taxonomy of Duckweeds (Lemnaceae), Potential New Crop Plants. *Emir. J. Food Agric.* **2016**, *28*, 291. [CrossRef]

20. Les, D.H.; Crawford, D.J.; Kimball, R.T.; Moody, M.L.; Landolt, E. Biogeography of Discontinuously Distributed Hydrophytes: A Molecular Appraisal of Intercontinental Disjunctions. *Int. J. Plant Sci.* **2003**, *164*, 917–932. [CrossRef]
21. Landolt, E. Lemnaceae. In *Flowering Plants Monocotyledons*; Kubitzki, K., Ed.; Springer: Berlin/Heidelberg, Germany, 1998; pp. 264–270. ISBN 978-3-642-08378-5.
22. Kimball, R.T.; Crawford, D.J.; Les, D.H.; Landolt, E. Out of Africa: Molecular Phylogenetics and Biogeography of Wolffiella (Lemnaceae): WOLFFIELLA PHYLOGENETICS and BIOGEOGRAPHY. *Biol. J. Linn. Soc.* **2003**, *79*, 565–576. [CrossRef]
23. Green, A.J. The Importance of Waterbirds as an Overlooked Pathway of Invasion for Alien Species. *Divers. Distrib.* **2016**, *22*, 239–247. [CrossRef]
24. Frankenberg, E. Will the Biogeographical Bridge Continue to Exist? *Isr. J. Zool.* **1999**, *45*, 65–74. [CrossRef]
25. Frumkin, R.; Pinshow, B.; Kleinhaus, S. A Review of Bird Migration over Israel. *J. Ornithol.* **1995**, *136*, 127–147. [CrossRef]
26. Kirby, J.S.; Stattersfield, A.J.; Butchart, S.H.M.; Evans, M.I.; Grimmett, R.F.A.; Jones, V.R.; O'Sullivan, J.; Tucker, G.M.; Newton, I. Key Conservation Issues for Migratory Land- and Waterbird Species on the World's Major Flyways. *Bird Conserv. Int.* **2008**, *18*, S49–S73. [CrossRef]
27. Reynolds, C.; Miranda, N.A.F.; Cumming, G.S. The Role of Waterbirds in the Dispersal of Aquatic Alien and Invasive Species. *Divers. Distrib.* **2015**, *21*, 744–754. [CrossRef]
28. Yan, Y.; Candreva, J.; Shi, H.; Ernst, E.; Martienssen, R.; Schwender, J.; Shanklin, J. Survey of the Total Fatty Acid and Triacylglycerol Composition and Content of 30 Duckweed Species and Cloning of a Δ6-Desaturase Responsible for the Production of γ-Linolenic and Stearidonic Acids in Lemna Gibba. *BMC Plant Biol.* **2013**, *13*, 201. [CrossRef]
29. Jones, D. *Factors for Converting Percentages of Nitrogen in Foods and Feeds into Percentages of Proteins Re*; US Department of Agriculture, USDA Publications: Washington, DC, USA, 1931.
30. Appenroth, K.-J.; Sree, K.S.; Bog, M.; Ecker, J.; Seeliger, C.; Böhm, V.; Lorkowski, S.; Sommer, K.; Vetter, W.; Tolzin-Banasch, K.; et al. Nutritional Value of the Duckweed Species of the Genus Wolffia (Lemnaceae) as Human Food. *Front. Chem.* **2018**, *6*, 483. [CrossRef]
31. Borisjuk, N.; Chu, P.; Gutierrez, R.; Zhang, H.; Acosta, K.; Friesen, N.; Sree, K.S.; Garcia, C.; Appenroth, K.J.; Lam, E. Assessment, Validation and Deployment Strategy of a Two-Barcode Protocol for Facile Genotyping of Duckweed Species. *Plant Biol. J.* **2015**, *17*, 42–49. [CrossRef]
32. Braglia, L.; Breviario, D.; Gianì, S.; Gavazzi, F.; De Gregori, J.; Morello, L. New Insights into Interspecific Hybridization in *Lemna* L. Sect. Lemna (Lemnaceae Martinov). *Plants* **2021**, *10*, 2767. [CrossRef]
33. Ho, E.K.H.; Bartkowska, M.; Wright, S.I.; Agrawal, A.F. Population Genomics of the Facultatively Asexual Duckweed *Spirodela Polyrhiza*. *New Phytol.* **2019**, *224*, 1361–1371. [CrossRef]
34. Xing, W.; Huang, W.; Liu, G. Effect of Excess Iron and Copper on Physiology of Aquatic Plant *Spirodela polyrrhiza* (L.) Schleid. *Environ. Toxicol.* **2009**. [CrossRef]
35. Caicedo, J. Effect of Total Ammonia Nitrogen Concentration and PH on Growth Rates of Duckweed (*Spirodela polyrrhiza*). *Water Res.* **2000**, *34*, 3829–3835. [CrossRef]
36. Ceschin, S.; Abati, S.; Leacche, I.; Zuccarello, V. Ecological Comparison between Duckweeds in Central Italy: The Invasive *Lemna Minuta* vs. the Native *L. Minor*. *Plant Biosyst. Int. J. Deal. All Asp. Plant Biol.* **2018**, *152*, 674–683. [CrossRef]
37. Yuan, J.-X.; Pan, J.; Wang, B.-S.; Zhang, D.-M. Genetic Differentiation of Wolffia Globosa in China. *J. Syst. Evol.* **2011**, *49*, 509–517. [CrossRef]
38. Lansdown, R.; Kitchener, G.; Jones, E. Wolffia Columbiana and W. Globosa (Araceae) New to Britain. *Br. Ir. Bot.* **2022**, *4*. [CrossRef]
39. Walsh, É.; Cialis, E.; Dillane, E.; Jansen, M.A.K. Lemnaceae Clones Collected from a Small Geographic Region Display Diverse Traits Relevant for the Remediation of Wastewater. *Environ. Technol. Innov.* **2022**, *28*, 102599. [CrossRef]
40. Fourounjian, P.; Fakhoorian, T.; Cao, X.H. Importance of Duckweeds in Basic Research and Their Industrial Applications. In *The Duckweed Genomes*; Cao, X.H., Fourounjian, P., Wang, W., Eds.; Compendium of Plant Genomes; Springer International Publishing: Cham, Switzerland, 2020; pp. 1–17. ISBN 978-3-030-11044-4.
41. O'Brien, A.M.; Yu, Z.H.; Luo, D.; Laurich, J.; Passeport, E.; Frederickson, M.E. Resilience to Multiple Stressors in an Aquatic Plant and Its Microbiome. *Am. J. Bot.* **2020**, *107*, 273–285. [CrossRef]

plants

MDPI

Communication

Chromosome Numbers and Genome Sizes of All 36 Duckweed Species (*Lemnaceae*)

Phuong T. N. Hoang †,‡, Jörg Fuchs †, Veit Schubert, Tram B. N. Tran ‡ and Ingo Schubert *

Leibniz Institute of Plant Genetics and Crop Plant Research (IPK), Gatersleben, D-06466 Stadt Seeland, Germany
* Correspondence: schubert@ipk-gatersleben.de
† These authors contributed equally.
‡ Permanent address: Biology Faculty, Dalat University, District 8, 66000 Dalat, Vietnam.

Abstract: Usually, chromosome sets (karyotypes) and genome sizes are rather stable for distinct species and therefore of diagnostic value for taxonomy. In combination with (cyto)genomics, both features provide essential cues for genome evolution and phylogenetic relationship studies within and between taxa above the species level. We present for the first time a survey on chromosome counts and genome size measurement for one or more accessions from all 36 duckweed species and discuss the evolutionary impact and peculiarities of both parameters in duckweeds.

Keywords: chromosome number; duckweeds; evolution; genome size; karyotype; *Lemnaceae*

1. Introduction

A review about the aquatic monocot model family of duckweeds and its emerging economic importance has been published recently [1]. However, at this time, not for all duckweed species chromosome counts and genome sizes were known. Here, we compile corresponding data for at least one accession (clone) for each species, for some of them for the first time (see Figure 1, Results and Discussion). For previous data see [2] and references therein.

check for updates

Citation: Hoang, P.T.N.; Fuchs, J.; Schubert, V.; Tran, T.B.N.; Schubert, I. Chromosome Numbers and Genome Sizes of All 36 Duckweed Species (*Lemnaceae*). *Plants* **2022**, *11*, 2674. https://doi.org/10.3390/plants11202674

Academic Editors: Viktor Oláh, Klaus-Jürgen Appenroth and K. Sowjanya Sree

Received: 28 June 2022
Accepted: 8 October 2022
Published: 11 October 2022

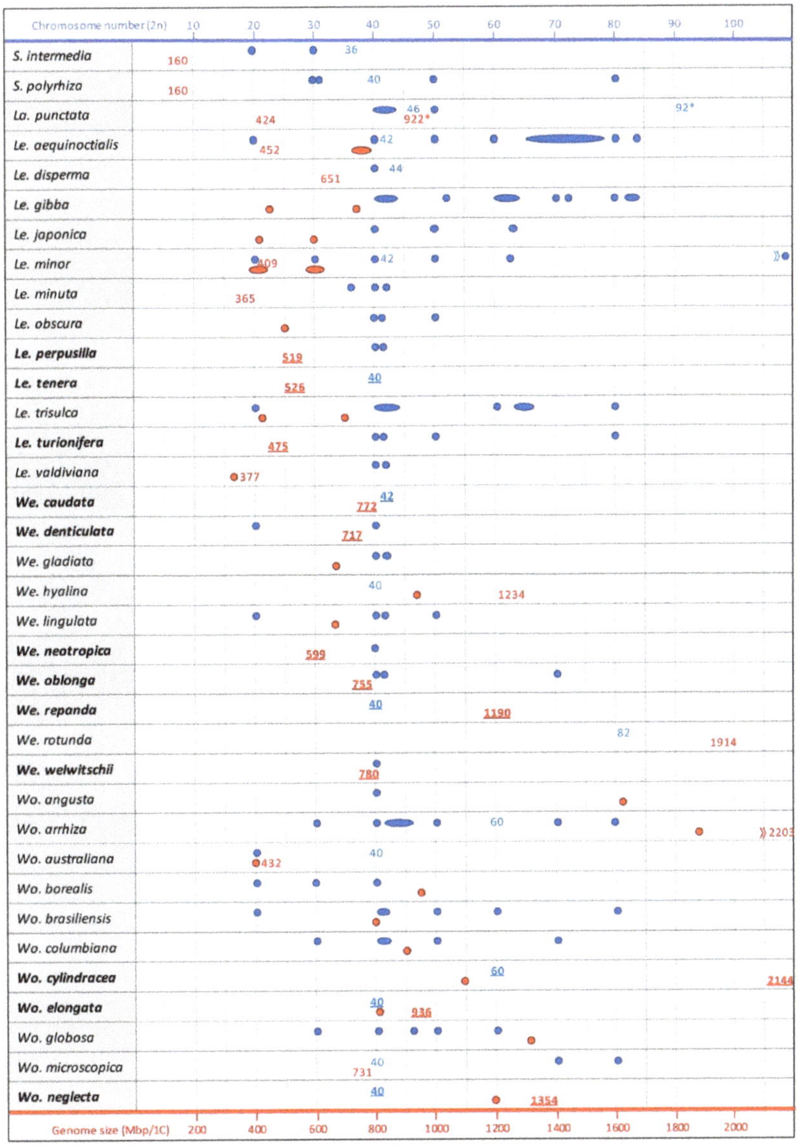

Figure 1. Chromosome number (blue) and genome size (red) of duckweeds. When chromosome numbers and genome sizes refer to newly investigated species or were taken from [2] or [3] (*Le. minuta*), they appear as numbers. For references to other chromosome counts (blue dots and ovals [for closely adjacent values]) see [2]. Other genome size estimations (red dots and ovals) were from [4]. The values for *Le. perpusilla* 8539 (519 Mbp), *Le. tenera* 9024 (526 Mbp), *Le. turionifera* 8693 (475 Mbp), *We. caudata* 9165 (772 Mbp); *We. denticulata* 8221 (717 Mbp), *We. neotropica* 8848 (599 Mbp), *We. oblonga* 9391 (755 Mbp), $2n = 42$), *We. repanda* 9062 (1190 Mbp, $2n = 40$), *We. welwitschii* 9469 (780 Mbp), *Wo. cylindracea* 9056 (2144 Mbp; $2n = 60$), *Wo. elongata* 9188 (936 Mbp; $2n = 40$), *Wo. neglecta* 9149 (1354 Mbp; $2n = 40$) were obtained by flow-cytometry and determined in this paper. Species for which we provide in this paper the first or new values are in bold and the corresponding values are underlined. * indicates the values for the colchicine-induced tetraploid *La. punctata* clone 5562.

2. Material and Methods

The duckweed accessions investigated in this paper are listed in the legend of Figure 1 and in Figure 2 and were kindly provided by Dr. K.-J. Appenroth, Friedrich-Schiller-University, Jena, Germany. For plant cultivation, chromosome preparation, super-resolution microscopy and genome size measurement, see [2].

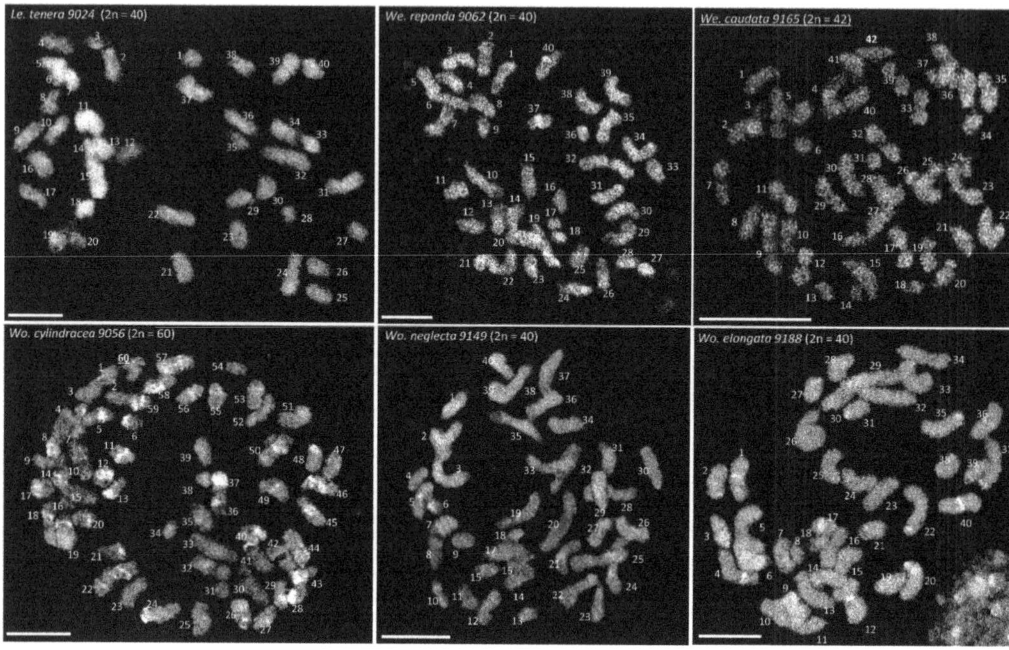

Figure 2. Metaphases with chromosome counts reported for the first time in this paper. To count clearly the chromosome number, image stacks from spatial super-resolution structured illumination microscopy (3D-SIM) [3] were used to identify overlapping chromosomes. Individual chromosomes were enumerated arbitrarily. Scale bar: 5 µm.

3. Results and Discussion

3.1. The Chromosomes

At least for one accession of all 36 duckweed species, chromosome counts are now available (for review see Figure 1 and [2]). For six species, *Lemna tenera* 9024 ($2n = 40$), *Wolffiella caudata* 9165 ($2n = 42$), *Wolffiella repanda* 9062 ($2n = 40$), *Wolffia cylindracea* 9056 ($2n = 60$), possibly a triploid accession, *Wolffia elongata* 9188 ($2n = 40$) and *Wolffia neglecta* 9149 ($2n = 40$), chromosome counts (Figure 2) are presented for the first time in this paper. Most frequent are chromosome counts around $2n = 40$. However, in some papers [2,5,6] (and in older references in [2,4]), highly variable chromosome numbers were reported for the same species, e.g., for *Lemna aequinoctialis* 20, 40, 50, 60 and 80 [5]; 40, 50, 66, 72, 78, 84 and 65–76 [7]; 42, 84 [6]; 40, 60, 80 [4]; and 42 [2]. Despite the predominantly asexual propagation of many duckweed species, such a high variability is strange. Moreover, in cases where previously variable chromosome numbers were reported and the corresponding clones are still available, deviating numbers could not be confirmed [2]. In all tested clones of *Spirodela polyrhiza* ($2n = 40$) and *Spirodela intermedia* ($2n = 36$), we constantly found the same chromosome number, even for clones for which deviating numbers were reported before (see [2]). Nevertheless, smaller deviations in chromosome number between asexual clones of the same species cannot be excluded due to the potential presence of B chromosomes. Some of the samples apparently indicate autotetraploidy (*Le. aequinoctialis*, clone 2018 $2n = 42$, and clone 6746 $2n = 84$, refs. [2,6]) which occurred spontaneously, or was induced [8] as in *Landoltia punctata* with $2n = 46$ and $2n = 92$ (clone 5562 [2]). However, the quite high number of rather small chromosomes, which barely show any morphological differentiation (primary constrictions = centromeres, and secondary constrictions = nucleolus organizing regions [NOR], are mostly not discernible), and which are difficult to prepare from the small meristems, can lead to counting errors. For instance, for *Wolffiella rotunda* clone 9072 ($2n = 82$), fluorescent in situ hybridization (FISH), with 45S rDNA as a probe, revealed chromosome satellites, distally flanking the NOR, which could have easily been miscounted as small separate chromosomes [2] (Figure S3 therein). Furthermore, small and poorly structured chromosomes often make it difficult to decide whether they are mono- or holocentric.

3.2. The Genome Sizes

There are several approaches to determining genome size, defined as pg/2C or Mbp/1C (C means the unreplicated haploid [or meiotically reduced] genome). Older papers applied Feulgen densitometry of nuclei, later ones mostly use flow cytometry [9] which allows for a much larger number of nuclei to be measured. These direct measurements require an internal reference standard with a known genome size close to that of the target species. Recently, based on sequenced genomes, k-mer distribution, as a bioinformatic approach, also served for genome size estimation. The accuracy of flow cytometry depends on the quality of the prepared nuclei suspension and the suitability of the used buffer and reference species. For precise genome size measurements, it is recommended that the coefficient of variation (CV) does not exceed 3% [9]. The k-mer size results provide a useful estimation, which should be complemented by direct measurements [10,11]. For all 36 duckweed species of the five genera, the genome size has been measured (Figure 1 and [2–4,12]). The values for *Lemna perpusilla* 8539 (519 Mbp), *Le. tenera* 9024 (526 Mbp), *Lemna turionifera* 8693 (475 Mbp), *We. caudata* 9165 (772 Mbp); *Wolffiella denticulata* 8221 (717 Mbp), *Wolffiella neotropica* 8848 (599 Mbp), *Wolffiella oblonga* 9391 (755 Mbp), $2n = 42$), *We. repanda* 9062 (1190 Mbp, $2n = 40$), *Wolffiella welwitschii* 9469 (780 Mbp), *Wo. cylindracea* 9056 (2144 Mbp; $2n = 60$), *Wo. elongata* 9188 (936 Mbp; $2n = 40$) and *Wo. neglecta* 9149 (1354 Mbp; $2n = 40$) were obtained by flow-cytometry in this paper. The results yielded a ~14-fold range from 160 Mbp in the phylogenetically oldest genus, comprising *S. polyrhiza* and *S. intermedia*, up to 2203 Mbp of *Wolffia arrhiza* of the phylogenetically youngest genus [2]. The largest intrageneric variation was found in the genus *Wolffia* with 432 Mbp for *Wolffia australiana* to 2203 Mbp for *Wo. arrhiza*. For 11 species representing all duckweed genera, genome size was positively correlated with nuclear and cell volume of guard cells, and with progressive organ reduction, but negatively with frond size and phylogenetic age. No correlation was observed with the number of chromosomes or the rDNA loci [2]. The larger genomes of phylogenetically younger genera do not necessarily indicate a general evolutionary genome expansion in duckweeds. The high ratio of soloLTRs to intact retroelements of ~8 in *S. polyrhiza* [13], and the respective value of 11–14 in *Wo. australiana* [14]—the species with the smallest genome in its genus—could rather be a hint that the oldest genus *Spirodela*, after early branching from the other duckweeds, experienced genome shrinking due to a deletion-biased DNA double-strand break repair pathway [15]. The same could be true for *Wo. australiana* within its genus. In these cases, the likely scenario would be a bi-directional evolution of genome size starting from an intermediate ancestral duckweed genome of ~400 to 800 Mbp. Studies on DNA double-strand break repair in species with a high ratio of soloLTRs to intact retroelements could test this assumption. Genome size estimations for *S. polyrhiza*, *S. intermedia* and *La. punctata* from different laboratories are rather similar [2–4]. The other genomes measured by Hoang et al. [2] are in general larger than those measured for the same species by Wang et al. [4]. This might be caused by the use of different internal reference standards or different equipment, but most likely by assuming a different genome size of *Arabidopsis thaliana* as a basis for calculation (157 Mbp in [2], *versus* 147 Mbp in [4]). An initially unexpected finding was the strong variation of genome size estimation for clones of the same species. In case of clones of *Le. aequinoctialis* and *La. punctata,* the explanation is spontaneous, respectively induced, whole genome duplication (WGD) [2] (Figure 1). The genome sizes for two groups of *Le. minor* clones (~400 Mbp *versus* ~600 Mbp), however, differ by >50% ([4]; M. Bog, J. Fuchs, K.-J. Appenroth unpublished). For clone 5500, a genome size of 481 Mbp was estimated [16] and 409 Mbp for clone 8623 with $2n = 42$ chromosomes [2]. For clone 8627, we counted 63 chromosomes and measured a genome size of >600 Mbp (PTNH & JF unpublished). Molecular fingerprinting for tubulin genes [17] suggested that clone 8627 belongs to *Lemna japonica* and is a hybrid of *Lemna minor* and *Le. turionifera.* Our data identified it as triploid. Further investigations are needed to uncover whether in other duckweed species with a high intraspecific variation of DNA content, clone-specific WGD with a fast subsequent genome size reduction occurred, or whether these groups of clones represent cryptic species or interspecific hybrids. The occurrence of interspecific hybrids in duckweeds is a surprise because they require sexual propagation—sometimes even involving unreduced gametes as in clone 8627 (see above)—which, according to laboratory observations, might occur very rarely in predominantly vegetatively propagating duckweed species. On the other hand, the maintenance of interspecific hybrids seems to be favored by asexual propagation.

3.3. *Evolutionary Impact of Genome Size and Karyotype Studies*

Chromosome numbers and genome size, varying in parallel by the same whole-number multiple, suggest a recent WGD (neopolyploidy) usually yielding autotetraploids in asexual clones. However, a nearly doubled genome size (or its absence) as well as a (nearly) doubled chromosome number (or its absence) alone are not sufficient to decide whether or not a WGD took place. Dysploid chromosome number reduction and/or fast reduction of genome size may blur real WGD. Therefore, in such cases, additional independent approaches are mandatory for arriving at conclusive statements. For instance, in Australian *Brassicaceae* with $n = 4$ to 6, evidence for mesopolyploidy (descendent dysploidy after WGD) was found through multicolor cross-FISH with bacterial artificial chromosome (BAC) pools representing the genome of *A. thaliana* [18]. All regions labeled by BAC pools from Arabidopsis, which represent the eight ancestral *Brassicaceae* chromosomes, appeared duplicated within the only four to six meiotic bivalents of the tested species. On the other hand, a genome can expand to double and more without WGD just by insertion-biased double-strand break repair (mainly retroelement insertion). In general, confirmed WGD leading to neopolyploidy seems to be more frequent between clones of distinct duckweed species than being responsible for genome size differences between species of the same or different genera. Of the natural species studied, only *We. rotunda* (clone 9072: $2n = 82$; 1914 Mbp/1C), *Wo. arrhiza* (clone 8872: $2n = 60$; 2203 Mbp/1C) and *Wo. cylindracea* (clone 9056: $2n = 60$; 2144 Mbp/1C) are, so far, candidates for neopolyploids.

Karyotype studies employing molecular cytogenetics may help to elucidate the evolutionary origin of polyploid species. Genomic in situ hybridization (GISH) can identify the parental species if species-specific repetitive sequences are present in the genomes of the suspected parental species. For 'quasi-diploid' or paleopolyploid species (in fact all land plants experienced at least one very remote WGD), FISH with anchored unique sequences is an independent and direct approach to confirm or correct genome assemblies which are based on probabilistic methods. In this way, a robust chromosome-scale genome map has been achieved in several steps for *S. polyrhiza* [19–21]. Cross-FISH with single copy sequences can uncover chromosome homeology and chromosome rearrangements between related species. The two *Spirodela* species (*S. polyrhiza*, $2n = 40$ and *S. intermedia*, $2n = 36$) were studied by sequential multicolor cross-FISH with different pools of 96 BACs anchored in the genome of *S. polyrhiza* [22] and compared with genome assemblies for both species [21]. Eight chromosome pairs did not reveal rearrangements between both species. The other twelve chromosome pairs of *S. polyrhiza* correspond to the remaining ten pairs of *S. intermedia* and display (in part multiple) rearrangements (Figure 3). The direction of the evolution cannot be determined with certainty, because the genus comprises only these two species. However, because most duckweed species tend to have 40 or more chromosomes, 40 might be the ancestral and 36 the derived chromosome number. No rearrangements were found between seven clones from different geographic origins of *S. polyrhiza* [20] and two clones of *S. intermedia* [21], respectively. These studies showed clearly that *S. polyrhiza* and *S. intermedia* are different species, despite similar genome size and overlap of most morphological features. In contrast to the situation described for the *Brassicaceae* family (e.g., ref. [23]), cross-FISH with *S. polyrhiza* BACs yielded only weak and dispersed signals but did not recognize chromosome homeology when applied to species of other duckweed genera (*La. punctata*, *Le. aequinoctialis*, *Wolffiella hyalina*, *Wo. arrhiza*), independently of stringency conditions [24]. So far, cross-FISH with oligo probes derived from chromosome assemblies of *S. polyrhiza* or *S. intermedia* also did not reveal chromosome homeology in other duckweed genera [25]. The probes yielded either dispersed signals in *La. punctata* and *Le. aequinoctialis* or no signals at all as in *We. hyalina* and *Wo. australiana*. Apparently, these genomes are too diverse and probe densities are too low to generate reliable signals across the duckweed genera. In future, homologous single copy sequences, selected assuming synteny between the genomes in question, should be designed for oligo-FISH to unassembled genomes [25].

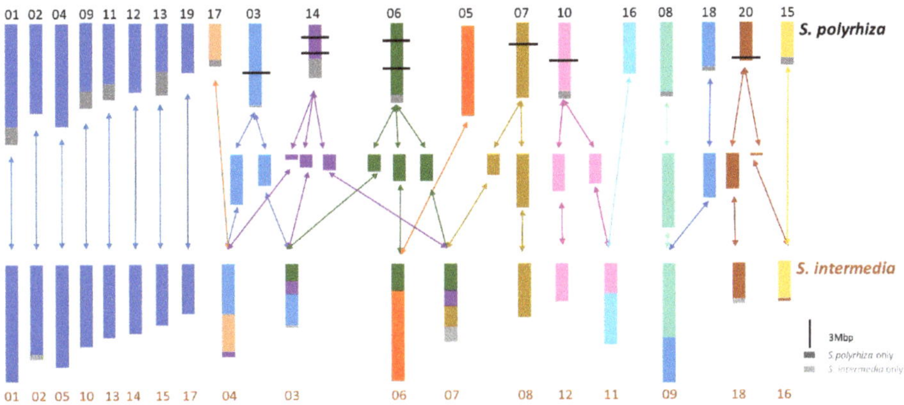

Figure 3. Chromosome homeology and rearrangements between *S. polyrhiza* (n = 20) and *S. intermedia* (n = 18) as revealed by cytogenomics, see [21,22]. The double arrows indicate that either direction of evolution could have happened. Dark blue are chromosomes not involved in rearrangements; other colors indicate rearranged chromosomes. Dark grey boxes: regions present only in *S. polyrhiza*; light grey boxes: regions present only in *S. intermedia*. Enumeration indicates distinct chromosomes of the respective species.

4. Conclusions

Having now compiled chromosome counts and genome sizes for at least one accession of all 36 duckweed species (for some species these are the first data), a basis is available upon which to investigate the evolution of duckweed genomes as well as the reasons for the apparent intraspecific variation of both parameters. A future combination of genomic data, genome size, chromosome number, GISH and FISH data will clarify the phylogenetic position and taxonomic status of intrageneric duckweed accessions which are difficult to assign to distinct species on the basis of morphological criteria.

Author Contributions: Conceptualization and funding aquisition, I.S.; data analysis P.T.N.H., J.F., V.S. and T.B.N.T., composition and writing, I.S., P.T.N.H., J.F. and V.S. All authors have read and agreed to the published version of the manuscript.

Funding: This work was supported in part by the German Research Foundation to IS (SCHU 951/18-1). PTNH was supported by the Vietnam National Foundation for Science and Technology Development (NAFOSTED) grant # 106.01-2020.33.

Data Availability Statement: Not applicable.

Conflicts of Interest: The authors declare no conflict of interest.

References

1. Acosta, K.; Appenroth, K.-J.; Borisjuk, L.; Edelman, M.; Heinig, U.; Jansen, M.; Oyama, T.; Pasaribu, B.; Schubert, I.; Sorrels, S.; et al. Return of the Lemnaceae: Duckweed as a model plant system in the genomics and post-genomics era. *Plant Cell* **2021**, *33*, 3207–3234. [CrossRef] [PubMed]
2. Hoang, P.T.N.; Schubert, V.; Meister, A.; Fuchs, J.; Schubert, I. Variation in genome size, cell and nucleus volume, chromosome number and rDNA loci among duckweeds. *Sci. Rep.* **2019**, *9*, 3234. [CrossRef] [PubMed]
3. Bog, M.; Sree, K.S.; Xu, S.; Himmelbach, A.; Brandt, R.; Fuchs, J.; Hoang, P.T.N.; Schubert, I.; Kuever, J.; Rabenstein, A.; et al. A taxonomic revision of the section *Uninerves*, genus *Lemna*. *Taxon* **2020**, *69*, 56–66. [CrossRef]
4. Wang, W.; Kerstetter, R.A.; Michael, T.P. Evolution of genome size in duckweeds (*Lemnaceae*). *J. Bot.* **2011**, *2011*, 570319. [CrossRef]
5. Urbanska, W.K. Cytological variation within the family of *Lemnaceae*. In *Veröffentlichungen des Geobotanischen Institutes der Eidg. Tech. Hochschule, Stiftung Rübel, in Zürich*; ETH Library: Zürich, Switzerland, 1980. [CrossRef]
6. Geber, G. Zur Karyosystematik der Lemnaceae. Ph.D. Thesis, University of Vienna, Vienna, Austria, 1989.
7. Beppu, T.; Takimoto, A. Geographical distribution and cytological variation of Lemna paucicostata Hegelm. in Japan. *Bot. Mag.* **1981**, *94*, 11–20. [CrossRef]
8. Vunsh, R.; Heinig, U.; Malitsky, S.; Aharoni, A.; Lerner, A.; Edelman, M. Manipulating duckweed through genome duplication. *Plant Biol.* **2014**, *17*, 115–119. [CrossRef] [PubMed]
9. Dolezel, J.; Greilhuber, J.; Suda, J. Estimation of nuclear DNA content in plants using flow cytometry. *Nat. Protoc.* **2007**, *2*, 2233–2244. [CrossRef]
10. Sun, H.; Ding, J.; Piednoel, M.; Schneeberger, K. *findGSE: Estimating* genome size variation within human and *Arabidopsis* using *k*-mer frequencies. *Bioinformatics* **2018**, *34*, 550–557. [CrossRef]
11. Pflug, J.M.; Holmes, V.R.; Burrus, C.; Johnston, J.S.; Maddison, D.R. Measuring genome sizes using read-depth, k-mers, and flow cytometry: Methodological comparisons in beetles (Coleoptera). *G3* **2020**, *10*, 3047–3060. [CrossRef]
12. Bog, M.; Lautenschläger, U.; Landrock, M.F.; Landolt, E.; Fuchs, J.; Sree, K.S.; Oberprieler, C.; Appenroth, K.J. Genetic characterization and barcoding of taxa in the genera Landoltia and Spirodela (*Lemnaceae*) by three plastidic markers and amplified fragment length polymorphism (AFLP). *Hydrobiologia* **2015**, *749*, 169–182. [CrossRef]
13. Michael, T.P.; Bryant, D.; Gutierrez, R.; Borisjuk, N.; Chu, P.; Zhang, H.; Xia, J.; Zhou, J.; Peng, H.; El Baidouri, M.; et al. Comprehensive definition of genome features in *Spirodela polyrhiza* by high-depth physical mapping and short-read DNA sequencing strategies. *Plant J.* **2017**, *89*, 617–635. [CrossRef]
14. Michael, T.P.; Ernst, E.; Hartwick, N.; Chu, P.; Bryant, D.; Gilbert, S.; Ortleb, S.; Baggs, E.L.; Sree, S.K.; Appenroth, K.J.; et al. Genome and time-of-day transcriptome of *Wolffia australiana* link morphological extreme minimization with gene loss and less growth control. *Genome Res.* **2021**, *31*, 225–238. [CrossRef]
15. Schubert, I.; Vu, G.T.H. Genome stability and evolution: Attempting a holistic view. *Trends Plant Sci.* **2016**, *21*, 749–757. [CrossRef] [PubMed]
16. Van Hoeck, A.; Horemans, N.; Monsieurs, P.; Cao, H.X.; Vandenhove, H.; Blust, R. The first draft genome of the aquatic model plant *Lemna minor* opens the route for future stress physiology research and biotechnological applications. *Biotechnol. Biofuels* **2015**, *8*, 188. [CrossRef]
17. Braglia, L.; Lauria, M.; Appenroth, K.-J.; Bog, M.; Breviario, D.; Grasso, A.; Gavazzi, F.; Morello, L.; Giani, S. Duckweed species genotyping and interspecific hybrid discovery by tubulin-based polymorphism fingerprinting. *Front. Plant Sci.* **2021**, *12*, 625670. [CrossRef]
18. Mandakova, T.; Joly, S.; Krzywinski, M.; Mummenhoff, K.; Lysak, M.A. Fast diploidization in close mesopolyploid relatives of *Arabidopsis*. *Plant Cell* **2010**, *22*, 2277–2290. [CrossRef] [PubMed]
19. Cao, H.X.; Vu, G.T.H.; Wang, W.; Appenroth, K.J.; Messing, J.; Schubert, I. The map-based genome sequence of *Spirodela polyrhiza* aligned with its chromosomes, a reference for karyotype evolution. *New Phytol.* **2016**, *209*, 354–363. [CrossRef] [PubMed]
20. Hoang, P.N.T.; Michael, T.P.; Gilbert, S.; Chu, P.; Motley, T.S.; Appenroth, K.J.; Schubert, I.; Lam, E. Generating a high-confidence reference genome map of the Greater Duckweed by integration of cytogenomic, optical mapping and Oxford Nanopore technologies. *Plant J.* **2018**, *96*, 670–684. [CrossRef]
21. Hoang, P.T.N.; Fiebig, A.; Novak, P.; Macas, J.; Cao, H.X.; Stepaneko, A.; Chen, G.; Borisjuk, N.; Scholz, U.; Schubert, I. Chromosome-scale genome assembly for the duckweed *Spirodela intermedia*, integrating cytogenetic maps, PacBio and Oxford Nanopore libraries. *Sci. Rep.* **2020**, *10*, 19230. [CrossRef]
22. Hoang, P.T.N.; Schubert, I. Reconstruction of chromosome rearrangements between the two most ancestral duckweed species *Spirodela polyrhiza* and *S. intermedia*. *Chromosoma* **2017**, *126*, 729–739. [CrossRef]

23. Lysak, M.A.; Berr, A.; Pecinka, A.; Schmidt, R.; McBreen, K.; Schubert, I. Mechanisms of chromosome number reduction in *Arabidopsis thaliana* and related *Brassicaceae* species. *Proc. Natl. Acad. Sci. USA* **2006**, *103*, 5224–5229. [CrossRef] [PubMed]
24. Hoang, P.T.N. Comparative Cytology and Cytogenomics for Representative Species of the Five Duckweed Genera. Ph.D. Thesis, Martin-Luther-Universität Halle-Wittenberg, Halle, Germany, 2019.
25. Hoang, P.T.N.; Macas, J.; Rouillard, J.-M.; Kubalová, I.; Schubert, V.; Schubert, I. Limitation of current probe design for oligo-cross FISH, exemplified by chromosome evolution studies in duckweeds. *Chromosoma* **2021**, *130*, 15–25. [CrossRef] [PubMed]
26. Weisshart, K.; Fuchs, J.; Schubert, V. Structured illumination microscopy(SIM) and photoactivated localization microscopy (PALM) to analyze the abundance and distribution of RNA polymerase II molecules on flow-sorted *Arabidopsis* nuclei. *Bio-Protocol* **2016**, *6*, e1725. [CrossRef]

plants

MDPI

Article

Characterisation of a Spontaneous Mutant of *Lemna gibba* G3 (Lemnaceae)

Lakshmi Pasricha Sarin [1,†,‡], K. Sowjanya Sree [2,‡], Károly Bóka [3], Áron Keresztes [3], Jörg Fuchs [4], Akhilesh K. Tyagi [1], Jitendra Paul Khurana [1,§] and Klaus-Juergen Appenroth [5,*]

1 Department of Plant Molecular Biology, University of Delhi South Campus, New Delhi 110021, India; lakshmi.sarin@rajguru.du.ac.in (L.P.S.); akhilesh@genomeindia.org (A.K.T.); khuranaj@genomeindia.org (J.P.K.)
2 Department of Environmental Science, Central University of Kerala, Periye 671320, India; ksowsree@gmail.com or ksowsree9@cukerala.ac.in
3 Department of Plant Anatomy, Eötvös Loránd University, H-1117 Budapest, Hungary; karolyboka@caesar.elte.hu (K.B.); keresztes.aron@ttk.elte.hu (Á.K.)
4 The Leibniz Institute of Plant Genetics and Crop Plant Research (IPK), 06466 Seeland, Germany; fuchs@ipk-gatersleben.de
5 Matthias Schleiden Institute—Plant Physiology, University of Jena, 07743 Jena, Germany
* Correspondence: klaus.appenroth@uni-jena.de
† Current address: Department of Biochemistry, Shaheed Rajguru College of Applied Sciences for Women, University of Delhi, Delhi 110096, India.
‡ These authors contributed equally to this work.
§ Professor Jitendra P. Khurana passed away on 27 October 2021.

Abstract: A spontaneous mutant of the duckweed *Lemna gibba* clone no. 7796 (known as strain G3, WT) was discovered. In this mutant clone, *L. gibba* clone no. 9602 (mt), the morphological parameters (frond length, frond width, root length, root diameter) indicated an enlarged size. A change in the frond shape was indicated by the decreased frond length/width ratio, which could have taxonomic consequences. Several different cell types in both the frond and the root were also enlarged. Flow cytometric measurements disclosed the genome size of the WT as 557 Mbp/1C and that of the mt strain as 1153 Mbp/1C. This represents the results of polyploidisation of a diploid clone to a tetraploid one. The mutant clone flowered under the influence of long day-treatment in half-strength Hutner's medium in striking contrast to the diploid WT. Low concentration of salicylic acid (<1 μM) induced flowering in the tetraploid mutant but not in the diploid plants. The transcript levels of nuclear-encoded genes of the photosynthetic apparatus (*CAB*, *RBCS*) showed higher abundance in light and less dramatic decline in darkness in the mt than in WT, while this was not the case with plastid-encoded genes (*RBCL*, *PSAA*, *PSBA*, *PSBC*).

Keywords: *Lemna gibba*; polyploidisation; spontaneous mutation

check for updates

Citation: Pasricha Sarin, L.; Sree, K.S.; Bóka, K.; Keresztes, Á.; Fuchs, J.; Tyagi, A.K.; Khurana, J.P.; Appenroth, K.-J. Characterisation of a Spontaneous Mutant of *Lemna gibba* G3 (Lemnaceae). *Plants* **2023**, *12*, 2525. https://doi.org/10.3390/plants12132525

Academic Editors: Richard R. C. Wang and Hisato Kunitake

Received: 18 April 2023
Revised: 17 June 2023
Accepted: 28 June 2023
Published: 2 July 2023

1. Introduction

Lemnaceae (waterlentils or duckweed) represent a small family of monocotyledonous, aquatic plants [1]. It consists of 36 species categorized into five genera, i.e., *Spirodela*, *Landoltia*, *Lemna*, *Wolffiella* and *Wolffia* [2]. Especially in the last decade [3] this plant family had unfolded a broad spectrum of potential practical applications. These range from phytoremediation of wastewater [4–6], to use for human nutrition and animal feeding [7–10], to the production of starch for bioalcohol conversion [11,12] and to biogas production [12]. Moreover, as a result of whole genome sequencing of increasing numbers of duckweed species and clones, e.g., in *Spirodela polyrhiza* [13,14], and the feasibility of application of other molecular methods such as genetic transformation using CRISPR/Cas9 [15], members of the Lemnaceae became "a model plant system in the genomics and postgenomics era" [16].

Although the members of Lemnaceae do undergo generative propagation through flowers and seeds, the most common mode of their growth and propagation is by vegetative means. Budding of daughter fronds from the vegetative pouch of the mother frond, producing clones of the mother is a common property of duckweeds [16]. On the way to becoming an unconventional crop plant, the natural variance of physiological and biochemical properties of these plants has to be investigated in order to prepare for biotechnological applications. Out of the several thousands of duckweed clonal accessions in the stock collections worldwide (International Steering Committee on Duckweed Research and Applications [17]), only a few dozens of them have been characterized concerning their growth rates or protein content. These investigations already suggest a huge natural variation between the clones or ecotypes of the same species. Apart from collecting a higher number of ecotypes of species, one other way to extend the available bioresource is to prospect for polyploid mutations in the already collected clones, either by scoring for spontaneous mutants or by artificially inducing mutations in the plants. The artificial induction of mutation was performed in *Landoltia punctata* (*Spirodela oligorrhiza*) which resulted in larger phenotypes with higher dry weight but without any changes in growth rates of the mutant plants when compared with the wild-type plants [18].

Lemna gibba G3 is a long-day strain, which has been extensively employed as a model system to artificially induce flowering [19–21]. In order to study the physiology of flowering, this strain was obtained from Charles F. Cleland (then at the Smithsonian Institution, Washington, DC, USA) in 1985–1986 by the late Jitendra P. Khurana (JPK) at the University of Delhi, South Campus and was maintained ever since under aseptic conditions. While rescuing the fronds of *L. gibba* G3 in late 1988 from an over-grown (3-month-old) agar culture, one of the colonies displayed an altered phenotype upon multiplication after being transferred to fresh liquid medium, i.e., it was considerably larger in size when compared to other colonies from the same culture. This putative mutant strain, which originated spontaneously, was transferred to fresh medium periodically and the progeny was found to retain the phenotype over several generations and has been maintained under axenic ("sterile") conditions ever since. Subsequently, this mutant was found to flower under certain conditions in which its parent strain *L. gibba* G3 did not flower normally.

In a previous study too, a mutant was resurrected from primary callus cultures of strain G3 of *L. gibba* [22], with an enlarged frond size. The self-pollination of cultures indicated that this mutant strain, jsR_1, contains essentially the same amount of DNA per nucleus as its parent line. However, a substantial increase in indole-3-acetic acid levels as estimated by GC-MS was recorded in the mutant in comparison to wild-type, and this tendency was observable at several stages of the culture raised under axenic conditions [22].

In the present work, we investigated the spontaneous mutant of *L. gibba* G3 (called throughout the manuscript as a mutant, mt) and compared it with the parental wild type (WT). As it turned out, the mt was generated by spontaneous polyploidisation, resulting in a tetraploid form. Morphological, anatomical, physiological and biochemical consequences of this genetic alteration were investigated, which included also the induction of flowering by salicylic acid and different daily light periods.

2. Results

The fronds of *L. gibba*, the parental wild type (WT, clone no. 7796) and the spontaneous mutant (mt, clone no. 9602) differed clearly in their morphology, strikingly visible concerning their frond size and frond shape. These differences are quite obvious at both the single colony and the population levels (Figure 1A–D). In Figure 1A,B, a representative colony of both WT and mt clones are shown. In each case, the colony depicts a mother frond, two daughter fronds and a granddaughter frond, connected by stipes. The biometric measurements showed that both the frond length and the frond width of mt clone were significantly larger than those of WT plants. These changes not only increased the size of the fronds in mt clone but also modified the length/width ratio of the frond. This ratio decreased in the mt clone as compared to the WT (Table 1). As a result, the shape of the frond was altered to being more obovate in mt clone than in WT clone. Differences were obvious concerning the root morphology as well. The mt possessed longer and thicker roots in comparison to those of WT (Table 1).

Figure 1. Fronds of the duckweed *Lemna gibba*. (**A**) Colony of the wild-type WT (clone number 7796) with mother frond, two daughter fronds and one granddaughter frond. (**B**) Colony of the mutant mt (clone number 9602), also with one mother frond, two daughter fronds, and one granddaughter frond. (**C**) Population of WT. (**D**) Population of mt, both in 9 cm Petri dishes. White circles focus on characteristic colonies.

Table 1. Biometry of morphological and anatomical parameters of the duckweed *Lemna gibba*, wild type (WT) and mutant (mt). * significance ($p < 0.05$). Data were given as Mean ± SD.

Parameter	Number of Samples	Wild Type	Mutant
Frond length (mm)	20	3.98 ± 0.11	6.31 ± 0.38 *
Frond width (mm)	20	3.12 ± 0.16	5.33 ± 0.37 *
Ratio frond length/width	20	1.28 ± 0.06	1.19 ± 0.06 *
Root length (cm)	20	1.20 ± 0.17	3.70 ± 0.59 *
Root diameter (µm)	43	161 ± 14	277 ± 30 *
Size of outer cortical cells of root (µm)	56	21.0 ± 4.9	40.8 ± 10.2 *

Anatomical observations were in accordance with the morphological biometric data. Light microscopic observations of the cross sections of fronds showed that the sub-epidermal cells or the mesophyll cells of the fronds were distinctly larger in mt than in the WT clone (Figure 2). The difference in cell size concerning the epidermal cells between the diploid and the tetraploid clones was less evident. This is mainly because of the large variability of these cells, as shown in the dorsal epidermis of the frond cross sections in both WT and mt (Figure 2B,D, respectively). The same is also shown in the light microscopic observations of the dorsal epidermal peels. The pavement cells of the dorsal epidermis were noted to be highly variable in shape and size in both clones (Figure 3A,B). However, the stomatal guard cells were more uniform and were significantly larger in the mt clone than in WT (Figure 3). Interesting observations were also made from the cross sections of roots of the WT and mt clones. It was noticed that the outer cortical cells in the root of mt were double the size of those in the root of WT (Table 1). Although the diameter of the root in clone mt was almost double that of clone WT (Table 1), the cross sections of the root showed that the number of root epidermal cells was equivalent at a similar position of the root (apical part) in both mt and WT clones. This indicates increased cell size of the root epidermal (atrichoblast) cells in mt in comparison to WT (Figure 4). The increase in cell size was also observed for all the endodermal cells in mt. In the root cross-sections, phloem and xylem cells also appeared to be larger in clone mt than in the WT clone (Figure 4).

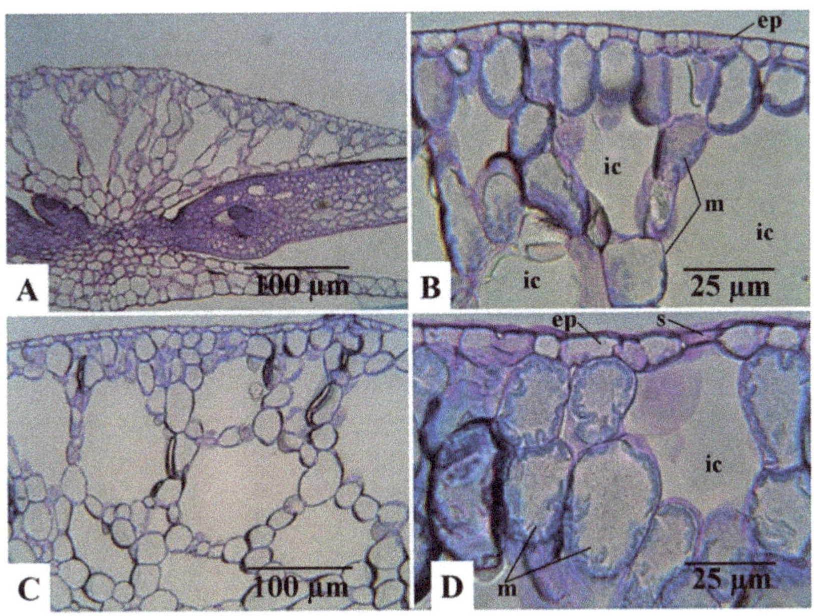

Figure 2. Micrographs of frond cross sections of *Lemna gibba* (**A**,**B**): WT 7796 and (**C**,**D**): mutant mt 9602 (ep: epidermis; ic: intercellular space; m: mesophyll cells; s: stoma).

Genome size measurements were carried out for *L. gibba* WT and mt by flow cytometry using internal reference standards (Figure 5). The genome size of WT was found to be 557 Mbp per 1C and that of mt was 1153 Mbp/1C. This indicated doubling of genome size in the mt when compared to WT. For comparison, three other clones of *L. gibba* collected from different geographic regions of the globe were also analysed. All these three clones of *L. gibba* had similar genome sizes as the WT clone (Table 2). The detailed measurement data can be found in Supplementary Table S1.

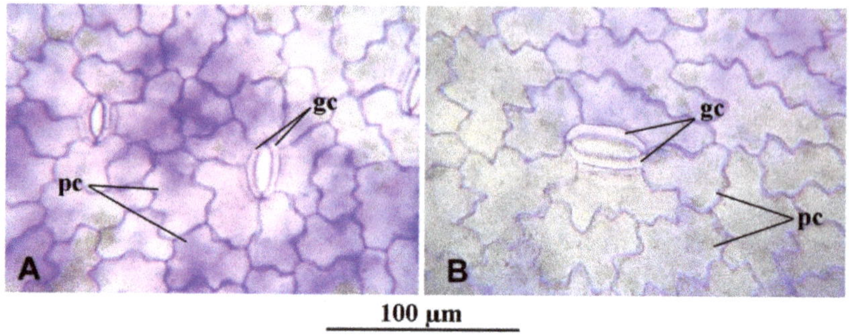

Figure 3. Micrographs of epidermal peels of dorsal epidermis of *Lemna gibba* showing stomatal guard cells (gc) and pavement cells (pc). (**A**): WT 7796, (**B**): mt 9602.

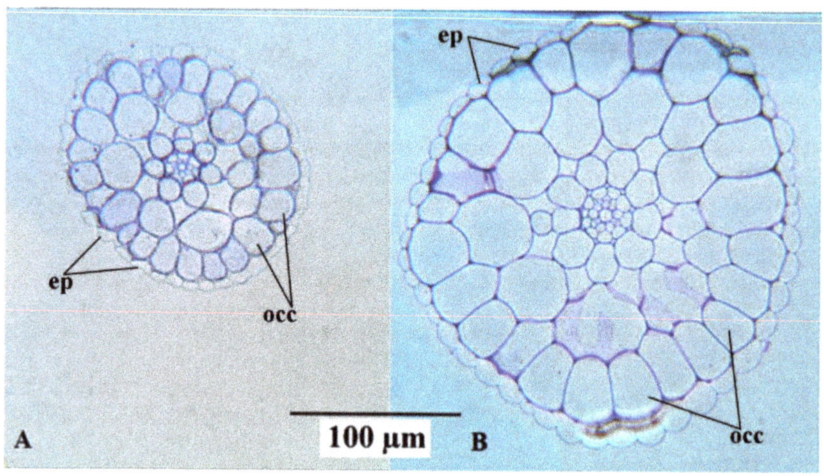

Figure 4. Micrographs of cross sections of roots of *Lemna gibba* (**A**): WT 7796, (**B**): mt 9602 (ep: epidermis; occ: outer cortical cells).

Table 2. Genome sizes of clones of *Lemna gibba*, clone WT 7796 and mt 9602 measured by flow cytometry. Genome sizes of three more clones of *L. gibba* collected from other regions were measured and presented for comparison. Data are given as Mean ± SD. For the detailed measurement data set see Supplementary Material, Table S1).

Clone ID	Origin	DNA Content (pg/C)	Genome Size (Mbp/1C)	Ploidy
WT (7796)	Italy	1.14 ± 0.02	557 ± 10	diploid
mt (9602)	Mutant of clone 7796	2.36 ± 0.05	1153 ± 26	tetraploid
		Other *L. gibba* clones		
7641	Israel	1.16 ± 0.01	568 ± 6	diploid
7922	Argentina	1.09 ± 0.02	535 ± 8	diploid
8682	Saudi Arabia	1.15 ± 0.01	565 ± 3	diploid

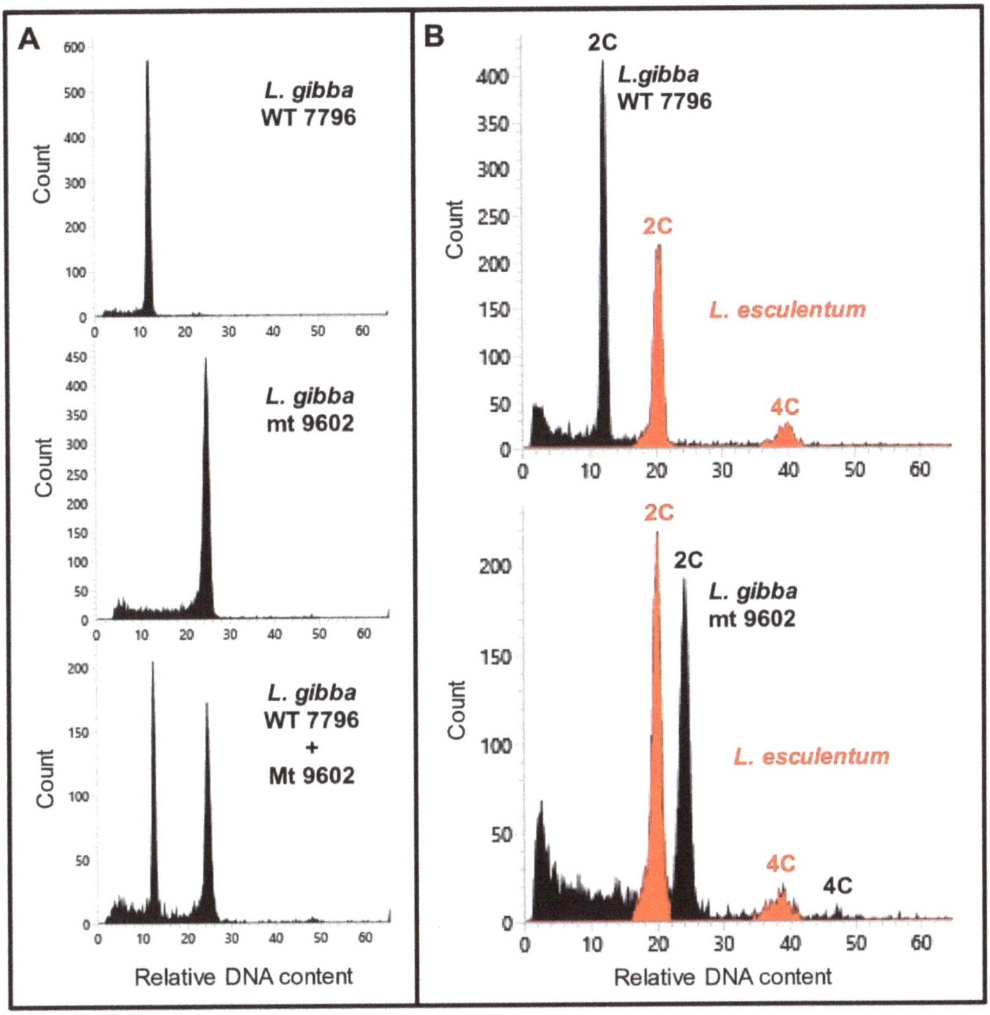

Figure 5. Flow cytometry confirmed the tetraploid status of the mutant clone 9602 in comparison to the diploid WT clone 7796. (**A**): Measurements of both clones separately or in combination clearly demonstrated different ploidy levels. (**B**): Genome size measurements using internal reference standards (exemplified here with *Lycopersicon esculentum* in red).

The physiological capacity of the mutant clone (mt) in comparison to the parent clone (WT) was investigated in terms of both vegetative propagation and generative propagation. The vegetative propagation was measured in terms of relative growth rates (RGR) and the generative propagation was scored by the flower induction capacity of the clones. The RGR was determined under standard growth conditions (following otherwise the ISO 20079 [23]) in two different nutrient media, i.e., N- and half-strength Hutner´s medium (see Supplementary Material, Table S2). Both the clones performed better in N-medium when compared to half-strength Hutner´s medium. However, it was found that the mt clone grows significantly slower than the parental strain in both nutrient media used in this study (Table 3).

Table 3. Growth rates of *Lemna gibba*, WT 7796 and mt 9602. Nutrient medium N and half-strength Hutner´s medium were used. *t* test (* significant different), *n* = 6.

Clone	RGR (d^{-1})	DT (d)	RY (g g^{-1} $week^{-1}$)
	N medium:		
WT	0.448 ± 0.004	1.56 ± 0.05	23.6 ± 2.9
mt	0.347 ± 0.040 *	2.03 ± 0.02	11.8 ± 3.2
	Half-strength Hutner´s medium:		
WT	0.385 ± 0.021	1.80 ± 0.09	14.8 ± 0.8
mt	0.324 ± 0.009 *	2.14 ± 0.01	9.62 ± 0.29

RGR = relative growth rate; DT = doubling time; RY = relative weekly yield. For more details, refer to Section 4.

In the following experiments, the influence of photoperiod on flower induction was investigated in detail. Flowering was not detected under continuous light in both clones. Clone mt flowered up to 20 and 40% under 14 and 18 h light periods per day, respectively, but not at shorter photoperiods. Clone WT, however, did not flower under any of the light conditions investigated (Figure 6A). It was demonstrated that eight days of long-day treatments were required to induce flowering in clone mt and the intensity of response increased further by increasing the number of days of long-day treatments to 14 or 15 days. The WT, however, did not respond at all with respect to flowering to any of these treatments (Figure 6B).

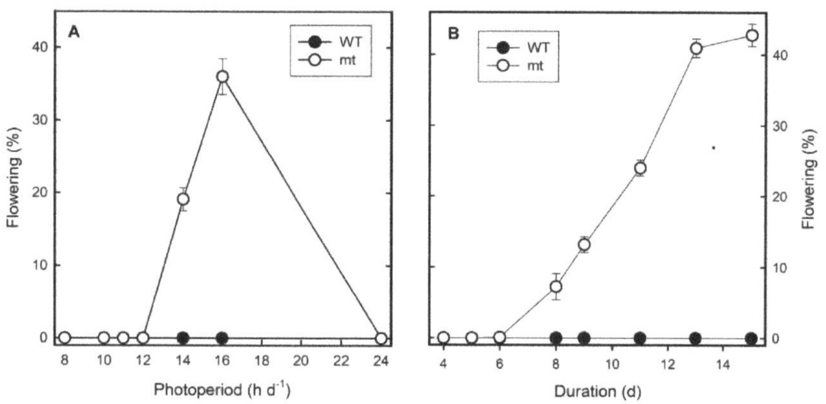

Figure 6. Influence of light periods on the flowering response of *Lemna gibba* WT 7796 and mt 9602. (**A**) Influence of photoperiod on flower induction. Cultures were pre-cultivated under non-inductive short days (8 h light:16 h dark) and thereafter subjected to appropriate photoperiods for 10 days, followed by 3 days of continuous light, and finally analysed for flowering. (**B**) Kinetics of induction of flowering in WT and mt under long days of 16 h light:8 h darkness. Single 4-frond colony was inoculated per flask and flowering was analysed on specified days. In all experiments, half-strength Hutner´s nutrient medium was used.

Flower induction was investigated under the influence of salicylic acid as has been investigated before with the same species and the same clone 7796 under long-day conditions (16 h light: 8 h darkness) [24]. At concentrations ≥1 µM of salicylic acid, both WT and mt, responded in a similar manner. However, there was a striking difference at lower salicylic acid concentrations (0.1 µM and 0.5 µM). The WT did not flower at all under these conditions, but the mt flowered to approximately 30%. At higher salicylic acid concentrations (10 µM), both the clones performed with ca. 80% flowering frequency (Figure 7).

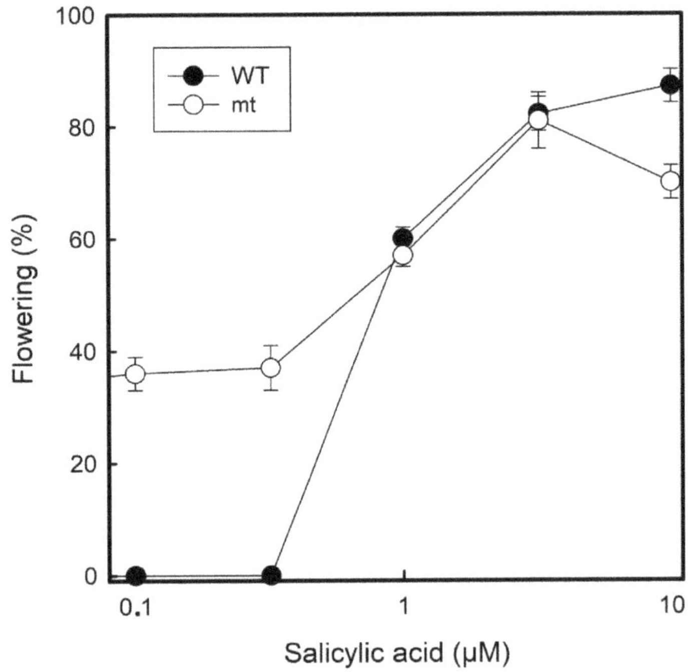

Figure 7. Effect of salicylic acid on flowering in *Lemna gibba* WT 7796 and mt 9602, grown in half-strength Hutner's medium. The experimental cultures were initiated in salicylic acid-containing medium and analysed for flowering after 11 long days (16 h light:8 h dark).

Biochemical properties were investigated and compared between both clones. Most of the parameters, chlorophyll a, carotenoid and protein content were lower in clone mt on a dry weight basis. However, dry weight and chlorophyll b content did not show any significant differences between the two clones (Table 4).

Table 4. Biochemical parameters of *Lemna gibba*, WT 7796 and mt 9602. Data are given per dry weight (DW). $n = 4$, means ± SD; * significant ($p < 0.05$).

Parameter	WT 7796	mt 9602
Dry weight (%)	5.30 ± 0.20	4.92 ± 0.29
Chlorophyll a (mg g^{-1} DW)	16.2 ± 1.82	8.78 ± 0.74 *
Chlorophyll b (mg g^{-1} DW)	6.05 ± 0.70	5.32 ± 0.91
Carotenoids (mg g^{-1} DW)	0.488 ± 0.060	0.203 ± 0.005 *
Protein content (% dry weight)	26.0 ± 0.1	22.8 ± 0.3 *

To examine the changes in expression of the genes of the photosynthetic apparatus, the changes in transcript levels of two nuclear-encoded (*CAB*, *RBCS*) and four plastid-encoded (*RBCL*, *PSBA*, *PSAA* and *PSBC*) genes were examined. An increase in steady-state transcript levels of the nuclear-encoded genes, *CAB* and *RBCS*, was observed in the mutant as compared to the wild-type and the expression was more under long-day conditions than short-day conditions (Figure 8) suggesting the influence of duration of light on the expression of these genes. Much less pronounced changes were observed for the plastid-encoded genes (*RBCL*, *PSAA*, *PSBA* and *PSBC*). The rDNA probe was used as a loading control and was found to be satisfactory as is evident from the autoradiogram presented in Figure 8.

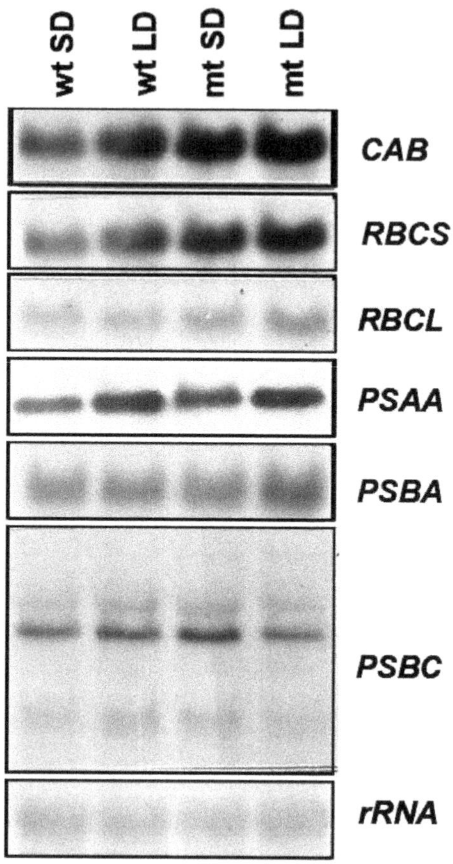

Figure 8. Changes in the steady-state transcript levels of nuclear (*CAB*, *RBCS*) and plastid (*RBCL*, *PSAA*, *PSBA*, *PSBC*) encoded genes of the photosynthetic apparatus in the wild-type and the mutant strain of *L. gibba* G3 grown in 1/2 Hutner's medium under long day and short day conditions. Note that the mutant strain was flowering under the LD conditions.

To find out if the lower content of chlorophyll in the mutant may be accounted for due to changes in expression pattern in the mt due to dark adaptation, the fronds of the wild-type and the mutant were initially grown under short day (so that both of them remain vegetative) for 15 cycles and then transferred to continuous light for two days before being subjected to dark adaptation. As is obvious from the autoradiogram in Figure 9, the transcript levels of the nuclear-encoded genes (*CAB* and *RBCS*) were somewhat higher in the mutant grown in continuous light and those of plastid-encoded genes showed no significant difference in the transcript abundance. On dark adaptation, the transcript levels of nuclear-encoded genes declined but remained somewhat steady up to 24 h and, in the following 24 h in darkness, a sharp decline was noted in the wild-type, whereas mutant showed relatively less drastic decline under similar conditions (Figure 9). In contrast to nuclear genes, plastid-encoded genes did not show much fluctuation in their transcript abundance, both in the mutant and the wild-type. However, most of the transcripts encoded by the *PSBC* operon did show a decline in dark adaptation (Figure 9). Strikingly, the transcript of *PSBC* operon displayed an increase in dark adaptation for 8 h or more, in both WT and mt.

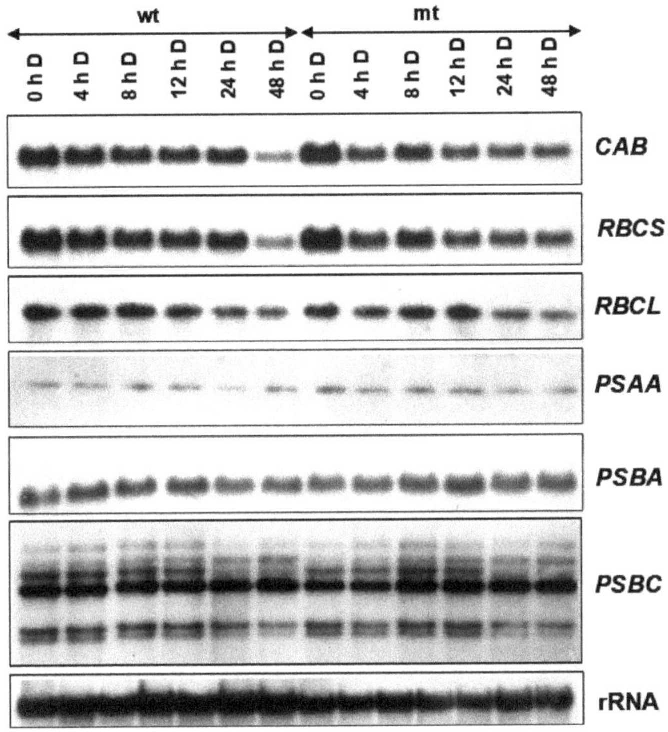

Figure 9. Dark adaptive changes in expression of nuclear (*CAB*, *RBCS*) and plastid (*RBCL*, *PSAA*, *PSBA*, *PSBC*) encoded genes of the photosynthetic apparatus in the wild-type and the mutant strain of *L. gibba* G3. Both the strains were initially grown under non-inductive short days for 15 cycles and then transferred to continuous light for 2 d, before subjecting them to dark adaptation for specified time (between 4 to 48 h). Total RNA was extracted, and northern blots were prepared. Northern analysis was performed using heterologous probes. For checking the quality and quantity of RNA, rRNA probe was used.

3. Discussion

The genome size of the two investigated clones of *L. gibba* showed that the WT 7796, also known as *L. gibba* G3, has a similar genome size as several other clones of this species (Table 2). Urbanska-Worytkiewicz [25] and Geber [26] investigated chromosome numbers of many duckweed clones including those belonging to *L. gibba* and reported values of 2n = 40, 42, 44, 50, 70, 80 and 84 for this species. These data were cited by Wang et al. [27] without critical evaluation. However, the technique was not yet much developed at that time and the results should be treated with care, considering the high chromosome number variation within the same species [25]. Comparing all available genome sizes and chromosome number measurements [25,26,28], the most probable chromosome number in *L. gibba* is 2n = 40 [28], defining the clone WT 7796 as a diploid. Consequently, the spontaneously formed mutant, mt 9602 is a tetraploid and might be an autotetraploid as the mutation proceeded in a closed Erlenmeyer flask of axenic ("sterile") cultures of the clone WT 7796. Theoretically, there are two mechanisms of how genome duplication could have happened. First of all, during vegetative propagation, the mitotic division between two S phases could have failed. Alternatively, two gametes of the same clone could be non-reduced. However, this second mechanism of self-fertilisation by spontaneously non-reduced gametes is very improbable considering also the low flowering frequency in this clone/species. To the best of our knowledge, this is the first report about spontaneous polyploidisation in the family of Lemnaceae. Vunsh et al. [18] reported polyploidisation in *L. punctata*, induced by treatment with colchicine that interferes with cell divisions [29]. Intraspecific fertilisation between different clones of a species has been reported in *Lemna aequinoctialis* (termed *Lemna paucicostata*; [30]) and in *L. gibba* [24]. However, no polyploidisation was observed in these cases.

Artificial polyploidisation is often used to improve plant quality, e.g., of ornamental plants in horticulture [29] and is generally used in plant breeding [31]. Moreover, polyploidisation (e.g., genome duplication) steps were important events during the evolution of plants. Often polyploidisation process results in plants with specific properties, e.g., larger leaves or fruits. In agreement with this general expectation, in the present work, the tetraploid plants were characterized by an increase in the size of several of the investigated morphological and anatomical parameters, such as frond and root size as well as the size of the cells in frond and root.

Morphologically, duckweeds are small in size, ranging from a centimetre to less than a millimetre. As a consequence, the morphological markers available for delineation of the different duckweed species, especially within a genus are scarce and often confusing for generalists. One of the features observed between the WT and mt clones investigated in this study was that the disproportionate increase in frond length and frond width had altered the shape of the frond in mt, thereby making it more obovate than the fronds of WT. Such changes in frond shape are noteworthy, especially in view of the fact that the above-mentioned biometric parameters play a significant role as morphological markers for the determination of duckweed species [2]. Hence it is suggested to use the morphological markers with care and caution and to use molecular markers for barcoding [16] wherever necessary for Lemnaceae species determination.

As a general physiological parameter, RGR was investigated. Unexpectedly, the RGR of the tetraploid clone 9602 was significantly lower than the diploid parent clone, both in N- and in half-strength Hutner´s medium. Notably, the decreased growth capacity correlated with the biochemical results: the chlorophyll a (but not chlorophyll b) and the carotenoid content were significantly lower in the tetraploid mt (clone 9602) than in the diploid WT parent clone (7796). Evidently, the photosynthetic capacity became partly impaired during polyploidisation. In accordance with these results, also the protein content, as a biochemical cross-parameter for the metabolic capacity, was lowered in the tetraploid clone.

The photoperiodic response of flower induction in *L. gibba* was already investigated by Kandeler [32]. He mainly used clone G1 but mentioned that clone G3 (the diploid clone investigated in the present paper) behaved very similarly. According to his results, *L. gibba* is a long-day plant with a critical day-length of 12 to 14 h. Cleland and Briggs [33,34] reported an increase in light sensitivity of flower induction in *L. gibba* G3 about 8 to 10 h after the start of the light period. In the present paper, both WT and mt *L. gibba* clones were investigated under different light regimes for 10 days. The mt did respond to these regimes in terms of flowering, unlike the WT. The critical day length was between 12 and 16 h. No flowering was observed in continuous light. Extending the duration of irradiation enhanced the effect on flowering until ca. 13 days and any further extension showed a saturated effect.

Exposure to salicylic acid is more effective in inducing flowering than different light periods alone. Here, the tetraploid mutant mt 9602 is more sensitive. While the wild type did not flower at low salicylic acid concentrations (<1 μM), the tetraploid mutant did flower to approximately 40%. Fu et al. [24] investigated the same WT clone of *L. gibba* (also clone G3, but registered in the duckweed stock collection of Elias Landolt in Zurich additionally as clone 7741) and found that application of salicylic acid is absolutely required to induce flowers and to produce seeds. However, these authors applied only concentrations of ≥ 10 μM salicylic acid. The whole genome of this clone has now been sequenced [35].

This difference in the response of WT and mt to flower induction is of interest for future studies on the probable sexually propagated generations of this mt clone of *L. gibba*. The genome sequencing of these clones can unravel more details on polyploidisation and can open further avenues.

The investigation of the transcript levels in WT and mt showed higher expression of the investigated nuclear-encoded genes (*CAB* and *RBCS*) in mt whereas the plastid-encoded genes (*RBCL*, *PSAA*, *PSBA* and *PSBC*) showed only minimal differences between both strains. During dark-adaption, the transcript levels of nuclear-encoded genes decreased even more dramatically in the mt whereas the plastidic-encoded transcripts showed no significant differences between the two strains. This does not explain immediately the lower chlorophyll content in mt. However, an imbalance between nuclear-encoded and plastid-encoded transcripts becomes evident and is even stronger in mt than in WT. It is difficult to explain the lower chlorophyll contents in mt than in WT considering the relative increase in abundance of the transcript, particularly of the nuclear-encoded genes, *RBCS* and *CAB*, and an altered rate of decline of transcripts during dark adaptation of mt, in comparison to its WT parent. It could be that the change in ploidy level in the mutant has caused a higher expression of nuclear-encoded genes in the mutant. However, this higher expression need not necessarily correlate with the level of the corresponding polypeptides that finally assemble in a definite ratio for establishing the photosynthetically competent chloroplast machinery.

4. Materials and Methods

4.1. Plant Material and Cultivation

Lemna gibba L., clone G3 was provided by Charles F. Cleland (then at the Smithsonian Institution, Washington, DC) in 1985–1986 to one of the authors (JPK) at the University of Delhi, South Campus, New Delhi, India. This clone is registered in many stock collections under several international IDs, e.g., 7796 at the University of Jena, Germany. In 1988, from a three-month-old culture, one of the authors (JPK) isolated a spontaneously formed mutant with a changed phenotype showing unusually large fronds and is now registered under the ID 9602. Throughout the manuscript, this mutant was indicated by "mt" whereas the parental, wild type is indicated by "WT". Three more clones of *L. gibba* were obtained from the stock collection of the University of Jena and used for genome size measurements as controls.

During the investigations, plants were either cultivated in N medium [36] or in half-strength Hutner´s medium [37]. For the composition of the media see Supplementary Material, Table S2. The temperature during all experiments was maintained at 25 °C. The plants were exposed to light conditions provided by fluorescence tubes at an intensity of 100 μmol m^{-2} s^{-1}.

4.2. Morphological-Anatomical Investigations

Clones 7796 (WT) and 9602 (mt) of *L. gibba* were investigated using a stereo microscope for morphological biometric studies. Anatomical investigations were carried out by light microscopic studies. Halved fronds and 1 cm long root pieces (apical part) were fixed in 4% formaldehyde solution (in 0.05 M phosphate buffer, pH 7.2). After thoroughly washing with the same buffer, samples were dehydrated in an increasing ethanol gradient and embedded in Technovit 7100 resin (Kulzer GmbH, Wehrheim, Germany). From 5-5 blocks, 1 μm thick sections were prepared with glass knives using a Microm HM 360 microtome (Microm International Ltd., Walldorf, Germany). Sections were stained with toluidine blue and observed in Nikon Eclipse 80i microscope (Nikon Corp., Tokyo, Japan).

The dorsal epidermal peels were stained with toluidine blue and observed using 40× magnifying objective under a light microscope.

4.3. Flow Cytometric Genome Size Estimation

Genome size measurements were performed using either a BD Influx cell sorter (BD Biosciences, San Jose, CA, USA) or a CyFlow Space flow cytometer (Partec GmbH, Münster, Germany) according to Dolezel et al. [38]. Fronds were chopped with a razor blade together with young leaf material of *Raphanus sativus* 'Voran' (IPK gene bank accession number RA 34; 2C = 1.11 pg; [39]), *Lycopersicon esculentum* Mill. convar. *infiniens* Lehm. var. *flammatum* Lehm. 'Stupicke Rane' (IPK gene bank accession number: LYC 418) or *Glycine max* cv. Cina 5202 'Voran' (IPK gene bank accession number SOJA 392; 2C = 2.23 pg; [40]) as internal reference standard.

For isolation of nuclei and their staining, either Galbraith's buffer [41] supplemented with DNAse-free RNase (50 µg/mL) and propidium iodide (50 µg/mL) or the DNA staining kit 'CyStain^R PI Absolute P' (Partec GmbH, Münster, Germany) was used. Usually, 10,000 nuclei per sample were analysed and at least four independent measurements per clone were performed. The absolute DNA contents (pg/2C) were calculated based on the mean values of the G1 peak and the corresponding genome sizes (Mbp/1C) according to Dolezel et al. [42].

4.4. Relative Growth Rate Measurements

Measuring growth rates on the basis of frond number and calculating the relative growth rates (RGR, d^{-1}), doubling time (DT, d) and relative weekly yield (RY, number of fronds after one week of cultivation starting with one gram fresh weight [43]) were performed as described earlier [44].

4.5. Induction of Flowering

Flowering was induced by different concentrations of salicylic acid applied directly to Hutner´s medium. During this treatment, long day conditions were applied (16 h light: 8 h dark) for 11 days. Thereafter, plants were de-stained in 70% ethanol overnight in glass tubes and flowering was evaluated under a stereo microscope. To investigate the influence of light periods, experimental cultures were initiated in half-strength Hutner's medium with the transfer of one 4-frond colony from cultures maintained in the same medium, under non-inductive short days (8 h light: 16 h dark). Cultures were then subjected to appropriate photoperiods for 10 days, followed by 3 days of continuous light, and analysed for flowering as described above. Finally, plants were treated under long-day conditions (16 h light: 8 h dark) and the duration of treatments was increased from 4 to 15 days.

4.6. Analytical Methods

Freshly harvested plant material was dried for 24 h at 105 °C and weighed before drying and thereafter. Chlorophyll and carotenoid content were investigated by homogenisation of 150 mg frozen fresh material in a mortar with ammonia-buffered acetone (800 mL acetone + 195 mL water + 5 mL 25% ammonia), and centrifugation at 15,000× *g* in a fixed-angle rotor of a bench-top centrifuge for 15 min. The absorbance of the supernatant was measured at 663 nm, 647 nm and 470 nm in a UV/Vis spectrophotometer and the pigment content was calculated according to Lichtenthaler [45]. The protein content was measured by homogenisation of 150 mg frozen fresh material in a mortar with 10 mM KOH and centrifuged at 15,000× *g* for 15 min. A fraction of the supernatant (100 µL) was mixed with 5 mL Bradford reagent and measured in a UV/Vis spectrophotometer at 595 nm and 465 nm. The ratio A_{595}/A_{465} was plotted against the concentration of BSA, and the protein content was calculated with the help of a linear regression line [46].

4.7. Isolation of Total RNA

Total RNA was isolated by the protocol described by Logemann et al. [47]. About two g of tissue was frozen in liquid nitrogen and homogenized to a fine powder with the help of a mortar and pestle. It was then transferred to SS 34 rotor tube, and suspended in two volumes of guanidine buffer (8 M guanidine hydrochloride, 20 mM MES, 20 mM

EDTA, 50 mM 2-mercaptoethanol, pH 7.0) immediately after the evaporation of liquid nitrogen. The extract was then centrifuged at 10,000 rpm for 15 min. The supernatant was filtered through one layer of mira cloth and extracted with one volume of phenol: chloroform: isoamyl alcohol (25:24:1, $v/v/v$) and centrifuged for 45 min at 10,000 rpm at room temperature to separate the organic and the aqueous phases. To the upper phase, 0.2 volume of 1 M acetic acid and 0.7 volume of chilled alcohol were added to precipitate the RNA. This was incubated at −20 °C overnight. The RNA was pelleted by centrifugation at 10,000 rpm for 10 min at 4 °C. The pellet was then washed twice with 3 M sodium acetate (pH 5.2) to dissolve the low molecular weight RNA and contaminating polysaccharides. Subsequently, it was washed with 70% ethanol and dissolved in DEPC-treated sterile-distilled water. The RNA was quantified spectrophotometrically, and its quality was checked on a mini-agarose gel.

All solutions (except phenol, chloroform, acetic acid, guanidine buffer and ethanol) and plastic ware and glassware were treated with 0.1% and 0.01% DEPC (diethylpyrocarbonate), respectively, and autoclaved.

4.8. Electrophoresis and Northern Transfer

For 350 mL of 1.2% gel solution, 4.2 g of agarose was added to 339.5 mL of 1× MOPS (20 mM MOPS, 5 mM sodium acetate, 1 mM EDTA, pH 7.0) and treated with 0.1% DEPC for 1 h at room temperature. Agarose was melted and cooled by keeping it in a water bath set at 60 °C. To this 10.5 mL of 37% formaldehyde was added. The gel was allowed to solidify and submerged in 1× gel running buffer (MOPS).

To the RNA samples (100 µg in sterile-distilled water), 3 volumes of denaturation mix (formamide 500 µL, formaldehyde (37%) 62 µL, 10× MOPS, 100 µL) was added and mixed well and incubated at 65 °C for 10 min. To each sample, 0.2 volume of gel loading buffer was added and immediately kept on ice.

The samples were electrophoresed in agarose-formaldehyde gel at 120 V (constant voltage). After a 4 h run, the gel was washed in sterile-distilled water to remove the formaldehyde.

For northern transfer, the gel was blotted onto nitrocellulose paper, by capillary blotting in 20× SSC for 24 h. Subsequently, the gel was lifted and rinsed in 3× SSC and air-dried for 1 h and finally baked at 80 °C for 2 h.

The nitrocellulose filters containing the bound RNA were pre-hybridized in closed, shallow plastic trays containing 200 $\mu L/cm^2$ prehybridisation solution (42% formamide, 5× SSC, 5× Denhardt's solution (5 mg/mL of polyvinylpyrrolidone, bovine serum albumin and ficoll 400), 50 mM sodium phosphate buffer, pH 6.5, 250 µg/mL sonicated and denatured herring sperm DNA) for 16 h at 37 °C for nuclear genes or 42 °C for chloroplast encoded genes, in an incubator shaker at 60 rpm. Various gene-specific DNA probes were prepared from recombinant plasmids containing DNA fragments, either using electroeluted fragments or low melting point agarose containing the fragment. These were then labelled with the help of multiprime DNA labelling or megaprime DNA labelling system (Amersham International, Amersham, UK) using alpha ^{32}P dATP, according to the manufacturer's specifications. Spinach chloroplast DNA gene probes were used as described by Kapoor et al. [48], and nuclear gene probes for *RBCS* and *CAB* from Coruzzi et al. [49]. 25S rRNA gene probe was used as a control [50]. Double-stranded radiolabelled probes were denatured by boiling in a water bath for 3 min, followed by immediate cooling on ice for the electroeluted fragments, and the probes were added directly to the hybridisation buffer (23% formamide, 5× SSC, 1× Denhardt's solution, 50 mM sodium phosphate, pH 6.5, 50 µg/mL sonicated and denatured herring sperm DNA). For the heterologous probes, 10% dextran sulphate was also added to the hybridisation mixture, the filters were then transferred to this after pre-hybridisation. Hybridisation was continued for 24–48 h in an incubator shaker at 37 °C or 42 °C, as per the requirement. The filters were first washed in solution 1 (42% formamide, 5× SSC, 0.1% SDS) at room temperature for 15 min. The second wash was given with 5× SSC and 0.1% SDS, at room temperature for 15 min, twice, and subsequently with 2× SSC and 0.1% SDS. The filters hybridized with chloroplast genes were washed with 2× SSC and 0.1% SDS at room temperature for 15 min, followed by 1× SSC and 0.1% SDS. When most of the non-specific radioactivity was removed, the filters were wrapped in cling film and exposed to the X-ray film (Konica, Tokyo, Japan) in a cassette containing an intensifying screen (Rege Chemicals, Delhi, India). The cassette was kept at −20 °C for exposure and the films developed after 24–72 h.

Supplementary Materials: The following supporting information can be downloaded at: https://www.mdpi.com/article/10.3390/plants12132525/s1, Table S1: List of individual genome size measurements (including the used cytometer and reference standard) performed for the *Lemna gibba* clones 7796, 9602, 7641, 7922 and 8682; Table S2: Nutrient media composition of N medium [36] and Hutner´s medium [37].

Author Contributions: The original concept was created by J.P.K. and A.K.T. and was extended by K.S.S. and K.-J.A., who prepared the first draft of the manuscript. The morphological-anatomical investigations were carried out by K.B., K.S.S. and Á.K., the genome size measurements were performed by J.F., Northern blots were performed by L.P.S., the physiological experiments and biochemical determinations were carried out by K.S.S., K.-J.A. and L.P.S. All authors have contributed to reviewing and revising the manuscript. All authors have read and agreed to the published version of the manuscript.

Funding: This research received no external funding.

Institutional Review Board Statement: Not applicable.

Informed Consent Statement: Not applicable.

Data Availability Statement: All data are available within the manuscript or Supplementary Material.

Conflicts of Interest: The authors declare no conflict of interest.

References

1. Tippery, N.P.; Les, D.H.; Appenroth, K.J.; Sree, K.S.; Crawford, D.J.; Bog, M. Lemnaceae and Orontiaceae are phylogenetically and morphologically distinct from Araceae. *Plants* **2021**, *10*, 2639. [CrossRef] [PubMed]
2. Bog, M.; Appenroth, K.J.; Sree, K.S. Key to the determination of taxa within the family of Lemnaceae: An update. *Nord. J. Bot.* **2020**, *38*, e02658. [CrossRef]
3. Zhao, H.; Appenroth, K.; Landesman, L.; Salesman, A.A.; Lam, E. Duckweed rising at Chengdu: Summary of the 1st International Conference on Duckweed Application and Research. *Plant Mol. Biol.* **2012**, *78*, 627–632. [CrossRef] [PubMed]
4. Ziegler, P.; Sree, K.S.; Appenroth, K.J. The uses of duckweed in relation to water remediation. *Desalination Water Treat.* **2017**, *63*, 327–342. [CrossRef]
5. Ansari, A.A.; Naeem, M.; Gill, S.S.; AlZuaibr, F.M. Phytoremediation of contaminated waters: An eco-friendly technology based on aquatic macrophytes application. *Egypt. J. Aq. Res.* **2020**, *46*, 371–376. [CrossRef]
6. Liu, Y.; Xu, H.; Yu, C.; Zhou, G. Multifaceted roles of duckweed in aquatic phytoremediation and bioproducts synthesis. *GCB Bioenergy* **2020**, *13*, 70–82. [CrossRef]
7. Cui, W.; Cheng, J.J. Growing duckweed for biofuel production: A review. *Plant Biol.* **2015**, *17*, 16–23. [CrossRef]
8. Appenroth, K.J.; Sree, K.S.; Boehm, V.; Hammann, S.; Vetter, W.; Leiterer, M.; Jahreis, G. Nutritional value of duckweeds (Lemnaceae) as human food. *Food Chem.* **2017**, *217*, 266–273. [CrossRef]
9. Appenroth, K.J.; Sree, K.S.; Bog, M.; Ecker, J.; Seeliger, C.; Boehm, V.; Lorkowski, S.; Sommer, K.; Vetter, W.; Tolzin-Banasch, K.; et al. Nutritional value of the duckweed species of the genus *Wolffia* (Lemnaceae) as human food. *Front. Chem.* **2018**, *6*, 483. [CrossRef]
10. Sońta, M.; Rekiel, A.; Batorska, M. Use of Duckweed (*Lemna* L.) in sustainable livestock production and aquaculture—A review. *Ann. Anim. Sci.* **2019**, *19*, 257–271. [CrossRef]
11. Appenroth, K.J.; Ziegler, P.; Sree, K.S. Accumulation of starch in duckweeds (Lemnaceae), potential energy plants. *Physiol. Mol. Biol. Plants* **2021**, *27*, 2621–2633. [CrossRef] [PubMed]
12. Chen, G.; Zhao, K.; Li, W.; Yan, B.; Yu, Y.; Li, J.; Zhang, Y.; Xia, S.; Cheng, Z.; Lin, F.; et al. A review on bioenergy production from duckweed. *Biomass Bioenergy* **2022**, *161*, 106468. [CrossRef]
13. Xu, S.; Stapley, J.; Gablenz, S.; Boyer, J.; Appenroth, K.J.; Sree, K.S.; Gershenzon, J.; Widmer, A.; Huber, M. Low genetic variation is associated with low mutation rate in the giant duckweed. *Nat. Commun.* **2019**, *10*, 1243. [CrossRef]
14. Michael, T.P.; Ernst, E.; Hartwick, N.; Chu, P.; Bryant, D.; Gilbert, S.; Ortleb, S.; Baggs, E.L.; Sree, K.S.; Appenroth, K.J.; et al. Genome and time-of-day transcriptome of *Wolffia australiana* link morphological minimization with gene loss and less growth control. *Genome Res.* **2021**, *31*, 225–238. [CrossRef]

15. Liu, Y.; Wang, Y.; Xu, S.; Tang, X.; Zhao, J.; Yu, C.; He, G.; Xu, H.; Wang, S.; Tang, Y.; et al. Efficient genetic transformation and CRISPR/Cas9-mediated genome editing in *Lemna aequinoctialis*. *Plant Biotechnol. J.* **2019**, *17*, 2143–2152. [CrossRef] [PubMed]
16. Acosta, K.; Appenroth, K.J.; Borisjuk, L.; Edelman, M.; Heinig, U.; Jansen, M.A.K.; Oyama, T.; Pasaribu, B.; Schubert, I.; Sorrels, S.; et al. Return of the Lemnaceae: Duckweed as a model plant system in the genomics and postgenomics era. *Plant Cell* **2021**, *33*, 3207–3234. [CrossRef] [PubMed]
17. International Steering Committee on Duckweed Research and Applications. Update: Registration and collections of duckweed clones/strains. *Duckweed Forum* **2022**, *10*, 21–22. Available online: http://www.ruduckweed.org/ (accessed on 29 June 2023).
18. Vunsh, R.; Heinig, U.; Malitsky, S.; Aharoni, A.; Lerner, A.; Edelman, M. Manipulating duckweed through genome duplication. *Plant Biol.* **2015**, *17*, 115–119. [CrossRef]
19. Kandeler, R. Flowering in the Lemna system. *Phyton* **1984**, *24*, 113–124.
20. Kandeler, R. Lemnaceae. In *Handbook of Flowering*; Halevy, A.H., Ed.; CRC Press: Boca Raton, FL, USA, 1985.
21. Khurana, J.P.; Cleland, C.F. Role of Salicylic Acid and Benzoic Acid in Flowering of a Photoperiod-Insensitive Strain, *Lemna paucicostata* LP6. *Plant Physiol.* **1992**, *100*, 1541–1546. [CrossRef]
22. Slovin, J.P.; Cohen, J.D. Levels of Indole-3-Acetic Acid in *Lemna gibba* G-3 and in a Large Lemna Mutant Regenerated from Tissue Culture. *Plant Physiol.* **1988**, *86*, 522–526. [CrossRef]
23. *ISO 20079*; Water Quality—Determination of the Toxic Effect of Water Constituents and Waste Water on Duckweed (Lemna minor)—Duckweed Growth Inhibition Test. ISO Central Secretariat: Geneva, Swizerland, 2005.
24. Fu, L.; Huang, M.; Han, B.; Sun, X.; Sree, K.S.; Appenroth, K.J.; Zhang, J. Flower induction, microscope-aided cross-pollination, and seed production in the duckweed *Lemna gibba* with discovery of a male-sterile clone. *Sci. Rep.* **2017**, *7*, 3047. [CrossRef] [PubMed]
25. Landolt, E.; Urbanska-Worytkiewicz, K. *List of the Studied Lemnaceae Samples: Origin and Chromosome Numbers*; Veröffentlichungen des Geobotanischen Institutes der ETH, Stiftung Rübel: Zürich, Switzerland, 1980; Volume 70, pp. 205–247.
26. Geber, G. Zur Karyosystematik der Lemnaceae. Ph.D. Thesis, University of Vienna, Vienna, Austria, 1989.
27. Wang, W.; Kerstetter, R.A.; Michael, T.P. Evolution of genome size in duckweeds (Lemnaceae). *J. Bot.* **2011**, *2011*, 570319. [CrossRef]
28. Hoang, P.T.N.; Fuchs, J.; Schubert, V.; Tran, T.B.N.; Schubert, I. Chromosome numbers and genome sizes of all 36 duckweed species (Lemnaceae). *Plants* **2022**, *11*, 2674. [CrossRef]
29. Eng, W.-H.; Ho, W.-S. Polyploidization using colchicine in horticultural plants: A review. *Sci. Hortic.* **2019**, *246*, 604–617. [CrossRef]
30. Beppo, T.; Takimoto, A. Flowering behaviour of the hybrids between strains 6746 and 371 of *Lemna paucicostata* Hegelm. *Aquat. Bot.* **1983**, *17*, 295–299. [CrossRef]
31. Sattler, M.C.; Carvalho, C.R.; Clarindo, W.R. The polyploidy and its key role in plant breeding. *Planta* **2016**, *243*, 281–296. [CrossRef]
32. Kandeler, R. Über die Blütenbildung bei *Lemna gibba* L. I. Kulturbedingungen und Tageslängenabhängigkeit. *Z. Bot.* **1955**, *43*, 61–67.
33. Cleland, C.F.; Briggs, W.R. Flowering Responses of the Long-day Plant *Lemna gibba* G3. *Plant Physiol.* **1967**, *42*, 1553–1561. [CrossRef]
34. Cleland, C.F.; Briggs, W.R. Effect of Low-Intensity Red and Far-Red Light and High-Intensity White Light on the Flowering Response of the Long-Day Plant *Lemna gibba* G3. *Plant Physiol.* **1968**, *43*, 157–162. [CrossRef]
35. Fu, L.; Ding, Z.; Kumpeangkeaw, A.; Tan, D.; Han, B.; Sun, X.; Zhang, J. De novo assembly, transcriptome characterization, and simple sequence repeat marker development in duckweed *Lemna gibba*. *Physiol. Mol. Biol. Plants* **2020**, *26*, 133–142. [CrossRef]
36. Appenroth, K.J.; Teller, S.; Horn, M. Photophysiology of turion formation and germination in *Spirodela polyrhiza*. *Biol. Plant.* **1996**, *38*, 95–106. [CrossRef]
37. Hutner, S.H. Comparative physiology of heterotrophic growth. In *Growth and Differentiation in Plants*; Iowa State University Digital Press: Ames, IA, USA, 1953; pp. 417–446.
38. Dolezel, J.; Greilhuber, J.; Suda, J. Estimation of nuclear DNA content in plants using flow cytometry. *Nat. Protoc.* **2007**, *2*, 2233–2244. [CrossRef]
39. Schmidt-Lebuhn, A.N.; Fuchs, J.; Hertel, D.; Hirsch, H.; Toivonen, J.; Kessler, M. An Andean radiation: Polyploidy in the tree genus Polylepis (Rosaceae, Sanguisorbeae). *Plant Biol.* **2010**, *12*, 917–926. [CrossRef]
40. Borchert, T.; Fuchs, J.; Winkelmann, T.; Hohe, A. Variable DNA content of *Cyclamen persicum* regenerated via somatic embryogenesis: Rethinking the concept of long-term callus and suspension cultures. *Plant Cell Tiss. Organ. Cult.* **2007**, *90*, 255–263. [CrossRef]
41. Galbraith, D.W.; Harkins, K.R.; Maddox, J.M.; Ayres, N.M.; Sharma, D.P.; Firoozabady, E. Rapid flow cytometric analysis of the cell cycle in intact plant tissues. *Science* **1983**, *220*, 1049–1051. [CrossRef]
42. Dolezel, J.; Bartos, J.; Voglmayr, H.; Greilhuber, J. Nuclear DNA content and genome size of trout and human. *Cytometry A* **2003**, *51*, 127–128. [CrossRef]
43. Ziegler, P.; Adelmann, K.; Zimmer, S.; Schmidt, C.; Appenroth, K.J. Relative in vitro growth rates of duckweeds (Lemnaceae)—The most rapidly growing higher plants. *Plant Biol.* **2015**, *17*, 33–41. [CrossRef]

44. Sree, K.S.; Sudakaran, S.; Appenroth, K.J. How fast can angiosperms grow? Species and clonal diversity of growth rates in the genus Wolffia (Lemnaceae). *Acta Physiol. Plant.* **2015**, *37*, 204. [CrossRef]
45. Lichtenthaler, H.K. Chlorophylls and carotenoids—Pigments of photosynthetic biomembranes. *Methods Enzymol.* **1987**, *148*, 350–382. [CrossRef]
46. Appenroth, K.J.; Augsten, H.; Liebermann, B.; Feist, H. Effects of light quality on amino acid composition of proteins in *Wolffia arrhiza* (L.) WIMM. using a specially modified Bradford method. *Biochem. Physiol. Pflanzen.* **1982**, *177*, 251–258. [CrossRef]
47. Logemann, J.; Schell, J.; Willmitzer, L. Improved method for the isolation of RNA from plant-tissue. *Anal. Biochem.* **1987**, *163*, 16–20. [CrossRef] [PubMed]
48. Kapoor, S.; Maheshwari, S.C.; Tyagi, A.K. Developmental and light-dependent cues interact to establish steady-state levels of transcripts for photosynthesis-related genes (psbA, psbD, psaA and rbcL) in rice (*Oryza sativa* L.). *Curr. Genet.* **1994**, *25*, 362–366. [CrossRef] [PubMed]
49. Coruzzi, G.; Broglie, R.; Cashmore, A.; Chua, N.H. Nucleotide-sequences of 2 Pea cDNA clones encoding the small subunit of Ribulose 1,5-biphosphate carboxylase and the major chlorophyll a/b-binding thylakoid polypeptide. *J. Biol. Chem.* **1983**, *258*, 1399–1402. [CrossRef]
50. Silverthorne, J.; Tobin, E.M. Demonstration of transcriptional regulation of specific genes by phytochrome action. *Proc. Natl. Acad. Sci. USA* **1984**, *81*, 1112–1116. [CrossRef]

plants

MDPI

Review

Duckweeds for Phytoremediation of Polluted Water

Yuzhen Zhou [1], Anton Stepanenko [2,3], Olena Kishchenko [2,3], Jianming Xu [1] and Nikolai Borisjuk [1,*]

1 School of Life Science, Huaiyin Normal University, Huai'an 223300, China
2 Leibniz Institute of Plant Genetics and Crop Plant Research (IPK), 06466 Gatersleben, Germany
3 Institute of Cell Biology and Genetic Engineering, National Academy of Sciences of Ukraine, 03143 Kyiv, Ukraine
* Correspondence: nborisjuk@hytc.edu.cn

Abstract: Tiny aquatic plants from the *Lemnaceae* family, commonly known as duckweeds, are often regarded as detrimental to the environment because of their ability to quickly populate and cover the surfaces of bodies of water. Due to their rapid vegetative propagation, duckweeds have one of the fastest growth rates among flowering plants and can accumulate large amounts of biomass in relatively short time periods. Due to the high yield of valuable biomass and ease of harvest, duckweeds can be used as feedstock for biofuels, animal feed, and other applications. Thanks to their efficient absorption of nitrogen- and phosphate-containing pollutants, duckweeds play an important role in the restorative ecology of water reservoirs. Moreover, compared to other species, duckweed species and ecotypes demonstrate exceptionally high adaptivity to a variety of environmental factors; indeed, duckweeds remove and convert many contaminants, such as nitrogen, into plant biomass. The global distribution of duckweeds and their tolerance of ammonia, heavy metals, other pollutants, and stresses are the major factors highlighting their potential for use in purifying agricultural, municipal, and some industrial wastewater. In summary, duckweeds are a powerful tool for bioremediation that can reduce anthropogenic pollution in aquatic ecosystems and prevent water eutrophication in a simple, inexpensive ecologically friendly way. Here we review the potential for using duckweeds in phytoremediation of several major water pollutants: mineral nitrogen and phosphorus, various organic chemicals, and heavy metals.

Keywords: duckweed; *Spirodela*; *Lemna*; water pollutants; nitrogen; phosphorus; heavy metals; agrochemicals wastewater remediation

check for updates

Citation: Zhou, Y.; Stepanenko, A.; Kishchenko, O.; Xu, J.; Borisjuk, N. Duckweeds for Phytoremediation of Polluted Water. *Plants* **2023**, *12*, 589. https://doi.org/10.3390/plants12030589

Academic Editors: Viktor Oláh, Klaus-Jürgen Appenroth and K. Sowjanya Sree

Received: 11 November 2022
Revised: 28 December 2022
Accepted: 19 January 2023
Published: 29 January 2023

1. Introduction

Pollution and shortages of potable water are two of the most serious problems facing humanity. In many Asian countries and elsewhere, the demand for potable water doubles every 10–15 years due to rising domestic and industrial consumption [1,2]. In addition, eutrophication, the nutrient enrichment of municipal, agricultural, and industrial water reservoirs due to human activities leading to stimulation of bacteria, algae, and plant growth and oxygen limitation, is a global concern and has been identified as a major environmental problem for water resource management.

The need to reduce anthropogenic nutrients in aquatic ecosystems to prevent water eutrophication has been widely recognized [3], and a number of physical, chemical, and biological methods for wastewater treatment have been tested [4]. The cultivation of aquatic plants is an attractive option for restoring eutrophic water bodies, offering an eco-friendly method for removing nutrients, bioaccumulating toxic nutrients and heavy metals for disposal, and regulating oxygen balance [5]. Various aquatic plants have been used for the bioremediation of wastewater with varying degrees of success [6], with duckweeds standing out because of their specific physiology, high growth rates, multiple options for biomass usage, simple maintenance, and easy harvesting [7].

Plants bioremediate pollutants by diverse mechanisms, depending on the pollutant nature. Simple nutrients such as N and P, which result from agricultural runoff and cause the eutrophication of water sources, can be used as nutrients to fuel plant growth. Organic compounds can be detoxified by cellular metabolism within the plant or by associated microbes; for successful bioremediation, the plant must tolerate the doses of these organic compounds present in the environment, take up the compound, and be able to metabolize it [8]. Some pollutants, such as industrial dyes, can also be removed by biosorption, in which the pollutant binds to the surface of the plant. For pollutants that cannot be broken down, such as heavy metals, bioremediation may involve uptake and sequestration of the pollutant, followed by removal and processing of the contaminated biomass [9,10]. Understanding the mechanism of bioremediation has key implications for selecting the species used and improving the ability of that species to bioremediate the pollutant in question.

Duckweed is a common name that unites a group of floating aquatic plants in the *Lemnaceae* that inhabit all continents except Antarctica [11,12]. Because of their rapid propagation, among the fastest growth rates of flowering plants [13], duckweeds play an important role in the ecology of water reservoirs worldwide. Often seen as detrimental to the environment due to their ability to quickly colonize and take over bodies of water, duckweeds have a long history of applications in medicine, the food chain, and rituals since ancient times, from the Chinese Han dynasty, to early Christians, to classic Mayan culture [14]. Since the dawn of modern molecular biochemistry, duckweeds have served as a model plant helping to reveal basic functions of proteins, nucleic acids, and hormones, and have provided insights into plants development, photosynthesis, nutrient turnover, and other key processes in plants [15–19]. With the search for new sources of renewable energy and biomaterials, the 2010s saw duckweeds reemerge in academic research and a wide spectrum of new practical applications [20–23]. Duckweeds have well-recognized potential uses as animal feed, biofuel feedstock, and human food because of their rapid accumulation of biomass and its high protein and starch contents [20,24–26]. Here we present a comprehensive summary highlighting recent applications of duckweeds for phytoremediation of major water pollutants: (i) mineral nitrogen and phosphorus, (ii) various organic chemicals, and (iii) heavy metals.

2. Duckweeds (*Lemnaceae*): Tiny Aquatic Plants with Unique Properties

Although duckweeds are often mistaken for algae because of their small size and reduced morphology, phylogenetically they are ancient monocotyledonous flowering plants represented by 36 currently recognized species grouped into five genera: *Spirodela*, *Landoltia*, *Lemna*, *Wolffiella* and *Wolffia* (Figure 1). Compared to the majority of plant species, leaves, and stems in duckweeds are merged into a simplified structure known as a frond, and roots, which are entirely lacking in two genera (*Wolffia* and *Wolffiella*). Species of the genus *Spirodela* have the largest fronds, up to 15 mm across, while those of *Wolffia* species are 2 mm or less in diameter with *Lemna* species are of intermediate size at 6–8 mm.

Because of the ancient origin of duckweeds about 100 million years ago, their tiny size, and their simple morphology, the phylogenic grouping of this clade remains a matter of debate [12], and recently has become more dependent on new molecular methods [27–29]. For example, analysis of chloroplast and nuclear DNA markers supported renaming *Spirodela punctata* to *Landoltia punctata* and separating the species into a new genus *Landoltia* [30], as well as a recent reduction in genus *Lemna* from 13 to 12 species [31].

Due to their very rapid vegetative propagation, duckweeds can produce a biomass mat capable of covering expansive water surfaces and formed by a single species or different species. With a doubling time of about 24 h for some duckweed species, they are among the fastest-growing flowering plant known and can reach an annual biomass productivity of 39–105 tons of dry weight per hectare per year [13]. For comparison, the productivity of *Miscanthus*, a major grass used for bioenergy production, is 5–44 tons of dry weight per hectare per year.

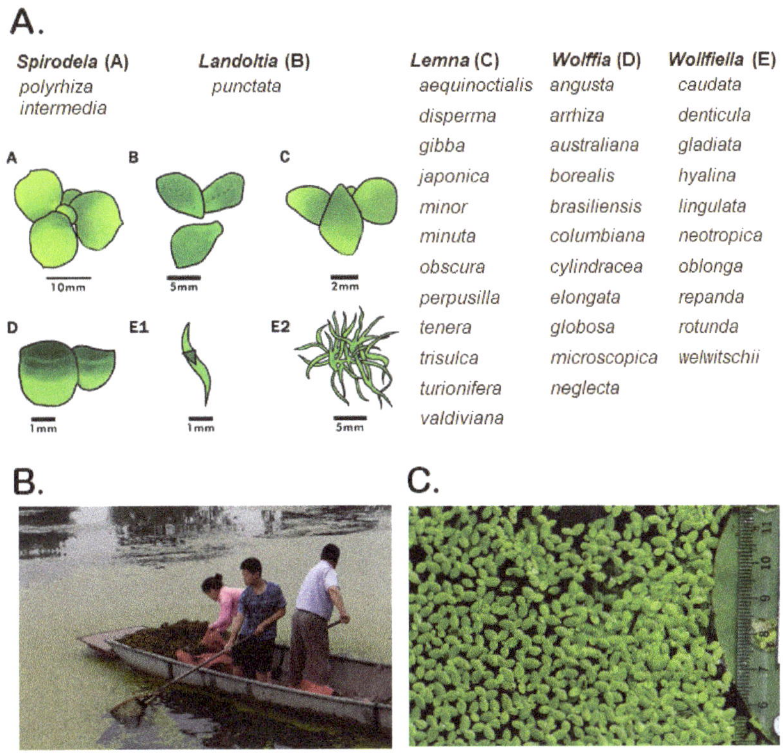

Figure 1. Duckweeds, the *Lemnaceae* plant family. (**A**) The *Lemnaceae* plant family contains 36 species grouped into five genera: *Spirodela*, *Landoltia*, *Lemna*, *Wolffia*, and *Wolffiella*. A, B, C, D, E1, E2 depict representative images of a species in the corresponding genera. The duckweed images are adapted from a drawing of Dr. K. Sowjanya Sree, Central University of Kerala, Periye, India [27]; (**B**) harvesting of duckweed covering a fishpond near Huai'an city, China; (**C**) *Lemna aequinoctialis* growing in the lake on the campus Huaiyin Normal University. Bar = 1 cm.

During biomass accumulation, duckweeds can very efficiently remediate different types of wastewater [32], and these traits can be further improved not only by selection of the best species or ecotypes [7,33,34], but also through optimization of plant's growth parameters such as ration of nutrients, light intensity, fronds density, etc. [35–38]. Biomass accumulation by plants in general strongly depends on efficient use of N, and duckweed plants are extremely efficient at assimilating N. Due to N remobilization and recycling by duckweeds, their nitrogen use efficiency is extremely high, reaching more than 68 kg biomass/kg of N under N limitation [39]. Simultaneously, duckweeds demonstrate relatively high tolerance to many water pollutants (e.g., ammonia, heavy metals, various organic compounds) and other environmental stresses when used for remediation of agricultural, municipal, and even industrial wastewater streams. These complementary features of water remediation and fast biomass accumulation have made duckweed a subject of intense academic research interest from the business community in recent years [20,40].

Recently, there has been significant progress in the areas of duckweed genomics, biochemistry, and developmental physiology [41]. We now have up-to-date, fully sequenced genomes of two ecotypes of *Spirodela polyrhiza* [42,43], *Spirodela intermedia* W. Koch [44], *Lemna minor* L. [45], *Lemna minuta* [46] and *Wolffia australiana* (Benth.) Hartog and Plas [47], as well as ongoing whole-genome sequencing projects for *Lemna gibba* L. [48]. Those efforts are further supported by establishment of duckweed collections hosting more than 2000 ecotypes [49]. The major world duckweed depository is hosted by Prof. E. Lam at the Rutgers Duckweed Stock Cooperative at Rutgers University (New Jersey, USA). There are also a number of live in vitro collections available in Canada, China, Germany, Hungary, India, Ireland, and Switzerland [50]. The broad geographical distribution of these duckweed collections clearly reflects the worldwide interest in promoting duckweed research and its exciting new applications.

3. Duckweeds for Remediating Water Contaminated with Nitrogen and Phosphorus

Excessive use and runoff of agrochemical fertilizers, primarily those containing nitrogen (N) and phosphorus (P), are considered to be the major causes of eutrophication [51]. To maximize crop yields, about 80 million tons of nitrogen fertilizers are applied globally per year [52]. It is likely that no more than 40% of this amount is taken up by crops [53], and the rest eventually ends up in freshwater reservoirs en route to the ocean. Aquaculture systems are another serious source of water pollutants, and the pollution of water bodies by aquaculture has increased by 2–4% per year over the last 20 years in the Yangtze River Basin and Zhujiang Delta Basin of China [54]. The main contaminants from aquaculture wastewater effluent are ammonium, organic N, and P [55]. Only about 15% of the N and 25% of the P from the feed used in aquaculture are consumed by fish and shrimp, with the unused part accumulating in the water or sediment [54].

Duckweeds have potential uses for low-cost wastewater treatment and efficient removal of excess N and P [34,56–58]. It has been estimated that duckweed can accumulate up to 9.1 t/ha/year of total N and 0.8 t/ha/year of total P in their biomass [58]. It has been demonstrated that after just 3 days of incubation of the duckweed *Lemna turionefera* in local municipal wastewater, the main nutrient concentrations (total N and total P) were lower than those in the effluent from a local wastewater treatment plant. In the same study of Zhou et al. [58], within 15 days of growth, four duckweed species removed more than 93% of total N and total P in local municipal wastewater. The final total N concentration was 1 mg/L, which is much lower than the national standard for treated wastewater (15 mg/L, China Standard GB 18918-2002) and is close to the total N level accepted for drinking water (1.5 mg/L, China Standard GB3838-2002). Similarly, high rates of removal were also demonstrated with duckweed growing on sewage water [59] and wastewater from a hog farm [60]. Moreover, 98% removal of N and P from pig-farm effluent has been achieved [61]. This was accompanied by a significant increase in the level of dissolved oxygen and the production of duckweed biomass with 35% crude protein.

Another advantage of duckweed is its tolerance of relatively high levels of ammonium ion (NH_4^+), which can be toxic to plants, animals, and humans at high concentrations [62]. The common duckweed (*L. minor*) has been reported to grow well at NH_4^+ concentrations of up to 84 mg/L [63]. The ability of duckweeds to take up and tolerate such high levels of NH_4^+ makes them particularly suited to the remediation of wastewater from domestic, agricultural, and especially aquaculture sources, which often contain considerable amounts of NH_4^+ due to the breakdown of urea in urine and runoff of NH_4^+-containing fertilizers. Moreover, unlike most plant species, duckweeds prefer NH_4^+ over nitrate (NO_3^-) as the source of N, as first demonstrated for dotted duckweed (*Landoltia punctata*) [64,65], and more recently confirmed for five other duckweed species representing the genera *Spirodela*, *Lemna*, and *Wolffia* [66].

To optimizing nutrient and fertilizer use and promote plant productivity, much recent work has focused on plant–microbe interactions in the rhizosphere [67–69]. To attract and feed root-associated microbes, plants invest a substantial part of their photosynthesized carbon into rhizosphere exudates [70,71]. In terrestrial [69,72] and aquatic plants [33,73,74], microbial-mediated denitrification limits nitrogen assimilation by reducing nitrate and nitrite ions to volatile NO, N_2O, and N_2 [75]. However, denitrification might accelerate bioremediation of wastewater containing high levels of N compounds. A detailed analysis of denitrifying bacteria interacting with the common duckweeds *S. polyrhiza* and *L. minor* showed that some derivatives of fatty acids and stigmasterol, previously revealed as the major components of the cuticule in duckweed [76], participate in plant–microbe interactions stimulating bacterial nitrogen metabolism by activating nitrate and nitrite reductases [77,78]. Moreover, the denitrifying rhizospheric bacterium *Pseudomonas* sp. RWX31 was able to modulate the chemical composition of root exudates in duckweed, specifically inducing the secretion of stigmasterol. In turn, stigmasterol appeared to alter the composition of the rhizosphere microbial community in favor of denitrifying bacteria [77].

4. Duckweeds for Remediating Water Contaminated with Organic Compounds

With the continuous development of agriculture, industry and economy, more and more organic pollutants are generated from agricultural irrigation; chemical, pharmaceutical, papermaking and other industries; and domestic sewage. Some organic pollutants do not biodegrade well and tend to accumulate in the environment, endangering the food chain. They are also often teratogenic, carcinogenic, and/or mutagenic to animals and humans, and this seriously threatens ecological environment security and human health. Therefore, it is important to seek efficient, low-cost and sustainable technologies to remove organic pollutants from water. Below we summarized the studies on duckweeds interaction with a variety of organic pollutants in water, such as agricultural chemicals, pharmaceuticals, and personal care products (PPCPs) and other industrial organic compounds.

4.1. Organic Agrochemicals

With the increasing demand for food and the development of agriculture and aquaculture, tons of toxic agrochemicals such as pesticides, herbicides, and fungicides are produced and applied annually. A considerable amount of these chemicals applied on farmlands and aquaculture ends up in the aquatic environment without treatment, causing substantial pressure on the environment. Aquatic non-targeted organisms are more likely to be exposed to herbicides in multiple pulse events than long continuous exposure. Therefore, the potential of an organism to recover between exposures has important effects on the overall toxicity. In addition, the organism used for bioremediation must be able to tolerate relevant concentrations of the compound while taking up some of the compound to metabolize it. Studies to test the toxicity to and uptake of agrochemicals by duckweed have primarily used *L. minor*.

Most agrochemicals are tolerated by duckweed at low concentrations but are toxic at higher concentrations. Wilson and Koch (2012) evaluated the effects and potential recovery of *L. minor* exposed to the herbicide norflurazon for 10 days under controlled conditions [79]. Duckweed was severely inhibited by norflurazon, but there was a rapid recovery for all norflurazon concentrations tested after the plant was removed from the media [79]. Varga et al. (2020) evaluated the growth patterns and recovery potential of duckweed between multiple exposures to the herbicide isoproturon [80]. Growth was significantly inhibited during each exposure phase with significant cumulative effects in subsequent treatment cycles, resulting in a cumulative decrease in biomass production. However, inhibitory effects were reversible upon transferring plants to a herbicide-free nutrient solution. These results indicate that *L. minor* plants have a high potential for recovery even after multiple exposures to isoproturon.

Burns et al. (2015) investigated the ability of two duckweed species (*L. minor* and *L. gibba*) to recover from a 7-day exposure to different concentrations (0.4–208 μg/L) of the herbicide diuron [81]. Diuron significantly inhibited duckweed growth and biomass production after the initial 7-day exposure. Following transfer to herbicide-free media, recovery was observed for all effects at concentrations ranging 60–111 μg/L for *L. minor* and 60–208 μg/L for *L. gibba*. These results suggest that recovery is possible for primary producers at environmentally relevant concentrations that are considered significant in ecological risk assessment. The herbicide glyphosate can induce oxidative stress in plants through H_2O_2 formation by targeting the mitochondrial electron transport chain and the deleterious effects of the herbicide, glyphosate, on duckweed photosynthesis, respiration, and pigment concentrations were related to glyphosate-induced oxidative stress through H_2O_2 accumulation [82].

Even though agrochemicals are toxic to duckweed, researchers showed that duckweed was able to remove agrochemicals from the environment, indicating that this aquatic plant can efficiently eliminate organic contaminants and may ultimately serve as phytoremediation agents in the natural environment. Dosnon-Olette studied the effect of two herbicides, isoproturon and glyphosate, on *L. minor* growth [83] and showed that 10 μg/L isoproturon and 80 μg/L glyphosate had little effect on the growth rate and chlorophyll fluorescence of *L. minor*, which was able to remove 25% and 8% of the isoproturon and glyphosate, respectively, after a four-day incubation. Mitsou et al. (2006) studied the toxicity of the rice herbicide propanil to *L. minor* and found that propanil, at a concentration of 1 mg/L, did not affect the growth of *L. minor*, and did not induce antioxidative defenses within the plant. In addition, *L. minor* accumulated and metabolized the propanil [84]. Prasertsup and Ariyakanon (2011) explored the potential of water lettuce (*Pistia stratiotes* L.) and duckweed (*L. minor*) to remove different concentrations of the herbicide chlorpyrifos under greenhouse conditions. Low concentrations (0.1 and 0.5 mg/L) of chlorpyrifos had no significant effect on the growth of *L. minor* and *P. stratiotes*, but a higher concentration (1 mg/L) inhibited their growth. The maximum removal of chlorpyrifos (initial culture concentration of 0.5 mg/L) by *P. stratiotes* and *L. minor* was 82% and 87%, respectively [85]. Olette et al. (2008) compared the ability of three aquatic plants to remove three pesticides and found that compared to two other aquatic plants (*Elodea canadensis* and *Elodea canadensis*), *L. minor* more efficiently removed the pesticides, causing reductions of 50%, 11.5% and 42% for copper sulfate, dimethomorph, and flazasulfuron, respectively [86].

Many organisms limit toxicity of environmental factors by not taking up these factors; however, successful bioremediation requires that the plant take up some of the compound and metabolize it into less-toxic byproducts. Dosnon-Olette et al. (2010) studied the factors affecting the rate of pesticide uptake by two duckweed species, *L. minor* and *S. polyrhiza* [87]. Increased sensitivity to the pesticide dimethomorph was observed with increasing duckweed population density, possibly explained by having less light due to crowding. Plant photosynthesis uses light as the energy source leading to the production of biochemical energy (e.g., ATP) and reducing power (NADPH), which in turn are used for carbon fixation. This light-dependent electron source contributes to the absobing and transformation pesticide dimethomorph. Panfili et al. (2019) showed that *L. minor* is suitable for cleaning water polluted with the herbicide terbuthylazine, and this potential can be successfully improved by treating the species with a biostimulant or a safener such as Megafol and benoxacor [88].

Tront and Saunders (2007) evaluated the uptake and accumulation of a fluorinated analog of 2,4-dichlorophenol, 4-chloro-2-fluorophenol (4-Cl-2-FP), by *L. minor* [89]. Time series data gathered from an experiment with an initial aqueous-phase concentration of 130 mM 4-Cl-2-FP showed that 4-Cl-2-FP was continuously removed from the aqueous phase and less than 2% of original 4-Cl-2-FP was detected in plant tissue within the time period of 77 h. An increasing amount of metabolites was detected in the plant tissue, comprising 18.9%, 28.6%, and 53.4% of original 4-Cl-2-FP at 10 h, 24 h, and 77 h, respectively. This means that over 95% of the initial compound accumulated by duckweed was broken down in the plant cells. Although many studies have focused on herbicides and their effects on aquatic plants, other agrochemicals also affect plants. For example, Yılmaz and Taş (2021) examined the effect of the synthetic pyrethroid insecticide zeta-cypermethrin on the growth and bioremediation of aquatic photosynthetic organisms and showed that *L. minor* used zeta-cypermethrin as a nutrient and increased its development in low zeta-cypermethrin concentration (150 μg/L) medium [90]. However, high concentrations (300–600 μg/L) were toxic and inhibited growth. In addition, *L. minor* removed 35.4–95.9% of zeta-cypermethrin, depending on the initial concentration.

4.2. Pharmaceuticals and Personal Care Products (PPCPs)

PPCPs, including antibiotics, painkillers, anti-inflammatory drugs, disinfectants, and aromatics, pose potential hazards to the environment and human health. These pollutants are becoming ubiquitous in the environment because they cannot be effectively removed by conventional wastewater treatment due to their toxic and recalcitrant nature. Though the PPCPs, and particularly pharmaceuticals are usually present in wastewaters at very low concentrations of nanograms per liter, their average annual world per capita is 15 g with 50–150 g in the most developed countries [8,91]. Considering that these compounds are often pretty stable, bioactive and bioaccumulative, they can present serious environmental and human health risks [92,93].

Toxicity to plants caused by pharmaceuticals is an important issue, and several plant species, including duckweed, have been considered for phytoremediation of pharmaceuticals in wetlands [94]. Like agrochemicals, most PPCPs are toxic to duckweeds. For example, three β-blockers, propranolo, atenolol and metoprolol, were found to be toxic to duckweed (*L. minor*), which was less sensitive than the arthropod *Daphnia magna* and the green alga *Desmodesmus subspicatus* [95]. Kaza et al. (2007) evaluated the toxicity of 13 pharmaceuticals, usually at ng/L to μg/L in the environment, to duckweed *L. minor* [96]. A total of 7 out of 13 drugs tested were toxic at concentrations below 100 mg/L. The antipsychotic drugs thioridazine and chlorpromazine were the most toxic substances, having effective concentrations (EC_{50}s) below 1 mg/L. Synthetic wastewater contaminated with the target compounds at 25 μg/L was prepared, and batch and continuous-flow experiments were conducted. Batch verification tests achieved removals of 98.8%, 96.4% and 95.4% for paracetamol, caffeine, and tricolsan, respectively. Overall removal of the PPCP contaminants was 97.7%, 98.0%, and 100% for paracetamol, caffeine, and tricolsan, respectively, by the constructed wetland system alone, while 97.5%, 98.2%, and 100%, respectively, were achieved by the lab-scale free water surface constructed wetland system [96].

In other work, Reinhold et al. (2010) tested the potential of both live and inactivated duckweed in removing pharmaceuticals in a microcosm wetland system [97]. Indeed, both live and inactivated duckweeds actively increased aqueous depletion of fluoxetine, ibuprofen, 2,4-dichlorophenoxyacetic acid, and the hand sanitizer triclosan. Some PPCPs can be used as a carbon source by duckweeds.

Amy-Sagers et al. (2017) conducted laboratory ecotoxicological assessments for a large range of concentrations of sucralose (an artificial sweetener) and fluoxetine (an antidepressant) on *L. minor* physiology and photosynthetic function [98]. Their results showed that, unlike humans who cannot break down and utilize sucralose, *L. minor* can use sucralose as a sugar substitute to increase its green leaf area and photosynthetic capacity. However, fluoxetine (323 nM) significantly decreased *L. minor* root growth, daily growth rate, and asexual reproduction.

4.2.1. Antibiotics

Although most antibiotics are toxic to duckweeds, they can tolerate and phytoremediate those compounds from the environment with different efficiency depending on particular types and concentrations of the antibiotic. Cascone et al. (2004) evaluated the phytotoxicity of the fluoroquinolone antibiotic flumequine on *L. minor* and plant drug uptake [99]. Flumequine, at all concentrations between 50 and 1000 μg/L tested, affected plant growth, but duckweed continued to grow over a five-week period. In media containing flumequine, a large proportion of the drug (about 96% at all concentrations tested) was degraded in the presence of *Lemna*. Gomes et al. (2017) studied the mechanism by which PPCPs affect duckweeds and found that in *L. minor*, high concentrations of the common antibiotic ciprofloxacin disrupted the normal electron flow in the respiratory electron transport chain and induced hydrogen peroxide production, thus changing the photosynthetic, respiratory pathway, and oxidative stress capacity of duckweed and affecting its ability to remove ciprofloxacin [100]. Therefore, when the concentration of antibiotics is high, the metabolism of duckweed changes, affecting its ability to remove the antibiotics. Singh et al. (2018) evaluated the potential toxicity of the antibiotic amoxicillin on the duckweed *S. polyrhiza* and found it was toxic, even at low concentrations [101]. Nonetheless, the duckweed contributed directly to the degradation of antibiotics in the water. In other study, the same group [102] estimated the phytotoxicity and degradation by *S. polyrhiza* of the antibiotic ofloxacin. The high concentrations of ofloxacin caused a reduction in biomass (4.8–41.3%), relative root growth, protein (4.16–11.28%) and photopigment contents. The fronds treated with ofloxacin showed an increased level of antioxidative enzymes (catalas, ascorbate peroxidase and superoxide dismutase) than control. The residual ofloxacin content in the medium was significantly reduced (93.73–98.36%) by day seven and phytodegradation was suggested to be the main mechanism for removal of this antibiotic.

The specific mechanism of PPCP removal by duckweeds depends on the type of PPCP and the duckweed species. Iatrou et al. (2017) explored the mechanism of removal effect of four kinds of antibiotics by *L. minor* [103]. The removal efficiencies of *L. minor* were 100% (cefadroxil), 96% (metronidazole), 59% (trimethoprim), and 73% (sulfamethoxazole), respectively. Plant uptake and biodegradation were the major mechanisms accounting for metronidazole removal; the most important mechanism for trimethoprim was plant uptake.

4.2.2. Analgesics and Anti-Inflammatory Drugs

The anti-inflammatory drugs and analgesics that do not require prescription in many countries, such as ibuprofen or paracetamol, are widely spread in the environment. Matamoros et al. (2012) found that caffeine and ibuprofen are removed by biodegradation and/or plant uptake by three aquatic plants, including the duckweed *L. minor*, and the removal rate was 83–99% in a microcosm wetland system [104]. In the presence of 1 mg/L ibuprofen, an increase in *L. gibba* frond number (+12%) and multiplication rate (+10%) was seen, while no variations in photosynthetic pigment content were observed [105]. Moreover, ibuprofen and 11 ibuprofen metabolites were detected in plants and in the growth medium, suggesting that *L. gibba* metabolizes ibuprofen. Li et al. (2017) studied the removal of four selected emerging PPCP compounds using greater duckweed (*S. polyrhiza)* in a laboratory-scale constructed wetland [106]. Di Baccio et al. (2017) explored the removal and metabolism of ibuprofen by *L. gibba* at high (0.20 and 1 mg/L) and environmentally relevant (0.02 mg/L) ibuprofen concentrations [107]. Ibuprofen uptake increased with increasing concentration, but the relative accumulation of ibuprofen and generation of hydroxy-ibuprofen was higher in the lower ibuprofen treatments. The main oxidized ibuprofen metabolites in humans (hydroxy-ibuprofen and carboxy-ibuprofen) were identified in the intact plants and in the growth solutions. Apart from a mean physical-chemical degradation of 8.2%, the ibuprofen removal by plants was highly efficient (89–92.5%) in all conditions tested.

4.3. Other Industrial Organic Compounds

Because of the efficient removal of pesticides and PPCPs by duckweed, researchers have explored whether duckweed can effectively remove other organic pollutants. In a recent study, the potential of *L. minor* for decolorization and degradation of malachite green (a triarylmethane dye) was investigated. The decolorization ability of the plant species was as high as 88%, and eight metabolic intermediate compounds were identified by gas chromatography-mass spectrometry [108]. Can-Terzi et al. (2021) [109] studied the phytoremediation potential of *L. minor* using methylene blue and showed that *L. minor* could effectively remove methylene blue from wastewater with the highest removal efficiency (98%) within 24 h. Fourier transform infrared spectroscopy (FTIR) and scanning electron microscopy (SEM) analyses indicated that dye removal was mainly by biosorption. Torbati (2019) evaluated the ability of *L. minor* to decolorize the acid Bordeaux B (ABB, an aminoazo benzene dye) [110]. Increased temperature and enhancement of initial plant weight increased the dye removal efficiency, but raising the initial dye concentration and pH reduced it. In optimum conditions, *L. minor* exhibited a considerable potential

(94% removal) for the phytoremediation of ABB. Seven intermediate ABB degradation products were identified using gas chromatography-mass spectrometry analysis, indicating biodegradation is one of the mechanisms of L. minor's removal and detoxification of ABB.

In a study of the fate of five benzotriazoles (used to inhibit the corrosion of copper) in a continuous-flow *L. minor* system, benzotriazole removal ranged between 26% and 72%. Plant uptake seemed to be the major mechanism governing the removal of benzotriazoles. When Zhang and Liang (2021) investigated the removal efficiency of 8 perfluoroalkyl acids by *L. minor* under aeration [111], they found that the removal efficiency of *L. minor* for long-chain perfluoroalkanes exceeded 95%, while the removal efficiency for short-chain perfluoroalkanes was marginal. The accumulation of perfluorooctane sulfonate in *L. minor* cells reached 14.4% after 2 weeks of exposure. Subsequently, the researchers further investigated the absorption and accumulation effect of *L. minor* on several intermediates of perfluoroalkyl compounds. The results showed that, after 14 days of exposure, *L. minor* accumulated 86.7 μG kg^{-1} and 1226 μG kg^{-1} for perfluorooctyl sulfonamide and fluorotelomere sulfonate, respectively [111]. In related work, Ekperusi et al. (2020) tested the potential of *Lemna paucicostata* (*Lemna aequinoctialis*, according to current classification) for removing petroleum hydrocarbons from crude oil-contaminated waters in a constructed wetland over a period of 120 days [112]. They found that *L. paucicostata* significantly ($F = 253.405$, $p < 0.05$) removed petroleum hydrocarbons from the wetland, reaching nearly 98% after 120 days, and estimated that about 97% of the petroleum hydrocarbons were biodegraded, because less than 1% bioaccumulated.

5. Duckweeds for Remediating Water Contaminated with Heavy Metals and Metalloids

5.1. Heavy Metals

Heavy metals are released into the environment from natural and anthropogenic sources, predominantly from mining and industrial activities. After entering the water environment, they accumulate in aquatic organisms, affecting their normal physiological and metabolic activities. Because they pose a threat to human health via the food chain and have serious impacts on the ecological environment, the removal of toxic pollutants is extremely important to minimize potential threats. Conventional techniques for the remediation of heavy metals are generally costly, time-consuming, and generate the problem of sludge disposal [113]. An environmentally friendly and economical treatment technology for the remediation of wastewater polluted with heavy metals is needed [114]. Duckweeds are relatively tolerant to heavy metals and able to take up many heavy metal ions, including those of cadmium, chromium, copper, iron, mercury, manganese, nickel, palladium, lead, and zinc [115–124]. Therefore, duckweed also has potential uses for monitoring and remediating heavy metals [125]. As a floating plant, duckweed can rapidly absorb heavy metals due to its special morphology and high growth rate [126]. In addition, duckweed can resist the toxicity of heavy metals through chelation and compartmentalization in vacuoles, effectively removing heavy metals in water through biological adsorption and intracellular accumulation [127].

A summary of studies of heavy metal uptake by duckweed species is shown in Table S1. Different duckweed species have different tolerances to various heavy metals, and their biomass, photosynthetic pigments, and antioxidant enzyme activities are significantly different. The toxic effect of heavy metals on duckweed is the main factor limiting the application of duckweed. Therefore, identifying duckweed species that can tolerate specific heavy metals, have suitable bioaccumulation ability, and have suitable resistance will help to improve the phytoremediation of heavy metals in polluted water by duckweed.

Some researchers found that mixing different species of duckweed and coculturing duckweed with microorganisms or other plants can affect the absorption of heavy metals. Due to differences in tolerance and accumulation ability of different duckweed species for various heavy metals, the coculture of different duckweed species can improve both biomass and antioxidant enzyme activity, reducing the toxicity of heavy metals to duckweed and thus aiding the removal of heavy metals from polluted water [128]. By coculturing

L. punctata and *L. minor* or individually in the medium with different concentrations of copper (Cu), Zhao (2015) found that coculturing produced better remediation effect than did single cultures at low Cu concentration; however, the single culture system was more effective at higher Cu concentration [129]. Duckweed can partly neutralize the toxic effect of high Cu concentrations by enhancing the activity of antioxidant enzymes, thus limiting the absorption of Cu.

The ability of duckweed to absorb heavy metals is also affected by the particular microorganisms symbiotically associated with the duckweed. Stout et al. (2010) showed that axenic duckweed, *L. minor*, accumulated slightly more Cd than did plants inoculated with bacterial isolates, suggesting that bacteria serve a phytoprotective role in their relationship with *L. minor* by preventing toxic Cd from entering plants [130].

Due to their ability to absorb heavy metals from the environment, duckweeds have been proposed for removing heavy metal contamination from wastewater. Bokhari et al. (2016) found that *L. minor* could effectively remediate both municipal and industrial wastewater [123], with removal rates of cadmium, copper, lead, and nickel all above 84% (Table S1). In addition, because dried duckweed power has a large specific surface area and high porosity, duckweed can also be processed into dry powder and used as a potential new adsorbent. Chen et al. (2013) found that the lead ion (Pb^{2+}) adsorption capacity of dried powder of *L. aequinoctialis* was more than 57 mg/g [131]. Nie et al. (2015) compared the removal rate of uranium ion (U^{4+}) by live *L. puntata* and its dry powder and found that the removal rate of 5 g/L U^{4+} was nearly 96% by 1.25 g/L dry powder at pH 5, which is higher than that (79%) by 2.5 g/L (FW, fresh weight) live *L. puntata* [132]. Li et al. (2017) studied the adsorption of cadmium ion (Cd^{2+}) in the aquatic environment by the dry powder of *S. polyrhiza* and *L. puntata* and found that the removal rates of Cd (50 mg/L) by the two kinds of dry powder of duckweed were 83% (*L. punctata*) and 96% (*S. polyrhiza*), respectively [133].

5.2. Metalloids: Boron and Arsenic

Boron (B) is an essential nutrient for plants but is toxic at high concentrations [134,135]. A study of the toxic effect of B (0.5–37 mg/L) on duckweed revealed that *S. polyrhiza* showed significantly reduced frond production and growth rates while significantly increasing the production of abnormal fronds. The authors concluded that *S. polyrhiza* could not remove significant amounts of B from the treatment solutions and, as a result, cannot be used as an effective component of B bioremediation systems [136]. Growing *L. gibba* at B concentrations of 0.3–10 mg/L showed no change in biomass production and a significant accumulation of B in fronds. At the same time, duckweed effectively reduced the B content in the environment in concentrations up to 2.0 mg/L [137]. A study of B toxicity using *L. minor* and *L. gibba*, with the aim of using them for phytoremediation and biomonitoring, revealed that significant inhibition of plant growth began at a B concentration of 16 mg/L. *L. minor* was more sensitive to B than *L. gibba*. The activity of the antioxidant enzymes superoxide dismutase, ascorbate peroxidase, and guaiacol peroxidase can serve as biomarkers for B-rich environments [138]. In another study, the combined use of *L. gibba* and chitosan beads effectively removed B from drinking water [139].

L. gibba showed the greatest potential to remove boron from irrigation water with B concentrations of 5.58–17.39 mg/L using a batch reactor. It was capable of removing 19–63% of the B from irrigation water, depending upon the level of contamination or initial concentration [140]. *L. gibba* and *L. minor* in the form of duckweed-based wastewater treatment systems coupled with microbial fuel cell reactor was shown to be an efficient method to simultaneously remove B from domestic wastewater/irrigation water and generate electricity [141,142]. In these studies, a monoculture of *L. gibba* showed the highest efficiency of B removal. Part of the research focused on the possibilities of B accumulation by duckweed under salt stress. Salt stress significantly affects the growth and B accumulation of L. minor. It was shown that only 7.9% to 15.5% of B was accumulated by L. minor during cultivation at NaCl concentration in a range of 0–200 mM. Finally, the authors concluded that L. minor is suitable for the accumulation of B when NaCl is below 100 mM [143]. Similar results were also shown for *S. polyrhiza* [144]. Thus, to date, information on the possibility of using duckweed for B removal is very limited, focusing on only three species, of which only *L. gibba* showed a sufficiently high potential for phytoremediation.

Arsenic (As) is present in the environment in inorganic and organic form and exists in four oxidation states—arsenate (As(V)), arsenite (As(III)), arsenic (As(0)), and arsine (As(-III)) [145]. Aquatic As phytoremediation approaches continue to be actively pursued [146,147]. Among 36 duckweed species, *L. gibba, L. minor, S. polyrhiza, W. globosa, W. australiana*, and *L. valdiviana* have been reported to remove As from water. The potential of duckweed for phytoremediation of As was first demonstrated in 2004 in waters from abandoned uranium mines. *L. gibba* revealed high arsenic bioaccumulation coefficients in wetlands of two former uranium mines in eastern Germany and under laboratory conditions. The potential extractions from mine surface waters using *L. gibba* were estimated to be 751.9 kg As/ha·yr [148]. In another study, *L. gibba* accumulated 10 times more As than background concentrations in the tailing waters of an abandoned uranium mine, reducing arsenic on average by 40.3% in the solutions [149].

L. minor showed high As accumulation (641 ± 21.3 nmol/g FW) when grown on As concentrations of 25–80 μM under laboratory conditions [150]. In another study, *L. minor* showed a removal rate of 140 mg As/ha·d, with a recovery of 5% when grown under a concentration of 0.15 mg/L [151]. The study of biological responses of *L. minor* revealed that both the duration of exposure and the concentration of inorganic As had a strong synergistic effect on antioxidant enzyme activity. *L. minor* showed a higher accumulation of As(III) compared to As(V) from polluted water [152]. A study of the accumulation of As by aquatic plants in running waters showed that *L. minor* is one of the top three studied species regarding arsenic accumulation (430 mg/kg DW). Higher values were observed only for *Callitriche lusitanica* and *Callitriche brutia* [153]. In hydroponics, *L. minor* revealed maximum removal of more than 70% As at a low concentration (0.5 mg/L) on day 15 of the experiment [154]. Another finding revealed that chelating agents had positive effects on As(III) or As(V) accumulation in *L. minor* [155].

For *L. valdiviana*, the As was only absorbed by the plant after a decline in the phosphate levels of the medium [156]. Concentrations greater than 1 mg/L As in the nutrient solution caused deleterious effects in *L. valdiviana* and compromised their phytoremediation capacity of water contaminated with As [156]. In addition, for *L. valdiviana*, As accumulation was dependent on pH. *L. valdiviana* accumulated 1190 mg/kg As (dry weight) from the aqueous media and reduced its initial concentration by 82% when cultivated between pH 6.3 and 7.0 [157].

At concentrations of 1.0, 2.0, and 4.0 μM As and dimethylarsinic acid, *S. polyrhiza* showed a significant level of As bioaccumulation, using different mechanisms for the degradation of arsenate vs. arsenite [158]. In addition, the uptake of inorganic arsenic (As (V) and As (III)) by *S. polyrhiza* was higher compared to the organic As sources, monomethylarsonic and dimethylarsinic acid. The addition of EDTA increased the uptake of inorganic As into the plant tissue, but the uptake of organic arsenic was not affected [159]. The study of the stability of *S. polyrhiza* at As (V) concentrations of 1, 5, 10, and 20 μM revealed an increase in the fresh biomass, photosynthetic pigments, and total protein contents of *S. polyrhiza* at lower concentrations of As (V) after 1 d of exposure, followed by a decrease in biomass with an increase in metal concentration [160]. In another study, *S. polyrhiza* showed the ability to survive in high concentrations of As (V) solution in hydroponics by decreasing As concentration, with a removal rate of 41% [161].

W. globosa accumulated 2–10 times more As than *S. polyrhiza/L. minor* and *Azolla* species [162]. At the low concentration range, the uptake rate was similar for arsenate and arsenite, but at the high concentration range, arsenite was taken up at a faster rate [162]. *W. globosa* was more resistant to external arsenate than arsenite but showed a similar degree of tolerance. A more detailed study of the mechanisms of As assimilation in *W. globosa* demonstrated an important role of phytochelatins in detoxifying As and enabling As accumulation [163]. A study conducted using *W. australiana* revealed the importance of microbial agglomerations for As assimilation. Sterile *W. australiana* did not oxidize As(III) in the growth medium or in plant tissue, whereas *W. australiana* with phyllosphere bacteria displayed substantial As(III) oxidation in the medium [164].

6. Conclusions and Perspectives

The beginning of the 21st century saw duckweed's revival as a model system for academic research and a wide spectrum of new applications boosted by growing concerns related to wastewater, renewable energy sources, and rising fossil fuel prices [41]. Researchers and entrepreneurs regard duckweed as a powerful tool for bioremediation that can reduce anthropogenic pollution in aquatic ecosystems and prevent water eutrophication in a simple, cheap, and environmentally friendly way. This is clearly reflected in the number of PubMed publications related to the search terms "duckweed remediation" compared to other popular categories such as "duckweed feed", "duckweed food", or "duckweed biofuel" (Figure 2).

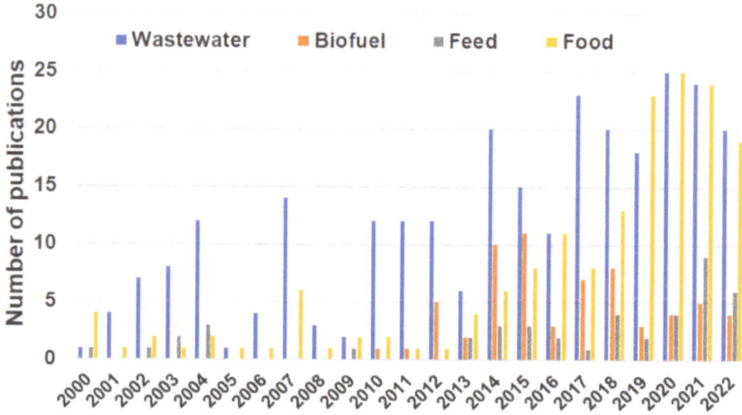

Figure 2. Publications related to duckweed research in the period 2000–2022. The total number of publications returned by the PubMed database (https://pubmed.ncbi.nlm.nih.gov, visited 3 November 2022) was 199 for "Wastewater", 57 for "Biofuel", 41 for "Feed", and 147 for "Food".

Testing different species and selecting duckweed varieties for more efficient phytoremediation remains an important and growing area of current research. Simultaneously, rapid advances in genome sequencing have revealed genes and metabolites responding to a growing number of specific pollutants and have provided valuable new information regarding the biochemical pathways underlying the uptake and assimilation of major pollutants by duckweed that are discussed in this review. This information, together with the optimized protocols for the genetic transformation of many duckweed species [165–168], should complement traditional selection efforts aimed at developing duckweed into an even more effective tool for protecting our environment and making our water as clean.

Supplementary Materials: The following supporting information can be downloaded at: https://www.mdpi.com/article/10.3390/plants12030589/s1. Table S1: Studies of heavy metals uptake by duckweeds. References [169–181] are cited in the supplementary materials.

Author Contributions: Conceptualization, Y.Z. and N.B.; investigation, Y.Z., A.S., O.K. and J.X.; resources, Y.Z., J.X. and N.B.; writing—original draft preparation, Y.Z., A.S. and N.B.; writing—review and editing, O.K. and N.B.; visualization, Y.Z. and N.B.; supervision and project administration, Y.Z., J.X. and N.B.; funding acquisition, Y.Z. and N.B. All authors have read and agreed to the published version of the manuscript.

Funding: This work was supported by an individual grant provided by Huaiyin Normal University (Huai'an, China) to N.B.

Institutional Review Board Statement: Not applicable.

Informed Consent Statement: Not applicable.

Data Availability Statement: Data are contained within the article.

Conflicts of Interest: The authors declare no conflict of interest.

References

1. Srivastava, J.; Gupta, A.; Chandra, H. Managing Water Quality with Aquatic Macrophytes. *Revis. Environ. Sci. Technol.* **2008**, *7*, 255–266. [CrossRef]
2. Priya, A.; Avishek, K.; Pathak, G. Assessing the Potentials of *Lemna minor* in the Treatment of Domestic Wastewater at Pilot Scale. *Environ. Monit. Assess.* **2012**, *184*, 4301–4307. [CrossRef]
3. Conley, D.J.; Paerl, H.W.; Howarth, R.W.; Boesch, D.F.; Seitzinger, S.P.; Havens, K.E.; Lancelot, C.; Likens, G.E. Controlling Eutrophication: Nitrogen and Phosphorus. *Science* **2009**, *323*, 1014–1015. [CrossRef]
4. González-González, R.B.; Flores-Contreras, E.A.; Parra-Saldívar, R.; Iqbal, H.M.N. Bio-Removal of Emerging Pollutants by Advanced Bioremediation Techniques. *Environ. Res.* **2022**, *214*, 113936. [CrossRef]
5. Dhote, S.; Dixit, S. Water Quality Improvement through Macrophytes—A Review. *Environ. Monit Assess.* **2009**, *152*, 149–153. [CrossRef]
6. Newete, S.W.; Byrne, M.J. The Capacity of Aquatic Macrophytes for Phytoremediation and Their Disposal with Specific Reference to Water Hyacinth. *Environ. Sci. Pollut. Res.* **2016**, *23*, 10630–10643. [CrossRef]
7. Ekperusi, A.O.; Sikoki, F.D.; Nwachukwu, E.O. Application of Common Duckweed (*Lemna minor*) in Phytoremediation of Chemicals in the Environment: State and Future Perspective. *Chemosphere* **2019**, *223*, 285–309. [CrossRef]
8. Becker, J.A.; Stefanakis, A.I. Pharmaceuticals and personal care products as emerging water contaminants. In *Pharmaceutical Sciences: Breakthroughs in Research and Practice*; Information Resources Management Association, Ed.; IGI Global: Hershey, PA, USA, 2017; pp. 1457–1475, ISBN 978-1-5225-1762-7. [CrossRef]
9. Skuza, L.; Szućko-Kociuba, I.; Filip, E.; Bożek, I. Natural Molecular Mechanisms of Plant Hyperaccumulation and Hypertolerance towards Heavy Metals. *Int. J. Mol. Sci.* **2022**, *23*, 9335. [CrossRef]
10. Noor, I.; Sohail, H.; Sun, J.; Nawaz, M.A.; Li, G.; Hasanuzzaman, M.; Liu, J. Heavy Metal and Metalloid Toxicity in Horticultural Plants: Tolerance Mechanism and Remediation Strategies. *Chemosphere* **2022**, *303*, 135196. [CrossRef]
11. Landolt, E. *Biosystematic Investigations in the Family of Duckweeds, Lemnaceae: The Family of Lemnaceae, a Monographic Study. Vol. 1: Morphology, Karyology, Ecology, Geographic Distribution, Systematic Position, Nomenclature, Descriptions*; Eidgenössischen Technischen Hochschule, Stiftung Rübel, Geobotanisches Institut: Zürich, Switzerland, 1986.
12. Tippery, N.P.; Les, D.H. Tiny Plants with Enormous Potential: Phylogeny and Evolution of Duckweeds. In *The Duckweed Genomes*; Cao, X.H., Fourounjian, P., Wang, W., Eds.; Springer International Publishing: Cham, Switzerland, 2020; pp. 19–38, ISBN 978-3-030-11044-4.
13. Ziegler, P.; Adelmann, K.; Zimmer, S.; Schmidt, C.; Appenroth, K.-J. Relative in Vitro Growth Rates of Duckweeds (*Lemnaceae*)—The Most Rapidly Growing Higher Plants. *Plant Biol.* **2015**, *17*, 33–41. [CrossRef]
14. Edelman, M.; Appenroth, K.-J.; Sree, K.S.; Oyama, T. Ethnobotanical History: Duckweeds in Different Civilizations. *Plants* **2022**, *11*, 2124. [CrossRef] [PubMed]
15. Trewavas, A. The Turnover of Nucleic Acids in *Lemna minor*. *Plant Physiol.* **1970**, *45*, 742–751. [CrossRef] [PubMed]
16. Frey, M.; Rail, S.; Roth, A.; Hemleben, V. Evidence for Uptake of Plamid DNA into Intact Plants (*Lemna perpusilla*) Proved by an E. Coli Transformation Assay. *Z. Nat. C* **1980**, *35*, 1104–1106. [CrossRef] [PubMed]
17. Mattoo, A.K.; Hoffman-Falk, H.; Marder, J.B.; Edelman, M. Regulation of Protein Metabolism: Coupling of Photosynthetic Electron Transport to in Vivo Degradation of the Rapidly Metabolized 32-Kilodalton Protein of the Chloroplast Membranes. *Proc. Natl. Acad. Sci. USA* **1984**, *81*, 1380–1384. [CrossRef]
18. Ben-Tal, Y.; Cleland, C.F. Uptake and Metabolism of [^{14}C] Salicylic Acid in *Lemna gibba* G3. *Plant Physiol.* **1982**, *70*, 291–296. [CrossRef] [PubMed]
19. Ingemarsson, B. Nitrogen Utilization in *Lemna*: I. Relations between Net Nitrate Flux, Nitrate Reduction, and in Vitro Activity and Stability of Nitrate Reductase. *Plant Physiol.* **1987**, *85*, 856–859. [CrossRef] [PubMed]
20. Appenroth, K.-J.; Sree, K.S.; Fakhoorian, T.; Lam, E. Resurgence of Duckweed Research and Applications: Report from the 3rd International Duckweed Conference. *Plant Mol. Biol.* **2015**, *89*, 647–654. [CrossRef]
21. An, D.; Li, C.; Zhou, Y.; Wu, Y.; Wang, W. Genomes and Transcriptomes of Duckweeds. *Front. Chem.* **2018**, *6*, 230. [CrossRef]
22. Baek, G.; Saeed, M.; Choi, H.-K. Duckweeds: Their Utilization, Metabolites and Cultivation. *Appl. Biol. Chem.* **2021**, *64*, 73. [CrossRef]
23. Yang, G.-L.; Feng, D.; Liu, Y.-T.; Lv, S.-M.; Zheng, M.-M.; Tan, A.-J. Research Progress of a Potential Bioreactor: Duckweed. *Biomolecules* **2021**, *11*, 93. [CrossRef]

24. Appenroth, K.-J.; Sree, K.S.; Bog, M.; Ecker, J.; Seeliger, C.; Böhm, V.; Lorkowski, S.; Sommer, K.; Vetter, W.; Tolzin-Banasch, K.; et al. Nutritional Value of the Duckweed Species of the Genus *Wolffia* (*Lemnaceae*) as Human Food. *Front. Chem.* **2018**, *6*, 483. [CrossRef]
25. Appenroth, K.-J.; Sree, K.S.; Böhm, V.; Hammann, S.; Vetter, W.; Leiterer, M.; Jahreis, G. Nutritional Value of Duckweeds (*Lemnaceae*) as Human Food. *Food Chem.* **2017**, *217*, 266–273. [CrossRef]
26. Appenroth, K.-J.; Ziegler, P.; Sree, K.S. Accumulation of Starch in Duckweeds (*Lemnaceae*), Potential Energy Plants. *Physiol. Mol. Biol. Plants* **2021**, *27*, 2621–2633. [CrossRef]
27. Appenroth, K.-J.; Borisjuk, N.; Lam, E. Telling Duckweed Apart: Genotyping Technologies for the *Lemnaceae*. *Chin. J. Appl. Environ. Biol.* **2013**, *19*, 1–10.
28. Borisjuk, N.; Chu, P.; Gutierrez, R.; Zhang, H.; Acosta, K.; Friesen, N.; Sree, K.S.; Garcia, C.; Appenroth, K.J.; Lam, E. Assessment, Validation and Deployment Strategy of a Two-Barcode Protocol for Facile Genotyping of Duckweed Species. *Plant Biol.* **2015**, *17*, 42–49. [CrossRef]
29. Braglia, L.; Lauria, M.; Appenroth, K.J.; Bog, M.; Breviario, D.; Grasso, A.; Gavazzi, F.; Morello, L. Duckweed Species Genotyping and Interspecific Hybrid Discovery by Tubulin-Based Polymorphism Fingerprinting. *Front. Plant Sci.* **2021**, *12*, 625670. [CrossRef]
30. Les, D.H.; Crawford, D.J. Landoltia (*Lemnaceae*), a new genus of duckweeds. *Novon* **1999**, *9*, 530–533. [CrossRef]
31. Bog, M.; Appenroth, K.-J.; Sree, K.S. Duckweed (*Lemnaceae*): Its Molecular Taxonomy. *Front. Sustain. Food Syst.* **2019**, *3*, 117. [CrossRef]
32. Zhou, Y.; Borisjuk, N. Small Aquatic Duckweed Plants with Big Potential for the Production of Valuable Biomass and Wastewater Remediation. *Int. J. Environ. Sci. Nat. Resour.* **2019**, *16*, 555942. [CrossRef]
33. Zhou, Y.R.; Yuan, X.Y.; Wang, J.R.; Bao, T.F.; Shi, W.M. Screening duckweed (*Lemnaceae*) species for efficient removal of waterbody nitrogen in the Tai Lake region and preliminary study on nitrogen removal mechanism. *Soils* **2010**, *42*, 390–397. (In Chinese)
34. Zhao, Y.; Fang, Y.; Jin, Y.; Huang, J.; Bao, S.; Fu, T.; He, Z.; Wang, F.; Zhao, H. Potential of Duckweed in the Conversion of Wastewater Nutrients to Valuable Biomass: A Pilot-Scale Comparison with Water Hyacinth. *Bioresour. Technol.* **2014**, *163*, 82–91. [CrossRef]
35. Kishchenko, O.; Stepanenko, A.; Straub, T.; Zhou, Y.; Neuhäuser, B.; Borisjuk, N. Ammonium Uptake, Mediated by Ammonium Transporters, Mitigates Manganese Toxicity in Duckweed, *Spirodela polyrhiza*. *Plants* **2023**, *12*, 208. [CrossRef]
36. Walsh, É.; Coughlan, N.E.; O'Brien, S.; Jansen, M.A.K.; Kuehnhold, H. Density Dependence Influences the Efficacy of Wastewater Remediation by *Lemna minor*. *Plants* **2021**, *10*, 1366. [CrossRef]
37. Walsh, É.; Kuehnhold, H.; O'Brien, S.; Coughlan, N.E.; Jansen, M.A.K. Light Intensity Alters the Phytoremediation Potential of *Lemna minor*. *Environ. Sci. Pollut. Res.* **2021**, *28*, 16394–16407. [CrossRef]
38. Walsh, É.; Paolacci, S.; Burnell, G.; Jansen, M.A.K. The Importance of the Calcium-to-Magnesium Ratio for Phytoremediation of Dairy Industry Wastewater Using the Aquatic Plant *Lemna minor* L. *Int. J. Phytoremediation* **2020**, *22*, 694–702. [CrossRef]
39. Guo, L.; Jin, Y.; Xiao, Y.; Tan, L.; Tian, X.; Ding, Y.; He, K.; Du, A.; Li, J.; Yi, Z.; et al. Energy-Efficient and Environmentally Friendly Production of Starch-Rich Duckweed Biomass Using Nitrogen-Limited Cultivation. *J. Clean. Produc.* **2020**, *251*, 119726. [CrossRef]
40. Lam, E.; Appenroth, K.J.; Michael, T.; Mori, K.; Fakhoorian, T. Duckweed in Bloom: The 2nd International Conference on Duckweed Research and Applications Heralds the Return of a Plant Model for Plant Biology. *Plant Mol. Biol.* **2014**, *84*, 737–742. [CrossRef]
41. Acosta, K.; Appenroth, K.J.; Borisjuk, L.; Edelman, M.; Heinig, U.; Jansen, M.A.K.; Oyama, T.; Pasaribu, B.; Schubert, I.; Sorrels, S.; et al. Return of the *Lemnaceae*: Duckweed as a Model Plant System in the Genomics and Post-Genomics Era. *Plant Cell* **2021**, *33*, 3207–3234. [CrossRef]
42. Wang, W.; Haberer, G.; Gundlach, H.; Gläßer, C.; Nussbaumer, T.; Luo, M.C.; Lomsadze, A.; Borodovsky, M.; Kerstetter, R.A.; Shanklin, J.; et al. The *Spirodela polyrhiza* Genome Reveals Insights into Its Neotenous Reduction Fast Growth and Aquatic Lifestyle. *Nat. Commun.* **2014**, *5*, 3311. [CrossRef]
43. Michael, T.P.; Bryant, D.; Gutierrez, R.; Borisjuk, N.; Chu, P.; Zhang, H.; Xia, J.; Zhou, J.; Peng, H.; El Baidouri, M.; et al. Comprehensive Definition of Genome Features in *Spirodela polyrhiza* by High-Depth Physical Mapping and Short-Read DNA Sequencing Strategies. *Plant J.* **2017**, *89*, 617–635. [CrossRef] [PubMed]
44. Hoang, P.T.N.; Fiebig, A.; Novák, P.; Macas, J.; Cao, H.X.; Stepanenko, A.; Chen, G.; Borisjuk, N.; Scholz, U.; Schubert, I. Chromosome-Scale Genome Assembly for the Duckweed *Spirodela intermedia*, Integrating Cytogenetic Maps, PacBio and Oxford Nanopore Libraries. *Sci. Rep.* **2020**, *10*, 19230. [CrossRef]
45. Van Hoeck, A.; Horemans, N.; Monsieurs, P.; Cao, H.X.; Vandenhove, H.; Blust, R. The First Draft Genome of the Aquatic Model Plant *Lemna minor* Opens the Route for Future Stress Physiology Research and Biotechnological Applications. *Biotechnol. Biofuels* **2015**, *8*, 188. [CrossRef]
46. Abramson, B.W.; Novotny, M.; Hartwick, N.T.; Colt, K.; Aevermann, B.D.; Scheuermann, R.H.; Michael, T.P. The genome and preliminary single-nuclei transcriptome of *Lemna minuta* reveals mechanisms of invasiveness. *Plant Physiol.* **2022**, *188*, 879–897. [CrossRef]
47. Michael, T.P.; Ernst, E.; Hartwick, N.; Chu, P.; Bryant, D.; Gilbert, S.; Ortleb, S.; Baggs, E.L.; Sree, K.S.; Appenroth, K.J.; et al. Genome and time-of-day transcriptome of *Wolffia australiana* link morphological minimization with gene loss and less growth control. *Genome Res.* **2020**, *31*, 225–238. [CrossRef]
48. Ernst, E. Status of the *Lemna gibba* 7742a and *Lemna minor* 8627 genomes. *ISCDRA* **2016**, *3*, 9–10.

49. Sree, K.S.; Appenroth, K.-J. Worldwide Genetic Resources of Duckweed: Stock Collections. In *The Duckweed Genomes*; Cao, X.H., Fourounjian, P., Wang, W., Eds.; Springer International Publishing: Cham, Switzerland, 2020; pp. 39–46. ISBN 978-3-030-11044-4.
50. Lam, E.; Appenroth, K.J.; Ma, Y.; Shoham, T.; Sree, K.S. Registration of Duckweed Clones/Strains-Future Approach. *Duckweed Forum* **2020**, *8*, 35–37.
51. Hilton, J.; O'Hare, M.; Bowes, M.J.; Jones, J.I. How Green Is My River? A New Paradigm of Eutrophication in Rivers. *Sci. Total Environ.* **2006**, *365*, 66–83. [CrossRef]
52. Food and Agriculture Organization (FAO). World Fertilizer Trends and Outlook to 2022. 2019. Available online: https://www.fao.org/3/ca6746en/CA6746EN.pdf?eloutlink=imf2fao (accessed on 20 October 2022).
53. Sylvester-Bradley, R.; Kindred, D.R. Analysing Nitrogen Responses of Cereals to Prioritize Routes to the Improvement of Nitrogen Use Efficiency. *J. Exp. Bot.* **2009**, *60*, 1939–1951. [CrossRef]
54. He, F.; Wu, Z. Application of Aquatic Plants in Sewage Treatment and Water Quality Improvement. *Chin. Bull. Bot.* **2003**, *20*, 641–647.
55. Cao, L.; Wang, W. Wastewater Management in Freshwater Pond Aquaculture in China. In *Sustainability in Food and Water: An Asian Perspective*; Sumi, A., Fukushi, K., Honda, R., Hassan, K.M., Eds.; Springer: Dordrecht, The Netherlands, 2010; pp. 181–190.
56. Zimmo, O.R.; van der Steen, N.P.; Gijzen, H.J. Nitrogen Mass Balance across Pilot-Scale Algae and Duckweed-Based Wastewater Stabilisation Ponds. *Water Res.* **2004**, *38*, 913–920. [CrossRef]
57. Cui, W.; Cheng, J.J. Growing Duckweed for Biofuel Production: A Review. *Plant Biol. J.* **2015**, *17*, 16–23. [CrossRef]
58. Zhou, Y.; Chen, G.; Peterson, A.; Zha, X.; Cheng, J.; Li, S.; Cui, D.; Zhu, H.; Kishchenko, O.; Borisjuk, N. Biodiversity of Duckweeds in Eastern China and Their Potential for Bioremediation of Municipal and Industrial Wastewater. *J. Geosci. Environ. Protect.* **2018**, *6*, 108–116. [CrossRef]
59. Yu, C.; Sun, C.; Yu, L.; Zhu, M.; Xu, H.; Zhao, J.; Ma, Y.; Zhou, G. Comparative Analysis of Duckweed Cultivation with Sewage Water and SH Media for Production of Fuel Ethanol. *PLoS ONE* **2014**, *9*, e115023. [CrossRef]
60. Cheng, J.J.; Stomp, A.-M. Growing Duckweed to Recover Nutrients from Wastewaters and for Production of Fuel Ethanol and Animal Feed. *Clean Soil Air Water* **2009**, *37*, 17–26. [CrossRef]
61. Mohedano, R.A.; Costa, R.H.R.; Tavares, F.A.; Belli Filho, P. High Nutrient Removal Rate from Swine Wastes and Protein Biomass Production by Full-Scale Duckweed Ponds. *Bioresour. Technol.* **2012**, *112*, 98–104. [CrossRef]
62. Britto, D.T.; Kronzucker, H.J. NH_4^+ Toxicity in Higher Plants: A Critical Review. *J. Plant Physiol.* **2002**, *159*, 567–584. [CrossRef]
63. Zhang, K.; Chen, Y.-P.; Zhang, T.-T.; Zhao, Y.; Shen, Y.; Huang, L.; Gao, X.; Guo, J.-S. The Logistic Growth of Duckweed (*Lemna minor*) and Kinetics of Ammonium Uptake. *Environ. Technol.* **2014**, *35*, 562–567. [CrossRef]
64. Fang, Y.Y.; Babourina, O.; Rengel, Z.; Yang, X.E.; Pu, P.M. Ammonium and Nitrate Uptake by the Floating Plant *Landoltia punctata*. *Ann. Bot.* **2007**, *99*, 365–370. [CrossRef] [PubMed]
65. Tian, X.; Fang, Y.; Jin, Y.; Yi, Z.; Li, J.; Du, A.; He, K.; Huang, Y.; Zhao, H. Ammonium Detoxification Mechanism of Ammonium-Tolerant Duckweed (*Landoltia punctata*) Revealed by Carbon and Nitrogen Metabolism under Ammonium Stress. *Environ. Pollut.* **2021**, *277*, 116834. [CrossRef]
66. Zhou, Y.; Kishchenko, O.; Stepanenko, A.; Chen, G.; Wang, W.; Zhou, J.; Pan, C.; Borisjuk, N. The Dynamics of NO_3^- and NH_4^+ Uptake in Duckweed Are Coordinated with the Expression of Major Nitrogen Assimilation Genes. *Plants* **2022**, *11*, 11. [CrossRef]
67. Feng, Z.; Li, X.; Lu, C.; Shen, Z.; Xu, F.; Chen, Y. Characterization of Pseudomonas Mendocina LR Capable of Removing Nitrogen from Various Nitrogen-Contaminated Water Samples When Cultivated with *Cyperus alternifolius* L. *J. Biosci. Bioeng.* **2012**, *114*, 182–187. [CrossRef]
68. Coskun, D.; Britto, D.T.; Shi, W.; Kronzucker, H.J. Nitrogen Transformations in Modern Agriculture and the Role of Biological Nitrification Inhibition. *Nat. Plants* **2017**, *3*, 17074. [CrossRef]
69. Gu, Y.; Dong, K.; Geisen, S.; Yang, W.; Yan, Y.; Gu, D.; Liu, N.; Borisjuk, N.; Luo, Y.; Friman, V.-P. The Effect of Microbial Inoculant Origin on the Rhizosphere Bacterial Community Composition and Plant Growth-Promotion. *Plant Soil* **2020**, *452*, 105–117. [CrossRef]
70. Bais, H.P.; Weir, T.L.; Perry, L.G.; Gilroy, S.; Vivanco, J.M. The Role of Root Exudates in Rhizosphere Interactions with Plants and Other Organisms. *Annu. Rev. Plant Biol.* **2006**, *57*, 233–266. [CrossRef]
71. Badri, D.V.; Vivanco, J.M. Regulation and Function of Root Exudates. *Plant Cell Environ.* **2009**, *32*, 666–681. [CrossRef]
72. Berendsen, R.L.; Pieterse, C.M.J.; Bakker, P.A.H.M. The Rhizosphere Microbiome and Plant Health. *Trends Plant Sci.* **2012**, *17*, 478–486. [CrossRef]
73. Bachand, P.A.M.; Horne, A.J. Denitrification in Constructed Free-Water Surface Wetlands: II. Effects of Vegetation and Temperature. *Ecol. Eng.* **1999**, *14*, 17–32. [CrossRef]
74. Zhou, Y.; Lu, Y.; Zhang, H.; Shi, W. Aerobic Denitrifying Characteristics of Duckweed Rhizosphere Bacterium RWX31. *Afr. J. Microbiol. Res.* **2013**, *7*, 211–219.
75. Ashok, V.; Hait, S. Remediation of Nitrate-Contaminated Water by Solid-Phase Denitrification Process—A Review. *Environ. Sci. Pollut. Res.* **2015**, *22*, 8075–8093. [CrossRef]
76. Borisjuk, N.; Peterson, A.A.; Lv, J.; Qu, G.; Luo, Q.; Shi, L.; Chen, G.; Kishchenko, O.; Zhou, Y.; Shi, J. Structural and Biochemical Properties of Duckweed Surface Cuticle. *Front. Chem.* **2018**, *6*, 317. [CrossRef]
77. Lu, Y.; Kronzucker, H.J.; Shi, W. Stigmasterol Root Exudation Arising from Pseudomonas Inoculation of the Duckweed Rhizosphere Enhances Nitrogen Removal from Polluted Waters. *Environ. Pollut* **2021**, *287*, 117587. [CrossRef]

78. Lu, Y.; Zhou, Y.; Nakai, S.; Hosomi, M.; Zhang, H.; Kronzucker, H.J.; Shi, W. Stimulation of Nitrogen Removal in the Rhizosphere of Aquatic Duckweed by Root Exudate Components. *Planta* **2014**, *239*, 591–603. [CrossRef] [PubMed]
79. Wilson, P.C.; Koch, R. Influence of Exposure Concentration and Duration on Effects and Recovery of *Lemna minor* Exposed to the Herbicide Norflurazon. *Arch. Environ. Contam. Toxicol.* **2013**, *64*, 228–234. [CrossRef] [PubMed]
80. Varga, M.; Žurga, P.; Brusić, I.; Horvatić, J.; Moslavac, M. Growth Inhibition and Recovery Patterns of Common Duckweed *Lemna minor* L. after Repeated Exposure to Isoproturon. *Ecotoxicology* **2020**, *29*, 1538–1551. [CrossRef]
81. Burns, M.; Hanson, M.L.; Prosser, R.S.; Crossan, A.N.; Kennedy, I.R. Growth Recovery of Lemna gibba and *Lemna minor* Following a 7-Day Exposure to the Herbicide Diuron. *Bull. Environ. Contam. Toxicol.* **2015**, *95*, 150–156. [CrossRef]
82. Gomes, M.P.; Juneau, P. Oxidative Stress in Duckweed (*Lemna minor* L.) Induced by Glyphosate: Is the Mitochondrial Electron Transport Chain a Target of This Herbicide? *Environ. Pollut.* **2016**, *218*, 402–409. [CrossRef] [PubMed]
83. Dosnon-Olette, R.; Couderchet, M.; Oturan, M.A.; Oturan, N.; Eullaffroy, P. Potential Use of *Lemna minor* for the Phytoremediation of Isoproturon and Glyphosate. *Int. J. Phytoremediation* **2011**, *13*, 601–612. [CrossRef]
84. Mitsou, K.; Koulianou, A.; Lambropoulou, D.; Pappas, P.; Albanis, T.; Lekka, M. Growth Rate Effects, Responses of Antioxidant Enzymes and Metabolic Fate of the Herbicide Propanil in the Aquatic Plant *Lemna minor*. *Chemosphere* **2006**, *62*, 275–284. [CrossRef]
85. Prasertsup, P.; Ariyakanon, N. Removal of Chlorpyrifos by Water Lettuce (*Pistia Stratiotes* L.) and Duckweed (*Lemna minor* L.). *Int. J. Phytoremediation* **2011**, *13*, 383–395. [CrossRef]
86. Olette, R.; Couderchet, M.; Biagianti, S.; Eullaffroy, P. Toxicity and Removal of Pesticides by Selected Aquatic Plants. *Chemosphere* **2008**, *70*, 1414–1421. [CrossRef]
87. Dosnon-Olette, R.; Couderchet, M.; El Arfaoui, A.; Sayen, S.; Eullaffroy, P. Influence of Initial Pesticide Concentrations and Plant Population Density on Dimethomorph Toxicity and Removal by Two Duckweed Species. *Sci. Total Environ.* **2010**, *408*, 2254–2259. [CrossRef]
88. Panfili, I.; Bartucca, M.L.; Del Buono, D. The Treatment of Duckweed with a Plant Biostimulant or a Safener Improves the Plant Capacity to Clean Water Polluted by Terbuthylazine. *Sci. Total Environ.* **2019**, *646*, 832–840. [CrossRef]
89. Tront, J.M.; Saunders, F.M. Sequestration of a Fluorinated Analog of 2,4-Dichlorophenol and Metabolic Products by *L. minor* as Evidenced by 19F NMR. *Environ. Pollut.* **2007**, *145*, 708–714. [CrossRef]
90. Yılmaz, Ö.; Taş, B. Feasibility and Assessment of the Phytoremediation Potential of Green Microalga and Duckweed for Zeta-Cypermethrin Insecticide Removal. *Desalination Water Treat.* **2021**, *209*, 131–143. [CrossRef]
91. Ortúzar, M.; Esterhuizen, M.; Olicón-Hernández, D.R.; González-López, J.; Aranda, E. Pharmaceutical Pollution in Aquatic Environments: A Concise Review of Environmental Impacts and Bioremediation Systems. *Front. Microbiol.* **2022**, *13*, 869332. [CrossRef]
92. Nassiri Koopaei, N.; Abdollahi, M. Health Risks Associated with the Pharmaceuticals in Wastewater. *DARU J. Pharm. Sci.* **2017**, *25*, 9. [CrossRef] [PubMed]
93. Chaves, M.d.J.S.; Kulzer, J.; Pujol de Lima, P.d.R.; Barbosa, S.C.; Primel, E.G. Updated Knowledge, Partitioning and Ecological Risk of Pharmaceuticals and Personal Care Products in Global Aquatic Environments. *Environ. Sci. Process. Impacts* **2022**, *24*, 1982–2008. [CrossRef]
94. Zhang, H.; Wang, X.C.; Zheng, Y.; Dzakpasu, M. Removal of Pharmaceutical Active Compounds in Wastewater by Constructed Wetlands: Performance and Mechanisms. *J. Environ. Manag.* **2023**, *325*, 116478. [CrossRef]
95. Cleuvers, M. Initial Risk Assessment for Three β-Blockers Found in the Aquatic Environment. *Chemosphere* **2005**, *59*, 199–205. [CrossRef]
96. Kaza, M.; Nalecz-Jawecki, G.; Sawicki, J. The Toxicity of Selected Pharmaceuticals to the Aquatic Plant *Lemn. Minor. Fresenius Environ. Bull.* **2007**, *16*, 524–531.
97. Reinhold, D.; Vishwanathan, S.; Park, J.J.; Oh, D.; Michael Saunders, F. Assessment of Plant-Driven Removal of Emerging Organic Pollutants by Duckweed. *Chemosphere* **2010**, *80*, 687–692. [CrossRef] [PubMed]
98. Amy-Sagers, C.; Reinhardt, K.; Larson, D.M. Ecotoxicological Assessments Show Sucralose and Fluoxetine Affect the Aquatic Plant, *Lemna minor*. *Aquat. Toxicol.* **2017**, *185*, 76–85. [CrossRef] [PubMed]
99. Cascone, A.; Forni, C.; Migliore, L. Flumequine Uptake and the Aquatic Duckweed, *Lemna minor* L. *Water Air Soil Pollut.* **2004**, *156*, 241–249. [CrossRef]
100. Gomes, M.P.; Gonçalves, C.A.; de Brito, J.C.M.; Souza, A.M.; da Silva Cruz, F.V.; Bicalho, E.M.; Figueredo, C.C.; Garcia, Q.S. Ciprofloxacin Induces Oxidative Stress in Duckweed (*Lemna minor* L.): Implications for Energy Metabolism and Antibiotic-Uptake Ability. *J. Hazard. Mater.* **2017**, *328*, 140–149. [CrossRef]
101. Singh, V.; Pandey, B.; Suthar, S. Phytotoxicity of Amoxicillin to the Duckweed *Spirodela polyrhiza*: Growth, Oxidative Stress, Biochemical Traits and Antibiotic Degradation. *Chemosphere* **2018**, *201*, 492–502. [CrossRef]
102. Singh, V.; Pandey, B.; Suthar, S. Phytotoxicity and Degradation of Antibiotic Ofloxacin in Duckweed (*Spirodela polyrhiza*) System. *Ecotoxicol. Environ. Saf.* **2019**, *179*, 88–95. [CrossRef]
103. Iatrou, E.I.; Gatidou, G.; Damalas, D.; Thomaidis, N.S.; Stasinakis, A.S. Fate of Antimicrobials in Duckweed *Lemna minor* Wastewater Treatment Systems. *J. Hazard. Mater.* **2017**, *330*, 116–126. [CrossRef]
104. Matamoros, V.; Nguyen, L.X.; Arias, C.A.; Salvadó, V.; Brix, H. Evaluation of Aquatic Plants for Removing Polar Microcontaminants: A Microcosm Experiment. *Chemosphere* **2012**, *88*, 1257–1264. [CrossRef]

105. Pietrini, F.; Di Baccio, D.; Aceña, J.; Pérez, S.; Barceló, D.; Zacchini, M. Ibuprofen Exposure in *Lemna gibba* L.: Evaluation of Growth and Phytotoxic Indicators, Detection of Ibuprofen and Identification of Its Metabolites in Plant and in the Medium. *J. Hazard. Mater.* **2015**, *300*, 189–193. [CrossRef]
106. Li, J.; Zhou, Q.; Campos, L.C. Removal of Selected Emerging PPCP Compounds Using Greater Duckweed (*Spirodela polyrhiza*) Based Lab-Scale Free Water Constructed Wetland. *Water Res.* **2017**, *126*, 252–261. [CrossRef]
107. Di Baccio, D.; Pietrini, F.; Bertolotto, P.; Pérez, S.; Barcelò, D.; Zacchini, M.; Donati, E. Response of *Lemna gibba* L. to High and Environmentally Relevant Concentrations of Ibuprofen: Removal, Metabolism and Morpho-Physiological Traits for Biomonitoring of Emerging Contaminants. *Sci. Total Environ.* **2017**, *584–585*, 363–373. [CrossRef] [PubMed]
108. Torbati, S. Feasibility and Assessment of the Phytoremediation Potential of Duckweed for Triarylmethane Dye Degradation with the Emphasis on Some Physiologicalresponses and Effect of Operational Parameters. *Turk. J. Biol.* **2015**, *39*, 438–446. [CrossRef]
109. Can-Terzi, B.; Goren, A.Y.; Okten, H.E.; Sofuoglu, S.C. Biosorption of Methylene Blue from Water by Live *Lemna minor*. *Environ. Technol. Innov.* **2021**, *22*, 101432. [CrossRef]
110. Torbati, S. Toxicological Risks of Acid Bordeaux B on Duckweed and the Plant Potential for Effective Remediation of Dye-Polluted Waters. *Environ. Sci. Pollut. Res.* **2019**, *26*, 27699–27711. [CrossRef] [PubMed]
111. Zhang, W.; Liang, L. Removal of eight perfluoroalkyl acids from aqueous solutions by aeration and duckweed. *Sci. Total Environ.* **2020**, *724*, 138357. [CrossRef]
112. Ekperusi, A.O.; Nwachukwu, E.O.; Sikoki, F.D. Assessing and Modelling the Efficacy of Lemna Paucicostata for the Phytoremediation of Petroleum Hydrocarbons in Crude Oil-Contaminated Wetlands. *Sci. Rep.* **2020**, *10*, 8489. [CrossRef]
113. Dhaliwal, S.S.; Singh, J.; Taneja, P.K.; Mandal, A. Remediation techniques for removal of heavy metals from the soil contaminated through different sources: A review. *Environ. Sci. Pollut. Res. Int.* **2020**, *27*, 1319–1333. [CrossRef] [PubMed]
114. Shahid, M.J.; Arslan, M.; Ali, S.; Siddique, M.; Afzal, M. Floating Wetlands: A Sustainable Tool for Wastewater Treatment. *Clean Soil Air Water* **2018**, *46*, 1800120. [CrossRef]
115. Uysal, Y.; Taner, F. Bioremoval of Cadmium by *Lemna minor* in Different Aquatic Conditions. *Clean Soil Air Water* **2010**, *38*, 370–377. [CrossRef]
116. Sekomo, C.B.; Rousseau, D.P.L.; Saleh, S.A.; Lens, P.N.L. Heavy Metal Removal in Duckweed and Algae Ponds as a Polishing Step for Textile Wastewater Treatment. *Ecol. Eng.* **2012**, *44*, 102–110. [CrossRef]
117. Parra, L.-M.M.; Torres, G.; Arenas, A.D.; Sánchez, E.; Rodríguez, K. Phytoremediation of Low Levels of Heavy Metals Using Duckweed (*Lemna minor*). In *Abiotic Stress Responses in Plants*; Springer: New York, NY, USA, 2012; pp. 451–463.minor). In *Abiotic Stress Responses in Plants*; Springer: New York, NY, USA, 2012; pp. 451–463.
118. Abdallah, M.A.M. Phytoremediation of Heavy Metals from Aqueous Solutions by Two Aquatic Macrophytes, *Ceratophyllum Demersum* and *Lemna gibba* L. *Environ. Technol.* **2012**, *33*, 1609–1614. [CrossRef]
119. Chaudhuri, D.; Majumder, A.; Misra, A.K.; Bandyopadhyay, K. Cadmium Removal by *Lemna minor* and *Spirodela polyrhiza*. *Int. J. Phytoremediation* **2014**, *16*, 1119–1132. [CrossRef] [PubMed]
120. Teixeira, S.; Vieira, M.N.; Marques, J.E.; Pereira, R. Bioremediation of an Iron-Rich Mine Effluent by *Lemna minor*. *Int. J. Phytoremediation* **2014**, *16*, 1228–1240. [CrossRef]
121. Miranda, A.F.; Muradov, N.; Gujar, A.; Stevenson, T.; Nugegoda, D.; Ball, A.S.; Mouradov, A. Application of Aquatic Plants for the Treatment of Selenium-Rich Mining Wastewater and Production of Renewable Fuels and Petrochemicals. *J. Sust. Bioenergy Syst.* **2014**, *04*, 97–112. [CrossRef]
122. Verma, R.; Suthar, S. Lead and Cadmium Removal from Water Using Duckweed—*Lemna gibba* L.: Impact of PH and Initial Metal Load. *Alex. Eng. J.* **2015**, *54*, 1297–1304. [CrossRef]
123. Bokhari, S.H.; Ahmad, I.; Mahmood-Ul-Hassan, M.; Mohammad, A. Phytoremediation Potential of *Lemna minor* L. for Heavy Metals. *Int. J. Phytoremediation* **2016**, *18*, 25–32. [CrossRef]
124. Zhou, Y.; Bai, T.; Kishchenko, O. Potential of Lemnoideae Species for Phytoremediation of Fresh Water with Elevated Manganese Concentration. *Innov. Biosyst. Bioeng.* **2019**, *3*, 232–238. [CrossRef]
125. Ziegler, P.; Sree, K.S.; Appenroth, K.-J. Duckweeds for Water Remediation and Toxicity Testing. *Toxicol. Environ. Chem.* **2016**, *98*, 1127–1154. [CrossRef]
126. Rezania, S.; Taib, S.M.; Md Din, M.F.; Dahalan, F.A.; Kamyab, H. Comprehensive Review on Phytotechnology: Heavy Metals Removal by Diverse Aquatic Plants Species from Wastewater. *J. Hazard. Mater.* **2016**, *318*, 587–599. [CrossRef]
127. Zhou, Q.; Lin, Y.; Li, X.; Yang, C.; Han, Z.; Zeng, G.; Lu, L.; He, S. Effect of Zinc Ions on Nutrient Removal and Growth of *Lemna Aequinoctialis* from Anaerobically Digested Swine Wastewater. *Bioresour. Technol.* **2018**, *249*, 457–463. [CrossRef]
128. Zhao, Z.; Shi, H.; Liu, C.; Kang, X.; Chen, L.; Liang, X.; Jin, L. Duckweed Diversity Decreases Heavy Metal Toxicity by Altering the Metabolic Function of Associated Microbial Communities. *Chemosphere* **2018**, *203*, 76–82. [CrossRef]
129. Zhao, Z.; Shi, H.; Duan, D.; Li, H.; Lei, T.; Wang, M.; Zhao, H.; Zhao, Y. The Influence of Duckweed Species Diversity on Ecophysiological Tolerance to Copper Exposure. *Aquat. Toxicol.* **2015**, *164*, 92–98. [CrossRef]
130. Stout, L.M.; Dodova, E.N.; Tyson, J.F.; Nüsslein, K. Phytoprotective Influence of Bacteria on Growth and Cadmium Accumulation in the Aquatic Plant *Lemna minor*. *Water Res.* **2010**, *44*, 4970–4979. [CrossRef]
131. Chen, L.C.; Fang, Y.; Jin, Y.L.; Chen, Q.; Zhao, Y.G.; Xiao, Y.; Zhao, H. Biosorption of Pb^{2+} by dried powder of duckweed (*Lemna aequinoctialis*). *Chin. J. Appl. Environ. Biol.* **2013**, *19*, 1046–1052. (In Chinese) [CrossRef]

132. Nie, X.Q.; Dong, F.Q.; Liu, N.; Zhang, D.; Liu, M.X.; Yang, J.; Zhang, W. Biosorption and biomineralization of uranium (VI) from aqueous solutions by *Landoltia puntata*. *Spectrosc. Spect. Anal.* **2015**, *35*, 2613–2619. (In Chinese)
133. Li, Y.; Yang, C.; Zhong, Y.; Tang, J. Adsorption properties of the dry powers of two duckweed species for Cd2+. *Jiangsu Agric. Sci.* **2017**, *45*, 248–254. [CrossRef]
134. Camacho-Cristóbal, J.J.; Rexach, J.; González-Fontes, A. Boron in Plants: Deficiency and Toxicity. *J. Integr. Plant Biol.* **2008**, *50*, 1247–1255. [CrossRef]
135. Kumar, V.; Pandita, S.; Kaur, R.; Kumar, A.; Bhardwaj, R. Biogeochemical Cycling, Tolerance Mechanism and Phytoremediation Strategies of Boron in Plants: A Critical Review. *Chemosphere* **2022**, *300*, 134505. [CrossRef]
136. Davis, S.M.; Drake, K.D.; Maier, K.J. Toxicity of Boron to the Duckweed, *Spirodella polyrrhiza*. *Chemosphere* **2002**, *48*, 615–620. [CrossRef]
137. Del-Campo Marín, C.M.; Oron, G. Boron Removal by the Duckweed Lemna Gibba: A Potential Method for the Remediation of Boron-Polluted Waters. *Water Res.* **2007**, *41*, 4579–4584. [CrossRef]
138. Gür, N.; Türker, O.C.; Böcük, H. Toxicity Assessment of Boron (B) by *Lemna minor* L. and *Lemna gibba* L. and Their Possible Use as Model Plants for Ecological Risk Assessment of Aquatic Ecosystems with Boron Pollution. *Chemosphere* **2016**, *157*, 1–9. [CrossRef] [PubMed]
139. Türker, O.C.; Baran, T. A Combination Method Based on Chitosan Adsorp-tion and Duckweed (*Lemna gibba* L.) Phytoremediation for Boron (B) Removal from Drinking Water. *Int. J. Phytoremediation* **2018**, *20*, 175–183. [CrossRef]
140. Türker, O.C.; Yakar, A.; Gür, N. Bioaccumulation and Toxicity Assessment of Irrigation Water Contaminated with Boron (B) Using Duckweed (*Lemna gibba* L.) in a Batch Reactor System. *J. Hazard. Mater.* **2017**, *324*, 151–159. [CrossRef] [PubMed]
141. Türker, O.C. Simultaneous Boron (B) Removal and Electricity Generation from Domestic Wastewater Using Duckweed-Based Wastewater Treatment Reactors Coupled with Microbial Fuel Cell. *J. Environ. Manag.* **2018**, *228*, 20–31. [CrossRef]
142. Türker, O.C.; Yakar, A.; Türe, C.; Saz, Ç. Boron (B) Removal and Bioelec-tricity Captured from Irrigation Water Using Engineered Duckweed-Microbial Fuel Cell: Effect of Plant Species and Vegetation Structure. *Environ. Sci. Pollut. Res. Int.* **2019**, *26*, 31522–31536. [CrossRef]
143. Liu, C.; Gu, W.; Dai, Z.; Li, J.; Jiang, H.; Zhang, Q. Boron Accumulation by *Lemna minor* L. under Salt Stress. *Sci. Rep.* **2018**, *8*, 8954. [CrossRef]
144. Uruc Parlak, K. Effects of Boron and NaCl on Antioxidant Defence Mechanisms in Duckweeds (*Spirodela polyrhiza* L.). *Pak. J. Biol. Sci.* **2021**, *24*, 989–996. [CrossRef]
145. Sharma, V.K.; Sohn, M. Aquatic Arsenic: Toxicity, Speciation, Transfor-mations, and Remediation. *Environ. Int.* **2009**, *35*, 743–759. [CrossRef]
146. Rahman, M.A.; Hasegawa, H. Aquatic Arsenic: Phytoremediation Using Floating Macrophytes. *Chemosphere* **2011**, *83*, 633–646. [CrossRef]
147. Khanna, K.; Kohli, S.K.; Kumar, P.; Ohri, P.; Bhardwaj, R.; Alam, P.; Ah-mad, P. Arsenic as Hazardous Pollutant: Perspectives on Engineering Remediation Tools. *Sci. Total Environ.* **2022**, *838*, 155870. [CrossRef]
148. Mkandawire, M.; Taubert, B.; Dudel, E.G. Capacity of *Lemna gibba* L. (Duckweed) for Uranium and Arsenic Phytoremediation in Mine Tailing Waters. *Int. J. Phytoremediation* **2004**, *6*, 347–362. [CrossRef] [PubMed]
149. Mkandawire, M.; Dudel, E.G. Accumulation of Arsenic in *Lemna gibba* L. (Duckweed) in Tailing Waters of Two Abandoned Uranium Mining Sites in Saxony, Germany. *Sci. Total Environ.* **2005**, *336*, 81–89. [CrossRef]
150. Charlier, H.A.J.; Albertson, C.; Thornock, C.; Warner, L.; Hurst, T.; Ellis, R. Comparison of the Effects of Arsenic (V), Cadmium (II), and Mercury (II) Single Metal and Mixed Metal Exposure in Radish, Raphanus Sativus, Fescue Grass, Festuca Ovina, and Duckweed, *Lemna minor*. *Bull. Environ. Contam. Toxicol.* **2005**, *75*, 474–481. [CrossRef] [PubMed]
151. Alvarado, S.; Guédez, M.; Lué-Merú, M.P.; Nelson, G.; Alvaro, A.; Jesús, A.C.; Gyula, Z. Arsenic Removal from Waters by Bioremediation with the Aquatic Plants Water Hyacinth (*Eichhornia crassipes*) and Lesser Duckweed (*Lemna minor*). *Bioresour. Technol.* **2008**, *99*, 8436–8440. [CrossRef]
152. Duman, F.; Ozturk, F.; Aydin, Z. Biological Responses of Duckweed (*Lemna minor* L.) Exposed to the Inorganic Arsenic Species As(III) and As(V): Ef-fects of Concentration and Duration of Exposure. *Ecotoxicology* **2010**, *19*, 983–993. [CrossRef]
153. Favas, P.J.C.; Pratas, J.; Prasad, M.N.V. Accumulation of Arsenic by Aquatic Plants in Large-Scale Field Conditions: Opportunities for Phytoremediation and Bioindication. *Sci. Total Environ.* **2012**, *433*, 390–397. [CrossRef]
154. Goswami, C.; Majumder, A.; Misra, A.K.; Bandyopadhyay, K. Arsenic Up-take by *Lemna minor* in Hydroponic System. *Int. J. Phytoremediation* **2014**, *16*, 1221–1227. [CrossRef]
155. Yang, G.-L.; Yang, M.-X.; Lv, S.-M.; Tan, A.-J. The Effect of Chelating Agents on Iron Plaques and Arsenic Accumulation in Duckweed (*Lemna minor*). *J. Hazard. Mater.* **2021**, *419*, 126410. [CrossRef]
156. De Souza, T.D.; Borges, A.C.; de Matos, A.T.; Veloso, R.W.; Braga, A.F. Kinetics of Arsenic Absorption by the Species Eichhornia crassipes and Lemna Valdiviana under Optimized Conditions. *Chemosphere* **2018**, *209*, 866–874. [CrossRef]
157. De Souza, T.D.; Borges, A.C.; Braga, A.F.; Veloso, R.W.; Teixeira de Matos, A. Phytoremediation of Arsenic-Contaminated Water by *Lemna Valdiviana*: An Optimization Study. *Chemosphere* **2019**, *234*, 402–408. [CrossRef]
158. Rahman, M.A.; Hasegawa, H.; Ueda, K.; Maki, T.; Okumura, C.; Rahman, M.M. Arsenic Accumulation in Duckweed (*Spirodela polyrhiza* L.): A Good Option for Phytoremediation. *Chemosphere* **2007**, *69*, 493–499. [CrossRef]

159. Rahman, M.A.; Hasegawa, H.; Ueda, K.; Maki, T.; Rahman, M.M. Influence of EDTA and Chemical Species on Arsenic Accumulation in *Spirodela polyrhiza* L. (Duckweed). *Ecotoxicol. Environ. Saf.* **2008**, *70*, 311–318. [CrossRef]
160. Seth, C.S.; Chaturvedi, P.K.; Misra, V. Toxic Effect of Arsenate and Cadmium Alone and in Combination on Giant Duckweed (*Spirodela polyrrhiza* L.) in Response to Its Accumulation. *Environ. Toxicol.* **2007**, *22*, 539–549. [CrossRef]
161. Zhang, X.; Hu, Y.; Liu, Y.; Chen, B. Arsenic Uptake, Accumulation and Phytofiltration by Duckweed (*Spirodela polyrhiza* L.). *J. Environ. Sci.* **2011**, *23*, 601–606. [CrossRef]
162. Zhang, X.; Zhao, F.-J.; Huang, Q.; Williams, P.N.; Sun, G.-X.; Zhu, Y.-G. Arsenic Uptake and Speciation in the Rootless Duckweed *Wolffia globosa*. *New Phytol.* **2009**, *182*, 421–428. [CrossRef]
163. Zhang, X.; Uroic, M.K.; Xie, W.-Y.; Zhu, Y.-G.; Chen, B.-D.; McGrath, S.P.; Feldmann, J.; Zhao, F.-J. Phytochelatins Play a Key Role in Arsenic Accumulation and Tolerance in the Aquatic Macrophyte *Wolffia globosa*. *Environ. Pollut.* **2012**, *165*, 18–24. [CrossRef]
164. Xie, W.-Y.; Su, J.-Q.; Zhu, Y.-G. Arsenite Oxidation by the Phyllosphere Bacterial Community Associated with *Wolffia Australiana*. *Environ. Sci. Technol.* **2014**, *48*, 9668–9674. [CrossRef] [PubMed]
165. Yang, G.L.; Fang, Y.; Xu, Y.L.; Tan, L.; Li, Q.; Liu, Y.; Lai, F.; Jin, Y.L.; Du, A.P.; He, K.Z.; et al. Frond transformation system mediated by *Agrobacterium tumefaciens* for *Lemna minor*. *Plant Mol. Biol.* **2018**, *98*, 319–331. [CrossRef]
166. Heenatigala, P.P.M.; Yang, J.; Bishopp, A.; Sun, Z.; Li, G.; Kumar, S.; Hu, S.; Wu, Z.; Lin, W.; Yao, L.; et al. Development of Efficient Protocols for Stable and Transient Gene Transformation for *Wolffia globosa* Using *Agrobacterium*. *Front. Chem.* **2018**, *6*, 227. [CrossRef]
167. Liu, Y.; Wang, Y.; Xu, S.; Tang, X.; Zhao, J.; Yu, C.; He, G.; Xu, H.; Wang, S.; Tang, Y.; et al. Efficient genetic transformation and CRISPR/Cas9-mediated genome editing in *Lemna aequinoctialis*. *Plant Biotechnol. J.* **2019**, *17*, 2143–2152. [CrossRef]
168. Peterson, A.; Kishchenko, O.; Zhou, Y.; Vasylenko, M.; Giritch, A.; Sun, J.; Borisjuk, N.; Kuchuk, M. Robust *Agrobacterium*-Mediated Transient Expression in Two Duckweed Species (*Lemnaceae*) Directed by Non-Replicating, Replicating, and Cell-to-Cell Spreading Vectors. *Front. Bioeng. Biotechnol.* **2021**, *9*, 761073. [CrossRef] [PubMed]
169. Megateli, S.; Semsari, S.; Couderchet, M. Toxicity and removal of heavy metals (cadmium, copper, and zinc) by Lemna gibba. *Ecotoxicol. Environ. Saf.* **2009**, *72*, 1774–1780. [CrossRef] [PubMed]
170. Yilmaz, D.D.; Akbulut, H. Effect of circulation on wastewater treatment by *Lemna gibba* and *Lemna minor* (Floating Aquatic Macrophytes). *Int. J. Phytoremediation* **2011**, *13*, 970–984. [CrossRef]
171. Ekta, C.; Praveen, S. Chromium and cadmium removal from wastewater using duckweed-*Lemna gibba* L. and ultrastructural deformation due to metal toxicity. *Int. J. Phytoremediation* **2019**, *21*, 279–286.
172. Khellaf, N.; Zerdaoui, M. Growth response of the duckweed *Lemna gibba* L. to copper and nickel phytoaccumulation. *Ecotoxicol. (Lond. Engl.)* **2010**, *19*, 1363–1368. [CrossRef] [PubMed]
173. Romero-Hernández, J.A.; Amaya-Chávez, A.; Balderas-Hernández, P.; Roa-Morales, G.; González-Rivas, N. Balderas-Plata, M.A. Tolerance and hyperaccumulation of a mixture of heavy metals (Cu, Pb, Hg, and Zn) by four aquatic macrophytes. *Int. J. Phytoremediation* **2017**, *19*, 239–245. [CrossRef] [PubMed]
174. Basile, A.; Sorbo, S.; Conte, B.; Cobianchi, R.C.; Trinchella, F.; Capasso, C.; Carginale, V. Toxicity, accumulation, and removal of heavy metals by three aquatic macrophytes, *Int. J. Phytoremediat.* **2012**, *14*, 374–387. [CrossRef]
175. Kaur, L.; Gadgil, K.; Sharma, S. Role of pH in the accumulation of lead and nickel by common duckweed (*Lemna minor*). *Int. J. Bioassays* **2012**, *12*, 191–195.
176. Azeez, N.M.; Sabbar, A.A. Efficiency of duckweed (*Lemna minor* L.) in phytotreatment of wastewater pollutants from Basrah oil refinery. *J. Appl. Phytotechnology Environ. Sanit.* **2012**, *1*, 163–172.
177. Daud, M.K.; Ali, S.; Abbas, Z.; Zaheer, I.E.; Riaz, M.A.; Malik, A.; Hussain, A.; Rizwan, M.; Rehman, M.Z.; Zhu, S.J. Potential of duckweed (*Lemna minor*) for the phytoremediation of landfill leachate. *J. Chem.* **2018**, *2018*, 1–9. [CrossRef]
178. Jafari, N.; Akhavan, M. Effect of pH and heavy metal concentration on phytoaccumulation of zinc by three duckweed species. *Am.-Eurasian J. Agric. Environ. Sci.* **2011**, *10*, 34–41.
179. Kumar, M.V.; Tripathi, B.D. Concurrent removal and accumulation of heavy metals by the three aquatic macrophytes. *Bioresour. Technol.* **2008**, *99*, 7091–7097. [CrossRef]
180. Al-Balawna, Z.A. Remove heavy metals (Cd and Pb) from irrigation water in Jordan valley by using duckweeds (*Spirodela polyrhiza*). *Acta Sci. Agric.* **2020**, *4*, 1–4. [CrossRef]
181. Loveson, A.; Sivalingam, R.; Syamkumar, R. Aquatic macrophyte *Spirodela polyrrhiza* as a phytoremediation tool in polluted wetland water from Eloor, Ernakulam district, Kerala. *IOSR J. Environ. Sci. Toxicol. Food Technol.* **2013**, *5*, 51–58. [CrossRef]

plants

MDPI

Article

Flow Rate and Water Depth Alters Biomass Production and Phytoremediation Capacity of *Lemna minor*

Neil E. Coughlan [1,2,*], Éamonn Walsh [1,2], Roger Ahern [1,2], Gavin Burnell [1,2], Rachel O'Mahoney [1,2], Holger Kuehnhold [3] and Marcel A. K. Jansen [1,2]

[1] School of Biological, Earth and Environmental Sciences, University College Cork, Distillery Fields, North Mall, T23 TK30 Cork, Ireland
[2] Environmental Research Institute, University College Cork, T23 XE10 Cork, Ireland
[3] Department of Ecology, Leibniz Centre for Tropical Marine Research (ZMT), 28359 Bremen, Germany
* Correspondence: neil.coughlan@ucc.ie

Abstract: Given its high biomass production, phytoremediation capacity and suitability as a feedstock for animal and human nutrition, duckweeds are valuable multipurpose plants that can underpin circular economy applications. In recent years, the use of duckweeds to mitigate environmental pollution and valorise wastewaters through the removal of excess nitrogen and phosphate from wastewaters has gained considerable scientific attention. However, quantitative data on optimisation of duckweed performance in phytoremediation systems remain scant. In particular, a mechanistical understanding of how physical flows affect duckweed growth and remediation capacity within vertical indoor multi-tiered bioreactors is unknown. Here, effects of flow rate (0.5, 1.5 or 3.0 L min^{-1}) and medium depth (25 mm or 50 mm) on *Lemna minor* biomass production and phytoremediation capacity were investigated. Results show that flow rates and water depths significantly affect both parameters. *L. minor* grew best at 1.5 L min^{-1} maintained at 50 mm, corresponding to a flow velocity of 0.0012 m s^{-1}. The data are interpreted to mean that flow velocities should be low enough not to physically disturb duckweed but still allow for adequate nutrient mixing. The data presented will considerably advance the optimisation of large-scale indoor (multi-tiered, stacked), as well as outdoor (pond, lagoon, canal), duckweed-based remediation of high nutrient wastewaters.

Keywords: biomass production; duckweed; Lemnaceae; nutrient recovery; phytoremediation

Citation: Coughlan, N.E.; Walsh, É.; Ahern, R.; Burnell, G.; O'Mahoney, R.; Kuehnhold, H.; Jansen, M.A.K. Flow Rate and Water Depth Alters Biomass Production and Phytoremediation Capacity of *Lemna minor*. *Plants* **2022**, *11*, 2170. https://doi.org/10.3390/plants11162170

Academic Editors: Viktor Oláh, Klaus-Jürgen Appenroth and K. Sowjanya Sree

Received: 19 July 2022
Accepted: 19 August 2022
Published: 21 August 2022

1. Introduction

The per capita availability of nutritional food and clean water is expected to substantially decrease in coming decades [1,2]. An anticipated global population growth of up to 9.8 billion people by 2050, coupled with the ongoing progression of climate change and increased levels of environmental pollution, will jeopardise food and water security worldwide [2,3]. Furthermore, given the array of environmental problems caused by fossil fuel consumption, there is an urgent need to identify and develop alternative energy sources such as sustainable biofuels, e.g. [4]. Accordingly, there is an urgent need to improve food, water and fuel security through the development of innovative novel crops and cropping systems that are sustainable and limit the consumption of finite resources, such as nutrients and water, and the generation of waste [3,5,6]. Closed-loop production systems can be used to minimise resource inputs and waste outputs in accordance with the principles of a circular economy approach, whereby the long-term retention and reuse of resources is prioritised over the addition of new raw materials [5,7]. For example, phytoremediation can be used to remove excess nutrients from wastewaters, facilitating re-use of water, and recycling of nutrients in a variety of industrial applications, e.g. [8,9].

As small floating aquatic plants, duckweed species (Lemnaceae) generally display rapid growth on nutrient-rich waters, as well as relatively high tolerances to a range of pollutants [10–12]. Further, the rapid growth of duckweed species is matched by a rapid uptake of plant nutrients from the medium e.g. [13,14]. Hence, duckweed species can be effectively used to remediate a variety of nutrient-rich wastewaters, including diluted livestock farm manure [9,14,15], aquaculture wastewater [16] or dairy processing wastewater [17,18]. Notably, duckweed biomass can be used to produce biofuel (e.g., ethanol and methane) [19]. Additionally, given their nutritionally desirable composition of both amino acids and poly-unsaturated fatty acids [10,20,21], duckweeds are also increasingly studied for their potential as a nutritious biomass for livestock and human consumption [10,22]. As such, duckweed-based water remediation can contribute to a recycling of plant nutrients in accordance with the principles of the circular economy [5]. Accordingly, innovative duckweed cropping systems could be used for environmental remediation and wastewater valorisation [23].

To date, duckweed biomass production, and phytoremediation of wastewaters, has generally occurred in ponds, lagoons, or canal-based systems [23,24]. These systems tend to be either outdoors or encapsulated within structures such as glasshouses or polytunnels [23]. In many regions, these systems can be relatively cheap to build and maintain and are often capable of accommodating 100s–1000s of litres of wastewater input per day. However, the large-scale cultivation of duckweed in pond-based systems can have a large spatial requirement [23]. As an alternative, indoor growth systems composed of multi-tiered, vertically stacked layers can be used to increase duckweed biomass production and phytoremediation capacity per unit area of land [23,25]. This may be especially pertinent in urban and semi-urban areas, where infrastructure is limited by horizontal rather than vertical space availability. Multi-tiered, indoor systems operated in controlled environments have additional potential benefits, such as optimised cultivation, improved year-round growth, and operation under sterile conditions, thus avoiding contamination by bacteria, viruses, fungi, algae, and invertebrates, e.g. [23,24,26].

There are considerable knowledge gaps concerning the basic operating parameters of both outdoor and indoor duckweed-based remediation systems. In particular, the effects of flow rate on duckweed biomass production and phytoremediation are largely unknown. To date, most studies have tended to assess performance of static or low flow systems with only periodic or extremely slow entry and exit of wastewater media (e.g., 1.0–1.5 L d^{-1}) [23] or have focused on wetland communities rather than strictly duckweed cultivation, but see [27]. Flow rates are critical determinants of the performance of duckweed-based systems, as they determine nutrient supply, mixing and the residence time of the medium. Sufficiently high flow rates relative to both the volume of the duckweed-based system and the duckweed surface cover, are required to avoid nutrient depletion. However, an adequate residence/retention time is required to facilitate optimal remediation. Despite this, an approach whereby flow rates and residence times are exclusively based on physico-chemical considerations is incomplete. In the case of duckweed bioreactors, plant biological aspects also need to be considered. Duckweeds are adapted to still or slow-flowing waters [28], however, the maximum flow rates tolerated by duckweeds are unknown. High flow rates may result in the formation of thick layers of piled-up duckweed and/or inhibit growth through physical disturbance. In this scenario, duckweed growth will be impeded, fronds will readily senesce and release nutrients back into the water column. Thus, it can be hypothesised that an optimal flow rate will exist that facilitates nutrient supply without impeding plant growth. Further, the water depth will co-determine the flow rate required to assure nutrient depletion. However, it is not clear whether the water depth will also directly impact on plant growth. Although duckweed has been observed to grow on just a few millimetres of water in the natural environment [29], in warm climates a sufficient depth of the water column may be required to prevent heat-stress in outdoor systems, e.g., >200 mm [30] and 500 mm [27].

In the present study, the effects of different flow rates (0.5, 1.5 or 3.0 L min^{-1}) and medium depths (25 mm or 50 mm) on duckweed biomass production, plant health, and nitrogen and phosphorous uptake were quantified. Plants were cultivated on a standardised nutrient-rich medium within a simple, indoor, vertically stacked recirculating system. We hypothesised that greater flow rates would reduce biomass production, decrease plant health, and curtail nutrient uptake by plants due to increased physical disturbance by medium (i.e., greater velocity). The data will improve operation of duckweed-based remediation, duckweed biomass production, and food-production systems that seek to use duckweed technologies to implement closed-loop production systems.

2. Methods

2.1. Stock Cultivation

The duckweed strain used in this study was *Lemna minor* L. (Blarney strain, number 5500 in the Rutgers Duckweed Stock Cooperative database: www.ruduckweed.org (accessed on 1 May 2022)). A non-axenic stock of *L. minor* was cultivated for the experiments. Stock plants were maintained on a nutrient-rich solution that consisted of tap water and commercially available nutrient additives: pH Perfect Grow (2 mL L^{-1}) and pH Perfect Micro (2 mL L^{-1}; Advanced Nutrients: Table S1). The stock culture was maintained indoors at an average light intensity of 150 μmol m^{-2} s^{-1} PAR (photosynthetically active radiation), at ~21 °C with a 16:8 hours light:dark photoperiod.

2.2. Experimental Design

The study of the effects of flow rate and water depth on *L. minor* was performed using an indoor, recirculating system. The system consisted of five vertically stacked trays (720 mm × 410 mm × 110 mm: length × width × height) and a sump tank (600 mm × 410 mm × 410 mm), operated at a combined capacity of 125 litres (Figure 1). Combined, the five trays had a surface area for duckweed growth of 1.48 m^2. The trays were suspended within a supporting stainless-steel framework. A nutrient-rich solution was continuously pumped from the sump tank at floor level to the top tray (Tray 1: Rio Pump 1700, TAAM Inc.), all other trays were gravity fed. Medium exited the final tray (i.e., tray 5) to be deposited back into the sump tank. All piping had an internal diameter of 20 mm. The nutrient-rich solution consisted of commercially available nutrient additives and was prepared using distilled water (FloraGrow, 0.25 mL L^{-1}; FloraMicro, 0.25 mL L^{-1}: General Hydroponics: Table S1). Plants were maintained within the stacked system at an average light intensity of 150 μmol m^{-2} s^{-1} PAR using LED strip lighting (Neonica Growy, Neonica Polska Sp. z o.o.) at ~ 21 °C with a 16:8 h light:dark photoperiod. The 10 mm-wide LED strips used were uniformly installed every 20 mm to ensure an even distribution of light. Upon completion of a duckweed growing cycle the system was drained and all surfaces were thoroughly cleaned with hot water (≥60 °C).

Growth of duckweed within the vertically stacked system was assessed in relation to three flow rates (0.5, 1.5 or 3.0 L min^{-1}) and two medium depths (25 mm or 50 mm). The highest flow rate (3.0 L min^{-1}) was selected based on a preliminary assessment of the cultivation system. Flow rates greater than 3.0 L min^{-1} tended to cause the formation of overlapping duckweed layers that inhibit growth. Lower flow rates were based on the partial incremental reduction of the maximum flow rate. Flow rates were modulated by diverting excess medium back to the sump prior to entering tray 1, this was achieved with the use of a control-valve that could be manually adjusted to alter medium flow rates. The lowest medium depth (25 mm) was selected based on a preliminary assessment of the cultivation system, and simply doubled for assessment of the greater depth (50 mm). Depths of less than 25 mm tended to cause the formation of overlapping duckweed layers that inhibit growth, as the plants did not tend to spread evenly across the medium surface, even for the lowest flow rate. Medium depths were modulated by raising or lowering tray exit pipes, which were inserted perpendicular to the base of the trays. Calculated surface flow velocities are shown in Table 1. All possible treatment combinations were assessed in a factorial design (treatment replication: $n = 3$; see Table 1). To commence the experiment, a 60% surface cover of *L. minor* was established for each tray. To achieve this, 75 g (fresh weight) of *L. minor* was added to each tray. Fresh weight was achieved by draining the duckweed of excess water through a fine mesh sieve, and then blotting the duckweed dry using highly absorbent paper towels. Preliminary work using the imaging software Easy Leaf Area established that 75 g consistently provided for 60% surface cover of the trays. Easy Leaf Area software distinguishes duckweed frond surface cover from non-duckweed covered surface area. Each treatment lasted seven days. However, to avoid overcrowding, an intermediate harvest was employed on the third day to return the surface cover to 60% [as per 8]. At the intermediate harvest, all duckweed was gently removed using a flat sieve and weighed, with 75 g being returned to each tray.

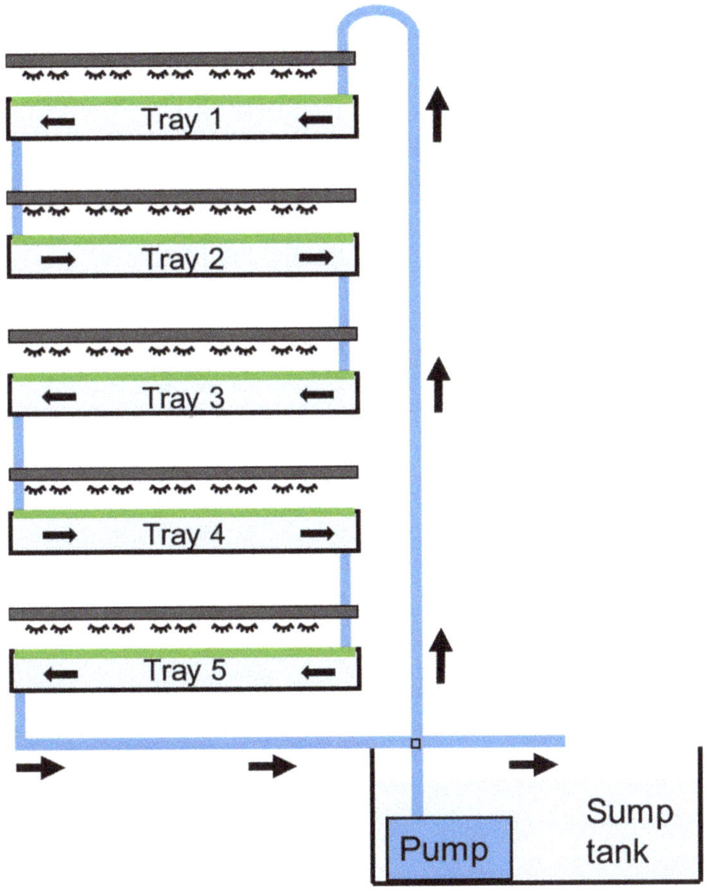

Figure 1. Schematic of the indoor, vertically stacked, 125-litre duckweed cultivation system.

Table 1. Treatment combinations of flow rates (0.5, 1.5 or 3.0 L min^{-1}) and medium depths (25 mm or 50 mm). Calculated system velocities are shown.

Treatment	Flow Rates (L min^{-1})	Medium Depth (mm)	Velocity (m s^{-1})
1	0.5	25	0.0008
2	1.5	25	0.0024
3	3.0	25	0.0048
4	0.5	50	0.0004
5	1.5	50	0.0012
6	3.0	50	0.0024

2.3. Data Collection

Total biomass yield (fresh weight) was determined through summation of biomass detected on the third and seventh days relative to the initial biomass (75 g) for each tray. Further, the relative growth rate of biomass yield was determined using the formula below [31]:

$$RGR = \frac{ln\frac{W_2}{W_1}}{\Delta T} \quad (1)$$

where ln is the natural log, W_1 is the initial biomass, W_2 is final biomass and ΔT is the length of the experiment. RGR was calculated from inoculation to intermediate harvest (day 0–3) and from the intermediate harvest to the final harvest (day 3–7). In addition, chlorophyll *a* fluorescence was measured using pulse amplitude modulated chlorophyll *a* fluorometry (WALZ Imaging fluorometer, Effeltrich, Germany). Chlorophyll *a* fluorescence was measured on days 0 and 7. Plants were adapted to dark conditions for 15 min immediately in advance of chlorophyll *a* fluorescence analysis. For assessment on day 0, three colonies (i.e., three plantlets of 3–4 fronds) were randomly selected from the stock of plants used to inoculate the stacked system. On day 7, three colonies were randomly taken from each tray. The measured values of these three colonies were averaged and treated as one replicate. The chlorophyll fluorescence analysis procedure used to measure fluorescence (F_0) and maximum fluorescence (Fm) of the dark-adapted plant, as well as the maximum fluorescence (Fm′) and steady-state fluorescence (Ft) under light-adapted conditions, was in accordance with Walsh et al. [8,18]. The actinic light intensity (i.e., photosynthetically active light) used was of 186 µmol m^{-2} s^{-1}. Subsequently, the maximum quantum efficiency of photosystem II (i.e., Fv/Fm), and the quantum efficiency of photosystem II under steady-state light conditions (i.e., Y(II)) were calculated according to the equations detailed by Maxwell and Johnson [32].

Plant biomass samples extracted from tray 1, 3 and 5 on day 7 were analysed for total nitrogen (TN) and total phosphate (TP) content (mg g^{-1}), as was a representative sample of the plants used to inoculate all trays on day 0. Plants were dried and milled, and then digested at 420 °C for one hour with concentrated sulphuric acid and Kjeltab Cu/3.5 in a TECATOR 2040 Digestor. Digested samples were diluted to a volume of 250 mL using deionised water. The total nitrogen was determined using QuickChem IC + FIA flow injection analyzer (8000 series: Lachat Instruments). For plant TP content, the samples were acid digested and analysed using the ammonium molybdate method [33]. Absorbance was measured using a UV-Visible Recording Spectrophotometer (UV-160A: SHIMADZU Corporation).

Samples of the liquid medium were taken for TN and TP analysis on days 0 and 7. Day 0 samples were taken from the sump tank, while day 7 samples were separately taken from trays 1, 3, and 5. Unfiltered samples were used for both TN and TP analysis. TN samples were digested with potassium persulfate and boric acid in alkaline conditions, while TP samples were digested with ammonium persulfate in acidic conditions. Both sets of samples were digested in an autoclave at 120 °C for 30 min. The resulting total oxidised nitrogen was analysed by automated cadmium reduction method using Lachat Quikchem

8000 by Zellweger Analytics, Inc. Milwaukee, WI, USA [34], while the resulting phosphate was analysed manually using the Murphy and Riley Method [33].

The measurements of the concentrations of TN and TP were used to assess the nett TN and TP depletion, i.e., the decrease in concentration, over the 7-day period, for the entire 125-litre system. Nett nutrient removal was calculated based on nutrient depletion and the volume of the system. Removal was also normalised for the amount of duckweed present, yielding estimations for milligrams of N and p removed per m^2 of *L. minor* per day (mg TN m^{-2} day^{-1}; mg TP m^{-2} day^{-1}) and milligrams of N and p removed per g of *L. minor* per day (mg TN g^{-1} day^{-1}; mg TP g^{-1} day^{-1}). For normalisation purposes, TN and TP removal was expressed relative to the starting amount of duckweed (i.e., 60% surface cover or 75 g per tray). The uptake of TN and TP by *L. minor* plants was also determined. To achieve this, the content of TN and TP per gram of dry duckweed biomass was converted to fresh weight using a conversion factor of 3.64% (% dry versus fresh weight). The conversion factor was determined on duckweed used to inoculate the 125-litre system by drying 50 g samples to a constant weight (60 °C; $n = 5$). The total amount of TN and TP taken up by plants was calculated by multiplying the plant TN and TP content by the total biomass present at the end of the 7-day cultivation period, and by subtracting the TN or TP content of the duckweed used to inoculate the system. An uptake rate was calculated by dividing by the duration of the experiment.

2.4. Data Analysis

Statistical analyses were conducted using R software (version R 4.1.2). All data were assessed for normality of residual distributions (Shapiro–Wilk test: library *psych*) and homoscedasticity of variances (Levene's test: library *car*). Where data were found to be normally distributed ($p > 0.05$) with homoscedastic residuals ($p > 0.05$), general linear models (ANOVA) were used to analyse differences in biomass yield, RGR and nutrient depletion of the media. Logistic regression in the form of generalised linear models (GLM: *car*) were employed for non-normal data and/or heteroscedastic residuals ($p < 0.05$). The effects of flow rate, medium depth, and tray position (1–5) were considered in all models. A stepwise depletion approach was used to remove non-significant terms, while overall model significance was determined using likelihood ratio tests in all cases (*lmtest*). Where p-values were significant ($\alpha < 0.05$), Tukey's LSD adjustments for multiple pairwise comparisons were used for post-hoc analysis (*emmeans*) [35].

3. Results

3.1. Duckweed Growth

The biomass yield of *L. minor* cultivated within the vertically stacked system was found to significantly differ across treatments of flow rates and medium depths (ANOVA: $\chi^2 = 70.881$, df = 3; $p < 0.001$: Figure 2A,B). As tray position did not have a significant effect on biomass yield ($p > 0.05$), data from different trays were pooled for a simplified model depiction (Figure 3). While both flow rate and medium depth had a significant effect on biomass yield (both $p < 0.001$), no interaction terms were detected amongst the model variables. Biomass yield was greatest for duckweed grown in a flow of 1.5 L min^{-1} with a medium depth of 50 mm compared to all other flow rate and depth combinations (all $p < 0.05$: see Figure 3).

Biomass yields were used to calculate RGR values to facilitate comparison with published literature. The RGR significantly differed for *L. minor* subjected to different treatments within the stacked system (ANOVA: $\chi^2 = 108.98$, df = 6; $p < 0.0001$: Figure 4A,B). Flow rate, medium depth, and assessment period (i.e., day 0–3 and day 3–7) had a significant effect on biomass yield (all $p < 0.0001$). RGR was generally greater for plants grown on a medium depth of 50 mm rather than 25 mm ($p < 0.0001$; Figure 4A,B). While RGR tended to be slightly lower for plants grown on a flow rate of 3.0 L min^{-1}, this was not always statistically apparent (Figure 4A,B). A significant interaction effect on RGR was detected for flow and assessment period ($p < 0.001$). Furthermore, although not always statistically apparent, the RGR tended to be greater for plants grown in the period of days 0–3 compared to day 3–7 (Figure 4A,B).

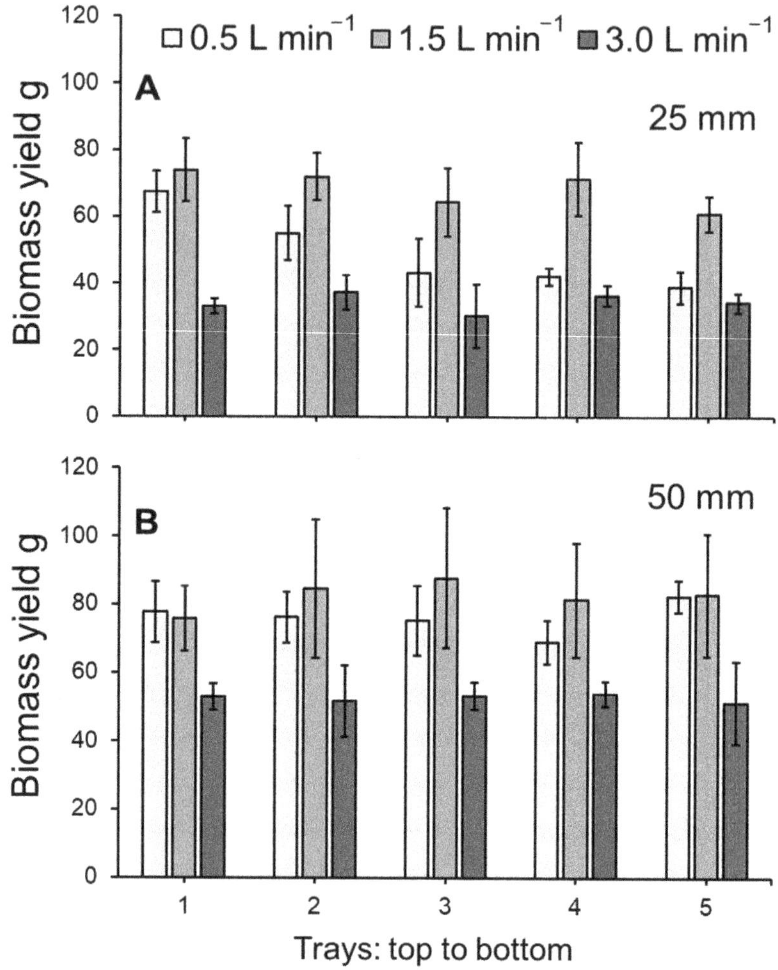

Figure 2. Mean (±SE) biomass yield (g) at medium depths of (**A**) 25 mm and (**B**) 50 mm for *Lemna minor* grown within an indoor, vertically stacked system. Plants were grown on hydroponics medium at three flow rates (0.5, 1.5 or 3.0 L min^{-1}) and two medium depths (25 or 50 mm) over 7 days. See Figure 3 for a simplified model depiction with post-hoc analysis.

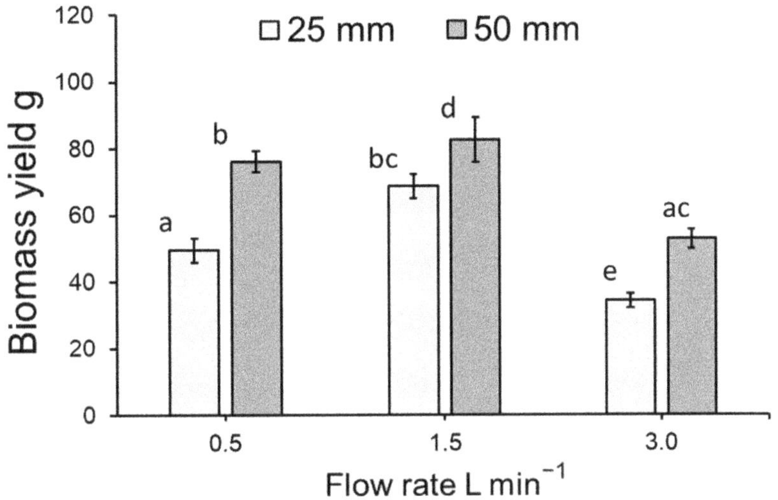

Figure 3. Mean (±SE) biomass yield (g) for *Lemna minor* grown within an indoor, vertically stacked system. As no effect of tray position was detected, trays have been pooled in this simplified model depiction. Letters show statistical similarity ($p > 0.05$). See Figure 2 for full model depiction.

3.2. *Chlorophyll a Fluorometry*

To determine the photosynthetic efficiencies of plants kept at different combinations of flow rate and water depth, chlorophyll *a* fluorescence was measured. In all cases, the maximum quantum efficiency (i.e., Fv/Fm) of photosystem II (PSII) for *L. minor* did not differ significantly for plants subjected to different treatments (GLM: $\chi^2 = 3.4032$, df = 17; $p > 0.05$). Mean Fv/Fm ranged from 0.63–0.77 across all treatments. Similarly, the quantum efficiency of PSII under steady-state light conditions (i.e., YII) for plants did not significantly differ (GLM: $\chi^2 = 20.003$, df = 17; $p > 0.05$). Mean YII ranged from 0.23–0.41 across all treatments.

Y(NPQ), the yield of regulated energy dissipation, was found to significantly vary (GLM: $\chi^2 = 8.6834$, df = 2; $p < 0.05$: Figure 5A). When pooled across depths and trays, a significantly lower Y(NPQ) was detected for *L. minor* cultivated at 3.0 L min^{-1} than for plants grown in a flow of 1.5 L min^{-1} ($p < 0.05$: Figure 5A). Finally, Y(NO), the yield of nonregulated energy dissipation, was also found to significantly vary (GLM: $\chi^2 = 8.4029$, df = 2; $p < 0.05$: Figure 5B). The combined Y(NO) of *L. minor* inoculated and cultivated at 3.0 L min^{-1} (pooled across depths and trays) was significantly greater than that of plants grown in a flow of 1.5 L min^{-1} ($p < 0.01$: Figure 5B).

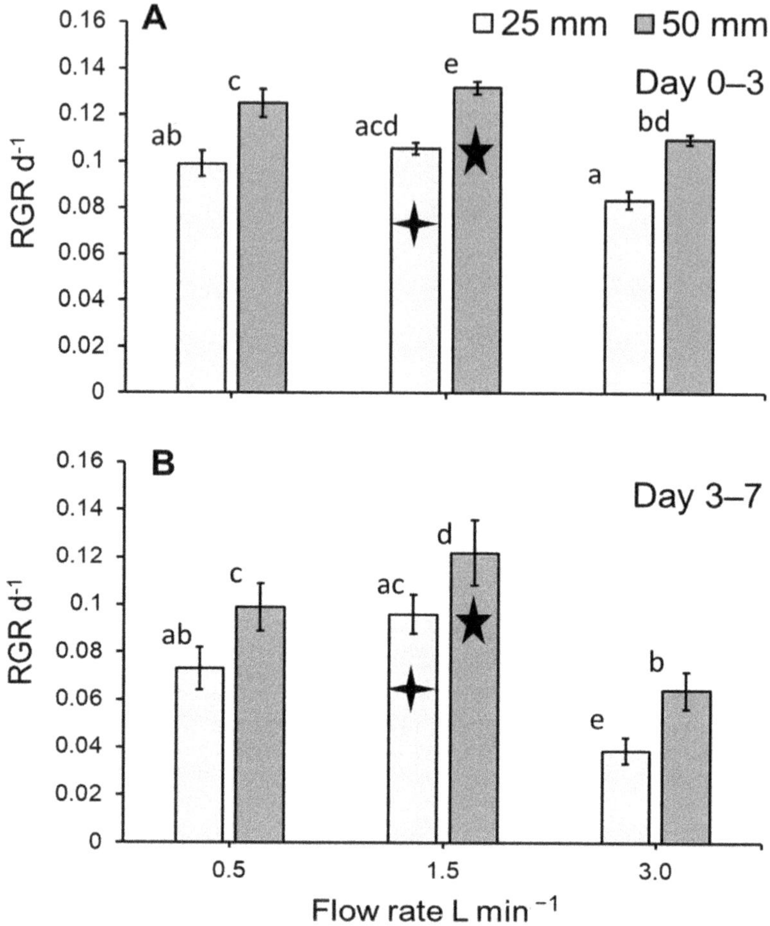

Figure 4. Mean (±SE) RGR (d^{-1}) for *Lemna minor* grown within an indoor, vertically stacked system. RGR between inoculation and the intermediate harvest (day 0–3; (**A**) and the final harvest (day 3–7; (**B**) are shown. As no effect of tray position was detected, trays have been pooled in this simplified model depiction. Letters show statistical similarity within each harvest period separately (i.e., day 0–3 or 3–7), while symbols indicate similarity amongst the harvest periods within treatments ($p > 0.05$). Similarity amongst harvest periods is considered for within each treatment only (see Table 1).

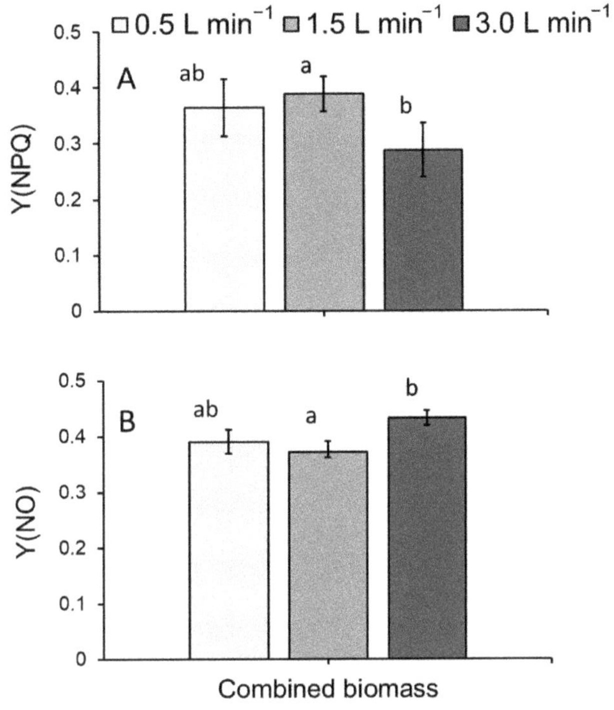

Figure 5. Mean (±SE) (**A**) Y(NPQ) and (**B**) Y(NO), for *Lemna minor* grown within an indoor, vertically stacked system. As no effect of depth or tray was detected, medium depths and tray positions have been pooled in this simplified model depiction. Letters show statistical similarity ($p > 0.05$).

3.3. Total Nitrogen Concentration of the Medium

The concentration of nitrogen (TN mg L^{-1}) remaining in the medium within the vertically stacked system significantly differed amongst treatments (GLM: $\chi^2 = 202.91$, df = 7; $p < 0.0001$: Figure 6A). Flow rate ($p < 0.05$), depth and time point (i.e., initial concentration day 0 vs. final concentration day 7) were found to have a significant effect on TN concentration in the medium (both $p < 0.0001$). Compared to the initial concentration of TN on day 0 (i.e., 22.42 ± 0.35 mg L^{-1}; mean ± SE), the concentration of nitrogen present in the medium was significantly reduced on day 7 for flow rates of 0.5 and 1.5 L min^{-1} (but not 3 L min^{-1}) maintained at a depth of 25 mm (both $p < 0.0001$; Figure 6A). Using a water depth of 50 mm, the concentration of TN was significantly less on day 7 than on day 0 for all flow rates (all $p < 0.0001$; Figure 6A). At individual flow rates, a greater reduction in the total nitrogen concentration occurred at the 50 mm depth than at the 25 mm depth (all $p < 0.001$). At the 50 mm depth, the flow rates that showed the lowest concentration of nitrogen in the medium on day 7 (i.e., the greatest removal) were 0.5 and 1.5 L min^{-1} (i.e., $p > 0.05$: Figure 6A). A significant interaction effect was detected for flow and day ($p < 0.001$), as well as depth and day ($p < 0.0001$).

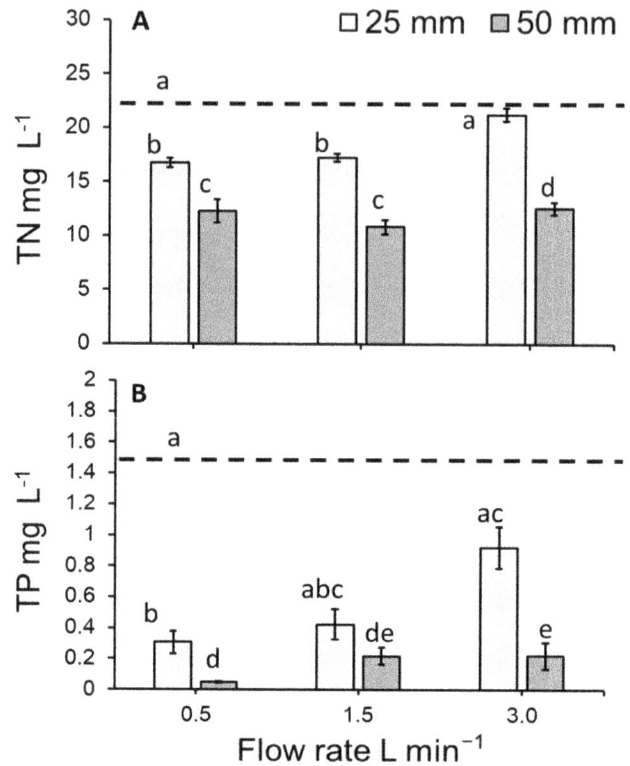

Figure 6. Mean (±SE) (**A**) total nitrogen concentration (TN mg L^{-1}) and (**B**) total phosphorous concentration (TP mg L^{-1}) of the hydroponics medium following the cultivate *Lemna minor* within an indoor, vertically stacked system. As no effect of tray was detected, trays have been pooled in this simplified model depiction. Letters show statistical similarity within each panel ($p > 0.05$). Dashed lines represent the initial concentration of TN (**A**) and TP (**B**).

3.4. Total Phosphorous Concentration of the Medium

The concentration of phosphorous (TP mg L^{-1}) in the medium within the vertically stacked system significantly differed amongst treatments (GLM: $\chi^2 = 135.88$, df = 7; $p < 0.0001$: Figure 6B). Time point was found to have a significant effect ($p < 0.0001$). Compared to the initial concentration of TP mg L^{-1} on day 0 (i.e., 1.46 ± 0.01 mg L^{-1}), the concentration of phosphorous present in the medium was significantly reduced on day 7 for the flow and depth combination of 0.5 L min^{-1} and 25 mm ($p < 0.0001$; Figure 6B). Similarly, the TP concentration on day 7 was significantly lower than the initial concentration of TP for all flow rates maintained at a depth of 50 mm (all $p < 0.0001$). Within individual flow rates, a greater reduction of phosphorous occurred at the 50 mm depth than at the 25 mm depth (all $p < 0.001$). At the 50 mm depth, flow rates of 0.5 and 1.5 L min^{-1} showed the lowest concentration of phosphorous on day 7 ($p > 0.05$: Figure 6B). Although an effect of flow or depth was not statistically apparent (i.e., $p > 0.05$), a significant interaction effect was detected for flow and day ($p < 0.001$), as well as depth and day ($p < 0.0001$).

3.5. Total Nitrogen and Total Phosphorous Removal Per m² of Lemna Minor

The total TN and TP removal from the system was calculated based on the depletion, i.e., decrease in the amount of TN and TP in the 125 litres of medium. Nutrient removal was normalised against the initial 60% surface cover or the initial inoculation mass of *L. minor* and recalculated as a rate by considering the duration of the treatment. The removal of mg TN m^{-2} day^{-1} from the medium significantly differed amongst treatments (GLM: $\chi^2 = 26.211$, df = 5; $p < 0.0001$: Figure 7A). Flow rate and depth (both $p < 0.01$) altered the rate of TN removal. A significant interaction effect was detected for flow and depth ($p < 0.001$). TN removal tended to not statistically differ among flow rates maintained at the same depth, nor between depths at a given flow rate. However, in comparison to all other treatment combinations, the removal of TN was significantly lower at a flow of 3 L min^{-1} when maintained at a depth of 25 mm (Figure 7A).

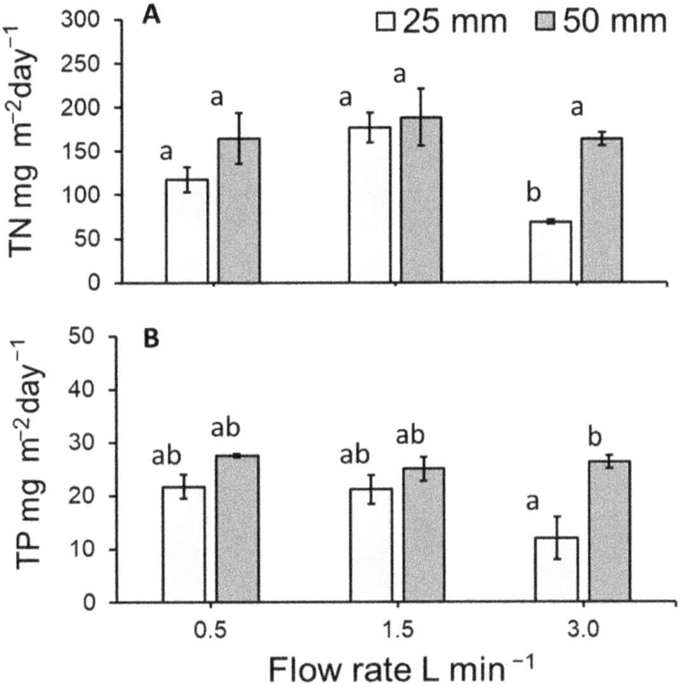

Figure 7. Mean (±SE) (**A**) TN removal per m² of *Lemna minor* per day (mg m^{-2} day^{-1}), (**B**) TP removal per m² of *L. minor* per day (mg m^{-2} day^{-1}) for plants cultivated on hydroponics medium at three flow rates and two medium depths over 7 days. Letters show statistical similarity within each panel ($p > 0.05$).

The removal of *p* (expressed as mg TP m^{-2} day^{-1}) from the medium significantly differed amongst treatments (GLM: $\chi^2 = 13.859$, df = 5; $p < 0.05$: Figure 7B). Whilst depth altered the rate of TP removal ($p < 0.05$), no other significant effects were detected. TP depletion was similar among flow rates maintained at the same depth. Further, TP depletion did not statistically differ between depths kept at the same flow rate with the exception of 3 L min^{-1}. The depletion of TP was significantly lower under the combined treatment of 3 L min^{-1} and 25 mm (Figure 7B).

3.6. Total Nitrogen and Total Phosphorous Uptake by Lemna Minor

As a function of dry-weight biomass, moderate TN values of 2.5–3.9% and low TP values of 0.2–0.6% were detected in the harvested duckweed biomass. The nitrogen and phosphorous content (mg g^{-1}) of dried *L. minor* biomass tended to be stable within treatments (Table S2). Calculated TN uptake rates (mg TN g^{-1} day^{-1}) for the dried *L. minor* plants cultivated in the stacked system significantly differed amongst treatments (ANOVA: χ^2 = 48.165, df = 3; $p < 0.0001$: Figure 8A). Flow rate and depth significantly altered the amount of TN taken up by plants (both $p < 0.0001$). No other significant effects were detected. The greatest TN uptake rate per gram of *L. minor* occurred at a depth of 50 mm for flows of 0.5 and 1.5 L min^{-1} (Figure 8A).

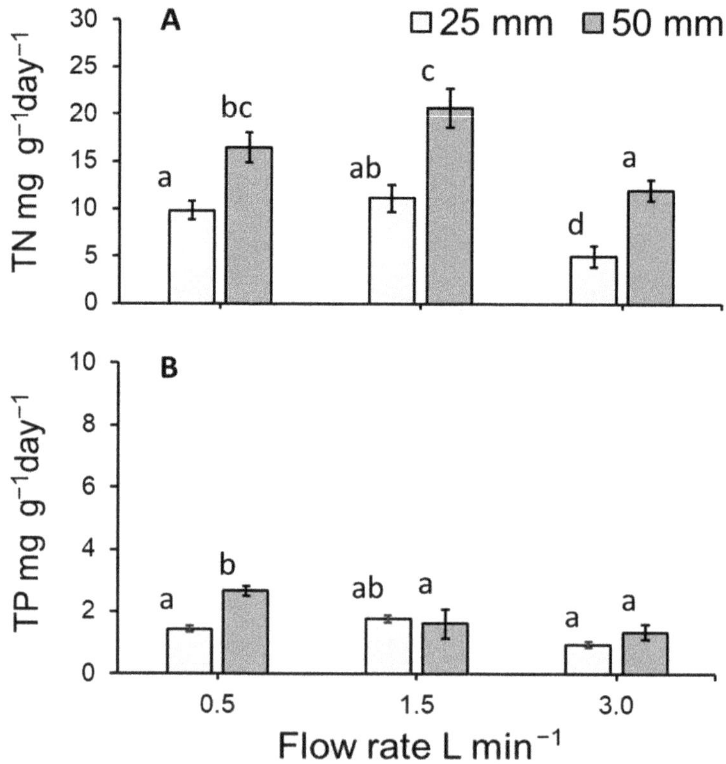

Figure 8. Mean (±SE) (**A**) TN and (**B**) TP uptake per gram of *Lemna minor* per day (mg g^{-1} day^{-1}) for the biomass of plants cultivated on hydroponics medium at three flow rates and two medium depths over 7 days. As no effect of tray was detected, trays have been pooled in this simplified model depiction. Letters show statistical similarity within each panel ($p > 0.05$).

The uptake of TP (mg TP g^{-1} day^{-1}) by plants cultivated within the stacked system significantly differed amongst treatments (ANOVA: χ^2 = 27.023, df = 5; $p < 0.0001$: Figure 8B). Flow rate ($p < 0.01$) and depth ($p < 0.05$) significantly affected the rate of TP uptake. An interaction effect between flow rate and depth was also detected ($p < 0.05$). Flows 0.5 L min^{-1} and 1.5 L min^{-1} maintained at depths of 25 mm and 50 mm respectively, both displayed the greatest uptake of TP per gram of *L. minor* cultivated within the system (Figure 8B).

3.7. Fate of Total Nitrogen and Total Phosphorous within the System

As the system was non-axenic, data on removal of nitrogen and phosphorous from the medium were compared with those concerning nutrient uptake by plants. The fate of TN differed amongst treatments (GLM: $\chi^2 = 49.634$, df = 7; $p < 0.0001$: Figure 9A,B). Flow rate and depth significantly altered the amounts of TN removed and taken up ($p < 0.01$ and $p < 0.001$, respectively). Interaction effects between TN fate (i.e., amount removed or taken up) and depth, as well as flow rate and depth were detected (both $p < 0.05$). In all cases, removal of TN from the medium was significantly greater than the amount taken up by plants cultivated at a depth of 25 mm (all $p < 0.0001$: Figure 9A), as well as 50 mm (all $p < 0.01$: Figure 9B). The amount of TN removed from the entire 125-litre system was the same amongst the flow rates within each depth category as was the amount of TN taken up by *L. minor*, with the exception of removal and depletion at 1.5 L min^{-1} and 3 L min^{-1} maintained at 25 mm ($p > 0.05$: Figure 9A,B). However, the amount of TN removed from the medium but not taken up by *L. minor* (i.e., the amount removed less the amount taken up) did not differ amongst treatments (GLM: $\chi^2 = 10.554$, df = 5; $p = 0.06$).

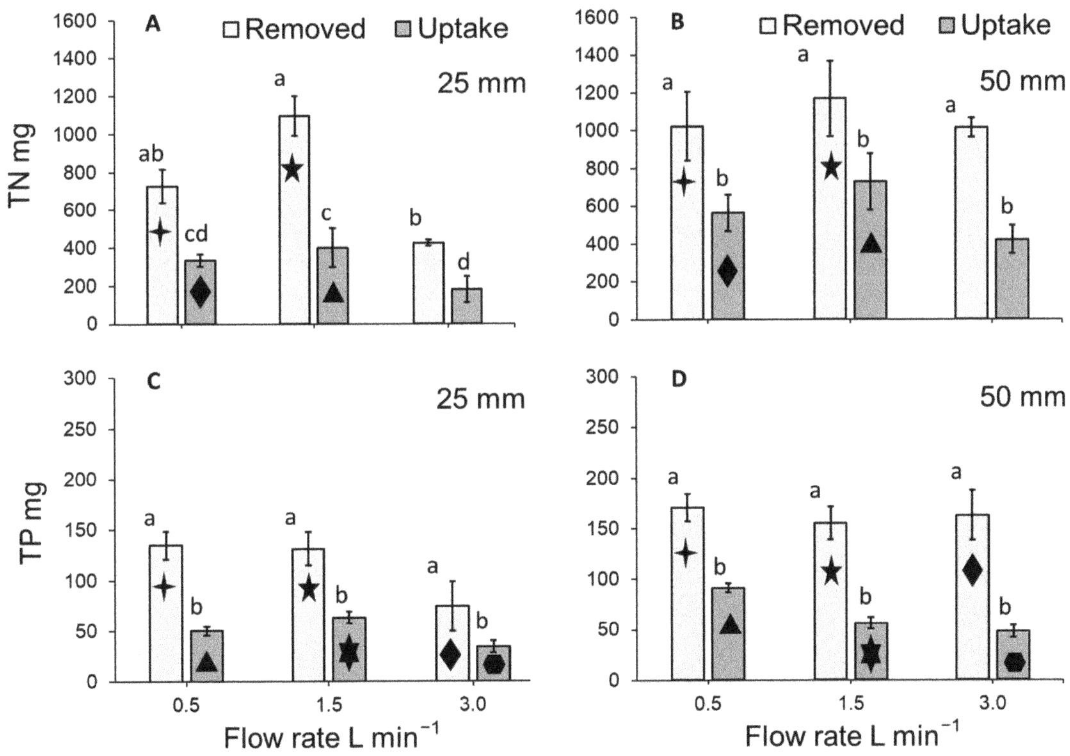

Figure 9. Mean (±SE) (**A**,**B**) total removal and total plant uptake of TN and (**C**,**D**) TP within the 125-litre stacked system. As no effect of tray was detected, trays have been pooled in this simplified model depiction. Letters show statistical similarity within each medium depth (i.e., within a panel), while symbols indicate similarity in relation to each flow rate across both medium depths ($p > 0.05$).

The fate of TP within the entire 125-litre system differed amongst treatments (GLM: $\chi^2 = 29.18$, df = 2; $p < 0.0001$: Figure 9C,D). Depth significantly altered the amount of TP removed and taken up ($p < 0.01$). No interactive effects were detected ($p > 0.05$). In all cases, removal of TP was significantly greater than the amount taken up by plants cultivated at either of the assessed depths (all $p < 0.0001$: Figure 9C,D). The amount of TP removed from the entire 125-litre system was the same for all flow rates maintained at either depth, as well as at a given flow rate between the assessed depths (Figure 9C,D). Similarly, TP uptake by *L. minor* cultivated within the stacked system was the same across all flow rates maintained at either depth, as well as between depths at a given flow rate (Figure 9C,D). The amount of TP removed from the medium but not taken up by *L. minor* (i.e., the amount removed less the amount taken up) did not differ amongst treatments (ANOVA: $\chi^2 = 6.9684$, df = 5; $p > 0.05$).

4. Discussion

The selection of flow rates and medium depths for duckweed biomass production and phytoremediation is a key challenge for the design and operation of both indoor and outdoor cultivation systems. Here, the growth, photosynthetic health, and nitrogen and phosphorous removal from the medium, and uptake by *L. minor* were quantified in relation to different combinations of flow rates and medium depths.

4.1. Duckweed Growth and Photosynthetic Health

The biomass yield and RGR of *L. minor* was reduced for plants maintained at a flow of 3 L min^{-1}, especially at the shallower depth of 25 mm. Overall, mean RGR values (0.04–0.13 d^{-1}) are lower than those reported in the literature for duckweed grown on an optimised medium (0.15–0.52 d^{-1}) [36]. Nonetheless, RGR values recorded by the present study are in the range reported for duckweed grown on wastewater (0.03–0.25 d^{-1}) [8,9,17]. Growth was fastest at the intermediate flow rate. The reduction in growth at the higher flow rate is conceivably caused by physical disturbance of the duckweed, especially in shallow water, with duckweed being forced away from the medium inlet and artificially overcrowded by the flow. Such physical disturbance is likely to be associated with water velocity, rather than flow rate. In this study, water velocity varied between 0.0004 and 0.0048 m s^{-1}, with the reduction in depth from 50 mm to 25 mm resulting in a doubling of water velocity for any flow rate (see Table 1). Thus, negative effects of flow rate in duckweed bioreactors can be avoided by increasing water depth. Overall, the data on slower growth at a higher flow rate are in agreement with the common perception that duckweeds prefer still and slow-flowing waters. Yet, few studies present actual quantified data on water velocities that facilitate duckweed growth. A report by Derksen and Zwart [37] refers to velocities less than 0.1 m s^{-1}, while a field study by Giblin et al. [33] indicated that a velocity of 0.095 m s^{-1} cannot be exceeded for a mixture of free-floating aquatic macrophytes, including various duckweed species: *L. minor*, *Spirodela polyrhiza*, and *Wolffia columbiana*. The high velocity observed by Giblin et al. [38], compared to the numbers reported in this study, may be explained by the focus on a maximal threshold for growth, as opposed to an optimal growth rate as aimed for in the current study. Indeed, in a field study by Janauer et al. [39], *L. minor* presence was associated with velocities of less than 0.05 m s^{-1}. It remains to be seen through what mechanism higher flow rates impede duckweed growth. One possible scenario is the formation of overlapping layers of duckweed. Overcrowding has been associated with decreases in growth [40]. However, in the current study no overlapping layers of duckweed were observed.

At a lower flow rate (0.5 L min^{-1}) we also observed small decreases in biomass accumulation and RGR, compared to growth at 1.5 L min^{-1}. This reduced yield could be indicative of poor nutrient mixing within the system, especially when operated at a relatively shallow depth. The duckweed literature contains extensive data on maximum residence times required to either achieve (or avoid) nutrient depletion. Hydraulic residence times depend on nutrient concentrations, removal and uptake kinetics, and duckweed surface-to-tank volume ratios and therefore will vary across applications. However, commonly used residence times range from just a few days to more than a week [41–43]. As a result, it can be surmised that in the case of nutrient depletion, the last tray in the multi-tiered system would show very slow growth. However, this was not the case, as no significant tray effect on growth was noted across the flow rate and water depth combinations. Accordingly, nutrient depletion is considered unlikely as a cause of growth inhibition at the lowest flow rate. Accordingly, future work should consider fluid dynamics in greater detail, to ensure that the diffusion of nutrients within the geometric design of the system is adequate. Furthermore, for improved accuracy and precision of duckweed growth assessments, future work should consider the use of dry-weight biomass to minimize the variability associated with wet-weight measurements.

Photosynthetic activity measured as Fv/Fm, the maximum quantum yield of photosystem II, and Y(II), the effective quantum yield of photosystem II, were not affected by flow rate or water depth. However, an intriguingly small decrease in Y(NPQ) of plants grown at 3.0 L min^{-1}, compared to 1.5 L min^{-1} was noted. This indicates an impairment in regulatory, non-photochemical quenching. In parallel, a small increase in Y(NO) is also noted. In combination with the decrease in Y(NPQ), this indicates a degree of plants stress [44]. One possible explanation for these data is that duckweed growth responded to the thigmotropic stimulus of water flow, which has been recorded for a range of aquatic plant species. Such a reorganisation of plant development may cause a decrease in growth [45].

4.2. Medium Concentration and Duckweed Removal (mg m^{-2} day^{-1}) of Total Nitrogen and Total Phosphorous

Compared to the initial concentration, nitrogen (TN mg L^{-1}) in the medium was similarly reduced at flow rates of 0.5 and 1.5 L min^{-1}. However, the decrease in TN concentration was substantially less when a flow rate of 3.0 L min^{-1} was used. This effect was particularly strong in case of the shallow water depth of 25 mm. When nutrient depletion was normalised versus duckweed surface cover, TN removal per m^2 of *L. minor* was broadly similar across all treatment combinations (mean: 117.22–188.39 mg TN m^{-2} day^{-1}), except for the combination of the highest flow rate (3.0 L min^{-1}) and shallow depth (25 mm) where nitrogen removal was substantially decreased. The observed TN removal rates were on the lower end of the wide range of values that can be found in the literature: 124–4400 mg TN m^{-2} d^{-1} [14,46–49]. Low TN removal rates may be due to the low TP concentration of the medium, which appears to slow both growth and TN removal. The lower TN removal rates at higher flow rates can be explained by relatively low duckweed growth rates observed under these conditions. Several authors have previously established links between growth and nutrient uptake, e.g. [8,13,18]. Thus, it appears that the negative effects of water velocity on plant growth are influencing plant phytoremediation potential.

The concentration of phosphorous (TP mg L^{-1}) in the medium was also substantially reduced at most flow and depth combinations (up to 96.7%). The reduction in TP concentration was lowest at a flow rate of 3.0 L min^{-1} and a water depth of 25 mm. Normalised against plant surface area, TP removal rates varied between 11.99 and 27.51 mg TP m^{-2} day^{-1}, with lowest removal at a flow of 3.0 L min^{-1} and a water depth of 25 mm. The observed TP removal rates were on the lower end of the wide range of values that can be found in the literature: 14–590 mg TP m^{-2} d^{-1} [14,46–49]. Low TP removal rates may be due to the low TP content of the medium. Furthermore, as noted for TN, low TP removal rates at higher flow rates can be explained by relatively low duckweed growth rates observed under these conditions. Although luxury uptake of phosphorous has been reported [50], in this study we noted a broad agreement between growth and phosphorous removal.

4.3. Duckweed Uptake (mg g^{-1} day^{-1}) of Total Nitrogen and Total Phosphorous

Duckweed grown on the 50 mm water depth tended to uptake more TN than plants grown under shallower conditions, and these data generally match data on plant yield and RGR. However, the slight increase in growth at 1.5 L min^{-1} at 25 mm is, apparently, not matched by a similar increase of plant TN uptake. In general, TN content of duckweed can range from 0.8–7.8% when expressed as a function of dry-weight biomass [51]. In the present study, moderate TN values of 2.5–3.9% were detected. For TP, literature indicates that content generally ranges from 0.03–2.8% [51]. In the present study, TP values tended to be low (0.2–0.6%) and this may in part be a consequence of the employed medium being low in TP. Although *L. minor* can absorb large amounts of excess phosphorous into its tissues [52], almost no variations in phosphorous uptake were observed across flow rates.

4.4. Fate of Total Nitrogen and Total Phosphorous within the System

The quantities of TN and TP removed from the medium were substantially greater than the amounts taken up by *L. minor*. Between 36.5–62.3% and 29.5–53.5% of TN and TP removed, respectively, was taken up by duckweed biomass. The discrepancy between nutrient removal from medium and uptake in the duckweed plants has previously been detailed, e.g. [15]. The removal of TN and TP that is not directly accounted for by the nutrient content of duckweed biomass likely reflects the establishment of a biofilm consisting of microorganisms on the submerged surfaces of the system over the duration of the seven-day growing cycle [15] (*per. obs.* NEC and RA). In the present study, the amount of TN and TP removed from the medium but not taken up by *L. minor* (i.e., the amount removed less the amount taken up) did not vary amongst treatments, which indicates that the formation of a biofilm was independent of flow rate and water depth. The creation of this biofilm was likely due to the non-axenic experimental conditions and is a typical component part of most duckweed cultivation and/or phytoremediation systems.

5. Conclusions

In general, little information concerning the combined effects of flow rates and medium depths on duckweed biomass production and phytoremediation is available within the literature, and this paucity of quantitative data has impeded the optimisation of duckweed cultivation systems. The present study shows that flow rates and water depths can alter biomass production and phytoremediation capacity of *L. minor*. Plants grew best at an intermediate flow rate, which is congruent with the commonly accepted view that duckweeds prefer still and slow-flowing waters. It appears that optimal growth will need to be supported by a sufficient flow rate and medium depth to enable adequate nutrient mixing, but without causing physical disruption of Lemnaceae cultures. Accordingly, future work should consider the growth and phytoremediation of duckweed species by integrating a detailed understanding of fluid dynamics, uptake kinetics, plant disturbance tolerance and biofilm formation.

Supplementary Materials: The following supporting information can be downloaded at: https://www.mdpi.com/article/10.3390/plants11162170/s1, Table S1: Water chemistry analysis for a single sample of the commercially available nutrient additives prepared using distilled water; Table S2: Mean proportional change of nitrogen (TN mg g^{-1}) and phosphorous (TP mg g^{-1}) within the dry-weight biomass of *Lemna minor* cultivated within an indoor, vertically stacked system.

Author Contributions: Conceptualization, N.E.C., É.W., H.K. and M.A.K.J.; Methodology, N.E.C., É.W., G.B. and M.A.K.J.; Validation, N.E.C., É.W. and M.A.K.J.; Formal Analysis, N.E.C.; Investigation, N.E.C., R.A. and R.O.; Resources, G.B. and M.A.K.J.; Data Curation, N.E.C., R.A. and R.O.; Writing—Original Draft Preparation, N.E.C.; Writing—Review & Editing, N.E.C., É.W., R.A., G.B., R.O., H.K. and M.A.K.J.; Visualization, N.E.C.; Supervision, N.E.C. and M.A.K.J.; Project Administration, G.B. and M.A.K.J.; Funding Acquisition, G.B. and M.A.K.J. All authors have read and agreed to the published version of the manuscript.

Funding: This study was funded under the EPA Research Programme 2014–2020 (grant 2016-W-LS-11). The EPA Research Programme is a Government of Ireland initiative funded by the Department of Communications, Climate Action and Environment. It is administered by the Environmental Protection Agency, which has the statutory function of co-ordinating and promoting environmental research. N.E.C., R.A., R.O., and M.A.K.J. gratefully acknowledge funding support from the European

Regional Development Fund through the Ireland Wales Cooperation Programme (82078), who part funded this study. M.A.K.J. also acknowledges support by WoB.

Institutional Review Board Statement: Not applicable.

Informed Consent Statement: Not applicable.

Data Availability Statement: The data presented in this study are available on request from the corresponding author.

Acknowledgments: The authors graciously thank Luke Harman for technical support.

Conflicts of Interest: The authors declare that they have no known competing financial interests or personal relationships that could have appeared to influence the work reported in this paper.

References

1. FAO. *The Future of Food and Agriculture—Alternative Pathways to 2050*, Summary version; FAO: Rome, Italy, 2018; p. 60.
2. Hamann, S. The global food system, agro-industrialization and governance: Alternative conceptions for sub-Saharan Africa. *Globalizations* **2020**, *8*, 1405–1420. [CrossRef]
3. Caine, R.S.; Yin, X.; Sloan, J.; Harrison, E.L.; Mohammed, U.; Fulton, T.; Biswal, A.K.; Dionora, J.; Chater, C.C.; Coe, R.A.; et al. Rice with reduced stomatal density conserves water and has improved drought tolerance under future climate conditions. *New Phytol.* **2019**, *221*, 371–384. [CrossRef] [PubMed]
4. Yin, Y.; Yu, C.; Yu, L.; Zhao, J.; Sun, C.; Ma, Y.; Zhou, G. The influence of light intensity and photoperiod on duckweed biomass and starch accumulation for bioethanol production. *Bioresour. Technol.* **2015**, *187*, 84–90. [CrossRef] [PubMed]
5. Rufí-Salís, M.; Petit-Boix, A.; Villalba, G.; Sanjuan-Delmás, D.; Parada, F.; Ercilla-Montserrat, M.; Arcas-Pilz, V.; Muñoz-Liesa, J.; Rieradevall, J.; Gabarrell, X. Recirculating water and nutrients in urban agriculture: An opportunity towards environmental sustainability and water use efficiency? *J. Clean. Prod.* **2020**, *261*, 121213. [CrossRef]
6. UN. *The Sustainable Development Goals Report 2021*; United Nations: New York, NY, USA, 2021.
7. Morseletto, P. Targets for a circular economy. *Resour. Conserv. Recycl.* **2020**, *153*, 104553. [CrossRef]
8. Walsh, É.; Coughlan, N.E.; O'Brien, S.; Jansen, M.A.K.; Kuehnhold, H. Density dependence influences the efficacy of wastewater remediation by Lemna minor. *Plants* **2021**, *10*, 1366. [CrossRef]
9. Dinh, T.T.U.; Soda, S.; Nguyen, T.A.H.; Nakajima, J.; Cao, T.H. Nutrient removal by duckweed from anaerobically treated swine wastewater in lab-scale stabilization ponds in Vietnam. *Sci. Total Environ.* **2020**, *722*, 137854. [CrossRef]
10. Cheng, J.J.; Stomp, A.-M. Growing duckweed to recover nutrients from wastewaters and for production of fuel ethanol and animal feed. *Clean Soil Air Water* **2009**, *37*, 17–26. [CrossRef]
11. Ziegler, P.; Sree, K.S.; Appenroth, K.J. Duckweeds for water remediation and toxicity testing. *Toxicol. Environ. Chem.* **2016**, *98*, 1127–1154. [CrossRef]
12. Ziegler, P.; Sree, K.S.; Appenroth, K.J. The uses of duckweed in relation to water remediation. *Desalination Water Treat.* **2017**, *63*, 327–342. [CrossRef]
13. Ceschin, S.; Crescenzi, M.; Iannelli, M.A. Phytoremediation potential of the duckweeds *Lemna minuta* and *Lemna minor* to remove nutrients from treated waters. *Environ. Sci. Pollut. Res.* **2020**, *27*, 15806–15814. [CrossRef] [PubMed]
14. Cheng, J.; Landesman, L.; Bergmann, B.A.; Classen, J.J.; Howard, J.W.; Yamamoto, Y.T. Nutrient removal from swine lagoon liquid by Lemna minor. *Trans. Am. Soc. Agric. Eng.* **2002**, *45*, 1003–1010. [CrossRef]
15. Chikuvire, T.J.; Muchaonyerwa, P.; Zengeni, R. Biomass, nitrogen uptake and content of *Wolffia arrhiza* depends on strength of swine lagoon water. *Water Environ. Res.* **2018**, *90*, 2066–2074. [CrossRef] [PubMed]
16. Paolacci, S.; Stejskal, V.; Jansen, M.A.K. Aquaculture international, estimation of the potential of *Lemna minor* for effluent remediation in integrated multi-trophic aquaculture using newly developed synthetic aquaculture wastewater. *Aquac. Int.* **2021**, *29*, 2101–2118. [CrossRef]
17. Walsh, É.; Kuehnhold, H.; O'Brien, S.; Coughlan, N.E.; Jansen, M.A.K. Light intensity alters the phytoremediation potential of Lemna minor. *Environ. Sci. Pollut. Res.* **2021**, *28*, 16394–16407. [CrossRef]
18. Walsh, É.; Paolacci, S.; Burnell, G.; Jansen, M.A.K. The importance of the calcium-to-magnesium ratio for phytoremediation of dairy industry wastewater using the aquatic plant *Lemna minor* L. *Int. J. Phytoremediation* **2020**, *22*, 694–702. [CrossRef]
19. Calicioglu, O.; Richard, T.L.; Brennan, R.A. Anaerobic bioprocessing of wastewater-derived duckweed: Maximizing product yields in a biorefinery value cascade. *Bioresour. Technol.* **2019**, *289*, 121716. [CrossRef]
20. Yan, Y.; Candreva, J.; Shi, H.; Ernst, E.; Martienssen, R.; Schwender, J.; Shanklin, J. Survey of the total fatty acid and triacylglycerol composition and content of 30 duckweed species and cloning of a D6-desaturase responsible for the production of c-linolenic and stearidonic acids in Lemna gibba. *BMC Plant Biol.* **2013**, *13*, 201. [CrossRef] [PubMed]
21. Appenroth, K.J.; Sree, K.S.; Böhm, V.; Hammann, S.; Vetter, W.; Leiterer, M.; Jahreis, G. Nutritional value of duckweeds (Lemnaceae) as human food. *Food Chem.* **2017**, *217*, 266–273. [CrossRef]

22. Fiordelmondo, E.; Ceschin, S.; Magi, G.E.; Mariotti, F.; Iaffaldano, N.; Galosi, L.; Roncarati, A. Effects of partial substitution of conventional protein sources with duckweed (*Lemna minor*) meal in the feeding of rainbow trout (Oncorhynchus mykiss) on growth performances and the quality product. *Plants* **2022**, *11*, 1220. [CrossRef]
23. Coughlan, N.E.; Walsh, É.; Bolger, P.; Burnell, G.; O'Leary, N.; O'Mahoney, M.; Paolacci, S.; Wall, D.; Jansen, M.A. Duckweed bioreactors: Challenges and opportunities for large-scale indoor cultivation of Lemnaceae. *J. Clean. Prod.* **2022**, *336*, 130285. [CrossRef]
24. Xu, J.; Cheng, J.J.; Stomp, A.M. Growing *Spirodela polyrrhiza* in Swine Wastewater for the Production of Animal Feed and Fuel Ethanol: A Pilot Study. *Clean Soil Air Water* **2012**, *40*, 760–765. [CrossRef]
25. Goddek, S.; Joyce, A.; Kotzen, B.; Burnell, G.M. *Aquaponics Food Production Systems*; Springer International Publishing: Cham, Switzerland, 2019. Available online: https://link.springer.com/book/10.1007%2F978-3-030-15943-6 (accessed on 1 May 2022).
26. Markou, G.; Wang, L.; Ye, J.; Unc, A. Using agro-industrial wastes for the cultivation of microalgae and duckweeds: Contamination risks and biomass safety concerns. *Biotechnol. Adv.* **2018**, *36*, 1238–1254. [CrossRef] [PubMed]
27. Zhao, Y.; Fang, Y.; Jin, Y.; Huang, J.; Bao, S.; He, Z.; Wang, F.; Zhao, H. Effects of operation parameters on nutrient removal from wastewater and high-protein biomass production in a duckweed-based (*Lemma aequinoctialis*) pilot-scale system. *Water Sci. Technol.* **2014**, *70*, 1195–1204. [CrossRef]
28. Landolt, E. Biosystematic investigations in the family of duckweeds (Lemnaceae). In *The Family of Lemnaceae—A Monographic Study*; Veröffentlichungen des Geobotanischen Institutes der ETH, Stiftung Rübel: Zürich, Switzerland, 1986; Volume 1.
29. Ceschin, S.; Abati, S.; Leacche, I.; Zuccarello, V. Ecological comparison between duckweeds in central Italy: The invasive *Lemna minuta* vs the native L. minor. *Plant Biosyst.* **2018**, *152*, 674–683. [CrossRef]
30. Skillicorn, P.; Spira, W.; Journey, W. *Duckweed Aquaculture*; The International Bank for Reconstruction and Development/The World Bank: Washington, DC, USA, 1993.
31. Connolly, J.; Wayne, P. Asymmetric competition between plant species. *Oecologia* **1996**, *108*, 311–320. [CrossRef]
32. Maxwell, K.; Johnson, G.N. Chlorophyll fluorescence—A practical guide. *J. Exp. Bot.* **2000**, *51*, 659–668. [CrossRef]
33. Rice, E.W.; Baird, R.B.; Eaton, A.D. *Standard Methods for the Examination of Water and Wastewater*; American Public Health Association (APHA): Washington, DC, USA; American Water Works Association (AWWA): Denver, CO, USA; Water Environmental Federation (WEF): Alexandria, VI, USA, 2005.
34. Grasshoff, K.; Kremling, K.; Ehrhardt, M. *Methods of Seawater Analysis*, 3rd ed.; John Wiley and Sons: Weinheim, Germany, 2009.
35. Lenth, R. Emmeans: Estimated Marginal Means, Aka Least-Squares Means. R Package Version 1.4.8. 2020. Available online: https://CRAN.R-project.org/package=emmeans (accessed on 1 May 2022).
36. Ziegler, P.; Adelmann, K.; Zimmer, S.; Schmidt, C.; Appenroth, K.J. Relative in vitro growth rates of duckweeds (Lemnaceae)—The most rapidly growing higher plants. *Plant Biol.* **2015**, *17*, 33–41. [CrossRef]
37. Derksen, H.; Zwart, L. *Eendenkroos Als Nieuw Eiwit-En Zetmeelgewas: Haalbaarheidsstudie November 2010, Eindrapport*; Stichting Sustenable Forum: Singapore, 2010.
38. Giblin, S.M.; Houser, J.N.; Sullivan, J.F.; Langrehr, H.A.; Rogala, J.T.; Campbell, B.D. Thresholds in the response of free-floating plant abundance to variation in hydraulic connectivity, nutrients, and macrophyte abundance in a large floodplain river. *Wetlands* **2014**, *34*, 413–425. [CrossRef]
39. Janauer, G.A.; Schmidt-Mumm, U.; Schmidt, B. Aquatic macrophytes and water current velocity in the Danube River. *Ecol. Eng.* **2010**, *36*, 1138–1145. [CrossRef]
40. Xu, J.; Cui, W.; Cheng, J.J.; Stomp, A.M. Production of high-starch duckweed and its conversion to bioethanol. *Biosyst. Eng.* **2011**, *110*, 67–72. [CrossRef]
41. Ran, N.; Agami, M.; Oron, G. A pilot study of constructed wetlands using duckweed (*Lemna gibba* L.) for treatment of domestic primary effluent in Israel. *Water Res.* **2004**, *38*, 2241–2248. [CrossRef] [PubMed]
42. Gatidou, G.; Oursouzidou, M.; Stefanatou, A.; Stasinakis, A.S. Removal mechanisms of benzotriazoles in duckweed *Lemna minor* wastewater treatment systems. *Sci. Total Environ.* **2017**, *596*, 12–17. [CrossRef]
43. Adhikari, U.; Harrigan, T.; Reinhold, D.M. Use of duckweed-based constructed wetlands for nutrient recovery and pollutant reduction from dairy wastewater. *Ecol. Eng.* **2015**, *78*, 6–14. [CrossRef]
44. Klughammer, C.; Schreiber, U. Complementary PS II quantum yields calculated from simple fluorescence parameters measured by PAM fluorometry and the Saturation Pulse method. *Pulse Amplitude Modul. Appl. Notes* **2008**, *1*, 27–35.
45. Schoelynck, J.; Puijalon, S.; Meire, P.; Struyf, E. Thigmomorphogenetic responses of an aquatic macrophyte to hydrodynamic stress. *Front. Plant Sci.* **2015**, *6*, 43. [CrossRef]
46. Körner, S.; Vermaat, J.E. The relative importance of *Lemna gibba* L., bacteria and algae for the nitrogen and phosphorus removal in duckweed-covered domestic wastewater. *Water Res.* **1998**, *32*, 3651–3661. [CrossRef]
47. Zimmo, O.R.; van der Steen, N.P.; Gijzen, H.J. Nitrogen mass balance across pilot-scale algae and duckweed-based wastewater stabilisation ponds. *Water Res.* **2004**, *38*, 913–920. [CrossRef]
48. Mohedano, R.A.; Costa, R.H.R.; Hofmann, S.M.; Belli Filho, P. Using full-scale duckweed ponds as the finish stage for swine waste treatment with a focus on organic matter degradation. *Water Sci. Technol.* **2014**, *69*, 2147–2154. [CrossRef]
49. Zhao, Y.; Fang, Y.; Jin, Y.; Huang, J.; Ma, X.; He, K.; He, Z.; Wang, F.; Zhao, H. Microbial community and removal of nitrogen via the addition of a carrier in a pilot-scale duckweed-based wastewater treatment system. *Bioresour. Technol.* **2015**, *179*, 549–558. [CrossRef]

50. Paolacci, S.; Harrison, S.; Jansen, M.A.K. A comparative study of the nutrient responses of the invasive duckweed *Lemna minuta*, and the native, co-generic species Lemna minor. *Aquat. Bot.* **2016**, *134*, 47–53. [CrossRef]
51. Landolt, E.; Kandeler, R. Biosystematic investigations in the family of duckweeds (Lemnaceae). In *The Family of Lemnaceae—A Monographic Study*; Veröffentlichungen des Geobotanischen Institutes der ETH, Stiftung Rübel: Zürich, Switzerland, 1987; Volume 2.
52. Lasfar, S.; Monette, F.; Millette, L.; Azzouz, A. Intrinsic growth rate: A new approach to evaluate the effects of temperature, photoperiod and phosphorus–nitrogen concentrations on duckweed growth under controlled eutrophication. *Water Res.* **2007**, *41*, 2333–2340. [CrossRef] [PubMed]

plants

Article

Lemna minor Cultivation for Treating Swine Manure and Providing Micronutrients for Animal Feed

Reindert Devlamynck [1,2,*], Marcella Fernandes de Souza [2], Jan Leenknegt [1], Liesbeth Jacxsens [3], Mia Eeckhout [3] and Erik Meers [2]

[1] Provincial Research and Advice Centre for Agriculture and Horticulture (Inagro vzw), Ieperseweg 87, 8800 Roeselare, Belgium; jan.leenknegt@inagro.be
[2] Department of Green Chemistry and Technology, Ghent University, Coupure Links 653, 9000 Ghent, Belgium; Marcella.FernandesDeSouza@UGent.be (M.F.d.S.); erik.meers@ugent.be (E.M.)
[3] Department of Food Technology, Safety and Health, Ghent University, Valentin Vaerwyckweg 1, 9000 Ghent, Belgium; liesbeth.jacxsens@ugent.be (L.J.); mia.eeckhout@ugent.be (M.E.)
* Correspondence: reindert.devlamynck@inagro.be; Tel.: +32-485-65-1931

Abstract: The potential of *Lemna minor* to valorise agricultural wastewater into a protein-rich feed component to meet the growing demand for animal feed protein and reduce the excess of nutrients in certain European regions was investigated. Three pilot-scale systems were monitored for nine weeks under outdoor conditions in Flanders. The systems were fed with a mixture of the liquid fraction and the biological effluent of a swine manure treatment system diluted with rainwater in order that the weekly N and P addition was equal to the N and P removal by the system. The design tested the accumulation of elements in a continuous recirculation system. Potassium, Cl, S, Ca, and Mg were abundantly available in the swine manure wastewaters and tended to accumulate, being a possible cause of concern for long-operating recirculation systems. The harvested duckweed was characterised for its mineral composition and protein content. In animal husbandry, trace elements are specifically added to animal feed as micronutrients and, thus, feedstuffs biofortified with essential trace elements can provide added value. Duckweed grown on the tested mixture of swine manure waste streams could be considered as a source of Mn, Zn, and Fe for swine feed, while it is not a source of Cu for swine feed. Moreover, it was observed that As, Cd, and Pb content were below the limits of the feed Directive 2002/32/EC in the duckweed grown on the tested medium. Overall, these results demonstrate that duckweed can effectively remove nutrients from agriculture wastewaters in a recirculated system while producing a feed source with a protein content of 35% DM.

Keywords: Lemnaceae; remediation; feed safety; mineral supplements; accumulation; agricultural wastewater; nutrient recovery

Citation: Devlamynck, R.; de Souza, M.F.; Leenknegt, J.; Jacxsens, L.; Eeckhout, M.; Meers, E. *Lemna minor* Cultivation for Treating Swine Manure and Providing Micronutrients for Animal Feed. *Plants* **2021**, *10*, 1124. https://doi.org/10.3390/plants10061124

Academic Editors: Viktor Oláh, Klaus-Jürgen Appenroth and K. Sowjanya Sree

Received: 20 April 2021
Accepted: 27 May 2021
Published: 1 June 2021

1. Introduction

The rising world population and the improving living standards have been driving the increase of animal-based food consumption [1]. Global livestock production has been estimated to expand by 21% between 2010 and 2025 [2]. In this context, proteins play a pivotal role in animal feed as a source of nitrogen and essential amino acids [3]. Crop production, land-use change, processing, and transport contribute to the overall greenhouse gas emission of livestock production [4]. Therefore, there is an urgent need to develop resource-efficient and innovative practices to locally produce protein-rich feed alternatives with high areal productivity.

On top of that, European pig production is characterised by its intensiveness, which results in a manure abundance in certain regions [5,6]. Hence, the treatment and nutrient recovery of these waste streams have an essential role in improving the sustainability of conventional agriculture.

Duckweed has been shown to grow on wastewaters and subsequently produce a protein-rich feed ingredient suitable for pig production, offering a possible solution for addressing both the protein scarcity and local nutrient abundancy [7–10]. These small floating macrophytes occur all over the world and are the most rapidly growing Angiosperms, following a quasi-exponential growth rate [11]. Doubling times of 1.34 to 4.54 days are reported in optimal conditions [12]; however, its productivity depends on the climatic variation, the length of the growing season, and the management of the system. The estimated production rate in Europe is between 7 and 22 tonnes dry weight (DW) ha^{-1} yr^{-1} [13]. Besides productivity, duckweed's key advantage is its high crude protein content of around 35% [14], and up to 45% DW [13]. In contrast to several major cultivated protein or starch crops, the entire plant is edible because of the lack of support tissues [15].

Although many studies have been conducted on the use of duckweed for the treatment of pig manure wastewaters, there are still several knowledge gaps. Most research has been carried out with either one large-scale pilot or replicates on a laboratory scale [10]. Phytoremediation aspects tend to focus on N, P, and biological oxygen demand (BOD); nevertheless, other elements are also present in wastewaters, which might gradually increase or decrease over time in a recirculated system. A gradual increase over time is considered as an accumulation, while a gradual decrease is considered as a depletion. Accumulation over time off, for example, Cl might eventually reach harmful levels and, hence, decrease the duckweed growth and environmental performance [16].

Besides duckweed cultivation, the produced biomass should also be suitable for animal consumption, and in this assessment, the mineral composition is generally overlooked. For example, Cu and Zn are frequently added to swine feed for improving feed efficiency [17,18], but these elements are only 10–20% absorbed by the animals. As a result, swine excrements have high concentrations of Cu and Zn [17]. Also, heavy metals like As, Cd, and Pb could have diverse toxicological health effects, including carcinogenesis, decreased reproductive ability, and damages to the nervous, skeletal, circulatory, endocrine, and immune systems of animals and humans [19]. Hence, these metals are regulated by European law for food and feedstuffs (respectively EC No 1881/2006 and Directive 2002/32/EC) [20].

To address these open questions, a pilot recirculation system was set up in triplicate at outdoor conditions and fed with a mixture of swine manure effluents. The elemental characterisation of the recirculated water and the produced duckweed was monitored over nine weeks to uncover trends in a continuous growing system and gather adequate data to compare with the mineral composition standards of animal feed. The aim was to identify if the elements K, Cl, S, Ca, Mg, Mn, Fe, Cu, Zn, As, Pb, and Cd would accumulate within a recirculated duckweed system using biological effluent and the liquid fraction of the swine manure treatment as nutrient sources. A second objective was to evaluate several mineral components (Mn, Cu, Fe, Zn, As, Cd, and Pb) in the duckweed biomass for potential nutritional and harmful effects in potential feed application.

2. Materials and Methods

2.1. Experimental Set-Up

2.1.1. Pilot

The growth was performed in 1000 L cubicontainers (BE COMPOSITE IBC, Mauser, Brühl, Germany), of which an area of 0.9 × 1.1 m was cut from the top. The sidewalls were covered with a black plastic foil to exclude light interference, which could trigger algae activity in the containers. A mesh (vidalXL, The Netherlands) with a pore size of 1.17 × 1.57 mm was placed over the cubicontainer to prevent that the insect *Cataclysta lemnata* would interfere with the experiment. On 22 August 2019, a pilot-scale growing experiment was inoculated with 500 g FW of *Lemna minor*. The identification of the duckweed species was performed using molecular barcoding based on plastidic markers prior to the experiment [21].

The experiment was conducted outdoors, and daily meteorological data, i.e., air temperature (°C), solar irradiance (W m^{-2}), daylength (h), precipitation (mm m^{-2}), and relative humidity (%), were received from the Belgian Royal Meteorological Institute for the complete growing season (Table A1). The solar irradiance was converted to light intensity, which is expressed in μmol m^{-2} s^{-1}, using a conversion factor of 4.6 [22].

2.1.2. Growing Medium

The starting medium was a mixture of rainwater, biological effluent (BE), and liquid fraction of pig manure (LF). These waste streams were sampled at the manure treatment facility of IVACO, Eernegem, Belgium. LF was obtained by separating the solid fraction from raw pig manure by centrifugation. This process reduces the P content in the LF [23]. Subsequently, the ammonia of the LF was nitrified and denitrified in separate tanks during the biological treatment, resulting in the BE stream with a reduced N content [23]. At Ivaco, the BE is partly further applied to arable land as a K-rich fertiliser, and partly treated to dischargeable water by a constructed wetland using reedbeds.

BE and LF were sampled on 22 August 2019. 600 L of BE was stored in a closed cubicontainer (BE COMPOSITE IBC, Mauser, Brühl, Germany), while 50 L of LF was stored in a closed 50-litre feed barrel (PV-50L-HA) during the experiment.

To determine the medium composition, a non-linear solver technique was performed using Microsoft Excel, and the results are given in Table 1. The aim was to maximise the LF composition within the following imposed constrictions:

1. the total mass of LF, BE, and RW equals 1000 kg, as this is the approximate limit of the cubicontainers,
2. all fractions of LF, BE, and RW are greater than zero,
3. the total N and total P content of the final mixture are below the limits proposed for *Lemna minor* [13],
4. the N/P ratio of the medium equals 3.0, as this is the ratio between the N removal and P removal determined in a duckweed system grown on diluted BE in outdoor systems [7].

Table 1. The total N, total P and the N to P ratio of LF, BE, and RW (with LF = liquid fraction raw swine manure, BE = Biological effluent of a pig manure treatment facility, and RW = rainwater), together with the calculated composition of the mixture after a non-linear solver technique maximising the LF composition within presented restrictions. The restrictions were found in literature [7,13].

	Total N	Total P	Ratio	Mass
	[mg kg^{-1}]	[mg/mg]		[kg]
LF	4800	225		4
BE	331	247		41
RW	0	0		955
Mixture	33	11	3.0	1000
Restrictions	350	11	3.0	

Considering the concentrations of the three streams, data was gathered prior to the experiment as follows. First, it was assumed that rainwater (RW) would have a negligible N and P concentration. Furthermore, LF's composition was extracted from a report on the valorisation of pig manure wastewaters in Flanders [24]. Finally, the concentration of the BE was extracted from an internal dataset containing T-N and T-P concentrations from the manure treatment facility that provided the LF and BE of this experiment. However, also samples were taken and frozen on the day of preparation and analysed afterwards to verify the theoretical composition.

First, the influent and cultivation cubicontainers were filled with the starting solution described in Table 1. During the growing period, the influent was transferred to the cultivation containers using a dosing pump (Etatron BT-MA/AD 50/3.0, Etatron, Italy) set at a debit of 800 L per week. Hence, via an overflow, the effluent container was filled (see Figure 1).

Figure 1. The three pilot-scale set-ups in which the influent is mixed and stored (**left**) before being pumped in the middle cubicontainer where duckweed is grown. With an overflow mechanism, the lower cubicontainer (**right**) is filled with effluent from the growing system.

The aim was to achieve a weekly loading rate of N and P that equalled the removal rate from the system. This was found to be 1.1 g N m^{-2} d^{-1} and 0.37 g P m^{-2} d^{-1} in duckweed cultivation systems on pig manure waste streams in Flanders [7]. Based on the concentrations mentioned in Table 1, BE and LF were added to the system in a ratio that allowed the added N and P to equal the removed. This resulted in the addition of 1 kg and 10 kg of respectively LF and BE. Furthermore, the influent cubicontainer was filled with the system effluent, achieving a recirculating system.

During the process, the volume of each cubicontainer was determined by measuring the height of the water table. In this way, the mass flows throughout the system could be determined without neglecting the effect of evaporation and precipitation on the medium concentrations.

2.1.3. Sampling

Each week, duckweed was harvested and drip-dried for 5 min before weighing. Subsequently, 500 g of the harvested biomass was inoculated back in the cultivation cubicontainers, while the rest was oven-dried for three days at 60 °C to determine a representative DW percentage (DW%). This fully dried duckweed was used for dry weight, protein, and heavy metal determination. Furthermore, a water sample of 1 L was taken from each container and directly stored and −18 °C. For the influent, this was performed after filling the cubicontainer with the affluent, BE and LF. For the affluent, this was done prior to the addition of BE and LF.

2.2. Plant Analysis

For the determination of the mineral composition (P, S, Ca, Mg, Na, K, Cu, Fe, Zn, As, Cd and Pb), an inductively coupled plasma optical emission spectrometry (ICP-OES, VISTA-MPX, Agilent Technologies, Santa Clara, CA, USA) was performed on the destructed plant samples. The dried plant samples were first homogenised using a Retch ZM-200 plant mill (Düsseldorf, Germany) with a 0.5 mm sieve. Subsequently, 0.25 g was weighted to digestion tubes containing Aqua regia (7.5 mL HCl with 2.5 mL HNO_3) and 50 µL of a pure gold (Au) solution of 1000 mg L^{-1}. Closed-vessel microwave destruction (MARS6, CEM, Matthews, NC, USA) was performed on these samples. Afterwards, the samples were diluted to 50 mL using Milli Q® (Merck, Belgium).

Kjeldahl nitrogen (Kj-N) was measured according to Van Ranst et al., (1999) [25], without the addition of a reduction agent, using a distiller (Büchi auto Kjeldahl Unit K-370, Büchi, Switzerland), destructor (Büchi digest automat K438, Büchi, Switzerland), sampler (Büchi Kjeldahl sampler type K-371, Büchi, Switzerland) and scrubber (Büchi scrubber B414, Büchi, Switzerland). This method measures organic and ammonium nitrogen. Additionally, Kj-N content was used to calculate the protein content by multiplying it with the factor 6.25 [26].

The total N content (T-N) was determined according to the procedure of Dumas using a CNS analyser (Primacs SNC-100, Skalar, The Netherlands), as described in the guideline NEN- EN16168:2012 presented by the Royal Dutch Normalisation Institute (NEN). In this method, 200 mg of dried plant material is combusted, and the produced N_2 is measured with a thermal conductivity sensor. As all nitrogen forms are combusted, this analysis gives the sum of organic, nitrate, nitrite, and ammonium nitrogen. With the T-N content, the N removal of the plant was calculated.

2.3. Medium Analysis

For water analysis, an open vessel microwave destruction using a MARS6 (CEM, Matthews, NC, USA) was performed. A centrifuge tube was filled with 5 mL of the sample, 3 mL of 65% HNO_3 (Chem-Lab, Belgium), and 3 mL of 30% H_2O_2 (Chem-lab, Belgium). For an optimal release, 50 µL of a 1000 mg L^{-1} Au solution was added. The mixture was subsequently destructed by a stepwise gradual increase of temperature to 100 °C. This temperature was held for 30 min before cooling to room temperature. Afterwards, the digested sample was diluted to 25 mL using Milli Q® (Merck, Belgium). Subsequently, inductively coupled plasma optical emission spectrometry (ICP-OES, VISTA-MPX, Agilent Technologies, Santa Clara, CA, USA) was performed on the destructed samples for analysing P, S, Ca, Mg, Na, K, Cu, Fe, Zn, As, Cd, and Pb.

Chloride, NO_3^-, PO_4^{2-}, and SO_4^{2-} were determined by liquid ion chromatography (850 Professional IC anion, Metrohm, Antwerpen, Belgium) in a 150 mm column (Metrosep A SUPP 5-150/4.0, Metrohm, Antwerpen, Belgium), following the ISO 10304-1:2007 method.

Total N content (T-N) was determined according to the procedure of Dumas using a CNS analyser (Primacs SNC-100, Skalar, Breda, The Netherlands).

Finally, pH and electric conductivity (EC) was measured with a pH-meter (ProfiLine pH 3110, WTW, Weilheim, Germany) and a conductivity tester, respectively (ProfiLine Cond 3110, WTW, Weilheim, Germany).

2.4. Calculations

First, biomass productivity or the linear growth rate (LGR) was calculated as follows [27]:

$$\text{LGR} = \frac{(\text{DW\%}_{\text{end}} * \text{FW}_{\text{end}} - \text{DW\%}_{\text{start}} * \text{FW}_{\text{start}})}{\text{time} * \text{surface}} \left[\text{g m}^{-2}\,\text{d}^{-1}\right] \quad (1)$$

Moreover, the N and P uptake by the plant and protein productivity were calculated by replacing the dry weight percentage (DW%) by, respectively, the T-N, T-P, and protein content and by replacing the fresh weight (FW) with the dry weight (DW) in Equation (1).

The removal of an element by a duckweed system is weekly determined with Equation (2). Here, 't' represents the conditions at the end of the week (before the addition of nutrients), and 't−1' represents the conditions at the beginning of the week (after the addition of nutrients). Furthermore, the volume and concentration of the influent (V_I

&V_I), cultivation cubicontainer (V_C &V_C), and effluent (V_E &V_E) are used for each specific parameter. This is subsequently divided by the surface area of the system (0.99 m^2) and duration between two measurements (7 days).

$$\text{Removal}_t = \frac{(V_I{*}C_I + V_c{*}C_c + V_E{*}C_E)_t - (V_I{*}C_I + V_c{*}C_c + V_E{*}C_E)_{t-1}}{\text{surface} * \text{duration}_{\text{week}}} \left[\text{mg m}^{-2}\ \text{d}^{-1}\right] \quad (2)$$

From the removed nutrients, only a part is taken up by the duckweed. The others remain in the growing medium or are removed by processes like sedimentation of particles containing nutrients or by conversion into gaseous forms like in volatilisation or denitrification. The contribution of duckweed in the nutrient removal is indicated with the relative uptake (%) and is calculated by dividing the nutrient uptake by the removal of the nutrient [28].

Although elements are removed in a duckweed system, each week nutrients are added by adding 10 kg BE and 1 kg LF. Therefore, the net balance between addition and removal can still be positive or negative, leading to a gradual increase or decrease over time. This phenomenon is respectively defined as an accumulation or depletion. The average daily increase or decrease of an element will be determined by linear regression using R. The slope of the regression is equal to the daily concentration increase. This is also defined as the accumulation rate in this manuscript. The significance of the increase or decrease over time was tested with a Pearson correlation function between the parameter and the duration of the experiment. The significance was evaluated on a 5% significance level ($p < 0.05$). Furthermore, the accumulation rate depends on both the plant performance as the system input. Therefore, this parameter only shows the gradual build-up or depletion of a respective element in the tested system. Finally, a negative result should be interpreted as depletion of a particular element in the system.

3. Results and Discussion

3.1. Medium Composition and Accumulation of Elements

During the experiment, the composition of the growing medium and the accumulation of elements was monitored. These conditions are summarised in Table 2, along with the optimal and maximal growing ranges for duckweed found in the literature [13,29]. When a certain component is within the optimal range, the growth will be optimal; when it is outside the maximal growing range, duckweed growth is theoretically impossible. In all other cases, duckweed growth is suboptimal.

Table 2. Average composition ± standard deviations of the cultivation cubicontainers at the start and end of the experiment; these compositions are compared with optimal and maximal growing ranges (where LOD means Limit of Detection).

	Start (10 September 2019)	End (28 October 2019)	Optimal Growing Ranges	Maximal Growing Ranges	Unit
pH	7.0 ± 0.1	6.8 ± 0.1	6.5–7.5 $^{\alpha}$	5.0–9.0 $^{\alpha}$	
EC	1.0 ± 0.1	1.1 ± 0.1	0.6–1.4 $^{\alpha}$	<10.9 $^{\alpha}$	mS/cm
T-N	19 ± 2	19 ± 1	2.8–350 $^{\alpha}$	<2100 $^{\alpha}$	mg L^{-1}
NO_3^-	0.2 ± 0.2	5.6 ± 0.3	2.8–350 $^{\alpha}$	<2100 $^{\alpha}$	mg L^{-1}
PO_4^{3-}	0.8 ± 0.0	0.5 ± 0	0.4–11 $^{\alpha}$	<55 $^{\alpha}$	mg L^{-1}
T-P	20 ± 0	21 ± 0	0.4–11 $^{\alpha}$	<55 $^{\alpha}$	mg L^{-1}
K	132 ± 9	200 ± 13	39–780 $^{\alpha}$	<2000 $^{\alpha}$	mg L^{-1}
Cl	16 ± 3	29 ± 3	0.4–36 $^{\alpha}$	<3500 $^{\alpha}$	mg L^{-1}
T-S	17 ± 1	23 ± 2	48–1900 $^{\alpha}$	<4800 $^{\alpha}$	mg L^{-1}
Ca	15 ± 0	17 ± 0	20–400 $^{\alpha}$	<2000 $^{\alpha}$	mg L^{-1}
Mg	5.9 ± 0.3	8.8 ± 0.7	5.0–97 $^{\alpha}$	<1200 $^{\alpha}$	mg L^{-1}
Fe	0.53 ± 0.07	0.51 ± 0.06	<27.9 $^{\beta}$	<100 $^{\beta}$	mg L^{-1}
Mn	16 ± 14	9 ± 7	<54,900 $^{\beta}$	<274,500 $^{\beta}$	µg L^{-1}
Cu	8.3 ± 6.3	3.4 ± 2.6	<3200 $^{\beta}$	<6300 $^{\beta}$	µg L^{-1}
Zn	28 ± 23	11 ± 8	<6500 $^{\beta}$	<65,300 $^{\beta}$	µg L^{-1}
As	<LOD	<LOD			
Cd	<LOD	<LOD			
Pd	<LOD	<LOD			

Source: $^{\alpha}$ [13], $^{\beta}$ [29].

Comparing the optimal and maximal growing ranges with the composition of the growing medium shows that the medium is suboptimal. EC, pH, T-N, K, Cl, Mg, and Fe are in optimal ranges, but a reduction of T-P and an increase of S and Ca theoretically improve the productivity.

It should be noted that the design of the experiment led to an optimal N and P content in the growing medium. These concentrations exceeded the discharge levels in Flanders for a manure treatment facility of 15 mg N L^{-1} and 2 mg P L^{-1} [30]. This effluent could be land applied or further treated to meet the discharge criteria. Treatment can be performed by the principle of a constructed wetland where several lagunes are placed in series. There is a various set of plants which can grow in the lagunes and remove the nutrients from the water. The processes in the system result in nutrient depletion, making the wastewater dischargeable without risk for eutrophication. However, it should be taken into account that the lower the nutrient content, also the lower the phytoremediation potential, and the protein composition of duckweed will be [7,28,31].

The average productivity of the period from 28/08 until 4/11 was 4.5 ± 1.7 g m^{-2} d^{-1}, and productivity peaked at the harvest of 24/09 with 6.6 ± 0.3 g m^{-2} d^{-1}. In that week, there was an average temperature of 15 °C, an average light intensity of 99 ± 15 µmol m^{-2} s^{-1}, and an average photoperiod of 8.34 h. The average productivity was lower than the one found in a previous study of duckweed grown on a diluted biological effluent from the same pig manure treatment facility, which resulted in an average productivity of 6.1 g m^{-2} d^{-1} and 10.7 t ha^{-1} yr^{-1} [7]. The latter was, however, reached in a full growing season including summer, which is a generally more productive season. The maximum duckweed growth rate in that study was 9.7 g m^{-2} d^{-1} at a weekly average temperature of 21 °C and light intensity of 115 µmol m^{-2} s^{-1} for 13.2 h [7]. A comparison of both studies at the same time period shows that current results are respective for its growing period (Figure A1). These results indicate that production levels would be higher if produced for a full growing season. An optimal temperature for duckweed growth depends on the species and ranges between 20 and 31 °C [32]. An optimal photoperiod at a light intensity of 371 µmol m^{-2} s^{-1}, amounts 12–13 h [32]. Therefore, it is concluded that, besides some suboptimal concentrations in the growing medium, temperature, light intensity, and photoperiod were suboptimal.

In Figure 2, the compositions show a sawtooth pattern because of the weekly addition of LF and BE, which contain nutrients that are removed afterwards by the combination of duckweed uptake and other removal mechanisms like sedimentation or denitrification. In the figure of total P and total N, the removal of nutrients is approximately equal to the addition however, variation occurs. This variation can be addressed to sampling, laboratory handling, and precipitation and evaporation in the system. In the figure of potassium, the removal by the system is smaller than the weekly addition (Figure 2). Hence, the K concentration in the growing medium increases gradually. Within this manuscript, this is defined as accumulation. If the concentration gradually decreases, this will be defined as depletion. In the case of T-N and T-P, it could be argued that there is a small depletion however, there was no significant correlation with the duration over time, meaning that the concentration was constant. In contrast, the correlation of K with the duration of the experiment was significant, meaning that the accumulation of K was significant ($p < 0.05$).

The accumulation rate of K can be estimated by the slope of the regression line, which is 1.0792 mg K L^{-1} d^{-1}. And indeed, after the experiment of 55 days, the K concentration should increase 55 times 1.0792 mg L^{-1} or 59.36 mg L^{-1}. The starting concentration was 132 ± 9, and the end concentration was 200 ± 13 (Table 2). The difference between start and end is, however, not exact 59.36 mg K L^{-1} but the figure also shows that the final concentration is slightly above the regression line (Figure 2). The correlation factor is a good way to estimate the average accumulation rate. This discussion can be done for each element and is summarised in Table 3.

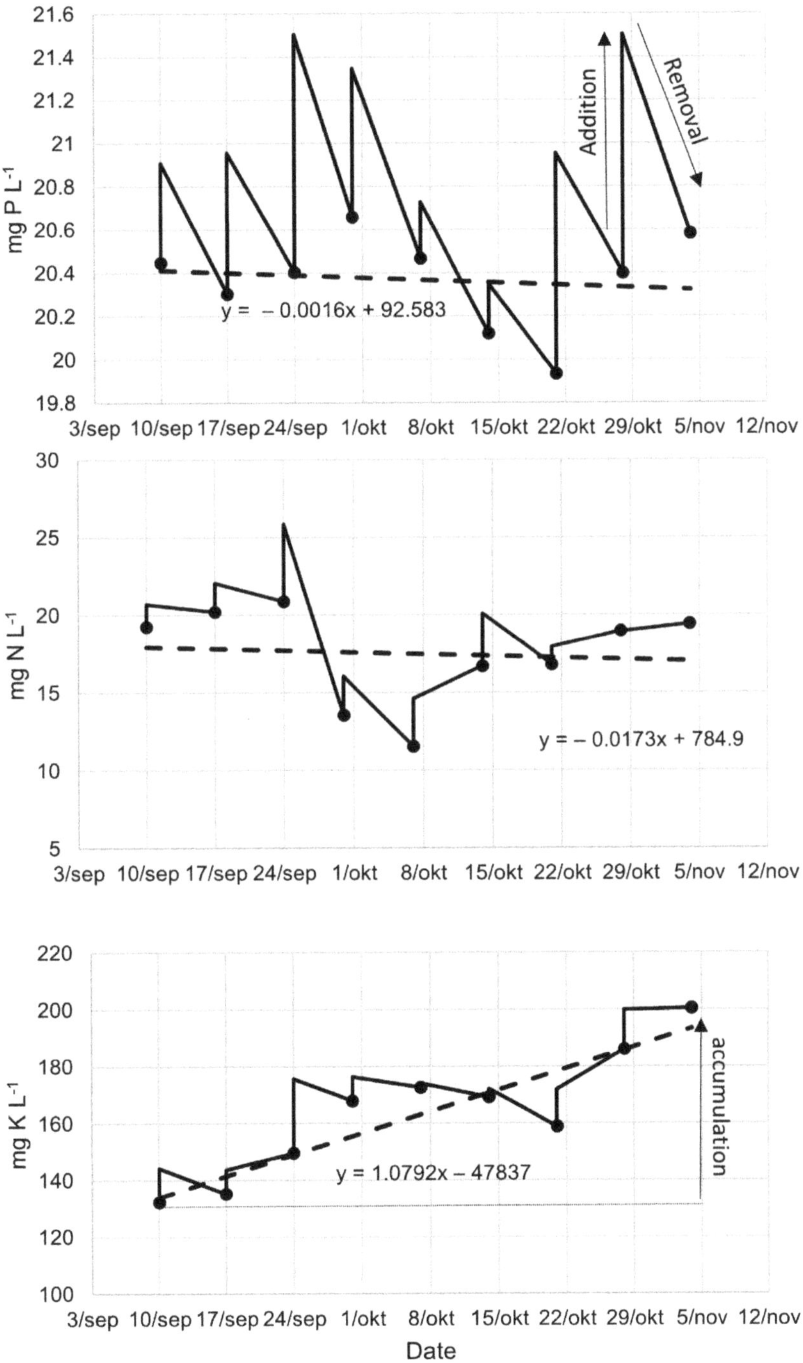

Figure 2. Average (**top**) total phosphorus, (**middle**) total nitrogen, and (**below**) potassium content in the growing media of the three cultivation systems (full line) with the linear regression of the concentration over time (dashed line).

Table 3. Summary of the daily accumulation rate of several parameters with their according significance value and interpretation.

Parameter	Accumulation Rate	Unit	*p*-Value	Interpretation
pH			0.432	Constant
EC	3.88	mS cm^{-1} d^{-1}	0.042	Accumulation
T-N			0.541	Constant
NO_3^-	0.12	mg L^{-1} d^{-1}	0.000	Accumulation
PO_4^{3-}	−0.004	mg L^{-1} d^{-1}	0.000	Depletion
T-P			0.685	Constant
K	1.09	mg L^{-1} d^{-1}	0.000	Accumulation
Cl	0.18	mg L^{-1} d^{-1}	0.000	Accumulation
T-S	0.11	mg L^{-1} d^{-1}	0.000	Accumulation
Ca	0.037	mg L^{-1} d^{-1}	0.0004	Accumulation
Mg	0.045	mg L^{-1} d^{-1}	0.000	Accumulation
Fe			0.363	Constant
Mn	−0.15	μg L^{-1} d^{-1}	0.022	Depletion
Cu			0.402	Constant
Zn			0.662	Constant
As	<LOD			<LOD
Cd	<LOD			<LOD
Pd	<LOD			<LOD

As the correlation of the T-N and T-P was not significant (Table 3), the goal of adding as much N and P to the system as it could remove was achieved, indicating that the non-linear solver technique predicted effectively the weekly required amount of LF and BE. Furthermore, analyses showed that BE and LF had an N content of respectively 301 ± 42 and 4665 ± 756 mg N L^{-1} and a P content of respectively 60 ± 3 and 206 ± 61 mg P L^{-1}. The observed compositions are in line with the average values found for LF and BE in Flanders [24]. Nevertheless, the starting conditions were not correctly predicted with the solver technique. The T-P content was twice as high as the result from the solver technique (20 instead of 10 mg L^{-1}), while the actual T-N of the starting condition was lower (20 instead of 33) (Tables 1 and 2). When preparing the growing medium, the BE is taken from a valve positioned at the bottom of the cubicontainer. Presumably, there is a sedimentation layer in the cubicontainer vessel of BE. The fraction of sediment was higher when preparing the growing medium. The solid fraction of pig manure contains generally a higher P and lower N content than the liquid fraction [33], which can explain the observed deviations.

Contrary to N and P, other elements tended to increase in the continuous cultivation (Tables 2 and 3). With the accumulation rates, a first extrapolation can be performed, although the removal rates might variate over different growing periods, and a determination over a full growing season and different years would give a more precise determination of the yearly accumulation rate. Electric conductivity, nitrate, K, Cl, T-S, Ca, and Mg rose steadily and would not have any impact on the growth performance of duckweed after one growing season of 175 days, which can be considered a feasible length of the growing season in Flanders [7]. For example, in the studied conditions, Cl content significantly increased, which might eventually induce salt stress. With the monitored accumulation rate of 0.18 mg L^{-1} d^{-1}, a discharge limit of 1000 mg L^{-1} would only be reached after 5466 days or 31 growing seasons of 175 days. This limit was imposed on pig manure treatment facilities in Flanders with a capacity of more than 60,000 tons of manure per year [30]. Even then, N and P removal by duckweed might not be affected, as this is only significantly reduced at a Cl concentration of 1772.5 ppm [34]. This should be a point of attention, as the chemical oxygen demand (COD) tends to increase at high salt concentrations [34]. Another important element to monitor is K, which had the highest accumulation rate in the system and would exceed the maximal growing range (2000 mg K L^{-1}) in 1713 days or 9.8 equivalent growing seasons.

In contrast, the differences in the concentration of Fe, Zn, and Cu were small to be statistically significant, while Mn showed a depletion over time. It was not possible, however, to draw a conclusion for potentially harmful elements like As, Cd, and Pb because all values found were below the limit of detection (LOD). These results indicate that more focus should be given to the accumulation effects on continuous recirculation cultivation at a large scale to determine at which point the water should be discharged, and the system should be restarted without losing productivity or treatment capacity.

3.2. Mineral Removal and Recovery Potential

Removal in a duckweed system is a combination of nutrient uptake by the plant and sedimentation. Except for carbon and nitrogen, which can also be removed in gaseous form by biodegradation [17]. The removal by the duckweed system and the contribution of the plant uptake are summarised in Table 4. For most of the monitored elements, duckweed's uptake was much lower than the system's removal, indicating that phytoremediation was only partly responsible for the wastewater treatment obtained.

Table 4. Average composition, uptake rate, removal rate and relative uptake rate of the removal by duckweed ± standard deviations (LOD stands for limit of detection).

	Content	Uptake	Removal	Relative Uptake
	[mg g^{-1} DW]	[mg m^{-2} d^{-1}]	[mg m^{-2} d^{-1}]	
N	60 ± 4	264 ± 123	1107 ± 715	24%
P	13 ± 1	58 ± 31	149 ± 150	39%
K	52 ± 10	233 ± 85	72 ± 3002	102%
S	6.1 ± 0.8	27 ± 15	47 ± 318	59%
Mg	2.8 ± 0.2	12 ± 5	32 ± 103	38%
Ca	11 ± 1	44 ± 18	242 ± 383	18%
Fe	1 ± 1	4.7 ± 14.5	44 ± 63	11%
Zn	0.25 ± 0.04	1.2 ± 0.7	2.4 ± 2.8	49%
	[μg g^{-1} DW]	[μg m^{-2} d^{-1}]	[μg m^{-2} d^{-1}]	
Cu	16 ± 15	82 ± 179	910 ± 1100	9%
As	0.32 ± 0.19	1.2 ± 2.0	<LOD	
Cd	0.1 ± 0.1	0.43 ± 1.05	<LOD	
Pb	3.0 ± 5.2	0.41 ± 11.36	<LOD	

The relative N uptake in this study was 24%, which is similar to the relative N uptake of 28% found in a duckweed pond fed with swine manure wastewater with a high nutrient load [28].

Remarkably, all K that was removed was taken up by the plant, which indicates that none was sedimented. Potassium is a more exchangeable element in wetland soils, making it highly mobile and explaining the observed result [35]. All removed K would re-enter the livestock system in the hypothesis that duckweed is used as a feed source.

The N and P removal are comparable to other duckweed treatment systems. Zimmo et al. found an N and P removal range of respectively 1.09–1.36 g N m^{-2} d^{-1} and 0.104–0.154 g P m^{-2} d^{-1} [36]. The N removal rate is comparable to a study on a reed-based constructed wetland which was monitored on a large scale in a temperate maritime climate, observing an N removal in summer of 1.22 g m^{-2} d^{-1} and 0.75 g m^{-2} d^{-1} in autumn [37].

With the removal rates found in the present study and considering a growing season of 175 days [7], it was extrapolated that 255 m^3 LF and 2550 m^3 BE per growing season would have been added and treated in a 1-ha duckweed pond with similar conditions to the experiment. The calculated treatment capacity was linked to the size of the farm and its pig places. A pig place is the average count of pigs present at the farm, taking into account that there are several production cycles in a year and that there are periods where no pigs are housed between two cycles. The average manure production was approximately

1.2 ton per year per pig place present at a farm in the period between 1998 and 2007 [38]. Furthermore, 0.83 m^3 of BE and 0.87 m^3 of LF are produced from 1 m^3 of raw manure with a density of 0.99-ton m^{-3} [23]. It should be noted that the extrapolation towards a hectare scale is a big step from the tested conditions. Therefore, an up-scaling validation is required to determine a more precise value. The manure generated by a farm with a housing capacity of 2805 pigs can be treated in the proposed duckweed system of 1 ha if the raw manure is pre-treated with a separation step followed by a biological treatment with an average performance. Hence, a considerable share could be treated of the manure produced in the manure facility which provided the wastewater, as this facility had a housing capacity of approximately 3000 pigs.

In this study, the swine wastewater was added to the pond after centrifugation and biological treatment (BE and LF) because this is a widespread treatment in Flanders [5]. Un-treated or anaerobically digested wastewater can also be treated using duckweed [9,10]. The N/K ratio will be different in these waste streams. As a result, the accumulation rate of K in a continuous recirculated system will be smaller. The same process holds for other accumulated elements. The downside is, however, that the areal need for processing the wastewater will be larger as these pretreatment techniques have a smaller reduction effect on the N and P concentration of the wastewater.

Finally, the removal rates might help to identify the potential accumulating elements for any given wastewater. Although, caution and validation over a full growing season at a large scale are advised, as the environmental performance depends on the growing medium and climate.

3.3. Feeding Value

After harvesting, duckweed could be used as a feed ingredient. This can be done fresh after harvest, or when it is dried, or when proteins are extracted. An ingredient is considered protein-rich by the EU protein balance sheet when the protein content is between 30–50 g per 100 g product [39]. When fully dried, duckweed contained 35 $\pm$ 2% DW and peaked at 10/09 with a protein content of 38 $\pm$ 2% DW (Figure A2). This is in line with the study of Zhao et al. [14]. This protein content allows classifying the produced duckweed as protein-rich biomass.

However, this holds for a fully dried product. The more moisture a product has, the more the contained elements are diluted. Duckweed should be almost fully dried before being considered a protein-rich product. The moisture content influences the mineral composition of a product. Therefore, all units are converted to a fully dried product in this manuscript to enable comparison with legal or suggested limits.

Copper, Fe, Zn, and Mn are known to have an essential role in metal-containing enzymes and lead to an improved immunity of livestock [40]. Conventional feeds are deficient in microelements, and for this reason, there is a need to supplement these constituents to livestock natural feeds [41]. This is also visible in the data presented in Table 5. Commodities like corn, oat, wheat, and steamed potatoes have Mn, Zn, and Fe levels below the feeding standard of both laying hens and swine (Table 5). No conventional feed ingredient can provide Zn levels that are close to 150 mg kg^{-1} DM. Duckweed grown in this study, however, has a much higher Zn, Mn, and Fe concentration than other commodities, while it has similar Cu values. It can be considered as a source of Mn, Zn, and Fe, and adding duckweed in the feed could act as a replacer for these elements in livestock feed. Nevertheless, duckweed cannot be unlimitedly mixed in feed, as its Zn content surpasses the proposed maximum of 150 mg Zn kg^{-1} DM in complete feed for piglets and sows [42].

Table 5. Recommended levels of micronutrients in livestock feed and the content of these elements in conventional feeds per kg dry matter (DM), adapted from Chojnacka (2008) [41], and duckweed (results from the present study ± standard deviations). The data of soybean meal was retrieved from Feedipedia [43].

Feeding Standard/Material	Mn	Zn	Cu	Fe
	[mg kg^{-1} DM]	[mg kg^{-1} DM]	[mg kg^{-1} DM]	[mg kg^{-1} DM]
Feeding standard for laying hens	60–70	50–60	5–6	60–70
Feeding standard for swine	30–40	20–165	70–150	90–100
Maximum limit piglets & sows		<150		
Corn (grain)	5	15	3.3	26
Oat (grain)	38	18	3.3	52
Wheat (grain)	26	23	2.7	43
Rye (grain)	58	26	3.3	60
Potatoes (steamed)	3	5	1.1	16
Fodder yeasts	14	9	12.6	90
Soybean meal	44	57	17	201
Duckweed	410 ± 60	250 ± 40	10 ± 2	372 ± 87

Additionally, the elements As, Cd, and Pb in feedstuffs are regulated by Directive 2002/32/EC of the European Parliament [20] and may not exceed the respective limits of 2, 1, and 10 mg kg^{-1} feedstuff. It should be noted that these limits do not hold for a completely dried product but a feedstuff with a water content of 12%. Converting the limits to the unit mg kg^{-1} dry weight results in the respective thresholds of 2.3, 1.1, and 11 mg kg^{-1} dry weight. These limits were not exceeded in the cultivation period between 10/09 and 4/11. Hence, duckweed can safely be used as a feed source when cultivated accordingly to this study, concerning As, Cd, and Pb (see Table 4 and Figure S2). It should be noted that there are several more aspects that should be monitored to guarantee feed safety, such as pathogens, viruses, and xenobiotics [19].

For all micronutrients analysed (Fe, Zn, Cu, As, Cd, and Pb), the contents do not seem to accumulate in the plant over time (see Figure S2). In contrast, the compositions sharply drop for Fe, Cu, Cd, and Pb, within the first three weeks, stabilizing afterwards. The comparison of the Fe and Zn concentrations with the suggested levels is complex. Therefore, it was chosen to compare the only concentrations from the stable period, being 30/09 to 4/11. This was not preferred for Cd and Pb, as safety should always be guaranteed.

The decreasing trend might suggest a reaction of the duckweed to the depletion of these nutrients within the growing medium. However, the Fe content in the medium does not sharply drop over time (Table 3). Arsenic, Cd, and Pb were all under the detection limit, suggesting nor a high content at the start nor the end of the experiment. Therefore, it is suggested that the high heavy metal concentrations of duckweed at the beginning of the experiment are caused by a historical accumulation at the sourcing location, which is subsequently thinned out by the continuous cultivation and harvest.

The observed high initial content is however, disadvantageous as the starting concentration of Pb is very high and close to the legal feed limit. Caution is advised at the start of the cultivation, and inoculating duckweed with As, Cd, and Pd concentrations that match the legal limits would minimise the risk in the first weeks.

4. Conclusions

With the help of a non-linear solver technique, a combination of liquid fraction and biological effluent from a swine manure treatment facility was treated with duckweed. The systems were fed with a mixture of the liquid fraction and the biological effluent of a swine manure treatment system diluted with rainwater in order that the weekly N and P addition was equal to the N and P removal by the system. The N and P concentrations in the system were high in order to have optimal duckweed growth. Potassium, Cl, S, Ca, and Mg showed an accumulation tendency within the wastewater. In a continuous recirculation growing system, these elements can eventually cause stress for duckweed cultivation. Nevertheless, N and P were removed from the growing medium, and protein-rich biomass was produced, with a content of 35 ± 2% dry weight. The mineral composition was rich in Mn, Zn, and Fe and can be seen as a source of these elements. The potential harmful heavy metals (As, Cd, and Pb) were monitored and were below the feed limits proposed by Directive 2002/32/EC. Hence, duckweed has the potential to be used to treat swine manure wastewater while producing a mineral- and protein-rich feed ingredient.

Supplementary Materials: The following are available online at https://www.mdpi.com/article/10.3390/plants10061124/s1, Figure S1: Comparing productivity, Figure S2: Protein.

Author Contributions: R.D.; formal analysis, R.D.; investigation, R.D., resources, J.L., M.E., E.M.; writing—original draft preparation, R.D.; writing—review and editing, M.F.d.S., L.J., M.E., E.M.; visualization, R.D.; supervision, J.L., L.J., M.E., E.M.; project administration, J.L., M.E., E.M.; funding acquisition, J.L., M.E., E.M. All authors have read and agreed to the published version of the manuscript.

Funding: This research was conducted within the framework of the H2020 project Nutri2Cycle, grant number: 773682. (https://www.nutri2cycle.eu/ accessed on 30 May 2021).

Institutional Review Board Statement: Not applicable.

Informed Consent Statement: Not applicable.

Data Availability Statement: Data is contained within the article and supplementary material.

Acknowledgments: Special thanks to the two master thesis students Emma Verhagen and Mathias Geerardyn who conducted several analyses and helped monitor the experimental set-up.

Conflicts of Interest: The authors declare no conflict of interest. The funders had no role in the design of the study; in the collection, analyses, or interpretation of data; in the writing of the manuscript, or in the decision to publish the results.

Appendix A

Table A1. Weather data.

Date	Average Temperature (°C)	Average Relative humidity (%)	Average Light Intensity (μmol m^{-2} s^{-1})	Average Photoperiod (h)		Total Precipitation (mm)
10/Sep	13 ± 2	82 ± 5	112 ± 14	6.4 ± 1.9		18.9
17/Sep	16 ± 2	83 ± 8	101 ± 16	9.5 ± 2.7		0.7
24/Sep	15 ± 0	72 ± 3	99 ± 15	8.4 ± 2.5		5.7
30/Sep	15 ± 2	91 ± 6	106 ± 20	5.9 ± 1.4		44
7/Oct	13 ± 2	86 ± 4	155 ± 64	3.9 ± 1.9		49.7
14/Oct	14 ± 2	89 ± 6	114 ± 36	4.6 ± 3.0		17.5
21/Oct	13 ± 2	90 ± 5	135 ± 69	3.7 ± 2.4		21.1
28/Oct	13 ± 3	90 ± 6	92 ± 30	4.0 ± 2.6		9.4
4/Nov	9 ± 3	90 ± 4	74 ± 9	6.1 ± 1.4		23.3
Average	14 ± 4	83 ± 9	109 ± 36	7.0 ± 3.5	Total	190.3

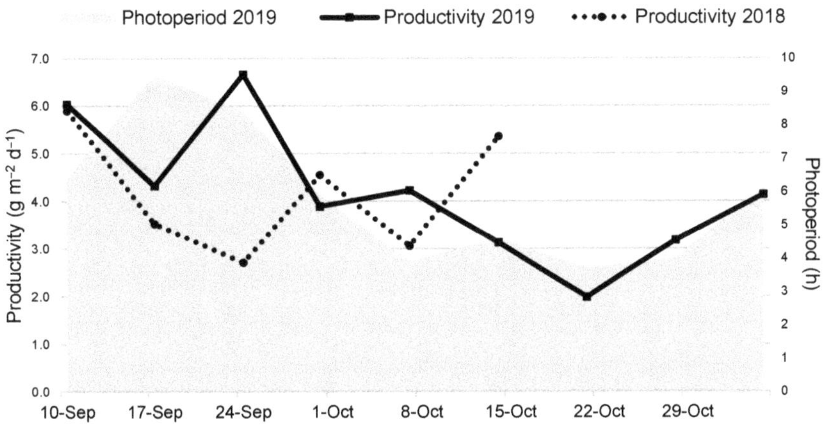

Figure A1. Duckweed productivity and the average daily photoperiod during the experiment compared with the productivity reached in a similar experiment using 22% of biological effluent from a pig manure treatment facility in 2018 [7].

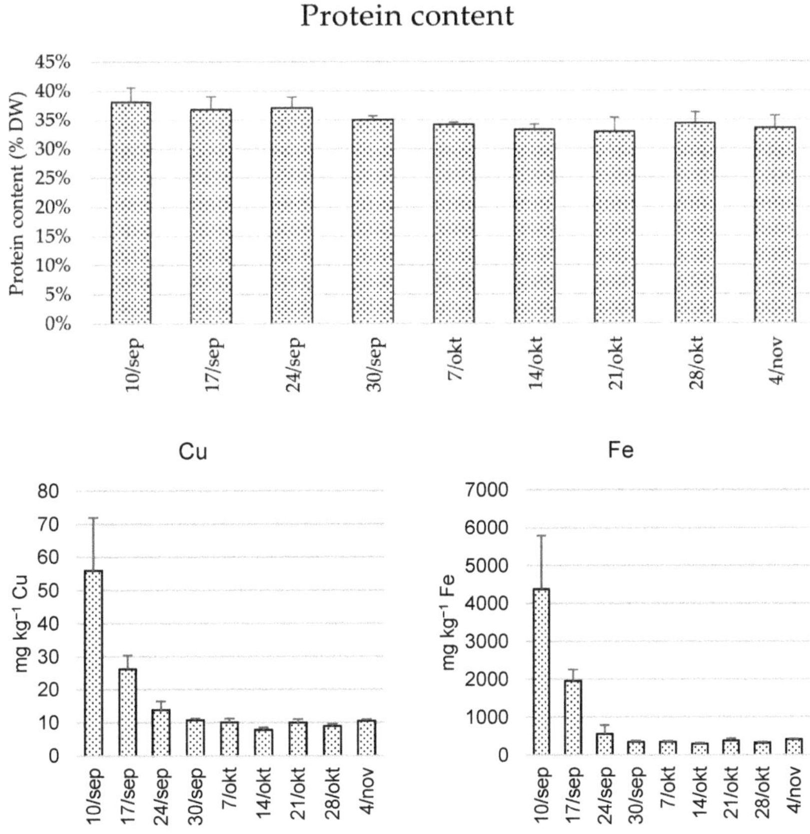

Figure A2. *Cont.*

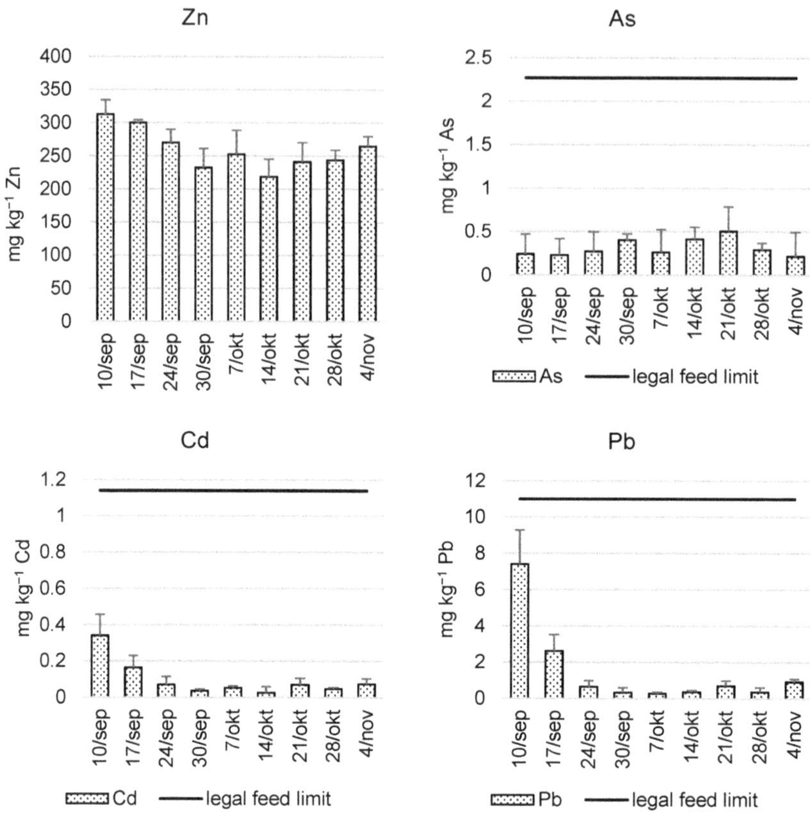

Figure A2. Protein, Fe, Zn, Cu, As, Cd, and Pb concentration with standard deviations (error flag) of the three duckweed Scheme 2019. *7*, 221–243, doi:10.1146/annurev-animal-030117-014838.

References

1. Kim, S.W.; Less, J.F.; Wang, L.; Yan, T.; Kiron, V.; Kaushik, S.J.; Lei, X.G. Meeting global feed protein demand: Challenge, Opportunity, and Strategy. *Annu. Rev. Anim. Biosci.* **2019**, *7*, 221–243. [CrossRef]
2. Mottet, A.; de Haan, C.; Falcucci, A.; Tempio, G.; Opio, C.; Gerber, P. Livestock: On our plates or eating at our table? A new analysis of the feed/food debate. *Glob. Food Sec.* **2017**, *14*, 1–8. [CrossRef]
3. Ritala, A.; Häkkinen, S.T.; Toivari, M.; Wiebe, M.G. Single cell protein-state-of-the-art, industrial landscape and patents 2001–2016. *Front. Microbiol.* **2017**, *8*, 2009. [CrossRef]
4. Meul, M.; Ginneberge, C.; Van Middelaar, C.E.; de Boer, I.J.M.; Fremaut, D.; Haesaert, G. Carbon footprint of five pig diets using three land use change accounting methods. *Livest. Sci.* **2012**, *149*, 215–223. [CrossRef]
5. Snauwaert, E.; Vannecke, T. *VCM-Enquête Operationele Stand van Zaken Mestverwerking in Vlaanderen*; VCM: Brugge, Belgium, 2017.
6. OECD. Nutrient Balance (Indicator). Available online: https://www.oecd-ilibrary.org/agriculture-and-food/oecd-compendium-of-agri-environmental-indicators_9789264186217-en (accessed on 14 May 2019).
7. Devlamynck, R.; de Souza, M.F.; Michels, E.; Sigurnjak, I.; Donoso, N.; Coudron, C.; Leenknegt, J.; Vermeir, P.; Eeckhout, M.; Meers, E. Agronomic and environmental performance of Lemna minor cultivated on agricultural wastewater Streams—A Practical Approach. *Sustainability* **2021**, *13*, 1570. [CrossRef]
8. Gwaze, F.; Mwale, M. The prospect of duckweed in pig nutrition: A Review. *J. Agric. Sci.* **2015**, *7*, 189. [CrossRef]
9. Zhang, M.; Luo, P.; Liu, F.; Li, H.; Zhang, S.; Xiao, R.; Yin, L.; Zhou, J.; Wu, J. Nitrogen removal and distribution of ammonia-oxidizing and denitrifying genes in an integrated constructed wetland for swine wastewater treatment. *Ecol. Eng.* **2017**, *104*, 30–38. [CrossRef]
10. Li, X.; Wu, S.; Yang, C.; Zeng, G. Microalgal and duckweed based constructed wetlands for swine wastewater treatment: A review. *Bioresour. Technol.* **2020**, *318*, 123858. [CrossRef] [PubMed]

11. Kutschera, U.; Niklas, K.J. Darwin-Wallace Demons: Survival of the fastest in populations of duckweeds and the evolutionary history of an enigmatic group of angiosperms. *Plant Biol.* **2015**, *17*, 24–32. [CrossRef] [PubMed]
12. Ziegler, P.; Adelmann, K.; Zimmer, S.; Schmidt, C.; Appenroth, K.-J. Relative in vitro growth rates of duckweeds (Lemnaceae)—The most rapidly growing higher plants. *Plant Biol.* **2015**, *17*, 33–41. [CrossRef]
13. Landolt, E.; Kandeler, R. *Biosystematic Investigations in the Family of Duckweeds (Lemnaceae), Vol. 4: The Family of LEMNACEAE-a Monographic Study, Vol. 2 Phytochemistry, Physiology*; Geobotanische Institut ETH: Zürich, Switzerland, 1987.
14. Zhao, Y.; Fang, Y.; Jin, Y.; Huang, J.; Bao, S.; Fu, T.; He, Z.; Wang, F.; Wang, M.; Zhao, H. Pilot-scale comparison of four duckweed strains from different genera for potential application in nutrient recovery from wastewater and valuable biomass production. *Plant Biol.* **2015**, *17*, 82–90. [CrossRef]
15. Skillicorn, P.; Spira, W.; Journey, W. *Duckweed Aquaculture—A New Aquatic Farming Fystem for Developing Countries*; The World Bank: Washington, DC, USA, 1993; ISBN 0-8213-2067-x.
16. Sikorski, Ł.; Piotrowicz-Cieślak, A.I.; Adomas, B. Phytotoxicity of sodium chloride towards common duckweed (*Lemna minor* L.) and yellow lupin (*Lupinus Luteus* L.). *Arch. Environ. Prot.* **2013**, *39*, 117–128. [CrossRef]
17. Suzuki, K.; Waki, M.; Yasuda, T.; Fukumoto, Y.; Kuroda, K.; Sakai, T.; Suzuki, N.; Suzuki, R.; Matsuba, K. Distribution of phosphorus, copper and zinc in activated sludge treatment process of swine wastewater. *Bioresour. Technol.* **2010**, *101*, 9399–9404. [CrossRef]
18. Nicholson, F.A.; Chambers, B.J.; Williams, J.R.; Unwin, R.J. Heavy metal contents of livestock feeds and animal manures in England and Wales. *Bioresour. Technol.* **1999**, *70*, 23–31. [CrossRef]
19. Markou, G.; Wang, L.; Ye, J.; Unc, A. Using agro-industrial wastes for the cultivation of microalgae and duckweeds: Contamination risks and biomass safety concerns. *Biotechnol. Adv.* **2018**, *36*, 1238–1254. [CrossRef]
20. European Commission. Directive 2002/32/EC of the European Parliament and of the Council of 7 May 2002 on undesirable substances in animal feed. *Off. J. Eur. Commun.* **2002**, *140*, 10–23.
21. Devlamynck, R.; Fernandes de Souza, M.; Bog, M.; Leenknegt, J.; Eeckhout, M.; Meers, E. Effect of the growth medium composition on nitrate accumulation in the novel protein crop Lemna minor. *Ecotoxicol. Environ. Saf.* **2020**, *206*, 111380. [CrossRef] [PubMed]
22. Sager, J.C.; Mc Farlane, J.C. 1. Radiation. In *Plant Growth Chamber Handbook*; Langhans, R.W., Tibbitts, T.W., Eds.; Iowa State University of Science and Technology: Ames, IA, USA, 1997; pp. 1–30.
23. Corbala-Robles, L.; Sastafiana, W.N.D.; van Linden, V.; Volcke, E.I.P.; Schaubroeck, T. Life cycle assessment of biological pig manure treatment versus direct land application—A trade-off story. *Resour. Conserv. Recycl.* **2018**, *131*, 86–98. [CrossRef]
24. Bodemkundige Dienst van België. *Eindrapport: Valorisatie van Resteffluenten Afkomstig van de Mestverwerking (Samenvatting)*; BDB: Leuven, Belgium, 2004.
25. Van Ranst, E.; Verloo, M.; Demeyer, A.; Pauwesl, J.M. *Manual for the Soil Chemistry and Fertility Laboratory: Analytical Methods for Soils and Plants Equipment and Management of Consumables*; NUGI 835; University of Ghent: Ghent, Belgium, 1999; ISBN 90-76603-01-4.
26. Casal, J.A.; Vermaat, J.E.; Wiegman, F. A test of two methods for plant protein determination using duckweed. *Aquat. Bot.* **2000**, *67*, 61–67. [CrossRef]
27. Xu, Y.-L.; Tan, L.; Guo, L.; Yang, G.-L.; Li, Q.; Lai, F.; He, K.-Z.; Jin, Y.; Du, A.; Fang, Y.; et al. Increasing starch productivity of Spirodela polyrhiza by precisely control the spectral composition and nutrients status. *Ind. Crops Prod.* **2019**, *134*, 284–291. [CrossRef]
28. Mohedano, R.A.; Costa, R.H.R.; Tavares, F.A.; Belli Filho, P. High nutrient removal rate from swine wastes and protein biomass production by full-scale duckweed ponds. *Bioresour. Technol.* **2012**, *112*, 98–104. [CrossRef]
29. Frick, H. Micronutrient tolerance and accumulation in the duckweed, Lemna. *J. Plant Nutr.* **1985**, *8*, 1131–1145. [CrossRef]
30. VLAREM II EMIS Navigator. Available online: https://navigator.emis.vito.be/mijn-navigator?woId=10112 (accessed on 19 February 2018).
31. Xu, J.; Cheng, J.J.; Stomp, A.-M. Growing Spirodela polyrrhiza in Swine Wastewater for the Production of Animal Feed and Fuel Ethanol: A Pilot Study. *CLEAN Soil Air Water* **2012**, *40*, 760–765. [CrossRef]
32. Lasfar, S.; Monette, F.; Millette, L.; Azzouz, A. Intrinsic growth rate: A new approach to evaluate the effects of temperature, photoperiod and phosphorus–nitrogen concentrations on duckweed growth under. *Water Res.* **2007**, *41*, 2333–2340. [CrossRef]
33. Hjorth, M.; Christensen, K.V.; Christensen, M.L.; Sommer, S.G. Solid-liquid separation of animal slurry in theory and practice. In *Sustainable Agriculture*; Springer Netherlands: Dordrecht, The Netherlands, 2009; Volume 2, pp. 953–986, ISBN 9789048126651.
34. Liu, C.; Dai, Z.; Sun, H. Potential of duckweed (*Lemna minor*) for removal of nitrogen and phosphorus from water under salt stress. *J. Environ. Manage.* **2017**, *187*, 497–503. [CrossRef]
35. Kadlec, R.H.; Wallace, S.D. Treatment Wetlands, 2nd ed.Lewis Publishers: New York, NY, USA, 1996; ISBN 0-87371-930-1.
36. Zimmo, O.R.O.; Steen, N.; van der Gijzen, H.J.H.; van der Steen, N.P.; Gijzen, H.J.H. Nitrogen mass balance across pilot-scale algae and duckweed-based wastewater stabilisation ponds. *Water Res.* **2004**, *38*, 913–920. [CrossRef]
37. Meers, E.; Tack, F.M.G.; Tolpe, I.; Michels, E. Application of a Full-scale Constructed Wetland for Tertiary Treatment of Piggery Manure: Monitoring Results. *Water. Air. Soil Pollut.* **2008**, *193*, 15–24. [CrossRef]
38. CBS. *Standardised Calculation Methods for Animal Manure and Nutrients*; Statistics: Den Haag, The Netherlands, 2012.
39. European Commission. *EU Feed Protein Balance Sheet*; EC: Brussels, Belgium, 2020.

40. Papadopoulos, G.A.; Maes, D.G.D.; Janssens, G.P.J. Mineral accretion in nursing piglets in relation to sow performance and mineral source. *Vet. Med.* **2009**, *54*, 41–46. [CrossRef]
41. Chojnacka, K. Using biosorption to enrich the biomass of seaweeds from the Baltic Sea with microelements to produce mineral feed supplement for livestock. *Biochem. Eng. J.* **2008**, *39*, 246–257. [CrossRef]
42. EFSA Scientific Opinion on the potential reduction of the currently authorised maximum zinc content in complete feed. *EFSA J.* **2014**, *12*. [CrossRef]
43. Heuzé, V.; Tran, G.; Kaushik, S. Soybean Meal; Feedipedia, a programme by INRAE, CIRAD, AFZ and FAO. Available online: https://www.feedipedia.org/node/674 (accessed on 14 May 2021).

plants

MDPI

Article

The Impact of Salt Accumulation on the Growth of Duckweed in a Continuous System for Pig Manure Treatment

Marie Lambert [1,2,*], Reindert Devlamynck [1,2], Marcella Fernandes de Souza [2], Jan Leenknegt [1], Katleen Raes [3], Mia Eeckhout [4] and Erik Meers [2]

1 Provincial Research and Advice Centre for Agriculture and Horticulture (Inagro vzw), Ieperseweg 87, 8800 Roeselare-Beitem, Belgium
2 Lab for Bioresource Recovery, Department of Green Chemistry and Technology, Ghent University, Coupure Links 653, 9000 Ghent, Belgium
3 Research Unit VEG-i-TEC, Department of Food Technology, Safety and Health, Ghent University, St-Martems Latemlaan 2B, 8500 Kortrijk, Belgium
4 Research Unit of Cereal and Feed Technology, Department of Food Technology, Safety and Health, Ghent University, Valentin Vaerwyckweg 1, 9000 Ghent, Belgium
* Correspondence: mmlamber.lambert@ugent.be; Tel.: +32-470550859

Abstract: Duckweed (*Lemna*) is a possible solution for the treatment of aqueous waste streams and the simultaneous provision of protein-rich biomass. Nitrification-Denitrification effluent (NDNE) from pig manure treatment has been previously used as a growing medium for duckweed. This study investigated the use of a continuous duckweed cultivation system to treat NDNE as a stand-alone technology. For this purpose, a system with a continuous supply of waste streams from the pig manure treatment, continuous biomass production, and continuous discharge that meets the legal standards in Flanders (Belgium) was simulated for a 175-day growing season. In this simulation, salt accumulation was taken into account. To prevent accumulating salts from reaching a toxic concentration and consequently inhibiting growth, the cultivation system must be buffered, which can be achieved by altering the depth of the system. To determine the minimum depth of such a system, a tray experiment was set up. For that, salt accumulation data obtained from previous research were used for simulating systems with different pond depths. It was found that a depth of at least 1 m is needed to prevent a significant relative growth inhibition at the end of the growing season compared to the start. This implies a high water consumption (5–10 times more than maize). As a response, a second cultivation system was investigated for the use of more concentrated NDNE. For this purpose, salt tolerance experiments were conducted on synthetic and biological media. Surprisingly, it was observed that duckweed grows better on diluted NDNE (to 75% NDNE, or EC of 8 mS/cm) than on a synthetic medium (EC of 1.5 mS/cm), indicating the potential of such a system.

Keywords: Lemnacea; alternative protein; agricultural wastewater; water recovery; accumulation; continuous systems

check for updates

Citation: Lambert, M.; Devlamynck, R.; Fernandes de Souza, M.; Leenknegt, J.; Raes, K.; Eeckhout, M.; Meers, E. The Impact of Salt Accumulation on the Growth of Duckweed in a Continuous System for Pig Manure Treatment. *Plants* **2022**, *11*, 3189. https://doi.org/10.3390/plants11233189

Academic Editors: Viktor Oláh, Klaus-Jürgen Appenroth and K. Sowjanya Sree

Received: 10 October 2022
Accepted: 16 November 2022
Published: 22 November 2022

Publisher's Note: MDPI stays neutral with regard to jurisdictional claims in published maps and institutional affiliations.

1. Introduction

Manure and agricultural wastewater treatment is a worldwide problem [1]. In some regions with intensive agriculture, such as Flanders (Belgium), manure application on land is limited and therefore its surplus needs to be treated to prevent eutrophication [2]. A considerable amount of nitrogen and phosphorus is therefore removed during the treatment of surplus manure and cannot be used for crop production [3]. At the same time, there is a significant import of fertilizers and proteins in these regions. Therefore, using this surplus manure can help to close the nutrient loop in the region.

A potential solution to close this cycle can be the cultivation of duckweed on wastewater. Duckweed is a general name for plants that belong to the family of Lemnaceae [4]. These small floating macrophytes can be found all over the world and belong to the most rapidly growing Angiosperms, following a quasi-exponential growth rate [5]. In Europe (Ireland), production rates of 37.89 tonnes of dry duckweed biomass have been reported [6]. However, higher productivity levels of 68 tonnes DW ha^{-1} $year^{-1}$ have also been reported [7]. Besides productivity, the primary advantage of duckweed is its high crude protein content of around 35% [8], and up to 45% DW [9]. Additionally, its moderate amount of fibres (5–15% on DW) makes it readily digestible for monogastric animals and many fishes [10,11].

Similar to the wetland technology developed to process pig manure [12], duckweed forms a floating mat that can remove N and P from wastewater. The advantage of duckweed in comparison to wetlands is the simultaneous provision of proteins for livestock production [13]. In this way, nutrients are recycled into a key feed ingredient. Duckweed has been shown to grow on dairy wastewater [14,15], pig manure wastewater [7,16] and aquaculture effluent [17,18].

Figure 1 presents an illustration of a manure treatment process and how duckweed can be added to that process. The first part of this scheme (up to NDNE) has already been applied at a manure treatment plant. The second part, the duckweed cultivation system, has not yet been applied in practice but was simulated in this study. This system was filled with a certain amount of NDNE and liquid fraction of pig manure (LF), and was diluted with rain water (RW) at the beginning of the growing season. Subsequently, a continuous supply of NDNE and LF was added so that the added N and P concentration was equal to the treatment capacity of the system, which equaled 1107 $\pm$ 715 $mg/m^2/d$ and 149 $\pm$ 150 $mg/m^2/d$, respectively [19]. In Flanders (Belgium), it is only feasible to grow duckweed for around 175 days each year, as was determined by Devlamynck et al. [13]. During this period, a continuous flow of dischargeable water is generated at the end of the system. After the growing season, the duckweed tank can be gradually emptied for the remaining days of the year.

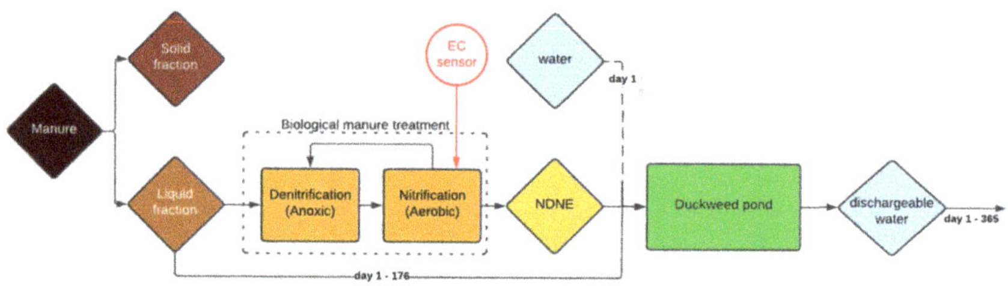

Figure 1. Schematic representation of the manure treatment process with centrifugal separation as the first step, biological treatment as the second step (with nitrification-denitrification = NDN) and a duckweed pond as the third and final treatment step. The duckweed pond in this scheme is not yet applied in practice, the other treatment steps are presented in the same order as they occur at IVACO, a pig manure treatment plant in Flanders (Belgium), Eernegem.

In the duckweed cultivation system, as presented in Figure 1, nutrients can accumulate over time when their incoming concentration is higher than the concentration that can be removed by the system [19]. It is, however, possible to reduce the accumulation rate by adjusting the system's volume or, in other words, the pond's depth. The deeper the pond, the lower the concentration of accumulating nutrients will be after one growing season. Yet, the depth must be minimized in order to reduce the water consumption of the system.

Therefore, this study investigated the possibility of using a continuous duckweed cultivation system to treat NDNE as a stand-alone approach (as represented in Figure 1). For this purpose, a system with a continuous supply of waste streams and a continuous discharge that meets the legal standards in Flanders (Belgium) was simulated for a 175-day growing season. Next, a tray experiment was carried out to determine the minimum depth of such a system. Furthermore, the impact of dilution on the suitability of NDNE as a growing medium for duckweed was also assessed. From salt tolerance experiments carried out by Landolt et al., an electrical conductivity (EC) of 10.9 mS/cm has been proposed as a maximal salt stress level for duckweed [9]. This indicates that undiluted NDNE cannot be used for duckweed cultivation as variations in EC between 7 and 24 mS/cm were observed [20]. However, EC does not distinguish between cations and anions or the share of beneficial and harmful elements. For example, Walsh et al., recently highlighted the Ca-to-Mg ratio as an extremely important parameter for duckweed growth [21]. Therefore, this study also investigated if the ratio of anions can play a role in the salt tolerance of duckweed.

2. Results and Discussion

2.1. *Simulation of a Continuous Duckweed Cultivation System (Tray Experiment 1)*

In a continuous system, the accumulation of ballast salts can occur [19], which can result in toxic salt levels and lead to a culture crash. Therefore, experiments mimicking continuous systems for obtaining more accurate results and assessing their water consumption are needed. In this study, a continuous system with a continuous supply of NDNE and LF and a continuous discharge that meets the discharge standards in Flanders (Belgium) that works for 175 days was simulated. The medium composition was calculated for systems with different depths (0.4 m, 0.7 m and 1 m). After simulating the system, a tray experiment was carried out with the calculated compositions. The relative growth (RGR) and chlorophyll inhibitions results, obtained from this experiment, are shown in Figure 2.

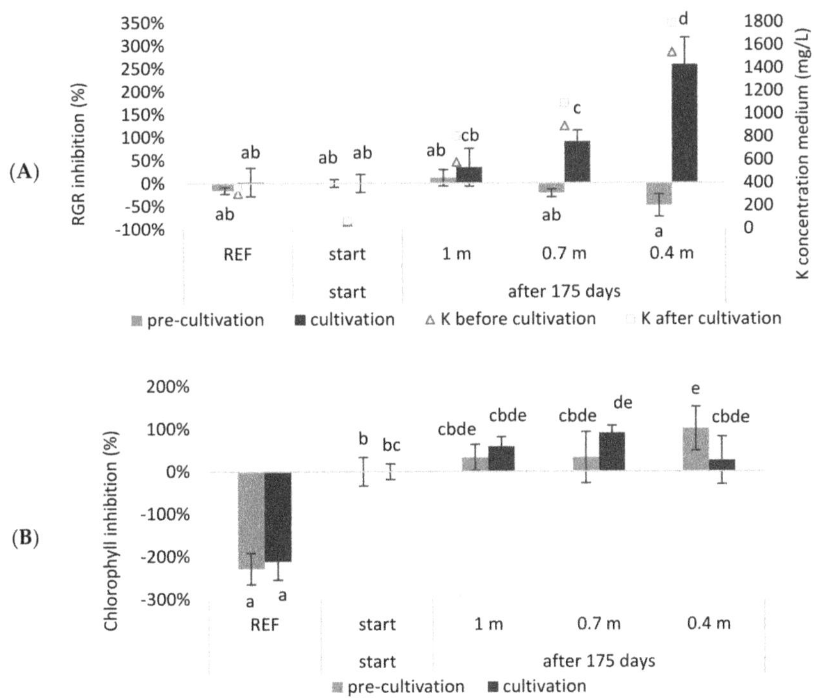

Figure 2. The relative growth rate inhibition (**A**), and the chlorophyll inhibition (**B**) of duckweed cultivated on a simulation of the concentration of wastewater of a continuous system, before and after 175 growing days, with different buffering capacities. Significant differences (n = 4 and $p < 0.05$) per graph are indicated by different letters. Error bars indicate standard deviations.

It was observed that the RGR inhibition of the continuous systems with a depth of 0.7 m or 0.4 m was significantly higher at the end of the growing season compared to the beginning of the season ($p = 3.98 \times 10^{-3}$ and 1.28×10^{-11}, respectively). Regarding chlorophyll inhibition, it was shown that the reference medium had a significantly higher chlorophyll content compared to the other treatments ($p < 8.99 \times 10^{-7}$). When comparing the chlorophyll concentrations at the start of the growing season with those at the end of the growing season, only the system with a depth of 0.7 m showed a significant inhibition ($p = 3.56 \times 10^{-2}$). For the 0.4 m system, the duckweed obtained in the pre-cultivation exhibited a significant chlorophyll inhibition ($p = 3.56 \times 10^{-2}$) but seemed to recover during the cultivation step.

As expected, the RGR of duckweed grown in a deeper system was less inhibited after 175 days than when it was growing in a shallower system. Accordingly, the deeper the system, or the higher the buffering capacity for the accumulating salts that enter the system, the lower the inhibition after one growing season. Roughly, a similar trend was shown for chlorophyll inhibition. The observation that the reference medium had a higher chlorophyll content can be explained by the low N and P concentrations of the wastewater-based growing media. Additionally, the pH of the reference medium was closer to the optimal pH for duckweed growth (Table A3) [9].

It can be concluded that, when duckweed is grown in a system with a continuous supply of NDNE and LF and a continuous discharge that meets the legal standards in Flanders (Belgium), a buffering capacity of at least 1 m depth is recommended. This would ensure that the relative growth inhibition would not be significantly higher at the end of the growing season compared to the start of the cultivation. However, in practice, this would mean that the continuous cultivation of duckweed would consume a significant amount of water. A buffering capacity of at least 1 m depth means that, for the cultivation of one hectare for one growing season, 10,000 m^3 water is needed. This is significantly higher compared to other crops such as maize, which has an irrigation water requirement of 900 to 1750 $m^3/ha/yr$ for optimal growth [22]. In reality, the water consumption itself may be even higher than calculated in this simulation, as the experiment did not take into account the naturally occurring evaporation of water.

The high water consumption is a result of the toxicity that occurs due to the accumulation of nutrients because of the continuous addition of biological waste streams. According to our simulation, this toxicity is mainly determined by potassium, since this element exceeds the toxicity limit according to Landolt et al., in the shortest time (see Table A3). However, if one considers the actual measured values before and after cultivation (Tables A1 and A2), it can be observed that the toxicity limit of potassium (2000 mg/L) was not exceeded. None of the growing media, neither before nor after cultivation, exceeded the toxicity limit for any element according to Landolt and Frick et al. [9,23]. Nevertheless, some elements, such as K and Cl, lie further outside their optimum range when a shallower system is simulated, which may be the reason for increased stress and therefore growth and chlorophyll inhibition.

Next to the high K concentration, a possible driving factor determining toxicity may be an adverse balance of nutrients. For example, it is known that a ratio which favors Mg over Ca negatively affects *L. minor* growth and its photosynthetic yield [21]. Therefore, a Ca:Mg ratio of 1:1.6 or greater is recommended for *L. minor* growth. In Figure 3 it is shown that, during cultivation, the Mg/Ca ratio of the reference and start medium was significantly lower than the maximum ratio of 1.6 ($p = 8.33 \times 10^{-5}$ and 1.42×10^{-7}, respectively). For the least buffered medium (0.4 m) it was shown that the Mg/Ca ratio was significantly higher than this maximum ratio ($p = 9.16 \times 10^{-5}$).

The shallower the system or the less buffering capacity the system has for accumulating salts, the higher the growth inhibition will be, and also, the more the nutrient uptake will decrease and will even become negative (Tables A7 and A8). A negative nutrient uptake indicates leaching, which is a clear signal of plant stress. A negative removal was observed for the elements Ca, P and Mn. This is an issue, especially for phosphorous, as the system was created/simulated in such a way that the added P concentrations should equal the removed P concentration. These measured nutrient removals also do not match those published by Devlamynck et al. [19], which were used to calculate the needed medium concentrations for the simulation. In addition, a constant removal was used to calculate all simulations for each nutrient, whereas we can see here that, for most nutrients, removal increases with increasing concentration. Even though more investigations are needed for continuous systems, this study indicates the importance of taking into account the salt accumulation in such setups.

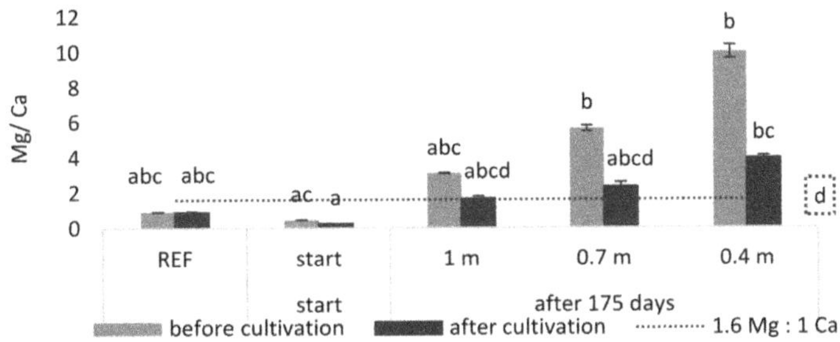

Figure 3. The Mg/Ca ratio in the medium of the simulated continuous system, before and after 175 growing days, with different buffering capacities. The dashed line indicates the minimum permitted Mg concentration in relation to the Ca concentration in the medium. Significant differences (n = 4 and $p < 0.05$) per graph are indicated by different letters. The letter in the dashed square belongs to the dashed line. Error bars indicate standard deviations.

Besides the high water consumption, a big disadvantage of the simulated system is that the growth medium must be discharged after 175 days, at the end of the growing season. Discharging this quantity of growth medium involves some practical problems, as the medium is too salty to be discharged into nature as a single flush. A possible solution is to opt for 175 days of duckweed growth in one year and then to gradually discharge for the rest of the year until the system is empty on the 365th day, as illustrated in Figure 1.

2.2. Duckweed Cultivation on NDNE (Tray Experiments 2 and 3)

2.2.1. NDNE as a Growing Medium

The simulation discussed in the previous section focuses on a system that treats NDNE and LF till a dischargeable effluent is obtained. However, it might be more interesting to treat a higher concentration of NDNE in a continuous system to achieve a higher biomass production and to reduce the water consumption needed for dilution. The purpose of the system is then changed from 'treating a waste stream as a stand-alone system to a dischargeable effluent' to 'production of protein-rich biomass with recycled nutrients'. In order to still obtain a dischargeable effluent at the end, the system can be combined with a constructed wetland using reedbeds as a final purification step.

To assess the maximum accepted waste stream concentration to minimize the needed dilution, a second tray experiment was done. Nine different dilutions/treatments of NDNE were tested (Table A2). It was investigated if the cultivation of duckweed on a biological medium (after dilution, addition of a salt solution or evaporation) resulted in better or worse growth compared to cultivation on a synthetic nutrient medium. In Figure 4, the RGR inhibition and the chlorophyll inhibition of the duckweed grown on different media are shown.

Surprisingly, the different dilutions and the undiluted NDNE showed no RGR inhibition compared to the reference medium. Even for the treatments where the EC was artificially increased there was only a significantly higher RGR inhibition after evaporation until an EC of 11.8 mS/cm was obtained ($p = 2.38 \times 10^{-9}$). However, for chlorophyll inhibition, a different trend was shown. There was a significantly higher chlorophyll inhibition for the undiluted NDNE and for most of the treatments where the EC was increased, indicating plant stress. For the diluted NDNE to an EC of 4, 6, or 8 mS/cm, the chlorophyll inhibitions were significantly lower than the reference ($p = 1.48 \times 10^{-6}$, 1.77×10^{-12} and

1.28×10^{-13}, respectively). The RGR and chlorophyll inhibitions of these dilutions were even below 0%.

(A)

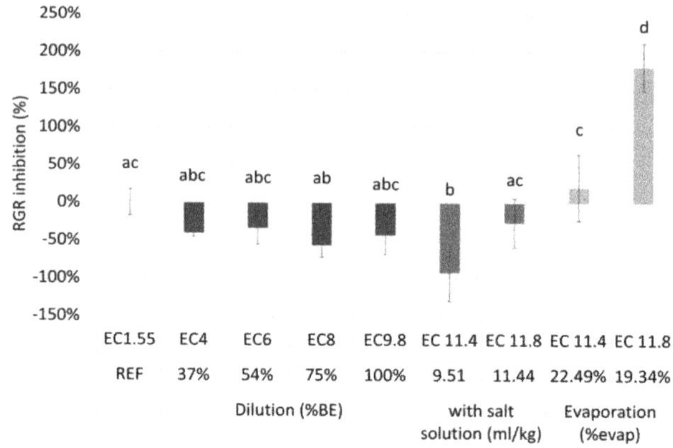

(B)

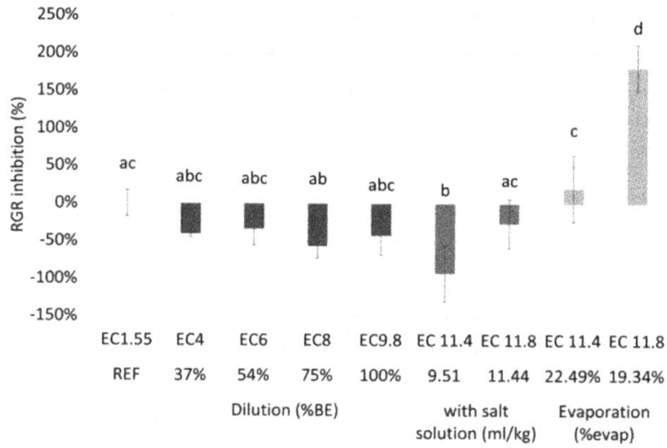

Figure 4. The relative growth rate inhibition (**A**) and the chlorophyll inhibition (**B**) of duckweed cultivated in different dilutions of NDNE, and NDNE to which a salt solution was added and in NDNE 'concentrated' after evaporation. On the x-axis, the EC values and respectively the percentage of dilution, the concentration of salt solution or the percentage of evaporation are given. Significant differences ($n = 4$ and $p < 0.05$) per graph are indicated by different letters. Error bars indicate standard deviations.

From the results shown in Figure 4, we can conclude that duckweed grew better on the diluted NDNE than on the reference medium. This is surprising, as the N medium is described in the literature as one of the best nutrient media used to support the fast growth of duckweed [24]. A possible explanation lies in the presence of organic components in the used waste streams. Organic matter contains humic substances that can be divided into three classes: fulvic acids (FA), humic acid (HA), and humin [25]. Humic substances might increase the uptake of both macro and micronutrients, such as N, P, K, Fe and Zn [26]. Additionally, they might also reduce the plant uptake of certain toxic metal ions, like Cd [27]. Thus, one might reason that the application of humic substances could improve plant response to salinity. However, in the literature, opinions are very much divided on this hypothesis. Liu et al., studied the influence of HA on the salt tolerance of hydroponically grown creeping bentgrass. They found out that, in general, the application of HA did not improve the salinity tolerance of the plant [28]. However, different findings were observed for other crops. Ghulam et al., studied the influence of HA on salt tolerance and nutrient uptake in wheat. They found out that a combined application of K and HA was promising for increasing wheat salt tolerance and nutrient uptake [29]. As both studies show very different results and worked with other crops than duckweed, it is difficult to make any concrete conclusions other than a possible indication that duckweed might behave like wheat and have higher salt tolerance in the presence of humic substances. Another explanation lies in the form in which nitrogen is present in the growth medium. Ammonium is the preferred nitrogen source of duckweed [30] and NDNE has more nitrogen present in the form of ammonium compared to the reference medium (Table A1), in which all the nitrogen is present in the form of nitrate.

In this study, it was shown that the duckweed grew better on the diluted NDNE media than on the reference medium. However, it should be considered that only data on growth and chlorophyll concentration was studied here. Osmotic stress can cause a reduction in protein concentration. Therefore, an important next step to this study should be biomass quality assessment of the duckweed grown on these different media.

2.2.2. Salt Tolerance of Duckweed

From the previous tray experiment, it seemed that duckweed is less salt-sensitive when grown on a biological medium compared to when grown on a synthetic medium. In order to better understand the salt tolerance of duckweed, an additional tray experiment was done. In this third experiment, a dose-response curve of *Lemna minor* was determined after adding different concentrations of NaCl to the synthetic reference medium. This approach is similar to how toxicity tests are usually done in literature [31]. However, this may not be the best way to determine salt tolerance as it is also important to take into account the ratio of certain elements. This was already demonstrated for the Mg/Ca ratio for example. Therefore, the results of this tray experiment were compared with another experiment where the same amount of Na^+ was added to a synthetic medium by adding a mixture of NaCl, Na_2SO_4 and K_2SO_4. This mixture had an SO_4^{2-}/Cl^- ratio of 0.25, similar to the ratio measured in NDNE in our previous research.

In Figure 5 both the RGR- and the chlorophyll inhibition are plotted as a function of the EC and the concentration of sodium in the mixture. It is shown that the EC is lower when Na^+ is added as NaCl than when the same concentration of Na^+ is added as a combination of NaCl, Na_2SO_4 and K_2SO_4. This is because SO_4^{2-} has twice as many charges as Cl^- and will therefore have a greater influence on the conductivity. Next, it was shown that the RGR inhibition was significantly higher at a certain EC when only NaCl was added compared to when a combination of NaCl, Na_2SO_4 and K_2SO_4 was added. A significant higher growth inhibition was observed after adding 46.6 mM Cl^- to the medium (up to EC = 6.51–6.65 mS/cm), compared to the reference medium (0 mM NaCl added—EC = 1.5 mS/cm) ($p = 2 \times 10^{-3}$ for pre-cultivation, $p < 1.08 \times 10^{-9}$ for cultivation).

Remarkably, when adding a combination of Cl^- and SO_4^{2-}, a significantly higher tolerance was observed in terms of growth inhibition in function of the EC, as total growth inhibition was only obtained after the addition of 37.3 mM Cl^- and 9.32 mM SO_4^{2-} to the medium (up to EC = 8.18–8.43 mS/cm) ($p = 4.88 \times 10^{-7}$ for cultivation). Presumably, total inhibition could have occurred even at lower concentrations than 46.6 mM NaCl. Therefore, it may be assumed that, in a synthetic medium, the duckweed was less inhibited when an equal amount of Cl^-/SO_4^{2-} was added than when only Cl^- was added.

The same trends are shown in Figure 5C,D for chlorophyll inhibition. However, it is possible to see significant differences at lower EC values or after the addition of lower Na^+ concentrations. A significantly higher chlorophyll inhibition was observed for pre-cultivation after adding 18 mM Cl^- to the medium (up to EC = 3.5 mS/cm), compared to the reference medium (0 mM NaCl added—EC = 1.5 mS/cm) ($p = 3.24 \times 10^{-4}$ for pre-cult). Remarkably, when adding a combination of Cl^- and SO_4^{2-} ions, a significantly higher tolerance was observed for the pre-cultivation, in terms of both growth inhibition in function of the EC and the Na^+ concentration. However, total chlorophyll inhibition was only obtained for the pre-cultivation after the addition of 25.6 mM Cl^- and 6.4 mM SO_4^{2-} to the medium (up to EC = 6.25 mS/cm) ($p = 2.19 \times 10^{-3}$). For the cultivation step, there was, for both experiments, a significant chlorophyll inhibition after the addition of 32 mM Na^+ (up to EC = 4.96–5.05 for Cl^- addition; EC = 6.25–6.44 for Cl^- and SO_4^{2-} addition) ($p = 1.02 \times 10^{-5}$ for Cl^- addition; $p = 5.63 \times 10^{-5}$ for Cl^- and SO_4^{2-} addition). The chlorophyll content of a plant can be used as a measure to assess oxidative damage in salt treatments. Oxidative stress is usual before a decrease in growth or die-off. In this sense, it is expected that chlorophyll inhibition already occurs at lower concentrations and EC values.

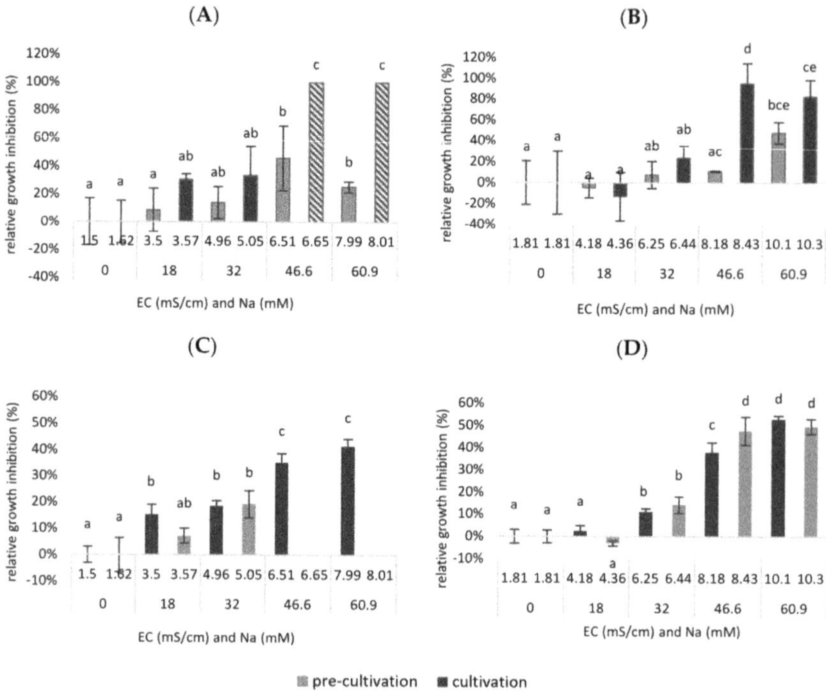

Figure 5. The relative growth rate inhibition of duckweed cultivated on a synthetic medium after NaCl addition (**A**) or NaCl, Na_2SO_4 and K_2SO_4 addition (**B**); the chlorophyll inhibition of duckweed cultivated on a synthetic medium after NaCl addition (**C**); or NaCl, Na_2SO_4 and K_2SO_4 addition (**D**). On the x-axis both the EC and the concentration of sodium in the medium is shown. A distinction is made between pre-cultivation and cultivation. For the inhibitions of the first experiment (**A**,**C**) there was no data obtained for the duckweed during cultivation grown on a medium with 46.6 and 60.9 mM Na due to die-off; therefore, in the RGRinh graph, a default value of 100% inhibition was taken (= no growth). Significant differences (n = 4 and $p < 0.05$) per graph are indicated by different letters. Error bars indicate standard deviations.

To conclude, in this experiment, duckweed was less inhibited when an equal amount of Cl^-/SO_4^{2-} was added than when only Cl^- was added. This proves that the composition of the anions plays a role in the salt tolerance of duckweed. As a result, the observation from the previous experiment that duckweed is less salt-sensitive when grown on a biological medium may be partly explained by the composition of the anions in this medium.

2.2.3. Variation in NDNE

From the second tray test, it could be concluded that the best medium for duckweed consists of a mixture of 75% NDNE and 25% demineralized water (EC = 8), both in terms of growth and chlorophyll concentration. However, the composition of the NDNE is not constant over time. In order to demonstrate that variation, the EC of NDNE was monitored in situ in a treatment facility. Over the same period, also the precipitation was monitored to calculate the dilution of the treatment system over its retention period (36 days).

Figure 6 shows that there is variation in the EC of NDNE over time. In fact, for 49% of the time, the measured EC was higher than 10.9 mS/cm, which is the maximal EC for duckweed survival according to Landolt et al. [9].

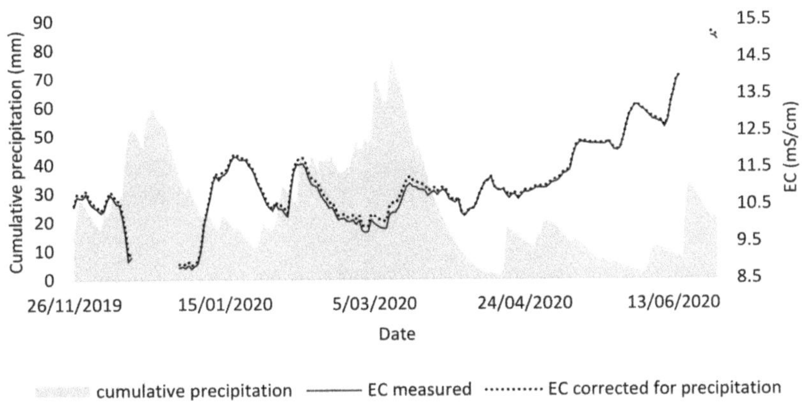

Figure 6. Electric conductivity (EC) measured in the aerobic tank (last tank) of a biological pig manure treatment system, together with the cumulative precipitation that could enter the system over a retention time of 36 days. Additionally, a corrected EC was calculated for the dilution from precipitation.

It was investigated whether the precipitation influenced the variation of the EC, as the rainwater that enters the system may dilute the stream and thus lower the EC, causing a higher variation. In this way, a lower EC could be expected in periods with high precipitation. To visualise this impact in the graph, a corrected EC was calculated by assuming that the rainwater entering the NDNE has an EC of 0. This leads to an underestimation but clearly shows that rainwater has no to little effect on the variation of the EC on NDNE.

More likely, the variable concentration of anions and cations in the NDNE is mainly caused by variations in the inlet stream and other process parameters. In this case, the monitored pig manure processing treatment has an inlet stream that consists of a combination of manure from fattening pigs and sows. The manure of sows is worth more for fertilisation and is preferred in times of high demand. Thus, around 10/05, the manure of sows was all diverted for fertilizing the lands. Hence, the manure from fattening pigs was fed to the treatment. The latter is thicker and has a higher conductivity. As a result, we can observe increasing conductivity from then on.

To conclude, the variation of the source material determines to a large extent the variation in the final growth medium. This variation can be reduced by the installation of buffering or storage lagoons, which would allow for a more stable effluent composition and an easier formulation of an adequate medium for duckweed growth on NDNE.

3. Materials and Methods

3.1. Tray Experiments

3.1.1. Experimental Conditions

Duckweed growth experiments were executed in PET containers (0.266 × 0.165 × 0.119 m) on a growth rack under laboratory conditions, as shown in Figure 7. The used containers were opaque to avoid light penetration through the walls and hence inhibit algae growth [32]. The total tray volume was 5.2 L, the surface area was 438.9 cm^2 and the trays were filled with 3 kg of growth medium for each experiment. The growth rack consisted of two operative levels. Each level could accommodate 10 containers, adding up to a total of maximally 20 containers per experiment.

Figure 7. Picture of the growth rack used for all tray experiments, with 10 PET containers and 4 parallel TL-lights per level.

Light was provided in a 16:8 h light-darkness regime by 4 parallel TL-light (TLD 36 w/86, Philips, the Netherlands) per level. Light intensity (PPFD, or photosynthetic photon flux density) ranged between 110–150 μmol/m^2/s and was measured at the respective duckweed mat level in the trays, in the centre of each tray. The simulation experiment took place in a climate chamber with a temperature of 25 ± 1 °C and air humidity of 70 ± 2%. The second tray experiment took place in the lab at a room temperature of 23 ± 2 °C.

To compensate for light asymmetry, the trays that were filled with the reference medium were placed in the centre of the rack. In this way, the reference was favoured over the other treatments. For the second experiment, where different dilutions and treatments of NDNE were used as a growing medium, a rotation scheme was designed to compensate for light asymmetry. A 5-period rotation was applied to keep each container in each row for the same time on each position. Besides, the water level was weight-adjusted with demineralized water to compensate for evaporation, solution mixing and to counter possible heat effects.

Plant material was sourced from a natural pond in Rumbeke-Beitem, Belgium. Visual determination according to [33] clarified that duckweed plants belonged to the *Lemna minor* species. The identification of the duckweed species was performed using molecular barcoding based on plastidic markers prior to the experiment [34].

Plant density was selected based on the work of Monette et al., to ensure total coverage of the water surface to minimize algae growth [35]. Since the overall goal was to obtain maximal biomass production, rather than maximal relative growth rate, a relatively high initial density of 30 g fresh weight (FW) was inoculated in the trays. Therefore, the density at the start of each cultivation was equal to 683.53 g/m^2.

3.1.2. Experimental Design

Three experiments were conducted where the dose-response curve of *Lemna minor* was determined. In the first experiment, waste streams from a pig manure treatment process at a pig farm in Pittem, Flanders (Belgium) were used to make the growth media. In the second experiment, NDNE from another pig manure treatment process in Flanders (Belgium) at IVACO, Eernegem was used. The latter was also the same manure treatment plant on which the EC of the NDNE was monitored. In the third experiment, duckweed was grown on a synthetic medium to which different concentrations of salt solutions were added.

The reference synthetic nutrient medium used in all experiments was the N medium as described in the ISCDRA newsletter [24]. The reference article for this medium was written by Appenroth et al. and these researchers stated that they "never found a nutrient media that supported a faster growth of duckweed than this one" [36]. The N medium was prepared as concentrated stock solutions (Table A1), of which 5 mL was taken for each litre of growth solution.

For each experiment, the pH of the reference medium was adjusted to 6–6.5 with 0.1 M NaOH. The other growth media used in the different experiments were also adjusted to the same pH, by adding either NaOH or HCl. For the first experiment, the pH of the growth media, other than the reference, was adjusted to the same pH as the start medium, which was 8.1 $\pm$ 1.

Tray Experiment 1: Simulation of a Continuous Duckweed Cultivation System

The growing medium of a continuous system with a depth of 1 m, 0.7 m or 0.4 m was simulated at the beginning and at the end of the growing season (175 days). In this tray experiment the depth of the simulated continuous system was determined. It should be stressed that the parameter of depth is a simulation using the assumption of a fixed nutrient accumulation during a growing period of 175 days. The experiment could also be interpreted as a pond with a fixed depth and where different periods of growth would be simulated if the pond would be placed in a region with a more suitable climate. For example, if the depth is fixed to 1 m, the treatments originally described as: '1 m', '0.7 m' and '0.4 m' would then be described as: 'after 175 days', '250 days' and '438 days', as calculated with the same simulation model using a non-linear solver technique.

To prepare the different growth media for this experiment, the N, P, K, S, Mg, Ca, Fe, Zn and Cl concentrations of the NDNE and the LF that would be added to the system were analysed. Next, the concentrations of LF, NDNE, and RW at the start of the growing season in the simulated system were determined using a non-linear solver technique. This was possible due to the following constrictions:

- All fractions of LF, NDNE, and RW are greater than zero;
- The sum of the fractions of LF, NDNE and RW equals 100%;
- The total N and P contents of the final mixture are below the discharge limits in Flanders (Belgium) [37];
- The N/P ratio of the medium equals 7.3, as this is the ratio between the N removal and P removal determined in an outdoor duckweed system, with diluted NDNE as the growing medium [19]

The most important restrictions are that the N and P concentrations of the effluent must not exceed the discharge limits in Belgium (resp. 15 and 2 mg/L) [37] and that the N and P removal of the system has to be the same as the N and P addition.

For the preparation of the different growth media, demineralized water (DW) was used instead of rainwater. The results of the non-linear solver technique, and thus the composition of the growing medium at the start of the growing season in the simulated continuous system, are given in Table 1.

Table 1. The total N, total P and the N to P ratio of the LF, NDNE and DW, together with the calculated composition of the mixture after a non-linear solver technique maximising the NDNE composition within presented restrictions.

	Total N (mg/kg)	Total P (mg/kg)	N:P Ratio	Mass Fraction (%)
LF	3440	191		0.1
NDNE	725	115		1.6
DW	0	0		98.3
Mixture	14.9	2	7.4	100
Restrictions	15	2	7.4	100

Next, the non-linear solver technique was used to calculate the accumulation coefficients of N, P, K, S, Mg, Ca, Fe, Zn and Cl. With these coefficients, it was possible to determine the concentrations of these elements in the growth medium of the system after one growing season, at a variable depth. For these calculations, it was assumed that the overall removal of the different nutrients by the system will be the same as the mean overall removal measured by Devlamynck et al. [19]. The calculated concentrations of the growth media are shown in Table 2.

Table 2. The simulated nutrient concentrations of the different growth media were calculated using a non-linear solver technique. 'Start' = the composition of the growing medium of a continuous system at the start of the growing season; '1 m'/'0.7 m'/'0.4 m' = the composition of the growing medium of a continuous system after 175 days of cultivation in a 1 m/0.7 m/0.4 m deep tank.

	Start	After One Growing Season		
Nutrient (mg/L)	(All Depths)	1 m	0.7 m	0.4 m
N	14.86	16.82	17.65	19.75
P	2.03	2.75	3.06	3.83
K	60.15	839.63	1173.70	2008.86
T-S	5.14	64.65	90.15	153.90
Mg	0.76	5.19	7.09	11.84
Ca	3.16	2.49	2.20	1.47
Fe	0.78	3.29	4.37	7.06
Zn	0.17	2.04	2.84	4.85
Cl	39.11	554.12	774.85	1326.65

Duckweed was grown on five different growth media. A reference medium (Table A1), a starting medium and three growing media with the same composition as at the end of the growing season depending on the depth of the tank. These last three growth media were prepared by adding extra nutrients via a specific salt solution in order to obtain the concentrations as shown in Table 2.

All growth media were prepared in volumes of 15 kg. Thus, for the starting medium, 15 g LF was mixed with 235 g NDNE and 14.749 kg DW. For the other growth media, 1 kg of the mass of the DW was replaced by 1 kg of a specific salt solution (Table A4). Afterwards, these growth media were divided into volumes of 3 kg for the tray experiment.

Cl^- was added by using NaCl and HCl. HCl was added via titration until a pH equal to that of the starting medium (8–8.2) was obtained. The necessary Cl^- concentration was then obtained by adding NaCl.

Tray Experiment 2: Different Dilutions/Treatments of NDNE

In the second tray experiment, it was determined to what extent NDNE is a suitable medium for duckweed cultivation. A dose-response curve with nine different growing media was set up (Table A2): (i) a reference medium (EC of 1.5 mS/cm); (ii, iii, and iv) NDNE diluted with demineralized water until an EC equal to 4, 6 and 8 mS/cm; and (v) undiluted NDNE (EC of 9.8 mS/cm).

In reality, the EC can even exceed 9.8 mS/cm. Therefore, these situations were mimicked by (vi and vii) spiking NDNE with a salt solution containing 237.4 mmol/L Na_2SO_4, 923.5 mmol/l KCl and 128 mmol/L NaCl until an EC of 11.4 and 11.8 mS/cm; and secondly by (viii and ix) heating NDNE until enough water was evaporated to reach an EC of 11.4 and 11.8 mS/cm.

Tray Experiment 3: Salt Tolerance Experiment

Two dose-response tests were conducted on synthetic media. For the first, only NaCl was added to the synthetic medium. In the second, the salt composition of NDNE was mimicked by adding a combination of salts in order to have a Cl/SO_4 ratio of 4:1 which is similar to the ratio found in the NDNE from previous analyses.

Therefore NaCl, Na_2SO_4 and K_2SO_4 were added as Na^+, K^+, Cl^-, $SO_4{}^{2-}$ with a molar ratio of 5:1:4:1 (Table 3). The amount of Na^+ ions were equal for both experiments.

Table 3. Salt concentrations added to the synthetic N medium during the first two salt experiments. The EC values were temperature corrected, and the interval gives the maximal range of measured values (at t0 and t1). Treatment n°1 of both experiments served as a reference.

	N°	NaCl (mM)	Na_2SO_4 (mM)	K_2SO_4 (mM)	EC (mS/cm)
Treatment NaCl	1	0.0	-	-	1.6–1.6
	2	18	-	-	3.5–3.6
	3	32	-	-	5.0–5.1
	4	46.6	-	-	6.5–6.7
	5	60.9	-	-	8.0–8.0
Treatment Cl/SO_4 (Ratio 4:1)	1	0.0	-	-	1.8–1.8
	2	14.4	1.8	1.8	4.2–4.4
	3	25.6	3.2	3.2	6.3–6.4
	4	37.3	4.7	4.7	8.2–8.4
	5	48.7	6.1	6.1	10.1–10.3

For both experiments, the pH was corrected by the addition of NaOH. Hence, this caused the addition of an amount of additional Na^+ in the different treatments; nevertheless, this was considered negligible. Water and biomass samples were taken before inoculation (t0), at the end of week 1 (t1) (pre-cultivation) and at the end of week 2 (t2) (cultivation).

3.1.3. Experimental Duration

The OECD guidelines describe a test period of 1 week as sufficient for toxicity tests [38]. However, phenomena such as luxury consumption have been described before [39]. As a result, a 'lag phase' in response curves might occur. Therefore a pre-cultivation period of 1 to 2 weeks was conducted on the same growing media to make sure that duckweed was adapted to the different conditions before the measurements were made. For experiments 2 and 3, the cultivation on the different growth media lasted for 1 week. However, for experiment 1, the cultivation period lasted for 3 days to prevent the depletion of N and P levels in the media.

3.1.4. Analytical Methods

Plant FW and DW Determination

Harvested duckweed material was measured both in terms of fresh (FW) and dry weight (DW). First, harvested fresh plant material was rinsed with tap water and drip-dried for 5 min in a fine mesh fishing net. Hereafter, the duckweed mass in the net was dried 2 times with a 5-folded paper towel for approx. 10 min. Afterwards, the duckweed pack was transferred from the net to aluminium cups and weighed on a balance (LA 320P, Sartorius Lab Instruments, Göttingen, Germany). Once the FW was determined and biomass for chlorophyll determination was separated, the samples in the aluminium cups were put in a drying oven at a low temperature (~60 °C) for a minimum of 3 days.

Plant Chlorophyll Content

To assess oxidative damage in the salt treatments, chlorophyll content was measured as an indicator via an ethanol extraction, as done by Liu et al. [40]. Therefore, 0.8 g FW was subsequently transferred into 40 mL of a 95% ethanol solution, stored for 5 days in the dark at room temperature, and centrifuged at 2790 rpm for 10 min (Centrifuge 5804 R, Eppendorf, Belgium) where after the supernatant was analysed with a UV-VIS spectrophotometer at 663 and 645 nm (Uvikon XL, Biotek Instruments, Santa Clara, CA, USA) to obtain chlorophyll contents. The concentrations of chlorophyll were calculated according to Huang et al. [41]:

$$C_a\left[\frac{\text{mg}}{\text{L}}\right] = 12.72A_{663} + 2.69A_{645} \tag{1}$$

$$C_b\left[\frac{\text{mg}}{\text{L}}\right] = 22.90A_{645} - 4.68A_{663} \tag{2}$$

$$C_{chl}\left[\frac{\text{mg}}{\text{L}}\right] = C_a + C_b \tag{3}$$

where C_a, C_b and C_{chl} represent the content of chlorophyll a, chlorophyll b, and total chlorophyll, respectively; A_{663} and A_{645} are the absorbances at 663 and 645 nm, respectively.

Compositional Analysis of Duckweed

The total N content (T-N) of the duckweed was determined before and after cultivation. A CN analyser (Primacs SNC-100, Skalar, Breda, The Netherlands) was used to determine the total nitrogen content in the duckweed. With the T-N content, the N removal and uptake of the plant were calculated.

For plant Ca, Mg, Na, K, P, S, Al, Cu, Fe, Mn and Zn content before and after cultivation, dried plant material, to which 65% HNO_3 was added, was first digested in a microwave (Milstone Ultrawave, SRC technology). Next, samples were accordingly diluted prior to elemental determination with inductively coupled plasma-optical emission spectrometry (ICP-OES) (Vista-MPX, Varian Inc., Palo Alto, CA, USA).

Analysis of NDNE, LF and Growing Media

For the determination of the T-N content of the undiluted NDNE and LF, the same CN analyser was used. However, as it was expected that the liquid samples contained nitrate and ammonia, 4 times the sample weight of sucrose and 1–2 drops of 20% o-Phosphoric acid solution were added to the crucibles before T-N analysis to maximize the N yield from NO_3^- and NH_4^+ in the sample.

For the determination of Ca, Mg, Na, K, P, S, Cu, Fe, Mn and Zn concentrations in the waste streams and the more diluted growth media before and after cultivation, the samples were first digested on a hot plate with a solution of H_2O_2 and HNO_3 (65%) in a 1:2 ratio for 30 min. After this, the sample was checked for its transparency. If the sample was still not transparent, then H_2O_2 and HNO_3 were again added at the same ratio used to continue the digestion until a transparent sample was obtained. Then the sample was filtered through a Whatman, grade 5 filter paper and diluted with milli-Q. Next, samples were accordingly diluted prior to elemental determination with ICP-OES.

The determination of Cl^-, NO_3^-, PO_4^{3-}, SO_4^{2-} in the water samples was done using ion chromatography (761 Compact Ion Chromatograph, Methrom, Herisau, Switzerland), preceded by 0.45 µm syringe filtration and dilution.

Finally, pH and electric conductivity (EC) was measured with a pH-meter (ProfiLine pH 3110, WTW, Weilheim, Germany) and a conductivity tester, respectively (ProfiLine Cond 3110, WTW, Weilheim, Germany).

3.2. *In Situ Monitoring of NDNE*

At the pig farm of IVACO, Eernegem, Belgium, pig manure is treated by a combination of centrifugation and subsequent biological treatment (Trevi, Ghent, Belgium) of the liquid fraction (LF). The loading rate is 20 m^3/day and the total retention time is 36 days. LF has a higher content of N and a higher N/P ratio than the solid fraction. NDNE is the result of the biological treatment process and was monitored from 26 November 2019, until 28 June 2020. For this, a solar power-driven 3798-S digital inductive electrical conductivity sensor (Hach, Belgium) was installed in the aeration tank in the last step of the process, as shown in Figure 1.

In order to assess the influence of weather conditions on conductivity, the monitored EC values were compared with climatological data obtained from the Royal Meteorological Institute. The climatological data also ran from 26 November 2019, to 28 June 2020. It was calculated how much water entered the manure treatment through precipitation. For this, the retention time of the system (36 days) and that of the different tanks individually were taken into account. For example, when one drop of rainwater enters the first step/tank of the treatment process, it will stay in the process until it leaves the last tank in the treatment (at day 36). Taking this into account, the cumulative precipitation was determined.

The biological manure treatment system of IVACO is divided into 4 tanks, 2 anaerobic and aerobic. For each tank, the cumulative precipitation was calculated as follows:

$$CP_k = \sum_{i=0}^{n} P_i * \frac{t_n - i}{t_n}, \tag{4}$$

with CP_k as the cumulative precipitation for tank k, P_i the precipitation on day i and with n the retention time (in days) of the specific tank. Next, the total cumulative precipitation for the whole system was calculated as follows:

$$CP_{tot} = \sum_{k=1}^{4} CP_k * \frac{A_k}{A_{tot}}, \tag{5}$$

with A_k as the surface area of tank k of the system and A_{tot} the total surface area of the system. With this total cumulative precipitation, the corrected EC was calculated as follows:

$$EC_{cor} = \frac{EC_{measured}}{1 - \left(\frac{\frac{CP_{tot}}{1000} * A_{tot}}{V_{tot}}\right)}, \tag{6}$$

with A_k and V_{tot} the total surface area and volume of the manure treatment system.

3.3. Calculations and Statistics

The relative growth rate was calculated based on the dry weight as follows [42]:

$$RGR\ [\text{g/g/d}] = \frac{\ln(DW_e) - \ln(DW_i)}{t_2 - t_1}, \tag{7}$$

with DW_e and DW_i representing respectively the dry weight of duckweed after (t_2) and before (t_1) cultivation. To make it possible to compare data from different experiments, the relative growth rate inhibition was also calculated:

$$RGRinhibition_i\ [\%] = 1 - \frac{RGR_i}{\text{mean}\left(RGR_{ref}\right)}, \tag{8}$$

with RGR_{ref} representing the RGR of the reference medium of the experiment. For experiment 1, where a continuous system was simulated, the start medium was chosen as the reference RGR. For experiments 2 and 3, the synthetic medium without the addition of extra salts was chosen as the reference RGR.

The same was done for the total chlorophyll concentration, the relative chlorophyll concentration was calculated as follows:

$$Chlorophyll\ inhibition_i\ [\%] = 1 - \frac{C_{Chl\ i}}{\text{mean}\left(C_{Chl\ ref}\right)} \tag{9}$$

with $C_{Chl\ ref}$, representing the total chlorophyll concentration of the reference medium of the experiment. The same references were chosen as for the calculation of the RGR inhibition.

Nutrient uptake/recovery rates were calculated considering the dry weight gain as well as the change in nutrient content in the biomass as follows:

$$Nutrient\ uptake\ \left[\text{g/m}^2/\text{d}\right] = \frac{C_i * DW_i - C_0 * DW_0}{area * t_2 - t_1} \tag{10}$$

with c_0 and c_i representing respectively the content of the specific nutrient in the plant before and after cultivation. For these concentrations, a correction was always made for the amount of water that evaporated during the cultivation step.

Nutrient removal by the system was calculated considering the nutrient concentrations of the growing media before and after cultivation as follows:

$$Nutrient\ removal\ \left[\text{g/m}^2/\text{d}\right] = \frac{c_0 - c_i}{area * (t_2 - t_1)}, \tag{11}$$

with c_0 and c_i representing respectively the content of the specific nutrient in the plant before and after cultivation. These concentrations were also corrected for the amount of water that evaporated during the cultivation step.

Microsoft Excel and R Statistical Software (v3.6.1, R Core Team 2019, Vienna, Austria) were used for statistical data processing and visual display. To show significant differences between the treatments in the experiments, parametric analyses such as one-way ANOVA, two-way ANOVA and Tukey's HSD tests were performed. These tests were only used if all conditions were met. The condition of whether the residuals are normally distributed or not was checked graphically by constructing qq-plots and numerically via the Shapiro-Wilk test. Homoscedasticity of the variations was checked graphically using box plots and numerically using the Modified Levene's-Test. If these conditions for these parametric analyses were not met, significant differences were demonstrated using the Kurskal-Wallis test followed by a post hoc Dunn's-Test. A significance value of 5% was used for all analyses, and in all cases the sample size (n) was equal to four.

4. Conclusions

This study showed that a continuous duckweed cultivation system can be used for the treatment of NDNE and the simultaneous provision of protein-rich biomass on a large scale. It was shown, however, that it is important to take into account that salts accumulate in long operating systems when their concentration added via the waste stream is higher than the concentration that can be removed by the system. This study concluded that, when duckweed is grown in a system with a continuous supply of NDNE and LF and a continuous discharge that meets the legal standards in Flanders (Belgium) (15 mg N/L and 2 mg P/L), a buffering capacity of at least 1 m depth is needed. This would ensure that the relative growth would not be inhibited at the end of the growing season compared to the start of the cultivation. However, in that case, this stand-alone treatment system would consume around 5 to 10 times more water than other crops such as maize. On the other hand, it was observed that duckweed grew better on NDNE (till 75% NDNE, or EC of 8 mS/cm) than on the reference N-medium, indicating that a lower dilution rate might be used if the duckweed system would be connected to a wetland for reaching the dischargeable legal limits. This study showed that the ratio of anions can partly explain this higher salt tolerance when grown on NDNE, but other mechanics remain uncovered. It is suggested that organic substances might have an effect on reducing salt stress. This observation increases the potential of using pig manure waste streams for duckweed cultivation. However, it was demonstrated by in-situ monitoring of the NDNE of a manure treatment plant that the composition of NDNE is not constant over time. Therefore, the optimal dilution found in the tray test is only temporarily valid. Further research on the operation of such a system, with the accumulation of elements taken into account, was shown to be relevant for the future application of this technology.

Author Contributions: Conceptualization, M.L., R.D., M.F.d.S. and E.M.; methodology, M.L. and R.D.; software, M.L.; validation, M.L., R.D., M.F.d.S. and E.M.; formal analysis, M.L.; investigation, M.L. and R.D.; resources, R.D., J.L., K.R., M.E. and E.M.; writing—original draft preparation, M.L. and R.D.; writing—review and editing, R.D., M.F.d.S. and E.M.; visualization, M.L.; supervision, R.D., M.F.d.S., J.L., K.R., M.E. and E.M.; project administration, R.D. and M.F.d.S.; funding acquisition, R.D., E.M., K.R. and M.E. All authors have read and agreed to the published version of the manuscript.

Funding: This research was performed in the framework of two projects, LemnaPro and Nutri2Cycle. LemnaPro is funded by Agentschap Innoveren & Ondernemen (www.vlaio.be, accessed on 9 October 2022) under grant agreement No HBC.2021.0173, with financial support of Flanders' FOOD. Nutri2Cycle is funded by the European Union's Horizon 2020 research and innovation programme under grant agreement No 773682.

Data Availability Statement: The data presented in this study are available on request from the corresponding author.

Conflicts of Interest: The authors declare that they have no known competing financial interests or personal relationships that could have appeared to influence the work reported in this paper.

Appendix A

Table A1. Composition of the N medium as described by the duckweed ISCDRA forum volume 3.5 [24].

Stock	Composition	Stock Concentration	Final Concentration
1	KH_2PO_4	0.2 M	1 mM
2	$Ca(NO_3)_2.4H_2O$	0.2 M	1 mM
3	KNO_3	1.6 M	8 mM
	H_3BO_3	1 mM	5 µM
	$MnCl_2.4H_2O$	2.6 mM	13 µM
	$Na_2MoO_4.2H_2O$	80 µM	0.4 µM
	$MgSO_4$.7H2O	0.2 M	1 mM
4	FeNaEDTA	5 mM	25 µM

Table A2. The different treatments used to test the salt tolerance in (adapted) NDNE. Treatment n°1: the reference medium; n°2,3,4 different dilutions of NDNE with demineralized water; n°5 undiluted NDNE; n°6,7 NDNE after evaporation; n° 8,9 NDNE with addition of a salt solution.

N°	EC (mS/cm)	Dilution (%NDNE)	Evaporated (%)	Salt Solution (mL/kg)
1	1.5	-	-	-
2	4	37	-	-
3	6	54	-	-
4	8	75	-	-
5	9.8	-	-	-
6	11.4	-	-	9.5
7	11.8	-	-	11.4
8	11.4	-	22.5	-
9	11.8	-	19.3	-

Table A3. Summary of the optimal and maximal growing ranges of *Lemna minor*, in which duckweed growth either is optimal or theoretically impossible [9,23].

	Optimal Growing Ranges	Maximal Growing Ranges	Unit
pH	6.5–7.5 $^{\alpha}$	5.0–9.0 $^{\alpha}$	
EC	0.6–1.4 $^{\alpha}$	0–10.9 $^{\alpha}$	mS/cm
NO_3_N	70–700 $^{\alpha}$	0–1400 $^{\alpha}$	mg/L
NO_2_N			mg/L
NH_4_N	45–90 $^{\alpha}$	9–1350 $^{\alpha}$	mg/L
T-DIN			mg/L
P	0.4–11 $^{\alpha}$	0–55 $^{\alpha}$	mg/L
K	39–780 $^{\alpha}$	0–2000 $^{\alpha}$	mg/L
Cl	0.4–36 $^{\alpha}$	0–3500 $^{\alpha}$	mg/L
SO_4^{2-}	48–1900 $^{\alpha}$	0–4800 $^{\alpha}$	mg/L
Ca	20–400 $^{\alpha}$	0–2000 $^{\alpha}$	mg/L
Mg	5.0–97 $^{\alpha}$	0–1200 $^{\alpha}$	mg/L
Na	120–230 $^{\alpha}$	0–3400 $^{\alpha}$	mg/L
H_2CO_3			mg/L
B	<17.3 $^{\beta}$	<86.5 $^{\beta}$	mg/L
Fe	<27.9 $^{\beta}$	<100 $^{\beta}$	mg/L
Mn	<54.9 $^{\beta}$	<274.5 $^{\beta}$	mg/L
Cu	<3.2 $^{\beta}$	<6.3 $^{\beta}$	mg/L
Zn	<6.5 $^{\beta}$	<65.3 $^{\beta}$	mg/L

$^{\alpha}$ [9]; $^{\beta}$ [23].

Table A4. Composition of the salt solutions that were added to the growth media that simulate the growing medium of a continuous system after 175 days of cultivation in a 1 m/0.7 m/0.4 m deep tank. DW = demi-water.

	1 m	0.7 m	0.4 m
	m%	m%	m%
K_2CO_3	1.735	2.478	4.337
K_2SO_4	0.418	0.597	1.045
$MgSO_4.7H_2O$	0.067	0.096	0.169
$FeSO_4.7H_2O$	0.019	0.027	0.047
$ZnSO_4.7H_2O$	0.012	0.018	0.031
NaCl	0.837	1.221	2.395
HCl	0.017	0.016	0.033
DW	96.640	95.195	91.484

Table A5. The measured composition of the simulated nutrient concentration of the different growth media before the cultivation step in the experiment ± standard deviations (LOQ stands for limit of quantification). 'Start' = the composition of the growing medium of a continuous system at the start of the growing season; '1 m'/'0.7 m'/'0.4 m' = the composition of the growing medium of a continuous system after 175 days of cultivation in a 1 m/0.7 m/0.4 m deep tank.

		Ref	Start	1 m	0.7 m	0.4 m
Ca	mg/L	38.24 ± 1.15	2.47 ± 0.08	2.33 ± 0.04	1.97 ± 0.07	1.89 ± 0.06
Mg	mg/L	21.35 ± 0.89	0.69 ± 0.02	4.36 ± 0.03	4.36 ± 0.03	11.46 ± 0.37
Na	mg/L	8.25 ± 0.47	19.35 ± 0.78	145.9 ± 1.37	145.9 ± 1.37	427.6 ± 2.96
K	mg/L	313.76 ± 3.56	64.28 ± 1.04	585.1 ± 6.6	585.1 ± 6.6	1539.86 ± 10.6
P	mg/L	31.05 ± 0.62	1.67 ± 0	1.81 ± 0	1.81 ± 0	2.13 ± 0
S	mg/L	29.22 ± 0.19	4.96 ± 0.29	56.94 ± 0.62	56.94 ± 0.62	155.83 ± 3.16
Cu	mg/L	0.02 ± 0	0.05 ± 0	0.06 ± 0	0.06 ± 0	0.06 ± 0.01
Fe	mg/L	1.19 ± 0.03	0.53 ± 0.04	2.46 ± 0.04	2.46 ± 0.04	6.36 ± 0.3
Mn	mg/L	0.67 ± 0.03	0.02 ± 0	0.01 ± 0	0.01 ± 0	0.01 ± 0
Zn	mg/L	0.02 ± 0.01	0.12 ± 0	1.63 ± 0.03	1.63 ± 0.03	4.28 ± 0.19
Cl	mg/L	208.27 ± 40.06	49.12 ± 1.85	671.4 ± 50.05	671.4 ± 50.05	1753.43 ± 86.35
NO_3	mg/L	608.79 ± 6.56	6.68 ± 1.21	51.96 ± 1.58	51.96 ± 1.58	<LOQ
PO_4	mg/L	99.71 ± 4.63	5.05 ± 0.34	<LOQ	<LOQ	<LOQ

Table A6. The measured composition of the simulated nutrient concentration of the different growth media after the cultivation step in the experiment ± standard deviations (LOQ stands for limit of quantification). 'Start' = the composition of the growing medium of a continuous system at the start of the growing season; '1 m'/'0.7 m'/'0.4 m' = the composition of the growing medium of a continuous system after 175 days of cultivation in a 1 m/0.7 m/0.4 m deep tank.

		Ref	Start	1 m	0.7 m	0.4 m
Ca	mg/L	50.63 ± 3.13	3.26 ± 0.3	6.1 ± 0.28	5.61 ± 0.85	1.89 ± 0.06
Mg	mg/L	27.58 ± 1.37	0.51 ± 0.04	6.12 ± 0.28	7.81 ± 0.37	11.46 ± 0.37
Na	mg/L	12.68 ± 0.34	23.79 ± 1.41	200.42 ± 3.11	317.07 ± 11.09	427.6 ± 2.96
K	mg/L	365.02 ± 26.02	71.98 ± 2.72	807.4 ± 17.85	1093.37 ± 61.69	1539.86 ± 10.6
P	mg/L	38.46 ± 2.82	1.36 ± 0	4.37 ± 0.34	4.51 ± 0.5	2.13 ± 0
S	mg/L	34.68 ± 1.97	6.32 ± 0.24	81.58 ± 2.94	108.1 ± 5.44	155.83 ± 3.16
Cu	mg/L	0.04 ± 0.01	0.1 ± 0.01	0.09 ± 0	0.02 ± 0	0.06 ± 0.01
Fe	mg/L	1.39 ± 0.07	0.57 ± 0.02	1.38 ± 0.08	1.7 ± 0.42	6.36 ± 0.3
Mn	mg/L	0.34 ± 0.11	0.1 ± 0.01	0.11 ± 0.04	0.09 ± 0.01	0.01 ± 0
Zn	mg/L	0.12 ± 0.01	0.12 ± 0.01	0.85 ± 0.08	1.1 ± 0.24	4.28 ± 0.19
Cl	mg/L	255.2 ± 96.19	53.16 ± 2.87	757.45 ± 23.04	1253.81 ± 120.31	1753.43 ± 86.35
NO_3	mg/L	655.78 ± 72.79	<LOQ	<LOQ	<LOQ	<LOQ
PO_4	mg/L	108.86 ± 10.27	3.72 ± 0.14	<LOQ	<LOQ	<LOQ

Table A7. The measured nutrient uptake by the duckweed in the simulated system, during cultivation ± standard deviations. 'Start' = the composition of the growing medium of a continuous system at the start of the growing season; '1 m'/'0.7 m'/'0.4 m' the composition of the growing medium of a continuous system after 175 days of cultivation in a 1 m/0.7 m/0.4 m deep tank.

		Ref	Start	1 m	0.7 m	0.4 m
T-N	mg/m²/d	141.07 ± 48.5	52.37 ± 33.81	33.11 ± 47.35	−21.31 ± 25.88	−154.92 ± 50.46
Ca	mg/m²/d	134.47 ± 40.32	106.15 ± 14.88	95.66 ± 24.5	−80.88 ± 29.27	−150.05 ± 17.39
Mg	mg/m²/d	27.69 ± 6.86	27.11 ± 2.94	43.16 ± 7.25	−6.67 ± 6.84	−38.55 ± 4.66
Na	mg/m²/d	47.01 ± 22.34	151.42 ± 18.76	177.5 ± 17.4	24.86 ± 42.92	−141.55 ± 40.18
K	mg/m²/d	736.47 ± 46.27	425.56 ± 74.55	467.68 ± 70.97	−321.33 ± 94.29	−627.37 ± 24.12
P	mg/m²/d	15.12 ± 60.27	17.3 ± 17.11	−59.09 ± 27.31	−127.31 ± 22.58	−165.53 ± 4.42
S	mg/m²/d	17.77 ± 13.08	7.06 ± 6.32	−9.37 ± 16.59	−51.78 ± 16.51	−79.73 ± 7.61
Al	mg/m²/d	−0.61 ± 0.96	0.2 ± 0.09	−0.69 ± 0.21	−3.50 ± 0.62	−1.44 ± 0.58
Cu	mg/m²/d	0.13 ± 0.03	0.33 ± 0.1	0.02 ± 0.17	−0.45 ± 0.15	−0.57 ± 0.06
Fe	mg/m²/d	1.03 ± 0.33	0.75 ± 0.47	2.11 ± 1.06	−1.13 ± 3.34	−0.83 ± 3.61
Mn	mg/m²/d	9.7 ± 2.4	−1.33 ± 2.42	−1.15 ± 2.07	−6.47 ± 1.85	−13.93 ± 3.05
Zn	mg/m²/d	−0.54 ± 0.16	3.32 ± 0.35	9.5 ± 1.37	5.56 ± 2.04	−1.60 ± 3.91

Table A8. The measured nutrient removal in the simulated system, during cultivation ± standard deviations (LOQ stands for limit of quantification). 'Start' = the composition of the growing medium of a continuous system at the start of the growing season; '1 m'/'0.7 m'/'0.4 m' = the composition of the growing medium of a continuous system after 175 days of cultivation in a 1 m/0.7 m/0.4 m deep tank.

		Ref	Start	1 m	0.7 m	0.4 m
Ca	mg/m²/d	73.84 ± 51.57	4.75 ± 3.67	−35.51 ± 2.89	−36.05 ± 10.74	−43.68 ± 5.43
Mg	mg/m²/d	50.34 ± 26.48	6.67 ± 0.64	3.91 ± 4.2	27.91 ± 7.55	31.83 ± 15.75
Na	mg/m²/d	−7.33 ± 9.5	61.62 ± 17.24	185.24 ± 63.54	753.74 ± 76.23	1580.68 ± 410.85
K	mg/m²/d	1275.63 ± 340.38	298.64 ± 31.13	694.56 ± 339.59	2993.52 ± 580.4	6270.47 ± 1816.01
P	mg/m²/d	95.05 ± 42.91	14.57 ± 0	−22.65 ± 4.55	−20.67 ± 6.59	−57.55 ± 3.3
S	mg/m²/d	109.56 ± 27.38	12.81 ± 6.29	27.66 ± 49.79	311.78 ± 73.45	546.73 ± 175.96
Cu	mg/m²/d	−0.12 ± 0	-0.32 ± 0.59	−0.12 ± 0.09	0.93 ± 0.36	0.75 ± 0.13
Fe	mg/m²/d	4.78 ± 1.26	2.74 ± 0.77	29.64 ± 0.92	42.45 ± 3.71	81.49 ± 6.16
Mn	mg/m²/d	8.66 ± 1.87	−1.05 ± 0.12	−1.18 ± 0.57	−1.10 ± 0.14	−5.87 ± 0.44
Zn	mg/m²/d	−1.22 ± 0.07	0.74 ± 0.18	20.6 ± 0.86	32.25 ± 2.71	53.93 ± 3.77
Cl	mg/m²/d	504.35 ± 901.86	271.64 ± 40.95	2986.44 ± 1264.25	4724.47 ± 2290.4	8063.42 ± 6035.16
NO_3	mg/m²/d	2800.82 ± 886.09	<LOQ	<LOQ	<LOQ	<LOQ
PO_4	mg/m²/d	518.28 ± 87.72	49.87 ± 8.33	<LOQ	<LOQ	<LOQ

References

1. Mateo-Sagasta, J.; Zadeh, S.M.; Turral, H. *More People, More Food, Worse Water? A Global Review of Water Pollution from Agriculture*; Food and Agriculture Organization of the United Nations: Roma, Italy, 2018.
2. OECD. *OECD Compendium of Agri-Environmental Indicators*; OECD Publishing: Paris, France, 2013. [CrossRef]
3. Snauwaert, E.; Vannecke, T. VCM Enquête Operationele Stand van Zaken Mestverwerking in Vlaanderen. Available online: https://www.vcm-mestverwerking.be/nl/kenniscentrum/5451/vcm-enquete (accessed on 8 October 2022).
4. Tippery, N.P.; Les, D.H.; Appenroth, K.J.; Sree, K.S.; Crawford, D.J.; Bog, M. Lemnaceae and Orontiaceae Are Phylogenetically and Morphologically Distinct from Araceae. *Plants* **2021**, *10*, 2639. [CrossRef]
5. Kutschera, U.; Niklas, K.J. *Darwin-Wallace Demons: Survival of the Fastest in Populations of Duckweeds and the Evolutionary History of an Enigmatic Group of Angiosperms*; Blackwell Publishing Ltd.: Hoboken, NJ, USA, 2015; Volume 17.
6. Paolacci, S.; Stejskal, V.; Toner, D.; Jansen, M.A.K. Wastewater Valorisation in an Integrated Multitrophic Aquaculture System; Assessing Nutrient Removal and Biomass Production by Duckweed Species. *Environ. Pollut.* **2022**, *302*, 119059. [CrossRef]
7. Mohedano, R.A.; Costa, R.H.R.; Tavares, F.A.; Belli Filho, P. High Nutrient Removal Rate from Swine Wastes and Protein Biomass Production by Full-Scale Duckweed Ponds. *Bioresour. Technol.* **2012**, *112*, 98–104. [CrossRef]
8. Zhao, Y.; Fang, Y.; Jin, Y.; Huang, J.; Bao, S.; Fu, T.; He, Z.; Wang, F.; Wang, M.; Zhao, H.; et al. Pilot-Scale Comparison of Four Duckweed Strains from Different Genera for Potential Application in Nutrient Recovery from Wastewater and Valuable Biomass Production. *Plant Biol.* **2014**, *17*, 82–90. [CrossRef]
9. Landolt, E.; Kandeler, R. *Biosystematic Investigations in the Family of Duckweeds ("Lemnaceae"). Volume 4: The Family of "Lemnaceae": A Monographic Study. Volume 2*; Veröffentlichungen des Geobotanischen Institutes ETH: Zurich, Switzerland, 1987; Volume 4.
10. Aslam, S.; Zuberi, A. Effect of Duckweed by Replacing Soybean in Fish Feed on Growth Performance of Grass Carp (Ctenopharyngodon Idella) and Silver Carp (Hypophthalmichthys Molitrix). *Int. J. Fish. Aquat. Stud.* **2017**, *5*, 278–282.
11. Leng, R.A.; Stambolie, J.H.; Bell, R. Duckweed—A Potential High-Protein Feed Resource for Domestic Animals and Fish. *Livest. Res. Rural. Dev.* **1995**, *7*, 36.
12. Meers, E.; Tack, F.M.G.; Tolpe, I.; Michels, E. *Application of a Full-Scale Constructed Wetland for Tertiary Treatment of Piggery Manure: Monitoring Results*; Springer: Berlin/Heidelberg, Germany, 2008. [CrossRef]
13. Devlamynck, R.; de Souza, M.F.; Michels, E.; Sigurnjak, I.; Donoso, N.; Coudron, C.; Leenknegt, J.; Vermeir, P.; Eeckhout, M.; Meers, E. Agronomic and Environmental Performance of *Lemna minor* Cultivated on Agricultural Wastewater Streams—A Practical Approach. *Sustainability* **2021**, *13*, 1570. [CrossRef]
14. Adhikari, U.; Harrigan, T.; Reinhold, D.M. Use of Duckweed-Based Constructed Wetlands for Nutrient Recovery and Pollutant Reduction from Dairy Wastewater. *Ecol. Eng.* **2015**, *78*, 6–14. [CrossRef]
15. Debusk, T.A.; Peterson, J.E.; Reddy, K.R. Use of Aquatic and Terrestrial Plants for Removing Phosphorus from Dairy Wastewaters. *Ecol. Eng.* **1995**, *5*, 371–390. [CrossRef]
16. Culley, D.D.; Epps, E.A. Use of Duckweed for Waste Treatment and Animal Feed. *Water Pollut. Control. Fed.* **1973**, *45*, 337–347.
17. Skillicorn, P.; Spira, W.; Journey, W. *Duckweed Aquaculture—A New Aquatic Farming System for Developing Countries*; The World Bank: Washington, DC, USA, 1993; ISBN 082132067X.
18. Stejskal, V.; Paolacci, S.; Toner, D.; Jansen, M.A.K. A Novel Multitrophic Concept for the Cultivation of Fish and Duckweed: A Technical Note. *J. Clean. Prod.* **2022**, *366*, 132881. [CrossRef]
19. Devlamynck, R.; de Souza, M.F.; Leenknegt, J.; Jacxsens, L.; Eeckhout, M.; Meers, E. *Lemna minor* Cultivation for Treating Swine Manure and Providing Micronutrients for Animal Feed. *Plants* **2021**, *10*, 1124. [CrossRef]
20. Verlinden, G.; Callens, D.; Demeulemester, K. *Valorisation of Effluents of Manure Processing*; Vlaamse Landmaatschappij, afdeling Mestbank: Leuven-Heverlee, Belgium, 2004.
21. Walsh, É.; Paolacci, S.; Burnell, G.; Jansen, M.A.K. The Importance of the Calcium-to-Magnesium Ratio for Phytoremediation of Dairy Industry Wastewater Using the Aquatic Plant *Lemna minor* L. *Int. J. Phytoremediation* **2020**, *22*, 694–702. [CrossRef]
22. Hoving, I.; Everts, H.; Alblas, J. *Beregenen Op Maat 1997—Toetsing Beregeningsplanner En-Wijzer in de Praktijk*; WUR: Wageningen, The Netherlands, 1997.
23. Frick, H. Micronutrient Tolerance and Accumulation in the Duckweed, Lemna. *J. Plant Nutr.* **1985**, *8*, 1131–1145. [CrossRef]
24. ISCDRA. *Newsletter for the Community of Duckweed Edited by the International Steering Committee on Duckweed Research and Applications*; ISCDRA: Rutgers, NJ, USA, 2015.
25. Vaughan, D.; Malcolm, R.E. *Soil Organic Matter and Biological Activity*; Springer: Dordrecht, The Netherlands, 1985.
26. Rauthan, B.S.; Schnitzer, M. Effects of a Soil Fulvic Acid on the Growth and Nutrient Content of Cucumber (Cucumis Sativus) Plants. *Plant Soil* **1981**, *63*, 491–495. [CrossRef]
27. Strickland, R.C.; Chaney, W.R.; Lamoreaux, R.J. *Organic Matter Influences Phytotoxicity of Cadmium to Soybeans*; Springer: Berlin/Heidelberg, Germany, 1979.
28. Liu, C.; Cooper, R. Humic Acid Application Does Not Improve Salt Tolerance of Hydroponically Grown Creeping Bentgrass. *J. Am. Soc. Hortic. Sci.* **2002**, *127*, 219–223. [CrossRef]
29. Abbas, G.; Rehman, S.; Siddiqui, M.H.; Ali, H.M.; Farooq, M.A.; Chen, Y. Potassium and Humic Acid Synergistically Increase Salt Tolerance and Nutrient Uptake in Contrasting Wheat Genotypes through Ionic Homeostasis and Activation of Antioxidant Enzymes. *Plants* **2022**, *11*, 263. [CrossRef]

30. Caicedo, J.R.; van der Steen, N.P.; Arce, O.; Gijzen, H.J. Effect of Total Ammonia Nitrogen Concentration and PH on Growth Rates of Duckweed (Spirodela Polyrrhiza). *Water Res.* **2000**, *34*, 3829–3835. [CrossRef]
31. Liu, C.; Dai, Z.; Sun, H. Potential of Duckweed (*Lemna minor*) for Removal of Nitrogen and Phosphorus from Water under Salt Stress. *J. Environ. Manag.* **2017**, *187*, 497–503. [CrossRef]
32. Lasfar, S.; Monette, F.; Millette, L.; Azzouz, A. Intrinsic Growth Rate: A New Approach to Evaluate the Effects of Temperature, Photoperiod and Phosphorus-Nitrogen Concentrations on Duckweed Growth under Controlled Eutrophication. *Water Res.* **2007**, *41*, 2333–2340. [CrossRef]
33. Van Landuyt, W. Herkenning van de Vier in België Voorkomende Drijvende Lemna-Soorten. *Dumortiera* **2007**, *91*, 16–20.
34. Devlamynck, R.; Fernandes de Souza, M.; Bog, M.; Leenknegt, J.; Eeckhout, M.; Meers, E. Effect of the Growth Medium Composition on Nitrate Accumulation in the Novel Protein Crop *Lemna minor*. *Ecotoxicol. Environ. Saf.* **2020**, *206*, 111380. [CrossRef]
35. Monette, F.; Lasfar, S.; Millette, L.; Azzouz, A. Comprehensive Modeling of Mat Density Effect on Duckweed (*Lemna minor*) Growth under Controlled Eutrophication. *Water Res.* **2006**, *40*, 2901–2910. [CrossRef]
36. Appenroth, K.J.; Teller, S.; Horn, M. Photophysiology of Turion Formation and Germination in Spirodela Polyrhiza. *Biol. Plant* **1996**, *38*, 95–106. [CrossRef]
37. VITO EMIS Navigator—VLAREM II. Available online: https://navigator.emis.vito.be/mijn-navigator?woId=74948 (accessed on 21 April 2022).
38. OECD. *OECD Guidelines for the Testing of Chemicals Revised Proposal for a New Guideline 221*; OECD: Paris, France, 2002.
39. Kufel, L.; Strzałek, M.; Wysokińska, U.; Biardzka, E.; Oknińska, S.; Ryś, K. Growth Rate of Duckweeds (Lemnaceae) in Relation to the Internal and Ambient Nutrient Concentrations—Testing the Droop and Monod Models. *Pol. J. Ecol.* **2012**, *60*, 241–249.
40. Liu, C.; Gu, W.; Dai, Z.; Li, J.; Jiang, H.; Zhang, Q. Boron Accumulation by *Lemna minor* L. under Salt Stress. *Sci. Rep.* **2018**, *8*, 8954. [CrossRef]
41. Huang, F.; Guo, Z.; Xu, Z. Determined Methods of Chlorophyll from Lemma Paucicostata. *Exp. Technol. Manag.* **2007**, *24*, 29–31.
42. Ziegler, P.; Adelmann, K.; Zimmer, S.; Schmidt, C.; Appenroth, K.J. Relative in Vitro Growth Rates of Duckweeds (Lemnaceae)—the Most Rapidly Growing Higher Plants. *Plant Biol.* **2015**, *17*, 33–41. [CrossRef]

plants

MDPI

Article

Density Dependence Influences the Efficacy of Wastewater Remediation by *Lemna minor*

Éamonn Walsh [1,2,*], Neil E. Coughlan [1,2], Seán O'Brien [1,2], Marcel A. K. Jansen [1,2] and Holger Kuehnhold [3]

1 School of Biological, Earth and Environmental Science, University College Cork, Distillery Fields, North Mall, T23N73K Cork, Ireland; neil.coughlan@ucc.ie (N.E.C.); 115449682@umail.ucc.ie (S.O.); m.jansen@ucc.ie (M.A.K.J.)
2 Environmental Research Institute, University College Cork, Lee Road, T23XE10 Cork, Ireland
3 Department of Ecology, Leibniz Centre for Tropical Marine Research (ZMT), 28359 Bremen, Germany; holger.kuehnhold@leibniz-zmt.de
* Correspondence: eamonnwalsh@umail.ucc.ie

Abstract: As part of a circular economy (CE) approach to food production systems, *Lemnaceae*, i.e., duckweed species, can be used to remediate wastewater due to rapid nutrient assimilation and tolerance of non-optimal growing conditions. Further, given rapid growth rates and high protein content, duckweed species are a valuable biomass. An important consideration for duckweed-mediated remediation is the density at which the plants grow on the surface of the wastewater, i.e., how much of the surface of the medium they cover. Higher duckweed density is known to have a negative effect on duckweed growth, which has implications for the development of duckweed-based remediation systems. In the present study, the effects of density (10–80% plant surface coverage) on *Lemna minor* growth, chlorophyll fluorescence and nutrient remediation of synthetic dairy processing wastewater were assessed in stationary (100 mL) and re-circulating non-axenic (11.7 L) remediation systems. Overall, *L. minor* growth, and TN and TP removal rates decreased as density increased. However, in the stationary system, absolute TN and TP removal were greater at higher densities (50–80% coverage). The exact cause of density related growth reduction in duckweed is unclear, especially at densities well below 100% surface coverage. A further experiment comparing duckweed grown at 'low' and 'high' density conditions with the same biomass and media volume conditions, showed that photosynthetic yield, Y(II), is reduced at high density despite the same nutrient availability at both densities, and arguably similar shading. The results demonstrate a negative effect of high density on duckweed growth and nutrient uptake, and point towards signals from neighbouring duckweed colonies as the possible cause.

Keywords: duckweed; wastewater; remediation; density; surface cover; circular economy; lemna

Citation: Walsh, É.; Coughlan, N.E.; O'Brien, S.; Jansen, M.A.K.; Kuehnhold, H. Density Dependence Influences the Efficacy of Wastewater Remediation by *Lemna minor*. *Plants* **2021**, *10*, 1366. https://doi.org/10.3390/plants10071366

Academic Editors: Viktor Oláh, Klaus-Jürgen Appenroth and K. Sowjanya Sree

Received: 25 May 2021
Accepted: 30 June 2021
Published: 3 July 2021

1. Introduction

Globally, the provision of nutritious food is a challenging endeavour [1,2]. Climate change, a reduction in the *per capita* availability of arable land, as well as soil erosion, chemical overuse and finite resources have decreased food security [3–6]. In recent years, the adoption of circular economy (CE) principles in food production systems has been suggested as a mechanism to improve resource-efficiency and the sustainability of food production [7]. In essence, CE promotes long-term retention and reuse of resources, as well as minimisation of waste generation, resulting in a reduced need for raw materials [8]. Thus, CE principles encourage the adoption of closed-loop production patterns, whereby waste is appropriated as a resource [8], reducing emissions and energy consumption in the process [9].

Dairy products are a major and important source of nutrition, employment and trade worldwide [10,11]. However, large volumes of wastewater are created as a consequence of dairy production and processing. It is estimated that up to 10 L of wastewater is created per litre of milk processed, making dairy processing waste one of the most significant waste streams in the food industry [12]. Dairy processing wastewaters tend to contain particularly high concentrations of organic matter, measured as chemical oxygen demand (COD): 2000–6000 mg L^{-1} COD [13]; 4420 mg L^{-1} COD [14]; 55,430–70,150 mg L^{-1} COD [15]. Moreover, these wastewaters generally contain high concentrations of nutrients, especially ammonium (64–270 mg L^{-1} NH_4-N), nitrate (9–30 mg L^{-1} NO_3-N) and phosphate (20–356 mg L^{-1} PO_4-P) [13,15,16]. The disposal of such wastewater often lacks value-capture in the treatment process [17]. For example, valuable nitrogen-containing nutrients such as nitrate and ammonium are commonly released as gaseous N_2 [12]. Phosphate is typically precipitated using aluminium chloride, lime and similar additives, to generate a precipitate sludge [18]. The resulting non-soluble form of phosphate has arguably limited further benefit as a fertiliser [19].

Phytoremediation has been proposed as a viable alternative to traditional wastewater treatments, as phytoremediation removes plant nutrients from wastewaters and also retains these elements in a chemical form suitable for further use [20,21]. Dairy processing wastewater is considered to be a good candidate for phytoremediation as it generally contains an abundance of essential plant nutrients, such as ammonium, nitrate and phosphate [22].

Duckweed, *Lemnaceae*, are a family of floating aquatic plants with excellent potential for phytoremediation due to a tolerance of wastewater conditions [23–25], fast growth rates [26] and high protein or starch content [27,28], as well as demonstrated use as feed, food and biofuel [29–31]. Thus, these plants can combine efficient wastewater remediation with the creation of a valuable plant biomass. To date, few studies have attempted to assess the suitability of duckweed for remediation of dairy processing wastewater. However, in principle duckweed has been shown to remediate dairy processing wastewater that lacks organic components, such as sugars and fats [32]. As a high proportion of these organic components are generally removed by existing microbial-based treatment technologies, such as sequential batch reactors or anaerobic digesters [33,34], the incorporation of duckweed into the remediation process is a realistic approach.

Wastewater remediation by duckweed is a surface process, whereby a layer of duckweed takes up nutrients from the underlying water column. In their natural habitats, most duckweed species grow in dense, floating mats [35]. Once mats have filled the available space, individual colonies begin to overlap and shade each other. Such highly crowded conditions negatively impact duckweed growth rates [36,37], and duckweed may even start to senesce and release nutrients back into the water column [38]. Conversely, a higher duckweed plant density can increase the potential for uptake of nitrogen and phosphorus [39]. Given the implications for biomass production and wastewater remediation [40], an improved understanding of the relationship between plant surface density and biomass yield, as well as net nutrient uptake is required. Earlier work has shown that a low growth rate at a high plant density does not necessarily imply a low biomass yield or low N and P removal [37]. Accordingly, to achieve effective phytoremediation, determination of optimal duckweed density for nutrient removal, plant growth and biomass yield per water surface area is required. In the present study, the effects of density on duckweed growth and remediation were quantified. This was done under axenic conditions, using stationary tanks containing either synthetic dairy processing wastewater or an optimal medium (half-strength Hutner's). Furthermore, with the aim of reproducing some of the conditions of large-scale duckweed phytoremediation systems, *L. minor* was cultivated on synthetic dairy wastewater using a larger scale non-axenic re-circulatory system. The results will inform management of duckweed-based remediation systems.

2. Materials and Methods

2.1. Stock Cultivation

The duckweed strain used in this study was *Lemna minor* L.–Blarney, strain number 5500 in the Rutgers Duckweed Stock Cooperative database [41]. A sterile stock of *L. minor* was cultivated on half-strength Hutner's medium [42] under an average light intensity of

50 µmol m^{-2} s^{-1} photosynthetically active radiation (PAR) in a controlled growth-room (22 °C, 14 h:10 h light:dark photoperiod).

2.2. Experimental Design

2.2.1. Synthetic Dairy Processing Wastewater

The synthetic dairy processing wastewater used in this study is based on the composition of real dairy processing wastewater found in dairy wastewater treatment facilities [43], with modifications as detailed in Walsh et al. [21]. The pH was reduced to, and maintained at, around 5.0 from a natural value of 8 with 1 M H_2SO_4 to facilitate optimal *L. minor* growth [44]. H_2SO_4 was chosen to decrease pH as SO_4-S has a wide 'optimal' range (0.5–20 mM) in which it does not cause adverse or beneficial effects towards duckweed, with a high maximum tolerated concentration of 60 mM, as per Walsh et al. [21].

2.2.2. Manipulation of Plant Density

In this paper, the term "plant density" is used to refer to the relative surface cover of the medium by *Lemna minor*, i.e., the proportional cover by duckweed as a fraction of the total available surface area. Plant density, i.e., surface cover, is linked to plant biomass per m^2. Plant biomass always refers to fresh duckweed biomass. Plant density was either measured directly, or estimated based on biomass per surface area. Direct density measurements were performed using the imaging software Easy Leaf Area [45] which distinguishes duckweed frond surface cover from non-duckweed covered surface area. This non-invasive technique could be used throughout the duration of an experiment. Alternatively, plant density, i.e., relative surface cover, was estimated based on biomass per m^2 of surface area. In this scenario, the latter values were calculated using a calibration curve for *L. minor* biomass versus surface area. To generate a calibration curve, a number of colonies were taken at random from a stock culture acclimated to the relevant medium. The total surface area and mass of these 'representative' colonies were measured and the area/mass ratio was calculated.

2.2.3. Stationary Remediation Experiment 1: Growth and Remediation at Variable Plant Densities

Two stationary experiments were conducted. In the first, scoping, experiment *L. minor* was grown on 100 mL of synthetic dairy wastewater for seven days (days 0–7) using a range of eight density conditions (10, 20, 30, 40, 50, 60, 70, 80% plant coverage of total surface area, $n = 6$; Figure 1). The corresponding biomass per container surface area (ranging from 21 to 154 g m^{-2}) was estimated based on a mass/area ratio of *L. minor* biomass. Plants were kept in Magenta vessels (GA-7, surface area (SA) 42.24 cm^2) in a controlled growth room (average light intensity 50 µmol m^{-2} s^{-1} PAR, 22 °C, 16 h:8 h light:dark photoperiod). To start experiments, *L. minor* colonies were taken at random from stock cultures that had been acclimated to synthetic wastewater for seven days. The range of density conditions was created by adding varying numbers of *L. minor* colonies and determining the total frond surface cover using Easy Leaf Area imaging. Plant densities were maintained at ±2% of target surface cover throughout the experiment by removing excess plant material every 2–3 days, and this process was guided by measurements of frond surface area, as determined by Easy Leaf Area. Excess plant biomass removed throughout the experiment was weighed and used to calculate specific growth rate (SGR) and relative growth rate (RGR) ($n = 6$, except for 40% where $n = 4$). Total nitrogen (TN) and total phosphorous (TP) were measured from medium samples taken on days 0 and 7 ($n = 6$, except for 40% where $n = 4$). Protein content was measured from plant samples taken on day 7 ($n = 6$, except for 40% where $n = 4$).

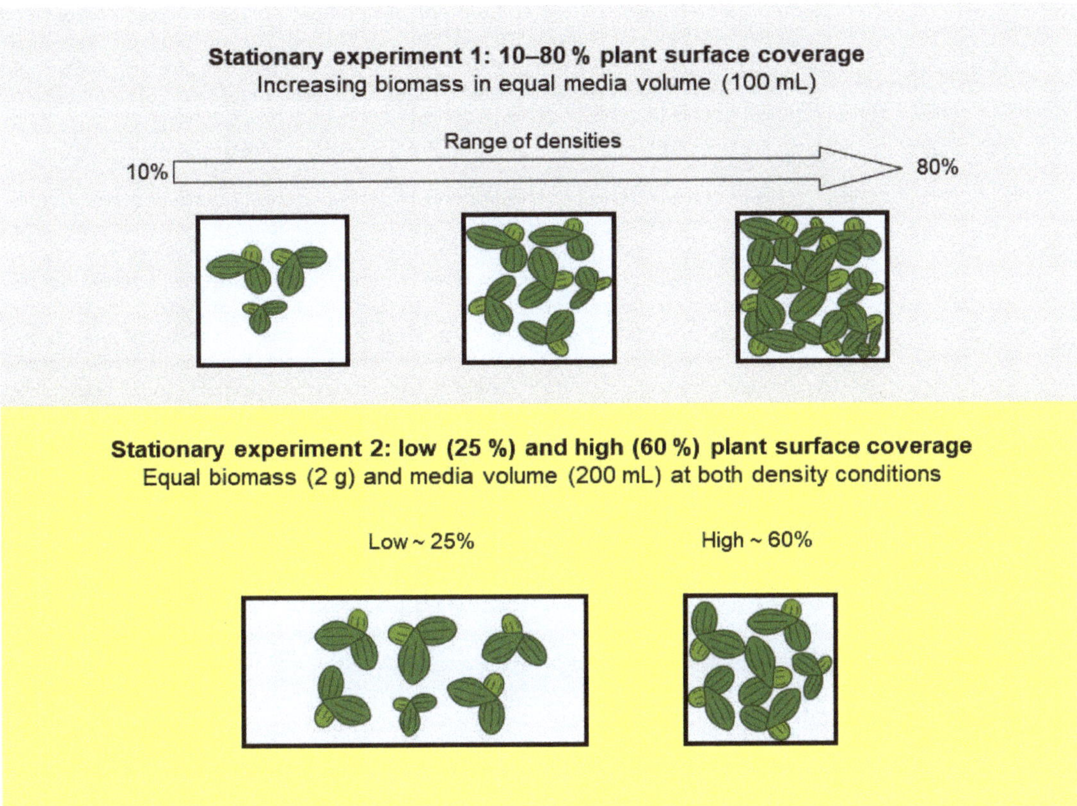

Figure 1. Set-up of stationary remediation experiments 1 and 2.

2.2.4. Stationary Remediation Experiment 2: Growth and Remediation at Low and High Density

In the second stationary experiment, the negative effect of higher duckweed density on growth was explored in greater detail. To achieve this, two density conditions were created by using two containers with different surface areas, but containing the same medium volume (200 mL) and initial plant biomass (2 g; Figure 1). For both density conditions, *L. minor* was grown for seven days (days 0–7) on either synthetic wastewater or half-strength Hutner's medium. The experiment was conducted in a controlled growth room (average light intensity 50 µmol m^{-2} s^{-1} PAR, 22 °C, 16 h:8 h light:dark photoperiod). The plant density conditions consisted of low (25%; 193 g m^{-2}) and high (60%; 476 g m^{-2}) plant coverage of the total surface area (n = 4 per experimental treatment). The % density was determined using Easy Leaf Area imaging, while the density in g m^{-2} was calculated based on the ratio of the weighted inoculum (2 g) and the container surface area. The two densities were created by using two types of growing containers (Magenta vessels with 42.25 cm^2 SA for 60% cover and larger circular glass containers with 103.87 cm^2 SA for 25% cover). In both cases, 2 g of colonies were selected randomly from stock cultures, which had already been acclimated to their respective media for seven days within the controlled growth room. Densities were maintained at ±2% of target surface cover throughout the experiment through the removal of excess plant material every 2–3 days, a process guided by measurements of frond surface area, as determined by Easy Leaf Area. Excess plant biomass removed throughout the experiment from each replicate was weighed and used to calculate SGR and RGR (n = 4). Chlorophyll *a* fluorescence measurements were taken on randomly selected plants on days 0 and 7 (n = 4). TN and TP were measured from medium samples taken on days 0 and 7 (n = 4). In both stationary experiments, any water loss due to evaporation was countered by adding deionised water to maintain original volumes.

2.2.5. Re-Circulating Remediation System: Growth and Remediation at Variable Plant Densities

To determine the effect of plant density on duckweed growth and remediation capacity under more realistic operating conditions, *L. minor* was grown in a non-axenic, re-circulating system containing 11.7 L of synthetic dairy wastewater, for five days (days 0–5), and at three densities (20, 50 and 80% plant coverage of total surface area of 600 cm^2, n = 4). Plant density was measured using Easy Leaf Area, whilst the corresponding biomass per container surface area (ranging from 50 to 187 g m^{-2}) was estimated based on a mass/area ratio of *L. minor* biomass. The experiment was conducted within a controlled environment room (300 µmol m^{-2} s^{-1} PAR, 16 h:8 h light:dark photoperiod). In this experimental system, synthetic wastewater was re-circulated between two tanks, an upper duckweed tank and a lower sump tank at a rate of 125 L per hour. The upper tanks of each replicate treatment were seeded at their respective plant surface densities on the initial day of the experiment, using stock plants acclimated to synthetic wastewater for seven days. Excess biomass grown over the course of the experiment was removed twice over the five-day experiment to maintain densities within ±2% of target surface coverage, as determined using Easy Leaf Area imaging. Excess plant biomass removed throughout the experiment from each replicate was weighed and used to calculate SGR (n = 4). TN and TP were measured from medium samples taken on days 0 and 5 (n = 4). Protein content was measured from plant samples taken on day 5 (n = 4).

2.3. Measured Parameters

2.3.1. Growth

All plant biomass was dried with absorbent tissue-paper to remove excess water and ensure reliable measurements before weight measurements. A specific growth rate (SGR), for growth comparisons within the present study, was calculated from estimations and measurements of fresh biomass using the formula [46]:

$$SGR = \frac{W_2/W_1}{\Delta T} \tag{1}$$

where W_1 is starting mass, W_2 is the increase in mass over the course of the entire experiment and ΔT is the length of the experiment. Except for stationary experiment 2, starting mass (W_1) was estimated rather than measured directly and this was guided by a calibration curve of biomass versus plant surface area.

For comparison with literature sources, a relative growth rate (RGR) was calculated from estimations and measurements of fresh biomass using the formula [47]:

$$RGR = \frac{In\frac{W_3}{W_1}}{\Delta T} \tag{2}$$

where ln is the natural log, W_1 is starting biomass, W_3 is total biomass on day 3 and ΔT is the length of time. As biomass was removed throughout each experiment to maintain a constant plant density, this formula was only used to calculate the RGR up to the first instance of removal (day 3). The total increase in mass over the course of the experiment is presented as the yield.

2.3.2. Chlorophyll *a* Fluorescence

Chlorophyll *a* fluorescence measurements were taken for randomly selected plants on days 0 and 7, using a pulse amplitude modulated fluorometer (WALZ Imaging fluorometer, Effeltrich, Germany). The procedure that was followed is detailed in Walsh et al. [32].

2.3.3. Total Nitrogen and Total Phosphorous Analysis

A sample of medium was taken for total nitrogen (TN) and total phosphorous (TP) analysis on the initial and final days of each experiment. For TN analysis, Hach test LCK138 was used with a Hach DR3900 spectrophotometer. Firstly, the sample was digested with peroxo-disulphate for one hour at 100 °C causing inorganically and organically bonded nitrogen to oxidise to nitrate (Koroleff digestion). The resulting oxidised nitrate was then analysed photometrically in a reaction with 2,6-dimethylphenol. For TP analysis, Hach test LCK348 was used. Firstly, the medium was digested using the persulphate digestion method for one hour at 100 °C. The resulting solution was then analysed photometrically through the ascorbic acid/phosphomolybdenum blue method.

2.3.4. Protein Analysis

Lemna minor samples, taken on the final day of experiments, were kept at −20 °C until used for protein extraction and analysis. Protein was extracted using 50 mM potassium phosphate buffer (pH 7, containing 0.1 mM polyvinylpyrrolidone (PVP) and 0.1 mM EDTA). Between 50–80 mg of fresh plant material was homogenised in cold potassium phosphate buffer (1 mL of buffer to 80 mg of plant sample). The homogenised sample was then centrifuged at 20,000× *g* for 30 min at 4 °C [48]. The resulting supernatant was used for protein analysis using the Bradford method with bovine serum albumin as a standard [49]. For absorbance measurements, 5 μL of sample was added to 1 mL of Bradford reagent in a cuvette and left for five minutes in dark conditions. Absorbance was measured at 595 nm using a spectrophotometer (UV-160A Shimadzu). In order to calculate the proportion of protein based on dry plant biomass, 4% dry weight content of fresh duckweed weight was used [28].

2.4. Data Analysis

Statistical analyses were conducted using R (version 3.4.3 [50]). One- and two-way ANOVAs were used to analyse differences between treatments for the measured parameters. Post hoc Tukey tests were used for pairwise comparisons of treatment groups. Normality was assessed through a graphical assessment of the distribution of the residual values for data points (i.e., histogram). Homoscedasticity was assessed with 'residuals vs. predicted values' plots as well as Fligner-Killeen and Levene's tests.

3. Results

3.1. Stationary Remediation Experiment 1: Growth and Remediation at Variable Plant Densities

The absolute plant biomass yield (g) did not significantly vary over the course of the experiment although the general trend of the average yield increased with increasing density (one-way ANOVA: $F(7) = 1.57$, $p = 0.174$; Figure 2a). SGR (d^{-1}) exhibited the opposite trend; rates decreased as density increased (one-way ANOVA: $F(7) = 8.357$, $p < 0.001$; Figure 2b). The overall removal of TN (mg) from synthetic dairy wastewater increased as plant density increased (one-way ANOVA: $F(7) = 2.574$, $p < 0.05$; Figure 2c). However, when TN removal was expressed per frond surface area (mg N m^{-2} day^{-1}), the rate decreased as density increased (one-way ANOVA: $F(7) = 9.287$, $p < 0.001$; Figure 2d). A similar pattern was found for TP removal in which the overall removal of TP (mg) from synthetic dairy wastewater increased as plant density increased (one-way ANOVA: $F(7) = 5.11$, $p < 0.001$; Figure 2e). While the TP removal rate per frond surface area (mg P m^{-2} day^{-1}) decreased as density increased (one-way ANOVA: $F(7) = 8.158$, $p < 0.001$; Figure 2f). There was no difference in protein content (% dry duckweed mass) detected in relation to plant density (one-way ANOVA: $F(7) = 0.334$, $p = 0.933$; Figure 2g).

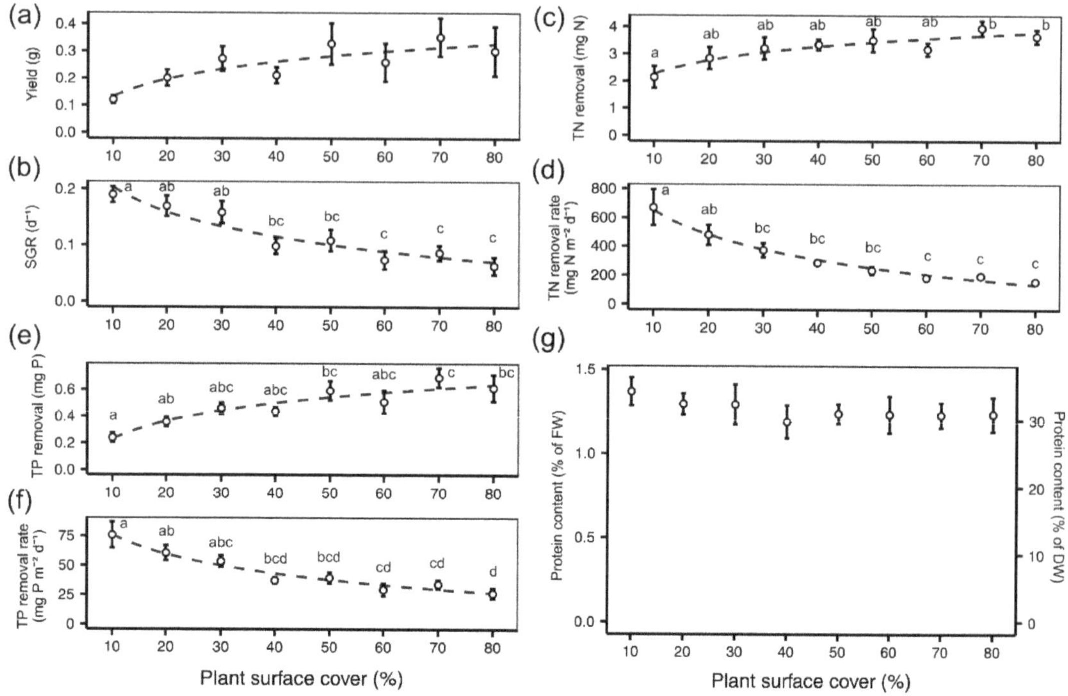

Figure 2. Mean (±SE) values with natural log trendline of (**a**) yield (g), (**b**) specific growth rate (SGR, d^{-1}), (**c**) total nitrogen (TN) removal (mg N), (**d**) TN removal rate per frond surface area (mg N m^{-2} d^{-1}), (**e**) total phosphorous (TP) removal (mg P), (**f**) TP removal rate per frond surface area (mg P m^{-2} d^{-1}) and (**g**) % protein content based on fresh duckweed biomass, FW, and dry duckweed biomass, DW, for *L. minor* grown on synthetic wastewater over 7 days under a range of plant surface covers (10–80%). Points that do not share the same letter significantly differ from one another, as per the Tukey post hoc test, $p < 0.05$.

3.2. Stationary Remediation Experiment 2: Growth and Remediation at Low and High Density

On both half-strength Hutner's and synthetic wastewater *L. minor* grown at a lower density (25% plant surface coverage) displayed a higher absolute yield (two-way ANOVA: $F(1) = 46.607$, $p < 0.001$; Figure 3a) than plants grown at the higher density condition (60% plant surface coverage). The same was found for SGR (two-way ANOVA: $F(1) = 45.994$, $p < 0.001$; Figure 3b). TN removal (mg) was not significantly affected by density (two-way ANOVA: $F(1) = 3.642$, $p = 0.0805$; Figure 3c), nor was TN removal rate per frond area (mg N m^{-2} d^{-1}) (two-way ANOVA: $F(1) = 3.154$, $p = 0.101$; Figure 3d), although average values were lower at the higher density. Density condition did not significantly affect TP removal (mg) (two-way ANOVA: $F(1) = 2.592$, $p = 0.136$; Figure 3e) or TP removal rate per frond area (mg P m^{-2} d^{-1}) (two-way ANOVA: $F(1) = 2.293$, $p = 0.158$; Figure 3f).

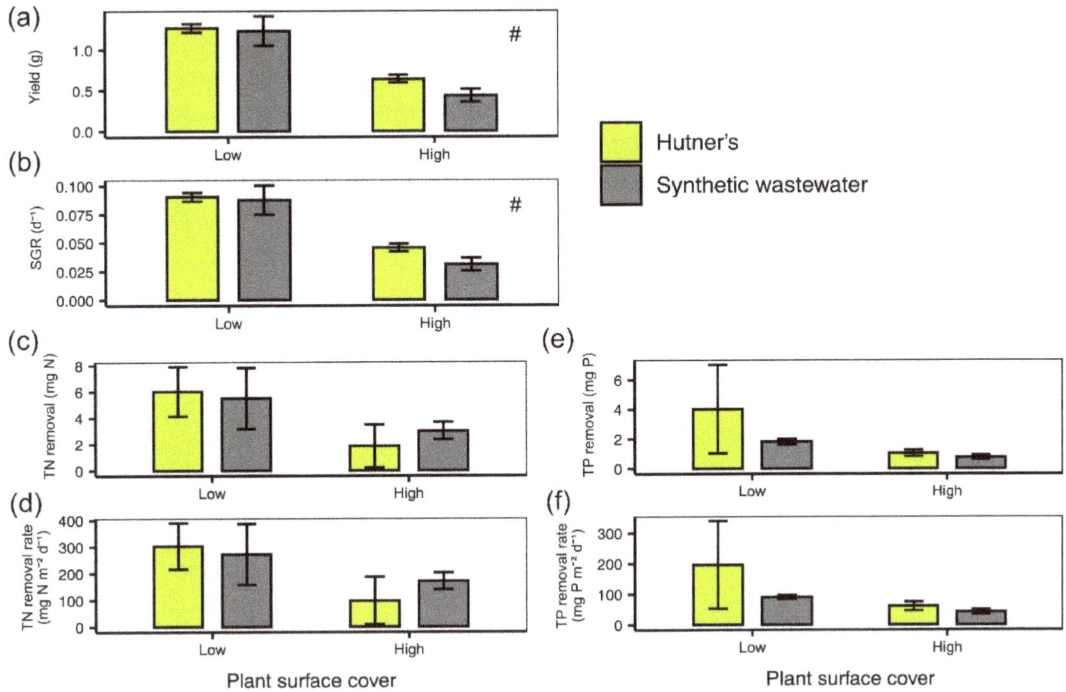

Figure 3. Mean (±SE) (**a**) yield (g), (**b**) SGR (d^{-1}), (**c**) TN removal (mg N), (**d**) TN removal rate per frond surface area (mg N m^{-2} d^{-1}), (**e**) TP removal (mg P) and (**f**) TP removal rate per frond surface area (mg P m^{-2} d^{-1}), for *L. minor* grown on synthetic wastewater or half-strength Hutner's medium at two plant surface covers (low, 25% and high, 60%) over 7 days. A hash symbol (#) denotes an effect of density for $p < 0.05$, as per the two-way ANOVA.

3.3. Chlorophyll a Fluorescence

Chlorophyll *a* fluorescence measurements were taken on the initial day of stationary experiment 2 (day 0, data not shown). They showed that plants grown in both half-strength Hutner's and synthetic wastewater displayed similar values for a range of chlorophyll fluorescence parameters: F_v/F_m, Y(II), Y(NPQ) and Y(NO) (one-way ANOVAs across all treatments: F(1) = 1.006, 0, 0.049, 0.076, p = 0.354, 0.986, 0.833, 0.792, respectively). Chlorophyll fluorescence measurements taken on the final day (day 7) of the experiment revealed some differences between treatments. Similar to day 0 values, mean F_v/F_m stayed largely constant between 0.68 and 0.8, and was not significantly affected by density (two-way ANOVA: F(1) = 2.038, p = 0.179; Figure 4a) or medium (two-way ANOVA: F(1) = 0.911, p = 0.359; Figure 4a). However, measurements taken on day seven showed that higher plant density in both media resulted in a lower Y(II) (two-way ANOVA: F(1) = 34.054, $p < 0.001$; Figure 4b). This effect was strongest in plants grown on synthetic wastewater (post hoc Tukey test low:high density in synthetic wastewater: $p < 0.001$; Figure 4b). Medium alone did not significantly affect Y(II) (two-way ANOVA: F(1) = 0.907, p = 0.360; Figure 4b). There was, however, a significant interaction between density and medium (two-way ANOVA interaction 'density*medium': F(1) = 7.369, $p < 0.05$; Figure 4b). Y(NPQ) was not significantly affected by density (two-way ANOVA: F(1) = 0.594, p = 0.456; Figure 4c) or medium (two-way ANOVA: F(1) = 1.953, p = 0.188; Figure 4c). Nor was Y(NO) significantly affected by density (two-way ANOVA: F(1) = 2.016, p = 0.181; Figure 4d) or medium (two-way ANOVA: F(1) = 2.631, p = 0.131; Figure 4d).

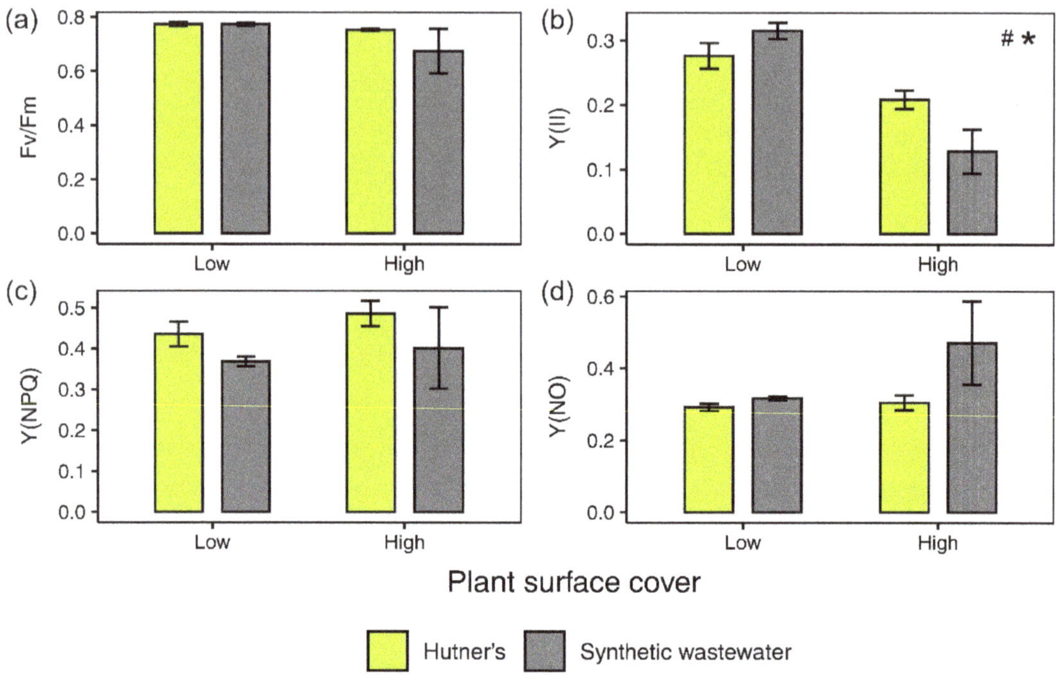

Figure 4. Mean (±SE) (**a**) F_v/F_m, (**b**) Y(II), (**c**) Y(NPQ) and (**d**) Y(NO), for *L. minor* grown on either synthetic wastewater or half-strength Hutner's medium at two plant surface covers (low, 25% and high, 60%) over 7 days. A hash symbol (#) denotes an effect of density for $p < 0.05$, and a star symbol (*) denotes an interactive effect between density and medium for $p < 0.05$, as per the two-way ANOVA.

3.4. Re-Circulating Remediation System: Growth and Remediation at Variable Plant Densities

The yield of *L. minor* in the re-circulating experiment was not significantly affected by the plant density (one-way ANOVA: F(2) = 3.238, $p = 0.087$; Figure 5a). However, the average biomass yield was lowest at the 20% plant density and increased to a plateau at 50% (post hoc Tukey: $p = 0.076$; Figure 5a). SGR (d^{-1}) steadily decreased with increasing density (one-way ANOVA: F(2) = 5.143, $p = 0.032$; post hoc Tukey 20–80: $p < 0.05$; Figure 5b). Overall, TN removal (mg) from synthetic wastewater was not significantly impacted by density (one-way ANOVA: F(2) = 0.698, $p = 0.525$; Figure 5c). However, TN removal rate (mg N m^{-2} day^{-1}) decreased significantly as density increased (one-way ANOVA: F(2) = 5.701, $p < 0.05$; Figure 5d). TN removal rate values dropped from around 2500 mg N m^{-2} day^{-1} at 20% plant surface cover to around 500 mg N m^{-2} day^{-1} at 80% (post hoc Tukey: $p < 0.05$; Figure 5d). Overall, mean TP removal remained at around 17 mg across the three plant densities (one-way ANOVA: F(2) = 0.419, $p = 0.670$; Figure 5e). TP removal rate (mg P m^{-2} day^{-1}) decreased with increasing density conditions (one-way ANOVA: F(2) = 29.240, $p < 0.001$; Figure 5f). Mean TP removal rate dropped from 300 mg P m^{-2} day^{-1} at 20% to 75 mg P m^{-2} day^{-1} at 80% (post hoc Tukey: $p < 0.001$; Figure 5f). Protein content (% protein of dry duckweed mass) did not vary between plant densities, with mean values remaining between 17–20% (one-way ANOVA: F(2) = 0.225, $p = 0.803$; Figure 5g).

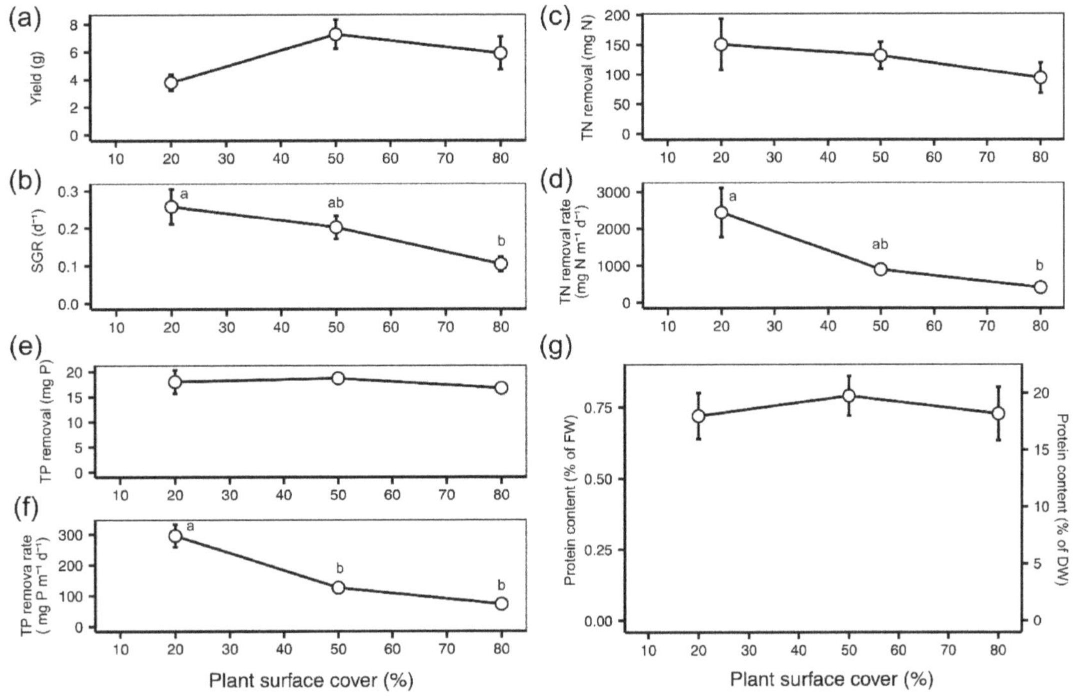

Figure 5. Mean (±SE) (**a**) yield (g), (**b**) SGR (d^{-1}), (**c**) TN removal (mg N), (**d**) TN removal rate per frond surface area (mg N m^{-2} d^{-1}), (**e**) TP removal (mg P), (**f**) TP removal rate per frond surface area (mg P m^{-2} d^{-1}) and (**g**) % protein content based on fresh duckweed biomass, FW, and dry duckweed biomass, DW, for *L. minor* grown on synthetic wastewater in a re-circulating system at three plant surface covers (20, 50 and 80 %) over 5 days. Points that do not share the same letter significantly differ from one another, as per the Tukey post hoc test, $p < 0.05$.

4. Discussion

4.1. L. minor Growth on Synthetic Wastewater

RGR values recorded in the stationary experiments (based on the first 3-days of growth; 0.03–0.15 d^{-1}) are at the lower end of the range found in the literature for duckweed grown on wastewater (0.04–0.3 d^{-1}) [24,51,52]. They are also lower than those found for duckweed grown on an optimised medium (0.153–0.519 d^{-1}) [26]. This may, in part, be due to the composition of the medium, as duckweed cultivated on wastewater often results in lower growth rates than obtained on optimised growing medium (compare growth rates in Ziegler et al. [26] with Al-Nozaily et al. [51]). As well as this, the relatively low light intensity of 50 μmol m^{-2} s^{-1} used for the stationary experiments may have also contributed to reduced growth rates [53]. Nevertheless, it has been shown that the light saturation point is around 50 μmol m^{-2} s^{-1} for *L. minor* grown on synthetic dairy wastewater [32], although the light saturation point for duckweed on more optimised media is 400–600 μmol m^{-2} s^{-1} [44,54]. Commonly used experimental light intensities range from 85 to 130 μmol m^{-2} s^{-1} [26,52,55], which are in line with OECD guidelines for duckweed toxicity growth inhibition tests [56]. Although, a higher light intensity of 300 μmol m^{-2} s^{-1} was used in the re-circulating experiment this only led to marginally improved growth rates. In both the stationary and re-circulating experiments, SGR decreased with increasing density, which corresponds with general trends noted within the literature [36,37].

4.2. Exploring the Mechanism Underlying Density Dependent Changes in Growth Using Chlorophyll Fluorometry

The negative effect of high density on growth may be a result of greater competition for nutrients [57], a quicker depletion of nutrients [58], or self-shading between colonies [59]. Yet, plant-plant competition has also been related to plant neighbour detection, including plant responses such as shade avoidance, root foraging, and use/induction of chemical defences [60,61]. Therefore, it is possible that *L. minor* senses the closeness of other plants and switches to a more defensive growth strategy, the trade-off of which is a reduction in growth rate. To explore the mechanism underlying the observed decrease in SGR with increasing plant surface density, plants were grown at lower and higher plant surface densities, but with identical biomass per medium volume (stationary experiment 2). This experiment confirmed the impediment of growth rate at higher plant surface densities. However, as plants at each surface density had access to the same volume of medium, the data imply that the growth impediment was due to factors other than nutrient depletion. Indeed, this point is further confirmed by the observation that similar growth impediments occurred at high plant density irrespective of whether nutrient rich Hutner's medium or more oligotrophic dairy wastewater was used.

Analysis of photosynthetic parameters in stationary experiment 2, measured after seven days of growth, showed that at high density the photosynthetic quantum yield of PSII, Y(II), was significantly depressed. A reduction in Y(II) for *L. minor* grown at high density means that the plants were using light energy less efficiently compared to those at a lower plant surface density [62]; an effect which was stronger for plants grown on synthetic wastewater. Y(II), together with Y(NPQ) and Y(NO) account for the partitioning of absorbed light energy in PSII [63], the sum of which equals 1 [64]. Accordingly, a reduction in Y(II) implies a concurrent increase in Y(NPQ) and/or Y(NO). The data reveal non-significant increases in Y(NPQ) at higher plant densities for both media, indicating minor increases in the amount of light energy dissipated in a regulated manner, i.e., through thermal dissipation [63], at these higher densities. Thus, a key finding is a density-dependent decrease in Y(II) which is not matched by clear significant parallel increases in Y(NO) and Y(NPQ), nor a clear effect on F_v/F_m.

Overall, the data indicate that duckweed density can affect aspects of the plant's metabolism (e.g., carbon assimilation or nitrogen metabolism), rather than having a direct effect on PSII activity (i.e., F_v/F_m). As such, this would indirectly reduce photosynthetic yield (Y(II)) and therefore biomass growth [65]. Previously, Kufel et al. [66] described complex changes in plant morphology in *L. minor* grown at different plant densities. While Zhang et al. [67] described morphological responses of *Spirodela polyrhiza* to population density that included decreased frond and root size, as well as increased frond thickness. It is possible that shading is a driver of these changes and induces a shade avoidance response in plants at higher densities [60,61]. Nevertheless, acclimation to shade typically results in increased Y(II) and decreased Y(NPQ) at low measuring light intensities [68,69]. However, the chlorophyll fluorescence data in this study show decreased Y(II) at higher plant densities where shading might potentially have been an issue. Therefore, neither a lack of nutrient nor light supply, two well-advocated explanations, adequately explain the high-density induced impediment of growth. Rather, the data point to plant neighbour detection between *Lemna*-colonies as the most likely explanation of this [60,61]. At present, touch, volatile organic compounds, chemical exudates and possibly even acoustic signals, have all been associated with neighbour detection [70]. Although it is not known to what extent these apply to duckweed, which tend to produce dense mats in natural habitats [35]. Jang et al. [71] reported that *Lemna japonica* may release interfering chemicals via its root systems, although the effects were interspecific. Similarly, Bich and Kato-Noguchi [72] reported on interspecific allelopathic signals from *Lemna minor*. It is less clear how allelopathic signals manifest between neighbouring duckweed colonies of the same species or clone. In some duckweed species high density, or 'overcrowding', has been associated with the production of ethylene, a possible early signal for an increasing lack of space which can inhibit growth [73]. Ethylene has been shown in other plant species to an inhibitor of plant growth [74]. Further studies have shown that the overcrowding-stimulated production of ethylene in duckweed is a Ca^{2+} and phytochrome-dependent process [75,76]. A transient increase in cytoplasmic Ca^{2+} is followed by an increase in ethylene production [75]. Nevertheless, further exploration is required to understand how plant signals, such as ethylene and other potentially unknown signals, influence *L. minor* growth and metabolism in high density conditions.

4.3. *L. minor* Biomass Yield and Protein Content under Variable Density and System Conditions

Given its high protein content [29,77], and usefulness as a biofuel and a source of phytochemicals [28,78], *Lemna minor* biomass is an important by-product of the wastewater remediation process. As such, the absolute biomass yield is an important parameter, as it relates directly to the amount of plant mass available for further use [27]. Previously, high duckweed densities of 60–80% plant coverage (around 160–280 g m^{-2}) have been reported to result in maximum yields, in combination with different harvesting regimes [39,79,80]. Data from stationary experiment 1 and the re-circulating experiment show that at higher plant densities the logarithmic relationship between plant density and yield will plateau. Consequently, the yield increment becomes smaller. However, significant differences between density treatments were not found. Therefore, a clear benefit of higher plant density for biomass yield was not detected. The second stationary experiment shows the opposite trend between density and yield. In this experiment the use of the same plant biomass, but with different container surface areas, led to both a higher growth rate and overall yield at low density. Thus, the use of a shallower/wider container for a volume of wastewater would improve duckweed yield, i.e., surface area space is an important factor in duckweed yield. However, consideration would have to be given to the impact of algae on this result, as less duckweed cover tends to increase the light availability to algae [81].

Duckweed density did not affect protein content in the stationary or re-circulating systems. Nevertheless, the protein content of 30–35% on a dry weight basis found in this study compares favourably with literature sources, where duckweed protein contents up to 45% of dry weight have been reported [44]. Although, more commonly reported values are between 20–35% [28,82]. Furthermore, this compares relatively well with the commonly used high-protein feed, soybean (33–49% [83]). It should be noted that the use of the Bradford assay with BSA as a standard can underestimate plant protein content when compared to other techniques [84,85], which may affect comparisons with literature sources.

The protein content detected for plants grown in the re-circulating system ranged from 17.5–20% of dry weight, which can be considered a low protein content relative to the published range. This demonstrates that scaling up, and the use of circulatory, non-axenic systems may have unexpected consequences. Starch content was not measured in this study, but, as has been observed in some studies, a lower protein content can result in, or be a result of, higher starch content [27]. If low protein content is a consistent problem for duckweed grown on a large-scale, there are established alternative uses for the biomass that do not depend on the protein content, such as biofuel production [86].

4.4. Remediation of TN and TP by *L. minor* from Synthetic Wastewater

The trend of decreasing growth with increasing density was reflected in the relationship between TN/TP removal rate and density. In both stationary and re-circulating experiments, the lower removal rates of TN and TP per plant surface area (g m^{-2} d^{-1}) at higher density conditions show that each duckweed colony is removing less nutrients at higher plant densities than those kept at lower densities. The TN and TP removal rates found in the stationary experiments were on the lower end of the wide range of values recorded in the literature: 124–4400 mg N m^{-2} d^{-1} and 14–590 mg P m^{-2} d^{-1} [57,82,87–90]. This can be explained by the lower growth rates observed in these experiments [91], which are likely to be in part due to the low light intensity used [32], the specific medium [23], as well as density effects on *L. minor* discussed previously. Both TN and TP removal rates from the non-axenic, re-circulating experiment compare well with values from non-axenic systems in literature sources [57,91]. Higher removal rates found in the re-circulating experiment compared to stationary may be mostly explained by the presence of algae and microorganisms, which can account for up to 50–70% of nutrient removal in non-axenic systems [57,88], but also by improved mixing of the medium.

Another important criterium for analysing the effectiveness of a remediation system is the absolute amount of nutrients removed from the remediated wastewater. In stationary experiment 1, a higher density of plants per surface area more effectively takes up nutrients, although the growth and nutrient removal rates per frond are significantly lower [36,37]. As the removal of nutrients started to plateau at higher densities of 50–80%, this suggests that 50–80% represents an optimal duckweed density for dairy wastewater phytoremediation.

There were additional issues encountered in a non-axenic re-circulating system, in which the absolute removal of TN and TP was relatively flat across the three densities. The exact cause of this is uncertain. However, microorganisms were present in substantial amounts at all plant densities in this non-axenic system. Previously, algae were shown to significantly contribute to nutrient removal [57,81], as well as having negative effects on duckweed such as increased decomposition rates [92]. At both high and low density, a high proportion of the absolute removal of TN and TP may be attributed to microorganisms, breaking the direct link between plant growth and nutrient removal (e.g., poor plant growth, but high nutrient removal). Considering wastewater remediation alone, whether remediation is fuelled primarily by duckweed or algae, nutrient removal is the desired outcome. However, if duckweed biomass is to be generated as a valuable by-product for further use, then strong competition for nutrients with non-utilised microorganisms is undesirable. Ideally, the duckweed should be taking up the majority of the nutrients. One way to achieve this is by maximising the duckweed surface density, thereby decreasing algal growth [93]. Overall, these results show that upscaling from laboratory-based, stationary systems to larger scale, recirculatory systems is complex, and that simple extrapolations are not necessarily correct. Accordingly, management of duckweed incubators will need to be informed by the effects of plant density on biomass yield, TN and TP removal, as well as competition with algal species.

5. Conclusions

Lemna minor can be successfully grown on synthetic dairy processing wastewater, opening the perspective to both remediate and valorise such waste, in accordance with the principles of the circular economy. *Lemna minor* has been shown to produce the best remediation at higher densities (50–80%), even though growth rates and nutrient uptake rates were slowest at these densities. The decrease in growth at high density was linked to a decrease in photosynthetic yield, rather than competition for light or nutrients, which points towards signals from neighbouring colonies as the potential cause of growth restrictions. However, in non-axenic, scaled-up conditions that better reflect an industrial duckweed-based remediation system, the benefits of high density were not as clear. High algal presence led to suppressed duckweed yield and static nutrient removal. Thus, despite the suitability of *L. minor* for valorisation of dairy processing waste, management of wastewater is subject to both interactions between plant density, yield and nutrient removal, as well as complex upscaling effects.

Author Contributions: É.W., H.K. and M.A.K.J. contributed to the study conception and design. Material preparation, data collection and analysis were performed by É.W., H.K. and S.O.; É.W., N.E.C., S.O., M.A.K.J. and H.K. contributed to the interpretation of results. The first draft of the manuscript was written by É.W.; É.W., N.E.C., S.O., M.A.K.J. and H.K. contributed to writing and editing the manuscript. All authors have read and agreed to the published version of the manuscript.

Funding: This project was funded under the EPA Research Programme 2014–2020 (grant 2016-W-LS-11). The EPA Research Programme is a Government of Ireland initiative funded by the Department of Communications, Climate Action and Environment. It is administered by the Environmental Protection Agency, which has the statutory function of co-ordinating and promoting environmental research.

Institutional Review Board Statement: Not applicable.

Informed Consent Statement: Not applicable.

Data Availability Statement: The datasets used in the current study are available from the corresponding author on reasonable request.

Acknowledgments: MAKJ acknowledges support by WoB.

Conflicts of Interest: The authors have no relevant financial or non-financial interests to disclose.

References

1. Mehrabi, Z.; Ellis, E.C.; Ramankutty, N. The challenge of feeding the world while conserving half the planet. *Nat. Sustain.* **2018**, *1*, 409–412. [CrossRef]
2. Ranganathan, J.; Vennard, D.; Waite, R.; Dumas, P.; Lipinski, B.; Searchinger, T. *Shifting Diets for a Sustainable Food Future*; World Resources Institute: Washington, DC, USA, 2016.
3. Nearing, M.A.; Xie, Y.; Liu, B.; Ye, Y. Natural and anthropogenic rates of soil erosion. *Int. Soil Water Conserv. Res.* **2017**, *5*, 77–84. [CrossRef]
4. Rosa, L.; Chiarelli, D.D.; Rulli, M.C.; Dell'Angelo, J.; D'Odorico, P. Global agricultural economic water scarcity. *Sci. Adv.* **2020**, *6*, eaaz6031. [CrossRef] [PubMed]
5. Pathak, T.B.; Maskey, M.L.; Dahlberg, J.A.; Kearns, F.; Bali, K.M.; Zaccaria, D. Climate Change Trends and Impacts on California Agriculture: A Detailed Review. *Agronomy* **2018**, *8*, 25. [CrossRef]
6. Obersteiner, M.; Peñuelas, J.; Ciais, P.; van der Velde, M.; Janssens, I.A. The phosphorus trilemma. *Nat. Geosci.* **2013**, *6*, 897–898. [CrossRef]
7. Del Borghi, A.; Moreschi, L.; Gallo, M. Circular economy approach to reduce water–energy–food nexus. *Curr. Opin. Environ. Sci. Health* **2020**, *13*, 23–28. [CrossRef]
8. Ghisellini, P.; Cialani, C.; Ulgiati, S. A review on circular economy: The expected transition to a balanced interplay of environmental and economic systems. *J. Clean. Prod.* **2016**, *114*, 11–32. [CrossRef]
9. Xue, Y.; Luan, W.; Wang, H.; Yang, Y. Environmental and economic benefits of carbon emission reduction in animal husbandry via the circular economy: Case study of pig farming in Liaoning, China. *J. Clean. Prod.* **2019**, *238*, 117968. [CrossRef]
10. Gaucheron, F. Milk and Dairy Products: A Unique Micronutrient Combination. *J. Am. Coll. Nutr.* **2011**, *30*, 400S–409S. [CrossRef]
11. OECD-FAO. Dairy and Dairy Products. In *OECD-FAO Agricultural Outlook 2020-2029*; OECD-FAO Agricultural Outlook; OECD: Rome, Italy, 2020; ISBN 9789264317673.
12. Wang, Y.; Serventi, L. Sustainability of dairy and soy processing: A review on wastewater recycling. *J. Clean. Prod.* **2019**, *237*, 117821. [CrossRef]
13. Ince, O. Performance of a two-phase anaerobic digestion system when treating dairy wastewater. *Water Res.* **1998**, *32*, 2707–2713. [CrossRef]
14. Demirel, B.; Yenigun, O. Anaerobic acidogenesis of dairy wastewater: The effects of variations in hydraulic retention time with no pH control. *J. Chem. Technol. Biotechnol.* **2004**, *79*, 755–760. [CrossRef]
15. Malaspina, F.; Cellamare, C.M.; Stante, L.; Tilche, A. Anaerobic treatment of cheese whey with a downflow-upflow hybrid reactor. *Bioresour. Technol.* **1996**, *55*, 131–139. [CrossRef]
16. Ghaly, A.E.; Singh, R.K. Pollution potential reduction of cheese whey through yeast fermentation. *Appl. Biochem. Biotechnol.* **1989**, *22*, 181–203. [CrossRef]
17. Ahmad, T.; Aadil, R.M.; Ahmed, H.; ur Rahman, U.; Soares, B.C.V.; Souza, S.L.Q.; Pimentel, T.C.; Scudino, H.; Guimarães, J.T.; Esmerino, E.A.; et al. Treatment and utilization of dairy industrial waste: A review. *Trends Food Sci. Technol.* **2019**, *88*, 361–372. [CrossRef]
18. Bunce, J.T.; Ndam, E.; Ofiteru, I.D.; Moore, A.; Graham, D.W. A Review of Phosphorus Removal Technologies and Their Applicability to Small-Scale Domestic Wastewater Treatment Systems. *Front. Environ. Sci.* **2018**, *6*, 8. [CrossRef]
19. López-Mosquera, M.E.; Moirón, C.; Carral, E. Use of dairy-industry sludge as fertiliser for grasslands in northwest Spain: Heavy metal levels in the soil and plants. *Resour. Conserv. Recycl.* **2000**, *30*, 95–109. [CrossRef]
20. Akansha, J.; Nidheesh, P.V.; Gopinath, A.; Anupama, K.V.; Kumar, M.S. Treatment of dairy industry wastewater by combined aerated electrocoagulation and phytoremediation process. *Chemosphere* **2020**, *253*, 126652. [CrossRef] [PubMed]
21. Walsh, É.; Paolacci, S.; Burnell, G.; Jansen, M.A.K. The importance of the calcium-to-magnesium ratio for phytoremediation of dairy industry wastewater using the aquatic plant *Lemna minor* L. *Int. J. Phytoremediat.* **2020**, *22*, 694–702. [CrossRef] [PubMed]
22. Carvalho, F.; Prazeres, A.R.; Rivas, J. Cheese whey wastewater: Characterization and treatment. *Sci. Total Environ.* **2013**, *445*, 385–396. [CrossRef]

23. Toyama, T.; Hanaoka, T.; Tanaka, Y.; Morikawa, M.; Mori, K. Comprehensive evaluation of nitrogen removal rate and biomass, ethanol, and methane production yields by combination of four major duckweeds and three types of wastewater effluent. *Bioresour. Technol.* **2018**, *250*, 464–473. [CrossRef] [PubMed]
24. Dinh, T.T.U.; Soda, S.; Nguyen, T.A.H.; Nakajima, J.; Cao, T.H. Nutrient removal by duckweed from anaerobically treated swine wastewater in lab-scale stabilization ponds in Vietnam. *Sci. Total Environ.* **2020**, *722*, 137854. [CrossRef]
25. Verma, R.; Suthar, S. Synchronized urban wastewater treatment and biomass production using duckweed *Lemna gibba* L. *Ecol. Eng.* **2014**, *64*, 337–343. [CrossRef]
26. Ziegler, P.; Adelmann, K.; Zimmer, S.; Schmidt, C.; Appenroth, K.J. Relative in vitro growth rates of duckweeds (*Lemnaceae*)—The most rapidly growing higher plants. *Plant Biol.* **2015**, *17*, 33–41. [CrossRef]
27. Xu, J.; Cui, W.; Cheng, J.J.; Stomp, A.-M. Production of high-starch duckweed and its conversion to bioethanol. *Biosyst. Eng.* **2011**, *110*, 67–72. [CrossRef]
28. Appenroth, K.J.; Sree, K.S.; Böhm, V.; Hammann, S.; Vetter, W.; Leiterer, M.; Jahreis, G. Nutritional value of duckweeds (*Lemnaceae*) as human food. *Food Chem.* **2017**, *217*, 266–273. [CrossRef] [PubMed]
29. Cheng, J.J.; Stomp, A.-M. Growing Duckweed to Recover Nutrients from Wastewaters and for Production of Fuel Ethanol and Animal Feed. *Clean Soil Air Water* **2009**, *37*, 17–26. [CrossRef]
30. de Beukelaar, M.F.A.; Zeinstra, G.G.; Mes, J.J.; Fischer, A.R.H. Duckweed as human food. The influence of meal context and information on duckweed acceptability of Dutch consumers. *Food Qual. Prefer.* **2019**, *71*, 76–86. [CrossRef]
31. Stadtlander, T.; Förster, S.; Rosskothen, D.; Leiber, F. Slurry-grown duckweed (*Spirodela polyrhiza*) as a means to recycle nitrogen into feed for rainbow trout fry. *J. Clean. Prod.* **2019**, *228*, 86–93. [CrossRef]
32. Walsh, É.; Kuehnhold, H.; O'Brien, S.; Coughlan, N.E.; Jansen, M.A.K. Light intensity alters the phytoremediation potential of *Lemna minor*. *Environ. Sci. Pollut. Res.* **2021**. [CrossRef] [PubMed]
33. Kushwaha, J.P.; Srivastava, V.C.; Mall, I.D. Sequential batch reactor for dairy wastewater treatment: Parametric optimization; kinetics and waste sludge disposal. *J. Environ. Chem. Eng.* **2013**, *1*, 1036–1043. [CrossRef]
34. Charalambous, P.; Shin, J.; Shin, S.G.; Vyrides, I. Anaerobic digestion of industrial dairy wastewater and cheese whey: Performance of internal circulation bioreactor and laboratory batch test at pH 5–6. *Renew. Energy* **2020**, *147*, 1–10. [CrossRef]
35. Landolt, E. *Biosystematic Investigations in the Family of Duckweeds (Lemnaceae). The Family of Lemnaceae—A Monographic Study, Volume 1*; Veröffentlichungen des Geobotanischen Institutes der ETH: Zürich, Switzerland, 1986.
36. Driever, S.M.; van Nes, E.H.; Roijackers, R.M.M. Growth limitation of *Lemna minor* due to high plant density. *Aquat. Bot.* **2005**, *81*, 245–251. [CrossRef]
37. Frédéric, M.; Samir, L.; Louise, M.; Abdelkrim, A. Comprehensive modeling of mat density effect on duckweed (*Lemna minor*) growth under controlled eutrophication. *Water Res.* **2006**, *40*, 2901–2910. [CrossRef] [PubMed]
38. Ceschin, S.; Sgambato, V.; Ellwood, N.T.W.; Zuccarello, V. Phytoremediation performance of *Lemna* communities in a constructed wetland system for wastewater treatment. *Environ. Exp. Bot.* **2019**, *162*, 67–71. [CrossRef]
39. Xu, J.; Shen, G. Growing duckweed in swine wastewater for nutrient recovery and biomass production. *Bioresour. Technol.* **2011**, *102*, 848–853. [CrossRef]
40. Xu, J.; Zhao, H.; Stomp, A.-M.; Cheng, J.J. The production of duckweed as a source of biofuels. *Biofuels* **2012**, *3*, 589–601. [CrossRef]
41. Lahive, E.; O'Halloran, J.; Jansen, M.A.K. Differential sensitivity of four *Lemnaceae* species to zinc sulphate. *Environ. Exp. Bot.* **2011**, *71*, 25–33. [CrossRef]
42. Hutner, S.H. Comparative physiology of heterotrophic growth in higher plants. In *Growth and Differentiation in Plants*; Iowa State College Press: Ames, IA, USA, 1953; pp. 417–447.
43. Tarpey, E. An Investigation into the Use of IASBRs for Treatment of Dairy Processing Wastewater. Master' Thesis, National University of Ireland, Galway, Ireland, 2016.
44. Landolt, E.; Kandeler, R. *Biosystematic Investigations in the Family of Duckweeds (Lemnaceae). The Family of Lemnaceae—A Monographic Study, Volume 2*; Veröffentlichungen des Geobotanischen Institutes der ETH: Zürich, Switzerland, 1987.
45. Easlon, H.M.; Bloom, A.J. Easy Leaf Area: Automated digital image analysis for rapid and accurate measurement of leaf area. *Appl. Plant Sci.* **2014**, *2*, 1400033. [CrossRef]
46. Chaiprapat, S.; Cheng, J.J.; Classen, J.J.; Liehr, S.K. Role of internal nutrient storage in duckweed growth for swine wastewater treatment. *Trans. ASAE* **2005**, *48*, 2247–2258. [CrossRef]
47. Connolly, J.; Wayne, P. Asymmetric competition between plant species. *Oecologia* **1996**, *108*, 311–320. [CrossRef] [PubMed]
48. Balen, B.; Tkalec, M.; Šikić, S.; Tolić, S.; Cvjetko, P.; Pavlica, M.; Vidaković-Cifrek, Ž. Biochemical responses of *Lemna minor* experimentally exposed to cadmium and zinc. *Ecotoxicology* **2011**, *20*, 815–826. [CrossRef]
49. Bradford, M.M. A rapid and sensitive method for the quantitation of microgram quantities of protein utilizing the principle of protein-dye binding. *Anal. Biochem.* **1976**, *72*, 248–254. [CrossRef]
50. R Core Team. *R: A Language and Environment for Statistical Computing*; R Foundation for Statistical Computing: Vienna, Austria, 2019.
51. Al-Nozaily, F.; Alaerts, G.; Veenstra, S. Performance of duckweed-covered sewage lagoons—II. Nitrogen and phosphorus balance and plant productivity. *Water Res.* **2000**, *34*, 2734–2741. [CrossRef]
52. Iatrou, E.I.; Stasinakis, A.S.; Aloupi, M. Cultivating duckweed *Lemna minor* in urine and treated domestic wastewater for simultaneous biomass production and removal of nutrients and antimicrobials. *Ecol. Eng.* **2015**, *84*, 632–639. [CrossRef]

53. Paolacci, S.; Harrison, S.; Jansen, M.A.K. The invasive duckweed *Lemna minuta* Kunth displays a different light utilisation strategy than native *Lemna minor* Linnaeus. *Aquat. Bot.* **2018**, *146*, 8–14. [CrossRef]
54. Wedge, R.M.; Burris, J.E. Effects of light and temperature on duckweed photosynthesis. *Aquat. Bot.* **1982**, *13*, 133–140. [CrossRef]
55. Al-Nozaily, F.; Alaerts, G.; Veenstra, S. Performance of duckweed-covered sewage lagoons—I. Oxygen balance and COD removal. *Water Res.* **2000**, *34*, 2727–2733. [CrossRef]
56. OECD. *Test No. 221: Lemna sp. Growth Inhibition Test*; OECD Guidelines for the Testing of Chemicals, Section 2; OECD: Paris, France, 2006.
57. Körner, S.; Vermaat, J.E. The relative importance of *Lemna gibba* L., bacteria and algae for the nitrogen and phosphorus removal in duckweed-covered domestic wastewater. *Water Res.* **1998**, *32*, 3651–3661. [CrossRef]
58. Porath, D.; Hepher, B.; Koton, A. Duckweed as an aquatic crop: Evaluation of clones for aquaculture. *Aquat. Bot.* **1979**, *7*, 273–278. [CrossRef]
59. Debusk, T.A.; Ryther, J.H.; Hanisak, M.D.; Williams, L.D. Effects of seasonality and plant density on the productivity of some freshwater macrophytes. *Aquat. Bot.* **1981**, *10*, 133–142. [CrossRef]
60. Kong, C.-H.; Zhang, S.-Z.; Li, Y.-H.; Xia, Z.-C.; Yang, X.-F.; Meiners, S.J.; Wang, P. Plant neighbor detection and allelochemical response are driven by root-secreted signaling chemicals. *Nat. Commun.* **2018**, *9*, 3867. [CrossRef]
61. Pierik, R.; Mommer, L.; Voesenek, L.A.C.J. Molecular mechanisms of plant competition: Neighbour detection and response strategies. *Funct. Ecol.* **2013**, *27*, 841–853. [CrossRef]
62. Murchie, E.H.; Lawson, T. Chlorophyll fluorescence analysis: A guide to good practice and understanding some new applications. *J. Exp. Bot.* **2013**, *64*, 3983–3998. [CrossRef] [PubMed]
63. Klughammer, C.; Schreiber, U. Complementary PS II quantum yields calculated from simple fluorescence parameters measured by PAM fluorometry and the Saturation Pulse method. *PAM Appl. Notes* **2008**, *1*, 27–35.
64. Kramer, D.M.; Johnson, G.; Kiirats, O.; Edwards, G.E. New fluorescence parameters for the determination of QA redox state and excitation energy fluxes. *Photosynth. Res.* **2004**, *79*, 209–218. [CrossRef] [PubMed]
65. Juneau, P.; Qiu, B.; Deblois, C.P. Use of chlorophyll fluorescence as a tool for determination of herbicide toxic effect: Review. *Toxicol. Environ. Chem.* **2007**, *89*, 609–625. [CrossRef]
66. Kufel, L.; Strzałek, M.; Przetakiewicz, A. Plant response to overcrowding–*Lemna minor* example. *Acta Oecologica* **2018**, *91*, 73–80. [CrossRef]
67. Zhang, L.-M.; Jin, Y.; Yao, S.-M.; Lei, N.-F.; Chen, J.-S.; Zhang, Q.; Yu, F.-H. Growth and Morphological Responses of Duckweed to Clonal Fragmentation, Nutrient Availability, and Population Density. *Front. Plant Sci.* **2020**, *11*, 618. [CrossRef] [PubMed]
68. Huang, D.; Wu, L.; Chen, J.R.; Dong, L. Morphological plasticity, photosynthesis and chlorophyll fluorescence of *Athyrium pachyphlebium* at different shade levels. *Photosynthetica* **2011**, *49*, 611–618. [CrossRef]
69. Hallik, L.; Niinemets, Ü.; Kull, O. Photosynthetic acclimation to light in woody and herbaceous species: A comparison of leaf structure, pigment content and chlorophyll fluorescence characteristics measured in the field. *Plant Biol.* **2012**, *14*, 88–99. [CrossRef] [PubMed]
70. Bilas, R.D.; Bretman, A.; Bennett, T. Friends, neighbours and enemies: An overview of the communal and social biology of plants. *Plant. Cell Environ.* **2021**, *44*, 997–1013. [CrossRef]
71. Jang, M.-H.; Ha, K.; Takamura, N. Reciprocal allelopathic responses between toxic cyanobacteria (*Microcystis aeruginosa*) and duckweed (*Lemna japonica*). *Toxicon* **2007**, *49*, 727–733. [CrossRef]
72. Bich, T.T.N.; Kato-Noguchi, H. Allelopathic potential of two aquatic plants, duckweed (*Lemna minor* L.) and water lettuce (*Pistia stratiotes* L.), on terrestrial plant species. *Aquat. Bot.* **2012**, *103*, 30–36. [CrossRef]
73. Färber, E.; Königshofer, H.; Kandeler, R. Ethylene Production and Overcrowding in *Lemnaceae*. *J. Plant Physiol.* **1986**, *124*, 379–384. [CrossRef]
74. Dubois, M.; Van den Broeck, L.; Inzé, D. The Pivotal Role of Ethylene in Plant Growth. *Trends Plant Sci.* **2018**, *23*, 311–323. [CrossRef] [PubMed]
75. Färber, E.; Kandeler, R. Significance of Calcium Ions in the Overcrowding Effect in *Spirodela Polyrrhiza* P 143. *J. Plant Physiol.* **1989**, *135*, 94–98. [CrossRef]
76. Färber, E.; Kandeler, R. Phytochrome effect on ethylene production after overcrowding in *Spirodela* (*Lemnaceae*). *Phyton Ann. Rei Bot. Austria* **1990**, *30*, 89–95.
77. Anderson, K.E.; Lowman, Z.; Stomp, A.M.; Chang, J. Duckweed as a Feed Ingredient in Laying Hen Diets and its Effect on Egg Production and Composition. *Int. J. Poult. Sci.* **2011**, *10*, 4–7. [CrossRef]
78. Ge, X.; Zhang, N.; Phillips, G.C.; Xu, J. Growing *Lemna minor* in agricultural wastewater and converting the duckweed biomass to ethanol. *Bioresour. Technol.* **2012**, *124*, 485–488. [CrossRef]
79. Verma, R.; Suthar, S. Impact of density loads on performance of duckweed bioreactor: A potential system for synchronized wastewater treatment and energy biomass production. *Environ. Prog. Sustain. Energy* **2015**, *34*, 1596–1604. [CrossRef]
80. Xu, J.; Shen, G. Effects of Harvest Regime and Water Depth on Nutrient Recovery from Swine Wastewater by Growing *Spirodela oligorrhiza*. *Water Environ. Res.* **2011**, *83*, 2049–2056. [CrossRef] [PubMed]
81. Roijackers, R.; Szabó, S.; Scheffer, M. Experimental analysis of the competition between algae and duckweed. *Arch. Hydrobiol.* **2004**, *160*, 401–412. [CrossRef]

82. Mohedano, R.A.; Costa, R.H.R.; Tavares, F.A.; Belli Filho, P. High nutrient removal rate from swine wastes and protein biomass production by full-scale duckweed ponds. *Bioresour. Technol.* **2012**, *112*, 98–104. [CrossRef]
83. Hymowitz, T.; Collins, F.I.; Panczner, J.; Walker, W.M. Relationship between the content of oil, protein, and sugar in soybean seed. *Agron. J.* **1972**, *64*, 613–616. [CrossRef]
84. Rekowski, A.; Langenkämper, G.; Dier, M.; Wimmer, M.A.; Scherf, K.A.; Zörb, C. Determination of soluble wheat protein fractions using the Bradford assay. *Cereal Chem.* **2021**. [CrossRef]
85. Mæhre, H.K.; Dalheim, L.; Edvinsen, G.K.; Elvevoll, E.O.; Jensen, I.-J. Protein Determination—Method Matters. *Foods* **2018**, *7*, 5. [CrossRef] [PubMed]
86. Cui, W.; Cheng, J.J. Growing duckweed for biofuel production: A review. *Plant Biol.* **2015**, *17*, 16–23. [CrossRef]
87. Zimmo, O.R.; van der Steen, N.P.; Gijzen, H.J. Nitrogen mass balance across pilot-scale algae and duckweed-based wastewater stabilisation ponds. *Water Res.* **2004**, *38*, 913–920. [CrossRef] [PubMed]
88. Zhao, Y.; Fang, Y.; Jin, Y.; Huang, J.; Ma, X.; He, K.; He, Z.; Wang, F.; Zhao, H. Microbial community and removal of nitrogen via the addition of a carrier in a pilot-scale duckweed-based wastewater treatment system. *Bioresour. Technol.* **2015**, *179*, 549–558. [CrossRef]
89. Benjawan, L.; Koottatep, T. Nitrogen removal in recirculated duckweed ponds system. *Water Sci. Technol.* **2007**, *55*, 103–110. [CrossRef] [PubMed]
90. Cheng, J.; Landesman, L.; Bergmann, B.; Classen, J.J.; Howard, J.W.; Yamamoto, Y.T. Nutrient Removal from swine lagoon liquid by *Lemna minor* 8627. *Trans. ASAE* **2002**, *45*, 1003–1010. [CrossRef]
91. Cheng, J.; Bergmann, B.A.; Classen, J.J.; Stomp, A.M.; Howard, J.W. Nutrient recovery from swine lagoon water by *Spirodela punctata*. *Bioresour. Technol.* **2002**, *81*, 81–85. [CrossRef]
92. Szabó, S.; Braun, M.; Nagy, P.; Balázsy, S.; Reisinger, O. Decomposition of duckweed (*Lemna gibba*) under axenic and microbial [-2pt] conditions: Flux of nutrients between litter water and sediment, the impact of leaching and microbial degradation. *Hydrobiologia* **2000**, *434*, 201–210. [CrossRef]
93. Szabó, S.; Braun, M.; Borics, G. Elemental flux between algae and duckweeds (*Lemna gibba*) during competition. *Arch. Hydrobiol.* **1999**, *146*, 355–367. [CrossRef]

plants

MDPI

Article

Cultivation of Lemna Minor on Industry-Derived, Anaerobically Digested, Dairy Processing Wastewater

Rachel O'Mahoney [1,2], Neil E. Coughlan [1,2], Éamonn Walsh [1,2] and Marcel A. K. Jansen [1,2,*]

1 School of Biological, Earth and Environmental Sciences, University College Cork, Distillery Fields, North Mall, T23 N73K Cork, Ireland

2 Environmental Research Institute, University College Cork, T23 N73K Cork, Ireland

* Correspondence: m.jansen@ucc.ie

Abstract: The growth and nutrient uptake capacity of a common duckweed (Lemnaceae) species, Lemna minor "Blarney", on dairy processing wastewater pre-treated by an anaerobic digester (AD-DPW) was explored. *L. minor* was cultivated in small stationary vessels in a controlled indoor environment, as well as in a semi-outdoor 35 L recirculatory system. The use of AD-DPW as a cultivation medium for *L. minor* offers a novel approach to dairy wastewater treatment, evolving from the current resource-intensive clean-up of wastewaters to duckweed-based valorisation, simultaneously generating valuable plant biomass and remediating the wastewater.

Keywords: anaerobic digestate; biomass generation; circular economy; Lemnaceae; nutrient recovery; phytoremediation

check for updates

Citation: O'Mahoney, R.; Coughlan, N.E.; Walsh, É.; Jansen, M.A.K. Cultivation of Lemna Minor on Industry-Derived, Anaerobically Digested, Dairy Processing Wastewater. *Plants* **2022**, *11*, 3027. https://doi.org/10.3390/plants11223027

Academic Editors: Viktor Oláh, Klaus-Jürgen Appenroth and K. Sowjanya Sree

Received: 26 August 2022
Accepted: 4 November 2022
Published: 9 November 2022

1. Introduction

Clean water is fundamental for life on Earth. However, by 2025 more than 40% of people worldwide will live under conditions of water insecurity [1]. Such a serious level of insecurity calls for advances in the efficiency of water use, for example, through water conservation and improved wastewater remediation [2]. Currently, food production industries account for a large portion of water use and wastewater production in many parts of the world [1,3,4]. Accordingly, there is considerable scope to develop circular economy approaches for improved water use efficiency and wastewater valorisation within many industry settings. For example, the dairy industry produces a considerable amount of wastewater during the production process and is in many countries considered the most significant source of wastewater [5]. It has been approximated that up to 11 litres of wastewater are produced for every one litre of milk processed, depending on the dairy product being made (e.g., drinking milk, cheese, butter, yogurt) [5,6]. Consequently, a dairy processing plant can produce upwards of 550 million litres of wastewater per year [6]. At present, dairy industries in most countries operate extensive wastewater treatment plants cleaning wastewater so that release on local surface waters is permissible [7]. This amounts to a considerable economic cost for the industry. Yet, nutrient-rich dairy processing wastewater (DPW) contains potentially valuable resources that can be valorised in line with the principles of circular economy [8].

A range of sequential biological and physicochemical techniques are used in DPW treatment facilities to minimise the concentrations of environmentally damaging pollutants [9], which include high biological oxygen demand (BOD) and chemical oxygen demand (COD) together with considerable amounts of nitrogen and phosphorus [5,10]. Anaerobic digestion is a widely employed method, in which microbial organisms, such as *Bacteria*, *Archaea*, and protozoa, metabolise organic compounds primarily into methane [11]. Anaerobic digestion can substantially decrease BOD and COD but is less effective in reducing nitrogen and phosphorus content [12]. Therefore, following biological treatment, physicochemical treatments serve to filter, precipitate, coagulate, flocculate or even electrochemically process nitrogen and phosphorus, as well as any remaining suspended fats, lactose, and other organic matter contents [13]. Typically, nitrogen is dissipated in the form of nitrogen–gas through the nitrification–denitrification cycle, whilst phosphorus is chemically precipitated, often using aluminium. While this remediates water and prevents eutrophication, the aforementioned nutrients are not reused. It has been argued that recycling of nutrients, particularly the phosphate, nitrate, and ammonium content of DPW, can help generate value from the waste and diminish the need to exploit limiting resources of rock phosphate as well as the need for fossil fuel-based urea production. Essentially, through the cultivation of plant biomass on wastewater, nutrients can be captured and recycled [14,15].

Duckweeds (Lemnaceae) are a family of small, free-floating aquatic macrophytes, of which there are five genera and 36 species dispersed throughout much of the world [16]. Lemnaceae are the fastest-growing angiosperms in the plant kingdom with growth often described as doubling times [17]. Lemnaceae are particularly apt for wastewater remediation due to their rapid growth, high nutrient uptake rate, broad-ranging tolerance of growing conditions, and overall ease of harvesting [17–19]. Lemnaceae preferentially take up ammonium over nitrate, which is paramount in wastewater remediation, in which ammonium is the dominant form of N [20,21]. Value from Lemnaceae's high biomass production can be found in use as biofuel, fertiliser, and/or feed [21,22]. In essence, Lemnaceae can clean water, while simultaneously producing a valuable protein-rich biomass that embodies the closed-loop, circular economy approach of retaining value through the recovery of resources from waste [21,22]. Thus, the integration of Lemnaceae cultivation into the wastewater purification processes can be advantageous to the long-term sustainability of the dairy industry.

In the present study, it was assessed to what extent a common Lemnaceae species, *Lemna minor*, can be grown on DPW pre-treated by an anaerobic digester (AD-DPW) with consideration of both biomass generation and wastewater remediation. The organic load of DPW is high and facilitates strong microbial growth, at the expense of duckweed growth. Therefore, pre-treatment of DPW is required to lower COD and BOD and to facilitate duckweed growth. Previous research has focussed on dairy wastewater pre-processed through small-batch microbial digesters, linked to the production of bioplastics [23]. Here, we pioneered *L. minor* growth on industry-derived, anaerobically digested, dairy processing wastewater (AD-DPW) in small, stationary vessels in a controlled indoor environment, as well as in a semi-outdoor, 35-litre recirculatory system. It was hypothesised that there would be a clear concentration dependence of growth, with lower concentrations of AD-DPW being sub-optimal, and higher concentrations of AD-DPW being super-optimal for key plant nutrients.

2. Results

2.1. Physicochemical Analysis of AD-DPW

Physicochemical analysis revealed the concentrations of compounds and elements present in 100% AD-DPW (Table 1). BOD and COD were present at 28.1 and 117 mg O_2 L^{-1}, respectively, while the concentration of total solids amounted to 3.29 g L^{-1}. The total nitrogen (TN) concentration was 105 mg L^{-1}, of which 95 mg N L^{-1} was ammonia and less than 0.01 mg L^{-1} was nitrate. Thus, the total ammonia concentration in AD-DPW was close to the maximum tolerated by duckweed, although at the applied pH (pH = 6) concentrations of toxic ammonia would have been very low. The total phosphorus (TP) concentration was 27.7 mg P L^{-1} of which 24.8 mg L^{-1} was orthophosphate. Thus, the orthophosphate concentration in the medium was close to optimal for duckweed growth. Other plant nutrients were also present in AD-DPW at concentrations close to optimal, such as potassium, calcium, magnesium, iron, zinc, and copper. Chloride is present at a high concentration, but within the limits typically tolerated by duckweed species.

Table 1. Concentration of compounds and elements present in AD-DPW [c], along with duckweed's required, tolerated, and optimal concentrations.

Parameters	100% AD-DPW [c]	Min. Required [a] ($mg\ L^{-1}$)	Max. Tolerated [a] ($mg\ L^{-1}$)	Optimal Range for Duckweed [a] ($mg\ L^{-1}$)
pH	7.9 [e]	-	-	-
BOD (mg O_2 L^{-1})	28.1	ND [b]	ND	ND
COD (mg O_2 L^{-1})	117	ND	ND	ND
Total Solids (g L^{-1})	3.29	ND	ND	ND
Total Nitrogen (mg N L^{-1})	105	0.07	2101	2.8–350
Ammonia (mg N L^{-1})	95	ND	98	20–50 [d]
Nitrate (mg N L^{-1})	<0.010	3	>1000	3–300
Nitrite (mg N L^{-1})	<0.001	ND	ND	ND
Total phosphorus (mg P L^{-1})	27.7	0.003	310	0.3–54.2
Orthophosphate (mg P L^{-1})	24.8	0.003	310	0.1–50
Chloride (mg Cl^- L^{-1})	1319	0.035	3545	0.035–350
Sulphate (mg SO_4^{2-} L^{-1})	ND	0.32	1924	16–641
Potassium (mg K L^{-1})	77	1.95	1564	20–782
Sodium (mg Na L^{-1})	1040	0	4600	0–230
Calcium (mg Ca L^{-1})	112	0.4	1600	8–800
Magnesium (mg Mg L^{-1})	11.6	0.1	800	1.2–240
Iron (mg Fe L^{-1})	0.42	0.06	56	0.06–11
Zinc (mg Zn L^{-1})	0.54	0.04	523	0.13–13
Copper (mg Cu L^{-1})	0.01	0.006	64	0.006–3.8
Manganese (mg Mn L^{-1})	0.03	0.005	55	0.05–5.5
Nickel (mg Ni L^{-1})	ND	0	1	0–0.1

[a] [15], [b] ND—not determined, [c] AD-DPW—anaerobically digested dairy processing wastewater, [d] see [24], [e] prior to experiments the pH of AD-DPW was reduced to pH 6.0.

2.2. The Effect of Varying AD-DPW Concentration on L. minor Growth

Plants displayed modest growth on all concentrations of AD-DPW, with RGR values reaching up to 0.13 d^{-1}. Overall, significant differences were observed between treatments (ANOVA: $\chi^2 = 12.788$, df = 6; $p = 0.03$: Figure 1) although, the post hoc analysis did not reveal any specific differences between concentrations. Plants cultivated on 10% AD-DPW had the highest RGR and appeared to have the healthiest colour and frond-to-colony ratio compared to all other AD-DPW concentrations. Frond size on 50% and 100% AD-DPW concentrations appeared reduced in comparison to 10% AD-DPW. Plants kept on 100% AD-DPW also displayed fragmented colonies and death (chlorosis).

2.3. Growth of L. minor on Varying Concentrations of AD-DPW over a Six-Week Period

For this experiment, the highest RGR value found was 0.35 d^{-1} (Figure 2). In some cases (50% and 100% AD-DPW) no growth was observed. The RGR differed significantly for *L. minor* grown on different AD-DPW concentrations over a six-week period (ANOVA: $\chi^2 = 303.94$, df = 35; $p < 0.0001$, Figure 2). Concentration had a significant effect on RGR ($p < 0.0001$), while the week had no effect ($p > 0.05$). A significant interaction effect on RGR was detected for concentration and week ($p < 0.0001$). For individual weeks, RGR was greater for plants grown on the control (half-strength Hutner's medium), 1%, 5%, and 10% concentrations than for duckweed cultivated on 50% and 100% concentrations. The 50% and 100% concentrations gave the least growth, with no survival occurring except for that of the 50% in week 1. Overall, as the concentration of AD-DPW increased, the

RGR decreased. However, the distribution of *L. minor* growth over the successive weeks followed a different trajectory pattern for each concentration.

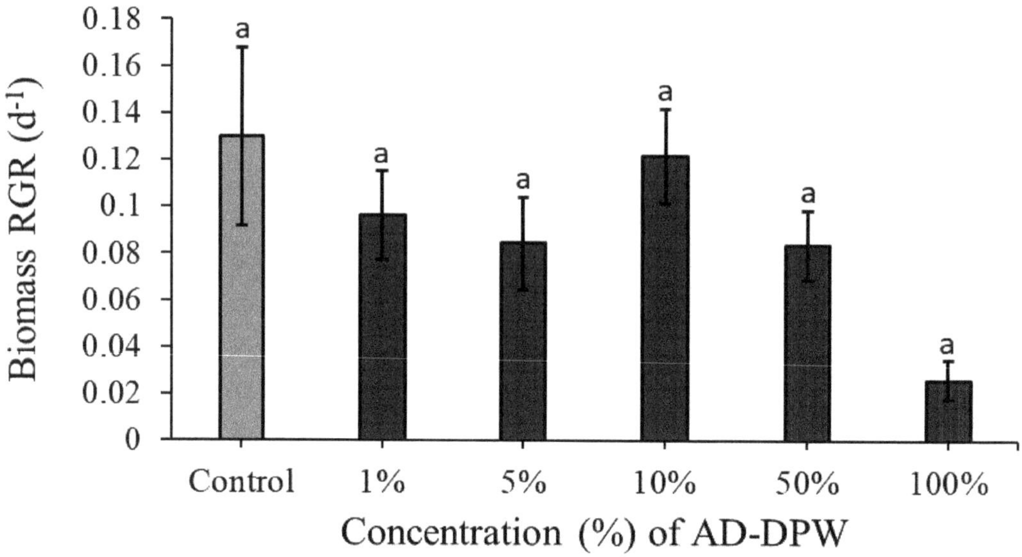

Figure 1. Mean (±SE) RGR (d^{-1}) of *Lemna minor* cultivated on AD-DPW at 1%, 5%, 10%, 50%, and 100% concentrations in one-week experiments, as well as a control treatment. "a": Denoting no significant difference was detected through post hoc analysis.

2.4. Growth of L. minor in Semi-Outdoor, Recirculating Systems

Plants displayed modest growth on both 5% and 10% AD-DPW, with RGR values ranging from 0.07 to 0.1 d^{-1}. The RGR significantly differed for *L. minor* cultivated on different concentrations of AD-DPW in the tiered tank system (ANOVA: $\chi^2 = 24.287$, df = 2; $p < 0.0001$, Figure 3). Concentration had a significant effect on RGR ($p < 0.0001$), while tank position in the tiered system had no effect ($p > 0.05$). No interaction effect on RGR was detected for concentration and tank ($p > 0.05$). Plants grown on 10% AD-DPW had a higher RGR than *L. minor* cultivated on either the control medium ($p < 0.0001$) or 5% AD-DPW ($p < 0.001$). The RGR (d^{-1}) of plants grown on 5% AD-DPW was similar to that of plants grown on the control medium ($p > 0.05$).

2.5. Nutrient Depletion and Removal Rates in the Semi-Outdoor, Recirculatory System

Total nitrogen (TN) and total phosphorus (TP) concentrations in the medium were determined at the start (day 0) and the end (day 7) of the 7-day experiments growing *L. minor* in the tiered tank system, using different media. Total nitrogen concentrations (TN) at day 7 were significantly lower than initial TN at day 0 for both 5% and 10% AD-DPW (both $p < 0.0001$).

TN was found to significantly differ between treatments and time-points (GLM: $\chi^2 = 85.757$, df = 3; $p < 0.0001$, Figure 4). The tested media (5% and 10% AD-DPW) and sampling day (day 0 or day 7) were found to have a significant effect on the TN concentration (both $p < 0.0001$). A significant interaction effect was also detected for the media and day ($p < 0.0001$).

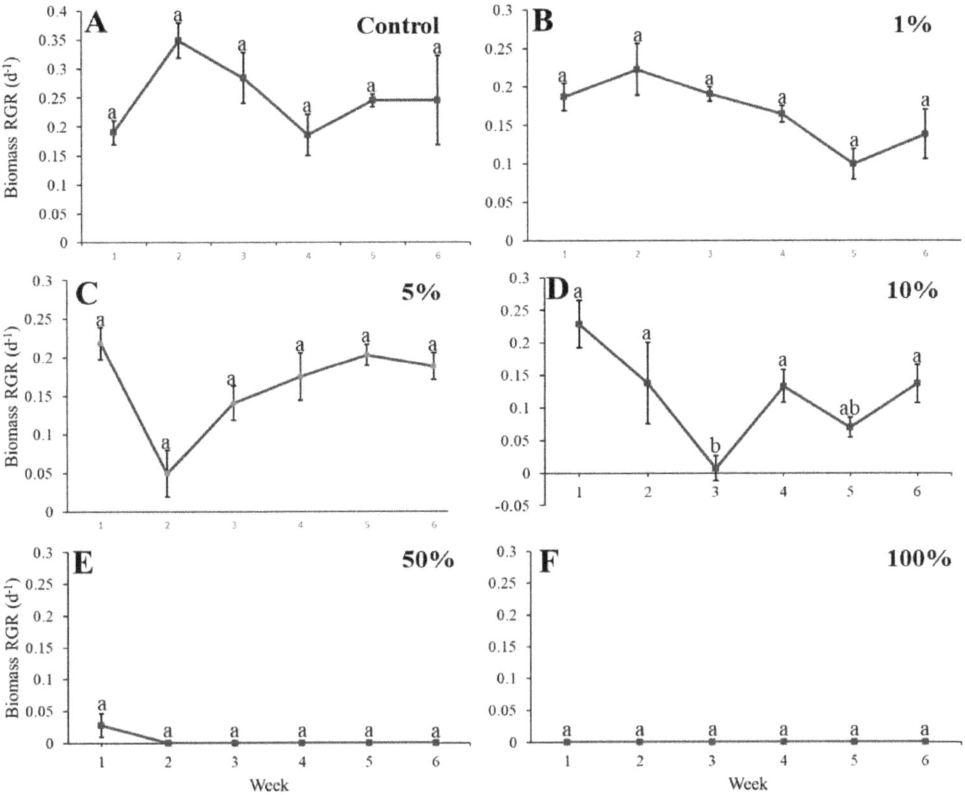

Figure 2. Mean (±SE) RGR (d^{-1}) of *Lemna minor* growth for the control treatment (panel (**A**)), as well as on 1%, 5%, 10%, 50%, and 100% AD-DPW (panel (**B–F**), respectively) over the course of the six successive weeks. Different letters within each panel denote statistical differences between weeks. For individual weeks, RGR was greater for plants grown on the control (half-strength Hutner's medium), 1%, 5%, and 10% concentrations than for duckweed cultivated on 50% and 100% concentrations: week 1 (all $p < 0.05$), week 2 (all $p < 0.05$), week 3 (all $p < 0.01$; except the 10% solution, $p > 0.05$), week 4 (all $p < 0.01$), week 5 (all $p < 0.01$), week 6 (all $p < 0.01$).

The concentration of total phosphorus (TP) within the tiered tank system was significantly reduced between day 0 and day 7, in both 5% and 10% AD-DPW (both $p < 0.001$). Similar to TN, TP was found to significantly differ across treatments and time-points (GLM: $\chi^2 = 29.233$, df = 3; $p < 0.0001$, Figure 4). The tested media (5% and 10% AD-DPW) and sampling day (day 0 or day 7) were found to have a significant effect on TP concentration ($p < 0.05$ and $p < 0.0001$, respectively). No significant interaction effect was detected.

The removal rate of TN and TP from AD-DPW by *L. minor* was determined by calculating the difference between the initial (day 0) and final (day 7) nutrient concentrations and expressing it per m^{-2} duckweed per day. The 5% AD-DPW had nutrient removal rates of 43.72 mg TN $m^{-2}d^{-1}$ and 53.76 mg TP $m^{-2}d^{-1}$ while 10% AD-DPW had higher nutrient removal rates of 248.46 mg TN $m^{-2}d^{-1}$ and 126.54 mg TP $m^{-2}d^{-1}$.

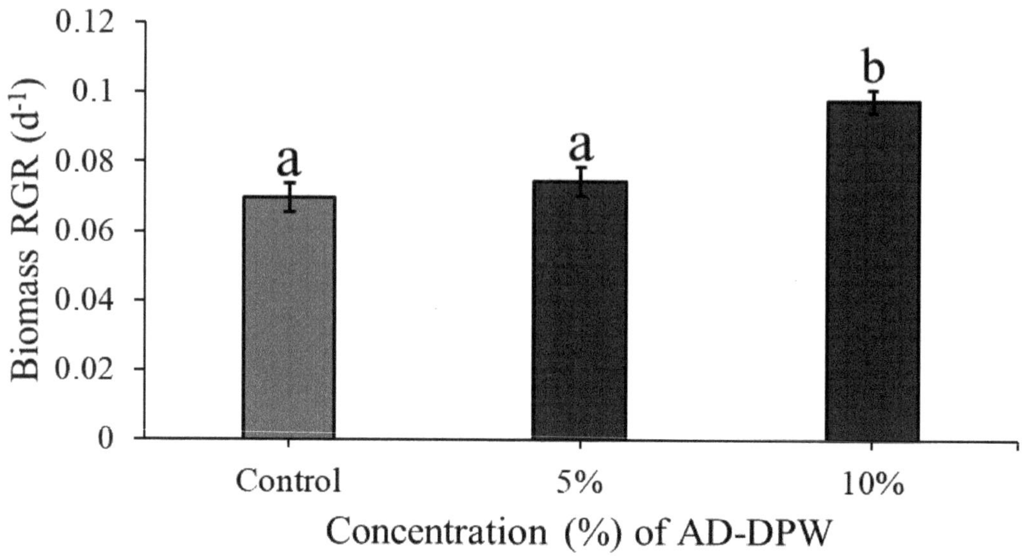

Figure 3. Mean (±SE) RGR (d^{-1}) for *Lemna minor* grown on the control medium (half-strength Hutner's medium), 5% and 10% AD-DPW in the semi-outdoor, recirculating system. Bars that do not share the same letter significantly differ from one another for $p < 0.001$, as per the post hoc test.

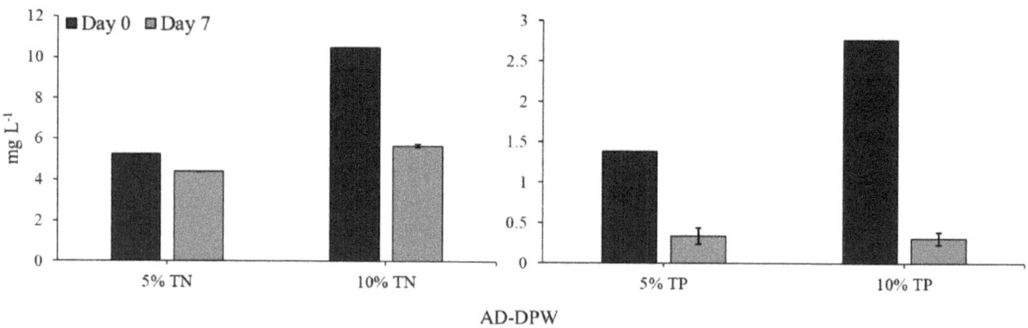

Figure 4. Comparison of the total nitrogen (TN) and the total phosphate (TP) concentration (mg L^{-1}) on day 0 and day 7 of *L. minor* growth, when cultivated on 5% or 10% AD-DPW (mean ± SE). For all treatments, the total nitrogen or total phosphate concentration was significantly lower on day 7 than on day 0 ($p < 0.05$).

3. Discussion

Multiple factors determine whether AD-DPW is a suitable growth medium for *L. minor*, and hence suitable for the process of phytoremediation. These include the presence of essential macro- and micronutrients in suitable quantities; low enough BOD and COD levels to avoid excessive microbial growth; a favourable combination of pH and ammonia for plant growth; and a suitable calcium-to-magnesium ratio for growth.

3.1. *Presence in Appropriate Quantities of All Essential Macro- and Micronutrients*

Various studies have shown that highly concentrated wastewater inhibits duckweed growth, which means that dilutions are necessary to prevent phytotoxic effects [20]. The present study shows that *L. minor* grows best on lower concentrations of AD-DPW, under both stationary indoor as well as recirculatory semi-outdoor conditions. The RGR of

L. minor was greater for plants cultivated on 10% AD-DPW compared to 5% which suggests that the nutrient composition of 5% AD-DPW is sub-optimal for growth, potentially with various nutrients becoming limiting. Conversely, reduced growth on 100% AD-DPW (compared to 10% AD-DPW) indicates some form of toxicity. To understand why *L. minor* grew slower or not at all at low (<5%) as well as high (>50%) concentrations of AD-DPW, nutrient requirements of *L. minor* were considered.

Duckweed prefers total nitrogen concentrations of between 2.8 and 350 N mg^{-1} L^{-1} and generally takes up N in the form of ammonium or nitrate (Gil-Pulido et al., 2018). The concentration of total ammonia on full-strength AD-DPW is above optimal, and this may contribute to the lower growth of *L. minor* on full-strength AD-DPW compared to dilutions [24]. However, there are also other elements present in the medium that may cause growth impediments at higher concentrations. For example, chloride (Cl) levels in 100% AD-DPW are at 1319 mg L^{-1}, well above the optimal range of 0.035–350 mg L^{-1} [25,26]. Similarly, sodium levels are higher than those tolerated by duckweed. Salinity-induced osmotic stress of *L. minor* inhibits vegetative growth [25,26]. Thus, it is concluded that to facilitate *L. minor* growth and remediation on AD-DPW, it needs to be substantially diluted. In contrast, concentrations of several cations (Mn, Zn, Fe, and K) are close to optimum in undiluted AD-DPW [26–28] and dilutions may cause deficiencies. Similarly, orthophosphate concentrations in undiluted AD-DPW are considered optimal, being between 0.3 and 54.2 P mg^{-1} L^{-1} [26,29]. So, phosphate will easily become a limiting factor at the lower concentrations tested here (i.e., 1% and 5% dilution), especially after a period of duckweed growth [30]. The present study shows that duckweed removed 46% of the TN and 90% of the TP present in the 10% concentration of AD-DPW, under semi-realistic conditions of medium velocity and outdoor light conditions. Thus, the low concentration of TP remaining in both 5% and 10% AD-DPW may in turn have hindered further growth and, thereby, uptake of TN, and explain both the relatively low RGR (d^{-1}) values and low TN removal. This observation underlines important issues concerning the management of wastewater and triggers the question of whether supplementation of wastewater-derived media with specific elements (i.e., phosphate in this case) is required to maintain sufficient *L. minor* growth and consequent parallel TN and TP removal.

3.2. Low BOD and COD Levels to Avoid Excessive Microbial Growth

Raw DPW contains a high load of BOD and COD and is not suitable for *L. minor* growth due to the rapid growth of a microbial scum on top of the lactose-rich medium (Walsh et al. in press). AD-DPW has already much of its organic components removed [31]. A physicochemical analysis of the raw non-diluted AD-DPW used by the present study (i.e., 100% concentration), confirms that both BOD and COD levels in 100% AD-DPW are unlikely to impede *L. minor* growth (see Table 1).

3.3. A Favourable Combination of pH and Ammonia for Plant Growth

The addition of H_2SO_4 to decrease the pH of the AD-DPW prevented the negative effects of a high pH on growth as observed in a pilot experiment. This is most likely due to avoidance of the combination of an alkaline pH with high ammonia concentrations. Total ammonia in solution consists of ionised ammonium (NH_4^+) and un-ionised ammonia (NH_3), with the relative proportion of each being pH-dependent. The main toxic form is un-ionised ammonia, with a more minor contribution of ammonium. So, a high total ammonia concentration is predominantly toxic at a higher pH [24,32]. For example, in the case of 100% AD-DPW, ammonia levels of 95 mg L^{-1} were present. At the native pH of 8, 3–4 percent of the total ammonia will have occurred as toxic un-ionised ammonia (i.e., ~3 mg L^{-1}). The latter concentration of un-ionised ammonia is reflected in the negative growth observed in the pilot experiment, in accordance with the isopleth maps generated by Körner et al. [24] for *L. gibba*. In contrast, at pH 6 less than 0.2 mg L^{-1} of un-ionised ammonia will have been present in 100% AD-DPW, but still, a modest growth inhibition can be predicted to occur due to ionised ammonium [24]. Körner et al. [24] found a maximum

RGR was achieved with 10 mg L^{-1} total ammonia at pH 6.8, resulting in a combination of 9.97 mg L^{-1} ionised ammonium (NH_4^+) and 0.03 mg L^{-1} un-ionised ammonia (NH_3). This explains why the best growth results were obtained using dilutions of around 10% of AD-DPW, in combination with a lowering of the pH to around 6.

3.4. A Suitable Calcium-to-Magnesium Ratio for Growth

The raw AD-DPW solution contained both Ca and Mg, both within the optimal range outlined for duckweed growth parameters. Recent studies have demonstrated the importance of the calcium-to-magnesium ratio for good growth of *L. minor* [15]. The 100% AD-DPW solution contained 112 mg L^{-1} of calcium and 11.6 mg L^{-1} of magnesium which translates to a molar Ca:Mg ratio of 5.6:1.0. This ratio aligns with the work demonstrated by Walsh et al. [15] which details the growth effect of a skewed Ca:Mg ratio in wastewater and the importance of a ratio which favours calcium over magnesium for *L. minor* growth.

It may be hypothesised that long-term cultivation of *L. minor* on AD-DPW will highlight both deficient and toxic phenomena, which may not be noticed in short-term experiments. However, over a six-week period, no negative effects were noted. It is concluded that diluted AD-DPW is suitable as a cultivation medium for *L. minor*. This leads to a subsequent question, namely, what is the nutrient uptake capacity of *L. minor* when grown on AD-DPW? TN and TP removal rates were assessed on 5% and 10% AD-DPW. At both dilutions decreases in TN and TP were measured. The nutrient depletion and removal rates of *L. minor* in the 10% conditions were greater in comparison with the lower concentration of 5%, which matches the established relationship between plant growth and nutrient uptake [33]. The TN removal rate achieved within the recirculating semi-outdoor setting was 248.46 mg TN $m^{-2}d^{-1}$ for the 10% concentration. This value is at the lower end of the removal rate range reported in the literature, which varies from 124 to 4400 mg TN $m^{-2}d^{-1}$ [34,35]. This may be explained by the relatively low growth rate, which may be associated with relatively low concentrations of phosphate, zinc, iron, potassium, and manganese at that dilution. The TP removal rate was 126.54 mg TP $m^{-2}d^{-1}$ for the 10% concentration which is also within the removal rate range commonly reported in the literature as between 14 and 590 mg TP m^{-2} d^{-1} [20,34]. The relatively low TN removal rate indicates that AD-DPW is not a particularly good medium for duckweed growth, as also indicated by the relatively low RGR values (Figures 1–3). Nevertheless, duckweed effectively lowered TN and TP concentrations in the AD-DPW. The EU Urban Wastewater Treatment Directive (98/15/EC) states TN and TP allowable limits for release into sensitive receiving waters as 10–15 mg L^{-1} and 1–2 mg L^{-1}, respectively. A comparison of the allowable release limits with the nutrient concentrations of AD-DPW shows that remediation of 10% AD-DPW lowered the mean TN content from 10.5 mg L^{-1} to 5.7 mg L^{-1} by day 7. Based on the measured nitrogen removal rate of 0.248 g TN m^{-2} d^{-1}, a system of six stacked recirculatory systems [36], each comprised of 10 m^2 duckweed per m^2 of the floor, would be capable of removing a total of 15 g TN d^{-1}, and clean 10% AD-DPW to releasable rates, thus indicating the potential for commercial applications. Nevertheless, expanding the scale beyond that of the present study requires critical consideration of scaling-up issues and adaption to operation within a commercial setting. Of particular interest is how the long-term operation of a remediation system will impact the concentrations of salts other than N and P: Will there be depletion, or rather a build-up of high concentrations of salts, such as manganese, iron, and zinc in the medium? The answer to this question is important to determine the viability of large-scale, long-term, duckweed flow-through systems.

4. Materials and Methods

4.1. Stock Cultivation and Anaerobically Digested DPW (AD-DPW) Source

The duckweed strain used in this study was *Lemna minor* L. "Blarney", number 5500 in the Rutgers Duckweed Stock Cooperative database. A stock of *L. minor* was maintained indoors at a light intensity of 50 µmol m^{-2} s^{-1} PAR (photosynthetically active radiation), at a temperature of 20 ± 2 °C with a 16:8 h light:dark photoperiod. Axenic stock cultures were maintained on a two-part commercial medium that consisted of pH Perfect Grow (2 mL L^{-1}) and pH Perfect Micro (2 mL L^{-1}) (Advanced Nutrients) [37].

The AD-DPW medium was sourced from a large-scale commercial dairy processing plant producing milk, cheese, and yogurt products in Ireland. In this dairy processing plant, wastewater is treated in an anaerobic digester-producing biogas from the organic load present in the DPW. The effluent released from the digester (i.e., AD-DPW) is low in BOD and COD and normally feeds a biological nutrient removal system [31].

For a full physicochemical assessment of the AD-DPW medium, samples of the wastewater were analysed by a GLP-certified laboratory (Aquatic Services Unit, Cork, Ireland). BOD, COD, total solids, total nitrogen (TN), and total phosphorus (TP) were measured on whole, unfiltered, wastewater samples, as per standard methods for wastewater analysis [38,39]. Wastewater was filtered (0.45 μm) to determine the dissolved concentrations of ammonia, nitrate, nitrite, and orthophosphate using the Lachat QuikChem 8000 by Zeilweger Analytics, Inc., Milwaukee, USA (QuikChem Methods 10-107-06-3-D, 10-107-04-1-C, 10-107-04-1-C, and 10-115-01-1-B, respectively). Sodium, potassium, calcium, magnesium, zinc, and iron were measured in filtered wastewater using a flame AAS (Varian Australia Ply Ltd., 1989). Copper and manganese were measured using a graphite furnace AAS (Varian Australia Ply Ltd., Mulgrave, Australia, 1989). Chloride was measured using the ferricyanide method on filtered wastewater [39].

4.2. *Experimental Designs*

4.2.1. The Effect of Varying AD-DPW Concentrations on *L. minor* Growth

Lemna minor was cultivated on AD-DPW at five different concentrations (1%, 5%, 10%, 50%, or 100%), in Magenta vessels with vented lids (GA-7, 7.7 cm length × 7.7 cm width × 9.7 cm height). The concentrations were obtained by diluting the AD-DPW with distilled water. Control treatment used half-strength Hutner's medium [40] since full-strength Hutner's medium is too concentrated for Lemnaceae [26]. At each treatment concentration, the initial pH was recorded using a pH meter (Hanna instruments). Typically, the pH of AD-DPW was pH 8. This was adjusted to pH 6 via the addition of sulfuric acid (H_2SO_4), thus reducing the pH to near-optimal values for the growth of *L. minor* [41]. A pilot experiment had shown that there was a significant effect of controlling the pH, with *L. minor* survival and growth occurring only on the pH-controlled medium. In each replicate experiment, three colonies of four fronds from the axenic *L. minor* stock culture were grown on 100 mL of AD-DPW in individual Magenta vessels. All replicates were kept under a light regime of 16:8 h light:dark photoperiod at a light intensity of 100 μmol m^{-2} s^{-1} PAR, and a temperature of 20 ± 2 °C for a period of seven days. The initial biomass of *L. minor* per Magenta was recorded based on an average of representative samples. At the end of the seven days, the fresh weight of each sample was determined. All treatments were replicated three times ($n = 3$).

4.2.2. Growth of *L. minor* on Varying Concentrations of AD-DPW over a Six-Week Period

Lemna minor was cultivated on five concentrations of AD-DPW (1%, 5%, 10%, 50%, or 100%) set up in Magenta vessels as described above. Initially, three colonies of four fronds from an axenic stock culture were added to each of the Magentas and an average representative initial biomass of *L. minor* was recorded ($n = 5$). At the end of each week, three colonies of four fronds were extracted from each Magenta vessel and placed on 100 mL of fresh AD-DPW media, at the same initial concentration as before. For the duration of the experiment, the media were based on the same AD-DPW stock. The experiment was maintained for six weeks.

4.2.3. Growth of *L. minor* in Semi-Outdoor Recirculating Systems

Lemna minor was cultivated in non-sterile, semi-outdoor, recirculating systems ($n = 3$). Each system consisted of five tiered tanks (21.5 cm length × 15.5 cm width × 11 cm height) and a lower sump tank (37 cm × 29 cm × 52 cm) with a capacity of 35 L per system

(Figure 5). The five tanks had a combined total surface area for duckweed growth of 0.165 m^2. The growing medium was pumped from the sump tank to the highest-tiered tank (1.1 L h^{-1}: JYC-2000 Jiayaocheng, Ltd., Foshan, China) at a slow pace to minimise plant disruption [37]. All other tanks were sequentially gravity fed from the top tank and effluent from the lowest tank was returned to the sump tank. Filters inserted into the outlet pipe of each tank prevented movement of *L. minor* between tanks, whilst allowing the continued circulation of the medium. A muslin cloth cover was suspended above the three recirculating systems to provide 50% shading from natural sunlight. A preliminary assessment indicated that *L. minor* grew better at a reduced light intensity rather than in direct sunlight. The mean ($\pm$ SE) light intensity experienced by *L. minor* between dawn and dusk was 139.9 ± 11.9 µmol m^{-2} s^{-1}, with a min–max of 2.9–463.2 µmol m^{-2} s^{-1} (measured using HOBO MX2202, Onset).

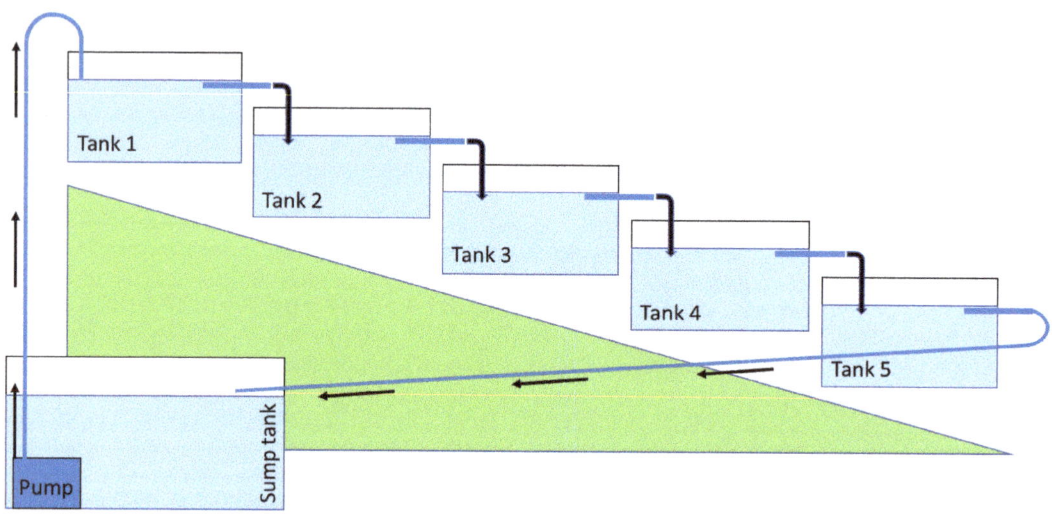

Figure 5. Schematic of the semi-outdoor, 35 L, recirculatory cultivation system.

In the recirculating system, *L. minor* was cultivated on 5% or 10% AD-DPW dilutions for 7 days. A two-part, non-axenic commercial hydroponics solution of FloraGrow (0.25 mL L^{-1}) and FloraMicro (0.25 mL L^{-1}) (GH Inc.) was used as a control medium ($n = 3$). The initial pH of the prepared solution in each sump tank was measured using a handheld pH meter and standardised to pH 6 by the addition of H_2SO_4.

At the start of the experiment, the five tanks of each replicate system (excluding the sump tank) were seeded to achieve 60% duckweed surface cover, using stock plants cultivated in the same semi-outdoor setting. Preliminary imaging analysis using Easy Leaf Area software established that 12.75 g (fresh weight) of *L. minor* per tank gave 60% plant surface cover. Each treatment was left for a seven-day growth period. At the end of each experiment, plant biomass was dried using absorbent tissue paper to remove excess water before being weighed. A representative, unfiltered, water sample was collected from the sump tank on day 0 and day 7 for each replicate and refrigerated at 4 °C until analysed.

4.3. Data Collection

For experiments in Magenta vessels, the total colony number and the total alive frond number were visually assessed and recorded. Where *L. minor* survived, the final fresh biomass was recorded. Total growth for each replicate was determined by subtracting the initial fresh biomass (W_1) from that of the final fresh biomass (W_2).

$$RGR = \frac{ln\frac{W_2}{W_1}}{\Delta T} \quad (1)$$

where ln is the natural log, W_1 is the initial biomass, W_2 is the final biomass and ΔT is the length of the experiment in days.

Total nitrogen (TN) and total phosphorus (TP) concentrations were measured in water samples taken on day 0 and day 7 of the recirculating system experiment. TN was determined using the LCK138 Laton testing kit (1–16 mg L^{-1} TN). TP was determined using the LCK348 Phosphate testing kit (0.5–5.0 mg L^{-1} PO_4-P). These concentration measurements were used to calculate the net TN and TP depletion per system, over the seven-day period and based on the 35 L volume of one individual system. Mean TN and TP removal rates were calculated based on the total available growing area per system. The calculated nutrient removal rates provide an estimate for N and P removed per m^2 of *L. minor* per day (mg TN m^{-2} d^{-1}; mg TP m^{-2} d^{-1}).

4.4. Data Analysis

Statistical analyses were conducted using R software (R Core Team 2021; R 4.1.2). All data were assessed for normality of residual distributions (Shapiro–Wilk test: library *psych*) and homoscedasticity of variances (Levene's test: library *car*). Where data were found to be normally distributed ($p > 0.05$) with homoscedastic residuals ($p > 0.05$), general linear models (ANOVA) were used to analyse differences in RGR and nutrient depletion of the media. Logistic regression in the form of generalised linear models (GLM: *car*) was employed for non-normal data and/or heteroscedastic residuals ($p < 0.05$). A stepwise depletion approach was used to remove non-significant terms if required, while the overall model significance was determined using likelihood ratio tests in all cases (*lmtest*). Where *p*-values were significant ($\alpha < 0.05$), a Tukey adjustment for the multiple pairwise comparison was used for post hoc analysis (*emmeans*; Lenth 2020).

5. Conclusions

AD-DPW can support the cultivation of *L. minor*. In particular, plants can be cultivated in semi-outdoor, recirculatory systems. Plant growth and nutrient removal rates from the medium are modest, but this is compensated by the double benefit of simultaneous water remediation and generation of protein-rich biomass. These data indicate a new perspective on AD-DPW treatment, whereby the emphasis moves from resource-intensive clean-up of AD-DPW to duckweed-based valorisation.

Author Contributions: Conceptualization, R.O., N.E.C., É.W. and M.A.K.J.; methodology, N.E.C., É.W. and M.A.K.J.; validation, R.O., N.E.C. and M.A.K.J.; formal analysis, R.O. and N.E.C.; investigation, R.O. and N.E.C.; resources, M.A.K.J.; data curation, N.E.C. and M.A.K.J.; writing—original draft preparation, R.O. and N.E.C.; writing—review and editing, R.O., N.E.C., É.W. and M.A.K.J.; supervision, N.E.C. and M.A.K.J.; project administration, M.A.K.J.; funding acquisition, M.A.K.J. All authors have read and agreed to the published version of the manuscript.

Funding: This research was funded by the European Regional Development Fund through the Ireland Wales Cooperation Programme (82078), and the Environmental Protection Agency (EPA)(2016-W-LS-11). The EPA Research Programme is a Government of Ireland initiative funded by the Department of Communications, Climate Action, and Environment. It is administered by the EPA, which has the statutory function of co-ordinating and promoting environmental research.

Institutional Review Board Statement: Not applicable.

Informed Consent Statement: Not applicable.

Data Availability Statement: The data presented in this study are available on request from the corresponding author.

Acknowledgments: M.A.K.J. acknowledges support from WoB.

Conflicts of Interest: The authors declare no conflict of interest.

References

1. United Nations. Water Scarcity. 2007. Available online: https://www.un.org/waterforlifedecade/scarcity.shtml (accessed on 20 May 2022).
2. Evans, R.G.; Sadler, E.J. Methods and technologies to improve efficiency of water use. *J. Water Resour. Res.* **2008**, *44*, W00E04. [CrossRef]
3. Wu, W.D.; Ju, X.; Li, C.; Xia, X.H.; Chen, G.Q. Water footprint of thermal power in China: Implications from the high amount of industrial water use by plant infrastructure of coal-fired generation system. *J. Energy Policy* **2019**, *132*, 452–461. [CrossRef]
4. FAO. *Progress on Change in Water-Use Efficiency: Global Status and Acceleration Needs for SDG Indicators 6.4.1*; FAO: Rome, Italy, 2021. [CrossRef]
5. Wang, Y.; Serventi, L. Sustainability of dairy and soy processing: A review on wastewater recycling. *Clean. Prod.* **2019**, *237*, 117821. [CrossRef]
6. Baskaran, K.; Palmowski, L.M.; Watson, B.M. Wastewater reuse and treatment options for the dairy industry. *Water Sci. Technol. Water Supply* **2003**, *3*, 85–91. [CrossRef]
7. Ahmad, T.; Aadil, R.M.; Ahmed, H.; Rahman, U.U.; Soares, B.C.V.; Souza, S.L.Q.; Pilmentel, T.C.; Scudino, H.; Guimaraes, J.T.; Esmerino, E.A.; et al. Treatment and utilization of dairy industrial waste: A review. *Trends Food Sci. Technol.* **2019**, *88*, 361–372. [CrossRef]
8. Furness, M.; Bello-Mendoza, R.; Dassonvalle, J.; Chamy-Maggi, R. Building the 'bio-factory': A bibliometric analysis of circular economies and life cycle sustainability assessment in wastewater treatment. *J. Clean. Prod.* **2021**, *323*, 129127. [CrossRef]
9. Bol, R.; Gruau, G.; Mellander, P.-E.; Dupas, R.; Bechmann, M.; Skarbovik, E.; Bieroza, M.; Djodijic, F.; Glendell, M.; Jordan, P.; et al. Challenges of Reducing Phosphorous Based Water Eutrophication in the Agricultural Landscapes of Northwest Europe. *Front. Mar. Sci.* **2018**, *5*, 276. [CrossRef]
10. Carvalho, F.; Prazeres, A.R.; Rivas, J. Cheese whey wastewater: Characterization and treatment. *Sci. Total Environ.* **2013**, *445*, 385–396. [CrossRef]
11. Guo, J.; Peng, Y.; Ni, B.J.; Han, X.; Fan, L.; Yuan, Z. Dissecting microbial community structure and methane-producing pathways of a full-scale anaerobic reactor digesting activated sludge from wastewater treatment by metagenomic sequencing. *Microb. Cell Factories* **2015**, *14*, 33. [CrossRef]
12. Campos, J.L.; Crutchik, D.; Franchi, Ó.; Pavissich, J.P.; Belmonte, M.; Pedrouso, A.; Mosquera-Corral, A.; Val del Río, Á. Nitrogen and Phosphorous Recovery from Anaerobically Pretreated Agro-Food Wastes: A Review. *Front. Sustain. Food Syst.* **2019**, *2*, 91. [CrossRef]
13. Yonar, T.; Sivrioglu, O.; Ozengin, N. *Physico-Chemical Treatment of Dairy Industry Wastewaters: A Review. Technological Approaches for Novel Applications in Dairy Processing*; IntechOpen: London, UK, 2018.
14. Elser, J.; Bennett, E. A broken biogeochemical cycle. *Nature* **2011**, *478*, 29–31. [CrossRef] [PubMed]
15. Walsh, É.; Paolacci, S.; Burnell, G.; Jansen, M.A.K. The importance of the calcium-to-magnesium ratio for phytoremediation of dairy industry wastewater using the aquatic plant *Lemna minor* L. *Int. J. Phytoremediation* **2020**, *22*, 694–702. [CrossRef] [PubMed]
16. Bog, M.; Appenroth, K.J.; Sree, S.K. Key to the determination of taxa of Lemnaceae: An update. *Nord. J. Bot.* **2020**, *38*. [CrossRef]
17. Ziegler, P.; Adelmann, K.; Zimmer, S.; Schmidt, C.; Appenroth, K.J. Relative in vitro growth rates of duckweeds (Lemnaceae)—The most rapidly growing higher plants. *Plant Biol.* **2015**, *17*, 33–41. [CrossRef]
18. Zhao, Z.; Shi, H.; Liu, Y.; Zhao, H.; Su, H.; Wang, M.; Zhao, Y. The influence of duckweed species diversity on biomass productivity and nutrient removal efficiency in swine wastewater. *Bioresour. Technol.* **2014**, *167*, 383–389. [CrossRef]
19. Calicioglu, O.; Richard, T.L.; Brennan, R.A. Anaerobic bioprocessing of wastewater-derived duckweed: Maximizing product yields in a biorefinery value cascade. *J. Bioresour. Technol.* **2019**, *289*, 121716. [CrossRef]
20. Cheng, J.; Landesman, L.; Bergmann, B.; Classen, J.J.; Howard, J.W.; Yamamoto, Y.T. Nutrient Removal from swine lagoon liquid by *Lemna minor* 8627. *Trans. Am. Soc. Agric. Eng.* **2002**, *45*, 1003–1010. [CrossRef]
21. Adhikari, U.; Harrigan, T.; Reinhold, D.M. Use of duckweed-based constructed wetlands for nutrient recovery and pollutant reduction from dairy wastewater. *Ecol. Eng.* **2015**, *78*, 6–14. [CrossRef]
22. Cheng, J.J.; Stomp, A.M. Growing Duckweed to Recover Nutrients from Wastewaters and for Production of Fuel Ethanol and Animal Feed. *CLEAN Soil Air Water* **2009**, *37*, 17–26. [CrossRef]
23. Walsh, É.; Margassery, L.; Coughlan, N.; Broughton, R.; Kühnhold, H.; Fricke, A.; Burnell, G.; O'Mahoney, M.; Wall, D.; Bolger, P.; et al. *Innovative Valorisation of Dairy Processing Wastewater Using a Circular Economy Approach (Newtrients)*; Report No. 441; EPA: Dublin, Ireland, 2022.
24. Körner, S.; Das, S.K.; Veenstra, S.; Vermaat, J.E. The effect of pH variation at the ammonium/ammonia equilibrium in wastewater and its toxicity to *Lemna gibba*. *J. Aquat. Bot.* **2001**, *71*, 71–78. [CrossRef]
25. Sree, K.S.; Adelmann, K.; Garcia, C.; Lam, E.; Appenroth, K.-J. Natural variance in salt tolerance and induction of starch accumulation in duckweeds. *Planta* **2015**, *241*, 1395–1404. [CrossRef] [PubMed]

26. Landolt, E.; Kandeler, R. *Biosystematic Investigations in the Family of Duckweeds (Lemnaceae), Vol. 4: The Family of Lemnaceae—A Monographic Study, Vol. 2. (Phytochemistry, Physiology, Application, Bibliography)*; ETH: Zürich, Switserland, 1987.
27. Wang, W. Toxicity tests of aquatic pollutants by using common duckweed. *Environ. Pollut. Ser. B Chem. Phys.* **1986**, *11*, 1–14. [CrossRef]
28. Lahive, E.; O'Halloran, J.; Jansen, M.A.K. Frond development gradients are a determinant of the impact of zinc on photosynthesis in three species of *Lemnaceae*. *J. Aquat. Bot.* **2012**, *101*, 55–63. [CrossRef]
29. Paolacci, S.; Harrison, S.; Jansen, M.A.K. A comparative study of the nutrient responses of the invasive duckweed *Lemna minuta*, and the native, co-generic species *Lemna minor*. *J. Aquat. Bot.* **2016**, *134*, 47–53. [CrossRef]
30. Chrismadha, T.; Suryono, T.; Magfiroh, M.; Mardiati, Y.; Mulyana, E. Phytoremediation of Maninjau Lake water using Minute Duckweed (*Lemna perpusilla* Torr.). *IOP Conf. Ser. Earth Environ. Sci.* **2019**, *308*, 012021. [CrossRef]
31. Gil-Pulido, B.; Tarpey, E.; Almeida, E.L.; Finnegan, W.; Zhan, X.; Dobson, A.D.W.; O'Leary, N. Evaluation of dairy processing wastewater in an IASBR system: Aeration rate impacts on performance and microbial ecology. *Biotechnol. Rep.* **2018**, *19*, e00263. [CrossRef]
32. Caicedo, J.R.; Van Der Steen, N.P.; Arce, O.; Gijzen, H.J. Effect of total ammonia nitrogen concentration and pH on growth rates of duckweed (*Spirodela polyrrhiza*). *J. Water Res.* **2000**, *34*, 3829–3835. [CrossRef]
33. Walsh, É.; Cialis, E.; Dillane, E.; Jansen, M.A.K. Lemnaceae clones collected from a small geographic region display diverse traits relevant for the remediation of wastewater. *J. Environ. Technol. Innov.* **2022**, *28*, 102599. [CrossRef]
34. Körner, S.; Vermaat, J.E. The relative importance of *Lemna gibba* L.; bacteria and algae for the nitrogen and phosphorus removal in duckweed-covered domestic wastewater. *Water Res.* **1998**, *32*, 3651–3661. [CrossRef]
35. Mohedano, R.A.; Costa, R.H.R.; Tavares, F.A.; Filho, P.B. High nutrient removal rate from swine wastes and protein biomass production by full-scale duckweed ponds. *Bioresour. Technol.* **2012**, *112*, 98–104. [CrossRef]
36. Coughlan, N.E.; Walsh, É.; Bolger, P.; Burnell, G.; O'Leary, N.; O'Mahoney, M.; Paolacci, S.; Wall, D.; Jansen, M.A.K. Duckweed bioreactors: Challenges and opportunities for large-scale indoor cultivation of Lemnaceae. *J. Clean. Prod.* **2022**, *336*, 130285. [CrossRef]
37. Coughlan, N.E.; Walsh, É.; Ahern, R.; Burnell, G.; O'Mahoney, R.; Kuehnhold, H.; Jansen, M.A.K. Flow Rate and Water Depth Alters Biomass Production and Phytoremediation Capacity of Lemna minor. *Plants* **2022**, *11*, 2170. [CrossRef] [PubMed]
38. Grasshoff, K.; Kremling, K.; Ehrhardt, M. *Methods of Seawater Analysis*, 3rd ed.; John Wiley and Sons: Weinheim, Germany, 2009.
39. Rice, E.W.; Baird, R.B.; Eaton, A.D. *Standard Methods for the Examination of Water and Wastewater*; American Public Health Association (APHA); American Water Works Association (AWWA); Water Environmental Federation (WEF): Washington, DC, USA, 2005.
40. Hutner, S.H. *Comparative Physiology of Heterotrophic Growth in Higher Plants: Growth and Differentiation in Plants*; Iowa State College Press: Ames, IA, USA, 1953; pp. 417–447.
41. McLay, C.L. The effect of pH on the population growth of three species of duckweed: *Spirodela oligorrhiza, Lemna minor* and *Wolffia arrhiza*. *Freshw. Biol.* **1976**, *6*, 125–136. [CrossRef]

plants

MDPI

Article

Integrated Multitrophic Aquaculture; Analysing Contributions of Different Biological Compartments to Nutrient Removal in a Duckweed-Based Water Remediation System

Simona Paolacci [1,*], Vlastimil Stejskal [1,2], Damien Toner [3] and Marcel A. K. Jansen [1,4]

1 School of Biological, Earth and Environmental Sciences, University College Cork, Distillery Fields, North Mall, T23 N73K Cork, Ireland
2 Faculty of Fisheries and Protection of Waters, South Bohemian Research Center of Aquaculture and Biodiversity of Hydrocenoses, Institute of Aquaculture, University of South Bohemia in Ceske Budejovice, Husova Tř. 458/102, 370 05 České Budějovice, Czech Republic
3 BIM, Ireland's Seafood Development Agency, Crofton Rd, Dun Laoghaire, A96 E5A0 Co. Dublin, Ireland
4 Environmental Research Institute, Lee Road, T23 N73K Cork, Ireland
* Correspondence: spaolacci@bmrs.ie

Abstract: Duckweed (Lemnaceae) can support the development of freshwater aquaculture if used as extractive species in Integrated MultiTrophic Aquaculture (IMTA) systems. These aquatic plants have the advantage of producing protein-rich biomass that has several potential uses. On the contrary, other biological compartments, such as microalgae and bacteria, present in the water and competing with duckweed for light and nutrients cannot be harvested easily from the water. Moreover, as phytoplankton cannot easily be harvested, nutrients are eventually re-released; hence, this compartment does not contribute to the overall water remediation process. In the present study, a mesocosm experiment was designed to quantify the portion of nutrients effectively removed by duckweed in a duckweed-based aquaculture wastewater remediation system. Three tanks were buried next to a pilot-scale IMTA system used for the production of rainbow trout and perch. The tanks received aquaculture effluents from the adjacent system, and 50% of their surface was covered by duckweed. Daily water analyses of samples at the inlet and outlet of the mesocosm allowed quantification of the amount of nutrients removed in total. The portion removed by duckweed was determined by examining the nutrient content in the initial and final biomass. The portion of nutrients removed by other compartments was similarly estimated. The results show that duckweed is responsible for the removal of 31% and 29% of N and P, respectively. Phytoplankton removed 33% and 38% of N and P, respectively, while the biofilm played no major role in nutrient removal. The remainder of the removed nutrients were probably assimilated by bacteria or sedimented. It is speculated that a higher initial duckweed density can limit phytoplankton growth and, therefore, increase the portion of nutrients removed by the duckweed compartment.

Keywords: Lemnaceae; aquaculture effluents; IMTA; RAS; phytoplankton; bacteria; phytoremediation

Citation: Paolacci, S.; Stejskal, V.; Toner, D.; Jansen, M.A.K. Integrated Multitrophic Aquaculture; Analysing Contributions of Different Biological Compartments to Nutrient Removal in a Duckweed-Based Water Remediation System. *Plants* **2022**, *11*, 3103. https://doi.org/10.3390/plants11223103

Academic Editors: Viktor Oláh, Klaus-Jürgen Appenroth and K. Sowjanya Sree

Received: 19 October 2022
Accepted: 12 November 2022
Published: 15 November 2022

1. Introduction

Demand for protein is rising sharply, with worldwide shortages of quality protein expected in the nearby future [1]. Aquaculture has the potential to contribute substantially to the production of protein required to feed an increasing world population [2]. Freshwater aquaculture can be a local source of protein, even in regions distant from the coast, where supply of marine seafood would involve food miles, and associated carbon emissions [3].

The development of freshwater aquaculture is sometimes impeded by concerns of negative environmental impacts. Consequently, the development of innovative, sustainable approaches to aquaculture is increasingly recognized as central to accelerate growth of the sector [4]. Such sustainable aquaculture should focus on high yields of quality produce, as well as minimise negative effects on the environment. Two negative effects of traditional freshwater aquaculture relate to water-use and eutrophication. Intensive aquaculture generates effluents rich in dissolved inorganic nutrients such as ammonia, and phosphate. If discharged without treatment, these effluents can have strong negative impacts on the water quality of receiving waterbodies. Eutrophication can lead to excessive growth of phytoplankton leading to algal blooms may, in turn, cause hypoxia which can affect a broad spectrum of organisms ranging from invertebrates to fish [5]. Moreover, the use of large volumes of freshwater for traditional aquaculture can exert pressure on local water resources in drier regions [6].

Recirculating Aquaculture Systems (RAS) can reduce the amount of water necessary to farm fish and prevent the negative effects of nutrients released by aquaculture on aquatic ecosystems [7]. In such recirculating systems, water is partially reused after undergoing remediation treatment using algae [8], bacteria [9] or aquatic plants [10]. Recently, several papers have described the use of duckweed as part of RAS system [11–13]. The use of duckweed in RAS has a promising future as it addresses two separate problems simultaneously; (1) it reduces the impact of aquaculture by capturing plant nutrients and (2) it converts a waste product into a resource that has an economic value [11].

The term duckweed refers to a group of freshwater free-floating plants belonging to the family of Lemnaceae. These plants are characterized by high growth rate and high protein content [14]. Thanks to their opportunistic nature, these plants thrive in eutrophic environments [15] and their use for the treatment of wastewater has been amply demonstrated, with early papers going back as far as 1973 [16]. In recent years, there has been a renewed interest in these species and their use for the treatment of a range of different agri-food wastewaters [17–19]. Rapid growth is associated with a high capacity to extract nutrients from water. Rapid growth similarly results in a high capacity to generate valuable biomass. Due to their high protein content, the use of these plants has been suggested as an alternative source of protein for livestock [20]. Alternatively, these plants can substitute conventional synthetic fertilizers [21] or be used as a biofuel [22].

A duckweed-based remediation system is characterized by a dynamic balance between the main compartments: duckweed, phytoplankton, biofilm, and other photosynthetic bacteria. Each group forms a functional compartment that affects the other two, and the co-existence of the three compartments determines the remediation efficiency of the whole system. The balance between duckweeds, algae and bacteria changes seasonally [12]. A reduced mat of duckweed in winter and spring is associated with a relative increase in phytoplankton. In contrast, during the summer, rapid growth of duckweed results in the formation of a conspicuous duckweed mat on the water surface, resulting in the shading of the underlying water column [12]. This, in turn, will impede algal photosynthesis, and growth, and ultimately the sequestration of nutrients from the water by algae [13,23]. Understanding of the relative proportion of nutrients removed by different taxa is important in order to make accurate models on water quality, and to inform IMTA management. Moreover, the removal of nutrients by taxa other than duckweed, will affect the portion of nutrients effectively recovered from the aquaculture system, as only duckweed can be readily harvested. In turn, this means that the competition for nutrients will also determine the yield of valuable biomass, and therefore the commercial outcome from a duckweed-based IMTA [23].

In the present study, an experiment was designed to quantify the relative uptake rate of nitrogen and phosphorus by duckweed species, algae and bacteria, in a duckweed-based aquaculture wastewater restoration system. The experiment aims to improve understanding of the balance between the different biological compartments in order to develop best practices for the management of these systems. The knowledge produced will help to optimize water restoration and maximize biomass production.

2. Materials and Methods

The experiment was carried out at an Integrated Multitrophic Aquaculture (IMTA) fish farm (Co. Offaly Ireland, coordinates 53.275555, −7.208392) where Oncorhynchus mykiss and Perca fluviatilis are farmed. The aquaculture effluents produced in four fish ponds are sent to a system of canals where the duckweed species *Lemna gibba* and *Lemna minor* are used to remove plant nutrients from the water. After treatment by duckweed, the water is returned to the fishponds. The technical details of this IMTA system have been extensively described by [13]. The experiment was carried out in August 2020, when suitable conditions for duckweed growth were present, and the effluent treatment canals were abundantly covered in *L. gibba* and, to a lesser extent, *L. minor*.

2.1. Experimental Set-Up

A small model of a duckweed-based effluent treatment system was set up next to the full-scale system. This mesocosm consisted of three independent tanks (i.e., three replicates) into which fresh aquaculture effluent was pumped.

Prior to the experiment, six tanks (area 35 × 60 cm, depth 45 cm) were submerged in one of the duckweed-covered canals for seven days (Figure 1) to establish a biofilm of microorganisms and sediment on the internal surface of the tanks. Out of these six tanks, three tanks were subsequently used as part of the experimental set-up, while a further three tanks were used to analyze the biofilm present at the start of the experiment. After having been submerged for seven days in a canal, the latter three tanks were air-dried for three days, and the dry biofilm was used to determine the organic fraction and Total Nitrogen (TN) and Total Phosphorous (TP) contained within the biofilm at the start of the experiment.

Figure 1. Tanks (circled in red) were submerged for a week in one of the duckweed-covered canals prior to being used for the mesocosm experiment. The pre-treatment resulted in the establishment of a biofilm on the internal surface of the tanks.

The experimental tanks were buried next to one of the IMTA treatment canals. Just 5 cm of the side of the tank was left above ground level. At the start of the experiment, the tanks were filled with 70 L of aquaculture effluent, and 40% of their surface area was covered with duckweed, both taken from the neighboring IMTA system. An equal amount of duckweed biomass was collected, weighed, dried at 60 °C for three days, and analyzed for TP and TN. These values of TP and TN represent the initial P and N content in duckweed biomass.

A small pump was placed inside each tank in order to generate internal effluent recirculation. The flow rate was 490 $L \cdot h^{-1}$, a value that is similar to the flow conditions in the IMTA canals (Figure 2.1). Every 12 h, a sample of water was collected from each replicate mesocosm tank, after which the tanks were gently drained without disruption of the biofilm or the duckweed biomass (Figure 2.2). The water removed was immediately replaced with fresh aquaculture effluent, pumped in from the adjoining canal (Figure 2.3). A sample of effluent pumped in from the canal was also collected for analysis in order to quantify the initial nutrient concentration in the water before being treated. The collected water samples were filtered to separate phytoplankton from the remainder of the water sample. TP and TN were quantified both in filtered water and in phytoplankton. Furthermore, algal chlorophyll content, cyanobacteria, and turbidity (expressed in Formazin Turbidity Unit, FTU) were measured both in the tanks and in the canal using an AlgaeTorch produced by Bbe-Modanke. A Seneye online system (Seneye Ltd., Norwich, UK) is present in the system to monitor pH and temperature.

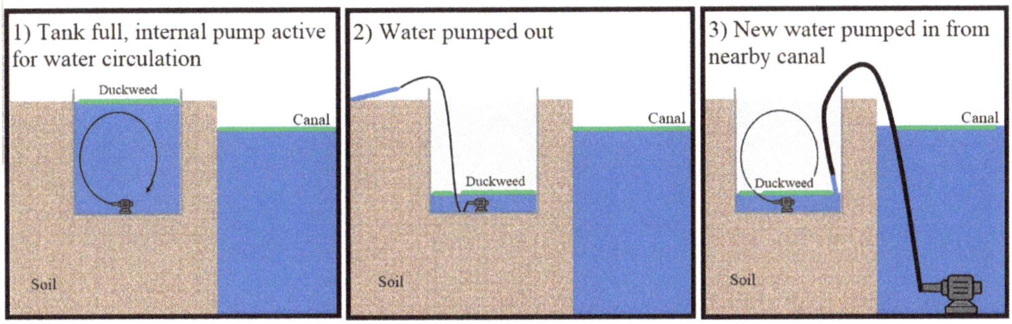

Figure 2. Experimental design. Tanks are buried next to a canal with aquaculture effluent. The tank is filled with aquaculture effluent (**1**). Twice a day, the tank is emptied without disrupting the biofilm or duckweed mat (**2**) and re-filled with fresh aquaculture effluent from the nearby canal (**3**).

The mesocosm experiment was terminated when the duckweed achieved a tank surface coverage close to 90%; this happened after eight days. At this stage, the duckweed biomass was harvested, dried at 60 °C for 3 days, and analyzed for TP and TN. The difference in the total amount of nutrients in the final duckweed biomass versus the initial duckweed biomass (i.e., plant concentration times biomass at start and finish) represents the TN or TP removed by duckweed.

Once the duckweed was removed, the tanks were emptied and air-dried for three days. The internal biofilm was gently removed and weighed. Half the biofilm biomass was used for the quantification of TP and TN, while the other half was retained to determine the organic and inorganic components. The difference between the total nutrient content in the final biofilm versus the initial biofilm (i.e., concentration times biomass at start and finish) represents the TN or TP removed from the water by bacteria, periphytic and epipelic algae, and other biofilm constituents. The experimental protocol is summarised in Table S1 of the Supplementary Material.

Half of the volume of water sampled was analyzed without filtering it, while the other half was filtered using a vacuum pump and a 5 μm filter. The phytoplankton biomass present in each sample was assessed by weighing the clean filter and then weighing it again after water sample filtration.

In the present paper, the term "duckweed" refers to the plant and its microbiome, and the term "biofilm" is used to indicate the mix of microorganisms (algae, protista, bacteria, and otherwise) and inorganic particles that build up on the internal surface of the tanks and contribute to nutrients sequestration. The term "phytoplankton" is used to indicate all the organisms with a diameter greater than 5 μm that live in the water column. Samples of biofilm and phytoplankton were observed under a Leica DM500 light microscope, and the microalgae were identified at the genus level using the dichotomic key presented in [24].

2.2. Analytical Methodology

TN in the unfiltered water sample was determined using an automated colorimetric method involving digestion of the unfiltered sample with potassium persulphate and boric acid in an alkaline solution in an autoclave at 121 °C for 30 min [25]. TP in the unfiltered water was determined using a modified Molybdate—Ascorbic Acid method following digestion of the unfiltered sample with persulphate and sulphuric acid in an autoclave at 121 °C for 30 min [26].

The other half of the samples were filtered and the dry weight of phytoplankton ($mg{\cdot}L^{-1}$) was determined. Total Dissolved Nitrogen (TDN) and Total Dissolved Phosphorus (TDP) were determined in the filtered effluent sample using the same method as detailed above. The difference between TN and TDN was taken as the total nitrogen present in phytoplankton. TP in phytoplankton was determined using the same approach. The amount of nutrients removed by phytoplankton was estimated by calculating the difference between TN and TP in the phytoplankton contained in the inlet and outlet water.

Plant samples, dried at 60 °C for 48 h and milled, were digested with concentrated sulphuric acid and Kjeltab Cu/3.5 in a TECATOR 2040 Digestor at 420 °C for 1 h. Digested samples were diluted to 250 mL using deionised water. The Total Kjeldahl Nitrogen was analysed using QuickChem IC + FIA flow injection analyzer (8000 series) manufactured by Lachat Instrument. For the determination of TP, the samples were acid digested and analysed using the ammonium molybdate method (Murphy and Riley, 1962). Absorbance was measured using a UV-Visible Recording Spectrophotometer manufactured by SHIMADZU Corporation (Model: SHIMADZU UV-160A). The same procedures were used for the analyses of TN and TP in the biofilm.

2.3. Duckweed Growth Analysis

A photograph of the duckweed mat was taken every morning and analysed using the imaging software ImageJ to determine the growth in the duckweed covered tank surface area. The Relative Growth Rate (RGR) was calculated using the formula by [27]:

$$RGR = \ln(Yf/Yi)/t$$

where Yi is the initial area of duckweed cover, Yf is the final area, t is the time in days and ln is the natural logarithm.

2.4. Calculations

The total amount of nitrogen removed from the water during the 8 days of experiments is expressed with the formula:

$$TNr = TNd + TNb + TNp + TNo$$

where:

TNr = Total Nitrogen removed from the wastewater
TNd = Total Nitrogen removed by duckweed
TNb = Total Nitrogen removed by the biofilm
TNp = Nitrogen removed by phytoplankton
TNo =Nitrogen removed by other processes/organisms
The variables of this equation were calculated as follows:

$$TNr = \sum_{day\ 1}^{day\ 8} (CN_i \times V) - (CN_o \times V)$$

CN_i = concentration of nitrogen in the inflow (filtered water pumped in from the canal) expressed in $mg{\cdot}L^{-1}$

CN_o = concentration of nitrogen in the outflow (water in the tank after 12 h) expressed in $mg{\cdot}L^{-1}$

V = volume of water treated in 12 h expressed in l

$$TNd = (CNd_f \times dwd_f) - (CNd_i \times dwd_i)$$

CNd_f = concentration nitrogen in final duckweed biomass
wd_f = final duckweed weight
CNd_i concentration of nitrogen in initial duckweed biomass
wd_i = initial duckweed weight

$$TNb = (CNb_f \times dwb_f) - (CNb_i \times dwb_i)$$

CNb_f = concentration of nitrogen in the final biofilm
dwb_f = final biofilm dry weight
CNb_i = concentration of nitrogen in the initial biofilm
Dwb_i = initial biofilm weight

$$TNp = \sum_{day\ 1}^{day\ 8} (CNp_f \times wp_f) - (CNp_i \times wp_i)$$

CNp_f = concentration of nitrogen in phytoplankton in the outflowing water
wp_f = final phytoplankton weight
CNp_i = concentration of nitrogen in phytoplankton in the inflowing effluent
wp_i = initial phytoplankton weight

After determining the variables described above, TNo was determined as a difference between the TNr and TN extracted by the other compartments:

$$TNo = TNr - (TNd + TNp + TNb)$$

The total amount of phosphorus removed from the water during the eight days of experiments was expressed with the same formula described above for the total amount of nitrogen removed.

2.5. Statistical Analyses

R studio version 4.1.1. was used for the statistical analysis. The correlation between the area of the duckweed mat and N and P removed from the tanks was analyzed with the non-parametric Kendall Tau test. A t-test was used to analyze differences in nutrient removal between night and day.

3. Results

3.1. Water Parameters

During the experiment, the water temperature varied between 11 and 19 °C.

The pH of the aquaculture effluent varied between 7.55 and 8.53 during the 8-day experiment, while the turbidity varied between 20.9 and 24.6 FTU (Figure 3).

The total dissolved nitrogen and phosphorus concentrations in the incoming aquaculture effluent were quantified twice a day, i.e., every time the water in the tanks was replaced. TN varied between 2.97 and 4.3 $mg \cdot L^{-1}$, while TP varied between 0.43 and 0.72 $mg \cdot L^{-1}$ (Figure 4).

The concentration of algal chlorophyll and cyanobacteria in the incoming effluent was also measured twice a day. The chlorophyll concentration in the water varied between 251.5 and 342.2 $\mu g \cdot L^{-1}$. The concentration of cyanobacteria was extremely high on the first day of the experiment (66.1 $\mu g \cdot L^{-1}$), but it decreased during the following days, reaching a low of 21.1 $\mu g \cdot L^{-1}$ toward the final days of the experiment (Figure 5).

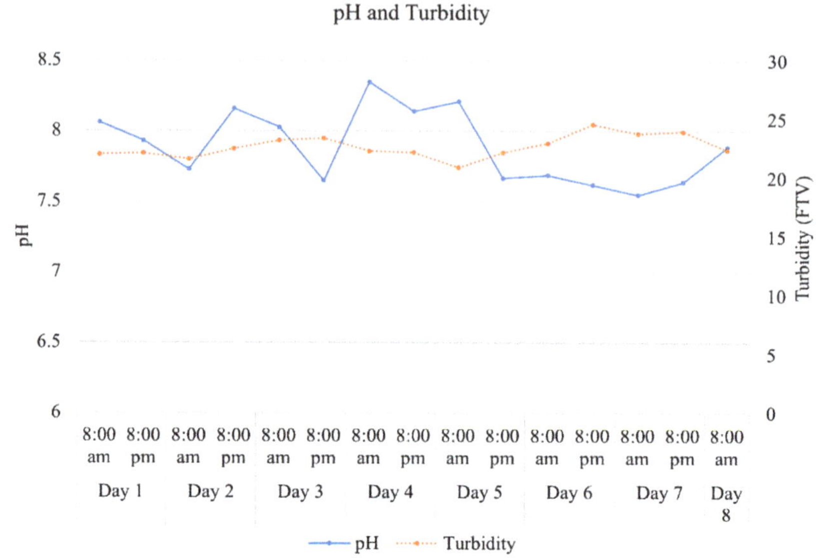

Figure 3. pH and turbidity in the aquaculture wastewater measured at 8:00 am and 8:00 pm during the eight days of the experiment. FTV = Formazin Turbidity Unit.

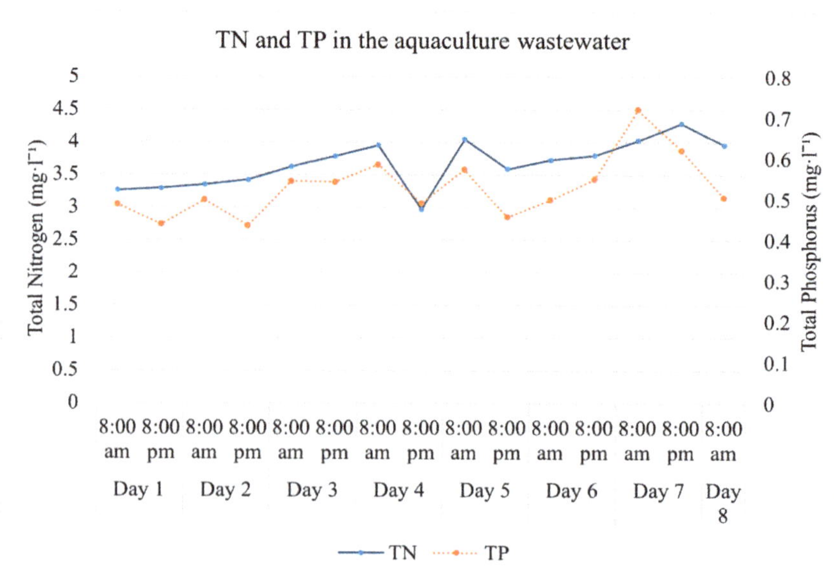

Figure 4. Total Nitrogen and Total Phosphorus concentrations in the aquaculture effluent measured at 8:00 am and 8:00 pm over the eight days of the experiment.

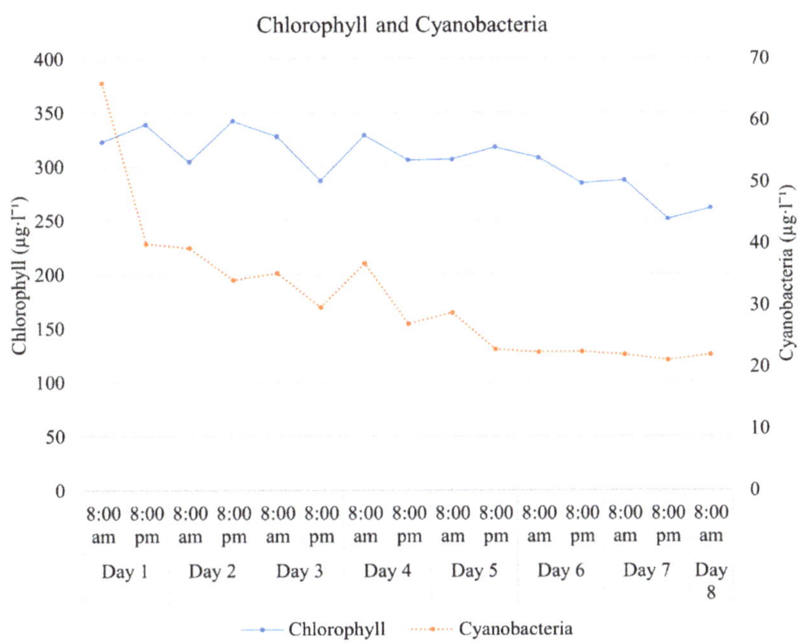

Figure 5. Algal chlorophyll and cyanobacteria concentration in the aquaculture wastewater measured at 8:00 am and 8:00 pm during the eight days of the experiment. N= night, D= day.

Samples of biofilm attached to the internal surface of the tanks were observed with a microscope. The genera identified are indicated in Table 1.

Table 1. Genera of microalgae identified in samples of biofilm and phytoplankton built in the internal surface of the tanks.

Genera Identified in the Phytoplankton and in the Biofilm
Bacillariophyta
Cyclotella sp.
Tabellaria sp.
Nitzschia sp.
Chlorophyta
Micractinium sp.
Scenedesmus sp.
Actinastrum sp.
Pediastrum sp.
Chlamydomonas sp.
Monoraphidium sp.
Charophyta
Closterium sp.

3.2. Nutrient Removal

The total volume of effluent treated by each tank in eight days was 1050 L. This volume, together with the cumulative differences in nutrient concentration between inflow and outflow water, allowed the calculation of the amount of dissolved N and P removed from the water.

A total of 689.93 mg of nitrogen (Figure 6) and 154.44 mg of phosphorus (Figure 7) were removed from the aquaculture wastewater. These numbers were estimated from the difference between inflow and outflow water in unfiltered samples. The N removal rate varied between −0.16 and 2.63 $mg{\cdot}L^{-1}{\cdot}d^{-1}$, while the P removal rate varied between 0.05 and 0.64 $mg{\cdot}L^{-1}{\cdot}d^{-1}$. There was no significant difference between N and P removed during the day and during the night.

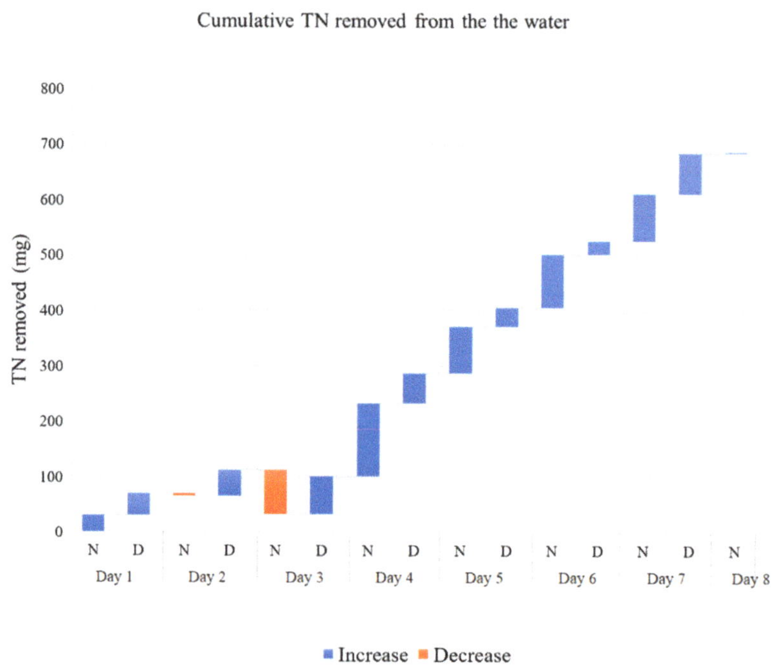

Figure 6. Cumulative Total Nitrogen removed from 70 L of aquaculture effluent every 12 h. N (Night) indicates the Total Nitrogen removed between 8:00 pm and 8:00 am; D (Day) indicates the Total Nitrogen removed between 8:00 am and 8:00 pm. The graph shows the average of three replicates.

3.3. Nutrients Accumulated by Phytoplankton

The phytoplankton contained in the inflow and outflow water was filtered and weighed. The filtered water samples (not containing phytoplankton) were analyzed for N and P content. The difference in nutrient concentration between filtered and unfiltered water samples represents the net nutrient removal by this biological compartment. The phytoplankton captured through filtering of the inlet and outlet water allowed the estimation that an average of 32.64 g of fresh weight phytoplankton (13.88 ± 1.22 during night hours and 18.66 ± 2.46 g during day hours) were produced, in each tank during the eight days of the experiment. The TN and TP concentrations in this biomass ranged between 14.82 and 35.2 $mg\ N{\cdot}g^{-1}$ and 5.72 and 10.09 $mg\ P{\cdot}g^{-1}$. The sums of TN and TP accumulated by phytoplankton are in total 230.06 mg (Figure 8) and 59.08 mg, respectively (Figure 9). Table 2 shows the details of phytoplankton produced and nutrients accumulated by this compartment.

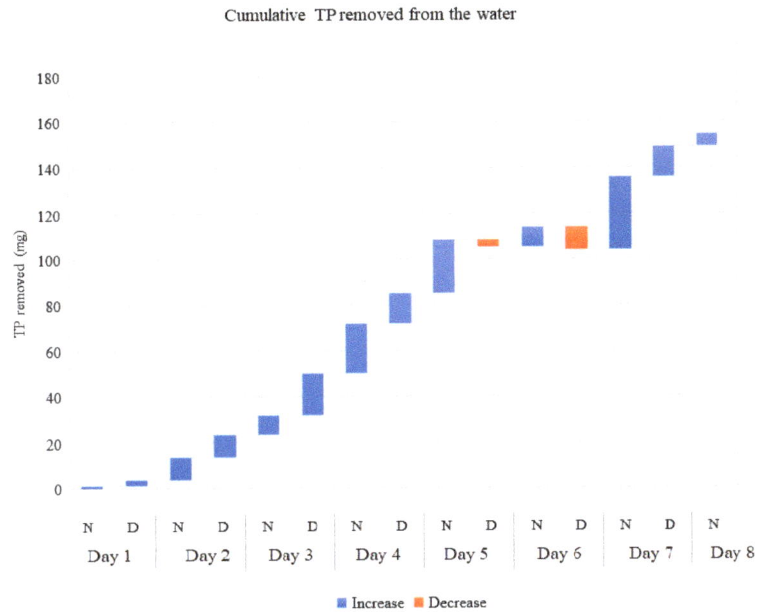

Figure 7. Cumulative Total Phosphorus removed from 70 L of aquaculture effluent every 12 h. N (Night) indicates the Total Phosphorus removed between 8:00 pm and 8:00 am; D (Day) indicates the Total Phosphorus removed between 8:00 am and 8:00 pm. The graph shows the average of three replicates.

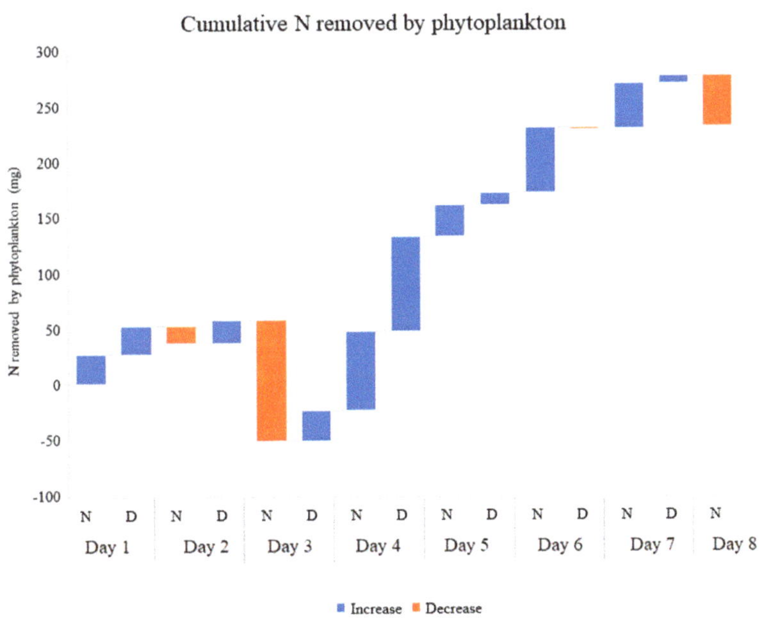

Figure 8. Cumulative Total Nitrogen removed by phytoplankton from 70 L of aquaculture effluent every 12 h. N (Night) indicates the Total Nitrogen removed between 8:00 pm and 8:00 am; D (Day) indicates the Total Nitrogen removed between 8:00 am and 8:00 pm. The graph shows the average of three replicates.

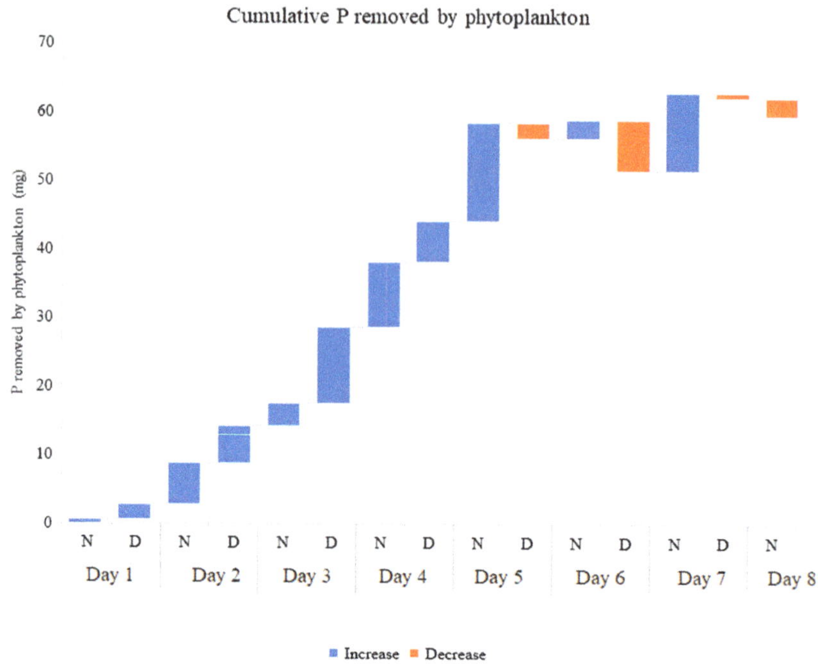

Figure 9. Cumulative total phosphorus removed from 70 L of aquaculture effluent every 12 h. N (Night) indicates the Total Phosphorus removed between 8:00 pm and 8:00 am; D (Day) indicates the Total Phosphorus removed between 8:00 am and 8:00 pm. The graph shows the average of three replicates.

Table 2. Biomass gained and nutrients accumulated by phytoplankton during the experiment. The total biomass produced was estimated from the phytoplankton filtered from the water samples collected twice a day from the inlet and outlet water.

Phytoplankton Biomass	
Total biomass that entered the system (g of fresh weight)	110.11
Total final biomass	142.75
N in tot. initial biomass (mg)	775.83
N in tot. final biomass (mg)	1005.89
N accumulated by phytoplankton (mg)	**230.06**
P in tot. initial biomass (mg)	199.3
P in tot. final biomass (mg)	258.38
P accumulated by phytoplankton (mg)	**59.08**

3.4. Duckweed Biomass

The area of the duckweed mat increased from 831.81 ± 23.03 cm^2 to 1936.47 ± 455.54 cm^2. The RGR calculated for the duration of the experiment was 0.11 ± 0.003 d^{-1}. Figure 10 shows the daily increase in duckweed surface. The plants grew slowly during the first days, while the absolute growth rate increased over the last two days.

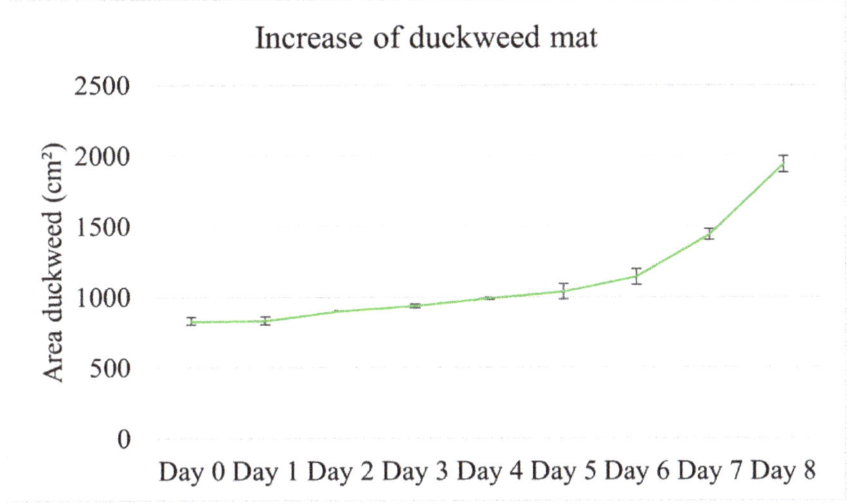

Figure 10. Increase in surface area covered by duckweed mat during the eight days of the experiment. Error bars are standard deviations.

The relationship between the increase in the area of the duckweed mat and the removal of nutrients was not linear (Supplementary Material); the non-parametric Kendall Tau test failed to identify a significant correlation between the duckweed area and the amount of N and P removed from the water.

The dry weight of duckweed, during the experiment, increased by 5.81 g. The content of N and P in duckweed biomass is indicated in Table 3. The aquatic plants removed, in 8 days, 210.1 mg of N and 44.5 mg of P from the entire water volume treated.

Table 3. Biomass gained and nutrients removed by duckweed during the experiment.

Duckweed Biomass	
Initial dry weight (g)	3.32 ± 0.41
Final dry weight (g)	9.13 ± 0.55
N in initial biomass (mg·g^{-1})	38.4 ± 0.36
N in final biomass (mg·g^{-1})	36.8 ± 4.33
N removed by duckweed (mg)	**210.1 ± 71.3**
P in initial biomass (mg·g^{-1})	8.61 ± 0.49
P in final biomass (mg·g^{-1})	7.99 ± 0.37
P removed by duckweed (mg)	**44.5 ± 6.2**

3.5. *Biofilm*

Most of the biofilm on the internal surface of the tanks constituted inorganic sediments. The organic matter represented only 15.6% of the biofilm at the start of the experiment and only 3.4% at the end of the experiment. The biofilm did not contribute substantially to the removal of nutrients from the wastewater. On the contrary, 11.2 mg of N and 4.9 mg of P were released from this compartment over the eight-day experimental period (Table 4).

Table 4. Biofilm weight, % of organic matter, and nutrient content at the start and at the end of the experiment.

Biofilm	
Initial dry weight (g)	4.57 ± 3.51
Organic fraction (%)	15.63
Final dry weight (g)	6.02 ± 0.97
Organic fraction (%)	3.39
N in initial biofilm (mg·g^{-1})	5.0 ± 1.6
N in final biofilm (mg·g^{-1})	1. 7 ± 0.6
N removed by biofilm (mg)	**−11.2 ± 13.2**
P in initial biofilm (mg·g^{-1})	2.0 ± 0.4
P in final biofilm (mg·g^{-1})	0.5 ± 0.1
P removed by biofilm (mg)	**−4.9 ± 3.7**

3.6. Nutrient Removal from Different Compartments

The calculations illustrated in Section 2.5. were used to estimate the relative nutrient uptake by the different biological compartments. Duckweed is responsible for the uptake from the wastewater of 31% of the total nitrogen removed, while phytoplankton is responsible for another 33% of nitrogen removal. Some 29% of total phosphorus removal can be attributed to duckweed, while 38% is removed by phytoplankton. The biofilm did not contribute to the removal of nutrients. The remaining 36% and 33% of N and P, respectively, were removed from the aquaculture effluent by other means, such as denitrification and volatilization (Figure 11).

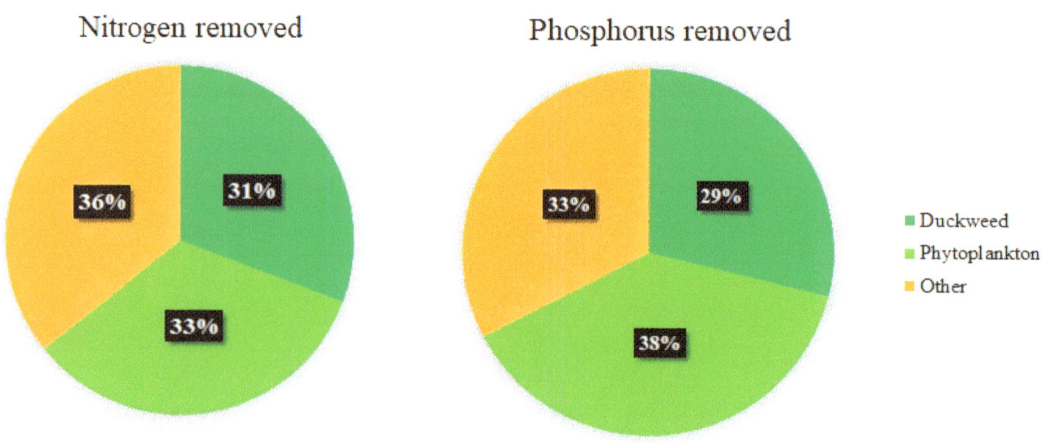

Figure 11. Percentages of total nitrogen and total phosphorus removed by the different compartments during the experiment.

4. Discussion

The present study assessed the amount of N and P removed from aquaculture wastewater by duckweed species, biofilm, and phytoplankton. The respective proportion of nutrients removed by these three compartments will inform system management and predict the portion of the nutrient load that can effectively be converted into valuable biomass as part of a circular economy approach.

4.1. Characteristics of the Effluent

The water quality parameters observed during the experiment are consistent with the parameters observed the year before by O'Neill et al. [28] during the same season, and at the same aquaculture farm. The nutrient concentration observed are also in the range previously identified by Paolacci et al. [29] The latter authors reviewed the characteristics of wastewater generated by rainbow trout and perch farms. They found that TN ranges between 0.5 and 70 mg·L^{-1}, while TP ranges between 0.42 and 15 mg·L^{-1}, depending mainly on the fish density.

The chlorophyll concentration is an indicator of the phytoplankton density in the water. Phytoplankton plays an important role in recirculating systems as it removes the NH4+-N from the effluents, contributing to the water restoration process [12]. The average chlorophyll concentration measured by [30] in a rainbow trout farm was 5 µg·L^{-1}, considerably lower than the concentrations observed in the effluents used for this experiment (between 251.5 and 342.2 µg·L^{-1}). However, Sen and Sonmez' data [30] refer to a flow through system. In an outdoor recirculating system the phytoplankton is never released into natural waters and it seems plausible that higher chlorophyll concentrations can accumulate in the water.

Water turbidity depends mainly on the concentration and the characteristics of the suspended particles which depend, amongst others, on the geological substrate [31]. The experimental site is located in a cutaway peatland and the values of turbidity measured are consistent with those measured by [32] in lakes present in the same area (Offaly, IE), and generated by flooding cutaway bogs without removing the residual peat (as it was done in the IMTA system of the present study). Turbidity affects light diffraction in the water column and can limit phytoplankton growth. Despite the high turbidity, the chlorophyll concentration measured suggests that this parameter was not suppressing phytoplankton. However, the turbidity can explain the low number of taxa observed in the biofilm and in the phytoplankton community [33].

4.2. Nutrient Removal from Water

The results show that the total amount of N removed from the water was nearly five times higher than the amount of P removed (684.9 g vs. 154.4 g). This proportion is consistent with the observations of other authors. For example, [34] reported that, in a duckweed-covered sewage lagoon, the removal rates for N and P were, respectively, 0.26 g·m^{-2}·d^{-1} and 0.05 g·m^{-2}·d^{-1}. This proportion reflects the N:P ratio in duckweed biomass [34] and also the N:P ratio observed in the biomass of other aquatic plants [35]. Similar N and P uptake data were also previously reported by [36] using dairy processing waste.

No significant difference was observed in nutrient removal during the night and the day. This can be explained by considering that the samples were taken at 8 am and 8 pm, in the summer, in Ireland. Between July and August (when the experiment was performed) this latitude experiences around 16 h of daylight. This means the 'night' uptake includes four hours of light during which duckweed and phytoplankton can perform photosynthesis and remove nutrients. Moreover, during night hours, sedimentation of nutrients is active and can contribute to N and P removal from the water column.

During the experiment, duckweed increased their surface density from 50% to >90%. The growth rate was consistent with the growth rates observed earlier for duckweed during the same time of the year in Ireland [37]. Around 30% of both N and P in the water was removed by duckweed, presumably to support growth, although some luxury uptake cannot be excluded. This percentage is consistent with the results of a small-scale experiment (1 L batches) performed by [38]. The latter authors determined the uptake rates of N and P in domestic wastewater by duckweed, algae and bacteria, at different initial densities of duckweed. They observed that duckweed was responsible for removing between 30% and 47% of N and up to 54% of P, depending on the initial plant density. The results of the present study are also in accordance with the observations of [23] who performed a competition experiment between phytoplankton and duckweed. When the duckweed mat was dense (like in the present study), the authors observed an increase in N and P content in the plant biomass, while at lower duckweed density the phytoplankton was competing with duckweed for nutrients, and this resulted in a decreased content of N and P in the biomass. The values reported by [23] for N and P content at high duckweed density are 36.65 and 10.95 $mg \cdot g^{-1}$ respectively, while the values observed in the present study are 36.8 (N) and 7.99 (P) $mg \cdot g^{-1}$ of biomass. The slightly lower value for P content is probably due to a lower concentration of this element in effluents used in the present study.

It is important to highlight that the removal of nutrients by duckweed measured in this study was determined by analysing N and P content in the initial and in the final biomass harvested at the end of the experiment. As a consequence, the values include the indirect contribution of algae and bacteria attached to, or incorporated into, the duckweed.

The phytoplankton removed on average 30 $\mu g \cdot L^{-1} \cdot d^{-1}$ of N and 7 $\mu g \cdot L^{-1} \cdot d^{-1}$ of P during the experiment. This constitutes around a third of the nutrients removed. Multiple studies have previously focused on the restoration ability of phytoplankton. For example, [39] observed that Pseudochlorella pringsheimii was able to remove 2.3 $mg \cdot L^{-1} \cdot d^{-1}$ of N and 1.3 $mg \cdot L^{-1} \cdot d^{-1}$ of P from aquaculture wastewater containing 34.8 and 18.6 $mg \cdot L^{-1}$ of N and P respectively. Ref. [40] grew Chlorella sp. in centrate wastewater containing between 150 and 340 $mg \cdot L^{-1}$ of N and between 90 and 300 $mg \cdot L^{-1}$ of P. The latter authors observed uptake rates between 14 and 20 $mg \cdot L^{-1} \cdot d^{-1}$ for N and of 2.8 $mg \cdot L^{-1} \cdot d^{-1}$ for P. In the present study, TN varied between 2.97 and 4.3 $mg \cdot L^{-1}$, while TP varied between 0.43 and 0.72 $mg \cdot L^{-1}$. The low nutrient concentrations will therefore have contributed to the reduced nutrient uptake rates. Furthermore, it is likely that the duckweed mat prevented light from entering the water column, inhibiting the phytoplankton growth and leading to a reduced nutrient uptake by this compartment.

The analysis of the biofilm revealed that this compartment released 11.2 mg of N and 4.9 mg of P (Table 3). The negative balance is probably associated with the reduction of the organic matter in the biofilm throughout the experiment.

The remaining third of nutrients removed from the water is probably linked to different processes such as sedimentation [41], denitrification and NH3-volatilisation [42]. Observed that, in diluted swine wastewater treated with the duckweed Spirodela oligorrhiza, 30% of TN present in the water is removed through ammonia volatilization [43]. Ammonia volatilization increases at pH values higher than 7 and a T close to 20 °C [44], two conditions consistent with the experimental conditions observed. Thus, it is possible that under the used experimental conditions part of the dissolved nitrogen was lost to the atmosphere.

Capturing the plant nutrients N and P using duckweed and using the plant biomass as part of a composite feed, can close the nutrient cycle and diminish the need for raw resources including mineable phosphate. However, the current experiment shows that under realistic conditions duckweed colonies capture just one-third of N and P removed from the medium. Thus, there is ample scope to improve the nutrient retention efficiency of duckweed systems, particularly by impeding phytoplankton growth and/or microbial activities.

5. Conclusions

The present study assessed the amounts of nutrients effectively recovered by duckweed from aquaculture effluents in a realistic, outdoor recirculating system. Phytoplankton in the water column removes a considerable amount of nutrients from the water. The extent of phytoplankton-mediated nutrient removal is likely to be particularly important in IMTA systems in seasons when duckweed cover is not present. However, in the long term, phytoplankton does not contribute to remediation as it cannot be easily removed from the water; hence, the nutrients will eventually be re-released into the water, nullifying the remediation process. In comparison, duckweed can be harvested, and if this is well managed, it can substantially contribute to the sequestration and removal of nutrients from aquaculture wastewater. This study confirms the phytoremediation ability of duckweed in aquaculture effluents; however, it also clearly shows that there is scope to further improve duckweed-based nutrient removal by impeding competing processes of algal growth and microbial activity.

Supplementary Materials: The following are available online at https://www.mdpi.com/article/10.3390/plants11223103/s1, Table S1: Summary of the experimental protocol.

Author Contributions: Conceptualization, S.P. and M.A.K.J.; methodology, S.P. and V.S.; formal analysis, S.P.; investigation, S.P., V.S. and M.A.K.J.; resources, D.T.; data curation, S.P.; writing—original draft preparation, S.P.; writing—review and editing, S.P., M.A.K.J., V.S. and D.T.; visualization, S.P.; supervision, M.A.K.J.; project administration, S.P. and M.A.K.J.; funding acquisition, M.A.K.J. All authors have read and agreed to the published version of the manuscript.

Funding: The study was supported by Bord Iascaigh Mhara through the Knowledge Gateway Scheme (co-funded by the Irish Government and the European Union as part of the European Maritime and Fisheries Fund 2014–2020) and project "CENAKVA" (LM2018099).

Informed Consent Statement: Not applicable.

Data Availability Statement: The data presented in this study are available on request from the corresponding author.

Acknowledgments: M.A.K.J. acknowledges support by WoB.

Conflicts of Interest: The authors declare no conflict of interest. The funders had no role in the design of the study; in the collection, analyses, or interpretation of data; in the writing of the manuscript, or in the decision to publish the results.

References

1. Aimutis, W.R. Plant-Based Proteins: The Good, Bad, and Ugly. *Annu. Rev. Food Sci. Technol.* **2022**, *13*, 1–17. [CrossRef] [PubMed]
2. Henriksson, P.J.G.; Troell, M.; Banks, L.K.; Belton, B.; Beveridge, M.C.M.; Klinger, D.H.; Pelletier, N.; Phillips, M.J.; Tran, N. Interventions for improving the productivity and environmental performance of global aquaculture for future food security. *One Earth* **2021**, *4*, 1220–1232. [CrossRef]
3. Zhong, S.; Li, A.; Wu, J. Eco-efficiency of freshwater aquaculture in China: An assessment considering the undesirable output of pollutant emissions. *Environ. Dev. Sustain.* **2022**, *2022*, 1–22. [CrossRef]
4. FAO. *The State of Fisheries and Aquaculture in the World 2018*; FAO: Rome, Italy, 2018.
5. Wang, J.; Beusen, A.H.W.; Liu, X.; Bouwman, A.F. Aquaculture Production is a Large, Spatially Concentrated Source of Nutrients in Chinese Freshwater and Coastal Seas. *Environ. Sci. Technol.* **2020**, *54*, 1464–1474. [CrossRef] [PubMed]
6. Verdegem, M.C.J.; Bosma, R.H. Water withdrawal for brackish and inland aquaculture, and options to produce more fish in ponds with present water use. *Water Policy* **2009**, *11*, 52–68. [CrossRef]
7. Martins, C.; Eding, E.; Verdegem, M.; Heinsbroek, L.; Schneider, O.; Blancheton, J.; D'Orbcastel, E.R.; Verreth, J. New developments in recirculating aquaculture systems in Europe: A perspective on environmental sustainability. *Aquac. Eng.* **2010**, *43*, 83–93. [CrossRef]
8. Craggs, R. Wastewater treatment by algal turf scrubbing. *Water Sci. Technol.* **2001**, *44*, 427–433. [CrossRef]
9. Ruiz, P.; Vidal, J.M.; Sepúlveda, D.; Torres, C.; Villouta, G.; Carrasco, C.; Aguilera, F.; Ruiz-Tagle, N.; Urrutia, H. Overview and future perspectives of nitrifying bacteria on biofilters for recirculating aquaculture systems. *Rev. Aquac.* **2020**, *12*, 1478–1494. [CrossRef]
10. Corpron, K.E.; Armstrong, D.A. Removal of nitrogen by an aquatic plant, *Elodea densa*, in recirculating Macrobrachium culture systems. *Aquaculture* **1983**, *32*, 347–360. [CrossRef]
11. Ahmed, N.; Turchini, G.M. Recirculating aquaculture systems (RAS): Environmental solution and climate change adaptation. *J. Clean. Prod.* **2021**, *297*, 126604. [CrossRef]
12. Paolacci, S.; Stejskal, V.; Toner, D.; Jansen, M.A. Wastewater valorisation in an integrated multitrophic aquaculture system; assessing nutrient removal and biomass production by duckweed species. *Environ. Pollut.* **2022**, *302*, 119059. [CrossRef]
13. Stejskjal, V.; Paolacci, S.; Toner, D.; Jansen, M.A.K. A novel multitrophic concept for the cultivation of fish and duckweed: A technical note. *J. Clean. Prod.* **2022**, *366*, 132881. [CrossRef]
14. Petersen, F.; Demann, J.; Restemeyer, D.; Ulbrich, A.; Olfs, H.-W.; Westendarp, H.; Appenroth, K.-J. Influence of the Nitrate-N to Ammonium-N Ratio on Relative Growth Rate and Crude Protein Content in the Duckweeds *Lemna minor* and *Wolffiella hyalina*. *Plants* **2021**, *10*, 1741. [CrossRef]
15. Paolacci, S.; Harrison, S.; Jansen, M.A. A comparative study of the nutrient responses of the invasive duckweed *Lemna minuta*, and the native, co-generic species *Lemna minor*. *Aquat. Bot.* **2016**, *134*, 47–53. [CrossRef]
16. Harvey, R.M.; Fox, J.L. Nutrient removal using *Lemna minor*. *J. Water Pollut. Control Fed.* **1973**, *45*, 1928–1938.
17. Devlamynck, R.; de Souza, M.F.; Michels, E.; Sigurnjak, I.; Donoso, N.; Coudron, C.; Leenknegt, J.; Vermeir, P.; Eeckhout, M.; Meers, E. Agronomic and Environmental Performance of *Lemna minor* Cultivated on Agricultural Wastewater Streams—A Practical Approach. *Sustainability* **2021**, *13*, 1570. [CrossRef]

18. Femeena, P.V.; House, G.R.; Brennan, R.A. Creating a Circular Nitrogen Bioeconomy in Agricultural Systems through Nutrient Recovery and Upcycling by Microalgae and Duckweed: Past Efforts and Future Trends. *J. ASABE* **2022**, *65*, 327–346. [CrossRef]
19. Walsh, É.; Paolacci, S.; Burnell, G.; Jansen, M.A.K. The importance of the calcium-to-magnesium ratio for phytoremediation of dairy industry wastewater using the aquatic plant *Lemna minor* L. *Int. J. Phytoremed.* **2020**, *22*, 694–702. [CrossRef]
20. Sońta, M.; Rekiel, A.; Batorska, M. Use of Duckweed (*Lemna* L.) in Sustainable Livestock Production and Aquaculture—A Review. *Ann. Anim. Sci.* **2019**, *19*, 257–271. [CrossRef]
21. Kreider, A.N.; Pulido, C.R.F.; Bruns, M.A.; Brennan, R.A. Duckweed as an Agricultural Amendment: Nitrogen Mineralization, Leaching, and Sorghum Uptake. *J. Environ. Qual.* **2019**, *48*, 469–475. [CrossRef]
22. Cui, W.; Cheng, J.J. Growing duckweed for biofuel production: A review. *Plant Biol.* **2015**, *17*, 16–23. [CrossRef] [PubMed]
23. Roijackers, R.; Szabó, S.; Scheffer, M. Experimental analysis of the competition between algae and duckweed. *Arch Hydrobiol.* **2004**, *160*, 401–412. [CrossRef]
24. Bellinger, E.G.; Sigee, D.C. A key to the more frequently occurring freshwater algae. In *Freshwater Algae, Identification and Use as Bioindicator*, 1st ed.; John Wiley & Sons Ltd.: Chichester, UK, 2010.
25. Grasshoff, K.; Ehrhardt, M.; Kremling, K. *Determination of oxygen. Methods of Seawater Analysis*; Wiley-VCH: Weinheim, Germany; New York, NY, USA; Chichester, UK; Brisbane, Australia; Singapore; Toronto, ON, Canada, 1983.
26. APHA-AWWA-WEF. *Standard Methods for the Examination of Water and Wastewater*, 21st ed.; American Public Health Association (APHA): Washington, DC, USA, 2005.
27. Connolly, J.; Wayne, P. Asymmetric competition between plant species. *Oecologia* **1996**, *108*, 311–320. [CrossRef] [PubMed]
28. O'Neill, E.A.; Stejskal, V.; Clifford, E.; Rowan, N.J. Novel use of peatlands as future locations for the sustainable intensification of freshwater aquaculture production—A case study from the Republic of Ireland. *Sci. Total Environ.* **2020**, *706*, 136044. [CrossRef] [PubMed]
29. Paolacci, S.; Stejskal, V.; Jansen, M.A.K. Estimation of the potential of *Lemna minor* for effluent remediation in integrated multi-trophic aquaculture using newly developed synthetic aquaculture wastewater. *Aquac. Int.* **2021**, *29*, 2101–2118. [CrossRef]
30. Sen, B.; Sonmez, F. A Study on the Algae in Fish Ponds and Their Seasonal Variations. *Int. J. Sci. Technol.* **2006**, *1*, 25–33.
31. Mukundan, R.; Pierson, D.; Schneiderman, E.; O'Donnell, D.; Pradhanang, S.; Zion, M.; Matonse, A. Factors affecting storm event turbidity in a New York City water supply stream. *Catena* **2013**, *107*, 80–88. [CrossRef]
32. Lally, H.; Higgins, T.; Colleran, E.; Gormally, M. Lakes: A new concept for wildlife conservation on Irish cutaway peatlands. In *After Wise Use–the Future of Peatlands, Proceedings of the 13th International Peat Congress, Tullamore, Ireland, 8–13 June 2008*; 2008. Available online: https://www.researchgate.net/profile/Heather-Lally/publication/297761882_LAKES_A_NEW_CONCEPT_FOR_WILDLIFE_CONSERVATION_ON_IRISH_CUTAWAY_PEATLANDS/links/56e34bfe08ae68afa10ca976/LAKES-A-NEW-CONCEPT-FOR-WILDLIFE-CONSERVATION-ON-IRISH-CUTAWAY-PEATLANDS.pdf (accessed on 11 November 2022).
33. Shen, P.-P.; Li, G.; Huang, L.-M.; Zhang, J.-L.; Tan, Y.-H. Spatio-temporal variability of phytoplankton assemblages in the Pearl River estuary, with special reference to the influence of turbidity and temperature. *Cont. Shelf Res.* **2011**, *31*, 1672–1681. [CrossRef]
34. Alaerts, G.; Mahbubar, R.; Kelderman, P. Performance analysis of a full-scale duckweed-covered sewage lagoon. *Water Res.* **1996**, *30*, 843–852. [CrossRef]
35. Hu, J.; Yu, H.; Li, Y.; Wang, J.; Lv, T.; Liu, C.; Yu, D. Variation in resource allocation strategies and environmental driving factors for different life-forms of aquatic plants in cold temperate zones. *J. Ecol.* **2021**, *109*, 3046–3059. [CrossRef]
36. Walsh, É.; Coughlan, N.; O'Brien, S.; Jansen, M.; Kuehnhold, H. Density Dependence Influences the Efficacy of Wastewater Remediation by *Lemna minor*. *Plants* **2021**, *10*, 1366. [CrossRef]
37. Paolacci, S.; Jansen, M.; Harrison, S. Competition Between *Lemna minuta*, *Lemna minor*, and *Azolla filiculoides*. Growing Fast or Being Steadfast? *Front. Chem.* **2018**, *6*, 207. [CrossRef]
38. Körner, S.; Vermaat, J. The relative importance of *Lemna gibba* L., bacteria and algae for the nitrogen and phosphorus removal in duckweed-covered domestic wastewater. *Water Res.* **1998**, *32*, 3651–3661. [CrossRef]
39. Kumar, V.; Jaiswal, K.K.; Verma, M.; Vlaskin, M.S.; Nanda, M.; Chauhan, P.K.; Singh, A.; Kim, H. Algae-based sustainable approach for simultaneous removal of micropollutants, and bacteria from urban wastewater and its real-time reuse for aquaculture. *Sci. Total Environ.* **2021**, *774*, 145556. [CrossRef]
40. Min, M.; Wang, L.; Li, Y.; Mohr, M.J.; Hu, B.; Zhou, W.; Chen, P.; Ruan, R. Cultivating Chlorella sp. in a Pilot-Scale Photobioreactor Using Centrate Wastewater for Microalgae Biomass Production and Wastewater Nutrient Removal. *Appl. Biochem. Biotechnol.* **2011**, *165*, 123–137. [CrossRef]
41. Nájera, A.F.; Serwecińska, L.; Szklarek, S.; Mankiewicz-Boczek, J. Characterization and comparison of microbial communities in sequential sedimentation-biofiltration systems for removal of nutrients in urban rivers. *Ecol. Eng.* **2020**, *149*, 105796. [CrossRef]
42. Chuong, T.; Plant, R.; Linquist, B.A. Fertilizer source and placement influence ammonia volatilization losses from water-seeded rice systems. *Soil Sci. Soc. Am. J.* **2020**, *84*, 784–797. [CrossRef]
43. Shen, G.X.; le Xu, J.; Hu, S.Q.; Zhao, Q.J.; di Liu, Y. Nitrogen removal pathways in shallow-water duckweed-based wastewater treatment systems. *J. Ecol. Rural. Environ.* **2006**, *22*, 42–47.
44. Liyuan, L.; Xiangqun, Z.; Chengfeng, P.; Junyi, L.; Yan, X. Driving forces and future trends on total nitrogen loss of planting in China. *Environ. Pollut.* **2020**, *267*, 115660. [CrossRef]

plants MDPI

Article

The Dynamics of NO_3^- and NH_4^+ Uptake in Duckweed Are Coordinated with the Expression of Major Nitrogen Assimilation Genes

Yuzhen Zhou, Olena Kishchenko, Anton Stepanenko, Guimin Chen, Wei Wang, Jie Zhou, Chaozhi Pan and Nikolai Borisjuk *

Jiangsu Key Laboratory for Eco-Agricultural Biotechnology around Hongze Lake, Jiangsu Collaborative Innovation Centre of Regional Modern Agriculture and Environmental Protection, Huaiyin Normal University, West Changjiang Road 111, Huai'an 223000, China; zyz@hytc.edu.cn (Y.Z.); o_kishchenko@hotmail.com (O.K.); stepanenko@hytc.edu.cn (A.S.); cgm@hytc.edu.cn (G.C.); ww376145775@163.com (W.W.); zhoujie554478400@163.com (J.Z.); pcz17878118183@163.com (C.P.)

* Correspondence: nborisjuk@hytc.edu.cn; Tel.: +86-0517-83525319

Abstract: Duckweed plants play important roles in aquatic ecosystems worldwide. They rapidly accumulate biomass and have potential uses in bioremediation of water polluted by fertilizer runoff or other chemicals. Here we studied the assimilation of two major sources of inorganic nitrogen, nitrate (NO_3^-) and ammonium (NH_4^+), in six duckweed species: *Spirodela polyrhiza, Landoltia punctata, Lemna aequinoctialis, Lemna turionifera, Lemna minor*, and *Wolffia globosa*. All six duckweed species preferred NH_4^+ over NO_3^- and started using NO_3^- only when NH_4^+ was depleted. Using the available genome sequence, we analyzed the molecular structure and expression of eight key nitrogen assimilation genes in *S. polyrhiza*. The expression of genes encoding nitrate reductase and nitrite reductase increased about 10-fold when NO_3^- was supplied and decreased when NH_4^+ was supplied. NO_3^- and NH_4^+ induced the glutamine synthetase (GS) genes *GS1;2* and the *GS2* by 2- to 5-fold, respectively, but repressed *GS1;1* and *GS1;3*. NH_4^+ and NO_3^- upregulated the genes encoding ferredoxin- and NADH-dependent glutamate synthases (Fd-GOGAT and NADH-GOGAT). A survey of nitrogen assimilation gene promoters suggested complex regulation, with major roles for NRE-like and GAATC/GATTC *cis*-elements, TATA-based enhancers, $(GA/CT)_n$ repeats, and G-quadruplex structures. These results will inform efforts to improve bioremediation and nitrogen use efficiency.

Keywords: duckweed; *Spirodela polyrhiza*; nitrogen assimilation; nitrate reductase; nitrite reductase; glutamine synthetase; GOGAT; gene expression

Citation: Zhou, Y.; Kishchenko, O.; Stepanenko, A.; Chen, G.; Wang, W.; Zhou, J.; Pan, C.; Borisjuk, N. The Dynamics of NO_3^- and NH_4^+ Uptake in Duckweed Are Coordinated with the Expression of Major Nitrogen Assimilation Genes. *Plants* **2022**, *11*, 11. https://doi.org/10.3390/plants11010011

Academic Editors: Viktor Oláh, Klaus-Jürgen Appenroth and K. Sowjanya Sree

Received: 10 November 2021
Accepted: 19 December 2021
Published: 21 December 2021

1. Introduction

The application of nitrogen (N) fertilizers produced substantial crop yield increases, but N fertilizers also cause serious environmental problems [1]. Plants only absorb about 50% of the N fertilizer applied in agriculture [2]; the remainder is mainly lost to the environment, leading to soil acidification, air pollution (ammonia and nitrogen oxides), and water eutrophication (mainly nitrate (NO_3^-) and ammonium (NH_4^+)) [3]. Agriculture is responsible for 59% of the current environmental N discharge, with the remaining 41% contributed by domestic and industrial waste [4]. Aquaculture and livestock wastewater also contribute to the eutrophication of water reservoirs [5]. Water eutrophication is a global concern, and a major environmental problem for water resource management. This is especially true in China, which has increased food crop production remarkably during recent decades, largely due to the extensive application of N fertilizers. In 2020, China accounted for over 30% of the 160 megatons of N fertilizer applied worldwide [6]. The resulting runoff has led to some regions substantially exceeding the surface-water quality standard of 1 mg N/L. Remedying these problems requires transformative changes to boost N recycling; implementing these changes was recently estimated to cost China \$18–29 billion per year [4].

Biological wastewater treatment using aquatic plants is a feasible, eco-friendly, and cost-effective approach [7–9]. For example, wetlands have been constructed worldwide to improve water quality for domestic reuse, irrigation, and environmental protection; the United States Department of Agriculture (USDA) alone has spent more than US \$4.2 billion on wetland restoration and protection, especially through the Conservation Reserve Program and the Wetland Reserve Program [10,11].

Floating aquatic macrophytes, including duckweeds (Lemnaceae), represented by 37 worldwide distributed species [12–14], have great potential for uses in sustainable wastewater recovery [1,15]. Duckweeds' applications rely on their capacity to efficiently take up the various contaminants responsible for eutrophication [16]. For example, about 98.0% of N and phosphorous (P) were absorbed in duckweed-populated wastewater reservoirs, with a simultaneous increase in dissolved oxygen [17,18]. Moreover, their exceptionally high propagation rates lead to fast accumulation of biomass rich in starch and protein and therefore, duckweed plants are considered a valuable feedstock for the production of biofuels [19], for livestock feed, and for human consumption [20].

Plant biomass accumulation is strongly associated with N utilization, and duckweed plants are extremely efficient at assimilating N. For example, duckweed nitrogen use efficiency (NUE) reached more than 68 kg biomass/kg N under N limitation due to N remobilization and recycling by the ubiquitin-proteasome system and autophagy [21]. However, despite intensive investigation of various duckweed species for remediation of wastewater and biomass production [22–25], studies of nutrient assimilation by duckweed species and the molecular mechanisms underlying duckweed's remarkable NUE remain limited to a few recent studies [26,27]. By contrast, the major enzymes and molecular aspects of N assimilation have been uncovered in other plant species, primarily *Arabidopsis thaliana* and rice (*Oryza sativa*) [28].

N mostly enters into plant tissues in inorganic form (NO_3^- and NH_4^+) by absorption from soil facilitated by nitrate transporters (NRTs) and ammonium transporters (AMTs) [29,30]. Inorganic N can be incorporated into cellular organic compounds only in the form of NH_4^+; therefore, NO_3^- is first reduced by cytosolic nitrate reductase (NR) to nitrite (NO_2^-), which is then imported into the plastid, where it is further reduced by nitrite reductase (NiR) to NH_4^+. The NH_4^+, whether taken directly from the environment or converted from NO_3^-, is assimilated by glutamine synthetase (GS) into glutamine, which provides N for virtually all cellular N-containing components directly or via glutamate (Figure S1).

Higher plants contain several GS isoenzymes, which are located in the cytosol (GS1) and in the plastids (GS2) and are encoded by a small multigene family [31]. Cytosolic GS1 plays a major role in primary NH_4^+ assimilation in roots and in re-assimilation of the NH_4^+ generated during protein degradation and amino acid catabolism; chloroplast GS2 is involved in assimilation of the NH_4^+ released during photorespiration or reduction of the NO_2^- generated by NO_3^- conversion. Glutamine-2-oxoglutarate aminotransferase (GOGAT) acts in tandem with GS2 to synthesize glutamate via the GS-GOGAT cycle. Plants have two different types of GOGAT enzymes: Fd-GOGAT (EC 1.4.7.1), which uses ferredoxin (Fd) as an electron donor, and NADH-GOGAT (EC 1.4.1.14), which uses NADH.

One important aspect of N assimilation is the plant's preference for NH_4^+ over NO_3^- as the source of N [32], a question that has attracted substantial attention because of its practical application in terms of the form of N supplied in fertilizer [2,29]. Most plants prefer NO_3^- to NH_4^+, although NO_3^- uptake requires more energy than NH_4^+, as absorption of NO_3^- works against a steep electrochemical gradient and NO_3^- must be reduced to NH_4^+ in the plant [33]. Moreover, NH_4^+ often triggers toxicity, manifested in leaves as chlorosis and a reduction of growth, but the threshold at which the symptoms become visible differs widely by species [34,35]. However, some species, such as rice [33], demonstrate a preference for NH_4^+. A similar bias for NH_4^+ over NO_3^- was shown for at least one duckweed species, dotted duckweed (*Landoltia punctata*) [36], which is also very tolerant to NH_4^+ stress [27].

Here, we explored the utilization of NO_3^- and NH_4^+ in six duckweed species representing four genera: *Spirodela* (*S. polyrhiza*), *Landoltia* (*L. punctata*), *Lemna* (*L. aequinoctialis*, *L. turionifera*, *L. minor*), and *Wolffia* (*W. globosa*). Taking advantage of the available genome sequence of great duckweed (*S. polyrhiza*) [37], we characterized the structure and expression profiles of the genes coding for eight key enzymes in N assimilation in *S. polyrhiza* grown in media supplied with NO_3^-, NH_4^+, or a combination.

2. Results

2.1. Identity of the Analyzed Species

The duckweeds used in this study include five species isolated in Eastern China (*Spirodela polyrhiza*, *Landoltia punctata*, *Lemna aequinoctialis*, *L. turionifera*, and *Wolffia globosa*) and *Lemna minor*, collected in Kazakhstan. Prior to the N assimilation experiments, the identity of all species grown in vitro from a single frond was confirmed by barcoding through sequencing the *atpF–atpH* (ATP) and *psbK–psbL* (PSB) intergenic spacers [38] and using BLAST searches against the NCBI sequence collection [39]. The obtained ATP and PSB sequences were deposited in GenBank with the sequence accession numbers listed in Figure 1.

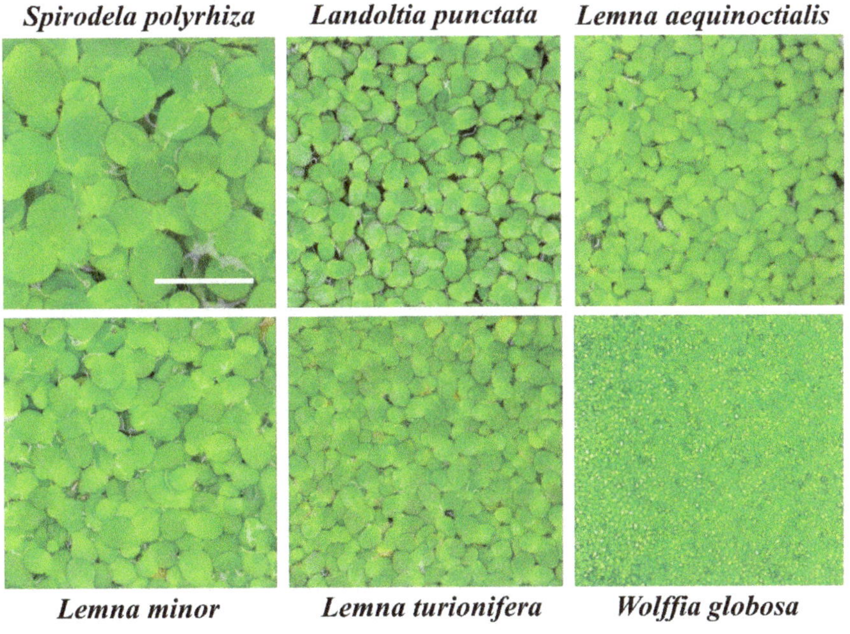

Figure 1. Images of the duckweed species used in the study at the exponential growth stage. All pictures were taken at the same magnification, bar corresponds to 1 cm. GenBank accession numbers for the *atpF–atpH* (ATP) and *psbK–psbL* (PSB) barcodes are: *S. polyrhiza* (NB5548), ATP MZ436185, PSB MZ436186; *L. punctata* (NB0031), ATP MZ436177, PSB MZ436178; *L. aequinoctialis* (NB0007), ATP MZ436181, PSB—MZ436182; *L. minor* (NB0020), ATP MZ436176; *L. turionifera* (NB0013), ATP MZ436179, PSB MZ436180; *W. globosa* (NB0015), ATP MZ436183, PSB MZ436184.

2.2. All Six Duckweed Species Demonstrate a Preference for NH_4^+ over NO_3

To estimate duckweed growth responses to NO_3^- and NH_4^+, all six duckweed species were cultivated under identical temperature and light conditions in 200 mL of liquid SH media. After a period of N starvation, the plants were supplied with 5 mM NO_3^-, 5 mM NH_4^+, or both (2.5 mM NO_3^- and 2.5 mM NH_4^+) and cultured for 12 days.

All six duckweed species grew well, showing no signs of chlorosis, when 5 mM NO_3^- was used as the sole N source, even though the medium pH went up to 6.4–6.9 during the 12-day cultivation. However, when 5 mM NH_4^+ was used as the sole N source, the duckweed plants showed noticeable growth defects at the late cultivation stages (Figure S2), and the medium pH dropped to 3.8–4.6 by the 4th day of cultivation (Figure S3). The first signs of chlorosis appeared on day 6 for *L. turionifera* and *L. minor*, day 8 for *S. polyrhiza* and *L. aequinoctialis*, and day 10 for *L. punctata* and *W. globosa*. On day 12, *L. turionifera* appeared to be the most damaged among the duckweed species, while *L. punctata* had the least number of fronds with chlorosis. The observed growth defects were less severe when the duckweeds were grown in medium with both NO_3^- and NH_4^+ (Figure S2).

To determine if the severe growth defects observed during the advanced cultivation stages on the duckweed plants grown in 5 mM NH_4^+ as the sole N source were due to the low pH, we adjusted the medium to the original pH of about 5.5 every other day in a second experiment. This pH correction, which more closely mirrored natural conditions in big, well-buffered water reservoirs, maintained duckweed growth for more than 2 weeks without any signs of chlorosis or depigmentation independent of the applied N source. This suggested that the growth defects and chlorosis were due to the low pH and not to the N supply.

We observed almost identical dynamics of N consumption by the duckweed species grown for 12 days in medium containing 5 mM of NO_3^- or NH_4^+ as the sole N source (Figure 2). Five species (excluding *W. globosa*), exhausted the N in the medium by day 8 independent of the source, with *S. polyrhiza*, *L. punctata*, and *L. turionifera* showing the most rapid consumption.

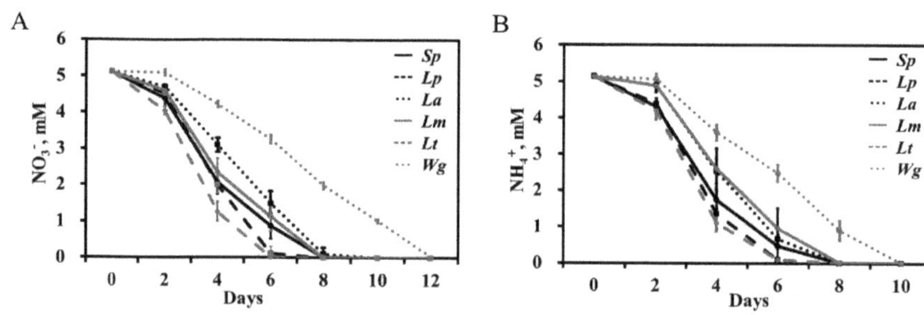

Figure 2. Dynamics of N uptake by six duckweed species grown in medium supplied with 5 mM NO_3^- (**A**) or NH_4^+ (**B**). The y-axis shows the concentration of NO_3^- or NH_4^+ remaining in the medium. *Sp*, *S. polyrhiza* (NB5548); *Lp*, *L. punctata* (NB0031); *La*, *L. aequinoctialis* (NB0007); *Lm*, *L. minor* (NB0020); *Lt*, *L. turionifera* (NB0013); *Wg*, *W. globosa* (NB0015).

When grown in medium supplied with both NH_4^+ and NO_3^-, the most common situation in the natural environment, all duckweed species demonstrated a clear preference for NH_4^+, with three species (*S. polyrhiza*, *L. punctata*, and *L. turionifera*) consuming almost all of the available NH_4^+ during the first four days of cultivation (Figure 3). The duckweeds started to utilize NO_3^- only when the concentration of NH_4^+ dropped below 0.5 mg/L (0.04 mM).

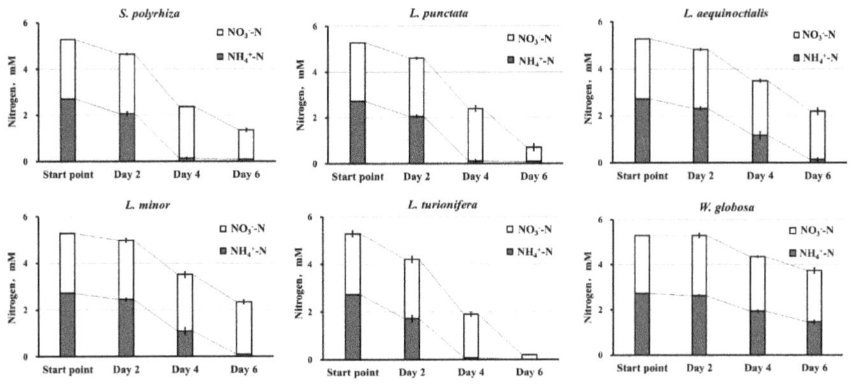

Figure 3. Relative dynamics of nitrate and ammonium uptake by duckweed species during six days of cultivation in medium supplied with equal amounts (2.5 mM) of NO_3^- and NH_4^+.

2.3. *Key Genes for N Assimilation in the Genome of S. polyrhiza*

We evaluated the key N assimilation genes encoding NR, NiR, GS, NADH-GOGAT, and Fd-GOGAT, which have been identified as the major players in the assimilation of inorganic N in many plant species (Figure S1), using *S. polyrhiza* as the representative species due to the availability of a well-characterized whole-genome sequence [37,40–42]. To validate the sequences available in the GenBank, we re-sequenced the cDNA clones prepared for the four *GS* genes, *NR*, *NiR*, *NADH-GOGAT* and *SpFd-GOGAT* for the *S. polyrhiza* ecotype NB5548 used in this study (the corresponding sequence accession IDs are: *SpGS1;1*-MZ605906, *SpGS1;2*-MZ605907, *SpGS1;3*-MZ605908, *SpGS2*-MZ605909, *SpNR*-OL421561, *SpNiR*-OL421562, *SpNADH-GOGAT*-OL421563, *SpFd-GOGAT*-MZ605910).

BLAST searches of the *S. polyrhiza* ecotype Sp9509 genome [37], available on the NCBI website (taxid: 29656, GCA_900492545.1), with rice protein queries revealed single genes coding for SpNR, SpNiR, SpNADH-GOGAT, and SpFd-GOGAT and four genes encoding GSs: SpGS1;1, SpGS1;2, and SpGS1;3 (which function in the cytoplasm) and SpGS2 (which is transported into chloroplasts). The exon/intron structures of the gene sequences deduced by their similarities with the corresponding rice sequences are represented in Figure 4.

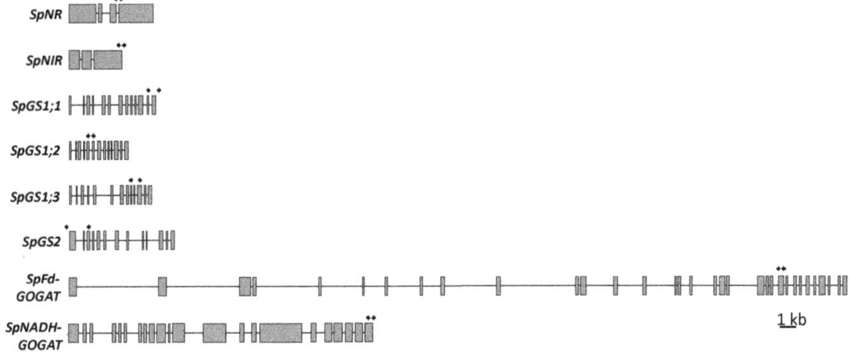

Figure 4. Structures of key *S. polyrhiza* genes involved in N assimilation. Exons are represented by grey boxes and lines represent introns. Arrows indicate the locations of primer binding sites used for gene expression analysis by RT-qPCR (primer sequences are listed in Table S2).

2.3.1. Nitrate and Nitrite Reductases

NR and NiR are encoded by single genes located on chromosome 18 of *S. polyrhiza*. For comparison, most terrestrial diploid plants have 2–3 *NR* genes and one *NiR* gene. *SpNR* is composed of four exons and three introns, typical for this plant lineage, whereas *SpNiR* has three exons and two introns (Figures 4 and S4) with exons 3 and 4 that are common in other plants fused into a single exon 3 in the genome of *S. polyrhiza* (Figure S6).

Multiple alignments of SpNR and SpNiR polypeptides with NR and NiR from other species showed a high level of protein conservation among flowering plants along the whole protein sequence, with the exception of a highly diverse N-terminal region (Figures S5A and S7A). The N-terminus of NiR proteins corresponds to the transit peptide for chloroplast targeting (Figure S7A). In the phylogenetic tree, *S. polyrhiza* NiR grouped with the monophyletic clade of NiRs from monocots, whereas SpNR was more closely related to the dicotyledonous plant NR clade (Figures S5B and S7B), which probably reflects the formation of the duckweed lineage around the time when the dicots and monocots diverged [13].

2.3.2. Glutamine Synthetases

Glutamine synthetases (GSs) are a family of enzymes involved in the primary incorporation of inorganic N in the form of NH_4^+ (absorbed directly, produced by NO_3^- conversion, or resulting from degradation of intracellular proteins and other organic compounds) into an organic form of glutamine. There are two major GS enzyme classes encoded in plant nuclear genomes: GS1, which is localized and functions in the cytoplasm, and GS2, which is transported to chloroplasts. Most plants have a small family of three to five genes encoding cytosolic GS1 isoforms and a single gene for GS2 [43,44].

The *GS* genes, searched from the two available *S. polyrhiza* genomes using the rice protein sequence of OsGS1;2 as a query, and showing no sequence variability between *S. polyrhiza* ecotypes 9509 and 7498, were classified as *SpGS1;1*, *SpGS1;2*, *SpGS1;3*, and *SpGS2* based on sequence similarities with the corresponding *GS* genes from rice, barley (*Hordeum vulgare*), and sorghum. All analyzed duckweed *GS* genes were composed of 13 exons and 12 introns, with no size variation of exons 1 through 12 (74, 40, 104, 49, 107, 88, 129, 75, 54, 38, 160, 61 bp, respectively) for *SpGS1;1-1;3* but some variation in the intron lengths. Compared to the three *SpGS1* genes, *SpGS2* is a bit larger in size due to longer introns, and longer exons 1 and 13, which contain a chloroplast signal peptide and short variable C-terminal extension peptides (Figures 4 and S8), the last one considered important for enzyme activity and do not take part in the import process to plastids [45].

Alignments of the amino acid sequences deduced from genomic DNA sequences of European and American ecotypes 9509 and 7894 and cDNA of Chinese ecotype NB5548, showed a very high degree of similarity between GSs in duckweed and other plants representing both monocot and dicot species (Figure S9A). The phylogenetic examination demonstrated that duckweed cytosolic GS1s and chloroplast GS2 form two sister groups (Figure S9B), consistent with previous studies of other plant taxa [46,47]. The separation of *GS1* and *GS2* is considered to have occurred due to a gene duplication that preceded the divergence of monocots and dicots. The degree of sequence conservation of *GS* genes can be used as a molecular clock in gene evolution studies [48]. SpGS2, SpGS1;1, and SpGS1;2 did not cluster with the respective GS sequences from monocots or dicots, while SpGS1;3 shared the highest sequence similarity with NnGS1;3 from lotus (*Nelumbo nucifera*), an aquatic dicot plant.

2.3.3. Fd-GOGAT and NADH-GOGAT

GOGATs and GS2 form a GS/GOGAT cycle in plant chloroplasts, where GS catalyzes the formation of Gln from Glu and NH_4^+, and Fd-GOGAT and NADH-GOGAT catalyze the transfer of an amide group from Gln to 2-oxoglutarate to produce two molecules of Glu (Figure S1). The genome of *S. polyrhiza* possesses a single gene for *Fd-GOGAT* and one for *NADH-GOGAT* (Figure 4). Similar to other characterized plant *Fd-GOGAT* genes, the duckweed homologue is composed of 33 exons and 32 introns with a total gene length of 29,677 bp. *SpNADH-GOGAT*, similar to its homologs from wheat (*Triticum aestivum*) [49] and rice [50], contains 22 exons and 21 introns with a total length of 11,391 bp (Figure 4, Figures S10 and S12). The presence of long introns is characteristic of *Fd-GOGAT* genes in many species; for example, lotus *NnFd-GOGAT* has 33 exons reaching almost 200 kb in size, while the *Fd-GOGAT* genes usually are of more than 330 kb in conifers [51].

Mature GOGAT proteins demonstrate high sequence conservation (Figures S11A and S13A). According to the phylogenetic tree shown in Figures S11B and S13B, both Fd-GOGAT and NADH-GOGAT of *S. polyrhiza* grouped with GOGAT proteins from dicots.

2.4. Expression of Key *S. polyrhiza* Genes Involved in N Assimilation

We measured gene expression based on the dynamics of N uptake observed in our experiment, with the most active N consumption occurring during the first 4 days of growth after the N source was added. Therefore, *S. polyrhiza* samples for RNA isolation were taken simultaneously with medium sampling for measurement of N as represented in Figures 2 and 3. All RT-qPCR reactions were performed in triplicate, normalized against the expression of two household genes (*β-actin* and *histone H3*), and related to the gene expression levels at the starting starvation point (day 0) (Figure 5).

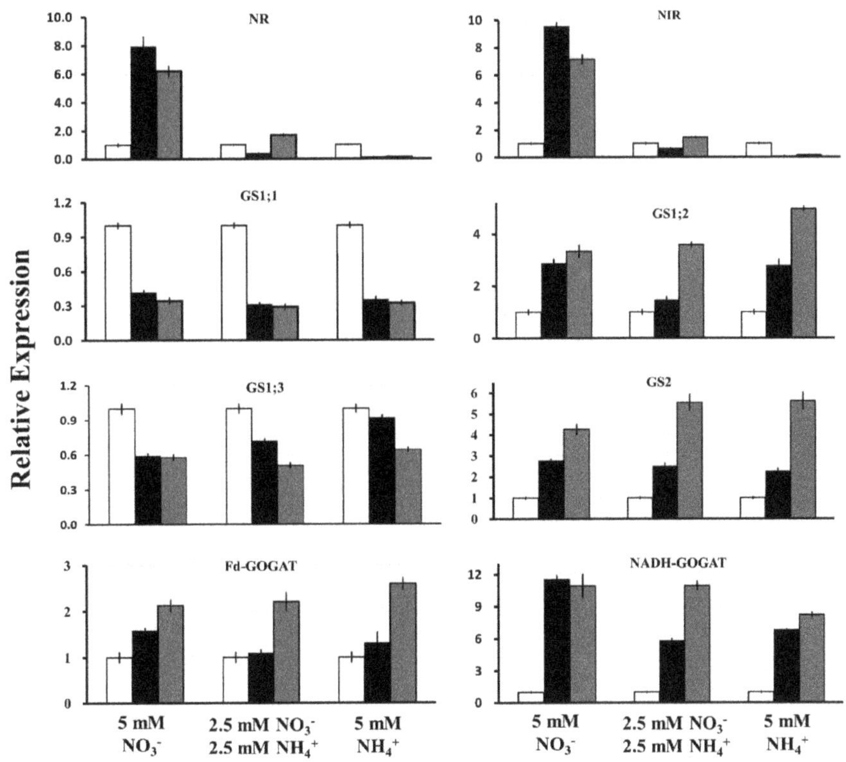

Figure 5. Relative expression of *S. polyrhiza NR, NiR, GS, Fd-*, and *NADH-GOGAT* genes in response to different forms of inorganic N as estimated by RT-qPCR. Gene expression levels are in relative units. Open bar: start point (day 0); black bar: day 2; gray bar: day 4. Error bars show ± SD of 3 replicates ($p < 0.05$).

The expression of *NR* and *NiR* was strongly induced in duckweed cultivated in medium with NO_3^- as the only N source. The highest expression was recorded for both genes at day 2, with an 8- and 9-fold increase for *NR* and *NiR*, respectively. The expression of *NR* and *NiR* then decreased slightly at day 4 following the drop in available NO_3^-. When both NO_3^- and NH_4^+ were supplied in the media, the relative expression of *NR* and *NiR* decreased about 2.5- and 1.5-fold on day 2, respectively, then the expression of both genes increased about 1.5-fold on day 4, indicating the start of their induction. Moreover, the relative expression of *NR* and *NiR* similarly decreased about 7- and 10-fold on day 2 and day 4, respectively, when NH_4^+ was used as the sole N source.

The relative expression of *GS1;2* and *GS2* gradually increased, with a more pronounced increase in samples grown in the presence of NH_4^+, reaching about 5-fold higher expression on day 4 for both genes compared to day 0. By contrast, the expression of *GS1;1* and *GS1;3* was suppressed by the addition of NO_3^- and/or NH_4^+.

Fd-GOGAT and *NADH-GOGAT* were induced by either NO_3^- or NH_4^+. The relative expression level of *Fd-GOGAT* increased 1.6- and 2.1-fold on day 2 and day 4, respectively, when 5 mM NO_3^- was used as the sole N source, and increased 1.3- and 2.6-fold on day 2 and day 4, respectively, when 5 mM NH_4^+ was used as the N source. A similar expression pattern was observed when the fronds were grown in the medium containing both NO_3^- and NH_4^+.

The relative expression of *NADH-GOGAT* increased 11.6- and 10.9-fold on day 2 and day 4, respectively, when 5 mM NO_3^- was used as the sole N source. Its expression increased 6.9- and 8.2-fold on day 2 and day 4, respectively, when 5 mM NH_4^+ was used as the only N source, and increased 5.8- and 10.9-fold on day 2 and day 4, respectively, when both NO_3^- and NH_4^+ were used.

2.5. Survey for Possible N-Responsive Promoter Cis-Elements in the N Assimilation Genes

To gain further insight into the transcriptional regulation of N assimilation genes in *S. polyrhiza*, we analyzed the gene promoter regions for the presence of possible regulatory *cis*-elements. The survey revealed the presence of *cis*-elements similar to the nitrate-responsive elements (NRE), first described for the *NiR* promoter in *A. thaliana* [52] and later characterized for many other NO_3^-—regulated genes [53–55], within the 1-kb DNA region upstream of the first ATG codon of all analyzed duckweed genes (Figure S14). While the NRE-like elements found in the promoters of *SpNR*, *SpNiR*, *SpGS1.1*, *SpGS1.2*, *SpGS1.3*, *SpGS2*, *SpNADH-GOGAT*, and *SpFd-GOGAT* showed some divergence from the canonical *A. thaliana* bipartite pseudo-palindromic sequence GACcCTT-N(10)-AAGagtcc, most of them aligned relatively well with the corresponding NREs found in *A. thaliana*, rice, sorghum, and maize (*Zea mays*) (Figure 6).

Moreover, *SpNiR* harbors four NRE-like copies positioned within the 242-bp promoter region upstream of the translation start site; *SpGS1.2* and *SpNADH-GOGAT*, which along with *SpNiR* demonstrated the highest upregulation by NO_3^- among the studied *S. polyrhiza* genes, possess three NRE-like elements each (Figure S14). Correlation between the number of NREs and the increase of nitrate-inducible expression was recently confirmed using synthetic promoters, which demonstrated that increasing the number of NREs in the promoter of rice *OsNiR*, which naturally has two NRE-like elements [56], led to a significant enhancement of N assimilation [57]. The NRE-like elements in *SpNADH-GOGAT* showed significant divergence from the canonic bipartite NRE sequence (Figure 6B), and their functionality remains to be tested.

Another relatively well-characterized molecular system for fine-tuning gene expression in response to the N supply is based on Nitrate-Inducible GARP-Type Transcriptional Repressor-1 (NIGT1) family proteins, first identified as transcriptional repressors in rice [58], and later studied in more detail in *A. thaliana* [59,60]. NIGT1 proteins demonstrate dual modes of promoter sequence recognition, binding to two types of *cis*-elements, GAATC or its reverse complement sequence GATTC, and GAATATTC [54,61]. A search for these elements in promoters of the duckweed N assimilation genes did not reveal the GAATATTC element, whereas multiple sites matching the GAATC/GATTC sequences were found in the promoters of *SpGS1.1* (4), *SpGS1.2* (6), *SpGS2* (4), and *SpNADH-GOGAT* (2), representing a potential opportunity for negative regulation by the SpNIGT1 homolog upon supply of NO_3^-. None of these *cis*-elements were found in the promoters of *SpNiR* and *SpGS1.3*, while *SpNR* and *SpFd-GOGAT* both contain a single copy of GAATC/GATTC (Table S3 and Figure S14).

A

Gene site		NRE-like motif	
AtNIR	-96	CTTGACCCTTTACATTG-CT-CAAGAGCTC	-69
OsNIRa	-224	TCTGCCCTTTTGAATTC-GC-CAAGAGCCA	-197
OsNIRb	-193	TCTGCCCTTTTGAATTC-GC-CAAGGGCCA	-166
SbNIR	-143	CCTGCCCTTTTGAACTC-CC-CAAGAGCCA	-116
ZmNIR2	-151	CCTGCCCCCTTGAACTC-CC-CAAGAGCCA	-124
SpNR	-261	CCTCGCCTTTCC--CTACCGCTGGGAGCTG	-234
SpNIRa	-242	TACGACCCTTCT--CTCGTC-CGTGAGTCC	-216
SpNIRb	-209	TCTGTCCTTTAGAGAGA-GC-GAAGAGCCC	-184
SpNIRc	-110	CGTCACCTCTCGAGGTG---TTAAGAACCC	-84
SpNIRd	-83	TGCGGCCTTCCCACTGCCCCGATGGAGTCG	-54
SpGS1;2a	-505	CCTCTGCCAGTACGCAG-CG-GAAGAGCCC	-478
SpGS1;2b	-217	CCTCCCTCTTCCTGTCGCAG-CAGGAGGAA	-189
SpGS1;3	-661	GGGATGCTTA-CCATGAAGC-TAAGAGCCC	-634
SpGS2	-888	GTTTTCCTCC-ACTCAAAGCGGAAGAACCC	-860
SpFd-GOGAT	-274	CTTTGGCGTTTCCTCTCCAC--GCGAGCCT	-247
SpNADH-GOGATa	-191	TTTTTTTTTTTTT-TAATTTTTGGGAGTTC	-161
SpNADH-GOGATb	-142	GATCTTCTTTTTCGCCGCCGGCGGTGGTTC	-113
Consensus		CCTGTCCTTTTGANNTC-CC-CAAGAGCCC	

B

NRE-like motifs logo

Figure 6. NRE-like sequences in promoters of *S. polyrhiza* N assimilation genes aligned with corresponding motifs in other plants. (**A**) Alignments of the NRE-like sequences identified in promoters of: *AtNiR* (AT2G15620, *Arabidopsis thaliana*); *OsNiR* (two motifs: *OsNiRa* and *OsNiRb*, LOC4326014, *Oryza sativa*); *SbNiR* (LOC8075200, *Sorghum bicolor*); *ZmNiR2* (LOC542264, *Zea mays*). Promoter sequences of the *SpNR*, *SpNiR*, *SpGS1;2*, *SpGS1;3*, *SpGS2*, *SpFd-GOGAT*, and *SpNADH-GOGAT* genes are from the genome of *S. polyrhiza* ecotype Sp9509, available in online databases (NCBI taxid: 29656; GCA_900492545.1). Numbers in front and at the end of the nucleotide motifs indicate their position relative to the gene translation start site. (**B**) The NRE-like motif logo displays the consensus sequences generated by CLC Main Workbench 7.6.1.

The presence of GAGA and/or complementary CTCT stretches, and G4-quadruplex structures, both of which are implicated in general regulation of gene transcription [62,63], is a prominent feature of the duckweed promoters analyzed in this study. All promoters of N assimilation genes, except for *SpGS1.1*, contained repetitive GAGA or CTCT stretches between −1070 and −330 nucleotides upstream of the ATG start site, with the *SpFd-GOGAT* promoter containing an exceptionally long GAGA region of 306 bp (positions −864 to −559) and a 22-nucleotide stretch of TCTC at −351 to −329 bp (Figure S14). Additionally, all analyzed promoters exhibited numerous TATA-like motifs, which represent not only important elements of a core promoter in many plant genes [64,65], but also may act as general transcriptional enhancers [66].

G-quadruplexes (G4) are secondary nucleotide structures found in guanidine-rich regions, which are implicated in various cellular processes in eukaryotic organisms [67]. The G4 structures formed along DNA or RNA strands by tetrads of guanine bases joined together via nonconventional hydrogen bonds are often located in gene promoters, introns, or 5′-untranslated regions (UTRs) and play important roles in regulation of gene functions [68,69]. The *pqsfinder* G4 prediction online tool [70] preset for the recommended scanning window of 100 bp [71] revealed characteristic patterns of G4-motif distribution in the analyzed duckweed promoters. In particular, *SpNR* and *SpNiR* promoters showed a strong G4 peak located at the same position between −265 and −290 bp relative to the translation start site, but on opposite DNA strands. This was the only G4 structure detected in the *SpNiR* promoter sequence, whereas the *SpNR* promoter had four more G4-motifs further upstream of the ATG start site (Figure S15). The promoters of *SpGS1.1*, *SpGS1.2*, and *SpGS1.3* each had one or two relatively strong G4 structures composed of more than four G4 stems situated at both DNA strands and in different locations along the 1-kb promoter sequence, while *SpGS2* possessed two relatively weak structures of three G4 units each. The *SpFd-GOGAT* and *SpNADF-GOGAT* promoters had a characteristic G4 peak directly adjacent to the gene translation start site, the 5′-UTR, which is the most common position of the G4 structures in genes of many monocot plants such as maize [72], rice [73], wheat [74], and barley [71].

3. Discussion

3.1. Duckweeds' Preference for NH_4^+ as a Source of N

The duckweed species investigated in this study (Figure 1) are the ones most commonly used for wastewater remediation [16,18,22,75]. All six species demonstrated an obvious preference for NH_4^+ over NO_3^- when given the choice between the two under our experimental growth conditions. A similar N source preference was previously shown for at least two representative species of duckweed, *Lemna gibba* [76] and *Landoltia punctata* [36], which are also among the most tolerant of NH_4^+ stress [27]. In our experiments the consumption of NO_3^- by the duckweeds did not start until the NH_4^+ was exhausted (Figure 3).

From the standpoint of cell metabolic economics, NH_4^+ is the obvious choice of N as it is the only N source that can be used for building various organic compounds, and NO_3^- must be converted into NH_4^+ to provide N for cellular metabolism. However, the majority of plant species prefer NO_3^- because NH_4^+ causes toxicity at certain concentrations, resulting in leaf chlorosis and a reduction of growth [34,35]. Duckweed also manifests symptoms of NH_4^+ stress [77], but its threshold is higher compared to other plants, which is probably one of the reasons why duckweed shows incredible adaptability and high growth rates.

3.2. NR and NiR Are Co-Ordinately Expressed, Stimulated by NO_3^- and Suppressed by NH_4^+

In accordance with their functional link in the step-by-step conversion of NO_3^- into NH_4^+, *SpNR* and *SpNiR* demonstrated very similar, almost identical, expression patterns in response to NH_4^+ and NO_3^- (Figure 5). The expression of both genes was clearly stimulated by addition of NO_3^- as the sole N source after a period of starvation. By contrast, NH_4^+ seemed to suppress *SpNR* and *SpNiR* expression, especially when it was supplied as the sole N source at a relatively high concentration of 5 mM. On the one hand, the stimulation of *NR* and *NiR* by NO_3^- has been well documented in many plant species [56,78]. On the other hand, to the best of our knowledge, we have demonstrated suppression of these genes in flowering plants at the transcriptional level by NH_4^+ for the first time, while the inhibition of NO_3^- uptake by NH_4^+ was previously shown for *Lemna* species [79], barley [80,81], and rice [82].

However, a similar expression switch of *NR* and *NiR* (upregulated by NO_3^- and downregulated by NH_4^+) was described for a wide range of algae, and has been successfully used to establish inducible systems for transgene expression [83–85].

3.3. *SpGS1;2* and *SpGS2* Are Regulated in a Very Similar Manner, Which Is Different from *SpGS1;1* and *SpGS1;3*

Our expression data showed drastic differences in the expression patterns between *SpGS1;1* and *SpGS1;3* and *SpGS1;2* and *SpGS2* (Figure 5). While the expression of *SpGS1;2* and *SpGS2* increased following addition of NO_3^- and/or NH_4^+, the expression of *SpGS1;1* and *SpGS1;3* decreased. This difference might be partially explained by the functional specializations of the GS isoenzymes [47,86].

GS1;2 transcripts are abundant in almost all plant tissues [87,88], especially as the dominant GS isoform in roots of monocot crops such as barley [89] and rice [50], where it is considered to play a pivotal role in the primary assimilation of NH_4^+. In green tissues, such as leaves and stems, GS1;2 complements GS2, which is a dominant enzyme in assimilating the NH_4^+ produced by photorespiration [90]. These genes are both upregulated by N, with a recent finding showing almost no difference in *GS2* expression patterns in response to NH_4^+ and NO_3^- in tea tree (*Camellia sinensis*) [91]. Moreover, GS1;2 and GS2 are essential for NUE, healthy development, and accumulation of vegetative biomass, as well as stress responses [86,92]. In duckweed, in which the uptake and assimilation functions of a root and a leaf are often combined in a single assembly of a frond, *GS1;2* and *GS2* likely play a central role in the plant's high growth rate and biomass accumulation.

In contrast to *SpGS1;2* and *SpGS2*, addition of fresh nutrient medium to an N-starved duckweed culture (start point in Figure 5) resulted in renewed vegetative growth and drastic downregulation of *SpGS1;1* and *SpGS1;3*. This is in agreement with a number of previous studies suggesting that cytosolic GS1;1 mainly functions in NH_4^+ remobilization from protein breakdown during starvation and/or senescence [93,94]. The cytosolic isoforms of GS in wheat (*TaGS1.1*) [31], barley (*HvGS1;1*) [89], and oilseed rape (*Brassica napus*; *BnaGLN1.1* and *BnaGLN1.4*) [95] are all upregulated in senescing leaves. In concert with GS1.3, GS1.1 is a major contributor to N supply during seed filling [31,89,96,97]. For example, a rice mutant lacking *OsGS1;1* showed a severe reduction in grain yield [96]; wheat with all three homologs of *TaGS1.1* knocked down by clustered regularly interspaced short palindromic repeats (CRISPR)/CRISPR-associated protein 9 (Cas9) had reduced N translocation efficiency and grain filling, fewer grains per spike, and significantly reduced yield compared to the wild type [98]. To the contrary, transgenic introduction of an extra copy of native *HvGS1;1* led to higher grain yields and NUE in barley [99]. Similarly, overexpression of *ZmGln1;3* in maize resulted in an increase in kernel number, ultimately leading to a higher yield in transgenic plants compared with controls [100].

Taking these observations together, it might be concluded that, similar to other plants [101], the contribution of *SpGS1;1* and *SpGS1;3* to primary N assimilation by duckweed is much lower than that of *SpGS1;2* and *SpGS2*. Accordingly, we suggest that *SpGS1;1* and *SpGS1;3* may play a role in filling of the turion with storage nutrients. Preliminary support for this assumption can be found in RNA-seq data related to turion formation in *S. polyrhiza* [102], showing 5-fold upregulation of *SpGS1;3* and simultaneous 5-fold downregulation of *SpGS1;2*.

3.4. *Fd-GOGAT* and *NADH-GOGAT* Have a Complex Exon-Intron Structure, and Are Upregulated by NO_3^- and NH_4^+

Approximately 95% of the NH_4^+ produced/assimilated in plant tissues is utilized through the GS-GOGAT cycle, facilitated by coordinated actions of glutamine synthetase and glutamine synthase (GOGAT), represented by Fd- and NADH-dependent glutamine:2-oxoglutarate amidotransferases (Fd-GOGAT and NADH-GOGAT) [103,104]. Our RT-qPCR data (Figure 5) demonstrated clear stimulation of *SpNADH-GOGAT* and *SpFd-GOGAT* expression by N added after starvation, where the relative expression of *Fd-GOGAT* increased about 2-fold, and the relative expression of *NADH-GOGAT* increased by more than 10-fold. While the GOGAT isozymes are both implicated in N assimilation processes, such as primary assimilation and in photorespiration [105,106], remobilization in senescing organs [50,107], and grain development [49,108], our data hint that SpNADH-

GOGAT is responsible for primarily N assimilation whereas SpFd-GOGAT is more active in N remobilization.

3.5. *Distinctive Promoter Elements in the N Assimilation Genes in S. polyrhiza*

A survey of the structural organization of promoter sequences of the analyzed genes revealed some common features, which might shed light on the role of N in gene regulation. The most intriguing finding is probably the signatures of NRE-like elements revealed in all characterized *S. polyrhiza* promoters. While some of these motifs show noticeable divergence from the canonic bipartite NRE, especially the 5′-halves of the NREs in the *SpNADH-GOGAT* promoter (Figures 6 and S14), their functionality remains to be tested.

The binding of $(GA/CT)_n$ repeats by a family of GAGA-binding transcription factors (GAFs) was first discovered and thoroughly characterized in *Drosophila*, where GAFs regulate numerous developmental genes in cooperation with chromatin remodeling factors [109,110]. The genomes of both *Spirodela* species, *S. polyrhiza* and *S. intermedia*, are unusually enriched in $(GA/CT)_n$ repeats as documented by whole-genome surveys [37,111]. However, although numerous transcription factors with GAGA-binding properties have been identified in a number of plant species [62,112], the function of GAF-regulated transcription in plants remains largely unknown [63]. Also, in the analyzed plants, G4 motifs were enriched in the first gene exons and first introns [71]. No such enrichments were found in the analyzed duckweed genes except for *SpGS1;1*, where a strong G4 structure was identified in the antisense strand of the first intron. Therefore, our analysis identified significant enrichment of G4 structures primarily in the promoter regions of the duckweed genes involved in N assimilation.

Overall, the array of potential regulatory DNA elements revealed in the *S. polyrhiza* promoters (summarized in Table S3) suggests complex regulation of the N assimilation genes with major roles attributed to the concerted actions involving multiple NRE-like and GAATC/GATTC *cis*-elements, TATA-based enhancers, $(GA/CT)_n$ repeats, and G-quadruplex structures, while the details of their individual roles and interaction will need to be uncovered in future research.

4. Materials and Methods

4.1. *Plant Materials*

The ecotypes used in this study were selected from the duckweed live in vitro collection recently established in the School of Life Sciences at Huaiyin Normal University, Huai'an, China (*S. polyrhiza* (collection ID: NB5548), *L. punctata* (NB0031), *L. turionifera* (NB0013), and *W. globosa* (NB0015)) were collected from small ponds and lakes at different locations between Huai'an city and Hongze lake in eastern China; *L. aequinoctialis* (NB0007) originated from the lake in a park next to People's Square in Shanghai; and *L. minor* (NB0020) was sampled in Nursultan, Kazakhstan. To propagate the samples under sterile conditions and form a stock of live material for further experiments, the collected duckweed samples were surface sterilized in a solution containing 0.5% sodium hypochlorite and 0.1% benzalkonium bromide, washed with autoclaved water, and single fronds were put on solid agar medium supplemented with SH salts [113]. The identity of the species was confirmed by double barcoding using primers specific for chloroplast DNA intergenic spacers *atpF–atpH* (ATP) and *psbK–psbL* (PSB) following the protocol described by Borisjuk et al. (2015) [38].

4.2. *Duckweed Cultivation Parameters and Determination of N Uptake*

To accumulate biomass, duckweed plants grown on solid agar medium were initially transferred into sterile liquid SH medium supplemented with 5 g/L sucrose and cultivated at 23 ± 1 °C with a photon flux density of 50–60 $\mu mol \cdot m^{-2} s^{-1}$ provided by cool white fluorescent bulbs in a 16-h light/8-h dark cycle. After four weeks of growth, the accumulated fronds were weighed and 0.5-g portions were inoculated into open 500-mL paper containers containing 200 mL of basic SH medium (no sugar) supplemented with three different formulations of N. The first medium was supplemented with NO_3^- as the sole N source (5 mM KNO_3, with 5 mM KH_2PO_4 replacing $NH_4H_2PO_4$ of the standard SH medium). The second formulation contained NH_4^+ (5 mM $NH_4H_2PO_4$ and 5 mM K_2SO_4 for the potassium salt), and the third formulation contained a mixture of NO_3^- and NH_4^+ (2.5 mM KNO_3, 2.5 mM KH_2PO_4, 2.5 mM $NH_4H_2PO_4$, and 2.5 mM K_2SO_4). The medium pH was originally adjusted to 5.5, and the cultivation medium was sampled to measure the acidity every two days. The cultivation medium was adjusted to the original pH of 5.5 every other day with HCl or NaOH during the 12-day experiments to measure N consumption.

The total nitrogen (TN) concentration was determined using standard alkaline potassium persulfate digestion followed by UV spectrophotometry as previously described [18]. The NO_3^- concentration in the growth media was measured spectrophotometrically as the difference in absorption between 220 and 275 nm [114]. The NH_4^+ concentration in the growth media was measured calorimetrically using the Nessler method [115].

4.3. Characterization of Major Duckweed Genes Related to N Assimilation

To access the duckweed orthologues of genes encoding critical enzymes involved in N assimilation, i.e., nitrate reductase (NR), nitrite reductase (NiR), glutamine synthetase (GS), NADH-dependent glutamate synthase (GLT, traditionally referred to as NADH-GOGAT), and ferredoxin-dependent glutamate synthase (Fd-GLT, also known as Fd-GOGAT), we searched the available duckweed genome sequences using the reference rice protein sequences OsNR (XP_015622710), OsNiR (XP_015641702), OsGS1;2 (XP_015631679.1), OsNADH-GOGAT (XP_015649242.1), and OsFd-GOGAT (XP_015646712.1) as the initial queries in tBLASTn searches.

The genes coding for SpNR, SpNiR, SpGS1;1, SpGS1;2, SpGS1;3, SpGS2, SpFd-GOGAT and SpNADH-GOGAT were validated by sequencing the PCR-amplified gene regions using cDNA prepared from local *S. polyrhiza*, ecotype NB5548, mRNA as a template. The PCR fragments were amplified with gene-specific primers designed according to the in silico sequence information available at NCBI (taxid: 29656, GCA_900492545.1) for *S. polyrhiza*, ecotype 9509 [37], cloned into the pMD19-T (Takara, China) vector following the manufacturer's instructions, and sent for custom sequencing to the Sangon Biotech (Shanghai, China). The specific primers used for gene amplification are listed in Table S1. The gene sequence assembly was carried out using the CLC Main Workbench 7.6.1.

The intron/exon structure of the duckweed genes was deduced based on similarities with the homologous genes of rice, *A. thaliana*, lotus (*Nelumbo nucifera*), and sorghum (*Sorghum bicolor*), available in GenBank [116], following the general rules of exon/intron prediction [117].

4.4. Phylogenetic Analysis

Mature protein sequences were compared with a selection of monocot and dicot sequences available in public databases [116,118–120]. Multiple alignments were generated in the CLC Main Workbench 7.6.1. Phylogenetic reconstructions were performed using the function "build" of ETE3 v3.1.1 [121]. The maximum-likelihood phylogenetic trees were constructed using RaxML v8.1.20 with model PROTGAMMAJTT and default parameters with branch supports of SH-like values [122].

4.5. Gene Expression Analysis by RT-qPCR

The transcript levels of eight target genes (*NR*, *NiR*, *GS1;1*, *GS1;2*, *GS1;3*, *GS2*, *Fd-GOGAT*, and *NADH-GOGAT*) were measured using reverse transcription quantitative PCR (RT-qPCR) with gene-specific primers designed based on the deduced exon sequences (Table S2, Figure 4). For gene expression analysis, total RNA was extracted from 100 μg of fresh *S. polyrhiza* fronds collected at 0, 48, and 96 h from the start of the experiment according to the protocol described in Box et al. (2011) [123]. The quality of isolated total RNA was estimated with a NanoDrop One C spectrophotometer (Thermo Fisher Scientific, USA) and 1% agarose gel electrophoresis. After DNAase treatment, 600 ng of total RNA was reverse transcribed using the Reverse Transcriptase cDNA synthesis kit (Takara, China), following the manufacturer's manual.

The qPCR reactions were performed using CFX Connect Real-Time detection system (Bio-Rad, Hercules, CA, USA) using the UltraSybr Mixture (High Rox) supplied by CWBio (Taizhou, China). The cycling conditions were as follows: initial denaturation at 95 °C for 10 min followed by 40 cycles of 30 s at 94 °C, and 20 s at the annealing temperature of the respective primers. The SYBR Green I fluorescence was monitored consecutively after the annealing step. The quality of products was checked by a thermal denaturation cycle. Only results providing a single peak were considered. The coefficient amplification efficiency for each pair of primers was determined by 10-fold serial dilutions. The level of relative expression was calculated by the $2^{-\Delta\Delta Ct}$ method [124]. Expression data for the target genes were normalized using the average expression of two *S. polyrhiza* ecotype NB5548 housekeeping genes, *histone H3* and *β-actin* (corresponding accession numbers, MZ605911 and MZ605912), according to the geNorm protocol [125]. Three replicates were performed for all samples. All data were analyzed using the program BIO-RAD CFX Manager 3.1 (Bio-Rad, USA) and Microsoft Excel 2016 software.

5. Conclusions

The present study demonstrated that duckweeds efficiently assimilate NO_3^- and NH_4^+, the main components of agricultural fertilizers and major contaminants of water reservoirs, with the specific regulation of key N-assimilation genes, in terms of changes in their expression levels in response to a supply of NO_3^- and NH_4^+. The activation of genes by NO_3^- or NH_4^+ as the sole N source, along with characterization of the promoter elements, details important features of N assimilation by plants.

The obtained data also provides valuable information for improvement of nitrogen assimilation efficiency, NUE, and phytoremediation of wastewater, potentially by up- or down-regulating certain genes [126], modulating the gene copy number [127], or using gene editing [128].

Supplementary Materials: The following supporting information can be downloaded at: https://www.mdpi.com/article/10.3390/plants11010011/s1, Figure S1: Simplified diagram of Nitrogen assimilation in plants, Figure S2: Photographic documentation of six duckweed species cultivated during 12 days on media supplied with 5 mM NO_3^-, 5 mM NH_4^+ as sole N source or 2.5 mM NO_3^- and 2.5 mM NH_4^+, Figure S3: Comparative dynamics of pH changes during the course of duckweed cultivation over 12 days period in the medium supplied with different sources of nitrogen, Figure S4: Comparison of the exon-intron structures between *NR* gene of *S. polyrhiza* and the homologues of some representative plant species, Figure S5: Sequence alignment of NR protein from *S. polyrhiza* with other representative species and the resulting proteins phylogenetic tree, Figure S6: Comparison of the exon-intron structures between *NIR* genes of *S. polyrhiza* and other representative plant species, Figure S7: Sequence alignment of NIR protein from *S. polyrhiza* with other representative species and the resulted phylogenetic tree, Figure S8: Comparison of the exon-intron structures between *GS* genes of *S. polyrhiza* and other representative plant species, Figure S9: Sequence alignment of GS proteins from *S.polyrhiza* with other representative species and the resulted phylogenetic tree, Figure S10: Diagrammatic representation of the structure of *Fd-GOGAT* genes, Figure S11: Sequence alignment of Fd-GOGAT protein from *S.polyrhiza* with other representative species and the resulted phylogenetic tree, Figure S12: Diagrammatic representation of the structure of *NADH-GOGAT* genes, Figure S13: Sequence alignment NADH-GOGAT protein from *S.polyrhiza* with other representative species and the resulted phylogenetic tree, Figure S14: Nucleotide sequences representing 1 kb promoter regions upstream of the starting ATG codon of six nitrogen assimilation genes of *S. polyrhiza* with marked locations of NRE-like *cis*-elements, GAGA and TCTC stretches and TATA-boxes, Figure S15: Patterns of G-quadruplex structures predicted for promoters of key *S. polyrhiza* genes involved in nitrogen assimilation; Table S1: List of RT-qPCR primers used for evaluating expression of the nitrogen assimilation genes in *S. polyrhiza* ecotype NB5548, Table S2: Distribution of potential regulatory DNA *cis*-elements along 1 kb promoter region upstream of the translation start of eight *S. polyrhiza* genes related to N assimilation, Table S3: List of primer used for cloning selected nitrogen assimilation genes of *S.polyrhiza* ecotype NB5548.

Author Contributions: Conceptualization, Y.Z. and N.B.; data curation, Y.Z., O.K., A.S. and N.B.; formal analysis, O.K. and G.C.; investigation, Y.Z., O.K., A.S., G.C., W.W., J.Z. and C.P.; methodology, A.S., W.W., J.Z. and C.P.; project administration, Y.Z. and N.B.; resources, Y.Z. and G.C.; software, O.K. and A.S.; supervision, Y.Z. and O.K.; validation, A.S.; writing—original draft, Y.Z., O.K. and A.S.; writing—review and editing, N.B. All authors have read and agreed to the published version of the manuscript.

Funding: This work was supported by an individual grant provided by Huaiyin Normal University (Huai'an, China) to NB.

Institutional Review Board Statement: Not applicable.

Informed Consent Statement: Not applicable.

Data Availability Statement: GenBank accession numbers for the *atpF–atpH* (ATP) and *psbK–psbL* (PSB) barcodes are: *S. polyrhiza* (NB5548), ATP MZ436185, PSB MZ436186; *L. punctata* (NB0031), ATP MZ436177, PSB MZ436178; *L. aequinoctialis* (NB0007), ATP MZ436181, PSB—MZ436182; *L. minor* (NB0020), ATP MZ436176; *L. turionifera* (NB0013), ATP MZ436179, PSB MZ436180; *W. globosa* (NB0015), ATP MZ436183, PSB MZ436184. GenBank accession numbers for *S. polyrhiza* (NB5548) genes are: *SpGS1;1*-MZ605906, *SpGS1;2*-MZ605907, *SpGS1;3*-MZ605908, *SpGS2*-MZ605909, *SpNR*-OL421561, *SpNiR*- OL421562, *SpNADH-GOGAT*-OL421563, *SpFd-GOGAT*-MZ605910, *histone H3*-MZ605911, and *β-actin*-MZ605912.

Conflicts of Interest: The authors declare no conflict of interest.

References

1. Liu, Y.; Xu, H.; Yu, C.; Zhou, G. Multifaceted Roles of Duckweed in Aquatic Phytoremediation and Bioproducts Synthesis. *GCB Bioenergy* **2021**, *13*, 70–82. [CrossRef]
2. Coskun, D.; Britto, D.T.; Shi, W.; Kronzucker, H.J. Nitrogen Transformations in Modern Agriculture and the Role of Biological Nitrification Inhibition. *Nat. Plants* **2017**, *3*, 17074. [CrossRef] [PubMed]
3. Camargo, J.A.; Alonso, Á. Ecological and Toxicological Effects of Inorganic Nitrogen Pollution in Aquatic Ecosystems: A Global Assessment. *Environ. Int.* **2006**, *32*, 831–849. [CrossRef] [PubMed]
4. Yu, C.; Huang, X.; Chen, H.; Godfray, H.C.J.; Wright, J.S.; Hall, J.W.; Gong, P.; Ni, S.; Qiao, S.; Huang, G.; et al. Managing Nitrogen to Restore Water Quality in China. *Nature* **2019**, *567*, 516–520. [CrossRef]
5. Smith, V.H. Eutrophication of Freshwater and Coastal Marine Ecosystems a Global Problem. *Environ. Sci. Pollut. Res.* **2003**, *10*, 126–139. [CrossRef]
6. Food and Agriculture Organization (FAO). World Fertilizer Trends and Outlook to 2022. 2019. Available online: https://www.fao.org/3/ca6746en/CA6746EN.pdf?eloutlink=imf2fao (accessed on 5 October 2021).
7. Rai, P.K. Heavy Metal Pollution in Aquatic Ecosystems and Its Phytoremediation Using Wetland Plants: An Ecosustainable Approach. *Int. J. Phytoremediation* **2008**, *10*, 131–158. [CrossRef]
8. Queiroz, R.C.S.; Lôbo, I.P.; Ribeiro, V.S.; Rodrigues, L.B.; Almeida Neto, J.A. Assessment of Autochthonous Aquatic Macrophytes with Phytoremediation Potential for Dairy Wastewater Treatment in Floating Constructed Wetlands. *Int. J. Phytoremediation* **2020**, *22*, 518–528. [CrossRef]
9. Kanwar, V.S.; Sharma, A.; Srivastav, A.L.; Rani, L. Phytoremediation of Toxic Metals Present in Soil and Water Environment: A Critical Review. *Environ. Sci. Pollut. Res. Int.* **2020**, *27*, 44835–44860. [CrossRef]
10. Hammer, D.A. *Constructed Wetlands for Wastewater Treatment: Municipal, Industrial, and Agricultural*, 1st ed.; Hammer, D.A., Ed.; CRC Press: Boca Raton, FL, USA, 2020. [CrossRef]
11. Koottatep, T.; Polprasert, C. Role of Plant Uptake on Nitrogen Removal in Constructed Wetlands Located in the Tropics. *Water Sci. Technol.* **1997**, *36*, 1–8. [CrossRef]
12. Landolt, E. *Biosystematic Investigations in the Family of Duckweeds (Lemnaceae) (Vol. 2). The family of Lemnaceae-a Monographic Study. vol. 1.*; Veroeffentlichungen des Geobotanischen Instituts der ETH, Stiftung Ruebel: Zurich, Switzerland, 1986.
13. Tippery, N.P.; Les, D.H. Tiny Plants with Enormous Potential: Phylogeny and Evolution of Duckweeds. In *The Duckweed Genomes*; Cao, X.H., Fourounjian, P., Wang, W., Eds.; Springer: Cham, Germany, 2020; pp. 19–38. [CrossRef]
14. Acosta, K.; Appenroth, K.J.; Borisjuk, L.; Edelman, M.; Heinig, U.; Jansen, M.A.K.; Oyama, T.; Pasaribu, B.; Schubert, I.; Sorrels, S.; et al. Return of the Lemnaceae: Duckweed as a Model Plant System in the Genomics and Post-Genomics Era. *Plant Cell* **2021**, *33*, 3207–3234. [CrossRef]
15. De Vasconcelos, V.M.; de Morais, E.R.C.; Faustino, S.J.B.; Hernandez, M.C.R.; Gaudêncio, H.R.S.C.; de Melo, R.R.; Bessa Junior, A.P. Floating Aquatic Macrophytes for the Treatment of Aquaculture Effluents. *Environ. Sci. Pollut. Res.* **2021**, *28*, 2600–2607. [CrossRef]

16. Zhou, Y.; Borisjuk, N. Small Aquatic Duckweed Plants with Big Potential for the Production of Valuable Biomass and Wastewater Remediation. *Int. J. Environ. Sci. Nat. Resour.* **2019**, *16*, 555942. [CrossRef]
17. Mohedano, R.A.; Costa, R.H.R.; Tavares, F.A.; Belli Filho, P. High Nutrient Removal Rate from Swine Wastes and Protein Biomass Production by Full-Scale Duckweed Ponds. *Bioresour. Technol.* **2012**, *112*, 98–104. [CrossRef]
18. Zhou, Y.; Chen, G.; Peterson, A.; Zha, X.; Cheng, J.; Li, S.; Cui, D.; Zhu, H.; Kishchenko, O.; Borisjuk, N. Biodiversity of Duckweeds in Eastern China and Their Potential for Bioremediation of Municipal and Industrial Wastewater. *J. Geosci. Environ. Protect.* **2018**, *6*, 108–116. [CrossRef]
19. Xu, S.; Stapley, J.; Gablenz, S.; Boyer, J.; Appenroth, K.J.; Sree, K.S.; Gershenzon, J.; Widmer, A.; Huber, M. Low Genetic Variation Is Associated with Low Mutation Rate in the Giant Duckweed. *Nat. Commun.* **2019**, *10*, 1243. [CrossRef]
20. Appenroth, K.-J.; Sree, K.S.; Böhm, V.; Hammann, S.; Vetter, W.; Leiterer, M.; Jahreis, G. Nutritional Value of Duckweeds (Lemnaceae) as Human Food. *Food Chem.* **2017**, *217*, 266–273. [CrossRef]
21. Guo, L.; Jin, Y.; Xiao, Y.; Tan, L.; Tian, X.; Ding, Y.; He, K.; Du, A.; Li, J.; Yi, Z.; et al. Energy-Efficient and Environmentally Friendly Production of Starch-Rich Duckweed Biomass Using Nitrogen-Limited Cultivation. *J. Clean. Produc.* **2020**, *251*, 119726. [CrossRef]
22. Ziegler, P.; Sree, K.S.; Appenroth, K.-J. Duckweeds for Water Remediation and Toxicity Testing. *Toxicol. Environ. Chem.* **2016**, *98*, 1127–1154. [CrossRef]
23. Ekperusi, A.O.; Sikoki, F.D.; Nwachukwu, E.O. Application of Common Duckweed (*Lemna minor*) in Phytoremediation of Chemicals in the Environment: State and Future Perspective. *Chemosphere* **2019**, *223*, 285–309. [CrossRef]
24. Walsh, É.; Paolacci, S.; Burnell, G.; Jansen, M.A.K. The Importance of the Calcium-to-Magnesium Ratio for Phytoremediation of Dairy Industry Wastewater Using the Aquatic Plant *Lemna minor* L. *Int. J. Phytoremediat.* **2020**, *22*, 694–702. [CrossRef]
25. Petersen, F.; Demann, J.; Restemeyer, D.; Ulbrich, A.; Olfs, H.-W.; Westendarp, H.; Appenroth, K.-J. Influence of the Nitrate-N to Ammonium-N Ratio on Relative Growth Rate and Crude Protein Content in the Duckweeds *Lemna minor* and *Wolffiella hyalina*. *Plants* **2021**, *10*, 1741. [CrossRef]
26. Wang, W.; Li, R.; Zhu, Q.; Tang, X.; Zhao, Q. Transcriptomic and physiological analysis of common duckweed *Lemna minor* responses to NH_4^+ toxicity. *BMC Plant Biol.* **2016**, *16*, 92. [CrossRef]
27. Tian, X.; Fang, Y.; Jin, Y.; Yi, Z.; Li, J.; Du, A.; He, K.; Huang, Y.; Zhao, H. Ammonium Detoxification Mechanism of Ammonium-Tolerant Duckweed (*Landoltia punctata*) Revealed by Carbon and Nitrogen Metabolism under Ammonium Stress. *Environ. Pollut.* **2021**, *277*, 116834. [CrossRef]
28. Li, H.; Hu, B.; Chu, C. Nitrogen Use Efficiency in Crops: Lessons from Arabidopsis and Rice. *J. Exp. Bot.* **2017**, *68*, 2477–2488. [CrossRef]
29. Masclaux-Daubresse, C.; Daniel-Vedele, F.; Dechorgnat, J.; Chardon, F.; Gaufichon, L.; Suzuki, A. Nitrogen Uptake, Assimilation and Remobilization in Plants: Challenges for Sustainable and Productive Agriculture. *Ann. Bot.* **2010**, *105*, 1141–1157. [CrossRef]
30. Krapp, A. Plant Nitrogen Assimilation and Its Regulation: A Complex Puzzle with Missing Pieces. *Curr. Opin. Plant Biol.* **2015**, *25*, 115–122. [CrossRef]
31. Bernard, S.M.; Møller, A.L.B.; Dionisio, G.; Kichey, T.; Jahn, T.P.; Dubois, F.; Baudo, M.; Lopes, M.S.; Tercé-Laforgue, T.; Foyer, C.H.; et al. Gene Expression, Cellular Localisation and Function of Glutamine Synthetase Isozymes in Wheat (*Triticum aestivum* L.). *Plant Molec. Biol.* **2008**, *67*, 89–105. [CrossRef]
32. Bloom, A.J. The Increasing Importance of Distinguishing among Plant Nitrogen Sources. *Curr. Opin. Plant Biol.* **2015**, *25*, 10–16. [CrossRef]
33. Britto, D.T.; Kronzucker, H.J. Ecological Significance and Complexity of N-Source Preference in Plants. *Ann. Bot.* **2013**, *112*, 957–963. [CrossRef]
34. Britto, D.T.; Kronzucker, H.J. NH_4^+ Toxicity in Higher Plants: A Critical Review. *J. Plant Physiol.* **2002**, *159*, 567–584. [CrossRef]
35. Esteban, R.; Ariz, I.; Cruz, C.; Moran, J.F. Review: Mechanisms of Ammonium Toxicity and the Quest for Tolerance. *Plant Sci.* **2016**, *248*, 92–101. [CrossRef] [PubMed]
36. Fang, Y.Y.; Babourina, O.; Rengel, Z.; Yang, X.E.; Pu, P.M. Ammonium and Nitrate Uptake by the Floating Plant *Landoltia punctata*. *Ann. Bot.* **2007**, *99*, 365–370. [CrossRef] [PubMed]
37. Michael, T.P.; Bryant, D.; Gutierrez, R.; Borisjuk, N.; Chu, P.; Zhang, H.; Xia, J.; Zhou, J.; Peng, H.; El Baidouri, M.; et al. Comprehensive Definition of Genome Features in *Spirodela polyrhiza* by High-Depth Physical Mapping and Short-Read DNA Sequencing Strategies. *Plant J.* **2017**, *89*, 617–635. [CrossRef] [PubMed]
38. Borisjuk, N.; Chu, P.; Gutierrez, R.; Zhang, H.; Acosta, K.; Friesen, N.; Sree, K.S.; Garcia, C.; Appenroth, K.J.; Lam, E. Assessment, Validation and Deployment Strategy of a Two-Barcode Protocol for Facile Genotyping of Duckweed Species. *Plant Biol.* **2015**, *17* (Suppl. S1), 42–49. [CrossRef]
39. BLAST: Basic Local Alignment Search Tool. Available online: https://blast.ncbi.nlm.nih.gov/Blast.cgi (accessed on 5 October 2021).
40. Wang, W.; Haberer, G.; Gundlach, H.; Glasser, C.; Nussbaumer, T.; Luo, M.C.; Lomsadze, A.; Borodovsky, M.; Kerstetter, R.A.; Shanklin, J.; et al. The *Spirodela polyrhiza* Genome Reveals Insights into Its Neotenous Reduction Fast Growth and Aquatic Lifestyle. *Nat Commun.* **2014**, *5*, 3311. [CrossRef]
41. An, D.; Zhou, Y.; Li, C.; Xiao, Q.; Wang, T.; Zhang, Y.; Wu, Y.; Li, Y.; Chao, D.-Y.; Messing, J.; et al. Plant evolution and environmental adaptation unveiled by long-read whole-genome sequencing of *Spirodela*. *Proc. Natl. Acad. Sci. USA* **2019**, *116*, 18893–18899. [CrossRef]

42. Harkess, A.; McLoughlin, F.; Bilkey, N.; Elliott, K.; Emenecker, R.; Mattoon, E.; Miller, K.; Czymmek, K.; Vierstra, R.D.; Meyers, B.C.; et al. Improved *Spirodela polyrhiza* Genome and Proteomic Analyses Reveal a Conserved Chromosomal Structure with High Abundance of Chloroplastic Proteins Favoring Energy Production. *J. Exp. Bot.* **2021**, *72*, 2491–2500. [CrossRef]
43. James, D.; Borphukan, B.; Fartyal, D.; Achary, V.M.M.; Reddy, M.K. Transgenic manipulation of glutamine synthetase: A target with untapped potential in various aspects of crop improvement. In *Biotechnologies of Crop Improvement, Volume 2*; Gosal, S.S., Wani, S.H., Eds.; Springer: Cham, Germany, 2018; pp. 367–416. [CrossRef]
44. Bernard, S.M.; Habash, D.Z. The Importance of Cytosolic Glutamine Synthetase in Nitrogen Assimilation and Recycling. *New Phytol.* **2009**, *182*, 608–620. [CrossRef]
45. Ferreira, M.J.; Vale, D.; Cunha, L.; Melo, P. Role of the C-Terminal Extension Peptide of Plastid Located Glutamine Synthetase from *Medicago truncatula*: Crucial for Enzyme Activity and Needless for Protein Import into the Plastids. *Plant Physiol. Biochem.* **2017**, *111*, 226–233. [CrossRef]
46. Biesiadka, J.; Legocki, A.B. Evolution of the Glutamine Synthetase Gene in Plants. *Plant Sci.* **1997**, *128*, 51–58. [CrossRef]
47. Thomsen, H.C.; Eriksson, D.; Møller, I.S.; Schjoerring, J.K. Cytosolic Glutamine Synthetase: A Target for Improvement of Crop Nitrogen Use Efficiency? *Trends Plant Sci.* **2014**, *19*, 656–663. [CrossRef]
48. Pesole, G.; Bozzetti, M.P.; Lanave, C.; Preparata, G.; Saccone, C. Glutamine Synthetase Gene Evolution: A Good Molecular Clock. *Proc. Natl. Acad. Sci. USA* **1991**, *88*, 522–526. [CrossRef]
49. Nigro, D.; Gu, Y.Q.; Huo, N.; Marcotuli, I.; Blanco, A.; Gadaleta, A.; Anderson, O.D. Structural Analysis of the Wheat Genes Encoding NADH-Dependent Glutamine-2-Oxoglutarate Amidotransferases and Correlation with Grain Protein Content. *PLoS ONE* **2013**, *8*, e73751. [CrossRef]
50. Yamaya, T.; Kusano, M. Evidence Supporting Distinct Functions of Three Cytosolic Glutamine Synthetases and Two NADH-Glutamate Synthases in Rice. *J. Exp. Bot.* **2014**, *65*, 5519–5525. [CrossRef]
51. García-Gutiérrez, Á.; Cánovas, F.M.; Ávila, C. Glutamate Synthases from Conifers: Gene Structure and Phylogenetic Studies. *BMC Genomics* **2018**, *19*, 65. [CrossRef]
52. Konishi, M.; Yanagisawa, S. Identification of a Nitrate-Responsive Cis-Element in the *Arabidopsis* NIR1 Promoter Defines the Presence of Multiple Cis-Regulatory Elements for Nitrogen Response: Nitrate-Responsive Element of *Arabidopsis* NIR1. *Plant J.* **2010**, *63*, 269–282. [CrossRef]
53. Feng, H.; Yan, M.; Fan, X.; Li, B.; Shen, Q.; Miller, A.J.; Xu, G. Spatial Expression and Regulation of Rice High-Affinity Nitrate Transporters by Nitrogen and Carbon Status. *J. Exp. Bot.* **2011**, *62*, 2319–2332. [CrossRef]
54. Konishi, M.; Yanagisawa, S. Emergence of a New Step towards Understanding the Molecular Mechanisms Underlying Nitrate-Regulated Gene Expression. *J. Exp. Bot.* **2014**, *65*, 5589–5600. [CrossRef]
55. Liu, X.; Feng, H.; Huang, D.; Song, M.; Fan, X.; Xu, G. Two Short Sequences in OsNAR2.1 Promoter Are Necessary for Fully Activating the Nitrate Induced Gene Expression in Rice Roots. *Sci. Rep.* **2015**, *5*, 11950. [CrossRef]
56. Konishi, M.; Yanagisawa, S. Roles of the Transcriptional Regulation Mediated by the Nitrate-Responsive Cis-Element in Higher Plants. *Biochem. Biophy. Res. Commun.* **2011**, *411*, 708–713. [CrossRef]
57. Yu, J.; Xuan, W.; Tian, Y.; Fan, L.; Sun, J.; Tang, W.; Chen, G.; Wang, B.; Liu, Y.; Wu, W.; et al. Enhanced OsNLP4-OsNiR Cascade Confers Nitrogen Use Efficiency by Promoting Tiller Number in Rice. *Plant Biotechnol. J.* **2021**, *19*, 167–176. [CrossRef]
58. Sawaki, N.; Tsujimoto, R.; Shigyo, M.; Konishi, M.; Toki, S.; Fujiwara, T.; Yanagisawa, S. A Nitrate-Inducible GARP Family Gene Encodes an Auto-Repressible Transcriptional Repressor in Rice. *Plant Cell Physiol.* **2013**, *54*, 506–517. [CrossRef]
59. Medici, A.; Marshall-Colon, A.; Ronzier, E.; Szponarski, W.; Wang, R.; Gojon, A.; Crawford, N.M.; Ruffel, S.; Coruzzi, G.M.; Krouk, G. AtNIGT1/HRS1 Integrates Nitrate and Phosphate Signals at the Arabidopsis Root Tip. *Nat. Commun.* **2015**, *6*, 6274. [CrossRef]
60. Ueda, Y.; Nosaki, S.; Sakuraba, Y.; Miyakawa, T.; Kiba, T.; Tanokura, M.; Yanagisawa, S. NIGT1 Family Proteins Exhibit Dual Mode DNA Recognition to Regulate Nutrient Response-Associated Genes in *Arabidopsis*. *PLoS Genet.* **2020**, *16*, e1009197. [CrossRef]
61. Yanagisawa, S. Characterization of a Nitrate-Inducible Transcriptional Repressor NIGT1 Provides New Insights into DNA Recognition by the GARP Family Proteins. *Plant Signal. Behav.* **2013**, *8*, e24447. [CrossRef]
62. Berger, N.; Dubreucq, B. Evolution Goes GAGA: GAGA Binding Proteins across Kingdoms. *Biochim. Biophys. Acta* **2012**, *1819*, 863–868. [CrossRef]
63. Gong, R.; Cao, H.; Zhang, J.; Xie, K.; Wang, D.; Yu, S. Divergent Functions of the GAGA-binding Transcription Factor Family in Rice. *Plant J.* **2018**, *94*, 32–47. [CrossRef]
64. Davison, B.L.; Egly, J.-M.; Mulvihill, E.R.; Chambon, P. Formation of Stable Preinitiation Complexes between Eukaryotic Class B Transcription Factors and Promoter Sequences. *Nature* **1983**, *301*, 680–686. [CrossRef]
65. Brázda, V.; Bartas, M.; Bowater, R.P. Evolution of Diverse Strategies for Promoter Regulation. *Trends Genet.* **2021**, *37*, 730–744. [CrossRef]
66. Komarnytsky, S.; Borisjuk, N. Functional Analysis of Promoter Elements in Plants. *Genet. Eng.* **2003**, *25*, 113–141. [CrossRef]
67. Rhodes, D.; Lipps, H.J. G-Quadruplexes and Their Regulatory Roles in Biology. *Nucleic Acids Res.* **2015**, *43*, 8627–8637. [CrossRef] [PubMed]
68. Du, Z.; Zhao, Y.; Li, N. Genome-Wide Colonization of Gene Regulatory Elements by G4 DNA Motifs. *Nucleic Acids Res.* **2009**, *37*, 6784–6798. [CrossRef] [PubMed]

69. Yadav, V.; Kim, N.; Tuteja, N.; Yadav, P. G Quadruplex in Plants: A Ubiquitous Regulatory Element and Its Biological Relevance. *Front. Plant Sci.* **2017**, *8*, 1163. [CrossRef] [PubMed]
70. Hon, J.; Martínek, T.; Zendulka, J.; Lexa, M. Pqsfinder: An Exhaustive and Imperfection-Tolerant Search Tool for Potential Quadruplex-Forming Sequences in R. *Bioinformatics* **2017**, *33*, 3373–3379. [CrossRef]
71. Cagirici, H.B.; Budak, H.; Sen, T.Z. Genome-Wide Discovery of G-Quadruplexes in Barley. *Sci. Rep.* **2021**, *11*, 7876. [CrossRef]
72. Andorf, C.M.; Kopylov, M.; Dobbs, D.; Koch, K.E.; Stroupe, M.E.; Lawrence, C.J.; Bass, H.W. G-Quadruplex (G4) Motifs in the Maize (*Zea mays* L.) Genome Are Enriched at Specific Locations in Thousands of Genes Coupled to Energy Status, Hypoxia, Low Sugar, and Nutrient Deprivation. *J. Genet. Genom.* **2014**, *41*, 627–647. [CrossRef]
73. Wang, Y.; Zhao, M.; Zhang, Q.; Zhu, G.-F.; Li, F.-F.; Du, L.-F. Genomic Distribution and Possible Functional Roles of Putative G-Quadruplex Motifs in Two Subspecies of *Oryza sativa*. *Comput. Biol. Chem.* **2015**, *56*, 122–130. [CrossRef]
74. Cagirici, H.B.; Sen, T.Z. Genome-Wide Discovery of G-Quadruplexes in Wheat: Distribution and Putative Functional Roles. *Sci. Rep.* **2020**, *10*, 2021–2032. [CrossRef]
75. Sekomo, C.B.; Rousseau, D.P.L.; Saleh, S.A.; Lens, P.N.L. Heavy Metal Removal in Duckweed and Algae Ponds as a Polishing Step for Textile Wastewater Treatment. *Ecol. Eng.* **2012**, *44*, 102–110. [CrossRef]
76. Porath, D.; Pollock, J. Ammonia Stripping by Duckweed and Its Feasibility in Circulating Aquaculture. *Aquat. Bot.* **1982**, *13*, 125–131. [CrossRef]
77. Huang, L.; Lu, Y.; Gao, X.; Du, G.; Ma, X.; Liu, M.; Guo, J.; Chen, Y. Ammonium-Induced Oxidative Stress on Plant Growth and Antioxidative Response of Duckweed (*Lemna minor* L.). *Ecol. Engineer.* **2013**, *58*, 355–362. [CrossRef]
78. Gowri, G.; Kenis, J.D.; Ingemarsson, B.; Redinbaugh, M.G.; Campbell, W.H. Nitrate Reductase Transcript Is Expressed in the Primary Response of Maize to Environmental Nitrate. *Plant Mol. Biol.* **1992**, *18*, 55–64. [CrossRef]
79. Ingemarsson, B.; Oscarson, P.; Af Ugglas, M.; Larsson, C.M. Nitrogen Utilization in *Lemna*: III. Short-Term Effects of Ammonium on Nitrate Uptake and Nitrate Reduction. *Plant Physiol.* **1987**, *85*, 865–867. [CrossRef]
80. Lee, R.B.; Drew, M.C. Rapid, Reversible Inhibition of Nitrate Influx in Barley by Ammonium. *J. Exp. Bot.* **1989**, *40*, 741–752. [CrossRef]
81. Kronzucker, H.J.; Glass, A.D.M.; Siddiqi, M.Y. Inhibition of Nitrate Uptake by Ammonium in Barley. Analysis of Component Fluxes. *Plant Physiol.* **1999**, *120*, 283–292. [CrossRef]
82. Kronzucker, H.J.; Siddiqi, M.Y.; Glass, A.D.M.; Kirk, G.J.D. Nitrate-Ammonium Synergism in Rice. a Subcellular Flux Analysis. *Plant Physiol.* **1999**, *119*, 1041–1046. [CrossRef]
83. Li, J.; Xue, L.; Yan, H.; Wang, L.; Liu, L.; Lu, Y.; Xie, H. The Nitrate Reductase Gene-Switch: A System for Regulated Expression in Transformed Cells of *Dunaliella salina*. *Gene* **2007**, *403*, 132–142. [CrossRef]
84. Von der Heyde, E.; Klein, B.; Abram, L.; Hallmann, A. The Inducible NitA Promoter Provides a Powerful Molecular Switch for Transgene Expression in *Volvox carteri*. *BMC Biotechnol.* **2015**, *15*, 5. [CrossRef]
85. Jackson, H.O.; Berepiki, A.; Baylay, A.J.; Terry, M.J.; Moore, C.M.; Bibby, T.S. An Inducible Expression System in the Alga *Nannochloropsis gaditana* Controlled by the Nitrate Reductase Promoter. *J. Appl. Phycol.* **2019**, *31*, 269–279. [CrossRef]
86. Ji, Y.; Li, Q.; Liu, G.; Selvaraj, G.; Zheng, Z.; Zou, J.; Wei, Y. Roles of Cytosolic Glutamine Synthetases in *Arabidopsis* Development and Stress Responses. *Plant Cell Physiol.* **2019**, *60*, 657–671. [CrossRef]
87. Lothier, J.; Gaufichon, L.; Sormani, R.; Lemaitre, T.; Azzopardi, M.; Morin, H.; Chardon, F.; Reisdorf-Cren, M.; Avice, J.-C.; Masclaux-Daubresse, C. The Cytosolic Glutamine Synthetase GLN1;2 Plays a Role in the Control of Plant Growth and Ammonium Homeostasis in *Arabidopsis* Rosettes When Nitrate Supply Is Not Limiting. *J. Exp. Bot.* **2011**, *62*, 1375–1390. [CrossRef]
88. Guan, M.; de Bang, T.C.; Pedersen, C.; Schjoerring, J.K. Cytosolic Glutamine Synthetase Gln1;2 Is the Main Isozyme Contributing to GS1 Activity and Can Be Up-Regulated to Relieve Ammonium Toxicity. *Plant Physiol.* **2016**, *171*, 1921–1933. [CrossRef]
89. Goodall, A.J.; Kumar, P.; Tobin, A.K. Identification and Expression Analyses of Cytosolic Glutamine Synthetase Genes in Barley (*Hordeum vulgare* L.). *Plant Cell Physiol.* **2013**, *54*, 492–505. [CrossRef]
90. Ferreira, S.; Moreira, E.; Amorim, I.; Santos, C.; Melo, P. *Arabidopsis thaliana* Mutants Devoid of Chloroplast Glutamine Synthetase (GS2) Have Non-Lethal Phenotype under Photorespiratory Conditions. *Plant Physiol. Biochem.* **2019**, *144*, 365–374. [CrossRef]
91. Tang, D.; Liu, M.-Y.; Zhang, Q.; Fan, K.; Ruan, J. Isolation and Characterization of Chloroplastic Glutamine Synthetase Gene (CsGS2) in Tea Plant *Camellia sinensis*. *Plant Physiol. Biochem.* **2020**, *155*, 321–329. [CrossRef]
92. Németh, E.; Nagy, Z.; Pécsváradi, A. Chloroplast Glutamine Synthetase, the Key Regulator of Nitrogen Metabolism in Wheat, Performs Its Role by Fine Regulation of Enzyme Activity via Negative Cooperativity of Its Subunits. *Front. Plant Sci.* **2018**, *9*, 191. [CrossRef]
93. Masclaux, C.; Valadier, M.-H.; Brugière, N.; Morot-Gaudry, J.-F.; Hirel, B. Characterization of the Sink/Source Transition in Tobacco (*Nicotiana tabacum* L.) Shoots in Relation to Nitrogen Management and Leaf Senescence. *Planta* **2000**, *211*, 510–518. [CrossRef]
94. Hörtensteiner, S.; Feller, U. Nitrogen Metabolism and Remobilization during Senescence. *J. Exp. Bot.* **2002**, *53*, 927–937. [CrossRef]
95. Orsel, M.; Moison, M.; Clouet, V.; Thomas, J.; Leprince, F.; Canoy, A.-S.; Just, J.; Chalhoub, B.; Masclaux-Daubresse, C. Sixteen Cytosolic Glutamine Synthetase Genes Identified in the *Brassica napus* L. Genome Are Differentially Regulated Depending on Nitrogen Regimes and Leaf Senescence. *J. Exp. Bot.* **2014**, *65*, 3927–3947. [CrossRef]

96. Tabuchi, M.; Sugiyama, K.; Ishiyama, K.; Inoue, E.; Sato, T.; Takahashi, H.; Yamaya, T. Severe Reduction in Growth Rate and Grain Filling of Rice Mutants Lacking OsGS1;1, a Cytosolic Glutamine Synthetase1;1: GS1;1-Knockout Mutant in Rice. *Plant J.* **2005**, *42*, 641–651. [CrossRef]
97. Gadaleta, A.; Nigro, D.; Marcotuli, I.; Giancaspro, A.; Giove, S.L.; Blanco, A. Isolation and Characterisation of Cytosolic Glutamine Synthetase (GSe) Genes and Association with Grain Protein Content in Durum Wheat. *Crop Pasture Sci.* **2014**, *65*, 38. [CrossRef]
98. Wang, Y.; Teng, W.; Wang, Y.; Ouyang, X.; Xue, H.; Zhao, X.; Gao, C.; Tong, Y. The Wheat Cytosolic Glutamine Synthetase GS1.1 Modulates N Assimilation and Spike Development by Characterizing CRISPR-Edited Mutants. *bioRxiv* **2020**. preprint. [CrossRef]
99. Gao, Y.; Bang, T.C.; Schjoerring, J.K. Cisgenic Overexpression of Cytosolic Glutamine Synthetase Improves Nitrogen Utilization Efficiency in Barley and Prevents Grain Protein Decline under Elevated CO_2. *J. Plant Biotechnol.* **2019**, *17*, 1209–1221. [CrossRef] [PubMed]
100. Martin, A.; Lee, J.; Kichey, T.; Gerentes, D.; Zivy, M.; Tatout, C.; Dubois, F.; Balliau, T.; Valot, B.; Davanture, M.; et al. Two Cytosolic Glutamine Synthetase Isoforms of Maize Are Specifically Involved in the Control of Grain Production. *Plant Cell* **2006**, *18*, 3252–3274. [CrossRef]
101. Konishi, N.; Ishiyama, K.; Beier, M.P.; Inoue, E.; Kanno, K.; Yamaya, T.; Takahashi, H.; Kojima, S. Contributions of Two Cytosolic Glutamine Synthetase Isozymes to Ammonium Assimilation in *Arabidopsis* Roots. *J. Exp. Bot.* **2017**, *68*, 613–625. [CrossRef]
102. Wang, W.; Wu, Y.; Messing, J. RNA-Seq Transcriptome Analysis of *Spirodela* Dormancy without Reproduction. *BMC Genom.* **2014**, *15*, 60. [CrossRef]
103. Lea, P.J.; Miflin, B.J. Alternative Route for Nitrogen Assimilation in Higher Plants. *Nature* **1974**, *251*, 614–616. [CrossRef]
104. Masclaux-Daubresse, C.; Reisdorf-Cren, M.; Pageau, K.; Lelandais, M.; Grandjean, O.; Kronenberger, J.; Valadier, M.-H.; Feraud, M.; Jouglet, T.; Suzuki, A. Glutamine Synthetase-Glutamate Synthase Pathway and Glutamate Dehydrogenase Play Distinct Roles in the Sink-Source Nitrogen Cycle in Tobacco. *Plant Physiol.* **2006**, *140*, 444–456. [CrossRef]
105. Lea, P.J.; Miflin, B.J. Glutamate Synthase and the Synthesis of Glutamate in Plants. *Plant Physiol. Biochem.* **2003**, *41*, 555–564. [CrossRef]
106. Suzuki, A.; Knaff, D.B. Glutamate Synthase: Structural, Mechanistic and Regulatory Properties, and Role in the Amino Acid Metabolism. *Photosynth. Res.* **2005**, *83*, 191–217. [CrossRef]
107. Zeng, D.-D.; Qin, R.; Li, M.; Alamin, M.; Jin, X.-L.; Liu, Y.; Shi, C.-H. The Ferredoxin-Dependent Glutamate Synthase (OsFd-GOGAT) Participates in Leaf Senescence and the Nitrogen Remobilization in Rice. *Mol. Genet. Genomics* **2017**, *292*, 385–395. [CrossRef]
108. Tamura, W.; Kojima, S.; Toyokawa, A.; Watanabe, H.; Tabuchi-Kobayashi, M.; Hayakawa, T.; Yamaya, T. Disruption of a Novel NADH-Glutamate Synthase 2 Gene Caused Marked Reduction in Spikelet Number of Rice. *Front. Plant Sci.* **2011**, *2*, 57. [CrossRef]
109. Tsukiyama, T.; Becker, P.B.; Wu, C. ATP-Dependent Nucleosome Disruption at a Heat-Shock Promoter Mediated by Binding of GAGA Transcription Factor. *Nature* **1994**, *367*, 525–532. [CrossRef]
110. Chetverina, D.; Erokhin, M.; Schedl, P. GAGA Factor: A Multifunctional Pioneering Chromatin Protein. *Cell. Mol. Life Sci.* **2021**, *78*, 4125–4141. [CrossRef]
111. Hoang, P.T.N.; Fiebig, A.; Novák, P.; Macas, J.; Cao, H.X.; Stepanenko, A.; Chen, G.; Borisjuk, N.; Scholz, U.; Schubert, I. Chromosome-Scale Genome Assembly for the Duckweed *Spirodela intermedia*, Integrating Cytogenetic Maps, PacBio and Oxford Nanopore Libraries. *Sci. Rep.* **2020**, *10*, 19230. [CrossRef]
112. Santi, L.; Wang, Y.; Stile, M.R.; Berendzen, K.; Wanke, D.; Roig, C.; Pozzi, C.; Müller, K.; Müller, J.; Rohde, W.; et al. The GA Octodinucleotide Repeat Binding Factor BBR Participates in the Transcriptional Regulation of the Homeobox Gene *Bkn3*. *Plant J.* **2003**, *34*, 813–826. [CrossRef]
113. Schenk, R.U.; Hildebrandt, A.C. Medium and Techniques for Induction and Growth of Monocotyledonous and Dicotyledonous Plant Cell Cultures. *Can. J. Bot.* **1972**, *50*, 199–204. [CrossRef]
114. Cedergreen, N.; Madsen, T.V. Nitrogen Uptake by the Floating Macrophyte *Lemna minor*. *New Phytol.* **2002**, *155*, 285–292. [CrossRef]
115. American Water Works Association (AWWA). *Standard Methods for the Examination of Water and Wastewater*, 21st ed.; AWWA: Washington, DC, USA, 2005.
116. National Center for Biotechnology Information (NCBI). Available online: https://www.ncbi.nlm.nih.gov (accessed on 5 October 2021).
117. Movassat, M.; Forouzmand, E.; Reese, F.; Hertel, K.J. Exon Size and Sequence Conservation Improves Identification of Splice-Altering Nucleotides. *RNA* **2019**, *25*, 1793–1805. [CrossRef]
118. Rice Genome Annotation Project. Available online: http://rice.plantbiology.msu.edu (accessed on 5 October 2021).
119. LEMNA.ORG. Available online: https://www.lemna.org/ (accessed on 5 October 2021).
120. The Arabidopsis Information Resource (TAIR). Available online: https://www.arabidopsis.org (accessed on 5 October 2021).
121. Huerta-Cepas, J.; Serra, F.; Bork, P. ETE 3: Reconstruction, Analysis, and Visualization of Phylogenomic Data. *Molecul. Biol. Evol.* **2016**, *33*, 1635–1638. [CrossRef]
122. Stamatakis, A. RAxML Version 8: A Tool for Phylogenetic Analysis and Post-Analysis of Large Phylogenies. *Bioinformatics* **2014**, *30*, 1312–1313. [CrossRef]
123. Box, M.S.; Coustham, V.; Dean, C.; Mylne, J.S. Protocol: A Simple Phenol-Based Method for 96-Well Extraction of High Quality RNA from Arabidopsis. *Plant Methods* **2011**, *7*, 7. [CrossRef]

124. Pfaffl, M.W. A New Mathematical Model for Relative Quantification in Real-Time RT-PCR. *Nucleic Acids Res.* **2001**, *29*, e45. [CrossRef]
125. Vandesompele, J.; De Preter, K.; Pattyn, F.; Poppe, B.; Van Roy, N.; De Paepe, A.; Speleman, F. Accurate Normalization of Real-Time Quantitative RT-PCR Data by Geometric Averaging of Multiple Internal Control Genes. *Genome Biol.* **2002**, *3*, research0034.1. [CrossRef]
126. López-Arredondo, D.L.; Leyva-González, M.A.; Alatorre-Cobos, F.; Herrera-Estrella, L. Biotechnology of Nutrient Uptake and Assimilation in Plants. *Int. J. Dev. Biol.* **2013**, *57*, 595–610. [CrossRef]
127. Borisjuk, N.; Borisjuk, L.; Komarnytsky, S.; Timeva, S.; Hemleben, V.; Gleba, Y.; Raskin, I. Tobacco Ribosomal DNA Spacer Element Stimulates Amplification and Expression of Heterologous Genes. *Nat. Biotechnol.* **2000**, *18*, 1303–1306. [CrossRef]
128. Liu, Y.; Wang, Y.; Xu, S.; Tang, X.; Zhao, J.; Yu, C.; He, G.; Xu, H.; Wang, S.; Tang, Y.; et al. Efficient Genetic Transformation and CRISPR/Cas9-mediated Genome Editing in *Lemna aequinoctialis*. *Plant Biotechnol. J.* **2019**, *17*, 2143–2152. [CrossRef]

plants

Article

Ammonium Uptake, Mediated by Ammonium Transporters, Mitigates Manganese Toxicity in Duckweed, *Spirodela polyrhiza*

Olena Kishchenko [1,2,3,†], Anton Stepanenko [1,2,3,†], Tatsiana Straub [4,†], Yuzhen Zhou [1], Benjamin Neuhäuser [4,*] and Nikolai Borisjuk [1,*]

1 Jiangsu Key Laboratory for Eco-Agricultural Biotechnology around Hongze Lake, Jiangsu Collaborative Innovation Centre of Regional Modern Agriculture and Environmental Protection, Huaiyin Normal University, West Changjiang Road 111, Huai'an 223000, China
2 Leibniz Institute of Plant Genetics and Crop Plant Research (IPK), 06466 Gatersleben, Germany
3 Institute of Cell Biology and Genetic Engineering, National Academy of Science of Ukraine, Acad. Zabolotnogo Str. 148, 03143 Kyiv, Ukraine
4 Institute of Crop Science, Nutritional Crop Physiology, University of Hohenheim, 70593 Stuttgart, Germany
* Correspondence: benjamin.neuhaeuser@uni-hohenheim.de (B.N.); nborisjuk@hytc.edu.cn (N.B.)
† These authors contributed equally to this work.

Abstract: Nitrogen is an essential nutrient that affects all aspects of the growth, development and metabolic responses of plants. Here we investigated the influence of the two major sources of inorganic nitrogen, nitrate and ammonium, on the toxicity caused by excess of Mn in great duckweed, *Spirodela polyrhiza*. The revealed alleviating effect of ammonium on Mn-mediated toxicity, was complemented by detailed molecular, biochemical and evolutionary characterization of the species ammonium transporters (AMTs). Four genes encoding AMTs in *S. polyrhiza*, were classified as *SpAMT1;1*, *SpAMT1;2*, *SpAMT1;3* and *SpAMT2*. Functional testing of the expressed proteins in yeast and *Xenopus* oocytes clearly demonstrated activity of SpAMT1;1 and SpAMT1;3 in transporting ammonium. Transcripts of all *SpAMT* genes were detected in duckweed fronds grown in cultivation medium, containing a physiological or 50-fold elevated concentration of Mn at the background of nitrogen or a mixture of nitrate and ammonium. Each gene demonstrated an individual expression pattern, revealed by RT-qPCR. Revealing the mitigating effect of ammonium uptake on manganese toxicity in aquatic duckweed *S. polyrhiza*, the study presents a comprehensive analysis of the transporters involved in the uptake of ammonium, shedding a new light on the interactions between the mechanisms of heavy metal toxicity and the regulation of the plant nitrogen metabolism.

Keywords: duckweed; *Spirodela polyrhiza*; manganese toxicity; ammonium transporter; gene expression; transcription factors

Citation: Kishchenko, O.; Stepanenko, A.; Straub, T.; Zhou, Y.; Neuhäuser, B.; Borisjuk, N. Ammonium Uptake, Mediated by Ammonium Transporters, Mitigates Manganese Toxicity in Duckweed, *Spirodela polyrhiza*. *Plants* **2023**, *12*, 208. https://doi.org/10.3390/plants12010208

Academic Editors: Viktor Oláh, Klaus-Jürgen Appenroth and K. Sowjanya Sree

Received: 14 November 2022
Revised: 27 December 2022
Accepted: 30 December 2022
Published: 3 January 2023

1. Introduction

Together with carbon, hydrogen and oxygen, nitrogen (N) is an essential nutrient that affects all aspects of the growth, development and metabolic responses of plants [1–3]. Nitrate (NO_3^-) and ammonium (NH_4^+) are the two major forms of N supply for plants [4]. With different plant species demonstrating preferences for either nitrate or ammonium as a preferential source of nitrogen [5–9], many studies revealed the dependence between nitrogen resource and the plants' responses to stresses [10,11]. A number of studies also demonstrated that the form of supplied nitrogen affects the response of different plant species to stress caused by heavy metals such as aluminum, cadmium, copper, iron or mercury [12–16].

For example, the alleviating effect of NH_4^+ compared to NO_3^- related to Al [15,17] and Cd toxicity [16] has been demonstrated in rice and wheat [18]. Similarly, increasing the ratio of NH_4^+ to NO_3^- slowed down Cd absorption and reduce Cd toxicity in tomato [19]. Ammonium was demonstrated to interact with iron homeostasis in *Brachypodium* [20], and alleviate iron deficiency in rice [21]. The NO_3^-/NH_4^+ ration influenced copper phytoextraction causing oxidative stress, and activating the antioxidant system in *Tanzania guinea* grass [22].

Manganese (Mn) plays a crucial role as a cofactor for numerous key enzymes [23] in all living organisms and occurs naturally in soil, sediment, air borne particulates and at relatively low levels in water reservoirs. It is of special importance for plants, where it is required for proper functioning and oxygen generation by photosystem II [24]. Despite its importance, Mn is required in relatively low amounts, usually between 20 and 40 µg/kg of the crop's dry biomass [11,23]. Deficiency of Mn may cause inhibition of growth, reduction in biomass [25] and could result in tissue necrosis due to decrease in Mn-dependent superoxide dismutase (MnSOD) and corresponding increase in free oxygen radicals [26]. On the other hand, excessive levels of Mn may be toxic for green vegetation. Plants are able to accumulate Mn in great access [27], which could lead to the Mn toxicity, usually manifested by symptoms of chlorosis, necrosis, deformation of young leaves [28,29] and characteristic brown spots on mature leaves [30]. High concentrations of this microelement can also prevent the uptake and translocation of other essential elements such as Ca, Mg and Fe inducing their deficiencies which is considered as an explanation for the inhibition of chlorophyll biosynthesis and growth [31,32]. Moreover, modern human activities such as mining, metal smelting and the application of fertilizers and fungicides to agricultural land have raised the Mn content in soil, sediments and groundwater [33–36]. These developments pollute the environment, negatively affect ecosystems and lead to accumulation of Mn in the food chain [37,38].

Duckweeds are a group of small fast growing aquatic plants, attracting increasing interest of the scientific community not only as a new model plant for various physiological, biochemical and molecular research [39], but also as a source of viable biomass [40] and a plant of choice for wastewater phytoremediation of many heavy metals [41]. Similar to rice growing on paddy fields [5,7], aquatic duckweeds have clear bias for ammonium as a source of nitrogen when given a choice between ammonium and nitrate [42]. In our previous study [43], we have demonstrated that great duckweed, *Spirodela polyrhiza*, is sensible to Mn toxicity and manifests characteristic symptoms of Mn stress when concentrations of Mn exceed 40 mg/L (0.73 mM).

In this study, we analysed the effects of 50-fold excessed Mn on various growth parameters of *S. polyrhiza* at the background of different combinations of NO_3^- and NH_4^+. The revealed physiological characteristics of biomass accumulation, N consumption and changes in cultivation medium pH, were related to comprehensive characterization of the molecular and biochemical properties of the four ammonium transporters SpAMT1;1, SpAMT1;2, SpAMT1;3 and SpAMT2, considered as the major entry pathways for NH_4^+ uptake.

2. Results

2.1. Parameters of the Experimental System: Dynamics of Biomass Accumulation, Nitrogen Uptake, Changes in Medium pH, during 12-Days of Duckweed Cultivation

The effects of Mn at 2.95 mM $MnSO_4$, 50-fold higher compared to the routinely used concentration in SH duckweed cultivation medium, were evaluated in dependence of two sources of inorganic nitrogen, supplied in form of nitrate (5 mM KNO_3) or a mixture of nitrate and ammonium (2.5 mM KNO_3 plus 2.5 mM $NH_4H_2PO_4$) during 12 days. No significant differences either in the biomass accumulation or plant appearance were observed up to four days of cultivation. However, by day 12 we observed significant, almost two-fold reduction in *S. polyrhiza* biomass accumulation in a medium supplied with nitrate and high Mn (N-Mn50) compared to nitrate medium with usual concentration of Mn (N-Mn1). Simultaneously, almost no difference in biomass accumulation was registered between duckweed grown on a mixture of nitrate and ammonium supplied with high (M-Mn50) of normal (M-Mn1) amounts of Mn (Figure 1).

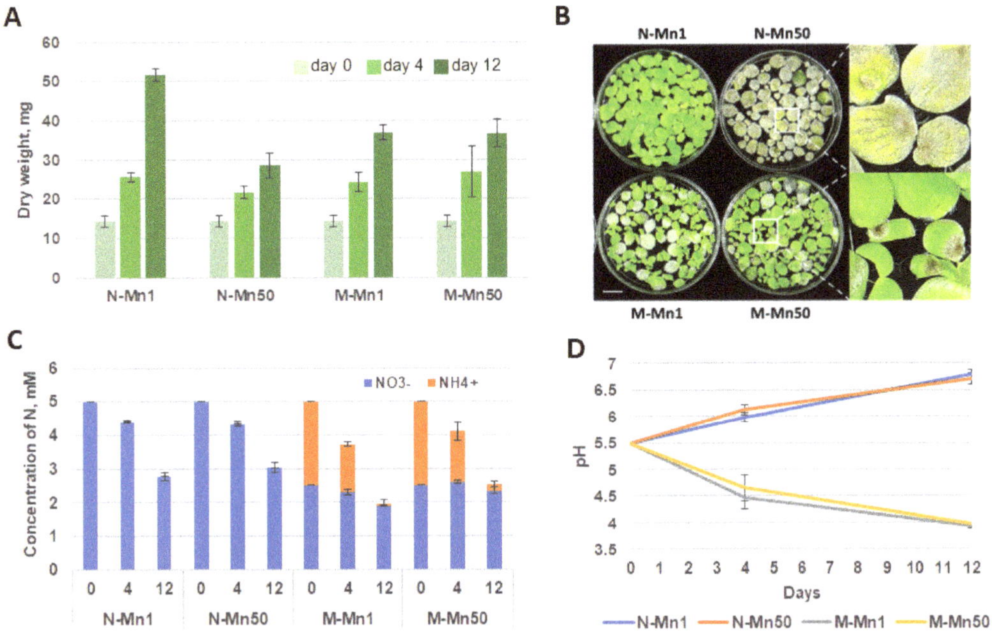

Figure 1. Dynamic parameters related to duckweed cultivation in media with different formulations of N and Mn: (**A**) Duckweed biomass at the beginning of the experiment, (light green) and following 4 days (green) and 12 days (dark green) cultivation in media supplied with different concentrations of nitrate, ammonium and Mn. N-Mn1—medium containing 5 mM NO_3^- and 0.059 mM Mn; N-Mn50—medium containing 5 mM NO_3^- and 2.95 mM Mn; M-Mn1—medium containing a mixture of 2.5 mM NO_3^-, 2.5 mM NH_4^+ and 0.059 mM Mn; M-Mn50—medium containing a mixture of 2.5 mM NO_3^-, 2.5 mM NH_4^+ and 2.95 mM Mn. (**B**) Images of *S. polyrhiza* fronds taken at day 12 following cultivation in media supplied with different concentrations of nitrate, ammonium and Mn; (**C**) Comparative dynamics of nitrogen uptake, supplied in the medium in form of nitrate (NO_3^-) or ammonium (NH_4^+); (**D**) Comparative dynamics of pH of four cultivation media in course of duckweed growth. Bar is 1 cm.

Suppression of biomass accumulation in the N-Mn50 medium was accompanied by severe chlorosis and death of whole duckweed fronds profoundly manifested at day 12. To the contrary, after 12 days cultivation, duckweed grown in the medium supplied with NH_4^+ (M-Mn50), mostly remained green with partial local browning of frond apexes, but no total chlorosis observed. In course of duckweed growth, the relevant changes in media such as consumption of nitrogen and pH have been monitored (Figure 1).

The dynamics of NO_3^- consumption by *S. polyrhiza* growing on N-Mn1 and N-Mn50 media were pretty similar, with the available nitrogen reduced by about a half by day 12. Duckweed cultivated in the media containing both NO_3^- and NH_4^+ (M-Mn1 and M-Mn50) preferentially utilized ammonium, with nitrate almost untouched in the M-Mn50 and only slightly consumed in M-Mn1 by day 12 (Figure 1C).

During the growth period, we observed contrasting dynamics of medium acidity between medium containing NO_3^-, as a soul source of nitrogen and a mixture of NO_3^- and NH_4^+. The original medium pH of 5.8 climbed up to 7.01–7.18 in the medium supplied with nitrate, whereas the pH gradually dropped to 3.94–4.02 on day 12 in the medium containing both nitrate and ammonium. There were no differences in pH dynamics in the pairs of media containing usual or high concentration of Mn, N-Mn1 and N-Mn50 or M-Mn1 and M-Mn50 (Figure 1D).

2.2. *Genome of S. polyrhiza Containes Four Genes Coding for Ammonium Transporters*

A blast search on the *S. polyrhiza* genome (ecotype Sp9509 [44]), available on the NCBI website (taxid: 29656, GCA_900492545.1), using the protein sequences of rice ammonium transporters in the queries revealed four genes potentially encoding ammonium transporters SpAMT1;1, SpAMT1;2, SpAMT1;3 and SpAMT2. The exon/intron structures of the gene sequences were deduced by their similarities with the corresponding rice sequences, showing three intron-free genes belonging to the *AMT1* subfamily and only *SpAMT2* containing two short introns. A similar arrangement of the *AMT* genes has been revealed also for the second species in the genus *Spirodela*, *S. intermedia* (Figure S1) by using BLAST against the corresponding genome at the NCBI website (taxid: 51605). To validate the sequences available in the GenBank, we sequenced the amplification products for the four *SpAMTs* derived from cDNA of ecotype NB5548 used in this study (the corresponding sequence accession numbers are: OP730321, OP730322, OP730323 and OP730324).

Analysis of the deduced polypeptide sequences using InterPro software (http://www.ebi.ac.uk/interpro/ accessed on 10 September 2022) revealed structural organization typical for ammonium transmembrane transporters with characteristic N-terminal non-cytoplasmic domains of different length (53 and 50aa for SpAMT1;1 and SpAMT1;2, and 15 and 23aa for SpAMT1;3 and SpAMT2) and cytoplasmic C-terminal domains ranging from 51aa for SpAMT1;3 to 62aa for SpAMT1;2 [45] (Figure S2).

Phylogenetic analysis of the deduced protein sequences showed that SpAMT1;1 and SpAMT1;2 are closely related to each other, clustering together with their homologues from *S. intermedia* and taro, *Colocasia esculenta*, a representative of Araceae, the sister family to Lemnaceae [46]. The sequences also demonstrate significant similarity to more remote monocotyledonous species (date palm, *Phoenix dactylifera;* Cocos, rice and wheat), as well as the representative of dicotyledonous, *Arabidopsis thaliana* (Figure 2). Similar phylogenetic distances relative to other monocotyledonous and dicotyledonous were revealed also for SpAMT1;3 and SpAMT2.

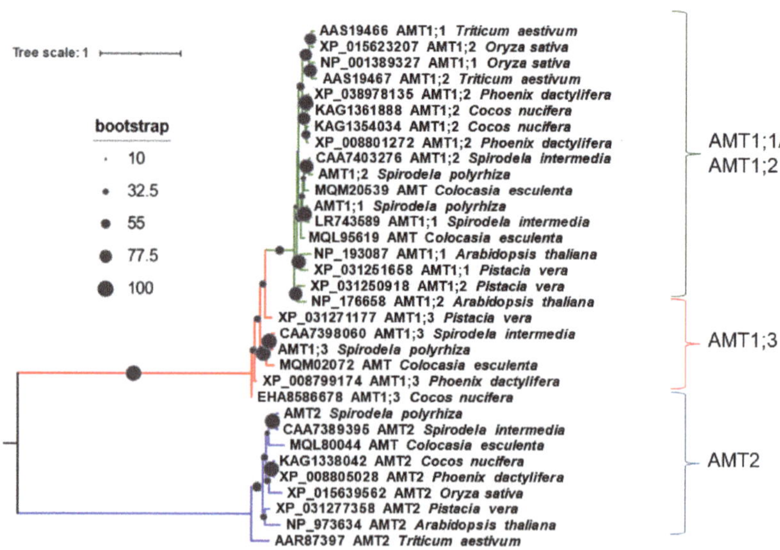

Figure 2. Clustering of AMT protein sequences from selected plant species. The phylograms shows

a maximum likelihood tree obtained using available AMT protein sequences from different plants. Multiple alignment and phylogenetic tree were generated in NGPhylogeny webservice (https://ngphylogeny.fr, accessed on 27 September 2022) [47] and iTOL (https://itol.embl.de, accessed on 29 September 2022) [48].

2.3. Functional Activity of the Ammonium Transporters in Yeast and Xenopus Oocytes Systems

To test general transport function of the identified potential SpAMTs we performed a complementation analysis by expressing the four duckweed ammonium transporter proteins from the pDR199 vector in a yeast strain deficient in AMT function (31019b, ΔΔΔmep; [49]). The yeast was grown on medium supplied with NH_4^+ as a sole source of Nitrogen at two concentrations, 0.3 and 1 mM and three different pH: 4.5, 5.5 and 6.5 (Figure 3). The highest complementation activity in all experimental settings, demonstrated the duckweed ammonium transporter SpAMT1;3, followed by the SpAMT1;1. The complementation efficiency of SpAMT1;3 was comparable to that of Arabidopsis AtAMT1;1, with the activity of SpAMT1;1 resembling that of AtAMT2. We were not able to detect any complementing activity of SpAMT1;2. The SpAMT2 showed some activity slightly above the pDR199 control only at 1 mM concentration of NH_4^+ and medium pH 6.5. In general, there was no pH effect on any of the transporters. When grown on 60 mM methylammonia (MeA), a toxic analog of NH_4^+, the active transporters conferred transport activity of MeA as well, which led to strongly reduced yeast growth.

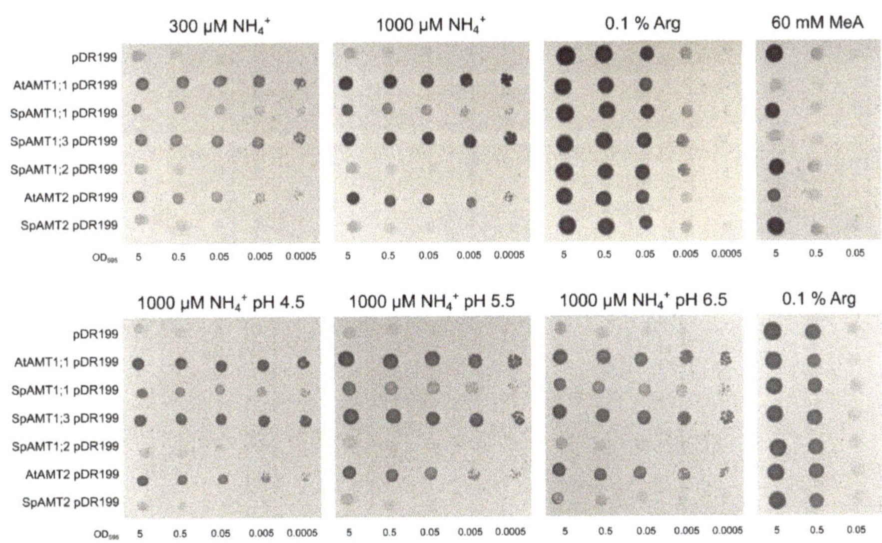

Figure 3. Transport activity of SpAMTs in yeast. Top panel represents growth on 0.3 or 1 mM NH_4Cl as sole nitrogen source, 0.1 % (*w*/*v*) Arginine was used as a positive control. Growth on 60 mM MeA was used as a toxicity test, with 0.1 % (*w*/*v*) Arginine as nitrogen source. The bottom panel shows growth on 1 mM NH4Cl when media was buffered at pH 4.5, 5.5 or 6.5 unbuffered media with 0.1 % (*w*/*v*) Arginine was used as dilution control. Spotted were five 10 times dilutions starting with OD_{595} = 5. Pictures are representatives of four independent repetitions.

To analyse the two functional duckweed AMTs in more detail, the proteins were expressed in *Xenopus leavis* oocytes and the ammonium induced currents where recorded by two electrode voltage clamp. Ammonium induced significant inward currents in the oocytes expressing SpAMT1;1 and SpAMT1;3 (Figure 4). As expected, the currents increased with more negative membrane potential (Figure 4A). Total currents by SpAMT1;3

were in the low nano ampere range and increased with increasing ammonium concentration with a low affinity K_m = 5.4 mM. The SpAMT1;1 mediated currents were about three times higher than the currents by SpAMT1;3 (Figure 4B) and saturated with a high affinity K_m = 116 µM tipical for plant AMTs (Figure 4C). No currect induction was detected in the oocytes transfected with SpAMT1;2 or SpAMT2 (data not shown). Based on the obtained results, we can characterise SpAMT1;1 as a tipical high affinity plant ammonium transporter while SpAMT1;3 showed uncommon low affinity for ammonium.

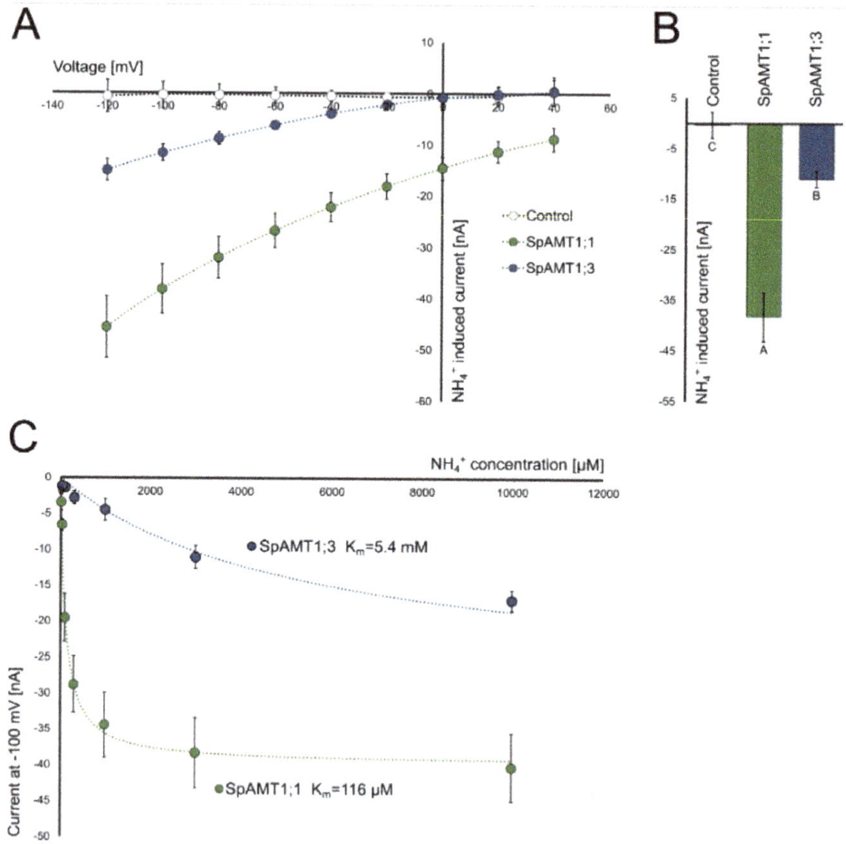

Figure 4. Functional characterization of SpAMT1;1 and SpAMT1;3 in *Xenopus laevis* oocytes: (**A**) Current/Voltage plot of ammonium induced currents in SpAMT1;1 and SpAMT1;3 expressing oocytes. Currents without ammonium were subtracted from currents with 3 mM ammonium; (**B**) For comparison of current intensities, ammonium induced (wash subtracted) currents by 3 mM and 100 mV are given as a bar chart; (**C**) Saturation kinetics of SpAMT1;1 and SpAMT1;3 mediated ammonium transport in oocytes using concentrations of ammonium from 0–10 mM NH_4Cl. Data show means +/− SEM; $n \geq 8$.

2.4. Expression Profiling of S. polyrhiza AMT Genes by qRT-PCR

The expression of the *SpAMT* genes was evaluated at the growth stage following 4 days of cultivation in the four media formulations mentioned above, using the primers represented in Table S1 according a previously described RT-qPCR protocol [42]. All RT-qPCR reactions were performed in triplicate, normalized against the expression of two housekeeping genes (*β-actin* and *histone H3*), and normalized to the gene expression levels at the starting point (SP). The expression of *S. polyrhiza* genes encoding AMT1;1 and AMT1;3, which demonstrated clear activity in oocytes and yeast tests, showed a complex and distinct regulation by Mn and the source of nitrogen. Adding of nitrogen in form of NO_3^- at the background of physiological concentration of Mn (N-Mn1) led to a slight activation of both *SpAMT1;1* and *SpAMT1;3*, whereas a mixture of NO_3^- and NH_4^+ stimulated expression of *SpAMT1;3* but did not *SpAMT1;1* (Figure 5). The 50-fold elevated concentration of Mn affected the expression of *SpAMT1;1* and *SpAMT1;3* in opposite directions, enhancing *SpAMT1;1* and suppressing *SpAMT1;3*, with these effects manifested much stronger in the presence of nitrate as a sole source of N, N-Mn50, compared to nitrate and ammonium mixture, M-Mn50.

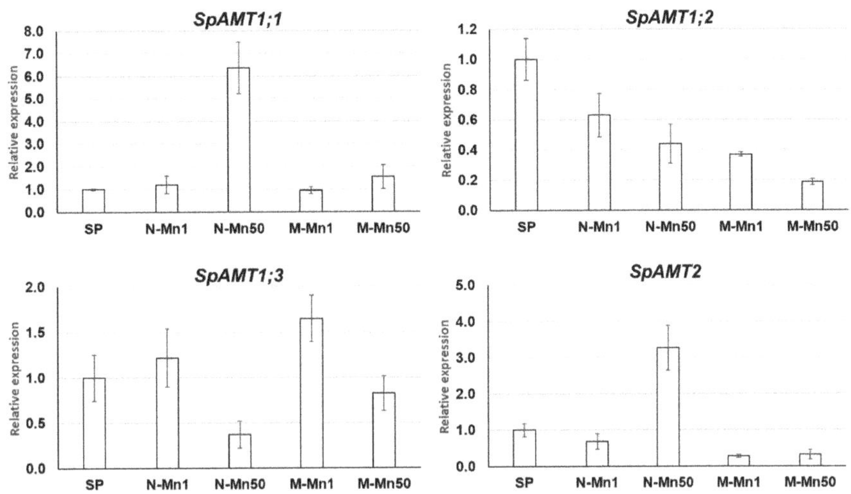

Figure 5. Expression patterns of the four *SpAMT* genes in duckweed cultivated in different media (N-Mn1, M-Mn1, N-Mn50 and M-Mn50). The gene expression was determined at cultivation day 4 by RT-qPCR relative to day 0, starting point (SP). Gene expression levels are in relative units. Error bars show $\pm$ SD of 3 replicates ($p < 0.05$).

The expression of *SpAMT1;2*, was downregulated in all experimental samples compared to SP. On the contrary, *SpAMT2*, which similar to *SpAMT1;2*, did not show a function either in oocyte or in the yeast test, demonstrated a 3.3-fold increased level of transcripts in the N-Mn50 sample, while expression was downregulated both in the M-Mn1 and the M-Mn50 samples compared to the starting point.

2.5. Regulatory Cis-Elements in the Promoters of Spirodela AMT Genes

To get further insights into the expression regulation of *S. polyrhiza AMT* genes, we conducted an in silico analysis of cis-elements in the gene promoters. The 1.5 kb long promoter sequences upstream from the protein translation starts, extracted from the available online genome sequence, version 3 of *S. polyrhiza*, strain 9509 (GCA_900492545.1) were analyzed using the New Place promoter analysis software (https://www.dna.affrc.go.jp/PLACE, accessed on 7 October 2022) [50]. In parallel, we conducted a similar analysis of corresponding gene promoters in a recently published genome of *S. intermedia* (taxid: 51605) a closely related duckweed species, with the aim to further highlight the evolutionary conserved regulatory elements (Figure 6).

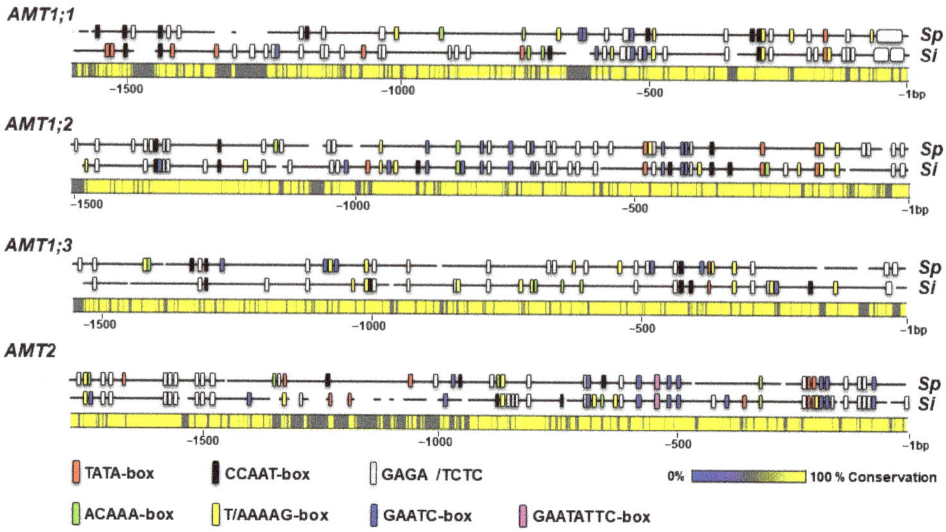

Figure 6. Schematic representation of promoters of the four *AMT* genes (*AMT1;1, AMT1;2, AMT1;3 and AMT2)* in the genomes of *Spirodela polyrhiza* (*Sp*) and *Spirodela intermedia* (*Si*). The negative numbers represent distance from the first ATG codon in nucleotides (bp). Colored boxes indicate locations of the corresponding promoter *cis*-elements. The conservation bar refers to the level of similarity between the homologous promoter sequences of *Spirodela polyrhiza* (*Sp*) and *Spirodela intermedia* (*Si*).

A prominent structural feature revealed in all analyzed promoters, was the abundant presence of GAGA motifs or its complement CTCT, involved in regulation of eukaryotic genomes through chromatin modulations [51–53]. Especially intriguing is the presence of a more than 50 bp long GAGA stretch in promoters of *AMT1;1* gene of *S. polyrhiza* and *S. intermedia*, closely adjacent to the ATG of protein translation start. Within the first 500 bp upstream of the translation start, all analyzed promoters feature one (*AMT1;1* and *AMT1;3*) or multiple (*AMT1;2* and *AMT2*) TATA motifs. The TATA-box is recognized as binding sites for a basic transcription factor TBP (TATA box-binding protein) involved in transcription initiation of many eukaryotic genes [54]. Moreover, inside 500 bp proximity to the ATG, all *AMT1* genes promoters contain general eukaryotic *cis*-element CCAAT-box [55] a target for transcription complex NF-Y (Nuclear Factor Y), which plays an essential role in many processes related to plant development, productivity and stress responses [56,57]. The *SpAMT2* and *SiAMT2* promoters have multiple CCAAT sites located between −1000 and −500 bp relating to the first ATG.

Transcription factors IDD (Indeterminate Domain) and DOF (DNA binding with one finger) were shown to be involved in regulation of the *AMT* genes in a number of different plants [58]. We located the recognition site for IDD, with the signature ACAAA, within the first 500 bp at the identical positions in the *AMT2* promoters of both *S. polyrhiza* and *S. intermedia.* In the promoters of the *AMT1* genes the site is scattered at various locations more remote from the translation start. As for the DNA motif recognized by DOF, T/AAAAG, multiple copies are found within the first 500 bp of all *AMT1* promoters, as well as scattered at more distant locations in promoters of both *AMT1* and *AMT2* genes.

The members of the NIGT1 (Nitrate-Inducible GARP-Type Transcriptional Repressor-1) subfamily in the superfamily of GARP transcription factors [59], participate in fine-tuning of gene expression in response to the plant nitrogen and phosphorus status [60]. The proteins bind to two types of *cis*-elements, GAATC and GAATATTC demonstrating dual modes of promoter recognition [61,62]. The GAATATTC site has been identified only in the

AMT2 promoters of *S. polyrhiza* and *S. intermedia* about 550 bp upstream of the first ATG, while the GAATC recognition site was located along the sequence of all promoters with the highest representation in the promoters of *AMT1;2* and *AMT2* genes.

3. Discussion

Our previous study demonstrated that the extend on Mn toxicity in four aquatic species of duckweed (Lemnaceae) depends on the source and concentration of nitrogen in the cultivation medium [43]. In particular, in great duckweed, *Spirodela polyrhiza*, the first characteristic symptoms of Mn stress appeared in form of characteristic brown spots when concentration of Mn was 40 mg/L (0.73 mM); however, the symptoms were largely reduced by the presence of ammonium in the medium in addition to nitrate. This observation, together with (i) increasing pollution of water resources with heavy metals [34–37], (ii) multiple previous reports linking the level of toxicity caused by various heavy metals to the nitrogen metabolism [12–16] and (iii) availability of the solid set of data on whole *Spirodela* genome sequences [45,63,64], encouraged us to take a closer look at Mn toxicity in *S. polyrhiza* in relation to the source of inorganic nitrogen with a special attention on the species's ammonium transporters (AMTs).

3.1. NH_4^+ Alleviated Mn Stress Symptoms for S. polyrhiza Compared with NO_3^-

Monitoring of *S. polyrhiza* fronds cultivated in the medium containing either a basic (0.059 mM) or 50-fold elevated (2.95 mM) concentration of Mn revealed profoundly different growth outcome, depending on the nitrogen source. Duckweed grown in the medium with basic Mn supplemented with NO_3^-, as a sole nitrogen source (N-Mn1), looked perfectly healthy and produced higher biomass compared to the medium with a mixture of a 50% of NO_3^- and 50% of NH_4^+ (Figure 1A). This observation is in accordance with the previously published data showing that growth rate of two duckweed species, *S. polyrrhiza* and *Lemna aequinoctialis* was lower in the medium containing ammonia compared to the medium with nitrate [65]. In addition to reduced biomass, selected fronds in the M-Mn1 medium showed some signs of chlorosis. These symptoms can be attributed to general ammonium toxicity [29–31], which we also observed in our previous study on six duckweed species, including *S. polyrhiza* [42].

The picture dramatically changed when the cultivation medium was supplied with elevated Mn. The fronds' appearance and biomass accumulation remained practically the same for duckweed in the medium with mixture of NO_3^- - and NH_4^+ (M-Mn50) compared to M-Mn1, except of occasional brown spots characteristic for Mn-mediated stress. In contrast, duckweed cultivated in the N-Mn50 medium with NO_3^- as a sole N source, manifests severe senescence, probably even death, and significant reduction in biomass accumulation compared not only to the N-Mn1 medium, but also to duckweed grown in medium containing the mixture of NO_3^- and NH_4^+ supplied with either normal (M-Mn1) or high concentration of Mn (M-Mn50). It is worth to notice that high concentration of Mn had no effect on the dynamics of NO_3^- and NH_4^+ uptake or the dynamics of pH during the cultivation period (Figure 1C,D). Confirming our previous study [42], utilization of NO_3^- led to a pH increase in the medium, whereas the uptake of NH_4^+ resulted in acidification of the nutrient medium, as has been shown earlier for a number of plants [42,66,67].

The lowering of pH in the media containing NH_4^+ might be a part of the mechanism of the observed alleviation of Mn toxicity mediated by NH_4^+ in *S. polyrhiza*. For example, NH_4^+ was shown to alleviate Mn stress symptoms in spruce and rice [68,69], and the decrease in Mn uptake by medium acidification was proposed as a mechanism underlying this phenomenon [69]. pH sensitivity has been reported as a general feature of transmembrane solute movement [70] as well as for many kinds of specific transporters in different organisms [71–74]. The reported data demonstrate that conformation and functionality of the plant transmembrane transporters are dependent upon cellular and intercellular pH. Dependence on pH, is especially well established in relation to the activity of aquaporins [75,76]. Aquaporins represent a large family of plant proteins with members in the model plant *Arabidopsis thaliana* [77] responsible not only for the plants water homeostasis, but for transporting many other molecules including ammonia and other small solutes [78]. Amounting evidence from across phyla suggests that some aquaporin types act as ion channels, and are modulated by divalent cations such as Ca^{2+} and Cd^{2+} [79]). Moreover, in *Arabidopsis,* in addition to pH, manganese was identified as potent inhibitors of aquaporin AtPIP2;1 [80].

In order to check the possibility that aquaporins could contribute to Mn stress in our system by responding to Mn and/or ammonium/pH, we have tested expression of two duckweed homologues of *Arabidopsis* aquaporins, *SpTIP2;1* and *SpPIP2;2* by quantitative RT-qPCR (Figure 7, Table S1). Indeed, both genes showed differential expression in response to the applied experimental conditions. The *SpTIP2;1* was significantly suppressed in the N-Mn50 environment but slightly upregulated in the M-Mn1 and M-Mn50, whereas *SpPIP2;2* was upregulated in all experimental medium, especially in the medium containing NH_4^+ (M-Mn1 and M-Mn50).

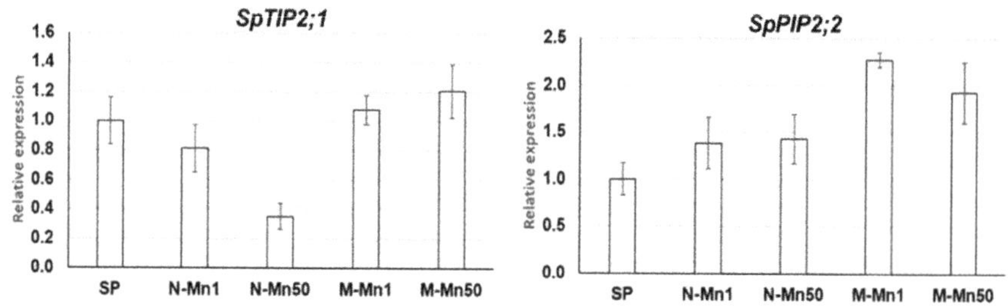

Figure 7. Expression patterns of two aquaporin genes, *SpTIP2;1* and *SpPIP2;2*, in duckweed cultivated in different media (N-Mn1, M-Mn1, N-Mn50 and M-Mn50). The genes expression was estimated at the cultivation day 4 by RT-qPCR relating to day 0, starting point (SP). Gene expression levels are in relative units. Error bars show ± SD of 3 replicates ($p < 0.05$).

We can assume that the acidification of the cultivation media and probably Mn itself may inactivate the transporters supporting inward Mn transport into the cytosol and lead to Mn exclusion from plant tissues. The presented data on differential expression of aquaporins may pave a way for future research in this area.

3.2. *Ammonium Transporter (AMT) Genes in the Genome of S. polyrhiza*

Ammonium transporters (AMTs) are transmembrane proteins, usually consisting of a N-terminal non-cytoplasmic domain, 11 transmembrane helices and C-terminal cytoplasmic domain [81,82], responsible for transporting NH_4^+ from the outside into the inside of the cell. Intensive studies on AMTs from different plant species revealed that this protein family is divided into two distinct subfamilies, the AMT1 and the AMT2, represented by different number of members in different species [58,83]. Compared to other plants, *S. polyrhiza* contains fewer genes coding for the AMTs. For example, *Arabidopsis thaliana* has five genes coding for AMT1 variants and one gene for AMT2 [83], and rice has three members of the AMT1 subfamily [84] and nine representatives of AMT2 [85]. Our analysis of the *S. polyrhiza* genome revealed three variants of AMT1: SpAMT1;1, SpAMT1;2, SpAMT1;3 and a single SpAMT2, all encoded by single copy genes located on different chromosomes. This finding further illustrates the reduced number of protein-coding genes in *S. polyrhiza* originally observed by the first genome sequencing [63], where a 28% reduction in genes compared to *Arabidopsis thaliana* and 50% reduction compared to rice was shown. This was later confirmed by Michael et al. (2017) [45] showing that most gene families in *Spirodela* have less members as compared with the model dicotyledonous plant *Arabidopsis* and monocotyledonous *Brachypodium*. Sequence analysis of the cloned cDNA from the Chinese ecotype NB5548 shows very high homology to the corresponding sequences of European ecotype 9509 (BlastN GCA_900492545.1) at the level of 99.6–99.8% similarity, with just a few observed C↔T or G↔A substitutes in the 1500 bp long sequences of the three *SpAMT1* genes, and 100% similarity for *SpAMT2*. This level of sequence conservation between ecotypes of *S. polyrhiza* is in a good agreement with the previously postulated extremely low rate of mutations revealed by comparing genomes of 68 worldwide geographic ecotypes of this species [86]. The *SpAMT* gene sequences also show similar arrangement, no introns in all three AMT1 gene sequences, two short introns within the AMT2 gene, and 95–97% sequences similarities with their homologues in the genome of the closely related *S. intermedia* [64], as represented in Figure S1.

The deduced SpAMT protein sequences demonstrate most close phylogenetic relationship to their counterparts from *S. intermedia*, followed by taro, *Colocasia esculenta* [87], a plant species which belongs to the group of early-branching monocotyledonous family Araceae [88], a sister family to Lemnaceae [46]. The SpAMT proteins also demonstrate significant levels of similarity to the homologues from other monocotyledonous banana, cocos, oil palm, rice and wheat, as well as dicotyledonous *Arabidopsis* (Figure 2), stressing strong evolutionary conservation of this important protein family along the plants [83,89]. Among *S. polyrhiza* AMTs, SpAMT1;1 and SpAMT1;2 share the highest homology, whereas the two proteins have certain differences in their C-termini (Figure S2). This offers possibilities for differential post-translational regulation since the proteins' C-terminal domain was reported as a target for environmental regulation through phosphorylation by protein kinases [82,90–92].

3.3. *SpAMT1;1 and SpAMT1;3 Demonstrate Clear Activity in Yeast and Oocytes*

From the four SpAMT proteins expressed in yeast and *Xenopus* oocytes, SpAMT1;1 and SpAMT1;3 demonstrated highly specific activity in parallel with the *Arabidopsis* controls (Figure 3). Interestingly, the homologues of these transporters in *Arabidopsis*, AtAMT1;1 and AtAMT1;3, have been reported for conferring 60% of total ammonium uptake combined [93,94]. To the contrary, SpAMT1;2 showed no activity in all of the experimental conditions and SpAMT2 was slightly activated at elevated pH in yeast complementation test. One might ponder about the silence of the SpAMT1;2, considering its phylogenetic closeness with the SpAMT1;1 (Figure 2) which shows a quite good activity in transporting NH_4^+. However, when looking on the structure of both proteins in more details, one can see that despite a high homology of the transmembrane part in the middle of the proteins, the non-cytoplasmic N-termini domains and the cytoplasmic C-termini domains show significant differences (Figure S2). For example, specific mutations of a conserved Gly in this domain was shown to inactivate AMTs from yeast and plants [49,95]. The mentioned mutation, however, cannot explain the functional differences between SpAMT1;1 and SpAMT1;2. The Gly is intact within the conserved first half of the C-terminal domain in both proteins. Another well-documented option of AMT regulation, is by post-translational phosphorylation of the Thr, located 4 residues downstream of the above mentioned Gly [96]. This position also does not differ between SpAMT1;1 and SpAMT1;2, as evident from the protein alignment in Figure S2. However, additional Thr residues uniquely positioned within both N-terminal and C-terminal domains of SpAMT1;2 (Figure S2), offer an opportunity of differential regulation of this protein compared to SpAMT1;1, through phosphorylation. These suggestions are purely speculative at the moment, and require an experimental verification.

3.4. *The Four S. polyrhiza AMT Genes Show Different Expression Patterns in Response to Nitrogen Source and Mn*

Our data show that both nitrogen source and elevated concentration of Mn affect expression of each of the *SpAMT* genes in a unique and in some instances opposite direction under the applied experimental conditions (Figure 3). Although NO_3^- slightly stimulated expression of *SpAMT1;1* and *SpAMT1;3*, it downregulated *SpAMT1;2* and *SpAMT2*. NH_4^+ was rather neutral on *SpAMT1;1*, upregulated *SpAMT1;3* and downregulated *SpAMT1;2* and *SpAMT2*. The revealed wide spectrum of *SpAMT* gene expression patterns in response to nitrogen is not too surprising. It is well in line with multiple previous reports showing diversity of AMT responses in other species such as *Arabidopsis* [97], rice [98], maize [99], poplar [14,100], apple [101] or tobacco [102]. The availability of multiple *AMT* genes with individual expression patterns, and multiple capacity of the expressed transporters for NH_4^+ uptake is considered to provide plants with great flexibility to respond differentially to varying environmental and nutritional conditions by fine tuning gene expression.

Simultaneous, presence of high concentrations of Mn in the cultivation media containing nitrate, N-Mn50, or a mixture of nitrate and ammonium, M-Mn50, dramatically changes the expression of *SpAMT* genes compared to the media with physiological concentration of Mn, N-Mn1 and M-Mn1. Thus, we observed drastic increase in expression of *SpAMT1;1* and *SpAMT2* in the medium containing NO_3^- as a sole source of nitrogen, N-Mn50, whereas the increase in the M-Mn50 medium was rather mild (Figure 5). Based on this specific expression dynamics, we are inclined to explain the observed expression boost not by direct influence of Mn, but rather by the process of senescence started by Mn toxicity and manifested in fronds cultivated in the N-Mn50 to a significantly greater extend, compared to the duckweed grown in the M-Mn50 (Figure 1). Observations in other plants, showing upregulation of the *SpAMT1;1* and *SpAMT2* homologues in old tissues, suggesting the role of these proteins in nitrogen remobilization [103], might be used in support of this notion. For example, van der Graaff et al., (2006) [104] revealed enhanced expression of *AtAMT1.1* in senescing leaves of *Arabidopsis*, and more recently, Liu et al., (2018) [102] showed that of nine AMT transporter genes identified in tobacco, only two, *NtAMT1.1* and *NtAMT2.1*, were strongly expressed in the old leaves, implying their dominant roles in N remobilization from senescent tissues.

Contrary to *SpAMT1;1* and *SpAMT2*, expression of the *SpAMT1;2* and *SpAMT1.3* seems to be influenced by Mn in a more direct manner. High Mn concentration suppressed expression of both genes independent of the: (i) supplied nitrogen formulation, N-Mn50 or M-Mn50 as, respectively, compared with N-Mn1 or M-Mn1 (Figure 5) and (ii) the differences in the acidity of the cultivation medium, correspondingly pH~6.5 for N-Mn1 and N-Mn50 and pH~ 4.0 for M-Mn1 and M-Mn50 (Figure 1). The question whether expression of *SpAMT1;2* and *SpAMT1.3* is directly controlled by Mn or through some Mn-mediated physiological changes such as hormonal status, generation of reactive oxygen species (ROS), or some other signaling pathway remains to be investigated.

3.5. Distinctive Promoter Elements in the AMT Genes in S. polyrhiza

Many environmental and physiological factors, nitrogen status, soil pH, photoperiod, developmental stage [58], have been recognized to influence ammonium transporter expression in plants. These factors regulate expression of *AMT* genes primarily via interaction of various transcription factors with specific *cis*-elements in the promoter of the genes [55]. Our analysis of *AMT* gene promoters in *Spirodela*, revealed characteristic *cis*-element patterns which might affect transcriptional regulation of ammonium uptake, and are distinctive from the genes encoding major nitrogen assimilation genes responsible to both NO_3^- and NH_4^+ [42].

The promoters of *AMT* genes are populated by $(GA/CT)_n$ repeats (Figure 6), targets for GAGA/TCTC-binding transcription factors (GAFs). Abundance of the GAGA in the promoters is not very surprising, considering that genomes of both *S. polyrhiza* and *S. intermedia*, are un-proportionally enriched in these motifs [45,64]. However, the contrast in distribution of the sequences in *AMTs* and nitrogen assimilation genes promoters is noticeable. In the analyzed *AMT* promoters, relatively short GAGA-stretches are more or less evenly distributed along the 1.5 kb length of the promoter, with a unique ~50 bp GAGA-stretch located just upstream of the ATG in the *AMT1;1* promoter (Figure 6), whereas in most of the promoters of nitrogen assimilation genes the $(GA/CT)_n$ repeats are represented by long stretches located outside of the 500 bp region adjacent to the genes' translation start.

Among prominent conservative *cis*-elements present within the 500 bp region upstream of translation start in all analyzed *Spirodela AMTs* are the TATA-box, a binding site for the general eukaryotic transcription factor TBP [54]. All *AMT* promoters also contain another general eukaryotic *cis*-element CCAAT-box [55], a binding site for NF-Y (nuclear factor Y) [105], involved in many important growth and developmental processes [106], including stress and hormone responses [107,108].

Overall, the range of *cis*-elements revealed in the *Sprirodela AMT* genes promoters in combination with the wide selection of corresponding transcription factors provide huge flexibility to regulate the *AMT* transcripts in response to various developmental, environmental and nutritional challenges faced by a plant during the life span.

4. Summary

Here we showed that the toxicity of manganese in duckweed plants could be mitigated by addition of ammonium as a nitrogen source. This mitigation effect might be pH related since the preferential uptake of ammonium by *S. polyrhiza* resulted in a pH decrease in the media which in turn might lead to a decreased activity of Mn uptake transporters e.g., aquaporins. The uptake of ammonium from the media could be mediated by four SpAMT proteins. The expression of the *AMT* genes was uniquely regulated by Mn and the available nitrogen source and two of the AMT proteins showed strong activity in heterologous systems. Their interesting unique transcriptional regulation is orchestrated by multiple transcription factor binding sites in the *AMT* promoters. The exact mechanism of Mn resistance by ammonium as well as the identification of transcription factors involved in *SpAMT* regulation are interesting questions which should be addressed in future studies.

5. Materials and Methods

5.1. Plant Material

The *S. polyrhiza* ecotype (collection ID: NB5548) used in this study was selected from the duckweed live in vitro collection recently established in the School of Life Sciences at Huaiyin Normal University, Huai'an, China. The NB5548 ecotype was originally collected in fall 2017 (N 33″17′40; E 118″49′45), surface sterilized and propagate on solid agar medium supplemented with SH salts [109] under sterile conditions.

5.2. Duckweed Cultivation Parameters

To accumulate biomass, duckweed plants grown on solid agar medium were initially transferred into 200 mL sterile liquid SH medium supplemented with 5 g/l sucrose contained in 500-mL flasks and cultivated at 23 ± 1 °C with a photon flux density of 50–60 $\mu mol \cdot m^{-2} s^{-1}$ provided by cool white fluorescent bulbs in a 16-h light/8-h dark cycle. After four weeks of growth, the accumulated fronds were washed 3 times with autoclaved water, and 200 mL sterile N- and sugar-free basal salt SH medium was added. After 3 days cultivation, 0.5-g portions were blotted and weighed in aseptic condition and then inoculated into 100-mL flasks containing 60 mL of autoclaved basic SH medium (no sugar) supplemented with two different formulations of N (N and M) and two concentrations of $MnSO_4$ (0.059 mM or 2.95 mM), namely N-Mn1, N-Mn50, M-Mn1 and M-Mn50. The basic medium labeled Mn1 contained 0.059 mM Mn that correspondent to SH medium formula, whereas experimental medium Mn50 contained 50-fold elevated concentration of Mn (2.95 mM). The media N-Mn1 and N-Mn50 contained NO_3^- as the sole N source (5 mM KNO_3, with 5 mM KH_2PO_4 replacing $NH_4H_2PO_4$ of the standard SH medium), and the media M-Mn1 and M-Mn50 had a mixture of NO_3^- and NH_4^+ (2.5 mM KNO_3, 2.5 mM KH_2PO_4, 2.5 mM $NH_4H_2PO_4$ and 2.5 mM K_2SO_4). The medium pH was adjusted to 5.5 and autoclaved. In total, 18 flasks with growing duckweed have been prepared for every variant of tested media. Three flasks (as 3 independent biological repeats) were randomly taken after 4 and 12 days for each of the analyses (RNA isolation for RT-qPCR, biomass measurement, determination of the total concentration of N). The duckweed biomass was also collected at the starting point (designated as SP), which corresponds to the moment before washing and 3-days cultivation on a medium containing no nitrogen (starvation).

5.3. Determination of N Concentration

The total nitrogen concentration was determined using standard alkaline potassium persulfate digestion followed by UV spectrophotometry as previously described [110]. The NO_3^- concentration in the growth media was measured spectrophotometrically as the difference in absorption between 220 and 275 nm [111]. The NH_4^+ concentration in the growth media was measured calorimetrically using the Nessler method [112].

5.4. Cloning and Analysis of *S. polyrhiza* Genes Encoding Ammonium Transporters

The *AMT1;1*, *AMT1;2*, *AMT1;3* and *AMT2* genes were obtained by sequencing and cloning the PCR-amplified gene regions using cDNA prepared from local *S. polyrhiza*, ecotype NB5548, mRNA as a template. The PCR fragments were amplified with gene-specific primers designed according to the in silico sequence information available at NCBI (taxid: 29656, GCA_900492545.1) for *S. polyrhiza*, ecotype 9509. The generated DNA fragments, cloned into the vector pMD19 (Takara, Dalian, China) following the manufacturer's instructions were custom sequenced (Sangon Biotech, Shanghai, China), and the obtained nucleotide sequences were analyzed using the CLC Main Workbench (Version 6.9.2, Qiagen) software. The specific primers used for gene amplification are listed in Table S2.

5.5. Phylogenetic Analysis

Protein sequences of AMT proteins were compared with a selection sequence available in public databases. The corresponding sequences available for monocotyledonous and dicotyledonous were extracted from the GenBank (National Center for Biotechnology Information (NCBI). Available online: https://www.ncbi.nlm.nih.gov (accessed on 15 August 2022) by blasting with the *S. polyrhiza* sequences as a query. The maximum-likelihood phylogenetic trees were constructed using NGPhylogeny webservice accessible through the https://ngphylogeny.fr [47] using MAFFT Multiple Sequence Alignment [113] and PhyML algorithm [114] with the Jones–Taylor–Thornton (JTT) model. Cleaning aligned sequences was made by utilizing BMGE tools [115]. Bootstrap support (BS) was estimated with 100 bootstrap replicates. iTOL (https://itol.embl.de) was used for displaying and annotating the generated phylogenetic trees [48].

5.6. Construction of Expression Vectors and Testing AMTs Functionality in Yeast and Xenopus oocytes

For testing in yeast and oocytes the coding sequences of *SpAMP1;1*, *SpAMP1;2*, *SpAMP1;3 and SpAMP2* were cut out from the original plasmid pMD19 and ligated into the vector pDR199 for expression in yeast [82] or the oocyte expression vector pOO2 [116]. Following verification by sequencing, the resulting vectors pDR199-SpAMP1;1, pDR199-SpAMP1;2, pDR199-SpAMP1;3 and pDR199-SpAMP2 were introduced into the AMT-deficient yeast strain 31019b [49]. The complementation analysis was performed according to the earlier described procedure [81], using the pDR199 vector as negative control and the pDR199 featuring *Arabidopsis thaliana* genes *AtAMT1;2* and *AtAMT2* as positive controls. Correspondingly, the four *S. polyrhiza AMT* genes were fitted into the pOO2 vector, and the resulting expression cassettes were used to test the proteins' abilities for transporting NH_4^+ in the oocyte according to an earlier established protocol [92]. Oocytes were ordered at EcoCyte Bioscience (Castrop-Rauxel, Germany), selected, washed and injected with 50 nl of linearized cRNA (0.4 μg/μL) of *SpAMTs*. Oocytes were kept in ND96 for 3 days at 18 °C and then placed in a recording chamber containing the recording solution (in mM): 110 CholineCl, 2 $CaCl_2$, $MgCl_2$, 5 N-morpholinoethane sulfonate (MES), pH adjusted to 5.5 with Tris(hydroxymethyl) aminomethane (TRIS). Ammonium was added as Cl salts. Oocytes were impaled with 3 M KCl-filled glass capillaries of around 0.8 Mohm resistance connected to a two-electrode voltage clamp amplifier (Dagan CA-1). Transport was measured at an array of NH_4^+ concentrations (10, 30, 100, 300, 1000, 3000 and 10,000 μM). Currents without ammonium were subtracted from currents with ammonium to give ammonium induced current.

5.7. Analysis of AMT Genes Expression in S. polyrhiza by RT-qPCR

After DNAase treatment, 600 ng of total RNA was reverse transcribed using Reverse Transcriptase cDNA synthesis kit (Takara, Beijing, China) following the manufacturer's manual. The qPCR reactions were performed using CFX Connect Real-Time detection system (Bio-Rad, Hercules, CA, USA) using the UltraSybr Mixture (High Rox) supplied by CWBio (Taizhou, China). The cycling conditions were as follows: initial denaturation at 95 °C for 10 min followed by 40 cycles of 30 s at 94 °C, and 20 s at the annealing temperature of the respective primers (Table S1). The SYBR Green I fluorescence was monitored consecutively after the annealing step. The quality of products was checked by a thermal denaturation cycle. Only results providing a single peak were considered. The coefficient amplification efficiency for each pair of primers was determined by 10-fold serial dilutions. The level of relative expression was calculated by the $2^{-\Delta\Delta Ct}$ method [117]. Expression data for the target genes were normalized using the average expression of two *S. polyrhiza* ecotype NB5548 housekeeping genes, *histone H3* (GenBank ID: MZ605911) and *β-actin* (GenBank ID: MZ605912), according to the geNorm protocol [118]. Three replicates were performed for all samples. Normalized expression was estimated using the ΔΔCq algorithm All data were analyzed using the program BIO-RAD CFX Manager 3.1 (Bio-Rad, USA) and Microsoft Excel 2016 software.

Supplementary Materials: The following supporting information can be downloaded at: https://www.mdpi.com/article/10.3390/plants12010208/s1, Figure S1: Schematic structure of AMT coding regions in genomes of *S. polyrhiza* and *S. intermedia*; Figure S2: Schematic structure of AMTs of *S. polyrhiza;* Figure S3: Sequence alignment of AMT1 proteins from *S. polyrhiza* and *A. thaliana*. Table S1: List of RT-qPCR primers used for evaluating expression of *AMTs* and *aquaporin* genes in *S. polyrhiza* line NB5548; Table S2: List of primer used for cloning *AMT* genes of *S.polyrhiza* ecotype NB5548.

Author Contributions: Conceptualization, O.K., B.N. and N.B.; data curation and formal analysis, O.K., A.S., B.N. and N.B.; investigation and methodology, O.K., A.S., Y.Z. and T.S.; project administration, Y.Z., B.N. and N.B.; resources, Y.Z., B.N. and N.B.; software, O.K., A.S. and T.S.; supervision, O.K., B.N. and N.B.; validation, O.K., A.S. and T.S.; writing—original draft, O.K., A.S., B.N. and N.B.; writing—review and editing, B.N and N.B. All authors have read and agreed to the published version of the manuscript.

Funding: This work was supported by an individual grant provided by the Huaiyin Normal University (Huai'an, China) to NB as well as general funding of BN by the University of Hohenheim (Stuttgart, Germany).

Data Availability Statement: GenBank accession numbers for *Spirodela polyrhiza* AMT genes are: *SpAMT1;1*—OP730321; *SpAMT1;2*—OP730322; *SpAMT1;3*—OP730323; *SpAMT2*—OP730324.

Conflicts of Interest: The authors declare no conflict of interest.

References

1. Scheible, W.-R.; Morcuende, R.; Czechowski, T.; Fritz, C.; Osuna, D.; Palacios-Rojas, N.; Schindelasch, D.; Thimm, O.; Udvardi, M.K.; Stitt, M. Genome-Wide Reprogramming of Primary and Secondary Metabolism, Protein Synthesis, Cellular Growth Processes, and the Regulatory Infrastructure of *Arabidopsis* in Response to Nitrogen. *Plant Physiol.* **2004**, *136*, 2483–2499. [CrossRef] [PubMed]
2. Xu, G.; Fan, X.; Miller, A.J. Plant Nitrogen Assimilation and Use Efficiency. *Annu. Rev. Plant Biol.* **2012**, *63*, 153–182. [CrossRef] [PubMed]
3. Saloner, A.; Bernstein, N. Nitrogen Source Matters: High NH_4/NO_3 Ratio Reduces Cannabinoids, Terpenoids, and Yield in Medical Cannabis. *Front. Plant Sci.* **2022**, *13*, 830224. [CrossRef]
4. Bloom, A.J. The Increasing Importance of Distinguishing among Plant Nitrogen Sources. *Curr. Opin. Plant Biol.* **2015**, *25*, 10–16. [CrossRef]

5. Chen, P.; Xu, J.; Zhang, Z.; Nie, T. "Preferential" Ammonium Uptake by Rice Does Not Always Turn into Higher N Recovery of Fertilizer Sources under Water-Saving Irrigation. *Agric. Water Manag.* **2022**, *272*, 107867. [CrossRef]
6. Cui, J.; Yu, C.; Qiao, N.; Xu, X.; Tian, Y.; Ouyang, H. Plant Preference for NH_4^+ versus NO_3^- at Different Growth Stages in an Alpine Agroecosystem. *Field Crops Res.* **2017**, *201*, 192–199. [CrossRef]
7. Duan, Y.H.; Zhang, Y.L.; Ye, L.T.; Fan, X.R.; Xu, G.H.; Shen, Q.R. Responses of Rice Cultivars with Different Nitrogen Use Efficiency to Partial Nitrate Nutrition. *Ann. Bot.* **2007**, *99*, 1153–1160. [CrossRef]
8. Hajari, E.; Snyman, S.J.; Watt, M.P. Nitrogen Use Efficiency of Sugarcane (*Saccharum* Spp.) Varieties under in Vitro Conditions with Varied N Supply. *Plant Cell Tiss Organ Cult.* **2015**, *122*, 21–29. [CrossRef]
9. Patterson, K.; Cakmak, T.; Cooper, A.; Lager, I.; Rasmusson, A.G.; Escobar, M.A. Distinct Signalling Pathways and Transcriptome Response Signatures Differentiate Ammonium- and Nitrate-Supplied Plants: Transcriptome Signatures of Ammonium and Nitrate Responses. *Plant Cell Environ.* **2010**, *33*, 1486–1501. [CrossRef]
10. Ding, L.; Lu, Z.; Gao, L.; Guo, S.; Shen, Q. Is Nitrogen a Key Determinant of Water Transport and Photosynthesis in Higher Plants Upon Drought Stress? *Front. Plant Sci.* **2018**, *9*, 1143. [CrossRef]
11. Jiang, W.Z. Mn Use Efficiency in Different Wheat Cultivars. *Environ. Exp. Bot.* **2006**, *57*, 41–50. [CrossRef]
12. Carrasco-Gil, S.; Estebaranz-Yubero, M.; Medel-Cuesta, D.; Millán, R.; Hernández, L.E. Influence of Nitrate Fertilization on Hg Uptake and Oxidative Stress Parameters in Alfalfa Plants Cultivated in a Hg-Polluted Soil. *Environ. Exp. Bot.* **2012**, *75*, 16–24. [CrossRef]
13. Huo, K.; Shangguan, X.; Xia, Y.; Shen, Z.; Chen, C. Excess Copper Inhibits the Growth of Rice Seedlings by Decreasing Uptake of Nitrate. *Ecotoxicol. Environ. Saf.* **2020**, *190*, 110105. [CrossRef]
14. Huang, J.; Wu, X.; Tian, F.; Chen, Q.; Luo, P.; Zhang, F.; Wan, X.; Zhong, Y.; Liu, Q.; Lin, T. Changes in Proteome and Protein Phosphorylation Reveal the Protective Roles of Exogenous Nitrogen in Alleviating Cadmium Toxicity in Poplar Plants. *Int. J. Mol. Sci.* **2019**, *21*, 278. [CrossRef] [PubMed]
15. Zhao, X.Q.; Guo, S.W.; Shinmachi, F.; Sunairi, M.; Noguchi, A.; Hasegawa, I.; Shen, R.F. Aluminium Tolerance in Rice Is Antagonistic with Nitrate Preference and Synergistic with Ammonium Preference. *Ann. Bot.* **2013**, *111*, 69–77. [CrossRef] [PubMed]
16. Zhu, C.Q.; Cao, X.C.; Zhu, L.F.; Hu, W.J.; Hu, A.Y.; Bai, Z.G.; Zhong, C.; Sun, L.M.; Liang, Q.D.; Huang, J.; et al. Ammonium Mitigates Cd Toxicity in Rice (*Oryza Sativa*) via Putrescine-Dependent Alterations of Cell Wall Composition. *Plant Physiol. Biochem.* **2018**, *132*, 189–201. [CrossRef]
17. Zhao, X.Q.; Shen, R.F. Aluminum–Nitrogen Interactions in the Soil–Plant System. *Front. Plant Sci.* **2018**, *9*, 807. [CrossRef]
18. Yu, L.; Wang, X.; Li, X.; Wang, Y.; Kang, H.; Chen, G.; Fan, X.; Sha, L.; Zhou, Y.; Zeng, J. Protective Effect of Different Forms of Nitrogen Application on Cadmium-Induced Toxicity in Wheat Seedlings. *Environ. Sci. Pollut. Res.* **2019**, *26*, 13085–13094. [CrossRef]
19. Afef, N.-H.; Chiraz, C.-H.; Houda, M.D.; Habib, G.; Houda, G. Substitution of NO_3^- by NH_4^+ Increases Ammonium-Assimilating Enzyme Activities and Reduces the Deleterious Effects of Cadmium on the Growth of Tomatoes. *Fresenius Environ. Bull.* **2012**, *21*, 665–671.
20. De la Peña, M.; Marín-Peña, A.J.; Urmeneta, L.; Coleto, I.; Castillo-González, J.; van Liempd, S.M.; Falcón-Pérez, J.M.; Álvarez-Fernández, A.; González-Moro, M.B.; Marino, D. Ammonium Nutrition Interacts with Iron Homeostasis in *Brachypodium Distachyon*. *J. Exp. Bot.* **2022**, *73*, 263–274. [CrossRef]
21. Zhang, X.; Liu, H.; Zhang, S.; Wang, J.; Wei, C. NH_4^+-N Alleviates Iron Deficiency in Rice Seedlings under Calcareous Conditions. *Sci. Rep.* **2019**, *9*, 12712. [CrossRef] [PubMed]
22. de Souza, J.C., Jr.; Nogueirol, R.C.; Monteiro, F.A. Nitrate and Ammonium Proportion Plays a Key Role in Copper Phytoextraction, Improving the Antioxidant Defense in Tanzania Guinea Grass. *Ecotoxicol. Environ. Saf.* **2019**, *171*, 823–832. [CrossRef] [PubMed]
23. Broadley, M.; Brown, P.; Cakmak, I.; Rengel, Z.; Zhao, F. Function of Nutrients: Micronutrients. In *Marschner's Mineral Nutrition of Higher Plants*; Elsevier: Amsterdam, The Netherlands, 2012; pp. 191–248. ISBN 978-0-12-384905-2.
24. Schmidt, S.B.; Husted, S. The Biochemical Properties of Manganese in Plants. *Plants* **2019**, *8*, 381. [CrossRef]
25. Schmidt, S.B.; Jensen, P.E.; Husted, S. Manganese Deficiency in Plants: The Impact on Photosystem II. *Trends Plant Sci.* **2016**, *21*, 622–632. [CrossRef] [PubMed]
26. Socha, A.L.; Guerinot, M.L. Mn-Euvering Manganese: The Role of Transporter Gene Family Members in Manganese Uptake and Mobilization in Plants. *Front. Plant Sci.* **2014**, *5*, 106. [CrossRef]
27. Pittman, J.K. Managing the Manganese: Molecular Mechanisms of Manganese Transport and Homeostasis. *New Phytol.* **2005**, *167*, 733–742. [CrossRef] [PubMed]
28. Foy, C.D.; Scott, B.J.; Fisher, J.A. Genetic Differences in Plant Tolerance to Manganese Toxicity. In *Manganese in Soils and Plants*; Developments in Plant and Soil Sciences; ©Kluwer Academic Publishers: Norwell, MA, USA, 1988; Volume 33, pp. 293–307. ISBN 978-94-010-7768-2.
29. Li, J.; Jia, Y.; Dong, R.; Huang, R.; Liu, P.; Li, X.; Wang, Z.; Liu, G.; Chen, Z. Advances in the Mechanisms of Plant Tolerance to Manganese Toxicity. *Int. J. Mol. Sci.* **2019**, *20*, 5096. [CrossRef]
30. Wissemeier, A.H.; Horst, W.J. Effect of Light Intensity on Manganese Toxicity Symptoms and Callose Formation in Cowpea (*Vigna Unguiculata* (L.) Walp.). *Plant Soil* **1992**, *143*, 299–309. [CrossRef]

31. Alejandro, S.; Höller, S.; Meier, B.; Peiter, E. Manganese in Plants: From Acquisition to Subcellular Allocation. *Front. Plant Sci.* **2020**, *11*, 300. [CrossRef]
32. Quartin, V.M.L.; Antunes, M.L.; Muralha, M.C.; Sousa, M.M.; Nunes, M.A. Mineral Imbalance Due to Manganese Excess in Triticales. *J. Plant Nutrit.* **2001**, *24*, 175–189. [CrossRef]
33. El Khalil, H.; El Hamiani, O.; Bitton, G.; Ouazzani, N.; Boularbah, A. Heavy Metal Contamination from Mining Sites in South Morocco: Monitoring Metal Content and Toxicity of Soil Runoff and Groundwater. *Environ. Monit. Assess.* **2008**, *136*, 147–160. [CrossRef]
34. Li, P.; Qian, H.; Howard, K.W.F.; Wu, J.; Lyu, X. Anthropogenic Pollution and Variability of Manganese in Alluvial Sediments of the Yellow River, Ningxia, Northwest China. *Environ. Monit. Assess.* **2014**, *186*, 1385–1398. [CrossRef] [PubMed]
35. Liu, Z.; Kuang, Y.; Lan, S.; Cao, W.; Yan, Z.; Chen, L.; Chen, Q.; Feng, Q.; Zhou, H. Pollution Distribution of Potentially Toxic Elements in a Karstic River Affected by Manganese Mining in Changyang, Western Hubei, Central China. *Int. J. Environ. Res. Public Helth* **2021**, *18*, 1870. [CrossRef] [PubMed]
36. Trueman, B.F.; Gregory, B.S.; McCormick, N.E.; Gao, Y.; Gora, S.; Anaviapik-Soucie, T.; L'Hérault, V.; Gagnon, G.A. Manganese Increases Lead Release to Drinking Water. *Environ. Sci. Technol.* **2019**, *53*, 4803–4812. [CrossRef]
37. Pinsino, A.; Matranga, V.; Roccheri, M.C. Manganese: A New Emerging Contaminant in the Environment. In *Environmental Contamination*; Srivastava, J., Ed.; IntechOpen: London, UK, 2012; pp. 17–37. ISBN 978-953-51-0120-8.
38. Queiroz, H.M.; Ying, S.C.; Abernathy, M.; Barcellos, D.; Gabriel, F.A.; Otero, X.L.; Nóbrega, G.N.; Bernardino, A.F.; Ferreira, T.O. Manganese: The Overlooked Contaminant in the World Largest Mine Tailings Dam Collapse. *Environ. Int.* **2021**, *146*, 106284. [CrossRef] [PubMed]
39. Acosta, K.; Appenroth, K.J.; Borisjuk, L.; Edelman, M.; Heinig, U.; Jansen, M.A.K.; Oyama, T.; Pasaribu, B.; Schubert, I.; Sorrels, S.; et al. Return of the Lemnaceae: Duckweed as a Model Plant System in the Genomics and Post-Genomics Era. *Plant Cell* **2021**, *33*, 3207–3234. [CrossRef]
40. Zhou, Y.; Borisjuk, N. Small Aquatic Duckweed Plants with Big Potential for the Production of Valuable Biomass and Wastewater Remediation. *Int. J. Environ. Sci. Nat. Resour.* **2019**, *16*, 555942. [CrossRef]
41. Ziegler, P.; Sree, K.S.; Appenroth, K.-J. Duckweeds for Water Remediation and Toxicity Testing. *Toxicol. Environ. Chem.* **2016**, *98*, 1127–1154. [CrossRef]
42. Zhou, Y.; Kishchenko, O.; Stepanenko, A.; Chen, G.; Wang, W.; Zhou, J.; Pan, C.; Borisjuk, N. The Dynamics of NO_3^- and NH_4^+ Uptake in Duckweed Are Coordinated with the Expression of Major Nitrogen Assimilation Genes. *Plants* **2022**, *11*, 11. [CrossRef]
43. Zhou, Y.; Bai, T.; Kishchenko, O. Potential of Lemnoideae Species for Phytoremediation of Fresh Water with Elevated Manganese Concentration. *Innov. Biosyst. Bioeng.* **2019**, *3*, 232–238. [CrossRef]
44. Michael, T.P.; Bryant, D.; Gutierrez, R.; Borisjuk, N.; Chu, P.; Zhang, H.; Xia, J.; Zhou, J.; Peng, H.; El Baidouri, M.; et al. Comprehensive Definition of Genome Features in *Spirodela Polyrhiza* by High-Depth Physical Mapping and Short-Read DNA Sequencing Strategies. *Plant J.* **2017**, *89*, 617–635. [CrossRef] [PubMed]
45. Omasits, U.; Ahrens, C.H.; Müller, S.; Wollscheid, B. Protter: Interactive Protein Feature Visualization and Integration with Experimental Proteomic Data. *Bioinformatics* **2014**, *30*, 884–886. [CrossRef] [PubMed]
46. Tippery, N.P.; Les, D.H. Tiny Plants with Enormous Potential: Phylogeny and Evolution of Duckweeds. In *The Duckweed Genomes*; Cao, X.H., Fourounjian, P., Wang, W., Eds.; Springer International Publishing: Cham, Switzerland, 2020; pp. 19–38. ISBN 978-3-030-11044-4.
47. Lemoine, F.; Correia, D.; Lefort, V.; Doppelt-Azeroual, O.; Mareuil, F.; Cohen-Boulakia, S.; Gascuel, O. NGPhylogeny.Fr: New Generation Phylogenetic Services for Non-Specialists. *Nucleic Acids Res.* **2019**, *47*, W260–W265. [CrossRef] [PubMed]
48. Letunic, I.; Bork, P. Interactive Tree of Life (ITOL) v5: An Online Tool for Phylogenetic Tree Display and Annotation. *Nucleic Acids Res.* **2021**, *49*, W293–W296. [CrossRef] [PubMed]
49. Marini, A.M.; Soussi-Boudekou, S.; Vissers, S.; Andre, B. A Family of Ammonium Transporters in Saccharomyces Cerevisiae. *Mol. Cell Biol.* **1997**, *17*, 4282–4293. [CrossRef]
50. Higo, K.; Ugawa, Y.; Iwamoto, M.; Korenaga, T. Plant Cis-Acting Regulatory DNA Elements (PLACE) Database: 1999. *Nucleic Acids Res.* **1999**, *27*, 297–300. [CrossRef]
51. Lehmann, M. Anything Else but GAGA: A Nonhistone Protein Complex Reshapes Chromatin Structure. *Trends Genet.* **2004**, *20*, 15–22. [CrossRef]
52. Hecker, A.; Brand, L.H.; Peter, S.; Simoncello, N.; Kilian, J.; Harter, K.; Gaudin, V.; Wanke, D. The Arabidopsis GAGA-Binding Factor BASIC PENTACYSTEINE6 Recruits the POLYCOMB-REPRESSIVE COMPLEX1 Component LIKE HETEROCHROMATIN PROTEIN1 to GAGA DNA Motifs. *Plant Physiol.* **2015**, *168*, 1013–1024. [CrossRef]
53. Tsai, S.-Y.; Chang, Y.-L.; Swamy, K.B.S.; Chiang, R.-L.; Huang, D.-H. GAGA Factor, a Positive Regulator of Global Gene Expression, Modulates Transcriptional Pausing and Organization of Upstream Nucleosomes. *Epigenet. Chromatin* **2016**, *9*, 32. [CrossRef]
54. Hernandez, N. TBP, a Universal Eukaryotic Transcription Factor? *Genes Dev.* **1993**, *7*, 1291–1308. [CrossRef]
55. Komarnytsky, S.; Borisjuk, N. Functional Analysis of Promoter Elements in Plants. *Genet. Eng.* **2003**, *25*, 113–141. [CrossRef]
56. Leyva-González, M.A.; Ibarra-Laclette, E.; Cruz-Ramírez, A.; Herrera-Estrella, L. Functional and Transcriptome Analysis Reveals an Acclimatization Strategy for Abiotic Stress Tolerance Mediated by Arabidopsis NF-YA Family Members. *PLoS ONE* **2012**, *7*, e48138. [CrossRef] [PubMed]

57. Petroni, K.; Kumimoto, R.W.; Gnesutta, N.; Calvenzani, V.; Fornari, M.; Tonelli, C.; Holt, B.F.; Mantovani, R. The Promiscuous Life of Plant NUCLEAR FACTOR Y Transcription Factors. *Plant Cell* **2013**, *24*, 4777–4792. [CrossRef] [PubMed]
58. Hao, D.-L.; Zhou, J.-Y.; Yang, S.-Y.; Qi, W.; Yang, K.-J.; Su, Y.-H. Function and Regulation of Ammonium Transporters in Plants. *Int. J. Mol. Sci.* **2020**, *21*, 3557. [CrossRef]
59. Zhao, X.; Yang, J.; Li, X.; Li, G.; Sun, Z.; Chen, Y.; Chen, Y.; Xia, M.; Li, Y.; Yao, L.; et al. Identification and Expression Analysis of GARP Superfamily Genes in Response to Nitrogen and Phosphorus Stress in *Spirodela Polyrhiza*. *BMC Plant Biol.* **2022**, *22*, 308. [CrossRef]
60. Wang, X.; Wang, H.-F.; Chen, Y.; Sun, M.-M.; Wang, Y.; Chen, Y.-F. The Transcription Factor NIGT1.2 Modulates Both Phosphate Uptake and Nitrate Influx during Phosphate Starvation in Arabidopsis and Maize. *Plant Cell* **2020**, *32*, 3519–3534. [CrossRef]
61. Yanagisawa, S. Characterization of a Nitrate-Inducible Transcriptional Repressor NIGT1 Provides New Insights into DNA Recognition by the GARP Family Proteins. *Plant Signal. Behav.* **2013**, *8*, e24447. [CrossRef]
62. Li, Q.; Zhou, L.; Li, Y.; Zhang, D.; Gao, Y. Plant NIGT1/HRS1/HHO Transcription Factors: Key Regulators with Multiple Roles in Plant Growth, Development, and Stress Responses. *Int. J. Mol. Sci.* **2021**, *22*, 8685. [CrossRef]
63. Wang, W.; Haberer, G.; Gundlach, H.; Gläßer, C.; Nussbaumer, T.; Luo, M.C.; Lomsadze, A.; Borodovsky, M.; Kerstetter, R.A.; Shanklin, J.; et al. The *Spirodela Polyrrhiza* Genome Reveals Insights into Its Neotenous Reduction Fast Growth and Aquatic Lifestyle. *Nat. Commun.* **2014**, *5*, 3311. [CrossRef]
64. Hoang, P.T.N.; Fiebig, A.; Novák, P.; Macas, J.; Cao, H.X.; Stepanenko, A.; Chen, G.; Borisjuk, N.; Scholz, U.; Schubert, I. Chromosome-Scale Genome Assembly for the Duckweed *Spirodela Intermedia*, Integrating Cytogenetic Maps, PacBio and Oxford Nanopore Libraries. *Sci. Rep.* **2020**, *10*, 19230. [CrossRef]
65. Chong, Y.X.; Hu, H.Y.; Qian, Y. Growth feature of biomass of *Lemna aequinoctialis* and *Spirodela polyrrhiza* in medium with nutrient character of wastewater. *Huan Jing Ke Xue* **2004**, *25*, 59–64. [PubMed]
66. Feng, H.; Fan, X.; Miller, A.J.; Xu, G. Plant Nitrogen Uptake and Assimilation: Regulation of Cellular PH Homeostasis. *J. Exp. Bot.* **2020**, *71*, 4380–4392. [CrossRef] [PubMed]
67. Hachiya, T.; Sakakibara, H. Interactions between Nitrate and Ammonium in Their Uptake, Allocation, Assimilation, and Signaling in Plants. *J. Exp. Bot.* **2016**, *68*, 2501–2512. [CrossRef] [PubMed]
68. Hu, A.Y.; Zheng, M.M.; Sun, L.M.; Zhao, X.Q.; Shen, R.F. Ammonium Alleviates Manganese Toxicity and Accumulation in Rice by Down-Regulating the Transporter Gene OsNramp5 Through Rhizosphere Acidification. *Front. Plant Sci.* **2019**, *10*, 1194. [CrossRef] [PubMed]
69. Langheinrich, U.; Tischner, R.; Godbold, D.L. Influence of a High Mn Supply on Norway Spruce (*Picea abies* (L.) Karst.) Seedlings in Relation to the Nitrogen Source. *Tree Physiol.* **1992**, *10*, 259–271. [CrossRef]
70. Gerbeau, P.; Amodeo, G.; Henzler, T.; Santoni, V.; Ripoche, P.; Maurel, C. The Water Permeability of Arabidopsis Plasma Membrane Is Regulated by Divalent Cations and PH. *Plant J.* **2002**, *30*, 71–81. [CrossRef] [PubMed]
71. Franz, M.C.; Pujol-Giménez, J.; Montalbetti, N.; Fernandez-Tenorio, M.; DeGrado, T.R.; Niggli, E.; Romero, M.F.; Hediger, M.A. Reassessment of the Transport Mechanism of the Human Zinc Transporter SLC39A2. *Biochemistry* **2018**, *57*, 3976–3986. [CrossRef]
72. Chauvigné, F.; Zapater, C.; Stavang, J.A.; Taranger, G.L.; Cerdà, J.; Finn, R.N. The PH Sensitivity of Aqp0 Channels in Tetraploid and Diploid Teleosts. *FASEB J.* **2015**, *29*, 2172–2184. [CrossRef]
73. Eide, D.J. Transcription Factors and Transporters in Zinc Homeostasis: Lessons Learned from Fungi. *Crit. Rev. Biochem. Mol. Biol.* **2020**, *55*, 88–110. [CrossRef]
74. Wang, J.; Yu, X.; Ding, Z.J.; Zhang, X.; Luo, Y.; Xu, X.; Xie, Y.; Li, X.; Yuan, T.; Zheng, S.J.; et al. Structural Basis of ALMT1-Mediated Aluminum Resistance in Arabidopsis. *Cell Res.* **2022**, *32*, 89–98. [CrossRef]
75. Tournaire-Roux, C.; Sutka, M.; Javot, H.; Gout, E.; Gerbeau, P.; Luu, D.-T.; Bligny, R.; Maurel, C. Cytosolic PH Regulates Root Water Transport during Anoxic Stress through Gating of Aquaporins. *Nature* **2003**, *425*, 393–397. [CrossRef] [PubMed]
76. Scochera, F.; Zerbetto De Palma, G.; Canessa Fortuna, A.; Chevriau, J.; Toriano, R.; Soto, G.; Zeida, A.; Alleva, K. PIP Aquaporin PH-sensing Is Regulated by the Length and Charge of the C-terminal Region. *FEBS J.* **2022**, *289*, 246–261. [CrossRef] [PubMed]
77. Soto, G.; Alleva, K.; Amodeo, G.; Muschietti, J.; Ayub, N.D. New Insight into the Evolution of Aquaporins from Flowering Plants and Vertebrates: Orthologous Identification and Functional Transfer Is Possible. *Gene* **2012**, *503*, 165–176. [CrossRef] [PubMed]
78. Maurel, C.; Boursiac, Y.; Luu, D.-T.; Santoni, V.; Shahzad, Z.; Verdoucq, L. Aquaporins in Plants. *Physiol. Rev.* **2015**, *95*, 1321–1358. [CrossRef]
79. Kourghi, M.; Nourmohammadi, S.; Pei, J.; Qiu, J.; McGaughey, S.; Tyerman, S.; Byrt, C.; Yool, A. Divalent Cations Regulate the Ion Conductance Properties of Diverse Classes of Aquaporins. *Int. J. Mol. Sci.* **2017**, *18*, 2323. [CrossRef]
80. Verdoucq, L.; Grondin, A.; Maurel, C. Structure–function Analysis of Plant Aquaporin *At*PIP2;1 Gating by Divalent Cations and Protons. *Biochem. J.* **2008**, *415*, 409–416. [CrossRef]
81. Schwacke, R.; Schneider, A.; van der Graaff, E.; Fischer, K.; Catoni, E.; Desimone, M.; Frommer, W.B.; Flügge, U.-I.; Kunze, R. ARAMEMNON, a Novel Database for Arabidopsis Integral Membrane Proteins. *Plant Physiol.* **2003**, *131*, 16–26. [CrossRef]
82. Neuhäuser, B.; Dynowski, M.; Mayer, M.; Ludewig, U. Regulation of NH4 + Transport by Essential Cross Talk between AMT Monomers through the Carboxyl Tails. *Plant Physiol.* **2007**, *143*, 1651–1659. [CrossRef]
83. Ludewig, U.; Neuhäuser, B.; Dynowski, M. Molecular Mechanisms of Ammonium Transport and Accumulation in Plants. *FEBS Lett.* **2007**, *581*, 2301–2308. [CrossRef]

84. Sonoda, Y.; Ikeda, A.; Saiki, S.; von Wirén, N.; Yamaya, T.; Yamaguchi, J. Distinct Expression and Function of Three Ammonium Transporter Genes (OsAMT1;1–1;3) in Rice. *Plant Cell Physiol.* **2003**, *44*, 726–734. [CrossRef]
85. Li, B.; Merrick, M.; Li, S.; Li, H.; Zhu, S.; Shi, W.; Su, Y. Molecular Basis and Regulation of Ammonium Transporter in Rice. *Rice Sci.* **2009**, *16*, 314–322. [CrossRef]
86. Xu, S.; Stapley, J.; Gablenz, S.; Boyer, J.; Appenroth, K.J.; Sree, K.S.; Gershenzon, J.; Widmer, A.; Huber, M. Low Genetic Variation Is Associated with Low Mutation Rate in the Giant Duckweed. *Nat. Commun.* **2019**, *10*, 1243. [CrossRef] [PubMed]
87. Bellinger, M.R.; Paudel, R.; Starnes, S.; Kambic, L.; Kantar, M.B.; Wolfgruber, T.; Lamour, K.; Geib, S.; Sim, S.; Miyasaka, S.C.; et al. Taro Genome Assembly and Linkage Map Reveal QTLs for Resistance to Taro Leaf Blight. *G3 Genes Genomes Genet.* **2020**, *10*, 2763–2775. [CrossRef] [PubMed]
88. Shi, T.; Huneau, C.; Zhang, Y.; Li, Y.; Chen, J.; Salse, J.; Wang, Q. The Slow-Evolving Acorus Tatarinowii Genome Sheds Light on Ancestral Monocot Evolution. *Nat. Plants* **2022**, *8*, 764–777. [CrossRef]
89. Wu, Z.; Gao, X.; Zhang, N.; Feng, X.; Huang, Y.; Zeng, Q.; Wu, J.; Zhang, J.; Qi, Y. Genome-Wide Identification and Transcriptional Analysis of Ammonium Transporters in Saccharum. *Genomics* **2021**, *113*, 1671–1680. [CrossRef]
90. Straub, T.; Ludewig, U.; Neuhäuser, B. The Kinase CIPK23 Inhibits Ammonium Transport in *Arabidopsis Thaliana*. *Plant Cell* **2017**, *29*, 409–422. [CrossRef]
91. Wu, X.; Liu, T.; Zhang, Y.; Duan, F.; Neuhäuser, B.; Ludewig, U.; Schulze, W.X.; Yuan, L. Ammonium and Nitrate Regulate NH_4^+ Uptake Activity of *Arabidopsis* Ammonium Transporter AtAMT1;3 via Phosphorylation at Multiple C-Terminal Sites. *J. Exp. Bot.* **2019**, *70*, 4919–4930. [CrossRef]
92. Ganz, P.; Porras-Murillo, R.; Ijato, T.; Menz, J.; Straub, T.; Stührwohldt, N.; Moradtalab, N.; Ludewig, U.; Neuhäuser, B. Abscisic Acid Influences Ammonium Transport via Regulation of Kinase CIPK23 and Ammonium Transporters. *Plant Physiol.* **2022**, *190*, 1275–1288. [CrossRef]
93. Kaiser, B.N. Functional Analysis of an Arabidopsis T-DNA "Knockout" of the High-Affinity NH4+ Transporter AtAMT1;1. *Plant Physiol.* **2002**, *130*, 1263–1275. [CrossRef]
94. Loqué, D.; Lalonde, S.; Looger, L.L.; von Wirén, N.; Frommer, W.B. A Cytosolic Trans-Activation Domain Essential for Ammonium Uptake. *Nature* **2007**, *446*, 195–198. [CrossRef]
95. Ludewig, U.; Wilken, S.; Wu, B.; Jost, W.; Obrdlik, P.; El Bakkoury, M.; Marini, A.-M.; André, B.; Hamacher, T.; Boles, E.; et al. Homo- and Hetero-Oligomerization of Ammonium Transporter-1 NH 4 + Uniporters. *J. Biol. Chem.* **2003**, *278*, 45603–45610. [CrossRef] [PubMed]
96. Nühse, T.S.; Stensballe, A.; Jensen, O.N.; Peck, S.C. Phosphoproteomics of the Arabidopsis Plasma Membrane and a New Phosphorylation Site Database. *Plant Cell* **2004**, *16*, 2394–2405. [CrossRef] [PubMed]
97. Gazzarrini, S.; Lejay, L.; Gojon, A.; Ninnemann, O.; Frommer, W.B.; von Wirén, N. Three Functional Transporters for Constitutive, Diurnally Regulated, and Starvation-Induced Uptake of *Ammonium* into *Arabidopsis* Roots. *Plant Cell* **1999**, *11*, 937–947. [CrossRef] [PubMed]
98. Kumar, A.; Silim, S.N.; Okamoto, M.; Siddiqi, M.Y.; Glass, A.D.M. Differential Expression of Three Members of the *AMT1* Gene Family Encoding Putative High-Affinity NH_4^+ Transporters in Roots of *Oryza Sativa* Subspecies *Indica*: Regulation of AMT1 Genes in Rice. *Plant Cell Environ.* **2003**, *26*, 907–914. [CrossRef] [PubMed]
99. Gu, R.; Duan, F.; An, X.; Zhang, F.; von Wirén, N.; Yuan, L. Characterization of AMT-Mediated High-Affinity Ammonium Uptake in Roots of Maize (*Zea mays* L.). *Plant Cell Physiol.* **2013**, *54*, 1515–1524. [CrossRef]
100. Wu, X.; Yang, H.; Qu, C.; Xu, Z.; Li, W.; Hao, B.; Yang, C.; Sun, G.; Liu, G. Sequence and Expression Analysis of the AMT Gene Family in Poplar. *Front. Plant Sci.* **2015**, *6*, 337. [CrossRef]
101. Li, H.; Yang, Q.; Liu, W.; Lin, J.; Chang, Y. The AMT1 Family Genes from Malus Robusta Display Differential Transcription Features and Ammonium Transport Abilities. *Mol. Biol. Rep.* **2017**, *44*, 379–390. [CrossRef]
102. Liu, L.-H.; Fan, T.-F.; Shi, D.-X.; Li, C.-J.; He, M.-J.; Chen, Y.-Y.; Zhang, L.; Yang, C.; Cheng, X.-Y.; Chen, X.; et al. Coding-Sequence Identification and Transcriptional Profiling of Nine AMTs and Four NRTs From Tobacco Revealed Their Differential Regulation by Developmental Stages, Nitrogen Nutrition, and Photoperiod. *Front. Plant Sci.* **2018**, *9*, 210. [CrossRef]
103. Masclaux-Daubresse, C.; Reisdorf-Cren, M.; Orsel, M. Leaf Nitrogen Remobilization for Plant Development and Grain Filling. *Plant Biol.* **2008**, *10*, 23–36. [CrossRef]
104. Van der Graaff, E.; Schwacke, R.; Schneider, A.; Desimone, M.; Flügge, U.-I.; Kunze, R. Transcription Analysis of *Arabidopsis* Membrane Transporters and Hormone Pathways during Developmental and Induced Leaf Senescence. *Plant Physiol.* **2006**, *141*, 776–792. [CrossRef]
105. Zhao, H.; Wu, D.; Kong, F.; Lin, K.; Zhang, H.; Li, G. The *Arabidopsis thaliana* Nuclear Factor Y Transcription Factors. *Front. Plant Sci.* **2017**, *7*, 2045. [CrossRef] [PubMed]
106. Myers, Z.A.; Holt, B.F. NUCLEAR FACTOR-Y: Still complex after all these years? *Curr. Opin. Plant Biol.* **2018**, *45*, 96–102. [CrossRef]
107. Swain, S.; Myers, Z.A.; Siriwardana, C.L.; Holt, B.F. The multifaceted roles of NUCLEAR FACTOR-Y in *Arabidopsis thaliana* development and stress responses. *Biochim. Biophys. Acta Gene Regul. Mech.* **2017**, *1860*, 636–644. [CrossRef]
108. Du, W.; Yang, J.; Li, Q.; He, C.; Pang, Y. Identification and Characterization of Abiotic Stress-Responsive NF-YB Family Genes in Medicago. *Int. J. Mol. Sci.* **2022**, *23*, 6906. [CrossRef] [PubMed]

109. Schenk, R.U.; Hildebrandt, A.C. Medium and Techniques for Induction and Growth of Monocotyledonous and Dicotyledonous Plant Cell Cultures. *Can. J. Bot.* **1972**, *50*, 199–204. [CrossRef]
110. Zhou, Y.; Chen, G.; Peterson, A.; Zha, X.; Cheng, J.; Li, S.; Cui, D.; Zhu, H.; Kishchenko, O.; Borisjuk, N. Biodiversity of Duckweeds in Eastern China and Their Potential for Bioremediation of Municipal and Industrial Wastewater. *J. Geosci. Environ. Protect.* **2018**, *06*, 108–116. [CrossRef]
111. Cedergreen, N.; Madsen, T.V. Nitrogen Uptake by the Floating Macrophyte Lemna Minor. *New Phytol.* **2002**, *155*, 285–292. [CrossRef]
112. American Water Works Association (AWWA). *Standard Methods for the Examination of Water and Wastewater*, 21st ed.; AWWA: Washington, DC, USA, 2005.
113. Katoh, K.; Standley, D.M. MAFFT Multiple Sequence Alignment Software Version 7: Improvements in Performance and Usability. *Mol. Biol. Evol.* **2013**, *30*, 772–780. [CrossRef]
114. Guindon, S.; Dufayard, J.-F.; Lefort, V.; Anisimova, M.; Hordijk, W.; Gascuel, O. New Algorithms and Methods to Estimate Maximum-Likelihood Phylogenies: Assessing the Performance of PhyML 3.0. *Syst. Biol.* **2010**, *59*, 307–321. [CrossRef] [PubMed]
115. Criscuolo, A.; Gribaldo, S. BMGE (Block Mapping and Gathering with Entropy): A New Software for Selection of Phylogenetic Informative Regions from Multiple Sequence Alignments. *BMC Evol. Biol.* **2010**, *10*, 210. [CrossRef]
116. Ludewig, U.; von Wirén, N.; Frommer, W.B. Uniport of NH 4 + by the Root Hair Plasma Membrane Ammonium Transporter LeAMT1;1. *J. Biol. Chem.* **2002**, *277*, 13548–13555. [CrossRef] [PubMed]
117. Pfaffl, M.W. A New Mathematical Model for Relative Quantification in Real-Time RT-PCR. *Nucleic Acids Res.* **2001**, *29*, e45. [CrossRef] [PubMed]
118. Vandesompele, J.; De Preter, K.; Pattyn, F.; Poppe, B.; Van Roy, N.; De Paepe, A.; Speleman, F. Accurate Normalization of Real-Time Quantitative RT-PCR Data by Geometric Averaging of Multiple Internal Control Genes. *Genome Biol.* **2002**, *3*, RESEARCH0034. [CrossRef] [PubMed]

plants

MDPI

Article

Examination of the Metallothionein Gene Family in Greater Duckweed *Spirodela polyrhiza*

Orathai Pakdee [1,2], Shomo Tshering [1,2], Prayad Pokethitiyook [1,2] and Metha Meetam [1,2,*]

1 Department of Biology, Faculty of Science, Mahidol University, Bangkok 10400, Thailand
2 Center of Excellence on Environmental Health and Toxicology (EHT), OPS, MHESI, Bangkok 10400, Thailand
* Correspondence: metha.mee@mahidol.ac.th

Abstract: Duckweeds are aquatic plants that proliferate rapidly in a wide range of freshwaters, and they are regarded as a potential source of sustainable biomass for various applications and the cost-effective bioremediation of heavy metal pollutants. To understand the cellular and molecular basis that underlies the high metal tolerance and accumulation capacity of duckweeds, we examined the forms and transcript profiles of the metallothionein (MT) gene family in the model duckweed *Spirodela polyrhiza*, whose genome has been completely sequenced. Four *S. polyrhiza* MT-like genes were identified and annotated as *SpMT2a*, *SpMT2b*, *SpMT3*, and *SpMT4*. All except *SpMT2b* showed high sequence homology including the conserved cysteine residues with the previously described MTs from flowering plants. The *S. polyrhiza* genome appears to lack the root-specific Type 1 MT. The transcripts of *SpMT2a*, *SpMT2b*, and *SpMT3* could be detected in the vegetative whole-plant tissues. The transcript abundance of *SpMT2a* was upregulated several-fold in response to cadmium stress, and the heterologous expression of *SpMT2a* conferred copper and cadmium tolerance to the metal-sensitive Δ*cup1* strain of *Saccharomyces cerevisiae*. Based on these results, we proposed that *SpMT2a* may play an important role in the metal detoxification mechanism of duckweed.

Keywords: duckweed; heavy metal; metallothionein

Citation: Pakdee, O.; Tshering, S.; Pokethitiyook, P.; Meetam, M. Examination of the Metallothionein Gene Family in Greater Duckweed *Spirodela polyrhiza*. *Plants* **2023**, 12, 125. https://doi.org/10.3390/plants12010125

Academic Editors: Klaus-Jürgen Appenroth, Viktor Oláh and K. Sowjanya Sree

Received: 11 October 2022
Revised: 20 December 2022
Accepted: 21 December 2022
Published: 27 December 2022

1. Introduction

Members of the metallothionein (MT) gene family encode low-molecular-weight, cysteine (Cys)-rich proteins and are believed to play various important roles in the homeostasis of metals and reactive oxygen species (ROS) in eukaryotic organisms and some bacteria. The Cys residues of MT proteins have been shown to coordinate metal ions in various configurations depending on the metal load, suggesting that MTs generally function as intracellular metal chelators for metal detoxification, distribution, and/or storage [1]. The Cys residues can also participate in the scavenging of ROS under oxidative stress [2]. Another unique characteristic of the MT proteins is the arrangement of the Cys residues that are highly conserved among the related MT homologs but differ between the MT gene lineages. In angiosperms, four lineages of MT genes, referred to as Types, have been described based on the conserved Cys arrangements of the gene products [3]. Studies on various flowering plants have shown that all four MT Types are present in the plants, but they differ in their expression patterns. For instance, Type 1 MTs are usually expressed in root tissues and Type 4 MTs are primarily expressed in seeds. The conservation in the amino acid sequences and expression patterns among the MT gene lineages clearly suggest that they inherit specific and indispensable functions [4]. However, the essential functions of each MT lineage in various organisms remain largely elusive.

Duckweeds are small aquatic flowering plants of the family *Lemnaceae*, with 36 plant species encompassing the following five genera: *Spirodela*, *Landoltia*, *Lemna*, *Wolffiella*, and *Wolffia* [5]. Due to their high proliferation rate and nutrient richness, duckweeds have recently attracted attention as sustainable sources of livestock feed, human nutrition, and renewable biomass for the biofuel production and green industries [6]. Another advantage of duckweeds over traditional crops is their ability to grow in eutrophic water and simultaneously remove undesirable pollutants [7]. Heavy metal contamination in water due to anthropogenic activities is a grave concern in the present century. The accumulation of heavy metals in water can have adverse effects on the health of all the aquatic organisms, as well as humans [8]. Some duckweeds are considerably metal-tolerant plant species, and many studies have indeed shown the high efficiency and cost-effectiveness of duckweeds in heavy metal removal applications [9]. Despite the great potential of duckweeds in heavy metal bioremediation, little is known about the cellular and molecular mechanisms that enable to duckweeds to efficiently take up, accumulate, and detoxify the heavy metals from the environment (for recent studies on the metal homeostasis network in duckweeds, see [10,11]).

In this study, the forms and potential functions of MTs were investigated in the greater duckweed *Spirodela polyrhiza*. The expression and transcriptional responses of the different *S. polyrhiza* MT-like genes to metal stress were examined. To test whether the protein products from the putative *S. polyrhiza* MT genes can serve as intracellular metal chelators, the effect of the heterologous expression of the *S. polyrhiza* MT homologs in the yeast *Saccharomyces cerevisiae* mutant Δ*cup1*, which lacked the major endogenous MT gene, was investigated.

2. Results

2.1. Identification and Annotation of Putative S. polyrhiza MT-like Genes

The MT-like genes in the genome of *S. polyrhiza* were first identified using a keyword search in the Phytozome database. The search resulted in four predicted transcripts: Spipo6G0071500, Spipo0G0175800, Spipo0G0112500, and Spipo14G0028700. The searches for additional *S. polyrhiza* MT-like genes were performed using the known sequences of the MT gene family from *Oryza sativa* and *Arabidopsis thaliana*, as well as some previously described MT sequences from representative monocot species, against the genomes of *S. polyrhiza* in the GenBank (NCBI) database, but no additional MT-like genes were identified. To annotate the *S. polyrhiza* MT-like genes, their predicted amino acid sequences were aligned against the known MTs from representative plant species (Figure 1a) and a neighbor-joining tree was constructed (Figure 1b). Based on these results, the *S. polyrhiza* MT-like genes were annotated (Table 1). Interestingly, no homolog of Type 1 MTs could be identified, suggesting that the *S. polyrhiza* genome lacks one of the typical MT lineages that are found in the genomes of flowering plants. Two Type 2 MT homologs were found in *S. polyrhiza*. The predicted amino acid sequence of Spipo0G0112500, annotated as *SpMT2a*, shared all of the fourteen Cys residues that are found in other flowering plants such as *O. sativa* and *A. thaliana*. In contrast, the predicted amino acid sequence of Spipo6G0071500, annotated as *SpMT2b*, only shared five of the fourteen Cys residues. As with SpMT2a, other Type 2 MT homologs from other duckweed species such as *Landolita punctata* and *Wolffia australiana* were also found to share all of the fourteen Cys residues. This suggests that *SpMT2b* does not encode a bona fide MT. Spipo14G0028700 and Spipo0G0175800 were found to encode single-member Type 3 and Type 4 MTs, respectively. The predicted amino acid sequences of *SpMT3* and *SpMT4* shared all of the critical Cys residues conserved among the flowering plant species. In addition, SpMT4 shared the two conserved histidine residues that are typically found in the middle Cys-rich domain of Type 4 MTs and which are potentially involved in metal coordination [12]. It should be noted that SpMT4 contains a stretch of nine non-Cys amino acids prior to the N-terminal Cys-rich domain. This extended N-terminal sequence is commonly found in the Type 4 MTs from dicot plants, but it is absent in OsMT4 and other Type 4 MTs from several monocots including wheat, maize, and barley [12]. This prompted us to investigate whether the extended N-terminal sequence found in SpMT4 is unique to the duckweed MT. A TBLASTN search using the SpMT4 amino acid sequence against the NCBI genome databases showed that SpMT4 was most closely related to a Type 4 MT from *Spirodela intermedia*, followed by Type 4 MTs from two other monocot plants—*Xerophyta humilis* and *Elaeis guineensis*—all of which contain an extended N-terminal sequence prior to the Cys-rich domain (data not shown). Therefore, SpMT4 is not the only monocot Type 4 MT that harbors the extended N-terminal sequence.

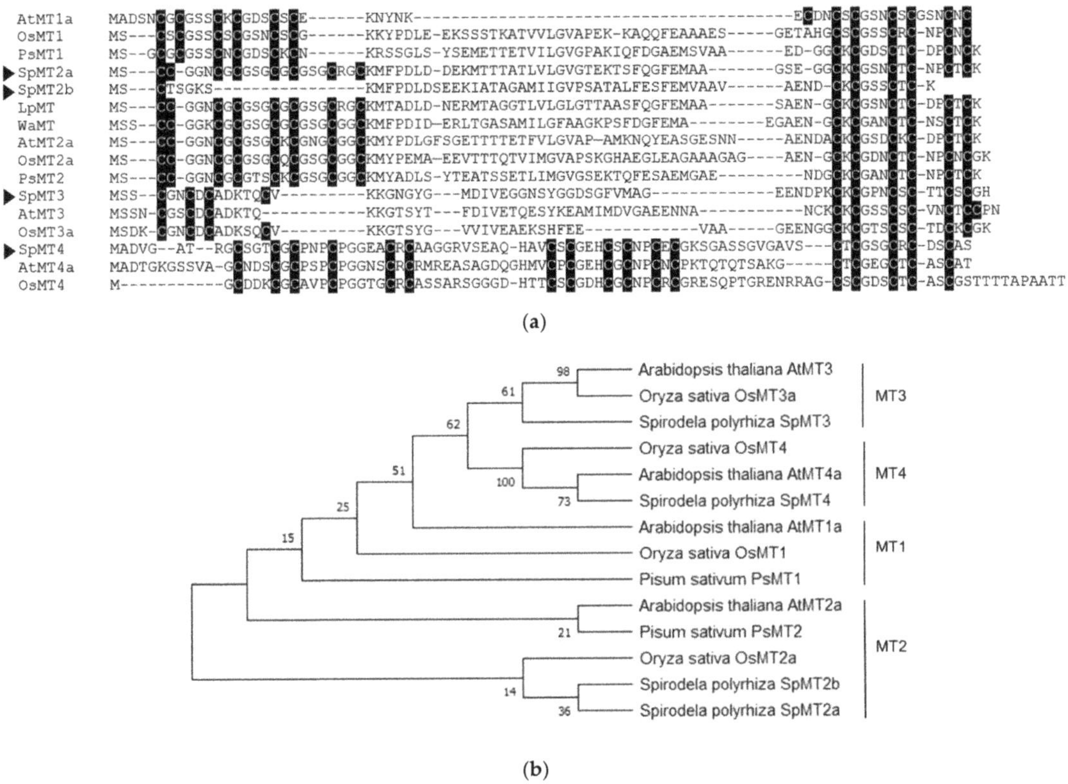

Figure 1. Identification of MT-like genes from *S. polyrhiza*: (**a**) alignment of predicted amino acid sequences of *S. polyrhiza* MT homologs against representative vascular-plant MTs; (**b**) a neighbor-joining tree based on the sequence alignment, with bootstrap values from 1000 iterations. Cysteine residues are highlighted in black. Arrows indicate the *S. polyrhiza* MT homologs. The accession numbers of the sequences used in the analysis are: AtMT1a (837273), OsMT1 (U43529), PsMT1 (BAD18382), LpMT (JZ977403.1), WaMT (JK990501.1), AtMT2a (820098), OsMT2a (P94029), PsMT2a (BAD18383), AtMT3 (820772), OsMT3a (A1YTM8), AtMT4a (818800), and OsMT4 (Q109B0).

Table 1. Annotation of *S. polyrhiza* MT-like genes.

Gene	Transcript	Location	No. of a.a.	No. of Cys
MT2a	Spipo0G0112500	Chr. 7	78	14
MT2b	Spipo6G0071500	Chr. 5	60	5
MT3	Spipo14G0028700	Chr. 9	65	10
MT4	Spipo0G0175800	Chr. 18	82	17

To confirm that the MT-like *S. polyrhiza* genes were expressed at the transcript level, the RT-PCR analysis was carried out using a total cDNA extract from the whole of *S. polyrhiza* plant. The transcripts of all three *S. polyrhiza* MT genes, but not of *SpMT4*, could be detected in the whole plant tissues, confirming that they are functionally transcribed (Figure 2). It is possible that the transcript of *SpMT4* was not detected in the whole plant tissues because the expression of Type 4 MTs is typically restricted to seeds [3]. However, the potential expression of *SpMT4* in the seeds of *S. polyrhiza* could not be investigated in this study due to the unavailability of the duckweed seeds at the time of the experiment.

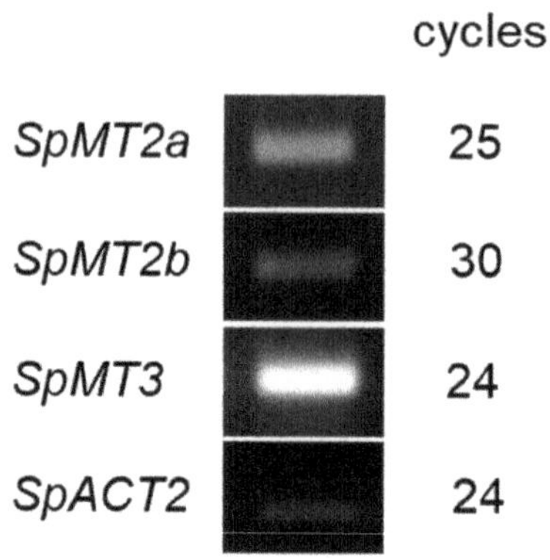

Figure 2. Semi-quantitative RT-PCR analysis of transcript abundance in the whole plant tissues of *S. polyrhiza* cultured in normal Hoagland solution. *SpACT2* was included as a standard control. The number of cycles used in the PCR reactions are indicated.

2.2. Transcriptional Response to Cadmium and Copper Stresses

The expression of plant *MT* genes is often induced by heavy metals, suggesting their role in heavy metal detoxification and tolerance. To investigate whether any of the *S. polyrhiza* MT genes might be functionally involved in the metal detoxification mechanism, the transcriptional responses of *SpMT2a* and *SpMT3* to heavy metal stress were examined. The *S. polyrhiza* plants were cultured for 24 h in medium containing various concentrations of $CuSO_4$ and $CdCl_2$ (Figure 3). The transcript abundance of *SpMT2b* was not investigated because it does not appear to encode a functional MT protein, whereas *SpMT4* was not expressed in the vegetative tissues, as demonstrated previously. The quantitative RT-PCR analysis showed that the transcript abundance of the *SpMT2a* genes was significantly up-regulated, more than two-fold ($p < 0.05$), by the $CdCl_2$ treatments. The $CuSO_4$ treatment slightly induced the *SpMT2a* expression, but not to a level that was statistically significant compared to the control. In contrast, the expression of *SpMT3* was not induced by the $CuSO_4$ or $CdCl_2$ treatments, and interestingly, it was partially down-regulated in response to high concentrations of $CuSO_4$ or $CdCl_2$. These results suggest that *SpMT2a* may play a role in the ability of *S. polyrhiza* to tolerate cadmium.

2.3. Copper and Cadmium Tolerance Conferred by the Heterologous Expression of SpMT2b in S. cerevisiae

We previously showed that the *S. polyrhiza* MT genes, except *SpMT2b*, shared with other plant MT homologs the conserved metal-binding motifs which may participate in metal chelation and contribute to the function of MT proteins in duckweed's metal homeostasis mechanism. To further test whether the protein products encoded by some of the *S. polyrhiza* MT genes could indeed function as metal chelators in vivo, we constitutively expressed *SpMT2a* and *SpMT3* in the *S. cerevisiae* copper-sensitive Δ*cup1* mutant, which lacked one of its two MT genes. The dilution spot assay showed that growth of the Δ*cup1* strain transformed with the empty p424-GPD vector was completely inhibited in medium supplemented with either 25 μM $CuSO_4$ or 25 μM $CdCl_2$ (Figure 4). For a positive control, the Δ*cup1* strain was transformed with the vector p424-GPD which harbored the yeast *CUP1* gene. The *CUP1* complement restored the copper and cadmium tolerance of the Δ*cup1* mutant. The heterologous expression of *SpMT2a* in the Δ*cup1* mutant could also confer growth tolerance to the cadmium and copper stress, although to a lower extent compared to the *CUP1* gene, indicating that the protein product of *SpMT2a* could function as a metal chelator *in vivo*. In contrast, the Δ*cup1* mutant that expressed *SpMT3* did not appear to grow under the metal stress under these conditions, although it should be noted that the inability of the *SpMT3* heterologous expression to impart metal tolerance in the yeast Δ*cup1* mutant could be attributed to several unforeseen reasons such as the failure of the duckweed protein to efficiently express inside the yeast cells.

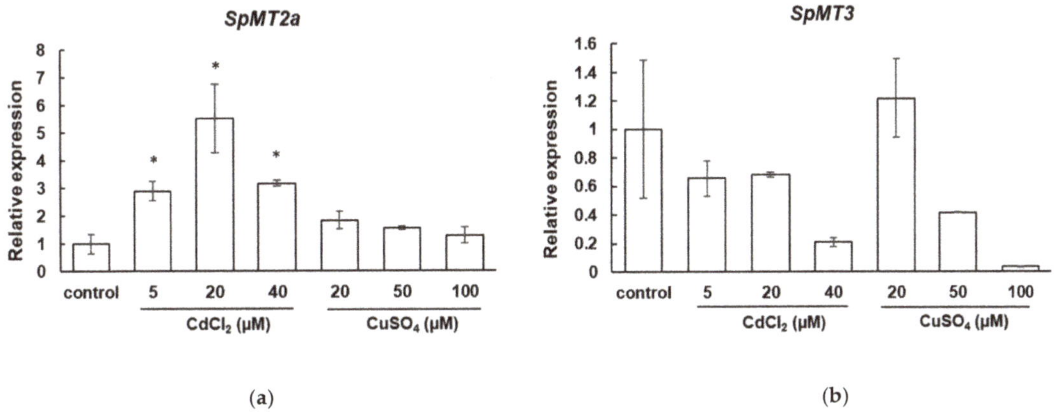

Figure 3. Transcript abundance of *S. polyrhiza* MT genes in whole plant tissues under metal stress: (**a**) *SpMT2a*. (**b**) *SpMT3*. The *S. polyrhiza* plants were cultured in Hoagland solution supplemented with the indicated concentrations of $CdCl_2$ or $CuSO_4$ for 24 h. The relative transcript abundance was quantified and normalized to that of *SpACT2*. The relative expression of the transcript abundance was adjusted to the control level. The asterisks indicate values that were significantly higher than the control condition and exceeded a fold change of two (one-tailed *t*-test, $p < 0.05$). The error bars represent SE (n = 3).

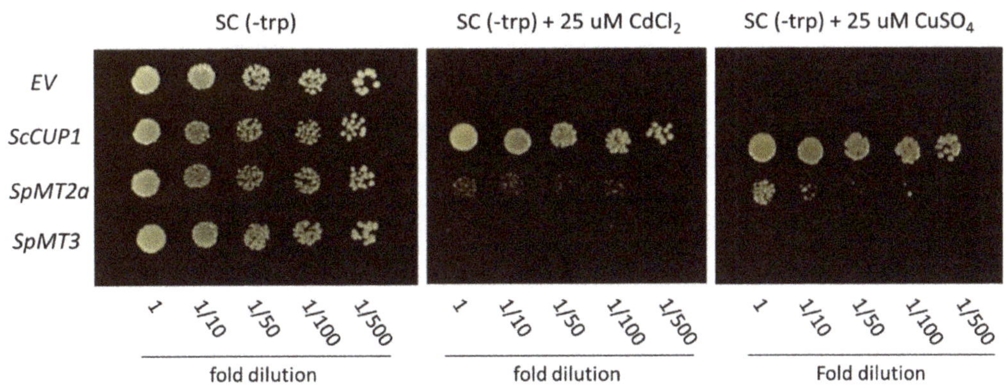

Figure 4. Metal sensitivity of the *S. cerevisiae* Δ*cup1* strain expressing the *SpMT2a* and *SpMT3* genes in comparison to the controls: empty vector (EV) or *S. cerevisiae CUP1* (Sc*CUP1*). Serially diluted yeast cultures were spotted on normal SC(-trp) agar or the medium supplemented with 25 μM $CdCl_2$ or 25 μM $CuSO_4$ for 3 days.

3. Discussion

Heavy metal contamination in aquatic environments is a global concern. The experimental application of various duckweeds in the phytoremediation of polluted water has been met with considerable success and improved the water parameters, including heavy metals reduction [7,13,14]. For instance, it has been reported that *Lemna gibba* was able to remove >90% of Ni, Pb, and Cd at industry-relevant metal concentrations [15]. The actual removal efficiency may depend on many factors, including duckweed species, metal species, initial metal concentration, and other water parameters. The duckweed byproduct can also be utilized in several applications including as biomass for biofuel and as biomaterial and in plant factories for the production of high-value bioproducts [15,16]. In general, duckweeds are considered promising candidates for the metal phytoremediation due to their high capacity for metal tolerance, accumulation, and bioconcentration factors [17–19]. Recently, Wang et al. [20] showed that the majority of cadmium taken up by *Landoltia punctata* was associated with the cell wall, whereas the remaining pool could be found in the soluble fraction and in the organellar fraction. The authors further showed that over 80% of the protein-bound cadmium pool was associated with the albumin and globumin classes of proteins. However, the exact identity of the major cadmium-associating proteins or other cellular ligands that contributed to the cadmium accumulation in the duckweed remained unknown [20]. MTs serve as a major intracellular metal-chelating protein in plants, and the high expression of MT genes has been shown to correlate with metal tolerance and/or accumulation [21]. Thus, a fundamental understanding of the MT family in duckweeds may help to improve their capacity in metal bioremediation applications, as well as in the fortification of duckweeds to provide animal or human nutrition.

Owing to its complete genome sequence and genome simplicity, the greater duckweed *S. polyrhiza* was used in this study to examine the duckweed MT gene family. The first genome assembly was based on the *S. polyrhiza* clone 7498 [22]. The genome assembly of *S. polyrhiza* clone 7498 was updated [23,24], and another genome sequence of *S. polyrhiza,* clone 9509, was later released [25,26]. The genome sequences of other duckweeds, including *Lemna gibba, Lemna minor, Spirodela intermedia,* and *Wolffia australiana,* have also become available and should permit more comprehensive investigations of the duckweed MT gene family [27,28].

A search in the genome of *S. intermedia* and the other available duckweed species also failed to identify a homolog of Type 1 MTs (data not shown). This suggests that Type 1 MTs were lost from the genome of the ancestral duckweed species. It is possible that the function of root-specific MTs is not needed for duckweeds, although whether the expression of Type 1 MTs in the roots is replaced by that of the other remaining MT genes remains to be tested. If this is true, the absence of MT gene expression in duckweed roots may be attributable to a different mechanism of metal transport, distribution, or homeostasis in the root tissues compared to other non-aquatic plants.

Two homologs were found for the Type 2 MTs, although the predicted amino acid sequence of MT2b contained Cys residues primarily in the C-terminal half, suggesting that it is not a bona fide MT. It should also be noted that the predicted amino acid sequence from the coding sequence of Spipo0G0112500 described in the Phytozome database appeared to be shortened compared to the typical Type 2 MTs, and so we searched and found an alternative start codon at the position −810 nucleotide upstream from the originally predicted start codon, and we used this to derive the amino acid sequence that is homologous to other Type 2 MTs, as shown in this study. A homolog of a Type 3 MT and another of a Type 4 MT were identified in the genome of *S. polyrhiza.* In flowering plants, Type 3 MTs are typically expressed in ripening fruits, but the expression of Type 3 MTs can also be located in the leaves and in other tissues in plants, such as *A. thaliana*, which do not bear fleshy fruits [29]. The expression of Type 4 MTs is largely restricted to the seeds of flowering plants [30]. Although the function of seed-specific MTs is not yet clear, it is believed that they function in the storage of essential nutrients such as zinc and copper for the embryo and the early seedling during germination [31]. As with most duckweeds, *S. polyrhiza* propagates primarily through asexual vegetative budding. During an unfavorable growth period, *S. polyrhiza* can also form turions, which are a form of dormant vegetative tissues and resume growth after the harsh period concludes. On rare occasions, *S. polyrhiza* can flower and produce seeds [32]. The mRNA expression of *SpMT4* was not found in the RT-PCR analysis of whole plant tissues, and the hypothetical expression of *SpMT4* in the seeds remains to be investigated in the future.

The qRT-PCR analysis showed that the expression level of *SpMT2a* transcripts was upregulated in response to the cadmium treatment, and it was only slightly upregulated by the copper treatment, suggesting that the *S. polyrhiza* MT may play a specific role in Cd tolerance and/or accumulation. In *A. thaliana* [29,33] and several flowering plants [34–36], the mRNA abundance of MT genes is upregulated primarily by copper stress. Thus, the transcriptional response of *SpMT2a* suggests that cadmium may pose a greater threat than copper to duckweed in its aquatic habitat. Alternatively, there may exist another mechanism that helps duckweed cope with excessive copper. A similar observation was made in the moss *Physciomitrella patents* whose MT genes also responded primarily to cadmium treatment, but not to copper [37]. The down-regulation of *SpMT3* in response to the cadmium and copper treatments was similar to the down-regulation of barley MTs during cadmium, copper, and zinc exposure [38]. While the purpose of the down-regulation of an MT gene is not yet clear, several explanations have been proposed, including a role of MTs in the homeostasis of intracellular ROS, e.g., to allow the transient accumulation of ROS to trigger a signaling cascade [39,40]. The hypothesis that *SpMT3* is not directly involved in the chelation of excess metal ions was further supported the finding that the heterologous expression of *SpMT2a*, but not of *SpMT3*, was able to increase the copper and cadmium tolerance of the yeast Δ*cup1* mutant. Nevertheless, since the expression levels of the SpMT genes in the transgenic yeast strains were not analyzed, it should be re-emphasized that the inability of SpMT3 to confer copper or cadmium tolerance could be attributed to the low protein stability and/or expression level in the yeast cells; this possibility should be investigated in the future. The specific contributions of individual duckweed MTs should also be tested using more concrete evidence, such as characterization of the loss-of-function mutants to verify their function as well as to investigate their additional roles under diverse physiological conditions.

The insights on the forms and potential functions of duckweed MTs gained from this and from future studies may contribute to a more efficient use of duckweeds in environmental cleanup, nutrient fortification, and other innovative applications. For instance, the expression levels of different MTs may be investigated for their correlation with the metal bioaccumulation levels among duckweed varieties to obtain varieties that have higher nutritional value or increased ability to remove toxic pollutants from the environment. On the other hand, the duckweed varieties that have lower MT expression and that hypo-accumulate toxic metals may be selected when duckweeds are cultivated for livestock feed or human food. As shown this study, the ability of SpMTs to confer metal tolerance when heterologously expressed in *S. cerevisiae* was considerably lower than that of CUP1, suggesting that duckweed MTs do not chelate metal ions as efficiently. Thus, it is possible to explore more efficient metal-chelating MT isoforms from different duckweed species or varieties. The use of genetically engineered duckweeds that overexpress a transgenic MT gene may also be investigated in the future.

4. Materials and Methods

4.1. Plant Material and Culture Conditions

The *S. polyrhiza* clone 5638 was used in this study. The *S. polyrhiza* clone was originally isolated from a local pond in Raleigh, North Carolina, USA, and it has previously been described [17]. The plant cultures were maintained in a 125 mL Erlenmeyer flask containing 30 mL sterile Hoagland solution under continuous white fluorescent light at approximately 50 μmol photon m^{-2} s^{-1} in a temperature-controlled growth room at 23–25 °C.

4.2. Identification and Annotation of Putative MT Genes

The *S. polyrhiza* clone 7498 genome database in Phytozome version 13 (JGI) was first examined using a keyword search for "metallothionein". To identify additional MT homologs, the *S. polyrhiza* genome databases in GenBank (NCBI) were examined using a TBLASTN search against amino acid sequences of the MT gene products from *O. sativa* and *A. thaliana*, as well as previously identified MT sequences from several monocot plants including *Zea mays* ZmMT1 (P30571), *Allium sativum* AsMT2 (AAV80430), *Posidonia oceanica* PoMT2 (CAF31414), *Wolffia arrhizal* WaMT2 (ADB85769), *Elaeis guineensis* EgMT3 (CAB52585), *Musa acuminate* MaMT3 (DN239297), *Triticum aestivum* TaMT4 (P30569.2), *Hordeum vulgare* HvMT4 (CAD88267.1), *Sorghum bicolor* SbMT4 (XP_002467575.1), and *Brachypodium distachyon* BdMT4 (XP_003572023.1). The putative MT-like sequences from *S. polyrhiza* were then aligned with the representative MT sequences obtained from the GenBank database using ClustalW (EMBL-EBI) and further adjusted manually using BioEdit 7.2 (MBIO-NCSU). The neighbor-joining tree was constructed using MEGA-X [41]. To find the chromosomal location of the genes, the coding sequences of the *S. polyrhiza* 7498 MT-like genes were used in a BLASTN search against *S. polyrhiza* 9509 in the GenBank (NCBI) database.

4.3. RT-PCR Analysis of S. polyrhiza MT Gene Expression

The *S. polyrhiza* clone WY001 plants, each containing 3–4 fronds, were cultured in sterile Hoagland solution for 3 days under continuous white fluorescent light. The whole plant tissues from five plants were transferred to a 1.5 mL microcentrifuge tube and frozen in liquid nitrogen before being ground with a plastic homogenizer pestle. For the analysis of the transcriptional responses to metal stress, $CuSO_4$ or $CdCl_2$ was added from 100 mM filter-sterile stock solutions to the indicated final concentration during the last 24 h. Triplicate cultures were included for each treatment. RNA extraction was performed using the TRIzol reagent (Invitrogen, Waltham, MA, USA) followed by using the TriRNA purification kit (Geneaid Biotech, New Taipei City, Taiwan ROC) according to the manufacturer's protocol. The cDNA was synthesized in a 20 μL reaction containing 0.5 μg of DNase-treated total RNA using the ImProm-II reverse transcriptase (Promega, Madison, WI, USA) and an oligo-dT primer. The PCR amplification was performed using gene-specific primers (Table 2), Taq DNA polymerase (Vivantis Technologies, Selangor Darul Ehsan, Malaysia), and 1 μL of the first-strand cDNA mixture. The PCR condition was as follows: initial denaturation at 95 °C for 2 min, followed by the indicated number of cycles of 95 °C for 15 s, 56 °C for 15 s, and 72 °C for 30 s, and the final extension period at 72 °C for 2 min. The sizes of the PCR amplicons were verified using 1.5% agarose gel electrophoresis (Supplemental Figure S1). For the quantitative RT-PCR analysis, 1 μL of the first-strand cDNA mixture was amplified in an Applied Biosystems 7500 real-time PCR system (Thermo Fisher Scientific, Waltham, MA, USA) with a KAPA SYBR® Fast ABI Prizm qPCR master mix (Kapa Biosystem, Potters Bar, UK) using the gene-specific primers. The expression levels were normalized to that of the *SpACT2* gene, which has been used as an internal standard in previous studies [36].

4.4. Heterologous Expression of SpMT2a and SpMT3

For the first-strand cDNA mixture, 1 μL, prepared as described above, was amplified using KOD high-fidelity DNA polymerase (Toyobo, Osaka, Japan) and the gene-specific primers that annealed to the 5′ and 3′ untranslated regions of *SpMT2a* and *SpMT3* (Table 2). The PCR condition was as follows: initial denaturation at 94 °C for 2 min, followed by 25 cycles of 94 °C for 15 s, 58 °C for 5 s, and 68 °C for 5 s, and the final extension period at 68 °C for 1 min. The PCR amplicons were purified using a PCR cleanup kit (Geneaid, New Taipei City, Taiwan ROC), incubated with Taq DNA polymerase (Vivantis Technologies, Selangor Darul Ehsan, Malaysia) to add a 3′-A overhang, and then ligated into the pGEMT-easy cloning vector (Promega, Madison, WI, USA). The sequence fidelity was confirmed by DNA sequencing. The full-length coding sequences were transferred to the p424-GPD expression vector [42] using the EcoRI and SpeI restriction sites between the *GPD* promoter and *CYC1* terminator. The construction of p424-GPD::*CUP1* was completed as previously described [37]. The wild type (DTY3) and Δ*cup1* (DTY4) strains of *S. cerevisiae* were kindly provided by Dr. Dennis J. Thiele of the Duke University School of Medicine, USA. All of the vectors including the p424-GPD empty vector were transformed into the yeast Δ*cup1* strain using a Frozen-EZ yeast transformation kit (Zymo Research, Irvine, CA, USA) according to the manufacturer's protocol. The yeast transformants were selected on synthetic complete agar medium lacking tryptophan (SC-Trp) from Himedia Laboratories (Maharashtra, India).

Table 2. Oligonucleotide primers used in the study.

Primers	Sequence (5′ → 3′)	Expected Amplicon Size (bp)	Annealing Temperature (°C)
Primers for RT-PCR Analysis			
SpMT2a-F	TGACGAGAAGATGACCACCA	148	56
SpMT2a-R	TCATTTGCAGGTGCAGGGAT		
SpMT2b-F	ATGTCTTGCACCAGCGGGAAA	134	56
SpMT2b-R	GCAGCGACCATCTCGAACGACTC		
SpMT3-F	ACCCAGTGCGTGAAGAAGGGCAA	162	56
SpMT3-R	TCAATGGCCGCAGGAGCAGGTGG		
SpMT4-F	GACGTAGGAGCGACGCGAGG	188	56
SpMT4-R	CCAACGCCAGAGGAAGCACC		
SpACT-F	GCGACATCAAGGAGAAGCTG	213	56
SpACT-R	AGTTGTAGGTGGTCTCGTGG		
Primers for heterologous expression			
SpMT2a-ORF-F	GAAGATGTCTTGCTGCGGAG	258	58
SpMT2a-ORF-R	CCTTCATACAGGAAGCGTCC		
SpMT3-ORF-F	CCATGTCGAGCTGCGGCAACT	203	58
SpMT3-ORF-R	CGCTCAATGGCCGCAGGA		

For the spot dilution assay, overnight *S. cerevisiae* cultures were inoculated in 50 mL of SC-Trp broth medium at the initial OD_{600} of 0.2 and then shaken at 30 °C until the OD_{600} reached 0.8–1.0. The cultures were diluted to an OD_{600} of 0.2 and then serially diluted in fresh SC-Trp broth. From the 1, 1/10, 1/50, 1/100, and 1/500 diluted cultures, samples of 3 µL were spotted on SC-Trp 2% agar medium or the medium supplemented with $CuSO_4$ or $CdCl_2$ at the indicated concentrations. After incubation at 30 °C for 2–3 days, the plates were photographed.

5. Conclusions

We showed that the *S. polyrhiza* genome contains four MT-like genes: *SpMT2a*, *SpMT2b*, *SpMT3*, and *SpMT4*, encompassing three of the four Types of MTs in flowering plants. Even though the *SpMT2b* gene is expressed in vegetative tissues, its predicted amino acid sequence lacks nine of the fourteen conserved Cys residues found in other Type 2 MTs, and it is unlikely to function as a bona fide MT. The transcript of *SpMT4* was not detected in the vegetative whole-plant tissues, which was in agreement with the typical seed-specific localization of Type 4 MTs. The transcript abundance of *SpMT2a*, but not of *SpMT3*, was upregulated in response to cadmium stress. The heterologous expression of *SpMT2a*, but not of *SpMT3*, conferred copper and cadmium tolerance in the *S. cerevisiae* Δ*cup1* mutant. Therefore, among the four *S. polyrhiza* MT-like genes, *SpMT2a* is a strong candidate for playing an important role in the metal tolerance and accumulation of duckweed. This hypothetical function should be further investigated. The application of this knowledge may contribute to a more efficient use of duckweed in the environmental cleanup of metal pollutants in the future.

Supplementary Materials: The following supporting information can be downloaded at: https://www.mdpi.com/article/10.3390/plants12010125/s1, Supplemental Figure S1. Electrophoretic analysis of the RT-PCR amplicons: (a) *SpMT2a*; (b) *SpMT2b*; (c) *SpMT3*; and (d) *SpACT2*. Total cDNA was synthesized from a whole-plant RNA extract using an oligo-dT primer, followed by the PCR amplification using gene-specific primers with the cDNA mixture (lane 1) or the water negative control (lane N). The PCR amplicons were analyzed under 1.5% agarose gel electrophoresis in comparison to a standard size marker (lane M).

Author Contributions: Conceptualization, methodology, and supervision: M.M. and P.P.; investigation, O.P. and S.T.; data curation and writing: M.M. and O.P. All authors have read and agreed to the published version of the manuscript.

Funding: This research was funded by a scholarship from the Thailand International Postgraduate Programme (TIPP), Thailand International Cooperation Agency (TICA), Ministry of Foreign Affairs, Thailand to S.T.

Data Availability Statement: The data presented in this study are available on request from the corresponding author.

Acknowledgments: We would like to thank the Thailand International Postgraduate Programme (TIPP), Thailand International Cooperation Agency (TICA), Ministry of Foreign Affairs, Thailand, for the graduate assistantship to S.T. and the Center of Excellence on Environmental Health and Toxicology (EHT), Office of the Permanent Secretary (OPS), Ministry of Higher Education, Science, Research and Innovation (MHESI), Thailand, for their technical support.

Conflicts of Interest: The authors declare no conflict of interest.

References

1. Nielson, K.B.; Atkin, C.L.; Winge, D.R. Distinct metal-binding configurations in metallothionein. *J. Biol. Chem.* **1985**, *260*, 5342–5350. [CrossRef]
2. Krężel, A.; Maret, W. The bioinorganic chemistry of mammalian metallothioneins. *Chem. Rev.* **2021**, *121*, 14594–14648. [CrossRef]
3. Cobbett, C.; Goldsbrough, P. Phytochelatins and metallothioneins: Roles in heavy metal detoxification and homeostasis. *Annu. Rev. Plant Biol.* **2002**, *53*, 159–182. [CrossRef] [PubMed]
4. Guo, W.J.; Meetam, M.; Goldsbrough, P.B. Examining the specific contributions of individual Arabidopsis metallothioneins to copper distribution and metal tolerance. *Plant Physiol.* **2008**, *146*, 1697–1706. [CrossRef]
5. Acosta, K.; Appenroth, K.J.; Borisjuk, L.; Edelman, M.; Heinig, U.; Jansen, M.A.K.; Oyama, T.; Pasaribu, B.; Schubert, I.; Sorrels, S.; et al. Return of the Lemnaceae: Duckweed as a model plant system in the genomics and postgenomics era. *Plant Cell* **2021**, *33*, 3207–3234. [CrossRef] [PubMed]
6. Vu, G.T.H.; Fourounjian, P.; Wang, W.; Cao, X.H. Future prospects of duckweed research and applications. In *The Duckweed Genomes*; Cao, X.H., Fourounjian, P., Wang, W., Eds.; Springer International Publishing: Cham, Switzerland, 2020; pp. 179–185.
7. Gupta, C.; Prakash, D. Duckweed: An effective tool for phyto-remediation. *Toxicol. Environ. Chem.* **2013**, *95*, 1256–1266. [CrossRef]
8. Vardhan, K.H.; Kumar, P.S.; Panda, R.C. A review on heavy metal pollution, toxicity and remedial measures: Current trends and future perspectives. *J. Mol. Liq.* **2019**, *290*, 111197. [CrossRef]
9. Ali, Z.; Waheed, H.; Kazi, A.G.; Hayat, A.; Ahmad, M. Chapter 16—Duckweed: An efficient hyperaccumulator of heavy metals in water bodies. In *Plant Metal Interaction*; Ahmad, P., Ed.; Elsevier: Amsterdam, The Netherlands, 2016; pp. 411–429.
10. Xu, H.; Yu, C.; Xia, X.; Li, M.; Li, H.; Wang, Y.; Wang, S.; Wang, C.; Ma, Y.; Zhou, G. Comparative transcriptome analysis of duckweed (*Landoltia punctata*) in response to cadmium provides insights into molecular mechanisms underlying hyperaccumulation. *Chemosphere* **2018**, *190*, 154–165. [CrossRef]
11. Chen, Y.; Zhao, X.; Li, G.; Kumar, S.; Sun, Z.; Li, Y.; Guo, W.; Yang, J.; Hou, H. Genome-wide identification of the nramp gene family in *Spirodela polyrhiza* and expression analysis under cadmium Stress. *Int. J. Mol. Sci.* **2021**, *22*, 6414. [CrossRef]
12. Chyan, C.L.; Lee, T.T.; Liu, C.P.; Yang, Y.C.; Tzen, J.T.; Chou, W.M. Cloning and expression of a seed-specific metallothionein-like protein from sesame. *Biosci. Biotechnol. Biochem.* **2005**, *69*, 2319–2325. [CrossRef]
13. Ekperusi, A.O.; Sikoki, F.D.; Nwachukwu, E.O. Application of common duckweed (*Lemna minor*) in phytoremediation of chemicals in the environment: State and future perspective. *Chemosphere* **2019**, *223*, 285–309. [CrossRef]
14. Ceschin, S.; Crescenzi, M.; Iannelli, M.A. Phytoremediation potential of the duckweeds *Lemna minuta* and *Lemna minor* to remove nutrients from treated waters. *Environ. Sci. Pollut. Res.* **2020**, *27*, 15806–15814. [CrossRef]
15. Liu, Y.; Xu, H.; Yu, C.; Zhou, G. Multifaceted roles of duckweed in aquatic phytoremediation and bioproducts synthesis. *GCB Bioenergy* **2021**, *13*, 70–82. [CrossRef]

16. Yoksan, R.; Boontanimitr, A.; Klompong, N.; Phothongsurakun, T. Poly(lactic acid)/thermoplastic cassava starch blends filled with duckweed biomass. *Int. J. Biol. Macromol.* **2022**, *203*, 369–378. [CrossRef]
17. Liu, Y.; Sanguanphun, T.; Yuan, W.; Cheng, J.J.; Meetam, M. The biological responses and metal phytoaccumulation of duckweed *Spirodela polyrhiza* to manganese and chromium. *Environ. Sci. Pollut. Res.* **2017**, *24*, 19104–19113. [CrossRef]
18. Hu, D.; Cheng, M.; Hu, K.; Zhang, W.; Yang, Y.; Xu, Q. Evaluation of cobalt hyperaccumulation and tolerance potential of the duckweed (*Lemna minor* L.). *Ecotoxicol. Environ. Saf.* **2019**, *179*, 79–87. [CrossRef]
19. Chaudhary, E.; Sharma, P. Chromium and cadmium removal from wastewater using duckweed—*Lemna gibba* L. and ultrastructural deformation due to metal toxicity. *Int. J. Phytoremediation* **2019**, *21*, 279–286. [CrossRef]
20. Wang, X.; Zhang, B.; Wu, D.; Hu, L.; Huang, T.; Gao, G.; Huang, S.; Wu, S. Chemical forms governing Cd tolerance and detoxification in duckweed (*Landoltia punctata*). *Ecotoxicol. Environ. Saf.* **2021**, *207*, 111553. [CrossRef]
21. Murphy, A.; Taiz, L. Comparison of metallothionein gene expression and nonprotein thiols in ten Arabidopsis ecotypes. Correlation with copper tolerance. *Plant Physiol.* **1995**, *109*, 945–954. [CrossRef]
22. Wang, W.; Haberer, G.; Gundlach, H.; Gläßer, C.; Nussbaumer, T.; Luo, M.C.; Lomsadze, A.; Borodovsky, M.; Kerstetter, R.A.; Shanklin, J.; et al. The *Spirodela polyrhiza* genome reveals insights into its neotenous reduction fast growth and aquatic lifestyle. *Nat. Commun.* **2014**, *5*, 3311. [CrossRef]
23. An, D.; Zhou, Y.; Li, C.; Xiao, Q.; Wang, T.; Zhang, Y.; Wu, Y.; Li, Y.; Chao, D.Y.; Messing, J.; et al. Plant evolution and environmental adaptation unveiled by long-read whole-genome sequencing of Spirodela. *Proc. Natl. Acad. Sci. USA* **2019**, *116*, 18893–18899. [CrossRef] [PubMed]
24. Harkess, A.; McLoughlin, F.; Bilkey, N.; Elliott, K.; Emenecker, R.; Mattoon, E.; Miller, K.; Czymmek, K.; Vierstra, R.D.; Meyers, B.C.; et al. Improved *Spirodela polyrhiza* genome and proteomic analyses reveal a conserved chromosomal structure with high abundance of chloroplastic proteins favoring energy production. *J. Exp. Bot.* **2021**, *72*, 2491–2500. [CrossRef] [PubMed]
25. Michael, T.P.; Bryant, D.; Gutierrez, R.; Borisjuk, N.; Chu, P.; Zhang, H.; Xia, J.; Zhou, J.; Peng, H.; El Baidouri, M.; et al. Comprehensive definition of genome features in *Spirodela polyrhiza* by high-depth physical mapping and short-read DNA sequencing strategies. *Plant J.* **2017**, *89*, 617–635. [CrossRef] [PubMed]
26. Hoang, P.N.T.; Michael, T.P.; Gilbert, S.; Chu, P.; Motley, S.T.; Appenroth, K.J.; Schubert, I.; Lam, E. Generating a high-confidence reference genome map of the Greater Duckweed by integration of cytogenomic, optical mapping, and Oxford Nanopore technologies. *Plant J.* **2018**, *96*, 670–684. [CrossRef]
27. Park, H.; Park, J.H.; Lee, Y.; Woo, D.U.; Jeon, H.H.; Sung, Y.W.; Shim, S.; Kim, S.H.; Lee, K.O.; Kim, J.-Y.; et al. Genome of the world's smallest flowering plant, *Wolffia australiana*, helps explain its specialized physiology and unique morphology. *Commun. Biol.* **2021**, *4*, 900. [CrossRef]
28. An, D.; Li, C.; Zhou, Y.; Wu, Y.; Wang, W. Genomes and transcriptomes of duckweeds. *Front. Chem.* **2018**, *6*, 230. [CrossRef]
29. Guo, W.-J.; Bundithya, W.; Goldsbrough, P.B. Characterization of the Arabidopsis metallothionein gene family: Tissue-specific expression and induction during senescence and in response to copper. *New Phytol.* **2003**, *159*, 369–381. [CrossRef]
30. Kawashima, I.; Kennedy, T.D.; Chino, M.; Lane, B.G. Wheat *Ec* metallothionein genes. Like mammalian Zn^{2+} metallothionein genes, wheat Zn^{2+} metallothionein genes are conspicuously expressed during embryogenesis. *Eur. J. Biochem.* **1992**, *209*, 971–976. [CrossRef]
31. Ren, Y.; Liu, Y.; Chen, H.; Li, G.; Zhang, X.; Zhao, J. Type 4 metallothionein genes are involved in regulating Zn ion accumulation in late embryo and in controlling early seedling growth in Arabidopsis. *Plant Cell Environ.* **2012**, *35*, 770–789. [CrossRef]
32. Fourounjian, P.; Slovin, J.; Messing, J. Flowering and seed production across the Lemnaceae. *Int. J. Mol. Sci.* **2021**, *22*, 2733. [CrossRef]
33. Zhou, J.; Goldsbrough, P.B. Structure, organization and expression of the metallothionein gene family in Arabidopsis. *Mol. Gen. Genet* **1995**, *248*, 318–328. [CrossRef] [PubMed]
34. Xu, X.; Duan, L.; Yu, J.; Su, C.; Li, J.; Chen, D.; Zhang, X.; Song, H.; Pan, Y. Characterization analysis and heavy metal-binding properties of CsMTL3 in *Escherichia coli*. *FEBS Open Bio* **2018**, *8*, 1820–1829. [CrossRef] [PubMed]
35. Schiller, M.; Hegelund, J.N.; Pedas, P.; Kichey, T.; Laursen, K.H.; Husted, S.; Schjoerring, J.K. Barley metallothioneins differ in ontogenetic pattern and response to metals. *Plant Cell Environ.* **2014**, *37*, 353–367. [CrossRef] [PubMed]
36. Hsieh, H.M.; Liu, W.K.; Huang, P.C. A novel stress-inducible metallothionein-like gene from rice. *Plant Mol. Biol.* **1995**, *28*, 381–389. [CrossRef]
37. Pakdee, O.; Songnuan, W.; Panvisavas, N.; Pokethitiyook, P.; Yokthongwattana, K.; Meetam, M. Functional characterization of metallothionein-like genes from *Physcomitrella patens*: Expression profiling, yeast heterologous expression, and disruption of *PpMT1.2a* gene. *Planta* **2019**, *250*, 427–443. [CrossRef]
38. Hegelund, J.N.; Schiller, M.; Kichey, T.; Hansen, T.H.; Pedas, P.; Husted, S.; Schjoerring, J.K. Barley metallothioneins: MT3 and MT4 are localized in the grain aleurone layer and show differential zinc binding. *Plant Physiol.* **2012**, *159*, 1125–1137. [CrossRef]
39. Wong, H.L.; Sakamoto, T.; Kawasaki, T.; Umemura, K.; Shimamoto, K. Down-regulation of metallothionein, a reactive oxygen scavenger, by the small GTPase OsRac1 in rice. *Plant Physiol.* **2004**, *135*, 1447–1456. [CrossRef]
40. Xue, T.; Li, X.; Zhu, W.; Wu, C.; Yang, G.; Zheng, C. Cotton metallothionein GhMT3a, a reactive oxygen species scavenger, increased tolerance against abiotic stress in transgenic tobacco and yeast. *J. Exp. Bot.* **2009**, *60*, 339–349. [CrossRef]

41. Kumar, S.; Stecher, G.; Li, M.; Knyaz, C.; Tamura, K. MEGA X: Molecular evolutionary genetics analysis across computing platforms. *Mol. Biol. Evol.* **2018**, *35*, 1547–1549. [CrossRef]
42. Mumberg, D.; Muller, R.; Funk, M. Yeast vectors for the controlled expression of heterologous proteins in different genetic backgrounds. *Gene* **1995**, *156*, 119–122. [CrossRef]

plants

Article

Species- and Metal-Specific Responses of the Ionome of Three Duckweed Species under Chromate and Nickel Treatments

Viktor Oláh [1,*], Muhammad Irfan [1], Zsuzsanna Barnáné Szabó [1], Zsófi Sajtos [2], Ágota Zsófia Ragyák [2,3], Boglárka Döncző [4], Marcel A. K. Jansen [5], Sándor Szabó [6] and Ilona Mészáros [1]

1 Department of Botany, Institute of Biology and Ecology, Faculty of Science and Technology, University of Debrecen, Egyetem Square 1, H-4032 Debrecen, Hungary
2 Atomic Spectroscopy Partner Laboratory, Department of Inorganic and Analytical Chemistry, Faculty of Science and Technology, University of Debrecen, Egyetem Square 1, H-4032 Debrecen, Hungary
3 Doctoral School of Chemistry, University of Debrecen, Egyetem Square 1, H-4032 Debrecen, Hungary
4 Institute for Nuclear Research (ATOMKI), Bem tér 18/c, H-4026 Debrecen, Hungary
5 School of Biological, Earth and Environmental Science, University College Cork, Distillery Fields, North Mall, T23N73K Cork, Ireland
6 Department of Biology, University of Nyiregyhaza, H-4401 Nyiregyhaza, Hungary
* Correspondence: olahviktor@unideb.hu

Abstract: In this study, growth and ionomic responses of three duckweed species were analyzed, namely *Lemna minor*, *Landoltia punctata*, and *Spirodela polyrhiza*, were exposed for short-term periods to hexavalent chromium or nickel under laboratory conditions. It was found that different duckweed species had distinct ionomic patterns that can change considerably due to metal treatments. The results also show that, because of the stress-induced increase in leaf mass-to-area ratio, the studied species showed different order of metal uptake efficiency if plant area was used as unit of reference instead of the traditional dry weight-based approach. Furthermore, this study revealed that µXRF is applicable in mapping elemental distributions in duckweed fronds. By using this method, we found that within-frond and within-colony compartmentation of metallic ions were strongly metal- and in part species-specific. Analysis of duckweed ionomics is a valuable approach in exploring factors that affect bioaccumulation of trace pollutants by these plants. Apart from remediating industrial effluents, this aspect will gain relevance in food and feed safety when duckweed biomass is produced for nutritional purposes.

Keywords: duckweed; metal accumulation; ionomics; ICP-OES; micro-XRF

check for updates

Citation: Oláh, V.; Irfan, M.; Szabó, Z.B.; Sajtos, Z.; Ragyák, Á.Z.; Döncző, B.; Jansen, M.A.K.; Szabó, S.; Mészáros, I. Species- and Metal-Specific Responses of the Ionome of Three Duckweed Species under Chromate and Nickel Treatments. *Plants* **2023**, *12*, 180. https://doi.org/10.3390/plants12010180

Academic Editor: Ferenc Fodor

Received: 22 November 2022
Revised: 25 December 2022
Accepted: 29 December 2022
Published: 1 January 2023

1. Introduction

Duckweeds (Lemnaceae) form a family of small aquatic monocots that usually inhabit the surface zone of slowly moving freshwaters. Their evolutionary adaptation to lentic habitats resulted in specific traits, such as a small, thallus-like body (frond) that displays a much-reduced anatomy [1]. Fronds lack a distinct shoot system and only have adventitious roots. It has long been argued that duckweed roots, and the root-to-shoot transport system, play less of a role in nutrient acquisition compared to most other plant species. Rather, fronds can absorb nutrients through their abaxial (lower) epidermis [2]. Another duckweed feature is the extremely rapid vegetative growth, with 2–3 days-long doubling time under suitable conditions [3]. The high growth rate involves efficient uptake and utilization of nutrients. Duckweeds have been reported to successfully remove different nitrogen forms and phosphorus from various types of waste waters [4,5]. The produced duckweed biomass has a favorable composition with e.g., high protein, starch and fiber content [6]. These traits make duckweeds potent candidate crops for use as biofuel, feed, or food in circular economic approaches [4,7]. In addition, duckweeds are also known to efficiently incorporate many trace elements, and this can potentially make these plants suitable for the remediation of industrial effluents. For example, duckweeds have high accumulation rates for arsenic, boron, cadmium, chromium, copper, manganese, nickel, and zinc [8–12].

When duckweed-based applications are considered, inherent traits of these plants should be taken into account. One of these traits is the high genetic diversity within the family. Thus, although these plants with their reduced morphology may look similar to one another, there is substantial interspecific diversity in traits such as growth potential, biomass composition, metal uptake rates, and/or stress tolerance [5,13,14]. Therefore, selection and matching of the most suitable species for any given application can be crucial. For example, pertaining to metal accumulation, the high affinity of a particular species to a particular trace element can be either advantageous in remediating polluted waters, e.g., [15], or a disadvantage if duckweed is produced for use in feed or food, e.g., [16]. Further, a high metal bioaccumulation rate may not involve high tolerance to the same element that potentially limit the usefulness of certain species in some applications. To even further complicate the situation, besides genotypic differences, mineral acquisition by plants is also influenced by interactions amongst different chemical elements in the medium that may lead to, for instance, uptake competition. Plant ionomics is quantitative and simultaneous analysis of multiple elements in the biomass and is an efficient approach to study the functional state of plants under different conditions [17]. This aspect has, however, been less considered in previous duckweed bioaccumulation studies.

Another consideration for bioremediation applications is that duckweeds are highly plastic plants that will respond to environmental stimuli by rapid phenotypic modulation. In response to changing ambient conditions, duckweeds can acclimate by altering e.g., the frond size or the concentration and ratio of photosynthetic pigments [18,19]. A typical plant response to suboptimal conditions is the increase of dry matter content and leaf (in case of duckweeds frond) mass-to-area ratio (LMA) [20–22]. In duckweeds, this response is rapid and may be attributed to several mechanisms, such as disturbed frond development and expansion [23], regulated modulation of frond development [20], or the rapid accumulation of starch in fronds [24]. Any of these mechanisms may trigger a stress-induced increase in frond LMA, and this will be inherently accompanied by slower growth in terms of the expansion of frond area. This can result in seemingly divergent growth rates depending on whether the growth was calculated on dry mass or surface area basis.

In the context of changing frond morphology, a further question is whether the acquired nutrients and/or metals are uniformly distributed within the fronds, or whether internal transport and redistribution lead to metal accumulation in particular frond regions. Previous reports showed that metal uptake was not uniform, and that the node-zone of fronds and roots showed higher uptake rates than other parts of the frond [25,26]. In the case of Cd-treated *Landoltia punctata*, this resulted in Cd accumulation primarily in the node and veins [27]. Thus far, studies have mostly analyzed metal accumulation at subcellular or tissue levels in duckweed fronds and roots [28,29]. Frond-level distribution is considered less often. Recently, various methods have become available to map elemental distributions in biological samples (e.g., histochemical assays, energy dispersive X-ray fluorescence microprobe—i.e., μXRF, and laser ablation plasma mass spectrometry—i.e., LA–ICP–MS). The thin, two-dimensional structure of fronds makes duckweeds particularly suitable for such analyses.

The aim of this study was to analyze the ionome of duckweed biomass in three species of Lemnaceae under either optimal growth conditions or following chromate or nickel treatments. More specifically, we tested (i) the species- and treatment-specific patterns in the ionome, (ii) whether the apparent order of species in metal accumulation efficiency was affected by the unit of reference (i.e., dry mass v. area), and (iii) if the within-frond distribution of the acquired Cr and Ni showed distinct patterns of accumulation or was rather uniform.

2. Materials and Methods

2.1. Culturing Conditions and Experimental Setup

Axenic stocks of the three studied duckweed species, *Spirodela polyrhiza* L. Schleid. (clone UD0401), *Landoltia punctata* (G. Meyer) Les & Crawford (Landolt clone #7760), and *Lemna minor* L. (clone UD0201) were maintained in 300 mL Erlenmeyer flasks on modified Steinberg medium (pH 6.0 ± 0.2, [30]) under constant temperature (24 ± 2 °C) and irradiation (PPFD 60 ± 10 µE m^{-2} s^{-1}, white) [31].

For the experimental work, healthy colonies of 7–8 days old stock cultures were used. The metal treatments were conducted in crystalizing dishes (80 mm diameter and 150 mL volume) and covered with plastic Petri dishes to reduce evaporative loss. Each vessel contained 100 mL of Steinberg medium either without supplemental metal (control), or with nominal added concentrations of 4 mg L^{-1} Cr(VI) ($K_2Cr_2O_7$ salt) or 2.5 mg L^{-1} Ni ($NiSO_4 \cdot 7\,H_2O$ salt), respectively. The applied metal concentrations were based on our previous work with the same *S. polyrhiza* clone [31] and aimed at resulting in significant growth inhibition without leading to frond mortality by the end of the experiments. For preparing the Steinberg medium and metal treatments, reagent grade chemicals and Type I ultrapure water were used.

The starting inoculum was 2.7 ± 0.6 cm^2 frond area per vessel for *S. polyrhiza;* 1.5 ± 0.4 cm^2 for *La. punctata* and 2.3 ± 0.7 cm^2 for *Le. minor*, corresponding to an initial biomass of 62.0 ± 21.9 mg FW (6.5 ± 1.3 mg DW) for *S. polyrhiza;* 58.1 ± 16.5 mg FW (5.0 ± 1.2 mg DW) for *La. punctata* and 71.1 ± 7.8 mg FW (6.3 ± 1.5 mg DW) for *Le. minor*.

The metal treatments lasted for 72 ± 2 h under ambient conditions identical to those used for stock culturing. Every treatment for each species was performed in triplicate and was repeated in two independent experiments.

2.2. Measurement of Growth

On the starting (0th) and final (3rd) day of experiments, digital images of each vessel were recorded by means of a custom-made photo hood and a PC-controlled camera (5 MP resolution) mounted in a perpendicular position to the surface of the cultures. Frond area (FA) in individual vessels was then determined from the obtained images by means of "Threshold colour" and "Analyze particles" functions of ImageJ image analysis software [32].

On the final day of experiments, the plants were harvested, thoroughly rinsed with Type I ultrapure water, and gently blotted dry with paper towels. The fresh (FW) and dry weight of plants (DW) in each vessel was determined using an analytical balance (Kern ABT 120-5DM). Prior to determination of DW, the plants were dried until constant weight (4 days at 65 °C).

We calculated the initial biomass using the frond areas in each vessel on the 0th day, and the average species-specific leaf mass-to-area ratios (LMA as mg DW cm^{-2}) and dry matter contents (DM%) of plants in the same stock cultures that were used for the experiments (n = 8 per species).

Based on the obtained biomass data, the following parameters were calculated:

$$\text{Dry matter content: DM\% (\%)} = (\text{DW}/\text{FW}) * 100$$

$$\text{Relative growth rate: } \text{RGR}_\text{x}\ (\text{day}^{-1}) = (\ln X_j - \ln X_i)/t$$

where X_i and X_j denote the respective parameter (FA, FW, DW) determined on the ith (0th) and jth (3rd) days of experiments, and t is the treatment duration (3) in days, according to [33].

$$\text{Growth inhibition: I\%} = ((\text{RGR}_\text{control} - \text{RGR}_\text{treated})/\text{RGR}_\text{control}) * 100$$

where $RGR_{treated}$ is the relative growth rate of a metal-treated culture and $RGR_{control}$ is the corresponding mean control relative growth rate, according to [33].

2.3. Determination of Metal Content in the Growth Medium and Biomass

For the analytical determination of metallic elements in the medium, 10 mL of medium from each vessel was filtered through a 100 μm pore size nylon mesh filter (SPL Life Sciences) on the 0th and 3rd days of the tests. On the 0th day, the samples were taken from the solutions prepared for control, Cr(VI) and Ni treatments in triplicates, while on the 3rd day from the media of each experimental vessel. The samples were immediately acidified with 2 drops of 65% (m/m) HNO_3 (reagent grade, Scharlau), and stored at room temperature prior to elemental analysis.

In analysis of biomass composition, 3–10 mg of the dried plant samples (see Section 2.2) was digested under atmospheric pressure in a mixture of 3.0 mL 65% (m/m) HNO_3 (reagent grade, Scharlau) and 1.0 mL 30% (m/m) H_2O_2 (reagent grade, Merck). Then, the digested extracts were transferred without loss into volume-calibrated plastic centrifuge tubes and diluted to a final volume of 5 mL with ultrapure water (Synergy UV Millipore). The solutions were stored at room temperature prior to further elemental analysis.

Metal concentration of medium and biomass samples were determined by means of inductively coupled plasma optical emission spectrometry (ICP-OES 5110 Vertical Dual View, Agilent Technologies, Santa Clara, CA, USA). An auto-sampler (Agilent SPS4), Meinhard® type nebulizer and double-pass spray chamber were used. The ICP-OES operating conditions and measurement parameters to determine the elemental concentrations of Al, B, Ba, Bi, Ca, Cd, Co, Cr, Cu, Fe, K, Li, Mg, Mn, Na, Ni, Pb, Sr, and Zn are provided in Tables S1 and S2. Standard solutions of the macro elements (Ca, K, Mg, and Na) were prepared from a mono-element spectroscopic standard of 1000 mg L^{-1} (Scharlau), while those of the micro elements (Al, B, Ba, Bi, Cd, Co, Cu, Cr, Fe, Li, Mn, Na, Ni, Pb, Sr, and Zn) were similarly prepared from the multi-element spectroscopic standard solution of 1000 mg L^{-1} (ICP IV, Merck). In both cases, a five-point calibration process was used, in which the standard solutions were diluted with 0.1 M HNO_3 prepared in ultrapure water.

Due to their low concentrations in both sample types (i.e., medium and biomass), Al, Ba, Bi, Cd, Co, Li, and Pb were excluded from the subsequent data processing.

The measured metal concentrations in the medium and biomass were used to calculate the metal- and species-specific bioconcentration factors (BCF) for the analyzed elements as follows:

$$BCF = Cc_{biomass}/Cc_{medium}$$

where $Cc_{biomass}$ and Cc_{medium} were the measured concentrations of a given element in the duckweed biomass (mg kg^{-1} DW) and in the medium (mg L^{-1}), respectively.

The measured Cr(VI) and Ni contents in the biomass were also transformed into metal content per unit frond area, as follows:

$$\text{metal content (mg m}^{-2}) = (Cc_{biomass} * DW)/(FA * 100)$$

where $Cc_{biomass}$ was the measured concentration of a given element in the duckweed biomass (mg kg^{-1} DW), DW was the harvested dry weight (mg), and FA is the frond area corresponding to the harvested biomass (cm^2)

2.4. mXRF Analyses

For the elemental mapping of metal distribution within fronds, the sample plants were subjected to the same Cr(VI) and Ni treatments and placed under the same ambient conditions as described in Section 2.1. Prior to μXRF scanning, the plants were transferred from metal-containing medium to 50 mL Type I ultrapure water for 10 min in order to remove excess metals adhering to the external surface of fronds and roots. Following that, the roots were carefully removed, and the plants were air-dried for 3 days, while being gently pressed to prevent deformation of fronds.

μXRF investigations were carried out using a Bruker M4 TORNADO Micro-XRF spectrometer (Bruker, Billerica, MA, USA) using a Rh-tube without any filter, at 50 kV accelerating voltage and 400 μA current. Characteristic X-ray lines were recorded by two energy dispersive detectors. Each of the two Be-window silicon drift detectors had a 30 mm^2 active area. The mapping was performed in 20 mbar vacuum. The beam diameter was focused to 20 μm by the built-in polycapillary lens. The recorded rectangular maps were acquired with 100 ms/pixel velocity and two accumulations. For QMap analysis, the M4 TORNADO software (version: 1.6.621.0) was used.

2.5. Data Processing and Statistics

To analyze species-specific and metal-induced growth and ionomic responses, respective data of the two independent experiments per species and per treatment were pooled, resulting in a sample size of $n = 6$ for each combination. Multiple comparisons were performed by means of Kruskal–Wallis test and, in case of significant differences, with *post hoc* Mann–Whitney pairwise comparisons of medians. The biomass ionomic dataset was also subjected to multivariate analysis by means of principal component analysis (PCA). All statistical analyses were performed by means of Past v4.0 [34], and for every analysis, <5% probability, that is, $p < 0.05$, was considered as statistically significant.

3. Results and Discussion

3.1. Metal-Induced Growth Inhibition

The observed trends indicate that, in general, the applied modified Steinberg medium can support normal growth of duckweeds. As expected, controls displayed good growth with RGR values ranging between 0.3 and 0.4 day^{-1} for every species and growth parameter. In general, the three species showed similar growth responses to Cr(VI) treatments, with a slightly larger frond area- and fresh weight-based growth inhibition for *S. polyrhiza* (I% = 70–77%) and a stronger dry weight-based growth inhibition (I% = 55%) for *Le. minor* (Figure 1). To Ni treatments, on the other hand, *S. polyrhiza* showed considerably greater sensitivity with almost arrested frond area (I% = 86%) and fresh weight growth (I% = 106%), while *La. punctata* proved to be slightly more tolerant (I% = 10–53% for the three growth parameters) than *Le. minor* (I% = 27–63% for the three growth parameters, Figure 1). It has previously been reported that different duckweed species may have different sensitivity to metals [35,36], and the same order of Ni-sensitivity amongst the three species was reported by Xyländer and Augsten [37]. Similarly, Appenroth et al. [38] found *S. polyrhiza* to be more Ni-sensitive than *Le. minor*. Thus, experimental data support our finding that *S. polyrhiza* is highly sensitive to Ni.

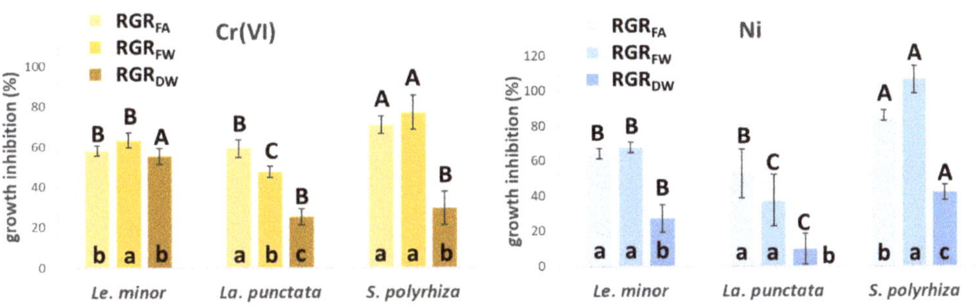

Figure 1. Cr(VI)- (**left**), and Ni-induced inhibition (**right**) of relative growth rates (RGR) derived from different growth parameters (frond area—FA, fresh weight—FW, dry weight—DW). The inhibition was compared to the respective species-specific control growth rates. Means ± SD of $n = 6$ samples; different lower cases at the bottom of bars denote significantly different ($p < 0.05$) inhibition of growth parameters with respect to the given species (*Le. minor*, *La. punctata* or *S. polyrhiza*) and treatment, while different capitals on top of the bars indicate significantly different ($p < 0.05$) growth inhibition of species based on the same growth parameter.

In general, RGR_{FA}- and RGR_{FW}-based growth inhibitions showed high similarity to each other (Figure 1). In contrast, I% based on RGR_{DW} indicated considerably lower growth decline in most cases than the former parameters. The reason for the diverging responses of RGR_{DW} versus RGR_{FA} and RGR_{FW} is in part the altered phenotype of newly formed fronds. Metal-induced disorders and morphogenic responses in frond expansion are known to increase LMA and DM% in duckweeds [23,38]. In *Le. minor*, DM% increased by 24 (Cr(VI) and 76% (Ni), in *La. punctata*, by 25 (Cr(VI) and 38% (Ni), and in *S. polyrhiza*, by 41 (Cr(VI) and 58% (Ni) in just 3 days (Table 1). This increase in DM%, to some extent, diminished the metal-induced decline in growth of dry matter of cultures. Apart from morphogenic modifications, elevated DM% can also be attributed to the accumulation of starch [24]. Rearranged carbon utilization to produce starch instead of new frond area is common amongst duckweeds when exposed to trace metals [13,39]. Ni proved to be particularly efficient in enhancing starch accumulation in duckweeds [40], and chromate is also known to have a similar effect [23].

Table 1. Dry matter content (DM%) of the studied duckweed species on the 3rd day of the applied metal treatments. Means ± SD of $n = 6$ samples; different capitals denote significantly different ($p < 0.05$) median DM% of species with respect to the given treatment (control, Cr(VI) and Ni), while different lower cases indicate significantly different ($p < 0.05$) median DM% of the treatments with respect to the given species.

	Control	Cr(VI)	Ni
Le. minor	6.70 ± 0.53 ^{Bc}	8.33 ± 0.19 ^{Cb}	11.79 ± 0.24 ^{Aa}
La. punctata	7.26 ± 0.24 ^{Bb}	9.11 ± 0.34 ^{Ba}	10.02 ± 0.35 ^{Ba}
S. polyrhiza	8.36 ± 1.01 ^{Ac}	11.78 ± 0.96 ^{Ab}	13.20 ± 1.09 ^{Aa}

3.2. Changes in the Composition of the Nutrient Medium

In general, the concentrations of most analyzed elements in the medium did not change much under control (no metal supplementation) conditions over the test duration (Figure 2, the actual concentrations are supplemented in Table S3). However, a substantial drop was found in the concentration of Mn, which had decreased by 44% (*S. polyrhiza*), 45% (*Le. minor*), and 55% (*La. punctata*) by the end of tests (Figure 2). In Cr(VI) treatments, the medium composition changed even less compared to that of the control (Figure 2). In fact, the concentration of most elements (B, Ca, Cu, K, Mg, and Sr) slightly increased, probably due to the evaporative loss. Similarly, chromium level stayed similar (101%) to its original concentration (Table S3). In Ni treatments, the concentration of B in the medium decreased strongly in the case of *Le minor* (−39%), while it stayed unchanged (99–101%) in case of the other two species (Figure 2). The other element that showed considerable depletion was Zn with a 27% (*S. polyrhiza*) to 46% (*La. punctata*) decrease in concentration compared to the initial content. The concentration of the supplied Ni decreased only marginally (by 2–3%) by the end of treatments (Table S3).

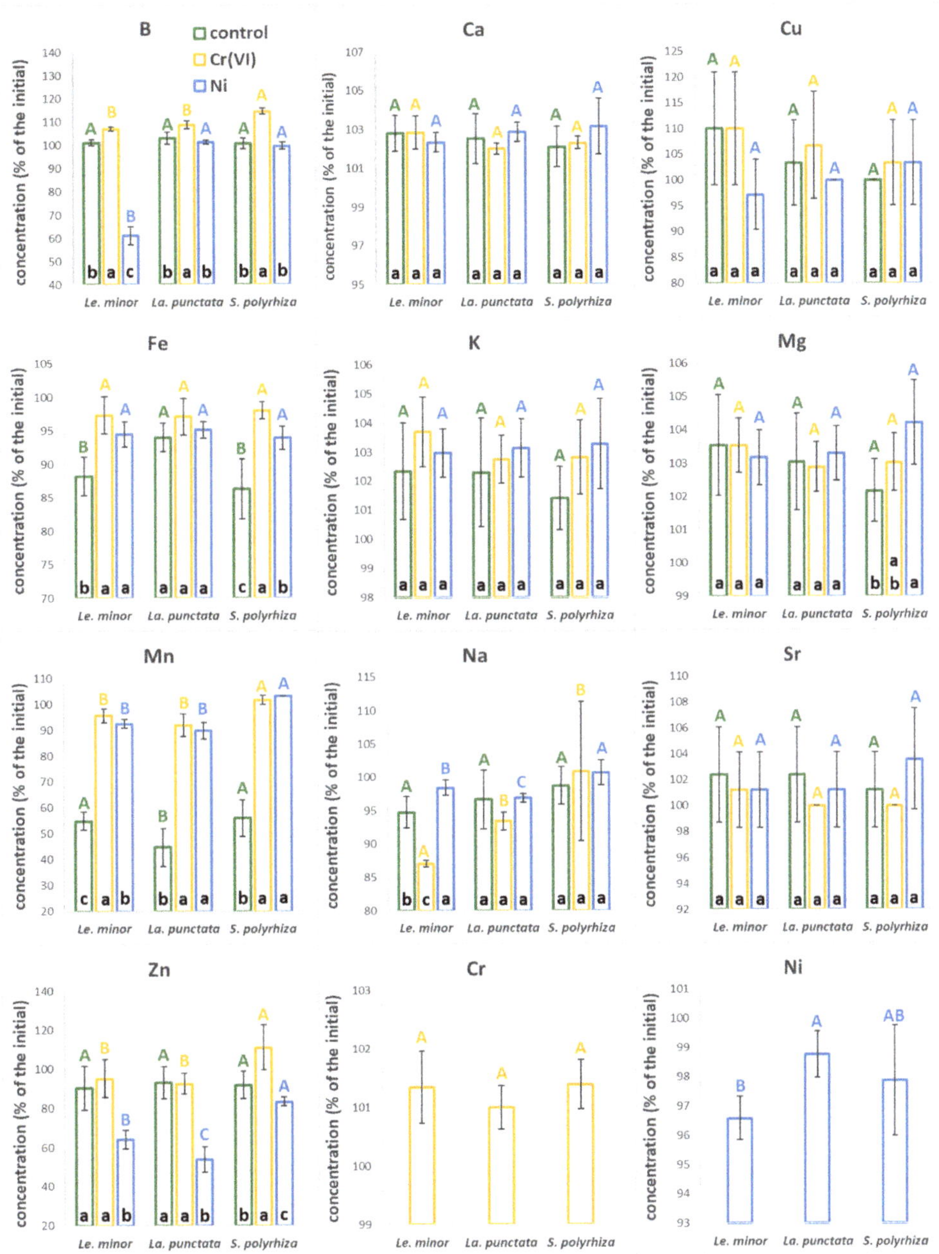

Figure 2. Ionic composition of the applied nutrient medium at the end (3rd day) of the experiments in control (green), Cr(VI) (yellow) and Ni treatments (blue). The data denote percentile concentrations as compared to the initial medium composition on the 0th day of the treatments. Means ± SD of $n = 6$ samples are shown. Different capitals with corresponding colors indicate significantly different ($p < 0.05$) medians of species across the same treatment. Different lower cases in the bottom of bars indicate significantly different ($p < 0.05$) medians of different treatments in case of the same species. The measured ionic concentrations are provided in Table S3.

Duckweeds have been reported as efficient accumulators of Mn [41]; and the fact that ~50% of this element was taken up by the control plants in just 3 days, points to the possible depletion from the medium during extended longer-term experiments under static conditions. Yet, Mn deficiencies are somewhat unlikely as duckweed colonies seem to have the ability to transport and redistribute metallic elements among and within fronds (see later in Section 3.5). Such an internal translocation may counterbalance the decreasing Mn supply from a depleting medium, even in Steinberg medium that represents the lower end in terms of Mn-content amongst the most frequently applied duckweed media [30]. Interestingly, both Cr(VI) and Ni treatments disrupted the enhanced uptake of Mn from the medium. Ni is known to affect Mn acquisition of both crops and hyperaccumulator plants by competing for uptake [42,43]. Cr(VI), however, forms oxyanions, and thus enters plants via different routes [44]. Previous literature is not consistent and report both synergistic and antagonistic effects of chromate treatments on Mn content of plants [45,46]. Park [44] suggested that Cr(III) can reduce Mn in the soil thus increasing its bioavailability. Reversing this logic, it can be speculated that Cr(VI) in turn might oxidize Mn, limiting its uptake by plants. Facilitated removal of B and Zn by *Le. minor* and *La. punctata* under Ni treatments indicates that wider nutrient uptake patterns of plants can be affected following metal treatment. A synergistic increase in the uptake of Zn in the presence of Ni was previously explained by up-regulated Zn-transporters in response to a virtual Zn-deficiency due to the substitution of this metal by Ni at functional binding sites, or by a disrupted sensing machinery for Zn uptake [47].

3.3. *Biomass Ionomic Composition*

The plant ionome is regulated by both taxonomic and environmental factors, and multi-elemental analyses can provide important additional information on the nutrient uptake preferences or functional state of the studied organisms [17]. Previously it has been demonstrated, as an example, that co-occurring, closely related Ericaceae species had distinct ionomic profiles that may be considered as a strategy to alleviate competition in densely populated habitats with high taxonomic diversity [48]. Considering duckweed mats as analogously "crowded" habitats, diverging nutrient preferences of species—or intraspecific genotypes—may thus be hypothesized supporting higher productivity of both natural and man-made (e.g., in constructed wetlands) duckweed polycultures.

According to the multi-elemental analysis, the first two principal components covered 55.4% of the overall variance in the ionomic dataset (Figure 3, the actual concentrations in the biomass and the comparative statistics are supplemented in Table S4). PC1 correlated strongly with Ni contents. Besides Ni, high positive correlations of this principal component were also found for B, Cu, and Zn (Table S5). Mg, Mn, and Na, in contrast, showed a weak negative correlation with PC1. Cr content of the biomass correlated positively with PC2, similarly to K (Figure 3). Ca and Sr, on the other hand, had negative correlations with that principal component (Table S5). The three studied duckweed species were also mostly separated along PC2 (Figure 3). This distribution suggests that genotype-specific differences in the biomass composition can be hypothesized even in these closely related species. From that aspect, *Le. minor* can be characterized as having higher K and Na contents than *La. punctata* and especially *S. polyrhiza* (Table S5). The latter species, on the other hand, had higher Ca and Sr contents. Being chemical analogues, the more efficient Ca uptake by *S. polyrhiza* compared to *La. punctata* and *Le. minor* can explain the higher Sr content in the biomass of this species as well [49].

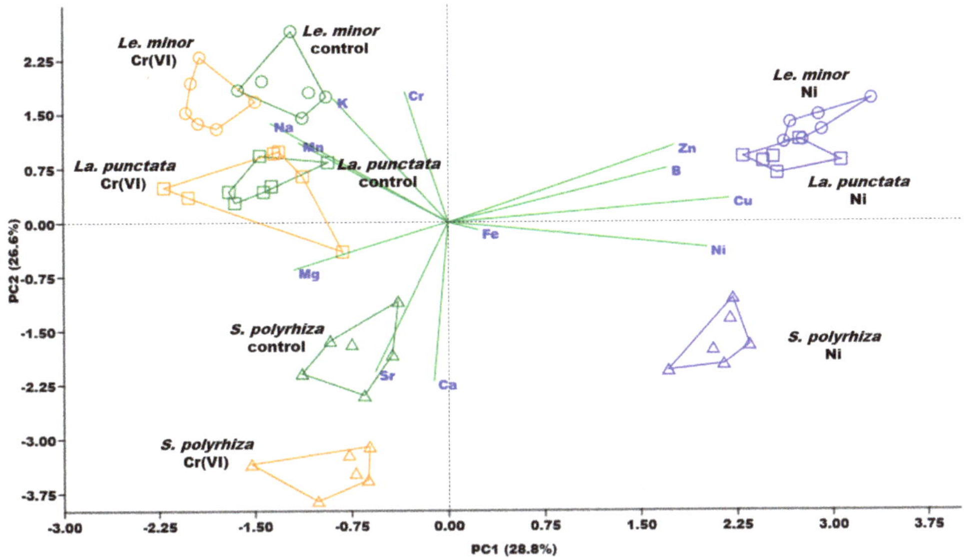

Figure 3. Overall patterns in the biomass ionomic composition of the studied duckweed species according to the first two principal components (PC1 and PC2, respectively) of the performed principal component analysis. Polygons delineate $n = 6$ samples of different treatments (control—green; 4 mg L^{-1} Cr(VI)—orange; or 2.5 mg L^{-1} Ni—blue) of the three species (*Le. minor*—circles, *La. punctata*—squares, *S. polyrhiza*—triangles).

The calculated bioconcentration factors (BCF) under control conditions indicate that the three tested duckweed species accumulated moderate concentrations of K, Na, Ca, and Mg, with BCF values in the range of ~130–1200 (Figure 4). The relatively high BCFs of the non-essential element Na can be explained by its low ambient concentration in the medium. However, the studied species seemed to accumulate Na differently (Table S4): *Le. minor* showed the highest BCF value (1159 ± 97), while *La. punctata* showed 24% lower (885 ± 89), and *S. polyrhiza* 67% showed lower (396 ± 118) bioaccumulation factors. Compared to Na, Ca, Mg, and K were more abundant in the medium and showed lower BCF values (~130–280 for Ca, ~170–230 for Mg and ~315–485 for K). As a result, those four elements comprised an average of 6.9 (*S. polyrhiza*), 7.2 (*La. punctata*), and 8.9% (*Le. minor*) of the dry biomass in the three species under control conditions (Table S4).

B, Cu, Fe, and Zn typically showed one order of magnitude higher range of BCF (~800–5700) compared to K, Na, Ca, and Mg, while the highest BCF in all the tested species was calculated for Mn (BCF = ~7500–15000, Figure 4). Duckweeds have long been known to efficiently accumulate various trace elements [2,50], and particularly high bioaccumulation capability was reported, amongst others, for B, Cu, Zn, and Mn [8,11,15,41,51–53]. Our data are in line with those observations and indicate that in the applied Steinberg medium, Mn was by far the most efficiently incorporated essential metal (Figure 4 and Table S4), explaining its rapid decline in the medium.

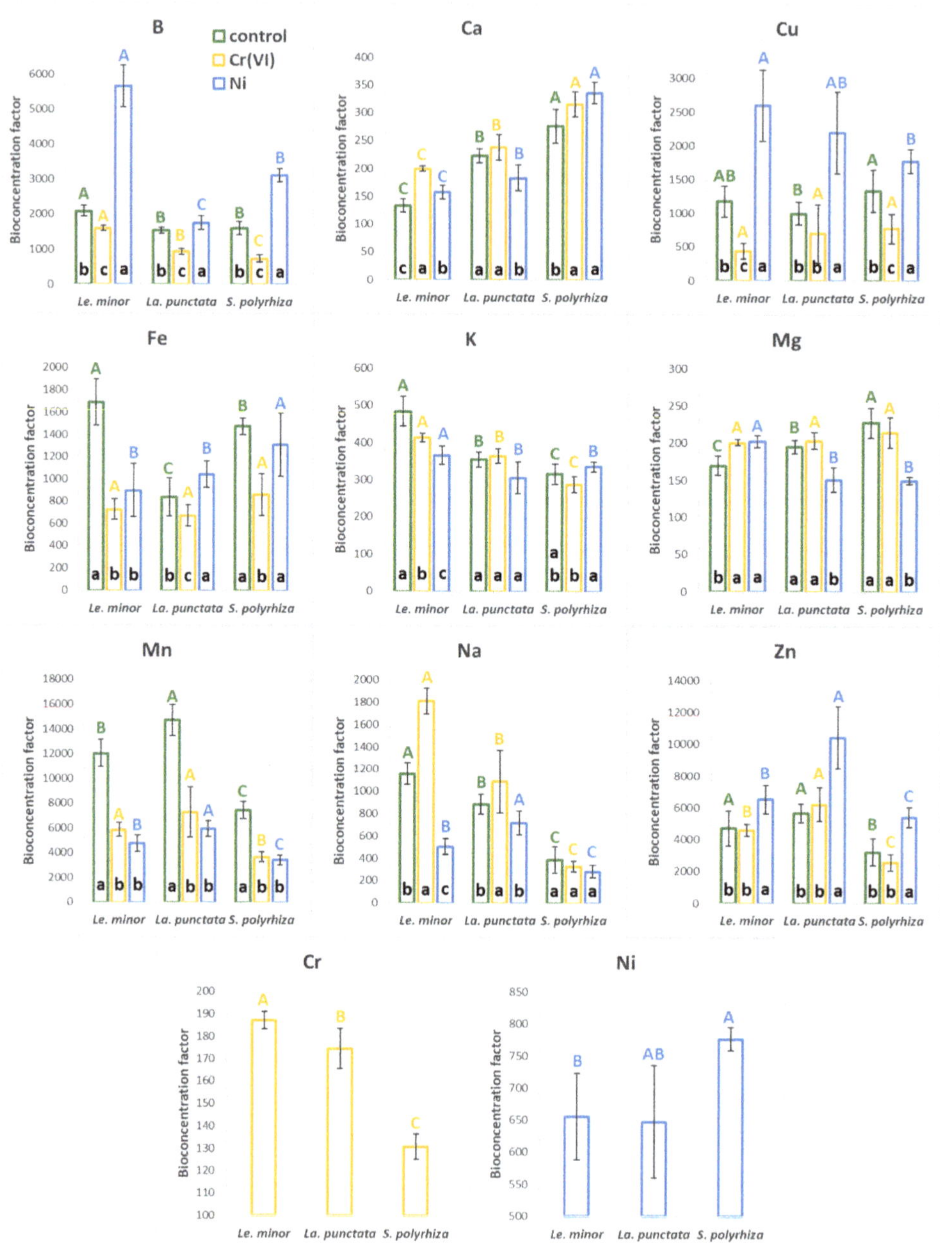

Figure 4. The calculated bioconcentration factors of the analyzed elements under control (green), Cr(VI) (orange), and Ni treatments (blue), in the three duckweed species. Means ± SD of $n = 6$ samples are shown. Different capitals with corresponding colors indicate significantly different ($p < 0.05$) medians of species across the same treatment. Different lower cases in the bottom of bars indicate significantly different ($p < 0.05$) medians of different treatments in case of the same species. Note: Results for Sr are not presented because the concentration of this element was below the detection limit in several samples. The original elemental concentrations have been supplemented in Table S4.

In the Cr(VI) treatment, Cr was accumulated with a relatively low efficiency; its BCF ranged from ~130 (*S. polyrhiza*) to ~190 (*Le. minor*). Similarly, in the Ni treatment, Ni had a fairly low BCF (~650–780), lower compared to e.g., Cu and Fe (Figure 4). Relatively low bioaccumulation factors can be explained by the high abundance of Cr(VI) and Ni in the medium. BCF in general tends to decrease with increasing ambient concentrations of the respective element due to homeostatic or protective regulation of uptake and cellular concentration [54,55]. The calculated BCFs fit well in the range reported previously for duckweeds under comparable ambient concentrations. Literature data delineate the typical BCF range in duckweeds as ~50–300 for Cr(VI) [9,56,57], and ~100–670 for Ni [10,52,58]. Thus, one can hypothesize that these BCFs are characteristic to duckweeds in general. Comparing the two applied metals, our results are in line with literature data and suggest higher affinity of duckweeds for Ni than for Cr(VI), and this can be partly explained by the essentiality of the former element.

As the PCA indicated, the exposure to and uptake of Cr(VI) and Ni altered the mineral composition of the biomass in terms of other elements. In Ni treatments, BCF of B, Cu, and Zn increased by 30–170% compared to the control, while that of Mn was reduced by 54–60% in all three species (Figure 4). In Cr(VI) treatments, on the other hand, BCF of B, Cu, Fe, and Mn showed a generally decreasing trend with a 20–72% drop. Meanwhile, BCF of Ca slightly increased (+7–50%) due to Cr(VI) treatment as compared to control. Changes in the bioaccumulation of Na were species-specific, rather than metal-specific (Figure 4).

It should be noted that even smaller changes in the biomass ionic concentrations of elements strongly affected the relative ratios of elements in the biomass of different species. For example, Cr(VI) treatment increased the Ca/Mg ratio from 4.2 ± 0.1 to 5.4 ± 0.2 (+27%) in *Le. minor* and from 6.5 ± 0.3 to 8.0 ± 0.5 (+22%) in *S. polyrhiza* compared to control, while in *La. punctata* it stayed similar (6.3 ± 0.5, i.e., +3%) to control (6.2 ± 0.1). The Ca/Mg ratio increased by 84% (to 12.0 ± 0.5) in *S. polyrhiza* due to Ni treatment, while it did not change significantly in the other two species (4.2 ± 0.3 and 6.5 ± 0.2 in *Le. minor* and *La. punctata*). In turn, this may have consequences for growth, as the Ca/Mg ratio is an important determinant of plant health [59]. K/Na ratios showed opposite trends in case of Cr(VI) and Ni, mostly due to the changes in Na-uptake (Figure 4). Under control conditions, this ratio was 107.3 ± 7.3 in *Le. minor*, 103.6 ± 9.2 in *La punctata*, and 226.9 ± 69.2 in *S. polyrhiza*. Treatments with Cr(VI) decreased the K/Na ratio to 36.4 ± 3.1 (-66%), to 54.8 ± 7.5 (-47%) and to 141.9 ± 19.5 (-37%) in *Le. minor*, *La. punctata* and *S. polyrhiza*, respectively. Ni, on the other hand, increased this ratio to 185.6 ± 16.4 (+73%) and to 311.5 ± 60.9 (+37%) in *Le. minor* and *S. polyrhiza*, while *La. punctata* was not affected significantly (109.4 ± 15.3, +5.5%, $p > 0.05$). Similarly to the Ca/Mg ratio, K/Na balance in plant tissues may play an important role, especially under saline conditions. Being the most abundant cation in plant cells, controlled uptake and accumulation of K^+ are crucial in many physiological processes, including osmoregulation, enzyme activities and membrane polarization [60].

3.4. Cr and Ni Accumulation

The dry mass-based metal contents indicated that most Cr was accumulated by *Le. minor* (Figure 5 and Table S4). Compared to this species, *La. punctata* and *S. polyrhiza* accumulated 7 and 30% less Cr after 3 days of exposure. The most Ni on a dry weight basis was accumulated by *S. polyrhiza*, while the two other duckweed species contained 16–17% less of this metal by the end of the Ni-treatments (Figure 5 and Table S4). The frond area-based metal contents showed a different order of the species. On this basis, *La. punctata* proved to have the highest Cr-and Ni-contents at the end of the metal treatments, though in the latter case it was not significantly higher than that of *S. polyrhiza* (Figure 5). *Le. minor* accumulated 28% less Cr and 15% less Ni on area basis, while *S. polyrhiza* accumulated 32 and 6% less Cr and Ni after 3 days of exposure, respectively. These results highlight that the apparent interspecific differences in metal accumulation can not only be attributed to the mere efficiency of uptake mechanisms (e.g., membrane transport, compartmentation), but also depend strongly on the biomass basis used for quantification.

This can be explained partly by inherent differences in the species-specific LMA. In addition, the observed interspecific differences suggest the influence of species-specific changes in DM% and LMA due to metal stress. This way, stress-induced morphogenic responses and increases in starch content can bias the apparent metal uptake efficiency. Therefore, it is advisable to extend the scope of ecotoxicological studies to the area-based metal accumulation capability, and this is particularly important when duckweeds are considered as bioremediating agents [12,53]. Such an approach is widely used when nutrient removal by duckweeds is studied [4,61], but typically lacks in studies that focus on potential reclamation of metal-contaminated waters.

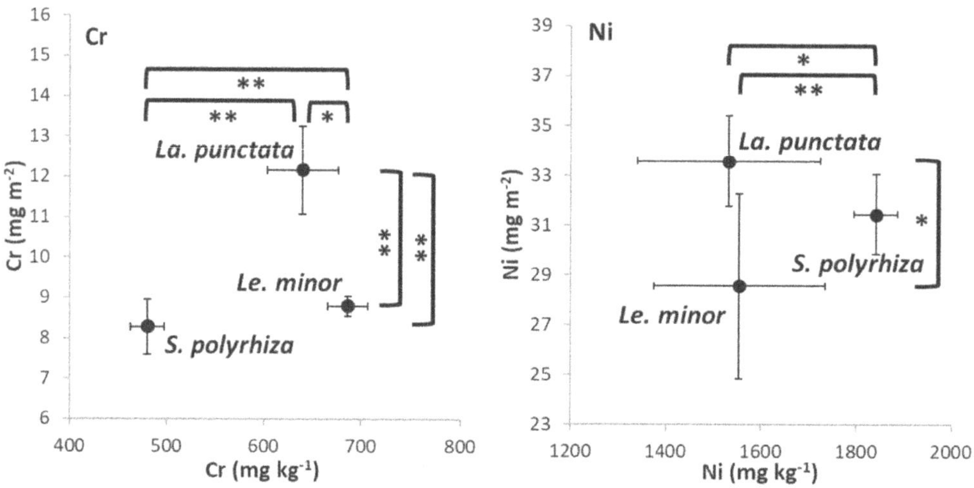

Figure 5. Cr (**left**) and Ni (**right**) contents in the biomass of the three tested duckweed species, expressed on dry weight- (x-axes) and area basis (y-axes). Means ± SD of $n = 6$ samples are shown. Asterisks denote significant differences between the respective species at * $p < 0.05$ and ** $p < 0.01$ probability levels according to the performed Kruskal–Wallis test and *post hoc* Mann–Whitney comparisons.

3.5. Within-Frond Distribution of Cr and Ni

The obtained elemental maps show that μXRF is suitable for the study of internal translocation and distribution of various metals in duckweeds (Figure 6). Considering the popularity of these plants in impact studies and water reclamation, this method may offer a better understanding of the interactions between the duckweed frond and its environment. The μXRF elemental maps revealed that the two metals studied distributed differently within the fronds. Cr primarily accumulated in the veins; the nodes contained particularly large amounts of this metal in all three species (Figure 6). Besides that, frond margins were areas of enhanced Cr accumulation too, especially in *Le. minor* and *S. polyrhiza*. In addition, it should be noted that in these two species, the mesophyll also contained detectable amounts of Cr. This was especially the case in *Le. minor*. In *La. punctata*, on the other hand, Cr accumulation was almost exclusively limited to the vascular tissues and the frond edges (Figure 6). In contrast with Cr, Ni distributed rather evenly in the mesophyll of fronds, with slightly higher concentrations in the frond margins. In *S. polyrhiza*, the node also contained a higher concentration of Ni. In *Le. minor* and *La. punctata*, on the other hand, basal regions of fronds, especially those of daughter fronds, contained elevated Ni contents (Figure 6).

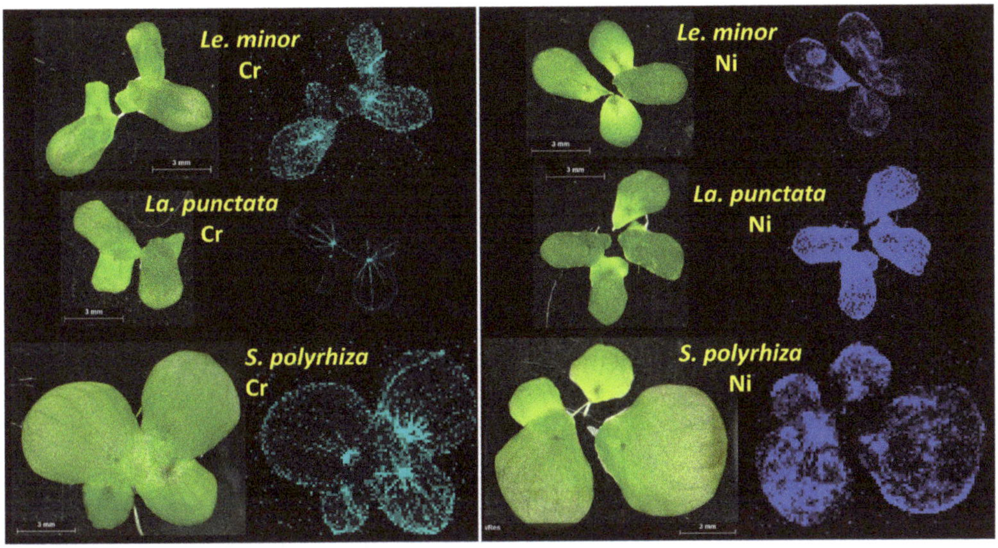

Figure 6. The distribution of Cr (**left**) and Ni (**right**) within the fronds of different duckweed species according to the μXRF maps. The color intensity is proportional to the elemental concentration of Cr (light blue) and Ni (blue). Note: the contrast in elemental maps was digitally enhanced after scanning.

Analogous patterns in terrestrial plants, that is, interveinal chlorosis, damaged leaf margins and tips [62,63], support our finding that vascular tissues play major role in internal translocation of Cr(VI) causing localized symptoms. Ni, on the other hand, is known to get easily transported via both the xylem and phloem, and to predominantly accumulate in the younger tissues [64]. This behavior explains well the observed accumulation of Ni in the basal parts of daughter fronds, where the cell maturation is still in process.

4. Conclusions

This study confirmed that different duckweed species have distinct ionomes, and that ionomic patterns can change considerably due to metal treatments. Duckweed ionomics, therefore, is a valuable approach in exploring factors that affect bioaccumulation of trace pollutants by these plants. Apart from remediating industrial effluents, this aspect may also gain relevance in food and feed safety when duckweed biomass is produced for dietary purposes.

The results also point to the importance of assessing metal accumulating potential on plant area basis. Stress-induced increases in dry matter content and the parallel decrease in horizontal growth can influence the calculated metal uptake efficiency that is traditionally calculated on a dry weight basis. Since the duckweed mat is a rather two-dimensional matrix for wastewater treatment, plant area-based metal accumulation can be at least as relevant as the biomass-based one.

In addition to the mere amount of trace metals removed by duckweed colonies, tracking the internal fate of metallic ions is also important. Our study revealed that within-frond and within-colony compartmentation of metal were strongly metal- and in part species-specific. These results also proved that μXRF can become a useful tool in mapping elemental distributions in duckweed fronds.

Supplementary Materials: The following is available online at https://www.mdpi.com/article/10.3390/plants12010180/s1, Table S1: ICP-OES settings I., Table S2: ICP-OES settings II., Table S3: Concentration of the analyzed elements in the Steinberg medium on the first and last day of heavy metal treatments, Table S4: Concentrations of the analyzed elements in the biomass of the three tested duckweed species, Table S5: Correlations of the corresponding elements with the principal components.

Author Contributions: Conceptualization, V.O.; methodology, V.O., Z.S., Á.Z.R. and B.D.; formal analysis, V.O.; investigation, M.I., Z.B.S., Z.S., Á.Z.R. and B.D.; writing—original draft preparation, V.O.; writing—review and editing, V.O., M.A.K.J., S.S. and I.M.; visualization, V.O. All authors have read and agreed to the published version of the manuscript.

Funding: This research was funded by the National Research, Development and Innovation Office —NKFIH—of the Hungarian Ministry for Innovation and Technology, grant number OTKA FK 134296. The μXRF measurements at ATOMKI were financed by the GINOP-2.3.3-15-2016-00029 'HSLab' project.

Institutional Review Board Statement: Not applicable.

Data Availability Statement: The datasets used in the present study are available from the corresponding author on reasonable request.

Acknowledgments: The authors are grateful to the Agora Science Centre (Debrecen, Hungary) for providing the opportunity to use the Imaging-PAM instrument. Muhammad Irfan is thankful for the support of the Tempus Public Foundation (Hungary) within the framework of the Stipendium Hungaricum Scholarship Programme.

Conflicts of Interest: The authors declare no conflict of interest.

References

1. Acosta, K.; Appenroth, K.J.; Borisjuk, L.; Edelman, M.; Heinig, U.; Jansen, M.A.K.; Oyama, T.; Pasaribu, B.; Schubert, I.; Sorrels, S.; et al. Return of the Lemnaceae: Duckweed as a Model Plant System in the Genomics and Postgenomics Era. *Plant Cell* **2021**, *33*, 3207–3234. [CrossRef] [PubMed]
2. Landolt, E.; Kandeler, R. *Biosystematic Investigations in the Family of Duckweeds (Lemnaceae), Vol. 4: The Family of Lemnaceae-a Monographic Study, Vol. 2 (Phytochemistry, Physiology, Application, Bibliography)*; Geobotanisches Institut der Eidgenoessischen Technischen Hochschule, Stiftung Ruebel: Zürich, Switzerland, 1987.
3. Ziegler, P.; Adelmann, K.; Zimmer, S.; Schmidt, C.; Appenroth, K.-J. Relative *in Vitro* Growth Rates of Duckweeds (Lemnaceae)-the Most Rapidly Growing Higher Plants. *Plant Biol. J.* **2015**, *17*, 33–41. [CrossRef] [PubMed]
4. Paolacci, S.; Stejskal, V.; Jansen, M.A.K. Estimation of the Potential of *Lemna Minor* for Effluent Remediation in Integrated Multi-Trophic Aquaculture Using Newly Developed Synthetic Aquaculture Wastewater. *Aquac. Int.* **2021**, *29*, 2101–2118. [CrossRef]
5. Ziegler, P.; Sree, K.S.; Appenroth, K.-J. Duckweeds for Water Remediation and Toxicity Testing. *Toxicol. Environ. Chem.* **2016**, *98*, 1127–1154. [CrossRef]
6. Pagliuso, D.; Grandis, A.; Fortirer, J.S.; Camargo, P.; Floh, E.I.S.; Buckeridge, M.S. Duckweeds as Promising Food Feedstocks Globally. *Agronomy* **2022**, *12*, 796. [CrossRef]
7. Petersen, F.; Demann, J.; Restemeyer, D.; Olfs, H.-W.; Westendarp, H.; Appenroth, K.-J.; Ulbrich, A. Influence of Light Intensity and Spectrum on Duckweed Growth and Proteins in a Small-Scale, Re-Circulating Indoor Vertical Farm. *Plants* **2022**, *11*, 1010. [CrossRef]
8. Abramson, B.W.; Novotny, M.; Hartwick, N.T.; Colt, K.; Aevermann, B.D.; Scheuermann, R.H.; Michael, T.P. The Genome and Preliminary Single-Nuclei Transcriptome of *Lemna minuta* Reveals Mechanisms of Invasiveness. *Plant Physiol.* **2022**, *188*, 879–897. [CrossRef]
9. Ater, M.; Ali, N.; Kasmi, H. Tolerance and accumulation of copper and chromium in two duckweed species: *Lemna minor* L. and *Lemna gibba* L. *Rev. Sci. L'eau/J. Water Sci.* **2006**, *19*, 57–67. [CrossRef]
10. Bokhari, S.H.; Mahmood-Ul-Hassan, M.; Ahmad, M. Phytoextraction of Ni, Pb and, Cd by Duckweeds. *Int. J. Phytoremediation* **2019**, *21*, 799–806. [CrossRef]
11. Khellaf, N.; Zerdaoui, M. Phytoaccumulation of Zinc Using the Duckweed *Lemna gibba* L.: Effect of Temperature, PH and Metal Source. *Desalination Water Treat.* **2013**, *51*, 5755–5760. [CrossRef]
12. Mkandawire, M.; Taubert, B.; Dudel, E.G. Capacity of *Lemna gibba* L. (Duckweed) for Uranium and Arsenic Phytoremediation in Mine Tailing Waters. *Int. J. Phytoremediation* **2004**, *6*, 347–362. [CrossRef] [PubMed]
13. Chen, D.; Zhang, H.; Wang, Q.; Shao, M.; Li, X.; Chen, D.; Zeng, R.; Song, Y. Intraspecific Variations in Cadmium Tolerance and Phytoaccumulation in Giant Duckweed (*Spirodela polyrhiza*). *J. Hazard. Mater.* **2020**, *395*, 122672. [CrossRef]

14. Sree, K.S.; Adelmann, K.; Garcia, C.; Lam, E.; Appenroth, K.-J. Natural Variance in Salt Tolerance and Induction of Starch Accumulation in Duckweeds. *Planta* **2015**, *241*, 1395–1404. [CrossRef] [PubMed]
15. Zhou, Y.; Bai, T.; Kishchenko, O. Potential of Lemnoideae species for phytoremediation of fresh water with elevated manganese concentration. *Innov. Biosyst. Bioeng.* **2019**, *3*, 232–238. [CrossRef]
16. van der Spiegel, M.; Noordam, M.y.; van der Fels-Klerx, H.j. Safety of Novel Protein Sources (Insects, Microalgae, Seaweed, Duckweed, and Rapeseed) and Legislative Aspects for Their Application in Food and Feed Production. *Compr. Rev. Food Sci. Food Saf.* **2013**, *12*, 662–678. [CrossRef] [PubMed]
17. Salt, D.E.; Baxter, I.; Lahner, B. Ionomics and the Study of the Plant Ionome. *Annu. Rev. Plant Biol.* **2008**, *59*, 709–733. [CrossRef]
18. Demmig-Adams, B.; López-Pozo, M.; Polutchko, S.K.; Fourounjian, P.; Stewart, J.J.; Zenir, M.C.; Adams, W.W. Growth and Nutritional Quality of Lemnaceae Viewed Comparatively in an Ecological and Evolutionary Context. *Plants* **2022**, *11*, 145. [CrossRef]
19. Kufel, L.; Strzałek, M.; Przetakiewicz, A. Plant Response to Overcrowding–*Lemna minor* Example. *Acta Oecologica* **2018**, *91*, 73–80. [CrossRef]
20. Potters, G.; Pasternak, T.P.; Guisez, Y.; Palme, K.J.; Jansen, M.A.K. Stress-Induced Morphogenic Responses: Growing out of Trouble? *Trends Plant Sci.* **2007**, *12*, 98–105. [CrossRef]
21. Szabó, S.; Koleszár, G.; Braun, M.; Nagy, Z.; Vicei, T.T.; Peeters, E.T.H.M. Submerged Rootless Macrophytes Sustain a Stable State against Free-Floating Plants. *Ecosystems* **2022**, *25*, 17–29. [CrossRef]
22. Szabó, S.; Koleszár, G.; Zavanyi, G.; Nagy, P.T.; Braun, M.; Hilt, S. Disentangling the Mechanisms Sustaining a Stable State of Submerged Macrophyte Dominance against Free-Floating Competitors. *Front. Plant Sci.* **2022**, *13*. [CrossRef] [PubMed]
23. Reale, L.; Ferranti, F.; Mantilacci, S.; Corboli, M.; Aversa, S.; Landucci, F.; Baldisserotto, C.; Ferroni, L.; Pancaldi, S.; Venanzoni, R. Cyto-Histological and Morpho-Physiological Responses of Common Duckweed (*Lemna minor* L.) to Chromium. *Chemosphere* **2016**, *145*, 98–105. [CrossRef] [PubMed]
24. Appenroth, K.-J.; Ziegler, P.; Sree, K.S. Accumulation of Starch in Duckweeds (Lemnaceae), Potential Energy Plants. *Physiol. Mol. Biol. Plants* **2021**, *27*, 2621–2633. [CrossRef] [PubMed]
25. Kocjan, G.; Samardakiewicz, S.; Woźny, A. Regions of Lead Uptake in *Lemna minor* Plants and Localization of This Metal within Selected Parts of the Root. *Biol. Plant.* **1996**, *38*, 107–117. [CrossRef]
26. Yue, L.; Zhao, J.; Yu, X.; Lv, K.; Wang, Z.; Xing, B. Interaction of CuO Nanoparticles with Duckweed (*Lemna minor* L.): Uptake, Distribution and ROS Production Sites. *Environ. Pollut.* **2018**, *243*, 543–552. [CrossRef]
27. Xu, H.; Yu, C.; Xia, X.; Li, M.; Li, H.; Wang, Y.; Wang, S.; Wang, C.; Ma, Y.; Zhou, G. Comparative Transcriptome Analysis of Duckweed (*Landoltia punctata*) in Response to Cadmium Provides Insights into Molecular Mechanisms Underlying Hyperaccumulation. *Chemosphere* **2018**, *190*, 154–165. [CrossRef]
28. Kanngießer, B.; Malzer, W.; Pagels, M.; Lühl, L.; Weseloh, G. Three-Dimensional Micro-XRF under Cryogenic Conditions: A Pilot Experiment for Spatially Resolved Trace Analysis in Biological Specimens. *Anal. Bioanal. Chem.* **2007**, *389*, 1171–1176. [CrossRef]
29. Morales-Barrera, L.; Flores-Ortiz, C.M.; Cristiani-Urbina, E. Single and Binary Equilibrium Studies for Ni^{2+} and Zn^{2+} Biosorption onto *Lemna gibba* from Aqueous Solutions. *Processes* **2020**, *8*, 1089. [CrossRef]
30. Environment Canada. *Biological Test Method-Test for Measuring the Inhibition of Growth Using the Freshwater Macrophyte Lemna minor*; Method Development and Applications Section, Environmental Technology Centre, Environment Canada: Ottawa, ON, Canada, 2007.
31. Oláh, V.; Hepp, A.; Irfan, M.; Mészáros, I. Chlorophyll Fluorescence Imaging-Based Duckweed Phenotyping to Assess Acute Phytotoxic Effects. *Plants* **2021**, *10*, 2763. [CrossRef]
32. Abràmoff, M.D.; Magalhães, P.J.; Ram, S.J. Image Processing with ImageJ. *Biophotonics Int.* **2004**, *11*, 32–64.
33. OECD. *Guidelines for the Testing of Chemicals, Revised Proposal for a New Guideline 221, Lemna Sp. Growth Inhibition Test*; OECD: Paris, France, 2006.
34. Hammer, Ø.; Harper, D.A.T.; Ryan, P.D. PAST: Paleontological Statistics Software Package for Education and Data Analysis. *Palaeontol. Electron.* **2001**, *4*, 9.
35. Lahive, E.; Halloran, J.O.; Jansen, M.A.K. Differential Sensitivity of Four Lemnaceae Species to Zinc Sulphate. *Environ. Exp. Bot.* **2011**, *71*, 25–33. [CrossRef]
36. Oláh, V.; Lakatos, G.; Bertók, C.; Kanalas, P.; Szőllősi, E.; Kis, J.; Mészáros, I. Short-Term Chromium(VI) Stress Induces Different Photosynthetic Responses in Two Duckweed Species, *Lemna gibba* L. and *Lemna minor* L. *Photosynthetica* **2010**, *48*, 513–520. [CrossRef]
37. Xyländer, M.; Augsten, H. Different Sensitivity of Some Lemnaceae to Nickel. *Beiträge Zur Biol. Der Pflanz.* **1992**, *67*, 89–99.
38. Appenroth, K.-J.; Krech, K.; Keresztes, Á.; Fischer, W.; Koloczek, H. Effects of Nickel on the Chloroplasts of the Duckweeds *Spirodela polyrhiza* and *Lemna minor* and Their Possible Use in Biomonitoring and Phytoremediation. *Chemosphere* **2010**, *78*, 216–223. [CrossRef]
39. Yang, J.; Li, G.; Bishopp, A.; Heenatigala, P.P.M.; Hu, S.; Chen, Y.; Wu, Z.; Kumar, S.; Duan, P.; Yao, L.; et al. A Comparison of Growth on Mercuric Chloride for Three Lemnaceae Species Reveals Differences in Growth Dynamics That Effect Their Suitability for Use in Either Monitoring or Remediating Ecosystems Contaminated with Mercury. *Front. Chem.* **2018**, *6*, 112. [CrossRef]
40. Shao, J.; Liu, Z.; Ding, Y.; Wang, J.; Li, X.; Yang, Y. Biosynthesis of the Starch Is Improved by the Supplement of Nickel (Ni^{2+}) in Duckweed (*Landoltia punctata*). *J. Plant Res.* **2020**, *133*, 587–596. [CrossRef]

41. Liu, Y.; Sanguanphun, T.; Yuan, W.; Cheng, J.J.; Meetam, M. The Biological Responses and Metal Phytoaccumulation of Duckweed *Spirodela polyrhiza* to Manganese and Chromium. *Environ. Sci. Pollut. Res.* **2017**, *24*, 19104–19113. [CrossRef]
42. Taylor, G.J.; Stadt, K.J. Interactive Effects of Cadmium, Copper, Manganese, Nickel, and Zinc on Root Growth of Wheat (*Triticum aestivum*) in Solution Culture. In *Plant Nutrition—Physiology and Applications: Proceedings of the Eleventh International Plant Nutrition Colloquium, Wageningen, The Netherlands, 30 July–4 August 1989*; van Beusichem, M.L., Ed.; Developments in Plant and Soil Sciences; Springer: Dordrecht, The Netherlands, 1990; pp. 317–322. ISBN 978-94-009-0585-6.
43. van der Ent, A.; Pillon, Y.; Fogliani, B.; Gei, V.; Jaffré, T.; Erskine, P.D.; Echevarria, G.; Spiers, K.M.; Paul, A.L.D.; Isnard, S. Contrasting Nickel and Manganese Accumulation and Localization in New Caledonian Cunoniaceae. *Plant Soil* **2022**, *475*, 515–534. [CrossRef]
44. Park, J.H. Contrasting Effects of Cr(III) and Cr(VI) on Lettuce Grown in Hydroponics and Soil: Chromium and Manganese Speciation. *Environ. Pollut.* **2020**, *266*, 115073. [CrossRef]
45. Dube, B.K.; Tewari, K.; Chatterjee, J.; Chatterjee, C. Excess Chromium Alters Uptake and Translocation of Certain Nutrients in Citrullus. *Chemosphere* **2003**, *53*, 1147–1153. [CrossRef]
46. Turner, M.A.; Rust, R.H. Effects of Chromium on Growth and Mineral Nutrition of Soybeans. *Soil Sci. Soc. Am. J.* **1971**, *35*, 755–758. [CrossRef]
47. Dalir, N.; Tandy, S.; Gramlich, A.; Khoshgoftarmanesh, A.; Schulin, R. Effects of Nickel on Zinc Uptake and Translocation in Two Wheat Cultivars Differing in Zinc Efficiency. *Environ. Exp. Bot.* **2017**, *134*, 96–101. [CrossRef]
48. Pillon, Y.; Petit, D.; Gady, C.; Soubrand, M.; Joussein, E.; Saladin, G. Ionomics Suggests Niche Differences between Sympatric Heathers (Ericaceae). *Plant Soil* **2019**, *434*, 481–489. [CrossRef]
49. Jovanović, P.; Rachmilevitch, S.; Roitman, N.; Erel, R. Strontium as a Tracer for Calcium: Uptake, Transport and Partitioning within Tomato Plants. *Plant Soil* **2021**, *466*, 303–316. [CrossRef]
50. Ali, Z.; Waheed, H.; Kazi, A.G.; Hayat, A.; Ahmad, M. Chapter 16-Duckweed: An Efficient Hyperaccumulator of Heavy Metals in Water Bodies. In *Plant Metal Interaction*; Ahmad, P., Ed.; Elsevier: Amsterdam, The Netherlands, 2016; pp. 411–429. ISBN 978-0-12-803158-2.
51. Dirilgen, N.; İnel, Y. Effects of Zinc and Copper on Growth and Metal Accumulation in Duckweed, *Lemna minor*. *Bull. Environ. Contam. Toxicol.* **1994**, *53*, 442–449. [CrossRef] [PubMed]
52. Khellaf, N.; Zerdaoui, M. Growth Response of the Duckweed *Lemna gibba* L. to Copper and Nickel Phytoaccumulation. *Ecotoxicology* **2010**, *19*, 1363–1368. [CrossRef] [PubMed]
53. Devlamynck, R.; de Souza, M.F.; Leenknegt, J.; Jacxsens, L.; Eeckhout, M.; Meers, E. *Lemna minor* Cultivation for Treating Swine Manure and Providing Micronutrients for Animal Feed. *Plants* **2021**, *10*, 1124. [CrossRef] [PubMed]
54. McGeer, J.C.; Brix, K.V.; Skeaff, J.M.; DeForest, D.K.; Brigham, S.I.; Adams, W.J.; Green, A. Inverse Relationship between Bioconcentration Factor and Exposure Concentration for Metals: Implications for Hazard Assessment of Metals in the Aquatic Environment. *Environ. Toxicol. Chem.* **2003**, *22*, 1017–1037. [CrossRef]
55. Regoli, L.; Van Tilborg, W.; Heijerick, D.; Stubblefield, W.; Carey, S. The Bioconcentration and Bioaccumulation Factors for Molybdenum in the Aquatic Environment from Natural Environmental Concentrations up to the Toxicity Boundary. *Sci. Total Environ.* **2012**, *435–436*, 96–106. [CrossRef]
56. Kalčíková, G.; Zupančič, M.; Jemec, A.; Žgajnar Gotvajn, A. The Impact of Humic Acid on Chromium Phytoextraction by Aquatic Macrophyte *Lemn minor*. *Chemosphere* **2016**, *147*, 311–317. [CrossRef]
57. Uysal, Y. Removal of Chromium Ions from Wastewater by Duckweed, *Lemna minor* L. by Using a Pilot System with Continuous Flow. *J. Hazard. Mater.* **2013**, *263*, 486–492. [CrossRef] [PubMed]
58. Yilmaz, D.D.; Parlak, K.U. Nickel-Induced Changes in Lipid Peroxidation, Antioxidative Enzymes, and Metal Accumulation in *Lemna gibba*. *Int. J. Phytoremediation* **2011**, *13*, 805–817. [CrossRef] [PubMed]
59. Walsh, É.; Paolacci, S.; Burnell, G.; Jansen, M.A.K. The Importance of the Calcium-to-Magnesium Ratio for Phytoremediation of Dairy Industry Wastewater Using the Aquatic Plant *Lemna minor* L. *Int. J. Phytoremediation* **2020**, *22*, 694–702. [CrossRef] [PubMed]
60. Matsuda, S.; Nagasawa, H.; Yamashiro, N.; Yasuno, N.; Watanabe, T.; Kitazawa, H.; Takano, S.; Tokuji, Y.; Tani, M.; Takamure, I.; et al. Rice RCN1/OsABCG5 Mutation Alters Accumulation of Essential and Nonessential Minerals and Causes a High Na/K Ratio, Resulting in a Salt-Sensitive Phenotype. *Plant Sci.* **2014**, *224*, 103–111. [CrossRef]
61. Walsh, É.; Coughlan, N.E.; O'Brien, S.; Jansen, M.A.K.; Kuehnhold, H. Density Dependence Influences the Efficacy of Wastewater Remediation by *Lemn minor*. *Plants* **2021**, *10*, 1366. [CrossRef]
62. Sharma, D.C.; Sharma, C.P.; Tripathi, R.D. Phytotoxic Lesions of Chromium in Maize. *Chemosphere* **2003**, *51*, 63–68. [CrossRef] [PubMed]
63. Singh, A.K. Effect of Trivalent and Hexavalent Chromium on Spinach (*Spinacea oleracea* L.). *Environ. Ecol.* **2001**, *19*, 807–810.
64. Hassan, M.U.; Chattha, M.U.; Khan, I.; Chattha, M.B.; Aamer, M.; Nawaz, M.; Ali, A.; Khan, M.A.U.; Khan, T.A. Nickel Toxicity in Plants: Reasons, Toxic Effects, Tolerance Mechanisms, and Remediation Possibilities—A Review. *Environ. Sci. Pollut. Res.* **2019**, *26*, 12673–12688. [CrossRef]

plants

MDPI

Article

Influence of the Nitrate-N to Ammonium-N Ratio on Relative Growth Rate and Crude Protein Content in the Duckweeds *Lemna minor* and *Wolffiella hyalina*

Finn Petersen [1,*], Johannes Demann [1], Dina Restemeyer [1], Andreas Ulbrich [1], Hans-Werner Olfs [1], Heiner Westendarp [1] and Klaus-Jürgen Appenroth [2]

[1] Faculty of Agricultural Sciences and Landscape Architecture, University of Applied Sciences Osnabrück, Am Krümpel 31, 49090 Osnabrück, Germany; johannes.demann@hs-osnabrueck.de (J.D.); dina.restemeyer@hs-osnabrueck.de (D.R.); a.ulbrich@hs-osnabrueck.de (A.U.); h-w.olfs@hs-osnabrueck.de (H.-W.O.); h.westendarp@hs-osnabrueck.de (H.W.)

[2] Matthias-Schleiden-Institute-Plant Physiology, University of Jena, Dornburger Str. 159, 07743 Jena, Germany; klaus.appenroth@uni-jena.de

* Correspondence: finn.petersen@hs-osnabrueck.de; Tel.: +49-5419695098

Abstract: In order to produce protein-rich duckweed for human and animal consumption, a stable cultivation process, including an optimal nutrient supply for each species, must be implemented. Modified nutrient media, based on the N-medium for duckweed cultivation, were tested on the relative growth rate (RGR) and crude protein content (CPC) of *Lemna minor* and *Wolffiella hyalina*, as well as the decrease of nitrate-N and ammonium-N in the media. Five different nitrate-N to ammonium-N molar ratios were diluted to 10% and 50% of the original N-medium concentration. The media mainly consisted of agricultural fertilizers. A ratio of 75% nitrate-N and 25% ammonium-N, with a dilution of 50%, yielded the best results for both species. Based on the dry weight (DW), *L. minor* achieved a RGR of $0.23 \pm 0.009\ d^{-1}$ and a CPC of $37.8 \pm 0.42\%$, while *W. hyalina*'s maximum RGR was $0.22 \pm 0.017\ d^{-1}$, with a CPC of $43.9 \pm 0.34\%$. The relative protein yield per week and m^2 was highest at this ratio and dilution, as well as the ammonium-N decrease in the corresponding medium. These results could be implemented in duckweed research and applications if a high protein content or protein yield is the aim.

Keywords: amino acids; biomass production; cultivation; Lemnaceae; nutrient medium; uptake; water lentils; yield

Citation: Petersen, F.; Demann, J.; Restemeyer, D.; Ulbrich, A.; Olfs, H.-W.; Westendarp, H.; Appenroth, K.-J. Influence of the Nitrate-N to Ammonium-N Ratio on Relative Growth Rate and Crude Protein Content in the Duckweeds *Lemna minor* and *Wolffiella hyalina*. *Plants* **2021**, *10*, 1741. https://doi.org/10.3390/plants10081741

Academic Editor: Mirza Hasanuzzaman

Received: 21 May 2021
Accepted: 19 August 2021
Published: 23 August 2021

1. Introduction

A growing world population, with an increasing demand for protein, will necessitate the efficient and increased production of food and animal feed. By 2050, the predicted global population is expected to increase to 9.5 billion, resulting in a rising global demand for protein of up to 78% under different scenarios [1]. In order to handle this challenge, cultivating plants with a high protein content is a promising option. One of the possible candidates is duckweed. However, for this purpose, duckweeds will need to be cultivated under standardized, large-scale conditions.

Duckweeds are an aquatic plant family (*Lemnaceae*), which have been gaining increased attention as an option for human nutrition and animal feeding. Several studies showed the potential of certain duckweed species as a nutrient source [2–4]. This is due to high relative growth rates (RGR) [5,6], a protein content comparable to soybeans, and, in accordance with WHO recommendations, an amino acid composition suitable for humans. The species *Wolffiella hyalina* and *Wolffia microscopica*, in particular, have been recommended for human nutrition [2,3]. Additionally, the nutritional values and proportions within the duckweeds can be modified by changing the cultivation conditions [2,7,8].

Several duckweed species have been tested as feed for animals, such as chickens, piglets and fish [9,10]. High contents of the essential amino acids lysine and methionine make some duckweed species, such as *W. hyalina*, an interesting substitute for today's foremost feed protein source soybean. Gwaze and Mwale [11] compiled several studies, which tested duckweed in pig nutrition. The replacement of soybean meal by 40% duckweed in the feeding rations of young piglets (0 to 10 days old) led to the highest average daily gain in body weight compared to the control of 100% soybean meal [12]. Nguyen and Ogle [13] showed that replacing 75% of roasted soybeans with *Lemna minor* in 5 to 15 week old Tau Vang chickens resulted in weight gain and a feed conversion optimum.

Such promising studies have led to the challenge of yielding high quantities of protein-rich duckweed biomass from a standardized, large-scale production process to incorporate duckweed in the food and feed industry. To economically operate such a system, inexpensive and easily available resources should be used. Moreover, an optimal nutrient supply for each duckweed species must be identified.

According to Appenroth et al. [14,15] no other medium supports faster growth of duckweeds than the N-medium. However, its only nitrogen source is nitrate, but different studies have emphasised the preferential uptake of ammonium over nitrate [16–19]. Little is known about the effect of different nitrate to ammonium ratios on the growth rate and nutritional components of duckweeds.

The aim of our research was to examine how five different nitrate-N to ammonium-N ratios in modified N-media [14,15] affected the RGR, crude protein content (CPC), and relative protein yield (RPY) of *L. minor* and *W. hyalina* (Figure 1). In order to minimize the inhibiting effect of algae on duckweed growth rate and biomass production, different steps of dilution were investigated. Additionally, the decrease of NO_3^--N and NH_4^+-N concentrations in the media, due to N-uptake by the duckweeds, was measured.

Figure 1. Investigated duckweed species. (**A**) *Lemna minor*, clone 9441. Mother frond (centre) is bearing two daughter fronds. (**B**) *Wolffiella hyalina*, clone 9525. Mother frond (bottom) is bearing a single daughter frond. Photos provided by Dr. K. Sowjanya Sree, Central University of Kerala, India.

2. Materials and Methods

2.1. Plant Material and Cultivation

Two duckweed species, *Lemna minor* L. (clone 9441; Germany) and *Wolffiella hyalina* Delile Monod (clone 9525; India), were used for the experiments due to their fast growth rates and high protein contents. The plant material was obtained from the Duckweed Stock Collection of the Department of Plant Physiology, University of Jena, Germany.

Experiments were carried out in a climate chamber (length × width × height: 4 × 3 × 3 m) at the campus of the University of Applied Sciences Osnabrück, Germany. The trials were conducted in black PE-vessels with a diameter of 24 cm, each containing 4 L of nutrient medium. All modified media used were based on the N-medium [14,15]. The following abbreviations for five different NO_3^--N to NH_4^+-N ratios are used throughout the manuscript: [100-0] = 100% NO_3^--N - 0% NH_4^+-N; [75-25] = 75% NO_3^--N - 25% NH_4^+-N; [50-50] = 50% NO_3^--N - 50% NH_4^+-N; [25-75] = 25% NO_3^--N - 75% NH_4^+-N; and [0-100] = 0% NO_3^--N - 100% NH_4^+-N. In preliminary experiments, it was observed that the growth of duckweed was disturbed by contamination of ubiquitous algae in the cultures. In order to minimize nutrient competition and growth inhibition of the duckweed due to algae and microorganism growth, several measures were implemented. Two different dilutions (10% and 50% of the original concentration) were used for all five NO_3^--N to NH_4^+-N ratios, indicated by "/10" and "/50" following the ratios. Dilutions of 1% and 5% (/1; /5) were used in initial experiments with 100-0, but omitted in later experiments because of poor results. The temperature was kept at 20.4 ± 1.3 °C. S4W LED elements (SANlight GmbH, Bludenz, Austria), with a photosynthetically active radiation of 350 µmol m^{-2} s^{-1}, were used as the light source. The photoperiod was set to 8 h of light and 16 h of darkness.

The pH showed a minor increase throughout the experimental period, rising from pH 6.6 to a maximum value of 7.0 in the 50% dilutions and from 7.2 to a maximum value of 8.0 in the 10% dilutions.

Six stock solutions, used for all five differently modified N-media, were mainly prepared with commercially available agricultural fertilizers and deionized water. The following products were used for preparing the stock solutions: Krista MKP, Calcinit, Krista K Plus (Yara GmbH and Co. KG, Dülmen, Germany), OCI Granular 2 (OCI NV, Amsterdam, Netherlands), potassium chloride, sodium molybdate dihydrate technical grade (AppliChem GmbH, Darmstadt, Germany), ammonium chloride p.a., calcium chloride dihydrate (Merck KGaA, Darmstadt, Germany), Borax, Mangaan (Horticoop, Bleiswijk, Netherlands), Epso Combitop (K+S AG, Kassel, Germany) and Ferty 72 (Planta Düngemittel GmbH, Regenstauf, Germany). The precise formulation for each stock solution is presented in Table 1.

Table 1. Formulation of seven stock solutions (g L^{-1}) for five different nitrate-N to ammonium-N ratios ([100-0], [75-25], [50-50], [25-75], and [0-100]), based on the N-medium.

Stock Solution	Product Name	Main Components	[100-0] (g L^{-1})	[75-25] (g L^{-1})	50-50 (g L^{-1})	[25-75] (g L^{-1})	[0-100] (g L^{-1})
1	Calcinit	NO_3^--N, NH_4^+-N, Ca^+	47.2	35.4	23.6	11.8	0
1	Krista K Plus	NO_3^--N, K^+	161.8	121.3	80.9	40.4	0
2	NH_4Cl	NH_4^+-N, Cl^-	0	0	26.7	53.5	80.2
3	OCI Granular 2	NH_4^+-N, SO_4^{2-}	0	33	33	33	33
4	KCl	K^+, Cl^-	0	29.8	59.6	89.5	119.3
4	$CaCl_2 \cdot 2\,H_20$	Ca^+, Cl^-	0	7.4	14.7	22.1	29.4
5	Krista MKP	PO_4^{3-}, K^+	27.2	27.2	27.2	27.2	27.2
6	Epso Combitop	Mg^{2+}, SO_4^{2-}, Mn^{2+}, Zn^{2+}	49.3	49.3	49.3	49.3	49.3
6	Borax	BO_3^{3-}	0.06	0.06	0.06	0.06	0.06
6	Mangaan	Mn^{2+}, SO_4^{2-}	0.44	0.44	0.44	0.44	0.44
6	$MoNa_2O_4 \cdot 2\,H_20$	MoO_4^{2-}, Na^+	0.02	0.02	0.02	0.02	0.02
7	Ferty 72	Fe^{3+}	2.2	2.2	2.2	2.2	2.2

The stock solutions were diluted with local tap water (see Table S1) to obtain the initial N-medium concentrations of 100%, as shown in Table 2.

Table 2. Nutrient composition (mM) of the modified N-media with five different NO_3^--N (light grey) to NH_4^+-N (dark grey) ratios at an initial concentration of 100%. These concentrations were diluted to the final concentrations of 10% and 50%, and, in some cases, to 1% and 5%.

NO_3^--N to NH_4^+-N Ratio Substance	[100-0] (mM)	[75-25] (mM)	[50-50] (mM)	[25-75] (mM)	[0-100] (mM)
NO_3^--N	10.1	7.6	5.1	2.6	0.1
NH_4^+-N	0	2.5	5	7.5	10
PO_4^{3-}	1	1	1	1	1
K^+	9.1	9.1	9.1	9.1	9.1
Mg^{2+}	1.3	1.3	1.3	1.3	1.3
SO_4^{2-}	2.0	3.2	3.2	3.2	3.2
Ca^+	2.2	2.2	2.2	2.2	2.2
Cl^-	0.9	3.4	8.4	13.4	18.4
Fe^{3+}	0.025	0.025	0.025	0.025	0.025
BO_3^{3-}	0.005	0.005	0.005	0.005	0.005
Mn^{2+}	0.013	0.013	0.013	0.013	0.013
Zn^{2+}	0.01	0.01	0.01	0.01	0.01
MoO_4^{2-}	0.0004	0.0004	0.0004	0.0004	0.0004
Na^+	0.8	0.8	0.8	0.8	0.8

Table 3 depicts the measured concentrations for nitrate-N and ammonium-N, as well as the corresponding electrical conductivity (EC), after dilution to the final concentrations of 1%, 5%, 10%, and 50% of the undiluted medium.

Table 3. Measured concentrations of NO_3^--N and NH_4^+-N (mg L^{-1}) and EC-values (mS cm^{-1}) at the start of the experiment in the different nutrient media. For the composition of the undiluted nutrient medium, see Table 2. The diluted nutrient media contain 1, 5, 10, or 50% of the undiluted medium.

Ratio	[100-0]				[75-25]		[50-50]		[25-75]		[0-100]	
Dilution (%)	1	5	10	50	10	50	10	50	10	50	10	50
NO_3^--N (mg L^{-1})	2.7	8.8	15.3	71.2	12.1	56.7	10.1	35.3	5.2	16.6	1.1	1.4
NH_4^+-N (mg L^{-1})	0.06	0.17	0.29	1.1	3.6	17.3	7.1	32.5	10.7	51.3	14.4	64.4
EC (mS cm^{-1})	0.43	0.46	0.53	1.15	0.6	1.33	0.66	1.58	0.65	1.64	0.64	1.84

Pre-cultivation occurred for three days within each of the differently diluted and modified nutrient media in order to avoid the lag-phase effect on the RGR data. Experiments lasted for seven days and were conducted under non-axenic growth conditions. The vessels were placed in a randomized block design within the climate chamber. In order to start with a similar surface coverage of 60%, the initial fresh weight biomass of 2 g for *L. minor* and 1.5 g for *W. hyalina* was placed in the above described vessels. At the end of the experiment, the duckweeds were harvested with a metal sieve, rinsed with fresh tap water to remove the attached nutrient solution, blotted with a paper towel to remove attached water, and weighed.

2.2. Analytical Methods

The dry weight (DW) was determined from fresh weight by oven drying at 65 °C for 72 h. At time 0, four samples per species of the same fresh weight as the starting material were used to determine the DW at the beginning of the experiments.

The RGR per day was calculated according to Equation (1) [5], using the values of the DW at the start (t0) and after seven days of cultivation (t7):

$$RGR = (\ln DW_{t7} - \ln DW_{t0})/(t7 - t0) \quad (1)$$

where RGR is the relative increase of the DW per unit time of 1 day (d^{-1}). The relative weekly yield (RY; g biomass obtained after one week of cultivation starting with 1 g) was calculated from the RGR using Equations (2) and (3):

$$\ln DW_{t7} = \ln DW_{t0} + RGR \cdot (t7 - t0) \quad (2)$$

$$RY = \exp(\ln DW_{t7}) \quad (3)$$

The RY (see Figure S1) was further used to calculate the RPY (g protein $week^{-1}$ m^{-2}) by multiplying it with the CPC and extrapolating it to one square metre, according to Equation (4):

$$RPY = RY \cdot CPC/(0.0452\ m^2 \cdot 100) \quad (4)$$

where 0.0452 m^2 is the cultivation area of the vessels used in the experiment.

Dried samples were ground and homogenized using a laboratory mill and stored for further analysis. The nitrogen content of the dried samples was determined by the Dumas method [20] using a FP628 (Leco, Saint Joseph, MI, USA), and was multiplied with the factor 6.25 to determine the CPC [2,21].

Nutrient solution samples were taken at the start (day 0) and end (day 7) of the experiment from each vessel, which were filtered (MN 619 EH, Machery Nagel GmbH and Co. KG, Düren, Germany) to remove particles and instantly frozen at −18 °C. The nitrate-N and ammonium-N concentrations in these samples were measured according to German standard methods [22,23] with a Lambda 25 UV/VIS Spectrometer (Perkin Elmer, Waltham, MA, USA).

2.3. Statistics

All the data is based on four replicates, which are given as mean ± standard deviations. The data were analysed statistically by one-way ANOVA and Tukey's post hoc test at 5% significance level, using the software program SPSS 25 (IBM, Armonk, NY, USA). Datasets fulfilled the one-way ANOVA postulates (including normal distribution and homogeneity of variance).

3. Results

3.1. Growth

The RGR was determined in dependence on the different nutrient media used (Figure 2). The highest value for *L. minor* was 0.23 ± 0.009 d^{-1} at [75-25]/50. The same RGR was determined for *W. hyalina* at [100-0]/50. The two most nitrate-rich ratios ([100-0] and [75-25]) showed an increasing RGR at higher nutrient concentrations (dilutions of 50%) compared to the 10% dilutions, which was significantly higher for *L. minor* in both ratios and for *W. hyalina* only in [100-0]. In these two ratios and dilutions, *W. hyalina* had a slightly higher RGR than *L. minor*, with the exception of [75-25]/50. This was contrary to when the ammonium concentration was increased. A significant drop of the RGR was observed for the ratios [50-50], [25-75], and [0-100] for *L. minor* compared to the 50% dilutions and for *W. hyalina* compared to the 10% and 50% dilutions, but it was more severe in the 50% dilutions. This decrease resulted in a minimum RGR of 0.09 ± 0.015 d^{-1} at [25-75]/50 for *W. hyalina*, while the decrease for *L. minor* (0.12 ± 0.002 d^{-1} at [0-100]/50) was less severe. *Lemna minor* achieved higher RGRs than *W. hyalina* at an overall lower level compared to the two most nitrate-rich ratios, except for [100-0]/1.

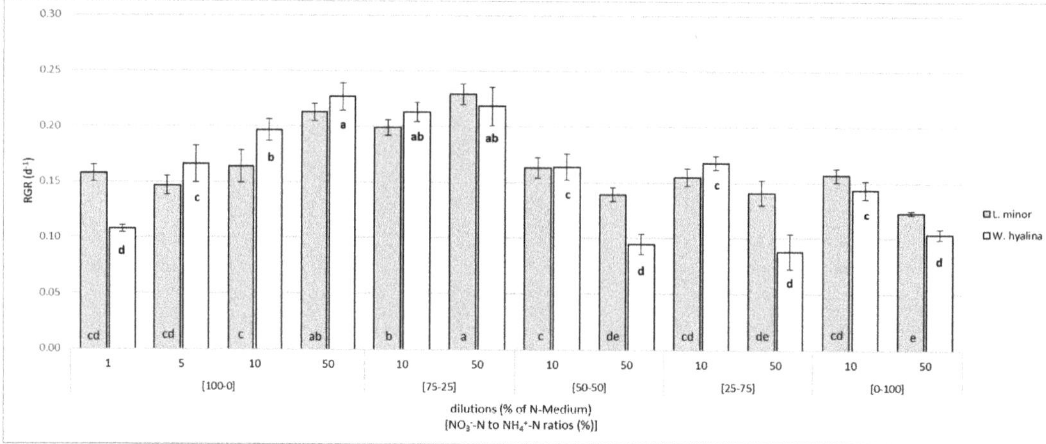

Figure 2. Relative growth rate (RGR; d^{-1}), based on dry weight, for *Lemna minor* (grey shaded columns) and *Wolffiella hyalina* (white columns). Plants were cultivated for seven days in nutrient solutions with varying nitrate-N to ammonium-N ratios (from [100-0] to [0-100]) in different dilutions (1, 5, 10, and 50% of the undiluted N-medium). For the different nutrient media see Tables 1–3. Number of parallel samples n = 4. Different letters indicate significances within a species, based on one-way ANOVA test, Tukey $p \leq 0.05$. Error bars indicate standard deviations.

3.2. Crude Protein Content and Protein Yield

The CPC increased in both duckweed species with increasing ammonium concentrations and a higher ammonium-N to nitrate-N ratio, but not significantly (Figure 3). The higher dilution of the nutrient media, i.e., lower nutrient concentrations, led to lower CPCs within each ratio. In general, *W. hyalina* had a higher CPC within each ratio and dilution compared to *L. minor*. The highest CPC in *L. minor* was reached at [0-100]/50 with 40.6 ± 0.48%, followed by 39.1 ± 0.43% at [0-100]/10. The maximum value measured for *W. hyalina* was 43.9 ± 0.34% at [75-25]/50, which is not significantly higher than the second highest CPC (43.0 ± 0.4%) at [0-100]/50. A minimum CPC of 21.1 ± 1.3% for *L. minor* and 30.3 ± 0.6% for *W. hyalina* were obtained in the ratio with the lowest concentration of nutrients available for the plants ([100-0]/1), which are significantly lower than the second lowest values for each species.

The highest RPY (g protein week^{-1} m^{-2}) was obtained at [75-25]/50 for both species, with a significant difference from the second highest value (Figure 4). A total of 41.6 ± 2.2 g week^{-1} m^{-2} were harvested from *L. minor* and 45.0 ± 5.7 g week^{-1} m^{-2} from *W. hyalina*. A higher nutrient concentration in the ratios [100-0] and [75-25] led to a higher protein yield. *W. hyalina* yielded more protein than *L. minor* under these conditions. The protein yield significantly decreased with an ammonium concentration of 50% and more compared to [100-0]/50 and [75-25]/50 for *L. minor* and [75-25]/50 for *W. hyalina*. At dilutions of 50%, *W. hyalina* performed worse than *L. minor*, while the RPY for both species was slightly higher at 10% dilutions at an overall low level of less than 30 g week^{-1} m^{-2}. The minimum RPYs of 14.1 ± 0.24 g week^{-1} m^{-2} for *L. minor* and 14.2 ± 0.28 g week^{-1} m^{-2} for *W. hyalina* were obtained in [100-0]/1.

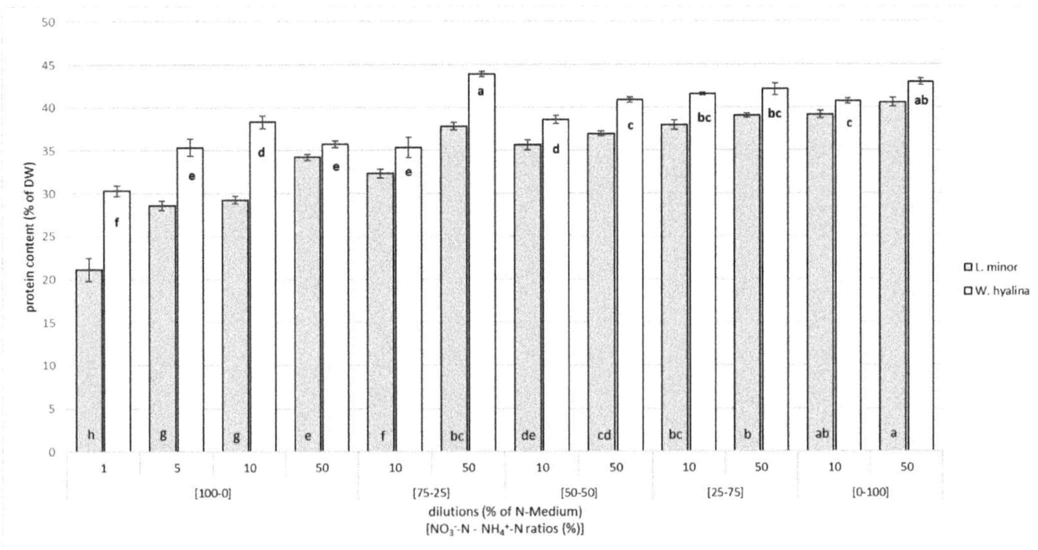

Figure 3. Crude protein content (% of DW) for *Lemna minor* (grey shaded columns) and *Wolffiella hyalina* (white columns), cultivated for seven days in nutrient solutions with different NO_3^--N to NH_4^+-N ratios in different dilutions based on N-medium. For further explanations, see Figure 2.

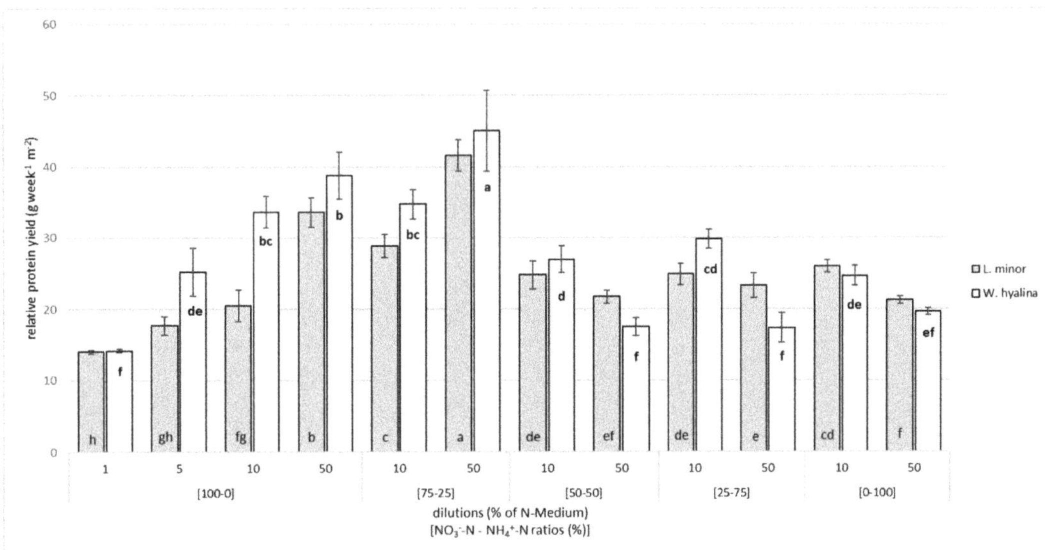

Figure 4. Relative protein yield, in grams per week and m^2 (based on DW), for *Lemna minor* (grey shaded columns) and *Wolffiella hyalina* (white columns), cultivated for seven days in nutrient solutions with different NO_3^--N to NH_4^+-N ratios in different dilutions based on N-medium. For further explanations, see Figure 2.

3.3. NO_3^--N and NH_4^+-N Reduction in the Media

Figure 5 shows the total reduction of NO_3^--N (mg L^{-1}) for each ratio and dilution at day seven. In the nitrate-only medium, [100-0], the higher NO_3^--N concentration led to a significantly higher absolute reduction in *L. minor*, but not in *W. hyalina*. The highest starting concentration of nitrate-N ([100-0]/50) led to the highest absolute decrease of nitrate-N, i.e., 6.5 mg L^{-1} and 6.9 mg L^{-1} for *L. minor* and *W. hyalina*, respectively. This corresponded to a relative reduction of 9.1% (*L. minor*) and 9.7% (*W. hyalina*). In general, higher nitrate-N concentrations resulted in a greater reduction, while an increasing NH_4^+-N concentration led to a decreasing nitrate-N reduction. These findings, however, were not significant. The maximal relative reduction of NO_3^--N for *L. minor* was found at [25-75]/10 with 29.2% and for *W. hyalina* at [75-25]/10 with 29.6%.

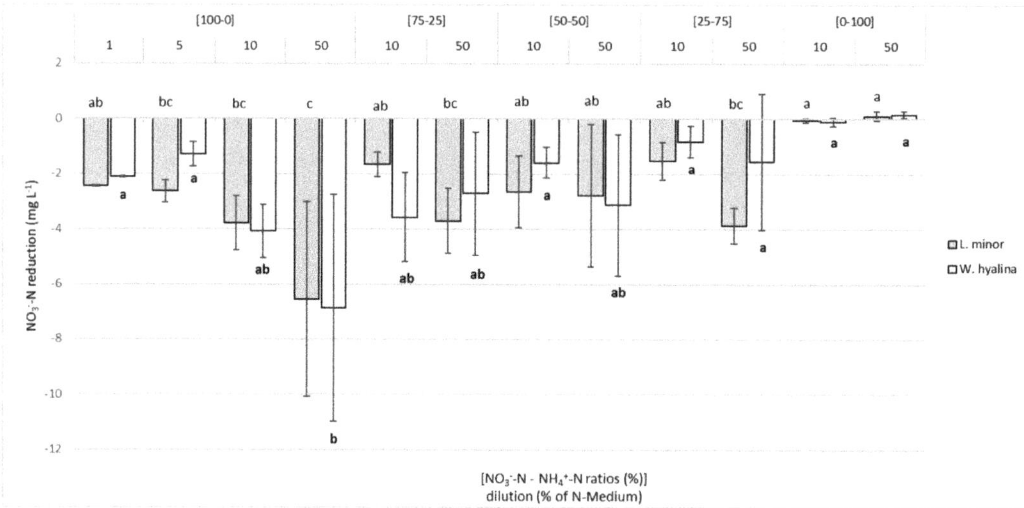

Figure 5. Decrease of the concentration of nitrate-N (mg L^{-1}) in the nutrient media for *Lemna minor* (grey shaded columns) and *Wolffiella hyalina* (white columns), cultivated for seven days in nutrient solutions with varying NO_3^--N to NH_4^+-N ratios in different dilutions. For further explanations, see Figure 2.

Figure 6 depicts the total reduction of NH_4^+-N (mg L^{-1}) for each ratio and dilution at day seven. In the [100-0] nutrient media, NH_4^+-N was present only in minor concentrations, which were decreased almost completely by both duckweed species. This also applied to [75-25]/10. The highest total reduction values were 8.1 ± 0.9 mg L^{-1} for *L. minor* and 7.2 ± 0.5 mg L^{-1} for *W. hyalina* in the [75-25]/50 treatments. This corresponded to a relative reduction of 46.8% (*L. minor*) and 41.6% (*W. hyalina*). The total, as well as the relative reduction, was slightly higher for *L. minor* than for *W. hyalina*. A significant drop was evident in the ammonium-only solutions, with the highest total NH_4^+-N concentration ([0-100]/50), compared to the same dilution in the ratios [75-25], [50-50], and [25-75]. The relative reduction for *L. minor* was 4.1% of the initially available NH_4^+-N, while for *W. hyalina* no reduction at all occurred. However, decreasing the total NH_4^+-N concentration, but keeping the ratio of the two N sources constant, i.e., [0-100]/10, resulted in much higher uptake rates for *L. minor* (7.8 ± 0.14 mg L^{-1}; 54% relative reduction) and *W. hyalina* (6.9 ± 0.27 mg L^{-1}; 48% relative reduction).

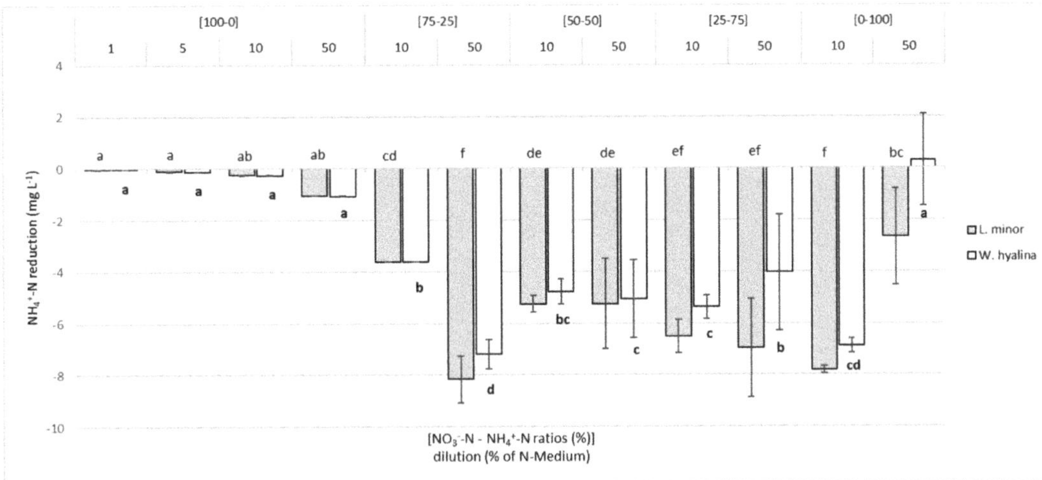

Figure 6. Decrease of the concentration of ammonium-N (mg L^{-1}) in the nutrient media for *Lemna minor* (grey shaded columns) and *Wolffiella hyalina* (white columns), cultivated for seven days in nutrient solutions with varying NO_3^--N to NH_4^+-N ratios in different dilutions. For further explanations, see Figure 2.

4. Discussion

The maximum RGR reached in this experiment was 0.23 d^{-1} for both species, which is lower compared to other studies. For *L. minor*, an RGR of 0.42 d^{-1} was reported, while *W. hyalina* had the highest RGR of all investigated species with a value of 0.519 d^{-1} [5]. The difference in the RGR was most likely caused by different growth conditions. Instead of an axenic in vitro set-up, both duckweed species in this study were cultivated under non-sterile conditions. The temperature was 5 °C lower and the photoperiod 16 h shorter, while the light intensity was 250 µmol m^{-2} s^{-1} higher compared to the experimental set-ups applied by Ziegler et al. [5]. These factors are possible explanations for the lower RGRs.

Iatrou et al. [24] achieved a maximum growth rate of 0.14 d^{-1} for *L. minor* at an ammonium-N concentration of 31.9 mg L^{-1}, using secondary treated wastewater. This was in agreement with our experimental results for the same species in the ratio of [50-50]/50 (RGR of 0.14 $\pm$ 0.009 d^{-1} at a NH_4^+-N concentration of 32.5 mg L^{-1}).

Caicedo et al. [18] observed that the highest RGR (0.3 d^{-1}) in *Spirodela polyrhiza* was obtained at the lowest total ammonium concentrations (3.5–20 mg L^{-1} N; equal to ca. 0.25–1.4 mM) and assumed an optimum NH_4^+-N concentration was below 20 mg L^{-1}. Zhang et al. [25] obtained the maximal RGR in *L. minor* at 3.5 mg L^{-1} ammonium-N. These data partly agree with our results concerning the total concentration of NH_4^+-N. High RGRs were obtained for the treatments [75-25]/10, with an initial NH_4^+-N concentration of 3.6 mg L^{-1}, and [75-25]/50, with a concentration of 17.3 mg L^{-1}. The 10% dilutions [50-50]/10, [25-75]/10, and [0-100]/10 had similar total NH_4^+-N starting values of 7.1, 10.7, and 14.4 mg L^{-1}, respectively. However, the RGR was significantly lower in these three treatments for both species. It can be assumed that other factors, such as the ratio of nitrate to ammonium, had a certain impact on the RGRs of *L. minor* and *W. hyalina*. This is in agreement with the investigations of Mehrer and Mohr [26] and Hecht and Mohr [27] on *Sinapis alba* seedlings. The explained the detrimental effects of higher ammonium concentrations by identifying that ammonium accumulation is not well regulated by plants. The stimulation of ammonium assimilation by simultaneously applied nitrate appears to explain the nitrate-mediated ammonium tolerance. A similar mechanism exists in duckweeds, as shown recently in *Landoltia punctata* [28]. A minor fraction of ammonium as the nitrogen source seemed to stimulate duckweed growth, while proportions of 50% and higher had a growth inhibiting effect.

Approximately six NO_3^- transporters and four NH_4^+ transporters are involved in the uptake of N for *Arabidopsis thaliana*. Nitrate acts as a signalling molecule that triggers changes in the expression of genes, metabolism, and growth. Plants have evolved several NO_3^- uptake systems to survive in the changing environment. While low affinity transporters are responsible for the uptake of a large amount of nitrate in the case of available high concentrations, high affinity transporters ensure plant survival in the presence of low nitrate concentrations [29]. Acquisition of ammonium from the aquatic environment is important, as this N source for plants may be the dominating form under certain conditions. While considerable progress was made in the last two decades, many aspects of the regulation of NH_4^+ uptake and metabolism are not yet well understood [30]. *Lemna minor* grown in an NH_4NO_3 (1:1 ratio between NH_4^+-N and NO_3^--N) containing nutrient solution preferentially took up ammonium over nitrate. It was discovered that both roots and fronds take up nitrate and ammonium from the medium. At low N concentrations, the root-to-frond biomass ratio increased. This is advantageous for the plant at a morphological level, wherein there is a lower biomass investment per unit surface area for roots than for fronds [19]. Fang et al. [16] reported a preference in NH_4^+ uptake compared to NO_3^- in *Landoltia punctata*. Turions of the duckweed *Spirodela polyrhiza* absorbed $^{15}NH_4^+$ much faster than $^{15}NO_3^-$ [31]. This was confirmed by our data, which showed that the average relative uptake rate of NH_4^+-N in almost all ratios and dilutions was higher than that of NO_3^--N.

A decrease of the ammonium concentration in nutrient media can be caused by plant uptake or by volatilization depending on the pH. With a pH value of 8 at 20 °C, less than 5% of the ammonium turns into NH_3 [32]. By looking at the pH in the present experiment, it can be concluded that the majority of the NH_4^+-N was taken up by the duckweeds.

The chloride concentrations in the experiment increased with increasing ammonium supplement because ammonium chloride was partly used to increase the NH_4^+-N concentrations. Liu et al. [33] recommended an NaCl concentration below 75 mM for *L. minor* for N and P removal from water. Concentrations of 50 mM and higher caused a decrease in the fresh weight and chlorophyll content of *L. minor*. The maximum Cl^- concentration used in the presented experiment was 9.2 mM in [0-100]/50. Therefore, the significantly reduced RGRs for both species in the ratios [50-50], [25-75], and [0-100], as compared to [75-25], could not be caused by the presence of chloride.

Duckweeds (species not identified) grown in irrigation ponds in Jordan yielded an average CPC of 26% [34]. Mohedano et al. [35] investigated the CPC of duckweed species grown in anaerobically digested swine manure in two consecutive ponds. The average CPC in the primary pond was 35% (based on dry matter) and decreased to 28% in the secondary pond, which had less nutrients available. The estimated productivity of both ponds was 24 t $year^{-1}$ ha^{-1} (ca. 46 g $week^{-1}$ m^{-2}). This value is slightly higher than our maximum value (45 g $week^{-1}$ m^{-2}). The lower CPC in their study was compensated for by a higher RGR (0.24 d^{-1}). Chakrabarti et al. [4] reported a yield of 703 kg $month^{-1}$ ha^{-1} *L. minor* (ca. 17.5 g $week^{-1}$ m^{-2}) with RGRs ranging between 0.073 d^{-1} and 0.422 d^{-1}. The duckweed was cultivated on different media containing organic manure or inorganic fertilizers. The final CPC was 36.07% and 27.12% for duckweeds grown in organic manure and in inorganic fertilizer, respectively.

The modified Schenk-Hildebrandt medium used by Appenroth et al. [2] had a nitrate-N to ammonium-N ratio of roughly [90-10]. The total ammonium-N concentration (1.3 mM) was about the same as in [75-25]/50 (1.24 mM) of the modified N-medium, while nitrate concentrations were higher in the modified Schenk-Hildebrand medium. The CPC in the presented investigation was above 25% for *L. minor* and above 35% for *W. hyalina* in almost all ratios and dilutions, which was also the result in Appenroth et al. [2] for both species. Only 100-0/1 showed a lower value of 21.1% and 30.3% for *L. minor* and *W.*

hyalina, respectively. The nitrogen availabilities in these two experiments were only slightly different, thereby confirming our own results.

If duckweed should be cultivated in an agrarian system in order to produce food and feed in the future, a standardized cultivation process needs to be implemented to yield a standardized product quality. Of high interest concerning a standardized non-axenic cultivation process is the growth of algae and microorganisms and how they influence duckweed growth and culture medium composition. The use of plant growth-promoting bacteria in particular may open up new opportunities [36]. Alongside the quality, the amount of biomass and protein yielded is of great importance. The variation of the initial biomass, hence surface coverage, could have an important impact on the productivity of a system. The higher the initial biomass, the higher the nutrient requirement over time. Therefore, highly diluted nutrient media result in low growth rates. An initial surface coverage of 20% seems optimal for a high RGR [37,38]. Such a low initial duckweed biomass, however, means less competition for other organisms competing for nutrients and light. Therefore, in these experiments, an initial surface coverage of 60% was selected. To avoid growth inhibition due to high densities ("overcrowding" [39]), a regular harvest interval should be defined. Regarding the protein yield, the RPY should be considered a good indicator for the productivity of a duckweed system.

5. Conclusions

Implementing conditions that increase the RGR and CPC, positively affect the RPY. One such condition is a suitable nutrient composition of standardized quality. The concentration of nutrients in the medium, as well as the ratio between nitrate-N and ammonium-N, influenced the RGR, CPC, and RPY in the duckweeds *L. minor* and *W. hyalina*. The modification of the promising N-medium, with a substitution of 25% nitrate-N by ammonium-N at 50% dilution, significantly increased the RPY for both species when compared to the nitrate-only ratio at the same dilution. *L. minor* yielded 41.6 ± 2.2 g week^{-1} m^{-2}, while *W. hyalina* reached 45.0 ± 5.7 g week^{-1} m^{-2}.

However, other abiotic factors, such as light intensity, light spectrum, photoperiod, temperature, water and duckweed movement, as well as biotic factors, such as the growth of algae and microorganisms and their effects on duckweed, should be closely investigated. A stable cultivation process is only possible if all the biotic and abiotic factors are complementary and optimized for the species of choice.

Supplementary Materials: The following are available online at https://www.mdpi.com/article/10.3390/plants10081741/s1, Figure S1: relative weekly yield (RY, week^{-1}) based on DW, for *L. minor* (grey shaded columns) and *W. hyalina* (white columns), cultivated for seven days in nutrient solutions with different NO_3^--N to NH_4^+-N ratios in different dilutions, based on N-medium. For further explanations, see Figure 1. Table S1: tap water analysis municipal utilities Osnabrueck Wittefeld.

Author Contributions: Conceptualization, A.U., J.D., D.R. and F.P.; methodology, A.U., J.D., D.R. and F.P.; validation, H.-W.O.; investigation, J.D., D.R. and F.P.; data curation, J.D. and F.P.; writing—original draft preparation, F.P.; writing—review and editing F.P., J.D., K.-J.A. and H.-W.O.; visualization, F.P.; supervision, A.U., K.-J.A. and H.W.; project administration, H.W.; funding acquisition, J.D., A.U., H.-W.O. and H.W. All authors have read and agreed to the published version of the manuscript.

Funding: This research was funded by Deutsche Bundesstiftung Umwelt (DBU), grant number 34223/01-46.

Institutional Review Board Statement: Not applicable.

Informed Consent Statement: Not applicable.

Conflicts of Interest: The authors declare no conflict of interest.

References

1. Henchion, M.; Hayes, M.; Mullen, A.M.; Fenelon, M.; Tiwari, B. Future protein supply and demand: Strategies and factors influencing a sustainable equilibrium. *Foods* **2017**, *6*, 53. [CrossRef]
2. Appenroth, K.J.; Sree, K.S.; Böhm, V.; Hammann, S.; Vetter, W.; Leiterer, M.; Jahreis, G. Nutritional value of duckweeds (Lemnaceae) as human food. *Food Chem.* **2017**, *217*, 266–273. [CrossRef]
3. Appenroth, K.J.; Sree, K.S.; Bog, M.; Ecker, J.; Seeliger, C.; Böhm, V.; Lorkowski, S.; Sommer, K.; Vetter, W.; Tolzin-Banasch, K.; et al. Nutritional value of the duckweed species of the genus Wolffia (Lemnaceae) as human food. *Front. Chem.* **2018**, *6*, 483. [CrossRef] [PubMed]
4. Chakrabarti, R.; Clark, W.D.; Sharma, J.G.; Goswami, R.K.; Shrivastav, A.K.; Tocher, D.R. Mass production of Lemna minor and its amino acid and fatty acid profiles. *Front. Chem.* **2018**, *6*, 479. [CrossRef] [PubMed]
5. Ziegler, P.; Adelmann, K.; Zimmer, S.; Schmidt, C.; Appenroth, K.J. Relative in vitro growth rates of duckweeds (Lemnaceae)—The most rapidly growing higher plants. *Plant Biol.* **2015**, *17*, 33–41. [CrossRef] [PubMed]
6. Sree, K.S.; Sudakaran, S.; Appenroth, K.J. How fast can angiosperms grow? Species and clonal diversity of growth rates in the genus Wolffia (Lemnaceae). *Acta Physiol. Plant* **2015**, *37*, 204. [CrossRef]
7. Xu, J.; Cui, W.; Cheng, J.J.; Stomp, A.M. Production of high-starch duckweed and its conversion to bioethanol. *Biosyst. Eng.* **2011**, *110*, 67–72. [CrossRef]
8. Xu, J.; Cheng, J.; Stomp, A.M. Growing Spirodela polyrhiza in swine wastewater for the production of animal feed and fuel ethanol: A Pilot Study. *CLEAN Soil Air Water* **2012**, *40*, 760–765. [CrossRef]
9. Landolt, E.; Kandeler, R. *Biosystematic Investigations in the Family of Duckweeds (Lemnaceae)—Vol. 4: The Family of Lemnaceae—A Monographic Study*; Geobotanisches Institut ETH: Zürich, Germany, 1987.
10. Sońta, M.; Rekiel, A.; Batorska, M. Use of duckweed (*Lemna* L.) in sustainable livestock production and aquaculture—A review. *Ann. Anim. Sci.* **2019**, *19*, 257–271. [CrossRef]
11. Gwaze, F.R.; Mwale, M. The Prospect of Duckweed in Pig Nutrition: A Review. *J. Agric. Sci.* **2015**, *7*, 189–199. [CrossRef]
12. Moss, M.E. Economics and Feed Value of Integrating Duckweed Production with a Swine Operation. Master's Thesis, Texas Tech University, Lubbock, TX, USA, 1999.
13. Nguyen, T.K.K.; Ogle, B. Effects of replacing roasted soya beans by broken rice and duckweed on performance of growing Tau Vang chickens confined on-station and scavenging on-farm. *Livest. Res. Rural Dev.* **2004**, *16*, 56.
14. Appenroth, K.J. Media for in vitro-cultivation of duckweed. *Duckweed Forum* **2015**, *3*, 180–186.
15. Appenroth, K.J.; Teller, S.; Horn, M. Photophysiology of turion formation and germination in *Spirodela polyrhiza*. *Biol. Plant* **1996**, *38*, 95–106. [CrossRef]
16. Fang, Y.Y.; Babourina, O.; Rengel, Z.; Yang, X.E.; Pu, P.M. Ammonium and nitrate uptake by the floating plant *Landoltia punctata*. *Ann. Bot.* **2007**, *99*, 365–370. [CrossRef] [PubMed]
17. Wang, W.; Yang, C.; Tang, X.; Gu, X.; Zhu, Q.; Pan, K.; Hu, Q.; Ma, D. Effects of high ammonium level on biomass accumulation of common duckweed *Lemna minor* L. *Environ. Sci. Pollut. Res. Int.* **2014**, *21*, 14202–14210. [CrossRef] [PubMed]
18. Caicedo, J.; van der Steen, N.P.; Arce, O.; Gijzen, H.J. Effect of total ammonia nitrogen concentration and pH on growth rates of duckweed (*Spirodela polyrhiza*). *Water Res.* **2000**, *34*, 3829–3835. [CrossRef]
19. Cedergreen, N.; Madsen, T.V. Nitrogen uptake by the floating macrophyte *Lemna minor*. *New Phytol.* **2002**, *155*, 285–292. [CrossRef]
20. Simonne, A.H.; Simonne, E.H.; Eitenmiller, R.R.; Mills, H.A.; Cresman, C.P. Could the Dumas method replace the Kjeldahl digestion for nitrogen and crude protein determinations in foods? *J. Sci. Food Agric.* **1997**, *73*, 39–45. [CrossRef]
21. Casal, J.A.; Vermaat, J.E.; Wiegman, F. A test of two methods for plant protein determination using duckweed. *Aquat. Bot.* **2000**, *67*, 61–67. [CrossRef]
22. VDLUFA. *Methodenbuch Band 1: Die Untersuchung von Böden, Methode A 6.1.4.1—Bestimmung von Mineralischem Stickstoff (Nitrat und Ammonium) in Bodenprofilen (Nmin-Labormethode)*; VDLUFA-Verlag: Darmstadt, Germany, 2012.
23. VDLUFA. *Methodenbuch Band 1: Die Untersuchung der Böden, Methode A 6.1.1.1—Bestimmung von Nitrat-Stickstoff durch UV-Absorption*; VDLUFA-Verlag: Darmstadt, Germany, 2012.
24. Iatrou, E.I.; Kora, E.; Stasinakis, A.S. Investigation of biomass production, crude protein and starch content in laboratory wastewater treatment systems planted with *Lemna minor* and *Lemna gibba*. *Environ. Technol.* **2019**, *40*, 2649–2656. [CrossRef]
25. Zhang, K.; Chen, Y.-P.; Zhang, T.-T.; Zhao, Y.; Shen, Y.; Huang, L.; Gao, X.; Guo, J.-S. The logistic growth of duckweed (*Lemna minor*) and kinetics of ammonium uptake. *Environ. Technol.* **2014**, *35*, 562–567. [CrossRef]
26. Mehrer, I.; Mohr, H. Ammonium toxicity: Description of the syndrome in *Sinapis alba* and the search for its causation. *Physiol. Plant.* **1989**, *77*, 545–554. [CrossRef]
27. Hecht, U.; Mohr, H. Factors controlling nitrate and ammonium accumulation in mustard (*Sinapis alba*) seedlings. *Physiol. Plant.* **1990**, *78*, 379–387. [CrossRef]
28. Tian, X.; Fang, Y.; Jin, Y.; Yi, Z.; Li, J.; Du, A.; He, K.; Huang, Y.; Zhao, H. Ammonium detoxification mechanism of ammonium-tolerant duckweed (*Landoltia punctata*) revealed by carbon and nitrogen metabolism under ammonium stress. *Environ. Pollut.* **2021**, *277*, 116834. [CrossRef] [PubMed]
29. Islam, S.; Islam, R.; Kandwal, P.; Khanam, S.; Proshad, R.; Kormoker, T.; Tusher, T.R. Nitrate transport and assimilation in plants: A potential review. *Arch. Agron. Soil Sci.* **2020**. [CrossRef]

30. Hao, D.-L.; Zhou, J.-Y.; Yang, S.-Y.; Qi, W.; Yang, K.-J.; Su, Y.-H. Function and Regulation of Ammonium Transporters in Plants. *Int. J. Mol. Sci.* **2020**, *21*, 3557. [CrossRef]
31. Appenroth, K.J.; Augsten, H.; Mattner, A.; Teller, S.; Döhler, G. Effect of UVB irradiation on enzymes of nitrogen metabolism in turions of *Spirodela polyrhiza* (L.) Schleiden. *J. Photochem. Photobiol. B* **1993**, *18*, 215–220. [CrossRef]
32. Emerson, K.; Russo, R.C.; Lund, R.E.; Thurston, R.V. Aqueous Ammonia Equilibrium Calculations: Effect of pH and Temperature. *J. Fish. Res. Board Can.* **1975**, *32*, 2379–2383. [CrossRef]
33. Liu, C.; Dai, Z.; Sun, H. Potential of duckweed (*Lemna minor*) for removal of nitrogen and phosphorus from water under salt stress. *J. Environ. Manag.* **2017**, *187*, 497–503. [CrossRef] [PubMed]
34. Shammout, M.W.; Zakaria, H. Water lentils (duckweed) in Jordan irrigation ponds as a natural water bioremediation agent and protein source for broilers. *J. Ecol. Eng.* **2015**, *83*, 71–77. [CrossRef]
35. Mohedano, R.A.; Costa, R.H.R.; Tavares, F.A.; Belli Filho, P. High nutrient removal rate from swine wastes and protein biomass production by full-scale duckweed ponds. *Bioresour. Technol.* **2012**, *112*, 98–104. [CrossRef] [PubMed]
36. Khairina, Y.; Jog, R.; Boonmak, C.; Toyama, T.; Oyama, T.; Morikawa, M. Indigenous bacteria, an excellent reservoir of functional plant growth promoters for enhancing duckweed biomass yield on site. *Chemosphere* **2021**, *268*, 129247. [CrossRef]
37. Hutabarat, R.C.S.M.; Indradewa, D. Effects of water flow rate and surface cover plant density on the growth of duckweed (*Lemna minor* L.). *Ilmu Pertan. Agric. Sci.* **2020**, *5*, 98–109. [CrossRef]
38. Verma, R.; Suthar, S. Impact of density loads on performance of duckweed bioreactor: A potential system for synchronized wastewater treatment and energy biomass production. *Environ. Prog. Sustain. Energy* **2015**, *34*, 1596–1604. [CrossRef]
39. Färber, E.; Königshofer, H.; Kandeler, R. Ethylene Production and Overcrowding in Lemnaceae. *J. Plant Physiol.* **1986**, *124*, 379–384. [CrossRef]

plants

MDPI

Article

Influence of Light Intensity and Spectrum on Duckweed Growth and Proteins in a Small-Scale, Re-Circulating Indoor Vertical Farm

Finn Petersen [1,*], Johannes Demann [1], Dina Restemeyer [1], Hans-Werner Olfs [1], Heiner Westendarp [1], Klaus-Juergen Appenroth [2] and Andreas Ulbrich [1]

[1] Faculty of Agricultural Sciences and Landscape Architecture, University of Applied Sciences Osnabrück, Am Krümpel 31, 49090 Osnabrück, Germany; johannes.demann@hs-osnabrueck.de (J.D.); dina.restemeyer@hs-osnabrueck.de (D.R.); h-w.olfs@hs-osnabrueck.de (H.-W.O.); h.westendarp@hs-osnabrueck.de (H.W.); a.ulbrich@hs-osnabrueck.de (A.U.)

[2] Matthias-Schleiden-Institute–Plant Physiology, University of Jena, Dornburger Str. 159, 07743 Jena, Germany; klaus.appenroth@uni-jena.de

* Correspondence: finn.petersen@hs-osnabrueck.de; Tel.: +49-54-1969-5098

Abstract: Duckweeds can be potentially used in human and animal nutrition, biotechnology or wastewater treatment. To cultivate large quantities of a defined product quality, a standardized production process is needed. A small-scale, re-circulating indoor vertical farm (IVF) with artificial lighting and a nutrient control and dosing system was used for this purpose. The influence of different light intensities (50, 100 and 150 µmol m^{-2} s^{-1}) and spectral distributions (red/blue ratios: 70/30, 50/50 and 30/70%) on relative growth rate (RGR), crude protein content (CPC), relative protein yield (RPY) and chlorophyll a of the duckweed species *Lemna minor* and *Wolffiella hyalina* were investigated. Increasing light intensity increased RGR (by 67% and 76%) and RPY (by 50% and 89%) and decreased chlorophyll a (by 27% and 32%) for *L. minor* and *W. hyalina*, respectively. The spectral distributions had no significant impact on any investigated parameter. *Wolffiella hyalina* achieved higher values in all investigated parameters compared to *L. minor*. This investigation proved the successful cultivation of duckweed in a small-scale, re-circulating IVF with artificial lighting.

Keywords: Lemnaceae; *Lemna minor*; *Wolffiella hyalina*; red/blue ratio; standardized production; yield; light quality; light quantity; controlled environment

check for updates

Citation: Petersen, F.; Demann, J.; Restemeyer, D.; Olfs, H.-W.; Westendarp, H.; Appenroth, K.-J.; Ulbrich, A. Influence of Light Intensity and Spectrum on Duckweed Growth and Proteins in a Small-Scale, Re-Circulating Indoor Vertical Farm. *Plants* **2022**, *11*, 1010. https://doi.org/10.3390/plants11081010

Academic Editor: Francesco Serio

Received: 25 February 2022
Accepted: 4 April 2022
Published: 7 April 2022

1. Introduction

The term duckweed comprises 36 species [1,2] of 5 genera, belonging to the family of Lemnaceae Martinov [3,4]. They are characterized, amongst other aspects, by their fast growth rate [5,6], high nutrient uptake capacity [7,8] as well as by their edibility [9,10] and variability of nutritional values influenced by cultivation conditions [11,12]. Those are key aspects for further use in human and animal nutrition, biotechnology or wastewater treatment.

In order to continuously produce large quantities of biomass with a defined quality (e.g., for human nutrition), a standardized cultivation process is necessary. One possible solution in the future might be the cultivation of duckweed in re-circulating (also described as closed) indoor vertical farms (IVF) with artificial lighting. By stacking several layers of cultivation areas above each other, the land utilization efficiency is increased [13,14]. When operating an IVF in a controlled environment, it is possible to regulate plant-relevant abiotic factors, e.g., nutrient composition and concentration, light intensity and spectrum, photoperiod, the temperature of water and air, water flow rate or humidity according to the grower's demand. Resources, such as nutrients, water and pesticides, can be used efficiently. This can positively affect the quantity and quality of the crops. Additionally, the use of IVFs will allow year-round crop production, even in areas with short growing seasons or unfavorable climatic conditions [13–16]. One shortcoming of this cultivation technology is the relatively high energy input, e.g., the production of one kg of curled lettuce required 7–9 kWh of electric energy [14].

However, closed hydroponic systems are already successfully used to cultivate different crops in large quantities. This includes tomatoes, cucumbers, peppers, different leafy greens, strawberries and even rice or maize [17]. The advantages of closed hydroponic systems compared to conventional farming are enormous, as up to 85% of fertilizers and 90% of water can be saved, while a productivity increase of up to 250% is possible [18]. The water and nutrient use efficiency of tomatoes cultivated in a closed hydroponic system was 23% higher compared to an open system in both cases [19]. The water use efficiency for tomatoes cultivated in a closed system in The Netherlands was 66 kg of yield per cubic meter of water applied [18]. Another study described zero discharge of nutrients and pesticides to the environment in the production of sweet peppers and autumn cucumber in a closed hydroponic system [20].

In order to also achieve an efficient system for duckweed cultivation, all necessary abiotic factors must be evaluated. Two of these abiotic factors are light intensity and the spectral light distribution. In nature, *Lemnaceae* grow in sunny as well as in shaded habitats, but the latter habitats are favorable due to lower light intensities and less extreme temperatures [21]. The plant's reaction to different light intensities is dependent on the species and abiotic factors, such as nutrients or temperature, while the light spectrum is another important parameter [22]. *Wolffia arrhiza* cultivated in steady-state conditions with blue light showed higher protein and chlorophyll contents compared to red light [22]. Increasing light intensities slightly increased the relative growth rates (RGRs) of *Lemna gibba* [23] and *Lemna aequinoctialis* [24]. Very high intensities, however, lead to light saturation. Light intensities above this point will not increase the photosynthetic activity of the plant and could lead to damages due to oxygen stress (photoinhibition). The light saturation point depends on factors such as temperature and varies for different duckweed species. A light saturation of 342 $\mu mol\ m^{-2}\ s^{-1}$ for *L. minor* [25] and of 400 $\mu mol\ m^{-2}\ s^{-1}$ for *L. minor* and *Lemna minuta* were observed [26], while *Landoltia punctata* (formerly *Spirodela punctata*) reached light saturation between 600 and 1200 $\mu mol\ m^{-2}\ s^{-1}$ at 30 °C [27]. Considering the cost of artificial lighting, an optimum of 110 $\mu mol\ m^{-2}\ s^{-1}$ was obtained for *L. aequinoctialis* [24].

The aim of our research was to evaluate the influence of different light intensities and spectral distributions on the RGR, crude protein content (CPC) and relative protein yield (RPY) in the duckweeds *L. minor* and *W. hyalina* when cultivated in a small-scale, re-circulating, aquatic IVF. Additionally, for both species, the chlorophyll a content was determined as a plant cultivation indicator. We selected clones of these two species because they showed good performance in earlier experiments concerning growth rates and protein contents [9].

2. Materials and Methods

2.1. Indoor Vertical Farm

Two duckweed species, *Lemna minor* L. (clone 9441; Germany) and *Wolffiella hyalina* Delile Monod (clone 9525; India), were chosen for the experiments due to their fast growth rates and high protein contents [28]. The plant material was obtained from the Duckweed Stock Collection of the Department of Plant Physiology, University of Jena, Germany.

Experiments were carried out in a container (length × width × height: 5 × 3 × 3 m) at the campus of the University of Applied Sciences Osnabrück, Germany. Trials were conducted in a re-circulating, aquatic IVF (Figures 1 and 2).

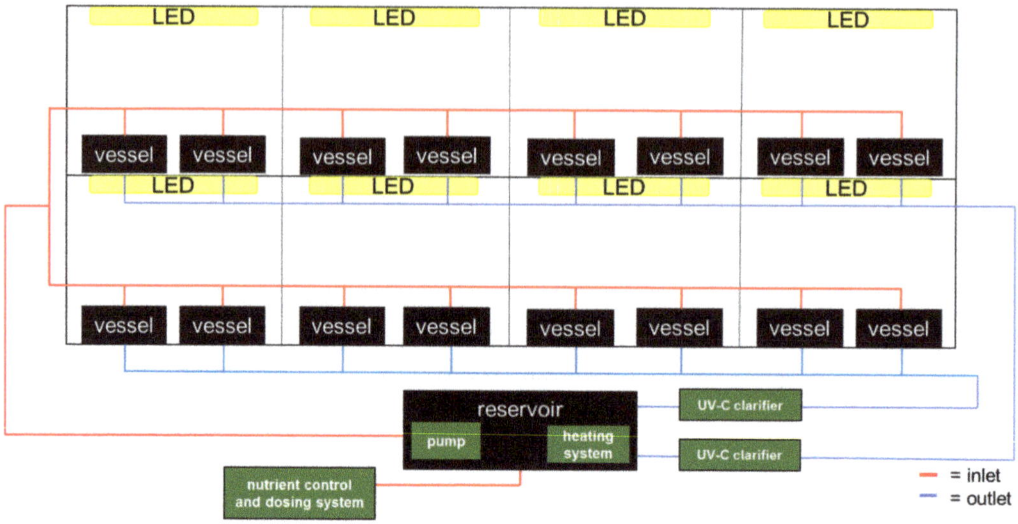

Figure 1. Scheme of the experimental set-up designed as indoor vertical farm (IVF). Black boxes depict the cultivation vessels for the duckweeds and the nutrient solution reservoir, yellow boxes depict the LEDs and green boxes depict the necessary technology to run the re-circulating system. Red lines indicate the nutrient solution inlet and blues lines the outlet.

It consisted of a 90 L reservoir for the nutrient solution connected to all duckweed cultivation vessels via flexible tubes. A submergible and adjustable pump (AquaForte DM-10000 Vario, SIBO BV, Veghel, The Netherlands) was installed at the bottom of the reservoir to create a continuous flow between reservoir and cultivation vessels. A nutrient control and dosing system (Pro Controller and PeriPods, Bluelab Corporation Ltd., Tauranga, New Zealand) added the required liquid fertilizers from stock solutions to the tap water in the reservoir. A heating system (Super Fish Smart Heater 500 W, Aquadistri BV, Klundert, The Netherlands) was installed at the bottom of the reservoir to keep a constant water temperature. The vessels (56 cm length × 37 cm width × 10 cm height) used for cultivation were positioned in a two-layer storage rack. On one side (width) of the cultivation vessel, the water inlet, a rectangular pipe leading the water inflow to the bottom of the vessel, was installed. On the opposite side, an outlet was located at 7 cm height. To guarantee no duckweed was lost from the vessel by flowing through the outlet, a wall was installed 7 cm before the outlet. The upper side of the wall was above water level, while the bottom side did not touch the ground of the vessel. This way, the nutrient solution could flow back into the reservoir, while the floating duckweed was hindered from passing the barrier. The net cultivation area per vessel decreased to 0.49×0.37 m = 0.1813 m^2 by applying this method. The unoccupied surface was covered with black PE in order to prevent algae growth in that area. The outlet solution from each of the two storage rack levels was led through UV-C clarifiers (OSAGA UVC36, Fischfarm Otto Schierhölter, Glandorf, Germany) in order to reduce the growth of ubiquitous algae and bacteria.

As light sources, dimmable LEDs with an adjustable spectrum (LED-LE1200-E03W-1-S, DH Licht GmbH, Wülfrath, Germany) were installed 34 cm above the water surface in the vessels. The settings were adjusted with the VisuSpectrum 3.0 software (DH Licht GmbH, Wülfrath, Germany and RAM GmbH Mess- und Regeltechnik, Herrsching, Germany).

Figure 2. Experimental set-up in the container at the Osnabrück University of Applied Sciences, Germany. The light colors of the different spectral treatments are visible.

2.2. Experimental Design

Three different light intensities (50, 100 and 150 µmol m^{-2} s^{-1}) were used for the experiments. All of the three spectral treatments contained 20% light at 6500 K (white light), and the remaining 80% were split according to the following red (660 nm)/blue (450 nm) ratio: 70/30, 50/50 and 30/70 (%). This resulted in eight different treatments (Table 1). Light intensities were controlled using a Light Meter LI-250A (LI-COR Biosciences, Lincoln, NE, USA). The photoperiod was set to 12 h of light and 12 h of darkness per day.

Table 1. Applied light intensities and red/blue ratios (spectral distributions) in the experiments as well as the corresponding treatment abbreviation, as used throughout the text.

Light Intensity	Red/Blue Ratios	Treatment Abbreviation
50	70/30	50–70/30
50	50/50	50–50/50
50	30/70	50–30/70
100	70/30	100–70/30
100	30/70	100–30/70
150	70/30	150–70/30
150	50/50	150–50/50
150	30/70	150–30/70

Pre-cultivation occurred for three days under above mentioned conditions. Experiments lasted for seven days and were conducted under non-axenic growth conditions. Vessels were placed in the storage rack based on a block design. This storage rack had eight compartments, each containing two LEDs and space for two experimental vessels. Eight treatments, with four replications for each of the two species, were investigated. In total, 16 vessels could be used at a time. Two replicates per light intensity and spectral distribution per species were investigated at the same time. To start with a similar surface coverage of ca. 80% in each vessel, 20 g of *L. minor* and 15 g of *W. hyalina* fresh weight (FW) biomass was placed in each vessel.

The nutrient medium applied mainly consisted of commercially available fertilizers (see Table S1). The nutrient dosing was set to an electrical conductivity (EC) value of 0.6 mS cm^{-1}, which corresponds to a nutrient solution of 75-25/10 with the following composition and concentrations (all given in mM): NO_3^--N: 0.76, NH_4^+-N: 0.25, PO_4^{3-}: 0.1, K^+: 0.91, Mg^{2+}: 0.13, SO_4^{2-}: 0.32, Ca^+: 0.22, Cl^-: 0.34, Fe^{3+}: 0.0025, BO_3^{3-}: 0.0005, Mn^{2+}: 0.0013, Zn^{2+}: 0.001 and Na^+: 0.08 [28]. When the EC dropped below target value in the time course of cultivation, additional nutrient solution was added until the target EC was reached again.

The pH at the beginning of the experiments was 7.6. The heating system was set to a target value of 24 °C, and the pump was adjusted to a flow rate of 2 L min^{-1}.

At the end of the experiments, duckweeds were harvested with a metal sieve, rinsed with tap water, spin-dried for three minutes with a Top Spin Compact (Chal-Tec GmbH, Berlin, Germany) to remove attached water and weighed.

2.3. *Analytical Methods*

2.3.1. Relative Growth Rate

Dry weight (DW) was determined from FW via oven drying at 65 °C for 72 h. At time 0, four samples per species of the same FW as the starting material were used to determine the DW at the beginning of the experiments.

Relative growth rates (RGRs) per day were calculated according to Equation (1) [6], using the values of the DW at the start (t0) and after seven days of cultivation (t7):

$$RGR = (\ln DW_{t7} - \ln DW_{t0})/(t7 - t0) \quad (1)$$

where RGR is the relative increase in the DW per day (d^{-1}).

2.3.2. Crude Protein Content and Relative Protein Yield

Dried samples were ground and homogenized using a laboratory mill and stored for further analysis. The nitrogen content of the dried samples was determined using the Dumas method [29] using an elemental analyzer (FP628, Leco, Saint Joseph, MI, USA), and CPC was calculated using the factor 6.25 [9,30].

The relative weekly yield (RY; g biomass obtained after one week of cultivation starting with 1 g) was calculated from the RGR using Equations (2) and (3):

$$\ln DW_{t7} = \ln DW_{t0} + RGR \cdot (t7 - t0) \quad (2)$$

$$RY = \exp(\ln DW_{t7}) \quad (3)$$

The RY was further used to calculate the relative protein yield (RPY; g protein $week^{-1}$ m^{-2}) by multiplying it with the crude protein content (CPC) and extrapolating it to one square meter, according to Equation (4):

$$RPY = RY \times CPC/(0.1813\ m^2 \times 100) \quad (4)$$

where 0.1813 m^2 is the cultivation area of the vessels used in the experiments.

2.3.3. Chlorophyll a

The chlorophyll a content was determined according to DIN 38409-60:2019-12 [31], using ethanol (ω(EtOH) = 90%) as a solvent. Four replicates of the starting biomass and four replicates of each treatment at the end of the experiments were analyzed. Laboratory analysis of the chlorophyll a content took place in the dark immediately after the samples were taken according to the following scheme: A net weight of 1.000 ± 0.005 g FW duckweed biomass was placed in 50 mL centrifuge tubes, filled with 10 mL of boiling solvent and homogenized for 60 s using an Ultra-Turrax. The resulting extract was cooled and treated in an ultrasonic bath for 30 min in the dark. Afterwards, the extract was filtered into a 100 mL volumetric flask, filled with ethanol to the calibration mark and homogenized again by shaking. The extract was placed into a glass cuvette. Of the remaining extract, 15 mL was put into a centrifuge tube, added with 100 µL of hydrochloric acid (2 M) and homogenized for the correction of phaeopigments. Both extracts and the pure solvent were finally put into different glass cuvettes and analyzed using a spectrophotometer (Specord 40, Analytik Jena AG, Jena, Germany) at 665 and 750 nm.

The following modified Equation (5) was applied to calculate the chlorophyll a content in the fresh duckweed biomass [31]:

$$\omega_{\text{Chlorophyll}-\text{a}} = ((A_{665v} - A_{750v}) - (A_{665n} - A_{750n}))\cdot\frac{R}{R-1}\cdot\frac{V_E}{m_p\cdot d\cdot\alpha\cdot 1000} \quad (5)$$

with

$\omega_{\text{Chlorophyll-a}}$: Chlorophyll a content (mg/g FW);
A_{665v}: Absorption of the extract before acidification, measured at 665 nm;
A_{750v}: Absorption of the extract before acidification, measured at 750 nm (for the correction of phaeopigments);
A_{665n}: Absorption of the extract after acidification, measured at 665 nm;
A_{750n}: Absorption of the extract after acidification, measured at 750 nm (for the correction of phaeopigments);
R: Ratio of A_{665v}/A_{665n} for pure Chlorophyll-a; R = 1.7;
V_E: Volume of the extract in milliliters (ml);
m_P: Net weight of the duckweed biomass sample (g);
d: Thickness of the cuvette (cm); d = 1.

Additionally, the dry matter content of each sample was determined by drying plant material at 105 °C until it reached a constant weight. The chlorophyll a FW content was then multiplied with the dry matter content to calculate the chlorophyll a DW content.

2.3.4. Nutrient Solution

A nutrient solution sample was taken at the start (day 0) and the end (day 7) of the experiments from the reservoir, filtered (MN 619 EH, Machery Nagel GmbH & Co. KG, Düren, Germany) to remove particles and instantly frozen at −18 °C. Nitrate-N and ammonium-N concentrations in these samples were measured according to German standard methods [32,33] with a Lambda 25 UV/VIS Spectrometer (Perkin Elmer, Waltham, MA, USA). Other nutrients were analyzed according to DIN EN ISO 11885:2009-09 with an ICP-OES (ICAP 7400, Thermo Fischer Scientific, Waltham, MA, USA) [34].

2.4. Statistics

All data are based on four replicates and are given as mean ± standard deviations. The data were analyzed statistically using one-way ANOVA and Tukey's post hoc test at 5% significance level, using the software program SPSS 25 (IBM, Armonk, NY, USA).

3. Results

3.1. Relative Growth Rate

Figure 3 shows the RGR based on DW. An increasing light intensity increased the RGR for both species. The highest RGR for *L. minor* was reached at 150–70/30 (0.13 ± 0.013 d^{-1}) and for *W. hyalina* at 150–50/50 (0.21 ± 0.01 d^{-1}). The minimum values were obtained at 50–30/70 for *L. minor* with an RGR of 0.078 ± 0.012 d^{-1} and at 50–50/50 for *W. hyalina* with an RGR of 0.119 ± 0.003 d^{-1}. The percentage increase from the lowest to the highest RGR was 67% for *L. minor* and 76% for *W. hyalina*. The results of all three *L. minor* treatments cultivated at 150 µmol m^{-2} s^{-1} were significantly higher compared to the 50 µmol m^{-2} s^{-1} treatments. *W. hyalina* cultivated at a light intensity of 150 µmol m^{-2} s^{-1} reached significantly higher RGRs than the 100 and 50 µmol m^{-2} s^{-1} treatments. The light spectrum showed no significant impact on the RGR in any treatment.

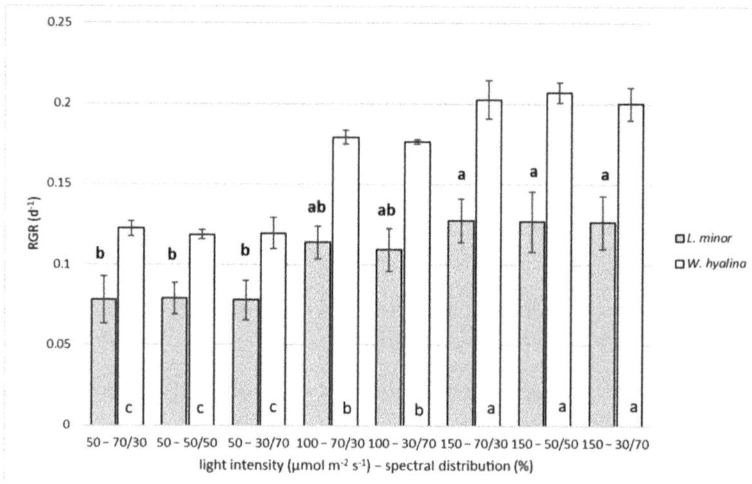

Figure 3. Relative growth rate (RGR; d^{-1}), based on dry weight, for Lemna minor (gray shaded columns) and Wolffiella hyalina (white columns). Plants were cultivated for seven days with different light intensities (50, 100 and 150 µmol m^{-2} s^{-1}) and spectral distributions (red/blue: 70/30, 50/50 and 30/70%). For the abbreviations used, see Table 1. Number of parallel samples n = 4. Different letters indicate significances within a species, based on one-way ANOVA test, Tukey $p \leq 0.05$. Error bars indicate standard deviations.

3.2. Crude Protein Content and Relative Protein Yield

The CPC, based on DW, varied in a narrow range between 31.8 ± 0.8% and 32.4 ± 1.2% for *L. minor* and between 39.3 ± 1.0% and 40.0 ± 0.8% for *W. hyalina* for the different treatments. No significant differences in the CPC for the different light intensities and spectral distributions within a species were detected.

The RPY in grams per week and m^2, based on DW, is presented in Figure 4. It ranged from 2.96 ± 0.30 to 4.44 ± 0.55 g $week^{-1}$ m^{-2} (50–70/30 and 150–50/50, respectively) for *L. minor*, while for *W. hyalina*, the range was from 5.01 ± 0.35 g $week^{-1}$ m^{-2} at 50–30/70 to 9.48 ± 0.39 g $week^{-1}$ m^{-2} at 150–50/50. The difference from the lowest to the highest value for *L. minor* was 50%, and for *W. hyalina*, it reached 89%. Higher light intensities resulted in higher relative protein yields. Overall, *W. hyalina* achieved higher RPYs in all treatments compared to *L. minor*. The higher the light intensity, the higher the difference between the species RPYs, meaning that at the highest light intensities (150 µmol m^{-2} s^{-1}), *W. hyalina* yielded more protein compared to *L. minor* than at the two lower light intensities. The treatments 50–70/30 and 50–30/70 were significantly lower compared to all other *L.*

minor treatments, except for 100–30/70. For *W. hyalina*, all treatments with a light intensity of 50 µmol m^{-2} s^{-1} (50–70/30, 50–50/50 and 50–30/70) were significantly lower compared to the other treatments with higher light intensities. No significant differences, in neither of the two duckweed species, were observed between the different spectral distributions.

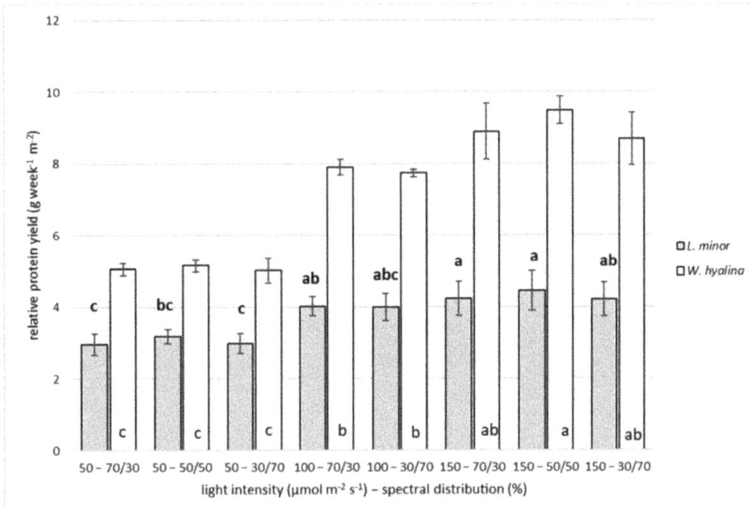

Figure 4. Relative protein yield (RPY; g $week^{-1}$ m^{-2}), based on dry weight, for *Lemna minor* (gray shaded columns) and *Wolffiella hyalina* (white columns). For further explanations, see Figure 3.

3.3. *Chlorophyll a*

The content of chlorophyll a for both species after seven days of experiments ranged between 5.32 ± 0.51 mg g^{-1} and 7.29 ± 0.39 mg g^{-1} for *L. minor* at 150–50/50 and 50–70/30, respectively (Figure 5). The maximum content for *W. hyalina* was 9.98 ± 1.01 mg g^{-1} chlorophyll a, achieved at 50–30/70, while the minimum content (6.83 ± 0.39 mg g^{-1}) was obtained at 150–30/70. This corresponded to a decrease of 27% for *L. minor* and 32% for *W. hyalina*.

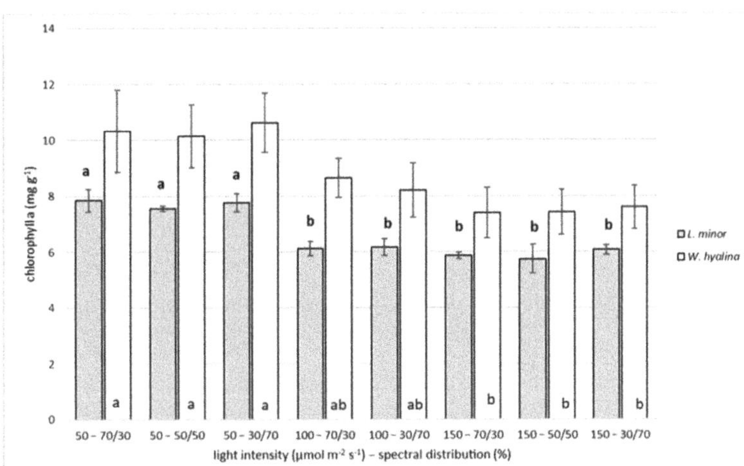

Figure 5. Chlorophyll a content, in mg g^{-1} (based on dry weight), for *Lemna minor* (gray shaded columns) and *Wolffiella hyalina* (white columns). For further explanations, see Figure 3.

A significant decline between the treatments of the lowest light intensity (50 µmol m^{-2} s^{-1}) and the two higher treatments (100 and 150 µmol m^{-2} s^{-1}) can be observed for *L. minor*. For *W. hyalina*, the 150 µmol m^{-2} s^{-1} treatments were significantly lower compared to the 50 µmol m^{-2} s^{-1} treatments. Different light spectra had no significant impact on the chlorophyll a content of both species.

3.4. Nutrients

In Table 2, the percentage reduction in different nutrient components in the nutrient medium after seven days of experiments compared to the initial concentration is presented. A percentage increase (shown as negative values) in certain substances was possible due to the EC-based nutrient dosing of the stock solutions.

Table 2. Percentage reduction in nutrient solution substances for *L. minor* and *W. hyalina*, based on one solution sample taken at the beginning and the end of experiments from the reservoir. Duckweeds were separately cultivated for seven days in a re-circulating, aquatic system. Negative values indicate an increase in the corresponding substance due to EC-based nutrient dosing.

Substance	*L. minor*	*W. hyalina*
NH_4^+-N	97.2	97.7
NO_3^--N	12.8	−6.6
PO_4^{3-}	52.8	26.6
K^+	7.9	−1.4
Mg^{2+}	−7.9	−22.6
SO_4^{2-}	−8.0	−11.7
Ca^+	−8.1	−10.1
Fe^{3+}	95.6	94.5
BO_3^{3-}	84.8	2.2
Mn^{2+}	80.3	98.6
Zn^{2+}	84.2	89.7
Na^+	29.1	23.7

A strong reduction of more than 80% can be seen for ammonium-N, iron, manganese, zinc, and in case of *L. minor*, also for boron. Nitrate-N was only slightly decreased for *L. minor* (12.8%) and showed a minor increase for *W. hyalina*. Similar results were also observed for potassium. An increase in magnesium, sulfur and calcium occurred for both species.

Compared to the start of experiments, the pH showed a minor increase with an average value of 7.8 for the *L. minor* experiments and 7.9 for the *W. hyalina* experiments.

4. Discussion

4.1. Relative Growth Rate

The RGR determined in our study differed between both investigated species and growth conditions. An increase in light intensity from 50 to 150 µmol m^{-2} s^{-1} significantly increased the RGR of *L. minor* and *W. hyalina*. Our data agree with other published investigations. Paolacci et al. [26] reported that increasing light intensities between 6 and 1000 µmol m^{-2} s^{-1} increased the RGR of *L. minor* and *L. minuta* cultivated in sterile growth rooms at 20 °C with a light:dark cycle of 16:8 h. At light intensities below 40 µmol m^{-2} s^{-1}, no significant differences were detected between the RGR of both species, while above 90 µmol m^{-2} s^{-1}, *L. minuta* had significantly higher RGRs than *L. minor*. The latter reached an RGR of 0.26 d^{-1} when grown at 150 µmol m^{-2} s^{-1}. This was higher compared to our result, but cultivation conditions varied, which might provide a possible explanation for this difference.

At comparatively low light intensities between 30 to 105 µmol m^{-2} s^{-1}, *L. aequinoctialis* reached an RGR of 0.19 d^{-1} at the highest light intensity, when cultivated in monoculture, while *L. punctata* and *Spirodela polyrhiza* reached 0.18 d^{-1} and 0.15 d^{-1} under the same growth conditions, respectively [35]. Increasing light intensity and photoperiod increased growth rate, biomass and starch production in *L. aequinoctialis*. Considering the costs for lighting, an optimum regarding those factors was reached at 110 µmol m^{-2} s^{-1} [24]. A sevenfold increase in light intensity (from 100 to 700 µmol m^{-2} s^{-1}) resulted in a 25% greater RGR of *L. gibba* [23]. This increase in RGR was lower compared to *L. minor's* RGR increase of 67% and *W. hyalina*'s increase of 76% at a 200% greater light input in our study.

The maximum obtained RGRs of 0.13 d^{-1} for *L. minor* and 0.21 d^{-1} for *W. hyalina* in the presented study are lower compared to the highest achieved values of 0.42 d^{-1} and 0.52 d^{-1} for the same clones, respectively, grown under sterile conditions in batch cultures [6]. However, under non-axenic conditions, certain cultivation adaptations due to inhibiting factors, such as algae or fungus growth, need to be considered [36,37]. A highly diluted growth medium, comparatively low light intensities and a moderate temperature were applied in our re-circulating IVF for non-axenic duckweed cultivation. Regarding the investigation of Petersen et al. [28], the same nutrient medium with a dilution of 10% resulted in an RGR of 0.21 d^{-1} for *W. hyalina*. This is in exact agreement with the results of the current study.

In contrast, other studies reported that different light intensities had no significant impact on the RGR of duckweed species. The RGR of *Lemna minor* grown on synthetic dairy wastewater did not increase with increasing light intensities between 50 and 850 µmol m^{-2} s^{-1} [38]. *Lemna gibba* reached constant high growth rates under different light intensities between 50 and 1000 µmol m^{-2} s^{-1}; however, higher intensities led to increasing zeaxanthin levels. This way, a large fraction of the absorbed light was dissipated non-photochemically [39].

The light spectra in the presented experiments had no significant impact on any investigated parameters for both species. However, it has to be kept in mind that in this study, pure red or blue light was never used. There was always a white light background of the light intensity of 20%, and the ratios between blue and red light were never higher than 70:30%.

Up to now, only a few investigations concerning this parameter have been carried out regarding duckweed RGR. *Landoltia punctata* cultivated under fluorescent white light, blue LED and white LED at 110 µmol m^{-2} s^{-1} showed no significant RGR differences [40]. There was also no significant difference in the RGR of *S. polyrhiza* when cultivated at 60 µmol m^{-2} s^{-1} using red and blue LEDs (660 and 460 nm, respectively) [41], which is in agreement with our results. Xu et al. [42] described that the application of red and blue light at the same time can be absorbed by plants more efficiently compared to other spectra and resulted in high photosynthetic efficiency. *Spirodela polyrhiza* cultured in eutrophic medium reached a significantly higher total biomass yield when a red:blue ratio of 2:1 or 4:1 at a light intensity of 110 µmol m^{-2} s^{-1} was applied compared to monochromatic (450, 630 or 660 nm) or fluorescent light sources at the same intensities.

4.2. Crude Protein Content and Relative Protein Yield

The presented crude protein contents for both species showed no significant difference between the tested light scenarios. This is in contrast to the results reported by Stewart et al. [39], who showed that the protein content of *L. gibba*, cultivated at 50 and 1000 µmol m^{-2} s^{-1}, increased from 25% to 46%, respectively. A protein content increase from 1.5% to 2% (based on FW) was observed for *L. minor* when cultivated on synthetic dairy wastewater at a light intensity of 850 µmol m^{-2} s^{-1} compared to 50 µmol m^{-2} s^{-1}. In C3 plants, such as duckweed, higher light intensities induce the increased production of Rubisco, a soluble protein [38]. A small increase in light intensity (from 200 to 400 µmol m^{-2} s^{-1}) only slightly increased the percentage of activated Rubisco in *S. polyrhiza* [43]. This could be an explanation for the relatively stable crude protein contents in our study, as the light intensity only slightly increased from 50 to

150 µmol m^{-2} s^{-1}. A more substantial increase in light intensity, as described above, will lead to rising protein contents.

The crude protein contents in the presented experiments were rather high considering the low nutrient concentration and the low light intensities, especially regarding *W. hyalina*. Appenroth et al. [9] reported a crude protein content of 35% for *W. hyalina* and 25% for *L. minor*. These duckweeds were cultivated with a modified Schenk–Hildebrandt medium at 100 µmol m^{-2} s^{-1} continuous white light. In another experiment, the highest values for crude protein of the three species *L. aequinoctialis*, *L. punctata* and *S. polyrhiza* (33.7, 32.3 and 36.8%, respectively), were reached at 105 µmol m^{-2} s^{-1} using a one-tenth strength Hoagland solution [35]. Petersen et al. [28] reached CPCs of 32.4% for *L. minor* and 35.3% for *W. hyalina* using a stationary system with the same nutrient solution as applied in these experiments. Wheeler at al. [44] assumed that a continuous supply of nitrogen caused higher protein levels in different crops (wheat, lettuce, potato and soybean) grown in a re-circulating hydroponic system compared to the same field-grown crops. Such a mechanism might also be responsible for the CPC increase in *W. hyalina*, cultivated in the re-circulating system compared to the stationary system.

A red:blue ratio of 1:2 can increase starch yield significantly, while a higher portion of the red spectrum under eutrophic conditions caused a strong inductive effect on turion formation in *S. polyrhiza* [42]. This is contrary to data reported by Zhong et al. [41], who detected an increased starch accumulation for the same species using red light, while blue light promoted protein accumulation. In *W. arrhiza*, using irradiation with wavelengths corresponding to white, red and blue light, no significant differences in amino acid concentrations of the soluble protein were detected [45]. These results fit to our findings that the spectral distribution as applied did not significantly influence CPC.

The protein productivity, given as RPY, was lower for *L. minor* compared to *W. hyalina*. The species *L. minor* reached a maximum of 4.44 ± 0.55 g $week^{-1}$ m^{-2} at 150–50/50 and *W. hyalina* of 9.48 ± 0.39 g $week^{-1}$ m^{-2} for the same treatment. This extrapolates to 2.31 and 4.93 t of pure protein per year and hectare, respectively. In the literature, a wide range of productivities are reported. For *L. minor* and *W. hyalina*, 28.8 and 34.7 g $week^{-1}$ m^{-2}, respectively, were reached using the same nutrient solution in a stationary system with smaller vessels [28]. Mohedano et al. [46] reported a protein productivity of 24 t $year^{-1}$ ha^{-1} (ca. 46 g $week^{-1}$ m^{-2}) for duckweeds. Chakrabarti et al. [47] reached a biomass yield of 703 kg $month^{-1}$ ha^{-1} (ca. 17.5 g $week^{-1}$ m^{-2}) for *L. minor*. Regarding protein content of 27.1% for duckweed grown on an inorganic fertilizer-based solution, the protein productivity resulted in 4.74 g $week^{-1}$ m^{-2}. Comparing these values to soybean with a yield of ca. 3 t $year^{-1}$ ha^{-1} and a protein content of 40% [48], the protein productivity of 1.2 t $year^{-1}$ ha^{-1} was considerably lower compared to any duckweed protein productivity projection.

4.3. Chlorophyll

The chlorophyll a content for both species was investigated as a parameter to indicate a possible color changes in the plants at different light conditions. It decreased with increasing light intensity. This negative correlation was also found for other duckweed species [23,26,38,39,49]. *L. minor* had higher total chlorophyll contents for all investigated light intensities (6 to 1000 µmol m^{-2} s^{-1}) than *L. minuta*, reaching up to ca. 1.4 mg g^{-1} of fresh biomass at the lowest light intensity [26]. *Lemna gibba* contained ca. 250 µmol m^{-2} of chlorophyll a and b at 50 µmol m^{-2} s^{-1} and ca. 300 µmol m^{-2} at 100 µmol m^{-2} s^{-1} [23,39]. The reduction in chlorophyll at high light intensities is an acclimation strategy, protecting the plant against light-induced damage due to photo oxidation [50]. Contrarily, high chlorophyll contents at low light intensities ensure maximal light absorption. Such plants are usually associated with shade tolerance [26].

The different investigated spectral distributions had no significant impact on both species' chlorophyll content. This has also been shown by Zhong et al. [41], who obtained no significant differences in *S. polyrhiza*, when cultivated under red, blue and white light. This missing effects of the light quality in our experiment might be also caused by the use of mixed light quality.

5. Conclusions

The duckweed cultivation system applied in our experiments was a small-scale, experimental prototype of a re-circulating, aquatic IVF and specifically designed and built for conducting scientific experiments. In the literature, only a theoretical approach [13], but no practical application of an IVF for duckweed cultivation has been described, neither on a small scale for experiments nor on a large scale for biomass production. This small-scale, re-circulating IVF for scientific experiments fits the criteria for a plant factory with artificial lighting regarding structure, functionality and operation goals in most aspects [16]. The results of the present study underline the idea that the cultivation of duckweeds in such a system under non-sterile conditions is feasible and might be up-scaled for mass production.

The applied system for nutrient control and dosing is based on EC values. When the actual EC values fell below the target EC, the dosing system pumped stock solution into the reservoir until the target value was reached again. This is a well-established system for nutrient dosing used in many different hydroponic applications [16,51]. However, when used in re-circulating systems, the disadvantages become obvious. An imbalance between nutrient composition of the stock solutions and actual nutrient uptake by the plants can cause increasing concentrations of certain substances in re-circulating systems, as happened in our experiments. The longer a re-circulating system runs, the greater the imbalances will become. A depletion of nutrients, such as ammonium, nitrate, sodium or magnesium, can cause reduced RGR, CPC or RPY in duckweed due to non-optimal nutrient ratios [28,52]. In the case of nitrogen, duckweeds preferentially take up ammonium over nitrate [53]. An adaptation of the stock solutions to the actual plants' demands is difficult due to plant physiological and technical reasons. Many crops have changing demands at different plant development stages. Additionally, the dosing pumps must work precisely, when dosing more than one stock solution, to keep the nutrient ratio at a given target level. The use of stationary, on-line, ion-selective sensors [54], ion-sensitive field-effect transistors [55] or mid-infrared sensors [56] might be options to solve the problem in the future, but to date, not all relevant nutrients for plant growth can be measured. Relevant aspects regarding the application in hydroponics are the frequency and complexity of sensor calibrations, lifespan and costs as well as the stability, selectivity and drift of these technologies [54,55,57]. The readiness levels of these technologies currently vary, but new components and membranes will improve the coming product generations [55].

To gain more data about the behavioral pattern of duckweed in re-circulating systems, longer-lasting experiments investigating a broad range of abiotic, and in the case of non-sterile experiments, also biotic, parameters are needed. Nonetheless, the findings and experiences of our study were already successfully implemented into the operation of a large scale, re-circulating, aquatic IVF for duckweed biomass cultivation (Figure 6).

Figure 6. Large scale, re-circulating indoor vertical farm (IVF) or duckweed biomass cultivation at the University of Applied Sciences Osnabrück, Germany.

Supplementary Materials: The following supporting information can be downloaded at: https://www.mdpi.com/article/10.3390/plants11081010/s1, Table S1. Formulation of seven stock solutions (g L^{-1}) for five different nitrate-N to ammonium-N ratios.

Author Contributions: Conceptualization, A.U., J.D., D.R. and F.P.; methodology, A.U., J.D., D.R. and F.P.; validation, H.-W.O.; investigation, J.D., D.R. and F.P.; data curation, D.R. and F.P.; writing—original draft preparation, F.P.; writing—review and editing F.P., J.D., K.-J.A. and H.-W.O.; visualization, F.P.; supervision, A.U., K.-J.A. and H.W.; project administration, H.W.; funding acquisition, J.D., A.U., H.-W.O. and H.W. All authors have read and agreed to the published version of the manuscript.

Funding: This research was funded by Deutsche Bundesstiftung Umwelt (DBU), grant number 34223/01-46 and the German Federal Ministry of Education and Research (BMBF), grant number 031B0728.

Institutional Review Board Statement: Not applicable.

Informed Consent Statement: Not applicable.

Data Availability Statement: All data are available within the manuscript or Supplementary Material.

Conflicts of Interest: The authors declare no conflict of interest.

References

1. Bog, M.; Sree, K.S.; Fuchs, J.; Hoang, P.T.; Schubert, I.; Kuever, J.; Rabenstein, A.; Paolacci, S.; Jansen, M.A.; Appenroth, K.-J. A taxonomic revision of *Lemna* sect. Uninerves (Lemnaceae). *TAXON* **2020**, *69*, 56–66. [CrossRef]
2. Bog, M.; Appenroth, K.-J.; Sree, K.S. Duckweed (Lemnaceae): Its Molecular Taxonomy. *Front. Sustain. Food Syst.* **2019**, *3*, 117. [CrossRef]
3. Tippery, N.P.; Les, D.H.; Appenroth, K.J.; Sree, K.S.; Crawford, D.J.; Bog, M. Lemnaceae and Orontiaceae Are Phylogenetically and Morphologically Distinct from Araceae. *Plants* **2021**, *10*, 2639. [CrossRef] [PubMed]
4. Martinov, I. *Techno-Botanical Dictionary (Техно-Ботанический Словарь)*; Pechashano v Imperatorskoĭ Tipografii: Saint Petersburg, Russia, 1820.

5. Sree, K.S.; Sudakaran, S.; Appenroth, K.J. How fast can angiosperms grow? Species and clonal diversity of growth rates in the genus *Wolffia* (Lemnaceae). *Acta Physiol. Plant.* **2015**, *37*, 204. [CrossRef]
6. Ziegler, P.; Adelmann, K.; Zimmer, S.; Schmidt, C.; Appenroth, K.J. Relative in vitro growth rates of duckweeds (Lemnaceae)—The most rapidly growing higher plants. *Plant Biol.* **2015**, *17*, 33–41. [CrossRef]
7. Xu, J.; Shen, G. Growing duckweed in swine wastewater for nutrient recovery and biomass production. *Bioresour. Technol.* **2011**, *102*, 848–853. [CrossRef]
8. Cedergreen, N.; Vindbæk Madsen, T. Nitrogen uptake by the floating macrophyte *Lemna minor*. *New Phytol.* **2002**, *155*, 285–292. [CrossRef]
9. Appenroth, K.J.; Sree, K.S.; Böhm, V.; Hammann, S.; Vetter, W.; Leiterer, M.; Jahreis, G. Nutritional value of duckweeds (Lemnaceae) as human food. *Food Chem.* **2017**, *217*, 266–273. [CrossRef]
10. Appenroth, K.J.; Sree, K.S.; Bog, M.; Ecker, J.; Seeliger, C.; Böhm, V.; Lorkowski, S.; Sommer, K.; Vetter, W.; Tolzin-Banasch, K. Nutritional value of the duckweed species of the genus *Wolffia* (Lemnaceae) as human food. *Front. Chem.* **2018**, *6*, 483. [CrossRef]
11. Xu, J.; Cheng, J.; Stomp, A.M. Growing *Spirodela polyrhiza* in swine wastewater for the production of animal feed and fuel ethanol: A Pilot Study. *Clean Soil Air Water* **2012**, *40*, 760–765. [CrossRef]
12. Xu, J.; Cui, W.; Cheng, J.J.; Stomp, A.M. Production of high-starch duckweed and its conversion to bioethanol. *Biosyst. Eng.* **2011**, *110*, 67–72. [CrossRef]
13. Coughlan, N.E.; Walsh, É.; Bolger, P.; Burnell, G.; O'Leary, N.; O'Mahoney, M.; Paolacci, S.; Wall, D.; Jansen, M.A. Duckweed bioreactors: Challenges and opportunities for large-scale indoor cultivation of Lemnaceae. *J. Clean. Prod.* **2022**, *336*, 130285. [CrossRef]
14. Keuter, V.; Deck, S.; Giesenkamp, H.; Gonglach, D.; Katayama, V.T.; Liesegang, S.; Petersen, F.; Schwindenhammer, S.; Steinmetz, H.; Ulbrich, A. Significance and Vision of Nutrient Recovery for Sustainable City Food Systems in Germany by 2050. *Sustainability* **2021**, *13*, 10772. [CrossRef]
15. Ragaveena, S.; Shirly Edward, A.; Surendran, U. Smart controlled environment agriculture methods: A holistic review. *Rev. Environ. Sci. Biotechnol.* **2021**, *20*, 887–913. [CrossRef]
16. Kozai, T.; Niu, G.; Takagaki, M. *Plant Factory: An Indoor Vertical Farming System for Efficient Quality Food Production*, 2nd ed.; Academic Press: London, UK, 2020.
17. Sharma, N.; Acharya, S.; Kumar, K.; Singh, N.; Chaurasia, O.P. Hydroponics as an advanced technique for vegetable production: An overview. *J. Soil Water. Conserv.* **2018**, *17*, 364. [CrossRef]
18. AlShrouf, A. Hydroponics, Aeroponic and Aquaponic as Compared with Conventional Farming. *Am. Sci. Res. J. Eng. Technol. Sci.* **2017**, *27*, 247–255.
19. La Rosa-Rodríguez, R.D.; Lara-Herrera, A.; Trejo-Téllez, L.I.; Padilla-Bernal, L.E.; Solis-Sánchez, L.O.; Ortiz-Rodríguez, J.M. Water and fertilizers use efficiency in two hydroponic systems for tomato production. *Hortic. Bras.* **2020**, *38*, 47–52. [CrossRef]
20. Van Os, E.A.; Beerling, E.; Blok, C.; Janse, J.; Leyh, R.; van Ruijven, J.; van der Staaij, M.; Kaarsemaker, R. Zero discharge of nutrients and pesticides to the environment in hydroponic production. *Acta Hortic.* **2019**, *1266*, 443–450. [CrossRef]
21. Landolt, E. *Biosystematic Investigations in the Family of Duckweeds (Lemnaceae), The Family of Lemnaceae—A Monographic Study*; Geobotanisches Institut ETH: Zürich, Switzerland, 1986; Volume 2.
22. Landolt, E.; Kandeler, R. *Biosystematic Investigations in the Family of Duckweeds (Lemnaceae), The Family of Lemnaceae—A Monographic Study*; Geobotanisches Institut ETH: Zürich, Switzerland, 1987; Volume 4.
23. Stewart, J.J.; Adams, W.W.; Escobar, C.M.; López-Pozo, M.; Demmig-Adams, B. Growth and Essential Carotenoid Micronutrients in *Lemna gibba* as a Function of Growth Light Intensity. *Front. Plant Sci.* **2020**, *11*, 480. [CrossRef]
24. Yin, Y.; Yu, C.; Yu, L.; Zhao, J.; Sun, C.; Ma, Y.; Zhou, G. The influence of light intensity and photoperiod on duckweed biomass and starch accumulation for bioethanol production. *Bioresour. Technol.* **2015**, *187*, 84–90. [CrossRef]
25. Lasfar, S.; Monette, F.; Millette, L.; Azzouz, A. Intrinsic growth rate: A new approach to evaluate the effects of temperature, photoperiod and phosphorus-nitrogen concentrations on duckweed growth under controlled eutrophication. *Water Res.* **2007**, *41*, 2333–2340. [CrossRef]
26. Paolacci, S.; Harrison, S.; Jansen, M.A.K. The invasive duckweed *Lemna minuta* Kunth displays a different light utilisation strategy than native *Lemna minor* Linnaeus. *Aquat. Bot.* **2018**, *146*, 8–14. [CrossRef]
27. Wedge, R.M.; Burris, J.E. Effects of light and temperature on duckweed photosynthesis. *Aquat. Bot.* **1982**, *13*, 133–140. [CrossRef]
28. Petersen, F.; Demann, J.; Restemeyer, D.; Ulbrich, A.; Olfs, H.-W.; Westendarp, H.; Appenroth, K.-J. Influence of the Nitrate-N to Ammonium-N Ratio on Relative Growth Rate and Crude Protein Content in the Duckweeds *Lemna minor* and *Wolffiella hyalina*. *Plants* **2021**, *10*, 1741. [CrossRef]
29. Simonne, A.H.; Simonne, E.H.; Eitenmiller, R.R.; Mills, H.A.; Cresman, C.P. Could the Dumas method replace the Kjeldahl digestion for nitrogen and crude protein determinations in foods? *J. Sci. Food Agric.* **1997**, *73*, 39–45. [CrossRef]
30. Casal, J.A.; Vermaat, J.E.; Wiegman, F. A test of two methods for plant protein determination using duckweed. *Aquat. Bot.* **2000**, *67*, 61–67. [CrossRef]
31. *DIN 38409-60:2019-12*; Deutsche Einheitsverfahren zur Wasser-, Abwasser- und Schlammuntersuchung—Summarische Wirkungs- und Stoffkenngrößen (Gruppe H) - Teil 60: Photometrische Bestimmung der Chlorophyll-a-Konzentration in Wasser (H 60). Beuth Verlag GmbH: Berlin, Germany, 2019.

32. VDLUFA. *Methodenbuch Band 1: Die Untersuchung von Böden, Methode A 6.1.4.1 Bestimmung von Mineralischem Stickstoff (Nitrat und Ammonium) in Bodenprofilen (Nmin-Labormethode)*; VDLUFA-Verlag: Darmstadt, Germany, 2012.
33. VDLUFA. *Methodenbuch Band 1 Die Untersuchung der Böden, Methode A 6.1.1.1 Bestimmung von Nitrat-Stickstoff Durch UV-Absorption*; VDLUFA-Verlag: Darmstadt, Germany, 2012.
34. *DIN EN ISO 11885:2009-09*; Wasserbeschaffenheit—Bestimmung von Ausgewählten Elementen durch Induktiv Gekoppelte Plasma-Atom-Emissionsspektrometrie (ICP-OES) (ISO_11885:2007); Deutsche Fassung EN_ISO_11885:2009. Beuth Verlag GmbH: Berlin, Germany, 2009.
35. Li, Y.; Zhang, F.; Daroch, M.; Tang, J. Positive effects of duckweed polycultures on starch and protein accumulation. *Biosci. Rep.* **2016**, *36*, e00380. [CrossRef]
36. Brand, T.; Petersen, F.; Demann, J.; Wichura, A. First report on *Pythium myriotylum* as pathogen on duckweed (*Lemna minor* L.) in hydroponic systems in Germany. *J. Cultiv. Plants* **2021**, *73*, 316–323.
37. Roijackers, R.; Szabó, S.; Scheffer, M. Experimental analysis of the competition between algae and duckweed. *Arch. Hydrobiol.* **2004**, *160*, 401–412. [CrossRef]
38. Walsh, É.; Kuehnhold, H.; O'Brien, S.; Coughlan, N.E.; Jansen, M.A.K. Light intensity alters the phytoremediation potential of *Lemna minor*. *Environ. Sci. Pollut. Res.* **2021**, *28*, 16394–16407. [CrossRef] [PubMed]
39. Stewart, J.J.; Adams, W.W.; López-Pozo, M.; Doherty Garcia, N.; McNamara, M.; Escobar, C.M.; Demmig-Adams, B. Features of the Duckweed *Lemna* That Support Rapid Growth under Extremes of Light Intensity. *Cells* **2021**, *10*, 1481. [CrossRef]
40. Gallego, L.M.; Chien, Y.-H.; Angeles Jr, I.P. Effects of light source and photoperiod on growth of duckweed *Landoltia punctata* and its water quality. *Aqua. Res.* **2021**, *53*, 398–408. [CrossRef]
41. Zhong, Y.; Wang, L.; Ma, Z.; Du, X. Physiological responses and transcriptome analysis of *Spirodela polyrhiza* under red, blue, and white light. *Planta* **2021**, *255*, 11. [CrossRef]
42. Xu, Y.-L.; Tan, L.; Guo, L.; Yang, G.-L.; Li, Q.; Lai, F.; He, K.-Z.; Jin, Y.; Du, A.; Fang, Y. Increasing starch productivity of *Spirodela polyrhiza* by precisely control the spectral composition and nutrients status. *Ind. Crops Prod.* **2019**, *134*, 284–291. [CrossRef]
43. Martindale, W.; Bowes, G. The effects of irradiance and CO_2 on the activity and activation of ribulose-1,5-bisphosphate carboxylase/oxygenase in the aquatic plant *Spirodela polyrhiza*. *J. Exp. Bot.* **1996**, *47*, 781–784. [CrossRef]
44. Wheeler, R.M.; Mackowiak, C.L.; Sager, J.C.; Knott, W.M.; Berry, W.L. Proximate composition of CELSS crops grown in NASA's biomass production chamber. *Adv. Space Res.* **1996**, *18*, 43–47. [CrossRef]
45. Appenroth, K.J.; Augsten, H.; Liebermann, B.; Feist, H. Effects of Light Quality on Amino Acid Composition of Proteins in *Wolffia arrhiza* (L.) Wimm. using a Specially Modified Bradford Method. *Biochem. Physiol. Pflanz.* **1982**, *177*, 251–258. [CrossRef]
46. Mohedano, R.A.; Costa, R.H.R.; Tavares, F.A.; Belli Filho, P. High nutrient removal rate from swine wastes and protein biomass production by full-scale duckweed ponds. *Bioresour. Technol.* **2012**, *112*, 98–104. [CrossRef]
47. Chakrabarti, R.; Clark, W.D.; Sharma, J.G.; Goswami, R.K.; Shrivastav, A.K.; Tocher, D.R. Mass production of *Lemna minor* and its amino acid and fatty acid profiles. *Front. Chem.* **2018**, *6*, 479. [CrossRef]
48. Helms, T.C.; Orf, J.H. Protein, Oil, and Yield of Soybean Lines Selected for Increased Protein. *Crop Sci.* **1998**, *38*, 707–711. [CrossRef]
49. Artetxe, U.; García-Plazaola, J.I.; Hernández, A.; Becerril, J.M. Low light grown duckweed plants are more protected against the toxicity induced by Zn and Cd. *Plant Physiol. Biochem.* **2002**, *40*, 859–863. [CrossRef]
50. Hendry, G.A.F.; Price, A.H. Stress Indicators: Chlorophylls and Carotenoids. In *Methods in Comparative Plant Ecology*; Hendry, G.A.F., Grime, J.P., Eds.; Chapman Hall: London, UK, 1993; pp. 148–152.
51. Hosseini, H.; Mozafari, V.; Roosta, H.R.; Shirani, H.; van de Vlasakker, P.C.H.; Farhangi, M. Nutrient Use in Vertical Farming: Optimal Electrical Conductivity of Nutrient Solution for Growth of Lettuce and Basil in Hydroponic Cultivation. *Horticulturae* **2021**, *7*, 283. [CrossRef]
52. Walsh, É.; Paolacci, S.; Burnell, G.; Jansen, M.A.K. The importance of the calcium-to-magnesium ratio for phytoremediation of dairy industry wastewater using the aquatic plant *Lemna minor* L. *Int. J. Phytoremediat.* **2020**, *22*, 694–702. [CrossRef]
53. Zhou, Y.; Kishchenko, O.; Stepanenko, A.; Chen, G.; Wang, W.; Zhou, J.; Pan, C.; Borisjuk, N. The Dynamics of NO_3^- and NH_4^+ Uptake in Duckweed Are Coordinated with the Expression of Major Nitrogen Assimilation Genes. *Plants* **2021**, *11*, 11. [CrossRef]
54. Richa, A.; Fizir, M.; Touil, S. Advanced monitoring of hydroponic solutions using ion-selective electrodes and the internet of things: A review. *Environ. Chem. Lett.* **2021**, *19*, 3445–3463. [CrossRef]
55. Bamsey, M.; Graham, T.; Thompson, C.; Berinstain, A.; Scott, A.; Dixon, M. Ion-specific nutrient management in closed systems: The necessity for ion-selective sensors in terrestrial and space-based agriculture and water management systems. *Sensors* **2012**, *12*, 13349–13392. [CrossRef]
56. Fan, R.; Yang, X.; Xie, H.; Reeb, M.-A. Determination of nutrients in hydroponic solutions using mid-infrared spectroscopy. *Sci. Hortic.* **2012**, *144*, 48–54. [CrossRef]
57. Jakobsen, Ø.; Schiefloe, M.; Mikkelsen, Ø.; Paille, C.; Jost, A. Real-time monitoring of chemical water quality in closed-loop hydroponics. *Acta Hortic.* **2020**, *1296*, 1005–1018. [CrossRef]

plants

MDPI

Review

The World Smallest Plants (*Wolffia* Sp.) as Potential Species for Bioregenerative Life Support Systems in Space

Leone Ermes Romano * and **Giovanna Aronne**

Department of Agricultural Sciences, University of Naples Federico II, 80055 Portici, Italy; aronne@unina.it
* Correspondence: leoneermes.romano@unina.it

Abstract: To colonise other planets, self-sufficiency of space missions is mandatory. To date, the most promising technology to support long-duration missions is the bioregenerative life support system (BLSS), in which plants as autotrophs play a crucial role in recycling wastes and producing food and oxygen. We reviewed the scientific literature on duckweed (Lemnaceae) and reported available information on plant biological traits, nutritional features, biomass production, and space applications, especially of the genus *Wolffia*. Results confirmed that the smallest existing higher plants are the best candidate for space BLSS. We discussed needs for further research before criticalities to be addressed to finalise the adoption of *Wolffia* species for space missions.

Keywords: duckweed; Lemnaceae; *Wolffia* sp.; space plant biology; astrobiology; bioregenerative life support system (BLSS); biomass

Citation: Romano, L.E.; Aronne, G. The World Smallest Plants (*Wolffia* Sp.) as Potential Species for Bioregenerative Life Support Systems in Space. *Plants* **2021**, *10*, 1896. https://doi.org/10.3390/plants10091896

Academic Editors: Viktor Oláh, Klaus-Jürgen Appenroth and K. Sowjanya Sree

Received: 3 August 2021
Accepted: 7 September 2021
Published: 13 September 2021

1. Introduction

To date, artificial ecosystems such as the bioregenerative life support system (BLSS) are the expected technology to support long interplanetary missions [1,2]. The BLSSs concept includes several interconnected compartments in which different types of organisms are used to recycle resources and making these available to other compartments of the system [3]. Within space BLSSs, the photoautotrophic compartment only requires light from outside of the system and recycles carbon dioxide, wastewater, and other wastes to produce edible biomass, oxygen, and water used by the astronauts. One such system is MELiSSA (Micro-Ecological Life Support System Alternative), a project of the European Space Agency for life support to space exploration [3]. The MELiSSA loop aims to produce food, oxygen and recycle water and carbon dioxide to sustain astronaut life in Space missions to reduce the initial payload and dependency on the Earth. Although a few common crop species have been successfully grown in space, a smooth transition from an experimental context to cultivation is not a given, and specific methodologies for selecting candidate plants for BLSS were thus proposed [4–6]. Space farming places more demands on plants than conventional agriculture due to the extreme condition of the space environment, which require plants to tolerate factors, such as cosmic radiation and the absence of gravity, while at the same time sustaining astronaut life. The best crop plants in space must produce edible biomass in a high-quality, fast and reliable way, without wasting resources on the production of non-edible biomass and while thus maximising resource utilisation [7]. Candidate species for space farming are increasing; plants at a different stage of development are considered; they include leafy greens, microgreens (e.g., *Brassica oleracea*, *Rumex acetosa*, *Lepidium bonariense*, *Coriandrum sativum*, *Amaranthus hypochondriacu*), fruit crops (e.g., *Fragaria vesca*, *Solanum lycopersicum*), and tuber crops (*Solanum tuberosum*) [8]. These candidate species were chosen from among crop species commonly used on Earth, and current efforts focus on improving these crops with respect to, e.g., small size (dwarf varieties), fast growth and optimal nutrient content for the astronaut diet [9]. A different approach focuses on alternative species not yet widely cultivated on Earth but possessing several attra ctive traits for food production in space. *Arthrospira platensis* (commonly referred to as Spirulina) is a current example. This filamentous cyanobacterium is the most widely cultivated photosynthetic prokaryote and is promoted as a candidate for BLSSs for space missions [3,10]. The present review focuses on additional new candidate crops, with attention to duckweeds (family Lemnaceae), and specifically the possible use of *Wolffia* species as plants to be cultivated in BLSSs in space.

Lemnaceae are flowering plants of the monocotyledon subdivision that populate freshwater ponds and slow-flowing water bodies all over the world. With the support of aerenchyma tissue, they float on water and may also be slightly submerged. Their diminutive size varies from 1.5 cm in diameter (*Spirodela polyrhiza*, Giant Duckweed) to less than 1 mm (*Wolffia angusta*, the smallest angiosperm). Flowering can occur among duckweeds, but most species exhibit vegetative reproduction as their primary propagation method. Vegetative reproduction occurs by budding from meristematic zones inside the frond pockets. Several generations of fronds exist at any given time inside the mother frond; this creates overcrowding and compression of the primordium daughter fronds. As soon as the oldest daughter frond detaches from its mother frond, the remaining daughters experience a phase of arrangement that pushes the oldest primordium towards the frond's pocket; this will generate a mature daughter frond. (For more details, see [11].)

Based on morphological markers, the family Lemnaceae is composed of two subfamilies, Wolffioideae (members lacking roots) and Lemnoideae (members with a variable number of roots) [12,13].

The family Lemnaceae comprises five genera (*Wolffia*, *Wolffiella*, *Spirodela*, *Lemna*, *Landoltia*) and currently includes 36 accepted species [14,15]. Scientific interest in these plants has been increasing in the last two decades, but the number of publications is not distributed uniformly among these genera. The trend of growth in interest focused on the genus *Lemna* (Figure 1).

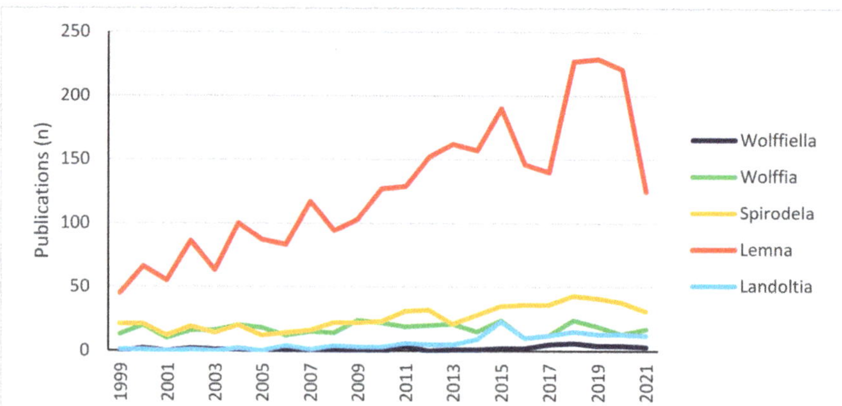

Figure 1. The number of scientific publications per genus of the family Lemnaceae from 1999 to date. Data show the discrepancy between the number of publications for the genus *Lemna* compared to others. Data are from https://www.scopus.com (last searched on 23 July 2021) using the name of each genus as keywords and including only results from 1999 to the date of the search.

Besides the nuanced morphological differences, important traits are common to all Lemnaceae and include small size, structural simplicity, exponential growth and genetical uniformity due to the predominant vegetative reproduction. Such traits have led scientists to suggest species of the family Lemnaceae as suitable candidates for space cultivation [16,17]. Despite the growing number of studies, the resulting knowledge cannot be directly applied to space farming and will need further testing in space-relevant environments. Moreover, and as mentioned above, most of the new scientific interest is focused on plants of the genus *Lemna*. The results do not represent all species in the family Lemnaceae due to a high degree of variability between and within species. Within the family Lemnaceae, the genus *Wolffia* includes eleven species and is well known for featuring both the fastest growing and the most diminutive flowering plants globally [18,19]. In addition to the peculiarity of these records, *Wolffia* species possess numerous traits that make them potentially suitable as a candidate crop for plant-based BLSSs. For instance, the average relative growth rate of *Wolffia* species is higher than that of most angiosperms and possibly even other Lemnaceae [19]. The rootless morphology maximises the harvest index [17]. The nutritional profile of *Wolffia* species has excellent characteristics [20] and is improved by the absence of toxic substances (e.g. oxalic acid) [13]. Moreover, *Wolffia* species are suitable for new biotechnological applications and nutrient removal from sewage water [21–25].

In addition, the positive buoyancy typically exhibited by these plants in water could facilitate the transition between Earth and the space environment where the effect of gravity is absent. We considered that all these traits could pave the way for *Wolffia* species to be suitable for space cultivation.

This work reviewed the primary scientific literature on plants in the family Lemnaceae, with an emphasis on *Wolffia* species. The objective was to synthesise the literature and identify possible bottlenecks for utilising *Wolffia* species in space cultivation in BLSS. We focused our attention on plant biological traits, nutritional features, biomass production, and space applications.

2. Plant Morphological and Reproductive Traits

Plants in the genus Wolffia have a cosmopolitan distribution, populating the lentic ecosystems in almost all the continents except the Antarctic and Arctic regions [11]. Wolffia is a genus of plants with 11 species and including both the fastest-growing angiosperm [19] and the smallest flowering plants [19] (Figure 2a). As is the case for other species in the family, plants in the genus Wolffia consist of a single physical unit termed frond, or thallus, and interpreted as a leaf and stem in an embryonic stage of development [13,26]. Fronds of Wolffia species have a globose, ovoid boat shape (Figure 2b). In each plant, the newly formed frond (daughter) develops from meristematic cells in a pocket of the older frond (mother) (Figure 2c).

Colonies of fronds consist of a large number of individuals forming the so-called "clusters". Colonies of Wolffia species rarely consist of more than two visible fronds per individual, while, anatomically, each mature plant is composed of multiple individual fronds at different stages of development (Figure 2c) [27,28]. Species of the genus Wolffia do not develop roots and are recounted to absorb water and nutrients through the underside of their main frond, making the root function redundant [27–29]. It is worth mentioning that this redundancy of roots functionalities also occurs in other Lemnaceae. More specifically, in Lemna species, plants under replete nutrient concentrations manage nutrient uptake mostly from leaves surface [30]. Vegetative propagation based on meristematic cells in the pocket of the fronds is by far the most recurrent way of plant reproduction [13]. However, under unfavourable conditions (temperatures below 15 °C or nutrient-depleted substrates), Wolffia species can produce perennating organs termed turions, the formation of which is an alternative strategy to normal frond development [31,32].

In Wolffia species, flowers develop inside a cavity that opens near the median line of the upper frond surface (Figure 2d). Flowers are composed of one stamen with the anther and a pistil within which there is one atropous ovule [13]. Recently, laboratory protocols were developed to control flowering in W. microscopica [33]. Nevertheless, several aspects of the reproductive biology and ecology of Wolffia species are still unexplored. One crucial point is to assess how these plants transfer pollen from one to another. Some authors have hypothesised that the main form of pollination in these plants is by fish, birds, or strong wind [13,34,35], but these dispersal models have not yet been verified. Such a lack of knowledge limits the possibility of increasing the genetic variability of plants for breeding and selection.

Figure 2. (**a**) Top view of plants of Wolffia arrhiza also showing a stainless-steel pin for size comparison; (**b**) close view of plants of Wolffia arrhiza. Some individual plants feature both the mother frond (mf) and the daughter frond (df); (**c**) schematic transversal section of a plant of Wolffia sp. vegetatively reproducing, mother frond (mf), meristem (m), daughter frond (df); (**d**) schematic transversal section of a plant of Wolffia sp. in full bloom, pistil(p), stamen (s). Red bars = 1 mm.

Wolffia species reproduce mainly by vegetative propagation. The average relative growth rate is noteworthy. It ranges from 0.155 to 0.559 (day^{-1}) but is variable among the species of the genera and different clones within a species [19].

Effective vegetative propagation increases biomass production rate, and cloning maintains genetic uniformity facilitating industrial applications. So far, only one genome has been sequenced (W. australiana) [36,37] further comparison among chloroplast genomes shows that the genus Wolffia possesses the most reduced genome size in Lemnaceae [38]. Genome size varied between 375 Mb in W. australiana to 1881 Mb in W. arrhiza [39]. Different genetic manipulation strategies have been conducted and tested on several Wolffia species (W. australiana, W. globosa, and W. columbiana) [40–42]. Such manipulations were not stable, except for those conducted by Khvatkov et al. [43] using W. arrhiza, a producer for recombinant human granulocyte colony-stimulating factor. The stability of transient transformation is crucial for being able to adopt Wolffia for mainstream commercial applications. Recently W. globosa was genetically transformed through transient genetic transformation protocols [44]. The advent of stable genetic modification protocols for the Wolffia species is now accelerating the application of these plants in biopharming, where recombinant plants produce complex molecules and proteins. One such example is hirudin production, a peptide produced by leeches saliva that has anticoagulant properties used by modern medicine to cure different types of thrombosis [45,46].

3. Human Nutrition

"Khai-nam" (eggs of the water) is the name for duckweed in many countries of southeast Asia. From samples retrieved in many local markets, these water eggs were mainly *W. globosa*, a species endemic to the region mentioned earlier [47,48]. Numerous studies and research projects have been conducted to deepen knowledge about *Wolffia* and the use of these hydrophytes in human and animal diets [29,49,50].

These tiny water plants attracted the interest of many researchers to find new, alternative sources of protein for human populations whose diets are based mainly on corn and starch. Plants of the genus *Wolffia* have a protein profile (both in terms of quality and quantity) necessary to supplement this type of diet [47,50]. *W. microscopica* showed a total protein content of more than 25% (dry weight) and the highest content of the 17 tested amino acids compared to other species of the family Lemnaceae [50]. According to the World Health Organization, the essential amino acid content is greater than the needs of preschool-aged children. The fat fraction comprises 80% polyunsaturated fatty acids, with a high proportion of n-3 polyunsaturated fatty acids; a balanced ratio of n-3 to n-6 in the human diet has been linked to reducing cardiovascular disease [51]. Plants of *Wolffia* species are rich in macro-and micronutrients, but those are strictly dependent on the cultivation substrate [18]. The nutritional values were extensively studied [48,52]. More recently, Appenroth et al. [20] showed that plants of the *Wolffia* species possess the needed and, in some cases, even more, essential amino acid content than those required by pre-school-aged children, with a protein content varying from 20 to 30% freeze-dried weight. Such results are consistent with those of previous research [53]. Starch, fat, and fibre content were also analysed and varied: 10–20% for starch, 1–5% for fat, and ~25% for fibre. Although the fat content was low in most species, the quantity of polyunsaturated fatty acid was around 80%, and among these, n-3 polyunsaturated fatty acids were more represented than n-6 polyunsaturated [54,55]. Micro and macro elements content were also analysed and reported to be dependent not only on the growth medium composition and the environmental parameters (light intensity) [56,57] but also on the genetic background of the species studied. Furthermore, carotenoid content has been linked to a stress response in *Lemna* species [56] further studies should also investigate this positive stress response in *Wolffia* species.

Due to the high nutrient content of the genus *Wolffia*, the selection of both species and ecotype requires attention to fine-tune cultivation to specific needs of astronauts in space. Among the eleven *Wolffia* species, *W. microscopica* is reported to possess the most significant potential for human nutrition since it has a fast reproduction time and an excellent nutritional profile [20].

Wolffia species grown under optimal conditions also resulted rich in phenolic substances [58], a class of bioactive compounds that plays a crucial role in the defence against biotic and abiotic stress in plants [59–61] is also well known to be essential for human wellbeing. *W. globose* plants were also reported to be rich in cobalamin (vitamin B12), in a form that is bioactive and well absorbed by the human body [62].

β-sitosterol and stigmasterol are important phytohormones that play a crucial role in plant resistance to pathogen infection and have been found in the extract of *W. arrhiza*. These compounds have high antimicrobial activity, and, have therefore boosted the interest from pharmaceutical and cosmetic companies as nutraceutical constituents [63].

In addition to the nutritional components, a randomised crossover control trial studied the effect of *W. globosa* on the postprandial and overnight glycaemic response [64] showing that patients with abdominal obesity fed a substitute meal shake made of *W. globosa* had a lower postprandial glucose peak than those provided with iso-carbohydrate/protein/calorie yoghurt shake. Moreover, it returned more quickly to baseline glucose levels regardless of the same carbohydrates content. The authors concluded that *W. globosa* might serve as a plant protein substitute with beneficial postprandial glycaemic effects.

When assessing the possibility of introducing plants of *Wolffia* species as a novel food into the human diet, one of the first steps is to study the possible toxic effects on the human body. The authors evaluated genotoxicity and repeated-dose toxicity of dried *W. globosa*, introducing different percentages of dried plants to the diets of rats. The results showed no harmful effect attributable to supplementation with different doses of *W. globosa* to the rat diet [65]. Further studies aimed at testing the possible cytotoxic effect of duckweed on human cells concluded that none of the duckweed extracts from five analysed genera (including *Wolffia*) possessed any detectable anti-proliferative or cytotoxic effect [66].

A further attractive feature for use in the human diet is that *Wolffia* species, unlike other genera in the family Lemnaceae (i.e., *Spirodela*, *Landoltia*, and *Lemna*), have the benefit of not containing any oxalate in the form of calcium oxalate, considering that this compound might be harmful to humans [67].

4. Biomass Production and Waste Recycling

Plants of *Wolffia* species can reach a productivity of 265 tons/ha (10.5 tons/ha of dry weight) of fresh weight calculated annually based on nine months in a non-controlled environment [47]. In controlled environments, clusters can reach a production rate of 86–160 g-wet/m^2/d for the vegetative fronds and 55–64 g-wet/m^2/d for the turions, where variations are nutrient-dependent [68].

Generally, for research purposes, duckweed growth is measured as the relative growth rate per day (RGR). Naumann et al. [69] describe the calculation of RGR for different bases, including fresh weight, frond number, chlorophyll content, or carotenoid content. The exponential growth rate of duckweeds is measured through doubling time (DT). Under controlled environmental conditions, the highest RGR (0.559/day) was observed in *W. globosa* and the fastest DT (29.3 h) in *W. microscopica*. It is worth noticing the profound variation in RGR and DT among locally adapted ecotypes of the same species, which complicates comparison among species (e.g. [19]). In a comparative study, the RGR of *Wolffia* species was not statistically different from the other genera of the family Lemnaceae [17]. The exponential growth that characterises these plants is presumably related to their high nutrient uptake of 2–6 g/day of nitrogen assimilation per kilo of fresh mass and 1.64–4.94 g/day of ammonium [70]. The natural capacity of assimilating high quantities of ammonium makes this group of plants ideal for recovering nutrients from waste products. *Wolffia* species can remove 70–80% of the nitrogen and phosphorus in secondary treated sewage effluent in 7 days [19]. More specifically, the possibility of using *Wolffia* species as a viable alternative to recover nutrients from wastewaters and converting the produced biomass to bioethanol has been investigated. Experiments conducted with *W. globosa* show that bioethanol production is achievable [71].

The high efficiency in taking nutrients has also been exploited in phytoremediation practice. Numerous experiments have been conducted in this framework, successfully removing pollutant such as arsenic, nitrate, and phosphate from water [23,24].

In addition to pollutants removal, *Wolffia* species also positively contributed to remediation in total suspended solids, chemical oxygen demand, and biological oxygen demand. Moreover, a 50–90% faecal coliform reduction was measured in waters with duckweed mats [72,73].

It is worth mentioning the symbiotic relationship that this group of plants establishes with microorganisms, such as plant-promoting bacteria, cyanobacteria, and microalgae [74]. These close relationships can boost these plants' nitrogen assimilation effectiveness to these organisms' capacity fixating nitrogen [75,76].

5. Space Application

The reduced size, clonal exponential growth, high yield, and biomass quality make Lemnaceae one of the most favourable candidate plants for use in BLSS [17,18]. The use of aquatic plants in a controlled ecological life support system (CELSS) is ideal [77], as demonstrated in various experiments [78,79]. In the 1990s, the first successful closed BLSS that operated in microgravity conditions was developed and autonomously grew different aquatic organisms. Among these organisms were flowering plants, including *Wolffia* species, animal organisms (invertebrates), algae, and microorganisms. The long duration of the experiment (4 months) made it possible even for the animals inhabiting the Autonomous Biological System to complete an entire life cycle in space [80]. These findings paved the way for the optimisation of food production in *W. arrhiza*-based bioreactors, combining intensive aquaculture systems and a closed regenerative food loop between fish and aquatic plants [22].

Experimental evidence is available for the feasibility of cultivating *Wolffia* in simulated microgravity, where plants exhibited enhanced growth [81]. In addition to the interaction with altered gravity, a study simulating the effect of cosmic radiation on plant growth showed that heavy ions increased the mortality rate of W. *arrhiza* [82]. Considering that the experiment was not conducted in space but on Earth with direct radiation generated under laboratory conditions, it would be worth verifying the effect of space radiation with experiments on orbital platforms such as the International Space Station (ISS). Studies on plants from the genus *Lemna* showed that radiation at low concentrations is mitigated by the production of antioxidant substances such as flavonoids [83]. The benefit of antioxidant substances in the human diet (especially in a space context) needs to be addressed [56,57]. Further studies would be desirable to verify if similar mitigation processes occur in *Wolffia* species.

In the possible cultivation of *Wolffia* species in BLSS, it is also noteworthy that, unlike other duckweeds, fronds of *Wolffia* species can thrive when either floating or submerged [84]. Such a trait could turn out to be helpful for the cultivation of plants in bioreactors instead of greenhouses for the maximisation of biomass production per volume rather than surface area.

6. General Discussion and Critical Insights

Plants of the genus *Wolffia* have been less studied than other species from the family Lemnaceae. Nevertheless, available information indicates that *Wolffia* species are attractive candidate species for BLSS and that future research is warranted to support the adoption of *Wolffia* species for space missions.

In particular, further work is needed on differences among species and locally adapted ecotypes in nutritional value, growth, genetics, and the effect of microgravity. The absence of buoyancy and sedimentation characterises the microgravity conditions, absence of convection, absence of hydrostatic pressure, and free-floating of the liquids in the air (container less float). Particular attention should be focused on understanding the physical and physiological reactions of plants *Wolffia* species in a microgravity context. The results of the new studies should be the base points to design appropriate cultivation hardware.

Wolffia species are rootless; consequently, they are different from other species already studied under space conditions. Therefore, to validate the adoption of duckweed based BLSS, specific studies should focus on the physiological aspects of nutrient uptakes under microgravity, hypergravity, and partial gravity levels (Moon and Mars gravities).

The vegetative reproduction that characterises most duckweed is fast and reliable to maintain genetic uniformity and maximise biomass production. However, to fast forward the commercial application on both Earth and space of *Wolffia* species and duckweeds in general, deepening knowledge on inducing flowering and achieving successful sexual reproduction is mandatory. The resulting increase of biodiversity will be essential to start the genetic improvement, employing breeding programs aimed at fine-tuning plants to the specific application requirements.

Plant cultivation in bioreactors could be beneficial for space exploration missions where batches of recombinant *Wolffia* species could be employed to produce complex molecules and nutraceutical constituents.

Additionally, as for other crop species, the symbiotic relationships that plants of the genus *Wolffia* can establish with bacteria and other microorganisms should be better investigated. Further investigation should focus on food safety issues as a possible hazard to crew health. However, the symbiotic relationship could maximise the efficiency in nutrient uptake and complex molecule degradation. These relationships could also be beneficial in building a BLSS, including micro-Angiosperm plants (duckweed) and microalgae as photoautotrophic organisms.

In a scenario of space greenhouses, productivity can be maximised by growing the plants in multiple stacked layers, allowing a vertical cultivation system. The possibility of optimising the whole volume of the growth chambers would be enhanced by exploring the potential to cultivate these plants in bioreactors as those currently employed for microalgae cultivation for space missions.

Finally, in a scenario of space cultivation with fully controlled growth systems, the nutritional content of plants of *Wolffia* species can be modulated by fine-tuning the environmental parameters. The protein and starch content can be adjusted to diet requirements by alternating the production of fronds (richer in protein) with turion (richer in starch).

In conclusion, *Wolffia* species hold many features that make them suitable candidate species for space BLSS. Further studies should focus on the effects of multiple environmental factors (including altered gravity, space radiation, light quality and intensity, nutrient management) on plant reproduction, propagation rate, and nutritional traits. Moreover, feasible applications would benefit from new concepts for efficient cultivation hardware.

Funding: This work was funded by the European Space Agency with a grant to the University of Naples Federico II–Department of Agricultural Sciences, ESA Contract No. 4000133778/21/NL/CBi.

Institutional Review Board Statement: Not applicable.

Informed Consent Statement: Not applicable.

Acknowledgments: The authors thank Nicola Carriero for plant images.

Conflicts of Interest: The authors declare no conflict of interest.

References

1. Tamponnet, C.; Savage, C. Closed ecological systems. *J. Biol. Educ.* **1994**, *28*, 167–174. [CrossRef]
2. De Micco, V.; Aronne, G.; Colla, G.; Fortezza, R.; De Pascale, S. Agro-biology for bioregenerative life support systems in long-term space missions: General constraints and the Italian efforts. *J. Plant Interactions* **2009**, *4*, 241–252. [CrossRef]
3. Gòdia, F.; Albiol, J.; Montesinos, J.; Pérez, J.; Creus, N.; Cabello, F.; Mengual, X.; Montras, A.; Lasseur, C. MELISSA: A loop of interconnected bioreactors to develop life support in space. *J. Biotechnol.* **2002**, *99*, 319–330. [CrossRef]
4. De Micco, V.; Buonomo, R.; Paradiso, R.; De Pascale, S.; Aronne, G. Soybean cultivar selection for Bioregenerative Life Support Systems (BLSS)—Theoretical selection. *Adv. Space Res.* **2012**, *49*, 1415–1421. [CrossRef]
5. Paradiso, R.; De Micco, V.; Buonomo, R.; Aronne, G.; Barbieri, G.; De Pascale, S. Soilless cultivation of soybean for Bioregenerative Life-Support Systems: A literature review and the experience of the MELiSSA Project—Food characterisation Phase I. *Plant Biol.* **2013**, *16*, 69–78. [CrossRef]
6. Wheeler, R.M. Agriculture for space: People and places paving the way. *Open Agric.* **2017**, *2*, 14–32. [CrossRef]
7. De Pascale, S.; Arena, C.; Aronne, G.; De Micco, V.; Pannico, A.; Paradiso, R.; Rouphael, Y. Rouphael, biology and crop production in Space environments: Challenges and opportunities. *Life Sci. Space Res.* **2021**, *29*, 30–37. [CrossRef] [PubMed]
8. Dueck, T.; Kempkes, F.; Meinen, E.; Stanghellini, C. Choosing crops for cultivation in space. *Contrib. Proc.* **2016**. [CrossRef]
9. Tuominen, L.K.; Levine, L.H.; Musgrave, M.E. Plant secondary metabolism in altered gravity. *Methods Mol. Biol.* **2009**, *547*, 373–386. [CrossRef]
10. Fahrion, J.; Mastroleo, F.; Dussap, C.-G.; Leys, N. Use of photobioreactors in regenerative life support systems for human space exploration. *Front. Microbiol.* **2021**, *12*. [CrossRef] [PubMed]
11. Sree, K.S.; Maheshwari, S.C.; Boka, K.; Khurana, J.P.; Keresztes, Á.; Appenroth, K.J. The duckweed *Wolffia* microscopica: A unique aquatic monocot. *Flora—Morphol. Distrib. Funct. Ecol. Plants* **2015**, *210*, 31–39. [CrossRef]
12. Landolt, E. Key to the determination of taxa within the family of "Lemnaceae". *Veröffentlichungen Geobot. Inst. Eidg. Tech. Hochsch. Stift Rübel* **1980**. [CrossRef]
13. Landolt, E. *The Family of Lemnaceae—A Monographic Study. Vol. 1. Biosystematic Investigations in the Family of Duckweeds;* Geobotanische Institut ETH: Zurich, Switzerland, 1986.
14. Bog, M.; Appenroth, K.-J.; Sree, K.S. Duckweed (Lemnaceae): Its molecular taxonomy. *Front. Sustain. Food Syst.* **2019**, *3*. [CrossRef]
15. Bog, M.; Appenroth, K.J.; Sree, K.S. Key to the determination of taxa of Lemnaceae: An update. *Nord. J. Bot.* **2020**, *38*. [CrossRef]
16. Tao, X.; Fang, Y.; Xiao, Y.; Jin, Y.-L.; Ma, X.-R.; Zhao, Y.; He, K.-Z.; Zhao, H.; Wang, H.-Y. Comparative transcriptome analysis to investigate the high starch accumulation of duckweed (Landoltia punctata) under nutrient starvation. *Biotechnol. Biofuels* **2013**, *6*, 72. [CrossRef]
17. Ziegler, P.; Adelmann, K.; Zimmer, S.; Schmidt, C.; Appenroth, K.-J. Appenroth, relative in vitro growth rates of duckweeds (Lemnaceae)—The most rapidly growing higher plants. *Plant Biol.* **2014**, *17*, 33–41. [CrossRef]
18. Bog, M.; Schneider, P.; Hellwig, F.; Sachse, S.; Kochieva, E.; Martyrosian, E.; Landolt, E.; Appenroth, K.-J. Genetic characterisation and barcoding of taxa in the genus *Wolffia* Horkel ex Schleid. (Lemnaceae) as revealed by two plastidic markers and amplified fragment length polymorphism (AFLP). *Planta* **2012**, *237*, 1–13. [CrossRef]
19. Sree, K.S.; Sudakaran, S.; Appenroth, K.-J. How fast can angiosperms grow? Species and clonal diversity of growth rates in the genus Wolffia (Lemnaceae). *Acta Physiol. Plant.* **2015**, *37*, 1–7. [CrossRef]

20. Appenroth, K.-J.; Sree, K.S.; Bog, M.; Ecker, J.; Seeliger, C.; Böhm, V.; Lorkowski, S.; Sommer, K.; Vetter, W.; Tolzin-Banasch, K.; et al. Nutritional value of the duckweed species of the genus wolffia (Lemnaceae) as human food. *Front. Chem.* **2018**, *6*, 483. [CrossRef]
21. Fujita, M.; Mori, K.; Kodera, T. Nutrient removal and starch production through cultivation of *Wolffia arrhiza*. *J. Biosci. Bioeng.* **1999**, *87*, 194–198. [CrossRef]
22. Bluem, V.; Paris, F. Novel aquatic modules for bioregenerative life-support systems based on the closed equilibrated biological aquatic system (C.E.B.A.S.). *Acta Astronaut.* **2002**, *50*, 775–785. [CrossRef]
23. Zhang, X.; Zhao, F.-J.; Huang, Q.; Williams, P.; Sun, G.; Zhu, Y. Arsenic uptake and speciation in the rootless duckweed *Wolffia globosa*. *New Phytol.* **2009**, *182*, 421–428. [CrossRef] [PubMed]
24. Gupta, C.; Prakash, D. Duckweed: An effective tool for phyto-remediation. *Toxicol. Environ. Chem.* **2013**, *95*, 1256–1266. [CrossRef]
25. Acosta, K.; Appenroth, K.J.; Borisjuk, L.; Edelman, M.; Heinig, U.; Jansen, M.A.K.; Oyama, T.; Pasaribu, B.; Schubert, I.; Sorrels, S.; et al. Return of the Lemnaceae: Duckweed as a model plant system in the genomics and postgenomics era. *Plant Cell* **2021**. [CrossRef]
26. Landolt, E. *Lemna yungensis*, a new duckweed species from rocks of the Andean Yungas in Bolivia. *Bull. Geobot. Inst. ETH* **1998**, *64*. [CrossRef]
27. Hillman, W.S. The lemnaceae, or duckweeds. *Bot. Rev.* **1961**, *27*, 221–287. [CrossRef]
28. Anderson, J.L.; Thomson, W.W.; Swader, J.A. Fine structure of Wolffia arrhiza. *Can. J. Bot.* **1973**, *51*, 1619–1622. [CrossRef]
29. Leng, R.A. *Duckweed: A Tiny Aquatic Plant with Enormous Potential for Agriculture and Environment*; FAO: Roma, Italy, 1999.
30. Cedergreen, N.; Madsen, T.V. Nitrogen uptake by the floating macrophyte *Lemna minor*. *New Phytol.* **2002**, *155*, 285–292. [CrossRef]
31. Takai, Y.; Mishima, D.; Kuniki, M.; Sei, K.; Soda, S.; Ike, M. Ethanol production from vegetative fronds and turions of *Wolffia arrhiza*. *Jpn. J. Water Treat. Biol.* **2014**, *50*, 133–140. [CrossRef]
32. Appenroth, K.J.; Teller, S.; Horn, M. Photophysiology of turion formation and germination in Spirodela polyrhiza. *Biol. Plant.* **1996**, *38*, 95–106. [CrossRef]
33. Fourounjian, P.; Slovin, J.; Messing, J. Flowering and seed production across the lemnaceae. *Int. J. Mol. Sci.* **2021**, *22*, 2733. [CrossRef]
34. Barrett, S.C.; Eckert, C.G.; Husband, B.C. Evolutionary processes in aquatic plant populations. *Aquat. Bot.* **1993**, *44*, 105–145. [CrossRef]
35. Crawford, D.J.; Landolt, E. Allozyme studies in Spirodela (Lemnaceae): Variation among conspecific clones and divergence among the species. *Syst. Bot.* **1993**, *18*, 389. [CrossRef]
36. Park, H.; Park, J.H.; Jeon, H.H.; Woo, D.U.; Lee, Y.; Kang, Y.J. Characterisation of the complete chloroplast genome sequence of *Wolffia globosa* (Lemnoideae) and its phylogenetic relationships to other Araceae family. *Mitochondrial DNA Part B* **2020**, *5*, 1905–1907. [CrossRef]
37. Park, H.; Park, J.H.; Lee, Y.; Woo, D.U.; Jeon, H.H.; Sung, Y.W.; Shim, S.; Kim, S.H.; Lee, K.O.; Kim, J.-Y.; et al. Genome of the world's smallest flowering plant, *Wolffia australiana*, helps explain its specialised physiology and unique morphology. *Commun. Biol.* **2021**, *4*, 900. [CrossRef] [PubMed]
38. Wang, W.; Messing, J. High-throughput sequencing of three lemnoideae (Duckweeds) chloroplast genomes from total DNA. *PLoS ONE* **2011**, *6*, e24670. [CrossRef] [PubMed]
39. Michael, T.P.; Ernst, E.; Hartwick, N.; Chu, P.; Bryant, D.; Gilbert, S.; Ortleb, S.; Baggs, E.L.; Sree, K.S.; Appenroth, K.J.; et al. Genome and time-of-day transcriptome of *Wolffia australiana* link morphological minimisation with gene loss and less growth control. *Genome Res.* **2020**, *31*, 225–238. [CrossRef]
40. Boehm, R.; Kruse, C.; Voeste, D.; Barth, S.; Schnabl, H. A transient transformation system for duckweed (*Wolffia* columbiana) using Agrobacterium-mediated gene transfer. *J. Appl. Bot.* **2001**, *75*, 107–111.
41. Kruse, C.; Boehm, R.; Voeste, D.; Barth, S.; Schnabl, H. Transient transformation of *Wolffia columbiana* by particle bombardment. *Aquat. Bot.* **2002**, *72*, 175–181. [CrossRef]
42. Pham, T.L.T.; Nguyen, H.A.; Pham, T.H.; Nguyen, T.H.; Le, H.H. Improvement of transformation procedure into duckweed (*Wolffia* sp.) via Agrobacterium tumefaciens. *Tapchi'Congnghe Sinhhoc* **2010**, *8*, 53–60.
43. Khvatkov, P.; Chernobrovkina, M.; Okuneva, A.; Pushin, A.; Dolgov, S. Transformation of *Wolffia arrhiza* (L.) Horkel ex Wimm. *Plant Cell Tissue Organ Cult. (PCTOC)* **2015**, *123*, 299–307. [CrossRef]
44. Heenatigala, P.P.M.; Yang, J.; Bishopp, A.; Sun, Z.; Li, G.; Kumar, S.; Hu, S.; Wu, Z.; Lin, W.; Yao, L.; et al. Development of efficient protocols for stable and transient gene transformation for *Wolffia globosa* using agrobacterium. *Front. Chem.* **2018**, *6*, 227. [CrossRef]
45. Khvatkov, P.; Firsov, A.; Shvedova, A.; Kozlov, O.; Chernobrovkina, M.; Pushin, A.; Shaloiko, L.; Dolgov, S. *Wolffia arrhiza* as a promising producer of recombinant hirudin. *3 Biotech* **2021**, *11*, 1–10. [CrossRef] [PubMed]
46. Mathew, R.; Mehta, R.L. Anticoagulation strategies for continuous renal replacement therapies. *Crit. Care Nephrol.* **2009**, 1342–1349. [CrossRef]
47. Bhanthumnavin, K.; McGarry, M.G. *Wolffia arrhiza* as a possible source of inexpensive protein. *Nature* **1971**, *232*, 495. [CrossRef] [PubMed]
48. Rusoff, L.L.; Blakeney, E.W.; Culley, D.D. Duckweeds (Lemnaceae family): A potential source of protein and amino acids. *J. Agric. Food Chem.* **1980**, *28*, 848–850. [CrossRef]

49. Skillicorn, P.; Spira, W.; Journey, W. *Duckweed Aquaculture: A New Aquatic Farming System for Developing Countries*; The World Bank: Washington, DC, USA, 1993.
50. Appenroth, K.-J.; Sree, K.S.; Böhm, V.; Hammann, S.; Vetter, W.; Leiterer, M.; Jahreis, G. Nutritional value of duckweeds (Lemnaceae) as human food. *Food Chem.* **2017**, *217*, 266–273. [CrossRef] [PubMed]
51. Simopoulos, A.P. Human requirement for N-3 polyunsaturated fatty acids. *Poult. Sci.* **2000**, *79*, 961–970. [CrossRef] [PubMed]
52. Mueller-Hiemeyer, R.; Amadò, R.; Marti, U. Proteingehalt, aminosäurezusammensetzung und neutralzuckergehalt von lemnaceen: Vorläufige mitteilung. *Veröffentlichungen Geobot. Instit. Eidg. Tech. Hochsch. Stift. Rübel Zürich* **1980**. [CrossRef]
53. Appenroth, K.-J.; Augsten, H.; Liebermann, B.; Feist, H. Feist, effects of light quality on amino acid composition of proteins in *Wolffia arrhiza* (L.) Wimm. using a specially modified bradford method. *Biochem. Physiol. Pflanz.* **1982**, *177*, 251–258. [CrossRef]
54. Tang, J.; Li, Y.; Ma, J.; Cheng, J.J. Cheng, survey of duckweed diversity in Lake Chao and total fatty acid, triacylglycerol, profiles of representative strains. *Plant Biol.* **2015**, *17*, 1066–1072. [CrossRef]
55. Yan, Y.; Candreva, J.; Shi, H.; Ernst, E.; Martienssen, R.; Schwender, J.; Shanklin, J. Shanklin, survey of the total fatty acid and triacylglycerol composition and content of 30 duckweed species and cloning of a Δ6-desaturase responsible for the production of γ-linolenic and stearidonic acids in *Lemna gibba*. *BMC Plant Biol.* **2013**, *13*, 201. [CrossRef] [PubMed]
56. Stewart, J.; Adams, W.W.I.; Escobar, C.M.; López-Pozo, M.; Demmig-Adams, B. Growth and essential carotenoid micronutrients in *Lemna gibba* as a function of growth light intensity. *Front. Plant Sci.* **2020**, *11*. [CrossRef] [PubMed]
57. Stewart, J.; Adams, W.; López-Pozo, M.; Garcia, N.D.; McNamara, M.; Escobar, C.; Demmig-Adams, B. Demmig-Adams, features of the duckweed *Lemna* that support rapid growth under extremes of light intensity. *Cells* **2021**, *10*, 1481. [CrossRef] [PubMed]
58. Pagliuso, D.; Jara, C.E.P.; Grandis, A.; Lam, E.; Ferreira, M.J.P.; Buckeridge, M.S. Flavonoids from duckweeds: Potential applications in the human diet. *RSC Adv.* **2020**, *10*, 44981–44988. [CrossRef]
59. Pichersky, E.; Gang, D.R. Genetics and biochemistry of secondary metabolites in plants: An evolutionary perspective. *Trends Plant Sci.* **2000**, *5*, 439–445. [CrossRef]
60. Plazonić, A.; Bucar, F.; Maleš, Ž.; Mornar, A.; Nigovic, B.; Kujundžić, N. Identification and quantification of flavonoids and phenolic acids in burr parsley (Caucalis platycarpos L.), using high-performance liquid chromatography with diode array detection and electrospray ionisation mass spectrometry. *Molecules* **2009**, *14*, 2466–2490. [CrossRef]
61. Panche, A.N.; Diwan, A.D.; Chandra, S.R. Flavonoids: An overview. *Science* **2016**, *5*, 1–15. [CrossRef]
62. Sela, I.; Meir, A.Y.; Brandis, A.; Krajmalnik-Brown, R.; Zeibich, L.; Chang, D.; Dirks, B.; Tsaban, G.; Kaplan, A.; Rinott, E.; et al. *Wolffia globosa*–Mankai plant-based protein contains bioactive vitamin B12 and is well absorbed in humans. *Nutrients* **2020**, *12*, 3067. [CrossRef] [PubMed]
63. Kotowska, U.; Piotrowska, A.; Isidorova, A.; Bajguz, A.; Isidorov, V. Gas chromatographic-mass spectrometric investigation of the chemical composition of the aquatic plant *Wolffia arrhiza* (Lemnaceae). *Oceanol. Hydrobiol. Stud.* **2013**, *42*, 181–187. [CrossRef]
64. Kaplan, A.; Zelicha, H.; Tsaban, G.; Meir, A.Y.; Rinott, E.; Kovsan, J.; Novack, L.; Thiery, J.; Ceglarek, U.; Burkhardt, R.; et al. Protein bioavailability of *Wolffia globosa* duckweed, a novel aquatic plant—A randomised controlled trial. *Clin. Nutr.* **2018**, *38*, 2576–2582. [CrossRef]
65. Kawamata, Y.; Shibui, Y.; Takumi, A.; Seki, T.; Shimada, T.; Hashimoto, M.; Inoue, N.; Kobayashi, H.; Narita, T. Genotoxicity and repeated-dose toxicity evaluation of dried *Wolffia globosa* Mankai. *Toxicol. Rep.* **2020**, *7*, 1233–1241. [CrossRef]
66. Sree, K.S.; Dahse, H.-M.; Chandran, J.N.; Schneider, B.; Jahreis, G.; Appenroth, K.J. Duckweed for human nutrition: No cytotoxic and no anti-proliferative effects on human cell lines. *Plant Foods Hum. Nutr.* **2019**, *74*, 223–224. [CrossRef]
67. Hejný, S. Ellias Landolt The family of Lemnaceae—A monographic study—Vol.1 biosystematic investigations in the family of Duckweeds (Lemnaceae) (vol. 2). *Folia Geobot. Phytotax* **1993**, *28*, 50. [CrossRef]
68. Soda, S.; Kawahata, Y.; Takai, Y.; Mishima, D.; Fujita, M.; Ike, M. Kinetics of nutrient removal and biomass production by duckweed *Wolffia arrhiza* in continuous-flow mesocosms. *Ecol. Eng.* **2013**, *57*, 210–215. [CrossRef]
69. Naumann, B.; Eberius, M.; Appenroth, K.-J. Growth rate based dose-response relationships and EC-values of ten heavy metals using the duckweed growth inhibition test (ISO 20079) with *Lemna minor* L. clone St. *J. Plant Physiol.* **2007**, *164*, 1656–1664. [CrossRef]
70. Sabliy, L.; Konontsev, S.; Grokhovska, J.; Widomski, M.K.; Łagód, G. Nitrogen removal from fish farms water by *Lemna minor* and *Wolffia arrhiza*. *Proc. ECOpole* **2016**, *10*. [CrossRef]
71. Fujita, T.; Nakao, E.; Takeuchi, M.; Tanimura, A.; Ando, A.; Kishino, S.; Kikukawa, H.; Shima, J.; Ogawa, J.; Shimizu, S. Characterisation of starch-accumulating duckweeds, *Wolffia globosa*, as renewable carbon source for bioethanol production. *Biocatal. Agric. Biotechnol.* **2016**, *6*, 123–127. [CrossRef]
72. Karpiscak, M.M.; Gerba, C.P.; Watt, P.M.; Foster, K.E.; Falabi, J.A. Multi-species plant systems for wastewater quality improvements and habitat enhancement. *Water Sci. Technol.* **1996**, *33*, 231–236. [CrossRef]
73. van der Steen, P.; Brenner, A.; Shabtai, Y.; Oron, G. Improved fecal coliform decay in integrated duckweed and algal ponds. *Water Sci. Technol.* **2000**, *42*, 363–370. [CrossRef]
74. Khairina, Y.; Jog, R.; Boonmak, C.; Toyama, T.; Morikawa, M. Indigenous bacteria, an excellent reservoir of functional plant growth promoters for enhancing duckweed biomass yield on site. *Chemosphere* **2020**, *268*, 129247. [CrossRef]
75. Duong, T.P.; Tiedje, J.M. Nitrogen fixation by naturally occurring duckweed–cyanobacterial associations. *Can. J. Microbiol.* **1985**, *31*, 327–330. [CrossRef]

76. Yoneda, Y.; Yamamoto, K.; Makino, A.; Tanaka, Y.; Meng, X.-Y.; Hashimoto, J.; Shin-Ya, K.; Satoh, N.; Fujie, M.; Toyama, T.; et al. Novel plant-associated acidobacteria promotes growth of common floating aquatic plants, duckweeds. *Microorganisms* **2021**, *9*, 1133. [CrossRef]
77. Gale, J.; Smernoff, D.; Macler, B.; MacElroy, R. Carbon balance and productivity of *Lemna gibba*, a candidate plant for CELSS. *Adv. Space Res.* **1989**, *9*, 43–52. [CrossRef]
78. Kutlakhmedov, L.A.; Sokirko, D.M.; Grodzinskii, A.L.; Mashinshii, G.S.; Nechitailo, N.I. Investigation of the effects of space flight factors on the emergence from the antibiotic state of turions of the great duckweed. *Aerosp. Med. Biol.* **1980**, *12*, 49–54.
79. Kutlakhmedov, L.A.; Sokirko, D.M.; Grodzinskii, G.S. The effects of simulated weightlessness on the reproductive capacity of the great duckweed in the norm and under irradiation. *Aerosp. Med. Biol.* **1980**, *12*, 55–58.
80. Maccallum, T.; Anderson, G.A.; Poynter, J.E.; Stodieck, L.S.; Klaus, D.M. Autonomous Biological System (ABS) experiments. *Biol. Sci. Space* **1998**, *12*, 363–365. [CrossRef] [PubMed]
81. Yuan, J.; Xu, K. Effects of simulated microgravity on the performance of the duckweeds *Lemna aequinoctialis* and *Wolffia globosa*. *Aquat. Bot.* **2017**, *137*, 65–71. [CrossRef]
82. Facius, R.; Scherer, K.; Strauch, W.; Nevzgodina, L.; Maximova, E.; Akatov, Y. Mortality and morphological anomalies related to the passage of cosmic heavy ions through the smallest flowering aquatic plant *Wolffia arrhiza*. *Adv. Space Res.* **1996**, *18*, 195–204. [CrossRef]
83. Van Hoeck, A.; Horemans, N.; Nauts, R.; Van Hees, M.; Vandenhove, H.; Blust, R. *Lemna minor* plants chronically exposed to ionising radiation: RNA-seq analysis indicates a dose rate dependent shift from acclimation to survival strategies. *Plant Sci.* **2017**, *257*, 84–95. [CrossRef] [PubMed]
84. Thompson, B. The maximisation of the productivity of aquatic plants for use in controlled ecological life support systems (CELSS). *Acta Astronaut.* **1989**, *19*, 269–273. [CrossRef]

plants

MDPI

Article

Effects of Partial Substitution of Conventional Protein Sources with Duckweed (*Lemna minor*) Meal in the Feeding of Rainbow Trout (*Oncorhynchus mykiss*) on Growth Performances and the Quality Product

Elisa Fiordelmondo [1], Simona Ceschin [2], Gian Enrico Magi [1], Francesca Mariotti [1], Nicolaia Iaffaldano [3], Livio Galosi [1] and Alessandra Roncarati [1,*]

[1] School of Biosciences and Veterinary Medicine, University of Camerino, Viale Circonvallazione 93–95, 62024 Matelica, Italy; elisa.fiordelmondo@unicam.it (E.F.); gianenrico.magi@unicam.it (G.E.M.); francesca.mariotti@unicam.it (F.M.); livio.galosi@unicam.it (L.G.)

[2] Laboratory of Systematic and Environmental Botany, Department of Sciences, Roma Tre University, Viale G. Marconi 446, 00146 Rome, Italy; simona.ceschin@uniroma3.it

[3] Department of Agricultural, Environmental and Food Sciences, University of Molise, Via De Sanctis s/n, 86100 Campobasso, Italy; nicolaia@unimol.it

* Correspondence: alessandra.roncarati@unicam.it; Tel.: +39-073-7403-448

check for updates

Citation: Fiordelmondo, E.; Ceschin, S.; Magi, G.E.; Mariotti, F.; Iaffaldano, N.; Galosi, L.; Roncarati, A. Effects of Partial Substitution of Conventional Protein Sources with Duckweed (*Lemna minor*) Meal in the Feeding of Rainbow Trout (*Oncorhynchus mykiss*) on Growth Performances and the Quality Product. *Plants* **2022**, *11*, 1220. https://doi.org/10.3390/plants11091220

Academic Editors: Viktor Oláh, Klaus-Jürgen Appenroth and K. Sowjanya Sree

Received: 7 February 2022
Accepted: 28 April 2022
Published: 30 April 2022

Publisher's Note: MDPI stays neutral with regard to jurisdictional claims in published maps and institutional affiliations.

Abstract: Duckweed (*Lemna minor*) meal was included in the formulation of three experimental feeds (L1, L2, L3) for rainbow trout at 10%, 20%, 28% of the protein source, respectively. Increasing the duckweed inclusion, the other protein sources were adjusted to get isonitrogenous (41%) and isolipidic (20%) diets, as the control diet (LC). 540 fish (mean body weight 124.5 ± 0.7 g) were randomly allocated in 12 tanks divided equally among the four different diets. After 90 days, fish were weighed and the most important productive performances, fillet quality and fatty acid profile were determined. The final body weight in L1 (340.53 g) and L2 (339.42 g) was not different from LC (348.80 g); L3 trout significantly ($p < 0.05$) exhibited the lowest one (302.16 g). Similar trends were found in final mean length, weight gain, specific growth rate, food conversion rate. Somatic indices were affected by duckweed inclusion. Diets had not significant effects on the proximate composition and fatty acids of the fillet in L1, L2, L3 respect to LC. Based on this study, duckweed meal derived from *Lemna minor* can be included in the feed for the rainbow trout without negative effects on the growth performances at 20% of the protein substitution.

Keywords: sustainable fish feed; aquatic plants; alternative proteins; on-growing phase

1. Introduction

According to Food and Agriculture Organization (FAO), fish consumption in the world has reached a new record of 20.5 kg per capita per year and it will increase further in the next decade [1]. In this context, sustainable development of aquaculture and effective management of fish resources are the key to support for this trend. The increase of fish production needs the use of new raw materials to be included in fish feeding, and the adoption of new technology and new strategies to produce greater quantities of fish in a sustainable way, thus avoiding natural resources exploitation. Nowadays, it is increasingly important to find alternative and innovative raw materials to be used in fish feeding, and to understand the level of the possible substitution without a negative influence in fish growth and fish meat quality, without forgetting the importance of the environmental sustainability. In particular, some non-conventional feeding sources are becoming strategically important from economic and environmental sustainability point of view; indeed, the use of expensive fish meal for fish feeding is leaving space for cheaper and less impactful alternative diets, based on protein sources able to replace this ingredient with others (insects, vegetables, algae, by-products from aquatic organisms) to combine fish growth and environmental sustainability [2,3]. The research of proteins of vegetable origin in the formulation of aquafeed has underlined that the crop-based agriculture can help the aquaculture to become more sustainable through expanding the variety of different plant sources [4,5]. The soybean meal is the conventional protein source mostly used as a complement to fish meal and consequently the demand for this feedstuff has significantly increased at the world level to skyrocket its price in these last five years; it is clear that this has made its use less and less sustainable both in economic terms, as many countries also have to import it, and in environmental terms. Therefore, the properties of other plants have been investigated to evaluate their potential use in fish feed as protein source [6–8]. Recently, the attention has been focused on small aquatic plants, known as duckweeds, appreciated for their ability to reduce nutrient concentrations in water absorbing nitrogen compounds [8,9] and for their nutritional properties [10].

Duckweeds (*Lemnaceae)* [11,12] are free-floating aquatic plants, occurring spontaneously in standing or slow-flowing waters. They grow very rapidly and widely in nature, showing to be one of the fastest growing higher plants [13,14], and their supply is easy to recover. These plants are characterized morphologically by a tiny (a few mm) leaf-shaped vegetative body (frond) in which the stem is not distinguishable from the leaves and with a root system consisting of a single root.

Concerning the cultivation, duckweeds can be produced quite easily and cheaply even without the necessity to use growth media and/or fertilizers since they are characterized by a high Relative Growth Rate (RGR) [14,15]. This means they are able to produce large quantities of biomass in a short time and in relatively small ponds filled with a few tens of centimeters of natural water (30–50 cm deep). Obviously, their productivity can increase the more the optimal ecological conditions for their growth are present. In optimal growth conditions, duckweeds show high concentrations of nutrients but pathogens, heavy metals and organic pollutants should be accumulated in the plants' tissues [16]. Controlling and monitoring the aquatic environment in which the plants grow is particularly important. Duckweeds productivity increases more if the optimal ecological conditions for growth are respected, which however are generally wide. These, while varying slightly from species to species, generally consist of moderately warm, sunny and nutrient-rich waters, as documented in ecological studies on some duckweed species of the *Lemna* genus [12,17,18]. However, a good productive performance of duckweeds can occur in a wide range of conditions with respect to some factors, such as temperature and pH [7,8] for example, which clearly points out how these plants can be easily cultivable in different habitats. It is interesting to consider the hypothesis of growing duckweeds in wastewater from aquaculture systems [19] that, generally rich in nutrients, could allow a production of them at low cost and eco-sustainable in line with the principles of the circular economy. This is important when considering the huge quantity of plant needed in large scale fish feeding. Other aspects such as the rapid growth and the composition of protein and polyunsaturated fatty acids [20,21] make duckweeds a good ingredient for feed applications.

Considering the composition of duckweeds, they provide a good source of protein (up to 45.5 g of crude protein [22]), lipid, and minerals [23–25]. For their good protein intake, duckweeds were also largely used for feeding ruminants [26,27], pigs [24,28] and poultry [29,30] and for making pet foods, as an alternative source of amino acids [20]. Duckweeds are also commonly consumed as food by people in some areas in different continents [31]. Amino acid profile and fatty acid profile have confirmed the suitability of these plants in the production of aquafeed [32]. Moreover, macronutrients and other compounds, such as β-carotene and xanthophyll, increase the importance of the duckweeds as a potential ingredient to be essayed in aquafeed [33] for warmwater fish species as rohu [34], carp [35] and tilapia [36]. However, in salmonids the dietary duckweeds meal content has been only evaluated during fry stage of the rainbow trout [22].

Based on these considerations, a trial was performed in order to evaluate the effects of duckweed meal as partial replacement of the main conventional protein sources (fish and soybean meal) in three different low fish meal diets on productive performances of rainbow trout reared during the on-growing phase and compared with coetaneous fish receiving a conventional feed.

2. Results

2.1. Water Physico-Chemical Characterization

Concerning the physico-chemical characteristics of the waters where the fish were reared, in all the groups the temperature ranged from 12 °C, at the beginning of the experiment, to 13.8 °C (L2) at the end (mean values: LC 11.05 ± 0.8 °C; L1 11.06 ± 0.9 °C; L2 11.04 ± 0.9 °C; L3 11.05 ± 0.9 °C). The pH ranged between 7.8 and 8.0 without notable variations among the groups (mean values: 7.9 ± 0.1). The dissolved oxygen was averagely always over 10 mg/L in all the groups (mean values: LC 10.9 ± 1.5 mg/L; L1 10.8 ± 1.6 mg/L; L2 10.9 ± 1.4 mg/L; L3 10.2 ± 1.2 mg/L). Water total nitrogen ammonia (TAN) was included between a minimum in LC and L3 tanks (0.11–0.12 mg/L) and a maximum in L2 and L3 (0.19–0.20 mg/L) (mean values: LC 0.16 ± 0.1 mg/L; L1 0.15 ± 0.05 mg/L; L2 0.15 ± 0.06 mg/L; L3 0.16 ± 0.06 mg/L). Nitrites (NO_2-N) ranged from 0.02 mg/L in LC to 0.03 mg/L in L3 tanks (mean values: LC 0.025 ± 0.004 mg/L; L1 0.029± 0.002 mg/L; L2 0.026 ± 0.003 mg/L; L3 0.028 ± 0.002 mg/L) while nitrates (NO_3-N) from 0.9 mg/L (L2) to 0.14 mg/L (L3) (mean values: LC 0.12 ± 0.2 mg/L; L1 0.12 ± 0.1 mg/L; L2 0.10 ± 0.1 mg/L; L3 0.12 ± 0.2 mg/L).

2.2. Productive Performances of *Oncorhynchus mykiss* under Different Experimental Diets

The productive parameters are reported in Table 1. The final mean body weight and length did not show significance differences among trout fed with L1 (340.5 g; 31.2 cm), L2 (339.4 g; 31.6 cm) and LC group (348.8 g; 31 cm) but were different from L3 group (302.16 g; 28.2 cm) that showed the lowest significantly final weight and size. Growth parameters, weight gain (WG) and specific growth rate (SGR), recorded similar performances among L1 (216 g; 1.26), L2 (214.9; 1.2) and LC (224.3; 1.29) and were significantly higher than L3 (177.66; 1.11). Food conversion rate (FCR) gave a favorable result without statistically differences among L1 (1.18), L2 (1.18) and LC (1.13) but significantly better than the one recorded in L3 group (1.37).

Table 1. Productive performances of rainbow trout fed with different experimental diets (mean ± standard deviation).

Parameters	LC	L1	L2	L3	*p*
Initial mean weight (g)	124.5 ± 0.7	124.5 ± 0.7	124.5 ± 0.7	124.5 ± 0.7	-
Initial mean length (cm)	20.0 ± 0.6	20.0 ± 0.6	20.0 ± 0.6	20.0 ± 0.6	-
Final mean weight (g)	348.80 ± 4.4 a	340.53 ± 4.3 a	339.42 ± 4.7 a	302.16 ± 2.2 b	<0.05
Final mean length (cm)	31.0 ± 1.2 a	31.2 ± 1.3 a	31.6 ± 1.5 a	28.2 ± 1.6 b	<0.05
WG (%)	224.3 ± 2.6 a	216.03 ± 2.8 a	214.92 ± 2.9 a	177.66 ± 2.7 b	<0.05
SGR (%/day)	1.29 ± 0.03 a	1.26 ± 0.04 a	1.25 ± 0.03 a	1.11 ± 0.01 b	<0.05
FCR (g/g)	1.13 ± 0.02 b	1.18 ± 0.02 b	1.18 ± 0.03 b	1.37 ± 0.02 a	<0.05
SR (%)	99 ± 0 a	98 ± 1 a	98 ± 1 a	98 ± 1 a	<0.05
Palatability	100 ± 0.0 a	99.6 ± 0.4 ab	98.8 ± 1 b	98.2 ± 1.1 b	<0.05
KI	1.17 ± 0.12 a	1.12 ± 0.13 a	1.08 ± 0.14 a	1.35 ± 0.22 b	<0.05
VSI	10.06 ± 0.41 c	10.28 ± 0.59 bc	11.57 ± 0.68 b	14.57 ± 0.54 a	<0.05
PFI	3.00 ± 0.36 b	2.91 ± 0.04 b	3.05 ± 0.12 b	3.68 ± 0.03 a	<0.05
HSI	1.05 ± 0.06 b	1.31 ± 0.08 a	1.35 ± 0.03 a	1.24 ± 0.06 a	<0.05

Different letters (a, b, c) on the same line show statistically significant differences ($p < 0.05$). WG: weight gain, SGR: specific growth rate, FCR: feed conversion rate, SR: survival rate, KI: condition index, VSI: viscerosomatic index, PFI: perivisceral fat index, HSI: hepato-somatic index.

The somatic indices, condition index (KI), viscerosomatic index (VSI), perivisceral fat index (PFI), hepatosomatic index (HSI), were significantly affected by diets with different levels of duckweed meal. KI exhibited differences when the inclusion was at the highest substitution (28%) in L3 group (1.35) respect to all the other groups (L1: 1.12; L2: 1.08; LC 1.17). VSI significantly increased from L1 (10.3), at an intermediate level between LC (10.06) and L2 (11.57), to L3 (14.57). PFI had not notable variations among L1 (2.9), L2 (3.05) and LC (3), that were all different from L3 (3.68). HSI appeared similar among the three experimental groups (L1: 1.31; L2: 1.35; L3: 1.24) but significantly higher than LC (1.05). No significant differences were observed in the survival rate (SR), with values ranging between 98 and 99% in all groups. Palatability of L1 (99.6) was at an intermediate level respect to LC (100) and L2 (98.8) and L3 (98.2); these last two experimental diets showed a similar acceptance.

The proximate composition of the fillets of rainbow trout fed with diets without (LC) or with different percentages of duckweed meal (L1, L2, L3) is reported in Table 2. For all the macronutrients considered (protein, fat, moisture and ash content) no significant differences were shown among the duckweed meal diets (L1, L2, L3) and the conventional control diet (LC).

Table 2. Proximate composition (% ww) (mean ± st.dev.) of the fillet of rainbow trout fed with the control diet (LC) and the three experimental diets (L1, L2, L3) at the end of the trial.

Parameters	LC	L1	L2	L3	*p*
Moisture	77.73 ± 1.4	77.76 ± 1.2	77.35 ± 1.1	77.41 ± 1.2	>0.05
Protein	19.44 ± 0.9	19.78 ± 1.0	18.46 ± 1.1	18.82 ± 0.8	>0.05
Fat	2.33 ± 1.1	2.54 ± 0.8	3.17 ± 0.9	3.31 ± 0.9	>0.05
Ash	1.37 ± 0.2	1.29 ± 0.1	1.25 ± 0.2	1.18 ± 0.2	>0.05

Due to no significant differences from the statistical point of view, no letter was reported among the parameters on the same line.

With regard to the main categories of fatty acids in the final composition of the fillet of rainbow trout, the saturated fatty acids (SFA) and monounsaturated fatty acids (MUFA) categories were similar among all the four groups without notable differences (Figure 1). Furthermore, the polyunsaturated fatty acids (PUFAs) (n-3 and n-6 series) did not show significant differences ($p < 0.05$) between the experimental (L1, L2, L3) and the control (LC) group (Figure 1).

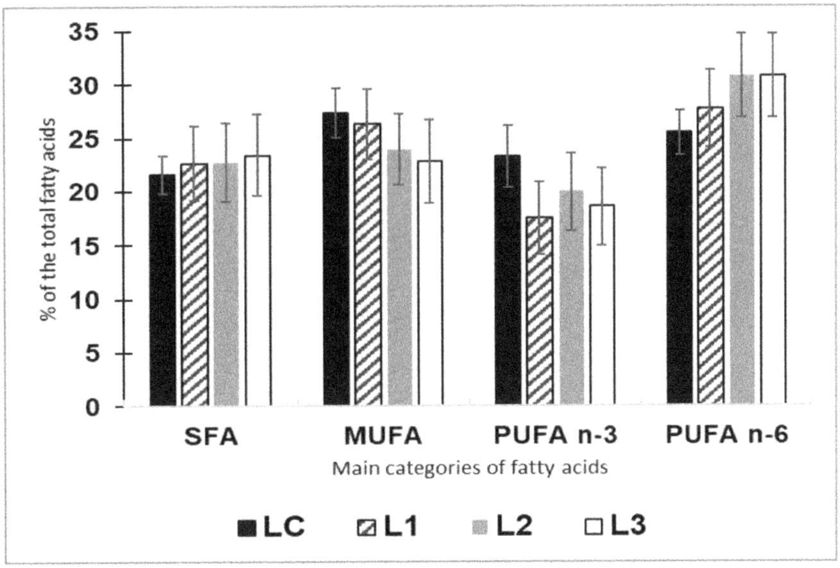

Figure 1. Categories of fatty acids (SFA: saturated fatty acids, MUFA: monounsaturated fatty acids,

PUFA: n-3, n-6 polyunsaturated fatty acids) determined in the fillet of trout fed with LC and L1, L2, L3 diets (% of the total fatty acids). Due to no significant difference from the statistical point of view, no letter was reported in the groups of column charts.

3. Discussion

For decades, duckweed has captured the interest of scientists [37] due to its organoleptic characteristics. It is still a current topic due to the urgent problem of finding new protein sources as alternatives to the standard ones to be used in aquafeed. In particular, the properties of duckweed meal have been studied for their protein content in freshwater and marine fish.

In this trial, duckweed meal was included in three experimental diets for rainbow trout as partial replacement of the two main protein feedstuffs, fish meal and soybean meal, in order to evaluate an alternative protein source less expensive and more sustainable than conventional ones. The substitution has mostly concerned soybean meal and secondly fish meal, because in the last years the strong increase of the production costs of aquafeeds has also concerned the price of feedstuffs of vegetable origin such as soybean meal. Besides, the maintenance of a fraction of fish meal was considered prudent by authors to assure good growth results for rainbow trout. To perform the study, we used duckweed collected in freshwater areas, strictly monitored in terms of quality as demonstrated by the analyzed physico-chemical parameters were within the range considered optimal for the rainbow trout species in all the fish groups tested [38]. The survival rate was very high in all the experimental groups.

The meal obtained by the common duckweed employed in the feeds had a protein content within the range specifically reported for that plant species [36]. The amount of essential amino acids was detected in a good quantity, except methionine. Because of that, DL-methionine was added in the formulation of all the three experimental diets used in the present trial, with the aim to administer balanced diets according to the specific requirements of the rainbow trout species [39]. The percentages of duckweed substitution were also evaluated and decided making a gradual increase of duckweed in the experimental diets. In duckweed meals used in the present study, the fat content (5%) was in agreement with the literature reporting ranges between 4% and 7% [24].

The productive results showed that no adverse effects were observed in mean body weight, weight gain, and final length of rainbow trout when fish receives diets including up to 20% of duckweed meal, whereas highest levels affected the growth performances and the FCR. These differences could be related to the diet palatability that slightly decreased at increasing the duckweed meal. As Table 1 shows, the decrease of the palatability of the duckweed meals corresponds to an up inclusion of *Lemna* meal. The rainbow trout receiving the feed with the highest duckweed meal content exhibited reduced productive performance and unfavorable FCR. In another study on the taste of various aquatic plants for tilapia [40], duckweed showed a low attractive effect due to the presence of flavonoids and triterpene compounds considered as not feed stimulating for fish. However, in common carp, diets with 20% of duckweed replacement gave results similar to conventional feeds in terms of growth performances [34]. In the Indian carp species (rohu *Labeo rohita*), the replacement of 30% of fish meal dietary with common duckweed did not affect the growth of fingerlings in comparison with coetaneous fed with other macrophytes [41]. Based on literature, studies on common duckweed in fish diets were mostly performed on cyprinids and tilapia, so fish with feeding (omnivorous) and living (warmwater) habits different respect to salmonids. In rainbow trout, a 4 weeks trial using another species of duckweed (*Spirodela polyrhiza*) at low and high (6.25–12.5% of feed) substitution for 4 weeks, had the same acceptance of the control diet although both the duckweed meal treatments resulted in 5 and 10% poorer growth traits [22]. As concerns the somatic indices, KI and PFI were affected by the highest percentage of duckweed inclusion followed the trend of growth increased, whereas VSI and HSI discriminated differences also among the three experimental diets as well as with the control feed. These variations in VSI and HSI could be associated to the carbohydrate fraction not used as energy source, and therefore accumulated in the liver and transformed in lipids and glycogen, resulting in an increase of this index as documented in rainbow trout fed with diets including alternative plant ingredients rich in indigestible carbohydrates, in the form of oligo- as well as polysaccharides [42]. To overcome this drawback, other works proposed to employ duckweed after a fermentation process that could considerably reduce the antinutritional factors and the crude fiber content [28]. Regarding the effects of dietary duckweed on the fish fillet quality, the proximate composition did not show notable differences among the macronutrients of the groups. In fact, as reported in Table 2, the proximate composition of the fillet of rainbow trout fed with LC, L1, L2 and L3 doesn't show significant differences from the statistical point of view. In terms of fatty acid categories in the meat, trout fed diets including *Lemna* appeared very similar among them and the control. It is well known that the final nutritional flesh quality is strongly affected by the diet composition administered to fish [43]. In the current study, the experimental feeds maintained the blend of fish and vegetable oil unchanged respect to the control diet with the aim to show the only effects of the duckweed meal ingredient. According to a study on nutritional value of different duckweed genera, *Lemna minor* is reported as a species with an intermediate proportion (27.99%) of SFA, a very low MUFA level (4.6%) and a very high PUFA n-3 rate (46%), surprisingly higher than content of PUFA n-6 (20%) [21]. In the current experiment, the fillet of rainbow trout had good content of fatty acids especially in terms of PUFA n-3, considered essentials of a balanced diet in humans providing beneficial effects on neural development [44] and in mitigating several pathological conditions [45].

4. Materials and Methods

4.1. Experimental Design and Fish Material

For the experiment, 12 concrete outdoor tanks were used, divided in 3 tanks for each of the four trout groups fed with different diets (LC, L1, L2, L3). Every tank had a length of 5 m, a width of 0.8 m, a depth of 0.5 m, and a volume of 2 m^3 and was filled with well water. During the experiment, the main water physico-chemical parameters (temperature, dissolved oxygen and pH) were daily recorded in every tank using portable electronic devices (YSI mod. 55 and 60, Yellow Springs, OH, USA). TAN, NO_2-N and NO_3-N were weekly analyzed following APHA standard methods [46]. All the basins were covered with an antifouling net in order to avoid algal development and to keep away ichthyophagous birds.

This experiment was performed during a standard zootechnical cycle, avoiding any animal suffering, and no sample was collected from live animals, according to the Italian Legislative Decree 26/2014. The farm applied an "Antibiotic-free Code of Prescription" to guarantee the quality of the product and respect the "antibiotic-free" approach. During spring season in 2021, a total of 540 rainbow trout (245 days old; mean body weight 124.5 $\pm$ 0.7 g) was randomly allocated in the 12 tanks, with 45 fish in each tank, at the initial stocking density of 6.2 kg/m^3. At the end of the experiment (90 days), fish were weighed, and their final length was recorded. Fish were fed by hand twice a day (8 a.m. and 3 p.m.) until the apparent satiation level; then the unconsumed feed was collected.

Palatability of the feeds was calculated according to the formula: ((ingested feed/administered feed) $\times$ 100) based on the index reported in previous studies [47,48]. The following zootechnical performances were evaluated in the four different groups: WG (%) = (final weight − initial weight) $\times$ 100/initial weight; SGR (%/day) = {Ln (final weight) − Ln (initial weight)/days} $\times$ 100; FCR = live weight gain (g)/feed administered (g); SR (%) = final number of fish/initial number of fish $\times$ 100 [49]. In addition, the condition factor KI = ([fish weight/fish length] $\times$ 100) [50] and the following biometric indices were calculated after having applied standard procedures for fish sampling [51]: VSI = ([weight viscera/whole body weight] $\times$ 100), PFI = ([perivisceral fat/body weight] $\times$ 100), and HSI = ([liver weight/body weight] $\times$ 100). In order to calculate the PFI and the VSI, the fat adherent to the digestive tract was accurately separated and individually weighed.

4.2. Plant Material and Experimental Diets

Fronds of a duckweed species of the genus *Lemna*, *L. minor* (common duckweed) were collected from ponds of a fish farm. We avoid the use of duckweeds coming from polluted areas or wastewater due to high sensitivity of this plant to a wide range of toxicants [52] which could threaten the fish health status [40]. Samples of duckweed were analyzed from the botanical point of view by means of stereomicroscope (mod. Stemi 305, Zeiss). The duckweed was submitted to washing, air exposure and oven-dried at 60 °C for 6 h, and finally milled modified after recent literature [30]. In order to employ the duckweed meal for the on-growing phase of rainbow trout, the proximate composition and the amino acid profile were determined on a duckweed frond sample (Table 3). The proximate composition (moisture, protein, lipid and ash) and the amino acid profile of duckweed meal were performed according to international methods. In particular, moisture and ash content were determined using the procedures described by the Official Analytical Chemists (AOAC) [53]. The protein content was determined using the standard Kjeldahl copper catalyst method. The aminoacid profile was determined by acid hydrolysis (6 N HCl for 24 hrs, at 110 °C) followed by ion ex-change chromatography with an amino-analyzer (L-8800 Auto-analyzer, HITACHI, Japan).

Table 3. Proximate composition (%) and amino acid profile (g/100 g) of duckweed meal used in the experiment.

Composition	%
Moisture	92.81
Crude protein	28.13
Crude lipid	5.10
Crude fibre	15.20
Ash	16.40
Amino acid	% of crude protein
Arginine	4.56
Histidine	3.28
Isoleucine	3.62
Leucine	6.41
Lysine	4.49
Methionine	1.74
Phenylalanine	4.25
Threonine	2.16
Tryptophan	3.89
Valine	3.53

Lemna minor meal was included in the formulation of three feeds (L1, L2, L3) at different rates (10%, 20%, 28%, respectively) of the protein source. At the increasing of the duckweed inclusion, soybean meal, fish meal, wheat flour and gluten wheat meal were reduced or adjusted in order to get similarly an isonitrogenous (41%) and isolipidic (20%) diets. A control diet (LC) was formulated with the same feedstuffs except duckweed meal (Table 4). This formulation aimed at saving the use of the various conventional protein sources, essaying a local feedstuff derived from duckweed plant; after the pandemic, also soybean meal and not only fish meal are becoming very expensive to get on the market.

Table 4. Formulation and proximate composition of the control diet without including duckweed meal (LC), and L1, L2 and L3 diets, with different inclusion of duckweed meal (10%, 20%, 28%, respectively).

	LC	L1	L2	L3
Ingredients (%)				
Duckweed meal	**0**	**10**	**20**	**28**
Soybean meal	22.3	21.8	11.0	7.0
Fish meal	21	20	18	17
Wheat flour	20	13	11	8
Haemoglobin meal	10	10	10	10
Gluten wheat meal	7.6	5.6	10.4	11.3
Fish oil	12	12	12	12
Soybean oil	5	5	5	5
L-Lysine	0.4	0.4	0.4	0.4
DL-Methionine	0.18	0.2	0.25	0.28
Vitamin-mineral mix	2	2	2	2
Proximate composition (%)				
Moisture	8.89	8.92	8.97	9.05
Protein	41.50	41.59	41.53	41.27
Lipid	20.00	20.12	20.30	20.00
Ash	6.81	7.28	7.10	7.26
Amino acid profile (% of crude protein)				
Arginine	5.24	4.51	4.00	3.92
Histidine	1.85	1.68	1.67	1.50
Isoleucine	1.39	1.37	1.31	1.30
Leucine	3.61	3.60	3.54	3.51
Lysine	4.99	3.90	3.56	3.01
Methionine	3.33	2.91	2.91	2.91
Phenylalanine	2.28	2.10	1.94	1.80
Threonine	2.06	1.53	1.45	1.42
Tryptophan	0.60	0.50	0.49	0.47
Valine	3.36	2.36	2.31	2.28

The feeds were manufactured in 3.5 mm size using a twin-screw extruder (100 rpm, 110 °C, 50 atm). After the coating, the feeds were stocked in buckets and kept in an aerated room. The proximate composition (moisture, protein, lipid and ash) and the amino acid profile of three samples of each feed type were performed according to the international methods reported for duckweed meal analysis [53]. For all the four diets, after determining total lipid content, using the procedure described by Folch et al. [54], fatty acids were converted to methyl esters following the method described by Christopherson and Glass [55]. The separation of fatty acids was carried out using a GC 3800 gas chromatography (Varian Strumentazione, Cernusco sul Naviglio, Italy) with a WP-4 Shimadzu integration system (Shimadzu Corporation, Tokyo, Japan), which was equipped with a Supelco SPTM—2340 capillary column (30 m × 0.25 mm i.d.; 0.25 μm film thickness; Supelco, Bellefonte, Pennsylvania, USA) and a flame ionization detector. The essential amino acid profile of the three feeds containing duckweed was obtained as previously indicated for duckweed meal.

4.3. Quality Traits of Fish Fillet

After 90 days of experiment, fish have reached the commercial size and are slaughtered by electrical stunning in an authorized slaughterhouse.

Fish were dissected as follow: after a check of the skin status on the whole body to be sure of the absence of skin lesions, the abdomen was opened with a cut starting from the anus to gills, and then a lateral line up the side of the fish allowed to check the status of gills and meat. From the anus the digestive system was tracked and removed; similarly, all the abdomen organs were checked and removed.

A portion of about 50 g of skinless dorsal left muscle from six fish casually selected for each diet group was collected, then homogenized and submitted to proximate composition analyses (moisture, protein, lipid, and ash content). The procedures adopted for these last analyses follow substantially the same methods indicated in the previous paragraph concerning the measurement of proximate composition of the different feedstuffs. The percentage of moisture was determined in duplicate according to the Association of Official Analytical Chemists procedure [53]. The protein content was determined using the standard Kjeldahl copper catalyst method. The ash content was determined using the procedure described by the AOAC [53]. Total lipids were measured using a modification of the chloroform:methanol procedure described by Folch et al. [54]. After determining the total lipid content, fatty acids were converted to methyl esters following the method described by Christopherson and Glass [55]. The separation of fatty acids was performed using a Carlo Erba HRGC 5160 gas chromatography (Carlo Erba Strumentazione, Rodano, MI, Italy) with a WP-4 Shimadzu integration system (Shimadzu Corporation, Tokyo, Japan) equipped with a Supelco SPTM-2340 capillary column (30 m$\times$ 0.32 mmi.d.; 0.20 μm film thickness; Supelco, Bellefonte, PA, USA) and a flame ionization detector. The operating conditions of the gas chromatography were as follows: the oven temperature was set at 170 °C for 15 min and subsequently increased to 190 °C at a rate of 1 °C/min, then increased to 220 °C at a rate of 5 °C/min and held at this temperature for 17 min. The concentration of individual fatty acid was calculated based on the relative proportion of each fatty acid compared with a known amount of the internal standard (17:0) added. The fatty acids were expressed as percentage of the total of fatty acids.

4.4. Statistical Data Analyses

Data collected (biometric parameters, final productive traits, fish fillet proximate composition, fatty acids categories) were subjected to one-way analysis of variance (ANOVA) using SPSS 25 [56] to check differences in productive performances and composition of fillet of rainbow trout fed with different experimental diets. Means and standard deviations were calculated. Means were considered significant with a value of $p < 0.05$ and compared using the Student-Newman-Keuls (SNK) test.

5. Conclusions

The current study aimed at evaluating the use of a local duckweed, collected in not contaminated waters, as protein source in partial substitution of the conventional feedstuffs (fish meal and soybean meal) to find out how could be used in animal feeding respecting the nature of the animal species. In particular, it provided useful information on the effects of duckweed meal diet on rainbow trout performances under on-growing phase and the quality fillet. This has great implication on responsible and sustainable aquaculture because it is essential to preserve fish biometric indices, rearing parameters and quality product, mainly when fish are reared under antibiotic-free approach. Through this study it has been possible to essay a protein source substitution with an alternative plant feedstuff, showing that the replacement should be done using *Lemna minor* at 20% of the protein sources, without negative consequences in the growth performance of the fish and quality fillet. At this rate of protein content, the feed with duckweed meal has shown satisfactory results. In this view, such study can represent a challenge for further investigations aimed to analyze possible variations of duckweed composition in different seasons, in different areas or considering different *Lemna* species occurring locally.

Author Contributions: Conceptualization, A.R. and E.F.; systematic classification of *L. minor* and contribution from the botanical point of view, S.C.; methodology, E.F. and A.R.; formal analysis, E.F., S.C., G.E.M., F.M., N.I., L.G., A.R.; data elaboration, E.F.; writing, review and editing, E.F., S.C., G.E.M., F.M., N.I., L.G., A.R.; funding acquisition, A.R. All authors have read and agreed to the published version of the manuscript.

Funding: This research was partially supported by the Eureka Project 2019–2020, Ph Doctoral Grant for Innovative Research in Life and Health Science, and by the School of Advanced Studies, University of Camerino (UNICAM), Italy.

Institutional Review Board Statement: Not applicable.

Informed Consent Statement: Not applicable.

Ethical Statement: This study was conducted according to the European Directive 2010/63/UE and the Italian Legislative Decree 26/2014.

Data Availability Statement: Data is contained within the current article.

Acknowledgments: We are grateful to the Erede Rossi Company for their valuable collaboration and technical assistance during the sampling activities.

Conflicts of Interest: The authors declare no conflict of interest. The funders had no role in the design of the study; in the collection, analyses or interpretation of data; in the writing of the manuscript; or in the decision to publish the results.

References

1. FAO. *The State of World Fisheries and Aquaculture 2020: Sustainability in Action*; FAO: Rome, Italy, 2020.
2. Gasco, L.; Acuti, G.; Bani, P.; Dalle Zotte, A.; Danieli, P.P.; De Angelis, A.; Fortina, R.; Marino, R.; Parisi, G.; Piccolo, G.; et al. Insect and fish by-products as sustainable alternatives to conventional animal proteins in animal nutrition. *Ital. J. Anim. Sci.* **2020**, *19*, 360–372. [CrossRef]
3. Parisi, G.; Tulli, F.; Fortina, R.; Marino, R.; Bani, P.; Dalle Zotte, A.; De Angelis, A.; Piccolo, G.; Pinotti, L.; Schiavone, A.; et al. Protein hunger of the feed sector: The alternatives offered by the plant world. *Ital. J. Anim. Sci.* **2020**, *19*, 1205–1227. [CrossRef]
4. Hua, K.; Cobcroft, J.M.; Cole, A.; Condon, K.; Jerry, D.R.; Mangott, A.; Praeger, C.; Vucko, M.J.; Zeng, C.; Zenger, K.; et al. The Future of Aquatic Protein: Implications for Protein Sources in Aquaculture Diets. *One Earth* **2019**, *1*, 316–329. [CrossRef]
5. Napier, J.A.; Haslam, R.P.; Olsen, R.-E.; Tocher, D.R.; Betancor, M.B. Agriculture can help aquaculture become greener. *Nat. Food* **2020**, *1*, 680–683. [CrossRef]
6. Dorothy, M.S.; Raman, S.; Nautiyal, V.; Singh, K.; Yogananda, T.; Kamei, M. Use of Potential Plant Leaves as Ingredient in Fish Feed–A Review. *Int. J. Curr. Microbiol. Appl. Sci.* **2018**, *7*, 112–125. [CrossRef]
7. Sudiarto, S.; Renggaman, A.; Choi, H.L. Floating aquatic plants for total nitrogen and phosphorus removal from treated swine wastewater and their biomass characteristics. *J. Environ. Manag.* **2019**, *23*, 763–769. [CrossRef]
8. Ceschin, S.; Crescenzi, M.; Iannelli, M.A. Phytoremediation potential of the duckweeds *Lemna minuta* and *Lemna minor* to remove nutrients from treated waters. *Environ. Sci. Pollut. Res.* **2020**, *27*, 15806–15814. [CrossRef]
9. Sońta, M.; Rekiel, A.; Batorska, M. Duckweed in sustainable livestock production and aquaculture. *Ann. Anim. Sci.* **2019**, *19*, 257–271. [CrossRef]
10. Bog, M.; Appenroth, K.J.; Sree, K.S. Duckweed (*Lemnaceae*): Its molecular taxonomy. *Front. Sustain. Food Syst.* **2019**, *3*, 117. [CrossRef]
11. Tippery, N.P.; Les, D.H.; Appenroth, K.J.; Sree, K.S.; Crawford, D.J.; Bog, M. *Lemnaceae* and *Orontiaceae* are phylogenetically and morphologically distinct from Araceae. *Plants* **2021**, *10*, 2639. [CrossRef]
12. Landolt, E. *The Family of Lemnaceae—A Monographyc Study*; Veröffentlichungen des Geobotanischen Institutes der ETH; Stiftung Rübel: Zürich, Switzerland, 1986; Volume 2, 567p.
13. Ziegler, P.; Adelmann, K.; Zimmer, S.; Schmidt, C.; Appenroth, K.-J.; Keurentjes, J. Relative in vitro growth rates of duckweeds (*Lemnaceae*)—The most rapidly growing higher plants. *Plant Biol.* **2015**, *17*, 33–41. [CrossRef] [PubMed]
14. Lemon, G.D.; Posluszny, U.; Husband, B.C. Potential and realized rates of vegetative reproduction in *Spirodela polyrhiza, Lemna minor*, and *Wolffia Boreal*. *Aquat. Bot.* **2001**, *70*, 79–87. [CrossRef]
15. Ceschin, S.; Della Bella, V.; Piccari, F.; Abati, S. Colonization dynamics of the alien macrophyte *Lemna minuta* Kunth: A case study from a semi-natural pond in Appia Antica Regional Park (Rome, Italy). *Fundam. Appl. Limnol.* **2016**, *188*, 93–101. [CrossRef]
16. Coughlan, N.E.; Walsh, E.; Bolger, P.; Burnell, G.; O'Leary, N.; O'Mahoney, M.; Paolacci, S.; Wall, D.; Jansen, M.A.L. Duckweed bioreactors: Challenges and opportunities for large-scale indoor cultivation of Lemnaceae. *J. Clean. Prod.* **2022**, *336*, 130285. [CrossRef]
17. Ceschin, S.; Abati, S.; Leacche, I.; Zuccarello, V. Ecological comparison between duckweeds in central Italy: The invasive *Lemna minuta* vs. the native *L. minor*. *Plant Biosyst.* **2018**, *152*, 674–683. [CrossRef]
18. Sharma, J.; Clark, W.D.; Shrivastav, A.K.; Goswami, R.K.; Tocher, D.R.; Chakrabarti, R. Production potential of greater duckweed *Spirodela polyrhiza* (L. Schleiden) and its biochemical composition evaluation. *Aquaculture* **2019**, *513*, 734419. [CrossRef]

19. Paolacci, S.; Stejskal, V.; Toner, D.; Jansen, M.A. Wastewater valorisation in an integrated multitrophic aquaculture system; assessing nutrient removal and biomass production by duckweed species. *Environ. Pollut.* **2022**, *302*, 119059. [CrossRef]
20. Yan, Y.; Candreva, J.; Shi, H.; Ernst, E.; Martienssen, R.; Schwender, J.; Shanklin, J. Survey of the total fatty acid and triacylglycerol composition and content of 30 duckweed species and cloning of a Δ6-desaturase responsible for the production of γ-linolenic and stearidonic acids in *Lemna gibba*. *BMC Plant Biol.* **2013**, *13*, 201. [CrossRef]
21. Appenroth, K.J.; Sree, K.S.; Böhm, V.; Hammann, S.; Vetter, W.; Leiterer, M.; Jahreis, G. Nutritional value of duckweeds (*Lemnaceae*) as human food. *Food Chem.* **2017**, *217*, 266–273. [CrossRef]
22. Stadtlander, T.; Förster, S.; Rosskothen, D.; Leiber, F. Slurry-grown duckweed (*Spirodela polyrhiza*) as a means to recycle nitrogen into feed for rainbow trout fry. *J. Clean. Prod.* **2019**, *228*, 86–93. [CrossRef]
23. Mbagwu, I.G.; Adeniji, H.A. The nutritional content of duckweed (*Lemna paucicostata* Hegelm.) in the Kainji lake area, Nigeria. *Aquat. Bot.* **1988**, *29*, 357–366. [CrossRef]
24. Leng, R.A.; Stambolie, J.H.; Bell, R. Duckweed—A potential high-protein feed resource for domestic animals and fish. *Livest. Res. Rural Dev.* **1995**, *7*, 36.
25. Men, B.X.; Ogle, B.; Lindberg, J.E. Use of duckweed as a protein supplement for growing ducks. *Asian Australas. J. Anim. Sci.* **2001**, *14*, 1741–1746. [CrossRef]
26. Huque, K.S.; Chowdhury, S.A.; Kibria, S.S. Study on the potentiality of Duckweeds as a feed for cattle. *Asian Australas. J. Anim. Sci.* **1996**, *9*, 133–138. [CrossRef]
27. Tanuwiria, U.H.; Mushawwir, A. Hematological and antioxidants responses of dairy cow fed with a combination of feed and duckweed (*Lemna minor*) as a mixture for improving milk biosynthesis. *Biodiversitas* **2020**, *21*, 4741–4746. [CrossRef]
28. Goopy, J.P.; Murray, P.J. Review on the Role of Duckweed in Nutrient Reclamation and as a Source of Animal Feed. *Asian Australas. J. Anim. Sci.* **2003**, *16*, 297–305. [CrossRef]
29. Mwale, M.; Gwaze, F.R. Characteristics of duckweed and its potential as feed source for chickens reared for meat production: A Review. *Sci. Res. Essays* **2013**, *8*, 689–697. [CrossRef]
30. Aderemi, F.A.; Alabi, O.M.; Agbaje, M.; Adeleke, A.G.; Ayoola, M.O. Utilization of Duckweed Meal as Replacement for Fish Meal by Broiler Chickens. *Insight Poult. Res.* **2018**, *8*, 1–9. [CrossRef]
31. McCusker, S.; Buff, P.R.; Yu, Z.; Fascetti, A.J. Amino acid content of selected plant, algae and insect species: A search for alternative protein sources for use in pet foods. *J. Nutr. Sci.* **2014**, *3*, e39. [CrossRef]
32. Beukelaar, M.F.A.; Zeinstraa, G.G.; Mes, J.J.; Fischer, A.R.H. Duckweed as human food. The influence of meal context and information on duckweed acceptability of Dutch consumers. *Food Qual. Prefer.* **2019**, *71*, 76–86. [CrossRef]
33. Chakrabarti, R.; Clark, W.D.; Sharma, J.G.; Goswami, R.K.; Shrivastav, A.K.; Tocher, D.R. Mass Production of *Lemna* spp. and Its Amino Acid and Fatty Acid Profiles. *Front Chem.* **2018**, *6*, 479. [CrossRef] [PubMed]
34. Bairagi, A.; Sarkar Ghosh, K.; Sen, S.K.; Ray, A.K. Duckweed (*Lemna polyrhiza*) leaf meal as a source of feedstuff in formulated diets for rohu (*Labeo rohita* Ham.) fingerlings after fermentation with a fish intestinal bacterium. *Bioresour. Technol.* **2002**, *85*, 17–24. [CrossRef]
35. Yilmaz, E.; Akyurt, I.; Günal, G. Use of Duckweed, *Lemna minor*, as a Protein Feedstuff in Practical Diets for Common Carp, *Cyprinus carpio*, Fry. *Turk. J. Fish. Aquat. Sci.* **2004**, *4*, 105–109.
36. Fasakin, E.A.; Balogun, A.M.; Fasuru, B.E. Use of duckweed, *Spirodela polyrrhiza* L. Schleiden, as a protein feedstuff in practical diets for tilapia, *Oreochromis niloticus* L. *Aquac. Res.* **1999**, *30*, 313–318. [CrossRef]
37. Van Dyke, J.M.; Sutton, D.L. Digestion of duckweed (*Lemna* spp.) by the grass carp (*Ctenopharyngodon idella*). *J. Fish Biol.* **1977**, *11*, 273–278. [CrossRef]
38. Fiordelmondo, E.; Magi, G.E.; Mariotti, F.; Bakiu, R.; Roncarati, A. Improvement of the Water Quality in Rainbow Trout Farming by Means of the Feeding Type and Management over 10 Years (2009–2019). *Animals* **2020**, *10*, 1541. [CrossRef]
39. National Research Council (NRC). *Nutrient Requirements of Fish and Shrimps*; The National Academic Press: Washington, DC, USA, 2011.
40. Vinogradskaya, M.I.; Kasumyan, A.O. Palatability of Water Organisms for Nile Tilapia *Oreochromis nilotocus* (Cichlidae). *J. Ichthyol.* **2019**, *59*, 3. [CrossRef]
41. Goswani, R.K.; Shrivastav, A.K.; Sharma, J.G.; Tocher, D.R.; Chakrabarti, R. Growth and digestive enzyme activities of rohu *Labeo rohita* fed diets containing macrophytes and almond oil-cake. *Anim. Feed Sci. Technol.* **2020**, *263*, 114456. [CrossRef]
42. Kaushik, S.; Panserat, S.; Schrama, J.W. Carbohydrates. In *Fish Nutrition*, 4th ed.; Hardy, R.W., Kaushik, S.J., Eds.; Elsevier: London, UK, 2022; pp. 555–591.
43. Sargent, J.R.; Bell, J.G.; McGhee, J.; McEvoy, J.; Webster, J.L. The nutritional value of fish. In *Farmed Fish Quality*; Fishing News Books; Kestin, S.C., Warriss, P.D., Eds.; Blackwell Science Ltd.: Oxford, UK, 2001; pp. 3–12.
44. Campoy, C.; Escolano-Margatt, V.; Anjos, T.; Szajewska, H.; Uauy, R. Omega 3 fatty acids on child growth, visual acuity and neurodevelopment. *Br. J. Nutr.* **2012**, *107*, S85–S106. [CrossRef]
45. Calder, P.C. Very long chain omega-3 (n-3) fatty acids and human health. *Eur. J. Lipid Sci. Technol.* **2014**, *116*, 1280–1300. [CrossRef]
46. APHA (American Water Works Association and Water Pollution Control Federation of American Public Health Association). *Standard Methods for the Examination of Water and Wastewater*, 19th ed.; American Public Health Association Inc.: New York, NY, USA, 1995.
47. Kasumyan, A.O. Gustatory reception and feeding behaviour in fish. *J. Ichthyol.* **1997**, *37*, 72–86.

48. Kasumyan, A.O.; Døving, K.B. Taste preferences in fish. *Fish Fish.* **2003**, *4*, 289–347. [CrossRef]
49. Steffens, W. *Principles of Fish Nutrition*; Ellis Horwood: Chichester, UK, 1989.
50. Bagenal, T.B.; Tesch, F.W. Methods of assessment of fish production in fresh waters. In *IBP Handbook N.3*, 3rd ed.; Blackwell Scientific: London, UK, 1978; pp. 101–136.
51. United States Environmental Protection Agency (U.S. EPA). *Guidance for Assessing Chemical Contaminant Data for Use in Fish Advisories*, 3rd ed.; Fish Sampling and Analysis; U.S. EPA: Washington, DC, USA, 2000.
52. Abdel-Gawad, F.K.; Khalil, W.K.; Bassem, S.M.; Kumar, V.; Parisi, C.; Inglese, S.; Temraz, T.A.; Nassar, H.F.; Guerriero, G. The Duckweed, *Lemna minor* Modulates Heavy Metal-Induced Oxidative Stress in the Nile Tilapia, *Oreochromis niloticus*. *Water* **2020**, *12*, 2983. [CrossRef]
53. AOAC, *Official Methods of Analysis*, 15th ed.; Association of Official Analytical Chemists: Washington, DC, USA, 1990; Volume 2.
54. Folch, J.; Lees, M.; Sloane Stanley, G.H. A Simple Method for the Isolation and Purification of Total Lipids from Animal Tissues. *J. Biol. Chem.* **1957**, *60*, 497–509. [CrossRef]
55. Christopherson, S.W.; Glass, R.L. Preparation of milk methyl esters by alcoholysis in an essentially nonalcoholic solution. *J. Dairy Sci.* **1969**, *52*, 1289–1290. [CrossRef]
56. IBM Corporation. Released. *IBM SPSS Statistics for Windows*; Version 25.0; IBM Corporation: Armonk, NY, USA, 2017.

plants

MDPI

Review

Growth and Nutritional Quality of Lemnaceae Viewed Comparatively in an Ecological and Evolutionary Context

Barbara Demmig-Adams [1,*], Marina López-Pozo [2], Stephanie K. Polutchko [1], Paul Fourounjian [1,3], Jared J. Stewart [1], Madeleine C. Zenir [1] and William W. Adams III [1]

1 Department of Ecology and Evolutionary Biology, University of Colorado, Boulder, CO 80309, USA; Stephanie.Polutchko@Colorado.EDU (S.K.P.); paul@internationallemnaassociation.org (P.F.); Jared.Stewart@Colorado.EDU (J.J.S.); Madeleine.Zenir@colorado.edu (M.C.Z.); William.Adams@colorado.edu (W.W.A.III)

2 Department of Plant Biology and Ecology, University of the Basque Country (UPV/EHU), 48049 Bilbao, Spain; marinalopezpozo@hotmail.com

3 International Lemna Association, Denville, NJ 07832, USA

* Correspondence: Barbara.Demmig-Adams@Colorado.EDU; Tel.: +1-303-492-5541

Abstract: This review focuses on recently characterized traits of the aquatic floating plant *Lemna* with an emphasis on its capacity to combine rapid growth with the accumulation of high levels of the essential human micronutrient zeaxanthin due to an unusual pigment composition not seen in other fast-growing plants. In addition, *Lemna*'s response to elevated CO_2 was evaluated in the context of the source–sink balance between plant sugar production and consumption. These and other traits of Lemnaceae are compared with those of other floating aquatic plants as well as terrestrial plants adapted to different environments. It was concluded that the unique features of aquatic plants reflect adaptations to the freshwater environment, including rapid growth, high productivity, and exceptionally strong accumulation of high-quality vegetative storage protein and human antioxidant micronutrients. It was further concluded that the insensitivity of growth rate to environmental conditions and plant source–sink imbalance may allow duckweeds to take advantage of elevated atmospheric CO_2 levels via particularly strong stimulation of biomass production and only minor declines in the growth of new tissue. It is proposed that declines in nutritional quality under elevated CO_2 (due to regulatory adjustments in photosynthetic metabolism) may be mitigated by plant–microbe interaction, for which duckweeds have a high propensity.

Keywords: chlorophyll fluorescence; electron transport chain; inflammation; lutein; photosystem; photosynthetic capacity; relative growth rate

Citation: Demmig-Adams, B.; López-Pozo, M.; Polutchko, S.K.; Fourounjian, P.; Stewart, J.J.; Zenir, M.C.; Adams, W.W., III. Growth and Nutritional Quality of Lemnaceae Viewed Comparatively in an Ecological and Evolutionary Context. *Plants* **2022**, *11*, 145. https://doi.org/10.3390/plants11020145

Academic Editors: Viktor Oláh, Klaus-Jürgen Appenroth and K. Sowjanya Sree

Received: 14 November 2021
Accepted: 4 January 2022
Published: 6 January 2022

1. Introduction

The smallest known flowering plants are found in the Lemnaceae family and are recognized (see recent comprehensive review by Acosta et al. [1]) for their attractive combination of extremely high growth rates [2,3] with high nutritional quality, including a high protein content, with all essential amino acids for humans, as well as a high content of essential human micronutrients [4]. The Lemnaceae, commonly known as duckweeds, water lentils, or water lenses, are comprised of five genera, including *Lemna* and *Wolffia* [1]. The high protein content of Lemnaceae is associated with their propensity for the efficient uptake and accumulation of nitrogen and other mineral nutrients, which makes them good at wastewater recycling [5–9] and contributing to a high nitrogen-use efficiency in agricultural contexts [10].

Additional traits of interest include duckweed's ability to accumulate high levels of starch as well as their relatively low susceptibility to the undesirable effects of elevated atmospheric carbon dioxide levels (for details, see sections dedicated to these topics below). Moreover, duckweed's rapid growth and diminutive size allow for a high volumetric yield in tight quarters such as greenhouses, urban rooftop growth facilities, and spacecrafts, where duckweed's insensitivity to microgravity is another boon [1,11]. Our review places the above traits, as well as additional traits recently described by our group, into the context of the ecology and evolution of Lemnaceae in comparison with terrestrial plants as well as other floating aquatic plants.

The traits of *Lemna* recently characterized by us, which are the focus of this review, include:

- The pronounced tolerance of a wide range of growth light intensities and a remarkable capacity to accumulate high levels of antioxidant micronutrients, such as the essential carotenoid zeaxanthin, due to an unusual pigment composition not seen in fast-growing land plants.
- The response to elevated CO_2 as evaluated in the context of plant metabolic regulation of the source–sink balance (balance between the plant's sugar production and consumption), carbon-to-nitrogen ratio, and redox homeostasis (balance of oxidants and antioxidants).

2. Exceptions to Common Trade-Offs: Araceae and Lemnaceae

2.1. *Trade-Off between the Ability to Grow in Deep Shade and Full Sun*

Many land plants exhibit a trade-off between the ability to tolerate deeply shaded growth environments on the one hand and extremely high light levels on the other. Specifically, fast-growing species are typically sun-loving and unable to grow in deep shade. Exceptions are found in the family Araceae, which belongs to the same order (Alismatales, water plantains) as the Lemnaceae. The Alismatales include many floating or submersed aquatic and wetland species found in marshy and marine habitats [12]. Araceae is the most species-rich family of the Alismatales and is remarkable for the highly diverse habitats in which its species thrive, ranging from open freshwater to deserts, and its diverse life forms, including hemi-epiphytes, epiphytes, terrestrial species, and aquatic plants [13]. It has been noted that some Araceae exhibit a high level of flexibility in the organization of the photosynthetic membrane [14] (see also below).

Plants can suffer damage from intense light unless they process absorbed light either in photosynthesis (which converts excitation energy to chemical energy) or via safe alternative processes (that, e.g., convert excitation energy to harmless thermal energy; see [15,16]). Many species in the Araceae can grow equally well in deep shade and in full sun (see [16,17]). For example, the hemi-epiphyte *Monstera deliciosa* (Araceae) germinates in deep shade on the rainforest floor, climbs the nearest tree, and eventually completes its life cycle in the sun-flooded forest canopy after shedding its connection to the soil. In full sun, *M. deliciosa* exhibits low rates of photochemical energy utilization but record rates of the alternative, non-photochemical dissipation of absorbed light as thermal energy [17]. This latter process is catalyzed by antioxidants with critical roles in fighting radiation damage in both plants and humans (for details, see sections below).

2.2. *Common Trade-Off between Fast Growth and Antioxidant Accumulation*

Terrestrial plants typically show a trade-off between growth rate and the accumulation of radiation-fighting antioxidants, especially the carotenoid zeaxanthin [16]. Fast-growing terrestrial species tend to accumulate less zeaxanthin, whereas slow-growing species tend to accumulate more zeaxanthin (Figure 1; [16,18]). This difference in response is the expected result of the above-described link to the fraction of absorbed light processed in photosynthesis. Fast-growing species use a greater fraction of the light they absorb at peak irradiance to support growth and accumulate less of the antioxidant zeaxanthin, which harmlessly removes light not usable for growth, i.e., excess light [16,18]. Conversely, slow-growing species (e.g., *M. deliciosa*) use a lesser fraction of absorbed light for growth and accumulate more zeaxanthin (Figure 1; [16,18]). The Araceae are thus no exception to the general rule of a trade-off between fast growth and accumulation of high levels of zeaxanthin [17,19].

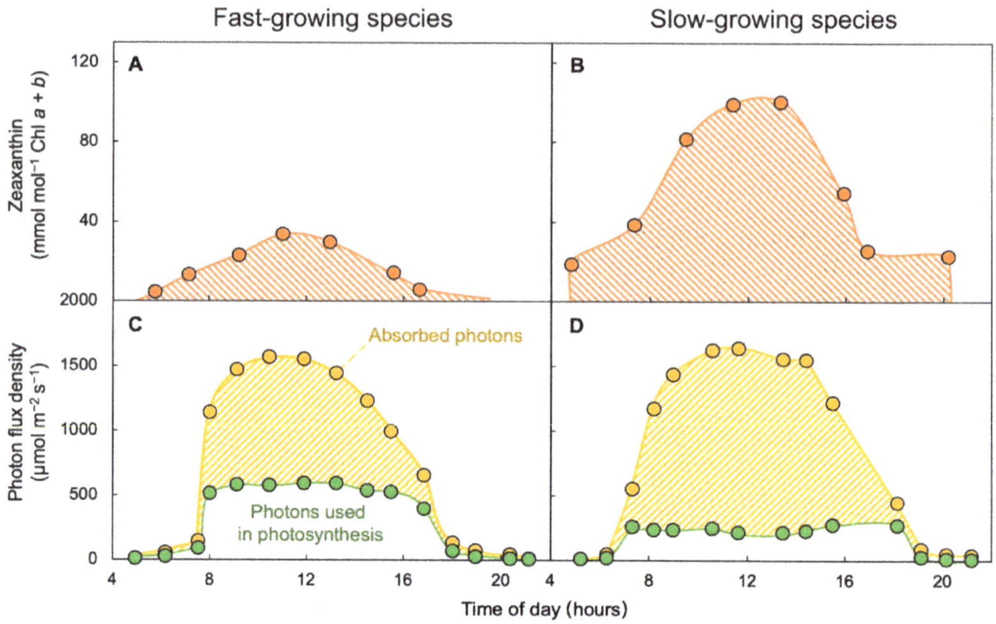

Figure 1. Diurnal time course of changes in (**A**,**B**) zeaxanthin level and (**C**,**D**) photons absorbed and utilized through photosynthetic electron transport in (**A**,**C**) the fast-growing annual crop *Helianthus annuus* (sunflower) versus (**B**,**D**) the slow-growing perennial *Vinca major*. Data from [18]; re-drawn from [16].

2.3. *Zeaxanthin—Essential Human Micronutrient and Hard to Come by in the Diet*

The essential micronutrient zeaxanthin is not easy to come by in the human diet for the very reason that leaves produce zeaxanthin to fight radiation damage only when necessary. The green parts of plants (such as leafy greens) only accumulate zeaxanthin when they need to actively dispose of excessive, potentially damaging light and quickly remove the dissipator zeaxanthin as soon as light levels drop. This dynamic process gives plants the advantage of allowing photosynthesis to return to highly efficient utilization of absorbed light (Figure 1; reviewed in [16]). Food sources of zeaxanthin, other than leafy greens, include egg yolks and yellow corn, which owe their color to the presence of zeaxanthin and the closely related xanthophyll lutein (see, e.g., [20]). However, neither eggs nor corn are particularly suitable for production in limited spaces. In particular, operation of a chicken farm or corn field is not feasible on a spacecraft and may also not be an attractive approach in an urban setting or a greenhouse.

The essential dietary carotenoid zeaxanthin is not only needed to fight radiation damage and system-wide inflammation in humans but also to support mental acuity in healthy young adults (Figure 2; [21,22]). Zeaxanthin enhances basic membrane function in the brain, which supports mental acuity and detoxifies free radicals and other oxidants, which lessens radiation damage and the system-wide inflammation linked to less-than-optimal function, disorders, and diseases. Additional essential micronutrients act synergistically with zeaxanthin in fighting radiation damage and inflammation; zeaxanthin thus works best in a total package—provided by whole foods—with additional plant antioxidants that recycle zeaxanthin for a longer lifetime in inflammation fighting (Figure 2; [22]). Figure 2 places antioxidants with overlapping essential functions in plants and humans into the cycle of oxygen and carbon dioxide exchange, organic nutrients (e.g., protein and antioxidants), and inorganic nutrients (human waste) between photosynthetic and non-photosynthetic organisms. Thereby, Figure 2 highlights that duckweed is an excellent plant component for a regenerative life support system on, e.g., a spacecraft (see also [23,24]).

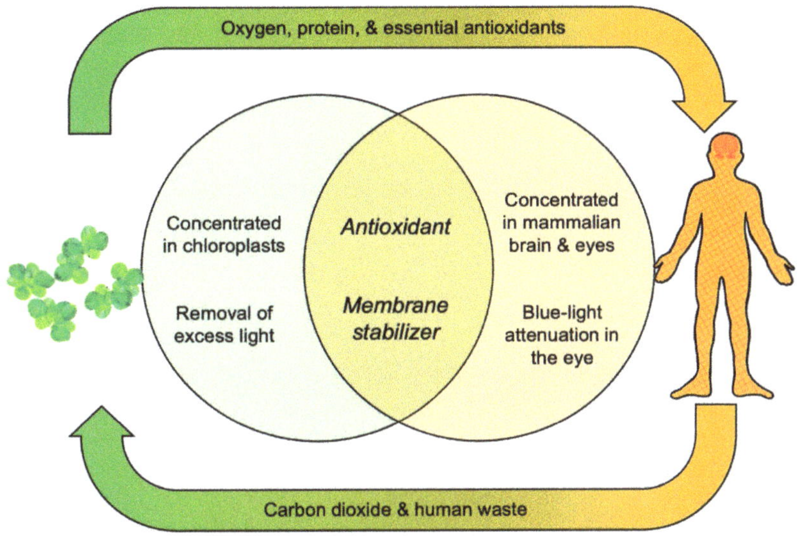

Figure 2. Schematic depiction of the functions of zeaxanthin in plants and humans with both overlapping and analogous aspects.

As reviewed in Demmig-Adams et al. [21], airline pilots (exposed to ionizing radiation) who reported eating more zeaxanthin exhibited less inflammation; similarly, healthy young subjects given zeaxanthin for six months showed less inflammation and enhanced cognitive performance in complex tasks, greater speed at completing those tasks, and better memory and attention. Sufficient dietary zeaxanthin would thus appear to be especially critical for long-duration human spaceflight. As further illustrated in the following section, duckweed provides a renewable food source that can be grown in limited space and can be exceptionally rich in zeaxanthin.

2.4. *Lemna: An Unusual, Fast-Growing Hyperaccumulator of Zeaxanthin*

Among land plants, it is the relatively slow-growing evergreens that typically exhibit the highest zeaxanthin concentrations [16,25]. An extreme example is evergreen conifers that arrest growth completely when they overwinter at high altitude where soil water freezes in the winter. While green needles can absorb a lot of light, they dissipate 100% of it as thermal energy via zeaxanthin as long as water lost from the needles during CO_2 uptake for photosynthesis cannot be replaced [26,27]. As stated above, fast-growing terrestrial annuals instead typically utilize a significant amount of the light they absorb in photosynthesis and do not dissipate as much. When subjected to environmental stresses that can curb their growth rate, such as limiting nutrient levels in the soil [28,29], annuals typically strongly downregulate chlorophyll levels (and thus light-harvesting capacity) and only exhibit moderate increases in zeaxanthin levels employed in energy dissipation. When grown under extreme light conditions (with respect to high intensity and long photoperiod) that resulted in a comparably low chlorophyll content per area, *L. gibba* accumulated significantly more zeaxanthin [23,24] than spinach grown under limiting nutrient supply [29,30].

Further characterization of *Lemna*'s exceptionally strong zeaxanthin accumulation led to the identification of a carotenoid composition that is unusual for fast-growing plants (Figure 3). A comparison of *Lemna* with several terrestrial species via principal component analysis showed that *Lemna* fell into the olive-green ellipse of slow-growing plants with high levels of zeaxanthin (and lutein) rather than the blue-green ellipse of other fast-growing plants (Figure 3; [24,25]). In other words, *Lemna* combines an exceptionally high growth rate with an unusual carotenoid composition and accumulation of as much zeaxanthin as seen in slow-growing land plants (see below).

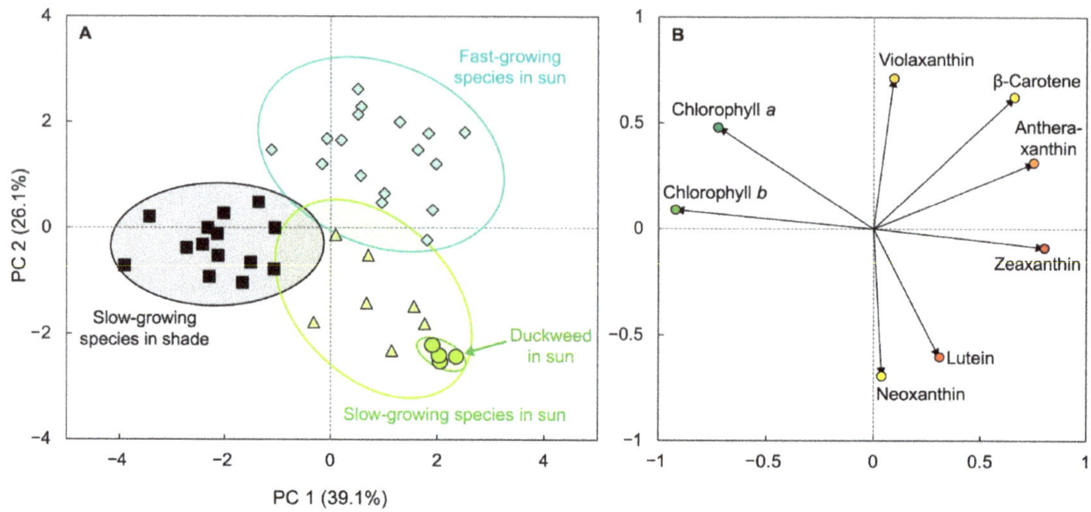

Figure 3. Principal component analysis showing (**A**) clusters of groups based on similarity (score plot) and (**B**) how each characteristic used influences a principal component, PC (loading plot). Characteristics used were the various leaf pigments of slow-growing terrestrial species either in shaded environments (black squares) or full sunlight (olive-green triangles), fast-growing (annual and biennial) species (turquoise diamonds) in full sunlight, and duckweed (*Lemna minor*) that was growing in full sunlight (olive-green circles) near Boulder, CO, USA (for details, see [24,25]). Data from [24,25].

Once again, the traits of duckweed are thus a mix of those seen in slow-growing terrestrial evergreens (with greater maximal zeaxanthin levels) and fast-growing terrestrial annual plants (with lower maximal zeaxanthin levels). Future research is needed to confirm that the high growth rate and the unusually pronounced accumulation of zeaxanthin in *Lemna* (Figure 4) is associated with thin leaves that likely cause a relatively greater portion of the frond cross-section to be exposed to high light when plants are grown under a high growth light intensity. Terrestrial plants often feature multi-layered canopies, of which only the top leaves receive unfiltered light. Additionally, terrestrial plants typically also feature multi-layered leaves, where only the top cell layer receives unfiltered light. When grown in full sun, leaves of terrestrial plants thus exhibit pronounced high-light acclimation [31], with the highest levels of xanthophylls, and especially zeaxanthin, in their top-most layers [32]. In contrast, duckweed and other floating plants consist of relatively thinner leaves/fronds that presumably experience less attenuation of light from top to bottom. This scenario is supported by the side-by-side comparison of a terrestrial weed with duckweed [24] growing in a high-light exposed location, where the terrestrial weed exhibited a much higher chlorophyll level per unit leaf area typical for that seen in leaves with multiple chloroplast-rich layers of mesophyll cells.

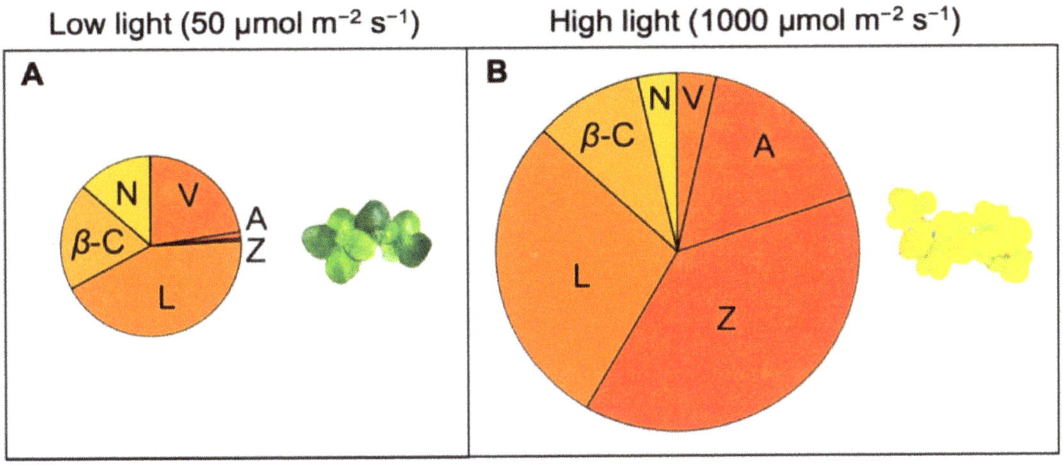

Figure 4. Carotenoid composition of *Lemna gibba* grown under continuous (**A**) low or (**B**) high light intensity. The size of the respective pies represents the molar ratio of the sum of all carotenoids to chlorophyll. Data from [24]; adapted from [16]. A, antheraxanthin, β-C, β-carotene, L, lutein, N, neoxanthin, V, violaxanthin, Z, zeaxanthin.

One can think of duckweed fronds as structures that allow a thin layer of chloroplasts to be uniformly exposed to unfiltered light, which may require unusually high levels of excess-light dissipation and zeaxanthin accumulation under exposure to high light intensities. Hypothetically, duckweed could simply lower its chlorophyll content (and antenna size) enough to avoid absorbing much excess light. However, the maintenance of sufficient chlorophyll levels, accompanied by strong protection from zeaxanthin as needed, may be advantageous to duckweeds that can rapidly shift between being at the top versus in a deeper layer when growing in dense mats.

In addition to being bound to specific designated sites in chlorophyll-binding light-harvesting complexes, additional zeaxanthin can be dissolved in the phospholipid bilayer of plants' photosynthetic membranes outside of chlorophyll-binding proteins (as first reported in [33] and recently reviewed in [34]). The report of a ratio greater than 0.5 (mol/mol) for zeaxanthin to chlorophyll $a + b$ [24] suggests that a substantial fraction of the zeaxanthin accumulated in golden-hued duckweed fronds (Figure 4) growing under continuous high light levels may be dissolved in these membranes rather than being bound to protein complexes. At the same time, this high zeaxanthin content is not associated with sustained lowering of photosystem II efficiency [23,24] as is seen in many plants (see [35] for *Arabidopsis thaliana* and [16] for evergreens) under environmental stresses. Such a scenario—high zeaxanthin levels uncoupled from sustained lowering of photochemical efficiency—is especially desirable for optimal human nutrition without plant productivity losses. Other aquatic floating flowering plant species (beyond Lemnaceae) have also been reported to exhibit very high levels of xanthophylls [36], although xanthophylls were not separated into zeaxanthin versus other di-hydroxy xanthophylls (like lutein) during these analyses. This possibility that other aquatic plants share the above-described features of duckweed is consistent with a report by Rice [37] that noted aquatic taxa possessed unique photosynthetic features, including pigment composition. Future research should assess these features of aquatic species, which could be of interest for human nutrition, e.g., with respect to zeaxanthin levels. It is noteworthy that many aquatic floating plants are described as edible and/or medicinal plants (see, e.g., [38]).

2.5. *Remote Sensing of Duckweed Zeaxanthin Content and Biomass Production*

Chloroplast-containing plant organs emit chlorophyll fluorescence, a signal that provides information about the fraction of absorbed light utilized in photosynthesis versus the fraction dissipated alternatively (non-photochemically) in the form of thermal energy. The fraction of absorbed light utilized in photosynthetic electron transport has been used, in conjunction with other parameters, to predict plant productivity (see, e.g., [39]). In terrestrial plants, this relationship is complex ([40,41]; see also the discussion in [24]). However, a simple, linear positive correlation exists in *Lemna* between this utilization of absorbed light as ascertained from chlorophyll fluorescence (as photosystem II activity per photons) and *Lemna* biomass production per photons (Figure 5A; [24]). In addition, the activity of non-photochemical, photoprotective energy dissipation assessed from chlorophyll fluorescence is positively and linearly correlated with xanthophyll cycle conversion to zeaxanthin (Figure 5B; [23,24]). Chlorophyll fluorescence measurements thus allow non-destructive, remote estimates of zeaxanthin level (see also [16]).

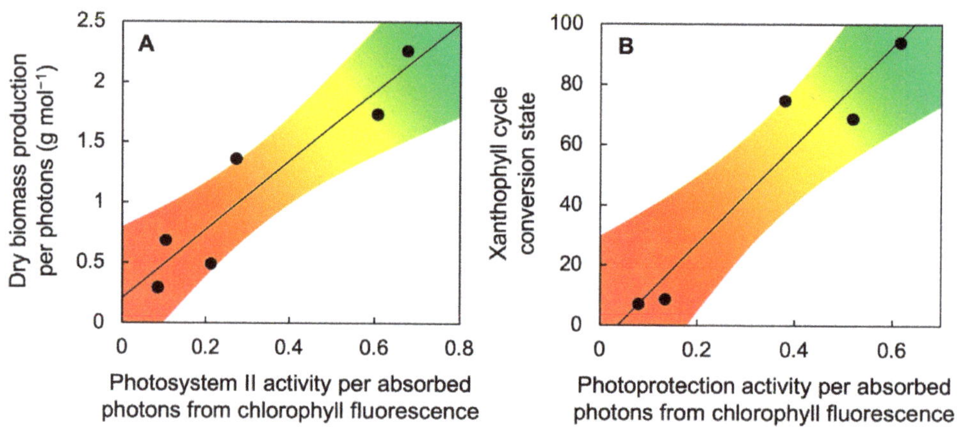

Figure 5. Relationship between (**A**) dry biomass production per photons and photosystem II activity per absorbed photons from chlorophyll fluorescence (as photosystem II efficiency from $F_v'/F_m' \times q_P$; see [18,24]) or (**B**) xanthophyll cycle conversion state, (Z + A)/(V + A + Z), and the activity of non-photochemical dissipation of excess absorbed light from chlorophyll fluorescence (as photoprotective thermal dissipation from $0.8 - F_v'/F_m'$; see [18,24]) under six different growth-light conditions over a range of light intensities from 50 to 1000 μmol photons m^{-2} s^{-1}. Data from [23,24]. A, antheraxanthin, V, violaxanthin, Z, zeaxanthin.

3. Comparative Evaluation of Adaptations in Aquatic and Terrestrial Plants

3.1. *Rapid Growth in Duckweeds and Other Aquatic Plants*

Lemna grows very fast [3] even under conditions of limited light supply [23,24]. We suggest that this unique feature may be associated with minimal self-shading within its relatively thin photosynthetic organs (fronds) and their single layer of fronds (no tiered structure) across the water surface, which should permit all chloroplasts to contribute significantly to sugar production. Other floating flowering plants also grow remarkably fast [42,43], which may be related to a high allocation of resources to photosynthetic tissue as discussed by Rice [37]. Terrestrial plants growing in soil must invest a sizeable fraction of their photosynthetically produced sugar into building (i) root structures to provide a stable anchor in the soil, for water (and nutrient) acquisition, and for the storage of carbohydrates and (ii) reinforced stem structures for support of the shoot system and transport of water, nutrients, and sugar between their photosynthetic organs and the roots. In contrast, floating plants have minimal non-photosynthetic tissue and can thus re-invest most of their photosynthate into more photosynthetic tissue. The main carbohydrate sink for duckweed is vegetative growth, its expanding daughter fronds that also contribute to photosynthetic productivity. Duckweed lacks typical stems or roots that would act as substantial sinks for photosynthate, and its roots can be green and contribute to photosynthesis. It has also been pointed out that Lemnaceae have undergone substantial genome reduction—including gene clusters that control growth [44] in response to the environment in terrestrial plants. Unabated growth of new tissue under a wide variety of environmental conditions is thus duckweed's main sink activity.

3.2. Continuum of Plant Adaptations to Water Availability

Another perspective from which to view the fast growth of aquatic plants is their rather continuous access to water. Variable access to water is one of the major environmental variables for terrestrial plants. Many, but not all, terrestrial species curb stomatal opening, and growth, under limiting water as a defense strategy [45]. Such a response would appear less necessary for species that float on water, and duckweeds indeed impose much less control on either stomata or growth rate [44]. A review by Dolferus [46] entitled "To grow or not to grow: A stressful decision for plants" concludes that highly responsive modulation of growth rate—with either slower or faster growth—is one of the major ways for terrestrial plants to respond to changes in water availability. Table 1 compares the response of aquatic floating plants with the continuum of responses of different groups of terrestrial plants.

Table 1. Water-acquisition strategies, growth patterns, and species examples that illustrate the continuum of plant adaptations to water availability.

Habitat	Water-Acquisition Strategy	Growth Pattern	Example
Aquatic plants	Continuous access to water body until it dries up	Fast growth	Lemnaceae
Terrestrial plants	Continuous access to water until source dries up	Very fast growth and life cycle completion	Desert ephemerals
	Continuous access to water, increased root volume, osmotic adjustment	Relatively fast growth throughout life cycle	Annuals and biennials
	Continuous access to water table	Steady growth throughout the seasons	Palm and mesquite
	Enhanced acquisition of minimal soil water via large root volume, osmotic adjustments	Slow growth despite minimal soil water	Desert shrub *Encelia*, creosote bush
	Storage of water in plants	Very slow growth	Succulents, cacti
	Tolerance of seasonal loss of water	Seasonal complete growth arrest	Conifers in frozen soil

The high growth rate seen in many aquatic plants [42,43] is reminiscent of the high growth rates of terrestrial plants that maintain constant access to water throughout their life cycle (Table 1). Desert ephemerals are an example of terrestrial plants with an extremely high growth rate that rapidly complete their accelerated life cycle after infrequent major rainfall events in arid environments [47]. Other examples for terrestrial plants with high growth rates are annuals and biennials that also complete their life cycle in a relatively short period of time and grow where they can maintain access to water. These species may employ evasive approaches to keep internal water content high, e.g., by performing osmotic adjustment to maintain metabolic activity [48] and increasing their investment in water-mining root structures [49] (Table 1). These terrestrial plants do not put the brakes on growth until water becomes scarce, and then switch to seed production—reminiscent of the production of vegetative storage forms (turions) in Lemnaceae. Yet, other terrestrial plants, such as palm trees or mesquite, have deep roots that access the ground-water table [50–54], which supports steady growth despite low rainfall or moisture level in the air. Pronounced osmotic adjustment [55] and expansion of root volume [56] is also employed by terrestrial plants that grow very slowly in areas with extremely low water availability (Table 1).

Those terrestrial plants that tolerate—rather than evade—internal water deficits exhibit trade-offs between growth rate and stress tolerance/antioxidant content. In extreme natural environments, growth is suspended entirely, such as in cold environments with seasonally frozen soils (Table 1), where the green needles of overwintering confers exhibit very high levels of zeaxanthin [26,27]. Many perennial species adapted to seasonally harsh environments grow incrementally over multiple years—by growing actively during favorable seasons and arresting growth during unfavorable seasons. In an agricultural context, unabated growth is desirable, and floating plants with their fast, continuous growth thus offer attractive opportunities that rival those of terrestrial annuals and biennials.

3.3. *Plant Source–Sink Balance and the Response to Light, CO_2, and Nutrient Supply*

While both CO_2 and light are necessary inputs for photosynthesis, the proverbial "too much of a good thing" with respect to CO_2 and/or light can have negative impacts on plant nutritional quality, growth, and plant lifespan. A report by Myers et al. [57] entitled "Increasing CO_2 threatens human nutrition" addressed possible adverse effects on plant protein as well as other nutrients. Whereas these impacts on plant nutritional quality for the consumer appear to be rather universal for C_3 species, other possible impacts of elevated CO_2 vary among species and with environmental conditions and include possible disruption of photosynthetic productivity as well as accelerated senescence (for a recent review, see [58]). In the following, we examine the possibility that duckweed may be less sensitive to the possible adverse effects of elevated CO_2 on plant productivity than many other species. We first discuss the effect of elevated CO_2 on protein (as well some micronutrients) in the context of plant source–sink balance. Photosynthetic organs are the plant's only source of newly formed carbohydrates, whereas all sugar-consuming tissues are sugar sinks. The maximal capacity of photosynthesis is regulated by the demand for sugar from the plant's sink tissues [30,59,60]. This regulation by demand is well described for terrestrial plants; excess carbohydrate build up in the leaves leads to downregulation of photosynthetic capacity, which can be the case under conditions of elevated atmospheric CO_2 and especially in combination with additional environmental factors that cause source–sink imbalances (see, e.g., [61]). *Lemna* exhibits some of the responses described for land plants with respect to the content of protein and micronutrients under very high light and/or CO_2 supply (see below).

Under moderate supply of light and earth-ambient CO_2, light drives photosynthetic electron transport and allows CO_2 to be fixed into sugar as the fuel for growth. A combination of high light supply and elevated CO_2 produces more sugar than is consumed by sink tissues, which leads to feedback inhibition (Figure 6) with build-up of carbohydrate in leaves, backed-up electrons in the photosynthetic electron transport chain, and formation of increased levels of reactive oxygen species (ROS) that are potent repressors of photosynthetic and other genes [62–65]. An extreme level of foliar carbohydrate build-up and ROS production under elevated CO_2 can curb the growth of new tissue and accelerate plant senescence [66].

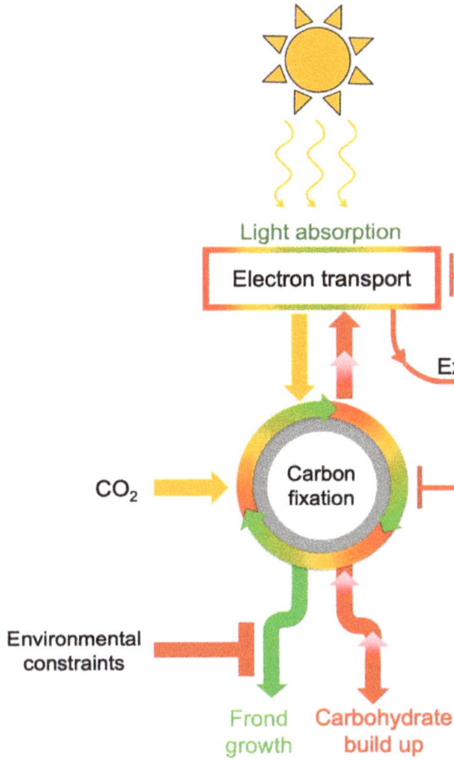

Figure 6. Schematic depiction of the effect of environmental inputs (light, CO_2, and factors that limit leaf/frond growth) in triggering feedback inhibition of photosynthesis via repression of photosynthetic genes when carbohydrate builds up in leaves/fronds because of limiting sink activity.

3.4. Comparative Evaluation of Plant Response to Light Supply

Unlike most terrestrial plants, *Lemna* exhibited the same growth rate in low and high light (Figure 7; [24]), which can be viewed as an absence of growth reductions in low light or as an absence of growth reductions in highly excessive light. In the example of Figure 7, *Lemna* exhibited the same relative growth rate, measured as daily area expansion of new tissue, when grown under low light (50 µmol photons m^{-2} s^{-1}) as when grown with a 20× greater light supply (1000 µmol photons m^{-2} s^{-1}) for 24 h a day (Figure 7), which exceeds the total daily light supply on the longest, brightest day on earth. However, plants did accumulate more dry biomass (presumably carbohydrate) per frond area under the high light level. Likewise, *Lemna aequinoctialis* fronds growing under continuous light (24 h photoperiod) accumulated more biomass than fronds growing under a 12 h photoperiod of an intermediate light intensity of 200 µmol photons m^{-2} s^{-1} [67].

At the same time, *Lemna* did show some downregulation of maximal photosynthetic capacity (as a correlate of the number of photosynthetic enzymes) as well as pronounced downregulation of light-harvesting capacity (chlorophyll content) under the latter very high light supply (Figure 7; [24]). This downregulation can be viewed as economy on part of a C_3 plant, whereby greater light exposure and more CO_2 are able to support the same growth rate (area expansion) and more biomass accumulation with a lesser investment in proteins for light collection and CO_2 fixation while releasing nitrogen from these photosynthetic proteins for use in area expansion [68]. A consequence of this downregulation is a lower nutritional quality in the form of a lower fraction of the accumulated biomass that consists of protein. The percentage of protein of dry biomass dropped from 46% in low light to under 24% in high light. It has been suggested [69] that the CO_2-fixing protein ribulose-bisphosphate carboxylase oxygenase may serve in a dual role as not only the CO_2-fixing enzyme in photosynthesis but also a vegetative storage protein in duckweed fronds. This dual role offers an explanation as to why duckweed vegetative protein would show a response to the demand for carbohydrate from sink tissues.

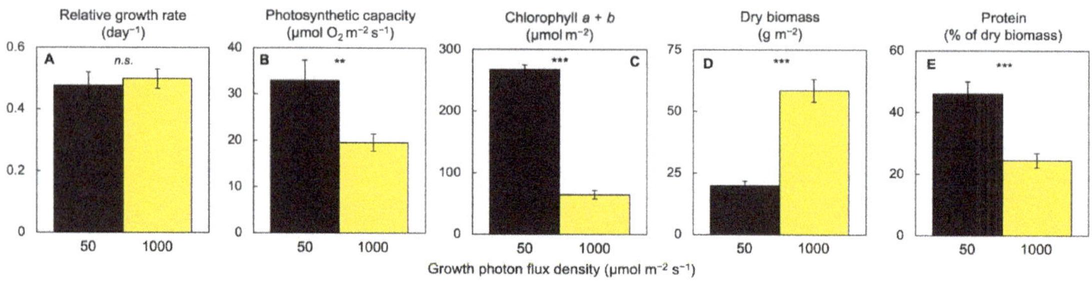

Figure 7. Effect of growth light intensity on (**A**) relative growth rate, (**B**) photosynthetic capacity, (**C**) chlorophyll content, and (**D**) biomass per frond area as well as (**E**) percentage of protein of this biomass for *Lemna gibba*. Data from [23]. *n.s.*, not significant. Asterisks indicate significant differences at $p < 0.01$ (**) or $p < 0.001$ (***).

Figures 8 and 9 illustrate responses of *Lemna* duckweed to a combination of high light and elevated CO_2 with respect to a suite of similar features related to plant productivity, photosynthesis, and nutritional quality for the consumer. Under an extremely high CO_2 level (0.7%) in the presence of high light (700 μmol photons m^{-2} s^{-1} for 24 h a day) and replete (sufficient) nutrient supply, duckweed grew new tissue at a relative growth rate of frond area expansion with a dry biomass content per area that were both similar under high versus ambient CO_2 (Figure 8A,B). This finding is consistent with the high growth rate, i.e., high sink activity in the form of new tissue growth, of duckweed, and supports the notion that duckweed is remarkably insensitive to the adverse effects of CO_2 on photosynthetic productivity. In contrast, there was—as expected—evidence for photosynthetic downregulation in the form of a somewhat lower photosynthetic capacity, much lower chlorophyll content, and much lower zeaxanthin content on a frond area basis (Figure 9) under high versus ambient CO_2. These findings are consistent with the principle that the plant can support growth and biomass accumulation with less CO_2-fixing protein in the presence of high CO_2 levels. Photosynthetic downregulation targets not only the CO_2-fixing enzymes but also several proteins involved in light collection and processing, including chlorophyll-binding proteins [60]. In turn, lower chlorophyll levels reduce light absorption and the need for the dissipation of excess excitation energy by zeaxanthin and other carotenoids. This downregulation of photosynthesis was thus associated with lower nutritional quality of the plant (Figures 8 and 9) in the form of a lower content of protein, zeaxanthin, lutein, and β-carotene (provitamin A). Lutein [70] and β-carotene [71] have additional roles in plant photoprotection and also act synergistically with zeaxanthin as membrane-soluble antioxidants that fight inflammation in humans [20–22].

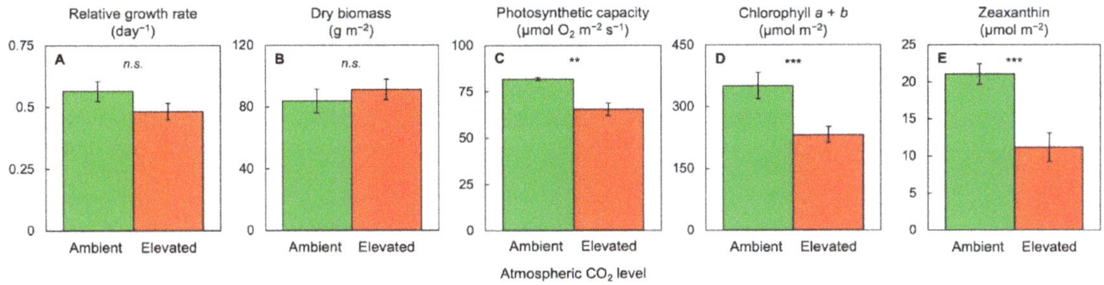

Figure 8. Effect of elevated compared to ambient atmospheric CO_2 level on (**A**) the relative growth rate (of frond area expansion) as well as (**B**) dry biomass, (**C**) photosynthetic capacity, (**D**) chlorophyll content, and (**E**) zeaxanthin content per frond area in *Lemna minor*. Elevated CO_2 level was 0.7%; light environment was continuous light of 700 μmol photons m^{-2} s^{-1}; nutrient medium was $\frac{1}{2}$ strength Schenk and Hildebrandt medium [72]. For other experimental conditions and methods, see [23]. *n.s.*, not significant. Asterisks indicate significant differences at $p < 0.01$ (**) or $p < 0.001$ (***).

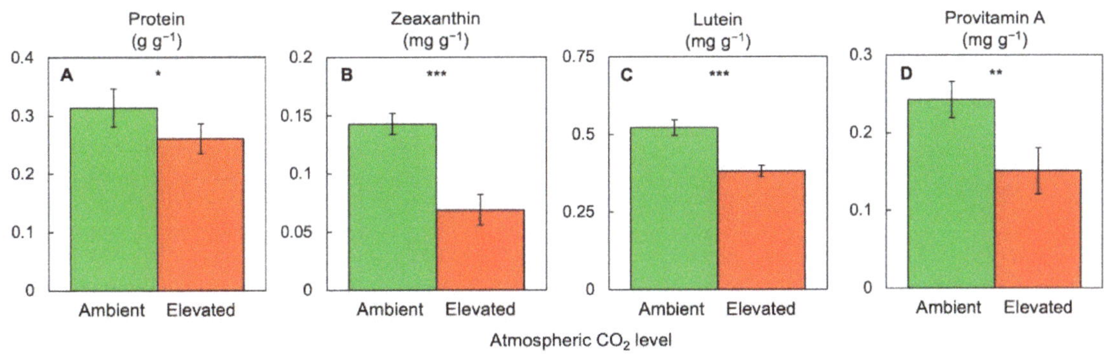

Figure 9. Effect of elevated compared to ambient atmospheric CO_2 level on various human nutrients including (**A**) protein, (**B**) zeaxanthin, (**C**) lutein, and (**D**) provitamin A (β-carotene) content per dry biomass in *Lemna minor*. CO_2, light, and nutrient conditions were as listed in the legend of Figure 8. For protein quantification, see [24]. For all other experimental conditions and carotenoid quantification, see [23]. Asterisks indicate significant differences at $p < 0.05$ (*), $p < 0.01$ (**), or $p < 0.001$ (***).

While *Lemna* is thus susceptible to effects of high CO_2 that are undesirable for the consumer, it exhibited no decline in the growth of new tissue under conditions of favorable nutrient supply. This relatively modest response to high light and CO_2 supply may, once again, be associated with the low responsiveness of duckweeds' growth rate to environmental conditions. Furthermore, we saw no difference in plant response to CO_2 over a wide range of CO_2 levels from 0.086% (2 × earth-ambient; not shown) up to 0.7%.

3.5. Comparative Evaluation of Plant Response to a Combination of High CO_2, High Light Supply, and Limiting Mineral Nutrient Supply

Source–sink imbalance is further exacerbated by a combination of elevated CO_2 and/or high light supply (which increase source strength) with additional environmental factors that decrease sink strength. An example of such an additional factor is a limited mineral nutrient supply [29,73]. The growth of sink tissues is highly sensitive to a shortage of mineral nutrients, especially nitrogen [74]. In terrestrial plants, the combination of elevated CO_2 and/or high light supply with low nitrogen supply exacerbates source–sink imbalance and downregulates the synthesis of photosynthetic protein [75,76], as well as other metabolic processes [61]. Growth under elevated CO_2 can also lead to premature leaf senescence when sugars accumulate to levels that inhibit photosynthetic gene expression [77] (see also [75,78–80]). Both growth declines and accelerated senescence can be viewed in the context of sugar signals controlling plant progression through the life cycle [81]. Carbohydrate accumulation can be a signal that sufficient resources are available for flowering and completion of the life cycle, and such signals are generated earlier in a plant's life cycle when limited mineral nutrients enhance carbohydrate accumulation [81]. In fact, it was proposed that some level of nitrogen fertilization of nutrient-limited natural plant communities may mitigate the negative effects of elevated CO_2 [82].

Duckweeds likewise respond with enhanced carbohydrate accumulation to a combination of high light supply and/or elevated CO_2 coupled with low nutrient levels in the medium. Pronounced starch accumulation in fronds growing in a nutrient-deficient medium [83–85] was further exacerbated by extended photoperiods [86]. However, several traits of duckweeds may lessen the overall impact of such combinations of environmental factors. Duckweeds exhibit a higher tolerance than terrestrial plants to carbohydrate build-up and can accumulate [5] considerable starch levels before showing photosynthetic downregulation [23,83,86,87]. In terrestrial plants, a limited capacity for nitrogen uptake and utilization caused enhanced sink limitation [66,88], whereas a high propensity for efficient nitrogen uptake alleviated sink limitation. Compared to terrestrial plants, *Lemna* exhibits highly efficient uptake [89,90] and conversion of nitrogen to amino acids [91] and accumulates much larger quantities of protein in its fronds [4,83].

Effective nutrient uptake and the propensity to accumulate high protein levels allows duckweed to reclaim polluted waters [6,8] and convert wastewater to high-quality animal feed [5,7,9,92]. The quantity of edible protein produced by duckweed per hectare of production area greatly exceeds that of soybean [5,93] since duckweeds accumulate high levels of vegetative storage protein in their fronds [4,93], whereas soybean accumulates high levels of protein only in its seeds, which represent a small fraction of the plant. Highly efficient nutrient uptake and accumulation of vegetative storage protein, resulting in exceptionally high plant protein contents, is also seen in other aquatic floating plants [94,95]. Many of these other aquatic plants are also edible; more than 70 wetland plants of India were identified as edible or medicinal plants by Swapna et al. ([38], e.g., ten species in the Asteraceae, six each in the Poaceae and Commelinaceae, and six in the order Alismatales, with four Araceae and two Hydrocharitaceae; see also [96]). From an ecological perspective, a combination of effective uptake and the storage of nitrogen as vegetative storage protein in aquatic environments may be advantageous in small- to mid-sized freshwater bodies that receive an intermittent influx of nutrients. Duckweeds were shown to continue growing in low-nutrient media for a certain amount of time by utilizing internal nutrient stocks [83]. Efficient nutrient uptake and storage, as vegetative protein, should enable the plants to take advantage of the pulses of nutrient availability in support of the continuous growth of new fronds until dense duckweed mats dramatically reduce the light available to algae and submerged aquatic plants that compete for nutrients. The properties shared by aquatic floating plants that set them apart from most terrestrial plants thus reflect adaptations to the unique aquatic environment with its inherent variability and opportunities.

3.6. Comparative Evaluation of Plant Response to Combinations of High CO_2, High Light Supply, and High Mineral Nutrient Supply

Remarkably, high nitrate supply can also decrease photosynthetic rates of terrestrial plants, especially when combined with conditions that result in carbohydrate build-up, such as extended photoperiods [97] or elevated CO_2 [98,99]. Under these conditions, excessive ROS production in the chloroplast is combined with the production of additional oxidants in other cell compartments that participate in nitrate metabolism. Specifically, a high activity of the nitrate-reducing enzyme nitrate reductase produces high levels of messengers (both ROS and reactive nitrogen species, RNS; [99]) that modulate nitrogen metabolism [100–102] and can also trigger plant senescence ([66,103]; see also [104,105]).

Such findings led to a suggestion that "in a future environment with high CO_2 levels the use of fertilizers containing low concentrations of nitrate could improve ... [nitrogen] assimilation" in terrestrial crops [106]. A similar warning and recommendations were issued by Bloom [107] in a communication entitled "As carbon dioxide rises, food quality will decline without careful nitrogen management." Ammonium metabolism does not have the same propensity as nitrate metabolism for the generation of ROS and RNS [108] and the resulting disruption of redox homeostasis. On the other hand, the accumulation of ammonium in plant tissue can disrupt cellular pH balance (and some other aspects of metabolism) in many terrestrial plants [109]. However, some plants, including those growing in marshes, are adept at using ammonium [107]. The preference of duckweeds (and other aquatic plants) for the uptake of ammonium over nitrate also has the potential to avoid high rates of nitrate reduction and its adverse effects, and duckweed's efficient conversion of ammonium to vegetative storage protein limits ammonium accumulation and resulting toxicity.

4. Plant–Microbe Interaction and the Abiotic Environment

Whereas adverse effects of high CO_2 in the absence of nutrient deficiency or other abiotic stresses can presumably be mitigated by avoiding high light intensities or long photoperiods, the additional presence of unfavorable nutrient conditions will likely require other measures. Plant–microbe interaction may offer the mitigation of adverse effects of combinations of environmental factors that exacerbate the source–sink imbalance. The extent of plant response to the presence of microorganisms depends on environmental conditions [110] such as CO_2 level (see, e.g., [111–114]) and nitrogen availability (see, e.g., [115,116]). *Lemna minor* exhibited a decline in growth rate, which was strongly exacerbated under elevated CO_2 levels (twice-ambient), starting several days following the transfer from replete nutrient medium to very low nutrient levels in Schenk and Hildebrandt [72] medium diluted by a factor of 1/20 [117].

While experimental manipulation of the plant microbiome is challenging in terrestrial plants growing in soil, some evidence is available for enhanced plant productivity in the presence of fungal partners of terrestrial plants [118,119]. Inoculation of terrestrial plant roots with arbuscular mycorrhizal fungi increased root volume and activity, triggering photosynthetic upregulation in some systems [118,120]. Conversely, the elimination of the mycorrhizal system of cucumber resulted in declining photosynthesis rates ([121]; see also [122]). Transcriptomic analysis of such systems revealed differential gene expression in pathways of photosynthesis, hormone metabolism, carbohydrate metabolism, amino acid metabolism, stress response, signal transduction, and antioxidation [119].

Legumes that feature symbioses with N_2-fixing bacteria had greater photosynthesis rates under elevated CO_2 and maintained higher protein content and higher overall ratios of nitrogen to carbon in their tissues [123]. In contrast, most non-leguminous C_3 crops exhibit increased carbon-to-nitrogen ratios under elevated atmospheric CO_2 [124,125]. Lemnaceae exhibit robust interaction with microorganisms [126–128], and duckweed photosynthesis rate as well as growth rate can be stimulated by the plant microbiome [127–132]. Whereas plant–microbe interaction can thus clearly benefit the plant, the existence of additional layers of complexity should be noted. Microorganisms may engage in competition with plants for mineral nutrients in certain environments; the same bacterial strains that strongly promoted duckweed growth under some conditions reduced plant growth under limiting levels of mineral nutrients other than nitrogen [127,130,133]. Duckweeds are an attractive model system for the study of plant–microbe interaction due to the ease of inoculation and manipulation of the rhizosphere as well as the fact that their small size and high growth rates favor multi-factorial design of growth experiments in different growth environments with plants fully acclimated to these conditions. Other aquatic floating plants also form alliances with microorganisms, especially potential N_2-fixing clades (see, e.g., [134]).

An integrative review of literature on (i) plant performance under high CO_2 and (ii) plant–microbe interaction and symbioses between photosynthetic and non-photosynthetic organisms [58] suggests that microorganisms can enhance plant photosynthetic productivity and nutritional quality (Figures 10 and 11) by (i) serving as an additional sugar sink that prevents carbohydrate build-up [130,132,135–138] and resulting excess ROS formation (see above), (ii) balancing nutrient supply (limiting or excess; see above) and producing growth factors [139], (iii) producing gene regulators that safely re-route electrons and, thereby, further counteract disruption of redox homeostasis under both limiting nutrient supply and very high nitrate supply (see, e.g., [108,140]). Alternative outlets for electrons include cyclic electron flow in the chloroplast [141–146]. Since mitochondria and cell membrane-associated processes also produce excess oxidants under elevated CO_2 levels, the maintenance of cellular redox homeostasis [64] requires coordination of electron flow in chloroplasts and mitochondria. Plant-specific mitochondrial alternative oxidase (AOX) is a key player in this coordination [147–151] and serves as a safe outlet for electrons when environmental shifts threaten to disrupt metabolism [147,152,153]. Plant AOX levels increased in response to high CO_2 and light supply [147,148,154] as well as under limiting nitrogen [155]. Furthermore, the plant rhizosphere microbiome interacts with plant AOX [140].

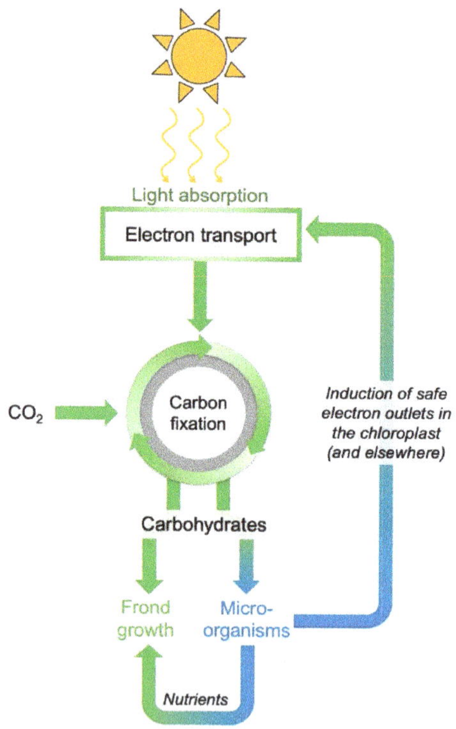

Figure 10. Schematic depiction of the effect of microorganisms in counteracting feedback inhibition of photosynthesis by (i) acting as a sink for carbohydrates, (ii) improving plant nitrogen content in support of new growth (as well as photosynthetic capacity), and (iii) indirectly and directly promoting safe electron outlets in chloroplasts and elsewhere.

Figure 11. Schematic depiction of a *Lemna* colony consisting of a mother frond as the source of photosynthate for expanding daughter fronds as well as for the rhizosphere microbiome as sinks of photosynthate.

Plant–microbe interaction may thereby allow plants to take advantage of high light and CO_2 for growth and biomass accumulation by maintaining a high photosynthetic capacity with high levels of photosynthetic protein, chlorophyll, and chlorophyll-protecting antioxidants. This suggestion is supported by recent findings that the inoculation of previously sanitized *L. minor* with pond water supporting populations of *L. minor* prevented the decline in relative growth rate seen upon transfer to 1/20 strength Schenk and Hildebrandt medium in un-inoculated controls [117]. Moreover, inoculated fronds accumulated greater levels of dry biomass with the same ratio of protein to biomass as uninoculated fronds under either ambient or twice-ambient CO_2 [117]. By supporting a sustained high photosynthetic capacity, microorganisms can thus favor the production of a steady stream of carbohydrate for consumption by both microorganisms and growing plant tissue and discourage metabolic imbalances that trigger photosynthetic downregulation and/or accelerated senescence.

5. Conclusions

The unique features of aquatic plants reflect adaptations to the freshwater environment and include rapid growth, high productivity, and strong accumulation of high-quality vegetative storage protein as well as essential human micronutrients. These micronutrients include carotenoids with wide-ranging roles in human health and a unique carotenoid composition of *Lemna*. Other aquatic plants may share this latter trait and should be further tested for edibility and nutritional quality. The relative insensitivity of duckweed's growth rate to environmental conditions and plant source–sink imbalance may allow duckweeds to take advantage of elevated atmospheric CO_2 levels via particularly strong stimulation of biomass production and relative insensitivity to declines in the growth of new tissue. It may be possible to mitigate the declines in nutritional quality under elevated CO_2 associated with regulatory adjustments in photosynthetic metabolism by plant–microbe interaction.

Author Contributions: Conceptualization, B.D.-A. and W.W.A.III; formal analysis, B.D.-A., W.W.A.III, M.L.-P., S.K.P. and J.J.S.; investigation, M.L.-P., W.W.A.III, J.J.S. and M.C.Z.; resources, B.D.-A. and W.W.A.III; writing—original draft preparation, B.D.-A., W.W.A.III and S.K.P.; writing—review and editing, J.J.S., M.L.-P., P.F. and M.C.Z.; visualization, B.D.-A., W.W.A.III, S.K.P. and J.J.S.; supervision and project administration, B.D.-A. and W.W.A.III; funding acquisition, B.D.-A., W.W.A.III and J.J.S. All authors have read and agreed to the published version of the manuscript.

Funding: This work was funded by the Translational Research Institute for Space Health through Cooperative Agreement NNX16AO69A, the National Science Foundation award number IOS-1907338, and the University of Colorado.

Institutional Review Board Statement: Not applicable.

Informed Consent Statement: Not applicable.

Data Availability Statement: Data are contained within the article.

Acknowledgments: We thank Christine and Adam Escobar for many stimulating discussions about duckweeds and their essential contributions to starting the work on duckweed in our group, including building photodiode arrays for lighting and improving CO_2 control of Conviron growth chambers. We thank Gabrielle Glime for assistance with lab maintenance.

Conflicts of Interest: The authors declare no conflict of interest.

References

1. Acosta, K.; Appenroth, K.J.; Borisjuk, L.; Edelman, M.; Heinig, U.; Jansen, M.A.K.; Oyama, T.; Pasaribu, B.; Schubert, I.; Sorrels, S.; et al. Return of the Lemnaceae: Duckweed as a model plant system in the genomics and postgenomics era. *Plant Cell* **2021**, *33*, 3207–3234. [CrossRef]
2. Sree, K.S.; Sudakaran, S.; Appenroth, K.-J. How fast can angiosperms grow? Species and clonal diversity of growth rates in the genus *Wolffia* (Lemnaceae). *Acta Physiol. Plant.* **2015**, *37*, 204. [CrossRef]
3. Ziegler, P.; Adelmann, K.; Zimmer, S.; Schmidt, C.; Appenroth, K.-J. Relative In Vitro growth rates of duckweeds (Lemnaceae)—The most rapidly growing higher plants. *Plant Biol.* **2015**, *17*, 33–41. [CrossRef]
4. Appenroth, K.-J.; Sree, K.S.; Böhm, V.; Hammann, S.; Vetter, W.; Leiterer, M.; Jahreis, G. Nutritional value of duckweeds (Lemnaceae) as human food. *Food Chem.* **2017**, *217*, 266–273. [CrossRef] [PubMed]
5. Skillicorn, P.; Spira, W.; Journey, W. *Duckweed Aquaculture: A New Aquatic Farming System for Developing Countries*; The World Bank: Washington, DC, USA, 1993.
6. Körner, S.; Vermaat, J.E. The relative importance of *Lemna gibba* L.; bacteria and algae for the nitrogen and phosphorus removal in duckweed-covered domestic wastewater. *Water Res.* **1998**, *32*, 3651–3661. [CrossRef]
7. Xu, J.; Shen, G. Effects of harvest regime and water depth on nutrient recovery from swine wastewater by growing *Spirodela oligorrhiza*. *Water Environ. Res.* **2011**, *83*, 2049–2056. [CrossRef]
8. Ekperusi, A.O.; Sikoki, F.D.; Nwachukwu, E.O. Application of common duckweed (*Lemna minor*) in phytoremediation of chemicals in the environment: State and future perspective. *Chemosphere* **2019**, *223*, 285–309. [CrossRef] [PubMed]
9. Roman, B.; Brennan, R.A. Coupling ecological wastewater treatment with the production of livestock feed and irrigation water provides net benefits to human health and the environment: A life cycle assessment. *J. Environ. Manag.* **2021**, *288*, 112361. [CrossRef] [PubMed]
10. Yao, Y.; Zhang, M.; Tian, Y.; Zhao, M.; Zhang, B.; Zhao, M.; Zeng, K.; Yin, B. Duckweed (*Spirodela polyrhiza*) as green manure for increasing yield and reducing nitrogen loss in rice production. *Field Crops Res.* **2017**, *214*, 273–282. [CrossRef]
11. Yuan, J.; Xu, K. Effects of simulated microgravity on the performance of the duckweeds *Lemna aequinoctialis* and *Wolffia globosa*. *Aquat. Bot.* **2017**, *137*, 65–71. [CrossRef]
12. Haynes, R.R.; Les, D.H. Alismatales (Water Plantains). In *eLS*; Wiley: Hoboken, NJ, USA, 2005. [CrossRef]
13. Ortiz, O.O.; Ibáñez, A.; Trujillo-Trujillo, E.; Croat, T.B. The emergent macrophyte *Montrichardia linifera* (Arruda) Schott (Alismatales: Araceae), a rekindled old friend from the Pacific Slope of lower Central America and western Colombia. *Nord. J. Bot.* **2020**, *38*, 1–10. [CrossRef]
14. Bonora, A.; Pancaldi, S.; Gualandri, R.; Fasulo, M.P. Carotenoid and ultrastructure variations is plastids of *Arum italicum* Miller fruit during maturation and ripening. *J. Exp. Bot.* **2000**, *51*, 873–884. [CrossRef] [PubMed]
15. Demmig-Adams, B.; Stewart, J.J.; Adams, W.W., III. Photoprotection and the trade-off between abiotic and biotic defense. In *Non-Photochemical Quenching and Energy Dissipation in Plants, Advances in Photosynthesis and Respiration*; Demmig-Adams, B., Garab, G., Adams, W.W., III, Govindjee, U., Eds.; Springer: Dordrecht, The Netherlands, 2014; Volume 40, pp. 631–643. [CrossRef]
16. Demmig-Adams, B.; Stewart, J.J.; López-Pozo, M.; Polutchko, S.K.; Adams, W.W., III. Zeaxanthin, a molecule for photoprotection in many different environments. *Molecules* **2020**, *25*, 5825. [CrossRef] [PubMed]
17. Demmig-Adams, B.; Ebbert, V.; Mellman, D.L.; Mueh, K.E.; Schaffer, L.; Funk, C.; Zarter, C.R.; Adamska, I.; Jansson, S.; Adams, W.W., III. Modulation of PsbS and flexible vs sustained energy dissipation by light environment in different species. *Physiol. Plant.* **2006**, *127*, 670–680. [CrossRef]
18. Demmig-Adams, B.; Adams, W.W., III; Barker, D.H.; Logan, B.A.; Verhoeven, A.S.; Bowling, D.R. Using chlorophyll fluorescence to assess the allocation of absorbed light to thermal dissipation of excess excitation. *Physiol. Plant.* **1996**, *98*, 253–264. [CrossRef]
19. Demmig-Adams, B.; Adams, W.W., III. Capacity for energy dissipation in the pigment bed in leaves with different xanthophyll cycle pools. *Funct. Plant Biol.* **1994**, *21*, 575–588. [CrossRef]

20. Polutchko, S.K.; Stewart, J.J.; Demmig-Adams, B. Integrative view of the nutrition of the eye. In *Nutraceuticals and Functional Foods in Human Health and Disease Prevention*; Bagchi, D., Preuss, H.G., Swaroop, A., Eds.; CRC Press/Taylor & Francis: Boca Raton, FL, USA, 2015; pp. 407–417. [CrossRef]
21. Demmig-Adams, B.; López-Pozo, M.; Stewart, J.J.; Adams, W.W., III. Zeaxanthin and lutein: Photoprotectors, anti-inflammatories, and brain food. *Molecules* **2020**, *25*, 3607. [CrossRef] [PubMed]
22. Polutchko, S.K.; Glime, G.N.E.; Demmig-Adams, B. Synergistic action of membrane-bound and water-soluble antioxidants in neuroprotection. *Molecules* **2021**, *26*, 5385. [CrossRef]
23. Stewart, J.J.; Adams, W.W., III; Escobar, C.M.; López-Pozo, M.; Demmig-Adams, B. Growth and essential carotenoid micronutrients in *Lemna gibba* as a function of growth light intensity. *Front. Plant Sci.* **2020**, *11*, 480. [CrossRef]
24. Stewart, J.J.; Adams, W.W., III; López-Pozo, M.; Doherty Garcia, N.; McNamara, M.; Escobar, C.M.; Demmig-Adams, B. Features of the duckweed *Lemna* that support rapid growth under extremes of light intensity. *Cells* **2021**, *10*, 1481. [CrossRef]
25. Demmig-Adams, B.; Adams, W.W. III. Carotenoid composition in sun and shade leaves of plants with different life forms. *Plant Cell Environ.* **1992**, *15*, 411–419. [CrossRef]
26. Zarter, C.R.; Adams, W.W., III; Ebbert, V.; Adamska, I.; Jansson, S.; Demmig-Adams, B. Winter acclimation of PsbS and related proteins in the evergreen *Arctostaphylos uva-ursi* as influenced by altitude and light environment. *Plant Cell Environ.* **2006**, *29*, 869–878. [CrossRef]
27. Zarter, C.R.; Adams, W.W., III; Ebbert, V.; Cuthbertson, D.J.; Adamska, I.; Demmig-Adams, B. Winter down-regulation of intrinsic photosynthetic capacity coupled with up-regulation of Elip-like proteins and persistent energy dissipation in a subalpine forest. *New Phytol.* **2006**, *172*, 272–282. [CrossRef]
28. Verhoeven, A.S.; Demmig-Adams, B.; Adams, W.W., III. Enhanced employment of the xanthophyll cycle and thermal energy dissipation in spinach exposed to high light and N stress. *Plant Physiol.* **1997**, *113*, 817–824. [CrossRef]
29. Logan, B.A.; Demmig-Adams, B.; Rosenstiel, T.N.; Adams, W.W., III. Effect of nitrogen limitation on foliar antioxidants in relationship to other metabolic characteristics. *Planta* **1999**, *209*, 213–220. [CrossRef]
30. Demmig-Adams, B.; Stewart, J.J.; Adams, W.W., III. Environmental regulation of intrinsic photosynthetic capacity: An integrated view. *Curr. Opin. Plant Biol.* **2017**, *37*, 34–41. [CrossRef] [PubMed]
31. Terashima, I. Productive structure of a leaf. In *Photosynthesis*; Briggs, W.R., Ed.; Alan R. Liss: New York, NY, USA, 1989; pp. 207–226.
32. Adams, W.W., III; Demmig-Adams, B.; Barker, D.H.; Kiley, S. Carotenoids and photosystem II characteristics of upper and lower halves of leaves acclimated to high light. *Aust. J. Plant Physiol.* **1996**, *23*, 669–677. [CrossRef]
33. Havaux, M.; Niyogi, K.K. The violaxanthin cycle protects plants from photooxidative damage by more than one mechanism. *Proc. Natl. Acad. Sci. USA* **1999**, *96*, 8762–8767. [CrossRef]
34. Havaux, M.; García-Plazaola, J.I. Beyond non-photochemical fluorescence quenching: The overlapping antioxidant functions of zeaxanthin and tocopherols. In *Non-Photochemical Quenching and Energy Dissipation in Plants, Advances in Photosynthesis and Respiration*; Demmig-Adams, B., Garab, G., Adams, W.W., III, Govindjee, U., Eds.; Springer: Dordrecht, The Netherlands, 2014; Volume 40, pp. 583–603. [CrossRef]
35. Nilkens, M.; Kress, E.; Lambrev, P.; Miloslavina, Y.; Müller, M.; Holzwarth, A.R.; Jahns, P. Identification of a slowly inducible zeaxanthin-dependent component of non-photochemical quenching of chlorophyll fluorescence generated under steady-state conditions in *Arabidopsis*. *Biochim. Biophys. Acta* **2010**, *1797*, 466–475. [CrossRef] [PubMed]
36. Muztar, J.A.; Slinger, S.J.; Burton, J.H. Chemical composition of aquatic macrophytes I. Investigation of organic constituents and nutritional potential. *Can. J. Plant Sci.* **1978**, *58*, 829–841. [CrossRef]
37. Rice, S.K. Patterns of allocation and growth in aquatic *Sphagnum* species. *Can. J. Bot.* **1995**, *73*, 349–359. [CrossRef]
38. Swapna, M.M.; Prakashkumar, R.; Anoop, K.P.; Manju, C.N.; Rajith, N.P. A review on the medicinal and edible aspects of aquatic and wetland plants of India. *J. Med. Plant Res.* **2011**, *5*, 7163–7176. [CrossRef]
39. Huang, M.; Shan, S.; Zhou, X.; Chen, J.; Cao, F.; Jiang, L.; Zou, Y. Leaf photosynthetic performance related to higher radiation use efficiency and grain yield in hybrid rice. *Field Crops Res.* **2016**, *193*, 87–93. [CrossRef]
40. Garbulsky, M.F.; Peñuelas, J.; Gamon, J.; Inoue, Y.; Filella, I. The photochemical reflectance index (PRI) and the remote sensing of leaf, canopy and ecosystem radiation use efficiencies: A review and meta-analysis. *Remote Sens. Environ.* **2011**, *115*, 281–297. [CrossRef]
41. Porcar-Castell, A.; Tyystjärvi, E.; Atherton, J.; van der Tol, C.; Flexas, J.; Pfündel, E.E.; Moreno, J.; Frankenberg, C.; Berry, J.A. Linking chlorophyll a fluorescence to photosynthesis for remote sensing applications: Mechanisms and challenges. *J. Exp. Bot.* **2014**, *65*, 4065–4095. [CrossRef]
42. Thouvenot, L.; Haury, J.; Thiebaut, G. A success story: Water primroses, aquatic plant pests. *Aquat. Conserv. Mar. Freshw. Ecosyst.* **2013**, *23*, 790–803. [CrossRef]
43. Chen, F.; Liu, X.; Yu, C.; Chen, Y.; Tang, H.; Zhang, L. Water lilies as emerging models for Darwin's abominable mystery. *Hortic. Res.* **2017**, *4*, 17051. [CrossRef]
44. Michael, T.P.; Ernst, E.; Hartwick, N.; Chu, P.; Bryant, D.; Gilbert, S.; Ortleb, S.; Baggs, E.L.; Sree, K.S.; Appenroth, K.J.; et al. Genome and time-of-day transcriptome of *Wolffia australiana* link morphological minimization with gene loss and less growth control. *Genome Res.* **2021**, *31*, 225–238. [CrossRef]
45. Hsiao, T.C. Plant responses to water stress. *Annu. Rev. Plant Physiol.* **1973**, *24*, 519–570. [CrossRef]

46. Dolferus, R. To grow or not to grow: A stressful decision for plants. *Plant Sci.* **2014**, *229*, 247–261. [CrossRef]
47. Patten, D.T. Productivity and production efficiency of an upper Sonoran Desert ephemeral community. *Am. J. Bot.* **1978**, *65*, 891–895. [CrossRef]
48. Parray, J.A.; Mir, M.Y.; Shameem, N. Stress management: Sustainable approach towards resilient agriculture. In *Sustainable Agriculture: Biotechniques in Plant Biology*; Parray, J.A., Mir, M.Y., Shameem, N., Eds.; Springer: Singapore, 2019; pp. 231–270. [CrossRef]
49. Beena, R.; Kirubakaran, S.; Nithya, N.; Manickavelu, A.; Sah, R.P.; Abida, P.S.; Sreekumar, J.; Jaslam, P.M.; Rejeth, R.; Jayalekshmy, V.G.; et al. Association mapping of drought tolerance and agronomic traits in rice (*Oryza sativa* L.) landraces. *BMC Plant Biol.* **2021**, *21*, 484. [CrossRef] [PubMed]
50. Nilsen, E.T.; Sharifi, M.R.; Rundel, P.W.; Jarrell, W.M.; Virginia, R.A. Diurnal and seasonal water relations of the desert phreatophyte *Prosopis glandulosa* (honey mesquite) in the Sonoran Desert of California. *Ecology* **1983**, *64*, 1381–1393. [CrossRef]
51. Nilsen, E.T.; Sharifi, M.R.; Rundel, P.W. Comparative water relations of phreatophytes in the Sonoran Desert of California. *Ecology* **1984**, *65*, 767–778. [CrossRef]
52. Sala, A.; Smith, S.D.; Devitt, D.A. Water use by *Tamarix ramosissima* and associated phreatophytes in a Mojave Desert floodplain. *Ecol. Appl.* **1996**, *6*, 888–898. [CrossRef]
53. Hultine, K.R.; Williams, D.G.; Burgess, S.S.O.; Keefer, T.O. Contrasting patterns of hydraulic redistribution in three desert phreatophytes. *Oecologia* **2003**, *135*, 167–175. [CrossRef]
54. Hultine, K.R.; Koepke, D.F.; Pockman, W.T.; Fravolini, A.; Sperry, J.S.; Williams, D.G. Influence of soil texture on hydraulic properties and water relations of a dominant warm-desert phreatophyte. *Tree Physiol.* **2006**, *26*, 313–323. [CrossRef]
55. Guo, J.S.; Gear, L.; Hultine, K.R.; Koch, G.W.; Ogle, K. Non-structural carbohydrate dynamics associated with antecedent stem water potential and air temperature in a dominant desert shrub. *Plant Cell Environ.* **2020**, *43*, 1467–1483. [CrossRef]
56. Carvajal, D.E.; Loayza, A.P.; Rios, R.S.; Gianoli, E.; Squeo, F.A. Population variation in drought-resistance strategies in a desert shrub along an aridity gradient: Interplay between phenotypic plasticity and ecotypic differentiation. *Perspect. Plant Ecol. Evol. Syst.* **2017**, *29*, 12–19. [CrossRef]
57. Myers, S.S.; Zanobetti, A.; Kloog, I.; Huybers, P.; Leakey, A.D.B.; Bloom, A.J.; Carlisle, E.; Dietterich, L.H.; Fitzgerald, G.; Hasegawa, T.; et al. Increasing CO_2 threatens human nutrition. *Nature* **2014**, *510*, 139–142. [CrossRef]
58. Demmig-Adams, B.; Polutchko, S.K.; Zenir, M.C.; Fourounjian, P.; Stewart, J.J.; López-Pozo, M.; Adams, W.W., III. Intersections: Photosynthesis, abiotic stress, and the plant microbiome. *Photosynthetica*, 2022; *in press*.
59. Körner, C. Paradigm shift in plant growth control. *Curr. Opin. Plant Biol.* **2015**, *25*, 107–114. [CrossRef]
60. Krapp, A.; Stitt, M. An evaluation of direct and indirect mechanisms for the "sink-regulation" of photosynthesis in spinach: Changes in gas exchange, carbohydrates, metabolites, enzyme activities and steady-state transcript levels after cold-girdling source leaves. *Planta* **1995**, *195*, 313–323. [CrossRef]
61. Tausz-Posch, S.; Tausz, M.; Bourgault, M. Elevated [CO_2] effects on crops: Advances in understanding acclimation, nitrogen dynamics and interactions with drought and other organisms. *Plant Biol.* **2020**, *22*, 38–51. [CrossRef]
62. Huang, H.; Ullah, F.; Zhou, D.X.; Yi, M.; Zhao, Y. Mechanisms of ROS regulation of plant development and stress responses. *Front. Plant Sci.* **2019**, *10*, 800. [CrossRef]
63. Foyer, C.H.; Noctor, G. Stress-triggered redox signalling: What's in pROSpect? *Plant Cell Environ.* **2016**, *39*, 951–964. [CrossRef]
64. Foyer, C.H.; Noctor, G. Redox homeostasis and signaling in a higher-CO_2 world. *Annu. Rev. Plant Biol.* **2020**, *71*, 157–182. [CrossRef] [PubMed]
65. Hasanuzzaman, M.; Bhuyan, M.H.M.B.; Parvin, K.; Bhuiyan, T.F.; Anee, T.I.; Nahar, K.; Hossen, M.S.; Zulfiqar, F.; Alam, M.M.; Fujita, M. Regulation of ros metabolism in plants under environmental stress: A review of recent experimental evidence. *Int. J. Mol. Sci.* **2020**, *21*, 8695. [CrossRef] [PubMed]
66. Padhan, B.K.; Sathee, L.; Meena, H.S.; Adavi, S.B.; Jha, S.K.; Chinnusamy, V. CO_2 Elevation accelerates phenology and alters carbon/nitrogen metabolism *vis-à-vis* ROS abundance in bread wheat. *Front. Plant Sci.* **2020**, *11*, 1061. [CrossRef]
67. Yin, Y.; Yu, C.; Yu, L.; Zhao, J.; Sun, C.; Ma, Y.; Zhou, G. The influence of light intensity and photoperiod on duckweed biomass and starch accumulation for bioethanol production. *Bioresour. Technol.* **2015**, *187*, 84–90. [CrossRef] [PubMed]
68. Paul, M.J.; Foyer, C.H. Sink regulation of photosynthesis. *J. Exp. Bot.* **2001**, *52*, 1383–1400. [CrossRef] [PubMed]
69. Martindale, W.; Bowes, G. The effects of irradiance and CO_2 on the activity and activation of ribulose-1,5-bisphosphate carboxylase/oxygenase in the aquatic plant *Spirodela polyrhiza*. *J. Exp. Bot.* **1996**, *47*, 781–784. [CrossRef]
70. Dall'Osto, L.; Lico, C.; Alric, J.; Giuliano, G.; Havaux, M.; Bassi, R. Lutein is needed for efficient chlorophyll triplet quenching in the major LHCII antenna complex of higher plants and effective photoprotection in vivo under strong light. *BMC Plant Biol.* **2006**, *6*, 32. [CrossRef] [PubMed]
71. Telfer, A. Singlet oxygen production by PSII under light stress: Mechanism, detection and the protective role of β-carotene. *Plant Cell Physiol.* **2014**, *55*, 1216–1223. [CrossRef] [PubMed]
72. Schenk, R.U.; Hildebrandt, A.C. Medium and techniques for induction and growth of monocotyledonous and dicotyledonous plant cell cultures. *Can. J. Bot.* **1972**, *50*, 199–204. [CrossRef]
73. Paul, M.J.; Driscoll, S.P. Sugar repression of photosynthesis: The role of carbohydrates in signalling nitrogen deficiency through source:sink imbalance. *Plant Cell Environ.* **1997**, *20*, 110–116. [CrossRef]

74. Burnett, A.C.; Rogers, A.; Rees, M.; Osborne, C.P. Nutrient sink limitation constrains growth in two barley species with contrasting growth strategies. *Plant Direct* **2018**, *2*, e00094. [CrossRef]
75. Moore, B.D.; Cheng, S.; Sims, D.; Seemann, J.R. The biochemical and molecular basis for photosynthetic acclimation to elevated atmospheric CO_2. *Plant Cell Environ.* **1999**, *22*, 567–582. [CrossRef]
76. Ainsworth, E.A.; Long, S.P. 30 years of free-air carbon dioxide enrichment (FACE): What have we learned about future crop productivity and its potential for adaptation? *Glob. Chang. Biol.* **2005**, *27*, 27–49. [CrossRef]
77. Prins, A.; Kunert, K.; Foyer, C.H. Atmospheric CO_2 signalling, cellular redox state and plant growth and development. In *Redox Metabolism and Longevity Relationships in Animals and Plants*; Foyer, C.H., Faragher, R., Thornalley, P., Eds.; Taylor & Francis: London, UK, 2008; Volume 62, pp. 229–252. [CrossRef]
78. Dai, N.; Schaffer, A.; Petreikov, M.; Shahak, Y.; Giller, Y.; Ratner, K.; Levine, A.; Granot, D. Overexpression of Arabidopsis hexokinase in tomato plants inhibits growth, reduces photosynthesis, and induces rapid senescence. *Plant Cell* **1999**, *11*, 1253–1266. [CrossRef]
79. Diaz, C.; Purdy, S.; Christ, A.; Morot-Gaudry, J.-F.; Wingler, A.; Masclaux-Daubresse, C. Characterization of markers to determine the extent and variability of leaf senescence in Arabidopsis. A metabolic profiling approach. *Plant Physiol.* **2005**, *138*, 898–908. [CrossRef]
80. Agüera, E.; De la Haba, P. Leaf senescence in response to elevated atmospheric CO_2 concentration and low nitrogen supply. *Biol. Plant.* **2018**, *62*, 401–408. [CrossRef]
81. Wingler, A.; Purdy, S.; MacLean, J.A.; Pourtau, N. The role of sugars in integrating environmental signals during the regulation of leaf senescence. *J. Exp. Bot.* **2005**, *57*, 391–399. [CrossRef]
82. Wang, S.; Zhang, Y.; Ju, W.; Chen, J.M.; Ciais, P.; Cescatti, A.; Sardans, J.; Janssens, I.A.; Wu, M.; Berry, J.A.; et al. Recent global decline of CO_2 fertilization effects on vegetation photosynthesis. *Science* **2020**, *370*, 1295–1300. [CrossRef] [PubMed]
83. Cheng, J.J.; Stomp, A.-M. Growing duckweed to recover nutrients from wastewaters and for production of fuel ethanol and animal feed. *Clean Soil Air Water* **2009**, *37*, 17–26. [CrossRef]
84. Xu, J.; Zhao, H.; Stomp, A.-M.; Cheng, J.J. The production of duckweed as a source of biofuels. *Biofuels* **2012**, *3*, 589–601. [CrossRef]
85. Tao, X.; Fang, Y.; Xiao, Y.; Jin, Y.L.; Ma, X.R.; Zhao, Y.; He, K.Z.; Zhao, H.; Wang, H.Y. Comparative transcriptome analysis to investigate the high starch accumulation of duckweed (*Landoltia punctata*) under nutrient starvation. *Biotechnol. Biofuels* **2013**, *6*, 72. [CrossRef]
86. Liu, Y.; Wang, X.; Fang, Y.; Huang, M.; Chen, X.; Zhang, Y.; Zhao, H. The effects of photoperiod and nutrition on duckweed (*Landoltia punctata*) growth and starch accumulation. *Ind. Crops Prod.* **2018**, *115*, 243–249. [CrossRef]
87. Appenroth, K.-J.; Sree, K.S.; Fakhoorian, T.; Lam, E. Resurgence of duckweed research and applications: Report from the 3rd International Duckweed Conference. *Plant Mol. Biol.* **2015**, *89*, 647–654. [CrossRef]
88. Adavi, S.B.; Sathee, L. Elevated CO_2 differentially regulates root nitrate transporter kinetics in a genotype and nitrate dose-dependent manner. *Plant Sci.* **2021**, *305*, 110807. [CrossRef] [PubMed]
89. Cedergreen, N.; Madsen, T.V. Nitrogen uptake by the floating macrophyte *Lemna minor*. *New Phytol.* **2002**, *155*, 285–292. [CrossRef]
90. Cedergreen, N.; Madsen, T.V. Light regulation of root and leaf NO_3-uptake and reduction in the floating macrophyte *Lemna minor*. *New Phytol.* **2004**, *161*, 449–457. [CrossRef]
91. Yu, C.; Zhao, X.; Qi, G.; Bai, Z.; Wang, Y.; Wang, S.; Ma, Y.; Liu, Q.; Hu, R.; Zhou, G. Integrated analysis of transcriptome and metabolites reveals an essential role of metabolic flux in starch accumulation under nitrogen starvation in duckweed. *Biotechnol. Biofuels* **2017**, *10*, 167. [CrossRef]
92. Zhao, Z.; Shi, H.; Liu, Y.; Zhao, H.; Su, H.; Wang, M.; Zhao, Y. The influence of duckweed species diversity on biomass productivity and nutrient removal efficiency in swine wastewater, *Bioresour. Technol.* **2014**, *167*, 383–389. [CrossRef]
93. Mohedano, R.A.; Costa, R.H.R.; Tavares, F.A.; Belli Filho, P. High nutrient removal rate from swine wastes and protein biomass production by full-scale duckweed ponds. *Bioresour. Technol.* **2012**, *112*, 98–104. [CrossRef] [PubMed]
94. Boyd, C.E. Fresh-water plants: A potential source of protein. *Econ. Bot.* **1968**, *22*, 359–368. [CrossRef]
95. Dewanji, A.; Chanda, S.; Si, L.; Barik, S.; Matai, S. Extractability and nutritional value of leaf protein from tropical aquatic plants. *Plant Foods Hum. Nutr.* **1997**, *50*, 349–357. [CrossRef]
96. Rai, U.N.; Sinha, S. Distribution of metals in aquatic edible plants: *Trapa natans* (Roxb.) Makino and *Ipomoea aquatica* Forsk. *Environ. Monit. Assess.* **2001**, *70*, 241–252. [CrossRef]
97. Zeng, J.; Sheng, H.; Liu, Y.; Wang, Y.; Wang, Y.; Kang, H.; Fan, X.; Sha, L.; Yuan, S.; Zhou, Y. High nitrogen supply induces physiological responsiveness to long photoperiod in barley. *Front. Plant Sci.* **2017**, *8*, 569. [CrossRef] [PubMed]
98. Du, S.; Zhang, Y.; Lin, X.; Wang, Y.; Tang, C. Regulation of nitrate reductase by nitric oxide in Chinese cabbage pakchoi (*Brassica chinensis* L.). *Plant Cell Environ.* **2008**, *31*, 195–204. [CrossRef]
99. Bian, Z.; Wang, Y.; Zhang, X.; Li, T.; Grundy, S.; Yang, Q.; Cheng, R. A review of environment effects on nitrate accumulation in leafy vegetables grown in controlled environments. *Foods* **2020**, *9*, 732. [CrossRef]
100. Shin, R.; Schachtman, D.P. Hydrogen peroxide mediates plant root cell response to nutrient deprivation. *Proc. Natl. Acad. Sci. USA* **2004**, *101*, 8827–8832. [CrossRef]
101. Kim, M.J.; Ciani, S.; Schachtman, D.P. A peroxidase contributes to ros production during *Arabidopsis* root response to potassium deficiency. *Mol. Plant* **2010**, *3*, 420–427. [CrossRef]

102. Wu, F.; Sun, X.; Hu, X.; Zou, B.; Lin, N.; Lin, J.; Ji, K. Response of nitrogen metabolism in masson pine needles to elevated CO_2. *Forests* **2020**, *11*, 390. [CrossRef]
103. Adavi, S.B.; Sathee, L. Elevated CO_2 alters tissue balance of nitrogen metabolism and downregulates nitrogen assimilation and signalling gene expression in wheat seedlings receiving high nitrate supply. *Protoplasma* **2020**, *258*, 219–233. [CrossRef] [PubMed]
104. Queval, G.; Issakidis-Bourguet, E.; Hoeberichts, F.A.; Vandorpe, M.; Gakière, B.; Vanacker, H.; Miginiac-Maslow, M.; Van Breusegem, F.; Noctor, G. Conditional oxidative stress responses in the Arabidopsis photorespiratory mutant *cat2* demonstrate that redox state is a key modulator of daylength-dependent gene expression, and define photoperiod as a crucial factor in the regulation of H_2O_2-induced cell death. *Plant J.* **2007**, *52*, 640–657. [CrossRef] [PubMed]
105. Krasensky-Wrzaczek, J.; Kangasjärvi, J. The role of reactive oxygen species in the integration of temperature and light signals. *J. Exp. Bot.* **2018**, *69*, 3347–3358. [CrossRef] [PubMed]
106. Begara-Morales, J.C. Nitric oxide signalling in a CO_2-enriched environment. *J. Exp. Bot.* **2016**, *67*, 560–561. [CrossRef] [PubMed]
107. Bloom, A.J. As carbon dioxide rises, food quality will decline without careful nitrogen management. *Calif. Agric.* **2009**, *63*, 67–72. [CrossRef]
108. de Sousa Leite, T.; Monteiro, F.A. Partial replacement of nitrate by ammonium increases photosynthesis and reduces oxidative stress in tanzania guinea grass exposed to cadmium. *Ecotoxicol. Environ. Saf.* **2019**, *174*, 592–600. [CrossRef]
109. Britto, D.T.; Kronzucker, H.J. NH_4^+ toxicity in higher plants: A critical review. *J. Plant Physiol.* **2002**, *159*, 567–584. [CrossRef]
110. Becklin, K.M.; Mullinix, G.W.R.; Ward, J.K. Host plant physiology and mycorrhizal functioning shift across a glacial through future [CO_2] gradient. *Plant Physiol.* **2016**, *172*, 789–801. [CrossRef]
111. Syvertsen, J.P.; Graham, J.H. Phosphorus supply and arbuscular mycorrhizas increase growth and net gas exchange responses of two *Citrus* spp. grown at elevated [CO_2]. *Plant Soil* **1999**, *208*, 209–219. [CrossRef]
112. Gavito, M.E.; Bruhn, D.; Jakobsen, I. Phosphorus uptake by arbuscular mycorrhizal hyphae does not increase when the host plant grows under atmospheric CO_2 enrichment. *New Phytol.* **2002**, *154*, 751–760. [CrossRef]
113. Jifon, J.L.; Wolfe, D.W. Photosynthetic acclimation to elevated CO_2 in *Phaseolus vulgaris* L. is altered by growth response to nitrogen supply. *Glob. Chang. Biol.* **2002**, *8*, 1018–1027. [CrossRef]
114. Nie, M.; Bell, C.; Wallenstein, M.D.; Pendall, E. Increased plant productivity and decreased microbial respiratory C loss by plant growth-promoting rhizobacteria under elevated CO_2. *Sci. Rep.* **2015**, *5*, 9212. [CrossRef] [PubMed]
115. Johnson, N.C.; Wilson, G.W.T.; Bowker, M.A.; Wilson, J.A.; Miller, R.M. Resource limitation is a driver of local adaptation in mycorrhizal symbioses. *Proc. Natl. Acad. Sci. USA* **2010**, *107*, 2093–2098. [CrossRef]
116. Weese, D.J.; Heath, K.D.; Dentinger, B.T.M.; Lau, J.A. Long-term nitrogen addition causes the evolution of less-cooperative mutualists. *Evolution* **2015**, *69*, 631–642. [CrossRef]
117. Zenir, M. Productivity and Nutritional Quality of *Lemna minor* in Different Physical and Biological Environments. Senior Honors Thesis, University of Colorado Boulder, Boulder, CO, USA, 2021.
118. Romero-Munar, A.; Del-Saz, N.F.; Ribas-Carbó, M.; Flexas, J.; Baraza, E.; Florez-Sarasa, I.; Fernie, A.R.; Gulías, J. Arbuscular mycorrhizal symbiosis with *Arundo donax* decreases root respiration and increases both photosynthesis and plant biomass accumulation. *Plant Cell Environ.* **2017**, *40*, 1115–1126. [CrossRef] [PubMed]
119. Yang, Y.; Guo, Y.; Zhong, J.; Zhang, T.; Li, D.; Ba, T.; Xu, T.; Chang, L.; Zhang, Q.; Sun, M. Root physiological traits and transcriptome analyses reveal that root zone water retention confers drought tolerance to *Opisthopappus taihangensis*. *Sci. Rep.* **2020**, *10*, 2627. [CrossRef]
120. Chen, J.; Zhang, H.; Zhang, X.; Tang, M. Arbuscular mycorrhizal symbiosis alleviates salt stress in black locust through improved photosynthesis, water status, and K^+/Na^+ homeostasis. *Front. Plant Sci.* **2017**, *8*, 1739. [CrossRef]
121. Gavito, M.E.; Jakobsen, I.; Mikkelsen, T.N.; Mora, F. Direct evidence for modulation of photosynthesis by an arbuscular mycorrhiza-induced carbon sink strength. *New Phytol.* **2019**, *223*, 896–907. [CrossRef]
122. Lamhamedi, M.S.; Godbout, C.; Fortin, J.A. Dependence of *Laccariabicolor* basidiome development on current photosynthesis of *Pinusstrobus* seedlings. *Can. J. For. Res.* **1994**, *24*, 1797–1804. [CrossRef]
123. Rogers, A.; Ainsworth, E.A.; Leakey, A.D.B. Will elevated carbon dioxide concentration amplify the benefits of nitrogen fixation in legumes? *Plant Physiol.* **2009**, *151*, 1009–1016. [CrossRef] [PubMed]
124. Leakey, A.D.B.; Ainsworth, E.A.; Bernacchi, C.J.; Rogers, A.; Long, S.P.; Ort, D.R. Elevated CO_2 effects on plant carbon, nitrogen, and water relations: Six important lessons from FACE. *J. Exp. Bot.* **2009**, *60*, 2859–2876. [CrossRef]
125. Bloom, A.J.; Asensio, J.S.R.; Randall, L.; Rachmilevitch, S.; Cousins, A.B.; Carlisle, E.A. CO_2 enrichment inhibits shoot nitrate assimilation in C_3 but not C_4 plants and slows growth under nitrate in C_3 plants. *Ecology* **2012**, *93*, 355–367. [CrossRef] [PubMed]
126. Acosta, K.; Xu, J.; Gilbert, S.; Denison, E.; Brinkman, T.; Lebeis, S.; Lam, E. Duckweed hosts a taxonomically similar bacterial assemblage as the terrestrial leaf microbiome. *PLoS ONE* **2020**, *15*, e0228560. [CrossRef] [PubMed]
127. Ishizawa, H.; Kuroda, M.; Inoue, D.; Morikawa, M.; Ike, M. Community dynamics of duckweed-associated bacteria upon inoculation of plant growth-promoting bacteria. *FEMS Microbiol. Ecol.* **2020**, *96*, fiaa101. [CrossRef]
128. O'Brien, A.M.; Laurich, J.; Lash, E.; Frederickson, M.E. Mutualistic outcomes across plant populations, microbes, and environments in the duckweed *Lemna minor*. *Microb. Ecol.* **2020**, *80*, 384–397. [CrossRef]
129. Zuberer, D.A. Nitrogen fixation (acetylene reduction) associated with duckweed (*Lemnaceae*) mats. *Appl. Environ. Microbiol.* **1982**, *43*, 823–828. [CrossRef]

130. Ishizawa, H.; Kuroda, M.; Morikawa, M.; Ike, M. Evaluation of environmental bacterial communities as a factor affecting the growth of duckweed *Lemna minor*. *Biotechnol. Biofuels* **2017**, *10*, 62. [CrossRef]
131. Toyama, T.; Kuroda, M.; Ogata, Y.; Hachiya, Y.; Quach, A.; Tokura, K.; Tanaka, Y.; Mori, K.; Morikawa, M.; Ike, M. Enhanced biomass production of duckweeds by inoculating a plant growth-promoting bacterium, *Acinetobacter calcoaceticus* P23, in sterile medium and non-sterile environmental waters. *Water Sci. Technol.* **2017**, *76*, 1418–1428. [CrossRef]
132. Yamakawa, Y.; Jog, R.; Morikawa, M. Effects of co-inoculation of two different plant growth-promoting bacteria on duckweed. *Plant Growth Regul.* **2018**, *86*, 287–296. [CrossRef]
133. Khairina, Y.; Jog, R.; Boonmak, C.; Toyama, T.; Oyama, T.; Morikawa, M. Indigenous bacteria, an excellent reservoir of functional plant growth promoters for enhancing duckweed biomass yield on site. *Chemosphere* **2021**, *268*, 129247. [CrossRef]
134. Quisehuatl-Tepexicuapan, E.; Ferrera-Cerrato, R.; Silva-Rojas, H.V.; Rodriguez-Zaragoza, S.; Alarcón, A.; Almaraz-Suárez, J.J. Free-living culturable bacteria and protozoa from the rhizoplanes of three floating aquatic plant species. *Plant Biosyst.* **2014**, *150*, 855–865. [CrossRef]
135. Grant, A.; Rémond, M.; Starke-Peterkovic, T.; Hinde, R. A cell signal from the coral *Plesiastrea versipora* reduces starch synthesis in its symbiotic alga, *Symbiodinium* sp. *Comp. Biochem. Physiol.-A Mol. Integr. Physiol.* **2006**, *144*, 458–463. [CrossRef]
136. Grant, A.; People, J.; Rémond, M.; Frankland, S.; Hinde, R. How a host cell signalling molecule modifies carbon metabolism in symbionts of the coral *Plesiastrea versipora*. *FEBS J.* **2013**, *280*, 2085–2096. [CrossRef] [PubMed]
137. Stefan, M.; Munteanu, N.; Stoleru, V.; Mihasan, M.; Hritcu, L. Seed inoculation with plant growth promoting rhizobacteria enhances photosynthesis and yield of runner bean (*Phaseolus coccineus* L.). *Sci. Hortic.* **2013**, *151*, 22–29. [CrossRef]
138. Adams, W.W., III; Stewart, J.J.; Polutchko, S.K.; Demmig-Adams, B. Leaf vasculature and the upper limit of photosynthesis. In *The Leaf: A Platform for Performing Photosynthesis, Advances in Photosynthesis and Respiration*; Adams, W.W., III, Terashima, I., Eds.; Springer: Cham, Switzerland, 2018; Volume 44, pp. 27–54. [CrossRef]
139. Vacheron, J.; Desbrosses, G.; Bouffaud, M.L.; Touraine, B.; Moënne-Loccoz, Y.; Muller, D.; Legendre, L.; Wisniewski-Dyé, F.; Prigent-Combaret, C. Plant growth-promoting rhizobacteria and root system functioning. *Front. Plant Sci.* **2013**, *4*, 356. [CrossRef]
140. Ortíz, J.; Sanhueza, C.; Romero-Munar, A.; Hidalgo-Castellanos, J.; Castro, C.; Bascuñán-godoy, L.; de la Peña, T.C.; López-Gómez, M.; Florez-Sarasa, I.; Del-Saz, N.F. In Vivo metabolic regulation of alternative oxidase under nutrient deficiency—interaction with arbuscular mycorrhizal fungi and *Rhizobium* bacteria. *Int. J. Mol. Sci.* **2020**, *21*, 4201. [CrossRef]
141. Kramer, D.M.; Avenson, T.J.; Edwards, G.E. Dynamic flexibility in the light reactions of photosynthesis governed by both electron and proton transfer reactions. *Trends Plant Sci.* **2004**, *9*, 349–357. [CrossRef] [PubMed]
142. Ivanov, A.G.; Rosso, D.; Savitch, L.V.; Stachula, P.; Rosembert, M.; Oquist, G.; Hurry, V.; Hüner, N.P.A. Implications of alternative electron sinks in increased resistance of PSII and PSI photochemistry to high light stress in cold-acclimated *Arabidopsis thaliana*. *Photosynth. Res.* **2012**, *113*, 191–206. [CrossRef]
143. Blanco, N.E.; Ceccoli, R.D.; Vía, M.V.D.; Voss, I.; Segretin, M.E.; Bravo-Almonacid, F.F.; Melzer, M.; Hajirezaei, M.-R.; Scheibe, R.; Hanke, G.T. Expression of the minor isoform pea ferredoxin in tobacco alters photosynthetic electron qartitioning and enhances cyclic electron flow. *Plant Physiol.* **2013**, *161*, 866–879. [CrossRef] [PubMed]
144. Courteille, A.; Vesa, S.; Sanz-Barrio, R.; Cazalé, A.-C.; Becuwe-Linka, N.; Farran, I.; Havaux, M.; Rey, P.; Rumeau, D. Thioredoxin m4 controls photosynthetic alternative electron pathways in Arabidopsis. *Plant Physiol.* **2013**, *161*, 508–520. [CrossRef]
145. Strand, D.D.; Fisher, N.; Kramer, D.M. Distinct energetics and regulatory functions of the two major cyclic electron flow pathways in chloroplasts. In *Chloroplasts: Current Research and Future Trends*; Kirchhoff, H., Ed.; Caister Academic Press: Norfolk, UK, 2016; pp. 89–100. [CrossRef]
146. Kawashima, R.; Sato, R.; Harada, K.; Masuda, S. Relative contributions of PGR5- and NDH-dependent photosystem I cyclic electron flow in the generation of a proton gradient in Arabidopsis chloroplasts. *Planta* **2017**, *246*, 1045–1050. [CrossRef]
147. Yoshida, K.; Terashima, I.; Noguchi, K. Up-regulation of mitochondrial alternative oxidase concomitant with chloroplast over-reduction by excess light. *Plant Cell Physiol.* **2007**, *48*, 606–614. [CrossRef] [PubMed]
148. Yoshida, K.; Watanabe, C.K.; Terashima, I.; Noguchi, K. Physiological impact of mitochondrial alternative oxidase on photosynthesis and growth in *Arabidopsis thaliana*. *Plant Cell Environ.* **2011**, *34*, 1890–1899. [CrossRef]
149. Vishwakarma, A.; Bashyam, L.; Senthilkumaran, B.; Scheibe, R.; Padmasree, K. Physiological role of AOX1a in photosynthesis and maintenance of cellular redox homeostasis under high light in *Arabidopsis thaliana*. *Plant Physiol. Biochem.* **2014**, *81*, 44–53. [CrossRef]
150. Vishwakarma, A.; Tetali, S.D.; Selinski, J.; Scheibe, R.; Padmasree, K. Importance of the alternative oxidase (AOX) pathway in regulating cellular redox and ROS homeostasis to optimize photosynthesis during restriction of the cytochrome oxidase pathway in *Arabidopsis thaliana*. *Ann. Bot.* **2015**, *116*, 555–569. [CrossRef]
151. Shapiguzov, A.; Vainonen, J.P.; Hunter, K.; Tossavainen, H.; Tiwari, A.; Järvi, S.; Hellman, M.; Aarabi, F.; Alseekh, S.; Wybouw, B.; et al. Arabidopsis RCD1 coordinates chloroplast and mitochondrial functions through interaction with ANAC transcription factors. *eLife* **2019**, *8*, e43284. [CrossRef]
152. Voss, I.; Sunil, B.; Scheibe, R.; Raghavendra, A.S. Emerging concept for the role of photorespiration as an important part of abiotic stress response. *Plant Biol.* **2013**, *15*, 713–722. [CrossRef] [PubMed]
153. Scheibe, R. Maintaining homeostasis by controlled alternatives for energy distribution in plant cells under changing conditions of supply and demand. *Photosynth. Res.* **2019**, *139*, 81–91. [CrossRef] [PubMed]

154. Wang, J.; Cheung, M.; Rasooli, L.; Amirsadeghi, S.; Vanlerberghe, G.C. Plant respiration in a high CO_2 world: How will alternative oxidase respond to future atmospheric and climatic conditions? *Can. J. Plant Sci.* **2014**, *94*, 1091–1101. [CrossRef]
155. Noguchi, K.O.; Terashima, I. Responses of spinach leaf mitochondria to low N availability. *Plant Cell Environ.* **2006**, *29*, 710–719. [CrossRef] [PubMed]

plants

MDPI

Article

Overexpression of the Phosphoserine Phosphatase-Encoding Gene (*AtPSP1*) Promotes Starch Accumulation in *Lemna turionifera* 5511 under Sulfur Deficiency

Lei Wang [1], Yingying Kuang [1], Siyu Zheng [1], Yana Tong [2,*], Yerong Zhu [1,*] and Yong Wang [1]

[1] College of Life Science, Nankai University, Tianjin 300071, China
[2] Tianjin Academy of Agricultural Sciences, Tianjin 300192, China
* Correspondence: tongyana03@163.com (Y.T.); zhuyr@nankai.edu.cn (Y.Z.)

Abstract: Duckweeds are well known for their high accumulation of starch under stress conditions, along with inhibited growth. The phosphorylation pathway of serine biosynthesis (PPSB) was reported as playing a vital role in linking the carbon, nitrogen, and sulfur metabolism in this plant. The overexpression of *AtPSP1*, the last key enzyme of the PPSB pathway in duckweed, was found to stimulate the accumulation of starch under sulfur-deficient conditions. The growth- and photosynthesis-related parameters were higher in the *AtPSP1* transgenic plants than in the WT. The transcriptional analysis showed that the expression of several genes in starch synthesis, TCA, and sulfur absorption, transportation, and assimilation was significantly up- or downregulated. The study suggests that *PSP* engineering could improve starch accumulation in *Lemna turionifera* 5511 by coordinating the carbon metabolism and sulfur assimilation under sulfur-deficient conditions.

Keywords: starch accumulation; *AtPSP1* overexpression; sulfur deficiency; *Lemna turionifera* 5511

Citation: Wang, L.; Kuang, Y.; Zheng, S.; Tong, Y.; Zhu, Y.; Wang, Y. Overexpression of the Phosphoserine Phosphatase-Encoding Gene (*AtPSP1*) Promotes Starch Accumulation in *Lemna turionifera* 5511 under Sulfur Deficiency. *Plants* **2023**, *12*, 1012. https://doi.org/10.3390/plants12051012

Academic Editors: Viktor Oláh, Klaus-Jürgen Appenroth and K. Sowjanya Sree

Received: 30 November 2022
Revised: 18 February 2023
Accepted: 20 February 2023
Published: 23 February 2023

1. Introduction

Duckweeds are the smallest flowering plant known to date and have severely reduced anatomies [1]. They are widely distributed all over the world, except in the Antarctic and Arctic regions [2]. There are 5 genera of duckweeds, namely *Spirodela*, *Landoltia*, *Lemna*, *Wolffiella*, and *Wolffia*, with a total of 36 species [3]. The biomass accumulation in duckweeds is much higher than the corn in dry weight (DW)/ha/year due to its asexual propagation and rapid growth [2]. The growth rate is attuned to the richness of the growth medium. It has been reported that duckweed can be converted into four different forms of energy, namely bio-oil, natural gas, bioethanol, and high-value-added industrial precursors, through different conversion technologies [4]. This makes duckweed a promising source of starch and a potential feedstock for the production of bioethanol and other biofuels [5]. The starch is mainly synthesized in the fronds of duckweeds, and 3–60% of dry weight can be accumulated when duckweeds are treated by growth regulators, heavy metals, nutrient deficiency, and salt stress [6–19]. However, these treatments always inhibit the growth of duckweed. The starch accumulation resulted from the decrease in duckweed biomass, while the sulfur deficiency was found to improve starch yields in duckweed without affecting its growth or biomass accumulation. Sulfur deprivation resulted in the highest starch yield, which was higher than the nitrogen or phosphorus limitation conditions. Previous research suggested that the cultivation of sulfur limitations is a potential strategy to prompt starch accumulation in duckweed [20].

In plants, three biosynthesis pathways of serine (Ser) have been described, which complicate the understanding of their metabolic regulation. One is the ethanoic acid pathway associated with photorespiration, and the other two alternative non-photorespiratory pathways of serine biosynthesis are the phosphorylation pathway and the glyceric acid pathway [21–23]. In most organisms, serine is synthesized via the phosphorylation pathway of serine biosynthesis (PPSB). Three enzymes are involved in the phosphorylation pathway of serine biosynthesis. First, 3PGA is oxidized to 3-phosphate hydroxypyruvate (3PHP) by 3-phosphoglycerate dehydrogenase (PGDH); then, 3PHP is converted to 3-phosphoserine via 3-phosphoserine aminotransferase (PSAT). Finally, 3-phosphoserine is dephosphorized to serine by 3-phosphoserine phosphatase (PSP) [24]. PSP was reported to be inhibited by high concentrations of serine; mutants lacking PSP in Arabidopsis are embryonically lethal, with altered pollen and chorion development, and *PSP* overexpression leads to increased nitrate reductase activity and photorespiration in leaves under light [25–28]. Metabolomic studies of *PSP1* overexpression and knockdown demonstrate that subtle changes in PPSB activity can modulate the glycolytic flux, affecting the TCA cycle and amino acid biosynthesis, which, in turn, affects glucose metabolism. These reports demonstrated that PSP plays a crucial role in plant metabolism and development, affecting glycolysis, amino acid synthesis, and the TCA cycle [28]. However, there are few studies on the relationship between PSP and starch biosynthesis in duckweed.

The sulfur element is an indispensable macronutrient to maintain normal growth in the plant. Sulfate is absorbed into root cells, transported to the plastids, and activated to form 5′-adenylylsulfate (APS) by ATP sulfurylase (ATPS); then, APS is reduced by APS reductase (APR) and sulfite reductase to sulfide; finally, sulfide and O-acetyl-L-serine (OAS) synthesize cysteine (Cys). Cys is the primary product of sulfur assimilation; its synthesis is dependent on the serine used to provide the carbon and nitrogen skeleton. Firstly, Ser is used with acetyl coenzyme A for the synthesis of O-acetyl serine (OAS) via serine acetyltransferase (SERAT). Then, O-acetyl serine (thiol) lyase (OAS-TL) replaces the activated acetyl portion of OAS with sulfide and releases cysteine [29,30]. Thus, Cys synthesis requires crosstalk between carbon and nitrogen metabolism and sulfur assimilation in the plant [31], while Ser is essential for the assimilation of sulfur.

PPSB synthesizes Ser in plastids from 3-phosphoglyceric acid (3-PGA). In duckweed, more 3-PGA was found to transfer to glucose and was used for starch biosynthesis after sulfur-deficiency treatment [20]. We are interested in whether there is a link between the phosphorylation pathway of serine (PPSB) and starch accumulation in duckweed. In this work, *AtPSP1*, encoding the key enzyme of PPSB in Arabidopsis, was overexpressed in duckweed when studying the effect on starch accumulation and growth under sulfur deficiency. The possible mechanism of starch accumulation was investigated in duckweed overexpressing *AtPSP1* under sulfur-deficient conditions.

2. Results

2.1. Generation of AtPSP1 Overexpressing Transgenic Lines

To explore the role that this pathway plays in starch accumulation and the interaction between PPSB and sulfur deficiency in duckweed, we constructed an *AtPSP1* overexpression vector based on pCAMBIA 1301, replacing the GUS coding sequence with the full-length CDS sequence (888 bp) of *AtPSP1* from Arabidopsis (Figure 1A). The vector was transformed into *Lemna turionifera* 5511 using *Agrobacterium tumefaciens EHA105*. The regenerated plants were screened with hygromycin to obtain resistant, transgenic plants. Nine transgenic lines with *AtPSP1* overexpression were confirmed via the PCR amplification of *AtPSP1* CDs (Figure 1B). Furthermore, all these lines were analyzed for semi-quantitative RT-PCR of the *AtPSP1* gene (Figure 1C), and three differentially expressed plants (Line 1/2/3, named PSP-1/PSP-2/PSP-3, respectively) were selected for quantitative RT-PCR assays; higher PSP enzyme activity was found in transgenic lines (Figure 1D,E).

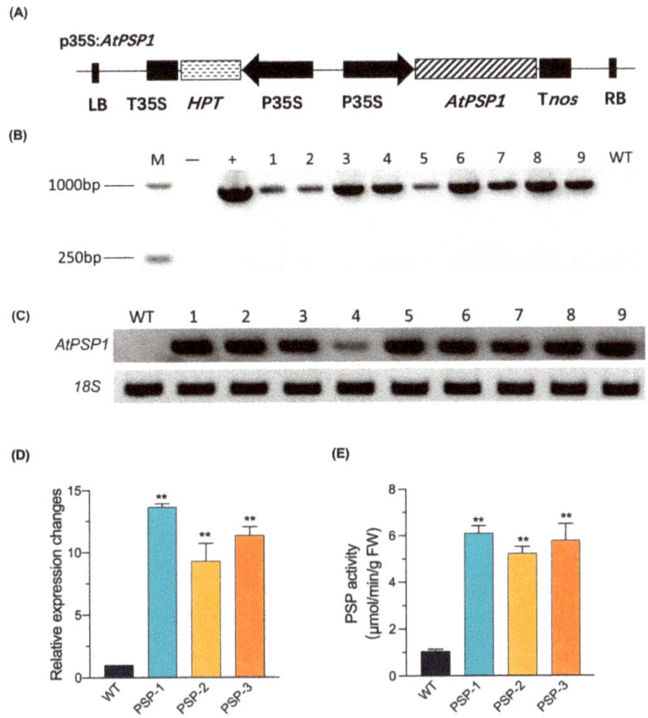

Figure 1. Construction of pCAMBIA1301-*AtPSP1* vector and identified the *AtPSP1* overexpressors lines: (**A**) T-DNA constructs designed for the overexpression of *AtPSP1*. LB, left border; *HPT*, encoding hygromycin-resistant gene; T35S and P35S, terminator and promoter of CaMV35S, respectively; T*nos*, terminator of *NOS*; RB, right border; (**B**) DNA-PCR identification of *AtPSP1* transformation, *AtPSP1* CDs was 888bp; (**C**) semi-quantitative RT-PCR analysis of *AtPSP1* transcripts with *18S* as reference gene; *18S* is *18S* ribosomal RNA of *Lemna turionifera* 5511; (**D**) real-time quantitative PCR analysis of *AtPSP1* transcript with *18S* as reference gene; (**E**) analysis of PSP enzyme activity. "M", DL15000 DNA ladder; "−", water as the PCR temple of negative control; "+", plastid as PCR temple of positive control; WT, wild-type; PSP-1, PSP-2, and PSP-3, the three different *AtPSP1* overexpressing lines. Values given in (**D**,**E**) are the means ± standard error of 3 independent experiments with 3 repeats. The double asterisks (**) indicate significant differences ($p < 0.01$) from controls based on one-way ANOVA.

2.2. Overexpression of AtPSP1 Increased Dry Weight under Sulfur Deficiency

Ten duckweed plants with similar growth conditions were selected from the wild-type or three transgenic lines under normal growth conditions, then cultured in a Datko (medium with the total nutrition) and sulfur-deficient medium. The phenotype of duckweed was observed and photos were taken on day 9. The results showed that there were no morphological differences between the transgenic lines and WT (Figures S1 and S2A–D). The statistical results of relative growth rate, fresh weight, and dry weight at 3d, 6d, 9d, 12d, and 15d also indicate that the overexpression of *AtPSP1* did not show any significant differences compared with the WT in the Datko medium (Figure S2E–G).

The fronds of duckweed only showed the etiolation phenotype after the duckweed was cultured under sulfur-deficient conditions, but there was no significant difference between the transgenic lines and WT (Figures 2A and S3). Although the relative growth rate (RGR) and fresh weight of duckweed gradually increased with treatment, the transgenic lines and WT showed the same growth trend (Figure 2B,C). The dry weight of the transgenic lines was progressively higher than that of the WT after 3 days and significantly higher than the WT at 9d and 12d (Figure 2D). The overexpression of *AtPSP1* in duckweed increased the dry weight under sulfur-deficient conditions.

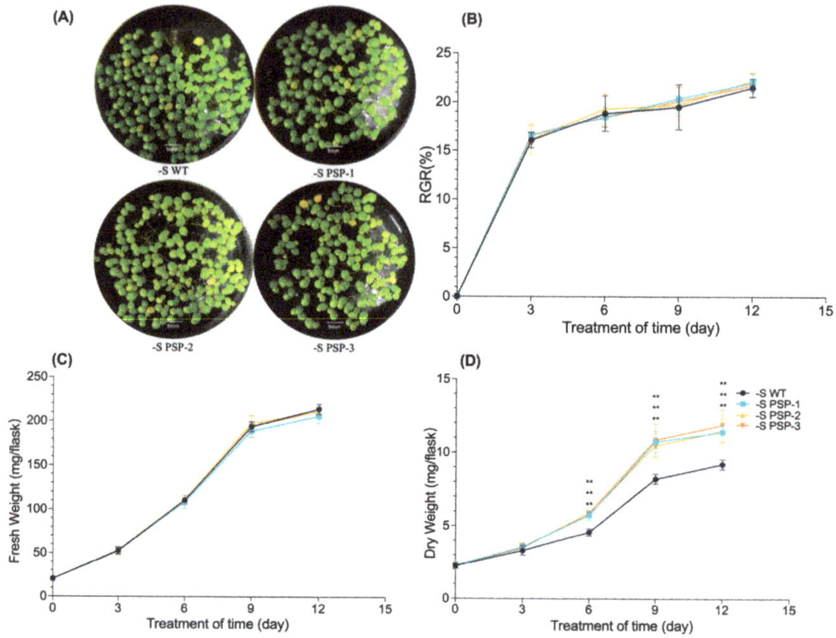

Figure 2. Analysis of phenotype, relative growth rate, fresh weight, and dry weight under sulfur-deficient conditions: (**A**) phenotypes of WT and three *AtPSP1* overexpressors; photos were taken after WT and *AtPSP1* transgenic plants were cultivated under sulfur-deficient conditions for 9 days; (**B**) analysis of relative growth rate (RGR); (**C**) comparison of fresh weight between WT and three *AtPSP1* overexpressors; (**D**) comparison of dry weight between WT and three *AtPSP1* overexpressors. The samples were harvested to analyze the growth rate, fresh weight, and dry weight after WT and *AtPSP1* transgenic plants were cultivated under sulfur-deficient conditions for 3, 6, 9, and 12 days, respectively. Each statistic is the mean ± standard error ($n = 9$, from 3 independent experiments with 3 repeats of each). The double asterisk symbol (**) indicates that the difference ($p < 0.01$), compared with the control, was significant, based on one-way ANOVA.

2.3. The Effect of the Overexpression of AtPSP1 on Photosynthetic Pigment and Chlorophyll Fluorescence Parameters under Sulfur Deficiency

The comparison of photosynthetic pigment content (chlorophyll a, chlorophyll b, total chlorophyll, and carotenoid) at 9d under Datko cultivation showed that there was no significant difference between the transgenic lines and WT (Figure S4A–D). The chlorophyll fluorescence parameters, after 6 days of cultivation in Datko, showed that Fv/Fm (PSII maximum light quantum production), the relative electron transfer rate (rETR), and Y(II) (actual photosynthetic efficiency of PS II) also had no significant differences (Figure S4E–G). Under sulfur-deficient conditions, the content of the photosynthetic pigment after 9 days of treatment and the chlorophyll fluorescence parameters after 6 days of treatment showed different trends in transgenic lines compared with WT (Figure 3). The photosynthetic pigment content (chlorophyll a, chlorophyll b, and total chlorophyll content) of the transgenic lines was significantly higher than that of the WT after 9 days of sulfur-deficient treatment, particularly PSP-2 (Figure 3A–C). The carotenoid content of the transgenic lines was also significantly higher than that of the WT, particularly *PSP-1* and *PSP-2* (Figure 3D). The value of Fv/Fm was significantly higher in the transgenic lines than in the WT. The value of Fv/Fm was the highest in *PSP-2* (Figure 3E). rETR and Y(II) of the transgenic lines were also significantly higher than those of the WT with the increase in light intensity (Figure 3F,G). The results indicated that the photosynthetic performance of the transgenic lines was significantly higher than that of the WT under sulfur-deficient conditions.

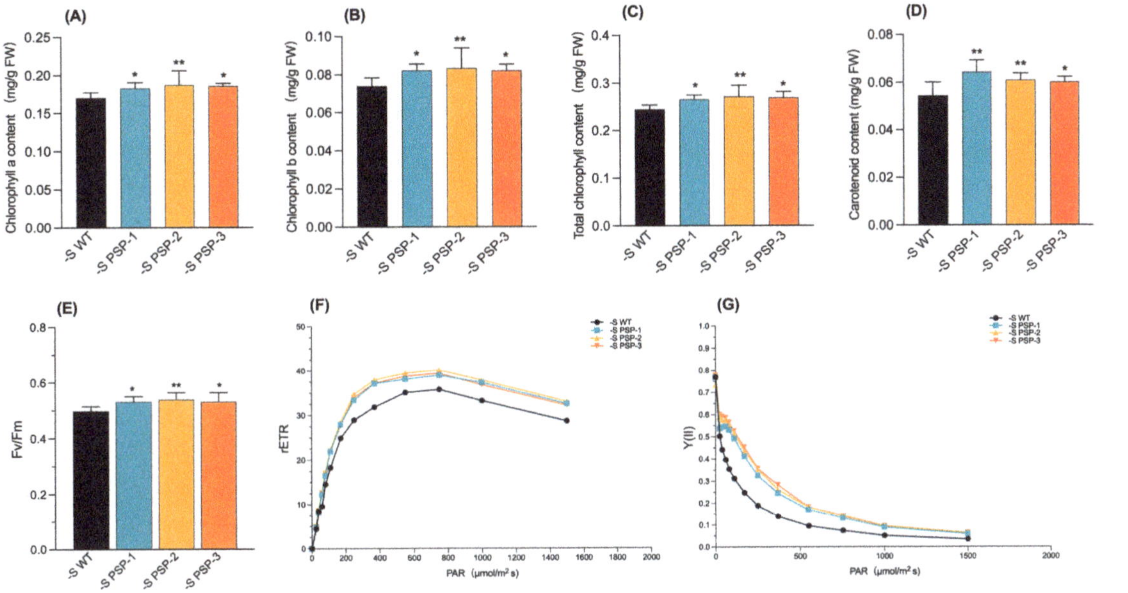

Figure 3. Analysis of photosynthetic pigment and chlorophyll fluorescence parameters in *AtPSP1*-transformed duckweed lines under sulfur-deficient cultivation: (**A**) chlorophyll a content; (**B**) chlorophyll b content; (**C**) total chlorophyll content; (**D**) carotenoid content. The photosynthetic pigment contents were measured after the samples were cultivated under sulfur deficiency for 9 days. (**E**), Fv/Fm; (**F**) rETR; (**G**) Y(II). Chlorophyll fluorescence parameters were measured after the samples were cultivated under sulfur deficiency for 6 days. Each statistic is the mean ± standard error ($n = 9$, from 3 independent experiments with 3 repeats of each, no average values in (**F**,**G**)). The asterisk symbol (*) represents statistically significant differences ($p < 0.05$); the double asterisk symbol (**) indicates that the difference ($p < 0.01$), compared with the control, was significant, based on one-way ANOVA.

2.4. The Effect of the Overexpression of AtPSP1 on Starch, Sugar and Soluble Protein Contents under Sulfur Deficiency

The starch content was measured after treatment for 3, 6, 9, and 12 days in the transgenic lines and WT under Datko medium and sulfur-deficient conditions. The starch content of the duckweed gradually increased with treatment, from about 25 mg/g DW at 3d to 45 mg/g DW at 12d (Figure S5A); the starch yield also gradually increased, from about 0.05 mg/flask at 3d to about 0.65 mg/flask at 12d in the Datko medium (Figure S5B). Both the starch content and yield showed no significant difference between the transgenic lines and WT.

The starch content was compared after the samples were treated for 3, 6, 9, and 12 days under sulfur-deficient conditions. The results showed that the starch content increased from about 35 mg/g DW at 3d to about 200 mg/g DW at 12d, which was significantly higher in the transgenic lines than that of the WT at 6d and 9d in Figure 4A. The starch yield also increased from about 0.1 mg/flask at 3d to more than 2.0 mg/flask at 12d, and the starch yield in the transgenic lines was significantly higher than that of the WT at both 6d and 9d (Figure 4B).

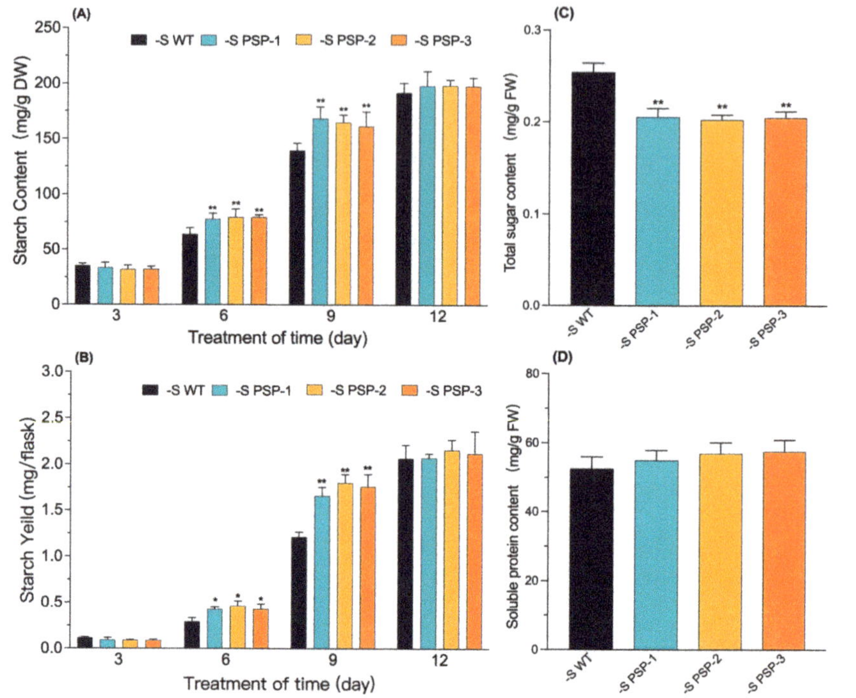

Figure 4. Effect on starch, total sugars, and soluble protein content of *AtPSP1*-transformed duckweed lines under sulfur deficiency, cultured for 9 days: (**A**) the starch content per gram dry weight; (**B**) the starch yield; (**C**) total sugar content; (**D**) soluble protein content. Each statistic is the mean ± standard error (n = 9, from 3 independent experiments with 3 repeats of each). The asterisk symbol (*) represents statistically significant differences ($p < 0.05$); the double asterisk symbol (**) indicates that the difference ($p < 0.01$), compared with the control, was significant based on one-way ANOVA.

The content of the total sugar and soluble protein was determined after 9 days under sulfur-deficient cultivations. The total sugar content was found to be 0.25 mg/g in the WT and 0.2 mg/g in the transgenic lines. The content of total sugar in the transgenic lines was significantly lower than that of the WT (Figure 4C). The soluble protein content was not significantly different between the transgenic lines and WT, with both ranging from 52 to 58 mg/g (Figure 4D). Starch accumulation was increased in the transgenic lines under sulfur-deficient conditions, while the total sugar content was reduced, possibly due to the conversion of sugars to starch.

2.5. The Effect of the Overexpression of AtPSP1 on the Expression of Sulfur Assimilation-Related Genes under Sulfur Deficiency

There are four SULTRs in Arabidopsis: *SULTR1* is responsible for the uptaking of sulfate from the rhizosphere, *LtSULTR2* and *LtSULTR3* play the role of a transition sulfate from root to shoot, and *LtSULTR4* is responsible for the transport of vascular sulfate [29–31]. Four genes were found to encode the sulfate transporter in duckweed: *LtSULTR1*, *LtSULTR2*, *LtSULTR3*, and *LtSULTR4*. The expression of *LtSULTR1*, *LtSULTR2*, *LtSULTR3*, and *LtSULTR4* in the transgenic lines was, on average, 14.5-, 3.2-, 5.65-, and 2.85-fold higher than that of the WT using qRT-PCR, respectively (Figure 5). The expression of *LtAPK1* (adenylyl sulfate kinase) was significantly upregulated 2.6–2.75-fold compared with the WT. The expression of *LtAPR1* (adenylyl sulfate reductase) was significantly higher than that of the WT, by about 3.35-fold; *LtSERAT1* (serine acetyltransferase coding gene) was also significantly upregulated in the transgenic lines, by 1.75–2.0-fold, compared with the WT. The expression of *LtSUOX1* (sulfite oxidase) was significantly upregulated by about 3.0-fold in all three transgenic lines compared with the WT. However, the expression of *LtSIR1* (sulfite reductase) was significantly lower than that of the WT, by about 0.3–0.4-fold; only the expression of *LtATPS1* (ATP sulfurylase) did not significantly differ between the transgenic lines and WT. The sulfur metabolic pathway was shown to be more active in the transgenic lines under sulfur-deficient conditions compared with normal growth conditions (Datko cultured condition).

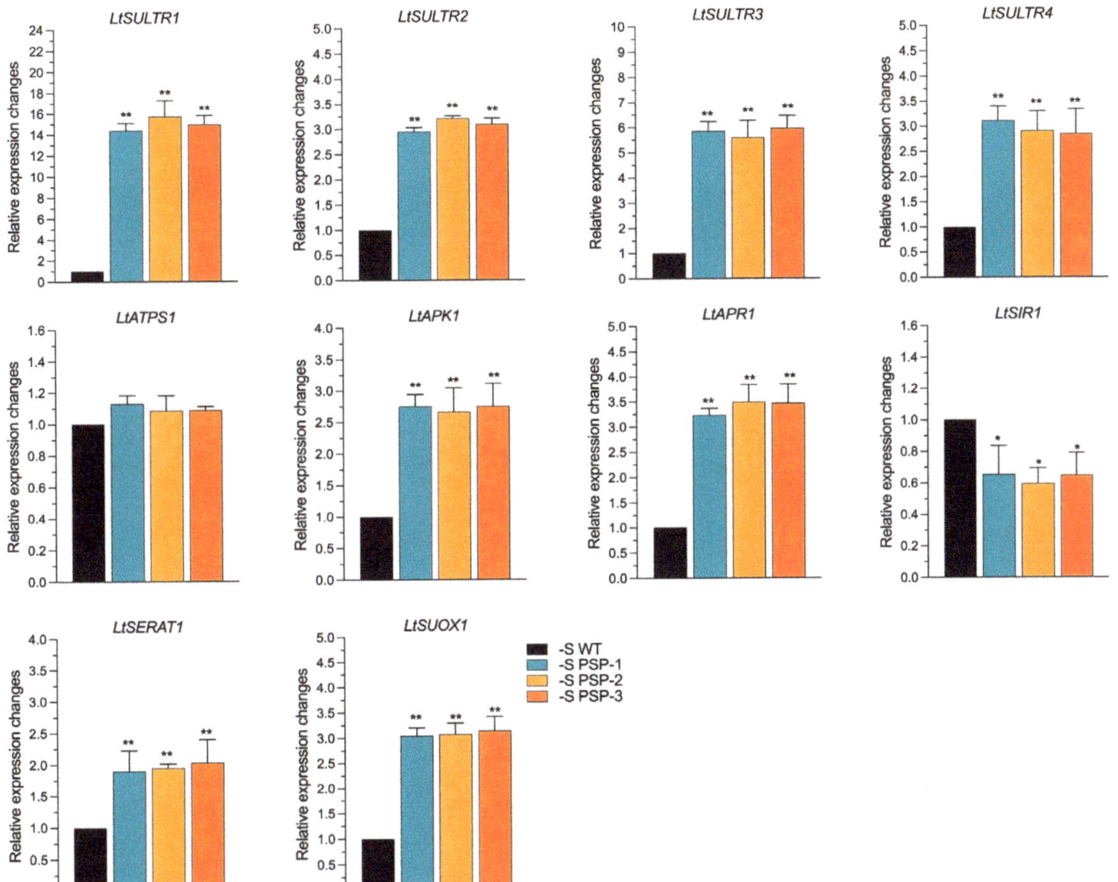

Figure 5. Effect on the expression of sulfur-assimilation-related genes in *AtPSP1*-transformed duckweed lines under sulfur-deficient conditions for 6 days. The relative expression level of the WT was normalized to 1. Each statistic is the mean ± standard error (n = 9, from 3 independent experiments with 3 repeats of each). The asterisk symbol (*) represents statistically significant differences ($p < 0.05$); the double asterisk symbol (**) indicates that the difference ($p < 0.01$), compared with the control, was significant, based on one-way ANOVA.

2.6. The Effect of the Overexpression of AtPSP1 on the Expression of Starch Metabolism-Related Genes under Sulfur Deficiency

The overexpression of *AtPSP1* promoted starch accumulation in duckweed under sulfur-deficient conditions. To study the expression of the genes related to starch metabolism, the transcriptional level of the genes was analyzed using qRT-PCR after sulfur-deficiency treatment for 6 days. The results showed that the expression of the two genes encoding ADP-glucose pyrophosphorylase, *LtAPS1* and *LtAPL1*, was significantly higher than that of the WT, by about 1.65-fold and 1.4-fold, respectively. The expression of *LtSSS1* (soluble starch synthase) was significantly higher in the three transgenic lines than in the WT, by about 4.5-, 5.2-, and 4.6-fold. The expression of *LtGBSS1*, the gene-encoded granule-bound starch synthase, was not significantly different compared with the WT, while the genes involved in the regulation of amylolytic metabolism showed different expression patterns. The expression of *LtISA1* (isoamylase) was significantly lower than that of the WT, by about 0.15-fold; the expression of *Ltα-Amy1*(α-amylase) was not significantly different compared with the WT; and *Ltβ-Amy1*(β-amylase) showed different expression patterns in the three transgenic lines—they were 0.7-fold, 0.85-fold, and 0.8-fold higher in the *PSP-1, PSP-2*, and *PSP-3* lines, respectively, than in the WT (Figure 6). The results showed that the expression of the genes involved in the starch synthesis pathway was upregulated, and the expression of the genes involved in the starch degradation pathway was downregulated.

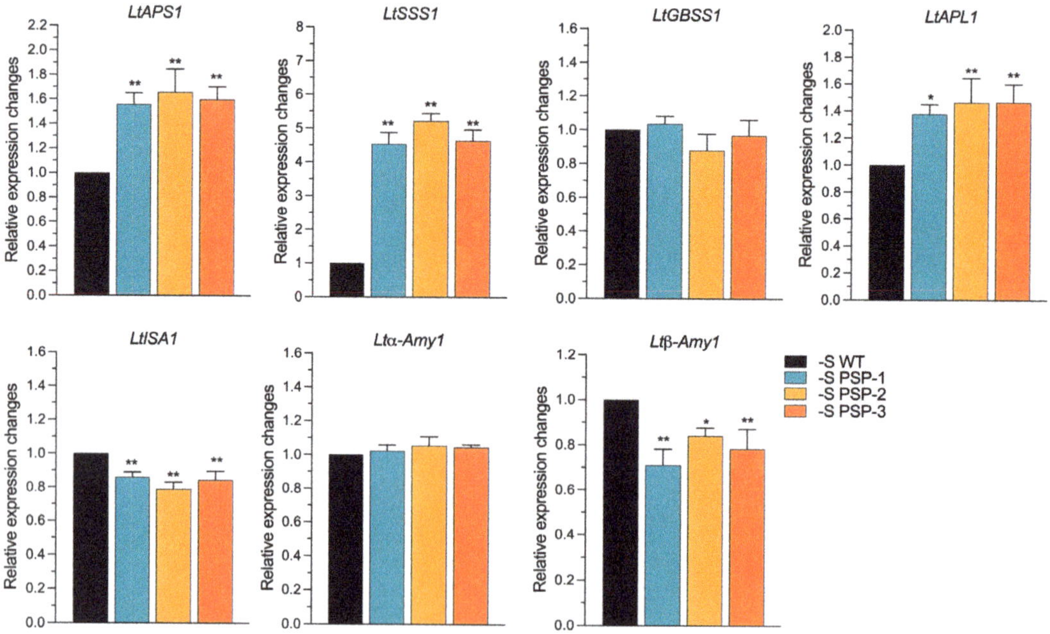

Figure 6. Effect on the expression of starch metabolism-related genes in *AtPSP1*-transformed duckweed lines under sulfur deficiency, cultured for 6 days. The relative expression level of the WT was normalized to 1. Each statistic is the mean ± standard error (n = 9, from three independent experiments with three repeats of each). The asterisk symbol (*) represents statistically significant differences ($p < 0.05$); the double asterisk symbol (**) indicates that the difference ($p < 0.01$), compared with the control, was significant based on one-way ANOVA.

2.7. The Effect of the Overexpression of AtPSP1 on the Expression of Carbon Metabolism-Related Genes under Sulfur Deficiency

The synthesis and degradation of plant starch are closely linked to carbon metabolism [32–34]; we examined the expression of the genes related to carbon metabolism in duckweed after 6 days of treatment under sulfur deficiency. According to the qRT-PCR results shown in Figure 7, the expression of *LtPFK1* (phosphofructokinase) in the glycolytic pathway was not significantly different between the transgenic lines and WT. In contrast, the expression of *LtFBA1* (fructose-diphosphate aldolase), *LtPK1* (pyruvate kinase encoding gene), and *LtPDC1* (pyruvate dehydrogenase complex) was significantly upregulated, by about 3.3-fold, 2.6-fold, and 2.4-fold, respectively, compared with WT. The expression patterns of the genes in the TCA cycle also differed. The expression of *LtACO1* (aconitase) was significantly upregulated in the transgenic lines, by about 2.4 times. The expression of *LtIDH1* (isocitrate dehydrogenase) was significantly higher than the WT, by about 1.2-fold. The expression of *Lt2OG-DH1* (2-OG dehydrogenase) was significantly higher in the three transgenic lines than in the WT, by about 2.6-fold, 2.25-fold, and 2.3-fold, respectively. The expression of *LtSDH1* (succinate dehydrogenase) in the three transgenic lines was, on average, 4.1-fold higher than in the WT. The expression of *LtFUM1* (fumarate) was also significantly higher than that of the WT, by about 3.3-fold, 2.75-fold, and 3.0-fold, in *PSP-1*, *PSP-2*, and *PSP-3*, respectively. The expression of *LtMDH1* (malate dehydrogenase) and *LtCS1* (citrate synthase) was upregulated compared with the WT but did not reach a significant difference. This indicated that the overexpression of *AtPSP1* increased the expression of carbon metabolism-related genes under sulfur-deficient conditions.

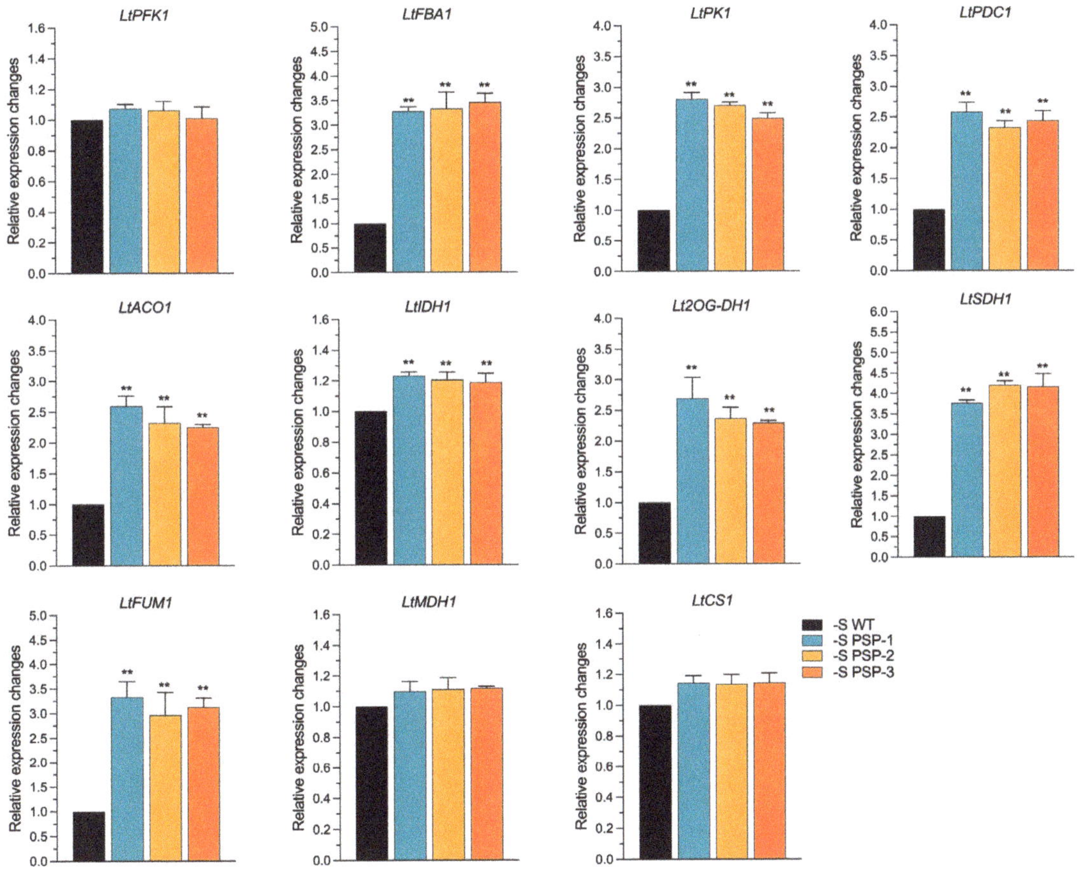

Figure 7. Effect on the expression of carbon metabolism-related genes in *AtPSP1*-transformed duckweed lines under sulfur deficiency, cultured for 6 days. The relative expression level of the WT was normalized to 1. Each statistic is the mean ± standard error ($n = 9$, from 3 independent experiments with 3 repeats of each). The asterisk symbol (*) represents statistically significant differences ($p < 0.05$); the double asterisk symbol (**) indicates that the difference ($p < 0.01$), compared with the control, was significant based on one-way ANOVA.

3. Discussion

3.1. The Overexpression of AtPSP1 Prompts Growth and Starch Accumulation in Duckweed under Sulfur Deficiency

Duckweed is rich in starch under normal growth conditions and can accumulate more starch when its growth is inhibited due to nutrient limitation, hormone treatments, etc. [6–15,35]. In recent years, much progress has been made regarding the mechanisms of starch accumulation in duckweed under nutrition stress conditions. It is believed that there is a tradeoff between starch accumulation and an increase in biomass or dry weight. In this work, the sulfur-deficient incubation did not cause morphological changes in the WT and *AtPSP1* expression lines. There was a reduction in the relative growth rate (RGR), fresh weight, and dry weight, as well as the content of chlorophyll and chlorophyll fluorescence parameters, in both the WT and transgenic duckweed, compared with normal growth conditions (Datko cultivation) (Figures 2 and S2). The above results indicated that sulfur deficiency affected the growth of the duckweed. The results are in accordance with former reports on the morphological observations and growth parameters after duckweed is cultivated under nutrient limitations [20,35–37].

Interestingly, although there were no significant differences in RGR and fresh weight between the transgenic lines and WT, the dry weight was significantly higher than that of the WT after 6 days under sulfur-deficient cultivation. The photosynthetic pigment content was also significantly higher than that of the WT, as well as the chlorophyll fluorescence parameters, indicating that the photosynthetic function was better in the transgenic lines than in the WT. This explains the accumulation of more dry matter, which is similar to the report on *Spirodela polyrhiza* under sulfur-deficient conditions [35]. Sulfur-deficient cultivation promoted starch accumulation in duckweed when compared to Datko cultivation (Figures 4A,B and S5), and the starch content was significantly higher in the transgenic lines than in the WT, indicating that the overexpression of *AtPSP1* could promote starch accumulation under sulfur-deficient conditions. The sugar content was lower than that of the WT in the transgenic duckweed, and there was almost no difference in protein content in both the transgenic duckweed and WT under sulfur limitation (Figure 4). There was a lower content of lipids and cellulose in duckweed. These results demonstrated that the dry weight increase was mainly caused by starch accumulation, which also suggested that there was no tradeoff between the accumulation of starch and dry weight under sulfur deficiency [20].

3.2. The Overexpression of AtPSP1 Coordinates Sulfur Assimilation, Starch Synthesis, and Carbon Metabolism under Sulfur Deficiency

How does the *AtPSP1* transgenic plant prompt starch accumulation under sulfur-deficient conditions? The transcriptional analysis of the genes related to sulfur assimilation showed that the overexpression of *AtPSP1* increased the expression of the genes involved in the process of sulfur assimilation, including the absorption, activation, and reduction of sulfur (Figure 5). The expression of four genes encoding the sulfate transporter, *LtSULTR1*, *LtSULTR2*, *LtSULTR3*, and *LtSULTR4*, was significantly upregulated. It has been reported that the plant could activate sulfate absorption from the rhizosphere by inducing the gene expression of high-affinity sulfate transporters (SULTR1) under sulfur limitations. Meanwhile, the plant could retrieve the sulfate stored in the vacuole by inducing the expression of SULTR4 and activate the translocation of sulfate from root to shoot by inducing the expression of SULTR2 and SULTR3 under sulfur limitation [31,35]. Our results indicated that the transgenic lines accelerated sulfate mobilization in duckweed to maintain the sulfur required for its growth. In addition to this, the increased expression of *LtSUOX1* could probably enhance the oxidation of endogenous sulfite to sulfate in the transgenic lines; the upregulation of the *LtAPR1* expression may increase the run-up of adenosine sulfate (APS) to sulfite; and the upregulation of *LtAPK1* expression presumably prompted the run-up of APS to PAPS (adenosine-5′-phosphoryl sulfate 3′-phosphate). With serine serving as a precursor of Cys biosynthesis, the increased expression of *LtSEART1* accelerated the conversion of serine to Cys (Figure 8). The expression mode of *LtSULTR1* and *LtSULTR4* is similar to the work on *Spirodela polyrhiza* under sulfur deficiency, as they were all upregulated [20], while the expression of the other genes involved in sulfur assimilation was between *Spirodela polyrhiza* and transgenic *AtPSP1* lines. Methionine (Met) and cysteine (Cys), which all contain sulfur, serve as essential compounds for plant growth and reproduction [38]. Serine is a precursor substance for their synthesis, and the mutation of *AtPSP1* was found to alter the sulfur metabolism in Arabidopsis [39]. This suggested that the overexpression of *AtPSP1* may play an important role in the prompting of sulfur assimilation, which could accelerate endogenous sulfur utilization to maintain duckweed growth.

The synthesis of starch is dependent on the genes involved in the starch synthesis pathway and degradation pathway. The expression of genes related to the starch synthesis and degradation pathway was examined under sulfur-deficient conditions (Figure 6). It was found that the expression of *LtAPS1*, *LtSSS1*, and *LtAPL1* was significantly higher in the starch synthesis pathway in the transgenic lines than in the WT, and the expression of *LtISA1* and *Ltβ-Amy1*, involved in the starch degradation pathway, was significantly lower than that of the WT. The expression pattern of the above genes is similar to the patterns found in *Lemna turionifera* 5511 under nitrogen starvation, which suggested that the accumulation of starch was probably a result of the upregulation of starch-synthesis-related genes and the downregulation of starch-degradation-related genes [15]. The synthesis of starch in plants is associated with the breakdown of substances such as sugars. The total sugar content in the transgenic lines was significantly lower than that of the WT under sulfur-deficient conditions (Figure 4C). This demonstrated that the increase in starch content was partly due to the conversion of sugars to starch, as previously found in *Lemna turionifera* 5511 under nitrogen starvation [15].

Plant starch biosynthesis is closely linked to carbon metabolic pathways, and the previous findings suggest that the process of starch accumulation affects the glycolytic and TCA cycle pathways [13]. The expression of the analyzed genes was involved in the glycolytic and TCA cycle pathways under sulfur deficiency (Figure 7). The glucose 6-phosphate is the main substance linking starch synthesis and the glycolytic pathway because the synthesis of starch requires the consumption of more glucose 6-phosphate. The expression of glycolysis-related genes, *LtFBA1*, *LtPK1*, and *LtPDC1*, was found to be significantly upregulated, which would increase the activity of the glycolytic pathway in the transgenic lines to maintain the balance of carbon metabolism in plants, including the accumulation of starch and the increase in the dry matter. The TCA cycle is one of the core pathways in plants; this is a downstream pathway linked to glycolysis and is closely related to the biosynthesis of many amino acids. In this work, the transcriptional level of TCA-cycle-related genes (mainly *LtACO1*, *LtIDH1*, *LtOG-DH1*, and *LtSDH1*) were significantly increased in *AtPSP1*-transformed duckweed under sulfur-deficient conditions, as observed in Arabidopsis when cultivated under a variety of stresses [40,41]. The results demonstrated that the *AtPSP1* overexpression prompted the link between the TCA cycle, carbon metabolism, and sulfur assimilation under sulfur-deficient conditions.

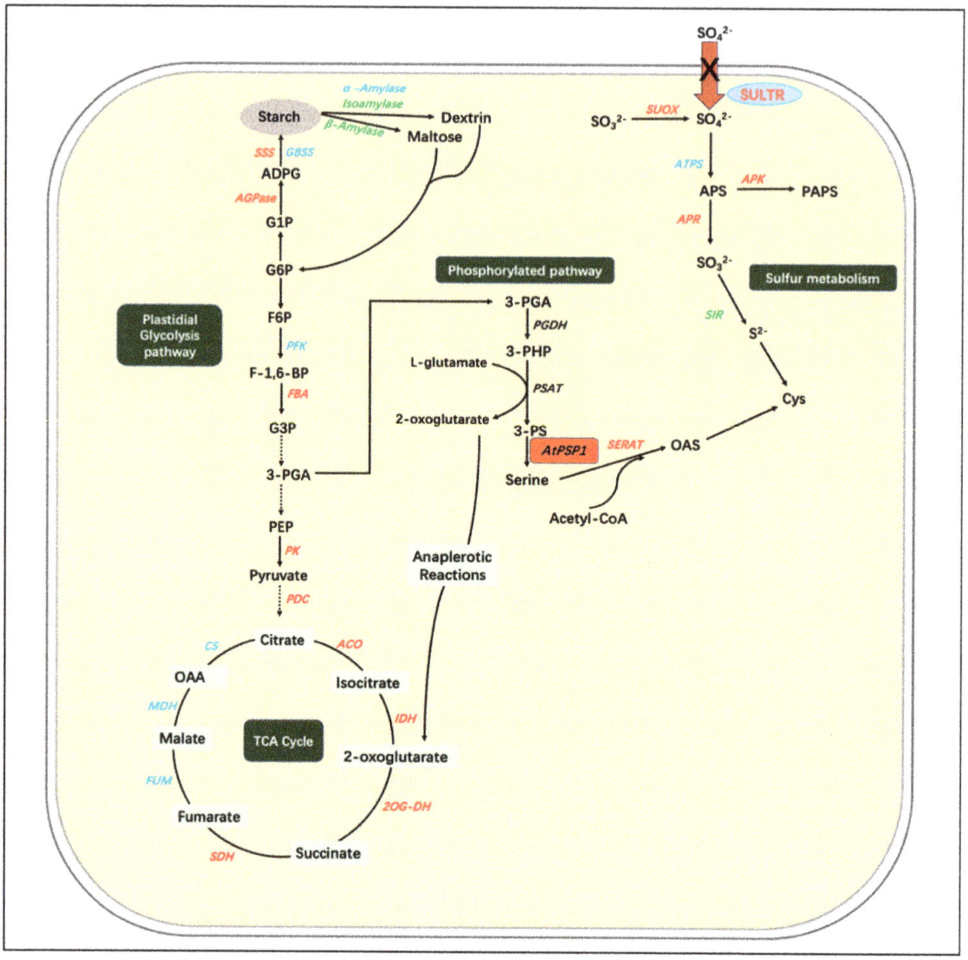

Figure 8. Expression pattern of genes involved in serine phosphorylated synthesis pathway, carbon metabolism, starch metabolism, and sulfur assimilation under sulfur deficiency in transgenic duckweed. The arrows indicate the directions of catalytic reactions or transportations. Red indicates upregulated expression, green indicates downregulated expression, and blue indicates no significant change. Dotted black arrows indicate omitted steps. AGPase, ADP−glucose pyrophosphorylase; ADPG, adenosine-5−diphosphoglucose; GBSS, granule−bound starch synthase; SSS, soluble starch synthase; G1P, glucose 1−phosphate; G6P, glucose 6−phosphate; F−1,6−BP, fructose1,6−bisphosphatase; F6P, fructose 6−phosphate; PFK, phosphofructokinase; FBA, fructose−bisphosphate aldolase; G3P, glyceraldehyde 3−phosphate; 3−PGA, 3−phosphate glycerate; PEP, phosphoenolpyruvate; PK, pyruvate kinase; PDC, pyruvate dehydrogenase complex; OAA, oxaloacetate; 2OG−DH, 2−OG dehydrogenase; ACO, aconitase; CS, citrate synthase; FUM, fumarase; IDH, isocitrate dehydrogenase; MDH, malate dehydrogenase; SDH, succinate dehydrogenase; TCA, tricarboxylic acid; 3−PHP, 3−phosphohydroxypyruvate; 3−PS, 3−phosphoserine; PGDH, 3−phosphoglycerate dehydrogenase; PSAT, 3−phosphoserine aminotransferase; *AtPSP1*, the gene encoding the Arabidopsis 3−phosphoserine phosphatase (PSP); APK, APS kinase; ATPS, ATP sulfurylase; SIR, sulfite reductase; SULTR, sulfate transporter; SUOX, sulfite oxidase; APR, adenylyl−sulfate reductase; SERAT, serine acetyltransferase; OAS, O−acetyl Ser; Cys, Cysteine.

4. Materials and Methods

4.1. Duckweed Culture and Sulfur-Deficiency Treatment

Lemna turionifera 5511 was grown in a sterile Datko medium, according to Yang [42]. Duckweed was cultured under long-day conditions (16 h light/8 h dark cycle) at 22 ± 2 °C and 100 μmol photons $m^{-2} \cdot s^{-1}$ light intensity. Initially, fresh duckweed was grown in the Datko medium for 7 days. Then, 9 plants were transferred to the sulfur-deficient Datko medium and fresh Datko medium in 100 mL conical flasks. Plants were cultivated under the same long-day conditions and temperature, and the horizontal light intensity of the plants was 100 μmol photons $m^{-2} \cdot s^{-1}$.

4.2. Vector Construction and Acquisition of Transgenic Duckweed

The known sequence of *AtPSP1* (the only gene encoding 3-phosphoserine phosphatase) was obtained from Arabidopsis cDNA using PCR. The pCAMBIA1301 plasmid containing the hygromycin resistance gene was inserted, the 5′ end was modified with the CaMV-35S promoter, and the 3′ end was the nopaline synthase gene terminator. The plant binary expression vector p*CAMBIA1301:35S:AtPSP1* was constructed. The vector was transformed into the Agrobacterium tumefaciens strain *EHA105* using the freeze–thaw method. The *AtPSP1* transgenic duckweed was obtained using the method of the callus transformation system described by Yang et al. [43]. DNA was isolated using the SDS method described by Yang et al. [43]. RNA was isolated using an Eastep Super Total RNA Extraction Kit (Promega, Shanghai, China) and reverse transcription to cDNA using a PrimeScriptTMRT reagent Kit with gDNA Eraser (TaKaRa, Beijing, China). The DNA and cDNA for *Lt18s* and *AtPSP1* were amplified using PCR and RT-PCR with specific primers (provided in Supplementary Materials Table S1) to identify the transgenic duckweed.

4.3. Measurement of the PSP Enzyme Activity

PSP activity was determined by measuring the phosphate release of 3-phosphoserine at 25 °C in 0.2 mL reaction mixture containing 80 mM Tris-HCl (pH 7.5), 10 mM 3-phosphoserine, 3 mM $MgCl_2$, and 160 μL of extract start the reaction. After 30 min, 0.25 mL of 15% trichloroacetic acid was added to stop the reaction. The concentration of free phosphate was determined, and one enzyme unit catalyzed the conversion of 1 mmol of substrate/min under specified conditions [27].

4.4. Biomass and Relative Growth Rate Determination

Ten duckweed lines were cultured in a 100 mL conical flask with the culture solution, and those treated for different growing days were pumped to dryness and weighed at fresh weight immediately, and then dried to constant weight in a hot-air-circulating oven at 80 °C and weighed as dry weight of the flask-cultured materials.

Duckweed's relative growth rate was calculated from the changes in fresh weight, and the 10 duckweed lines were weighed at the beginning growth day (d_0 = 0 d) and at the end (d_x = 3, 6, 9, and 12 d) of the experimental period. The relative growth rate (RGR) was calculated according to Su [44].

$$\text{RGR} = (\ln Nj - \ln Ni)/\Delta t \quad (1)$$

ln represents the natural logarithm, Nj and Ni represent the fresh weight at time j and time i, respectively, and Δt is the growth day instead of $d_x - d_0$.

4.5. Determination of Photosynthetic Pigments and Chlorophyll Fluorescence Parameters

A total of 0.1 g of duckweed material was treated for 9 days in 1 mL of 95% ethanol and placed in an incubator at 28 °C until the duckweed turned completely white. After taking 200 μL of extract, the OD values at 663 nm, 645 nm, and 440 nm were measured under a plate reader, and the content of photosynthetic pigments was calculated according to the Arnon method [45]. Chlorophyll fluorescence parameters were determined using Mini-

PAM-II (WALZ, Effeltrith, Germany). The values of Fv/Fm, rETR, and Y(II) of photosystem II were measured using the WALZ protocol.

4.6. Starch, Total Sugar, and Soluble Protein Determination

A starch extraction and measurement kit was used to extract and measure starch content according to the manufacturer's protocol (BC0700, Solarbio Biological and Technology Co., Ltd., Beijing, China). The total sugar was extracted and measured using a total sugar extraction and measurement kit according to the manufacturer's protocol (BC2715, Solarbio Biological and Technology Co., Ltd., Beijing, China). The soluble protein content extraction was measured by using 30 mM Tris-HCl (pH 7.0) according to the manufacturer's protocol (P0006C, Beyotime Biotechnology Co., Ltd., Shanghai, China).

4.7. DNA and RNA Extraction and qRT-PCR

RNA was isolated using an Eastep Super Total RNA Extraction Kit (Promega, Shanghai, China). After reverse transcription, cDNAs for *Lt18s*, *AtPSP1*, starch metabolism genes, sulfur metabolism genes, and carbon metabolism genes were amplified using qRT-PCR and the specific primers designed from the duckweed genome sequence obtained through RNA-Seq (in Supplementary File Table S1). qRT-PCR was performed on an iCycler thermal cycler (Bio-Rad iQ5, Hercules, CA, USA) using TB Green Premix Ex TaqII (code RR420A: TB Green Premix Ex Taq TM, Dalian Bao, Dalian, China) according to the manufacturer's protocol. The reaction mixture was heated to 95 °C for 30 s, and then 40 PCR cycles were carried out, at 95 °C for 5 s, 58 °C for 30 s, and 72 °C for 30 s. After normalizing the data to *Lt18s*, differences in the relative expression levels of the detected genes were calculated using the $2^{-\Delta\Delta CT}$ method. All values are shown as the mean ± standard error of the mean using at least three biological replicates.

4.8. Data Analysis

All data were measured in at least three biological replicates, and the experiments were repeated at least three times independently. Data were organized using Excel, statistically analyzed using SPSS 22.0 (IBM, Chicago, IL, USA), analyzed for significance using one-way ANOVA (* $p < 0.05$ and ** $p < 0.01$), and graphed using GraphPad 9.2 software (GraphPad Software, San Diego, CA, USA).

5. Conclusions

Duckweed is a promising feedstock for bioenergy. We found that the overexpression of *AtPSP1*, the last key enzyme of PPSB, can increase the dry weight and starch accumulation in duckweed, accompanied by a higher photosynthetic capacity under sulfur-deficient conditions. The transcriptional analysis demonstrated that better growth and starch accumulation resulted from coordinating the expression of the key enzymes involved in sulfur assimilation, starch synthesis, and metabolism, as well as the glycolytic pathways and TCA cycle in *AtPSP1*-overexpressing lines under sulfur deficiency. This study provides a prospective pathway for starch accumulation through genetic engineering and sheds light on the potential mechanism of PPSB-promoting starch accumulation under sulfur-deficient conditions.

Supplementary Materials: The following supporting information can be downloaded at: https://www.mdpi.com/article/10.3390/plants12051012/s1, Figure S1. Phenotype of WT and three *AtPSP1* transgenic lines. The photos were taken after WT and *AtPSP1* transgenic lines were cultivated after full-nutrition condition (Datko) for 9 days; Figure S2. Analysis of phenotype, relative growth rate (RGR), fresh weight, and dry weight in WT and three *AtPSP1* transgenic lines under Datko condition; Figure S3. Phenotype of WT and three *AtPSP1* transgenic lines under sulfur-deficient conditions; Figure S4. The analysis of photosynthetic pigment and chlorophyll fluorescence parameters in duckweed under full-nutrition condition (Datko); Figure S5. The content of starch, yield, and soluble protein in WT and three AtPSP1 transgenic lines under full-nutrition condition in different treatment times; Table S1. The primer sequences in the study.

Author Contributions: L.W. performed most of the experiments and analyzed the data; Y.K., S.Z., Y.T., Y.Z. and Y.W. were involved in the experiments; Y.W. and Y.Z. designed the experiments; L.W. wrote the manuscript. All authors have read and agreed to the published version of the manuscript.

Funding: This work was supported by the Natural Science Foundation of China (NSFC, No. 31970365) and Tianjin Academy of Agricultural Sciences Youth Science and Technology Innovation General Project (No.2022006).

Institutional Review Board Statement: Not applicable.

Informed Consent Statement: Not applicable.

Data Availability Statement: Not applicable.

Acknowledgments: The authors would like to thank Xue-Qin Shi from Nankai University College of Life Science for providing technical support in Chlorophyll Fluorometer (Mini-PAM-II) acquisition.

Conflicts of Interest: The authors declare no conflict of interest.

References

1. Les, D.H.; Crawford, D.J.; Landolt, E.; Gabel, J.D.; Rettig, J.H. Phylogeny and Systematics of Lemnaceae, the Duckweed Family. *Syst. Bot.* **2002**, *27*, 221–240. [CrossRef]
2. Stomp, A.M. The duckweeds: A valuable plant for biomanufacturing. *Biotechnol. Annu. Rev.* **2005**, *11*, 69–99. [CrossRef]
3. Bog, M.; Appenroth, K.J.; Sree, K.S. Key to the determination of taxa of Lemnaceae: An update. *Nord. J. Bot.* **2020**, *38*. [CrossRef]
4. Chen, G.; Zhao, K.; Li, W.; Yan, B.; Yu, Y.; Li, J.; Zhang, Y.; Xia, S.; Cheng, Z.; Lin, F.; et al. A review on bioenergy production from duckweed. *Biomass Bioenergy* **2022**, *161*, 1064–1068. [CrossRef]
5. Cui, W.; Cheng, J.J.; Appenroth, K. Growing duckweed for biofuel production: A review. *Plant Biol.* **2015**, *17*, 16–23. [CrossRef]
6. Zhu, Y.; Li, X.; Gao, X.; Sun, J.; Ji, X.; Feng, G.; Shen, G.; Xiang, B.; Wang, Y. Molecular mechanism underlying the effect of maleic hydrazide treatment on starch accumulation in *S. polyrrhiza* 7498 fronds. *Biotechnol. Biofuels* **2021**, *14*, 99. [CrossRef]
7. Wang, X.; Cui, W.; Hu, W.; Feng, C. Abscisic acid-enhanced starch accumulation of bioenergy crop duckweed (*Spirodela polyrrhiza*). *RSC Adv.* **2020**, *10*, 10394–10401. [CrossRef]
8. Faizal, A.; Putra, R. Uniconazole increases starch content in duckweed (*Lemna aequinoctialis* Welw). *3bio J. Biol. Sci. Technol. Manag.* **2019**, *1*, 1. [CrossRef]
9. Kurepa, J.; Shull, T.E.; Smalle, J.A. Cytokinin-induced growth in the duckweeds Lemna gibba and *Spirodela polyrhiza*. *Plant Growth Regul.* **2018**, *86*, 477–486. [CrossRef]
10. de Morais, M.B.; Barbosa-Neto, A.G.; Willadino, L.; Ulisses, C.; Calsa Junior, T. Salt Stress Induces Increase in Starch Accumulation in Duckweed (*Lemna aequinoctialis*, Lemnaceae): Biochemical and Physiological Aspects. *J. Plant Growth Regul.* **2019**, *38*, 683–700. [CrossRef]
11. Chen, D.; Zhang, H.; Wang, Q.; Shao, M.; Li, X.; Chen, D.; Zeng, R.; Song, Y. Intraspecific variations in cadmium tolerance and phytoaccumulation in giant duckweed (*Spirodela polyrhiza*). *J. Hazard. Mater.* **2020**, *395*, 122672. [CrossRef]
12. Shao, J.; Liu, Z.; Ding, Y.; Wang, J.; Li, X.; Yang, Y. Biosynthesis of the starch is improved by the supplement of nickel (Ni^{2+}) in duckweed (*Landoltia punctata*). *J. Plant Res.* **2020**, *133*, 587–596. [CrossRef]
13. Yu, C.; Xiaowen, Z.; Qi, G.; Bai, Z.; Wang, Y.; Wang, S.; Ma, Y.; Liu, Q.; Hu, R.; Zhou, G. Integrated analysis of transcriptome and metabolites reveals an essential role of metabolic flux in starch accumulation under nitrogen starvation in duckweed. *Biotechnol. Biofuels* **2017**, *10*, 167. [CrossRef]
14. Guo, L.; Jin, Y.; Xiao, Y.; Tan, L.; Xueping, T.; Ding, Y.; He, K.-Z.; Du, A.; Li, J.; Yi, Z.; et al. Energy-efficient and environmentally friendly production of starch-rich duckweed biomass using nitrogen-limited cultivation. *J. Clean. Prod.* **2019**, *251*, 119726. [CrossRef]
15. Wang, L.; Li, S.; Sun, L.; Tong, Y.; Yang, L.; Zhu, Y.; Wang, Y. Over-Expression of Phosphoserine Aminotransferase-Encoding Gene (AtPSAT1) Prompts Starch Accumulation in *L. turionifera* under Nitrogen Starvation. *Int. J. Mol. Sci.* **2022**, *23*, 11563. [CrossRef]
16. Huang, M.; Fang, Y.; Liu, Y.; Jin, Y.; Sun, J.; Tao, X.; Ma, X.; He, K.; Zhao, H. Using proteomic analysis to investigate uniconazole-induced phytohormone variation and starch accumulation in duckweed (*Landoltia punctata*). *BMC Biotechnol.* **2015**, *15*, 81. [CrossRef]
17. Sree, K.S.; Adelmann, K.; Garcia, C.; Lam, E.; Appenroth, K.-J. Natural variance in salt tolerance and induction of starch accumulation in duckweeds. *Planta* **2015**, *241*, 1395–1404. [CrossRef]
18. Wang, W.; Messing, J. Analysis of ADP-glucose pyrophosphorylase expression during turion formation induced by abscisic acid in *Spirodela polyrhiza* (greater duckweed). *BMC Plant Biol.* **2012**, *12*, 5. [CrossRef]
19. Guo, L.; Ding, Y.; Xu, Y.; Li, Z.; Jin, Y.; He, K.; Fang, Y.; Zhao, H. Responses of *Landoltia punctata* to cobalt and nickel: Removal, growth, photosynthesis, antioxidant system and starch metabolism. *Aquat. Toxicol.* **2017**, *190*, 87–93. [CrossRef]
20. Sun, Z.; Guo, W.; Zhao, X.; Chen, Y.; Yang, J.; Xu, S.; Hou, H. Sulfur limitation boosts more starch accumulation than nitrogen or phosphorus limitation in duckweed (*Spirodela polyrhiza*). *Ind. Crops Prod.* **2022**, *185*, 115098. [CrossRef]
21. Bauwe, H.; Hagemann, M.; Fernie, A. Photorespiration: Players, partners and origin. *Trends Plant Sci.* **2010**, *15*, 330–336. [CrossRef]

22. Douce, R.; Bourguignon, J.; Neuburger, M.; Rébeillé, F. The glycine decarboxylase system: A fascinating complex. *Trends Plant Sci.* **2001**, *6*, 167–176. [CrossRef]
23. Kleczkowski, L.; Givan, C. Serine Formation in Leaves by Mechanisms other than the Glycolate Pathway. *J. Plant Physiol.* **1988**, *132*, 641–652. [CrossRef]
24. Ros, R.; Muñoz-Bertomeu, J.; Krueger, S. Serine in plants: Biosynthesis, metabolism, and functions. *Trends Plant Sci.* **2014**, *19*, 564–569. [CrossRef]
25. Benstein, R.; Ludewig, K.; Wulfert, S.; Wittek, S.; Gigolashvili, T.; Frerigmann, H.; Gierth, M.; Flügge, U.-I.; Krueger, S. Arabidopsis Phosphoglycerate Dehydrogenase1 of the Phosphoserine Pathway Is Essential for Development and Required for Ammonium Assimilation and Tryptophan Biosynthesis. *Plant Cell* **2013**, *25*, 5011–5029. [CrossRef]
26. Flores-Tornero, M.; Anoman, A.; Rosa Téllez, S.; Ros, R. Lack of phosphoserine phosphatase activity alters pollen and tapetum development in Arabidopsis thaliana. *Plant Sci.* **2015**, *235*, 81–88. [CrossRef]
27. Han, X.; Peng, K.; Wu, H.; Song, S.; Zhu, Y.; Bai, Y.; Wang, Y. Overexpression of PSP1 enhances growth of transgenic Arabidopsis plants under ambient air conditions. *Plant Mol. Biol.* **2017**, *94*, 419–431. [CrossRef]
28. Ros, R.; Cascales-Miñana, B.; Segura, J.; Anoman, A.; Toujani, W.; Flores-Tornero, M.; Rosa Téllez, S.; Muñoz-Bertomeu, J. Serine biosynthesis by photorespiratory and non-photorespiratory pathways: An interesting interplay with unknown regulatory networks. *Plant Biol.* **2012**, *15*, 707–712. [CrossRef]
29. Takahashi, H.; Kopriva, S.; Giordano, M.; Saito, K.; Hell, R. Sulfur Assimilation in Photosynthetic Organisms: Molecular Functions and Regulations of Transporters and Assimilatory Enzymes. *Annu. Rev. Plant Biol.* **2010**, *62*, 157–184. [CrossRef]
30. Rennenberg, H.; Herschbach, P.D.C. A detailed view on sulphur metabolism at the cellular and whole-plant level illustrates challenges in metabolite flux analyses. *J. Exp. Bot.* **2014**, *65*, 5711–5724. [CrossRef]
31. Kopriva, S.; Rennenberg, H. Control of sulphate assimilation and glutathione synthesis: Interaction with N and C metabolism. *J. Exp. Bot.* **2004**, *55*, 1831–1842. [CrossRef]
32. Smith, A.M.; Zeeman, S.C. Starch: A Flexible, Adaptable Carbon Store Coupled to Plant Growth. *Annu. Rev. Plant Biol.* **2020**, *71*, 217–245. [CrossRef]
33. Streb, S.; Zeeman, S. Starch Metabolism in Arabidopsis. *Arab. Book Am. Soc. Plant Biol.* **2012**, *10*, e0160. [CrossRef]
34. Smith, A.M. Starch in the Arabidopsis plant. *Starch—Stärke* **2012**, *64*, 421–434. [CrossRef]
35. Sun, Z.; Guo, W.; Zhao, X.; Yang, J.; Duan, P.; Xu, S.; Hou, H. Sulfur limitation increases duckweed starch accumulation without compromising growth. *bioRxiv* **2021**, *185*, 115098. [CrossRef]
36. Tao, X.; Fang, Y.; Xiao, Y.; Jin, Y.-L.; Ma, X.-R.; Zhao, Y.; He, K.-Z.; Zhao, H.; Wang, H.-Y. Comparative transcriptome analysis to investigate the high starch accumulation of duckweed (*Landoltia punctata*) under nutrient starvation. *Biotechnol. Biofuels* **2013**, *6*, 72. [CrossRef]
37. Liu, Y.; Wang, X.; Fang, Y.; Huang, M.; Chen, X.; Zhang, Y.; Zhao, H. The effects of photoperiod and nutrition on duckweed (*Landoltia punctata*) growth and starch accumulation. *Ind. Crops Prod.* **2018**, *115*, 243–249. [CrossRef]
38. Maruyama-Nakashita, A. Metabolic changes sustain the plant life in low-sulfur environments. *Curr. Opin. Plant Biol.* **2017**, *39*, 144–151. [CrossRef]
39. Samuilov, S.; Rademacher, N.; Brilhaus, D.; Kurz, S.; Arab, L.; Kopriva, S.; Weber, A.; Mettler-Altmann, T.; Rennenberg, H. Knock-Down of the Phosphoserine Phosphatase Gene Effects Rather N- Than S-Metabolism in *Arabidopsis thaliana*. *Front. Plant Sci.* **2018**, *9*, 1830. [CrossRef]
40. Niehaus, T.D. Phosphatases are involved in modulating the TCA cycle in plants. *Mol. Plant* **2021**, *14*, 1036–1037. [CrossRef]
41. Lehmann, M.; Schwarzländer, M.; Obata, T.; Sirikantaramas, S.; Burow, M.; Olsen, C.E.; Tohge, T.; Fricker, M.D.; Møller, B.L.; Fernie, A.R.; et al. The Metabolic Response of Arabidopsis Roots to Oxidative Stress is Distinct from that of Heterotrophic Cells in Culture and Highlights a Complex Relationship between the Levels of Transcripts, Metabolites, and Flux. *Mol. Plant* **2009**, *2*, 390–406. [CrossRef]
42. Yang, L.; Han, Y.; Wu, D.; Yong, W.; Liu, M.; Wang, S.; Liu, W.; Lu, M.; Wei, Y.; Sun, J. Salt and cadmium stress tolerance caused by overexpression of the Glycine Max Na^+/H^+ Antiporter (*GmNHX1*) gene in duckweed (*Lemna turionifera* 5511). *Aquat. Toxicol.* **2017**, *192*, 127–135. [CrossRef]
43. Yang, L.; Han, H.; Liu, M.; Zuo, Z.; Zhou, K.; Lü, J.; Zhu, Y.; Bai, Y.; Wang, Y. Overexpression of the Arabidopsis photorespiratory pathway gene, serine: Glyoxylate aminotransferase (*AtAGT1*), leads to salt stress tolerance in transgenic duckweed (*Lemna minor*). *Plant Cell Tissue Organ Cult. (PCTOC)* **2013**, *113*, 407–416. [CrossRef]
44. Su, C.; Jiang, Y.; Yang, Y.; Zhang, W.; Xu, Q. Responses of duckweed (*Lemna minor* L.) to aluminum stress: Physiological and proteomics analyses. *Ecotoxicol. Environ. Saf.* **2019**, *170*, 127–140. [CrossRef]
45. Arnon, D.I. Copper Enzymes in Isolated Chloroplasts. Polyphenoloxidase in Beta Vulgaris. *Plant Physiol.* **1949**, *24*, 1–15. [CrossRef]

plants

Article

Optimization of Molecular Methods for Detecting Duckweed-Associated Bacteria

Kenneth Acosta [1], Shawn Sorrels [1], William Chrisler [2], Weijuan Huang [3], Sarah Gilbert [1], Thomas Brinkman [1], Todd P. Michael [4], Sarah L. Lebeis [5,6,7] and Eric Lam [1,*]

1 Department of Plant Biology, Rutgers the State University of New Jersey, New Brunswick, NJ 08901, USA
2 Environmental Molecular Sciences Laboratory (EMSL), Pacific Northwest National Laboratory (PNNL), Richland, WA 99354, USA
3 Institute of Nanfan & Seed Industry, Guangdong Academy of Sciences, Guangzhou 510316, China
4 The Plant Molecular and Cellular Biology Laboratory, The Salk Institute for Biological Studies, La Jolla, CA 92037, USA
5 Department of Plant, Soil and Microbial Sciences, Michigan State University, East Lansing, MI 48824, USA
6 Department of Microbiology and Molecular Genetics, Michigan State University, East Lansing, MI 48824, USA
7 Plant Resilience Institute, Michigan State University, East Lansing, MI 48824, USA
* Correspondence: eric.lam@rutgers.edu

Abstract: The bacterial colonization dynamics of plants can differ between phylogenetically similar bacterial strains and in the context of complex bacterial communities. Quantitative methods that can resolve closely related bacteria within complex communities can lead to a better understanding of plant–microbe interactions. However, current methods often lack the specificity to differentiate phylogenetically similar bacterial strains. In this study, we describe molecular strategies to study duckweed–associated bacteria. We first systematically optimized a bead-beating protocol to co-isolate nucleic acids simultaneously from duckweed and bacteria. We then developed a generic fingerprinting assay to detect bacteria present in duckweed samples. To detect specific duckweed–bacterium associations, we developed a genomics-based computational pipeline to generate bacterial strain-specific primers. These strain-specific primers differentiated bacterial strains from the same genus and enabled the detection of specific duckweed–bacterium associations present in a community context. Moreover, we used these strain-specific primers to quantify the bacterial colonization of duckweed by normalization to a plant reference gene and revealed differences in colonization levels between strains from the same genus. Lastly, confocal microscopy of inoculated duckweed further supported our PCR results and showed bacterial colonization of the duckweed root–frond interface and root interior. The molecular methods introduced in this work should enable the tracking and quantification of specific plant-microbe associations within plant-microbial communities.

Keywords: plant-microbe interactions; plant-bacteria associations; bacterial colonization; duckweed; duckweed-associated bacteria; RISA; *Azospirillum brasilense* Sp7; *Azospirillum brasilense* Sp245; strain-specific primers; bead-beating

check for updates

Citation: Acosta, K.; Sorrels, S.; Chrisler, W.; Huang, W.; Gilbert, S.; Brinkman, T.; Michael, T.P.; Lebeis, S.L.; Lam, E. Optimization of Molecular Methods for Detecting Duckweed-Associated Bacteria. *Plants* **2023**, *12*, 872. https://doi.org/10.3390/plants12040872

Academic Editor: Viktor Oláh

Received: 31 December 2022
Revised: 5 February 2023
Accepted: 8 February 2023
Published: 15 February 2023

1. Introduction

Lemnaceae, commonly known as duckweed, is a family of freshwater aquatic plants [1]. Their small size, short doubling time, growth habitat, and reduced morphology put forth duckweed as a model system to study plant–microbe interactions. Indeed, many similarities can be found between the structuring of duckweed-associated bacterial (DAB) communities and terrestrial plant bacterial communities. For example, both terrestrial plants and duckweed host distinct bacterial communities when compared to the host environment, demonstrating that selection strongly shapes the bacterial communities of both terrestrial plants and duckweed [2–5]. Moreover, similar bacterial taxa are found among terrestrial plant bacterial communities and DAB communities, suggesting bacterial adaptation to these plant habitats [3]. Therefore, studying duckweed–bacteria interactions may help reveal conserved mechanisms involved in plant–bacteria interactions.

The study of plant–bacteria interactions is complicated by many factors. One factor is the functional diversity found among phylogenetically similar bacteria associated with plants. Despite their similar phylogeny, these related bacteria can interact differently with plant hosts and may serve diverse roles in plant microbial communities. Community surveys of plant bacterial communities show that bacteria of the same genus can have different colonization dynamics in different plant tissues and developmental stages [6]. Other community surveys show bacteria of the same family can have distinct plant host preferences [7]. In support of these community surveys, functional studies show bacterial strains from the same genus can colonize plants at different concentrations and protect against disease to different degrees [8,9]. Another factor that adds complexity to plant–microbe interactions is the presence of microbe–microbe interactions within microbial communities [10]. Bacteria may readily colonize plants when no other microbes are present, but could fail to stably colonize plants in a community context [11]. The presence of microbe–microbe interactions in microbial communities is a major reason why many bacteria display plant-growth-promoting behavior in the laboratory in mono-association studies but not in field trials when natural microbial communities are present [12]. Thus, differentiating phylogenetically similar bacteria in diverse contexts will be important to unravel the complexity of plant–bacteria interactions.

Different methods can be used to detect plant-bacteria associations, including culture-dependent methods, microscopy, and molecular approaches [13,14]. Together, these methods differ in the information they provide and in the context in which they can be applied. The colony-forming units (CFU) assay is a culture-dependent method used to quantify live bacteria. In the context of plant-bacteria interactions, this method has typically been used to quantify the colonization of plants by single bacterial isolates [15–18], but with the implementation of selective culture media, members in a small community of plants and bacteria can also be distinguished [19]. Still, this method can be laborious, imprecise, and cannot be used to quantify bacteria found in complex microbial communities [14]. In contrast to the CFU assay, microscopy is a qualitative approach used to observe the spatial and temporal colonization dynamics of bacteria on plants [20]. Its application has revealed the presence of colonization hotspots on plants and different colonization patterns between bacteria when applied individually to plant tissues [21–24]. However, microscopy techniques commonly use generic fluorescent stains or oligonucleotide probes that cannot detect specific interactions within a bacterial community [13,25,26]. An alternative microscopy approach involves labeling and monitoring a bacterial strain of interest with an in vivo reporter gene, such as GFP or GUS, but this application can also be laborious and is dependent on the transformability of the bacterium of interest [27–29]. Thus, the CFU assay and microscopy methods are often used to study bipartite plant–bacterium interactions, since they lack the specificity required to study the interactions between plants and specific members in complex bacterial communities.

The most common method to detect bacteria in complex communities is 16S rDNA amplicon sequencing, in which variable regions of the 16S rRNA gene are selectively amplified and commonly sequenced by short-read sequencing technologies [30–32]. Initially, 16S rDNA amplicon sequencing provided the relative abundance of bacteria within communities, but recent innovations allow for the absolute abundance of community members to be quantified [33–35]. Other recent innovations include the application of long-read sequencing technologies to sequence full-length 16S rRNA gene sequences, since full-length sequences demonstrate higher resolution than partial sequences [36–38]. Initially, the high error rate of long-read sequencing technologies limited their use in 16S rDNA amplicon sequencing, but this error rate has diminished with the implementation of circular consensus sequencing and denoising algorithms tailored to long-read technologies [37,39]. Despite these innovations, this molecular approach is still limited by the extent of polymorphisms in the 16S rRNA gene, which can distinguish between bacterial families and genera but often lacks resolution between strains of the same bacterial species and between some closely related bacterial species [40–42]. This is also exemplified in community surveys using full-length 16S rRNA gene sequences, where some reads cannot be assigned at the species level and some closely related bacterial species cannot be resolved [37,38]. In addition to low resolution, some bacterial taxa contain multiple non-identical copies of the 16S rRNA gene, further complicating the differentiation of closely related bacteria with this molecular approach [43,44]. For these reasons, a different approach, not reliant on the sequencing of the 16S rRNA gene, is needed to differentiate phylogenetically similar bacteria in community scenarios.

One molecular approach that can track and quantify specific bacterial isolates of interest within complex communities is real-time PCR (qPCR) [13,14,45]. In this context, strain-specific primers are designed for qPCR using unique sequences from the bacterium of interest. Initially, sequences from the 16S rRNA gene or sequence-characterized amplified regions were used to design strain-specific primers [46,47], but with current technologies entire bacterial genomes can be examined for strain-specific sequences [48–50]. As a result, the qPCR approach can provide high specificity and sensitivity, demonstrated by its ability to detect specific bacterial strains in non-sterile soils [51–53]. However, results from this molecular method are dependent on the efficiency of the method used to extract DNA and the resulting sample quality. Furthermore, reported strategies to design primers either involve custom protocols [51,53] or fragmenting bacterial genomes into smaller sequences to search against public databases [48,49]. These approaches can be inefficient if strain-specific sequences are desired for many bacteria. Therefore, a broadly applicable and efficient strategy for designing strain-specific primers is preferred to fully realize the potential of qPCR for studying specific plant–bacterium associations.

In this study, we optimized and applied molecular methods to detect duckweed-associated bacteria in simple (i.e., binary) or community contexts. To apply these molecular methods, we first systematically optimized a bead-beating protocol to co-isolate nucleic acids simultaneously from both duckweed and bacteria. Second, we combined ribosomal intergenic spacer analysis (RISA) and PCR of a plant-specific marker to detect bacteria associated with duckweed. Third, we used publicly available computational tools and developed an accessible computational pipeline to design strain-specific primers for different bacteria. These primers were able to detect and quantify the associations of duckweed with the respective bacteria, either alone or in the presence of ambient microbial communities from wastewater samples. Lastly, we used confocal microscopy as a complementary approach to describe bacterial colonization dynamics on the duckweed *Lemna minor*. The molecular methods introduced in this work could enable high-resolution, quantitative studies of duckweed–associated bacteria in diverse contexts.

2. Results

2.1. Selection of Duckweed Strain and Bacterial Isolates

Duckweed and bacteria were obtained to study the colonization dynamics of bacteria on duckweed. The duckweed strain *Lemna minor* 5576 (Lm5576) was acquired from the Rutgers Duckweed Stock Cooperative (RDSC; New Brunswick, NJ, USA). This duckweed strain has been previously used to study duckweed-associated bacterial communities [3]. Bacteria originating from different hosts were acquired (File S1). One of the duckweed-associated bacterial (DAB) isolates, *Microbacterium* sp. RU370.1 (DAB 1A), was isolated from Lm5576 and can produce the phytohormone indole-3-acetic acid (IAA) as well as colonize and affect the root development of *Arabidopsis thaliana* [54,55]. Another bacterial isolate was retrieved from the seaweed *Ulva fasciata*. This seaweed-associated bacterium, *Bacillus simplex* RUG2-6 (G2-6), was hypothesized to be a weak colonizer of duckweed due to the large evolutionary divergence between macroalgae and angiosperms. Two bacterial isolates of well-characterized plant colonizers, an epiphyte *Azospirillum brasilense* Sp7 (Sp7) and a known endophyte *Azospirillum baldaniorum* Sp245 (Sp245), were acquired to act as positive colonization controls [56]. Both these strains (Sp7 and Sp245) contain similar 16S rRNA gene sequences (97–99.9% identity) and were initially classified as *Azospirillum brasilense*, but recent phylogenomic analyses show Sp245 belongs to the novel species named *Azospirillum baldaniorum* [57]. Together, these bacteria were used to inoculate duckweed to study their colonization dynamics of axenic Lm5576.

2.2. Systematic Optimization of a Nucleic Acid Extraction Method for Duckweed and Bacteria

To characterize bacterial colonization of duckweed using molecular methods, an optimized protocol for efficiently isolating nucleic acid simultaneously from duckweed and different bacteria was developed. First, nucleic acid extraction was compared between bead-beating and homogenization with a mortar and pestle using a modified CTAB protocol [58] (Figure S1). While the mortar and pestle extracted more nucleic acids from Lm5576 than bead-beating, only bead-beating was able to extract both duckweed and bacterial nucleic acids. Therefore, bead-beating was selected as the physical lysis method for nucleic acid extraction. Various parameters of the bead-beating protocol were then modified to improve duckweed and bacteria nucleic acid extraction. First, three different sizes of beads were compared for their ability to extract plant and bacterial nucleic acids (Figures S2 and S3). These tests showed 1.7 mm zirconium beads were the most effective at homogenizing duckweed tissues and extracting duckweed nucleic acids, and 100 μm silica beads were the best for extracting bacterial nucleic acids. Furthermore, a combination of different-sized beads effectively extracted nucleic acids from both duckweed and bacteria. Therefore, a combination of different-sized beads was used for the bead-beating protocol. Chloroform and a heating step at 65 °C were then added to the lysis step to test their ability to improve nucleic acid extraction (Figure S4A). Both chloroform and heating improved nucleic acid extraction from duckweed. Bead-beating was also performed at different temperatures with and without the addition of a reducing agent to test for improvement of nucleic acid extraction (Figure S4B,C). All conditions resulted in high yields of intact nucleic acids, but bead-beating at room temperature without a reducing agent showed a slightly higher yield of nucleic acids and with DNA higher in apparent molecular weight. Lastly, nucleic acid extractions were performed using various incubation times in the lysis buffer with different bacteria, including isolates from monoderm (Firmicutes and Actinobacteria) and diderm (Proteobacteria) bacterial phyla (Figure S5). Nucleic acids were extracted from both monoderm and diderm bacteria, and the nucleic acid yield increased with longer incubation times in the lysis buffer for some isolates. From these experiments, an optimized bead-beating protocol was developed that could be used to effectively extract nucleic acids from duckweed inoculated with diverse bacteria types.

2.3. Establishment of a PCR-Based DNA Fingerprinting Assay for Detecting Duckweed–Associated Bacteria

rRNA intergenic spacer analysis (RISA) is a PCR-based method that amplifies the intergenic spacer region between 16S and 23S rRNA genes. This region can vary in copy number, sequence, and length among bacterial species. As a result, RISA can be used to estimate bacterial community composition by generating community fingerprints [59] and for rapid, universal bacterial typing [60]. In this study, RISA was applied as a simple molecular approach to detect duckweed-associated bacteria. Different RISA primer sets were tested for their ability to amplify DNA from different bacterial species and duckweed (File S2, Figure S6). 16S-e1390f and 23S-e130r primers produced distinct fingerprints between bacterial species while importantly they did not show significant amplification products from Lm5576 DNA under our conditions. Therefore, these primers were selected for detecting bacterial colonization of Lm5576. In addition to RISA, a plant-specific marker was used to compare the relative amount of Lm5576 genomic DNA between samples and to control for sample quality. Primers were designed for detecting the single-copy *Lemna minor* ortholog of the plant-specific *LEAFY* gene (*LmLFY*), which is a master transcription factor for flowering control (File S3). PCR using *LmLFY* primers specifically detected and allowed visual estimation of the relative quantity of Lm5576 DNA between samples (Figure S7).

Both RISA PCR and *LmLFY* PCR were used in concert to monitor bacterial colonization of Lm5576. This strategy is subsequently referred to as "attachment PCR".

2.4. Standardization and Validation of Attachment PCR Assay for Detecting Duckweed–Associated Bacteria

Attachment PCR was used to detect and compare the colonization of Lm5576 by G2-6, DAB 1A, Sp7, and Sp245 bacterial strains (Figure 1A). Axenic Lm5576 plants were inoculated with bacteria for seven days. After seven days, inoculated Lm5576 tissue was collected, rinsed with sterile water, and nucleic acids were isolated using the bead-beating protocol described above. Isolated DNA from pure bacterial cultures and sterile Lm5576 were used as controls to compare RISA fingerprints from inoculated Lm5576 samples. RISA PCR did not generate any banding pattern from axenic Lm5576 DNA, whose sterility was verified by culturing on LB agar plates (Section 5.1). RISA PCR of bacterial DNA controls produced distinct fingerprints between G2-6, DAB 1A, and *Azospirillum* strains (Sp7 and Sp245). *LmLFY* PCR of DNA controls only produced a PCR product from the axenic Lm5576 DNA control and none from bacterial DNA controls. All inoculated Lm5576 samples showed *LmLFY* PCR products, indicating good sample quality and the presence of Lm5576 DNA for reference. RISA PCR of Lm5576 inoculated with G2-6 sample did not generate any bacterial fingerprint, suggesting G2-6 was either not able to colonize Lm5576 or colonized Lm5576 at a low concentration not detectable by RISA PCR. RISA PCR of Lm5576 inoculated with DAB 1A produced a fingerprint consisting of a single PCR band that matched the fingerprint of DAB 1A DNA, demonstrating that DAB 1A colonized Lm5576. RISA PCR of Lm5576 inoculated with Sp7 or Sp245 produced a similar fingerprint consisting of two major PCR bands. These two bands were the most prominent PCR bands found in RISA PCR of Sp7 and Sp245 DNA controls indicating Sp7 and Sp245 were both able to colonize Lm5576. The higher concentrations of DNA used for Sp7 and Sp245 DNA controls may explain why the additional PCR bands were not clearly observed in Lm5576 inoculated with Sp7 or Sp245. Overall, attachment PCR showed that DAB 1A, Sp7, and Sp245 could colonize Lm5576 at detectable levels. The exact matching of RISA PCR fingerprints between inoculated Lm5576 samples and DNA controls confirmed the colonization of Lm5576 by the respective bacteria. In addition, this exact matching suggested no contaminating bacteria were present. Therefore, fingerprint matching between RISA PCR of inoculated duckweed and DNA controls can be used to confirm what bacteria are present in duckweed samples.

2.5. Computational Pipeline for Primer Design to Detect and Quantify Specific Duckweed-Associated Bacteria

Attachment PCR using RISA and *LmLFY* primer sets detected the colonization of Lm5576 by different bacteria, but it was unable to differentiate strains of the same genus (i.e., Sp7 and Sp245). To distinguish between closely related bacteria, a genomics-enabled approach was taken where strain-specific primers for end-point PCR were designed for each bacterium using available computational pipelines (File S2). For this approach, genomes of G2-6 and DAB 1A were sequenced and the genomes of Sp7 and Sp245 were retrieved from public databases (File S1). The strain-specific primers designed from this pipeline were used for PCR of DNA controls to validate their specificity (Figure 1A). As expected, strain-specific PCR of DNA controls uniquely detected the target bacteria and differentiated Sp7 and Sp245 strains. Strain-specific PCR of inoculated Lm5576 samples showed DAB 1A, Sp7, and Sp245 significantly colonized Lm5576. G2-6-specific PCR showed a faint amplification product in the Lm5576 sample inoculated with G2-6, in contrast to RISA PCR results, suggesting G2-6 may be able to attach to Lm5576 tissues at low concentrations not detectable by RISA PCR. These results show that PCR using bacterial strain-specific primers can distinguish between phylogenetically similar bacterial strains and can be used to detect specific duckweed–bacterium associations.

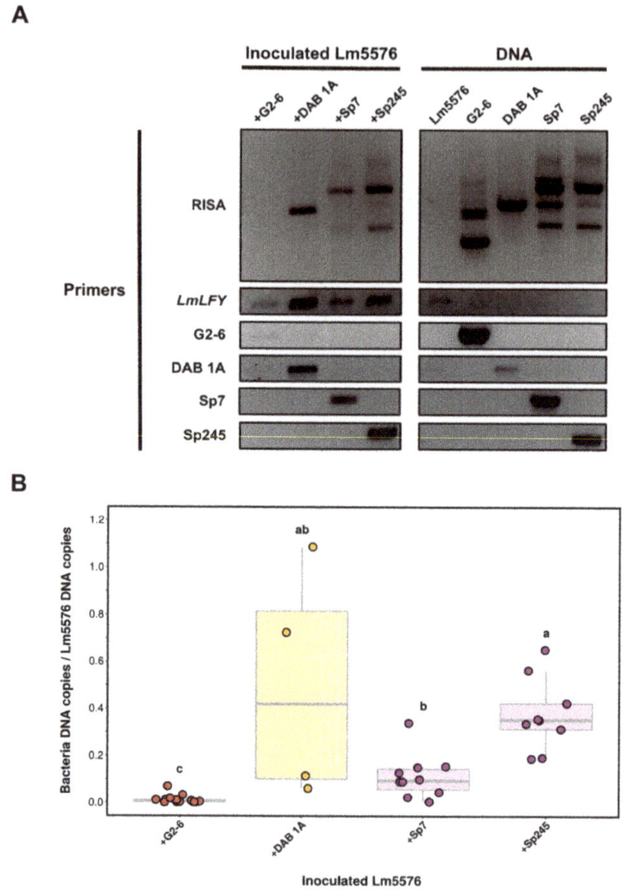

Figure 1. Molecular detection of duckweed-associated bacteria. Axenic Lm5576 was inoculated with different bacteria in 0.5X SH media. After seven days, inoculated Lm5576 tissue was collected, washed with sterile water, and nucleic acids were isolated for analysis. (**A**) Representative gel electrophoresis results of end-point PCR using RISA, *LmLFY*, and strain-specific primers (File S2). RISA PCR fingerprints from inoculated Lm5576 samples were compared to the respective DNA controls from Lm5576 and bacteria alone. Samples: +G2-6 = Lm5576 inoculated with *Bacillus simplex* RUG2-6, +DAB 1A = Lm5576 inoculated with *Microbacterium* sp. RU370.1, +Sp7 = Lm5576 inoculated with *Azospirillum brasilense* Sp7, +Sp245 = Lm5576 inoculated with *Azospirillum baldaniorum* Sp245. Primers: RISA = PCR using 16S-1390f and 23S-e130r primers, *LmLFY* = PCR using LmLFY-F and LmLFY-R primers specific to Lm5576, DAB 1A = PCR using AmRU370.1-F and AmRU370.1-R primers specific to *Microbacterium* sp. RU370.1, G2-6 = PCR using BsRUG2.6-F and BsRUG2.6-R primers specific to *Bacillus simplex* RUG2-6, Sp7 = PCR using AbSp7-F and AbSp7-R primers specific to *Azospirillum brasilense* Sp7, Sp245 = PCR using AbSp245-F and AbSp245-R primers specific to *Azospirillum baldaniorum* Sp245. (**B**) Bacterial association with Lm5576 was quantified using real-time PCR. Bacterial DNA copy number was determined for each inoculated Lm5576 sample and normalized to Lm5576 DNA copy number. Different colors were used for the different bacterial genera. Each data point represents an experimental repeat except for +G2-6, for which each sample was measured twice. A significant difference was found in colonization levels between bacteria (Kruskal–Wallis, p-value = 4.58×10^{-6}). Pairwise comparisons were performed using Dunn's test, and results are displayed as compact letters (a–c). Bacteria with significantly different colonization levels from each other, according to Dunn's test (FDR < 0.05), do not share any letters.

While end-point PCR with strain-specific primers detected specific duckweed–bacterium associations, it could not be used to accurately quantify bacterial colonization levels. To quantify bacterial colonization of Lm5576, bacterial strain-specific primers were designed for real-time PCR (qPCR) assays using a computational pipeline developed in this study called UniAmp (Section 5.8, Figure S8, File S4). For this computational pipeline, unique genomic sequences were identified and retrieved for each bacterial genome. These unique sequences were then used for optimal primer design. Bacterial strain-specific primers designed by this pipeline were then used to determine bacterial DNA copy number in respective inoculated Lm5576 samples. Bacteria DNA copy number was then normalized to the Lm5576 DNA copy number of the inoculated Lm5576 sample using Lm5576-specific primers. This approach is referred to as "attachment qPCR". Attachment qPCR showed a significant difference (Kruskal–Wallis, p-value = 4.58×10^{-6}) in the colonization of Lm5576 between the bacteria tested (Figure 1B). Attachment qPCR showed G2-6 associated with Lm5576 in significantly lower concentrations compared to the other bacteria tested (Dunn's test, p-value < 0.005 for all comparisons), similarly to what was found qualitatively by end-point strain-specific PCR (Figure 1A). DAB 1A and Sp245 had the highest colonization levels on Lm5576. However, DAB 1A displayed higher variability between experimental repeat samples, so no significant difference was established compared to Sp7 or Sp245. Sp7 colonized Lm5576 at significantly lower concentrations than Sp245 (Dunn's test, p-value < 0.05). In conclusion, attachment qPCR revealed a significant difference in colonization levels between bacterial isolates from plants compared to the bacterial isolate from seaweed and detected significant differences in colonization levels between phylogenetically similar bacteria. These findings demonstrate attachment qPCR can be used to quantify colonization levels of bacteria on plants with high resolution.

2.6. Bacterial Colonization of *Lemna minor* Visualized by Confocal Microscopy

As a complementary approach to the PCR-based methods described above, confocal microscopy was performed on inoculated Lm5576 samples to qualitatively describe bacterial colonization patterns (Figure 2, File S5). Attachment PCR was performed on all microscopy samples and confirmed the colonization of Lm5576 by the respective bacteria and the absence of contaminating bacteria (File S6). All the bacteria tested were found to colonize the surfaces of Lm5576 fronds (Figure 2). G2-6, DAB 1A, and Sp245 were spread over the surfaces of Lm5576 fronds in smaller colonies, and Sp7 was mostly localized to the root–frond interface in aggregates. No bacteria were observed to colonize the inside of Lm5576 fronds in these experiments. Bacteria displayed different colonization patterns of Lm5576 roots (Figure 2, File S5). G2-6 was found on the surfaces of Lm5576 roots at a low density. As mentioned above, Sp7 aggregates were mostly located on the surfaces of Lm5576 roots near the root–frond interface. DAB 1A and Sp245 were also found mostly at the root–frond interface on the surfaces of Lm5576 roots. Microscopy showed DAB 1A was present in higher concentrations at the root–frond interfaces than the other bacteria tested. This correlates with the high colonization levels observed in attachment qPCR experiments for some samples (Figure 1B). Interestingly, Sp245 was found inside Lm5576 roots, within the endodermis, at high concentrations. This also agreed with the attachment qPCR results that revealed a significantly high colonization level for Sp245 and shows it is an endophyte for Lm5576. DAB 1A and Sp7 were also sporadically found inside Lm5576 roots but at much lower frequencies and concentrations. Together, confocal microscopy of inoculated Lm5576 samples revealed that the bacteria tested were able to colonize Lm5576 fronds and roots to various extents, and that the root–frond interface is a hotspot for bacterial colonization. In addition, these results support the main conclusions of the attachment qPCR experiments in this study (Figure 1B) while contributing to the understanding of spatial colonization dynamics of bacteria on duckweed tissues.

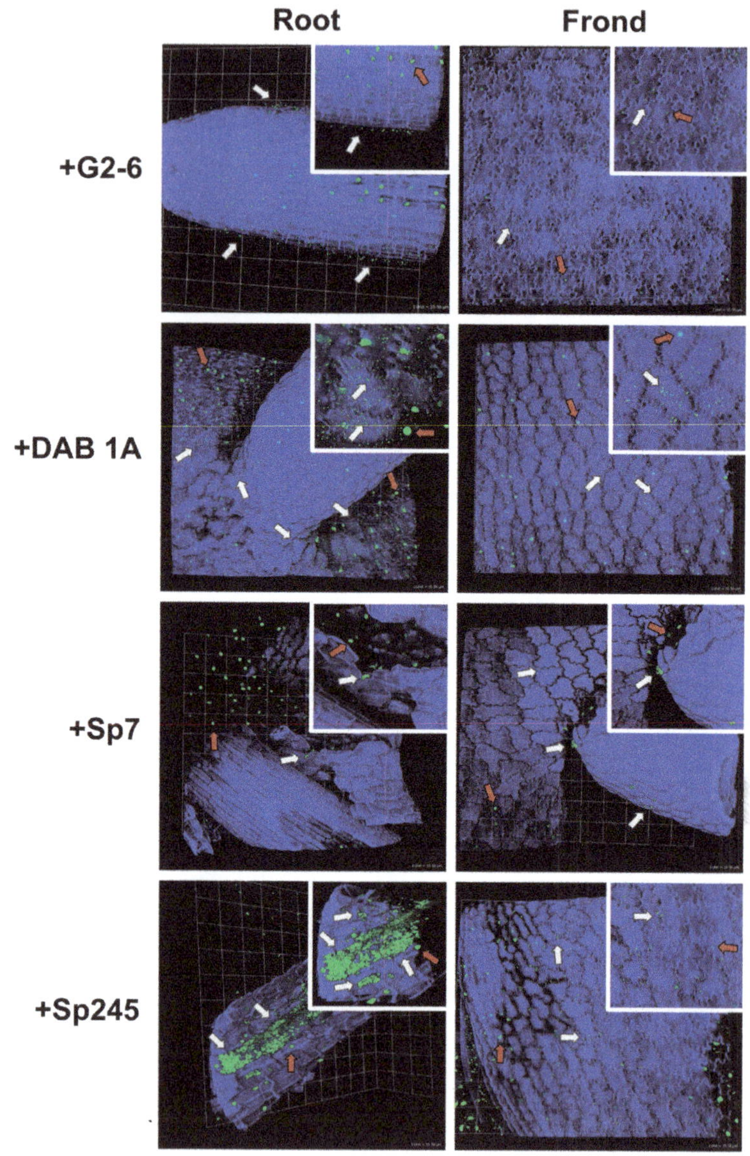

Figure 2. Two-dimensional confocal microscopy of inoculated duckweed samples. Confocal microscopy (40X/1.1 objective) was performed on inoculated Lm5576 in 0.5X SH media to characterize the spatial colonization dynamics of bacteria on duckweed. Calcofluor white was used to stain Lm5576 cellulose and visualized with the blue channel, whereas SYBR Gold was used to stain DNA and visualized with the green channel. Both bacterial and Lm5576 DNA were stained by SYBR Gold and pictured in green, but bacterial DNA (indicated by white arrows) is smaller in size than Lm5576 nuclei (indicated by red arrows), and these bacteria are often found in clustered colonies. For each main image, 1 unit of the grid scale is equal to 35.56 μm and zoomed-in images are pictured in the top-right corner. +G2-6 = Lm5576 inoculated with *Bacillus simplex* RUG2-6, +DAB 1A = Lm5576 inoculated with *Microbacterium* sp. RU370.1, +Sp7 = Lm5576 inoculated with *Azospirillum brasilense* Sp7, +Sp245 = Lm5576 inoculated with *Azospirillum baldaniorum* Sp245.

2.7. Strain-Specific Monitoring of Duckweed–Associated Bacteria in a Community Context

To further illustrate the efficacy of attachment PCR assay using bacterial strain-specific primers, this molecular approach was used to detect specific duckweed–bacterium associations in the presence of other bacterial isolates and in the presence of microbes in wastewater (Figure 3). Attachment PCR was also tested using another duckweed strain obtained from the RDSC, *Spirodela polyrhiza* strain 9509 (dw9509), whose genome was sequenced to reference quality [61,62]. For these experiments, axenic dw9509 was inoculated with DAB 1A, *Bacillus* sp. RU9509-4 (DAB 3D), Sp245, or with wastewater containing microbes (NS) for five days. In addition, Sp245 was co-inoculated onto dw9509 with DAB 1A, DAB 3D, or wastewater containing microbes (NS) to test bacterial colonization in the presence of other microbes. After five days, inoculated duckweed tissue was collected, rinsed with sterile water, and nucleic acids were isolated. *SpLFY* PCR generated a PCR product for all inoculated dw9509 samples, ensuring good sample quality. RISA PCR and strain-specific PCR did not generate any signals for axenic dw9509, confirming its sterility. RISA PCR showed DAB 1A, DAB 3D, Sp245, and wastewater microbes (NS) colonized dw9509. Additionally, strain-specific PCR confirmed DAB 1A, DAB 3D, and Sp245 colonized dw9509 while no amplification products were obtained with wastewater containing microbes sample (NS). RISA PCR and Sp245 strain-specific PCR demonstrated Sp245 was able to colonize dw9509 in the presence of DAB 1A, DAB 3D, and a wastewater microbial community, indicating robust colonization ability by Sp245 in diverse contexts. While both DAB 1A and DAB 3D were able to colonize dw9509 in the presence of Sp245, DAB 3D strain-specific PCR showed a lower amplification signal in the dw9509 sample co-inoculated with DAB 3D and Sp245 compared to the dw9509 sample inoculated only with DAB 3D, suggesting DAB 3D colonization, but not DAB 1A, of dw9509 was reduced in the presence of Sp245. These experiments illustrate the efficacy of attachment PCR and strain-specific PCR to detect specific duckweed–associated bacteria in a community context. In addition, quantitative effects on bacterial colonization of the plant host resulting from distinct microbe-microbe interactions can be revealed.

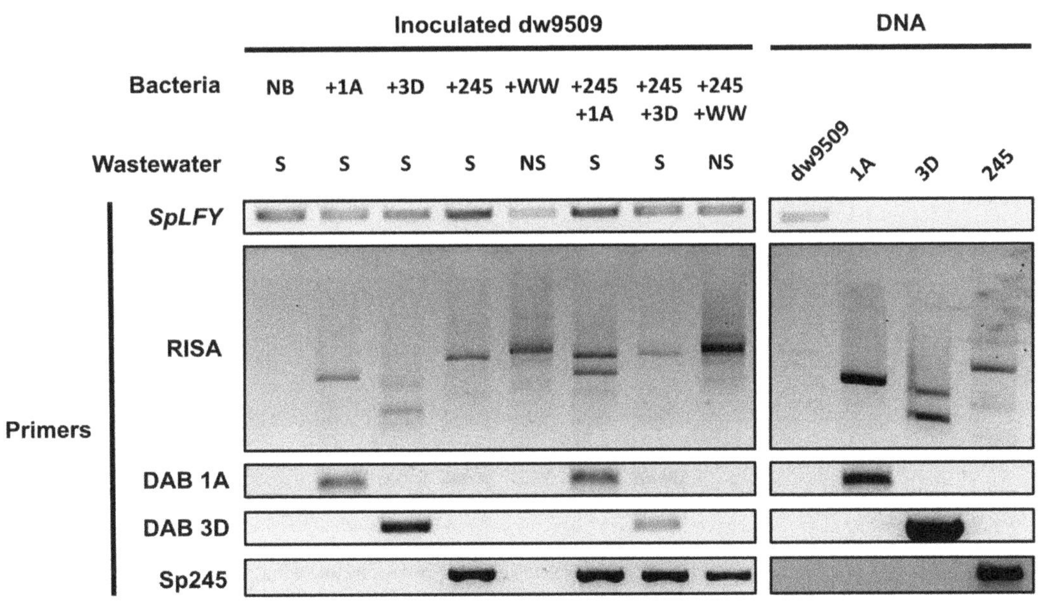

Figure 3. Molecular detection of duckweed-associated bacteria in a community context. Axenic dw9509 was inoculated with different bacteria in wastewater with (NS) or without (S) microbes.

In addition, Sp245 was co-inoculated onto dw9509 with DAB isolates or non-sterile wastewater-containing microbes. For co-inoculated samples, bacteria were mixed at a 1:1 ratio based on OD_{600}. After five days, dw9509 tissue was collected, washed with sterile water, and nucleic acids were isolated for end-point PCR using RISA, *SpLFY*, and strain-specific primers (File S2). RISA PCR fingerprints from inoculated dw9509 samples were compared to DNA controls from dw9509 and bacteria alone. Wastewater: S = filter-sterilized wastewater not containing microbes, NS = non-sterile wastewater containing microbes; Bacteria: NB = axenic dw9509, +1A = dw9509 inoculated with *Microbacterium* sp. RU370.1, +3D = dw9509 inoculated with *Bacillus* sp. RU9509.4, +Sp245 = dw9509 inoculated with *Azospirillum baldaniorum* Sp245, +WW = dw9509 inoculated with non-sterile wastewater containing microbes; Primers: *SpLFY* = PCR using SpLFY-F and SpLFY-R primers specific to dw9509, RISA = PCR using 16S-1390f and 23S-e130r primers, DAB 1A = PCR using AmRU370.1-F and AmRU370.1-R primers specific to *Microbacterium* sp. RU370.1, DAB 3D = PCR using BsRU9509.4-F and BsRU9509.4-R primers specific to *Bacillus* sp. RU9509.4, Sp245 = PCR using AbSp245-F and AbSp245-R primers specific to *Azospirillum baldaniorum* Sp245.

3. Discussion

3.1. Localization of Bacteria on Duckweed

In terrestrial plants, bacteria have been shown to consistently colonize certain areas of plants termed colonization hot spots [21,22], which include root cracks where lateral roots emerge from the main root [23]. One explanation for this bacterial colonization pattern is that root cracks may release cell lysates and exudates that could help attract bacteria and other microorganisms [63]. In duckweed, a few studies have described bacterial colonization patterns on these aquatic plants. In one study, duckweed collected from chalk streams was found to have higher densities of bacteria on the submerged abaxial surface of duckweed fronds compared to the aerial adaxial surface [64]. In another study on *L. minor*, the plant-growth-promoting bacterium, *Bacillus amyloliquefaciens* FZB42, was found to initially colonize *L. minor* at the root tip and root–frond interface, and later the grooves between root epidermal cells and concavities of the abaxial frond surface [29]. For a rootless duckweed, *Wolffia australiana*, bacteria present in the surrounding greenhouse environment were found to colonize near reproductive pockets, where mother and daughter fronds attach, and the stomata [4]. In the present study, we performed high-resolution confocal microscopy on inoculated *L. minor* samples to further study bacterial colonization patterns of duckweed (Figure 2). All bacteria in this study were able to associate with the abaxial surfaces of duckweed fronds and roots to varying extents. Like previous reports, bacterial strains used in this study also showed a preference for the root–frond interface. Together, these studies show that bacteria readily colonize the abaxial surface of duckweed fronds, at least for duckweeds with roots. One possible explanation for this observation is that the abaxial side of duckweed fronds is in direct contact with the microbial inoculum present in the surrounding water environment. Another explanation is that the surface composition, such as that of the cuticle, is distinct between the abaxial and adaxial surfaces of fronds [65] and may play a role in the differential attachment of microbes. Although duckweeds do not make lateral roots [66], the root–frond interface in duckweed may serve as a hotspot akin to the root cracks in terrestrial plants, where cell contents are released or secreted to attract microbes.

While the surface colonization patterns of bacteria on duckweed have been described, there is no description of whether bacteria can colonize the inside of duckweed tissues. In this study, Sp245 colonized Lm5576 at the root–frond interface and the inside of duckweed roots, within the endodermis [66], at high densities (Figure 2). In terrestrial plants, colonization hot spots, such as root cracks, can be used by bacteria to enter the roots of terrestrial plants [23,67,68]. Likewise in duckweed, one possibility could be that Sp245 entered through cracks at the root–frond interface of Lm5576 and proceeded to colonize the interior of Lm5576 roots. Recently, we studied the Sp245 interaction with the model plant *Arabidopsis thaliana* and revealed its potential interaction with guard cells in leaf tissues as a means for entering the endosphere [55]. Strikingly, this interaction and targeting to the guard cells by Sp245 are abolished in the pleiotropic *axr1* mutant of *A. thaliana*, suggesting specific involvement of this gene in the signaling between plants and certain microbes. However, our microscopy studies with Lm5576 failed to observe this guard-cell colonization of Sp245 on duckweed fronds, indicating this mode of interaction could be lost or modified in duckweed. Sp245 was originally isolated from wheat experiments in Brazil [69] and colonizes the inside of wheat roots at high densities [24,26]. In contrast to the endophytic colonization pattern of Sp245, Sp7, originally isolated from *Digitaria* in Brazil [70], is an epiphyte shown to colonize only the surfaces of plant roots such as wheat and corn [24,26,71]. In particular, Sp7 has been found to form aggregates on the surface of these plant roots. Likewise, we found Sp7 mostly colonized the surface of Lm5576 roots near the root–frond interface, also in aggregates. These Sp7 and Sp245 colonization experiments on Lm5576 demonstrated that plant-associated bacteria can have similar colonization patterns with both terrestrial plants, such as wheat, and aquatic monocots, such as duckweed. This suggests a likely conservation of bacterial mechanisms for associating with duckweed and other higher plants.

3.2. Molecular Methods for Detection of Duckweed-Associated Bacteria

Classical microbiology techniques, such as the CFU assay, have been used to monitor bacterial colonization of duckweed [72]. As mentioned above, this CFU assay lacks the specificity to differentiate bacteria within a community context. More recent studies have demonstrated the ability of a molecular approach using bacterial primers and real-time PCR to quantify bacterial colonization levels on duckweed in the presence of other microbes [73,74]. In this study, we built on this molecular approach and developed different PCR-based strategies to detect duckweed-associated bacteria. A prerequisite for using such approaches is a protocol capable of efficiently isolating nucleic acids. Thus far, there have been no published attempts to systematically develop a protocol capable of efficiently isolating nucleic acids from both duckweed and bacteria. A working protocol for isolating nucleic acids is critical for studying duckweed–bacteria associations, since different DNA extraction protocols can introduce significant biases toward what bacteria are detected and in what quantities [75,76]. These differences can be partly explained by the inability of certain protocols to efficiently lyse monoderm bacteria, Gram-positive bacteria consisting of a thick peptidoglycan layer. However, nucleic-acid-isolation protocols implementing physical lysis methods such as bead-beating can efficiently lyse monoderm bacteria, especially when longer bead-beating times are used [77–79]. Bead-beating protocols are also reproducible [76,80], yield high concentrations of nucleic acid [76,81], and produce relatively accurate community profiles [82]. In addition, combining bead-beating and chemical lysis, such as phenol or chloroform, can dramatically increase DNA extraction efficiency and quality [76,81]. For these reasons, bead-beating is recommended for nucleic acid extraction protocols [83]. Here, we implemented and optimized a bead-beating protocol to simultaneously co-isolate duckweed and bacterial nucleic acids. By combining different bead sizes and a CTAB/chloroform lysis buffer, this bead-beating protocol produced high yields of intact nucleic acids from duckweed and different bacteria, including monoderm and diderm bacteria (Figures S2–S5). Furthermore, through various testing of the bead-beating protocol, we observed increases in nucleic acid yields with longer incubation time periods in the lysis buffer (Figure S5) and with longer alcohol precipitation time periods. However, the ability of this bead-beating protocol to generate representative profiles of duckweed colonized by complex bacterial communities remains to be validated. This could be tested by isolating nucleic acid from mixtures of bacteria in known concentrations, known as mock communities [84,85]. This will be an important validation step for applying this bead-beating protocol to study the interactions between duckweed and complex microbial communities in the future. Lastly, this bead-beating protocol can be modified to isolate only DNA or RNA from duckweed or bacteria by adding an RNase or DNase treatment step, respectively.

rRNA intergenic spacer analysis (RISA) has been commonly used for community fingerprinting [59] and bacterial typing [60]. However, RISA has also been used to study plant–bacteria associations. For example, an automated version of RISA (ARISA) that applies fluorescently tagged primers and detects fluorescent PCR fragments [86] was used to monitor changes in the composition of synthetic bacterial communities [87]. Here, we used RISA to detect bacterial colonization of duckweed by comparing fingerprints of inoculated duckweed samples to DNA controls of duckweed and the respective bacteria (Figures 1A and 3). This molecular approach serves many purposes. First, RISA can be used to determine the axenic status of *L. minor* and *S. polyrhiza* plants used in experiments, since RISA PCR does not produce any amplicons from sterile Lm5576 and dw9509. This is worth highlighting since difficulties can be encountered in obtaining sterile duckweed [88]. As we optimized RISA PCR for use with *L. minor* and *S. polyrhiza* in this study, RISA PCR may need to be optimized for use with other duckweed species by using different RISA primer sets and/or PCR conditions. Second, RISA can be used to determine the colonization of duckweed by different bacterial genera, since RISA PCR can produce distinct fingerprints between bacterial genera. This is particularly useful when no sequence information is available to design primers for the bacteria being studied. Third, because fingerprints generated from inoculated duckweed samples are compared to DNA controls of the organisms being studied, RISA can also reveal non-matching fingerprints that are due to contaminating or exogenous bacteria during the experiment. In addition to RISA, we included a duckweed-specific marker to control for sample quality and to provide a reference for the relative quantity of duckweed DNA between samples. We termed this approach of combining RISA PCR and PCR of a plant-specific marker as attachment PCR. Attachment PCR was recently used in our laboratory to detect interactions between bacteria isolated from rice and duckweed [89]. Under laboratory conditions, this method showed that *Pantoea* isolates from rice were able to colonize duckweed such as Lm5576, despite the low representation of *Pantoea* bacteria in community surveys of duckweed-associated bacterial (DAB) communities from the same rice paddies. This suggests that microbe-microbe interactions or environmental factors could be responsible for the low representation of *Pantoea* in DAB communities, but not in rice tissues, in this context. These case studies demonstrate the utility of RISA in general and its application for attachment PCR specifically.

Despite its advantages, we found RISA was unable to distinguish between strains from the same genus (Figure 1A). As mentioned previously, plant-associated bacteria from the same genus can be functionally diverse. Thus, it is important to identify methods that can differentiate closely related bacteria isolated from plants. Previous studies have used strain-specific primers to monitor bacteria associated with duckweed [73,74]. However, these studies did not explicitly define the primer design strategy and did not verify whether the primers used could differentiate phylogenetically similar bacteria. Here, we used available bioinformatic tools and developed a computational pipeline (UniAmp) to generate strain-specific primers, leveraging the large and growing databases for bacteria (Figure S8, Files S2 and S4). Strain-specific primers for end-point PCR were able to clearly differentiate strains from the same genus (Figure 1A) and detected specific bacteria even in the presence of other microbes (Figure 3). Strain-specific primers were also designed for real-time PCR to determine colonization levels of specific bacteria on duckweed by normalization to a duckweed-specific reference marker (Figure 1B). This "attachment qPCR" approach, of normalizing bacterial abundance to an internal plant marker gene, has been applied in previous studies to quantify maize root colonization by plant-growth-promoting bacteria (PGPB) [48], *Petunia* root colonization by arbuscular mycorrhizal fungi [90], and *Arabidopsis* root colonization by bacteria belonging to the Rhizobiales family [91]. For all these uses of strain-specific primers to detect bacterial colonization of different plants, many studies do not provide an accessible primer design strategy. In a few of the remaining studies, a primer design strategy is used that splits bacterial genome sequences into smaller sequence fragments that are then used for a BLAST similarity search against public databases to find unique sequences for primer design [48,49]. Yet, this primer design strategy can be inefficient if many strain-specific primers are desired or if large genomes are being used. Here, we developed a straightforward computational pipeline (UniAmp) to design strain-specific primers for any bacterial strain with a sequenced genome. In this pipeline, pairwise genome alignments are performed between a reference genome for the bacterial strain of interest and a selected set of query genomes. By retrieving a set of query genomes from public databases that are highly similar to the reference genome, unique sequences to the reference genome can be determined and used for primer design. Due to the efficiency of pairwise genome alignments [92], this approach can be readily applied to design strain-specific primers for many genomes or for large genomes. While the UniAmp pipeline was used in this study to design primers to detect bacteria associated with duckweed, the pipeline could be readily applied to other host systems and many other scenarios.

3.3. Bacterial Adaptation to Plant Habitats and Colonization Dynamics

Selection is a major driver in structuring plant bacterial communities [2]. As a result of this selection, certain bacteria have adapted to occupy different plant habitats [93,94]. For example, genomic analyses have shown that plant-associated bacterial genomes are enriched in certain functions, such as chemotaxis, motility, and carbohydrate metabolism [95,96]. In support of these analyses, genome-wide functional screens using transposon mutagenesis sequencing in both terrestrial plants and duckweed confirm the involvement of chemotaxis, motility, and carbon metabolism in bacterial colonization of plants [97,98]. In addition to these functions, many plant-associated bacteria can produce phytohormones, such as auxins, which can have either beneficial or detrimental effects on plant hosts [99]. Most studies on bacterial auxin production have focused on the effects on plant growth, but one recent study investigated the role of bacterial auxin production in plant colonization [100]. This study showed that auxin production is necessary for some bacteria to efficiently colonize plant roots and revealed a feedback loop between auxin produced by bacteria and reactive oxygen species (ROS) produced by the plant host. Together, these studies describe some of the functions that may have evolved in bacteria to help colonize plants.

In this study, we explored the colonization levels among a non-plant-associated bacterial isolate and several plant-associated bacterial isolates using attachment qPCR. G2-6 was isolated from seaweed, a macroalga from salt water, and likely has not adapted or evolved to colonize freshwater macrophytes such as duckweed. We thus expected G2-6 to colonize duckweed at very low levels, if at all. Indeed, attachment qPCR showed G2-6 associated with Lm5576 at significantly lower concentrations compared to all the plant-associated bacterial isolates tested (Figure 1B). DAB 1A was originally isolated from Lm5576 and produces high levels of the auxin indole-3-acetic acid (IAA) that can affect the root development of *Arabidopsis thaliana* [54,55]. Therefore, we expected DAB 1A to re-colonize Lm5576 in this study. Confocal microscopy confirmed these expectations and showed high levels of DAB 1A near the root–frond interface of Lm5576 (Figure 2). Attachment qPCR showed variable colonization levels, where DAB 1A colonized Lm5576 at high levels in some samples (Figure 1B). Extending these results, confocal microscopy of *A. thaliana* inoculated with DAB 1A showed high concentrations of DAB 1A present on the root surface [55]. Interestingly, this same study showed another DAB isolate, DAB 33B, was not able to colonize *A. thaliana*, even though it belonged to the same genus, *Microbacterium*, as DAB 1A. In addition to this inability to colonize *A. thaliana*, DAB 33B was shown to produce significantly lower levels of IAA compared to DAB 1A. Together, one explanation for the different colonization dynamics between these phylogenetically similar strains (DAB 1A and DAB 33B) could be that high levels of auxin production facilitate DAB 1A colonization of plants. Members of the *Azospirillum* genus are well-known plant colonizers and have been shown to fix nitrogen and produce phytohormones, such as IAA, which may promote plant growth [56]. Interestingly, *Azospirillum* taxa have also been detected in surveys of DAB communities [3] and isolated from duckweed tissues [101]. Therefore, we hypothesized that the *Azospirillum* strains Sp7 and Sp245 would be able to colonize Lm5576 to some extent. Attachment qPCR showed Sp245 colonized Lm5576 at significantly higher levels than Sp7 (Figure 1B). This was further supported by confocal microscopy, which showed high levels of Sp245 within duckweed roots (Figure 2). These results are similar to a previous study that found Sp245 colonized the root endosphere of wheat and contained higher overall colonization levels compared to Sp7 [24].

Together, these attachment qPCR results raise several implications about the bacterial colonization of plants. For one, these data suggest that bacteria adapted to plants may display significantly higher colonization levels compared to non-adapted bacteria (Figure 1B). If so, then attachment qPCR can be used to screen for bacteria adapted to colonizing plant habitats. This kind of experiment may help to discover novel traits necessary for the successful bacterial colonization of plants. Secondly, the plant-associated bacterial isolates examined in the present work showed different colonization levels. This raises the question of what traits determine the colonization levels of bacteria on plants? As mentioned above, auxin production is necessary for some bacteria to colonize plants [100]. Interestingly, the plant-associated bacterial isolates DAB 1A, Sp7, and Sp245 all produce significant levels of auxin [54,55]. Future work could use attachment qPCR to examine the relationship between the level of bacterial auxin produced and the effect on bacterial colonization level of the plant host. Results from this study also showed significantly higher colonization levels for the endophyte Sp245 compared to the epiphyte Sp7 (Figure 1B). This also raises the question of what is the relationship between bacterial colonization levels and bacterial colonization patterns? Do all endophytes display high colonization levels? If not, what controls the colonization levels of different endophytes? To answer this, attachment qPCR experiments could be performed to quantify the colonization levels of different bacterial endophytes. In summary, quantitative studies using attachment qPCR could lead to an improved understanding of traits underlying bacterial colonization of plants.

3.4. *Molecular Detection of Specific Duckweed–Associated Bacteria within a Community Context*

Similar to findings with terrestrial plant bacterial communities, microbe–microbe interactions likely play an important role in the bacterial colonization of duckweed. One study reported that a PGPB and two different plant-growth-inhibiting bacteria (PGIB) showed stable colonization levels of duckweed when inoculated separately [73]. However, when the PGPB and PGIB were co-inoculated together, the PGPB strain completely excluded one of the PGIB from colonizing duckweed. In another study, the same PGPB strain slowly decreased in abundance over time on duckweed in the presence of different bacterial communities [74]. Thus, the ability to distinguish between phylogenetically similar microbes in both mono-associations and within a community context will be important for studying the interactions between plants and complex microbial communities. In our work, strain-specific primers were shown to detect specific duckweed–associated bacteria within a community context (Figure 3). The specificity demonstrated by strain-specific PCR has a pertinent application in the synthetic ecology approach used to study plant–microbe interactions [102]. In this approach, synthetic bacterial communities (SynComs) are constructed from bacterial isolates that are representative of members found in wild plant bacterial communities. In contrast to wild bacterial communities, SynComs are experimentally amenable and tractable, allowing causal relationships to be determined. As a constructed community, SynComs can capture the complexity of plant bacterial communities found in nature while providing a means to decipher mechanisms underlying community dynamics and functions [20]. However, SynComs are limited by methods commonly used to track member presence and abundance, such as 16S rDNA amplicon sequencing. Since 16S rDNA amplicon sequencing cannot distinguish between many phylogenetically similar bacteria, SynComs have to be carefully designed in a way to select distinguishable members [103]. As a result, this can severely limit the diversity and representativeness of SynComs that can be used to effectively study the colonization dynamics of plant microbial communities. However, using strain-specific primers will allow closely related bacteria to be included and monitored in SynCom experiments. Moreover, attachment qPCR can be used to quantify member abundance in SynCom experiments. The strategy used in attachment qPCR where bacteria load is normalized to the quantity of host DNA is similar to traditional qPCR used in RNA-sequencing experiments to validate gene expression, where a target gene is normalized to a housekeeping gene. In an analogous fashion, attachment qPCR could be used to compare member abundance generated from 16S rDNA amplicon sequencing in SynCom experiments, since both approaches are DNA-based. Moreover, attachment qPCR could allow phylogenetically similar bacteria with different colonization dynamics and functional traits to be used in SynComs. Such comparisons could facilitate the assignment of the different phenotypes observed in SynCom experiments to specific molecular features. Together, these kinds of experiments should facilitate a mechanistic understanding of the interactions between plant hosts and their associated microbes.

4. Conclusions

In conclusion, the PCR-based approaches introduced in this study have been shown to be effective for studying duckweed–associated bacteria. Attachment PCR with generic RISA primers can be used to reveal the bacteria associated with duckweed, and PCR using strain-specific primers can be used to differentiate specific duckweed–bacterium associations. Additionally, the attachment qPCR approach can be used to determine colonization levels of bacteria under binary or community contexts. While these molecular approaches were used to study duckweed–associated bacteria in this study, they should be easily adopted for use with other host-microbe systems. Together, these strain-specific approaches overcome the limitations of current methods used to detect plant–bacteria associations and enable the detection and quantification of specific plant–microbe associations under diverse scenarios.

5. Materials and Methods

5.1. Duckweed Sterilization and Propagation

Cultures of *Lemna minor* 5576 (Lm5576) and *Spirodela polyrhiza* (dw9509) were obtained from the Rutgers Duckweed Stock Cooperative (RDSC; Rutgers University, New Brunswick, NJ, USA). Duckweed cultures were sterilized using a modified protocol from a previously described procedure [88]. For this procedure, duckweed plants were transferred to 1.7 mL microcentrifuge tubes and washed with 500 µL of salt and detergent solution (1% Triton-X 100, 137 mM NaCl, 2.7 mM KCl, 10 mM Na_2HPO, 1.8 mM KH_2PO_4, 0.5 mM $MgSO_4$, 1 mM $CaCl_2$, pH 7.4) to facilitate surface sterilization. Duckweed plants were then surface sterilized using 5–10% (*v/v*) household bleach (0.5–1% sodium hypochlorite). Duckweed plants were surface sterilized until most frond tissues turned white, and only the meristematic regions retained chlorophyll and remained green. Following surface sterilization, 2% (*w/v*) of sodium thiosulfate was added to help neutralize residual sodium hypochlorite [104]. Surface-sterilized duckweed plants were then rinsed with sterile water and aseptically transferred to 0.8% (*w/v*) agar (BD, Franklin Lakes, NJ, USA; catalog #214530) plates with 1.6 g/L of Schenk and Hildebrandt basal salt mixture (0.5X SH) media (Phytotechnology Laboratories, Lenexa, KS, USA; catalog #S816) containing 0.5% sucrose and 100 µg/mL cefotaxime (GoldBio, St. Louis, MO, USA; catalog #C-104-25) at pH 6.5–7.0. In addition, surface-sterilized duckweed plants were transferred to 1.5% (*w/v*) agar plates with Miller's lysogeny (or Luria) broth (LB) (10 g/L tryptone, 5 g/L yeast extract, 10 g/L NaCl). Surface-sterilized duckweed plants were propagated on 0.5X SH agar plates and a LB agar plate. The LB agar plate was used to detect any signs of microbial growth. If microbial growth was observed on duckweed plants growing on the LB agar plate, then the surface-sterilization procedure was repeated on the surface-sterilized duckweed plants growing on the 0.5X SH agar plates.

Once axenic duckweed plants were obtained, stock cultures and working cultures of axenic duckweed were generated. Stock cultures of axenic duckweed were stored at 15 °C and only used when required. Axenic working cultures of duckweed were generated by transferring a few duckweed plants to a 0.5X SH agar plate with 0.5% (*w/v*) sucrose and an LB agar plate after each experiment. If no microbial growth was observed on the LB agar plate, then duckweed plants on the 0.5X SH agar media were propagated for experiments. If microbial growth was observed, then a stock culture of axenic duckweed was retrieved from storage and propagated for experiments.

Axenic duckweed plants were propagated in a growth chamber on 0.5X SH agar media with 0.5% (*w/v*) sucrose (pH 6.5–7.0) at 25 °C under a photoperiod of 16 h light and 8 h dark for 2–4 weeks. Duckweed plants from the agar plate were then transferred to 100 mL of liquid culture of 0.5X SH with 0.1% (*w/v*) sucrose and propagated for 1–2 weeks under the same growth chamber conditions. Axenic duckweed plants from these liquid cultures were then transferred for experiments. Duckweed sterility was confirmed between transfers by plating duckweed plants on LB agar plates and checking for microbial growth.

5.2. Isolation and Identification of Bacteria

To inoculate duckweed with bacteria for experiments, bacteria were isolated from different duckweed samples and the seaweed *Ulva fasciata* by washing tissues before homogenization and plating on LB agar or tryptic soy agar (TSA; BD, Franklin Lakes, NJ, USA; catalog #236950) plates at 28 °C for up to three days (File S1). For some bacterial isolates, plant host tissues were surface sterilized, using the procedure described above, before isolation. Pure cultures for these bacterial isolates were generated by picking single colonies from LB agar or TSA plates and inoculating liquid LB or tryptic soy broth (TSB; Hardy Diagnostics, Santa Maria, CA, USA; catalog #C7141) for up to two days at 28 °C. Glycerol stocks were then generated for each isolate and stored at –80 °C as stock cultures until further use. Cultures of *Azospirillum* strains, *A. brasilense* Sp7, and *A. baldaniorum* Sp245 (formerly *A. brasilense*) [57] were obtained from S. Lebeis (MSU) and stored as glycerol stocks.

Bacterial isolates from duckweed and seaweed were previously identified using the following procedure [54]. The 16S rRNA gene fragment was amplified with polymerase chain reaction (PCR) using the primers 16S-e9f and 16S-e926r (File S2) [32]. PCR reactions were prepared in a total volume of 25 uL consisting of: 1X PCR buffer with Mg^{2+} (1.5 mM $MgCl_2$, 10 mM KCl, 8 mM $(NH_4)_2SO_4$, 10 mM Tris-HCl, pH 9.0, 0.05 % NP-40; Thomas Scientific, Swedesboro, NJ, USA), 0.4 μM forward primer, 0.4 μM reverse primer, 0.2 mM dNTPs, 2 units of Choice-Taq DNA polymerase (Thomas Scientific, Swedesboro, NJ, USA; catalog #CB4050-2), and 1 μL of either bacterial nucleic acid (100 ng/μL), bacterial DNA (5 ng/μL), or bacterial liquid culture. PCR reactions were performed using the following 3-stage thermocycler program: (1) denaturation stage of 95 °C for 5 min; (2) 25 cycles consisting of 95 °C for 30 s, 50 °C for 30 s, and 72 °C for 1 min; and (3) a final extension stage of 72 °C for 5 min. PCR products were cleaned using ExoSAP-It PCR Product Cleanup Reagent (ThermoFisher Scientific, Waltham, MA, USA; catalog #78200.200.UL) or DNA Clean and Concentrator-5 kit (Zymo Research, Irvine, CA, USA; catalog #D4003). PCR products were sent to Genewiz (South Plainfield, NJ, USA) for sequencing using both 16S-e9f and 16S-e926r primers.

For each isolate, the resulting chromatograms for both forward and reverse sequences were analyzed, and poor-quality sequences at both 5′ and 3′ ends were cropped using FinchTV v1.3.0 (Geospiza, Inc.) (www.digitalworldbiology.com). Forward and reverse sequences were aligned using SerialCloner v2.6.1 (http://serialbasics.free.fr/Serial_Cloner.html) to generate a consensus sequence. Gaps and mismatches were corrected in the consensus sequence using the chromatograms of the raw sequences. The consensus sequence was cropped 216 bp downstream and 385 bp upstream of the conserved U515 (5′-GTGCCAGCAGCCGCGGTAA-3′) sequence [32] to generate a 620 bp processed sequence. Processed sequences were annotated using the RDP classifier v2.13 with the 16S rRNA training set 18 [105].

5.3. Genome Sequencing of Bacterial Isolates

Draft genomes were generated at the DOE Joint Genome Institute (JGI) for duckweed-associated bacterial (DAB) isolates DAB 1A and DAB 3D and the seaweed bacterial isolate G2-6 (File S1). Standard 300 bp Illumina shotgun libraries were constructed for all isolates.

DAB 1A (*Microbacterium* sp. RU370.1) and DAB 3D (*Bacillus* sp. RU9509.4) libraries were sequenced with the Illumina HiSeq 2000 platform. Raw reads were filtered for artifacts using BBDUK (Bushnell B., sourceforge.net/projects/bbmap/). Filtered reads were assembled using Velvet v1.2.07 [106] with the following parameters for velveth: 63, -shortPaired and the following parameters for velvetg: -very_clean yes, -exportFiltered yes, -min_contig_lgth 500, -scaffolding no, -cov_cutoff 10. Velvet contigs were then used to create 1–3 kb simulated paired end reads using wgsim v0.3.0 (https://github.com/lh3/wgsim) with the following parameters: -e 0, -1 100, -2 100, -r 0, -R 0, -X 0. Simulated read pairs were then used to assemble Illumina reads using Allpaths-LG version r46642 [107] with the following parameters: PrepareAllpathsINputs, RunAllpathsLG. Assembly of 16S rRNA genes (percent 16S rRNA sequence covered in assembly is $\geq$80% or length $\geq$ 1000 bp) was performed using filtered Illumina reads and non-duplicated sequences were merged into Allpaths assembly.

G2-6 (*Bacillus simplex* RUG2-6) libraries were sequenced with the Illumina HiSeq-2500 1TB platform. Read were processed using the BBTools suite at JGI (BBMap—Bushnell B.—sourceforge.net/projects/bbmap/). Raw reads were filtered for artifacts using BBDUK based on the following criteria: more than one N, quality scores on average less than 8 (before trimming), or reads shorter than 51 bp (after trimming). Reads were then mapped to masked versions of human, cat, and dog references and discarded if identity was greater than 95% using BBMap. Reads were then masked using BBMask. Processed reads were assembled using SPAdes v3.6.2 [108] with the following parameters: –cov-cutoff auto, –phred-offset 33, -t 8, -m 40, –careful, -k 25,55,95, –12. Assembly contigs less than 1 kbp were discarded.

5.4. Inoculation of Duckweed with Bacteria

To study the bacterial colonization of duckweed, axenic duckweed was inoculated with the bacterial isolates described above. To inoculate duckweed with bacteria, a glycerol stock for the respective bacterium was used to inoculate a 5 mL liquid culture of LB or TSB and grown for up to two days at 28 °C by shaking on a rotating platform at 220 rpm. A volume of 500 µL from the 5 mL liquid culture was then used to inoculate a 50 mL liquid culture of LB or TSB and grown for up to two days at 28 °C at 220 rpm. The 50 mL bacterial culture was spun at 8000 rpm for 5 min at 4 °C. The supernatant was then decanted, and the bacterial pellet was resuspended and washed with 0.5X SH. The sample was centrifuged as mentioned above. The resulting bacterial pellet was resuspended in 0.5X SH media and diluted to an optical density (600 nm) of 0.2 in a final volume of 50 mL in a glass plant tissue culture vessel (Phytotechnology Laboratories, Lenexa, KS, USA; catalog #C1770). Duckweed was then transferred to this 50 mL bacterial culture to cover the entire surface of the 50 mL bacterial culture. Inoculated duckweed was then incubated in a growth chamber under the same conditions used for duckweed propagation described above.

Wastewater samples were used to examine the colonization of duckweed by bacterial isolates in the presence of a microbial community. Wastewater samples were collected from the United Water Princeton Meadows wastewater treatment facility (Plainsboro, NJ, USA) after secondary clarification. For wastewater experiments, duckweed was inoculated as described above in 50 mL of non-sterile or filter-sterilized wastewater using 0.2 µm polyethersulfone filters.

5.5. Nucleic Acid Isolation from Duckweed and Bacteria

A bead-beating protocol was used to isolate nucleic acid from duckweed and bacteria. A combination of one 4 mm glass bead (OPS Diagnostics, Lebanon, NJ, USA; catalog #BAWG 4000-200-18), 0.5 g of 1.7 mm zirconium beads (OPS Diagnostics, Lebanon, NJ, USA; catalog #BAWG 1700-300-22), and 0.5 g of 100 µm silica beads (OPS Diagnostics, Lebanon, NJ, USA; catalog #BAWG 100-200-10) was used for bead-beating to lyse samples. The lysis buffer consisted of 300 µL of high salt CTAB buffer (100 mM Tris-HCl pH 8, 2.0 M NaCl, 20 mM EDTA, 2% CTAB) and 300 µL chloroform. Duckweed, bacteria, or inoculated duckweed samples were transferred to bead-beating tubes with beads and lysis buffer and homogenized using an HT6 benchtop homogenizer from OPS Diagnostics (Lebanon, NJ, USA) for 5 min (10 cycles of 30-s homogenization and 10-s pause) at 4000 rpm at room temperature or 4°C. The remaining steps were carried out at room temperature. After homogenization, samples were then centrifuged at 16,000× *g* for 5–10 min. Supernatants were transferred to new tubes and washed with 1× volume of 24:1 chloroform:isoamyl alcohol to remove protein precipitate and centrifuged at 16,000× *g* for 5–10 min. This wash step was repeated. Supernatants were then transferred to new tubes and 0.5× volume of 7.5 M ammonium acetate and 2.5× volume of 95% chilled ethanol were added [109]. Samples were centrifuged at 16,000× *g* for 30 min to pellet the precipitated nucleic acid. The resulting sample pellets were washed with 70% chilled ethanol and centrifuged at 16,000× *g* for 5–10 min. This step was repeated. Sample pellets were then air-dried and resuspended in 20 μL of sterile water or TE buffer. Nucleic acid concentration of samples were measured with a NanoDrop microvolume spectrophotometer (ThermoFisher Scientific, Waltham, MA, USA).

5.6. Molecular Detection of Duckweed-Associated Bacteria by rDNA Intergenic Spacer Analysis (RISA)

Primers for rRNA intergenic spacer analysis (RISA) were designed using previous reports (File S2) [32,59,60,86]. The primers 16S-e1390f and 23S-e130r were selected to detect bacterial colonization of Lm5576 (File S2). RISA PCR reactions were prepared in a total volume of 25 μL consisting of: 0.5 mM $MgCl_2$, 1X PCR buffer with Mg^{2+} (1.5 mM $MgCl_2$, 10 mM KCl, 8 mM $(NH_4)_2SO_4$, 10 mM Tris-HCl, pH 9.0, 0.05% NP-40; Thomas Scientific, Swedesboro, NJ, USA), 0.2 mM dNTPs, 0.8 μM forward primer, 0.8 μM reverse primer, 2.5 units of ChoiceTaq DNA polymerase (Thomas Scientific, Swedesboro, NJ, USA; catalog # CB4050–2), and 2 μL of either nucleic acids from inoculated duckweed (~100 ng/μL) or bacterial DNA (~5 ng/μL). RISA PCR reactions were executed using the following 3-stage thermocycler program: (1) denaturation stage of 95 °C for 5 min; (2) 30 cycles consisting of 95 °C for 15 s, 60 °C for 30 s, and 72 °C for 1 min 30 s; and (3) a final extension stage of 72 °C for 5 min. RISA PCR products were visualized on a 1.0% (*w/v*) agarose gel stained with ethidium bromide.

To verify sample quality and the relative amount of duckweed DNA in samples, primers were designed for the single-copy, plant-specific *LEAFY* gene (File S2). *LEAFY* gene (*LFY*) primers were designed for dw9509 [62] and *L. minor* 5500 (Lm5500) [110]. Assembly and annotation files were retrieved from CoGe (https://genomevolution.org/coge/) for dw9509 (id 51364) and Lm5500 (id 27408). The LEAFY protein from *Arabidopsis thaliana* (NP_200993.1) was searched against the proteomes of dw9509 and Lm5500 using BLASTP v2.10.0+ [111]. Gene sequences were retrieved for top hits and a pairwise global alignment was generated using MUSCLE v3.8 [112]. The primers LmLFY-F and LmLFY-R were used to amplify the *LEAFY* gene from Lm5576 for endpoint PCR (*LmLFY* PCR) (File S2). The primers SpLFY-F and SpLFY-R were used to amplify the *LEAFY* gene from dw9509 for endpoint PCR (*LmLFY* PCR) (File S2). The primers qLFY-F and qLFY-R were used to amplify the *LEAFY* gene from Lm5576 for real-time PCR (File S2). *LmLFY* and *SpLFY* PCR reactions were prepared in a total volume of 25 μL consisting of: 1× PCR buffer with Mg^{2+} (1.5 mM $MgCl_2$, 10 mM KCl, 8 mM $(NH_4)_2SO_4$, 10 mM Tris-HCl, pH 9.0, 0.05% NP-40; Thomas Scientific, Swedesboro, NJ, USA), 0.2 mM dNTPs, 0.4 μM forward primer, 0.4 μM reverse primer, 2 units of ChoiceTaq DNA polymerase (Thomas Scientific, Swedesboro, NJ; catalog # CB4050-2), and 2 μL of either nucleic acids from inoculated duckweed (~100 ng/μL) or duckweed DNA (~5 ng/μL). *LmLFY* PCR reactions were executed using the following 3-stage thermocycler program: (1) denaturation stage of 95 °C for 5 min; (2) 28 cycles consisting of 95 °C for 15 s, 60 °C for 15 s, and 72 °C for 45 s; and (3) a final extension stage of 72 °C for 5 min. *SpLFY* PCR reactions were executed using the following 3-stage thermocycler program: (1) denaturation stage of 95 °C for 5 min; (2) 26 cycles consisting of 95 °C for 15 s, 60 °C for 15 s, and 72 °C for 30 s; and (3) a final extension stage of 72 °C for 5 min. *LmLFY* and *SpLFY* PCR products were visualized on a 1.0% (*w/v*) agarose gel stained with ethidium bromide.

5.7. Confocal Microscopy of Lm5576 Colonized by Bacteria

Lm5576 was inoculated with bacteria, as described above. After seven days, inoculated Lm5576 tissue was harvested, washed with sterile H_2O, and fixed in 1 mL of 4% paraformaldehyde at room temperature in the dark overnight. The following day, the fixative solution was decanted, and the fixed tissue was washed with 1 mL of 1× phosphate-buffered saline (PBS) twice. Fixed tissue was then stored at 4 °C in 1 mL 1× PBS until further processing.

For confocal microscopy, paraformaldehyde fixed Lm5576 plants were gently washed in 1× PBS and stained for DNA 16 h at 4 °C with SYBR Gold nucleic acid stain (ThermoFisher Scientific, Waltham, MA, USA) diluted 1000 times in 1× PBS. Samples were then washed with 1× PBS and stained with 0.5 mg/mL calcofluor white stain (Sigma-Aldrich, St. Louis, MO, USA) for cellulose for 10 min at 22 °C. Confocal images were acquired at 1 μm z-steps on a Zeiss LSM 710 (Carl Zeiss MicroImaging GmbH, Jena, Germany) scanning head confocal microscope with a Zeiss plan apo 40X/1.1 objective. Excitation lasers were 405, 488, and 633 nm for the blue, green, and red emission channels, respectively. The calcofluor white fluorescence was detected at 410–551 nm, SYBR Gold fluorescence was detected at 533–572 nm, and chlorophyll autofluorescence was detected at 680–721 nm. Laser dwell times were 2.55 μs for all channels. Image analysis (2D and 3D) was conducted using Zen (Zeiss) or Volocity (PerkinElmer, Waltham, MA, USA). Both duckweed DNA and bacteria DNA were stained by SYBR Gold. Two-dimensional confocal Z-plane image stacks were visually analyzed to distinguish Lm5576 nuclei from stained bacteria cells based on their sizes and patterns of localization.

5.8. Strain-Specific Primer Design and End-Point PCR

Strain-specific primers were designed to detect the colonization of duckweed by specific bacterial strains (File S2). Two approaches were used to design primers; both approaches required sequenced genomes of the respective bacterial strains. The first approach used Panseq v3.2.1 [113] to find unique sequences for primer design for endpoint PCR. The following configuration settings were used: minimumNovelRegionSize 500, novelRegionFinderMode unique, fragmentationSize 1000, percentIdentityCutoff 85, and coreGenomeThreshold 2, runMode novel. The resulting unique sequences were then used for primer design. Primers were designed using the Primer3Plus web interface [114] with the following general settings: Primer Size Min 18, Primer Size Opt 20, Primer Size Max 25, Primer Tm Min 57, Primer Tm Opt 60, Primer Tm Max 63, Primer GC% Min 40, Primer GC% Opt 50, Primer GC% Max 60.

In the second approach, a computational pipeline (UniAmP) composed of wrapper scripts was developed to design strain-specific primers for bacteria to use in real-time PCR. To accomplish this: (1) unique sequences to a reference genome were determined and (2) these unique sequences were then used for primer design. To find unique sequences in a reference genome, query genomes with a high sequence similarity to the reference genome were retrieved. The Genome Taxonomy Database Toolkit (GTDB-tk) v1.7.0 was used to retrieve closely related query genomes from the Genome Taxonomy Database (GTDB) (release 202) [115]. Additionally, the GenBank and RefSeq databases from the National Center for Biotechnology Information (NCBI) were remotely searched using the datasets v10.25.0 command line tool (https://github.com/ncbi/datasets). For this search, all genomes pertaining to the same genus as the reference genome were downloaded. RNAmmer v1.2 [116] was then used to extract the 16S rRNA gene sequences from these genomes. Only genomes whose 16S rRNA gene was >97% identical to the 16S rRNA gene from the reference genome were used as queries. Second, pairwise genome alignments were performed between each query genome and the reference genome using nucmer v3.1 [92]. Unique sequences in the reference genome, not found in any of the query genomes, were extracted using BedTools v2.25.0 [117]. Only unique reference genome sequences that were 150–250 bp long and contained GC content of 40–60% were selected for further processing. As one last step to confirm sequences were unique to the reference genome, pairwise local alignments were performed between each unique sequence and query sequences from the same genus in the GenBank nucleotide database. Query sequences, from the same genus, were retrieved using the e-utilities from NCBI and compared using BLASTN v2.10.0+ [111]. Only the most unique reference sequence was used for primer design. To design primers, the unique reference sequence was used in a Primer-BLAST [118] search using the specified parameters: PCR product size Min 100, PCR product size Max 200, # of primers to return 500, Database nr, Organism bacteria (taxid: 2), Primer must have at least 5 total mismatches to unintended targets, including at least 2 mismatches within the last 3 bps at the 3′ end, Primer Size Min 18, Primer Size Opt 22, Primer Size Max 26, Primer GC content (%) Min 40, Primer GC content (%) Max 60. Primer-BLAST results were saved as a HTML file and parsed using a custom Python script. In silico PCR was then performed using USEARCH v11.0.667 [119] to determine the number of amplicons in the reference genome and in a selected set of query genomes. For each bacterial strain, primers with the fewest number of non-target amplicons found in the Primer-BLAST search, only 1 reference amplicon generated from in silico PCR, and the lowest primer pair complementarity based on Primer-BLAST results were used for real-time PCR experiments. Primers were also subjected to PCR suitability tests using the PCR Primer Stats function of the online Sequence Manipulation Suite (https://www.bioinformatics.org/sms2/index.html) [120].

Strain-specific PCR reactions were prepared in a total volume of 25 μL consisting of 1× PCR buffer with Mg^{2+} (1.5 mM $MgCl_2$, 10 mM KCl, 8 mM $(NH_4)_2SO_4$, 10 mM Tris-HCl, pH 9.0, 0.05% NP-40; Thomas Scientific, Swedesboro, NJ, USA), 0.2 mM dNTPs, 0.4 μM forward primer, 0.4 μM reverse primer, 2 units of ChoiceTaq DNA polymerase (Thomas Scientific, Swedesboro, NJ, USA; catalog # CB4050-2), and 2 μL of either nucleic acids from inoculated duckweed (~100 ng/μL) or bacteria DNA (5 ng/μL). For Sp7 and DAB 1A-specific PCR, 2% and 10% DMSO were added, respectively, to end-point PCR reactions to avoid non-specific amplification due to high genome GC content (File S1). PCR reactions were executed using the following 3-stage thermocycler program: (1) a denaturation stage of 95 °C for 5 min; (2) 30 cycles consisting of 95 °C for 15 s, 60 °C for 15 s, and 72 °C for 30 s; and (3) a final extension stage of 72 °C for 5 min. Strain-specific PCR products were visualized on a 1.0% (*w/v*) agarose gel stained with ethidium bromide.

5.9. *Quantification of Bacteria Colonization Levels on Duckweed*

The colonization levels of bacteria associated with duckweed were determined by real-time PCR (qPCR) using bacterial strain-specific and *qLFY* primers (File S2). Bacterial strain-specific primers were designed by the UniAmp computational pipeline, and *qLFY* primers were designed to be complementary to the single-copy, plant-specific *LEAFY* gene (File S3). For each inoculated Lm5576 sample, bacterial strain-specific primers were used to determine bacteria DNA copy number and *qLFY* primers were used to determine Lm5576 DNA copy number. To calculate bacterial colonization levels, bacteria DNA copy number was divided by Lm5576 DNA copy number for each inoculated Lm5576 sample [48].

For each qPCR reaction, a total volume of 20 μL was used and consisted of: 250 nM forward primer, 250 nM reverse primer, 1× Power SYBR Green PCR Master Mix (Thermo Fisher Scientific, Waltham, MA, USA; catalog # 4367659), and 5 μL of either nucleic acids from inoculated duckweed (~100 ng/μL) or 5 μL of DNA from Lm5576 or bacteria alone. qPCR reactions were executed and analyzed using the StepOnePlus Real-Time PCR System from Applied Biosystems with StepOne software v2.2.2 (ThermoFisher Scientific, Waltham, MA, USA). The following settings were used: standard curve experiment; method with a holding stage of 10 min at 95 °C and cycling stage of 40 cycles consisting of 95 °C for 15 s and 60 °C for 1 min.

To determine the DNA copy number in bacteria and Lm5576, standard curves were generated using bacteria and Lm5576 DNA [48]. DNA was isolated using the nucleic acid protocol described in this study with an additional RNase treatment at 37 °C for 30–60 min prior to the chloroform washes. DNA was quantified with a NanoDrop microvolume spectrophotometer (ThermoFisher Scientific, Waltham, MA, USA). Tenfold serial dilutions were performed to generate the following DNA standards: 5, 0.5, 0.05, 0.005, 0.0005 ng/µL for bacteria DNA and 50, 5, 0.5, 0.05, 0.005 ng/µL for Lm5576 DNA. The DNA copy number was then determined for each standard using the following equation:

$$DNA\ copy\ number = \frac{DNA\ quantity\ (\mathrm{ng}) \times Avogrado's\ constant}{Genome\ size\ (\mathrm{bp}) \times 10^9 \times MW\ of\ DNA}$$

where *Avogadro's constant* is 6.022×10^{23}, and the *MW of DNA* is 660 Da per 1 bp of double-stranded DNA. The following genome sizes were used: 6.07 Mbp for G2-6 (File S1), 3.39 Mbp for DAB 1A (File S1), 7.03 Mbp for Sp7 (JGI IMG genome id 2597490356), 7.53 Mbp for Sp245 (GenBank GCA_000237365.1), and 398.93 Mbp for Lm5576 (unpublished genome draft). The copy number of the DNA standards ranged from 10^5 to 10^1 copies for G2-6, Sp7, Sp245, and Lm5576, whereas a range of 10^6 to 10^2 copies was produced for DAB 1A. qPCR reactions were then executed for DNA standards to generate standard curves of bacteria and Lm5576 DNA copy numbers. These standards curves were then used by the StepOne software to calculate the bacteria and Lm5576 DNA copy number for inoculated Lm5576 samples. All standard curves contained an $R^2 > 0.97$ and efficiencies that ranged between 82–106%. Reactions with a Ct > 30 were not considered to be positive for amplification. Bacterial strain-specific primers did not amplify Lm5576 DNA, and *qLFY* primers did not amplify any bacteria DNA. No amplification was observed in any no template controls.

Supplementary Materials: The following supporting information can be downloaded at: https://www.mdpi.com/article/10.3390/plants12040872/s1. Figure S1. Nucleic acid isolation between mortar & pestle and bead-beating. Figure S2. Nucleic acid isolation from Lm5576 with bead-beating. Figure S3. Nucleic acid isolation from bacteria with bead-beating. Figure S4. Optimization of nucleic acid extraction using a bead-beating protocol. Figure S5. Nucleic acid isolation from different bacteria. Figure S6. Amplification of bacteria DNA using RISA primers. Figure S7. Optimization of *LEAFY* gene PCR. Figure S8. Overview of UniAmp computational pipeline to design strain-specific primers. File S1. Metadata of bacterial isolates used in this study. File S2. Information on primers used in this study. File S3. Design of duckweed *LEAFY* gene primers. File S4. Strain-specific primers generated using UniAmp computational pipeline. File S5. Three-dimensional confocal microscopy of inoculated Lm5576 samples. File S6. Attachment PCR results of confocal microscopy samples.

Author Contributions: Conceptualization: K.A., E.L. and S.L.L.; methodology: K.A., E.L., S.S., W.H. and W.C.; software: K.A. and T.P.M.; validation: K.A., S.S., W.H. and W.C.; formal analysis: K.A.; investigation: K.A., S.S., W.C., W.H., S.G. and T.B.; resources: K.A., E.L., W.C., S.G., T.B. and S.L.L.; Data Curation: K.A.; writing—original draft preparation: K.A.; writing—review editing: All authors; visualization: K.A. and W.C.; supervision: E.L.; project administration: E.L. and S.L.L.; funding acquisition: E.L. and S.L.L. All authors have read and agreed to the published version of the manuscript.

Funding: Duckweed research at the Lam laboratory was supported in part by a grant from the Department of Energy (DE-SC0018244). The Lam lab was also supported by a Hatch project (#12116), and a Multi-State Capacity project (#NJ12710) from the New Jersey Agricultural Experiment Station at Rutgers University during this work. Contributions by the Facilities Integrating Collaborations for User Science (FICUS) initiative, and under contract numbers DE732 AC02-05CH11231 (JGI) and DE-AC05-76RL01830 (EMSL), for the characterization of the duckweed microbiome, are also gratefully acknowledged.

Data Availability Statement: Raw experimental data, bioinformatic analyses, and protocols used in this study can be found on figshare (https://figshare.com/projects/Acosta_2023_Plants/155327). Protocols can also be found on figshare (https://figshare.com/projects/duckweed_microbiome/155330) and protocols.io (https://www.protocols.io/workspaces/duckweed_microbiome). The UniAmp pipeline is available at: https://github.com/kenscripts/UniAmp. Identifiers for 16S rRNA gene sequences and genomes generated in this study can be found in File S1. For the *Azospirillum brasilense* Sp7 genome, the JGI assembly with IMG genome id 2597490356 was used. For the *Azospirillum baldaniorum* Sp245 genome, the GenBank assembly GCA_000237365.1 was used.

Acknowledgments: A portion of this research was performed at the Environmental Molecular Sciences Laboratory, a national scientific user facility sponsored by the Department of Energy's Office of Biological and Environmental Research and located at Pacific Northwest National Laboratory.

Conflicts of Interest: The authors declare no conflict of interest.

References

1. Acosta, K.; Appenroth, K.J.; Borisjuk, L.; Edelman, M.; Heinig, U.; Jansen, M.A.K.; Oyama, T.; Pasaribu, B.; Schubert, I.; Sorrels, S.; et al. Return of the Lemnaceae: Duckweed as a Model Plant System in the Genomics and Postgenomics Era. *Plant Cell* **2021**, *33*, 3207–3234. [CrossRef] [PubMed]
2. Fitzpatrick, C.R.; Salas-González, I.; Conway, J.M.; Finkel, O.M.; Gilbert, S.; Russ, D.; Teixeira, P.J.P.L.; Dangl, J.L. The Plant Microbiome: From Ecology to Reductionism and Beyond. *Annu. Rev. Microbiol.* **2020**, *74*, 81–100. [CrossRef] [PubMed]
3. Acosta, K.; Xu, J.; Gilbert, S.; Denison, E.; Brinkman, T.; Lebeis, S.; Lam, E. Duckweed Hosts a Taxonomically Similar Bacterial Assemblage as the Terrestrial Leaf Microbiome. *PLoS ONE* **2020**, *15*, e0228560. [CrossRef]
4. Xie, W.-Y.; Su, J.-Q.; Zhu, Y.-G. Phyllosphere Bacterial Community of Floating Macrophytes in Paddy Soil Environments as Revealed by Illumina High-Throughput Sequencing. *Appl. Environ. Microbiol.* **2015**, *81*, 522–532. [CrossRef] [PubMed]
5. Inoue, D.; Hiroshima, N.; Ishizawa, H.; Ike, M. Whole Structures, Core Taxa, and Functional Properties of Duckweed Microbiomes. *Bioresour. Technol. Rep.* **2022**, *18*, 101060. [CrossRef]
6. Beilsmith, K.; Perisin, M.; Bergelson, J. Natural Bacterial Assemblages in *Arabidopsis thaliana* Tissues Become More Distinguishable and Diverse during Host Development. *mBio* **2021**, *12*, e02723-20. [CrossRef]
7. Hacquard, S.; Garrido-Oter, R.; González, A.; Spaepen, S.; Ackermann, G.; Lebeis, S.; McHardy, A.C.; Dangl, J.L.; Knight, R.; Ley, R.; et al. Microbiota and Host Nutrition across Plant and Animal Kingdoms. *Cell Host Microbe* **2015**, *17*, 603–616. [CrossRef] [PubMed]
8. Wang, N.R.; Wiesmann, C.L.; Melnyk, R.A.; Hossain, S.S.; Chi, M.-H.; Martens, K.; Craven, K.; Haney, C.H. Commensal Pseudomonas Fluorescens Strains Protect *Arabidopsis* from Closely Related Pseudomonas Pathogens in a Colonization-Dependent Manner. *mBio* **2022**, *13*, e02892-21. [CrossRef]
9. Innerebner, G.; Knief, C.; Vorholt, J.A. Protection of *Arabidopsis thaliana* against Leaf-Pathogenic *Pseudomonas syringae* by *Sphingomonas* Strains in a Controlled Model System. *Appl. Environ. Microbiol.* **2011**, *77*, 3202–3210. [CrossRef]
10. Hassani, M.A.; Amine Hassani, M.; Durán, P.; Hacquard, S. Microbial Interactions within the Plant Holobiont. *Microbiome* **2018**, *6*, 58. [CrossRef]
11. Wippel, K.; Tao, K.; Niu, Y.; Zgadzaj, R.; Kiel, N.; Guan, R.; Dahms, E.; Zhang, P.; Jensen, D.B.; Logemann, E.; et al. Host Preference and Invasiveness of Commensal Bacteria in the Lotus and Arabidopsis Root Microbiota. *Nat. Microbiol.* **2021**, *6*, 1150–1162. [CrossRef] [PubMed]
12. Van Veen, J.A.; van Overbeek, L.S.; van Elsas, J.D. Fate and Activity of Microorganisms Introduced into Soil. *Microbiol. Mol. Biol. Rev.* **1997**, *61*, 121–135. [PubMed]
13. Gamalero, E.; Lingua, G.; Berta, G.; Lemanceau, P. Methods for Studying Root Colonization by Introduced Beneficial Bacteria. *Sustain. Agric.* **2009**, 601–615. [CrossRef]
14. Romano, I.; Ventorino, V.; Pepe, O. Effectiveness of Plant Beneficial Microbes: Overview of the Methodological Approaches for the Assessment of Root Colonization and Persistence. *Front. Plant Sci.* **2020**, *11*, 6. [CrossRef] [PubMed]
15. Katagiri, F.; Thilmony, R.; He, S.Y. The Arabidopsis Thaliana-Pseudomonas Syringae Interaction. *Arabidopsis Book Am. Soc. Plant Biol.* **2002**, *1*, e0039. [CrossRef] [PubMed]
16. Tornero, P.; Dangl, J.L. A High-Throughput Method for Quantifying Growth of Phytopathogenic Bacteria in *Arabidopsis thaliana*. *Plant J.* **2002**, *28*, 475–481. [CrossRef]
17. Haney, C.H.; Samuel, B.S.; Bush, J.; Ausubel, F.M. Associations with Rhizosphere Bacteria Can Confer an Adaptive Advantage to Plants. *Nat. Plants* **2015**, *1*, 15051. [CrossRef]
18. Zinniel, D.K.; Lambrecht, P.; Harris, N.B.; Feng, Z.; Kuczmarski, D.; Higley, P.; Ishimaru, C.A.; Arunakumari, A.; Barletta, R.G.; Vidaver, A.K. Isolation and Characterization of Endophytic Colonizing Bacteria from Agronomic Crops and Prairie Plants. *Appl. Environ. Microbiol.* **2002**, *68*, 2198–2208. [CrossRef]
19. Niu, B.; Paulson, J.N.; Zheng, X.; Kolter, R. Simplified and Representative Bacterial Community of Maize Roots. *Proc. Natl. Acad. Sci. USA* **2017**, *114*, E2450–E2459. [CrossRef]
20. O'Banion, B.S.; O'Neal, L.; Alexandre, G.; Lebeis, S.L. Bridging the Gap Between Single-Strain and Community-Level Plant-Microbe Chemical Interactions. *Mol. Plant. Microbe Interact.* **2020**, *33*, 124–134. [CrossRef]
21. Reinhold-Hurek, B.; Hurek, T. Life in Grasses: Diazotrophic Endophytes. *Trends Microbiol.* **1998**, *6*, 139–144. [CrossRef]
22. Kandel, S.L.; Joubert, P.M.; Doty, S.L. Bacterial Endophyte Colonization and Distribution within Plants. *Microorganisms* **2017**, *5*, 77. [CrossRef] [PubMed]
23. James, E.K.; Gyaneshwar, P.; Mathan, N.; Barraquio, W.L.; Reddy, P.M.; Iannetta, P.P.M.; Olivares, F.L.; Ladha, J.K. Infection and Colonization of Rice Seedlings by the Plant Growth-Promoting Bacterium *Herbaspirillum seropedicae* Z67. *Mol. Plant. Microbe Interact.* **2002**, *15*, 894–906. [CrossRef] [PubMed]

24. Schloter, M.; Hartmann, A. Endophytic and Surface Colonization of Wheat Roots (*Triticum aestivum*) by Different *Azospirillum brasilense* Strains Studied with Strain-Specific Monoclonal Antibodies. *Symbiosis* **1998**, *25*, 159–179.
25. Yamaga, F.; Washio, K.; Morikawa, M. Sustainable Biodegradation of Phenol by *Acinetobacter calcoaceticus* P23 Isolated from the Rhizosphere of Duckweed *Lemna aoukikusa*. *Environ. Sci. Technol.* **2010**, *44*, 6470–6474. [CrossRef] [PubMed]
26. Assmus, B.; Hutzler, P.; Kirchhof, G.; Amann, R.; Lawrence, J.R.; Hartmann, A. In Situ Localization of *Azospirillum brasilense* in the Rhizosphere of Wheat with Fluorescently Labeled, rRNA-Targeted Oligonucleotide Probes and Scanning Confocal Laser Microscopy. *Appl. Environ. Microbiol.* **1995**, *61*, 1013–1019. [CrossRef]
27. Chelius, M.K.; Triplett, E.W. Immunolocalization of Dinitrogenase Reductase Produced by *Klebsiella pneumoniae* in Association with *Zea mays* L. *Appl. Environ. Microbiol.* **2000**, *66*, 783–787. [CrossRef]
28. Hurek, T.; Reinhold-Hurek, B.; Van Montagu, M.; Kellenberger, E. Root Colonization and Systemic Spreading of *Azoarcus* sp. Strain BH72 in Grasses. *J. Bacteriol.* **1994**, *176*, 1913–1923. [CrossRef]
29. Fan, B.; Borriss, R.; Bleiss, W.; Wu, X. Gram-Positive Rhizobacterium Bacillus amyloliquefaciens FZB42 Colonizes Three Types of Plants in Different Patterns. *J. Microbiol.* **2012**, *50*, 38–44. [CrossRef]
30. Tringe, S.G.; Hugenholtz, P. A Renaissance for the Pioneering 16S rRNA Gene. *Curr. Opin. Microbiol.* **2008**, *11*, 442–446. [CrossRef] [PubMed]
31. Thompson, L.R.; Sanders, J.G.; McDonald, D.; Amir, A.; Ladau, J.; Locey, K.J.; Prill, R.J.; Tripathi, A.; Gibbons, S.M.; Ackermann, G.; et al. A Communal Catalogue Reveals Earth's Multiscale Microbial Diversity. *Nature* **2017**, *551*, 457–463. [CrossRef]
32. Baker, G.C.; Smith, J.J.; Cowan, D.A. Review and Re-Analysis of Domain-Specific 16S Primers. *J. Microbiol. Methods* **2003**, *55*, 541–555. [CrossRef]
33. Tkacz, A.; Hortala, M.; Poole, P.S. Absolute Quantitation of Microbiota Abundance in Environmental Samples. *Microbiome* **2018**, *6*, 110. [CrossRef]
34. Guo, X.; Zhang, X.; Qin, Y.; Liu, Y.-X.; Zhang, J.; Zhang, N.; Wu, K.; Qu, B.; He, Z.; Wang, X.; et al. Host-Associated Quantitative Abundance Profiling Reveals the Microbial Load Variation of Root Microbiome. *Plant Commun.* **2020**, *1*, 100003. [CrossRef] [PubMed]
35. Lundberg, D.S.; Pramoj Na Ayutthaya, P.; Strauß, A.; Shirsekar, G.; Lo, W.-S.; Lahaye, T.; Weigel, D. Host-Associated Microbe PCR (hamPCR) Enables Convenient Measurement of Both Microbial Load and Community Composition. *Elife* **2021**, *10*, e66186. [CrossRef] [PubMed]
36. Kim, M.; Morrison, M.; Yu, Z. Evaluation of Different Partial 16S rRNA Gene Sequence Regions for Phylogenetic Analysis of Microbiomes. *J. Microbiol. Methods* **2011**, *84*, 81–87. [CrossRef] [PubMed]
37. Johnson, J.S.; Spakowicz, D.J.; Hong, B.-Y.; Petersen, L.M.; Demkowicz, P.; Chen, L.; Leopold, S.R.; Hanson, B.M.; Agresta, H.O.; Gerstein, M.; et al. Evaluation of 16S rRNA Gene Sequencing for Species and Strain-Level Microbiome Analysis. *Nat. Commun.* **2019**, *10*, 5029. [CrossRef]
38. Matsuo, Y.; Komiya, S.; Yasumizu, Y.; Yasuoka, Y.; Mizushima, K.; Takagi, T.; Kryukov, K.; Fukuda, A.; Morimoto, Y.; Naito, Y.; et al. Full-Length 16S rRNA Gene Amplicon Analysis of Human Gut Microbiota Using MinION^{TM} Nanopore Sequencing Confers Species-Level Resolution. *BMC Microbiol.* **2021**, *21*, 35. [CrossRef]
39. Callahan, B.J.; Wong, J.; Heiner, C.; Oh, S.; Theriot, C.M.; Gulati, A.S.; McGill, S.K.; Dougherty, M.K. High-Throughput Amplicon Sequencing of the Full-Length 16S rRNA Gene with Single-Nucleotide Resolution. *Nucleic Acids Res.* **2019**, *47*, e103. [CrossRef]
40. Janda, J.M.; Abbott, S.L. 16S rRNA Gene Sequencing for Bacterial Identification in the Diagnostic Laboratory: Pluses, Perils, and Pitfalls. *J. Clin. Microbiol.* **2007**, *45*, 2761–2764. [CrossRef]
41. Mignard, S.; Flandrois, J.P. 16S rRNA Sequencing in Routine Bacterial Identification: A 30-Month Experiment. *J. Microbiol. Methods* **2006**, *67*, 574–581. [CrossRef]
42. Yang, B.; Wang, Y.; Qian, P.-Y. Sensitivity and Correlation of Hypervariable Regions in 16S rRNA Genes in Phylogenetic Analysis. *BMC Bioinform.* **2016**, *17*, 135. [CrossRef] [PubMed]
43. Pei, A.Y.; Oberdorf, W.E.; Nossa, C.W.; Agarwal, A.; Chokshi, P.; Gerz, E.A.; Jin, Z.; Lee, P.; Yang, L.; Poles, M.; et al. Diversity of 16S rRNA Genes within Individual Prokaryotic Genomes. *Appl. Environ. Microbiol.* **2010**, *76*, 3886–3897. [CrossRef] [PubMed]
44. Maroniche, G.A.; García, J.E.; Salcedo, F.; Creus, C.M. Molecular Identification of Azospirillum Spp.: Limitations of 16S rRNA and Qualities of rpoD as Genetic Markers. *Microbiol. Res.* **2017**, *195*, 1–10. [CrossRef]
45. Rilling, J.I.; Acuña, J.J.; Nannipieri, P.; Cassan, F.; Maruyama, F.; Jorquera, M.A. Current opinion and perspectives on the methods for tracking and monitoring plant growth–promoting bacteria. *Soil Biol. Biochem.* **2019**, *130*, 205–219. [CrossRef]
46. Ruppel, S.; Rühlmann, J.; Merbach, W. Quantification and localization of bacteria in plant tissues using quantitative real-time PCR and online emission fingerprinting. *Plant Soil* **2006**, *286*, 21–35. [CrossRef]
47. Couillerot, O.; Poirier, M.-A.; Prigent-Combaret, C.; Mavingui, P.; Caballero-Mellado, J.; Moënne-Loccoz, Y. Assessment of SCAR Markers to Design Real-Time PCR Primers for Rhizosphere Quantification of *Azospirillum brasilense* Phytostimulatory Inoculants of Maize. *J. Appl. Microbiol.* **2010**, *109*, 528–538. [CrossRef]
48. Pereira, T.P.; do Amaral, F.P.; Dall'Asta, P.; Brod, F.C.A.; Arisi, A.C.M. Real-Time PCR Quantification of the Plant Growth Promoting Bacteria *Herbaspirillum seropedicae* Strain SmR1 in Maize Roots. *Mol. Biotechnol.* **2014**, *56*, 660–670. [CrossRef]
49. Stets, M.I.; Alqueres, S.M.C.; Souza, E.M.; Pedrosa, F.d.O.; Schmid, M.; Hartmann, A.; Cruz, L.M. Quantification of *Azospirillum brasilense* FP2 Bacteria in Wheat Roots by Strain-Specific Quantitative PCR. *Appl. Environ. Microbiol.* **2015**, *81*, 6700–6709. [CrossRef]

50. Zhang, Y.; Gao, X.; Wang, S.; Zhu, C.; Li, R.; Shen, Q. Application of *Bacillus velezensis* NJAU-Z9 Enhanced Plant Growth Associated with Efficient Rhizospheric Colonization Monitored by qPCR with Primers Designed from the Whole Genome Sequence. *Curr. Microbiol.* **2018**, *75*, 1574–1583. [CrossRef]
51. Mendis, H.C.; Thomas, V.P.; Schwientek, P.; Salamzade, R.; Chien, J.-T.; Waidyarathne, P.; Kloepper, J.; De La Fuente, L. Strain-Specific Quantification of Root Colonization by Plant Growth Promoting Rhizobacteria *Bacillus firmus* I-1582 and *Bacillus amyloliquefaciens* QST713 in Non-Sterile Soil and Field Conditions. *PLoS ONE* **2018**, *13*, e0193119. [CrossRef]
52. Urrea-Valencia, S.; Etto, R.M.; Takahashi, W.Y.; Caires, E.F.; Bini, A.R.; Ayub, R.A.; Stets, M.I.; Cruz, L.M.; Galvao, C.W. Detection of *Azospirillum brasilense* by qPCR throughout a maize field trial. *Appl. Soil Ecol.* **2021**, *160*, 103849. [CrossRef]
53. Kaminsky, L.M.; Bell, T.H. Novel primers for quantification of *Priestia megaterium* populations in soil using qPCR. *Appl. Soil Ecol.* **2022**, *180*, 104628. [CrossRef]
54. Gilbert, S.; Xu, J.; Acosta, K.; Poulev, A.; Lebeis, S.; Lam, E. Bacterial Production of Indole Related Compounds Reveals Their Role in Association Between Duckweeds and Endophytes. *Front. Chem.* **2018**, *6*, 265. [CrossRef] [PubMed]
55. Gilbert, S.; Poulev, A.; Chrisler, W.; Acosta, K.; Orr, G.; Lebeis, S.; Lam, E. Auxin-Producing Bacteria from Duckweeds Have Different Colonization Patterns and Effects on Plant Morphology. *Plants* **2022**, *11*, 721. [CrossRef] [PubMed]
56. Steenhoudt, O.; Vanderleyden, J. Azospirillum, a Free-Living Nitrogen-Fixing Bacterium Closely Associated with Grasses: Genetic, Biochemical and Ecological Aspects. *FEMS Microbiol. Rev.* **2000**, *24*, 487–506. [CrossRef] [PubMed]
57. Ferreira, N.d.S.; dos Santos Ferreira, N.; Anna, F.H.S.; Reis, V.M.; Ambrosini, A.; Volpiano, C.G.; Rothballer, M.; Schwab, S.; Baura, V.A.; Balsanelli, E.; et al. Genome-Based Reclassification of *Azospirillum brasilense* Sp245 as the Type Strain of *Azospirillum baldaniorum* Sp. Nov. *Int. J. Syst. Evol. Microbiol.* **2020**, *70*, 6203–6212. [CrossRef]
58. Borisjuk, N.; Chu, P.; Gutierrez, R.; Zhang, H.; Acosta, K.; Friesen, N.; Sree, K.S.; Garcia, C.; Appenroth, K.J.; Lam, E. Assessment, Validation and Deployment Strategy of a Two-Barcode Protocol for Facile Genotyping of Duckweed Species. *Plant Biol.* **2015**, *17* (Suppl. S1), 42–49. [CrossRef] [PubMed]
59. García-Martínez, J.; Acinas, S.G.; Antón, A.I.; Rodríguez-Valera, F. Use of the 16S–23S Ribosomal Genes Spacer Region in Studies of Prokaryotic Diversity. *J. Microbiol. Methods* **1999**, *36*, 55–64. [CrossRef]
60. Gürtler, V.; Stanisich, V.A. New Approaches to Typing and Identification of Bacteria Using the 16S-23S rDNA Spacer Region. *Microbiology* **1996**, *142*, 3–16. [CrossRef]
61. Michael, T.P.; Bryant, D.; Gutierrez, R.; Borisjuk, N.; Chu, P.; Zhang, H.; Xia, J.; Zhou, J.; Peng, H.; El Baidouri, M.; et al. Comprehensive Definition of Genome Features in *Spirodela polyrhiza* by High-Depth Physical Mapping and Short-Read DNA Sequencing Strategies. *Plant J.* **2017**, *89*, 617–635. [CrossRef] [PubMed]
62. Hoang, P.N.T.; Michael, T.P.; Gilbert, S.; Chu, P.; Motley, S.T.; Appenroth, K.J.; Schubert, I.; Lam, E. Generating a High-Confidence Reference Genome Map of the Greater Duckweed by Integration of Cytogenomic, Optical Mapping, and Oxford Nanopore Technologies. *Plant J.* **2018**, *96*, 670–684. [CrossRef] [PubMed]
63. Dennis, P.G.; Miller, A.J.; Hirsch, P.R. Are Root Exudates More Important than Other Sources of Rhizodeposits in Structuring Rhizosphere Bacterial Communities? *FEMS Microbiol. Ecol.* **2010**, *72*, 313–327. [CrossRef] [PubMed]
64. Baker, J.H.; Orr, D.R. Distribution of Epiphytic Bacteria on Freshwater Plants. *J. Ecol.* **1986**, *74*, 155–165. [CrossRef]
65. Borisjuk, N.; Peterson, A.A.; Lv, J.; Qu, G.; Luo, Q.; Shi, L.; Chen, G.; Kishchenko, O.; Zhou, Y.; Shi, J. Structural and Biochemical Properties of Duckweed Surface Cuticle. *Front. Chem.* **2018**, *6*, 317. [CrossRef] [PubMed]
66. Ware, A.; Jones, D.H.; Flis, P.; Smith, K.; Kümpers, B.; Yant, L.; Atkinson, J.A.; Wells, D.M.; Bishopp, A. Duckweed Roots Are Dispensable and Are on a Trajectory toward Vestigiality. *bioRxiv* **2022**, *1*, 475062. [CrossRef]
67. Chaintreuil, C.; Giraud, E.; Prin, Y.; Lorquin, J.; Bâ, A.; Gillis, M.; de Lajudie, P.; Dreyfus, B. Photosynthetic Bradyrhizobia Are Natural Endophytes of the African Wild Rice *Oryza breviligulata*. *Appl. Environ. Microbiol.* **2000**, *66*, 5437–5447. [CrossRef]
68. Gopalaswamy, G.; Kannaiyan, S.; O'Callaghan, K.J.; Davey, M.R.; Cocking, E.C. The Xylem of Rice (*Oryza Sativa*) Is Colonized by *Azorhizobium caulinodans*. *Proc. R. Soc. Lond. Ser. B Biol. Sci.* **2000**, *267*, 103–107. [CrossRef]
69. Baldani, V.L.D.; Alvarez, M.A.; Baldani, J.I.; Döbereiner, J. Establishment of Inoculated *Azospirillum* spp. in the Rhizosphere and in Roots of Field Grown Wheat and Sorghum. *Plant Soil* **1986**, *90*, 35–46. [CrossRef]
70. Tarrand, J.J.; Krieg, N.R.; Döbereiner, J. A Taxonomic Study of the *Spirillum lipoferum* Group, with Descriptions of a New Genus, Azospirillum Gen. Nov. and Two Species, *Azospirillum lipoferum* (Beijerinck) Comb. Nov. and *Azospirillum brasilense* Sp. Nov. *Can. J. Microbiol.* **1978**, *24*, 967–980. [CrossRef]
71. Gafny, R.; Okon, Y.; Kapulnik, Y.; Fischer, M. Adsorption of *Azospirillum brasilense* to Corn Roots. *Soil Biol. Biochem.* **1986**, *18*, 69–75. [CrossRef]
72. Mori, K.; Toyama, T.; Sei, K. Surfactants Degrading Activities in the Rhizosphere of Giant Duckweed (*Spirodela polyrhiza*). *Jpn. J. Water Treat. Biol.* **2005**, *41*, 129–140. [CrossRef]
73. Ishizawa, H.; Kuroda, M.; Inoue, K.; Inoue, D.; Morikawa, M.; Ike, M. Colonization and Competition Dynamics of Plant Growth-Promoting/Inhibiting Bacteria in the Phytosphere of the Duckweed *Lemna minor*. *Microb. Ecol.* **2019**, *77*, 440–450. [CrossRef]
74. Ishizawa, H.; Kuroda, M.; Inoue, D.; Morikawa, M.; Ike, M. Community Dynamics of Duckweed-Associated Bacteria upon Inoculation of Plant Growth-Promoting Bacteria. *FEMS Microbiol. Ecol.* **2020**, *96*. [CrossRef] [PubMed]

75. Haro, C.; Anguita-Maeso, M.; Metsis, M.; Navas-Cortés, J.A.; Landa, B.B. Evaluation of Established Methods for DNA Extraction and Primer Pairs Targeting 16S rRNA Gene for Bacterial Microbiota Profiling of Olive Xylem Sap. *Front. Plant Sci.* **2021**, *12*, 640829. [CrossRef]
76. Henderson, G.; Cox, F.; Kittelmann, S.; Miri, V.H.; Zethof, M.; Noel, S.J.; Waghorn, G.C.; Janssen, P.H. Effect of DNA Extraction Methods and Sampling Techniques on the Apparent Structure of Cow and Sheep Rumen Microbial Communities. *PLoS ONE* **2013**, *8*, e74787. [CrossRef]
77. Walker, A.W.; Martin, J.C.; Scott, P.; Parkhill, J.; Flint, H.J.; Scott, K.P. 16S rRNA Gene-Based Profiling of the Human Infant Gut Microbiota Is Strongly Influenced by Sample Processing and PCR Primer Choice. *Microbiome* **2015**, *3*, 26. [CrossRef]
78. Maukonen, J.; Simões, C.; Saarela, M. The Currently Used Commercial DNA-Extraction Methods Give Different Results of Clostridial and Actinobacterial Populations Derived from Human Fecal Samples. *FEMS Microbiol. Ecol.* **2012**, *79*, 697–708. [CrossRef]
79. Salonen, A.; Nikkilä, J.; Jalanka-Tuovinen, J.; Immonen, O.; Rajilić-Stojanović, M.; Kekkonen, R.A.; Palva, A.; de Vos, W.M. Comparative Analysis of Fecal DNA Extraction Methods with Phylogenetic Microarray: Effective Recovery of Bacterial and Archaeal DNA Using Mechanical Cell Lysis. *J. Microbiol. Methods* **2010**, *81*, 127–134. [CrossRef]
80. Leite, G.M.; Magan, N.; Medina, Á. Comparison of Different Bead-Beating RNA Extraction Strategies: An Optimized Method for Filamentous Fungi. *J. Microbiol. Methods* **2012**, *88*, 413–418. [CrossRef]
81. Miller, D.N.; Bryant, J.E.; Madsen, E.L.; Ghiorse, W.C. Evaluation and Optimization of DNA Extraction and Purification Procedures for Soil and Sediment Samples. *Appl. Environ. Microbiol.* **1999**, *65*, 4715–4724. [CrossRef] [PubMed]
82. Yuan, S.; Cohen, D.B.; Ravel, J.; Abdo, Z.; Forney, L.J. Evaluation of Methods for the Extraction and Purification of DNA from the Human Microbiome. *PLoS ONE* **2012**, *7*, e33865. [CrossRef] [PubMed]
83. Pollock, J.; Glendinning, L.; Wisedchanwet, T.; Watson, M. The Madness of Microbiome: Attempting to Find Consensus "Best Practice" for 16S Microbiome Studies. *Appl. Environ. Microbiol.* **2018**, *84*, e02627-17. [CrossRef] [PubMed]
84. Fouhy, F.; Clooney, A.G.; Stanton, C.; Claesson, M.J.; Cotter, P.D. 16S rRNA Gene Sequencing of Mock Microbial Populations-Impact of DNA Extraction Method, Primer Choice and Sequencing Platform. *BMC Microbiol.* **2016**, *16*, 123. [CrossRef]
85. Brooks, J.P.; Edwards, D.J.; Harwich, M.D., Jr.; Rivera, M.C.; Fettweis, J.M.; Serrano, M.G.; Reris, R.A.; Sheth, N.U.; Huang, B.; Girerd, P.; et al. The Truth about Metagenomics: Quantifying and Counteracting Bias in 16S rRNA Studies. *BMC Microbiol.* **2015**, *15*, 66. [CrossRef] [PubMed]
86. Fisher, M.M.; Triplett, E.W. Automated Approach for Ribosomal Intergenic Spacer Analysis of Microbial Diversity and Its Application to Freshwater Bacterial Communities. *Appl. Environ. Microbiol.* **1999**, *65*, 4630–4636. [CrossRef] [PubMed]
87. Bodenhausen, N.; Bortfeld-Miller, M.; Ackermann, M.; Vorholt, J.A. A Synthetic Community Approach Reveals Plant Genotypes Affecting the Phyllosphere Microbiota. *PLoS Genet.* **2014**, *10*, e1004283. [CrossRef] [PubMed]
88. Bowker, D.W.; Duffield, A.N.; Denny, P. Methods for the Isolation, Sterilization and Cultivation of Lemnaceae. *Freshw. Biol.* **1980**, *10*, 385–388. [CrossRef]
89. Huang, W.; Gilbert, S.; Poulev, A.; Acosta, K.; Lebeis, S.; Long, C.; Lam, E. Host-Specific and Tissue-Dependent Orchestration of Microbiome Community Structure in Traditional Rice Paddy Ecosystems. *Plant Soil* **2020**, *452*, 379–395. [CrossRef]
90. Bodenhausen, N.; Deslandes-Hérold, G.; Waelchli, J.; Held, A.; van der Heijden, M.G.A.; Schlaeppi, K. Relative qPCR to Quantify Colonization of Plant Roots by Arbuscular Mycorrhizal Fungi. *Mycorrhiza* **2021**, *31*, 137–148. [CrossRef]
91. Garrido-Oter, R.; Nakano, R.T.; Dombrowski, N.; Ma, K.-W.; AgBiome Team; McHardy, A.C.; Schulze-Lefert, P. Modular Traits of the Rhizobiales Root Microbiota and Their Evolutionary Relationship with Symbiotic Rhizobia. *Cell Host Microbe* **2018**, *24*, 155–167.e5. [CrossRef] [PubMed]
92. Kurtz, S.; Phillippy, A.; Delcher, A.L.; Smoot, M.; Shumway, M.; Antonescu, C.; Salzberg, S.L. Versatile and Open Software for Comparing Large Genomes. *Genome Biol.* **2004**, *5*, R12. [CrossRef] [PubMed]
93. Müller, D.B.; Vogel, C.; Bai, Y.; Vorholt, J.A. The Plant Microbiota: Systems-Level Insights and Perspectives. *Annu. Rev. Genet.* **2016**, *50*, 211–234. [CrossRef]
94. Trivedi, P.; Leach, J.E.; Tringe, S.G.; Sa, T.; Singh, B.K. Plant–microbiome Interactions: From Community Assembly to Plant Health. *Nat. Rev. Microbiol.* **2020**, *18*, 607–621. [CrossRef] [PubMed]
95. Levy, A.; Salas Gonzalez, I.; Mittelviefhaus, M.; Clingenpeel, S.; Herrera Paredes, S.; Miao, J.; Wang, K.; Devescovi, G.; Stillman, K.; Monteiro, F.; et al. Genomic Features of Bacterial Adaptation to Plants. *Nat. Genet.* **2017**, *50*, 138–150. [CrossRef]
96. Hardoim, P.R.; van Overbeek, L.S.; van Elsas, J.D. Properties of Bacterial Endophytes and Their Proposed Role in Plant Growth. *Trends Microbiol.* **2008**, *16*, 463–471. [CrossRef]
97. Cole, B.J.; Feltcher, M.E.; Waters, R.J.; Wetmore, K.M.; Mucyn, T.S.; Ryan, E.M.; Wang, G.; Ul-Hasan, S.; McDonald, M.; Yoshikuni, Y.; et al. Genome-Wide Identification of Bacterial Plant Colonization Genes. *PLoS Biol.* **2017**, *15*, e2002860. [CrossRef]
98. Ishizawa, H.; Kuroda, M.; Inoue, D.; Ike, M. Genome-Wide Identification of Bacterial Colonization and Fitness Determinants on the Floating Macrophyte, Duckweed. *Commun. Biol.* **2022**, *5*, 68. [CrossRef]
99. Duca, D.; Lorv, J.; Patten, C.L.; Rose, D.; Glick, B.R. Indole-3-Acetic Acid in Plant–microbe Interactions. *Antonie Van Leeuwenhoek* **2014**, *106*, 85–125. [CrossRef]
100. Tzipilevich, E.; Russ, D.; Dangl, J.L.; Benfey, P.N. Plant Immune System Activation Is Necessary for Efficient Root Colonization by Auxin-Secreting Beneficial Bacteria. *Cell Host Microbe* **2021**, *29*, 1507–1520.e4. [CrossRef]

101. Ishizawa, H.; Kuroda, M.; Morikawa, M.; Ike, M. Evaluation of Environmental Bacterial Communities as a Factor Affecting the Growth of Duckweed *Lemna Minor*. *Biotechnol. Biofuels* **2017**, *10*, 62. [CrossRef]
102. Vorholt, J.A.; Vogel, C.; Carlström, C.I.; Müller, D.B. Establishing Causality: Opportunities of Synthetic Communities for Plant Microbiome Research. *Cell Host Microbe* **2017**, *22*, 142–155. [CrossRef] [PubMed]
103. Carper, D.L.; Lawrence, T.J.; Carrell, A.A.; Pelletier, D.A.; Weston, D.J. DISCo-Microbe: Design of an Identifiable Synthetic Community of Microbes. *PeerJ* **2020**, *8*, e8534. [CrossRef] [PubMed]
104. Miché, L.; Balandreau, J. Effects of Rice Seed Surface Sterilization with Hypochlorite on Inoculated *Burkholderia vietnamiensis*. *Appl. Environ. Microbiol.* **2001**, *67*, 3046–3052. [CrossRef]
105. Wang, Q.; Garrity, G.M.; Tiedje, J.M.; Cole, J.R. Naive Bayesian Classifier for Rapid Assignment of rRNA Sequences into the New Bacterial Taxonomy. *Appl. Environ. Microbiol.* **2007**, *73*, 5261–5267. [CrossRef] [PubMed]
106. Zerbino, D.R.; Birney, E. Velvet: Algorithms for de Novo Short Read Assembly Using de Bruijn Graphs. *Genome Res.* **2008**, *18*, 821–829. [CrossRef]
107. Butler, J.; MacCallum, I.; Kleber, M.; Shlyakhter, I.A.; Belmonte, M.K.; Lander, E.S.; Nusbaum, C.; Jaffe, D.B. ALLPATHS: De Novo Assembly of Whole-Genome Shotgun Microreads. *Genome Res.* **2008**, *18*, 810–820. [CrossRef]
108. Bankevich, A.; Nurk, S.; Antipov, D.; Gurevich, A.A.; Dvorkin, M.; Kulikov, A.S.; Lesin, V.M.; Nikolenko, S.I.; Pham, S.; Prjibelski, A.D.; et al. SPAdes: A New Genome Assembly Algorithm and Its Applications to Single-Cell Sequencing. *J. Comput. Biol.* **2012**, *19*, 455–477. [CrossRef]
109. Crouse, J.; Amorese, D. Ethanol Precipitation: Ammonium Acetate as an Alternative to Sodium Acetate. *Focus* **1996**, *19*, 17–20.
110. Van Hoeck, A.; Horemans, N.; Monsieurs, P.; Cao, H.X.; Vandenhove, H.; Blust, R. The First Draft Genome of the Aquatic Model Plant *Lemna minor* Opens the Route for Future Stress Physiology Research and Biotechnological Applications. *Biotechnol. Biofuels* **2015**, *8*, 188. [CrossRef]
111. Camacho, C.; Coulouris, G.; Avagyan, V.; Ma, N.; Papadopoulos, J.; Bealer, K.; Madden, T.L. BLAST+: Architecture and Applications. *BMC Bioinform.* **2009**, *10*, 421. [CrossRef] [PubMed]
112. Edgar, R.C. MUSCLE: Multiple Sequence Alignment with High Accuracy and High Throughput. *Nucleic Acids Res.* **2004**, *32*, 1792–1797. [CrossRef] [PubMed]
113. Laing, C.; Buchanan, C.; Taboada, E.N.; Zhang, Y.; Kropinski, A.; Villegas, A.; Thomas, J.E.; Gannon, V.P.J. Pan-Genome Sequence Analysis Using Panseq: An Online Tool for the Rapid Analysis of Core and Accessory Genomic Regions. *BMC Bioinform.* **2010**, *11*, 461. [CrossRef] [PubMed]
114. Untergasser, A.; Nijveen, H.; Rao, X.; Bisseling, T.; Geurts, R.; Leunissen, J.A.M. Primer3Plus, an Enhanced Web Interface to Primer3. *Nucleic Acids Res.* **2007**, *35*, W71–W74. [CrossRef] [PubMed]
115. Chaumeil, P.-A.; Mussig, A.J.; Hugenholtz, P.; Parks, D.H. GTDB-Tk: A Toolkit to Classify Genomes with the Genome Taxonomy Database. *Bioinformatics* **2019**, *36*, 1925–1927. [CrossRef]
116. Lagesen, K.; Hallin, P.; Rødland, E.A.; Staerfeldt, H.-H.; Rognes, T.; Ussery, D.W. RNAmmer: Consistent and Rapid Annotation of Ribosomal RNA Genes. *Nucleic Acids Res.* **2007**, *35*, 3100–3108. [CrossRef]
117. Quinlan, A.R.; Hall, I.M. BEDTools: A Flexible Suite of Utilities for Comparing Genomic Features. *Bioinformatics* **2010**, *26*, 841–842. [CrossRef]
118. Ye, J.; Coulouris, G.; Zaretskaya, I.; Cutcutache, I.; Rozen, S.; Madden, T.L. Primer-BLAST: A tool to design target-specific primers for polymerase chain reaction. *BMC bioinform.* **2012**, *13*, 134. [CrossRef]
119. Edgar, R.C. Search and Clustering Orders of Magnitude Faster than BLAST. *Bioinformatics* **2010**, *26*, 2460–2461. [CrossRef]
120. Stothard, P. The Sequence Manipulation Suite: JavaScript Programs for Analyzing and Formatting Protein and DNA Sequences. *Biotechniques* **2000**, *28*, 1102–1104. [CrossRef]

plants

MDPI

Article

Auxin-Producing Bacteria from Duckweeds Have Different Colonization Patterns and Effects on Plant Morphology

Sarah Gilbert [1], Alexander Poulev [1], William Chrisler [2], Kenneth Acosta [1], Galya Orr [2], Sarah Lebeis [3] and Eric Lam [1,*]

[1] Department of Plant Biology, Rutgers University, New Brunswick, NJ 08901, USA; sarahg19@email.unc.edu (S.G.); apoulev@sebs.rutgers.edu (A.P.); ikennethacosta@gmail.com (K.A.)
[2] Environmental Molecular Sciences Laboratory, Pacific Northwest National Laboratory, Richland, WA 99352, USA; william.chrisler@pnnl.gov (W.C.); galya.orr@pnnl.gov (G.O.)
[3] Department of Microbiology and Molecular Genetics, Michigan State University, East Lansing, MI 48824, USA; lebeis.sarah@gmail.com
* Correspondence: ericl89@hotmail.com or eric.lam@rutgers.edu

Abstract: The role of auxin in plant–microbe interaction has primarily been studied using indole-3-acetic acid (IAA)-producing pathogenic or plant-growth-promoting bacteria. However, the IAA biosynthesis pathway in bacteria involves indole-related compounds (IRCs) and intermediates with less known functions. Here, we seek to understand changes in plant response to multiple plant-associated bacteria taxa and strains that differ in their ability to produce IRCs. We had previously studied 47 bacterial strains isolated from several duckweed species and determined that 79% of these strains produced IRCs in culture, such as IAA, indole lactic acid (ILA), and indole. Using *Arabidopsis thaliana* as our model plant with excellent genetic tools, we performed binary association assays on a subset of these strains to evaluate morphological responses in the plant host and the mode of bacterial colonization. Of the 21 tested strains, only four high-quantity IAA-producing *Microbacterium* strains caused an auxin root phenotype. Compared to the commonly used colorimetric Salkowski assay, auxin concentration determined by LC–MS was a superior indicator of a bacteria's ability to cause an auxin root phenotype. Studies with the auxin response mutant *axr1-3* provided further genetic support for the role of auxin signaling in mediating the root morphology response to IAA-producing bacteria strains. Interestingly, our microscopy results also revealed new evidence for the role of the conserved *AXR1* gene in endophytic colonization of IAA-producing *Azospirillum baldaniorum* Sp245 via the guard cells.

Keywords: duckweed-associated bacteria; *Microbacterium*; *Azospirillum*; auxin; *AXR1*; Arabidopsis

check for updates

Citation: Gilbert, S.; Poulev, A.; Chrisler, W.; Acosta, K.; Orr, G.; Lebeis, S.; Lam, E. Auxin-Producing Bacteria from Duckweeds Have Different Colonization Patterns and Effects on Plant Morphology. *Plants* **2022**, *11*, 721. https://doi.org/10.3390/plants11060721

Academic Editors: Viktor Oláh, Klaus-Jürgen Appenroth, Sowjanya Sree and Petronia Carillo

Received: 18 December 2021
Accepted: 20 February 2022
Published: 8 March 2022

1. Introduction

The phytohormone indole-3-acetic acid (IAA) is the most commonly occurring auxin found in nature and is produced by both plants and bacteria through a similar biosynthetic pathway [1–3]. In addition to its role in gravitropism and cell elongation, IAA can alter plant root architecture to increase the efficiency of nutrient acquisition or its action may be downregulated by the plant to optimize defense against pathogens [4–7]. Homeostasis of auxin activities through biosynthesis, conjugation, oxidation, and transport is important for plants to maintain a balance between defense response and growth [8]. According to the "cry for help" hypothesis, when a plant detects a pathogen, it alters its root exudation profile to recruit and assemble a beneficial microbiome [9,10]. A recent study showed that elevated reactive oxygen species (ROS) levels in Arabidopsis activated IAA production by *Bacillus velezensis* FZB42, which is necessary for its colonization [11]. Microbes may modulate plant defense or growth by manipulation of the auxin pathway in the host by directly producing IAA themselves or altering levels of endogenous IAA levels through effects on plant auxin synthesis and/or conjugation pathways [1,12–19].

With access to a large collection of aquatic plants in the duckweed family, we previously carried out an initial survey of auxin-producing bacteria within the duckweed microbiome of diverse species and genera [20]. In the bacterial supernatant from a subset of strains that were studied, a variety of IRCs were detected, such as IAA, indole-3-lactic acid, and indole. We sought to examine the specificity of active auxins by coculturing the strains with the model plant *Arabidopsis thaliana* (hereafter Arabidopsis). We chose to use Arabidopsis as the host plant in our studies because of the vast resource of characterized mutant lines and high-quality genomic resources available for this species. In this study, we used the auxin response mutant *axr1-3* for comparison to wild type as both backgrounds have a similar root phenotype grown under sterile conditions. The highly conserved *Auxin Resistant-1* (*AXR1*) gene is involved in downstream auxin signaling through canonical auxin sensor F-box proteins, such as the TIR1 (transport inhibitor response 1) receptor [21,22]. AXR1 acts via conjugation of the small, ubiquitin-related protein NEDD8 (also called RUB in plants and yeast) to CULLIN1-containing E3 ligase SCF^{TIR1} [23] and enhances their ubiquitylation of target IAA/Aux repressor proteins, such as AXR3, for rapid turnover via the proteasome [24]. The AXR1 protein is structurally related to a ubiquitin-activating E1 enzyme, and the *axr1-3* mutant shows reduced sensitivity to IAA in the roots along with various other tissues [25,26].

In addition to the lack of characterized mutants in auxin response, duckweeds reproduce asexually, with some species doubling themselves in one to two days, making it challenging to observe the effects of plant-growth-promoting bacteria on an individual plant [27]. Some duckweed species lack roots or have multiple roots, which also makes it difficult to observe a bacteria-induced auxin root phenotype [28]. Previously, we found that most of our isolates from bleach-treated duckweed were of the phyla Proteobacteria, Firmicutes, and Actinobacteria [20], which is similar to reports from model land plants, such as the dicot Arabidopsis [29–32]. In addition, epiphytic and endophytic *Azospirillum* strains of PGPB isolated from wheat behaved similarly when cocultivated with duckweed [33], suggesting likely conservation of plant–microbe association mechanisms. A more recent global survey of duckweed-associated bacteria (DAB) community structure provided robust support for the core microbiome of duckweeds being very similar to those found in leaf tissues of monocots and dicots, indicating a highly conserved mode of selection for many bacteria taxa across plant species [34].

As we had previously found that strongly associated bacteria from duckweed tissues produce different IRCs, we sought to determine whether these strains have plant-growth-promoting abilities. We hypothesized that strains producing different IRCs, such as indole-3-acetic acid, indole lactic acid, and indole, may have different colonization patterns and morphological effects on plants. In this work, 21 bacterial strains capable of producing IRCs were individually inoculated onto gnotobiotically grown Arabidopsis seedlings from wild type and the auxin response mutant line *axr1-3*. Plant morphological responses and the pattern of bacterial colonization were assessed to evaluate the effects of bacteria on root development in these genetic backgrounds.

2. Results

2.1. The Salkowski Assay Is Insufficient as a Proxy for Auxin Production by Bacteria

Upon addition of the Salkowski reagent to bacterial supernatant, a color change from yellow to red can indicate that an IRC, such as IAA, is present. With its simplicity and low cost, this method is commonly used in a high-throughput format to screen for bacteria capable of producing auxin, which refers to the well-known ability of IAA to produce a stereotypical root phenotype. Using the Salkowski assay, we previously screened a collection of 47 bacterial isolates from 16 duckweed ecotypes for their ability to produce IRCs in vitro [20]. These duckweed-associated bacteria (DABs) were classified as "pink-type" or "brown-type" depending on the color change of their supernatant when the Salkowski reagent was added [20]). Using a combination of synthetic standards for various IRCs and liquid chromatography–mass spectrometry (LC–MS), we determined that indole-

3-acetic acid (IAA) results in a pink color change and indole results in a brown color change. Our work thus demonstrated that using optical density at a single wavelength (typically at 530–535 nm) with the Salkowski reagent, as is commonly done, would not be sufficient to accurately identify IAA-producing bacteria due to high occurrence of false positives. Through LC–MS, we demonstrated the production of indole lactic acid in addition to IAA from one strain of DAB, *Herbaspirillum* RU5E [20]. Our results show that more than one type of IRC can be produced by a single DAB and that indole producers can be commonly found among Salkowski-positive bacteria strains.

In this study, we tested whether strains that were able to produce IRCs in vitro as determined by the Salkowski assay were also able to alter the physiology of a host plant, such as the production of a short root phenotype when inoculated onto Arabidopsis seedlings. This phenotype is indicative of an auxin response that results in decreased primary root length while increasing lateral root number and root hairs [35]. As positive controls, we used *Azospirillum* strains originally isolated from wheat, Sp7 and Sp245, which are well-studied PGPBs that can produce IAA and affect growth in various plant species, including Arabidopsis [36]. Of the 21 screened IAA-producing and/or indole-producing DABs, only four IAA-producing strains caused a short root phenotype (Figure 1). The strains that inhibited primary root length—*Microbacterium* sp. RU1A, *Microbacterium* sp. RU1D, *Microbacterium* sp. RU19A, and *Microbacterium* sp. RU19B—were derived from the duckweed genus *Lemna*. Only one other bacterial strain of the 21 tested was of the genus *Microbacterium*, and this strain, *Microbacterium* sp. RU33B, which was isolated from a duckweed in the genus *Wolffia*, did not inhibit primary root length in Arabidopsis. Under brightfield microscopy, we observed that primary root length inhibition was accompanied by an increase in root hairs, as demonstrated in wild-type Arabidopsis roots cocultivated with RU1A (Figure S1). This phenotype is thus indicative of auxin response in the plant by the bacteria treatment.

None of the brown-type strains that produced a significant amount of indole and a small but detectable amount of IAA [20] caused a short root phenotype in Arabidopsis. Moreover, strains that turned the darkest shade of red by the Salkowski assay and were first suspected to be high producers of IRCs did not produce a short root phenotype (Figure 1). In our assays, exogenous tryptophan, a precursor for a common pathway of IAA biosynthesis in plants and bacteria, was not added to the bacteria growth medium before inoculation onto the plant. Exogenous tryptophan would thus need to be supplied by the plant if any was taken up by the bacterial strains. Of the five strains that tested positive in the Salkowski assay without exogenous L-tryptophan [20], only one strain, *Microbacterium* RU1D, caused a short root phenotype (Figure 1). Therefore, the ability to produce IRCs, including IAA, without exogenous L-tryptophan is insufficient for the bacteria to cause a short root phenotype.

Figure 1. Bacterial strains that have a positive Salkowski assay result do not always result in primary root length inhibition. Representative image of wild-type root development after 7 days of treatment with bacteria or IAA. Out of 21 DAB strains producing a positive Salkowski assay result, only four *Microbacterium* strains—RU19B, RU19A, RU1A, and RU1D—caused an auxin root phenotype. *Azospirillum baldaniorum* strains Sp7 and Sp245 are known auxin-producing, plant-growth-promoting strains derived from wheat. *Bacillus* RU3D produced a negative Salkowski assay result and *Microbacterium* RU33B, *Rhizobium* RU20A, *Rhizobium* RU33A, *Herbaspirillum* RU5E, and *Azospirillum* RU37A produced a positive Salkowski assay result yet did not cause an auxin root phenotype.

2.2. Comparison of IAA Quantification Methods

We next asked whether the short root phenotype caused by *Microbacterium* strains may be quantitatively related to their ability to produce higher levels of IAA by the bacteria in vitro. LC–MS was used to identify and quantify the amount of free IAA in the supernatant of various bacterial strains that tested positive in our Salkowski assay. The molecular weight of free IAA is 175 g/mol, with positive ionization resulting in a molecular ion at a m/z value of 176 [M + H] and a fragment at m/z of 130, as previously determined [20]. The retention time of free IAA in our LC–MS system was determined to be approximately 9.7 min from our previous work [20]. A free IAA standard was used to determine the HPLC UV absorbance signal at 280 nm for quantification. The resulting standard curve equation was generated: $y = 5722x - 193.47$ with an R^2 value of 1.00. Using three biological replicates of 1 μL injections each, we calculated the % recovery for free IAA in our extraction with 5 ng/μL spike samples. The free IAA spike in the LB medium was 2.408 ng/μL ± 0.173 ng/μL (48% recovery), and the amount of free IAA spike in the TSB medium was 2.750 ng/μL ± 0.184 ng/μL (55% recovery).

We screened the Salkowski-negative control strain *Bacillus* RU3D, two Salkowski-positive control strains *Azospirillum* Sp7 and Sp245, two *Microbacterium* strains RU1A and RU19A that caused a short root phenotype, *Microbacterium* RU33B that did not cause a short root phenotype, as well as four additional strains that do not produce a short root phenotype and yet were top producers of IRCs based on the Salkowski assay (RU5E, RU20A, RU33A, and RU37A). The strains incapable of causing a short root phenotype in Arabidopsis seedlings all produced lower than 1 ng/µL of free IAA (Figure 2). Positive control strain Sp245 produced a similar level of IAA as previously reported [37]. In sum, our comparative analysis across these 10 strains of plant-associated bacteria indicates a requirement of higher levels (>1 ng/µL in the culture media) of IAA production by the particular strain for their ability to alter root development in Arabidopsis. By comparing the Salkowski assay to LC–MS, we determined that LC–MS is clearly a more accurate method for predicting an auxin root phenotype as the former cannot resolve various IRCs, many of which do not function as auxins.

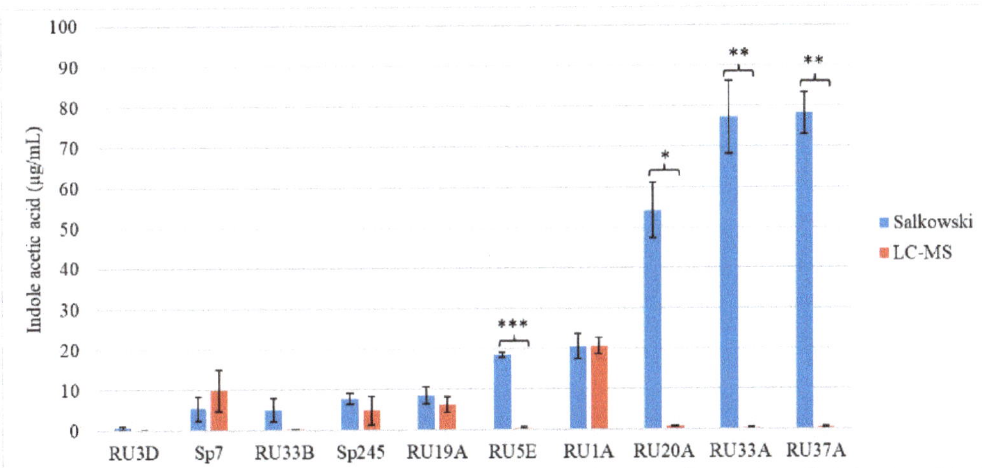

Figure 2. LC–MS and the Salkowski colorimetric assay quantify different concentrations of bacterial-derived indole acetic acid. Bars represent background subtracted mean values and standard deviation. Student's T-test (* = $p < 0.05$, ** = $p < 0.005$, *** = $p < 0.0005$) was performed with n = 3.

2.3. *Inoculation of Bacteria on Auxin Response Mutant Plants*

To further confirm that the short root phenotype we observed with the bacterial strains that can produce high levels of IAA is indeed mediated through the auxin response pathway, we tested a subset of bacteria on a characterized Arabidopsis auxin response mutant at the *AXR1* locus to determine whether their effect on root length would be suppressed. As expected, exogenously applied 1 µM IAA no longer inhibited root length in this mutant background in comparison to wild-type seedlings, thus verifying that the IAA-induced short root phenotype requires this known auxin response mediator (Figure 3). DAB RU1A also failed to inhibit root length in *axr1-3* seedlings (Figure 3). Similarly, the positive control IAA-producing strain Sp245 no longer inhibited root length in *axr1-3* (Figure 3). In summary, these results indicate that the *AXR1* gene is involved in the root response to IAA-producing bacteria, such as RU1A and Sp245. Coupled with the lack of any root response in the various strains of DABs tested, which showed little to no IAA production, our data supports the hypothesis that the auxin produced by these plant-associated bacteria, when produced at sufficiently high levels, can mediate the physiological changes in the roots of host plants via their phytohormone pathways.

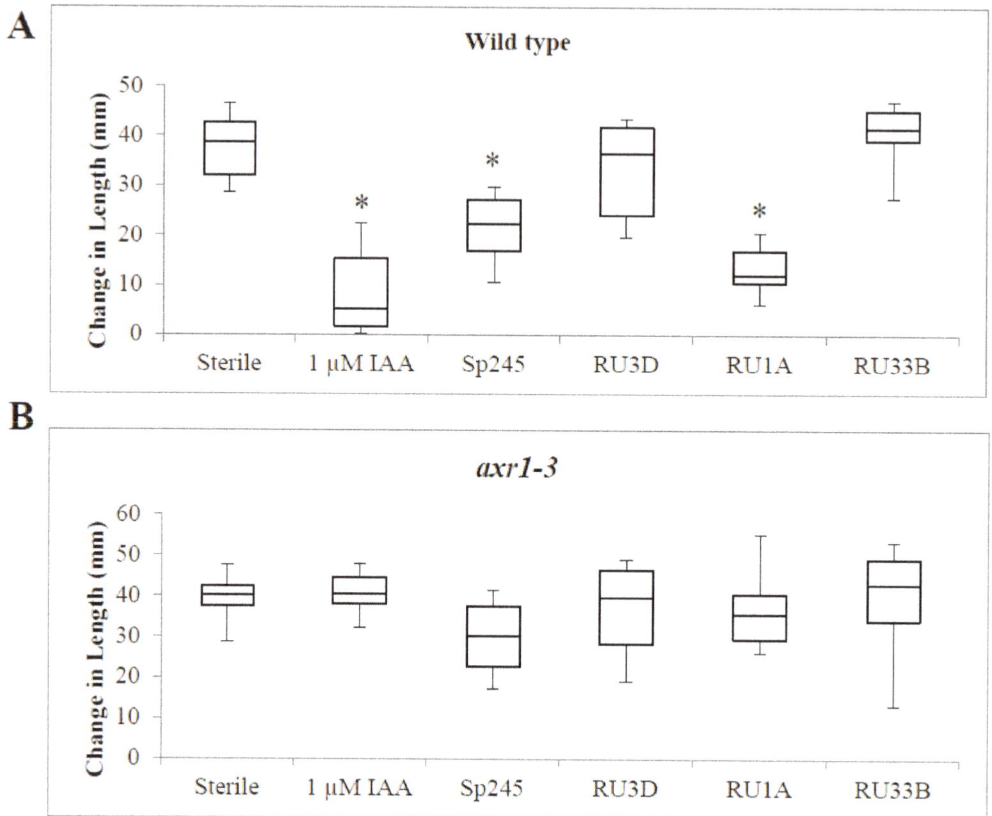

Figure 3. Auxin response gene *AXR1* is necessary for primary root length inhibition by IAA-producing strains *Azospirillum* Sp245 and *Microbacterium* RU1A. Change in primary root length after 7 days in the genetic backgrounds: (**A**) wild type and (**B**) auxin response mutant *axr1-3*. For box plots, horizontal lines represent the median, with the box representing the 25th and 75th percentiles, and the whiskers representing the minimum and maximum. For positive control, 1 μM IAA was used. Student's T-test ($p < 0.05$) was performed (n = 18), and an asterisk indicates significant difference compared to the sterile control.

2.4. Colonization of Auxin-Producing Bacteria on Wild-Type Plants

We also investigated how colonization patterns of IAA-producing bacteria on plant roots may vary considering their different abilities to inhibit root length. We compared negative control *Bacillus* strain RU3D, which does not produce detectable IAA, to strains that produce a short root phenotype (IAA-producing *Microbacterium* RU1A and *Azospirillum baldaniorum* Sp245) and low IAA-producing strains that do not produce a short root phenotype (*Microbacterium* RU33B and *Herbaspirillum* RU5E). After treatment of Arabidopsis seedlings with each of the bacteria separately for 7 days, we used high-resolution 3D confocal microscopy with nucleic acid binding dyes to observe localization of the bacteria on inoculated gnotobiotic plant tissues from these seedlings. While nuclear DNA is also stained by these dyes, the size and morphology of the stained bodies readily distinguish them from the stained bacteria colonies. RU1A and Sp245 were found to be more abundant on the root surface than RU33B and RU5E (Figure 4). Imaging leaf tissues revealed that RU33B is more abundant on the leaves than the roots (Figure S2). In contrast, our data indicated that RU1A associated more strongly with root than leaf tissues of Arabidopsis seedlings (Figure S2). The lack of a short root phenotype by RU33B may thus result from low production of IAA as well as less efficient bacterial attachment and epiphytic colonization on Arabidopsis tissues, especially roots.

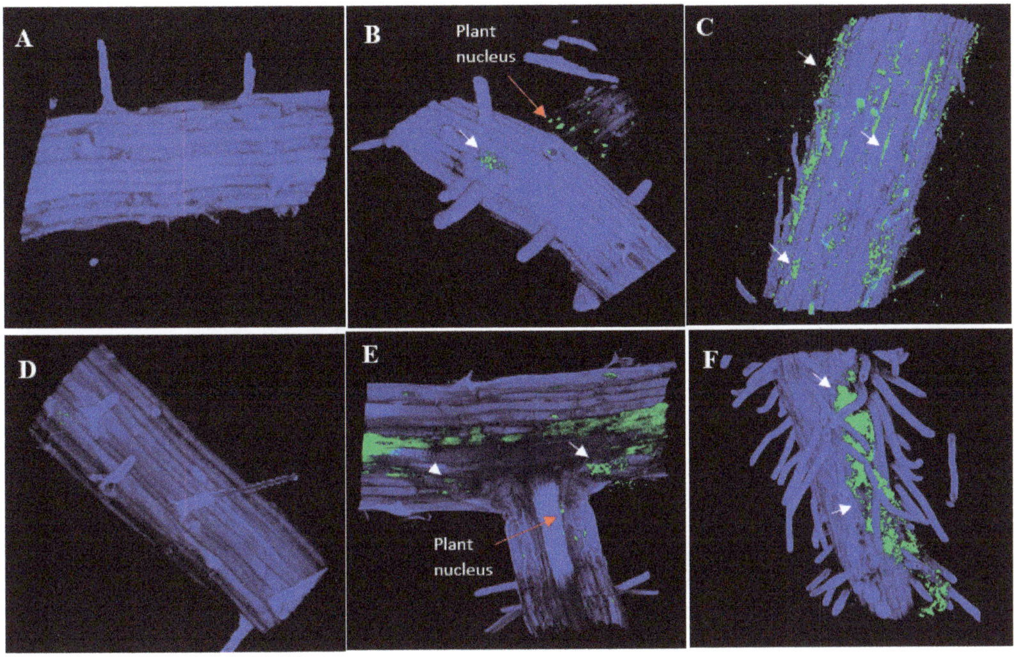

Figure 4. IAA-producing strains that inhibit root length are abundant on wild-type root tissue. A 3D reconstruction from confocal microscopy of wild-type Arabidopsis roots upon treatment with (**A**) no bacteria, (**B**) non-IAA-producing *Bacillus* RU3D, (**C**) IAA-producing *Microbacterium* RU1A, (**D**) IAA-producing *Microbacterium* RU33B, (**E**) IAA-producing *Herbaspirillum* RU5E, and (**F**) IAA-producing *Azospirillum baldaniorum* Sp245. The microscopy channels are blue (calcofluor white used to stain the cell wall) and green (SYBR Gold DNA used to stain the nuclei and bacteria). White arrows indicate bacteria location based on the size of the DNA-stained spots. Bacteria are shown as green spots that are smaller in size to plant nuclei.

2.5. Colonization of Auxin-Producing Bacteria on Auxin Response Mutant Plants

Although RU5E was not highly abundant on the root surface and did not cause a short root phenotype, we observed detectable colonization under the root epidermis (Figure 5). Similarly, RU1A and Sp245 could also colonize the intercellular space beneath the root epidermis, suggesting that these could be endophytic bacteria (Figure 5). The pattern of RU5E and RU1A colonization did not change in the roots of the auxin response mutant *axr1-3*; however, Sp245 became unable to colonize the root epidermis of *axr1-3* plants and was instead more abundant on the root surface (Figure 5). Interestingly, on wild-type leaf tissues, Sp245 appeared to often target and accumulate inside the open stomata, which are pores located on the leaf surface and used for gas exchange and water transpiration (Figure 6). Strikingly, the leaf surface of *axr1-3* mutants showed no targeting of Sp245 at the stomata and were more randomly aggregated at the intercellular grooves (Figure 6). This suggests a potential role of *AXR1* in mediating endophytic colonization of IAA-producing Sp245 by targeting the stomatal pore as a point of entry, perhaps via a guard-cell-specific signaling pathway.

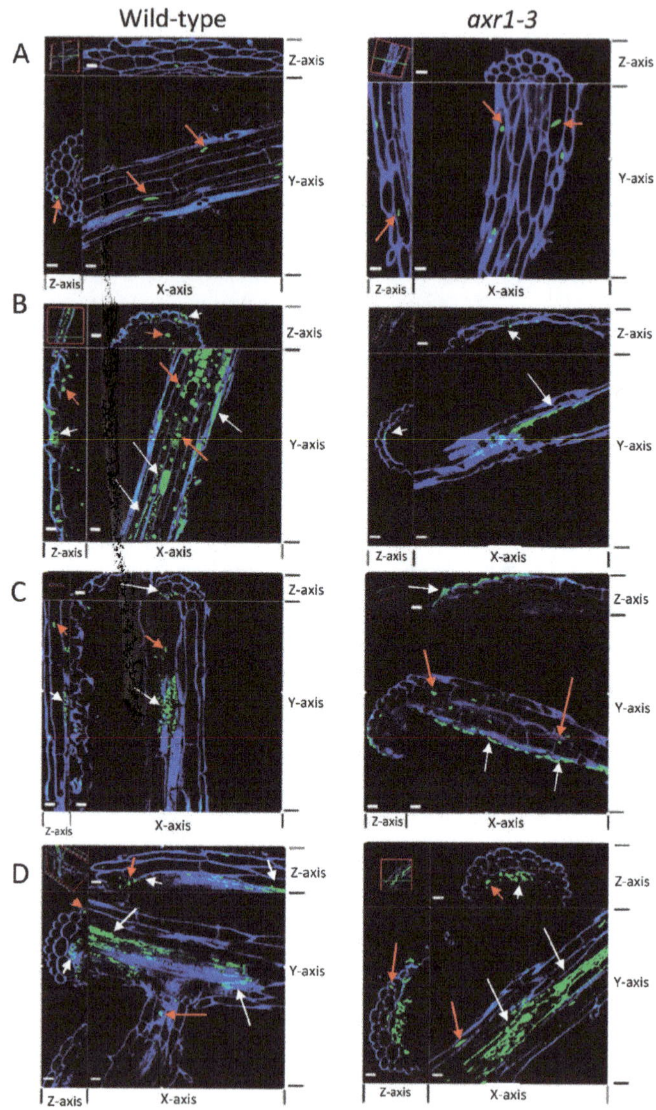

Figure 5. IAA-producing strains that inhibit root length differentially colonize *axr1-3* root tissue. Confocal microscopy showing orthogonal view of wild-type (left panels) and *axr1-3* (right panels) roots upon treatment with (**A**) no bacteria, (**B**) IAA-producing *Microbacterium* RU1A, (**C**) IAA-producing *Azospirillum baldaniorum* Sp245, and (**D**) low IAA-producing *Herbaspirillum* RU5E that does not inhibit root length. The microscopy channels for each dye are blue (calcofluor white used to stain cell wall) and green (SYBR Gold DNA used to stain the nucleus and bacteria). White arrows indicate bacteria locations based on the size and morphology of the DNA-stained spots. Bacteria are shown as green spots that are smaller in size to plant nuclei (shown with red arrows) and tend to form clusters. The size bar in white represents 20 μm on each panel. The 3D images are rotated at the z-axis at two different locations of the tissue shown (top and left sections of each panel as shown by the cross-hair in the upper left corner image) to illustrate transverse views at the location of the stained spots and demonstrate either epiphytic or endophytic locations.

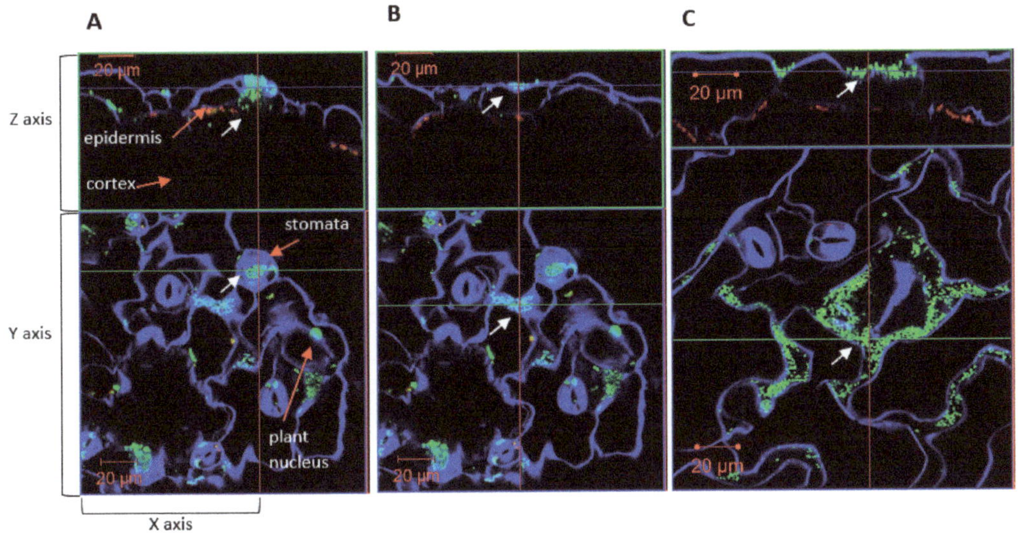

Figure 6. IAA-producing strain *Azospirillum baldaniorum* Sp245 colonizes leaf tissue through stomata. Confocal microscopy showing orthogonal view of (**A**,**B**) wild-type and (**C**) *axr1-3* leaf tissue inoculated with *Azospirillum baldaniorum* Sp245. Note that panels A and B present the z-axis inside the stomata and at the cell surface, respectively, for comparison. The microscopy channels are blue (calcofluor white), green (SYBR Gold DNA), and red (chlorophyll autofluorescence). White arrows indicate bacteria location based on the size of the DNA-stained spots. Bacteria are shown as green spots that are smaller in size to plant nuclei.

3. Discussion

3.1. Limitations of the Salkowski Assay as a Screen for PGPB

To utilize auxin-producing bacteria for agricultural applications, such as with synthetic bacterial communities, it is important to elucidate the role and mechanism of auxin signaling in the context of the plant microbiome [1,5,17,38]. Out of the 21 DAB strains capable of producing IRCs in our previous study [20], we identified only four *Microbacterium* strains that caused a short root phenotype in Arabidopsis. While the *Microbacterium* RU33B strain produced a positive Salkowski assay result indicative for synthesis of IRCs, it did not cause a short root phenotype in Arabidopsis seedlings. Using LC–MS to accurately quantify IAA in a collection of plant-associated bacteria isolates, we found that high levels of IAA (>1 ng/µL) in the bacteria's growth medium correlated with the strain's ability to cause a short root phenotype in Arabidopsis. Strains that were top producers of IRCs based on the Salkowski assay results (but, in many cases, apparently did not correspond to IAA) were not able to cause a short root phenotype. Thus, this commonly used colorimetric assay for detecting auxin-producing strains can often result in false positives [39–42]. Kuźniar et al. [43] detected IAA and IAA conjugates from endophytic bacteria isolated from winter wheat species using a combination of the Salkowski assay and LC–MS. They further tested bioactivity of the bacterial supernatant on wheat coleoptile segments and found the conjugates had lower biological activity in comparison to IAA. Our results highlight the importance of using LC–MS in combination with the Salkowski assay to screen for PGPB across plant species and identify bona fide auxin-producing bacteria strains.

While the correlation between higher levels of IAA-producing capability in the bacteria strain and its ability to modify root development of Arabidopsis seedlings is striking in this study (compare Figures 1 and 2), the sample number in terms of different genera and strains of bacteria tested is likely too low in our current dataset to make a general statement about the threshold of IAA production needed to be effective in planta. Further testing of additional plant-associated bacteria with varying capacity for IAA production would be necessary to build on this initial work to define the threshold level(s) of auxin biosynthesis by the bacteria and its ability to modify host root morphology. As a recent example, microbial-community-derived auxin was posited to play a possible role in increasing *Lemna minor* fitness as measured by the increased number of plants, although this work relied on only using the Salkowski assay to infer auxin production by the bacteria [44]. Similarly, *Bacillus safensis* strains were screened for IAA-producing capability solely using the Salkowski assay, and their function in inducing Cd stress tolerance and promoting plant growth was partly based on the strains' ability to produce auxins [41,42]. Confirmation of this type of results by applying the more definitive LC–MS method to determine the identity and quantity of auxin(s) that are being produced would be important. Future work to quantify the concentration of DAB-derived IAA produced in vivo will be invaluable for creating synthetic DAB communities and understanding how they can be deployed to improve duckweed growth [45].

3.2. Colonization of Bacteria That Produce Different Indole-Containing Compounds

In this study, we found that DAB strains that caused a short root phenotype in Arabidopsis were abundant on the root surface. For example, *Microbacterium* RU1A appeared to be more abundant on the root than the leaf tissues. This contrasts with *Microbacterium* RU33B, which was more abundant on the leaf surface than on the root and did not cause a short root phenotype. In contrast to these *Microbacteria* isolates, *Herbaspirillum* RU5E produced a higher concentration of indole lactic acid than indole-3-acetic acid in vitro. Despite it being endophytic in the root, albeit at low abundance, RU5E did not cause a short root phenotype. This suggests that the duckweed microbiome can produce different indole-containing compounds at various concentrations, with strains having unique colonization patterns and potentially occupying different niches [46–48]. Whether these colonization patterns of different DAB strains may be altered in the presence of other microbes will need to be examined in future synthetic community studies to further define the rules governing the ecological interactions that give rise to the microbiome's structure on host plants.

How plants select for beneficial bacteria while defending against pathogens is not yet well understood [49,50]. Over the past decade, the complex roles that guard cells can play in plant responses to biotic and abiotic stresses have been revealed [51]. In addition to the well-established function of regulating gas exchange and transpiration, these specialized pores also play critical roles in microbial defense through their regulation via the phytohormone abscisic acid as well as others, such as salicylic acid and jasmonic acid [52]. By regulating closure of the aperture between the guard cells, these phytohormones can control the physical barrier that often allow entry of microbes into the intercellular space of plant aerial tissues. Bacterial phytotoxin, such as coronatin, has been demonstrated to be an important virulence determinant through its ability to maintain the stomata in the open state, while common molecular patterns of bacteria, such as the flagellar peptide flg22 that induce basal immunity functions, are known to induce closure of the stomata. In this study, we used *Azospirillum baldaniorum* Sp245 as a positive control since it is a well-characterized IAA-producing endophyte. Including this control in our work led to the unexpected finding that the *AXR1* gene, which is known to be involved in downstream auxin signaling, is necessary for endophytic root colonization of Sp245. Our microscopy results also uncovered the potential role of the guard cells in mediating bacterial entry for this strain, as shown by the remarkable concentration of bacteria inside the open stomata of wild-type plants but not in the *axr1-3* mutant background. In contrast, endophytic colonization of *Microbacterium* RU1A and *Herbaspirillum* RU5E were not altered in the *axr1-3* mutant. These observations suggest that plants have multiple mechanisms to regulate endophytic colonization by different IAA-producing bacteria, one of which requires guard-cell-specific signaling in an AXR-1-dependent manner.

4. Materials and Methods

4.1. Bacterial Strains and Media

Bacterial strains were previously isolated from surface-sterilized duckweed ecotypes as described in Gilbert et al. [20] using either a salt/detergent solution alone or with a bleach wash. Well-characterized IAA-producing *Azospirillum* strains Sp7 [53] and Sp245 [54] isolated from wheat tissue were used as controls. Bacterial strains were stored at −80 °C in LB (Miller's) from IBI Scientific (Dubuque, IA, USA) or tryptic soy broth (TSB) (Hardy Diagnostics, Springboro, OH, USA) depending on the medium of isolation, and supplemented with 40% (*v*/*v*) sterilized glycerol. To isolate single colonies, bacteria from a glycerol stock was spread onto an agar plate (LB or TSB depending on the medium of isolation) and then stored at 28 °C for 2 days or until single colonies were grown. Next, 6 mL liquid cultures of LB or TSB broth were made from a single colony and grown for 1 day at 28 °C and shaken at 240 rpm except for RU33B cultures, which were grown for 2 days at the same temperature and rpm due to slower growth. Bacteria 16S rRNA gene sequence data are available at NCBI GenBank under accession numbers MH217512–MH217560.

4.2. Colorimetric Detection of Indole-Related Compounds

For each strain, a single colony was used to inoculate 6 mL of liquid LB medium with 5 mM L-tryptophan. For DAB 33B, liquid TSB with 5 mM L-tryptophan was used instead due to difficulty growing on LB medium. After 48 h of growing at 28 °C with shaking at 240 rpm, 1 mL of culture was centrifuged for 5 min at 14,000× *g* rpm to collect the supernatant. The original Salkowski assay based on the Gordon and Weber protocol was adapted for a 96-well format [39]. In a Corning 96-well clear-bottom white plate, 100 µL of the supernatant was added to 200 µL of Salkowski reagent (10 mM $FeCl_3$, 97% reagent grade, and 34.3% perchloric acid, ACS grade) in duplicate. After incubating samples with the Salkowski reagent at room temperature for 30 min, the color change was recorded. A BioTek Synergy HT microplate reader was used to determine the absorbance (O.D.) at a single wavelength of 530 nm. To estimate the amount of indole-related compounds at 530 nm, an IAA standard curve was generated by suspending IAA (Gibco Laboratories, Life Technologies, Inc., New York, NY, USA) in 100% acetonitrile at a concentration of 1 mg/mL and diluting in LB medium or TSB to a concentration of 100, 50, 20, 10, 5, and 0 µg/mL. Sterile LB medium with 5 mM L-tryptophan and sterile TSB with 5 mM L-tryptophan were used as controls. The concentration of IRCs at 530 nm of the sterile control sample, either LB or TSB depending on the bacterial medium used, was subtracted from the concentration of indole-related compounds at 530 nm of the bacterial samples to obtain a background-subtracted concentration.

4.3. Extraction of IAA

From glycerol stocks, bacterial strains were streaked onto an LB or TSA (for DAB 33B) agar plate and grown at 28 °C. A single colony was used to inoculate a starter culture of 6 mL liquid LB medium, supplemented with 5 mM L-tryptophan (Sigma-Aldrich, St. Louis, MO, USA), and grown at 28 °C and 240 rpm. After 24 h, the starter culture was used to make a 60 mL culture of liquid LB medium, supplemented with 5 mM L-tryptophan, at OD_{600} 0.01. The cultures were grown at 28 °C and 240 rpm for 24 h. The supernatant was collected at 8000× *g* at 4 °C. For IAA spike samples, 300 µg of IAA was added to the culture by first generating a 1 mg/mL IAA solution in 100% acetonitrile and diluting to 100 µg/mL IAA solution in LB medium, supplemented with 5 mM L-tryptophan. Samples were then acidified with 1N HCl to a pH of 3.0. The samples were then separated into 20 mL aliquots for biological triplicates.

A Sep-Pak C18 cartridge (360 mg sorbent, 55–105 µm particle size) was prepared for each sample by washing with 10 mL of 100% acetonitrile followed by 10 mL of water. The acidified supernatant was passed through the C18 cartridge. The C18 cartridge was then washed with 10 mL of water and eluted with 5 mL of 80% (*v*/*v*) acetonitrile. The eluate was centrifuged at 12,000× *g* rpm for 5 min at 4 °C to remove solid particles. A

20 ng/µL solution of IAA was suspended into 100% acetonitrile for use as a standard in mass spectrometry. Acetonitrile of HPLC grade and HCl of ACS grade were used for the experiment, and water was prepared from Millipore Synergy 185.

4.4. LC–MS

Samples were separated and analyzed by a UPLC/MS system with the Dionex® UltiMate 3000 RSLC ultrahigh-pressure liquid chromatography system consisting of a workstation with ThermoFisher Scientific's Xcalibur v. 4.0 software package combined with Dionex®'s SII LC control software, solvent rack/degasser SRD-3400, pulseless chromatography pump HPG-3400RS, autosampler WPS-3000RS, column compartment TCC-3000RS, and photodiode array detector DAD-3000RS. After the photodiode array detector, the eluent flow was guided to a Q Exactive Plus Orbitrap high-resolution high-mass-accuracy mass spectrometer (MS). Mass detection was full MS scan with low-energy collision-induced dissociation (CID) from 100 to 1000 m/z in positive ionization mode with electrospray (ESI) interface. Sheath gas flow rate was 30 arbitrary units, auxiliary gas flow rate was 7, and sweep gas flow rate was 1. The spray voltage was 3500 volts (−3500 for negative ESI) with a capillary temperature of 275 °C. The mass resolution was 140,000, and the isolation window was 4.0 mDa. Substances were separated on a Phenomenex™ Kinetex C8 reverse-phase column, size 100 × 2 mm, particle size 2.6 mm, pore size 100 Å. The mobile phase consisted of two components: solvent A (0.5% ACS grade acetic acid in LCMS grade water, pH 3–3.5) and solvent B (100% acetonitrile, LCMS grade). The mobile phase flow was 0.20 mL/min, and a gradient mode was used for all analyses. The initial conditions of the gradient were 95% A and 5% B. After 30 min, the proportion reached 5% A and 95% B, which was kept for the next 8 min. During the following 4 min, the ratio was brought to initial conditions. An 8 min equilibration interval was included between subsequent injections. The average pump pressure using these parameters was typically around 3900 psi for the initial conditions.

Putative formulas of IAA metabolites were determined by performing isotope abundance analysis on the high-resolution mass spectral data with Xcalibur v. 4.0 software and reporting the best fitting empirical formula. Database searches were performed using reaxys.com (RELX Intellectual Properties SA, Neuchatel, Switzerland) and SciFinder (American Chemical Society, Washington, DC, USA). Using the external standard of IAA with concentrations of 2.5, 5, 50, and 100 ng/µL with 0.2 µL injections, we calculated the concentration of free IAA in the samples using the peak area in UV chromatograms at 280 nm. To calculate the concentration in the original culture, the concentration was then divided by four to account for the original culture volume being 20 µL and the final elution volume being 5 µL. The concentration of IAA in the LB or TSB medium control sample was then subtracted to obtain the final concentration of IAA produced by the bacteria.

4.5. *Arabidopsis Growth Assay*

The Arabidopsis growth assay was performed in a similar manner for observation of root lengths and microscopy. For each assay, 200 *Arabidopsis thaliana* (Col-0 ecotype) seeds (wild type or *axr1-3* genotype) were sterilized using 50% (v/v) bleach solution (0.3% sodium hypochlorite) in a 1.5 mL microcentrifuge tube for 4 min with continuous shaking using a vortex (Fisher Genie 2) shake setting of 6. The bleach solution was removed, and the seeds were washed four times in 1 mL of sterile water. After removing the water, the seeds were suspended in 0.1% (w/v) Difco agar granulated (Becton Dickinson, Sparks, MD, USA). Seeds were placed onto circular 100 × 15 mm plates containing 0.5× Murashige and Skoog (MS) modified basal medium with Gamborg vitamins (PhytoTech Laboratories, Lenexa, KS, USA), 1% sucrose, pH 5.7, 0.25% phytagel (Sigma-Aldrich, St. Louis, MO, USA). The seeds were vernalized at 4 °C in the dark for 2 days and then stored vertically in a growth chamber at 22 °C under 100 µmol m^{-2} s^{-1} of 12 h light. After 6 days, previously grown bacterial cultures were prepared by taking 1 mL of culture and centrifuging at 14,000× g rpm for 5 min. The supernatant was discarded, and the bacterial pellet was resuspended in sterile water to an OD_{600}/cm of 0.7 (1.58×10^7 CFU/mL was measured for RU1A). Bacterial cultures for heat-killed samples were autoclaved and centrifuged at 8000× *g* rpm for 5 min, and the pellet was diluted to an OD_{600}/cm of 0.7 before plating 100 μL onto an LB plate to check vitality. Next, 100 μL of heat-killed or living bacterial solution was spread onto square 100 × 15 mm plates containing 0.5× MS, pH 5.7, 0.5% gellan gum powder (PhytoTech Laboratories, Lenexa, KS, USA). Media containing 1 μM IAA (Gibco Laboratories, Grand Island, NY, USA) was previously prepared by adding IAA dissolved in DMSO (Sigma-Aldrich, St. Louis, MO, USA) directly to the media before pouring and solidifying. Then, 6–12 seedlings (depending on the assay) were transferred onto each plate, which were then sealed with a self-adherent wrap (3M micropore surgical tape; Coban, St. Paul, MN, USA). Plates were then placed in the same growth chamber under the same conditions as previously described for 7 days until processing for all subsequent experiments. Pictures of plants were taken with a Nikon D5200 camera, and roots were measured using ImageJ. Water was prepared from Millipore Synergy 185 and sterilized using a 0.2 micron polyethersulfone syringe filter.

4.6. Confocal Microscopy

Five whole seedlings of sterile or bacteria-treated, from wild-type and *axr1-3* genotypes, were fixed in 1 mL of 4% paraformaldehyde overnight at room temperature. The solution was removed followed by washing twice with 1 mL of sterile phosphate buffer saline (1.37 M NaCl, 26 mM KCl, 10 mM $Na_2HPO_4 \cdot 7H_2O$, 17.6 mM KH_2PO_4, pH 7.4) and then storing at 4 °C. Images were acquired by EMSL (Richland, WA, USA) using a Zeiss LSM 710 scanning confocal microscope. The channels used were blue (calcofluor white), green (SYBR Gold DNA), red (chlorophyll autofluorescence), and gray (transmitted light).

Supplementary Materials: The following are available online at https://www.mdpi.com/article/10.3390/plants11060721/s1, Figure S1: Effect of *Microbacterium* RU1A on Arabidopsis root hairs. Figure S2: IAA-producing strains differentially colonize wild-type Arabidopsis leaf tissue.

Author Contributions: E.L. conceived the project; S.G. performed most of the experiments; A.P. performed LC–MS; W.C. performed confocal microscopy; K.A., G.O., and S.L. advised the experiments; S.G. wrote the manuscript; and E.L. assisted in data analysis, interpretation, and finalizing the presentation. All authors have read and agreed to the published version of the manuscript.

Funding: Duckweed research at the Lam laboratory is provided by a Hatch project (#12116) from the New Jersey Agricultural Experiment Station at Rutgers University and in part by the U.S. Department of Energy, Office of Science, Office of Biological and Environmental Research program under award number DE-SC0018244. S.G. was also supported in part by a Robert and Lillian White-Stevens Graduate Fellowship (between 2015 and 2017) during part of this work. Contribution by the Facilities Integrating Collaborations for User Science (FICUS) initiative and under contract numbers DE732 AC02-05CH11231 (JGI) and DE-AC05-76RL01830 (EMSL) to the characterization of the duckweed microbiome is also gratefully acknowledged.

Acknowledgments: A portion of this research was performed at the Environmental Molecular Sciences Laboratory, a national scientific user facility sponsored by the Department of Energy's Office of Biological and Environmental Research and located at Pacific Northwest National Laboratory. We thank Mark Estelle for providing *axr1-3* seeds, Yi Zhang (UC San Diego) and Jocelyn Malamy (U. Chicago) for advice on auxin response mutants, and Ilya Raskin for providing access to the mass spectrometry instrument in his laboratory as well as providing LC–MS materials. We thank Henry Qu for technical assistance.

Conflicts of Interest: The authors declare no conflict of interest.

References

1. Spaepen, S.; Vanderleyden, J. Auxin and Plant-Microbe Interactions. *Cold Spring Harb. Perspect. Biol.* **2011**, *3*, a001438. [CrossRef]
2. Mano, Y.; Nemoto, K. The pathway of auxin biosynthesis in plants. *J. Exp. Bot.* **2012**, *63*, 2853–2872. [CrossRef] [PubMed]
3. Patten, C.L.; Blakney, A.J.C.; Coulson, T.J.D. Activity, distribution and function of indole-3-acetic acid biosynthetic pathways in bacteria. *Crit. Rev. Microbiol.* **2012**, *39*, 395–415. [CrossRef] [PubMed]

4. Navarro, L.; Dunoyer, P.; Jay, F.; Arnold, B.; Dharmasiri, N.; Estelle, M.; Voinnet, O.; Jones, J.D.G. A Plant miRNA Contributes to Antibacterial Resistance by Repressing Auxin Signaling. *Science* **2006**, *312*, 436–439. [CrossRef] [PubMed]
5. Duca, D.; Lorv, J.; Patten, C.L.; Rose, D.; Glick, B.R. Indole-3-acetic acid in plant–microbe interactions. *Antonie Leeuwenhoek* **2014**, *106*, 85–125. [CrossRef] [PubMed]
6. Fitzpatrick, C.R.; Salas-González, I.; Conway, J.M.; Finkel, O.M.; Gilbert, S.; Russ, D.; Teixeira, P.J.; Dangl, J.L. The Plant Microbiome: From Ecology to Reductionism and Beyond. *Annu. Rev. Microbiol.* **2020**, *74*, 81–100. [CrossRef]
7. Eichmann, R.; Richards, L.; Schäfer, P. Hormones as go-betweens in plant microbiome assembly. *Plant J.* **2020**, *105*, 518–541. [CrossRef]
8. Rosquete, M.R.; Barbez, E.; Kleine-Vehn, J. Cellular Auxin Homeostasis: Gatekeeping Is Housekeeping. *Mol. Plant* **2012**, *5*, 772–786. [CrossRef]
9. Rolfe, S.A.; Griffiths, J.; Ton, J. Crying out for help with root exudates: Adaptive mechanisms by which stressed plants assemble health-promoting soil microbiomes. *Curr. Opin. Microbiol.* **2019**, *49*, 73–82. [CrossRef]
10. Rolli, E.; Vergani, L.; Ghitti, E.; Patania, G.; Mapelli, F.; Borin, S. "Cry-for-help" in contaminated soil: A dialogue among plants and soil microbiome to survive in hostile conditions. *Environ. Microbiol.* **2021**, *23*, 5690–5703. [CrossRef]
11. Tzipilevich, E.; Russ, D.; Dangl, J.L.; Benfey, P.N. Plant immune system activation is necessary for efficient root colonization by auxin-secreting beneficial bacteria. *Cell Host Microbe* **2021**, *29*, 1507–1520.e4. [CrossRef] [PubMed]
12. Chen, Z.; Agnew, J.L.; Cohen, J.; He, P.; Shan, L.; Sheen, J.; Kunkel, B.N. *Pseudomonas syringae* type III effector AvrRpt2 alters *Arabidopsis thaliana* auxin physiology. *Proc. Natl. Acad. Sci. USA* **2007**, *104*, 20131–20136. [CrossRef] [PubMed]
13. Grunewald, W.; van Noorden, G.; Van Isterdael, G.; Beeckman, T.; Gheysen, G.; Mathesius, U. Manipulation of Auxin Transport in Plant Roots during Rhizobium Symbiosis and Nematode Parasitism. *Plant Cell* **2009**, *21*, 2553–2562. [CrossRef]
14. Evangelisti, E.; Govetto, B.; Minet-Kebdani, N.; Kuhn, M.L.; Attard, A.; Ponchet, M.; Panabières, F.; Gourgues, M. The *Phytophthora parasitica* RXLR effector Penetration-Specific Effector 1 favours *Arabidopsis thaliana* infection by interfering with auxin physiology. *New Phytol.* **2013**, *199*, 476–489. [CrossRef]
15. Ludwig-Müller, J. Bacteria and fungi controlling plant growth by manipulating auxin: Balance between development and defense. *J. Plant Physiol.* **2015**, *172*, 4–12. [CrossRef]
16. Ma, K.-W.; Ma, W. Phytohormone pathways as targets of pathogens to facilitate infection. *Plant Mol. Biol.* **2016**, *91*, 713–725. [CrossRef] [PubMed]
17. Kunkel, B.N.; Harper, C. The roles of auxin during interactions between bacterial plant pathogens and their hosts. *J. Exp. Bot.* **2017**, *69*, 245–254. [CrossRef] [PubMed]
18. Lu, T.; Ke, M.; Lavoie, M.; Jin, Y.; Fan, X.; Zhang, Z.; Fu, Z.; Sun, L.; Gillings, M.; Peñuelas, J.; et al. Rhizosphere microorganisms can influence the timing of plant flowering. *Microbiome* **2018**, *6*, 231. [CrossRef]
19. Finkel, O.M.; Salas-González, I.; Castrillo, G.; Conway, J.M.; Law, T.F.; Teixeira, P.J.P.L.; Wilson, E.D.; Fitzpatrick, C.R.; Jones, C.D.; Dangl, J.L. A single bacterial genus maintains root growth in a complex microbiome. *Nature* **2020**, *587*, 103–108. [CrossRef]
20. Gilbert, S.; Xu, J.; Acosta, K.; Poulev, A.; Lebeis, S.; Lam, E. Bacterial Production of Indole Related Compounds Reveals Their Role in Association Between Duckweeds and Endophytes. *Front. Chem.* **2018**, *6*, 265. [CrossRef]
21. Ruegger, M.; Dewey, E.; Gray, W.M.; Hobbie, L.; Turner, J.; Estelle, M. The TIR1 protein of *Arabidopsis* functions in auxin response and is related to human SKP2 and yeast Grr1p. *Genes Dev.* **1998**, *12*, 198–207. [CrossRef] [PubMed]
22. Trenner, J.; Poeschl, Y.; Grau, J.; Gogol-Döring, A.; Quint, M.; Delker, C. Auxin-induced expression divergence between *Arabidopsis* species may originate within the TIR1/AFB–AUX/IAA–ARF module. *J. Exp. Bot.* **2017**, *68*, 539–552. [PubMed]
23. Del Pozo, J.C.; Timpte, C.; Tan, S.; Callis, J.; Estelle, M. The Ubiquitin-Related Protein RUB1 and Auxin Response in *Arabidopsis*. *Science* **1998**, *280*, 1760–1763. [CrossRef] [PubMed]
24. Schwechheimer, C. NEDD8—Its role in the regulation of Cullin-RING ligases. *Curr. Opin. Plant Biol.* **2018**, *45*, 112–119. [CrossRef]
25. Lincoln, C.; Britton, J.H.; Estelle, M. Growth and development of the *axr1* mutants of *Arabidopsis*. *Plant Cell* **1990**, *2*, 1071–1080.
26. Leyser, H.M.O.; Lincoln, C.A.; Timpte, C.; Lammer, D.; Turner, J.; Estelle, M. *Arabidopsis* auxin-resistance gene AXR1 encodes a protein related to ubiquitin-activating enzyme E1. *Nature* **1993**, *364*, 161–164. [CrossRef]
27. Ziegler, P.; Adelmann, K.; Zimmer, S.; Schmidt, C.; Appenroth, K.J. Relative in vitro growth rates of duckweeds (Lemnaceae)-the most rapidly growing higher plants. *Plant Biol.* **2018**, *17*, 33–41. [CrossRef]
28. Utami, D.; Kawahata, A.; Sugawara, M.; Jog, R.N.; Miwa, K.; Morikawa, M. Effect of Exogenous General Plant Growth Regulators on the Growth of the Duckweed Lemna minor. *Front. Chem.* **2018**, *6*, 251. [CrossRef]
29. Bulgarelli, D.; Rott, M.; Schlaeppi, K.; Van Themaat, E.V.L.; Ahmadinejad, N.; Assenza, F.; Rauf, P.; Huettel, B.; Reinhardt, R.; Schmelzer, E.; et al. Revealing structure and assembly cues for *Arabidopsis* root-inhabiting bacterial microbiota. *Nature* **2012**, *488*, 91–95. [CrossRef]
30. Lundberg, D.S.; Lebeis, S.L.; Paredes, S.H.; Yourstone, S.; Gehring, J.; Malfatti, S.; Tremblay, J.; Engelbrektson, A.; Kunin, V.; Del Rio, T.G.; et al. Defining the core *Arabidopsis thaliana* root microbiome. *Nature* **2012**, *488*, 86–90. [CrossRef]
31. Bodenhausen, N.; Horton, M.; Bergelson, J. Bacterial Communities Associated with the Leaves and the Roots of *Arabidopsis thaliana*. *PLoS ONE* **2013**, *8*, e56329. [CrossRef] [PubMed]
32. Schlaeppi, K.; Dombrowski, N.; Oter, R.G.; van Themaat, E.V.L.; Schulze-Lefert, P. Quantitative divergence of the bacterial root microbiota in *Arabidopsis thaliana* relatives. *Proc. Natl. Acad. Sci. USA* **2013**, *111*, 585–592. [CrossRef] [PubMed]

33. Huang, W.; Gilbert, S.; Poulev, A.; Acosta, K.; Lebeis, S.; Long, C.; Lam, E. Host-specific and tissue-dependent orchestration of microbiome community structure in traditional rice paddy ecosystems. *Plant Soil* **2020**, *452*, 379–395. [CrossRef]
34. Acosta, K.; Xu, J.; Gilbert, S.; Denison, E.; Brinkman, T.; Lebeis, S.; Lam, E. Duckweed hosts a taxonomically similar bacterial assemblage as the terrestrial leaf microbiome. *PLoS ONE* **2020**, *15*, e0228560. [CrossRef]
35. Vissenberg, K.; Claeijs, N.; Balcerowicz, D.; Schoenaers, S. Hormonal regulation of root hair growth and responses to the environment in *Arabidopsis*. *J. Exp. Bot.* **2020**, *71*, 2412–2427. [CrossRef]
36. Spaepen, S.; Bossuyt, S.; Engelen, K.; Marchal, K.; Vanderleyden, J. Phenotypical and molecular responses of Arabidopsis thaliana roots as a result of inoculation with the auxin-producing bacterium *Azospirillum brasilense*. *New Phytol.* **2014**, *201*, 850–861. [CrossRef]
37. Ona, O.; Van Impe, J.; Prinsen, E.; Vanderleyden, J. Growth and indole-3-acetic acid biosynthesis of *Azospirillum brasilense* Sp245 is environmentally controlled. *FEMS Microbiol. Lett.* **2005**, *246*, 125–132. [CrossRef]
38. Ishizawa, H.; Kuroda, M.; Morikawa, M.; Ike, M. Evaluation of environmental bacterial communities as a factor affecting the growth of duckweed Lemna minor. *Biotechnol. Biofuels* **2017**, *10*, 62. [CrossRef]
39. Gordon, S.A.; Weber, R.P. Colorimetric estimation of indoleacetic acid. *Plant Physiol.* **1951**, *26*, 192–195. [CrossRef]
40. Glickmann, E.; Dessaux, Y. A critical examination of the specificity of the salkowski reagent for indolic compounds produced by phytopathogenic bacteria. *Appl. Environ. Microbiol.* **1995**, *61*, 793–796. [CrossRef]
41. Nazli, F.; Jamil, M.; Hussain, A.; Hussain, T. Exopolysaccharides and indole-3-acetic acid producing *Bacillus safensis* strain FN13 potential candidate for phytostabilization of heavy metals. *Environ. Monit. Assess.* **2020**, *192*, 438. [CrossRef] [PubMed]
42. Nazli, F.; Wang, X.; Ahmad, M.; Hussain, A.; Bushra; Dar, A.; Nasim, M.; Jamil, M.; Panpluem, N.; Mustafa, A. Efficacy of Indole Acetic Acid and Exopolysaccharides-Producing *Bacillus safensis* Strain FN13 for Inducing Cd-Stress Tolerance and Plant Growth Promotion in *Brassica juncea* (L.). *Appl. Sci.* **2021**, *11*, 4160. [CrossRef]
43. Kuźniar, A.; Włodarczyk, K.; Sadok, I.; Staniszewska, M.; Woźniak, M.; Furtak, K.; Grządziel, J.; Gałązka, A.; Skórzyńska-Polit, E.; Wolińska, A. A Comprehensive Analysis Using Colorimetry, Liquid Chromatography-Tandem Mass Spectrometry and Bioassays for the Assessment of Indole Related Compounds Produced by Endophytes of Selected Wheat Cultivars. *Molecules* **2021**, *26*, 1394. [CrossRef] [PubMed]
44. Tan, J.; Kerstetter, J.E.; Turcotte, M.M. Eco-evolutionary interaction between microbiome presence and rapid biofilm evolution determines plant host fitness. *Nat. Ecol. Evol.* **2021**, *5*, 670–676. [CrossRef]
45. Marín, O.; González, B.; Poupin, M.J. From Microbial Dynamics to Functionality in the Rhizosphere: A Systematic Review of the Opportunities with Synthetic Microbial Communities. *Front. Plant Sci.* **2021**, *12*, 650609. [CrossRef] [PubMed]
46. Goldford, J.E.; Lu, N.; Bajić, D.; Estrela, S.; Tikhonov, M.; Sanchez-Gorostiaga, A.; Segrè, D.; Mehta, P.; Sanchez, A. Emergent simplicity in microbial community assembly. *Science* **2018**, *361*, 469–474. [CrossRef]
47. Plett, J.M.; Martin, F.M. Know your enemy, embrace your friend: Using omics to understand how plants respond differently to pathogenic and mutualistic microorganisms. *Plant J.* **2018**, *93*, 729–746. [CrossRef]
48. Ishizawa, H.; Tada, M.; Kuroda, M.; Inoue, D.; Futamata, H.; Ike, M. Synthetic Bacterial Community of Duckweed: A Simple and Stable System to Study Plant-microbe Interactions. *Microbes Environ.* **2020**, *35*, ME20112. [CrossRef]
49. Yu, K.; Pieterse, C.M.; Bakker, P.A.; Berendsen, R.L. Beneficial microbes going underground of root immunity. *Plant Cell Environ.* **2019**, *42*, 2860–2870. [CrossRef]
50. Thoms, D.; Liang, Y.; Haney, C.H. Maintaining Symbiotic Homeostasis: How Do Plants Engage With Beneficial Microorganisms While at the Same Time Restricting Pathogens? *Mol. Plant Microbe Interact.* **2021**, *34*, 462–469. [CrossRef]
51. Melotto, M.; Zhang, L.; Oblessuc, P.R.; He, S.Y. Stomatal Defense a Decade Later. *Plant Physiol.* **2017**, *174*, 561–571. [CrossRef] [PubMed]
52. Bharath, P.; Gahir, S.; Raghavendra, A.S. Abscisic Acid-Induced Stomatal Closure: An Important Component of Plant Defense against Abiotic and Biotic Stress. *Front. Plant Sci.* **2021**, *12*, 615114. [CrossRef] [PubMed]
53. Tarrand, J.J.; Krieg, N.R.; Döbereiner, J. A taxonomic study of the *Spirillum lipoferum* group, with descriptions of a new genus, *Azospirillum* gen. nov. and two species, *Azospirillum lipoferum* (Beijerinck) comb. nov. and *Azospirillum brasilense* sp. nov. *Can. J. Microbiol.* **1978**, *24*, 967–980. [CrossRef] [PubMed]
54. Baldani, V.L.D.; Alvarez, M.A.D.B.; Baldani, J.I.; Döbereiner, J. Establishment of inoculated *Azospirillum* spp. in the rhizosphere and in roots of field grown wheat and sorghum. *Plant Soil* **1986**, *90*, 35–46. [CrossRef]

plants

MDPI

Article

Dynamic Alteration of Microbial Communities of Duckweeds from Nature to Nutrient-Deficient Condition

Chakrit Bunyoo [1,2,3], Peerapat Roongsattham [2,3], Sirikorn Khumwan [2,3], Juthaporn Phonmakham [2,3], Passorn Wonnapinij [2,3,4] and Arinthip Thamchaipenet [2,3,4,*]

1 Interdisciplinary Graduate Program in Bioscience, Faculty of Science, Kasetsart University, Bangkok 10900, Thailand
2 Department of Genetics, Faculty of Science, Kasetsart University, Bangkok 10900, Thailand
3 Duckweed Holobiont Resource & Research Center (DHbRC), Kasetsart University, Bangkok 10900, Thailand
4 Omics Center for Agriculture, Bioresource, Food and Health Kasetsart University (OmiKU), Bangkok 10900, Thailand
* Correspondence: arinthip.t@ku.ac.th

Abstract: Duckweeds live with complex assemblages of microbes as holobionts that play an important role in duckweed growth and phytoremediation ability. In this study, the structure and diversity of duckweed-associated bacteria (DAB) among four duckweed subtypes under natural and nutrient-deficient conditions were investigated using V3-V4 16S rRNA amplicon sequencing. High throughput sequencing analysis indicated that phylum Proteobacteria was predominant in across duckweed samples. A total of 24 microbial genera were identified as a core microbiome that presented in high abundance with consistent proportions across all duckweed subtypes. The most abundant microbes belonged to the genus *Rhodobacter*, followed by other common DAB, including *Acinetobacter*, *Allorhizobium-Neorhizobium-Pararhizobium-Rhizobium*, and *Pseudomonas*. After nutrient-deficient stress, diversity of microbial communities was significantly deceased. However, the relative abundance of *Allorhizobium-Neorhizobium-Pararhizobium-Rhizobium*, *Pelomonas*, *Roseateles* and *Novosphingobium* were significantly enhanced in stressed duckweeds. Functional prediction of the metagenome data displayed the relative abundance of essential pathways involved in DAB colonization, such as bacterial motility and biofilm formation, as well as biodegradable ability, such as benzoate degradation and nitrogen metabolism, were significantly enriched under stress condition. The findings improve the understanding of the complexity of duckweed microbiomes and facilitate the establishment of a stable microbiome used for co-cultivation with duckweeds for enhancement of biomass and phytoremediation under environmental stress.

Keywords: duckweed; microbiome; 16S rRNA; metagenome; stress

Citation: Bunyoo, C.; Roongsattham, P.; Khumwan, S.; Phonmakham, J.; Wonnapinij, P.; Thamchaipenet, A. Dynamic Alteration of Microbial Communities of Duckweeds from Nature to Nutrient-Deficient Condition. *Plants* **2022**, *11*, 2915. https://doi.org/10.3390/plants11212915

Academic Editors: Viktor Oláh, Klaus-Jürgen Appenroth and K. Sowjanya Sree

Received: 10 October 2022
Accepted: 25 October 2022
Published: 29 October 2022

1. Introduction

Duckweeds, tiny flowering aquatic plants, belong to the family *Lemnaceae*, consisting of five genera; *Spirodela*, *Landoltia*, *Lemna*, *Wolffia*, and *Wolffiella*. Presently, 36 species of duckweed have been identified worldwide [1]. Duckweeds have been intensively studied in terms of aquatic plant models, animal feed, human food, biofuel production, and wastewater treatment due to their richness of nutrition, as well as their capability for phytoremediation [2–6].

Recently, several potential plant-growth-promoting bacteria (PGPB) have been isolated from duckweeds [7,8]. For instance, *Acinetobacter calcoaceticus* P23 has been proven to benefit *Lemna aoukikusa* by promoting biomass and facilitating phytoremediation [7,9]. Co-cultivation of two PGPB strains, *Ac. calcoaceticus* P23 and *Pseudomonas* sp. Ps6, enhanced growth of *Lemna minor* [10]. An indigenous wastewater bacteria, *Chryseobacterium* sp. 27AL, promoted biomass production of *Lemna gibba* under N-rich wastewater and limited-N conditions [11]. Using next-generation sequencing, the microbial community associated with natural growing duckweeds has been identified to consist of members in phyla Proteobacteria, Bacteroidetes, Firmicutes, and Actinobacteria [12]. However, the application of those PGPB was limited by the competitive indigenous community. For example, inoculation of *Aquitalea magnusonii* H3 to *L. minor* promoted duckweed biomass at the beginning of the association, but gradually lost its benefit due to the indigenous community competition [13]. Moreover, the duckweed-inoculated microbial community was changed over time during phytoremediation process [5]. Environmental abiotic factors, such as salinity, also altered the bacterial community of *L. minor* [14]. Thus, a successful PGPB inoculant should be able to invade and persist against both indigenous bacteria and variable abiotic stresses. Understanding the interaction between duckweeds and PGPB towards a dynamic change of the associated microbes over various environmental conditions is necessary to improve PGPB application [15].

In this work, the duckweed-associated bacterial community of four subtypes of natural growing duckweeds, *Spirodela*, *Landoltia*, *Lemna*, and *Wolffia*, in Thailand, were investigated using a metagenomic approach. Microbiome profiles of natural duckweeds before and after growing in an extreme nutrient-deficient condition were examined, and a set of "core" microbiomes were identified. The findings in this study will enhance the understanding of duckweed-microbial communities for the establishment of a stable PGPB community used for duckweed applications.

2. Results

2.1. Microbial Diversity and Composition

Duckweeds in natural conditions (NC) were identified as *Landoltia punctata*, *Lemna aequinoctialis*, *Spirodela polyrhiza*, and *Wolffia globose*, based on two-barcode approaches (data not shown). Chemical composition of ambient (surrounding) water (AW) composed of 35.53 mg/L of total N and 0.71 mg/L of NO_3-N, at pH 7.53 with 0.227 dS/m of EC. No P, K, As, or Cd were detected. To evaluate the dynamic change of duckweed microbial communities under nutrient-deficient conditions (stress condition; SC), NC duckweeds were grown in sterilized distilled water at 25 °C under a 12-h photoperiod corresponding to the ambient temperature and daylight hours of NC. This condition was adopted to minimize the effect of environmental conditions that may alter bacterial communities. After cultivation under SC for two weeks, the growth of the NC duckweeds was retarded, and half of them turned yellowish and pale (data not shown).

Microbiomes of NC ($n = 20$) and SC ($n = 20$) duckweeds, as well as AW ($n = 5$), were determined using V3–V4 region of 16S rRNA gene amplicon sequencing. Chimera, chloroplast, mitochondria, and low frequency ASVs were removed from the total 5,932,394 reads of 50 data sets to obtain 3,123,655 reads (Table S1). The number of processed reads per sample ranged between 27,601 to 95,416 with a median of 59,800 (Table S2). To minimize bias introduced by the magnitude of sample depth, all samples were rarefied to an even number of 27,601 reads prior to diversity analysis (Figure S2). The number of detected ASVs and the calculated diversity index across samples are listed in Table S3. The highest number of ASVs was detected in NC *Spirodela* (1,345 ASVs), while the lowest number was that of SC *Wolffia* (449 ASVs) (Table S3).

Within sample diversity (alpha diversity) based on the Shannon index, NC *Landoltia*, *Lemna*, and *Spirodela* harbored microbial diversity higher than that of AW (Figure 1A; p-vaule < 0.05); whereas NC *Wolffia* revealed the smallest degree of microbial diversity across all duckweeds. NC duckweeds harbored microbial communities with a similar degree of diversity, with a Shannon index ranging from 7.59–8.57 for *Landoltia*, *Lemna*, and *Spirodela*, and except *Wolffia*, which displayed a significantly smaller Shannon index (6.92–7.50) (Table S3). There was no significant difference observed in the Shannon index between NC *Wolffia* and AW (Figure 1A; p-value = 0.07). After a 2-week nutrient starvation, most of the duckweeds, except *Wolffia*, loosened their microbial diversity (Figure 1B). However, the reduction of bacterial diversity was observed in SC *Wolffia* without statistical significance (Figure 1B; p-value = 0.67).

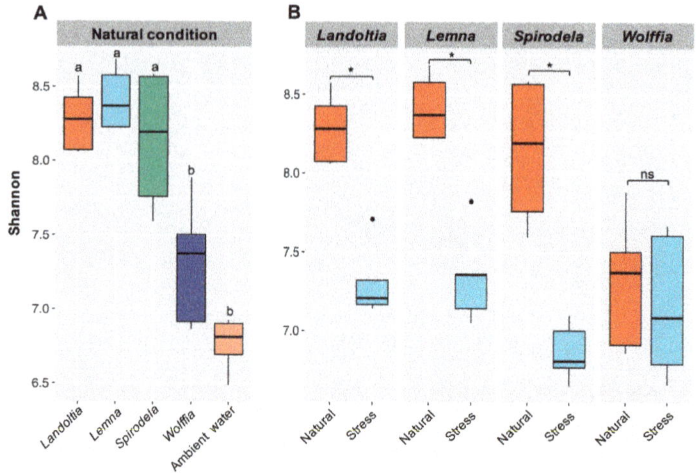

Figure 1. Comparison of alpha diversity index (Shannon). (**A**) Between natural duckweed subtypes and ambient water; and (**B**) between natural and nutrient-deficient (stress) conditions. Asterisks (*) indicate significant difference based on Kruskal-Wallis test, p-value < 0.05; ns, no significant difference.

Diversity between samples (beta diversity) based on the Bray-Curtis dissimilarity matrix was calculated to estimate the effect of environmental conditions and duckweed subtypes in shaping the microbiome community. The microbial communities associated with the same environmental conditions were likely to be clustered. Conversely, the microbial communities from different environmental conditions were clearly distinguished on the nMDS ordination (Figure 2). PERMANOVA analysis strongly supported the nMDS ordination result. The microbial communities detected in NC, SC, and AW were significantly different from one another (p-value < 0.001).

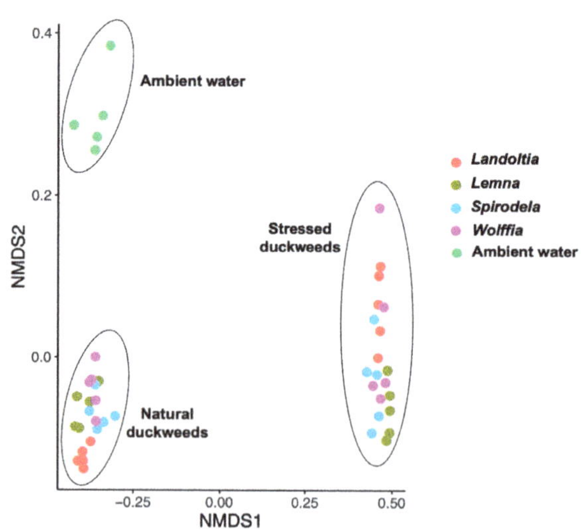

Figure 2. Non-metric multidimensional scaling (nMDS) analysis of the Bray-Curtis dissimilarity index between natural, stressed duckweeds (nutrient deficiency) and ambient water. All conditions displayed significant difference based on PERMANOVA analysis (p-value < 0.01).

Microbial taxa composition of NC and SC duckweeds, as well as AW, were classified into taxonomic levels. Proteobacteria was the most prominent phylum across samples (67.5%, 71.4%, and 49.3% median relative abundance of NC, SC, and AW, respectively; Figure 3). For NC duckweeds, the highly abundant phyla consisted of Bacteroidota (10.3%) and Acidobacteriota (5.3%). Whereas, Firmicutes, Bacteroidota, and Actinobacteria were prevalent in SC duckweeds with median relative abundances of 12.9%, 10.7%, and 2.4%, respectively (Figure 3). Conversely, AW harbored Bacteroidota and Actinobacteria with median relative abundances of 33.3% and 19.4%, respectively (Figure 3). Although Proteobacteria were predominant throughout the samples, the median relative abundance of Proteobacteria in AW was significantly lower than those of NC and SC duckweeds (Figure 3). However, Bacteroidota and Actinobacteria detected in AW displayed a median relative abundance greater than that of NC and SC duckweeds. Furthermore, Firmicutes were detected in SC duckweed with a median relative abundance significantly higher than that of NC duckweed and AW (Figure 3).

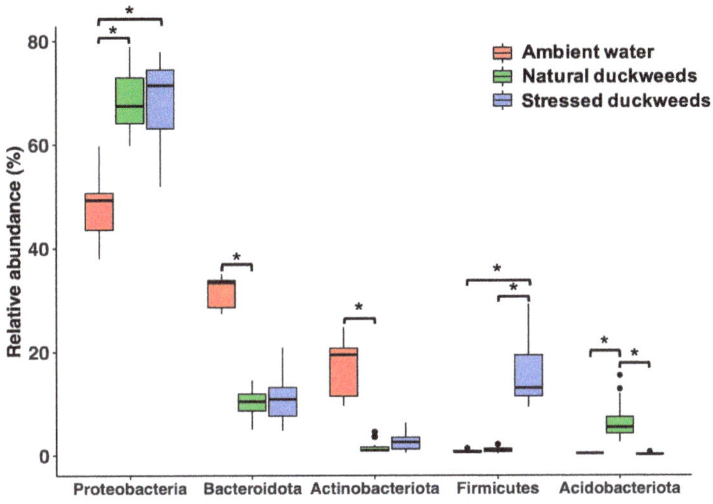

Figure 3. Taxonomic composition of microbial community associated with natural, stressed duckweeds (nutrient deficiency) and ambient water. The most abundance phyla (>5% median relative abundance) were plotted. Asterisks (*) indicate significant difference based on Wilcoxon test, p-value < 0.05. Dots indicate potential outliers.

2.2. Core Microbiomes of Natural Duckweed

To evaluate the core bacterial community associated with NC duckweeds, all ASVs were analyzed at the genus level. A sum of 315 putative core microbiomes were defined by the bacterial genera that presented in every biological replicate of each duckweed subtype, disregarding their relative abundances (Table S4). Almost half of overall genus candidates (148 of 315; 46.8%) were consistently found among the four duckweed species (Figure 4A). There were bacterial genera that were exclusively detected in *Landoltia*, *Lemna*, *Spirodela*, and *Wolffia* at 6.0% (19 of 315), 6.9% (22 of 315), 3.8% (7 of 315), and 8.8% (12 of 315), respectively (Figure 4A). However, these unique genera presented rather low relative abundances ranging between 0.01% to 0.60%; whereas those of conserved core genera varied between 0.01% to 20.2% across the four duckweed species. In order to define low or high abundance taxa, the counts or taxa abundances were transformed into centered log ratio (clr) where the abundance counts were compared to their geometric mean. The taxa carrying clr values close to 0 indicated their average abundance. By this criterion, the majority of the conserved genera (141 of 148; 95.2%) displayed abundance above average (clr > 0; Table S5), while all unique genera showed low abundance (clr < 0; Table S5). The results indicated that the moderate to high abundances of microbial community were shared across duckweed subtypes.

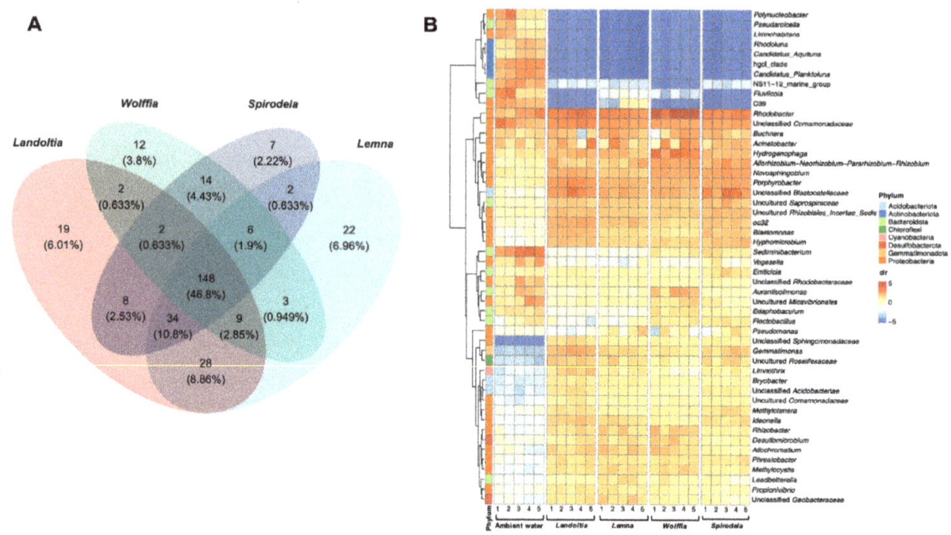

Figure 4. Candidate core microbiomes defined by bacterial genera that presented in every biological replicate of duckweed subtypes and ambient water, disregarding their relative abundance using (**A**) Venn diagram, and (**B**) heatmap representing the top 50 abundant core genera.

Rhodobacter was predominant in almost duckweed subtypes, including *Landoltia*, *Lemna*, and *Wolffia*, with average relative abundances of 8.9%, 8.9%, and 20.2%, respectively; whereas the unclassified genus of the family *Blastocatellaceae* was prominent in *Spirodela*, followed by *Rhodobacter*, with average relative abundances of 9.4% and 9.2%, respectively (Table S6). There were 24 genera presented in high abundance (>1% relative abundance; clr > 3) with consistent proportion across all duckweed subtypes (Table S5). The most abundant were *Acinetobacter*, *Allorhizobium-Neorhizobium-Pararhizobium-Rhizobium*, *Hydrogenophaga*, *Novosphingobium*, *Porphyrobacter*, and *Rhodobacter* (Figure 4B). For ambient water, a total of 156 genera were defined as putative core microbiomes of which *Sediminibacterium* was predominant (12.8% relative abundance), followed by hgcl clade, an unclassified genus of the families *Comamonadaceae*, *Fluviicola*, and *Rhodobacter*, with relative abundances of 6.8%, 6.8%, 6.3%, and 5.7%, respectively (Table S5). Of those, *Rhodobacter*, an unclassified genus of the families *Comamonadaceae*, *Buchnera*, and *Acinetobacter*, were also found in NC and SC duckweeds (Figure 4B).

2.3. Nutrient-Deficient Condition Altered Duckweed Core Microbiomes

After treating four subtypes of NC duckweeds under stress of nutrient starvation, 174 bacterial genera were observed as putative core microbiomes. Approximately, 20% of putative core genera (38 of 174) were conserved across all duckweed subtypes (Figure S3). Members of the putative core genera identified in SC duckweeds were less than those of NC duckweeds (174 vs. 315 genera). The results indicated that most duckweeds lost their core microbial community under nutrient-deficient conditions, which was clearly supported by alpha diversity analysis (Figure 1B).

The most abundant core microbiome exhibited in *Landoltia* was a member of genus *Allorhizobium-Neorhizobium-Pararhizobium-Rhizobium*, with a relative abundance of 17.8%, followed by unclassified genera of the families *Comamonadaceae*, *Pelomonas*, *Roseateles*, and *Novosphingobium*, with relative abundances of 13.3%, 12.7%, 10%, 10.6%, and 6.4%, respectively (Table S7). Similarly, *Allorhizobium-Neorhizobium-Pararhizobium-Rhizobium* was the most frequent core genus found in *Wolffia*, with a relative abundance of 14.5%, followed by *Pelomonas*, *Novosphingobium*, *Roseateles*, and an unclassified genus of the family *Comamonadaceae*, with relative abundances of 12.3%, 8.2%, 7.9%, and 7.8%, respectively (Table S7). *Roseateles* was the predominant core genus detected in *Lemna*, with a relative abundance of 12.4%, followed by *Allorhizobium-Neorhizobium-Pararhizobium-Rhizobium*, *Novosphingobium*, *Pelomonas*, and *Curvibacter*, with relative abundances of 11.1%, 6.8%, 5.3%, and 4.7%, respectively (Table S7). *Spirodela* harbored a majority of *Allorhizobium-Neorhizobium-Pararhizobium-Rhizobium*, with a proportion of 14.9%, followed by unclassified genera of the families *Comamonadaceae*, *Roseateles*, *Novosphingobium*, and *Pelomonas*, with proportions of 14.7%, 8.5%, 8.4%, and 7.1%, respectively (Table S7).

Differential abundance of core microbial communities between NC and SC duckweeds displayed around 34% of the core genera (147 of 427) with significant differences (Table S8). Under nutrient-deficient stress, the core microbiomes were dynamically changed (Figure 5). Of those, the prominent genera consistently found in NC duckweeds, such as *Rhodobacter* and *Acinetobacter*, were significantly diminished under nutrient-deficient stress (Figure 5, Table S8). Interestingly, rare core microbiomes in the phylum *Proteobacteria*, including *Roseateles*, *Sphingomonas*, and *Pelomonas*, detected in NC duckweeds were greatly enhanced under SC treatments, followed by members in the phylum *Firmicutes*, such as *Lactobacillus* and *Romboutsia* (Figure 5, Table S8). In addition, the genus *Allorhizobium-Neorhizobium-Pararhizobium-Rhizobium*, which presented with high abundance in NC duckweeds, were increased under SC treatments.

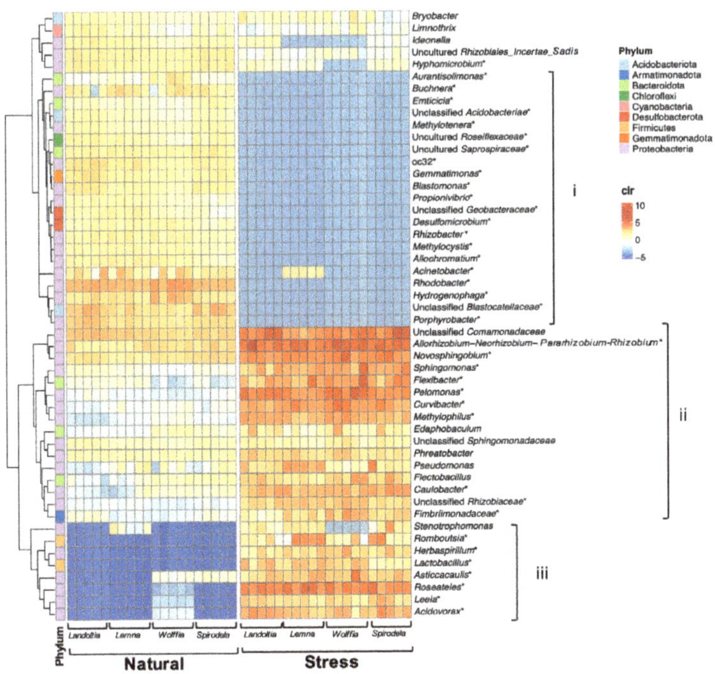

Figure 5. Top 50 abundant core genera of natural and stressed duckweeds (nutrient deficiency). The alteration of core microbiomes was categorized into: **(i)** high abundance in natural duckweeds but diminished under stress; **(ii)** high abundance in natural duckweeds and enriched under stress; and **(iii)** low abundance in natural duckweeds but highly enriched under stress. Asterisks (*) indicate significant difference based on ALDEx2, p-value < 0.05.

2.4. Functional Prediction of Microbial Communities of Duckweeds

Based on KEGG orthologs, 84 functional pathways of duckweed microbial communities were predicted (Table S9). Of those, 73 pathways were categorized into metabolism, environmental information processing, cellular processes, genetic information processing, and drug resistance that were significantly different between NC and SC duckweed microbial communities (Figure 6). Under stress conditions, the relative abundance of four pathways in cellular processes were significantly enriched, including bacterial chemotaxis, biofilm formation, flagellar assembly, and quorum sensing. The pathways involved in environmental information processing also displayed relative abundance enrichment, such as ABC transporters, bacterial secretion systems, and two-component systems. Furthermore, relative abundance of nitrogen metabolism related to plant growth, promoting function and benzoate degradation involved in biodegradation, were significantly increased (Figure 6). Conversely, the pathways mainly related to amino acid metabolism displayed significantly lower relative abundance in SC duckweed microbial communities (Figure 6).

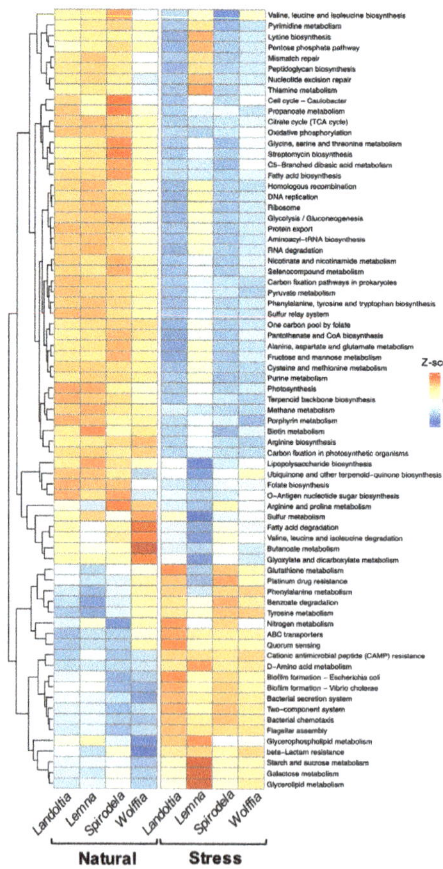

Figure 6. Functional prediction of bacterial communities of duckweeds predicted by PICRUSt2. Relative abundance of pathways that are significantly different (p-value < 0.05) between natural and stressed duckweeds are shown.

3. Discussion

Duckweeds are known to be associated with beneficial PGPB as holobionts [8]. These associated bacteria help promote duckweed growth and phytoremediation performance [7,9].

However, utilization of these PGPB in the real environment is limited, since understanding of the interaction between host and microbes in various environmental conditions is required [15]. Here, we investigated a duckweed-associated microbial community which has not yet been reported in this region using 16S rRNA amplicon metagenome strategy. The "core" microbiomes of the four subtypes of duckweeds, both in natural and stress conditions, were identified.

The natural condition (NC) duckweeds from the same location harbored microbial communities with similar degrees of diversity, except *Wolffia*. This may suggest that the rootless morphology of *Wolffia* and its physiology affected the microbial diversity. The microbial richness (observed ASVs) of NC duckweeds in this study, ranging from 769–1345 ASVs (Table S3), was greater than those of duckweeds collected from ponds in the U.S. [12]. The divergence of the microbial diversity among different studies was possibly caused by the distinction of geographic locations, environmental conditions or variation of methodologies in those studies, such as sample collection approaches, DNA preparation protocols, sequencing depths, and 16S rRNA regions.

The microbial diversity of NC, SC duckweeds, and the ambient water (AW) were clearly distinguished. The bacterial communities of most subtypes of NC duckweeds in this study were higher than those of AW, which agrees with the richness of bacterial diversity of duckweeds collected in Japan that was higher than that in their surrounding water [16]. Conversely, the microbial diversity of duckweeds collected in the U.S. was lower than in their surrounding water [12]. These findings indicate that the floating duckweeds directly interacted with water act as a microbial shelter in aquatic environments, and the associated bacterial communities under the same environmental conditions are likely to be clustered. The different sources of water influenced the microbial assemblage of duckweeds that changed the microbial communities [17]. The results of this study also suggested that environment conditions have a forceful consequence on duckweed microbial composition rather than the duckweed subtypes. In land plants, soil components were an important factor that significantly influenced the soil microbiome [18,19]. However, further investigation is required to determine the chemical composition of water that possibly affects the duckweed bacterial assembly. When the four subtypes of NC duckweeds were cultivated in nutrient-deficient environments, most of the duckweeds significantly lost their bacterial diversity. The results agree with the diversity of *L. minor* microbiome, which was reduced under salinity stress [14].

Microbiomes of NC and SC duckweeds, and of AW, displayed Proteobacteria as the most predominant phylum, similar to results outlined in previous reports [12,16,20]. Members of Proteobacteria, followed by Bacteroidota, were the major phyla found in NC duckweeds, which was supported by previous duckweed microbiome studies [12,16]. In general, Proteobacteria and Firmicutes were dominant phyla in plant endospheres, while Proteobacteria and Bacteroidota were mainly composed in phyllospheres [21]. Pangenomic analysis of Proteobacteria isolated from roots of *Brassicaceae*, poplar, and maize revealed a higher number of substrate transporters that could export/import a board range of compounds [22]. This may explain their high abundance in plant environments and may enable fast-growing characteristics during nutrient-deficient conditions [21]. Moreover, the proportion of Firmicutes was significantly enhanced in SC duckweeds compared to that of NC. The Firmicutes were previously enriched during a period of drought stress due to their thicker cell walls, which promoted their stress tolerance [23,24]. However, most plant-microbe interaction studies have been performed on terrestrial plants, and so the nature of duckweed holobionts in an aquatic lifestyle remains unclear.

To shade more light on duckweed-associated microbes, the microbial community profiles were classified at the genus level. The term "core" microbiome aims to identify a group of potential microbes that are consistently present in duckweed hosts [12]. In this study, half of the core microbiomes were shared across the four subtypes of duckweed collected from the same natural site. The data suggested that duckweeds growing in the same location harbored a remarkable conserved core microbiome across duckweed subtypes. Core genera exclusively found in one subtype but not in the others, presented in small relative abundance (0.01–0.60%), which did not represent specific taxa, but likely occurred due to non-captured sequencing.

Several dominant genera of Proteobacteria, such as *Acinetobacter*, *Allorhizobium-Neorhizobium-Pararhizobium-Rhizobium*, *Hydrogenophaga*, *Novosphingobium*, and *Rhodobacter*, were found to be universal "core" microbiomes of NC duckweeds, which were similarly observed as duckweed microbiomes [12,16]. These stable genera suggested that the core microbiomes tentatively act as the universal duckweed-associated bacteria (DAB). *Acinetobacter calcoaceticus* P23 has been proven to promote duckweed biomass and facilitate phytoremediation through phenol degradation [7,9]. Members of the genus *Rhizobium* are well-known as a typical symbiosis of leguminous plants, as well as the other plants [25]. Several species of *Novosphingobium* promoted plant growth by the production of indole-3-acetic acid (IAA) [26–28]; while members of *Rhodobacter* were recognized as plant growth-promoting bacteria [29,30]. Although many potential DAB have been successfully inocculated to duckweeds through culture-dependent [7,8,16] and culture-independent methods [12,16], associations occurred for a short period and then vanished [13]. Therefore, high potential DAB suitable for long-term applications should be selected from "stable" core microbiomes. In addition, some common DAB found in NC duckweeds, such as *Rhodobacter* and *Acinetobacter*, are also present in ambient water. The results suggest that those core genera, originally exhibited in the surrounding water, were exclusively recruited by the duckweed host [12,17].

The "stable" core microbiomes of four subtypes of duckweeds under nutrient-deficient conditions were investigated. The candidate-beneficial DAB, such as *Acinetobacter* and *Rhodobacter*, found highly abundantly in NC duckweed, were significantly diminished; whereas *Allorhizobium-Neorhizobium-Pararhizobium-Rhizobium* and *Novosphingobium* were persistently enhanced. The results suggested that these potential duckweed-associated microbiomes experience a dynamic change in response to the environment or stressor. The disappearing scenario was observed when a beneficial DAB, *Aquitalea magnusonii* H3, was inoculated to *L. minor*; it could promote growth in just a week and vanished after growing under several conditions [13]. Members of the genus *Rhizobium* could promote drought tolerance in both leguminous [31,32] and non-leguminous plants [33]. Genome analysis of *Rhizobium* strains revealed a set of genes that are involved in plant-growth-promoting and stress-tolerant traits, including phosphate solubilization, production of IAA, exopolysaccharide, siderophores, and 1-aminocyclopropane-1-carboxylate (ACC) deaminase [32]. A *Novosphingobium* strain was reported to increase salinity tolerance and induce accumulation of IAA in plants [34]. Interestingly, several rare core microbiomes belonged to the phylum Proteobacteria, such as *Pelomonas*, *Roseateles*, and *Sphingomonas*, and phylum Firmicutes, such as *Lactobacillus*, were also detected in NC duckweeds and were greatly enhanced under nutrient deficiency. These genera have been reported as PGPB; for instance, *Pelomonas* sp. MRB3 has been recently proven as a DAB by root colonization and growth promotion of *L. minor* [35]. An endophytic *Sphigomonas* was reported to promote growth of tomatoes by the production of phytohormones, IAA, and gibberellins [26]. Genome analysis of *Sphingomonas*-determined genes related to adaptation to extreme oligotrophic environments [36]. Similarly, *Lactobacillus*, associated with plants, displayed plant growth promoting traits, such as IAA production, phosphate solubilization, and anti-phytopathogenic activity [37,38] which could be applied as biofertilizer in a variety of plants, such as wheat, tomato, pepper, and cucumber [37]. Apart from PGP traits, *Roseateles depolymerans* TB-87, isolated from fresh water, was reported to be able to decompose various bioplastics that may be useful for bioremediation [39].

Functional predictions of microbial communities in NC and SC duckweeds displayed alterations in the relative abundance of the pathways. Under stress conditions, pathways involved in bacterial motility, biofilm formation, chemotaxis, flagellar assembly, and two-component systems were significantly enhanced. The findings are comparable to those functional predictions of natural water-obtained DAB, co-cultivated with several duckweed species, including *Spirodela*, *Landoltia*, *Lemna*, *Wolffiella*, and *Wolffia* [13,17]. These enhanced pathways may explain essential steps for DAB colonization. A recent study on duckweed illuminated these functions, particularly flagellar motility and cell surface structures such as lipopolysaccharide and type-IV pili synthesis, were essential for colonization and fitness regulation of DAB, *A. magnusonii* H3, to *L. minor* surfaces [40]. Comparable to those of terrestrial plants, *Arabidopsis thaliana* attracted beneficial *Bacillus subtilis* via root exudates, while flagellar motility and chemotaxis machinery mediated *B. subtilis* contact and settled on the roots before forming into biofilm for long-term colonization [41]. Two component system signal transduction is the key pathway involved in differentiation of bacterial cells to biofilm-producing cells [42]. Similarly, biofilm formation contributed to corn root colonization and seed adhesion of plant-beneficial *Pseudomonas putida* KT2440 [43,44]. Additionally, the relative abundance of benzoate degradation and nitrogen metabolism pathways are significantly increased in the microbial communities of SC duckweeds, which may suggest the enhancement of the bioremediation ability of those DAB, such as phenolic compound degradation and nitrogen removal [13,17]. The functional prediction results suggested that those persistently presented bacteria were likely to be a real DAB, and have a positive effect on duckweed fitness under stress conditions.

Altogether, *Lactobacillus*, *Novosphingobium*, *Pelomonas*, *Rhizobium*, *Roseateles*, and *Sphingomonas* are proposed to be "stable" DAB of duckweeds; potential candidates for duckweed utilization under stress environments. Further investigation is required to understand the actual DAB traits and their interaction with the duckweed host.

4. Materials and Methods

4.1. Sample Collection

Four duckweed genera (*Spirodela*, *Landoltia*, *Lemna*, and *Wolffia*) and ambient (surrounding) water were collected from drainage ditches in Nakorn Pathom, Thailand (14°00′34.7″ N 99°58′13.3″ E) in June 2021 (Figure S1) in five replicates (duckweeds, $n = 20$; ambient water, $n = 5$). Duckweed samples were rinsed three times in sterilized water, transferred to 5 mL centrifuge tubes containing 3 mL of DNA/RNA shield™ (Zymo Research Corp, Irvine, CA, USA), and immediately stored at −80 °C until used. An amount of 500 milliliters of water samples were passed through sterilized Whatman filter paper, grade 4 (20–25 μm), to get rid of impurities before being filtering through Whatman WME membrane (0.2 μm) to capture microbial communities. The filters were excised into small pieces and then transferred to 5 mL centrifuge tubes containing 3 mL of DNA/RNA shield™ (Zymo Research Corp, Irvine, CA, USA) and immediately stored at −80 °C until used.

EC and chemical composition of the ambient water samples were analyzed for total N, P, K, As, Cd, and NO_3-N at the Central Laboratory and Greenhouse Complex, Kasetsart University, Kamphaeng Saen Campus, Thailand.

4.2. Experimental Design for Nutrient-Deficient Condition

Approximately 5 g fresh weight of natural duckweeds ($n = 20$) were grown in sterilized distilled water in clean glass containers (length × width × height: 31 × 18.5 × 19 cm) at 25 °C under a photo-period of 12 h with a light intensity of 50 μmol m^{-2} s^{-1}. After 14-day cultivation, duckweed and water were harvested and preserved as described above.

4.3. Duckweed Genotyping

Duckweed samples were identified by two barcodes: *atp*F-*atp*H (5′-ACTCGCACACAC TCCCTTTCC-3′ and 5′-GCTTTTATGGAAGCTTTAACAAT-3′) and *psb*K-*psb*I (5′-TTAGCAT TTGTTTGGCAAG-3′ and 5′-AAAGTTTGAGAGTAAGCAT-3′), using PCR conditions as previously described [45].

4.4. DNA Extraction and Sequencing

Approximately 250 mg of each duckweed sample obtained from natural and nutrient-deficient experiments were homogenized in liquid nitrogen. DNA from duckweed and water filtrates was isolated using a ZymoBIOMICS™ DNA Miniprep Kit (Zymo Research Corp, USA) according to the manufacturer's instructions. All DNA samples were quantified using a NanoDrop™ One/OneC Microvolume UV-Vis Spectrophotometer (Thermo Fisher Scientific, Waltham, MA, USA). Library was prepared using V3-V4 region of 16S rRNA amplification with primers 341F (5-CCTAYGGGRBGCASCAG-3) and 806R (5-GGACTACNNGGGTATCTAAT-3) [46]. Finally, 2×250 bp pair-end sequencing was performed using an Illumina Novaseq 6000 platform at NovogeneAIT Genomics Singapore Pte. ZymoBIOMICS™ Microbial Community DNA Standard (Mock; Zymo Research Corp, CA, USA), and was used as a control.

4.5. Data Processing and Metagenome Analysis

The raw reads were pre-processed by the removal of adaptors and primers performed by NovogeneAIT. The pair-end reads were denoised, dereplicated, and chimeras-filtered using a dada2 plugin [47] under QIIME2 (q2) version 2021.8 [48]. The amplicon sequence variances (ASVs) were classified using a pre-formatted SILVA version 138 reference database [49] and q2-feature-classifier classify-sklearn. The suspected background contamination features were subtracted using Microdecon [50]. The final feature table was filtered for chloroplast, mitochondria, and low frequency ASVs (<10 reads across all samples).

Rarefaction analysis was conducted based on the feature table with a random sampling to the minimal read number (27,601) of all samples. Alpha diversity analysis was calculated using a q2-diversity alpha plugin. Non-parametric Kruskal-Wallis tests [51] were performed to compare the alpha diversity index. For beta diversity analysis, Bray-Curtis's dissimilarity [52] was calculated using a q2-diversity core-metrics plugin. Permutation multivariate analysis of variance (PERMANOVA) [53] was used to compare community composition between groups of samples. Non-metric multidimensional scaling (nMDS) was calculated from Bray-Curtis's dissimilarity matrix using the vegan package (https://CRAN.R-project.org/package=vegan (accessed on 24 February 2022)) and were plotted using the ggplot2 package [54] in R version 4.1.1. Differential abundance comparisons were performed using ALDEx2 [55]. The p-value obtained from multiple pairwise testing was adjusted by Holm's sequential Bonferroni method [56]. Adjusted p-value < 0.05 was considered as statistically significant. Data visualization was conducted by ggplot2, ComplexHeatmap [57], and the VennDiagram package (https://CRAN.R-project.org/package=VennDiagram (accessed on 24 February 2022)) in R version 4.1.1.

4.6. Functional Metagenome Prediction

The obtained ASVs table was subjected to PICRUSt2 [58] to predict functional profiles of metagenome data. The predicted functional table was categorized into pathways based on the Kyoto Encyclopedia of Genes and Genomes (KEGG) Orthologs [59]. Those predicted pathways with low mean relative abundance (<0.4%) were filtered. The differential abundance of predicted pathways between natural and nutrient-deficient conditions was evaluated by the Wilcoxon (Mann-Whitney U) test [60].

Supplementary Materials: The following supporting information can be downloaded at: https://www.mdpi.com/article/10.3390/plants11212915/s1, Table S1 ASVs information and sequence reads for duckweeds under natural, nutrient-deficient conditions, and ambient water; Table S2 Number of sequences obtained from all samples; Table S3 Alpha diversity analysis of all samples with observed ASVs and Shannon diversity index; Table S4 Lists and abundance of core genera of duckweeds collected from natural site; Table S5 Lists of core microbiome obtained from four duckweed subtypes collected from natural site and ambient water; Table S6 High relative abundance (>1%) of core genera detected in duckweeds collected from natural site; Table S7 High relative abundance (>1%) of core genera detected in duckweeds under nutrient-deficient condition; Table S8 Differential abundance testing of bacterial communities between duckweeds in natural and in nutrient-deficient condition using ALDEx2; Table S9 Functional prediction of metagenome data using PICRUSt2; Figure S1, Duckweed samples collected from Kasetsart University Kamphaeng

Saen campus, Nakhon Pathom, Thailand during June 2021; Figure S2, The rarefaction curve between observed ASVs and sample depth displayed nearly plateau at 27,601 reads; Figure S3, Candidate core microbiomes from duckweeds under nutrient-deficient condition.

Author Contributions: C.B., P.R., S.K. and J.P. collected samples and performed experiments; C.B. and A.T. analyzed, interpreted the data, and writing the original draft manuscript; P.W. supported metagenome analysis; A.T. conceived the project, designed the methodology, supervised the research, and revised the manuscript. All authors have read and agreed to the published version of the manuscript.

Funding: C.B. has been awarded PhD scholarship from Chulabhorn Royal Academy. J.P. has been awarded a post-master fellowship under Kasetsart University Reinventing University Program 2022 supported by the Ministry of Higher Education, Science, Research and Innovation and the Thailand Science Research and Innovation. This work was funded by Kasetsart University Research and Development Institute (KURDI), grant number FF(KU)4.64, and the Science and Technology Research Partnership for Sustainable Development (SATREPS), JICA, Japan.

Data Availability Statement: The 16S rRNA metagenome data were deposited in the GenBank NCBI Bioproject number PRJNA888649.

Acknowledgments: We would like to thank Metha Meetam, Mahidol University for his suggestion in surveying duckweeds.

Conflicts of Interest: The authors declare no conflict of interest.

References

1. Bog, M.; Appenroth, K.-J.; Sree, K.S. Duckweed (Lemnaceae): Its molecular taxonomy. *Front. Sustain. Food Syst.* **2019**, *3*, 117. [CrossRef]
2. Van der Spiegel, M.; Noordam, M.Y.; van der Fels-Klerx, H.J. Safety of novel protein sources (insects, microalgae, seaweed, duckweed, and rapeseed) and legislative aspects for their application in food and feed production. *Compr. Rev. Food Sci. Food Saf.* **2013**, *12*, 662–678. [CrossRef]
3. Cui, W.; Cheng, J.J. Growing duckweed for biofuel production: A review. *Plant Biol.* **2015**, *17* (Suppl S1), 16–23. [CrossRef] [PubMed]
4. Appenroth, K.J.; Sree, K.S.; Bog, M.; Ecker, J.; Seeliger, C.; Bohm, V.; Lorkowski, S.; Sommer, K.; Vetter, W.; Tolzin-Banasch, K.; et al. Nutritional value of the duckweed species of the genus *Wolffia* (Lemnaceae) as human food. *Front. Chem.* **2018**, *6*, 483. [CrossRef]
5. Chen, G.; Huang, J.; Fang, Y.; Zhao, Y.; Tian, X.; Jin, Y.; Zhao, H. Microbial community succession and pollutants removal of a novel carriers enhanced duckweed treatment system for rural wastewater in Dianchi Lake basin. *Bioresour. Technol.* **2019**, *276*, 8–17. [CrossRef] [PubMed]
6. Acosta, K.; Appenroth, K.J.; Borisjuk, L.; Edelman, M.; Heinig, U.; Jansen, M.A.K.; Oyama, T.; Pasaribu, B.; Schubert, I.; Sorrels, S.; et al. Return of the Lemnaceae: Duckweed as a model plant system in the genomics and postgenomics era. *Plant Cell* **2021**, *33*, 3207–3234. [CrossRef]
7. Yamaga, F.; Washio, K.; Morikawa, M. Sustainable biodegradation of phenol by *Acinetobacter calcoaceticus* P23 isolated from the rhizosphere of duckweed *Lemna aoukikusa*. *Environ. Sci. Technol.* **2010**, *44*, 6470–6474. [CrossRef] [PubMed]
8. Gilbert, S.; Xu, J.; Acosta, K.; Poulev, A.; Lebeis, S.; Lam, E. Bacterial production of indole related compounds reveals their role in association between duckweeds and endophytes. *Front. Chem.* **2018**, *6*, 265. [CrossRef]
9. Suzuki, W.; Sugawara, M.; Miwa, K.; Morikawa, M. Plant growth-promoting bacterium *Acinetobacter calcoaceticus* P23 increases the chlorophyll content of the monocot *Lemna minor* (duckweed) and the dicot *Lactuca sativa* (lettuce). *J. Biosci. Bioeng.* **2014**, *118*, 41–44. [CrossRef] [PubMed]
10. Yamakawa, Y.; Jog, R.; Morikawa, M. Effects of co-inoculation of two different plant growth-promoting bacteria on duckweed. *Plant Growth Regul.* **2018**, *86*, 287–296. [CrossRef]
11. Khairina, Y.; Jog, R.; Boonmak, C.; Toyama, T.; Oyama, T.; Morikawa, M. Indigenous bacteria, an excellent reservoir of functional plant growth promoters for enhancing duckweed biomass yield on site. *Chemosphere* **2021**, *268*, 129247. [CrossRef]
12. Acosta, K.; Xu, J.; Gilbert, S.; Denison, E.; Brinkman, T.; Lebeis, S.; Lam, E. Duckweed hosts a taxonomically similar bacterial assemblage as the terrestrial leaf microbiome. *PLoS ONE* **2020**, *15*, e0228560. [CrossRef] [PubMed]
13. Ishizawa, H.; Kuroda, M.; Inoue, D.; Morikawa, M.; Ike, M. Community dynamics of duckweed-associated bacteria upon inoculation of plant growth-promoting bacteria. *FEMS Microbiol. Ecol.* **2020**, *96*, fiaa101. [CrossRef] [PubMed]
14. O'Brien, A.M.; Yu, Z.H.; Luo, D.Y.; Laurich, J.; Passeport, E.; Frederickson, M.E. Resilience to multiple stressors in an aquatic plant and its microbiome. *Am. J. Bot.* **2020**, *107*, 273–285. [CrossRef] [PubMed]
15. Finkel, O.M.; Castrillo, G.; Herrera Paredes, S.; Salas Gonzalez, I.; Dangl, J.L. Understanding and exploiting plant beneficial microbes. *Curr. Opin. Plant Biol.* **2017**, *38*, 155–163. [CrossRef] [PubMed]

16. Iwashita, T.; Tanaka, Y.; Tamaki, H.; Yoneda, Y.; Makino, A.; Tateno, Y.; Li, Y.; Toyama, T.; Kamagata, Y.; Mori, K. Comparative analysis of microbial communities in fronds and roots of three duckweed species: *Spirodela polyrhiza*, *Lemna minor*, and *Lemna aequinoctialis*. *Microbes Environ.* **2020**, *35*, ME20081. [CrossRef] [PubMed]
17. Inoue, D.; Hiroshima, N.; Ishizawa, H.; Ike, M. Whole structures, core taxa, and functional properties of duckweed microbiomes. *Bioresour. Technol. Rep.* **2022**, *18*, 101060. [CrossRef]
18. Fierer, N.; Lauber, C.L.; Ramirez, K.S.; Zaneveld, J.; Bradford, M.A.; Knight, R. Comparative metagenomic, phylogenetic and physiological analyses of soil microbial communities across nitrogen gradients. *ISME J.* **2012**, *6*, 1007–1017. [CrossRef]
19. Sheldrake, M.; Rosenstock, N.P.; Mangan, S.; Revillini, D.; Sayer, E.J.; Olsson, P.A.; Verbruggen, E.; Tanner, E.V.J.; Turner, B.L.; Wright, S.J. Responses of arbuscular mycorrhizal fungi to long-term inorganic and organic nutrient addition in a lowland tropical forest. *ISME J.* **2018**, *12*, 2433–2445. [CrossRef] [PubMed]
20. Huang, W.; Gilbert, S.; Poulev, A.; Acosta, K.; Lebeis, S.; Long, C.; Lam, E. Host-specific and tissue-dependent orchestration of microbiome community structure in traditional rice paddy ecosystems. *Plant Soil* **2020**, *452*, 379–395. [CrossRef]
21. Trivedi, P.; Leach, J.E.; Tringe, S.G.; Sa, T.; Singh, B.K. Plant-microbiome interactions: From community assembly to plant health. *Nat. Rev. Microbiol.* **2020**, *18*, 607–621. [CrossRef]
22. Levy, A.; Salas Gonzalez, I.; Mittelviefhaus, M.; Clingenpeel, S.; Herrera Paredes, S.; Miao, J.; Wang, K.; Devescovi, G.; Stillman, K.; Monteiro, F.; et al. Genomic features of bacterial adaptation to plants. *Nat. Genet.* **2017**, *50*, 138–150. [CrossRef]
23. Xu, L.; Naylor, D.; Dong, Z.; Simmons, T.; Pierroz, G.; Hixson, K.K.; Kim, Y.M.; Zink, E.M.; Engbrecht, K.M.; Wang, Y.; et al. Drought delays development of the sorghum root microbiome and enriches for monoderm bacteria. *Proc. Natl. Acad. Sci. USA* **2018**, *115*, E4284–E4293. [CrossRef]
24. Xu, L.; Coleman-Derr, D. Causes and consequences of a conserved bacterial root microbiome response to drought stress. *Curr. Opin. Microbiol.* **2019**, *49*, 1–6. [CrossRef] [PubMed]
25. Jaiswal, S.K.; Mohammed, M.; Ibny, F.Y.I.; Dakora, F.D. Rhizobia as a source of plant growth-promoting molecules: Potential applications and possible operational mechanisms. *Front. Sustain. Food Syst.* **2021**, *4*, 619676. [CrossRef]
26. Khan, A.L.; Waqas, M.; Kang, S.M.; Al-Harrasi, A.; Hussain, J.; Al-Rawahi, A.; Al-Khiziri, S.; Ullah, I.; Ali, L.; Jung, H.Y.; et al. Bacterial endophyte *Sphingomonas* sp. LK11 produces gibberellins and IAA and promotes tomato plant growth. *J. Microbiol.* **2014**, *52*, 689–695. [CrossRef]
27. Rangjaroen, C.; Rerkasem, B.; Teaumroong, N.; Noisangiam, R.; Lumyong, S. Promoting plant growth in a commercial rice cultivar by endophytic diazotrophic bacteria isolated from rice landraces. *Ann. Microbiol.* **2015**, *65*, 253–266. [CrossRef]
28. Banik, A.; Mukhopadhaya, S.K.; Dangar, T.K. Characterization of N2-fixing plant growth promoting endophytic and epiphytic bacterial community of Indian cultivated and wild rice (*Oryza* spp.) genotypes. *Planta* **2016**, *243*, 799–812. [CrossRef] [PubMed]
29. Gamal-Eldin, H.; Elbanna, K. Field evidence for the potential of *Rhodobacter capsulatus* as biofertilizer for flooded rice. *Curr. Microbiol.* **2011**, *62*, 391–395. [CrossRef] [PubMed]
30. Hayashi, S.; Iwamoto, Y.; Hirakawa, Y.; Mori, K.; Yamada, N.; Maki, T.; Yamamoto, S.; Miyasaka, H. Plant-growth-promoting effect by cell components of purple non-sulfur photosynthetic bacteria. *Microorganisms* **2022**, *10*, 771. [CrossRef]
31. Staudinger, C.; Mehmeti-Tershani, V.; Gil-Quintana, E.; Gonzalez, E.M.; Hofhansl, F.; Bachmann, G.; Wienkoop, S. Evidence for a rhizobia-induced drought stress response strategy in *Medicago truncatula*. *J. Proteom.* **2016**, *136*, 202–213. [CrossRef] [PubMed]
32. Igiehon, N.O.; Babalola, O.O.; Aremu, B.R. Genomic insights into plant growth promoting rhizobia capable of enhancing soybean germination under drought stress. *BMC Microbiol.* **2019**, *19*, 159. [CrossRef] [PubMed]
33. Tulumello, J.; Chabert, N.; Rodriguez, J.; Long, J.; Nalin, R.; Achouak, W.; Heulin, T. *Rhizobium alamii* improves water stress tolerance in a non-legume. *Sci. Total Environ.* **2021**, *797*, 148895. [CrossRef] [PubMed]
34. Vives-Peris, V.; Gómez-Cadenas, A.; Pérez-Clemente, R.M. Salt stress alleviation in citrus plants by plant growth-promoting rhizobacteria *Pseudomonas putida* and *Novosphingobium* sp. *Plant Cell Rep.* **2018**, *37*, 1557–1569. [CrossRef]
35. Makino, A.; Nakai, R.; Yoneda, Y.; Toyama, T.; Tanaka, Y.; Meng, X.-Y.; Mori, K.; Ike, M.; Morikawa, M.; Kamagata, Y.; et al. Isolation of aquatic plant growth-promoting bacteria for the floating plant duckweed (*Lemna minor*). *Microorganisms* **2022**, *10*, 1564. [CrossRef]
36. Aylward, F.O.; McDonald, B.R.; Adams, S.M.; Valenzuela, A.; Schmidt, R.A.; Goodwin, L.A.; Woyke, T.; Currie, C.R.; Suen, G.; Poulsen, M. Comparison of 26 sphingomonad genomes reveals diverse environmental adaptations and biodegradative capabilities. *Appl. Environ. Microbiol.* **2013**, *79*, 3724–3733. [CrossRef]
37. Lamont, J.R.; Wilkins, O.; Bywater-Ekegärd, M.; Smith, D.L. From yogurt to yield: Potential applications of lactic acid bacteria in plant production. *Soil Biol. Biochem.* **2017**, *111*, 1–9. [CrossRef]
38. Quattrini, M.; Bernardi, C.; Stuknyte, M.; Masotti, F.; Passera, A.; Ricci, G.; Vallone, L.; De Noni, I.; Brasca, M.; Fortina, M.G. Functional characterization of *Lactobacillus plantarum* ITEM 17215: A potential biocontrol agent of fungi with plant growth promoting traits, able to enhance the nutritional value of cereal products. *Food Res. Int.* **2018**, *106*, 936–944. [CrossRef]
39. Shah, A.A.; Eguchi, T.; Mayumi, D.; Kato, S.; Shintani, N.; Kamini, N.R.; Nakajima-Kambe, T. Purification and properties of novel aliphatic-aromatic co-polyesters degrading enzymes from newly isolated *Roseateles depolymerans* strain TB-87. *Polym. Degrad. Stab.* **2013**, *98*, 609–618. [CrossRef]
40. Ishizawa, H.; Kuroda, M.; Inoue, D.; Ike, M. Genome-wide identification of bacterial colonization and fitness determinants on the floating macrophyte, duckweed. *Commun. Biol.* **2022**, *5*, 68. [CrossRef]

41. Allard-Massicotte, R.; Tessier, L.; Lécuyer, F.; Lakshmanan, V.; Lucier, J.-F.; Garneau, D.; Caudwell, L.; Vlamakis, H.; Bais, H.P.; Beauregard, P.B. *Bacillus subtilis* early colonization of *Arabidopsis thaliana* roots involves multiple chemotaxis receptors. *mBio* **2016**, *7*, e01664-16. [CrossRef] [PubMed]
42. Liu, C.; Sun, D.; Zhu, J.; Liu, W. Two-component signal transduction systems: A major strategy for connecting input stimuli to biofilm formation. *Front. Microbiol.* **2019**, *9*, 3279. [CrossRef]
43. Espinosa-Urgel, M.; Salido, A.; Ramos, J.L. Genetic analysis of functions involved in adhesion of *Pseudomonas putida* to seeds. *J. Bacteriol.* **2000**, *182*, 2363–2369. [CrossRef] [PubMed]
44. Martínez-Gil, M.; Yousef-Coronado, F.; Espinosa-Urgel, M. LapF, the second largest *Pseudomonas putida* protein, contributes to plant root colonization and determines biofilm architecture. *Mol. Microbiol.* **2010**, *77*, 549–561. [CrossRef] [PubMed]
45. Wang, W.; Wu, Y.; Yan, Y.; Ermakova, M.; Kerstetter, R.; Messing, J. DNA barcoding of the Lemnaceae, a family of aquatic monocots. *BMC Plant Biol.* **2010**, *10*, 205. [CrossRef]
46. Yu, Y.; Lee, C.; Kim, J.; Hwang, S. Group-specific primer and probe sets to detect methanogenic communities using quantitative real-time polymerase chain reaction. *Biotechnol. Bioeng.* **2005**, *89*, 670–679. [CrossRef]
47. Callahan, B.J.; McMurdie, P.J.; Rosen, M.J.; Han, A.W.; Johnson, A.J.; Holmes, S.P. DADA2: High-resolution sample inference from Illumina amplicon data. *Nat. Methods.* **2016**, *13*, 581–583. [CrossRef]
48. Bolyen, E.; Rideout, J.R.; Dillon, M.R.; Bokulich, N.A.; Abnet, C.C.; Al-Ghalith, G.A.; Alexander, H.; Alm, E.J.; Arumugam, M.; Asnicar, F.; et al. Reproducible, interactive, scalable and extensible microbiome data science using QIIME 2. *Nat. Biotechnol.* **2019**, *37*, 852–857. [CrossRef]
49. Robeson, M.S., 2nd; O'Rourke, D.R.; Kaehler, B.D.; Ziemski, M.; Dillon, M.R.; Foster, J.T.; Bokulich, N.A. RESCRIPt: Reproducible sequence taxonomy reference database management. *PLoS Comput. Biol.* **2021**, *17*, e1009581. [CrossRef]
50. McKnight, D.T.; Huerlimann, R.; Bower, D.S.; Schwarzkopf, L.; Alford, R.A.; Zenger, K.R. microDecon: A highly accurate read-subtraction tool for the post-sequencing removal of contamination in metabarcoding studies. *Environ. DNA* **2019**, *1*, 14–25. [CrossRef]
51. Kruskal, W.H.; Wallis, W.A. Use of ranks in one-criterion variance analysis. *J. Am. Stat. Assoc.* **1952**, *47*, 583–621. [CrossRef]
52. Sorensen, T.A. A method of establishing groups of equal amplitude in plant sociology based on similarity of species content and its application to analyses of the vegetation on Danish commons. *Biol. Skar.* **1948**, *5*, 1–34.
53. Anderson, M.J. Permutational multivariate analysis of variance (PERMANOVA). In *Wiley StatsRef: Statistics Reference Online*; Wiley: Hoboken, NJ, USA, 2017; pp. 1–5. [CrossRef]
54. Wickham, H. *Package 'ggplot2': Elegant Graphics for Data Analysis*; Springer: New York, NY, USA, 2016; Volume 10, pp. 970–978.
55. Fernandes, A.D.; Reid, J.N.; Macklaim, J.M.; McMurrough, T.A.; Edgell, D.R.; Gloor, G.B. Unifying the analysis of high-throughput sequencing datasets: Characterizing RNA-seq, 16S rRNA gene sequencing and selective growth experiments by compositional data analysis. *Microbiome* **2014**, *2*, 15. [CrossRef]
56. Holm, S. A simple sequentially rejective multiple test procedure. *Scand. J. Stat.* **1979**, *6*, 65–70.
57. Gu, Z.; Eils, R.; Schlesner, M. Complex heatmaps reveal patterns and correlations in multidimensional genomic data. *Bioinformatics* **2016**, *32*, 2847–2849. [CrossRef]
58. Douglas, G.M.; Maffei, V.J.; Zaneveld, J.R.; Yurgel, S.N.; Brown, J.R.; Taylor, C.M.; Huttenhower, C.; Langille, M.G.I. PICRUSt2 for prediction of metagenome functions. *Nat. Biotechnol.* **2020**, *38*, 685–688. [CrossRef]
59. Kanehisa, M.; Goto, S. KEGG: Kyoto encyclopedia of genes and genomes. *Nucleic Acids Res.* **2000**, *28*, 27–30. [CrossRef]
60. Wilcoxon, F. Individual comparisons by ranking methods. *Biom. Bull.* **1945**, *1*, 80–83. [CrossRef]

plants

MDPI

Communication

The Effects of Microbiota on the Herbivory Resistance of the Giant Duckweed Are Plant Genotype-Dependent

Martin Schäfer [1,2] and Shuqing Xu [1,2,*]

1 Institute for Evolution and Biodiversity, University of Münster, 48149 Münster, Germany
2 Institute of Organismic and Molecular Evolution, Johannes Gutenberg University Mainz, 55128 Mainz, Germany
* Correspondence: shuqing.xu@uni-mainz.de

Abstract: In nature, all plants live with microbes, which can directly affect their host plants' physiology and metabolism, as well as their interacting partners, such as herbivores. However, to what extent the microbiota shapes the adaptive evolution to herbivory is unclear. To address this challenge, it is essential to quantify the intra-specific variations of microbiota effects on plant fitness. Here, we quantified the fitness effects of microbiota on the growth, tolerance, and resistance to herbivory among six genotypes of the giant duckweed, *Spirodela polyrhiza.* We found that the plant genotypes differed in their intrinsic growth rate and tolerance, but not in their resistance to a native herbivore, the great pond snail. Inoculation with microbiota associated with *S. polyrhiza* growing outdoors reduced the growth rate and tolerance in all genotypes. Additionally, the microbiota treatment altered the herbivory resistance in a genotype-specific manner. Together, these data show the potential of microbiota in shaping the adaptive evolution of plants.

Keywords: *Spirodela polyrhiza*; giant duckweed; *Lymnaea stagnalis*; great pond snail; microbiota; evolution; adaptation; herbivory; tolerance; resistance

check for updates

Citation: Schäfer, M.; Xu, S. The Effects of Microbiota on the Herbivory Resistance of the Giant Duckweed Are Plant Genotype-Dependent. *Plants* **2022**, *11*, 3317. https://doi.org/10.3390/plants11233317

Academic Editors: Viktor Oláh, Klaus-Jürgen Appenroth and K. Sowjanya Sree

Received: 10 October 2022
Accepted: 26 November 2022
Published: 1 December 2022

1. Introduction

All plants in nature live with microbes that populate their close proximity, live on their surface, or even live directly within them [1,2]. As plant microbiota can profoundly change the metabolism, physiology, and fitness of the host plants, the intimate interaction between plants and their microbiota can also affect their interaction with herbivores [3] and thus potentially also the evolutionary trajectory of adaptation to herbivory. While increasing studies suggest that the plant microbiota alters plant growth and defenses [3–5], they have largely focused on single genotypes. However, to understand whether and how the plant microbiota affects plant evolution, it is essential to quantify the effect of microbiota on plant fitness using different plant genotypes (fitness landscape).

Here, we address this challenge using the hydrophyte, *Spirodela polyrhiza* (Araceae, Lemnoideae), one of the fastest-growing angiosperms, in which the microbiota can be manipulated to assess the microbe-dependent plant fitness effects [6,7]. For example, inoculation of *S. polyrhiza* with *Ensifer sp.* strain SP4 can promote plant growth [8], and *S. polyrhiza*-associated microorganisms can contribute to the degradation of microcystins, which are a group of cyanotoxins [9]. Using six globally distributed *S. polyrhiza* genotypes, we showed that the microbiota reduced plant growth and tolerance to herbivory by freshwater snails, while it changed the resistance to herbivory in a genotype-specific manner, indicating the importance of microbiota in shaping the evolutionary trajectory of plants in nature.

2. Results

2.1. *Different Genotypes Varied in Intrinsic Growth Rate, but Not Resistance to Herbivory by Snails*

To investigate the genotype-dependent fitness in *S. polyrhiza*, we first analyzed the growth rate of six different *S. polyrhiza* genotypes originating from Asia, North America, Australia, and Europe. We calculated the relative growth rate (RGR) based on the changes in frond numbers within 7 days. As each frond represents an individual plant, the frond number is strongly correlated with fitness in *S. polyrhiza*, although additional measurements of biomass, surface area, and chlorophyll content could further improve our mechanistic understanding. Under the greenhouse conditions, the six genotypes showed different RGR (Figure 1A, genotype effects: $p = 2.2 \times 10^{-16}$, Kruskal–Wallis rank sum test). While genotype SP05 showed the highest growth rate, genotype SP30 had the lowest growth rate.

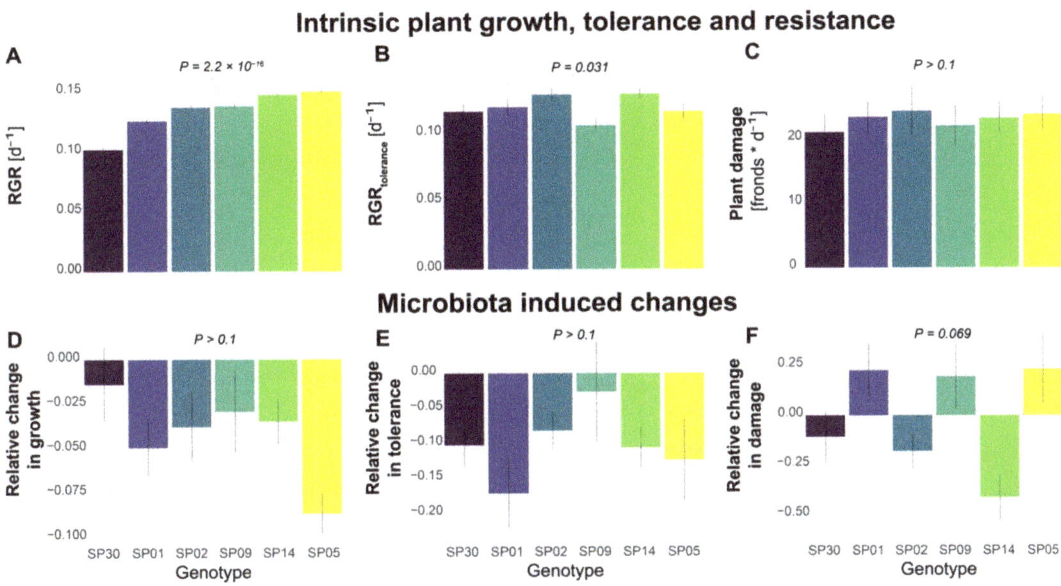

Figure 1. Intra-specific variations of microbiota effects on plant growth, tolerance, and resistance to the snail herbivory in *S. polyrhiza*. (**A**) Relative growth rate (RGR) based on the number of fronds grown within a week. (**B**) Tolerance as RGR within 5 days after snail herbivory ($RGR_{tolerance}$). (**C**) Plant damage, which was calculated as the number of fronds consumed or damaged by one *L. stagnalis* within 24 h and represents the level of plant resistance to herbivory, with high damage levels indicating low resistance. (**D–F**) The effects of plant microbiota on growth rate, tolerance, and resistance to snail herbivory, estimated as the percentage of the growth/tolerance/damage rate that was altered by the pre-treatment with a microbe inoculum. Positive and negative values refer to increased or decreased growth/tolerance/resistance levels. Columns represent mean values and error bars show the standard error. Each column represents a genotype with N > 16 for (**A**,**D**) and N > 11 for (**B**,**C**,**E**,**F**). *p*-values refer to genotype effects.

We then performed bioassays to quantify the tolerance after snail herbivory, based on the RGR within 5 days after snail exposure ($RGR_{tolerance}$). During snail exposure, snails were allowed to feed on the duckweed populations for 24 h. The $RGR_{tolerance}$ was calculated using the remaining fronds after snail exposure, which still possessed at least one of the reproductive pouches (including damaged fronds) and therefore were expected to still have the potential to further reproduce, as a starting point. The six genotypes showed different $RGR_{tolerance}$ (Figure 1B, genotype effects: $p = 0.031$, Kruskal–Wallis rank sum test). While genotype SP14 showed the highest $RGR_{tolerance}$, genotype SP02 had the lowest $RGR_{tolerance}$.

Additionally, we quantified the plant resistance to snail herbivory based on the number of consumed or damaged fronds by one snail within 24 h. High levels of plant damage would represent a low plant resistance to herbivory. On average, each snail consumed approximately 20–25 fronds. However, no statistical difference was found among the genotypes (Figure 1C, genotype effects: $p > 0.1$, one-way ANOVA).

2.2. *Microbiota Inoculation Altered Tolerance and Resistance to Herbivory in a Genotype-Specific Manner*

To examine whether the microbiota affects the growth rate, tolerance, and herbivory resistance, we inoculated six genotypes with microbiota collected from a natural population of *S. polyrhiza*. To prevent algae overgrowth, the inoculum was filtered through a 3 μm filter, which is expected to also remove most fungal cells. Therefore, the inoculum should mainly consist of the bacteria associated with duckweed and its environment. It still might have included some microalgae such as the prokaryotic cyanobacteria, but no excessive algae growth was visually observed. Interestingly, inoculation of microbiota reduced the growth rate (Figure 1D, $p = 1.8 \times 10^{-6}$) and tolerance (Figure 1E, $p = 6.4 \times 10^{-5}$) of all genotypes, but no statistical differences were found among the genotypes ($p > 0.1$).

We then quantified the effects of the microbiota on herbivore resistance. Among the six genotypes, the effects of microbiota on resistance differed (Figure 1F, $p = 0.069$). The microbiota inoculation decreased the resistance of genotype SP05 to herbivory by 23%, while it had an opposite effect on genotype SP14 by increasing its resistance by 41% ($p = 0.014$).

3. Discussion

In nature, plants interact with complex biotic interaction partners, such as beneficial or detrimental microbes, and herbivores. The evolutionary trajectory of a plant therefore not only depends on one interaction partner, but the combination of different interacting members in the community. Consistent to this prediction, we found that the interaction of both an herbivore and the microbiota might alter the fitness landscape among different *S. polyrhiza* genotypes.

We found different *S. polyrhiza* genotypes differed in their intrinsic growth rate and tolerance. Interestingly, microbiota inoculation reduced the growth rate and tolerance in all genotypes, and no significant differences were found among the genotypes. Many previous studies reported the microbe-induced growth promotion of duckweed [8,10,11], but also negative growth effects have been reported before [11,12]. Overall, we observed a relatively low growth rate, which could be caused by non-optimal growth conditions in the greenhouse and the comparably low nutrient concentration of the medium. However, in nature, the growth conditions for *S. polyrhiza* are also often suboptimal and dependent on the season, the geographic location, and the surrounding vegetation, while nutrient levels vary between different water bodies and can also change over time [13].

Mechanistically, the observed effects of the microbiota on plant growth, tolerance, and resistance could be caused by at least three different non-exclusive mechanisms. First, the microbiota might have induced changes in signaling pathways either via microbial elicitors, such as flagellin, or via microbe-produced bioactive compounds, such as phytohormones [11,14], that could directly affect the plants physiology and metabolisms and/or its defense response to herbivory. Second, the microbiota might have altered the nutrient supply of the *S. polyrhiza* plants. For example, some duckweed-associated microbes can solubilize insoluble phosphate, produce siderophores, or improve nitrogen uptake [8,11]. In contrast, other microorganisms, such as algae can compete with duckweed for the available nutrients in the medium [12]. The nutrient availability can further affect the plants primary and secondary metabolisms [15], which are key components of plant defenses. Third, the microbiota might have produced specific toxins that directly affected the plant or deterred the feeding from the snails. For example, the endophytic fungus, *Acremonium coenophialu*, in tall fescue (*Festuca arundinace*) produces alkaloids that deter herbivores on its host [16] and *Lemna minor*-associated microbes are able to produce hydrogen cyanide that can inhibit the plant growth [11]. Furthermore, the genotypes might assemble different microbe communities despite the treatment with the same inoculum [17] and therefore differentially affect the plants. Although most of the available literature in this field is based on terrestrial plants, the conserved structure of the duckweed microbiome and the terrestrial leaf microbiome indicate similar mechanisms underlying the assembly of the associated microbe community [18]. In the future, the combination of multi-omics profiling, synthetic microbe communities, and genetic manipulation approaches in *S. polyrhiza* will shed light on the mechanisms involved in plant–microbe–herbivore interactions.

A major limitation in studying plant–microbe–herbivore interactions is the efficiency and stability of microbiota inoculation. The microbiota consists of different living microbes that can dynamically change their composition depending on the experimental conditions, such as the pH, temperature, and chemical composition of the media used for extraction, inoculation, and plant cultivation. Therefore, we currently do not know to what extent the inoculation of microbiota reassembled the natural microbiota in our experiments. Differences between a field microbiome and the respective inoculum have been found before [10], although fewer differences might be expected from protocols without a microbe cultivation step, e.g., by direct inoculation with freshly isolated microbes. Additionally, during growth under semi-sterile conditions, the control plants might also assemble a specific microbiome and during some steps, such as the addition of the non-sterile snails, further changes in the microbiome can occur. Sequencing the microbiota using meta-barcoding might offer a glimpse into the differences between natural and re-inoculated microbiota. Additionally, a more natural way of inoculation is to grow sterile genotypes together with outdoor growing *S. polyrhiza* for several generations. Consistently, we also found genotype-dependent microbiota effects on plant resistance in a current study using such a natural inoculation approach in a long-term experimental evolution experiment [19].

In summary, this study demonstrates the potential of microbiota in shaping the evolutionary trajectory of plants in nature, highlighting the need to consider microbiota in studying plant evolution and plant–herbivore interactions.

4. Methods

4.1. Duckweed Growth with and without Microbiota-Inoculation

Spirodela polyrhiza plants of the genotypes SP1 (ID 0040, China), SP2 (ID 6613, USA), SP5 (ID 7551, Australia), SP9 (ID 8756, Ethiopia), SP14 (ID 9509, Germany), and SP30 (ID 8442, India) were pre-cultivated in Erlenmeyer flasks under sterile conditions for 3 weeks with high nutrient availability in N-medium (KH_2PO_4 150 µM, KNO_3 8 mM, $Ca(NO_3)_2$ 1 mM, H_3BO_3 5 µM, $MnCl_2$ 13 µM, Na_2MoO_4 0.4 µM, $MgSO_4$ 1 mM, FeNaEDTA 25 µM) [20] to generate sufficient fronds. Subsequently plants were transferred to Hoagland medium (NaH_2PO_4 32 µM, KNO_3 357 µM, H_3BO_3 15 µM, $MnCl_2$ 1.97 µM, H_2MoO_4 1.09 µM, $MgSO_4$ 418 µM, K_2SO_4 1.68 mM, $CuSO_4$ 120 nM, $ZnSO_4$ 278 nM, $FeSO_4$ 12 µM, $CaCl_2$ 1 mM; pH 7.0) [11] with reduced nutrient availability to acclimate for 3 days before the microbe inoculum or a buffer control was added. The plants were then incubated for 5 days in the medium supplemented with the microbe inoculum. Inoculation was conducted for all plants of each genotype together and all of the genotypes were treated with a subfraction of the same microbe isolate. At the beginning of the experiment (t0) for each replicate, 20 fronds were transferred under semi-sterile conditions to plastic cups (PP, transparent, round, 250 mL, Plastikbecher.de GmbH) containing 150 mL of fresh Hoagland medium. The plastic cups were closed with a perforated lid (PP, transparent, round, 101 mm, Plastikbecher.de GmbH). The plants originate from our in-house stock collection, which is kept at 18 °C, with a day/night period of 12:12. With the start of the pre-cultivation step, the plants were grown under glasshouse conditions between August and September 2018 without light supplementation or temperature regulation. Their growth was measured after 7 days (t7), by counting the frond number (FN), to calculate the relative growth rate with the formula RGR = $(LnFN_{t7} - LnFN_{t0})/(t7 - t0)$ [21]. An overview of the experimental procedure is shown in Figure 2. The RGR values were largely affected by the direct offspring of the starting fronds. Genotype-dependent microbiota effects were calculated based on the relative difference between the plants treated with and without microbe inoculum.

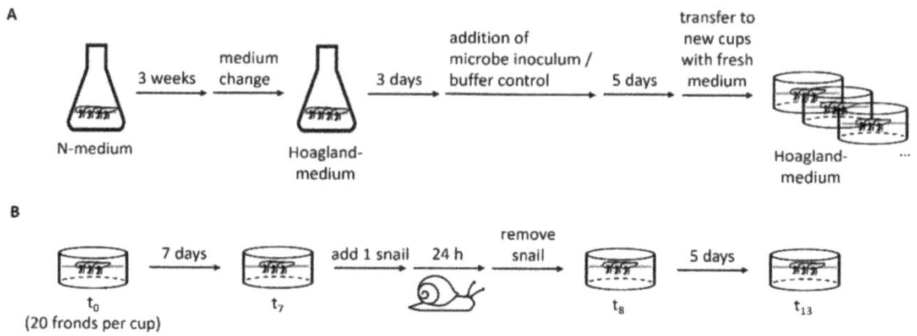

Figure 2. Overview of the experimental procedure. (**A**) Scheme of the plant pre-treatment, including plant propagation on N-medium, adaptation to Hoagland medium, microbe inoculation, and incubation before the plants were divided to the cups for conducting the experiment. For each genotype, one Erlenmeyer flask was used per treatment. (**B**) Experimental procedure to determine the plant growth before (t_0–t_7) and after herbivory attack (t_8–t_{13}) as well as the herbivore damage within 24 h.

4.2. Microbe Inoculum

Microbes were extracted in a similar way to the method as described by Ishizawa et al. (2017) [11]. In brief, *S. polyrhiza* plants from a population that was under herbivory pressure outdoors and water from these ponds were used as starting materials for the microbe inoculum. An approximately similar number of fronds was used as the number of fronds supposed to be treated. Duckweed plants were homogenized in 5 mg/L sodium tripolyphosphate (Supelco) using a kitchen blender. Subsequently, the water and plant homogenate were filtered with a coffee filter and then through a 3.0 μm filter (MCE membrane, MF-Millipore, Merck, Germany). The flow-through was centrifuged (10 min, 10,000× *g*, 4 °C) to collect the microbes. The pellet was washed twice with Hoagland medium and subsequently re-suspended in Hoagland medium. The resulting microbe suspension was used for inoculation and pure Hoagland medium as a buffer control.

4.3. Herbivore Assay

At the end of the growth assay (t7), one pond snail, *Lymnaea stagnalis*, was added per cup and the number of remaining fronds was counted after 24 h (t8). The snails were removed and 5 days later (t13), the fronds were counted again. Tolerance was defined as the RGR observed after the snail feeding $RGR_{tolerance} = (LnFN_{t13} - LnFN_{t8})/(t13 - t8)$ including intact and damaged fronds. Herbivore resistance was determined based on the number of fronds consumed or damaged by the snail, which was calculated as the difference between the observed undamaged fronds and the number of expected fronds for day 8 ($FN_{t8expected}$–FN_{t8}). $FN_{t8expected}$ was calculated based on the frond number on day 7 and the RGR between t0 and t7. An overview of the experimental procedure is shown in Figure 2B. The snails were collected from an outdoor experimental pond that was covered by duckweed (*L. minor*), cleaned with tap water, and starved for one day before the start of the experiment. The snails used for the experiment were sorted by size and the snails with a length of approximately 2 cm were used. The snails were randomly added to the cups.

4.4. Statistics and Data Analysis

All data were analyzed in R v4.2.1 [22]. Differences in growth rate and resistance among genotypes were tested using linear models. Non-parametric tests were used when the data were not normally distributed. The microbiota effects were respectively normalized to the trait of each genotype. The original data and data analysis R-scripts are provided in a Supplementary File. The cups that were used for the growth, herbivore, and tolerance assay were treated as experimental units and the number of replicates is provided in the figure legend. For the growth assay, the genotypes SP1, SP2, SP5, SP14 and SP30 had 20 replicates, and SP9 had 16 replicates. For the tolerance and herbivore assay, SP1, SP2, SP5, SP14 and SP30 had 15 replicates, and SP9 had 11 replicates.

Supplementary Materials: The following supporting information can be downloaded at: https://www.mdpi.com/article/10.3390/plants11233317/s1, Supplementary data 1_original data and R-scripts.

Author Contributions: Conceptualization, M.S. and S.X.; methodology, M.S.; software, S.X.; validation, M.S. and S.X.; formal analysis, M.S. and S.X.; investigation, M.S.; resources, S.X.; data curation, M.S. and S.X.; writing—original draft preparation, M.S. and S.X.; writing—review and editing, M.S. and S.X.; visualization, M.S. and S.X.; supervision, M.S. and S.X.; project administration, S.X.; funding acquisition, S.X. All authors have read and agreed to the published version of the manuscript.

Funding: The duckweed projects in the group Xu are supported by the German Research Foundation (DFG project number 438887884), the University of Münster, and the Johannes Gutenberg University Mainz. This research was inspired by discussions with the members of the CRC TRR 212 (NC^3) – Project number 316099922.

Data Availability Statement: The data presented in this study are available in the Supplementary Materials.

Acknowledgments: We thank K. Sowjanya Sree for the invitation to contribute to this special issue and for providing a publication fee waiver for this manuscript.

Conflicts of Interest: The authors declare no conflict of interest.

Ethical Statements: *Spirodela polyrhiza* plants were used from an in-house duckweed collection. Their original sampling can be traced based on the 4-digit number provided in the method section. *Lymnaea stagnalis* were collected from outdoor ponds near our institute in Münster, Germany. During the experiments, the snails were used as herbivores and placed in plastic cups with *S. polyrhiza* for 24 h. The snails also live together and feed on duckweed in nature. After the experiments, the snails were placed back in the ponds from which they were collected.

References

1. Müller, D.B.; Vogel, C.; Bai, Y.; Vorholt, J.A. The plant microbiota: Systems-level insights and perspectives. *Annu. Rev. Genet.* **2016**, *50*, 211–234. [CrossRef]
2. Trivedi, P.; Leach, J.E.; Tringe, S.G.; Sa, T.; Singh, B.K. Plant–microbiome interactions: From community assembly to plant health. *Nat. Rev. Microbiol.* **2020**, *18*, 607–621. [CrossRef] [PubMed]
3. Shikano, I.; Rosa, C.; Tan, C.-W.; Felton, G.W. Tritrophic interactions: Microbe-mediated plant effects on insect herbivores. *Annu. Rev. Phytopathol.* **2017**, *55*, 313–331. [CrossRef] [PubMed]
4. Hu, L.; Robert, C.A.M.; Cadot, S.; Zhang, X.; Ye, M.; Li, B.; Manzo, D.; Chervet, N.; Steinger, T.; van der Heijden, M.G.A.; et al. Root exudate metabolites drive plant-soil feedbacks on growth and defense by shaping the rhizosphere microbiota. *Nat. Commun.* **2018**, *9*, 2738. [CrossRef] [PubMed]
5. Friman, J.; Karssemeijer, P.N.; Haller, J.; de Kreek, K.; van Loon, J.J.A.; Dicke, M. Shoot and root insect herbivory change the plant rhizosphere microbiome and affects cabbage–insect interactions through plant–soil feedback. *New Phytol.* **2021**, *232*, 2475–2490. [CrossRef]
6. Toyama, T.; Yu, N.; Kumada, H.; Sei, K.; Ike, M.; Fujita, M. Accelerated aromatic compounds degradation in aquatic environment by use of interaction between *Spirodela polyrrhiza* and bacteria in its rhizosphere. *J. Biosci. Bioeng.* **2006**, *101*, 346–353. [CrossRef]
7. Hoang, H.; Yu, N.; Toyama, T.; Inoue, D.; Sei, K.; Ike, M. Accelerated degradation of a variety of aromatic compounds by *Spirodela polyrrhiza*-bacterial associations and contribution of root exudates released from *S. polyrrhiza*. *J. Environ. Sci.* **2010**, *22*, 494–499. [CrossRef] [PubMed]
8. Toyama, T.; Mori, K.; Tanaka, Y.; Ike, M.; Morikawa, M. Growth promotion of giant duckweed *Spirodela polyrhiza* (Lemnaceae) by *Ensifer* sp. SP4 through enhancement of nitrogen metabolism and photosynthesis. *Mol. Plant Microbe. Interact.* **2021**, *35*, 28–38. [CrossRef] [PubMed]
9. Toporowska, M. Degradation of three microcystin variants in the presence of the macrophyte *Spirodela polyrhiza* and the associated microbial communities. *Int. J. Environ. Res. Public Health* **2022**, *19*, 6086. [CrossRef] [PubMed]
10. O'Brien, A.M.; Laurich, J.; Lash, E.; Frederickson, M.E. Mutualistic outcomes across plant populations, microbes, and environments in the duckweed *Lemna minor*. *Microb. Ecol.* **2020**, *80*, 384–397. [CrossRef] [PubMed]

11. Ishizawa, H.; Kuroda, M.; Morikawa, M.; Ike, M. Evaluation of environmental bacterial communities as a factor affecting the growth of duckweed *Lemna minor*. *Biotechnol. Biofuels* **2017**, *10*, 62. [CrossRef] [PubMed]
12. Roijackers, R.; Szabó, S.; Scheffer, M. Experimental analysis of the competition between algae and duckweed. *Arch. Hydrobiol.* **2004**, *160*, 401–412. [CrossRef]
13. Appenroth, K.-J.; Nickel, G. Turion formation in *Spirodela polyrhiza*: The environmental signals that induce the developmental process in nature. *Physiol. Plant.* **2010**, *138*, 312–320. [CrossRef] [PubMed]
14. Gilbert, S.; Xu, J.; Acosta, K.; Poulev, A.; Lebeis, S.L.; Lam, E. Bacterial production of indole related compounds reveals their role in association between duckweeds and endophytes. *Front. Chem.* **2018**, *6*, 265. [CrossRef]
15. Kumar, V.; Sharma, S.S. Nutrient deficiency-dependent anthocyanin development in *Spirodela polyrhiza* L. Schleid. *Biol. Plant.* **1999**, *42*, 621–624. [CrossRef]
16. Bultman, T.L.; Borowicz, K.L.; Schneble, R.M.; Coudron, T.A.; Bush, L.P. Effect of a fungal endophyte on the growth and survival of two *Euplectrus* parasitoids. *Oikos* **1997**, *78*, 170–176. [CrossRef]
17. Bodenhausen, N.; Bortfeld-Miller, M.; Ackermann, M.; Vorholt, J.A. A synthetic community approach reveals plant genotypes affecting the phyllosphere microbiota. *PLoS Genet.* **2014**, *10*, e1004283. [CrossRef]
18. Acosta, K.; Xu, J.; Gilbert, S.; Denison, E.; Brinkman, T.; Lebeis, S.; Lam, E. Duckweed hosts a taxonomically similar bacterial assemblage as the terrestrial leaf microbiome. *PLoS ONE* **2020**, *15*, e0228560. [CrossRef]
19. Malacrinò, A.; Böttner, L.; Nouere, S.; Huber, M.; Schäfer, M.; Xu, S. Induced responses contribute to rapid plant adaptation to herbivory. *bioRxiv* **2022**. [CrossRef]
20. Appenroth, K. Media for in vitro-cultivation of duckweed. *Duckweed Forum* **2015**, *3*, 180–186.
21. Ziegler, P.; Adelmann, K.; Zimmer, S.; Schmidt, C.; Appenroth, K.-J. Relative in vitro growth rates of duckweeds (Lemnaceae)–the most rapidly growing higher plants. *Plant Biol.* **2015**, *17*, 33–41. [CrossRef] [PubMed]
22. R Core Team. *R: A Language and Environment for Statistical Computing*; R Foundation for Statistical Computing: Vienna, Austria, 2022. Available online: https://www.R-project.org/ (accessed on 1 October 2022).

plants

Article

Brassinolide Enhances the Level of Brassinosteroids, Protein, Pigments, and Monosaccharides in *Wolffia arrhiza* Treated with Brassinazole

Magdalena Chmur and Andrzej Bajguz *

Department of Biology and Plant Ecology, Faculty of Biology, University of Bialystok, Ciolkowskiego 1J, 15-245 Bialystok, Poland; m.chmur@uwb.edu.pl
* Correspondence: abajguz@uwb.edu.pl; Tel.: +48-85-7388361

Abstract: Brassinolide (BL) represents brassinosteroids (BRs)—a group of phytohormones that are essential for plant growth and development. Brassinazole (Brz) is as a synthetic inhibitor of BRs' biosynthesis. In the present study, the responses of *Wolffia arrhiza* to the treatment with BL, Brz, and the combination of BL with Brz were analyzed. The analysis of BRs and Brz was performed using LC-MS/MS. The photosynthetic pigments (chlorophylls, carotenes, and xanthophylls) levels were determined using HPLC, but protein and monosaccharides level using spectrophotometric methods. The obtained results indicated that BL and Brz influence *W. arrhiza* cultures in a concentration-dependent manner. The most stimulatory effects on the growth, level of BRs (BL, 24-epibrassinolide, 28-homobrassinolide, 28-norbrassinolide, catasterone, castasterone, 24-epicastasterone, typhasterol, and 6-deoxytyphasterol), and the content of pigments, protein, and monosaccharides, were observed in plants treated with 0.1 μM BL. Whereas the application of 1 μM and 10 μM Brz caused a significant decrease in duckweed weight and level of targeted compounds. Application of BL caused the mitigation of the Brz inhibitory effect and enhanced the BR level in duckweed treated with Brz. The level of BRs was reported for the first time in duckweed treated with BL and/or Brz.

Keywords: biosynthesis inhibitor; duckweed; occurrence; overcome; phytohormones

Citation: Chmur, M.; Bajguz, A. Brassinolide Enhances the Level of Brassinosteroids, Protein, Pigments, and Monosaccharides in *Wolffia arrhiza* Treated with Brassinazole. *Plants* **2021**, *10*, 1311. https://doi.org/10.3390/plants10071311

Academic Editors: Klaus-Jürgen Appenroth, K. Sowjanya Sree and Viktor Oláh

Received: 6 June 2021
Accepted: 25 June 2021
Published: 28 June 2021

1. Introduction

Brassinosteroids (BRs) are a class of steroid phytohormones represented by the above 70 compounds as free compounds or fatty acid and glucose conjugates. They are widely distributed in plants; their presence has been confirmed in algae, mosses, and vascular plants [1–5]. Brassinolide (BL) is the first discovered, simultaneously the most widespread and active representative of BRs [5,6]. BRs regulate many physiological processes in plants, including cell division and elongation, vessel differentiation, reproductive development, seed germination, flowering, pollen development, modulation of gene expression, maturation, and aging of the plant. Moreover, BRs significantly improve the efficiency of transpiration and cause the increase of chlorophylls, carbohydrates, and protein contents [7–11]. Besides, BRs participate in plants' tolerance to various abiotic stresses, such as hypoxia, heavy metals, drought, salinity, and oxidative stress. Moreover, BRs are involved in plant protection against pathogen attacks [12–15].

Depending on the number of carbon atoms in a molecule, these sterols are divided into C_{27}-,C_{28}- or C_{29}-type of BRs, among which, the most widespread are C_{28}-BRs [5]. Their biosynthesis occurs through the early or late C6 oxidation pathway with a precursory compound named campestanol (CN). During the late C6 pathway, CN is hydroxylated in C-22 position into the 6-deoxocatasterone (6dCT), which is hydroxylated in C-23 position into the 6-deoxoteasterone (6dTE). The resulting compound is converted into the 3-dehydro-6-dTE, which is converted to 6-deoxotyphasterol (6dTY). Then, 6-deoxoTY is hydroxylated to 6-deoxocastasterone (6dCS), which is converted to castasterone (CS). In contrast, the early C6 oxidation pathway begins from the oxidation of CN to 6-oxoCN. Then, 6-oxoCN is converted to catasterone (CT), teasterone (TE), 3-dehydroteasterone, typhasterol (TY), CS, and BL, respectively [3,16,17] BRs' biosynthesis can be blocked through the application of specific inhibitors [18]. One of them is brassinazole (Brz), a synthetic triazole-type inhibitor of BRs' synthesis, which blocks the hydroxylation of CN to the 6dCT and hydroxylation of 6-deoxoCT to the 6dTE during the late C6 pathway, and analogously, it inhibits the conversion of 6-oxocampestanol (6-oxoCN) to CT and conversion of CT to the TE during the early C6 pathway [19,20]. Plants treated with Brz indicate the phenotype changes, primarily manifested in dwarfism or growth inhibition. However, exogenously applied BRs can alleviate these unfavorable morphological alterations. Brz application aims at confirmation of BRs' biological activity and role in plants [21–26]. Another type of BR biosynthesis inhibitor is Brz2001, a modified form of Brz containing an allyl moiety instead of the methyl group [16,18].

The duckweeds belonging to the *Wolffia* genus are the smallest angiosperms with strongly reduced organs; they do not create a stem, leaves, and roots system. The body size of *W. arrhiza* reaches about 1 mm in diameter. Rarely does flowering occur; therefore, most commonly, it reproduces vegetatively. Despite a simple body structure, *W. arrhiza* has an excellent adaptation for living in the aquatic environment. In organic-rich water, it changes the way of feeding on photoautotrophic into mixotrophic or heterotrophic. Moreover, *W. arrhiza* can accumulate xenobiotics, e.g., metals, radionuclides, nanoparticles, pesticides, and pharmaceuticals, from polluted water. Therefore, the fast reproductions, good absorption ability of organic compounds, and simple breeding methods of *W. arrhiza* provide this duckweed to be a good plant for practical usage in wastewater treatment [27–30].

The present work is a continuation of the research performed by Bajguz and Asami [31], in which the growth and level of chlorophylls, carotenoids, monosaccharides, and proteins under the influence of EBL and/or Brz2001 in *W. arrhiza* cultures was studied. The present study's main aim was to determine the endogenous level of Brz and BRs in the duckweed treated with BL and/or Brz. Besides hormone analysis, the duckweed growth rate and level of photosynthetic pigments, proteins, and monosaccharides under the influence of Brz and BL were also analyzed. For this, the following hypotheses were tested: (1) BL has a stimulatory effect on the growth and endogenous level of BRs, photosynthetic pigments, proteins, and monosaccharides in *W. arrhiza*; (2) Brz decreases the fresh weight and content of BRs and primary metabolites in the duckweed; (3) Brz is absorbed by *W. arrhiza* cultures; (4) BL effectively overcomes the repressive impact of Brz on the growth rate and level of targeted compounds; (5) the effect of BL and Brz in *W. arrhiza* differs in a concentration-dependent manner. LC-MS/MS system was used to identify and quantify BRs as well as Brz. The LC-MS/MS with a multiple reaction monitoring (MRM) mode is one of the most predominant analytical methods of BRs' analysis due to the high sensitivity and efficient separation performance [5,32]. For the analysis of pigments belonging to the chlorophylls, carotenes, and xanthophylls, the HPLC method was used. Obtained results show the presence and essential role of BRs in *W. arrhiza* growth.

2. Results

2.1. Growth Rate of W. arrhiza

The effect of BL, Brz, and the combination of BL with Brz on the *W. arrhiza* growth rate is presented in Figure 1. The initial weight of duckweed in each variant was 1 g. After 7 days of cultivation, the biomass of untreated plants increased up to 1.37 g. Application of BL in range of concentration 0.001–0.1 µM caused the promotive effect on *W. arrhiza* growth. The slight increases were observed in plants treated with 0.001 µM BL, while the most stimulatory effect was noted in plants exposed to the 0.1 µM BL. The cultivation with the addition of 0.1 µM BL caused a considerable increase of duckweed weight, up to 1.65 g, and this is a 20% rise compared to the control plants. Whereas treatment with 1 µM BL had a slight inhibitory effect on the growth rate, causing a 4% decline of biomass in relation to the untreated plant.

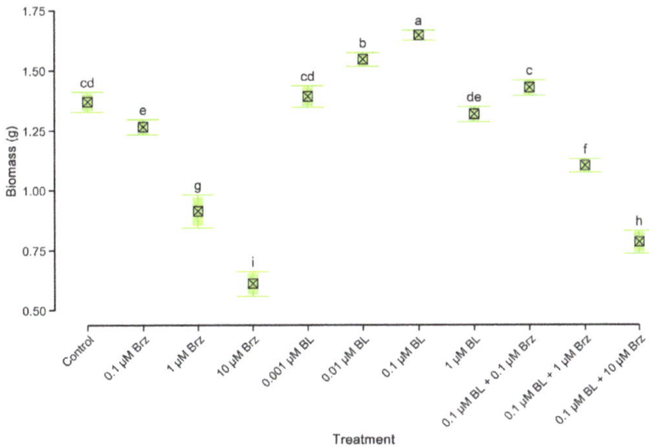

Figure 1. The effect of BL and/or Brz on *W. arrhiza* biomass after 7 days of cultivation. The crossed square shows the mean (n = 5). The lower and upper hinges correspond to the lower and upper bounds of 95% confidence interval of the mean. The lower and upper whiskers extend from the hinge to mean ± standard deviation. Means with the same letters are not significantly different ($p \geq 0.05$) according to Tukey's post-hoc test.

An opposite effect on duckweed growth was observed in plants treated with BR biosynthesis inhibitor. Exogenously applied Brz led to the decrease of *W. arrhiza* weight in a concentration-dependent manner. Thus, the application of 0.1 μM Brz had a slight inhibitory effect on plant growth, whereas under the influence of 1–10 μM Brz, the fresh weight of duckweed was considerably reduced, reaching the highest of a 55% decrease in plant treated with 10 μM Brz.

The inhibitory effect of Brz on *W. arrhiza* growth was reversed through the application of 0.1 μM BL. Combination of 0.1 μM BL with 0.1 μM Brz caused a statistically inconsiderable enhance of plant weight compared to the control. The mixtures of 0.1 μM BL with 1–10 μM Brz also positively affected duckweed weight, and caused the 20% and 28% increases to plant treated with 1–10 μM Brz alone, respectively. However, the biomass values of duckweed exposed on 0.1 μM BL with 1–10 μM Brz was lower than the control group, by 20% and 43%, respectively.

2.2. Brassinazole and Brassinosteroids' Content in W. arrhiza

Table 1 presents the endogenous level of BRs and Brz in *W. arrhiza* treated with BL and/or Brz. The LC-MS/MS analysis indicated the occurrence of nine BRs, i.e., BL, 24-epibrassinolide (EBL), 28-homobrassinolide (HBL), 28-norbrassinolide (norBL), CT, CS, 24-epicastasterone (ECS), TY, and 6dTY.

Table 1. The endogenous level (ng/g fresh weight) of Brz and BRs in *W. arrhiza* treated with Brz and/or BL after 7 days of cultivation. Data present the mean (n = 5) ± standard deviation. The range in square brackets corresponds to the 95% confidence interval of the mean. Means with the same letters are not significantly different ($p \geq 0.05$) according to Tukey's post hoc test.

Treatment	Brz	BL	EBL	HBL	norBL
Control	0	1.79 ± 0.31 ^{d} [1.52–2.06]	0.17 ± 0.04 ^{d} [0.13–0.21]	0.63 ± 0.15 ^{c,d} [0.5–0.76]	0.12 ± 0.04 ^{c,d} [0.09–0.15]
0.1 μM Brz	911.04 ± 81.21 ^{c} [839.86–982.22]	1.19 ± 0.38 ^{d} [0.85–1.52]	0.12 ± 0.03 ^{d} [0.1–0.14]	0.49 ± 0.08 ^{d,e} [0.42–0.56]	0.09 ± 0.02 ^{c,d} [0.07–0.11]
1 μM Brz	6987.57 ± 263.03 ^{b} [6757.01−7218.13]	0.44 ± 0.09 ^{d} [0.36–0.51]	0.06 ± 0.02 ^{d} [0.05–0.08]	0.13 ± 0.04 ^{e} [0.1–0.17]	0.05 ± 0.01 ^{c,d} [0.04–0.06]
10 μM Brz	9381.43 ± 1007.15 ^{a} [8498.62–10264.23]	0.13 ± 0.04 ^{d} [0.1–0.17]	0.02 ± 0.02 ^{d} [0–0.04]	0.06 ± 0.02 ^{e} [0.05–0.07]	0.02 ± 0.01 ^{d} [0.01–0.02]
0.001 μM BL	0	2 ± 0.23 ^{d} [1.8–2.2]	0.21 ± 0.06 ^{d} [0.16–0.26]	0.68 ± 0.14 ^{c,d} [0.56–0.81]	0.16 ± 0.04 ^{c} [0.12–0.2]
0.01 μM BL	0	94.27 ± 19.58 ^{d} [77.11–111.43]	1.23 ± 0.17 ^{d} [1.08–1.38]	1.17 ± 0.38 ^{b} [0.84–1.5]	0.29 ± 0.06 ^{b} [0.24–0.35]
0.1 μM BL	0	447.18 ± 40.86 ^{b} [411.37–482.99]	16.08 ± 3.2 ^{a} [13.27–18.89]	1.72 ± 0.3 ^{a} [1.45–1.98]	0.49 ± 0.11 ^{a} [0.4–0.59]
1 μM BL	0	682.61 ± 82.16 ^{a} [610.6–754.63]	14.36 ± 1.73 ^{a,b} [12.84–15.88]	1.03 ± 0.18 ^{b,c} [0.88–1.19]	0.37 ± 0.08 ^{b} [0.3–0.45]
0.1 μM BL + 0.1 μM Brz	719.6 ± 104.76 ^{c} [627.78–811.43]	433.03 ± 52.71 ^{b} [386.83–479.24]	13.37 ± 2.68 ^{a,b} [11.02–15.72]	1.21 ± 0.36 ^{b} [0.9–1.53]	0.34 ± 0.06 ^{b} [0.28–0.39]
0.1 μM BL + 1 μM Brz	6303.29 ± 1051.26 ^{b} [5381.82–7224.76]	362.98 ± 100.02 ^{b} [275.31–450.65]	11.68 ± 1.93 ^{b} [9.98–13.37]	0.37 ± 0.13 ^{d,e} [0.26–0.48]	0.08 ± 0.01 ^{c,d} [0.07–0.08]
0.1 μM BL + 10 μM Brz	10677.36 ± 2183.66 ^{a} [8763.29–12591.42]	259.68 ± 42 ^{c} [222.87–296.5]	6.51 ± 1.56 ^{c} [5.14–7.87]	0.1 ± 0.02 ^{e} [0.08–0.12]	0.03 ± 0.01 ^{d} [0.02–0.03]

	CT	CS	ECS	TY	6dTY
Control	1.96 ± 0.61 ^{b,c,d} [1.42–2.49]	2.77 ± 0.29 ^{c,d} [2.52–3.03]	0.03 ± 0.01 ^{d,e} [0.02–0.03]	0.18 ± 0.05 ^{d,e,f} [0.14–0.22]	0.03 ± 0.01 ^{b,c,d} [0.02–0.04]
0.1 μM Brz	1.6 ± 0.15 ^{c,d,e} [1.46–1.73]	2.2 ± 0.27 ^{d} [1.96–2.43]	0.02 ± 0 ^{e,f} [0.01–0.02]	0.14 ± 0.03 ^{e,f,g} [0.12–0.16]	0.02 ± 0.01 ^{d} [0.01–0.03]
1 μM Brz	0.94 ± 0.16 ^{d,e,f} [0.8–1.08]	1.12 ± 0.11 ^{e,f} [1.02–1.22]	0.01 ± 0 ^{f} [0.01–0.01]	0.06 ± 0.01 ^{f,g} [0.05–0.07]	0
10 μM Brz	0.52 ± 0.13 ^{f} [0.4–0.63]	0.73 ± 0.16 ^{f} [0.59–0.87]	0.01 ± 0.01 ^{f} [0–0.01]	0.02 ± 0 ^{g} [0.02–0.03]	0
0.001 μM BL	2.08 ± 0.15 ^{b,c} [1.94–2.21]	2.95 ± 0.17 ^{b,c} [2.8–3.1]	0.03 ± 0.01 ^{c,d} [0.03–0.04]	0.31 ± 0.05 ^{c,d} [0.27–0.35]	0.04 ± 0.01 ^{b} [0.04–0.04]
0.01 μM BL	2.76 ± 0.52 ^{a,b} [2.31–3.22]	3.53 ± 0.31 ^{a,b} [3.26–3.8]	0.04 ± 0.01 ^{b,c} [0.04–0.05]	0.49 ± 0.1 ^{a,b} [0.4–0.58]	0.05 ± 0.01 ^{b} [0.04–0.05]
0.1 μM BL	3.47 ± 0.56 ^{a} [2.98–3.96]	4.02 ± 0.26 ^{a} [3.79–4.24]	0.05 ± 0.01 ^{a} [0.05–0.06]	0.57 ± 0.12 ^{a} [0.47–0.67]	0.06 ± 0.01 ^{a} [0.05–0.07]
1 μM BL	2.03 ± 0.8 ^{b,c} [1.33–2.74]	2.75 ± 0.13 ^{c,d} [2.63–2.87]	0.03 ± 0 ^{d,e} [0.03–0.03]	0.42 ± 0.08 ^{b,c} [0.35–0.48]	0.04 ± 0.01 ^{b} [0.03–0.05]
0.1 μM BL + 0.1 μM Brz	3.01 ± 1.02 ^{a,b} [2.12–3.9]	3.54 ± 0.44 ^{a} [3.15–3.92]	0.05 ± 0 ^{a,b} [0.04–0.05]	0.25 ± 0.14 ^{d,e} [0.14–0.37]	0.04 ± 0.01 ^{b,c} [0.03–0.04]
0.1 μM BL + 1 μM Brz	1.25 ± 0.13 ^{c,d,e,f} [1.14–1.37]	1.6 ± 0.44 ^{e} [1.21–1.99]	0.01 ± 0 ^{f} [0.01–0.02]	0.03 ± 0.01 ^{g} [0.02–0.04]	0.02 ± 0.01 ^{c,d} [0.02–0.03]
0.1 μM BL + 10 μM Brz	0.7 ± 0.11 ^{e,f} [0.61–0.79]	0.87 ± 0.13 ^{f} [0.76–0.98]	0.01 ± 0 ^{f} [0.01–0.02]	0	0

In untreated plants, the highest content of CS, CT, and BL was noted (2.77 ng/g, 1.96 ng/g, and 1.79 ng/g, respectively). Simultaneously, the lowest amount of ECS and 6dTY was observed (0.03 ng/g for both of them). The total content of BRs in the control group was 7.68 ng/g. Furthermore, the effect of treatment with BL, Brz, and mixture of BL with Brz on BRs' content was studied. Application of 0.001–0.1 μM BL had a stimulatory effect on the content of all detected BRs. The addition of 0.001 μM BL caused a slight increase of BRs level, whereas exposition on 0.01–1 μM BL led to the significant enhancement of BRs' content in *W. arrhiza*. For instance, the endogenous CS and CT level increased sequentially by 70% and 45% in plants treated with 0.1 μM BL to the control. Additionally, the endogenous level of BL was considerably higher in plant exposed on BL, that is related with the absorption of the exogenously applied hormone by *W. arrhiza*. Furthermore, the content of EBL also was significantly higher in plants treated with BL. The amount of BL and EBL enhanced gradually with the increasing concentrations of BL. Whereas the overall content of BRs, except BL and EBL, was 10.38 ng/g in plant exposed on 0.1 μM BL. Thus, this is an almost 2-fold increase of BRs' content to the control group, in which total BRs' content, without BL and EBL, was 5.72 ng/g. In duckweed exposed on 1 μM BL, the overall content of HBL, norBL, CT, CS, ECS, TY, and 6dTY was 6.67 ng/g. To summarize, the treatment with 0.1 μM BL had the most stimulating effect on the content of BRs in *W. arrhiza*.

An opposite effect was observed in a group of plants treated with 0.1–10 μM Brz. The endogenous level of BRs decreased proportionally to the increase of Brz concentration. In duckweed exposed on 10 μM Brz, TY and 6dTY were not detected, whereas the BL level decreased by 92% in the untreated plants. The total BRs' content was 1.49 ng/g, 81% lower than in control. The strong inhibitory action is probably connected with a huge absorption of Brz from the medium. In the plants exposed to 1 μM Brz, the endogenous inhibitor level was about 7000 ng/g. Therefore, 11% of Brz was absorbed by 1 g of duckweed.

The inhibitory effect of Brz on the BRs' synthesis was reduced by the application of 0.1 μM BL. In plants treated with the combination of 0.1 μM BL with 0.1 μM Brz, the level of BRs was slightly higher than in control. The application of 0.1 μM BL with 1 μM Brz caused a significant increase of BRs' content to plant treated with 1 μM Brz alone, except for ECS and TY. However, the amount of BRs was lower by 40% compared to the control, while BL and EBL levels were lower than in plants treated with BL alone. The least restored effect was observed in duckweed treated with a mixture of 0.1 μM BL with 10 μM Brz. The total inhibition of TY and 6dTY synthesis under the influence of 10 μM of the addition of BL did not reverse Brz. Whereas the endogenous level of BL and EBL was considerably lower than the plant treated with BL alone. The content of remaining BRs was slightly higher than in plant exposed only with 10 μM Brz. For example, the content of CS increased from 0.73 to 0.87 ng/g, and the level of CT enhanced from 0.52 to 0.7 ng/g. However, in plants exposed to 0.1 μM BL with 10 μM Brz, the endogenous level of BRs was three times lower compared to the control. The addition of μM 0.1 BL also resulted in a reduction of Brz accumulation in duckweed treated with 0.1 μM and 1 μM Brz; however, in the plant exposed with 10 μM Brz, the endogenous level of inhibitor was slightly higher compared with plant treated with 10 μM Brz alone.

2.3. Photosynthetic Pigments' Content in W. arrhiza

The level of photosynthetic pigments in duckweed *W. arrhiza* exposed on Brz and/or BL is shown in Table 2. In the control group, the total content of chlorophylls, carotenes, and xanthophylls was 204.35, 3.07, and 9.89 ng/g, respectively. The treatment with BL enhanced the pigment synthesis in a concentration-dependent manner. Application of 0.001 μM BL caused an inconsiderable increase in the level of chlorophylls, carotenes and xanthophylls compared with untreated duckweed. The treatment with 0.1 μM BL had the most stimulatory effect on pigment content. The endogenous level of chlorophyll *a* and *b* increased by 39% in the plant treated with 0.1 μM BL to the control. Similarly, carotenoids' content increased by 69% and 48%, respectively, compared with the plant without hormone addition.

Table 2. The endogenous level of chlorophylls, carotenes, and xanthophylls (µg/g fresh weight) in *W. arrhiza* treated with Brz and/or BL after 7 days of cultivation. Data present the mean (n = 5) ± standard deviation. The range in square brackets corresponds to the 95% confidence interval of the mean. Means with the same letters are not significantly different ($p \geq 0.05$) according to Tukey's post hoc test.

Treatment	Chlorophyll *a*	Chlorophyll *b*	α-Carotene	β-Carotene	Neoxanthin
Control	162.16 ± 2.69 ^{e,f} [159.8–164.51]	42.19 ± 2.2 ^{d,e} [40.26–44.13]	1.29 ± 0.01 ^{e} [1.28–1.3]	1.78 ± 0.07 ^{c} [1.72–1.84]	0.92 ± 0.04 ^{e} [0.88–0.95]
0.1 µM Brz	157.7 ± 1.53 ^{f} [156.36–159.04]	40.63 ± 1.68 ^{e} [39.15–42.1]	1.17 ± 0.01 ^{f} [1.15–1.18]	1.52 ± 0.07 ^{d} [1.47–1.58]	0.85 ± 0.01 ^{e} [0.84–0.86]
1 µM Brz	120.71 ± 1.52 ^{h} [119.38–122.04]	29.77 ± 1.01 ^{g} [28.88–30.65]	0.91 ± 0.05 ^{g} [0.87–0.95]	1.2 ± 0.05 ^{e} [1.16–1.25]	0.52 ± 0.04 ^{g} [0.48–0.55]
10 µM Brz	87.77 ± 2.41 ^{i} [85.66–89.89]	20.35 ± 0.89 ^{i} [19.57–21.13]	0.61 ± 0.02 ^{h} [0.59–0.63]	0.78 ± 0.1 ^{g} [0.69–0.86]	0.35 ± 0.02 ^{h} [0.33–0.36]
0.001 µM BL	163.83 ± 2.73 ^{e} [161.44–166.22]	44.64 ± 3.02 ^{d} [41.99–47.28]	1.49 ± 0.03 ^{d} [1.46–1.52]	1.83 ± 0.03 ^{c} [1.8–1.86]	1.14 ± 0.08 ^{d} [1.07–1.2]
0.01 µM BL	192.25 ± 2.52 ^{c} [190.04–194.45]	50.91 ± 2.27 ^{c} [48.92–52.9]	1.73 ± 0.06 ^{c} [1.68–1.78]	2.24 ± 0.08 ^{b} [2.17–2.31]	1.31 ± 0.02 ^{b} [1.29–1.33]
0.1 µM BL	222.8 ± 1.73 ^{a} [221.28–224.32]	61.64 ± 1.23 ^{a} [60.57–62.72]	2.4 ± 0.05 ^{a} [2.36–2.44]	2.79 ± 0.12 ^{a} [2.69–2.9]	1.59 ± 0.04 ^{a} [1.56–1.62]
1 µM BL	172.23 ± 2.69 ^{d} [169.87–174.59]	49.07 ± 1.33 ^{c} [47.9–50.23]	1.74 ± 0.07 ^{c} [1.68–1.8]	2.29 ± 0.06 ^{b} [2.24–2.34]	1.22 ± 0.04 ^{c} [1.18–1.25]
0.1 µM BL + 0.1 µM Brz	202.95 ± 2.26 ^{b} [200.97–204.93]	55.45 ± 1.14 ^{b} [54.46–56.45]	2.12 ± 0.04 ^{b} [2.09–2.15]	2.68 ± 0.07 ^{a} [2.62–2.74]	1.27 ± 0.03 ^{b,c} [1.25–1.3]
0.1 µM BL + 1 µM Brz	152.32 ± 1.44 ^{g} [151.06–153.58]	35.01 ± 1.7 ^{f} [33.52–36.5]	1.38 ± 0.04 ^{e} [1.34–1.42]	1.7 ± 0.01 ^{c} [1.69–1.71]	0.73 ± 0.01 ^{f} [0.72–0.74]
0.1 µM BL + 10 µM Brz	122.16 ± 4.24 ^{h} [118.44–125.88]	25.23 ± 1.05 ^{h} [24.31–26.15]	0.81 ± 0.09 ^{g} [0.72–0.89]	1.04 ± 0.04 ^{f} [1.01–1.08]	0.45 ± 0.01 ^{g} [0.44–0.46]

	Violaxanthin	Astaxanthin	Zeaxanthin	Cryptoxanthin	Lutein
Control	0.7 ± 0.01 ^{f} [0.69–0.71]	0.35 ± 0.02 ^{d,e} [0.33–0.37]	3.21 ± 0.06 ^{d} [3.16–3.26]	4.24 ± 0.08 ^{e} [4.17–4.31]	0.47 ± 0.01 ^{d} [0.47–0.48]
0.1 µM Brz	0.6 ± 0.01 ^{h} [0.6–0.61]	0.32 ± 0.01 ^{e,f} [0.31–0.33]	2.73 ± 0.1 ^{e} [2.64–2.82]	3.96 ± 0.15 ^{f} [3.83–4.09]	0.42 ± 0.02 ^{e} [0.4–0.44]
1 µM Brz	0.45 ± 0.01 ^{i} [0.45–0.46]	0.18 ± 0.01 ^{g} [0.17–0.19]	2.09 ± 0.14 ^{f} [1.98–2.21]	2.58 ± 0.1 ^{g} [2.49–2.66]	0.33 ± 0.02 ^{f} [0.31–0.35]
10 µM Brz	0.25 ± 0.01 ^{k} [0.25–0.26]	0.08 ± 0.01 ^{h} [0.07–0.09]	0.94 ± 0.07 ^{h} [0.88–0.99]	2.15 ± 0.07 ^{h} [2.09–2.21]	0.25 ± 0.02 ^{h} [0.22–0.27]
0.001 µM BL	0.73 ± 0.01 ^{e} [0.72–0.74]	0.38 ± 0.01 ^{d} [0.38–0.39]	3.27 ± 0.06 ^{d} [3.22–3.32]	4.42 ± 0.13 ^{e} [4.31–4.54]	0.51 ± 0.01 ^{c} [0.5–0.53]
0.01 µM BL	0.84 ± 0.01 ^{c} [0.83–0.85]	0.51 ± 0.01 ^{b} [0.5–0.52]	3.77 ± 0.08 ^{c} [3.69–3.84]	5.21 ± 0.13 ^{c} [5.09–5.32]	0.61 ± 0.01 ^{b} [0.6–0.62]
0.1 µM BL	0.96 ± 0.01 ^{a} [0.95–0.97]	0.58 ± 0.01 ^{a} [0.58–0.59]	4.61 ± 0.09 ^{a} [4.53–4.69]	6.25 ± 0.08 ^{a} [6.19–6.32]	0.72 ± 0.01 ^{a} [0.71–0.73]
1 µM BL	0.78 ± 0.01 ^{d} [0.76–0.79]	0.44 ± 0.01 ^{c} [0.43–0.45]	3.65 ± 0.06 ^{c} [3.59–3.7]	4.93 ± 0.16 ^{d} [4.79–5.07]	0.6 ± 0.01 ^{b} [0.59–0.61]
0.1 µM BL + 0.1 µM Brz	0.91 ± 0.01 ^{b} [0.91–0.92]	0.55 ± 0.01 ^{a,b} [0.54–0.55]	4.23 ± 0.08 ^{b} [4.16–4.29]	5.92 ± 0.05 ^{b} [5.87–5.97]	0.62 ± 0.01 ^{b} [0.6–0.63]
0.1 µM BL + 1 µM Brz	0.65 ± 0.01 ^{g} [0.64–0.66]	0.28 ± 0.03 ^{f} [0.26–0.31]	2.72 ± 0.11 ^{e} [2.62–2.82]	3.73 ± 0.12 ^{f} [3.62–3.83]	0.47 ± 0.01 ^{d} [0.46–0.48]
0.1 µM BL + 10 µM Brz	0.33 ± 0.01 ^{j} [0.32–0.34]	0.16 ± 0.04 ^{g} [0.13–0.19]	1.3 ± 0.08 ^{g} [1.23–1.37]	2.68 ± 0.16 ^{g} [2.54–2.82]	0.3 ± 0.01 ^{g} [0.28–0.31]

The application of Brz resulted in a decrease of synthesis in all detected pigments. The overall pigment content decreased proportionally with the increasing inhibitor concentration, reaching the most inhibitory effect in duckweed treated with 10 µM Brz. For example, the level of chlorophyll *a* in a plant with 10 µM Brz was 54% lower than in unexposed plants. The treatment with 0.1 µM BL mitigated the inhibitory effect of Brz on pigment synthesis. In duckweed exposed on simultaneous action of 0.1 µM BL and 0.1 µM Brz, their overall content was 20% higher than in the control group. Whereas the combinatory effect of 0.1 μM BL with 1 μM Brz or 10 μM Brz resulted in a decline of pigment content to the untreated plants, but their level was higher than in plants treated with 1 μM Brz and 10 μM alone.

2.4. Protein and Monosaccharides' Content in *W. arrhiza*

Application of BL at the range of concentrations of 0.01–1 μM had a stimulatory effect on the protein (Figure 2) and monosaccharides' (Figure 3) content. The greatest increase of these compounds was observed under the influence of 0.1 μM BL. Thus, the total content of protein enhanced by 25%, and the total level of monosaccharides increased by 22% to the untreated plant.

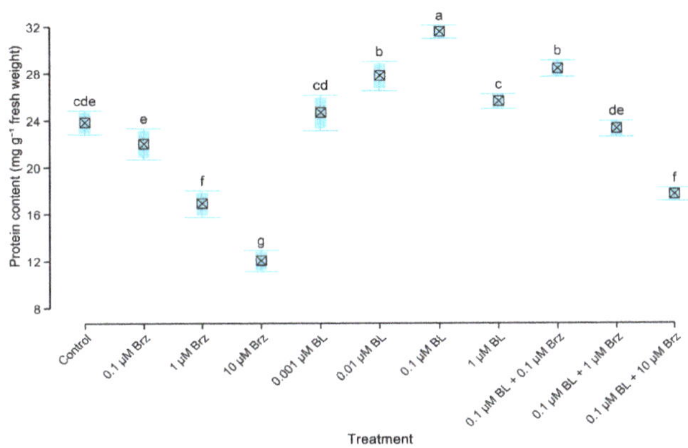

Figure 2. The effect of BL and/or Brz on protein content in *W. arrhiza* after 7 days of cultivation. The crossed square shows the mean ($n = 5$). The lower and upper hinges correspond to the lower and upper bounds of 95% confidence interval of the mean. The lower and upper whiskers extend from the hinge to mean ± standard deviation. Means with the same letters are not significantly different ($p \geq 0.05$) according to Tukey's post hoc test.

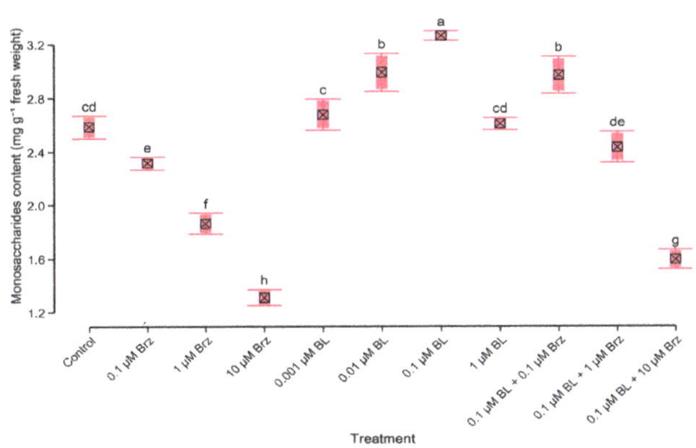

Figure 3. The effect of BL and/or Brz on content of monosaccharides in *W. arrhiza* after 7 days of cultivation. The crossed square shows the mean ($n = 5$). The lower and upper hinges correspond to the lower and upper bounds of 95% confidence interval of the mean. The lower and upper whiskers extend from the hinge to mean ± standard deviation. Means with the same letters are not significantly different ($p \geq 0.05$) according to Tukey's post hoc test.

In opposition to the BL, the treatment with inhibitor caused a considerable decrease in the contents of protein (Figure 2) and monosaccharides (Figure 3). Plant exposition on 0.1 μM Brz had a slight inhibitory effect on amount of monosaccharides while protein did not change significantly. However, in duckweed exposed to the 10 μM Brz, the level of protein and monosaccharides decreased by 50% and 52% compared to the control, respectively. Moreover, the effect of 0.1 μM Brz on the protein level and monosaccharides was suppressed by the application of 0.1 μM BL. The use of 0.1 μM BL with 0.1 μM Brz led to the 15% increase of the amount of protein and monosaccharides to the control. Whereas the amount of these compounds in duckweed treated with 0.1 μM BL with 1 μM Brz, or 10 μM Brz was lower than in control but higher than in the group treated with 1 μM Brz or 10 μM Brz alone.

3. Discussion

3.1. *Exogenous Brassinolide Improves the Growth Rate and Content of Endogenous Hormones and Primary Metabolites*

Phytohormones promote plant growth and biosynthesis of targeted compounds in a concentration-dependent manner. An activity of exogenously applied hormones depends on their absorption efficiency by plants. Exogenous BRs are uptaken and transported in dependence on the application method. In plants, the most efficient absorption of BRs occurs by the roots due to their function which is to uptake substances from the soil [33]. While in water, deprived of stem and roots of plants, absorption of BRs and other molecules takes place through the endocytosis process [29]. As one of the most active BRs, BL plays a vital role in plant growth and regulation of physiological processes [34,35]. In the present study, the enhancement of growth rate and level of detected BRs in the duckweed *W. arrhiza* exposed on BL was observed. Moreover, the content of primary metabolites, such as photosynthetic pigments, monosaccharides, and protein, increased in the presence of BL.

The various hormone concentrations may have opposite effects on plant growth and metabolism. BL in the range of 0.001–0.1 μM had the stimulatory effect on the *W. arrhiza* growth; however, treatment with 1 μM BL caused a slight decrease of duckweed growth. Plant growth is dependent upon the synthesis of nucleic acids and protein. It was demonstrated that the activation of the growth of plant tissue and higher levels of DNA and RNA polymerase is manifested by the increase of the content of nucleic acids and protein. It may be concerned directly or indirectly, with growth promotion induced by BRs that enhancement of DNA and RNA polymerase activities may be a result of regulation of gene expression. Moreover, the high doses of BRs may act antagonistically—in this case, the rate of growth is decreased. Suppression of plant growth is a result of mitosis blocking, DNA replication inhibition, the breakdown of cell integrity and cell membrane, and the degradation of cell wall polysaccharides induced by high doses of BRs [36–39]. Jiroutova et al. [40] showed that high EBL concentration (above 0.5 μM) considerably enhanced ethylene production, which induces plant senescence. Furthermore, high doses of BRs cause gibberellin inactivation, which is correlated with the inhibition of rice (*Oryza sativa*) cell elongation [41]. Other studies demonstrated the stimulatory effect of BRs; however, the growth rate under the influence of various BR concentrations differed among plant species. For instance, in the green alga *Acutodesmus obliquus*, the most promotive effect on the number of algal cells was reached at a concentration of 1 μM EBL [42], while the greatest increases of the *Chlorella vulgaris* growth was observed in alga treated with 0.01 μM BL [43]. In another study, the most promotive effect on the weight of *Brassica junceae* shoots and roots was noted after treatment with 0.5 μM BL [44]. The present study shows that 0.1 μM BL stimulated duckweed growth the most. Summarizing, the explanation of interactions between BRs and other phytohormones on plant growth and development requires further analysis.

The most important part of the presented work was confirmation of the occurrence and determination of the endogenous level of BRs under the influence of exogenously applied BL with Brz in duckweed *W. arrhiza*. Although the presence of BRs has been evidenced both in angiosperms and gymnosperms plants, *W. arrhiza* remains the only species among

Lemnoideae in which these hormones were identified [27]. In our study, the content of BRs is in the range of 0.03–2.77 ng/g in the control group of duckweed. The presence of BRs belonging to the early and late C-6 oxidation pathway was reported. In addition, the occurrence of norBL, which possesses 27 atoms of carbon, and HBL with 29 carbon atoms was also identified. Among all detected BRs, CS, CT, and BL dominate, while ECS and 6dTY occur in a trace amount. In turn, the presence of 6dCT was not detected. These results suggest that the early C-6 oxidation pathway is predominant. Additionally, the presence of EBL and norBL in *W. arrhiza* was noted for the first time. In other reports, the levels and the profiles of BRs varied in dependence on family, species, or environment. Among freshwater plants, eight various BRs were detected in *A. obliquus* [45], while the presence of the most active BL and CS was confirmed in many algal strains [46,47]. Whereas, among freshwater angiosperms, the level of BRs was not studied.

The duckweed treatment with BL had a positive effect on the biosynthesis of all detected BRs. Similarly to the growth rate, the most increase of BRs' level was reported under the influence of 0.1 μM BL, while during the treatment with 1 μM BL, the content of BRs declined. These BRs' level alterations may be associated with the direct effect of exogenously applied BRs on their biosynthetic pathways [48]. Furthermore, the application of BL to hydroponically cultured duckweed influenced the considerable absorption of BL manifested in the increases of its endogenous level. Moreover, the endogenous level of EBL was considerably enhanced in duckweed treated with BL. It suggests that in abundant conditions of BL, BL can be converted directly into EBL. These results are similar to the studies of Janeczko and Swaczynova [49], who demonstrated the increases of endogenous content of BL and CS in wheat (*Triticum aestivum*) seedlings treated with 0.1 μM EBL. The stimulatory effect of 0.1 μM EBL on the level of BRs in barley (*Hordeum vulgare*) was also reported [48].

As is known, chlorophylls are the necessary photosynthetic pigments, and their major role is light-absorbing from solar power for ATP synthesis. In plants, two types of chlorophylls are presented, predominantly chlorophyll *a* and, in less amount, chlorophyll *b*. Phytohormones and environmental conditions determine their biosynthesis and degradation. Determination of the chlorophyll content is used to analyze the photosynthesis rate in plants. As the second group of photosynthetic pigments, carotenoids possess antioxidant activity and protect chlorophylls against photo-oxidative destruction [8]. Carotenoids are divided into carotenes, and their oxygen derivatives mean xanthophylls. Whereas xanthophylls are divided into the oxygen-poor compounds containing one or two atoms of oxygen in the molecule, e.g., citranaxanthin, zeaxanthin, lutein, rhodoxanthin, canthaxanthin, cryptoxanthin, and oxygen-rich molecules containing 3 or 4 oxygen atoms, e.g., astaxanthin, antheraxanthin, neoxanthin, flavaxanthin, and violaxanthin [50,51]. Both chlorophylls and carotenoids are essential to the light phase of photosynthesis [52]. The presence of photosynthetic pigments with the specific individual xanthophyll was confirmed mainly in algae [42,53] and useless agricultural plants, e.g., maize (*Zea mays*) [54], wheat [55], or soybean (*Glycine max*) [56]. While in the present studies, the occurrence of two chlorophylls and eight carotenoids was also confirmed (Table 2). The increased level of pigments in duckweed treated with BL shows that exogenously, BL contributes to photosynthesis efficiency improvements. The direct function of BRs in photosynthesis is the increase of plants' light-capturing efficiency and induction of the activity of chlorophyll biosynthesis enzymes [8]. Referring to the previous studies, EBL caused the increase of pigment content in alga *A. obliquus* [42], *C. vulgaris* [57], and wheat [58]. In another study, Maity and Bera [59] observed the promotive effect of BL on the total chlorophylls' content in the mung bean (*Vigna radiata*).

Sugars, as signaling molecules, are a great reservoir of carbon and energy necessary for all biochemical changes in plants. Monosaccharides, mainly glucose, are produced during a photosynthesis process [60]. The effective role of BRs in photosynthesis is the improvement of the CO_2 assimilation in the Calvin cycle, which leads to an increase of Rubisco activity, resulting in enhanced content of monosaccharides [26,31]. Therefore, the rise of monosaccharides' level in the presence of BL is associated with the increase of photosynthesis activity, resulting from the increases of synthesis of chlorophylls in *W. arrhiza*. These results are in agreement with the literature data. In earlier studies focused on *W. arrhiza*, the level of monosaccharides in the presence of EBL was enhanced [31]. Whereas the analysis of Yu et al. [61] indicated the increase of sugar content in cucumber (*Cucumis savitus*) treated with EBL.

Similar to the sugars, the endogenous content of soluble protein increased in the presence of BL. Moreover, it was evident that the enhanced synthesis of nucleic acids and proteins contributes to the increase of plant growth rate. Treatment with BL causes the acceleration of metabolic processes that favors the accumulation of soluble proteins. Therefore, it suggests that the BL application also affects the level of nucleic acids and, consequently, contributes to the rise in the translation process rate [38,57]. According to the earlier reports, exogenously applied EBL positively affected the protein content in *W. arrhiza* [31]. In another study, the stimulating effect of BL on the level of soluble protein was noted in *V. radiata* [59].

3.2. *Exogenous Brassinolide Overcame the Negative Effect of Brassinazole*

The results presented above show the positive effect of exogenously applied BL on the *W. arrhiza* growth and level of targeted compounds but do not provide certain information about the direct role of endogenous BRs in duckweed development. Therefore, there are two major ways to analyze the functions of endogenous hormones in plants. The first method is to create BRs' deficient mutants through the mutation of genes encoding key enzymes in BRs' biosynthesis. In model plants, e.g., *Arabidopsis thaliana*, mutants' growth is drastically inhibited, the length of hypocotyl is reduced, roots and petioles are shorter, while leaves are smaller and discolored. Moreover, the leaf petiole and blade lengths are significantly reduced, while the stamen filament elongation is impaired, leading to male sterility [62–64]. However, the more universal and accessible method is using the BR biosynthesis inhibitors. The plant responses on the treatment with inhibitors led to recognize the importance of BRs in many physiological and biochemical processes in plants. Brz, as a specific inhibitor of BRs, blocks the conversion of 6-oxoCN to CT and conversion of CT to the TE, and analogously, it blocks conversion CN to the 6-deoxoCT and 6-deoxoCT to the 6-deoxoTE. The chemical structure of Brz contains a triazole ring and methyl residue attached to the carbon atom, which includes a hydroxy group. The methyl group is required for the activity of Brz because the compound without these groups does not indicate any inhibiting activity of BRs. The measurement of plant growth rate under the influence of various Brz concentrations was performed. In plants, often with the Brassicaceae family, the hypocotyl elongation in seedlings treated with Brz to untreated plants is compared. The hypocotyl length of cress (*Lepidium sativum*) was considerably lower in cress exposed to 1 μM and 10 μM Brz. Additionally, in plants treated with 0.1 μM Brz, a slight decrease of hypocotyl length was observed [18]. The foliar application of Brz caused a growth decrease in soybean [25]. Besides growth reduction, Brz-treated cress exhibited curled, dark-green leaves [21]. The inhibitory effect of Brz on the hypocotyl elongation in seedlings of barley [48], cucumber, tomato (*Solanum lycopersicum*), and tobacco (*Nicotiana tabacum*) [20], or *A. thaliana* [19] was also reported. Thus, the phenotype modifications of Brz-treated plants are similar to the genetically modified BR-deficient mutants. Whereas in small, deprived-of-the-stem-and-roots freshwater *W. arrhiza*, the growth rate analysis was performed by comparing plant fresh weight. The present study revealed the strong negative effect of 1 μM and 10 μM Brz on duckweed growth. The reproduction of duckweed was inhibited. The weight of cultures exposed on 10 μM Brz was 2-fold reduced after 7 days of treatment. The phenotype of the whole plant was lightened and withered.

Discoloration of duckweed is connected with a considerable decrease of chlorophylls and other pigments. Similar to the pigments, the endogenous amount of monosaccharides and soluble proteins was significantly lower. The decrease of content of monosaccharides, observed in *W. arrhiza* treated with Brz, was caused by the degradation of photosynthetic pigments contributing to decreased photosynthesis rate and monosaccharides' accumulation. The inhibited growth and reduced concentrations of targeted metabolites are associated with the inhibition of BRs' biosynthesis. Thus, in the present study, application of Brz resulted in BR-deficiency phenotypes through the decrease of endogenous BR levels. Besides BRs, the endogenous level of Brz in *W. arrhiza* treated with inhibitor was also determined. Interestingly, this is the first report about the content of Brz in duckweed tissues. The previous studies indicated the effect of Brz on the morphological processes in plants but did not provide any data about concentrations of this inhibitor in plant organs. The exogenous treatment with BL reversed the inhibitory effect of Brz on the BRs' biosynthesis in duckweed. Interestingly, simultaneous application of 0.1 µM BL with 0.1 µM of Brz caused an enhanced level of BRs compared with untreated plants. The co-application of BL with Brz also had a stimulatory effect on the growth and metabolites' content to the group exposed on Brz alone. Concerning the literature data, exogenous BRs' influence on their accumulation in plants treated with Brz was exceptionally rarely studied. It was reported that exogenous EBL application overcame the inhibition of BRs' synthesis caused by treatment with Brz [48]. Thus, the previous research on *W. arrhiza* showed the reduction of fresh weight, proteins, and monosaccharides' content in cultures treated with Brz2001 and their restoration after application of EBL [31]. Referring to the land plants, BL reversed the dwarf phenotype of *A. thaliana* seedling treated with Brz [20]. While, in plants treated with Brz2001, the recovery effects of treatment with EBL on the length of soybean seedling [65] and algal growth (expressed as a number of cells) [26] were also observed.

4. Materials and Methods

4.1. Plant Material and Growth Conditions

The culture of *W. arrhiza* (L.) Horkel ex Wimm. was obtained from the Faculty of Biology of the University of Bialystok. The breeding of *W. arrhiza* was conducted under controlled conditions at 22.0 ± 0.5 °C, 16 h photoperiod (photon flux of 100 µmol/m/s), and a relative humidity of 65%. One gram of plant was placed in a sterile, glass vessel containing 200 mL of Hunter medium with the following composition: 500 mg/L EDTA, 500 mg/L $MgSO_4{\cdot}7H_2O$, 400 mg/L KH_2PO_4, 354 mg/L $Ca(NO_3)_2{\cdot}4H_2O$, 200 mg/L KOH, 200 mg/L NH_4VO_3, 65.9 mg/L $ZnSO_4{\cdot}7H_2O$, 25.2 mg/L $Na_2MoO_4 \cdot 2H_2O$, 24.9 mg/L $FeSO_4{\cdot}7H_2O$, 17.9 mg/L $MnCl_2$ $4H_2O$, 14.2 mg/L H_3BO_3, 3.95 mg/L $CuSO_4{\cdot}5$ H_2O, 0.2 mg/L $Co(NO_3)_2{\cdot}6H_2O$ [66].

The whole experiment was divided into two major steps. In the first stage of the experiment, the cultures of *W. arrhiza* were treated with Brz in the range of concentrations 0.1–10 µM, or BL in the range of concentrations 0.01–1 µM. The varied solutions were prepared through the diluting of hormone and inhibitor in Hunter medium. After 7 days of treatment, the duckweed biomass was separated from the medium by filtration using a vacuum pump (KNF Neuberger, Inc., Trenton, NJ, USA). Then, the collected plant was weighed and homogenized in liquid nitrogen using a mortar and pestle. The obtained powder was used in further analysis. In the second experiment, the cultures of *W. arrhiza* were treated with a mixture of 0.1 µM BL with 0.1, 1, and 10 µM Brz. The concentration of 0.1 M µM BL was selected because it indicated the greatest effect on the growth and the content of analyzed compounds in duckweed. The subsequent parts of the sample preparation were analogs to stage one.

After 7 days of treatment, the analysis of growth and selected biochemical parameters was performed. The measurement of biomass, protein, pigments, and monosaccharides is necessary for each research involving plants because those parameters constitute a basic indicator of plant responses.

4.2. Chemicals

The eleven standards of BRs (BL, EBL, HBL, norBL, CS, ECS, 6dCS, CT, 6dCT, TY, and 6dTY, as well as stable isotope-labeled standards of [$^2H_3$]BL and [$^2H_3$]CS, were purchased from OlChemIm (Olomouc, Czech Republic). Brz was purchased from TCI Europe N.V. (Zwijndrecht, Belgium). The initial solution of the inhibitor was prepared by the dissolution of Brz powder in dimethyl sulfoxide (DMSO), while the BL was dissolved in 70% ethanol (EtOH). The final amount of DMSO and EtOH added to the medium did not affect the *W. arrhiza* growth. Bovine albumin standard, all chemicals for Hunter's medium, Bradford and Somogyi reagents were purchased from Sigma-Aldrich (St. Louis, MO, USA). Methanol (MeOH), EtOH, acetonitrile (ACN), water (LC-MS purity), formic acid (FA), potassium hydroxide (KOH), and 4-(dimethylamino)phenylboronic acid (DMAPBA), and DMSO were purchased from Merck KGaA (Darmstadt, Germany). Ten standards of photosynthesis pigments (chlorophyll *a*, chlorophyll *b*, α-carotene, β-carotene, neoxanthin, violaxanthin, astaxanthin, zeaxanthin, cryptoxanthin, and lutein) were purchased from DHI (Horsholm, Denmark).

4.3. Quantification of Brassinazole and Brassinosteroids

The measurement of the endogenous level of Brz and BRs was performed using the LC-MS method. For this purpose, 200 mg of plant powders were placed into 2 mL Eppendorf tubes, suspended in 1 mL 95% (*v*/*v*) MeOH, and homogenized in a bead mill (50 Hz, 10 min, TissueLyser LT; Qiagen GmbH, Düsseldorf, Germany) using three 2 mm tungsten balls. Then, the homogenates were centrifuged (2800× *g*, 10 min; MPW-55 Med. Instruments, Gliwice, Poland), and supernatants were transferred to the glass flasks with a flat bottom. The remaining precipitates were suspended in MeOH and centrifuged again. This procedure was repeated five times. The final volume of supernatants (5 mL) was mixed (90 rpm, Laboratory shaker LC-350, Pol-Eko-Aparatura, Wodzisław Śląski, Poland) in temperature of 5 °C for 12 h. For quantification of BRs, $[^2H_3]$BL (2 ng) and $[^2H_3]$CS (2 ng) were added into the mixture, followed by extraction with MeOH as internal standards. The sample purification was performed according to Xin et al.'s [67] method. Therefore, the obtained supernatant (5 mL) was purified from pigments and other pollutions using solid-phase extraction (SPE) MAX cartridge (6 mL, 500 mg, Waters Corporation, Milford, MA, USA), which was activated and equilibrated with 99.9% MeOH, H_2O, 1 M KOH, 10% (*v*/*v*) MeOH, and 95% (*v*/*v*) MeOH, respectively. Purified extracts were dried up using a centrifugal vacuum concentrator, reconstructed in 10% (*v*/*v*) MeOH, and passed through Waters SPE MCX cartridge to remove ion contamination. Cartridges were previously activated and equilibrated with 5% (*v*/*v*) FA in 5% (*v*/*v*) MeOH, 5% (*v*/*v*) MeOH, 5% (*v*/*v*) NH_4OH in 5% (*v*/*v*) MeOH, and 5% (*v*/*v*) MeOH, respectively. Then, samples were eluted using 80% (*v*/*v*) MeOH. Eluents were dried up using a centrifugal vacuum concentrator, suspended in 98 µL of 96% (*v*/*v*) EtOH, and derivatized (45 °C, 1 h), using 2 µL of DMAPBA reagent.

Detection and quantitative analysis of Brz and BR-DMAPBA were performed using the Shimadzu LC-MS-MS-8050 system consisting of pump, degasser, autosampler, column oven, and mass spectrometer with triple quadrupole (Shimadzu Corporation, Kyoto, Japan); 10 µL of each sample was injected on the Waters XBridge C_{18} column (250 mm × 4.6 mm, 1.7 µm, Waters Corporation, Milford, MA, USA). The temperature of the column oven was 25 °C. Mobile phase A was 0.01% (*v*/*v*) FA in H_20 and phase B was 0.01% (*v*/*v*) FA in ACN. The gradient of phase A was 25% from 0 to 14 min, 5% from 14 to 25 min, 20% from 25 to 25.1 min, and 25% from 25.1 to 30 min. The flow was 1 mL/min. Chromatographic parameters of detected compounds are presented in Table 3. Analytical data were analyzed using Shimadzu Browser Workstation Software For LC/MS.

Table 3. Chromatographic properties of Brz and BRs in positive ionization scan mode of ESI-LC-MS.

Compound	Precursor *m/z*	Product *m/z*	CE	Retention Time (min)
Brz	327.8	70.05	−25	3.976
BL	610.0	176.05	−43	11.201
EBL	610.0	190.2	−42	10.720
HBL	624.0	190.1	−43	13.118
norBL	596.0	190.0	−44	9.350
CT	433.0	415.2	−21	1.796
CS	594.1	175.95	−36	13.254
ECS	594.1	175.95	−38	13.552
6dCS	580.1	562.2	−40	23.778
TY	578.1	560.15	−37	18765
6dTY	564.2	176.15	−55	18.957

4.4. Quantification of Photosynthetic Pigments

The endogenous level of photosynthetic pigments in *W. arrhiza* was determined using HPLC method [68]. Therefore, 0.5 g of duckweed powder was suspended in MeOH and homogenized in a bead mill (50 Hz, 10 min, TissueLyser LT; Qiagen GmbH, Düsseldorf, Germany) for cell disruption. The obtained homogenate was left in the fridge for 12 h for pigment extraction. Afterward, samples were centrifuged (2800× *g*, 10 min), and the resulting extract was analyzed.

For pigment separation and analysis, the Agilent 1260 Infinity Series HPLC apparat (Agilent Technologies, Inc., Santa Clara, CA, USA) with quaternary pump with an in-line vacuum degasser, photo-diode array detector set to monitor 350 and 700 nm, refrigerated autosampler with autoinjector sample loop, and thermostatic Eclipse XDB C_8 column (150 mm × 4.6 mm; Agilent Technologies, Inc., Santa Clara, CA, USA) kept at 25 °C was used. The injection volume was 500 µL. The flow was 1 mL/min. The total analysis time was 40 min. Eluent A of mobile phase was MeOH/ACN/0.25 M aqueous pyridine (pH 5.0) in proportion 50/25/25 (*v*/*v*/*v*), while eluent B was MeOH/ACN/acetone in proportion 20/60/20 (*v*/*v*/*v*). The linear gradient of solvent A was as follows: from 100% in the 1st min to 60% in the 22nd min, from 60% in the 22nd min to 5% in the 38th min, from 5% in the 38th min to 100% in the 40th min. Analytical data were integrated using ChemStation software for LC systems (Agilent Technologies, Inc., Santa Clara, CA, USA).

4.5. Determination of Proteins' Content

The content of soluble proteins in *W. arrhiza* was determined spectrophotometrically (Hitachi U-5100 UV-Vis spectrophotometer; Hitachi High-Tech Science Corporation, Tokyo, Japan) by the Bradford [69] method. This method is based on the ability to create ionic and hydrophobic bonds between the protein and the Coomassie Brilliant Blue G-250 dye. For the Bradford reagent preparation, 500 mg of targeted dye was dissolved in 250 mL of 95% (*v*/*v*) EtOH. Then, this mixture was filtered, and then 85% (*w*/*v*) orthophosphate acid was added. The obtained reagent was filled up with water to 1000 mL. Albumin standard was prepared through the dissolved 30 mg bovine albumin with 100 mL of distilled water. For the analytic sample preparation, the harvested and filtered duckweed was extracted in the $\frac{1}{4}$ dilution of the Bradford reagent. The blind sample was distilled water with Bradford reagent, while the standard sample was albumin with Bradford reagent. Finally, 3 mL of distilled water was added to all samples. The measurement of absorbance was performed at 595 nm 60 min after sample preparation.

4.6. Determination of Monosaccharides' Content

The monosaccharides' content was estimated spectrophotometrically (Hitachi U-5100 UV-Vis spectrophotometer; Hitachi High-Tech Science Corporation, Tokyo, Japan) according to the Somogyi-Nelson method [70,71]. A sample of *W. arrhiza* (0.5 g) was extracted in 5 mL of 62.5% (*v*/*v*) MeOH in a water bath for 30 min at 60 °C [72]. The standard sample was glucose, which was prepared by the dissolving of 30 mg glucose with 62.5% (*v*/*v*) MeOH. All samples (0.5 mL) were treated with 0.5 mL of copper reagent and placed in a boiling water bath for 20 min. Next, the 0.5 mL of arsenomolybdate reagent was added, then, after 5 min, the extract was diluted in 3.5 mL of water and mixed. The value of absorbance was read at 540 nm.

4.7. Statistical Treatment

The R software was used to perform statistical analyses [73]. Basic descriptive statistics were calculated for the data dataset grouped by treatment (n = 5) using 'summaryBy' function from 'doBy' package [74]. Then, the dataset was subjected to the one-way ANOVA (Table S1) ('aov' function from 'stats' package) followed by Tukey's post hoc test ('LTukey' function from 'laercio' package [75]). Differences were considered significant for $p < 0.05$. The Shapiro–Wilk normality test ('shapiro.test' function from 'stats' package) and Bartlett's test of homogeneity of variances ('bartlett.test' function from 'stats' package) were used to verify ANOVA assumptions with $\alpha = 0.05$. Results from ANOVA were visualized as plots (with the help of 'ggplot2' package [76]) and tables.

5. Conclusions

The obtained results of performed experiments showed that the application of BL and/or Brz effects on the *W. arrhiza* cultures in a concentration-dependent manner. Treatment with BL in concentrations of 0.01 µM and 0.1 µM caused statistically significant increases in the growth rate, as well as the content of BRs, photosynthetic pigments, soluble proteins, and monosaccharides in duckweed after 7 days of cultivation. Otherwise, the influence of 0.001 µM BL was statistically insignificant, while the exposition on the 1 µM BL had a statistically important effect depending on the analyzed parameter. Additionally, the inhibitory effect of Brz (1 µM and 10 µM) on the growth rate and level of BRs as well as primary metabolites was verified. However, the inhibitory effect of Brz was effectively mitigated by the application of 0.1 µM BL. The level of nine BRs (BL, EBL, HBL, norBL, CT, CS, ECS, TY, and 6dTY) was reported for the first time in duckweed treated with BL and/or Brz. Additionally, the presence of EBL in *W. arrhiza* was noted for the first time.

Supplementary Materials: The following are available online at https://www.mdpi.com/article/10.3390/plants10071311/s1, Table S1: One-way ANOVA results.

Author Contributions: Conceptualization: A.B.; Methodology: A.B. and M.C.; Validation: M.C.; Formal analysis: A.B.; Investigation: M.C.; Resources: M.C.; Data curation: A.B.; Writing—original draft preparation: M.C.; Writing—review and editing: A.B.; Visualization: A.B.; Supervision: A.B.; Project administration: A.B.; Funding acquisition: A.B. Both authors have read and agreed to the published version of the manuscript.

Funding: This work was funded by the Ministry of Science and Higher Education as part of subsidies for maintaining research potential awarded to the Faculty of Biology of the University of Bialystok (No. SWB-3).

Institutional Review Board Statement: Not applicable.

Informed Consent Statement: Not applicable.

Data Availability Statement: Data is contained within the current article and Supplementary Materials.

Acknowledgments: We would like to thank Andrzej Reszka and Marcin Gawryś from "SHIM-POL A.M. Borzymowski" E. Borzymowska-Reszka, A. Reszka Spółka Jawna for Shimadzu LC-MS-MS-8050 renting. A.B. thanks Adam Bajguz for programming consultations.

Conflicts of Interest: The authors declare no conflict of interest.

References

1. Zullo, M.A.T.; Bajguz, A. The brassinosteroids family—structural diversity of natural compounds and their precursors. In *Brassinosteroids: Plant Growth and Development*; Hayat, S., Yusuf, M., Bhardwaj, R., Bajguz, A., Eds.; Springer: Singapore, 2019; pp. 1–44. [CrossRef]
2. Bajguz, A. Brassinosteroids in microalgae: Application for growth improvement and protection against abiotic stresses. In *Brassinosteroids: Plant Growth and Development*; Hayat, S., Yusuf, M., Bhardwaj, R., Bajguz, A., Eds.; Springer: Singapore, 2019; pp. 45–58. [CrossRef]
3. Tarkowska, D.; Strnad, M. Isoprenoid-derived plant signaling molecules: Biosynthesis and biological importance. *Planta* **2018**, *247*, 1051–1066. [CrossRef]
4. Bajguz, A.; Tretyn, A. The chemical characteristic and distribution of brassinosteroids in plants. *Phytochemistry* **2003**, *62*, 1027–1046. [CrossRef]
5. Kanwar, M.K.; Bajguz, A.; Zhou, J.; Bhardwaj, R. Analysis of brassinosteroids in plants. *J. Plant Growth Regul.* **2017**, *36*, 1002–1030. [CrossRef]
6. Grove, M.D.; Spencer, G.F.; Rohwedder, W.K.; Mandava, N.; Worley, J.F.; Warthen, J.D., Jr.; Steffens, G.L.; Flippen-Anderson, J.L.; Cook, J.C., Jr. Brassinolide, a plant growth-promoting steroid isolated from *Brassica napus* pollen. *Nature* **1979**, *281*, 216–217. [CrossRef]
7. Nolan, T.M.; Vukašinović, N.; Liu, D.; Russinova, E.; Yin, Y. Brassinosteroids: Multidimensional regulators of plant growth, development, and stress responses. *Plant Cell* **2020**, *32*, 295–318. [CrossRef] [PubMed]
8. Siddiqui, H.; Hayat, S.; Bajguz, A. Regulation of photosynthesis by brassinosteroids in plants. *Acta Physiol. Plant.* **2018**, *40*, 59. [CrossRef]
9. Hayat, S.; Mori, M.; Fariduddin, Q.; Bajguz, A.; Ahmad, A. Physiological role of brassinosteroids: An update. *Indian J. Plant Physiol.* **2010**, *15*, 99–109.
10. Li, Z.; He, Y. Roles of brassinosteroids in plant reproduction. *Int. J. Mol. Sci.* **2020**, *21*, 872. [CrossRef]
11. Hussain, M.A.; Fahad, S.; Sharif, R.; Jan, M.F.; Mujtaba, M.; Ali, Q.; Ahmad, A.; Ahmad, H.; Amin, N.; Ajayo, B.S.; et al. Multifunctional role of brassinosteroid and its analogues in plants. *Plant Growth Regul.* **2020**, *92*, 141–156. [CrossRef]
12. Bajguz, A.; Hayat, S. Effects of brassinosteroids on the plant responses to environmental stresses. *Plant Physiol. Biochem.* **2009**, *47*, 1–8. [CrossRef]
13. Rajewska, I.; Talarek, M.; Bajguz, A. Brassinosteroids and response of plants to heavy metals action. *Front. Plant Sci.* **2016**, *7*, 629. [CrossRef]
14. Ahanger, M.A.; Ashraf, M.; Bajguz, A.; Ahmad, P. Brassinosteroids regulate growth in plants under stressful environments and crosstalk with other potential phytohormones. *J. Plant Growth Regul.* **2018**, *37*, 1007–1024. [CrossRef]
15. Li, S.; Zheng, H.; Lin, L.; Wang, F.; Sui, N. Roles of brassinosteroids in plant growth and abiotic stress response. *Plant Growth Regul.* **2020**, *93*, 29–38. [CrossRef]
16. Bajguz, A.; Chmur, M.; Gruszka, D. Comprehensive overview of the brassinosteroid biosynthesis pathways: Substrates, products, inhibitors, and connections. *Front. Plant Sci.* **2020**, *11*, 1034. [CrossRef]
17. Wei, Z.; Li, J. Regulation of brassinosteroid homeostasis in higher plants. *Front. Plant Sci.* **2020**, *11*, 583622. [CrossRef]
18. Rozhon, W.; Akter, S.; Fernandez, A.; Poppenberger, B. Inhibitors of brassinosteroid biosynthesis and signal transduction. *Molecules* **2019**, *24*, 4372. [CrossRef] [PubMed]
19. Asami, T.; Min, Y.K.; Nagata, N.; Yamagishi, K.; Takatsuto, S.; Fujioka, S.; Murofushi, N.; Yamaguchi, I.; Yoshida, S. Characterization of brassinazole, a triazole-type brassinosteroid biosynthesis inhibitor. *Plant Physiol.* **2000**, *123*, 93–99. [CrossRef] [PubMed]
20. Asami, T.; Yoshida, S. Brassinosteroid biosynthesis inhibitors. *Trends Plant Sci.* **1999**, *4*, 348–353. [CrossRef]
21. Nagata, N.; Asami, T.; Yoshida, S. Brassinazole, an inhibitor of brassinosteroid biosynthesis, inhibits development of secondary xylem in cress plants (*Lepidium sativum*). *Plant Cell Physiol.* **2001**, *42*, 1006–1011. [CrossRef] [PubMed]
22. Nagata, N.; Min, Y.K.; Nakano, T.; Asami, T.; Yoshida, S. Treatment of dark-grown *Arabidopsis thaliana* with a brassinosteroid-biosynthesis inhibitor, brassinazole, induces some characteristics of light-grown plants. *Planta* **2000**, *211*, 781–790. [CrossRef] [PubMed]
23. Pereira-Netto, A.B.; Roessner, U.; Fujioka, S.; Bacic, A.; Asami, T.; Yoshida, S.; Clouse, S.D. Shooting control by brassinosteroids: Metabolomic analysis and effect of brassinazole on *Malus prunifolia,* the Marubakaido apple rootstock. *Tree Physiol.* **2009**, *29*, 607–620. [CrossRef]
24. Pramod, S.; Anju, M.; Rajesh, H.; Thulaseedharan, A.; Rao, K.S. Effect of exogenously applied 24-epibrassinolide and brassinazole on xylogenesis and microdistribution of cell wall polymers in *Leucaena leucocephala* (Lam) De Wit. *J. Plant Growth Regul.* **2021**, in press. [CrossRef]
25. Terakado, J.; Fujihara, S.; Goto, S.; Kuratani, R.; Suzuki, Y.; Yoshida, S.; Yoneyama, T. Systemic effect of a brassinosteroid on root nodule formation in soybean as revealed by the application of brassinolide and brassinazole. *Soil Sci. Plant Nutr.* **2005**, *51*, 389–395. [CrossRef]
26. Bajguz, A.; Asami, T. Effects of brassinazole, an inhibitor of brassinosteroid biosynthesis, on light- and dark-grown *Chlorella vulgaris*. *Planta* **2004**, *218*, 869–877. [CrossRef] [PubMed]

27. Chmur, M.; Bajguz, A.; Piotrowska-Niczyporuk, A. Effect of cadmium on the level of isoprenoid-derived phytohormones in duckweed *Wolffia arrhiza*. *J. Plant Growth Regul.* **2020**, *39*, 1518–1530. [CrossRef]
28. Kotowska, U.; Karpinska, J.; Kapelewska, J.; Kowejsza, E.M.; Piotrowska-Niczyporuk, A.; Piekutin, J.; Kotowski, A. Removal of phthalates and other contaminants from municipal wastewater during cultivation of *Wolffia arrhiza*. *Process Saf. Environ. Prot.* **2018**, *120*, 268–277. [CrossRef]
29. Piotrowska, A.; Bajguz, A.; Czerpak, R.; Kot, K. Changes in the growth, chemical composition, and antioxidant activity in the aquatic plant *Wolffia arrhiza* (L.) Wimm. (Lemnaceae) exposed to jasmonic acid. *J. Plant Growth Regul.* **2010**, *29*, 53–62. [CrossRef]
30. Forni, C.; Tommasi, F. Duckweed: A tool for ecotoxicology and a candidate for phytoremediation. *Curr. Biotechnol.* **2016**, *5*, 2–10. [CrossRef]
31. Bajguz, A.; Asami, T. Suppression of *Wolffia arrhiza* growth by brassinazole, an inhibitor of brassinosteroid biosynthesis and its restoration by endogenous 24-epibrassinolide. *Phytochemistry* **2005**, *66*, 1787–1796. [CrossRef]
32. Li, Y.X.; Deng, T.; Duan, C.F.; Ni, L.X.; Wang, N.; Guan, Y.F. Dispersive matrix solid-phase extraction method coupled with high performance liquid chromatography-tandem mass spectrometry for ultrasensitive quantification of endogenous brassinosteroids in minute plants and its application for geographical distribution study. *J. Agric. Food Chem.* **2019**, *67*, 3037–3045. [CrossRef] [PubMed]
33. Symons, G.M.; Reid, J.B. Brassinosteroids do not undergo long-distance transport in pea. Implications for the regulation of endogenous brassinosteroid levels. *Plant Physiol.* **2004**, *135*, 2196–2206. [CrossRef] [PubMed]
34. Roh, J.; Yeom, H.S.; Jang, H.; Kim, S.; Youn, J.H.; Kim, S.K. Identification and biosynthesis of C-24 ethylidene brassinosteroids in *Arabidopsis thaliana*. *J. Plant Biol.* **2017**, *60*, 533–538. [CrossRef]
35. Ackerman-Lavert, M.; Savaldi-Goldstein, S. Growth models from a brassinosteroid perspective. *Curr. Opin. Plant Biol.* **2020**, *53*, 90–97. [CrossRef] [PubMed]
36. Kalinich, J.F.; Mandava, N.B.; Todhunter, J.A. Relationship of nucleic acid metabolism to brassinolide-induced responses in beans. *J. Plant Physiol.* **1986**, *125*, 345–353. [CrossRef]
37. Mandava, N.B.; Thompson, M.J.; Yopp, J.H. Effects of selected inhibitors of RNA and protein synthesis on brassinosteroid-induced responses in mung bean epicotyls. *J. Plant Physiol.* **1987**, *128*, 53–65. [CrossRef]
38. Bajguz, A. Effect of brassinosteroids on nucleic acids and protein content in cultured cells of *Chlorella vulgaris*. *Plant Physiol. Biochem.* **2000**, *38*, 209–215. [CrossRef]
39. Mussig, C. Brassinosteroid-promoted growth. *Plant Biol.* **2005**, *7*, 110–117. [CrossRef] [PubMed]
40. Jiroutova, P.; Mikulik, J.; Novak, O.; Strnad, M.; Oklestkova, J. Brassinosteroids induce strong, dose-dependent inhibition of etiolated pea seedling growth correlated with ethylene production. *Biomolecules* **2019**, *9*, 849. [CrossRef]
41. Tong, H.N.; Xiao, Y.H.; Liu, D.P.; Gao, S.P.; Liu, L.C.; Yin, Y.H.; Jin, Y.; Qian, Q.; Chu, C.C. Brassinosteroid regulates cell elongation by modulating gibberellin metabolism in rice. *Plant Cell* **2014**, *26*, 4376–4393. [CrossRef]
42. Talarek-Karwel, M.; Bajguz, A.; Piotrowska-Niczyporuk, A.; Rajewska, I. The effect of 24-epibrassinolide on the green alga *Acutodesmus obliquus* (Chlorophyceae). *Plant Physiol. Biochem.* **2018**, *124*, 175–183. [CrossRef]
43. Bajguz, A. An enhancing effect of exogenous brassinolide on the growth and antioxidant activity in *Chlorella vulgaris* cultures under heavy metals stress. *Environ. Exp. Bot.* **2010**, *68*, 175–179. [CrossRef]
44. Latha, P.; Vardhini, V. Effect of brassinolide on the growth of mustard crops grown in semi-arid tropics of nizamabad. *Int. J. Plant Soil Sci.* **2016**, *9*, 1–5. [CrossRef]
45. Talarek-Karwel, M.; Bajguz, A.; Piotrowska-Niczyporuk, A. Hormonal response of *Acutodesmus obliquus* exposed to combined treatment with 24-epibrassinolide and lead. *J. Appl. Phycol.* **2020**, *32*, 2903–2914. [CrossRef]
46. Stirk, W.A.; Balint, P.; Tarkowska, D.; Novak, O.; Strnad, M.; Ordog, V.; Van Staden, J. Hormone profiles in microalgae: Gibberellins and brassinosteroids. *Plant Physiol. Biochem.* **2013**, *70*, 348–353. [CrossRef] [PubMed]
47. Bajguz, A. Isolation and characterization of brassinosteroids from algal cultures of *Chlorella vulgaris* Beijerinck (Trebouxiophyceae). *J. Plant Physiol.* **2009**, *166*, 1946–1949. [CrossRef] [PubMed]
48. Bajguz, A.; Orczyk, W.; Golebiewska, A.; Chmur, M.; Piotrowska-Niczyporuk, A. Occurrence of brassinosteroids and influence of 24-epibrassinolide with brassinazole on their content in the leaves and roots of *Hordeum vulgare* L. cv. Golden Promise. *Planta* **2019**, *249*, 123–137. [CrossRef]
49. Janeczko, A.; Swaczynova, J. Endogenous brassinosteroids in wheat treated with 24-epibrassinolide. *Biol. Plant.* **2010**, *54*, 477–482. [CrossRef]
50. Patias, L.D.; Fernandes, A.S.; Petry, F.C.; Mercadante, A.Z.; Jacob-Lopes, E.; Zepka, L.Q. Carotenoid profile of three microalgae/cyanobacteria species with peroxyl radical scavenger capacity. *Food Res. Int.* **2017**, *100*, 260–266. [CrossRef] [PubMed]
51. Poojary, M.M.; Barba, F.J.; Aliakbarian, B.; Donsì, F.; Pataro, G.; Dias, D.A.; Juliano, P. Innovative alternative technologies to extract carotenoids from microalgae and seaweeds. *Mar. Drugs* **2016**, *14*, 214. [CrossRef] [PubMed]
52. Czerpak, R.; Dobrzyń, P.; Krotke, A.; Kicińska, E. The effect of auxins and salicylic acid on chlorophyll and carotenoid contents in *Wolffia arrhiza* (L.) Wimm. (Lemnaceae) growing on media of various trophicities. *Pol. J. Environ. Stud.* **2002**, *11*, 231–235.
53. Piotrowska-Niczyporuk, A.; Bajguz, A.; Kotowska, U.; Bralska, M.; Talarek-Karwel, M. Growth, metabolite profile, oxidative status, and phytohormone levels in the green alga *Acutodesmus obliquus* exposed to exogenous auxins and cytokinins. *J. Plant Growth Regul.* **2018**, *37*, 1159–1174. [CrossRef]

54. Ortiz, D.; Ponrajan, A.; Bonnet, J.P.; Rocheford, T.; Ferruzzi, M.G. Carotenoid stability during dry milling, storage, and extrusion processing of biofortified maize genotypes. *J. Agric. Food Chem.* **2018**, *66*, 4683–4691. [CrossRef] [PubMed]
55. Paznocht, L.; Burešová, B.; Kotíková, Z.; Martinek, P. Carotenoid content of extruded and puffed products made of colored-grain wheats. *Food Chem.* **2021**, *340*, 127951. [CrossRef] [PubMed]
56. Alamu, E.O.; Olatunde, G.O.; Adegunwa, M.O.; Adebanjo, L.A.; Awoyinfa, O.C.; Soyoye, J.B. Carotenoid profile and functional properties of flour blends from biofortified maize and improved soybean varieties for product developments. *Cogent Food Agric.* **2021**, *7*, 1868665. [CrossRef]
57. Bajguz, A.; Czerpak, R. Physiological and biochemical role of brassinosteroids and their structure-activity relationship in the green alga *Chlorella vulgaris* Beijerinck (Chlorophyceae). *J. Plant Growth Regul.* **1998**, *17*, 131–139. [CrossRef]
58. Dong, Y.J.; Wang, W.W.; Hu, G.Q.; Chen, W.F.; Zhuge, Y.P.; Wang, Z.L.; He, M.R. Role of exogenous 24-epibrassinolide in enhancing the salt tolerance of wheat seedlings. *J. Soil Sci. Plant Nutr.* **2017**, *17*, 554–569. [CrossRef]
59. Maity, U.; Bera, A. Effect of exogenous application of brassinolide and salicylic acid on certain physiological and biochemical aspects of green gram (*Vigna radiata* L. Wilczek). *Indian J. Agric. Res.* **2009**, *43*, 194–199.
60. Eckstein, A.; Zięba, P.; Gabryś, H. Sugar and light effects on the condition of the photosynthetic apparatus of *Arabidopsis thaliana* cultured in vitro. *J. Plant Growth Regul.* **2012**, *31*, 90–101. [CrossRef]
61. Yu, J.Q.; Huang, L.F.; Hu, W.H.; Zhou, Y.H.; Mao, W.H.; Ye, S.F.; Nogués, S. A role for brassinosteroids in the regulation of photosynthesis in *Cucumis sativus*. *J. Exp. Bot.* **2004**, *55*, 1135–1143. [CrossRef]
62. Ren, C.M.; Han, C.Y.; Peng, W.; Huang, Y.; Peng, Z.H.; Xiong, X.Y.; Zhu, Q.; Gao, B.D.; Xie, D.X. A leaky mutation in DWARF4 reveals an antagonistic role of brassinosteroid in the inhibition of root growth by jasmonate in *Arabidopsis*. *Plant Physiol.* **2009**, *151*, 1412–1420. [CrossRef] [PubMed]
63. Wang, M.J.; Liu, X.Y.; Wang, R.; Li, W.C.; Rodermel, S.; Yu, F. Overexpression of a putative *Arabidopsis* BAHD acyltransferase causes dwarfism that can be rescued by brassinosteroid. *J. Exp. Bot.* **2012**, *63*, 5787–5801. [CrossRef]
64. Shahnejat-Bushehri, S.; Tarkowska, D.; Sakuraba, Y.; Balazadeh, S. *Arabidopsis* NAC transcription factor JUB1 regulates GA/BR metabolism and signalling. *Nat. Plants* **2016**, *2*, 16013. [CrossRef] [PubMed]
65. Mazorra, L.M.; Nunez, M.; Napoles, M.C.; Yoshida, S.; Robaina, C.; Coll, F.; Asami, T. Effects of structural analogs of brassinosteroids on the recovery of growth inhibition by a specific brassinosteroid biosynthesis inhibitor. *Plant Growth Regul.* **2004**, *44*, 183–185. [CrossRef]
66. Hutner, S.H. Comparative physiology of heterotrophic growth in plants. In *Growth and Differentiation in Plants*; Loomis, W.E., Ed.; Iowa State College Press: Ames, IA, USA, 1953; pp. 417–446.
67. Xin, P.Y.; Yan, J.J.; Fan, J.S.; Chu, J.F.; Yan, C.Y. An improved simplified high-sensitivity quantification method for determining brassinosteroids in different tissues of rice and *Arabidopsis*. *Plant Physiol.* **2013**, *162*, 2056–2066. [CrossRef]
68. Zapata, M.; Rodríguez, F.; Garrido, J.L. Separation of chlorophylls and carotenoids from marine phytoplankton: A new HPLC method using a reversed phase C_8 column and pyridine-containing mobile phases. *Mar. Ecol. Prog. Ser.* **2000**, *195*, 29–45. [CrossRef]
69. Bradford, M.M. A rapid and sensitive method for the quantitation of microgram quantities of protein utilizing the principle of protein-dye binding. *Anal. Biochem.* **1976**, *72*, 248–254. [CrossRef]
70. Somogyi, M. Notes on sugar determination. *J. Biol. Chem.* **1952**, *195*, 19–23. [CrossRef]
71. Nelson, N. A photometric adaptation of the Somogyi method for the determination of glucose. *J. Biol. Chem.* **1944**, *153*, 375–380. [CrossRef]
72. Blunden, C.A.; Wilson, M.F. A specific method for the determination of soluble sugars in plant extracts using enzymatic analysis and its application to the sugar content of developing pear fruit buds. *Anal. Biochem.* **1985**, *151*, 403–408. [CrossRef]
73. R Core Team. *R: A Language and Environment for Statistical Computing (R Version 4.0.2, Taking Off Again)*; R Foundation for Statistical Computing: Vienna, Austria, 2020; Available online: https://www.R-project.org/ (accessed on 22 June 2020).
74. Hojsgaard, S.; Halekoh, U. doBy: Groupwise Statistics, LSmeans, Linear Contrasts, Utilities (R Package Version 4.6.8). 2020. Available online: https://CRAN.R-project.org/package=doBy (accessed on 15 February 2020).
75. Da Silva, L.J. Laercio-Package: Duncan Test, Tukey Test and Scott-Knott Test. 2010. Available online: https://CRAN.R-project.org/package=laercio (accessed on 20 February 2015).
76. Wickham, H. *ggplot2: Elegant Graphics for Data Analysis*; Springer: New York, NY, USA, 2016.

plants

MDPI

Article

Effect of 17β-Estradiol on Growth and Biosynthesis of Microalgae *Scenedesmus quadricauda* (CPCC-158) and Duckweed *Lemna minor* (CPCC-490) Grown in Three Different Media

Tatiana A. Kozlova [1,2,*] and David B. Levin [3]

1 Laboratory of Controlled Photobiosynthesis, Timiryazev Institute of Plant Physiology of RAS, Botanicheskaya 35, 127276 Moscow, Russia
2 Laboratory of Ecology, Institute of Natural and Technical Systems of RAS, Teatralnaya 8A, 354000 Sochi, Russia
3 Department of BioSystems Engineering, University of Manitoba, Room E2-370, 75A Chancellor's Circle, Winnipeg, MB R3T 2N2, Canada; david.levin@umanitoba.ca
* Correspondence: kozlova@ifr.moscow; Tel.: +7-977-595-77-18

Abstract: As fish farm wastewaters have detectable levels of fish hormones, such as 17β-estradiol (E2), an understanding of the influence of fish steroids on algal (*Scenedesmus quadricauda*) and duckweed (*Lemna minor*) physiology is relevant to the potential use of fishery wastewaters for microalgae and plant biomass production. The study was conducted using three types of media: Bold Basal Medium (BBM), natural fishery wastewater (FWW), and reconstituted fishery wastewater (RFWW) with the nutrient composition adjusted to mimic FWW. During the experiment, the media were aerated and changes in the pH and conductivity of the water were closely monitored. E2 promoted the growth of *S. quadricauda* and *L. minor*, with significant accumulation of high-value biomolecules at very low steroid concentrations. However, clear differences in growth performance were observed in both test cultures, *S. quadricauda* and *L. minor*, grown in different media, and the most effective hormone concentrations were evidently different for the algae and the plant.

Keywords: aquaponics; steroid 17β-estradiol; *Scenedesmus quadricauda*; *Lemna minor*; fish wastewater; population growth; biomolecule synthesis

Citation: Kozlova, T.A.; Levin, D.B. Effect of 17β-Estradiol on Growth and Biosynthesis of Microalgae *Scenedesmus quadricauda* (CPCC-158) and Duckweed *Lemna minor* (CPCC-490) Grown in Three Different Media. *Plants* **2022**, *11*, 1669. https://doi.org/10.3390/plants11131669

Academic Editors: Viktor Oláh, Klaus-Jürgen Appenroth and K. Sowjanya Sree

Received: 25 April 2022
Accepted: 19 June 2022
Published: 24 June 2022

1. Introduction

There is a great difference in the ability of different organic compounds to bind to and cross cell membranes under conditions that promote cell growth. Most organic molecules bind strongly to ions dissolved in water, which normally reduces membrane permeability, and thus, the bioavailability of both molecules [1,2]. Consequently, the concentrations of dissolved ions (hardness, conductivity), pH, light quality and quantity, and temperature, are among the major factors that influence the impact of organic compounds on the growth and biosynthesis of microalgae and aquatic plants [3–5].

Numerous reports provide the range of residual 17β-estradiol (E2) in surface and ground waters as 0.1 to 200 ng/L E2 in North America and Europe [5–9]. In fish-farming wastewaters, the concentration of E2 is much higher, reaching 100–200 μg/L [10–12]. With respect to the fate of estradiol in aqueous environments and aquaponic systems, three major processes must be considered. The first process is the "reverse uptake" of steroid hormones into fish, although the percentage of E2 reabsorbed by fish is small [13], and in the case of water flowing out of fish tanks to microalgae and/or plant production units, this possibility can be ignored. The second consideration is abiotic degradation (e.g., photodegradation) of E2. It has been established that between 2% and 10% of spiked E2 concentrations may be photodegraded in 48 h [14]. The third process is biotic degradation of the steroid by the microbial community (bacteria and/or microalgae) of the system. Most exogenous estrogens undergo biotic degradation, which can be simply described as an oxidative process beginning with hydroxylation that increases the solubility of estrogen, followed by glycosylation and methylation. During this process, E2 is converted to estrone (E1), then to estriol (E3) and hydroxyestrone [15] plus low concentrations of an unknown metabolite [3,5,16]. A recent paper by Liu et al. [5] reported the ratio of extracellular (adsorbed) to intracellular (absorbed) E2, as well as residual E2 in the water, after four days of exposure, which appeared to be dependent on the initial concentration of E2, time, and water chemistry. Duckweed was shown as a more powerful agent for removing estrogens from water when compared to microalgae [3]. However, in this work, the authors report that algae and duckweed equally adsorb only 5% of dissolved estrogen at a given time and that the process of absorption and further biodegradation of estrogens occurs quickly.

Different algae species have been shown to remove and degrade estradiols from aquatic environments via biodegradation or biotransformation processes, rather than simple adsorption and accumulation in the cells [5,17]. The ratio of adsorbed to biodegraded estradiols varies among the studies available in the literature but is often reported more or less as half and half [15]. However, some studies suggest that most of the adsorbed estradiol becomes absorbed and biodegraded [5,18] and, importantly, does not accumulate intracellularly, at least not when the concentrations of estradiol were below 100 ng/L [18].

It has been reported that mammalian sex steroids can be synthesized by plants. About 70–80% of plant species tested were found to synthesize progesterone (including 17, 20β-P) and androgens. Estrogens were found to be synthesized by about 50% of the plant species tested [19,20]. The metabolic pathways of sex steroids in plants are suggested to be very similar to those in animals and humans [21–23] and that the same enzymes carry out these reactions in animal cells (for example, 17β-hydroxysteroid dehydrogenase). It is likely also true for microalgae species, but research on mammal steroid synthesis by algae cells is not as well documented.

The waterborne hormone, E2, has been shown to promote plant and microalgae growth and biosynthesis. Increases in the biomass of plant shoots and roots have been observed in *Medicago sativa* L. plants at 5 to 500 ng/L, while 50 μg/L E2 and higher was toxic for this plant [24]. Other studies confirmed the positive effect of E2 on flowering in *Cichorium intybus* and *Arabidopsis thaliana* [25,26], as well as an improvement in the development of reproductive organs in *Salvia splendens* [27]. Thus, the processes targeted by estrogens in plants are primarily the early stages of plant development (shoots and young roots) and the formation of the sexual reproduction organs [19,28,29]. In aquatic plants, and particularly in the members of the family Lemnaceae, E2 was reported to increase chlorophyll and total carotenoid content [30], nucleic acids (DNA and RNA), soluble protein content, and reducing sugars content [31]. Stimulation of *L. minor* flowering by E2 was also observed [32].

The growing interest in microalgae biomass production for biofuels, pharmaceuticals, and the agricultural industry has motivated investigations of the influence of steroids, which could be found in fishery wastewater in nano-concentrations, on algal growth and biosynthesis, although the initial studies on the subject of steroid influences on microalgae were conducted more than 45 years ago [33]. It was demonstrated that E2 increases growth and biomass production, as well as the biosynthesis of chlorophyll, carotenoids, proteins, and sugars in *Ch. vulgaris* cells [34–37] when concentrations between 0.3 to 3000 ng/L exogenous E2 were applied.

Our particular interest in investigating the effect of steroids, commonly detected in fishery wastewater, on the physiology of potential objects for biomass production is dictated by the objectives of a joint scientific and industrial project between the University of Manitoba and Myera Nu-Agrinomics Group Canada Inc. In our previous study on the influence of the estrogen E2, and the progesterone 17, 20β-P, on the microalgal species, *S. quadricauda*, we observed a positive effect of both steroids on the rate of cell growth, the final cell density, and the biosynthesis and accumulation of chlorophyll-a, total carotenoids, and lipids [38]. In the present study, the general goal was to evaluate the ability of low concentrations of the common fishery wastewater steroid E2 to induct growth and commercially valuable biomolecule accumulation in two aquatic organisms proposed for biomass production. Thus, growth parameters and soluble protein production of both *S. quadricauda* and the duckweed species, *L. minor*, as well as pigment (chlorophyll-a and total carotenoids) accumulation in *S. quadricauda* were investigated. An important subject of the study was the evaluation of the steroid influence on aquatic organisms in three different media. Therefore, our experiments were conducted in three different media—Bolt Basal Medium (BBM), fishery wastewater from a hatchery farm (FWW), and reconstituted fishery wastewater (RFWW)—over an 11-day period. RFWW was designed based on the analyzed chemical parameters of fishery wastewater from the trout farms of Manitoba, MB. Changes in water chemistry (pH and conductivity) were assessed, and the concentrations of the major macro- and micronutrients of FWW were analyzed (Tables S1 and S2).

2. Materials and Methods

2.1. Strains and Culturing Conditions

The test organisms, *Scenedesmus quadricauda* (CPCC-158) and *Lemna minor* (CPCC-140), were obtained from the Canadian Phycological Culture Centre (CPCC; Waterloo, ON, Canada). *S. quadricauda* was chosen as it is known to grow well on wastewaters, exhibiting high biomass, lipid, and carotenoid production [39–41]. *L. minor* was selected as a model plant in this study due to its high growth rate and its ability to accumulate high concentrations of protein, carotenoids, and lipids [42–44].

A stock culture of algae was grown aseptically in standard Bold's Basal Medium (BBM) [45] with some modifications (NH_4Cl was added to 4.5 ppm) for at least two months prior to the experiment at 23 °C, pH 6.9 ± 0.1, with Plant Grow fluorescent lighting at 4000 lux (120 $\mu mol/m^2/s$) measured with a CalRight CI-1010 digital lux-meter (CalRight Instrument Inc., San Diego, US) in a chamber shaker Innova-44R (New Brunswick Scientific, Enfield, US). A stock culture of duckweed was grown aseptically in standard 30% Hoagland's medium (30% HG) in the same chamber shaker.

Prior to the experiments with altered ambient conditions (e.g., lighting in the walk-in growth chamber, aeration) all cultures were acclimated to the experimental conditions for a minimum of one month. Before starting the experimental work, the status of the cultures was checked using the health criteria identified by Environment Canada [46,47]. This ensures the quality and relevance of the obtained data. The seeding density of *S. quadricauda* cells in all trials was $6.0 \pm 1.0 \times 10^4$ cell/mL. The initial density of the *L. minor* culture was ten plants with three fronds on each.

All experiments were conducted over an 11-day period. Aeration was applied aseptically to 250 mL Erlenmeyer flasks containing *S. quadricauda* and 300 mL glass bowls containing *L. minor* (Figure 1A,B). The air bubbled through the cultures was dispersed through the plastic discs with 2 mm pores situated at the bottom of the culture bowls (Figure 1A) and Erlenmeyer flasks (Figure 1B). This experiment was conducted in a walk-in growth chamber with controlled lighting (Plant Growing fluorescent Lamp at 4000 lux (120 $\mu mol/m^2/s$) and temperature (23 °C ± 1) (Figure 1A,B).

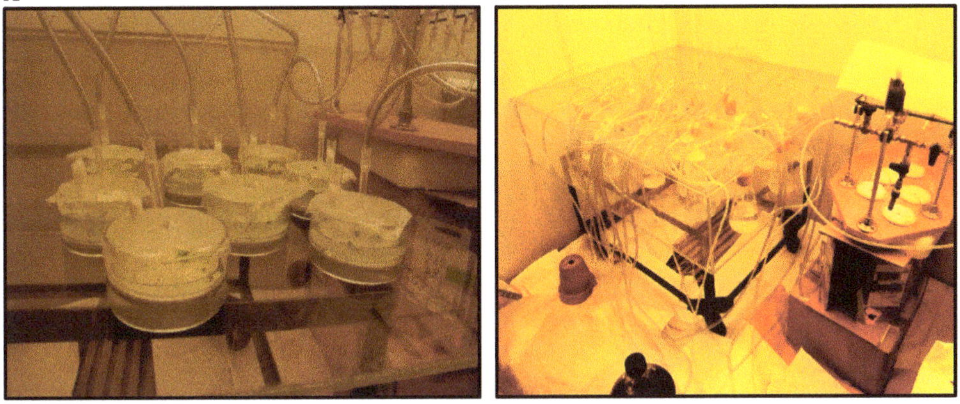

Figure 1. The experimental setup for duckweed in the 300 mL glass bowls (**A**) and microalgae in 250 mL Erlenmeyer flasks (**B**) grown with aeration through plastic discs with 2 mm pores and a flow control manifold (**A**,**B**) in a walk-in chamber. Photo by Kozlova T.A.

2.2. Water Chemistry

The monitoring of general water chemistry and ambient conditions was performed daily: temperature, pH, and conductivity were measured using an Orion pH-meter 420 (ThermoFisher Scientific Ltd., Waltham, WA, US). The composition of dissolved nutrients (phosphate, nitrate, nitrite, and ammonium ions) and their depletion were monitored using a HACH DR-900 colorimeter (HACH Co., London, ON, Canada). Analysis of water ions (Ca, Mg, K, P, and metals) was conducted by inductively coupled plasma mass spectrometry (ICP-MS) at the Manitoba Chemical Analysis Laboratory (MCAL) of the University of Manitoba. Water samples for ion analysis were acidified with 16N HNO_3 (1% acidification).

The experimental framework was completed on three different media. The RFWW was designed taking into account both BBM and natural fish wastewater (FWW) physico-chemical parameters. To eliminate the possible influence of pH on *S. quadricauda* and *L. minor* physiology and steroid performance, the pH of all the media tested was adjusted to 7 with 1 N HCl or 1 N NaOH solutions. All pH-adjusted media were equilibrated for 1 day prior to being used in the tests.

2.3. The Range of E2 Concentrations

The range of E2 concentrations used in these experiments was based on the previously reported steroid loading in hatchery wastewater, surface waters, and groundwater in North America [48,49] and the reported impact of E2 on vascular plants and algae [15,18]. The range of E2 concentrations tested was 0.25, 0.75, 2.50, 5.0, and 15.0 ng/L; the highest tested concentration is approximately 2×10^5 fold lower than EC50 96 h for microalgae (European Chemicals Agency, ECHA (https://echa.europa.eu/information-on-chemicals; accessed on 14 October 2020) and the United States Environmental Protection Agency, EPA US EcoTox Databases (https://cfpub.epa.gov/ecotox/search; accessed on 24 October 2020) and duckweed species [3,50,51]). The same hormone concentrations were tested at the same time on RFWW, FWW, and BBM. Water-soluble estradiol was purchased from Sigma Aldrich Inc., Oakville, ON, Canada (β-Estradiol Watersoluable, Sigma-Aldrich E4389-100mg Lot # SLBF2466V). E2 stock solutions were prepared on distillate water (based on E2 concentration as 5% of chemical powder). A control-blank (no hormone) was assessed in each trial.

2.4. Sampling and Analytical Methods

2.4.1. Algal Cell Counting

Cell density was determined by counting with a Bright-Line haemacytometer (Hausser Scientific Com., Horsham, PA, US) using a Nikon-Eclipse-T*ι* microscope (Nikon Instrument Inc., Melville, NY, US) taking 7 to 10 technical replicates. Cell size was assessed using the NIS-Elements-D3.1 software program of the microscope. The cell's wellness was evaluated by a criterion of the degree of plasmolysis using the same NIS-Elements-D3.1 software program, where critical plasmolysis was at 50% of protoplasm shrank.

As the seeding density in each trial could not be absolutely equal, the proportional rate of increase between each time point was used to assess cell growth, using the following formula:

$$N_a = (N_t - N_o)/t,$$

where, N_a = the calculated cell density, N_o = initial cell density, and N_t = cell density on day t.

2.4.2. Assessment of Biomass Production

To determine the total dry mass weight (drw), a 20 mL aliquot of well-shaken culture was taken at the end of the experiment (day 11) from each flask into the centrifuge tubes. The cells were pelleted using a Thermo Scientific^{TM} Sorvall^{TM} RC 6 Plus centrifuge at 4500 rpm ($2000\times g$) for 15 min. The supernatant was then discharged and the cells were washed with 0.9% NaCl, pelleted again, dewatered, and oven-dried at 60 °C until the weight was stabilized.

2.4.3. Assessment of Chlorophyll-a and Total Carotenoids Concentrations

To determine concentrations of chlorophyll-a and carotenoids, samples were collected according to the sampling schedule, placed on ice, centrifuged, resuspended in 99.8% methanol (Sigma Aldrich Inc.), and stored in the dark for a minimum of 24 h. Aliquots of each extract were transferred into wells of a 96-well plate and measured for optical density using a microplate spectrophotometer (BioTek Synergy 4-Hybrid). Chlorophyll-a and total carotenoids were calculated using the method for methanol extraction described by [52]. Chlorophyll and carotenoid concentrations per cell (ng/cell) were calculated using the following equation:

$$X = C/N \times 1000,$$

where, X = pigment concentration per cell in ng/cell, C = pigment concentration in μg/mL of extract, and N = number of cells per mL.

2.4.4. Duckweed Assessment

Growth parameters of *L. minor* were assessed according to the methods EPS 1/RM/37 [47] and ISO/FDIS/20079 [53] described by Environmental Canada. A minor adjustment was accomplished for root growth evaluation by taking into account the method described by Greenberg et al. [54]. Soluble proteins of dry duckweed biomass were assessed with a Bradford assay.

2.5. Statistical Analyses

To determine the statistical relevance of the data, all trials were conducted using four to five independent biological replicates (n = 4 or 5) with four to fifteen technical replicates which were then analyzed for significant differences ($p < 0.05$) against the control (zero hormone concentration) using the Analysis of Variance test (ANOVA). The significance of measured differences in cell size and biomass was assessed with the Mann–Whitney Rank Sum Test using SigmaStat software (version 3.5, Systat Software, Inc., Palo Alto, CA, US).

3. Results

3.1. Water Chemistry

The FWW from a hatchery fish farm was screened for major inorganic and organic compounds, and it was found that concentrations of key macro- and micronutrients in the wastewater were sufficient for algae and duckweed biomass production (See Tables S1 and S2 in the Supplementary Materials). Concentrations of copper (Cu) were found to be above LOEC, which is about 1 to 3 μg/L for Cu [55,56]. Reconstituted fishery wastewater was used to evaluate of influence of macro- and micronutrients in FWW, eliminating the possible co-influence of unknown dissolved inorganic and/or organic compounds in the water derived from an actual fish hatchery farm.

Over the 11 days of experiments with *S. quadricauda*, the pH of the growth media shifted to alkaline conditions, but the extent of the pH change was different for each of the three media tested (see Figure S1A–C in the Supplementary Materials). In both artificial media (BBM and RFFW), pH increases occurred to a lesser degree compared with FWW, and it appears that the E2 presence did not influence pH changes in these media at the tested E2 concentrations. However, a change in pH was observed in the FWW medium at E2 concentrations of 0.25 to 2.5 ng/L. The pH increase was significantly less than that observed in the control cultures (with no steroid present) and cultures with 5 and 15 ng/L E2 (Figure S1A). Interestingly, the pH of the FWW medium was more alkaline at all concentrations of E2 tested, compared with the equivalent E2 concentrations in the two artificial media (Figure S1A–C). The conductivity of the media did not change significantly over the 11 days of the experiment (data are not shown). However, the conductivity of RFWW was about 1.3-fold higher than the conductivity observed in BBM and 2.2-fold higher than the conductivity observed in FWW.

3.2. Effect of E2 on S. quadricauda and Duckweed (Lemna minor) Growth and Biomass Production

3.2.1. Growth and Biomass Production in *S. quadricauda* under Different E2 Concentrations

The growth of *S. quadricauda* in the presence of E2 varied with the different types of media (Figure 2A–C). All E2 concentrations tested showed significant stimulation of *S. quadricauda* growth in FWW and BBM. The greatest (3.5-fold) effect was observed in the FFW medium at 0.75 ng/L E2 (Figure 2A). Ranking the growth-stimulating concentrations revealed that 0.75 ng/L > 0.25 ng/L > 2.5 ng/L ≈ 5 ng/L in both FWW and BBM mediums. An E2 concentration of 0.75 ng/L significantly induced greater growth in both FWW and BBM compared to higher tested hormone concentrations.

In RFWW, E2 inhibited the growth of *S. quadricauda* at concentrations between 0.25 and 2.5 ng/L not only compared to the control but also compared to the higher concentrations ($p < 0.05$). In addition, a high percentage of unhealthy cells was observed in all cultures containing E2 (Figure 2B). E2 concentrations of 5 and 15 ng/L increased total cell density, although the percent of healthy cells was lower at these concentrations compared to the control. *S. quadricauda* grown in control (no E2 added) in FWW and BBM at the end of the experiment increased cell density to a similar degree (Figure 2A,C). However, in the RFWW trial, the numbers of healthy, unhealthy, and total cell density were significantly greater in the control (2.0–2.4 times) compared to these numbers in the other two media (Figure 2A–C).

Biomass yields of *S. quadricauda* were significantly higher in FFW and BMM at all E2 concentrations tested compared with the control cultures (Table 1). The greatest increase in biomass production (4.5-fold) was observed in the FWW medium at 0.75 ng/L E2. No increase in biomass production was observed in the RFWW medium. Moreover, concentrations from 0.25 to 2.5 ng/L E2 demonstrated adverse effects on biomass yield on day 11 of the experiment. When comparing biomass production in the controls, the highest value was obtained in RFWW. Biomass yields of *L. minor* were positively affected by most tested E2 concentrations in all three media (the only exception was 15 ng/L E2 in BBM where no effect was observed). The greatest increase in biomass production was in the RFWW medium with a maximum increase of 2.4-fold at 0.75 ng/L despite the fact, that the highest biomass yield in control was in the FWW medium (Table 1).

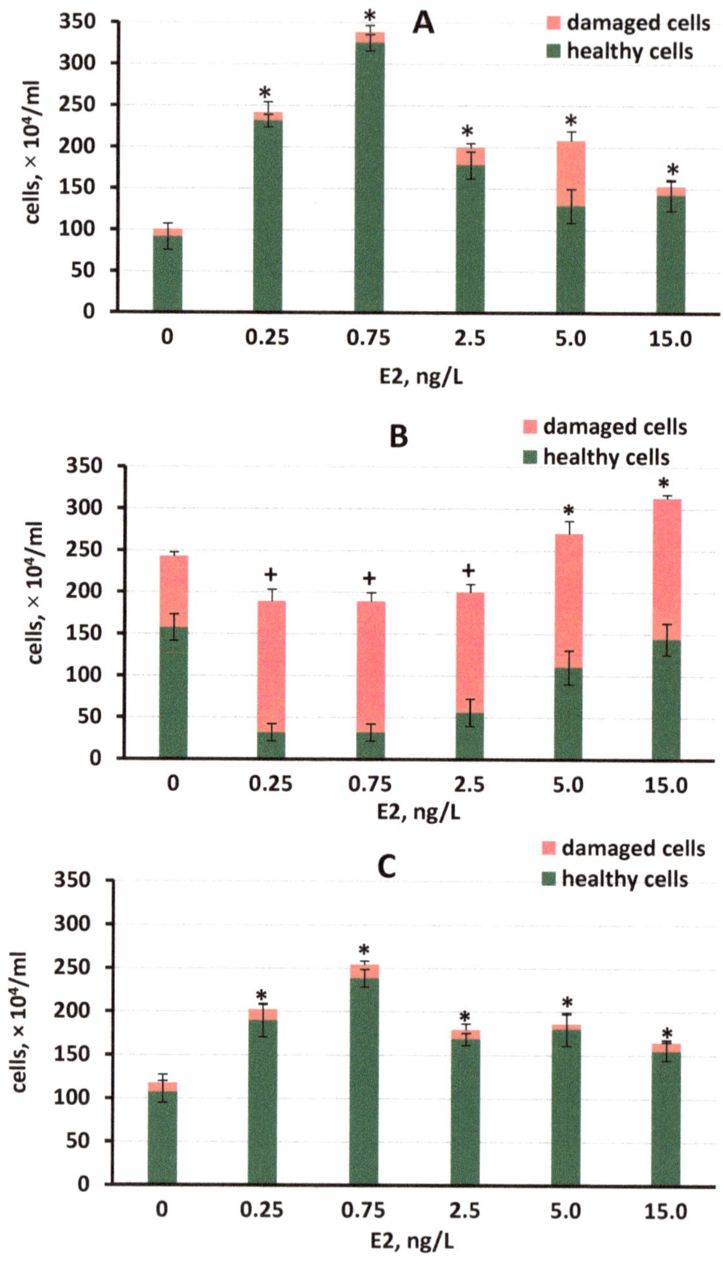

Figure 2. Cell density of *S. quadricauda* cultured in the three media containing E2 concentrations of 0.25 ng/L to 15 ng/L on day 11 of the experiment: (**A**) FFW; (**B**) RFWW; and (**C**) BBM. Statistically significant differences from the controls ($p < 0.05$) are shown for the total cell density, where *—significantly greater than the control; +—significantly lower than the control. Data are shown as the mean ± SD, n = 15. Control cultures did not contain E2, indicated by "0".

Table 1. Changes in growth parameters of *Scenedesmus quadricauda* and *Lemna minor* when cultured in the three growth media, on day 11 of the experiment (the day on which the experiment was terminated).

	S. quadricauda			*L. minor*
E2 Concentration, (ng/L)	**Cell Size (µm)**	**Biomass (g drw/L)**	**Cell Density ($\times 10^4$/mL)**	**Biomass (g drw/L)**
		FWW trial		
Control	44.77 ± 3.26	0.40 ± 0.07	100.07 ± 5.23	3.40 ± 0.05
0.25	46.65 ± 4.73 ^{a}	1.15 ± 0.11 ^{*a}	241.43 ± 3.87 ^{*a}	5.05 ± 0.09 ^{*a}
0.75	46.87 ± 2.31 ^{a}	1.79 ± 0.05 ^{*b}	338.34 ± 3.18 ^{*b}	4.94 ± 0.06 ^{*a}
2.5	42.05 ± 1.45 ^{a}	0.88 ± 0.10 ^{*c}	199.61 ± 4.93 ^{*c}	4.86 ± 0.08 ^{*a}
5	37.23 ± 1.85 ^{b}	0.86 ± 0.14 ^{*c}	208.12 ± 8.89 ^{*d}	4.89 ± 0.09 ^{*a}
15	31.65 ± 2.22 $^{+c}$	0.78 ± 0.08 ^{*c}	152.63 ± 5.47 ^{*e}	5.59 ± 0.08 ^{*b}
		RFWW trial		
Control	43.77 ± 4.25	0.69 ± 0.09	242.67 ± 4.45	2.49 ± 0.07
0.25	47.14 ± 3.43 ^{a}	0.44 ± 0.05 $^{+a}$	188.81 ± 3.22 $^{+a}$	5.45 ± 0.07 ^{*a}
0.75	48.25 ± 2.62 ^{a}	0.46 ± 0.04 $^{+a}$	191.52 ± 3.42 $^{+a}$	5.98 ± 0.08 ^{*b}
2.5	39.84 ± 5.66 ^{a}	0.41 ± 0.07 $^{+a}$	200.21 ± 5.25 $^{+ab}$	4.71 ± 0.09 ^{*c}
5	35.92 ± 4.32 ^{ab}	0.72 ± 0.11 ^{b}	270.42 ± 8.61 ^{*c}	4.86 ± 0.07 ^{*c}
15	26.52 ± 3.65 $^{+c}$	0.76 ± 0.06 ^{b}	312.71 ± 6.05 ^{*d}	4.66 ± 0.04 ^{*c}
		BBM trial		
Control	44.76 ± 0.92	0.44 ± 0.11	128.51 ± 10.66	2.68 ± 0.07
0.25	52.67 ± 1.01 ^{*a}	0.85 ± 0.09 ^{*a}	202.12 ± 12.54 ^{*a}	3.35 ± 0.08 ^{*a}
0.75	56.72 ± 1.25 ^{*b}	1.24 ± 0.04 ^{*b}	254.24 ± 8.42 ^{*b}	4.14 ± 0.06 ^{*b}
2.5	46.74 ± 0.63 ^{*c}	0.79 ± 0.08 ^{*a}	179.13 ± 5.57 ^{*c}	3.89 ± 0.09 ^{*b}
5	46.84 ± 1.05 ^{c}	0.81 ± 0.08 ^{*a}	186.02 ± 14.05 ^{*ac}	3.84 ± 0.06 ^{*bc}
15	42.95 ± 1.48 ^{d}	0.75± 0.06 ^{*a}	164.64 ± 7.33 ^{*ac}	2.95± 0.09 ^{d}

*—significantly greater than control ($p < 0.05$); +—significantly lower than control ($p < 0.05$). Within a trial, and for each column, values labeled with the same letter are not significantly different from each other ($p > 0.05$) and values labeled with two letters indicate a significant difference from other letters, but no statistical difference from other equal letters. Data are shown as the mean ± SD (for cell size n = 15; for biomass n = 4; for cell density n = 12).

S. quadricauda cell size was affected differently by E2 in all three media. In FWW and RFWW, cell size was not affected, except at the highest tested E2 concentration of 15 ng/L, at which an adverse effect was observed (Table 1). In BBM medium, concentrations of 0.25 to 2.5 ng/L E2 resulted in significantly increased cell size, with no effect being observed at 5 and 15 ng/L. Higher diversity in cell size was noted in the RFWW medium at all E2 concentrations tested (Table 1), although the cell size in the controls prepared on FWW and RFWW media was diverse too. No change in cell size was observed in the control cultures of all three media during the 11 days of the experiment. The ability of E2 to induce growth and biosynthesis of both tested organisms depends on the tested concentration and the parameter, sometimes showing a greater effect at a smaller concentration. (Table 1).

3.2.2. *L. minor* Growth under Different E2 Concentrations in Three Media

The influence of E2 on *L. minor* growth was assessed by measuring the development of the roots and fronds. All E2 concentrations tested significantly stimulated the growth of *L. minor* fronds, in all three media, except in BBM at 0.25 and 15 ng/L E2 (Figure 3A), with the greatest effect observed in RFWW at 0.25 and 0.75 ng/L E2. The influence of E2 on frond development was comparable in FWW and BBM media, while the stimulation of *L. minor* fronds was greatest in the RFWW medium (Figure 3A). Stimulation of *L. minor* root development was observed in FWW at 0.25 and 15 ng/L E2 and in RFWW at 0.75 and 5 ng/L E2. However, no significant influence on root development was observed at any E2 concentration in the BBM trial (Figure 3B). The variation in root length was significant in all three media, although the SD was lower in FFW compared to the other media.

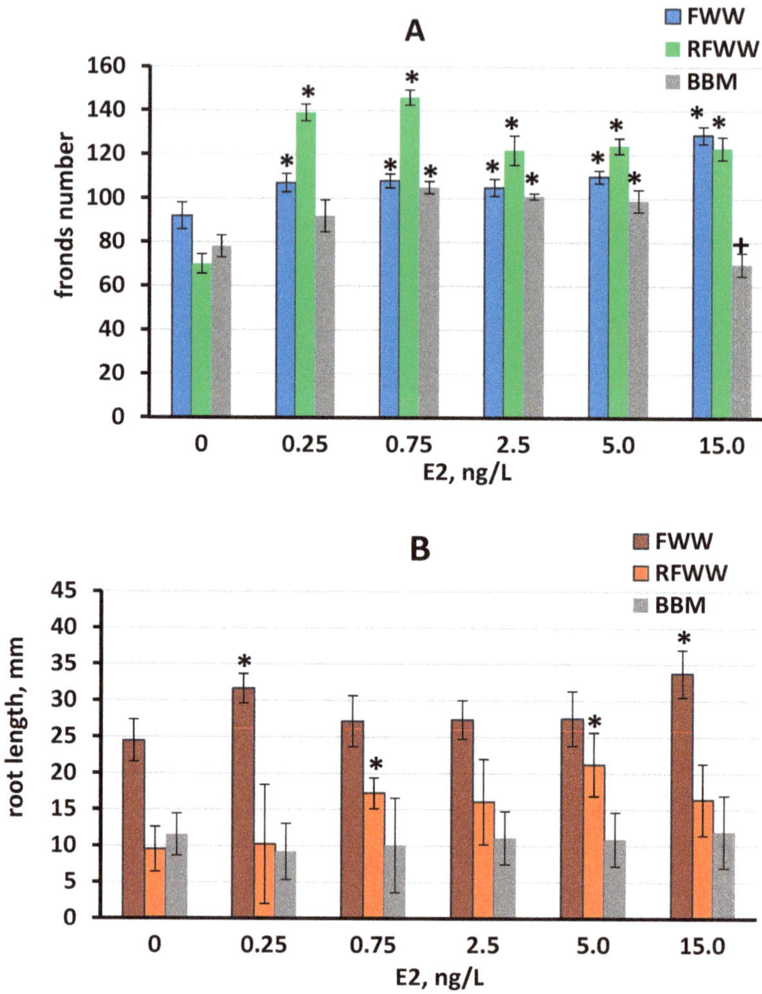

Figure 3. Growth of *L. minor* cultured in the three media containing E2 concentrations of 0.25 ng/L to 15 ng/L, on day 11 of the experiment: (**A**) foliage development in FFW, RFWW, and BBM media; (**B**) root length in FFW, RFWW, and BBM media. Statistically significant differences from the controls ($p < 0.05$) are indicated, where *—significantly greater than the control; +—significantly lower than the control. Data are shown as the mean ± SD, n = 15. Control cultures did not contain E2, indicated by "0".

3.3. Effect of E2 on Chlorophyll-a and Total Carotenoid Production by S. quadricauda Cells

Significant increases in chlorophyll-a and total carotenoids were observed at all E2 concentrations tested in FWW and BBM (Figure 4A,C). In contrast, no increase in either pigment was observed in RFWW (Figure 4B). Concentrations of both chlorophyll-a and total carotenoids increased in *S. quadricauda* cells to higher levels than in untreated control cells in the FWW medium. A maximum chlorophyll-a concentration of 174.1 ng/cell was obtained in FWW at 0.75 ng/L E2, although at E2 concentrations of 0.25 to 5 ng/L, chlorophyll-a concentrations increased by 1.8- to 2.1-fold compared to the untreated control cultures.

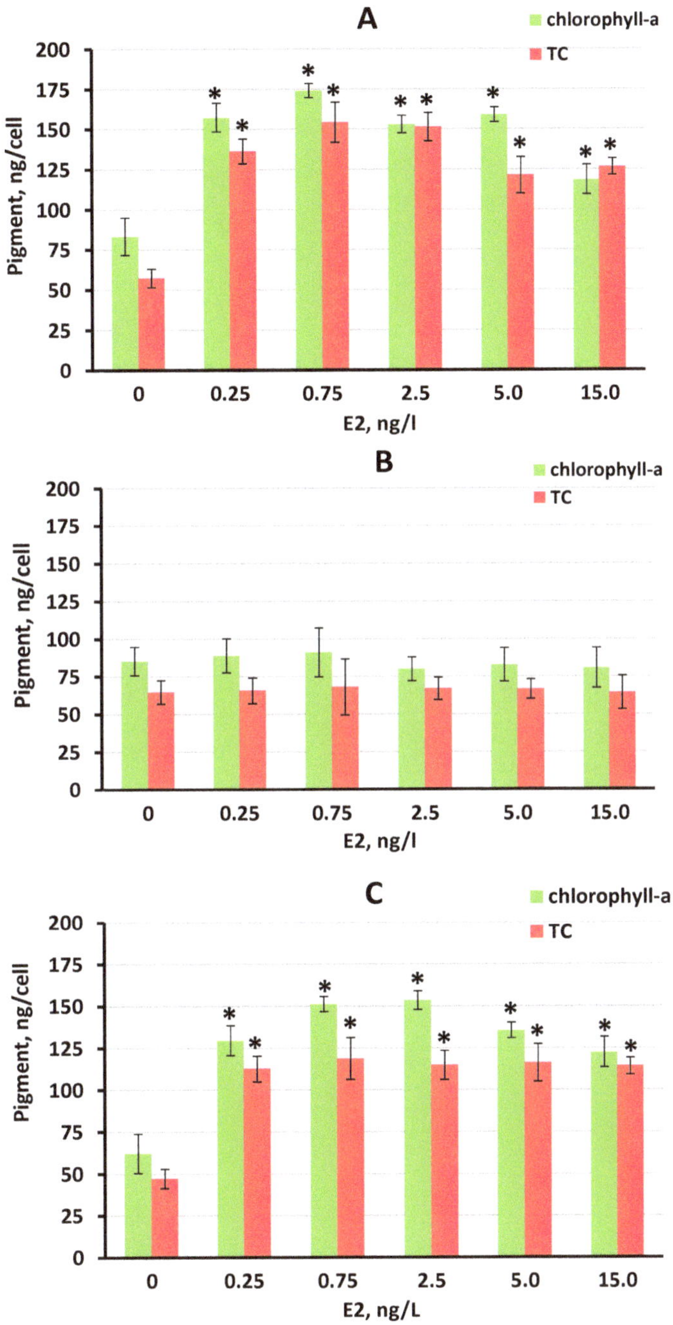

Figure 4. Chlorophyll-a and total carotenoid (TC) content in *S. quadricauda* cells cultured in the three media containing E2 concentrations of 0.25 ng/L to 15 ng/L on day 11 of the experiment: (**A**) FWW; (**B**) RFWW; and (**C**) BBM. Statistically significant differences from the controls ($p < 0.05$) are shown for total cell density, where *—significantly greater than the control. Data are shown as the mean ± SD, n = 6. Control cultures did not contain E2, indicated by "0".

In this trial, the maximum concentration of total carotenoids was nearly equal at 0.75 and 2.5 ng/L E2 (154.2 and 151.7, respectively), with an increase of 2.7-fold over the concentrations of total carotenoids observed in the untreated control cultures (Figure 4A). Ranking the efficacy of stimulation of chlorophyll-a by E2 in FWW, we determined that: 0.75 ng/L > 0.25 ng/L ≥ 5 ng/L ≥ 2.5 ng/L > 15 ng/L. Ranking the efficacy of stimulation of total carotenoids by E2 in FWW, we determined that: 0.75 ng/L ≈ 2.5 ng/L > 0.25 ng/L ≥ 5 ng/L ≥ 15 ng/L.

In BBM medium, maximum values of 153.4 and 151.2 ng/cell of chlorophyll-a were observed at 2.5 and 0.75 ng/L E2, respectively (Figure 4C). The accumulation of chlorophyll-a was stimulated to a lesser extent at 0.25, 5, and 15 ng/L E2. Total carotenoid accumulation in *S. quadricauda* cells was stimulated 2.4- to 2.5-fold at all E2 concentrations tested in BBM medium (Figure 4C). Ranking the efficacy of stimulation of chlorophyll-a in BBM medium, we found that 2.5 ng/L ≈ 0.75 ng/L > 5 ng/L ≥ 2.5 ng/L ≥ 15 ng/L. Ranking the efficacy of stimulation of total carotenoids in BBM medium, we found that 0.75 ng/L ≥ 2.5 g/L ≈ 5 ng/L ≈ 15 ng/L ≥ 0.25 ng/L. Between the tested E2 concentrations, the stimulatory effect was not significantly different, except for the highest tested E2 concentration (15 ng/L) in FWW where both pigments were reduced compared to other E2 concentrations.

3.4. Soluble Protein Production by L. minor and S. quadricauda

Accumulation of water-soluble protein in *L. minor* and *S. quadricauda* biomass was different when the test organisms were grown in three media containing the range concentrations of E2, compared to untreated control cultures (Table 2). In the untreated control cultures, both *L. minor* and *S. quadricauda* cultured in the FWW medium produced the highest concentration of soluble proteins (36.12 and 30.11 mg/g frw, respectively), compared with the untreated controls cultured in RFWW and BBM. In the presence of E2, elevated protein concentrations (compared with the untreated control cultures) were observed in *L. minor* and *S. quadricauda* cultured in FWW at concentrations of E2 from 0.25 to 15 ng/L, although the greatest stimulation of protein accumulation in both cultures was at 2.5 ng/L. In RFWW, elevated protein concentrations were observed in cultures containing 0.25, 0.75, and 2.5 ng/L E2 for *L. minor* and in *S. quadricauda* cells at any tested concentration (compared with the untreated control cultures). In BBM medium, protein accumulation was different for the two tested cultures. For *L. minor* tissue, elevated concentrations of protein compared with the untreated control cultures were only detected in a medium containing 15 ng/L. *S. quadricauda* cells reacted stronger in the presence of E2. The concentrations from 0.75 to 15 ng/L E2 were effective in elevating the algae protein content (Table 2). Overall, E2 was more effective in the induction of soluble protein accumulation in the algae compared to duckweed. Between the tested E2 concentrations, the greatest effect (2.0 fold induction) demonstrated 15 ng/L E2 in BBM in *S. quadricauda*, while the weakest difference in induction was observed in *L. minor* grown in BBM (Table 2).

Table 2. Protein (mg/g frw) accumulation in *L. minor* grown in three different media on the day of the test termination (day 11).

E2, ng/L	Control	0.25	0.75	2.5	5	15
			Lemna minor			
FWW	36.12 ± 1.06	34.77 ± 0.76 [a]	38.95 ± 0.48 *[b]	40.41 ± 0.57 *[b]	38.92 ± 0.91 *[b]	39.68 ± 0.66 *[b]
RFWW	31.73 ± 0.72	40.74 ± 1.21 *[a]	41.12 ± 0.43 *[a]	36.02 ± 0.79 *[b]	32.82 ± 1.04 [c]	32.51 ± 0.72 [c]
BBM	24.75 ± 0.66	25.34 ± 0.26 [a]	25.28 ± 1.01 [a]	26.25 ± 0.76 [a]	25.78 ± 0.86 [a]	27.78 ± 0.58 *[b]
			Scenedesmus quadricauda			
FWW	30.11 ± 0.56	37.22 ± 1.03 *[a]	43.35 ± 0.19 *[b]	46.76 ± 0.63 *[c]	44.52 ± 1.52 *[bc]	40.14 ± 1.06 *[d]
RFWW	25.88 ± 0.69	29.94 ± 0.21 *[a]	36.02 ± 0.48 *[b]	39.93 ± 0.55 *[c]	43.76 ± 1.04 *[d]	29.57 ± 1.48 *[a]
BBM	20.38 ± 0.21	21.84 ± 0.75 [a]	31.08 ± 0.61 *[b]	36.25 ± 0.65 *[c]	38.78 ± 0.34 *[d]	41.88 ± 0.71 *[e]

*—significantly greater than control ($p < 0.05$). Within a trial, and for each row, values labeled with the same letter are not significantly different from each other ($p > 0.05$), and values labeled with two letters indicate significant difference from other letters, but no statistical difference from other equal letters. Data are shown as the mean ± SD, n = 6.

4. Discussion

The range of E2 concentrations examined in this experiment was consistent with the concentrations typically found in Canadian surface and ground waters, and fish-farming wastewaters [57–59]. Moreover, a reduced range of steroid concentrations was chosen for investigation, considering two additional important points: (a) toxicity of the hormones to plants and algae; and (b) in an attempt to avoid the possibility of hormone accumulation in the algal cells. Toxicity of E2 to microalgae is relatively low, normally more than 3.2 mg/L 96 h EC50 and [5], but is from 0.2 mg/L in some reports (ECOSAR v 2.2) depending on the species and water chemistry. NOEC of E2 is in the range of 10 to 100 μg/L with the lower toxicity of natural estradiol (EPA US EcoTox Database).

The chemical compound E2 that we used in this study contains cyclodextrin (β-CD) as a concomitant that makes E2 water-soluble. This may raise the question of the co-influence of β-CD on the effect of E2 or the bioavailability of steroids. However, according to the information on β-CD provided in the Material Safety Data Sheet (MSDS, Sigma-Aldrich) and ECHA and the EPA US EcoTox Databases, EC50 and even NOEC, for β-CD, demonstrated very low toxicity or no impact of the compound on the physiology of microalgae species and other aquatic organisms (LC50 > 100 mg/L and up to 7500 mg/L; NOEC 30–40 mg/L). Thus, the highest expected β-CD concentration in our experiment is 285 ng/L, which is about 10^5 lower than NOEC concentrations for green algae species. One would not expect any co-influence of β-CD in such concentrations, especially in conditions of intensive stirring during the preparation of the stock solutions and subsequent aeration of the test media. The matter of solvent concentration was clearly demonstrated in our previous study on phytohormones' influence on microalgae physiology [60]. The influence of dissolved compounds on each other as well as on a test organism is highly dependent on the concentration of each compound in the solution and lesser on the ratio of two concomitant substances (in our case E2 and β-CD) [50,61]. Numerous studies, which used β-Estradiol water-soluble (Sigma-Aldrich E4389) did not conduct additional tests on the influence of β-CD, when much higher E2 concentrations, compared to our study (the difference is about 1×10^3 7×10^5 fold), were examined [62–65]. However, an additional test on β-CD influence was not conducted in these studies, likely due to the significantly lower concentration used, compared to EC50 48h β-CD (ECHA and EPA US EcoTox Databases).

E2 is one of the frequently detected hormones in fishery wastewaters, which is why it may be used as a beneficial component of growth media in combined fish-algae-duckweed integrated production systems. In our previous study [38], E2 demonstrated a powerful ability to simultaneously induce the growth and biosynthesis of valuable bio-molecules in *S. quadricauda* cells. The current study was conducted to clarify whether the influence of

E2 on the algae could be affected by physicochemical differences in three different growth media, including RFWW, which mimics the concentrations of macro- and micronutrients in real fish wastewaters (FWW). This mimicry was made for the possible replacement of FWW for further biomass production in pilot-scale facilities. Moreover, this study compares the effect of E2 on *S. quadricauda* physiology, with the steroid influence on *L. minor*, when these two aquatic organisms were grown under the same conditions.

Although ambient conditions in our previous study had some dissimilarities from this experimental work (the light intensity was 20% higher and constant aeration of the growth mediums was not provided during the first experiment), the results of E2 influence on *S. quadricauda* growth parameters and biosynthesis of pigments were similar, comparing the values on days 10 and 11 of the experiments conducted in BBM medium.

However, clear dissimilarities in the algae population growth and biosynthetic activity were observed in three media tested. Remarkably, the dissimilarities in algae performance were found between the controls of three trials as well as between the two test cultures containing equal concentrations of E2. Furthermore, the pattern of dissimilarity between the controls varied with the type of medium used. For instance, cell density was about two times higher in untreated control cultures in RFWW media compared to the control cultures with BBM and FWW media. However, when E2 was present in the cultures, the effect was the opposite: in RFWW, cell density and biomass yield of *S. quadricauda* were negatively affected by the three lowest E2 concentrations (0.25, 0.75, and 2.5 ng/L), while in FWW and BBM, these concentrations stimulated an increase in cell density and biomass yield.

The RFWW medium was prepared according to the results of chemical analyses of actual FWW from two trout farms. The macro- and micronutrients of FWW were incorporated into the design. The main metals, which could cause toxicity of FWW, were also analyzed and were found to be under LOEC for each metal [66,67], except Cu, which was slightly higher (Table S2). Why E2 negatively affected cell growth in the RFWW and why RFWW compromised cell health needs further investigation. A possible reason may be related to the buffering capacity of the medium, where sodium and potassium carbonates were used to stabilize the pH of RFWW. This buffering strategy decreased the shifting of the pH to alkaline conditions and increased the dissolved carbon value, which could directly affect the binding capacity of E2, and/or membrane permeability [3]. It is also possible that the period of time for cells' acclimatization to the RFWW media was insufficient for the algae to stabilize its physiology prior to the test. When considering the effect of water chemistry on chlorophyll-a and carotenoids, it may be that the stimulatory effect of E2 observed in the FWW and BBM media was abolished in the RFWW medium.

In contrast to RFWW, the positive effect of E2 on *S. quadricauda* growth and biosynthesis in natural FWW was the most pronounced. The effect of water chemistry on the promoting efficiency of an induction factor (such as E2) has been highlighted previously by research conducted on aquatic plants [68–70] and microalgal species [71–73]. It was demonstrated that a change in water chemistry could activate more than one defense pathway at the same time in algae cells. The combination of these pathways and dominance of one of them depends on the factor altered, whether it is a media chemistry parameter or an additional stressor, such as a hormone, or both [71,74].

Remarkably, the influence of RFWW on *L. minor* growth was opposite to its influence on *S. quadricauda* growth. The growth of foliage, root development, and protein accumulation in the duckweed was stimulated by greater E2 in the RFWW medium, compared with the effect of E2 in the other two media, although FWW was a significantly better medium for *L. minor* growth when E2 was not present. In addition, it is not clear if the greater effect of E2 on protein content in RFWW was due to the altered water chemistry. Soluble protein concentrations in *S. quadricauda* cells in all the tested mediums were significantly lower than in *L. minor* tissue. However, E2 was more effective in the induction of soluble protein accumulation in the algae compared to duckweed. Coupled with the fact that induction of cell growth and population density by E2 was stronger for *S. quadricauda* compared to the effect of E2 on *L. minor* foliage growth (maximum biomass increase as 4.5-fold and 2.4-fold induction, respectively), we can conclude that E2 is a more effective inductor for *S. quadricauda* than for *L. minor* at the E2 concentrations and experimental conditions tested.

Our results on the influence of E2 on the soluble protein accumulation by *L. minor* are consistent with research conducted on another duckweed species, *Wolffia arrhizal* [31]. However, we observed the maximum effect at much lower E2 concentrations in both RFWW (0.25 and 0.75 ng/L E2) and FWW (between 0.75 and 15 ng/L E2) media, compared to the effect on *W. arrhizal*, where the concentrations of 272, 2720, and 27.2 μg/L E2 (in that order) were most stimulatory. In the study by Szamrej and Czerpak [30,31], soluble proteins were induced to a greater degree (1.8-fold increase) compared to maximum induction in our experiment (1.3-fold increase). The species-specific differences (e.g., *W. arrhizal* does not have roots) and growth media chemistry are likely responsible for the difference in results.

Lemnaceae species are broadly used for wastewater treatment, including domestic and pharmaceutical effluences, which are the prime resources of animal steroids in surface water. Some previous research indicates that duckweed species are one of the best agents for removing hormones from water [3,75,76]. The low E2 waterborne concentrations were likely inactivated and biotransformed by *L. minor* shortly after the exposure. E2 and other animal steroids have also been shown to induce biosynthesis of valuable molecules in Lemnaceae plants, when applied in relatively low concentrations in the range of 10^{-5} to 10^{-7} M. For example, corticosteroids significantly stimulated the accumulation of pigments (chlorophylls and carotenoids), nucleic acids, soluble proteins, and sugar biosynthesis in *W. arrhizal* tissues, although the extent of stimulation was less than the effect of E2 [30,31]. To the best of our knowledge, no study has measured duckweed growth performance in the presence of known concentrations of waterborne steroid hormones. The effect of *L. minor* on buffering and stabilizing the pH of the growth media has also not previously been reported. However, efforts were made for the development of a mathematical model for a better understanding of the influence of growth medium alteration on Lemnaceae performance in vitro and in situ [77–79].

Our results on the influence of E2 on *S. quadricauda* growth and biomass production (cell size and density) and the accumulation of pigments (chlorophyll-a and total carotenoids) in BBM medium over 11 days were similar to that obtained in the previous 20-day experiment [38]. However, the stimulatory effects in this study were more pronounced than the effects observed on days 10 to 15 in the previous experiment. The better results are likely due to the constant aeration condition applied in this (11-day) experiment.

5. Conclusions

We have compared the influence of an animal steroid, E2 on the growth and physiology of two primary producers: the microalga *S. quadricauda* and the aquatic plant *L. minor*. The results of this study may be organized into two groups. First, it was confirmed that E2 in nano-concentrations has a strong positive effect on the cell growth (cell size and final cell density) and biosynthesis of valuable molecules of *S. quadricauda*, as was previously demonstrated in a 20-day non-aerated experiment. Moreover, we have demonstrated that E2 is capable of increasing foliage growth, root growth, and protein biosynthesis in *L. minor*, although root development was stimulated by the presence of E2 to a lesser extent. Importantly, E2 exhibited a stronger stimulatory effect on *S. quadricauda* compared to *L. minor* when population growth and protein content were assessed. Second, we have compared the influence of three growth media (FWW, RFWW, and BBM) on the two test organisms, both in the absence and presence of E2. The results of this comparison clearly show the importance of the physicochemical parameters of the growth medium on the physiology and metabolism of aquatic organisms, as well as on the extent to which E2 affected the physiology of *S. quadricauda* and *L. minor*. Our results highlight the need for further investigations of the relationships between the physicochemical parameters of growth media, the particular organism to be cultured, and any additional stressor or chemical inducer that may be applied to stimulate biomass production and biosynthesis of the molecules of interest.

Supplementary Materials: The following supporting information can be downloaded at: https://www.mdpi.com/article/10.3390/plants11131669/s1. Table S1: Macronutrients and buffering capacity of standard synthetic Bold's Basal Medium (BBM), 30% Hoagland's medium (30% HG), natural fishery wastewater (FWW), and reconstituted fishery wastewater (RFWW) used in the experiment, ppm; Table S2: Micronutrients in standard synthetic Bold's Basal Medium (BBM), 30% Hoagland's medium (30% HG), natural fishery wastewater (FWW), and reconstituted fishery wastewater (RFWW) used in the experiment, ppm.; Figure S1: pH of three culture media on day 11 of the experiment. Micronutrients in standard synthetic Bold's Basal Medium (BBM), 30% Hoagland's medium (30% HG), natural fishery wastewater (FWW), and reconstituted fishery wastewater (RFWW) used in the experiment, ppm.

Author Contributions: T.A.K.—experimental design, conduction of the experiment, analytical methods and data analyses, writing—original draft preparation. D.B.L.—writing—review and editing, provided reagents, facilities and analytical tools, supervision, and project administration. All authors have read and agreed to the published version of the manuscript.

Funding: This work was supported by the Natural Sciences and Engineering Research Council of Canada (NSERC) through an NSERC Discovery grant (RGPIN-04945-2017).

Institutional Review Board Statement: Not applicable.

Informed Consent Statement: Not applicable.

Data Availability Statement: Data are contained within the article and Supplementary Materials. Raw data were generated at the Department of Biosystems Engineering, University of Manitoba and are available from the corresponding author (initials) upon request.

Acknowledgments: The authors would like to extend sincere thanks to all the people from Levin's laboratory for their support and priceless collaboration. The authors would like to extend sincere thanks to Claudio Stasolla and the management of the greenhouse (the Department of Plant Science, University of Manitoba) for their continuous help, useful suggestions in the experimental setup, and for their hospitality and patience in providing facilities for the experimental work. The authors greatly thank the Natural Sciences and Engineering Research Council of Canada (NSERC) for its invaluable support of this research.

Conflicts of Interest: The authors have no conflict of interest to disclose. No conflict, informed consent, or human or animal rights are applicable.

References

1. Smith, D.S.; Kramer, J.R.; Bell, R.A. Metal speciation with organic matter. *Comp. Biochem. Physiol.* **2002**, *133*, 65–74.
2. Richards, J.C.; Curtis, P.J.; Bumison, B.K.; Playle, R. Effect of natural organic matter sources on reducing metal toxicity to rainbow trout (*Oncorhynchus mykiss*) and on metal binding to their gills. *Environ. Toxicol. Chem.* **2001**, *20*, 1159–1166. [CrossRef] [PubMed]
3. Shi, W.; Wang, L.; Rousseau, D.P.L.; Lens, P.N.L. Removal of estrone, 17α-ethinylestradiol, and 17ß-estradiol in algae and Duckweed-based wastewater treatment systems. *Environ. Sci. Pollut. Res.* **2010**, *17*, 824–833. [CrossRef] [PubMed]
4. Xin, L.; Hu, H.Y.; Gan, K.; Sun, Y.X. Effects of different nitrogen and phosphorus concentrations on the growth, nutrient uptake, and lipid accumulation of a freshwater microalga *Scenedesmus* sp. *Bioresour. Technol.* **2010**, *101*, 5494. [CrossRef] [PubMed]
5. Liu, W.; Chen, Q.; He, N.; Sun, K.; Sun, D.; Wu, X.; Duan, S. Removal and biodegradation of 17β-Estradiol and diethylstilbestrol by the freshwater microalgae *Raphidocelis subcapitata*. *Int. J. Environ. Res. Public Health* **2018**, *15*, 452. [CrossRef] [PubMed]
6. Pojana, G.; Gomiero, A.; Jonkers, N.; Marcomini, A. Natural and synthetic endocrine disrupting compounds (EDCs) in water, sediment, and biota of a coastal lagoon. *Environ. Int.* **2007**, *33*, 929–936. [CrossRef]
7. Zhou, W.; Li, Y.; Min, M.; Hu, B.; Chen, P.; Ruan, R. Local bioprospecting for high-lipid producing microalgal strains to be grown on concentrated municipal wastewater for biofuel production. *Bioresour. Technol.* **2011**, *102*, 6909–6919. [CrossRef]
8. Zhang, X.; Gao, Y.; Li, Q.; Li, G.; Guo, Q.; Yan, C. Estrogenic compounds and estrogenicity in surface water, sediments, and organisms from Yundang lagoon in Xiamen, China. *Arch. Environ. Contam. Toxicol.* **2011**, *61*, 93–100. [CrossRef]
9. Esteban, S.; Gorga, M.; Petrovic, M.; Gonzalez-Alonso, S.; Barcelo, D.; Valcarcel, Y. Analysis and occurrence of endocrine-disrupting compounds and estrogenic activity in the surface waters of central Spain. *Sci. Total Environ.* **2014**, *466–467*, 939–951. [CrossRef]

10. Barry, T.P.; Riebe, J.D.; Parrish, J.J.; Malison, J.A. Effects of 17α, 20β-Dihydroxy-4-pregnen-3-one on cortisol production by Rainbow Trout interrenal tissue in vitro. *Gen. Comp. Endocrinol.* **1997**, *107*, 172–181. [CrossRef]
11. Kidd, C.E.; Kidd, M.R.; Hofmann, H.A. Measuring multiple hormones from a single water sample using enzyme immunoassays. *Gen. Comp. Endocrinol.* **2010**, *165*, 277–285. [CrossRef]
12. Hala, D.; Petersen, L.H.; Martinovic, D.; Huggett, D.B. Constraints-based stoichiometric analysis of hypoxic stress on steroidogenesis in fathead minnows, *Pimephales promelas*. *J. Exp. Biol.* **2012**, *215*, 1753–1765. [CrossRef] [PubMed]
13. Scott, A.P.; Ellis, T. Measurement of fish steroids in water—A review. Minireview. *Gen. Comp. Endocrinol.* **2007**, *153*, 392–400. [CrossRef] [PubMed]
14. Norvill, Z.N.; Toledocervantes, A.; Blanco, S.; Shilton, A.; Guieysse, B.; Muñoz, R. Photodegradation and sorption govern tetracycline removal during wastewater treatment in algal ponds. *Bioresour. Technol.* **2017**, *232*, 35–43. [CrossRef] [PubMed]
15. Lai, K.M.; Scrimshaw, M.D.; Lester, J.N. Prediction of the bioaccumulation factors and body burden of natural and synthetic estrogens in aquatic organisms in the river systems. *Sci. Total Environ.* **2002**, *289*, 159–168. [CrossRef]
16. Wang, P.; Wong, Y.S.; Tam, F.Y. Green microalgae in removal and biotransformation of estradiol and ethinylestradiol. *J. Appl. Phycol.* **2017**, *29*, 263–273. [CrossRef]
17. Hom-Diaz, A.; Llorca, M.; Rodríguez-Mozaz, S.; Vicent, T.; Barceló, D.; Blánquez, P. Microalgae cultivation on wastewater digestate: β-estradiol and 17α-ethynylestradiol degradation and transformation products identification. *J. Environ. Manag.* **2015**, *155*, 106–113. [CrossRef]
18. Beecher, L. Assessment of 17β-Estradiol Removal from Wastewater via Abiotic and Biotic Routes and Potential Effects on Food Chain Pathways. Ph.D. Thesis, Clemson University, Clemson, SC, USA, 2013; 152p.
19. Zhang, J.S.; Yang, Z.H.; Tsao, T.H. The occurrence of estrogens in relation to reproductive processes in flowering plants. *Sex. Plant Reprod.* **1991**, *4*, 193–196. [CrossRef]
20. Janeczko, A.; Skoczowski, A. Mammalian sex hormones in plants. *Folia Histochemica et cytobiologica* **2005**, *43*, 71–79.
21. Itagaki, E.; Iwaya, T. Purification and characterization of 17β-hydroxysteroid dehydrogenase from *Cylindrocarpon Radicicola*. *J. Biochem.* **1988**, *103*, 1039–1044. [CrossRef]
22. Rosati, F.; Danza, G.; Guarna, A.; Cini, N.; Racchi, M.L.; Serio, M. New evidence of similarity between human and plant steroid metabolism: 5α-reductase activity in *Solanum malacoxylon*. *Endocrinology* **2003**, *144*, 220–229. [CrossRef] [PubMed]
23. Milanesi, L.; Monje, P.; Boland, R. Presence of estrogens and estrogen receptor-like proteins in *Solanum glaucophyllum*. *Biochem. Biophys. Res. Commun.* **2001**, *289*, 1175–1179. [CrossRef] [PubMed]
24. Shore, L.S.; Kapulnik, Y.; Ben-Dor, B.; Fridman, Y.; Wininger, S.; Shemesh, M. Effects of estrone and 17β-estradiol on vegetative growth of *Medicago sativa*. *Physiol. Plant.* **1992**, *84*, 217–222. [CrossRef]
25. Janeczko, A.; Filek, W. Stimulation of generative development in partly vernalized winter wheat by animal sex hormones. *Acta Physiol. Plant.* **2002**, *24*, 291–295. [CrossRef]
26. Kopcewicz, J. Influence of estrogens on the auxins content in plants. *Naturwissenschaffen* **1970**, *57*, 48. [CrossRef] [PubMed]
27. Kopcewicz, J.; Porazinski, Z. Effect of growth regulators, steroids and estrogen fraction from sage plants on flowering of a long-day plant, *Salvia splendens*, grown under non-inductive light conditions. *Biol. Plant.* **1974**, *16*, 132–135. [CrossRef]
28. Geuns, J.M.C. Steroid hormones and plant growth and development. *Phytochemistry* **1978**, *17*, 1–14. [CrossRef]
29. Zhong-Han, Y.; Yin, T.; Zong-Xun, C.; Tsao, T.H. The changes of steroidal sex hormone–testosterone contents in reproductive organs of *Lilium davidii* Duch. *Acta Bot. Sin.* **1994**, *36*, 215–220.
30. Czerpak, R.; Szamrej, I.K. The Effect of β-estradiol and corticosteroids on chlorophylls and carotenoids content in *Wolffia arrhiza* (L.) Wimm. (Lemnaceae) Growing in Municipal Bialystok Tap Water. *Pol. J. Environ. Stud.* **2003**, *12*, 677–684.
31. Szamrej, I.K.; Czerpak, R. The effect of sex steroids and corticosteroids on the content of soluble proteins, nucleic acids and reducing sugars in *Wolffia arrhiza* (L.) Wimm. (Lemnaceae). *Pol. J. Environ. Stud.* **2004**, *13*, 565–571.
32. Czygan, J.C. Blütenbildung bei *Lemna minor* nach Zusatz von Oestrogen. *Naturwissenschaften* **1962**, *49*, 285–286. [CrossRef]
33. Sayegh, A.; Greppin, H. *Chlorella rubescens* essai de synchronisation et mise en evidence de rythmes endogens. *Arch. Sci. Geneve* **1973**, *8*, 6–18.
34. Bajguz, A.; Czerpak, R. Metabolic activity of estradiol in *Chlorella vulgaris* Beijerinck (*Chlorophytaceae*) Part 1. Content of photosynthetic pigments. *Pol. Arch. Hydrobiol.* **1996**, *43*, 421–426.
35. Bajguz, A.; Czerpak, R. Metabolic activity of estradiol in *Chlorella vulgaris* Beijerinck (Chlorophyceae) Part II. Content of the cellular sugar and protein accumulation. *Pol. Arch. Hydrobiol.* **1996**, *43*, 427–430.
36. Czerpak, R.; Szamrej, I.K. Metabolitic activity of 11-deoxycorticosterone and prednisolone in the alga *Chlorella vulgaris* Beijerinck. *Acta Soc. Bot. Pol.* **2000**, *69*, 25. [CrossRef]
37. Czerpak, R.; Bajguz, A.; Iwaniuk, D. Comparison of the influence of hydrocortisone and progesterone on the content of protein and sugar in the green alga *Chlorella vulgaris* Beijerinck. *Ecohydrol. Hydrobiol.* **2001**, *1*, 473.
38. Kozlova, T.A.; Hardy, B.P.; Levin, D.B. Effect of Fish Steroids 17β-estradiol and 17,20β-dihydroxy-4-pregnen-3-one on Growth, Accumulation of Pigments, and Fatty Acid Profiles in the Microalgae *Scenedesmus quadricauda* (CPCC-158). *Renew. Energy* **2019**, *148*, 798–806. [CrossRef]
39. Kim, M.K.; Park, J.W.; Park, C.S.; Kim, S.J.; Jeune, K.H.; Chang, M.U.; Acreman, J. Enhanced production of *Scenedesmus* spp. (green microalgae) using a new medium containing fermented swine wastewater. *Bioresour. Technol.* **2007**, *98*, 2220–2228. [CrossRef]

40. Goswami, R.D.; Kalita, M.C. *Scenedesmus dimorphus* and *Scenedesmus quadricauda*: Two potent indigenous microalgae strains for biomass production and CO_2 mitigation—A study on their growth behavior and lipid productivity under different concentration of urea as nitrogen source. *J. Algal Biomass Utln.* **2011**, *2*, 42–49.
41. Mata, T.; Meloa, A.C.; Meireles, S.; Mendes, A.M.; Martins, A.A.; Caetano, N.C. Potential of microalgae *Scenedesmus obliquus* grown in brewery wastewater for biodiesel production. *Chem. Eng. Trans.* **2013**, *32*, 901–907.
42. Leng, R.A.; Stambolie, J.H.; Bell, R. Duckweed—A potential high-protein feed resource for domestic animals and fish. *Livest. Res. Rural Dev.* **1995**, *7*, 36.
43. Landolt, E.; Kandeler, R. The family of Lemnaceae—A monographic study, Phytochemistry, physiology, application and bibliography. In *Biosystematic Investigations in the Family of Duckweeds (Lemnaceae)*; Geobotanischen Instutites der ETH, Stiftung Rubel: Zurich, Switzerland, 1987; Volume 4, p. 638. [CrossRef]
44. Guimaraes, F.P.; Aguiar, R.; Oliveira, J.A.; Silva, J.A.A.; Karam, D. Potential of macrophyte for removing arsenic from aqueous solution. *Planta Daninha* **2012**, *30*, 112–118. [CrossRef]
45. Stein, J. (Ed.) *Handbook of Phycological Methods. Culture Methods and Growth Measurements*; Cambridge University Press: Cambridge, UK, 1973; 448p.
46. EC. Biological Test Method: Growth Inhibition Test Using a Freshwater Alga, EPS1/RM/25. 2007. Available online: https://www.canada.ca/en/environment-climate-change/services/wildlife-research-landscape-science/biological-test-method-publications/growth-inhibition-test-freshwater-alga.html (accessed on 10 April 2018).
47. EC. Biological Test Method: Test for Measuring the Inhibition of Growth Using the Freshwater Macrophyte, Lemna minor, EPS1/RM/37. 2014. Available online: https://www.canada.ca/en/environment-climate-change/services/wildlife-research-landscape-science/biological-test-method-publications/inhibition-growth-freshwater-macrophyte/chapter-2.html (accessed on 24 April 2018).
48. Silva, C.P.; Otero, M.; Esteves, V. Processes for the elimination of estrogenic steroid hormones from water: A review. *Environ. Pollut.* **2012**, *165*, 38–58. [CrossRef] [PubMed]
49. Mota, V.C.; Martins, C.I.M.; Eding, E.H.; Canário, A.V.M.; Verreth, J.A.J. Steroids accumulate in the rearing water of commercial recirculating aquaculture systems. *Aquac. Eng.* **2014**, *62*, 9–16. [CrossRef]
50. Markovic, M.; Neale, P.A.; Nidumolu, B.; Kumar, R. Combined toxicity of therapeutic pharmaceuticals to duckweed, *Lemna minor*. *Ecotoxicol. Environ.Saf.* **2021**, *208*, 111428. [CrossRef]
51. Santos, A.N.; Fachini, A.A.; Pena, A.; Delerue-Matos, C.; Montenegro, M.C.B.S.M. Ecotoxicological aspects related to the presence of pharmaceuticals in the aquatic Environment. *J. Hazard. Mater.* **2010**, *175*, 45–95. [CrossRef]
52. Eaton, A.D.; Franson, M.A.H. *American Public Health, American Water Works, and Water Environment: Standard Methods for the Examination of Water & Wastewater*; American Public Health Association: Washington, DC, USA, 2005.
53. *International Standard ISO/FDIS/20079*; Water Quality—Determination of Toxic Effect of Water Constituents and Wastewater to Duckweed (Lemna minor)-Duckweed Growth Inhibition Test. International Organization for Standardization: Geneva, Switzerland, 2005.
54. Greenberg, B.M.; Huang, X.D.; Dixon, D.G. Applications of the aquatic higher plant *Lemna gibba* for Ecotoxicological Assessment. *J. Aquat. Ecosyst. Health* **1992**, *1*, 147–155. [CrossRef]
55. Franklin, N.M.; Stauber, J.L.; Markich, S.J.; Lim, R.P. pH-dependent toxicity of copper and uranium to a tropical freshwater alga (*Chlorella* sp.). *Aquat. Toxicol.* **2000**, *48*, 275–289. [CrossRef]
56. Wilde, K.L.; Stauber, J.L.; Markich, S.J.; Franklin, N.M.; Brown, P.L. The effect of pH on the uptake and toxicity of copper and zinc in a tropical freshwater alga (*Chlorella* sp.). *Arch. Environ. Contam. Toxicol.* **2006**, *51*, 174–185. [CrossRef]
57. Nichols, D.J.; Daniel, T.C.; Edwards, D.R.; Moore, P.A., Jr.; Pote, D.H. Use of grass filter strips to reduce 17 beta-estradiol in runoff from fescue-applied poultry litter. *J. Soil Water Conserv.* **1998**, *53*, 74–77.
58. Peterson, E.W.; Davis, R.K.; Orndorff, H.A. 17h-Estradiol as an indicator of animal waste contamination in mantled karst aquifers. *J. Environ. Qual.* **2001**, *29*, 826–834. [CrossRef]
59. Ying, G.G.; Kookanaa, R.S.; Ru, Y.J. Occurrence and fate of hormone steroids in the environment. *Environ. Int.* **2002**, *28*, 545–551. [CrossRef]
60. Patel, M.; Kumar, R.; Kishor, K.; Mlsna, T.; Pittman, C.U., Jr.; Mohan, D. Pharmaceuticals of Emerging Concern in Aquatic Systems: Chemistry, Occurrence, Effects, and Removal Methods. *Chem. Rev.* **2019**, *119*, 3510–3673. [CrossRef] [PubMed]
61. ECETOC (European Centre for Ecotoxicology and Toxicology of Chemicals). *Environmental Risk Assessment of Difficult Substances*; Technical Report No. 88; ECETOC: Brussels, Belgium, 2003; pp. 17–33.
62. De Bree, L.C.J.; Janssen, R.; Aaby, P.; van Crevel, R.; Joosten, L.A.B.; Stabell Benn, C.; Netea, M.G. The impact of sex hormones on BCG-induced trained immunity. *J. Leukoc. Biol.* **2018**, *104*, 573–578. [CrossRef]
63. Escribese, M.M.; Kraus, T.; Rhee, E.; Fernandez-Sesma, A.; López, C.B.; Moran, T.M. Estrogen inhibits dendritic cell maturation to RNA viruses. *Blood* **2008**, *112*, 4574–4584. [CrossRef]
64. Saraceno, G.E.; Bellini, M.J.; Garcia-Segura, L.M.; Capani, F. Estradiol activates PI3K/Akt/GSK3 pathway under chronic neurodegenerative conditions triggered by perinatal asphyxia. *Front. Pharmacol.* **2018**, *9*, 1–10. [CrossRef]
65. Verma, M.; Pandey, S.; Bhat, I.A.; Mukesh, B.; Anand, J.; Chandra, V.; Sharma, G.T. Impact of l-carnitine on lipid content and post thaw survivability of buffalo embryos produced in vitro. *Cryobiology* **2018**, *82*, 99–105. [CrossRef]

66. Fujiwara, K.; Matsumoto, Y.; Kawakami, H.; Aoki, M.; Tuzuki, M. Evaluation of metal toxicity in *Chlorella kessleri* from the perspective of the Periodic Table. *Bull. Chem. Soc. Jpn.* **2008**, *81*, 478–488. [CrossRef]
67. Morris, J.M.; Lipton, J.; Brinkman, S. Copper toxicity to Rainbow Trout and Fathead Minnows in Low-Hardness Waters: Comparisons of BLM Predictions of Toxicity to Bioassay Results Using Laboratory Water and Site-Collected Water from Upper Talarik Creek in Bristol Bay, Southwest Alaska Salmon Science Workshop, Anchorage, AK, 2013, December 4–5. Available online: http://www.southwestsalmon.org/wpcontent/uploads/2014/01/Morris_Stratus_CuTox_RBT_FHM_LowHardnessH2O.pdf/ (accessed on 12 January 2021).
68. Agami, M.; Reddy, R. Inter-relationships between *Salvinia rotundifolia* and *Spirodela polyrhiza* at various interaction stages. *J. Aquat. Plant Manag.* **1989**, *27*, 96–102.
69. Davies, P.J. *Plant Hormones;* Kluwer Academic Publishers: Alphen aan den Rijn, The Netherlands, 2004; 716p. [CrossRef]
70. Borker, A.R.; Mane, A.V.; Saratale, G.D.; Pathade, G.R. Phytoremediation potential of *Eichhornia crassipes* for the treatment of cadmium in relation with biochemical and water parameters. *Emir. Emir. J. Food Agric.* **2013**, *25*, 443–456. [CrossRef]
71. Wang, S.B.; Hu, Q.; Sommerfeld, M.; Chen, F. Cell wall proteomics of the green alga *Haematococcus pluvialis* (Chlorophyceae). *Proteomics* **2004**, *4*, 692–708. [CrossRef] [PubMed]
72. Li, Q.; Du, W.; Liu, D. Perspectives of microbial oils for biodiesel production. *Appl. Microbiol. Biotechnol.* **2008**, *80*, 749–756. [CrossRef] [PubMed]
73. Damiani, M.C.; Popovich, C.A.; Constenla, D.; Leonardi, P.I. Lipid analysis in *Haematococcus pluvialis* to assess its potential use as a biodiesel feedstock. *Bioresour. Technol.* **2010**, *101*, 3801–3807. [CrossRef] [PubMed]
74. Lemoine, Y.; Schoefs, B. Secondary ketocarotenoid astaxanthin biosynthesis in algae: A multifunctional response to stress. *Photosynth. Res.* **2010**, *106*, 155–177. [CrossRef] [PubMed]
75. Joss, A.; Andersen, H.; Ternes, T.; Richle, P.R.; Siegrist, H. Removal of estrogens in municipal wastewater treatment under aerobic and anaerobic conditions: Consequences for Plant Optimization. *Environ. Sci. Technol.* **2004**, *38*, 3047–3055. [CrossRef]
76. Muradov, N.; Fidalgo, B.; Gujara, A.C.; T-Raissi, A. Pyrolysis of fast-growing aquatic biomass—*Lemna minor* (Duckweed): Characterization of pyrolysis products. *Bioresour. Technol.* **2014**, *101*, 8424–8428. [CrossRef]
77. Landesman, L.; Parker, N.C.; Fedler, C.B.; Konikoff, M. Modeling duckweed growth in wastewater treatment systems. *Livest. Res. Rural Dev.* **2005**, *17*, 2005.
78. Frédéric, M.; Samir, L.; Louise, M.; Abdelkrim, A. Comprehensive modeling of mat density effect on duckweed (*Lemna minor*) growth under controlled eutrophication. *Water Res.* **2006**, *40*, 2901–2910. [CrossRef]
79. Khvatkov, P.; Chernobrovkina, M.; Okuneva, A.; Dolgov, S. Creation of culture media for efficient Duckweeds micropropagation (*Wolffia arrhiza* and *Lemna minor*) using artificial mathematical optimization models. *Plant Cell Tissue Organ Cult.* **2018**, *139*, 85–100. [CrossRef]

plants

Article

The Response of Duckweed *Lemna minor* to Microplastics and Its Potential Use as a Bioindicator of Microplastic Pollution

Ula Rozman and Gabriela Kalčíková *

Faculty of Chemistry and Chemical Technology, University of Ljubljana, 113 Večna pot, SI-1000 Ljubljana, Slovenia
* Correspondence: gabriela.kalcikova@fkkt.uni-lj.si

Abstract: Biomonitoring has become an indispensable tool for detecting various environmental pollutants, but microplastics have been greatly neglected in this context. They are currently monitored using multistep physico-chemical methods that are time-consuming and expensive, making the search for new monitoring options of great interest. In this context, the aim of this study was to investigate the possibility of using an aquatic macrophyte as a bioindicator of microplastic pollution in freshwaters. Therefore, the effects and adhesion of three types of microplastics (polyethylene microbeads, tire wear particles, and polyethylene terephthalate fibers) and two types of natural particles (wood dust and cellulose particles) to duckweed *Lemna minor* were investigated. The results showed that fibers and natural particles had no effect on the specific growth rate, chlorophyll *a* content, and root length of duckweed, while a significant reduction in the latter was observed when duckweed was exposed to microbeads and tire wear particles. The percentage of adhered particles was ten times higher for polyethylene microbeads than for other microplastics and natural particles, suggesting that the adhesion of polyethylene microbeads to duckweed is specific. Because the majority of microplastics in freshwaters are made of polyethylene, the use of duckweed for their biomonitoring could provide important information on microplastic pollution in freshwaters.

Keywords: adhesion; aquatic; biomonitoring; microplastics; microbeads; phytoremediation

Citation: Rozman, U.; Kalčíková, G. The Response of Duckweed *Lemna minor* to Microplastics and Its Potential Use as a Bioindicator of Microplastic Pollution. *Plants* **2022**, *11*, 2953. https://doi.org/10.3390/plants11212953

Academic Editors: Viktor Oláh, Klaus-Jürgen Appenroth and K. Sowjanya Sree

Received: 9 October 2022
Accepted: 1 November 2022
Published: 2 November 2022

1. Introduction

Plastic pollution has become one of the most important environmental issues of the last decades, with microplastics (MPs, pieces of plastic from 1 to 1000 μm [1]) being of increasing public and scientific concern due to their widespread occurrence [2]. They enter the environment through many pathways, including wastewaters [3], runoff [4], and atmospheric deposition [5]. MPs are also formed directly in the environment by the fragmentation of plastic items [6], but are degraded very slowly under natural conditions and therefore remain in aquatic ecosystems for a long time [7].

Traditionally, the presence and abundance of MPs in the aquatic ecosystem is monitored using physico-chemical methods [2]. However, such monitoring is challenging as MPs are not uniformly distributed in the water phase. Most of them float on the water surface because they are composed of low density polymers [8]. There, they interact with microorganisms that form a biofilm on their surface, resulting in increased size and density, and MPs can then sink deeper into the water body [9]. Further monitoring is even more difficult because MPs move dynamically from one environmental compartment to another, interact with biota, and become incorporated into sediment after settling [10,11]. However, reliable monitoring of MPs in the aquatic environment is crucial to identify the sources of MPs and to establish regulatory limits and measures to reduce them in the environment [12].

Recent research has shown that MPs interact with different organisms, so the use of biological indicators (bioindicators) could provide an alternative to traditional monitoring methods [13]. Ideally, the bioindicator should accumulate a high concentration of pollutant with low impact, be widespread and abundant, and be sessile to represent the local population [14]. Animal species such as fish and invertebrates may not meet the requirements for a bioindicator because the bioaccumulation of MPs may be low; many studies indicated that ingested MPs are also rapidly excreted [15,16]. On the other hand, recent studies showed that MPs can attach to biotic surfaces such as biofilms [17] or aquatic macrophytes [18]. The latter showed great potential to interact with MPs under laboratory conditions [19,20] and in field studies [11,21]. Macrophytes are minimally affected by environmentally relevant MP concentrations [22,23], and floating macrophytes are among the first organisms with which MPs interact when they enter the aquatic environment, as they collectively occupy the water surface. Therefore, floating macrophytes could be appropriate organisms for the biomonitoring of MPs in freshwaters.

In this context, the aim of this study was to investigate the potential use of the floating macrophyte duckweed *Lemna minor* as a bioindicator of MPs in freshwaters. Duckweed is widely used for toxicity testing [24] and as a bioindicator (e.g., for metal pollution [25]). It is tolerant to the presence of MPs [23], grows wild in European regions, and plays an important role as food for other organisms and habitat for various aquatic organisms [26]. For this purpose, the effect of various MPs and natural particles was tested to evaluate a specific response of duckweed to MPs exposure. The selected endpoints represent three main areas where MPs can affect plants: specific growth rate is an indicator of leaf damage, measurement of root length shows the effects on roots, and measurement of photosynthetic pigment content is a sensitive biomarker often used to detect adverse effects on photosynthesis [27,28]. Furthermore, the number of MPs and natural particles adhered to the duckweed biomass was monitored to evaluate the efficiency of duckweed in capturing MPs and thus its potential application for monitoring MP pollution. To our knowledge, this is the first study to consider floating aquatic macrophytes as potential bioindicators of MP pollution, and likely one of the first to monitor particulate matter in the aquatic environment.

2. Results

2.1. Characterization of Microplastics and Natural Particles

The characteristics of microplastics (MPs) and natural particles are shown in Figure 1 and Table 1. Microbeads, tire wear particles, and wood dust had irregular shapes, while fibers were smooth and uniform. Cellulose particles had the shape of beads, but with irregularities on the surface (Figure 1).

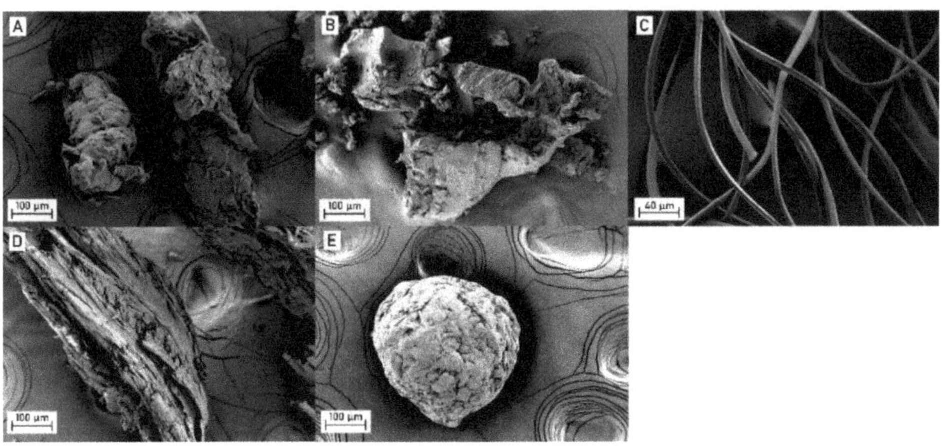

Figure 1. Microplastics and natural particles under the field-emission scanning electron microscope: (**A**) microbeads, (**B**) tire wear particles, (**C**) fibers, (**D**) wood dust, and (**E**) cellulose particles.

Table 1. Characteristics of microplastics (microbeads, tire wear particles, fibers) and natural particles (wood dust, cellulose particles).

	Size (Mean ± SD) (μm)	Number of Particles per Mass (particles/mg)	Chemical Composition
Microbeads	149 ± 75	68	Low density Polyethylene (PE)
Tire wear particles	47 ± 22	445	Rubber
Fibers	Length: 5362 ± 1082 Diameter: 9.6 ± 3.5	581	Polyethylene Terephthalate (PET)
Wood dust	253 ± 142	44	/
Cellulose particles	296 ± 45	48	/

/—not determined.

Microbeads, wood dust, and cellulose particles had a similar number of particles per mass, while fibers and tire wear particles had a higher number of particles per mass (Table 1). The mean size of MPs and natural particles was different, ranging from 47 ± 22 μm to 296 ± 45 μm, while for fibers the mean length and diameter were determined separately due to large length-to-diameter ratio (Table 1). The chemical characterization of MPs was previously performed by Rozman et al. and, based on the results of FTIR analysis, the microbeads and fibers were pure low-density polyethylene and polyethylene terephthalate, respectively, while the FTIR spectrum of tire wear particles confirmed that particles were derived from rubber tires [23].

2.2. *Effects of Microplastics and Natural Particles on Duckweed*

All particles introduced into the test floated on the water surface and were thus in contact with duckweed. After exposure to MPs and natural particles, the specific growth of duckweed *Lemna minor* was not significantly affected (Table 2). The average specific growth rates in the control, microbeads, tire wear particles, fibers, wood dust, and cellulose particles treatments were 0.324 ± 0.008 day^{-1}, 0.313 ± 0.015 day^{-1}, 0.302 ± 0.022 day^{-1}, 0.302 ± 0.017 day^{-1}, 0.310 ± 0.026 day^{-1}, and 0.306 ± 0.033 day^{-1}, respectively. No significant reduction in chlorophyll *a* content was observed in all treatments compared to the control treatment. The chlorophyll *a* content was 0.503 ± 0.027 mg/g, 0.518 ± 0.027 mg/g, 0.535 ± 0.031 mg/g, 0.450 ± 0.038 mg/g, 0.505 ± 0.063 mg/g, and 0.581 ± 0.185 mg/g in the control, microbeads, tire wear particles, fibers, wood dust, and cellulose particles treatments, respectively. On the other hand, microbeads (DF = 6, p = 0.00868) and tire wear particles (DF = 6, p = 0.000039) caused a significant reduction in the length of duckweed roots (mean root length of microbeads and tire wear particles were 23.3 ± 2.6 mm and 21.8 ± 2.0 mm, respectively). The effects of fibers on the roots of duckweed were comparable to those of natural particles, as the mean root length in the control, fibers, wood dust, and cellulose particles treatments was 29.2 ± 1.6 mm, 27.6 ± 0.6 mm, 28.2 ± 1.1 mm, and 29.3 ± 3.3 mm, respectively.

Table 2. Inhibition of specific growth rate, root length, and chlorophyll *a* content after exposure to microplastics (microbeads, tire wear particles, fibers) and natural particles (wood dust, cellulose particles).

	Inhibition (%)		
	Specific Growth Rate	Root Length	Chlorophyll *a*
Microbeads	3.4	20.2 *	0
Tire wear particles	6.8	25.3 *	0
Fibers	6.8	5.5	10.5
Wood dust	4.3	3.4	0
Cellulose particles	5.6	0	0

* Statistical significance compared to control ($p < 0.05$).

2.3. Adhesion of Particles to Duckweed

The percentage of adhered particles to duckweed is shown in Figure 2. The adhesion of tire wear particles, fibers, wood dust, and cellulose particles was similar (approximately 1%); however, the percentage of adhered microbeads was 10-times higher. Due to the high number of adhered microbeads (Figure 3), the effect of gentle shaking was further investigated, and the number of adhered microbeads with shaking was slightly higher than the number of adhered microbeads without shaking (13.0 ± 3.5% and 10.0 ± 5.1%, respectively); however, it was not statistically significant (U = 11.5, p = 0.3350).

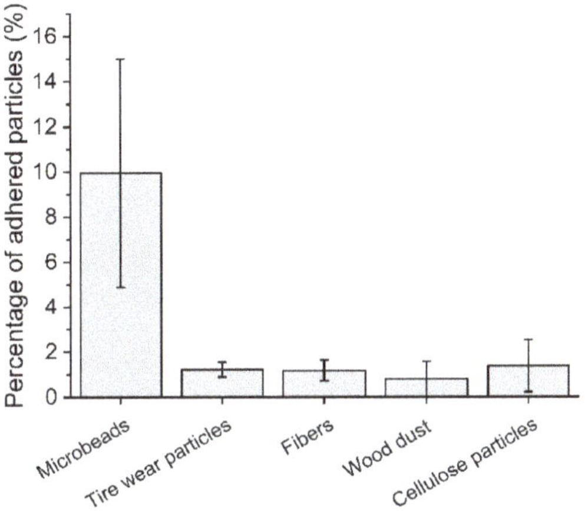

Figure 2. The percentage of adhered microplastics and natural particles to duckweed.

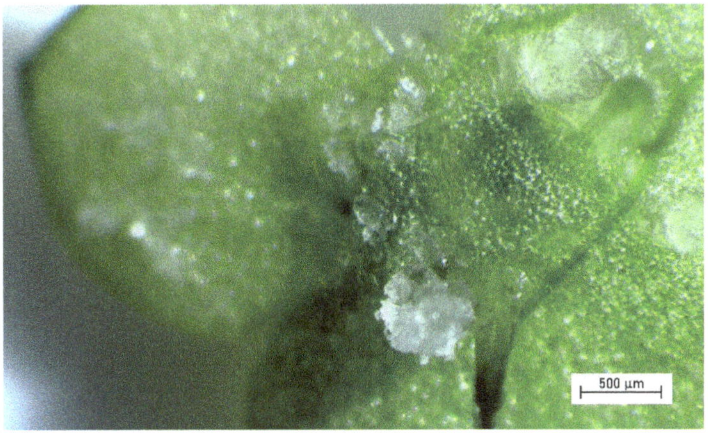

Figure 3. Microbeads adhered to fronds of duckweed.

3. Discussion

Large quantities of MPs continuously enter aquatic ecosystems, but their monitoring is difficult, as MP concentrations are highly heterogeneous [2], vary over time [29], and fluctuate even within a site [30]. Therefore, the use of biomonitoring could be beneficial because organisms live at the site for a long period of time and are thus in long-term contact with MPs. The use of organisms as bioindicators of MPs is still in its early stages, with a focus on the marine environment and the use of animal species [31,32]. Therefore, in this study, we focused for the first time on the use of the floating macrophyte duckweed *Lemna minor* as a potential bioindicator of MP pollution in freshwaters.

In the first part of this study, the specific response of duckweed to the presence of MPs was studied, because it is possible to monitor pollutants based on the effect they trigger (e.g., use of biomarkers) [33]. Results showed that MPs did not elicit a negative response on duckweed specific growth rate, including chlorophyll *a* content. Microbeads and tire wear particles caused a reduction in root length, while fibers and natural particles did not. In general, MPs can induce an adverse response by mechanical and/or chemical stress. The latter can occur through the leaching of additives [34], but this was not the case here because all MPs were previously tested for possible leaching and their leachates had no effect on duckweed [23]. It is very unlikely that the MPs used in this study penetrated the roots because they are too large to pass through the cell wall. This is in agreement with the results of Dovidat et al. [34], who investigated the uptake of micro- and nano-plastics by confocal microscopy in *Spirodela polyrhiza* and found that plastic particles adhered only externally, while no micro- or even nano-plastics were allocated inside the roots. Therefore, it is plausible that root length was affected due to mechanical stress (external abrasion of roots), and the effect may be related to the surface morphology of the MPs, as particles with sharp edges could affect root length and root cell viability [27]; microbeads and tire wear particles had sharp edges, while fibers were perfectly smooth. Although wood particles also had sharp edges, they did not affect duckweed, most probably due to the softness of wood [35]. In previous studies, long-term exposure of duckweed to microbeads also showed no effects on carbohydrates, lipids, proteins, the activity of electron transport system, and antioxidant capacity [20], and thus the most sensitive endpoint following exposure of duckweed to MPs appear to be the root length. However, in the aquatic ecosystem, MPs are inhabited by microorganisms that can form a biofilm [36] that covers their surface and mitigates the effect of sharp MPs [37]. Therefore, the MP-induced reduction in duckweed root length may not occur in the environment, making monitoring of this endpoint in duckweed impractical for MP biomonitoring.

Furthermore, we focused on the monitoring of MPs adhered to duckweed biomass as previous studies have shown that MPs can interact with organisms via bio-adhesion [38,39]. The results showed that only microbeads (made of polyethylene) were adhered to a greater extent and that the percentage of adhered tire wear particles, fibers, wood dust, and cellulose particles was comparable. It is plausible that the initial interactions between MPs and duckweed are electrostatic in nature, as negatively charged plant biomass attracts positively charged MPs [40]. This would also explain the low adhesion of natural particles made of wood and cellulose as they carry the same charge as plant biomass ([41] and [42], respectively). Similarly, used rubber tires had a negative charge as they obtain an excess of electrons from the road [43], while polyethylene microbeads can be positively charged [44] (no data on the charge of PET fibers were found). The interaction between microbeads and duckweed can be rather strong as water movement did not affect the percentage of adhered microbeads, and thus natural water flow may not have a significant impact on MPs adhesion.

From the results, it is apparent that polyethylene microbeads are the only particles that adhered to plant biomass to a higher extend. These microbeads originated from a cosmetic product but they are also similar to the MPs sampled in freshwaters; they had the shape of irregular fragments and a similar particle size (~150 μm) to MPs detected in rivers and lakes [45,46], and the concentration used here (100 mg/L = 6 800 MPs/L, calculated according to Table 1) can also be considered relevant to MP hotspots, as Pivokonsky et al. detected up to 3 605 MPs/L in freshwater lakes [47]. Thus, the environmental relevance of this study is undisputed; however, it should be noted that many physico-chemical and biological processes occur in the environment and that further research, including initial field studies, is needed for the successful use of duckweeds as bioindicators of MP pollution. Currently, we can support our conclusion only with the results of monitoring studies in the marine environment, where the strong interactions between MPs and plant biomass were also confirmed. For example, Goss et al. monitored MPs adhered to seagrass *Thalassia testudinum* and found a number of MPs attached to the blades and overgrown by periphyton [18]. Huang et al. monitored the abundance and diversity of MPs in seagrass *Enhalus acodoides* and found that vegetated sites had up to 2.9 times more MPs than bare sites, with polyethylene MPs being the most abundant MP type.

The use of duckweed as a bioindicator for polyethylene MPs seems promising, not only because of the intensive adhesion to plant biomass, but also because polyethylene MPs float on the water surface and can be immediately captured by duckweed before they begin to sink or are transported further. Therefore, duckweed-derived MPs can indicate relatively fresh/immediate MP contamination, and the biomonitoring could identify nearby sources of MPs. In addition to the use for biomonitoring, the use of duckweed to collect MPs could also be used for phytoremediation. Removal of contaminated biomass could reduce the number of MPs by preventing them from spreading further into aquatic ecosystems where their removal is currently impossible.

4. Materials and Methods

4.1. Microplastics and Natural Particles

Three different types of microplastics (MPs) and two types of natural particles were used in this study. Microbeads and cellulose particles were extracted from two different facial scrubs where they serve as abrasives. Both were extracted using the same methods described in Kalčíková et al. [27]. Briefly, 50 mL of a facial scrub was dissolved in a warm deionized water under a stirring condition (400 rpm). The solution was filtered through filter paper (pore size 12–25 μm, Macherey-Nagel, Düren, Germany), and both types of particles were washed three times with deionized water and dried at 40 ± 2 °C overnight. Tire wear particles were obtained from a local car repair service, and particles were prepared by cutting pieces of old used tires. Particles were separated by size using sieves with mesh sizes of 355 and 125 μm, and the middle fraction (between 125 and 355 μm) was further used in this study. Fibers were obtained from synthetic clothing by milling at 8000 rpm for 6 min (Tube Mill 100 control, IKA, Staufen, Germany) [23]. The wood dust was from beech (*Fagus* sp.), and the particles were prepared by drilling holes in a beech wood slab, and the resulting sawdust was sieved (800 μm) to obtain a smaller size fraction of the dust.

MPs and natural particles were characterized in terms of their size, number of particles per mass, shape, morphology, and chemical composition, as described in Rozman et al. [23]. Size was determined by a laser diffraction analyzer (S3500 Bluewave, Microtrac, Haan/Duesseldorf, Germany) using a dry unit. The measurement was repeated three times, and the results were expressed as the number of particle size distribution. To determine the number of particles per mass, an amount of 1–2 mg of particles was weighed and counted using a stereo microscope (SMZ-171, Motic, Xiamen, China). The procedure was repeated ten-times to include at least 1000 particles in the analysis. The number of particles per mass of fibers was determined by measuring the length and diameter of numerous fibers under an optical microscope (Imager.Z2m, Zeiss, Oberkochen, Germany). The mass of each particle was calculated based on their density [48] and the volume of the fiber, resulting in the calculation of the mean number of particles per mg. The shape and morphology of the particles were examined using a field-emission scanning electron microscope (FE-SEM, Ultra plus, Zeiss, Oberkochen, Germany) at an accelerating voltage of 2 kV using a secondary detector. Before analysis, MPs and natural particles were coated with a thin Au/Pb layer. Chemical composition was determined by a Fourier transform infrared (FTIR) spectrometer (Spectrum Two FT-IR, PerkinElmer, Beaconsfield, UK) in the wavelength range from 4000 cm^{-1} to 400 cm^{-1} (resolution 2 cm^{-1}, 10 scans). The background and ATR correlation of the spectra was performed [23].

4.2. Duckweed Lemna minor

The duckweed *Lemna minor* used in this study originated from a permanent laboratory culture. Plants were grown in 2-litre rectangular vessels in Steinberg medium [49] at 24 ± 1 °C and a 16/8-h photoperiod (light/dark) with a light intensity of 3500 ± 500 lx.

The medium was changed weekly, and overgrown biomass was removed, leaving at least 2/3 of the surface free for further growth.

4.3. Ecotoxicity Test

The ecotoxicity test largely followed the OECD Guidelines No. 221 [49], with some minor modification. The concentration of 100 mg/L of MPs and natural particles was used for the experiment as it is recommended as a limit concentration by the OECD [50] and used in other MPs studies [51,52]. MPs and natural particles were directly weighted into 100 mL glass beakers and, afterwards, 50 mL of Steinberg medium was added into each glass beaker. In all treatments, the roots of duckweed were removed before exposure, and randomly selected plants with a total of ten fronds were placed into each beaker. They were incubated at the same temperature and photoperiod as in the permanent laboratory culture with a high light intensity of 7000 $\pm$ 500 lx and humidity of >70% to minimize evaporation of the medium from the test vessels. Each treatment was replicated four times. After seven days of incubation, the number of fronds was counted and the specific growth rate was calculated based on the OECD Guidelines No. 221 [49]. The root length of ten randomly selected plants in each test vessel was measured using millimeter paper [35]. To determine chlorophyll *a* content, approximately 15 mg of the fresh plant was homogenized in cold 95% (v/v) ethanol. After 24 h incubation in a freezer at -18 ± 2 °C, the absorbance of the supernatant was measured at 664.2 nm and 648.6 nm using a spectrophotometer (Cary 50 UV-Vis spectrophotometer, Agilent Technologies, Santa Clara, CA, USA) [53], and chlorophyll *a* content (mg/g) was calculated according to Lichtenthaler [54].

4.4. Adhesion of Particles to Duckweed

Based on our preliminary experiments, the maximum number of adhered polyethylene MPs to duckweed was reached after 24 h (not yet published); therefore, incubation time for adhesion experiments was the same. The experiment was set up in a similar way to the ecotoxicity test: the concentration of MPs or natural particles was 100 mg/L, the test was performed in test vessels containing 50 mL of Steinberg medium, and ten fronds were added to each test vessel, but in this case the roots were not removed. The test vessels were incubated for 24 h under the same conditions as in the ecotoxicity test (16/8 h, 7000 $\pm$ 500 lx, 24 $\pm$ 1 °C, humidity of >70%). After incubation, the number of adhered MPs and natural particles was determined. Plant biomass from each test vessel was washed with deionized water, and the washing water was filtered (S-Pak filter, pore size 0.22 µm, Merk Millipore, Burlington, MA, USA). The filters with retained particles were dried at room temperature for 24 h, and the particles were counted using the stereo microscope. The plant biomass was then weighed and digested by Fenton oxidation, as described in Rozman et al. [20]. Briefly, 2 mL of 0.015 g/mL $Fe_2SO_4 \cdot 7H_2O$ (with 3 mL/L of H_2SO_4 (97% v/v)) and 2 mL of 30% (w/w) H_2O_2 [55] were added to each test tube containing a previously weighed plant. The digestion process lasted 24 h at room temperature (22 $\pm$ 2 °C). The digestate was filtered (S-Pak filter, pore size 0.22 µm, Merk Millipore, Burlington, MA, USA), the filters were dried at room temperature for 24 h, and the number of particles on the filter paper was counted under the stereo microscope. The total number of adhered particles to duckweed was the sum of the washed particles and the particles remaining in the digestate. Each treatment was replicated four times, and the results were expressed as the number of particles per fresh weight of duckweed [20].

Due to the extensive adhesion of polyethylene microbeads to duckweed, an additional experiment was conducted to investigate the effects of water movement. The experiment was performed under the same conditions as described above, except that it was slightly shaken (70 rpm) on an orbital shaker (Orbit 19000, Labnet, Edison, NJ, USA).

The plant biomass in the control treatment was processed in the same way to monitor for possible airborne contamination, but no particles were detected. MPs and natural particles only (excluding plant tissue) were also subjected to Fenton oxidation under the same conditions as described above (each replicated five times) to evaluate the effects of the process (filtration, digestion, etc.) on particle loss. The mass was reduced by 2.8 ± 1.7%, 6.8 ± 1.2%, 3.6 ± 3.7%, 7.5 ± 2.7%, and 2.3 ± 1.4% for microbeads, tire wear particles, fibers, wood dust, and cellulose particles, respectively, so the effect of the procedure was considered to be of minor importance.

The percentage of MPs or natural particles adhered to plant biomass was calculated by comparing the total number of MPs or natural particles in the test vessel (calculated based on the number of particles per mass, Table 1) and their number adhered to plant biomass:

$$P = (n_0 - n_x)/n_0 \cdot 100 \quad (1)$$

where P (%) is the percentage of adhered MPs or natural particles to plant biomass, n_0 (/) is the number of MPs or natural particles introduced into the vessels, and n_x (/) is the number of MPs or natural particles adhered to plant biomass [20].

4.5. Data Analysis

To analyze statistically significant differences between treated and control groups when testing the effects of MPs and natural particles on duckweed, normality was tested using the Shapiro–Wilk test and homogeneity of variances with Levene's test. When normality and/or homogeneity of variances were not achieved, statistical differences compared to control were tested with the Mann–Whitney U test, but when the data were considered as normal and homogeneous, Student's t-test was used. Differences were considered statistically significant if $p < 0.05$. All data analysis was preformed using OriginPro 2021b software (OriginLab Corp., Northampton, MA, USA).

5. Conclusions

The biomonitoring of various environmental pollutants is a widely used strategy, and bioindicators provide important information about the quality of the surrounding environment. So far, bioindicators have been used to monitor organic and inorganic substances, but particulate matter has remained unnoticed, and thus the biomonitoring of MPs is also in its infancy. This study presents a proof of concept for the monitoring of MPs in freshwaters by using duckweed *Lemna minor*. The results showed that polyethylene microbeads adhere to the plant biomass to a significant extent compared to other MPs or natural particles. Because MPs can change their properties and disperse in the water column over time, it is plausible to consider duckweed as a bioindicator of fresh MP pollution when MPs are still associated with the water-air interface and can be in close contact with the floating macrophyte. Because biomonitoring and the use of bioindicators are an important way to monitor pollutants as complex as MPs, both a systematic survey and a field study under environmentally relevant conditions are recommended.

Author Contributions: Conceptualization, U.R. and G.K.; methodology, U.R. and G.K.; formal analysis, U.R. and G.K.; investigation, U.R.; writing-original draft preparation, U.R. and G.K.; project administration, G.K., funding acquisition, G.K. All authors have read and agreed to the published version of the manuscript.

Funding: This research was funded by the Slovenian Research Agency (Research programs P2-0191 and the project Plasti-C-Wetland (J2-2491) (https://planterastics.fkkt.uni-lj.si/plasti-c-wetland, accessed on 8 October 2022).

Data Availability Statement: The data presented in this study are available on request from the corresponding author.

Acknowledgments: The authors are thankful to Tina Skalar for help with particle size measurements. This article is based upon work from COST Action CA20101 Plastics monitoRIng detectiOn RemedIa-Tion recoverY—PRIORITY, supported by COST (European Cooperation in Science and Technology, www.cost.eu, accessed on 8 October 2022).

Conflicts of Interest: The authors declare no conflict of interest.

References

1. Hartmann, N.B.; Hüffer, T.; Thompson, R.C.; Hassellöv, M.; Verschoor, A.; Daugaard, A.E.; Rist, S.; Karlsson, T.; Brennholt, N.; Cole, M.; et al. Are We Speaking the Same Language? Recommendations for a Definition and Categorization Framework for Plastic Debris. *Environ. Sci. Technol.* **2019**, *53*, 1039–1047. [CrossRef] [PubMed]
2. Rezania, S.; Park, J.; Md Din, M.F.; Mat Taib, S.; Talaiekhozani, A.; Kumar Yadav, K.; Kamyab, H. Microplastics pollution in different aquatic environments and biota: A review of recent studies. *Mar. Pollut. Bull.* **2018**, *133*, 191–208. [CrossRef] [PubMed]
3. Murphy, F.; Ewins, C.; Carbonnier, F.; Quinn, B. Wastewater Treatment Works (WwTW) as a Source of Microplastics in the Aquatic Environment. *Environ. Sci. Technol.* **2016**, *50*, 5800–5808. [CrossRef] [PubMed]
4. Piñon-Colin, T.d.J.; Rodriguez-Jimenez, R.; Rogel-Hernandez, E.; Alvarez-Andrade, A.; Wakida, F.T. Microplastics in stormwater runoff in a semiarid region, Tijuana, Mexico. *Sci. Total Environ.* **2020**, *704*, 135411. [CrossRef] [PubMed]
5. Ross, P.S.; Chastain, S.; Vassilenko, E.; Etemadifar, A.; Zimmermann, S.; Quesnel, S.-A.; Eert, J.; Solomon, E.; Patankar, S.; Posacka, A.M.; et al. Pervasive distribution of polyester fibres in the Arctic Ocean is driven by Atlantic inputs. *Nat. Commun.* **2021**, *12*, 106. [CrossRef] [PubMed]
6. Julienne, F.; Delorme, N.; Lagarde, F. From macroplastics to microplastics: Role of water in the fragmentation of polyethylene. *Chemosphere* **2019**, *236*, 124409. [CrossRef]
7. Priya, K.; Renjith, K.R.; Joseph, C.J.; Indu, M.S.; Srinivas, R.; Haddout, S. Fate, transport and degradation pathway of microplastics in aquatic environment—A critical review. *Reg. Stud. Mar. Sci.* **2022**, *56*, 102647. [CrossRef]
8. Eibes, P.M.; Gabel, F. Floating microplastic debris in a rural river in Germany: Distribution, types and potential sources and sinks. *Sci. Total Environ.* **2022**, *816*, 151641. [CrossRef]
9. Kalčíková, G.; Bundschuh, M. Aquatic Biofilms—Sink or Source of Microplastics? A Critical Reflection on Current Knowledge. *Environ. Toxicol. Chem.* **2021**, *41*, 838–843. [CrossRef]
10. Di, M.; Wang, J. Microplastics in surface waters and sediments of the Three Gorges Reservoir, China. *Sci. Total Environ.* **2018**, *616*, 1620–1627. [CrossRef]
11. Huang, Y.; Xiao, X.; Xu, C.; Perianen, Y.D.; Hu, J.; Holmer, M. Seagrass beds acting as a trap of microplastics-Emerging hotspot in the coastal region? *Environ. Pollut.* **2020**, *257*, 113450. [CrossRef] [PubMed]
12. Artiola, J.F.; Brusseau, M.L. Chapter 10-The Role of Environmental Monitoring in Pollution Science. In *Environmental and Pollution Science*, 3rd ed.; Brusseau, M.L., Pepper, I.L., Gerba, C.P., Eds.; Academic Press: Cambridge, UK, 2019; pp. 149–162.
13. Sumudumali, R.G.I.; Jayawardana, J.M.C.K. A Review of Biological Monitoring of Aquatic Ecosystems Approaches: With Special Reference to Macroinvertebrates and Pesticide Pollution. *Environ. Manag.* **2021**, *67*, 263–276. [CrossRef]
14. Zhou, Q.; Zhang, J.; Fu, J.; Shi, J.; Jiang, G. Biomonitoring: An appealing tool for assessment of metal pollution in the aquatic ecosystem. *Anal. Chim. Acta* **2008**, *606*, 135–150. [CrossRef]
15. Ogonowski, M.; Schür, C.; Jarsén, Å.; Gorokhova, E. The Effects of Natural and Anthropogenic Microparticles on Individual Fitness in Daphnia magna. *PLoS ONE* **2016**, *11*, e0155063. [CrossRef]
16. Woods, M.N.; Stack, M.E.; Fields, D.M.; Shaw, S.D.; Matrai, P.A. Microplastic fiber uptake, ingestion, and egestion rates in the blue mussel (*Mytilus edulis*). *Mar. Pollut. Bull.* **2018**, *137*, 638–645. [CrossRef] [PubMed]
17. Huang, S.; Peng, C.; Wang, Z.; Xiong, X.; Bi, Y.; Liu, Y.; Li, D. Spatiotemporal distribution of microplastics in surface water, biofilms, and sediments in the world's largest drinking water diversion project. *Sci. Total Environ.* **2021**, *789*, 148001. [CrossRef] [PubMed]
18. Goss, H.; Jaskiel, J.; Rotjan, R. *Thalassia testudinum* as a potential vector for incorporating microplastics into benthic marine food webs. *Mar. Pollut. Bull.* **2018**, *135*, 1085–1089. [CrossRef]
19. Mateos-Cárdenas, A.; Scott, D.T.; Seitmaganbetova, G.; Pelt Frank, N.A.M.; John, O.H.; Marcel, A.K.J. Polyethylene microplastics adhere to *Lemna minor* (L.), yet have no effects on plant growth or feeding by *Gammarus duebeni* (Lillj.). *Sci. Total Environ.* **2019**, *689*, 413–421. [CrossRef]
20. Rozman, U.; Jemec Kokalj, A.; Dolar, A.; Drobne, D.; Kalčíková, G. Long-term interactions between microplastics and floating macrophyte *Lemna minor*: The potential for phytoremediation of microplastics in the aquatic environment. *Sci. Total Environ.* **2022**, *831*, 154866. [CrossRef]
21. Jones, K.L.; Hartl, M.G.J.; Bell, M.C.; Capper, A. Microplastic accumulation in a *Zostera marina* L. bed at Deerness Sound, Orkney, Scotland. *Mar. Pollut. Bull.* **2020**, *152*, 110883. [CrossRef]
22. van Weert, S.; Redondo-Hasselerharm, P.E.; Diepens, N.J.; Koelmans, A.A. Effects of nanoplastics and microplastics on the growth of sediment-rooted macrophytes. *Sci Total Environ.* **2019**, *654*, 1040–1047. [CrossRef] [PubMed]
23. Rozman, U.; Turk, T.; Skalar, T.; Zupančič, M.; Čelan Korošin, N.; Marinšek, M.; Olivero-Verbel, J.; Kalčíková, G. An extensive characterization of various environmentally relevant microplastics–Material properties, leaching and ecotoxicity testing. *Sci. Total Environ.* **2021**, *773*, 145576. [CrossRef] [PubMed]
24. Ziegler, P.; Sree, K.S.; Appenroth, K.J. Duckweeds for water remediation and toxicity testing. *Toxicol. Environ. Chem.* **2016**, *98*, 1127–1154. [CrossRef]
25. Appenroth, K.J.; Krech, K.; Keresztes, Á.; Fischer, W.; Koloczek, H. Effects of nickel on the chloroplasts of the duckweeds *Spirodela polyrhiza* and *Lemna minor* and their possible use in biomonitoring and phytoremediation. *Chemosphere* **2010**, *78*, 216–223. [CrossRef]

26. Ekperusi, A.O.; Sikoki, F.D.; Nwachukwu, E.O. Application of common duckweed (*Lemna minor*) in phytoremediation of chemicals in the environment: State and future perspective. *Chemosphere* **2019**, *223*, 285–309. [CrossRef]
27. Kalčíková, G.; Žgajnar Gotvajn, A.; Kladnik, A.; Jemec, A. Impact of polyethylene microbeads on the floating freshwater plant duckweed Lemna minor. *Environ. Pollut.* **2017**, *230*, 1108–1115. [CrossRef]
28. Malec, P.; Maleva, M.G.; Prasad, M.N.V.; Strzałka, K. Responses of *Lemna trisulca* L. (Duckweed) exposed to low doses of cadmium: Thiols, metal binding complexes, and photosynthetic pigments as sensitive biomarkers of ecotoxicity. *Protoplasma* **2010**, *240*, 69–74. [CrossRef]
29. Rodrigues, M.O.; Abrantes, N.; Gonçalves, F.J.M.; Nogueira, H.; Marques, J.C.; Gonçalves, A.M.M. Spatial and temporal distribution of microplastics in water and sediments of a freshwater system (Antuã River, Portugal). *Sci. Total Environ.* **2018**, *633*, 1549–1559. [CrossRef]
30. Dris, R.; Gasperi, J.; Rocher, V.; Saad, M.; Renault, N.; Tassin, B. Microplastic contamination in an urban area: A case study in Greater Paris. *Environ. Chem.* **2015**, *12*, 592–599. [CrossRef]
31. Bonanno, G.; Orlando-Bonaca, M. Perspectives on using marine species as bioindicators of plastic pollution. *Mar. Pollut. Bull.* **2018**, *137*, 209–221. [CrossRef]
32. Li, J.; Lusher, A.L.; Rotchell, J.M.; Deudero, S.; Turra, A.; Bråte, I.L.N.; Sun, C.; Shahadat Hossain, M.; Li, Q.; Kolandhasamy, P.; et al. Using mussel as a global bioindicator of coastal microplastic pollution. *Environ. Pollut.* **2019**, *244*, 522–533. [CrossRef]
33. de Souza, C.P.; Guedes, T.d.A.; Fontanetti, C.S. Evaluation of herbicides action on plant bioindicators by genetic biomarkers: A review. *Environ. Monit. Assess.* **2016**, *188*, 694. [CrossRef]
34. Dovidat, L.C.; Brinkmann, B.W.; Vijver, M.G.; Bosker, T. Plastic particles adsorb to the roots of freshwater vascular plant *Spirodela polyrhiza* but do not impair growth. *Limnol. Oceanogr. Lett.* **2020**, *5*, 37–45. [CrossRef]
35. Kalčíková, G.; Skalar, T.; Marolt, G.; Jemec Kokalj, A. An environmental concentration of aged microplastics with adsorbed silver significantly affects aquatic organisms. *Water Res.* **2020**, *175*, 115644. [CrossRef]
36. Parrish, K.; Fahrenfeld, N.L. Microplastic biofilm in fresh- and wastewater as a function of microparticle type and size class. *Environ. Sci. Water Res. Technol.* **2019**, *5*, 495–505. [CrossRef]
37. Jemec Kokalj, A.; Kuehnel, D.; Puntar, B.; Žgajnar Gotvajn, A.; Kalčikova, G. An exploratory ecotoxicity study of primary microplastics versus aged in natural waters and wastewaters. *Environ. Pollut.* **2019**, *254*, 112980. [CrossRef]
38. Klun, B.; Rozman, U.; Ogrizek, M.; Kalčíková, G. The first plastic produced, but the latest studied in microplastics research: The assessment of leaching, ecotoxicity and bioadhesion of Bakelite microplastics. *Environ. Pollut.* **2022**, *307*, 119454. [CrossRef]
39. Martin, C.; Corona, E.; Mahadik, G.A.; Duarte, C.M. Adhesion to coral surface as a potential sink for marine microplastics. *Environ. Pollut.* **2019**, *255*, 113281. [CrossRef]
40. Kalčíková, G. Aquatic vascular plants–A forgotten piece of nature in microplastic research. *Environ. Pollut.* **2020**, *262*, 114354. [CrossRef]
41. Anton, P. The Surface Charge of Wood: Application Report. Available online: https://s3-eu-central-1.amazonaws.com/centaur-wp/theengineer/prod/content/uploads/2017/01/25142553/Surpass-%E2%80%93-The-Surface-Charge-of-Wood.pdf (accessed on 7 October 2022).
42. Maurer, H.W. Chapter 18-Starch in the Paper Industry. In *Starch*, 3rd ed.; BeMiller, J., Whistler, R., Eds.; Academic Press: San Diego, CA, USA, 2009; pp. 657–713.
43. Amato-Lourenço, L.F.; Carvalho-Oliveira, R.; Júnior, G.R.; dos Santos Galvão, L.; Ando, R.A.; Mauad, T. Presence of airborne microplastics in human lung tissue. *J. Hazard. Mater.* **2021**, *416*, 126124. [CrossRef]
44. Kalčíková, G.; Alič, B.; Skalar, T.; Bundschuh, M.; Gotvajn, A.Ž. Wastewater treatment plant effluents as source of cosmetic polyethylene microbeads to freshwater. *Chemosphere* **2017**, *188*, 25–31. [CrossRef]
45. Kameda, Y.; Yamada, N.; Fujita, E. Source- and polymer-specific size distributions of fine microplastics in surface water in an urban river. *Environ. Pollut.* **2021**, *284*, 117516. [CrossRef]
46. Lu, H.-C.; Ziajahromi, S.; Neale, P.A.; Leusch, F.D.L. A systematic review of freshwater microplastics in water and sediments: Recommendations for harmonisation to enhance future study comparisons. *Sci. Total Environ.* **2021**, *781*, 146693. [CrossRef]
47. Pivokonsky, M.; Cermakova, L.; Novotna, K.; Peer, P.; Cajthaml, T.; Janda, V. Occurrence of microplastics in raw and treated drinking water. *Sci. Total Environ.* **2018**, *643*, 1644–1651. [CrossRef]
48. Hidalgo-Ruz, V.; Gutow, L.; Thompson, R.C.; Thiel, M. Microplastics in the Marine Environment: A Review of the Methods Used for Identification and Quantification. *Environ. Sci. Technol.* **2012**, *46*, 3060–3075. [CrossRef]
49. OECD. *Test No. 221: Lemna sp. Growth Inhibition Test*; OECD Publishing: Berlin, Germany, 2006.
50. OECD. *Test No. 203: Fish, Acute Toxicity Test*; OECD Publishing: Berlin, Germany, 2019. [CrossRef]
51. Kokalj, A.J.; Kunej, U.; Skalar, T. Screening study of four environmentally relevant microplastic pollutants: Uptake and effects on *Daphnia magna* and *Artemia franciscana*. *Chemosphere* **2018**, *208*, 522–529. [CrossRef]
52. Zong, X.; Zhang, J.; Zhu, J.; Zhang, L.; Jiang, L.; Yin, Y.; Guo, H. Effects of polystyrene microplastic on uptake and toxicity of copper and cadmium in hydroponic wheat seedlings (*Triticum aestivum* L.). *Ecotoxicol. Environ. Saf.* **2021**, *217*, 112217. [CrossRef]
53. Kalčíková, G.; Zupančič, M.; Jemec, A.; Žgajnar Gotvajn, A. The impact of humic acid on chromium phytoextraction by aquatic macrophyte *Lemna minor*. *Chemosphere* **2016**, *147*, 311–317. [CrossRef]

54. Lichtenthaler, H.K. 34 Chlorophylls and carotenoids: Pigments of photosynthetic biomembranes. *Methods Enzymol.* **1987**, *148*, 350–382. [CrossRef]
55. Prata, J.C.; da Costa, J.P.; Girão, A.V.; Lopes, I.; Duarte, A.C.; Rocha-Santos, T. Identifying a quick and efficient method of removing organic matter without damaging microplastic samples. *Sci. Total Environ.* **2019**, *686*, 131–139. [CrossRef]

plants

Article

Effects of Microplastic Contamination on the Aquatic Plant *Lemna minuta* (Least Duckweed)

Simona Ceschin [1,*], Flaminia Mariani [1,*], Dario Di Lernia [1], Iole Venditti [1], Emanuele Pelella [1] and Maria Adelaide Iannelli [2]

[1] Department of Sciences, University of Roma Tre, Viale G. Marconi 446, 00146 Rome, Italy
[2] Institute of Agricultural Biology and Biotechnology—National Research Council (IBBA-CNR), Via Salaria Km 29.300, Monterotondo Scalo, 00015 Rome, Italy
* Correspondence: simona.ceschin@uniroma3.it (S.C.); flaminia.mariani@uniroma3.it (F.M.); Tel.: +39-06-57336324 (S.C.)

Abstract: Microplastics are widely spread in aquatic environments. Although they are considered among the most alarming contaminants, toxic effects on organisms are unclear, particularly on freshwater plants. In this study, the duckweed *Lemna minuta* was grown on different concentrations (50, 100 mg/L) of poly(styrene-co-methyl methacrylate) microplastics (MP) and exposure times (T0, T7, T14, T28 days). The phytotoxic effects of MP were investigated by analyzing several plant morphological and biochemical parameters (frond and root size, plant growth, chlorophyll, and malondialdehyde content). Observations by scanning electron microscope revealed MP adsorption on plant surfaces. Exposition to MP adversely affected plant growth and chlorophyll content with respect to both MP concentrations and exposure times. Conversely, malondialdehyde measurements did not indicate an alteration of oxidative lipid damage in plant tissue. The presence of MP induced root elongation when compared to the control plants. The effects of MP on *L. minuta* plants were more evident at T28. These results contribute to a better understanding of MP's impact on aquatic plants and highlight that MP contamination manifests with chronic-type effects, which are thus detectable at longer exposure times of 7 days than those traditionally used in phytotoxicology tests on duckweeds.

Keywords: poly(styrene-co-methylmethacrylate); free-floating plant; freshwater; microplastic adsorption; phytotoxic effect; chronic impact

Citation: Ceschin, S.; Mariani, F.; Di Lernia, D.; Venditti, I.; Pelella, E.; Iannelli, M.A. Effects of Microplastic Contamination on the Aquatic Plant *Lemna minuta* (Least Duckweed). *Plants* **2023**, *12*, 207. https://doi.org/10.3390/plants12010207

Academic Editors: Viktor Oláh, Klaus-Jürgen Appenroth and K. Sowjanya Sree

Received: 28 October 2022
Revised: 2 December 2022
Accepted: 30 December 2022
Published: 3 January 2023

1. Introduction

Plastics are synthetic organic polymers that are mainly derived from fossil fuel-based chemicals like natural gas or petroleum [1]. Certain characteristics of plastics have caused their wide use, such as their lightweight nature, versatility, strength, longevity, hygiene, food compatibility, and washability. Because of that, plastics production increased from 15 million tons in 1964 to more than 368 million tons produced in the year 2019, of which 114 were produced in China alone and 59 in Europe, and its capacity is expected to double by 2040 [2].

In short, plastics have been and continue to be one of the most widely used synthetic materials in the world, in daily life products as well as in agriculture and industry. Despite the benefits of using plastic products, the release of large amounts of this material into the environment has become a cause for increasing global concern, being considered the second most alarming environmental problem after global warming [2].

Unauthorized discharges and inadequate waste management lead to the massive release of plastics into the environment that accumulates in various environmental matrices, taking many years to degrade. In addition, plastic waste debris exposed to weathering gradually fragments into smaller pieces that are dispersed into aquatic and terrestrial environments [3,4], increasing their potential for adsorption, ingestion, and accumulation by living organisms [5].

The polymers most commonly produced as plastics are polystyrene (PS), polyurethane (PUR), polyvinyl chloride (PVC), polypropylene (PP), polyethylene (PE), polyethylene terephthalate (PET), and their copolymeric compounds, which together account for about 80 percent of total plastic production [6].

Plastics can be classified according to their size in: macroplastics (>25 mm), mesoplastics (5–25 mm), microplastics (0.1–5 mm), and nanoplastics (<100 nm) [2]. In aquatic environments, the main sources of microplastics are direct discharges from water treatment plants, industrial and agricultural wastewater, and spontaneous degradation of macro- and mesoplastics in water [7]. When microplastics enter the aquatic environment, some remain suspended in the body of water, others float to the surface, and still others with a higher density settle to the bottom. Animal organisms belonging to different trophic levels can easily ingest microplastics, and more and more cases are highlighting the worrying phenomenon of biomagnification of plastics along the food chain [7].

Although most of the research has been focused primarily on the ecological problem of plastic contamination in marine ecosystems [1,8–11], some recent investigations have highlighted that plastics are an equally serious source of environmental risk to freshwater ecosystems [12–16].

Furthermore, investigations of microplastics and their impact on aquatic communities have mainly focused on animal organisms, largely neglecting plant organisms, even though plants play a central role in ecosystems both trophically, structurally and functionally [17,18]. Furthermore, plants, the first interface between the abiotic and biotic components of an ecosystem, assume the important role of early warning systems, essential for early intercepting contamination and, therefore, limiting biomagnification processes both along the food chain and in the environment [19].

Although aquatic and bank plants are often exposed to plastic pollution, studies verifying the effects of microplastics on freshwater plants are still scarce and concern very few species [19], including microalgae of the genus *Scenedesmus* and *Chlorella* and some flowering plants, such as *Lemna minor* L. and *Myriophyllum spicatum* L. [18,20,21]. The available literature points out that the main phytotoxic effects of plastics are inhibition of photosynthesis and limitation of shoot and root growth. These effects would appear to be due to microplastic particles adsorbing on the outer plant tissues and forming physical blockages to light and air by hindering photosynthesis and respiration activities [7,20,22–25]. However, some of these studies showed that generally, plant species are only affected when the concentrations of microplastics are higher than those occurring in nature [20,24].

Low-density microplastics with specific densities < 1 g cm^{-3}, such as microspheres of polyethylene (PE), polystyrene (PS), and polypropylene (PP), are distributed in the upper layers of slow-flowing waters [26,27]. Here, microplastics frequently encounter pleustophytes, that are aquatic plants free-floating on the water surface, which have roots and the lower surface of vegetative body in direct contact with the water. Very common and widespread pleustophytes are the duckweeds (Lemnaceae family) that, although characterized by a very tiny vegetative body (frond), are able to produce a floating plant mat [28], which can easily trap plastic material in slow water of lakes, ponds, canals, lentic stretches of rivers, or smaller water basins [22]. Some field observations have shown how some duckweed species of the *Lemna* genus, such as *Lemna minuta* Kunth, retain a large amount of surface floating pollutants, including microplastics. The scarcity of studies on the interactions and effects of microplastics on duckweeds makes it necessary to acquire more information about it, as the duckweeds play a major role in aquatic ecosystems. Indeed, they serve both as habitats for many animal species, providing protection from predators or sites for larval spawning, and as a food source for many insects, fish, and waterfowl [29], becoming the basis of many food chains in aquatic environments. Based on the above-mentioned characteristics, duckweed species could be used for the removal of plastic material from the aquatic environment by phytostabilization. Phytostabilization involves stabilizing and binding the contaminant by adsorption on the leaves and roots, reducing its dispersion in water. Recently, it has been shown that, in addition to dissolved contaminants [30] and nanoplastics [31], microplastics may be adsorbed on the surface and accumulated by vascular plants [20,32]. Although they have not yet been studied consistently, several mechanisms have been suggested to explain this phenomenon of the adsorption of plastic material by aquatic macrophytes. The electrostatic forces of the cellulosic constituents of plant cells can attract microplastics, and their adsorption is facilitated by the roughness of plant surfaces, which provide many binding sites for plastic particles [33]. The surface morphology of the plant organisms can also play an important role in microplastic-plant interactions; in fact, for micro- and macroalgae, the more complex the algal thallus structure is on the surface, the more it can trap microplastics [34–36]. In addition, if a periphyton layer (e.g., composed by microalgae) is present on the plant surfaces, it creates a higher viscosity that increases the retention of microplastics [37].

Polystyrene (PS) and their copolymeric compounds, due to their insulating properties and extreme lightness, are among the most widely used plastic materials in construction and beyond. They can be found on the market in the common version (hard and rigid) or in the form of an expanded product (commonly called polystyrene), depending on their functions. Expanded PS is mainly used for packaging, and thermal and electrical insulation, while common PS is used for many disposable items (e.g., cutlery, razors, CD cases), furniture items, tableware, toys, lining of household appliances, and for many other items. Especially, polymethyl methacrylate is widely used in medicine for bone cements, contact and intraocular lenses, screw fixation in bone, filler for bone cavities and skull defects, vertebral stabilization in osteoporotic patients, and for packaging of medical devices [38]. In this research, poly(styrene-co-methyl methacrylate) microplastics (MP) were analyzed due to their widespread use and the large amount of waste generated by this low-density material.

Namely, this study aimed to assess (i) the ability of the duckweed *L. minuta* to adsorb MP from the water medium and (ii) the phytotoxic effects of MP on this aquatic plant. The results obtained can contribute to better understanding the type of impact that MP may have on aquatic plant organisms as well as further investigating the adsorption mechanisms of these contaminants, knowledge of which can be relevant to safeguarding the health not only of the plant community but also of the entire aquatic ecosystem.

2. Results and Discussion

2.1. Water Chemical and Physical Parameters

Water chemical- and physical parameters were measured at T0, T7, T14, and T28 in the control tests and in two treatments with MP (MP50 and MP100) (Table S1).

In all tests, the dissolved oxygen concentration (DO) of the water increased overall over time, but in both microplastic treatments the values were significantly higher than in the control ($p < 0.001$; increase of about 20%) (Figure 1; Table S2). This result is justifiable by considering the positive correlation between DO and amount of Lemna biomass recorded during the experiment (rho = 0.855; p < 0.001) (Figure 2). In fact, in the MP50 and MP100 treatments, the amount of biomass, although increasing over time and positively correlated with DO (rho = 0.855; $p < 0.001$), was significantly lower at T28 (a decrease of about 45%) than in the control, forming thinner floating mats that did not cover the entire water surface, therefore, allowing gaseous exchanges between air and water. On the contrary, in the control test, the greater amount of biomass produced limited gas exchanges between air and water, causing a higher DO reduction compared to the treatments. The negative influence of *L. minuta* floating mats on DO concentrations as a function of their thickness confirms findings from previous laboratory and field studies on the impact of this duckweed on the water chemical and physical components in aquatic ecosystems [39,40].

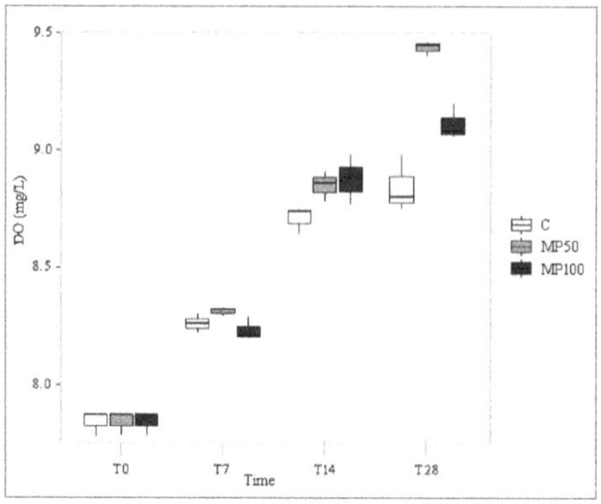

Figure 1. Dissolved oxygen (DO) in the control (C) and microplastic treatments (MP50 and MP100) at different exposure times (T0, T7, T14, and T28).

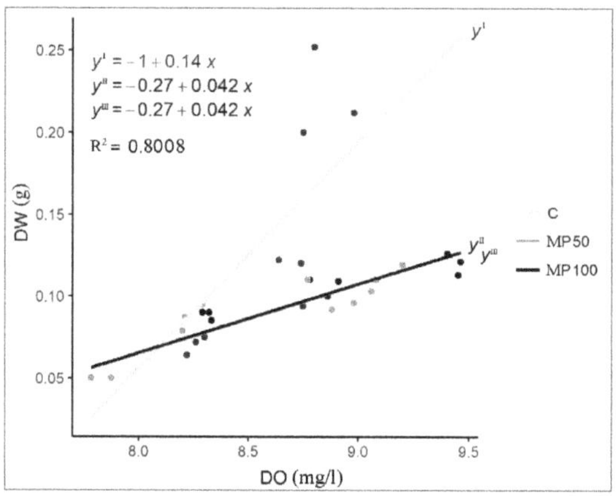

Figure 2. Regression model between DW and DO in control and microplastic treatments (MP50 and MP100). The regression lines of MP50 and MP100 are overlapping.

The mean water temperature values in the two MP treatments did not differ significantly from the control ($p > 0.05$) (Tables S1 and S2), varying in function of the air temperature recorded in the laboratory. The mean pH values increased progressively throughout the experiment in all tests, increasing by more than one unit in both treatments compared to T0 (Table S1). However, pH was significantly different only during MP100 treatment compared to the control (Table S2).

Conductivity, a factor dependent on the water ionic components, and salinity, related to the dissolved salt content of the water, decreased in all tests from T0 to T7, due to the presence of the starting *Lemna* populations, which absorbed these solutes for their vegetative growth. However, from T7 to T28, conductivity and salinity showed a steady increase (Table S1), which is most likely due to weekly refills of mineral water rich in ions and salts, whose utilization by *Lemna* plants failed to compensate for the continuous inputs.

2.2. *Effects of Microplastics on Plant Growth Parameters*

In general, the amount of biomass (FW and DW) increased linearly in all tests during the experiment. However, up to T14, no significant differences ($p > 0.05$) were observed between the tests, whereas at T28, the difference was highly significant ($p < 0.001$) between the control (an increase of 130%) and treatments, as was consequently the difference recorded for the relative growth rate (RGR) ($p < 0.001$) (Figures 3 and 4; Table S3). These results show that exposure of *L. minuta* populations to MP induces a negative effect on plant growth and that this effect occurs at both MP concentrations tested but after a prolonged period of exposure. These findings would differ from previous studies conducted on the related species *L. minor*, exposed for seven days to microplastics [20,22], in which no significant effects on biomass and RGR were recorded.

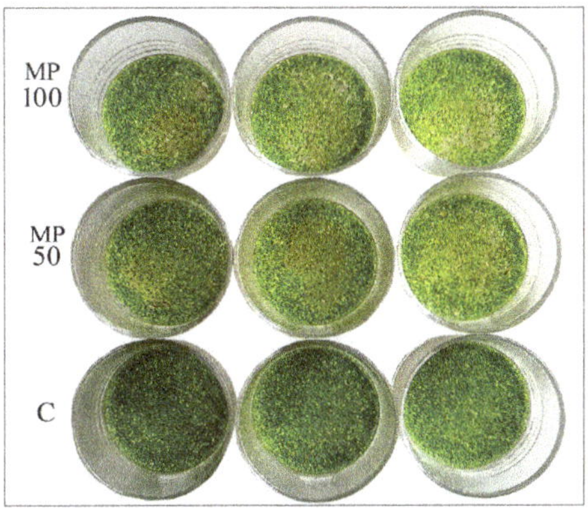

Figure 3. Biomass of *L. minuta* in the control (C) and microplastic treatments (MP50 and MP100) at T28. For each group, the three replications are shown.

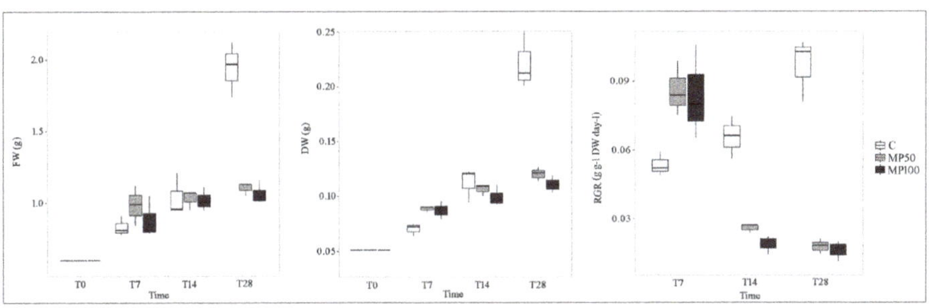

Figure 4. Fresh weight (FW), dry weight (DW), and relative growth rate (RGR) of *L. minuta* in control (C) and microplastic treatments (MP50 and MP100) at different exposure times (T0, T7, T14, and T28).

Lemna samples observed by SEM showed MP microspheres adsorbed on roots and both upper and lower frond surfaces (Figure 5). This suggests the hypothesis that the adsorption of microparticles on roots and frond surfaces physically impedes the smooth passage of light and oxygen and the uptake of nutrients, consequently limiting the regular growth of the plant.

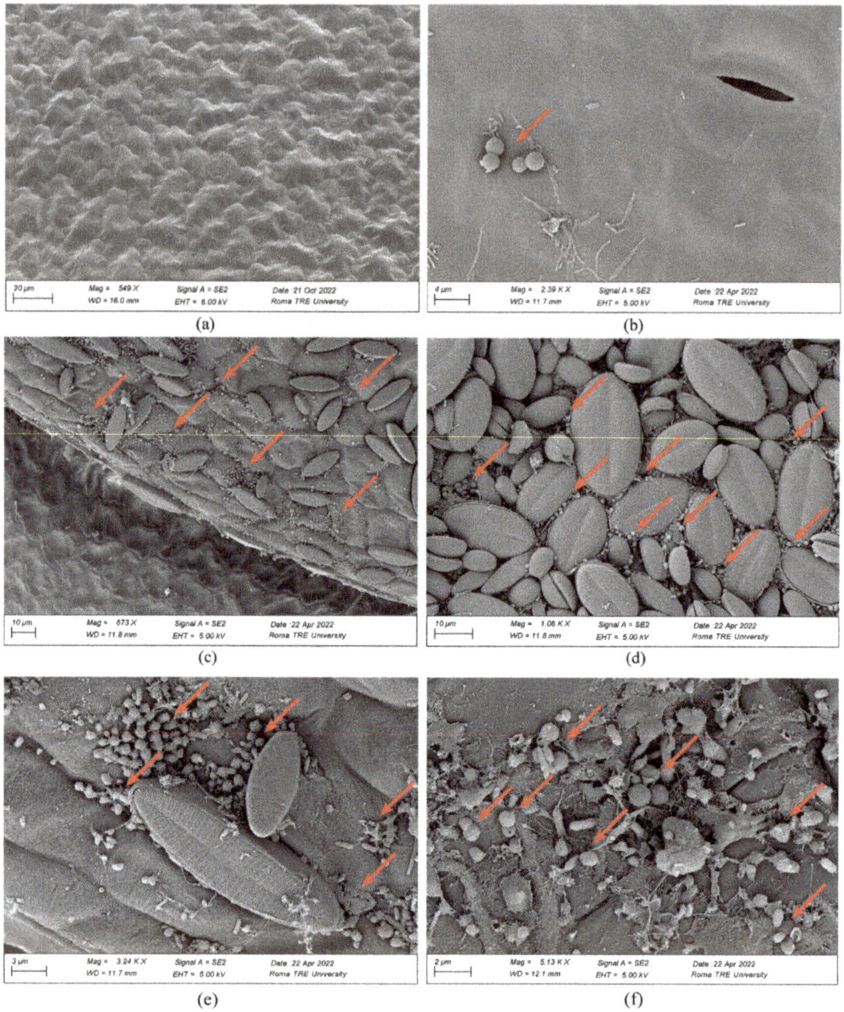

Figure 5. Scanning Electron Microscopy (SEM) images of *L. minuta* fronds grown in aqueous medium without (**a**) and with MP particles (arrows) (**b**–**f**). In detail: adaxial frond surface in control test at T28 (**a**), MP microspheres adsorbed onto the adaxial surface in PS50 treatment at T7 (**b**), and on abaxial surfaces in PS50 at T7 (**c**), in PS50 at T28 (**d**), and in PS100 at T28 (**e**–**f**). Evident aggregates of MPs with pennate diatoms (**c**–**e**).

Frond length and width and root width showed no significant differences between controls and treatments ($p > 0.05$) in the presence of MP. In contrast, root length increased significantly from T14 to T28 during exposure to the two different MP concentrations (MP50, 85%; MP100, 50%) compared with the control (Figure 6; Table S3). Already, Kalčíková et al. [22] highlighted that depending on composition, concentration, and size, microplastics show different impacts on plant growth, such as a re-direction of growth between root and frond or between root thickness and root elongation. However, while they recorded that microplastics cause mechanical stress that hinders root growth, in contrast, root elongation occurred in this study. The increase in root length observed in both treatments might suggest that treated *Lemna* samples tend to elongate their roots to reach those portions of the water column furthest from the surface; here, in fact, there are fewer MP microparticles in suspension due to their low specific density [26,27], and thus the plant might have more surface area available to take up water and nutrients without physical obstructions.

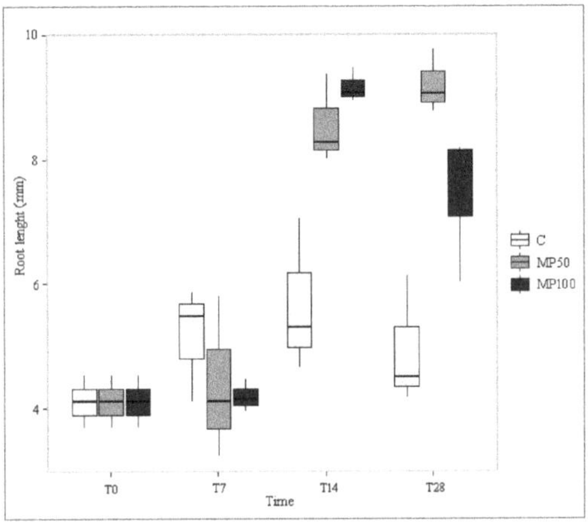

Figure 6. Root length (mm) of *L. minuta*, measured in control (C) and microplastic treatments (MP50, MP100) at different exposure times (T0, T7, T14, T28).

2.3. Effects of MP on Biochemical Parameters

Total chlorophyll content (Chl_{tot}) was not affected by the exposition to both MP concentrations (MP50 and MP100) until T7, and these results agree with other studies carried out on L. minor over shorter experimental times [20–22]. Differently, in the MP100 treatments both at T14 and T28, Chl_{tot} was significantly lower than in the control ($p < 0.001$) and the reduction was about 20% (Figure 7; Table S3), while for MP50 the pigment content is slightly decreased 10% only at T28. Long-term exposure to high microplastic concentrations led to chlorosis effects on *L. minuta* plants, as it is observable in Figure 3. Conversely, some studies highlighted that microplastics had no significant effect on *L. minor* chlorophyll content over both shorter [20–22] and longer experimental times [21].

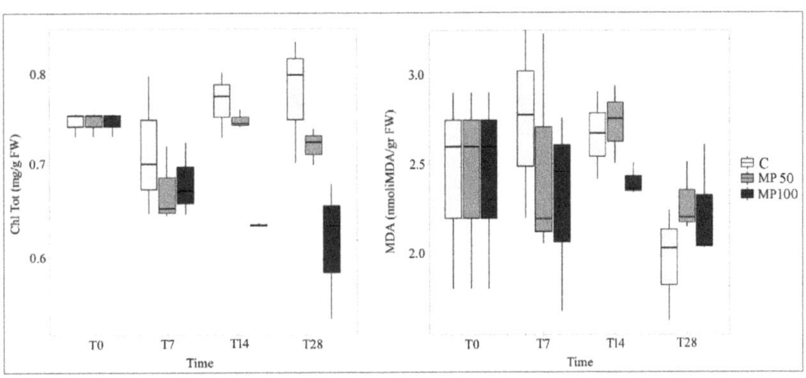

Figure 7. Total chlorophyll (Chl_{tot}) and malondialdehyde (MDA) content in *L. minuta* samples in control (C) and microplastic treatments (MP50, MP100) at different exposure times (T0, T7, T14, T28).

Malondialdehyde (MDA) content showed no changes during the experiment, and no significant difference was observed between the treatments MP50 and M100 compared to the control (Figure 7; Table S3). MDA measurement is closely related to lipid peroxidation activity occurring at the subcellular level, and its increase implies the induction of an oxidative stress condition. Therefore, these results on MDA content suggest that the MP used were adsorbed, but not absorbed by *Lemna*, which then did not internalize them.

2.4. Adsorption of MP on Plants

Quali-quantitative SEM observation of single *Lemna* samples from the two treatments revealed that MP were primarily adsorbed on the portions of the plant most exposed to the contaminant, namely the abaxial surface of the fronds, secondarily, the adaxial surface (Figure 5). In addition to MP, the presence of many pinnate diatoms (unicellular microalgae with silicified plant walls) was noted, especially on the abaxial surfaces (Figure 5). Preliminary observations of the *Lemna* samples collected in the field highlighted that these microalgae were already associated with *Lemna* in nature, and, thus, their presence was not related to a later contamination occurring in the laboratory during the experiment. An aspect that relates diatoms and microplastics has been pointed out in some recent studies where plastic seems to behave as a "rigid" substrate on which microalgae can more easily adhere and grow according to the phenomenon of biofouling [41,42]. The cause of such aggregation is not yet clear, but specific properties of plastics may help microalgae aggregation and growth [43]. Adsorption of MP on *Lemna* samples may have facilitated increased adhesion of microalgae on the surfaces of the treated *Lemna* samples, potentially amplifying the phytotoxic effects of microplastics.

3. Materials and Methods

3.1. Production of Microplastics

Microplastics (MP) used to investigate the effect on *L. minuta* plants were obtained from pellets of poly(styrene-co-methyl methacrylate) [P(S-co-MMA)] (Aldrich 462896, pellets average Mw 100,000–150,000 pellets, styrene 40%) following the OBM method reported in a previous work [44]. In particular, 300 mg of P(S-co-MMA) were dissolved in 10 mL of acetone (C_3H_6O, technical grade, Merck) and stirred for 24 h, then an aliquot of 7 mL was transferred into a dialysis cellulose membrane (width 10 mm, Sigma Aldrich D9277-100FT) and further immersed into 200 mL of distilled water for 5 days at constant temperature (T = 24 °C). MP were observed using a Gemini 300 field emission SEM system (Carl Zeiss AG, Jena, Germany), and their mean diameter was verified on SEM images by ImageJ software vers. 1.53t (National Institutes of Health, Bethesda, MD, USA). The mean diameter ± SE was calculated to be 2.60 μm ± 1.54 (Figure 8).

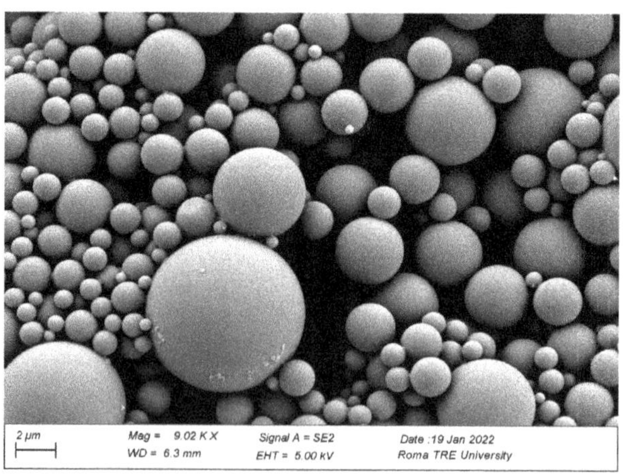

Figure 8. Scanning electron microscope (SEM) image of stock suspension of MP particles.

Stock MP suspensions were prepared at two different concentrations, 50 (MP50) and 100 (MP100) mg/L. These concentrations, on average higher than those recorded in nature, were used to stress the system with the aim of obtaining more evident biological responses and comparing the effects with other similar studies [20–22].

3.2. *Plant Material and Experimental Set-Up*

Samples of *L. minuta* were collected from a natural pond within the Appia Antica Regional Park (Rome, Italy), and then transported to the laboratory in containers containing local water. *Lemna minuta* samples were acclimated for seven days in mineral water of known chemical composition (Table S4). The mineral profile of the chosen water was the closest match to the one in ponds where *L. minuta* grows spontaneously in nature [28].

100 mL of MP stock suspensions were transferred into cylindrical glass containers (7 × 7 mm, 240 mL), and an amount of 0.6 g of *L. minuta* fronds was added, corresponding to about 80 percent coverage of the water surface. In parallel, a control set was arranged with *L. minuta* and water without MP. Three replicates (n = 3) were set up for the two MP concentrations (MP50, MP100) and control. Plants in the control and treatment tests were grown for 28 days and were sampled at different time points: 0 (T0), 7 (T7), 14 (T14), and 28 (T28) days (Figure 3). The choice of longer exposure times than those used in similar studies [20–22] is based on the assumption that microparticles mainly cause a chronic, rather than acute, toxic effect [45].

Every 7 days, to restore the water level to 100 mL, each container was refilled using the same water medium in which the plant samples were grown.

3.3. *Determination of Water Chemical and Physical Parameters*

The measurement of water chemical and physical parameters, such as temperature (T, °C), pH (pH, pH values), conductivity (C, μS/cm), salinity (S, ‰), and dissolved oxygen concentration (DO, mg/L) was done using a multiparameter immersion probe (Hach-Lange HQ40d) at different time points T0, T7, T14 and T28 days.

3.4. *Plant Growth and Morphological Measurements*

3.4.1. Plant Growth Analysis

The biomass amount of *L. minuta* was measured at each experimental time (T0, T7, T14, and T28) to quantify any changes during plant growth. *Lemna minuta* biomass was collected with a fine-mesh metal sieve and, after having dried for one minute on blotting paper, fresh weight (FW) was measured using a precision scale (AS2001, Digital Scale, Ascher). Then, plant biomass was completely dried for 72 h at 60 °C to determine dry weight (DW). Thereafter, Relative Growth Rate (RGR, $g^{-1}day^{-1}$) was calculated using the following formula [46]:

$$RGR = (\ln DW_f - \ln DW_i)/(T_f - T_i)$$

where: DW_f = final dry weight (g), DW_i = initial dry weight (g), T_f = total incubation period (day), and T_i = initial time (day) at each experimental time.

3.4.2. Morphological Analysis

To examine possible variations in frond and root sizes, five individuals of *L. minuta* were taken from each replicate and placed on graph paper to be observed and photographed under a stereomicroscope (Stemi 305, ZEISS). Specifically, length and width of both fronds and roots were measured for each individual, using ImageJ software vers. 1.53t (National Institutes of Health, Bethesda, MD, USA).

3.5. *Analysis of Biochemical Parameters*

At each experimental time, biochemical parameters such as total chlorophyll (Chl_{tot}) and malondialdehyde (MDA) content were measured to analyze the plant physiological performance in response to MP treatments.

3.5.1. Determination of Chlorophyll Content

Total chlorophyll content in *Lemna* fronds was measured as an indicator of the physiological status of the plant. Fresh *Lemna* fronds (0.2–0.5 g) were soaked in 10 mL of 95% (*v*/*v*) ethanol for 3 days in a stoppered tube at room temperature and in the dark. Samples were centrifuged at 3000× *g* for 10 min, and the absorbance of the supernatant was measured at 663 and 645 nm [47]. Chlorophyll concentration was calculated following the equations described by Huang et al. [46]:

$$Chl_a = 12.72\ A_{663} - 2.69\ A_{645} \quad (1)$$

$$Chl_b = 22.90\ A_{645} - 4.68\ A_{663} \quad (2)$$

$$Chl_{Tot} = Chl_a + Chl_b \quad (3)$$

where Chl_a, Chl_b, and Chl_{Tot} represent chlorophyll *a*, chlorophyll *b*, and total chlorophyll contents, respectively. A_{663} and A_{645} are the absorbances at 663 and 645 nm, respectively. Results are expressed as mg of total chlorophyll per gram of fresh weight plant tissue (mg/g FW).

3.5.2. Determination of Malondialdehyde Content

Lipid peroxidation was measured by spectrophotometric methods by estimating malondialdehyde (MDA) content, which is considered a biomarker of oxidative damage in plant tissue and thus an indicator of plant stress conditions.

Frozen samples were homogenized in a precooled mortar and pestle with two volumes of ice-cold 0.1% (*w*/*v*) trichloroacetic acid (TCA), and 1 mM Ethylenediamine tetraacetic acid (EDTA) and centrifuged for 15 min at 16,000× *g*. A mixture containing 1 mL of supernatant and 2 mL of 0.5% (*w*/*v*) thiobarbituric acid (TBA) in 20% (*w*/*v*) TCA was heated to 95 °C for 30 min and then rapidly cooled in an ice bath.

After centrifugation (16,000× *g* for 10 min at 4 °C), the absorbance of the supernatant was read at 532 nm, and the values corresponding to non-specific adsorption at 600 nm were subtracted. The concentration of MDA was calculated using the extinction coefficient (ε = 155 mM/cm).

3.6. SEM Observations of Microplastics

At the end of each experimental time (T0, T7, T14, and T28), from each control and treatment test, 50 fronds of *Lemna* were randomly taken and dehydrated through EtOH baths in series at increasing concentrations (10, 30, 50, 70, 90, and 100%). Then, they were dried at the critical point (Bal-Tec CPD 030), mounted on a stub (using self-adhesive carbon discs), gold sputter coated (Emitech k550), and observed by scanning electron microscope (SEM) (Gemini 300, Carl Zeiss AG, Jena, Germany).

On selected acquired SEM images, the adsorption of MP on *L. minuta* was verified considering all plant surfaces, thus both the adaxial and abaxial surfaces of the frond as well as the entire root surface.

3.7. Statistical Analyses

Multiple two-way ANOVA tests were conducted to compare changes in chemical and physical water parameters and plant physiological measurements in the separate tests (C, MP50, MP100) and over different exposure times (T0, T7, T14, e T28). A post-hoc analysis (Tukey's Test) was conducted for each ANOVA test. Assumptions of normality and homoscedasticity were tested both prior to ANOVA and on the model residuals. The correlation between biotic and abiotic parameters that were significantly different between treatments was investigated by calculating Spearman's rank correlation coefficient. Where there was a strong and significant correlation, further analyses were conducted via analysis of covariance (ANCOVA). Where assumptions of homoskedasticity were not met, a logarithmic transformation of the dependent variable was performed. Non-significant interaction terms were removed from the ANCOVA models via stepwise selection. Graphs were made using the ggplot2 and sjPlot packages [48,49]. All statistical analyses were conducted using R software vers. 4.2.1 [50].

4. Conclusions

This study demonstrates that the aquatic plant *L. minuta* can adsorb MP large 1–5 μm, whose adsorption occurs mainly on plant surfaces in direct contact with the contaminated suspension (i.e., abaxial frond surface). The amount of adsorbed MP on *L. minuta* fronds was dose- and time-dependent. Indeed, MP most affected *L. minuta* growth (biomass, RGR) at the highest concentration (MP100) and after 28 days of exposure (T28). Simultaneously, reduction of chlorophyll content was evident, indicating that the plant exposition to MP contamination has reduced its photosynthetic capacity. Anyway, long-term monitoring of the effects of MP on the growth and biochemical parameters of *L. minuta* pointed out that this plant can tolerate high MP concentrations. Thus, free-floating mats of *L. minuta*, thanks to that tolerance and ability of phytostabilizing MP particles, it could be exploited in the phytoremediation of water contaminated by microplastics. It should be noted that the ability of this plant to capture microplastics can depend by different environmental conditions; for example, the natural presence of periphyton on the plant tissue may increase the number of microplastics adsorbed by the plants.

As a whole, these results contribute to a better understanding of microplastics impact on aquatic plants and highlight that MP contamination manifests with chronic-type effects, thus observable at longer exposure times than those traditionally used of 7 days in phytotoxicity tests on duckweeds [51,52]. Further investigations on the adsorption mechanisms of microplastics by this duckweed will be relevant to verifying the actual possibility of using this type of plant organisms in the phytoremediation of freshwaters contaminated with microplastics, which would currently seem to be one of the most promising biological approaches to remove microparticles in situ and then to safeguard the health of the plant community and the entire aquatic ecosystem.

Supplementary Materials: The following supporting information can be downloaded at: https://www.mdpi.com/article/10.3390/plants12010207/s1, Table S1: Mean values of chemico-physical water parameters measured in control (C) and microplastic treatments (MP50, MP100) at different exposure times (T0, T7, T14, T28); Table S2: ANOVA results related to chemico-physical water parameters; Table S3: ANOVA results related to plant parameters; Table S4: Chemico-physical composition of mineral water used as an aqueous medium for experiments.

Author Contributions: Conceptualization, S.C. and F.M.; methodology, S.C., F.M., D.D.L., I.V. and M.A.I.; validation, S.C., F.M., I.V. and M.A.I.; formal analysis, D.D.L., E.P. and M.A.I.; investigation, S.C., F.M., D.D.L. and M.A.I.; resources, S.C., I.V. and M.A.I.; data curation S.C., F.M., D.D.L. and M.A.I.; writing—original draft preparation, S.C. and D.D.L.; writing—review and editing, S.C., F.M., I.V., E.P. and M.A.I.; visualization, S.C. and D.D.L.; supervision, S.C. and M.A.I.; project administration, S.C.; funding acquisition, S.C., I.V. and M.A.I. All authors have read and agreed to the published version of the manuscript.

Funding: This research did not receive any specific funding but was supported by the Grant to the Department of Science, Roma Tre University (MIUR-Italy Departments of Excellence, Article 1, Comma 314–337, Low 232/2016) and by Project "Biomemory", FOE-2021 (MIUR), Ministerial Decree n. 844 of the 16–7–2021.

Data Availability Statement: Not applicable.

Acknowledgments: The authors are very grateful to Valerio Renzoni for his support during experimentation and Andrea Di Giulio for his contribution to the observations of the samples by scanning electron microscopy at the LIME, the Interdepartmental Laboratory for Electron Microscopy at University of Roma TRE.

Conflicts of Interest: The authors declare no conflict of interest.

References

1. Cole, M.; Lindeque, P.; Halsband, C.; Galloway, T.S. Microplastics as contaminants in the marine environment: A review. *Mar. Pollut. Bull.* **2011**, *62*, 2588–2597. [CrossRef] [PubMed]
2. Azeem, I.; Adeel, M.; Ahmad, M.A.; Shakoor, N.; Jiangcuo, G.D.; Azeem, K.; Ishfaq, M.; Shakoor, A.; Ayaz, M.; Xu, M.; et al. Uptake and Accumulation of Nano/Microplastics in Plants: A Critical Review. *Nanomaterials* **2021**, *11*, 2935. [CrossRef] [PubMed]
3. Thompson, R.C.; Olsen, Y.; Mitchell, R.P.; Davis, A.; Rowland, S.J.; John, A.W.G.; McGonigle, D.; Russel, A.E. Lost at sea: Where is all the plastic? *Science* **2004**, *304*, 838. [CrossRef] [PubMed]
4. Ryan, P.G.; Moore, C.J.; Van Franeker, J.A.; Moloney, C.L. Monitoring the abundance of plastic debris in the marine environment. *Philosoph. Trans. R. Soc. B Biol. Sci.* **2009**, *364*, 1999–2012. [CrossRef] [PubMed]
5. Pereao, O.; Opeolu, B.; Fatoki, O. Microplastics in aquatic environment: Characterization, ecotoxicological effect, implications for ecosystems and developments in South Africa. *Environ. Sci. Pollut. Res.* **2020**, *27*, 22271–22291. [CrossRef]
6. Almroth, C.B.; Eggert, H. Marine plastic pollution: Sources, impacts, and policy issues. *Rev. Env. Econ. Policy* **2019**, *13*, 317–326. [CrossRef]
7. Ma, P.; Wei Wang, M.; Liu, H.; Feng Chen, Y.; Xia, J. Research on ecotoxicology of microplastics on freshwater aquatic organisms. *Env. Poll. Bioavailab.* **2019**, *31*, 131–137. [CrossRef]
8. Barnes, D.K.A.; Galgani, F.; Thompson, R.C.; Barlaz, M. Accumulation and fragmentation of plastic debris in global environments. *Philosoph. Trans. R. Soc. B Biol. Sci.* **2009**, *364*, 1985–1998. [CrossRef]
9. Browne, M.A.; Crump, P.; Niven, S.J.; Teuten, E.; Tonkin, A.; Galloway, T.; Thompson, R. Accumulation of microplastic on shorelines worldwide: Sources and sinks. *Environ. Sci. Technol.* **2011**, *45*, 9175–9179. [CrossRef]
10. Arossa, S.; Martin, C.; Rossbach, S.; Duarte, C.M. Microplastic removal by Red Sea giant clam (*Tridacna maxima*). *Environ. Pollut.* **2019**, *252*, 1257–1266. [CrossRef]
11. Horton, A.A.; Walton, A.; Spurgeon, D.J.; Lahive, E.; Svendsen, C. Microplastics in freshwater and terrestrial environments: Evaluating the current understanding to identify the knowledge gaps and future research priorities. *Sci. Total Environ.* **2017**, *586*, 127–141. [CrossRef]
12. Zbyszewski, M.; Corcoran, P.L. Distribution and degradation of freshwater plastic particles along the beaches of Lake Huron, Canada. *Water Air Soil Pollut.* **2011**, *220*, 365–372. [CrossRef]
13. Zbyszewski, M.; Corcoran, P.L.; Hockin, A. Comparison of the distribution and degradation of plastic debris along shorelines of the Great Lakes, North America. *J. Great Lakes Res.* **2014**, *40*, 288–299. [CrossRef]
14. Koelmans, A.A.; Besseling, E.; Shim, W.J. Nanoplastics in the aquatic environment. Critical review. In *Marine Anthropogenic Litter*; Bergmann, M., Gutow, L., Klages, M., Eds.; Springer: Cham, Switzerland, 2015; pp. 325–340.
15. Mattsson, K.; Hansson, L.A.; Cedervall, T. Nano-plastics in the aquatic environment. *Environ. Sci. Process. Impacts* **2015**, *17*, 1712–1721. [CrossRef]
16. Van Sebille, E.; Wilcox, C.; Lebreton, L.; Maximenko, N.; Hardesty, B.D.; Van Franeker, J.A.; Eriksen, M.; Siegel, D.; Galgani, F.; Law, K.L. A global inventory of small floating plastic debris. *Environ. Res. Lett.* **2015**, *10*, 124006. [CrossRef]
17. Rillig, M.C.; Lehmann, A.; de Souza Machado, A.A.; Yang, G. Microplastic effects on plants. *New Phytol.* **2019**, *223*, 1066–1070. [CrossRef]
18. Kalčíková, G. Aquatic vascular plants—A forgotten piece of nature in microplastic research. *Environ. Pollut.* **2020**, *262*, 114354. [CrossRef]
19. Ceschin, S.; Bellini, A.; Scalici, M. Aquatic plants and ecotoxicological assessment in freshwater ecosystems: A review. *Environ. Sci. Pollut. Res.* **2021**, *28*, 4975–4988. [CrossRef]
20. Mateos-Cárdenas, A.; Scott, D.T.; Seitmaganbetova, G.; Van Pelt, F.N.A.M.; O'Halloran, J.; Jansen, M.A.K. Polyethylene microplastics adhere to *Lemna minor* (L.) yet have no effects on plant growth or feeding by *Gammarus duebeni* (Lillj.). *Sci. Total Environ.* **2019**, *689*, 413–421. [CrossRef]
21. Rozman, U.; Kokalj, A.J.; Dolar, A.; Drobne, D.; Kalčíková, G. Long-term interactions between microplastics and floating macrophyte *Lemna minor*: The potential for phytoremediation of microplastics in the aquatic environment. *Sci. Total Environ.* **2022**, *831*, 154866. [CrossRef]
22. Kalčíková, G.; Žgajnar Gotvajn, A.; Kladnik, A.; Jemec, A. Impact of polyethylene microbeads on the floating freshwater plant duckweed *Lemna minor*. *Environ. Pollut.* **2017**, *230*, 1108–1115. [CrossRef] [PubMed]
23. Bosker, T.; Bouwman, L.J.; Brun, N.R.; Behrens, P.; Vijver, M.G. Microplastics accumulate on pores in seed capsule and delay germination and root growth of the terrestrial vascular plant *Lepidium sativum*. *Chemosphere* **2019**, *226*, 774–781. [CrossRef] [PubMed]
24. Van Weert, S.; Redondo-Hasselerharm, P.E.; Diepens, N.J.; Koelmans, A.A. Effects of nanoplastics and microplastics on the growth of sediment-rooted macrophytes. *Sci. Total Environ.* **2019**, *654*, 1040–1047. [CrossRef] [PubMed]
25. Yi, X.; Chi, T.; Li, Z.; Wang, J.; Yu, M.; Zhou, H. Combined effect of polystyrene plastics and triphenyltin chloride on the green algae *Chlorella pyrenoidosa*. *Environ. Sci. Pollut. Res.* **2019**, *26*, 15011–15018. [CrossRef] [PubMed]
26. Eerkes-Medrano, D.; Thompson, R.C.; Aldridge, D.C. Microplastics in freshwater systems: A review of the emerging threats, identification of knowledge gaps and prioritisation of research needs. *Water Res.* **2015**, *75*, 63–82. [CrossRef]
27. Marchetto, D.; Latella, A.; Pojana, G. Le microplastiche nell'ambiente marino. *Chim. Ambiente* **2017**, *1*, 18–25.

28. Ceschin, S.; Della Bella, V.; Piccari, F.; Abati, S. Colonization dynamics of the alien macrophyte *Lemna minuta* Kunth: A case study from a semi-natural pond in Appia Antica Regional Park (Rome, Italy). *Fundam. Appl. Limnol.* **2016**, *188*, 93–101. [CrossRef]
29. Van Hoeck, A.; Horemans, N.; Monsieurs, P.; Cao, H.X.; Vandenhove, H.; Blust, R. The first draft genome of the aquatic model plant *Lemna minor* opens the route for future stress physiology research and biotechnological applications. *Biotechnol. Biofuels* **2015**, *8*, 188. [CrossRef]
30. Rezania, S.; Taib, S.M.; Md Din, M.F.; Dahalan, F.A.; Kamyab, H. Comprehensive review on phytotechnology: Heavy metals removal by diverse aquatic plants species from wastewater. *J. Hazard. Mater.* **2016**, *318*, 587–599. [CrossRef]
31. Ekperusi, A.O.; Sikoki, F.D.; Nwachukwu, E.O. Application of common duckweed (*Lemna minor*) in phytoremediation of chemicals in the environment: State and future perspective. *Chemosphere* **2019**, *223*, 285–309. [CrossRef]
32. Dovidat, L.C.; Brinkmann, B.W.; Vijver, M.G.; Bosker, T. Plastic particles adsorb to the roots of freshwater vascular plant *Spirodela polyrhiza* but do not impair growth. *Limnol. Oceanogr. Lett.* **2020**, *5*, 37–45. [CrossRef]
33. Bhattacharya, P.; Lin, S.; Turner, J.P.; Ke, P.C. Physical adsorption of charged plastic nanoparticles affects algal photosynthesis. *J. Phys. Chem. C* **2010**, *114*, 16556–16561. [CrossRef]
34. Schaffelke, B. Particulate organic matter as an alternative nutrient source for tropical *sargassum* species (Fucales, phaeophyceae). *J. Phycol.* **1999**, *35*, 1150–1157. [CrossRef]
35. Gutow, L.; Eckerlebe, A.; Gim_enez, L.; Saborowski, R. Experimental evaluation of seaweeds as a vector for microplastics into marine food webs. Environ. *Sci. Technol.* **2016**, *50*, 915–923. [CrossRef]
36. Savio, S.; Farrotti, S.; Di Giulio, A.; De Santis, S.; Ellwood, N.T.W.; Ceschin, S.; Congestri, R. Functionalization of frustules of the diatom *Staurosirella pinnata* for nickel (Ni) adsorption from contaminated aqueous solutions. *Front. Mar. Sci.* **2022**, *9*, 889832. [CrossRef]
37. Goss, H.; Jaskiel, J.; Rotjan, R. *Thalassia testudinum* as a potential vector for incorporating microplastics into benthic marine food webs. *Mar. Pollut. Bull.* **2018**, *135*, 1085–1089. [CrossRef]
38. Frazer, R.Q.; Byron, R.T.; Osborne, P.B.; West, K.P. PMMA: An essential material in medicine and dentistry. *J. Long. Term. Eff. Med. Implants* **2005**, *15*, 629–639. [CrossRef]
39. Ceschin, S.; Abati, S.; Traversetti, L.; Spani, F.; Del Grosso, F.; Scalici, M. Effects of the invasive duckweed *Lemna minuta* on aquatic animals: Evidence from an indoor experiment. *Plant Biosyst.* **2019**, *153*, 749–755. [CrossRef]
40. Ceschin, S.; Ellwood, N.T.W.; Ferrante, G.; Mariani, F.; Traversetti, L. Habitat change and alteration of plant and invertebrate communities in waterbodies dominated by the invasive alien macrophyte *Lemna minuta* Kunth. *Biol. Invasions* **2020**, *22*, 1325–1337. [CrossRef]
41. Prata, J.C.; Da Costa, J.P.; Lopes, I.; Duarte, A.C.; Rocha-Santos, T. Effects of microplastics on microalgae populations: A critical review. *Sci. Total Environ.* **2019**, *665*, 400–405. [CrossRef]
42. Mateos-Cárdenas, A.; van Pelt, F.N.; O'Halloran, J.; Jansen, M.A. Adsorption, uptake and toxicity of micro-and nanoplastics: Effects on terrestrial plants and aquatic macrophytes. *Environ. Pollut.* **2021**, *284*, 117183. [CrossRef] [PubMed]
43. Nava, V.; Leoni, B. A critical review of interactions between microplastics, microalgae and aquatic ecosystem function. *Water Res.* **2021**, *188*, 116476. [CrossRef] [PubMed]
44. Chronopoulou, L.; Fratoddi, I.; Palocci, C.; Venditti, I.; Russo, M.V. Osmosis based method drives the self-assembly of polymeric chains into micro-and nanostructures. *Langmuir* **2009**, *25*, 11940–11946. [CrossRef] [PubMed]
45. Sharma, S.; Chatterjee, S. Microplastic pollution, a threat to marine ecosystem and human health: A short review. *Environ. Sci. Pollut. Res.* **2017**, *24*, 21530–21547. [CrossRef]
46. Radford, P.J. Growth analysis formulae—Their use and abuse. *Crop. Sci.* **1967**, *7*, 171–175. [CrossRef]
47. Huang, F.; Guo, Z.; Xu, Z. Determined methods of chlorophyll from *Lemma paucicostata*. *Exp. Technol. Manag.* **2007**, *24*, 29–31.
48. Kassambara, A. *R Package*, Version 0.4.0. ggpubr: 'ggplot2' Based Publication Ready Plots. R Foundation for Statistical Computing: Vienna, Austria, 2020.
49. Lüdecke, D. *R Package*, Version 2.8.11. sjPlot: Data Visualization for Statistics in Social Science. CERN, European Organization for Nuclear Research: Genève, Switzerland, 2022.
50. R Core Team. *R: A Language and Environment for Statistical Computing*; R Foundation for Statistical Computing: Vienna, Austria, 2022.
51. ISO. *ISO 20079*. Water Quality—Determination of the Toxic Effect of Water Constituents and Wastewater on Duckweed (Lemna minor)—Duckweed Growth Inhibition Test. International Organization for Standardization: Geneve, Switzerland, 2005; pp. 1–23.
52. OECD. *OECD Test No. 221: Lemna sp. Growth Inhibition Test*; OECD Guidelines for the Testing of Chemicals, Section 2; OECD Publishing: Paris, France, 2006; pp. 1–22.

plants

MDPI

Article

Chlorophyll Fluorescence Imaging-Based Duckweed Phenotyping to Assess Acute Phytotoxic Effects

Viktor Oláh *, Anna Hepp, Muhammad Irfan and Ilona Mészáros

Department of Botany, Faculty of Science and Technology, University of Debrecen, Egyetem tér 1, H-4032 Debrecen, Hungary; pitypang666@gmail.com (A.H.); muhadirfan6167@gmail.com (M.I.); immeszaros@unideb.hu (I.M.)

* Correspondence: olahviktor@unideb.hu

Abstract: Duckweeds (Lemnaceae species) are extensively used models in ecotoxicology, and chlorophyll fluorescence imaging offers a sensitive and high throughput platform for phytotoxicity assays with these tiny plants. However, the vast number of potentially applicable chlorophyll fluorescence-based test endpoints makes comparison and generalization of results hard among different studies. The present study aimed to jointly measure and compare the sensitivity of various chlorophyll fluorescence parameters in *Spirodela polyrhiza* (giant duckweed) plants exposed to nickel, chromate (hexavalent chromium) and sodium chloride for 72 h, respectively. The photochemistry of Photosystem II in both dark- and light-adapted states of plants was assessed via *in vivo* chlorophyll fluorescence imaging method. Our results indicated that the studied parameters responded with very divergent sensitivity, highlighting the importance of parallelly assessing several chlorophyll fluorescence parameters. Generally, the light-adapted parameters were more sensitive than the dark-adapted ones. Thus, the former ones might be the preferred endpoints in phytotoxicity assays. Fv/Fm, i.e., the most extensively reported parameter literature-wise, proved to be the least sensitive endpoint; therefore, future studies might also consider reporting Fv/Fo, as its more responsive analogue. The tested toxicants induced different trends in the basic chlorophyll fluorescence parameters and, at least partly, in relative proportions of different quenching processes, suggesting that a basic distinction of water pollutants with different modes of action might be achievable by this method. We found definite hormetic patterns in responses to several endpoints. Hormesis occurred in the concentration ranges where the applied toxicants resulted in strong growth inhibition in longer-term exposures of the same duckweed clone in previous studies. These findings indicate that changes in the photochemical efficiency of plants do not necessarily go hand in hand with growth responses, and care should be taken when one exclusively interprets chlorophyll fluorescence-based endpoints as general proxies for phytotoxic effects.

Keywords: chlorophyll fluorescence imaging; PAM fluorometry; duckweed test; *Spirodela polyrhiza*; phytotoxicity; hormesis

Citation: Oláh, V.; Hepp, A.; Irfan, M.; Mészáros, I. Chlorophyll Fluorescence Imaging-Based Duckweed Phenotyping to Assess Acute Phytotoxic Effects. *Plants* **2021**, *10*, 2763. https://doi.org/10.3390/plants10122763

Academic Editor: Miguel Pedro Mourato

Received: 2 October 2021
Accepted: 9 December 2021
Published: 14 December 2021

1. Introduction

Duckweed species (members of the Lemnaceae family) are widely used test objects in ecotoxicology research. In fact, phytotoxic effects of potential toxicants on aquatic macrophytes are presently modelled predominantly by using duckweeds (*Lemna minor* L., *L. gibba* L. and *Spirodela polyrhiza* L. Scleid.) in standardized ecotoxicity tests (e.g., OECD Guideline 221, ISO No. 20079 and ISO/NP No. 20227, respectively). They owe this popularity due to their small size, fast and predominantly vegetative reproduction, simple anatomy and sensitivity to various aquatic toxicants [1,2]. Duckweed ramets—usually referred to as fronds—basically consist of an upper and lower epidermis, a spongy mesophyll and one or two meristematic regions, which differentiate new fronds. The size of the fronds is in the mm–cm range and the doubling time of their cultures can be as fast as two days under suitable growth conditions [3]. In duckweed-based phytotoxicity assays, toxic effects have been usually quantified via the inhibition of biomass growth in cultures, in terms of frond number, frond area, fresh or dry mass change, respectively [4,5]. In the course of the methodological development, other test endpoints, such as chlorophyll- or N-content [6], colony disintegration [7], or root growth [8] have also been applied, but due to their simplicity and straightforward interpretation, the biomass-based toxicity endpoints became the most common parameters in *Lemna*-tests. In order to properly measure biomass growth inhibition, it is inevitable to ensure the test cultures grow for several frond generations. The most common ISO [4] and OECD [5] test protocols use seven-day-long exposures and require a control biomass doubling time shorter than 2.5 days in order to ensure adequate sensitivity of tests.

As a promising alternative to the biomass-based methods in duckweed phytotoxicity-tests, the applicability of *in vivo* chlorophyll fluorescence (ChlF) has been long studied [9]. ChlF predominantly originates from Photosystem II (PSII) as the fraction of absorbed light energy that has not been utilized in photochemical energy conversion nor released as heat. Photosynthesis provides the basis for plant growth and is a complex process with several sensitive components. Thus, ChlF supposedly can be used as an early-warning proxy of adverse effects, offering fast and sensitive phytotoxicity test protocols [10,11]. Pulse-amplitude modulated (PAM) fluorescence imaging is an emerging tool to assess phytotoxic effects [12]. It is based on acquiring fluorescence signals from 2 or three-dimensional objects via a CCD camera, and thus enables analyses on spatial heterogeneities in photosynthetic properties. This technique also allows parallel phenotyping of several samples under identical conditions, e.g., in multiwell plates.

Imaging PAM fluorometry has already been proven to be applicable in assessing phytotoxic effects in algae [13,14], seagrass [15,16], or mosses [17]. Duckweeds are apparently ideal objects for ChlF imaging-based phytotoxicity assays: they are easy to grow in genetically homogenous, aseptic cultures and are small enough to fit into multiwell plates. In addition, the flat fronds of most species float on the water surface facing towards the CCD camera while taking up toxicants directly to the assimilating tissue. The internationally standardized protocols and the already published, extensive duckweed phytotoxicity literature provides solid grounds for further methodological development. An increasing body of reports indicates that ChlF imaging-based duckweed phenotyping has been successfully applied in assessing the effects of various stressors (Table S1). In the studies listed in Table S1, ChlF was predominantly assessed by means of two commercially available platforms: the Imaging-PAM (Walz GmbH, Effeltrich, Germany, 22 out of 29 papers) and FluorCam (Photon System Instruments, Brno, Czech Republic, 6 out of 29 papers) instruments. The ChlF parameters were used either solely or in combination with other test endpoints derived from, e.g., plant growth, photosynthetic pigment content, or oxidative stress markers, and proved to be a robust proxy for phytotoxic effects in several cases.

PAM chlorophyll fluorescence measurements generate an excessive number of various parameters that are basically derived from five mutually independent chlorophyll fluorescence levels, that is ground (Fo) and maximal fluorescence yields (Fm) in the dark-adapted state; and steady-state (Fs), ground (F'o) and maximal fluorescence yields (F'm) in light-acclimated state of samples, respectively. The referred 29 papers in Figure 1 (for details, see Table S1) report a total of 18 ChlF-derived test endpoints -with a median of 4 parameters per study- of which Fv/Fm (90% of papers), Y(II) (55% of papers) and NPQ (45% of papers) were the most common ones. This diversity of potentially applicable test endpoints makes it difficult to compare ChlF-based phytotoxicity data from different sources. Additionally, most papers report data for single toxicants or well-defined groups of toxicants with similar modes of action (e.g., PSII inhibiting herbicides), but apply various exposure conditions. Thus, it is hard to generalize plant responses in terms of ChlF parameters, or to assess the specificity of particular ChlF endpoints to different toxicants. It can also be noted that, though most of the reviewed studies jointly used various types of chlorophyll fluorescence parameters, a significant part (~30%) of them relied exclusively on either dark- or light-adapted photochemical efficiency indices (Figure 1).

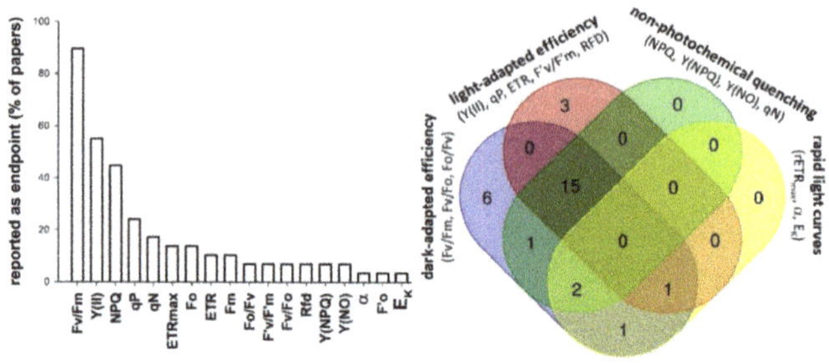

Figure 1. Reported application of various chlorophyll fluorescence induction-derived phytotoxicity test endpoints in scientific studies assessing duckweeds by means of chlorophyll fluorescence imaging (**left**), and Venn diagram denoting the number of studies using different types of chlorophyll fluorescence parameters as test endpoints (**right**). Details of the reviewed 29 papers are summarized in Supplementary Table S1.

In the present study, we assessed the applicability of a ChlF imaging-based duckweed test method performed in multiwell tissue culturing plates as an alternative to the standard duckweed growth inhibition tests. In order to do this, we used three different toxicants, namely nickel (Ni), hexavalent chromium -Cr(VI)- and sodium chloride (NaCl), and parallelly analyzed concentration-response curves of several ChlF parameters. The applied toxicants have different modes of action, and they have also been suggested as reference toxicants in standard duckweed growth inhibition tests [1,5]. Ni is essential for plants as a cofactor for enzymes involved in N-metabolism. Ni thus promotes plant growth at low concentrations, but in high doses, it induces disorders in nutrient uptake, N-metabolism, and water balance, leading to growth inhibition and declining fitness in the longer term [18]. Ni at higher concentrations affects photosynthesis by decreasing chlorophyll content, altering chloroplast ultrastructure, displacing Mg in chlorophyll and RuBisCo, inhibiting the photosynthetic electron transport chain, and inactivating PSII and Photosystem I (PSI), respectively [19–21]. In aqueous environments, its prevalent oxidation form is Ni^{2+} ion, and it disturbs duckweed growth starting from the 10^{-4} g L^{-1} concentration range [22,23]. Chromium has no proven beneficial physiological role in plants [24]. Hexavalent form of chromium occurs in waters as $CrO_4{}^{2-}$ and $Cr_2O_7{}^{2-}$ oxyanions, and those forms can be actively taken up by plants via the sulfate transport system [25]. Cr(VI) toxicity can lead to impaired nutrient balance, e.g., because of the inhibition of plasma membrane H^+-ATPase, nitrate reductase, and Fe(III)-reductase enzymes. Cr(VI) can react with macromolecules in cells, and thus triggers oxidative stress while it gets reduced to Cr(III) [18]. At the photosynthesis level, Cr(VI) decreases chlorophyll content, inhibits the photosynthetic electron transport, inactivates enzymes of the Calvin–Benson-cycle, and disorganizes chloroplast ultrastructure, respectively [25]. According to Naumann et al. [22] and Oláh et al. [23], Cr(VI) inhibits duckweed growth in the 10^{-3}–10^{-4} g L^{-1} range. NaCl, besides inducing osmotic stress, impairs plant growth by Na^+ and Cl^- ion toxicity. Salinity stress disturbs mineral nutrition and water relations, alters protein conformation and triggers the formation of reactive oxygen species (ROS) [26]. NaCl was reported to reduce chlorophyll content, inhibit electron transport and change chloroplast ultrastructure [27]. It inhibits the growth of various duckweed species in the 10^0–10^1 g L^{-1} concentration range [28].

By applying the above toxicants, our specific aims were (i) to assess the concentration-response relationships of different ChlF-based test endpoints, (ii) to find the most responsive ones, and (iii) to test if different ChlF-parameters show general or rather toxicant-specific patterns in responses.

2. Materials and Methods

2.1. Plant Material and Treatments

The tests were performed with the giant duckweed (*Spirodela polyrhiza* (L.) Schleid.) clone UD0401 (RDSC #5501) [29]. Test plants were obtained from seven-day-old stock cultures maintained on 100 mL of Steinberg medium ([1] pH 6.0 ± 0.2) in 300 mL Erlenmeyer flasks under constant temperature (24 ± 2 °C) and irradiation (white, 58 ± 4 μE m^{-2} s^{-1}).

The reference toxicants were applied in the following final concentrations:

Ni (as $NiSO_4 \times 7H_2O$) or Cr(VI) (as $K_2Cr_2O_7$): 0; 0.039; 0.078; 0.156; 0.313; 0.625; 1.25; 2.5; 5.0 and 10.0 mg L^{-1}, respectively.

NaCl: 0; 2; 4; 6; 8; 10; 12; 14 and 16 g L^{-1}, respectively.

The exposures were performed in 12-well tissue culture plates, and each well contained 4 mL test solution, i.e., Steinberg medium containing either toxicant at an above-listed concentration. At the beginning of tests, 1-1 healthy *S. polyrhiza* colony with 4–5 fronds were inoculated in each well. The exposures lasted for three days (72 ± 4 h). This duration was based on preliminary experiments, which indicated that ChlF parameters were not affected by nutrient depletion or crowding while frond numbers had doubled under control conditions.

2.2. Chlorophyll Fluorescence Imaging

After three days of toxicant treatments, the ChlF parameters of test plants were measured by means of Maxi Imaging PAM (Heinz Walz GmbH, Effeltrich, Germany) equipped with blue Imag-Max/L LED-array Illumination Unit (wavelength λ = 450 nm) and IMAG-K6 CCD camera (2/3" chip with 1392 × 1040 pixels resolution) using the following protocol (for example fluorescence charts see also Supplementary Figure S1):

1. First, the test plants in plate were dark-adapted for 20 min in order to reach a fully oxidized state of PSII photochemistry (all PSII reaction centers were "open").
2. In dark-adapted state ground fluorescence yield (Fo) of plants was determined using weak, non-inductive measuring light (blue, 450 nm, intensity: 2, frequency: 1).
3. A single saturating pulse (blue, 450 nm, intensity: 9, i.e., ~4000 μE m^{-2} s^{-1} with the applied instrumental setup, length 720 ms) was used to saturate PSII photochemistry (all PSII reaction centers became "closed") and to determine maximal fluorescence yield (Fm) of plants.
4. After the saturation pulse, a continuous actinic irradiation (blue, 450 nm) with similar intensity (intensity: 5, equivalent to ~77 μE m^{-2} s^{-1} in the applied instrumental setup) to that of the plants' ambient light environment was switched on for 10 min to induce a light-adapted state of plants (Supplementary Figure S1). After 10 min, steady-state chlorophyll fluorescence yield (Fs) of plants under actinic illumination was determined and a second saturating pulse—with the same settings as for determining Fm—was applied to measure their light-adapted maximal fluorescence yield (F'm). Due to technical reasons, F'o, which is the ground fluorescence yield in a light-acclimated state, was not directly measured by the instrument, but calculated using the formula [30]:

$$F'o = Fo/(Fv/Fm + Fo/F'm).$$

Using the above five, mutually independent ChlF yields, ChlF parameters were calculated according to Table 1.

Table 1. The calculated ChlF parameters in dark-adapted and light-adapted states of *S. polyrhiza* test plants, measured after 20 min of dark-adaption and 10 min of non-saturating actinic irradiation, respectively.

Parameter	Calculation	Description	Reference
Fv/Fm	(Fm − Fo)/Fm	Maximum quantum yield of PSII photochemistry	[31]
Fv/Fo	(Fm − Fo)/Fo	Maximum ratio of quantum yields of photochemical and concurrent non-photochemical processes in PSII in dark-adapted state	[31]
Y(II)	ΔF/F′m	The effective quantum yield of photochemical energy conversion in PSII	[31]
Y(NPQ)	(Fs/F′m) − (Fs/Fm)	Quantum yield of regulated non-photochemical energy loss in PSII	[32]
Y(NO)	Fs/Fm	Quantum yield of non-regulated heat dissipation and fluorescence emission	[32]
qP	1 − ((Fs − F′o)/F′v) = ΔF/F′v = (F′m − Fs)/(F′m − F′o)	Photochemical quenching of variable ChlF, i.e., the fraction of open PSII reaction centers in light-adapted state	[31]
F′v/F′m	(F′m − F′o)/F′m	Effective quantum yield of PSII photochemistry in light-adapted state	[31]
F′v/F′o	(F′m − F′o)/F′o	Effective ratio of quantum yields of photochemical and concurrent non-photochemical processes in PSII related to light-adapted state	[31]
Rfd	(Fm − Fs)/Fs	Chlorophyll fluorescence decrease ratio proportional to net photosynthesis; a.k.a. "vitality index"	[33]

Immediately after the termination of the above measuring sequence, a so-called rapid light curve (RLC) of PSII photochemistry was performed with 15 consecutively increasing irradiation steps (PPFD: 0-530 E m^{-2} s^{-1}, each light step lasted for 30 s), using the built-in light curve protocol of the instrument. At each step, the electron transport rate (ETR) of plants was calculated according to the formula [30]:

$$\text{ETR } (\mu\text{mol e}^- \text{ m}^{-2} \text{ s}^{-1}) = (\Delta F/F'm) \times \text{PPFD} \times 0.5 \times 0.84 \quad (1)$$

where "ΔF/F′m" was the effective quantum yield of photochemical energy conversion in PSII [i.e., Y(II)] at a given light intensity; "PPFD" is the nominal incident photon flux density in terms of $\mu E\ m^{-2}\ s^{-1}$; "0.5" referred to an assumed stoichiometric distribution of absorbed light energy between the two Photosystems, and "0.84" denoted an empirical 84% absorption efficiency of incident light in higher plants [12,15,34,35].

By plotting the obtained ETR values as a function of the corresponding actinic irradiation intensities, a saturation curve was obtained that allowed the calculation of maximal photon use efficiency (α) and maximal ETR (ETR_{max}) under the applied ambient conditions as follows:

α was calculated as the initial slope of the curve at the first light step, i.e.:

$$\alpha = \text{ETR}/11\ \mu\text{E m}^{-2}\text{ s}^{-1} \quad (2)$$

ETR_{max} was considered as the highest calculated ETR value of the curve.

2.3. Data Processing and Statistical Analyses

Each concentration was applied in four parallel treatments (wells) and the experiments were repeated three times with the same experimental design. During data processing, means of the four parallels within an experiment were normalized to their respective control means, resulting in a total of n = 3 data per toxicant concentration. Concentration-dependent responses of Fv/Fm, Fv/Fo, Y(II), qP, F'v/F'm, F'v/F'o, Rfd, α and ETR_{max} were analyzed by fitting non-linear regression models to the data. Based on the assumption that all the above endpoints reflect the efficiency of PSII photochemistry in their specific way, and, theoretically, can decrease to zero, the concentration-dependence of ChlF parameters were assessed by means of log-logistic models fixing the lower limit of the functions to 0. Model fittings were performed by means of the "drc"-package (version 2.5-12) [36], in R statistical environment (version 3.2) [37], using "LL.3" (three-parameter log-logistic) function of the package.

Additionally, plant responses were also modelled by the modified four-parameter log-logistic functions "BC.4" and "CRS.4" of the "drc"-package, respectively. These models were developed by [38] ("BC"-model) and by [39] ("CRS"-model) in order to include a hormetic component at the lower concentration range. After visually checking the model fittings, the goodness of fit for different regression models was compared based on the Akaike information criterion (AIC), using the "mselect" function of the "drc"-package. In those cases, where the hormetic component did not contain additional information based on the AIC score, the three-parameter log-logistic model was applied to the given endpoint. In those cases, on the other hand, when the hormetic models described the concentration-response relationships better, the model selection was based on the estimated residual standard error of the models by means of the "mselect" function. The selected model for a given endpoint was also checked by the lack-of-fit test using the "modelFit" function and tested by means of pseudo-R^2 using the "cor" function of the "drc"-package, respectively. In the latter procedure, correlation between the predicted and observed responses in terms of each ChlF parameter was compared at the applied toxicant concentrations.

Based on the fitted regression models, the effective concentrations resulting in 20 and 50% inhibition of the given endpoint (EC_{20} and EC_{50}, respectively) were estimated by means of the "ED" function of the "drc"-package. When hormetic models were used, the "No Observable Adverse Effect Concentration" (NOAEC), i.e., the toxicant concentration at which the hormetic effect diminishes; the maximum of the hormetic response ("max"), and the corresponding concentration at which the maximal hormetic response occurs (EC_{max}) were also estimated using the "MAX" function of the "drc"-package.

3. Results

After 72 h-long exposures, all three toxicants affected the assessed ChlF parameters in the applied concentration range (Figure 2). In general, Ni has an inhibiting effect (>5%) from 1.25 mg L^{-1}. From that concentration, the assessed indices of PSII efficiency decreased in a concentration-dependent manner (Figure 2, the summary statistics of the original data are supplemented in Table S2). On the other hand, below 1.25 mg L^{-1} Ni proved to have hormetic effect in the case of most parameters (Table 2, for the plotted model fittings, refer to Supplementary Figure S2). Though the maximal response did not exceed 105% of control in some cases, concentration-response relationships were better described by those non-linear models that included the hormetic component, with the only exception of that of Fv/Fm. The maximal hormetic response exceeded the control by 10% in the case of maximal electron transport rate (ETR_{max}). In general, the NOAEC ranged between 0.6–0.7 mg L^{-1} and the Ni concentration resulting in maximal hormetic response was 20–40% of that concentration (0.1–0.3 mg L^{-1}, Table 2). By comparing the obtained EC_{20} and EC_{50} values, we found considerable differences amongst different parameters. The calculated EC_{20} values were in the 1.0–3.7 mg L^{-1} range while EC_{50} values ranged from 1.8 to 20.6 mg L^{-1}, and four of the assessed nine parameters had higher calculated EC_{50} than the maximal applied Ni-concentration (Table 2). ETR_{max} and the effective quantum yield of photochemical energy conversion in a light-adapted state [Y(II)] responded with the highest sensitivity (i.e., lowest EC_{20} and EC_{50}) while the maximum quantum yield of PSII photochemistry (Fv/Fm) and the relative fluorescence decrease (Rfd) proved to be the least responsive parameters, respectively.

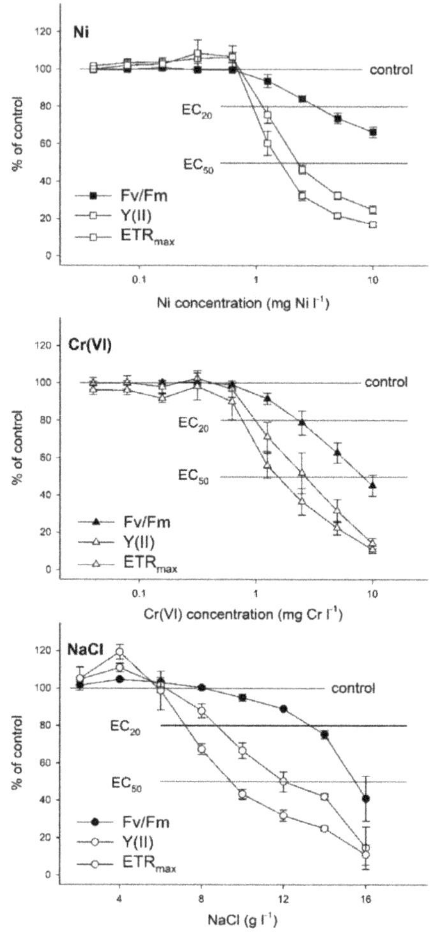

Figure 2. Concentration-response relationships of the dark-adapted (Fv/Fm) and light-adapted photochemical energy conversion -Y(II)- and of the maximal electron transport rate (ETR_{max}) for *S. polyrhiza* UD0401 test plants exposed to Ni, Cr(VI) or NaCl for 72 h, respectively. Symbols denote grand means (means of means) ± SE of three experiments (n = 3), as a percentage of respective control data. For summary statistics of every measured parameter, please refer to Supplementary Tables S3, S5 and S7, respectively.

Cr(VI)-induced inhibition started from 1.25 mg L^{-1} for most ChlF parameters, though ETR_{max} and the ratio of quantum yields of photochemical and concurrent non-photochemical processes in PSII in the light-adapted state (F'v/F'o) showed ~10% decrease even at 0.625 mg L^{-1} (Figure 2, the summary statistics of the original data are supplemented in Table S3). In contrast with Ni, Cr(VI) did not result in hormetic response (Table 3). Though concentration-response relationships for most ChlF parameters were better described by models including a hormetic component than by the three-parameters log-logistic model, the calculated maxima did not exceed 102% of the control with the only exception of Rfd (Table 3, for the plotted model fittings, refer to (Supplementary Figure S3). The calculated EC_{20} ranged from 0.7 to 2.5 mg L^{-1} while EC_{50}s were in the 1.8–9.0 mg L^{-1} range. Similar to Ni, the lowest EC_{20} and EC_{50} was calculated for ETR_{max}, while the highest effective concentrations were found in cases of Rfd and Fv/Fm (Table 3).

Table 2. Results of non-linear regression model fittings to the assessed ChlF parameters after 72 h of Ni-treatments. Models "LL.3", "BC.4" and "CRS.4" denote three-parameter log-logistic, 4-parameter Brain-Cousens and 4-parameter Cedergreen-Ritz-Streibig models, respectively. In the case of hormetic models, "max" denotes the highest calculated hormetic response as percent of the respective control data, "EC_{max}" denote the Ni-concentration resulting in maximal hormetic response, and "NOAEC" denotes the "No Adversible Effect Concentration", that is the Ni-concentration at which the hormetic effect diminishes. Bold numbers highlight maximal hormetic responses higher than 105%. EC_{20} and EC_{50} denote Ni-concentrations (as mg L^{-1} ± 95% confidence interval) resulting in 20 and 50% inhibition of the respective parameter. RSS and pseudo-R^2 indicate the residual sum of squares and the correlation between the predicted and observed responses at the applied concentrations, respectively.

Parameter	Model	DF	RSS	F-Value	Pseudo-R^2	Max	EC_{max}	NOAEC (95% CI)	EC_{20} (95% CI)	EC_{50} (95% CI)
Fv/Fm	LL.3	27	323.56	1.7206	0.9286	NA	NA	NA	3.72 (2.87–4.57)	18.6 * (13.8–23.4)
Fv/Fo	BC.4	26	914.79	0.5439	0.9522	102.5	0.20	0.52 (0.16–0.88)	1.41 (1.10–1.72)	4.20 (3.36–5.03)
Y(II)	BC.4	26	679	3.6103	0.9775	**108.3**	**0.26**	**0.67 (0.49–0.84)**	1.22 (1.05–1.38)	2.68 (2.35–3.01)
qP	CRS.4c	26	1652	1.0886	0.9151	**105.8**	**0.13**	**0.61 (0.09–1.14)**	1.74 (1.14–2.35)	5.19 (3.81–6.56)
F′v/F′m	BC.4	26	404.32	1.4081	0.9600	102.6	0.21	0.61 (0.30–0.91)	2.03 (1.65–2.41)	10.8 * (7.55–14.0)
F′v/F′o	BC.4	26	1051.19	1.3835	0.9554	**105.6**	**0.23**	**0.64 (0.38–0.89)**	1.3 (1.04–1.56)	3.40 (2.74–4.07)
Rfd	BC.4	26	302.72	3.8371	0.9602	103.4	0.13	0.38 (0.05–0.71)	2.13 (1.69–2.57)	20.59 * (12.0–29.2)
α	BC.4	26	747.14	0.3612	0.9267	103.3	0.23	0.68 (0.25–1.11)	2.20 (1.66–2.75)	11.4 * (6.70–16.1)
ETR_{max}	BC.4	26	1671.24	2.5494	0.9600	**110.0**	**0.28**	**0.66 (0.47–0.85)**	1.02 (0.85–1.19)	1.84 (1.56–2.13)

* Denote extrapolated effective concentrations, exceeding the highest applied Ni concentration (10 mg L^{-1}). NA indicates that the respective parameter is not applicable due to a lack of hormetic response. The summary statistics of the original data and the fitted regression models are supplemented in Table S2 and Figure S2.

Table 3. Results of non-linear regression model fittings to the assessed ChlF parameters after 72 h of Cr(VI)-treatments. Models "LL.3", "BC.4" and "CRS.4" denote three-parameter log-logistic, 4-parameter Brain-Cousens and 4-parameter Cedergreen-Ritz-Streibig models, respectively. In the case of hormetic models, "max" denote the highest calculated hormetic response as a percent of the respective control data, "EC_{max}" denote the Cr(VI)-concentration resulting in maximal hormetic response, and "NOAEC" denote the "No Adversible Effect Concentration", that is the Cr(VI)-concentration at which the hormetic effect diminishes. Bold numbers highlight maximal hormetic responses higher than 105%. EC_{20} and EC_{50} denote Cr(VI)-concentrations (as mg L^{-1} ± 95% confidence interval) resulting in 20 and 50% inhibition of the respective parameter. RSS and pseudo-R^2 indicate the residual sum of squares and the correlation between the predicted and observed responses at the applied concentrations, respectively.

Parameter	Model	DF	RSS	F-Value	Pseudo-R^2	Max	EC_{max}	NOAEC (95% CI)	EC_{20} (95% CI)	EC_{50} (95% CI)
Fv/Fm	BC.4	26	727.5	0.0317	0.9336	100.5	0.18	0.45 (−0.58–1.47)	2.46 (1.89–3.03)	8.26 (6.25–10.3)
Fv/Fo	BC.4	26	1783.9	0.0571	0.9382	102.1	0.21	0.49 (−0.02–1.00)	1.21 (0.89–1.53)	2.77 (2.24–3.30)
Y(II)	CRS.4b	26	1873.8	0.2508	0.9395	102.2	0.24	0.4 (−0.03–0.83)	1.06 (0.72–1.40)	2.54 (2.00–3.09)
qP	LL.3	27	5191.1	0.3166	0.8252	NA	NA	NA	2.21 (1.13–3.23)	4.47 (3.26–5.68)
F′v/F′m	BC.4	26	984.35	0.066	0.9426	100.3	0.13	0.32 (−0.36–1.00)	1.48 (1.09–1.87)	4.70 (3.74–5.67)

Table 3. *Cont.*

Parameter	Model	DF	RSS	F-Value	Pseudo-R^2	Max	EC_{max}	NOAEC (95% CI)	EC_{20} (95% CI)	EC_{50} (95% CI)
F'v/F'o	BC.4	26	2127.9	0.1227	0.9307	100.9	0.14	0.33 (−0.16–0.82)	0.93 (0.63–1.23)	2.25 (1.75–2.74)
Rfd	BC.4	26	1537.2	0.2505	0.8784	**105.3**	**0.34**	**0.98 (0.37–1.58)**	2.47 (1.77–3.17)	9.02 (5.12–12.9)
α	LL.3	27	918.75	0.2719	0.9357	NA	NA	NA	1.88 (1.34–2.43)	6.13 (5.11–7.15)
ETR_{max}	LL.3	27	2303.1	1.0996	0.9344	NA	NA	NA	0.72 (0.47–0.98)	1.85 (1.45–2.24)

NA indicates that the respective parameter is not applicable due to a lack of hormetic response. The summary statistics of the original data and the fitted regression models are supplemented in Table S3 and Figure S3.

Different ChlF parameters showed diverging responses to NaCl-treatments too. Photochemical quenching (qP) and Rfd started to decrease from 6 g L^{-1} NaCl, Y(II) and ETR_{max} from 8 g L^{-1}, while Fv/Fm, Fv/Fo, F'v/F'm (i.e., the effective quantum yield of PSII photochemistry in the light-adapted state), F'v/F'o and the maximal photon use efficiency (i.e., α) decreased from 10 g L^{-1} NaCl, after 72 h of exposure (Figure 2, the summary statistics of the original data are supplemented in Table S4). Several parameters showed distinct hormetic responses to lower NaCl concentrations, including Fv/Fo, Y(II), F'v/F'o, and ETR_{max}, respectively (Figure 2 and Figure S4, Table 4). The maximal stimulation ranged from 108 to 125% of respective control data with calculated EC_{max} of 3.2–4.6 g L^{-1} that is 50–57% of the calculated NOAEC. It should also be noted that even those parameters which were better described by three-parameter log-logistic functions increased by 2−7% at lower (2−4 g L^{-1}) concentrations (data not shown). The calculated EC_{20} concentrations were in the 7.4−13.2 g L^{-1} range, while EC_{50} was between 9.7 and 15.5 g L^{-1} for the assessed endpoints (Table 4). Similar to the heavy metal treatments, ETR_{max} showed the highest sensitivity, and Fv/Fm and Rfd were the least responsive parameters.

In order to compare the overall sensitivity of the assessed tests endpoints, we used mean ranks based on their calculated EC_{20} and EC_{50} values for the three toxicants, respectively. The ranks indicated a consistent order of sensitivity with the lowest effective concentrations calculated for ETR_{max}, Y(II) and F'v/F'o while Fv/Fm proved to be the least sensitive parameter (Figure 3a,b). Even Fv/Fo, which is based on the same two basic ChlF yields, i.e., ground (Fo) and maximal (Fm) fluorescence yields, resulted in lower effective concentrations as compared to Fv/Fm, or the light acclimated test endpoints qP, F'v/F'm, Rfd and α. The concentration ranges for the obtained EC_{20} and EC_{50} values were considerably wider in the case of the two tested heavy metals, while effective concentrations scattered in a much narrower range when NaCl was analyzed (Figure 3c,d).

Concentration-response relationships of the basic ChlF parameters determining the variable fluorescence Fv in dark-adapted (Fo, Fm) and ΔF in the light-adapted state (Fs, F'm) indicated different patterns for heavy metals as compared to NaCl (Figure 4, the summary statistics of the original data are supplemented in Table S5). Ni and Cr(VI) reduced Fv and ΔF via a parallel increase in Fo and Fs and a decrease in Fm and F'm, respectively. It should also be noted that Fo showed a considerably larger percentile increase due to heavy metal treatments (151 and 186% of control at 10 mg L^{-1} Ni and Cr(VI) compared to Fs (111 and 124% of control at 10 mg L^{-1} Ni and Cr(VI), Figure 4). Fm and F'm, on the other hand, were similarly affected: 10 mg L^{-1} Ni and Cr(VI) decreased Fm by 20 and 24% compared to control, while F'm was lowered by 28 and 25%, respectively (Figure 4). Contrary to the tested heavy metals, NaCl decreased all ChlF yields (Fo, Fm, Fs and F'm) in a concentration-dependent manner, and inhibited the calculated quantum efficiencies through reducing maximal fluorescence parameters (i.e., Fm and F'm) stronger as compared to Fo and Fs (Figure 4).

Table 4. Results of non-linear regression model fittings to the assessed ChlF parameters after 72 h of NaCl treatments. Models "LL.3", "BC.4" and "CRS.4" denote three-parameter log-logistic, 4-parameter Brain-Cousens and 4-parameter Cedergreen-Ritz-Streibig models, respectively. In the case of hormetic models, "max" denotes the highest calculated hormetic response as a percent of the respective control data, "EC_{max}" denote the NaCl-concentration resulting in maximal hormetic response, and "NOAEC" denotes the "No Adversible Effect Concentration", that is the NaCl-concentration at which the hormetic effect diminishes. Bold numbers highlight maximal hormetic responses higher than 105%. EC_{20} and EC_{50} denote NaCl concentrations (as g L^{-1} ± 95% confidence interval), resulting in 20 and 50% inhibition of the respective parameter. RSS and pseudo-R^2 indicate the residual sum of squares and the correlation between the predicted and observed responses at the applied concentrations, respectively.

Parameter	Model	Df	Rss	F-Value	Pseudo-R^2	Max	EC_{max}	NOAEC (95% CI)	EC_{20} (95% CI)	EC_{50} (95% CI)
Fv/Fm	LL.3	24	1109.59	0.5767	0.8996	NA	NA	NA	13.2 (12.4–14.0)	15.5 (15.0–15.9)
Fv/Fo	CRS.4a	23	1273.78	2.8564	0.9554	**113.2**	**4.55**	**8.78 (7.47–10.1)**	10.7 (9.96–11.5)	13.2 (12.6–13.8)
Y(II)	CRS.4a	23	1771.4	1.3314	0.9381	**108.2**	**3.20**	**6.48 (4.34–8.61)**	8.95 (7.87–10.0)	12.0 (11.1–12.8)
qP	LL.3	24	2319.2	1.4362	0.8816	NA	NA	NA	9.65 (8.13–11.2)	13.3 (12.3–14.3)
F′v/F′m	LL.3	24	1546.45	1.983	0.8844	NA	NA	NA	11.8 (10.7–13.0)	15.1 (14.4–15.9)
F′v/F′o	BC.4	23	2031	3.5422	0.9333	**124.6**	**4.63**	**8.57 (7.52–9.61)**	10.2 (9.37–11.1)	13.3 (12.2–14.3)
Rfd	LL.3	24	1206.4	1.4771	0.9374	NA	NA	NA	10.1 (9.09–11.2)	13.4 (12.8–14.0)
α	LL.3	24	1479	0.8038	0.9053	NA	NA	NA	11.9 (10.9–12.9)	14.6 (14.1–15.2)
ETR_{max}	BC.4	23	1632.4	0.9605	0.9595	**115.8**	**3.45**	**6.06 (5.12–7.00)**	7.44 (6.76–8.11)	9.71 (9.06–10.4)

NA indicates that the respective parameter is not applicable due to a lack of hormetic response. The summary statistics of the original data and the fitted regression models are supplemented in Table S4 and Figure S4.

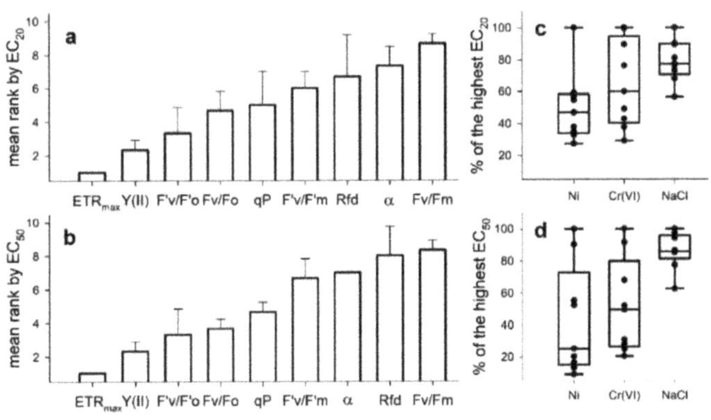

Figure 3. Responsivity of the assessed ChlF-based test endpoints as their mean relative sensitivity ranks (± SE, n = 3, subfigures (**a**) for EC_{20} and (**b**) for EC_{50}). Relative ranks were based on the calculated effective Ni, Cr(VI) and NaCl concentrations resulting in 20 (EC_{20}) and 50% (EC_{50}) inhibition of the respective parameter after 72 h long exposures. Variability in the calculated EC_{20} and EC_{50} concentrations for the assessed ChlF parameters (subfigures (**c**) for EC_{20}, (**d**) for EC_{50}) were normalized to the highest respective calculated effective concentration in the case of each applied toxicant.

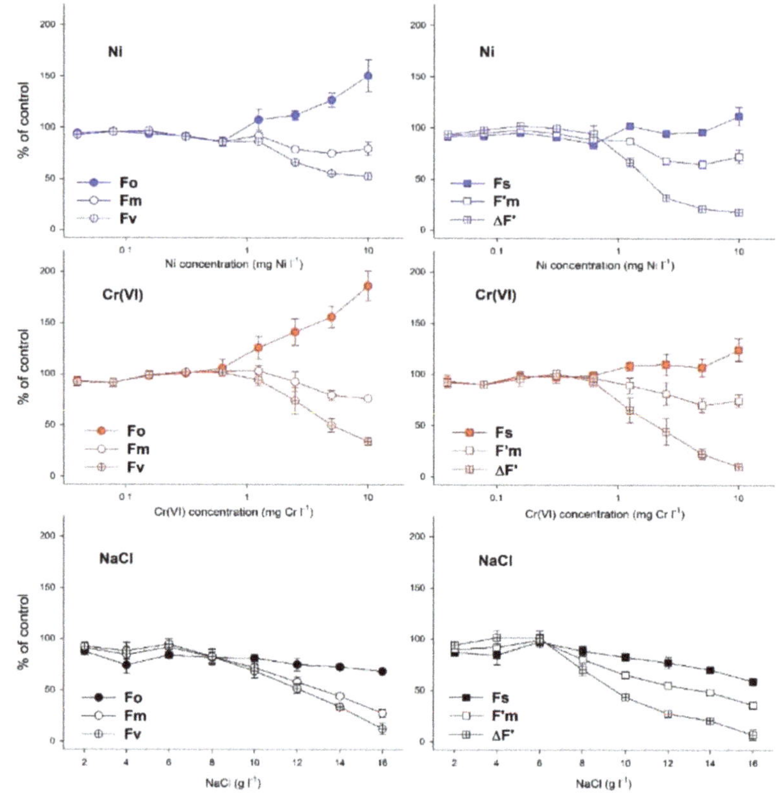

Figure 4. Changes in the dark-adapted (Fo, Fm, **left**) and light-adapted (Fs, F′m, **right**) basic ChlF yields and the derived variable fluorescence parameters (Fv, ΔF′) after 72 h long exposures to Ni, Cr(VI) and NaCl, respectively. Symbols denote grand means (means of means) ± SE of three experiments (n = 3), as a percentage of respective control data. For summary statistics of every measured parameter, please refer to Supplementary Table S5.

When proportions of chlorophyll fluorescence quenching processes were compared, similar response patterns were found for different toxicants in the lower concentration ranges (Figure 5). Compared to the control, low concentrations did not change the ratio of the effective quantum yield of PSII photochemical energy conversion [Y(II)], the quantum yield of regulated PSII non-photochemical energy loss (Y(NPQ)) and the quantum yield of non-regulated heat dissipation and fluorescence emission (Y(NO)); or even resulted in a slight increase in Y(II), as a hormetic response, with a parallel decrease in Y(NPQ). Increasing concentrations of the two heavy metals, however, resulted in a gradual decline in Y(II), in parallel with an increase in Y(NPQ), and also slightly elevated Y(NO) at the highest concentrations (Figure 5). Contrary to them, NaCl-induced increase in Y(NPQ) was only intermittent (up to 12 g L^{-1}) and was followed by a decline in that parameter with parallel increase in Y(NO) (Figure 5).

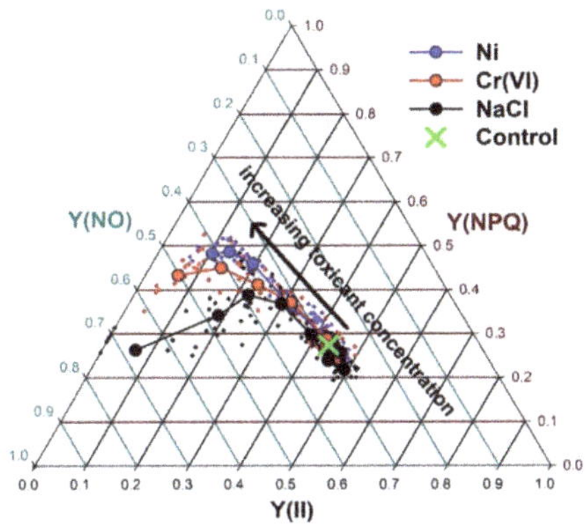

Figure 5. Changes in the relative ratios of photochemical -Y(II), regulated non-photochemical -Y(NPQ), and non-regulated non-photochemical -Y(NO)- quenching along the increasing Ni, Cr(VI) and NaCl concentrations, respectively. Small dots denote original data, large symbols with corresponding colors denote means of 12 samples at a given toxicant concentration.

4. Discussion

Our results indicate that acute phytotoxicity of various chemicals can be efficiently characterized via the ChlF imaging-based duckweed phenotyping method. One of the crucial points is the proper choice of the most responsive ChlF parameters as test endpoints [35]. The sensitivity of the assessed ChlF parameters was found to be highly variable with respect to a given toxicant used in our study. The difference between the endpoints with the lowest and highest calculated EC_{50}-s was 11-fold, 5-fold and 1.6-fold in the case of Ni, Cr(VI) and NaCl, respectively. In general, Fv/Fm and Rfd proved to be the least sensitive endpoints. Rfd has been applied widely in plant ecophysiology studies, while Fv/Fm is by far the most basically reported ChlF parameter literature-wise. Our results indicate that one should be cautious with relying exclusively on the latter parameter in toxicological studies. Fv/Fo resulted in considerably lower effective concentrations, though the calculation of this parameter uses the same basic ChlF levels, i.e., Fo and Fm, as Fv/Fm. Due to the different mathematical basis, Fv/Fo has higher values and performs a larger dynamic range than Fv/Fm [33]. Consequently, the relationship between these two endpoints is non-linear and results in higher sensitivity of Fv/Fo in the physiologically near-optimum range [31,40–43].

Differences in the photochemical efficiency of plants can be better distinguished if actinic irradiation is applied, as the limiting steps of photochemistry are put under pressure in this way [10,44–48]. Our results confirmed those previous observations: Y(II) was a very responsive endpoint, and both F'v/F'm and F'v/F'o exhibited lower calculated effective concentrations as compared to Fv/Fm and Fv/Fo, respectively. ETR_{max} had also been reported as a very sensitive endpoint in detecting herbicide and heavy metal toxicity in many previous duckweed tests [34,42,44,49–51]. From this aspect, the more pressure seems to be the better, as maximum electron transport rate (ETR_{max}) proved to be the most sensitive parameter to each toxicant in our study. ChlF imaging-based assays generate detailed datasets both in terms of spatial resolution and different aspects of photosynthetic processes [11]. Besides the above indices for photochemical efficiency, basic ChlF parameters (Fo, Fm, F'o and F'm) or indicators of non-photochemical quenching (Y(NPQ), Y(NO), NPQ, and qN) can also be useful to characterize particular effects of various toxicants [44,45,52]. Comparing the effects of the two heavy metals and NaCl, we found different patterns of changes in ChlF parameters along with increasing toxicant concentrations. These results underline the importance of analyzing additional ChlF parameters besides the most commonly reported Fv/Fm and Y(II). Changes in non-photochemical quenching can be seen as a sensitive indicator of physiological changes as Y(II), and may provide important additional data for understanding the phytotoxic effects of the applied toxicants [32,48,53].

ChlF parameters are considered to have comparable sensitivity to growth-based endpoints in aquatic plants [9,10]. In duckweed tests conducted with herbicide and phenol treatments for seven and three days, respectively, Fv/Fm and ETR_{max} showed lower calculated effective concentrations than the relative growth rates [44,52]. Our results did not support those former observations. Irrespective of their relatively high sensitivity, the calculated EC_{50}s for ETR_{max} and Y(II) was considerably higher in the present study as compared to those calculated in growth inhibition tests using the same *S. polyrhiza* clone (clone UD0401, Figure 6). In the case of Ni and Cr(VI), the calculated EC_{50}-s for frond area- and frond number-based growth rates were in the 0.18–0.2 mg L^{-1} range [23], which is one order of magnitude lower than the respective concentrations for ETR_{max}.

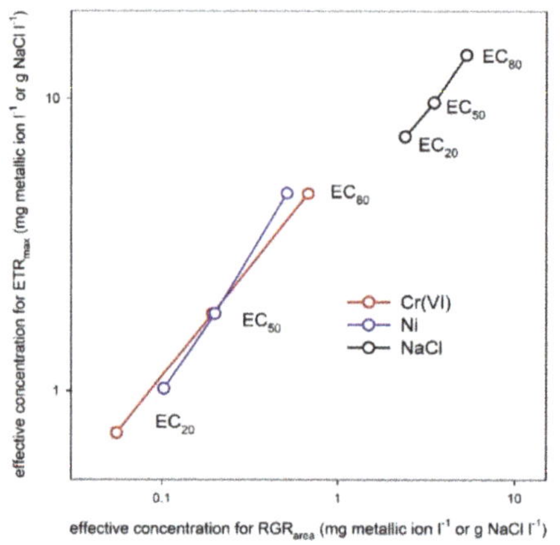

Figure 6. Correlations of the 20, 50 and 80% inhibiting concentrations (EC_{20}, EC_{50} and EC_{80}) of Cr(VI), Ni and NaCl were calculated for frond area-based relative growth rates (RGR_{area}) in seven-day-long duckweed growth tests, and for maximal electron transport rates (ETR_{max}) measured in three-day-long multiwell-based phytotoxicity tests with the *S. polyrhiza* clone UD0401, respectively. The effective concentrations for frond area growth inhibition were based on the data reported by Oláh et al. [23] and Hepp et al. [54], respectively.

In NaCl treatments, the calculated EC_{50}s were 3.45 and 4.51 g L^{-1} for frond area and frond number growth rates, respectively [54], which is one-half to one-third of that of ETR_{max} in the present study. It should be noted, however, that in the present study, relative growth rates were not determined, but obtained from a different experimental setup, that is, the OECD-conform [5] duckweed growth inhibition test. Thus, a possible reason for such diverging sensitivities can be the different treatment volume and duration: the growth inhibition tests were performed in 100 mL volume for seven days, while toxicant exposures in the present study were conducted in smaller medium volume (4 mL) and shorter exposure time (three days). This latter experimental setup thus may result in exposure to smaller doses on a biomass basis at the same nominal toxicant concentrations, and for a shorter duration as compared to the growth tests. Extended exposure times can drastically increase the sensitivity of phytotoxicity assays: in Cd-treated cultures of the same *S. polyrhiza* clone and under the same experimental conditions, three-day-long exposures resulted in three times higher calculated EC_{50}-s of growth rate compared to the seven-day-long treatments [55].

Another explanation of the higher effective concentrations can be that growth rates reflect the overall performance of test plants under toxicant treatments. ChlF-based endpoints, on the other hand, reflect the operability of certain physiological processes, and if those are not affected directly, their response can be weaker and/or delayed [9,56]. Growth conditions might also be of crucial importance: test plants exposed to toxicants under sub-saturating irradiance—as was the case in the present study—have a better chance to up-regulate PSII repair systems and ROS scavenging than plants grown under higher irradiance levels [16,57]. This factor can further reduce the sensitivity of ChlF-based test endpoints compared to those derived from growth parameters.

Hormesis has been discussed extensively as a potentially general response to moderate environmental stress [58,59]. Hormetic-type responses were reported in duckweed toxicity tests conducted with organic and inorganic pollutants [39,60–63]. On the other hand, hormesis in photosynthetic responses of duckweeds is still scarcely studied [64]. Our results indicate that hormesis might be a frequent response of ChlF-based endpoints, at least in the shorter term. Based on the tested regression models, Ni and NaCl had a definite stimulating effect in case of several ChlF parameters in the low concentration ranges. The results suggest that, in general, the more sensitive parameters (Fv/Fo, Y(II), qP, F'v/F'o, ETR_{max}) were more likely to indicate hormesis induced by Ni and NaCl treatments. For Cr(VI), hormetic models also described the concentration-response relationships better for several ChlF parameters. However, the maximal stimulation generally stayed below 3% in these cases, which is lower than the 5% limit, commonly defined as a criterion for hormetic responses [60]. The only exception was Rfd, which is the least sensitive endpoint in Cr(VI) treatments—which increased by a maximum of 5% as compared to control.

The concentrations needed for the maximal hormetic response (EC_{max}) were in the range of 2–15% of that of EC_{50} for Ni and 25–35% for NaCl, respectively. This observation was similar to that of [39], who found that stimulating concentrations usually fell in the 20–25% range of the EC_{50} concentrations in aquatic macrophytes treated with the herbicide terbuthylazine. The maximal response ranged in the 105–110% (Ni) and 110–125% (NaCl) of the respective control, similarly to earlier observations on plant hormetic responses [58,64]. Ni is essential for plants and Steinberg's medium does not contain directly added Ni. Thus, Ni-induced stimulation of plant metabolism can be explained by its physiological role [18]. Similarly, NaCl was reported to enhance plant photosynthesis at low concentrations [26]. More interestingly, we found that stimulating Ni and NaCl concentrations for ChlF parameters in the present study were in the range of growth-inhibiting EC_{50} concentrations calculated in seven-day-long growth tests with the same *S. polyrhiza* clone [23,54]. These differences indicate that ChlF- and growth rate-based test endpoints do not implicitly describe the same phytotoxic pattern within the same concentration range, and one should be cautious when relying exclusively on ChlF-based endpoints as a general proxy for phytotoxic effects.

5. Conclusions

Chlorophyll fluorescence imaging-based duckweed phenotyping offers a non-destructive, fast, and easy way to assess the potential phytotoxicity of water pollutants, resulting in high throughput systems. Our results indicated that different chlorophyll fluorescence induction parameters responded with very different sensitivity to the applied treatments. In general, parameters measured in the dark-adapted state of test plants proved to be less sensitive than those measured in the light-adapted state. Thus, the latter ones might be the preferred endpoints in phytotoxicity assays. On the other hand, we also found that dark-adapted chlorophyll fluorescence parameters can result in different calculated effective concentrations, and Fv/Fo showed considerably higher responsivity than Fv/Fm.

In future studies, it would be important to take this aspect into account and to report information on the Fv/Fo parameter as well. This might also be encouraged by the manufacturers by including this parameter in the default ones calculated by their instruments, as presently it's missing from the repertoire of the most widely used Imaging-PAM platform. These results highlight the importance of parallel assessment of as many chlorophyll fluorescence parameters as possible, besides the most commonly reported Fv/Fm, Y(II), or NPQ. This finding was also supported by the observation that the tested toxicants induced different trends in the basic chlorophyll fluorescence parameters and, at least partly, in relative proportions of different quenching processes. Based on those differences, a basic distinction of water pollutants with different modes of action seems to be achievable by using this method. In the applied test protocol, the calculated effective concentrations proved to be higher than those calculated in growth inhibition tests performed with the same duckweed clone. Moreover, definite hormetic trends were found in responses of several endpoints in those concentration ranges where the applied toxicants resulted in strong growth inhibition in longer-term exposures. These differences suggest that changes in the photochemical efficiency of plants do not necessarily go hand in hand with growth responses, and care should be taken when one interprets exclusively ChlF-based endpoints as general proxies for phytotoxic effects.

Supplementary Materials: The following is available online at https://www.mdpi.com/article/10.3390/plants10122763/s1, Table S1: Literature reports on the application of chlorophyll fluorescence imaging method in duckweed ecotoxicology and ecophysiology studies [65–81]. Table S2: Summary statistics of the assessed chlorophyll fluorescence induction parameters after 72 h-long Ni-treatments of the S. polyrhiza UD0401 clone. The table summarizes minimums (Min), maximums (Max), arithmetic means (Mean), standard deviations (SD) and coefficients of variation (CV) of pooled data, expressed as percentage of their respective control means, from 3 independent experiments with 4-4 parallel treatments at each applied Ni concentraiton (n = 12). Different upper cases indicate significantly ($p < 0.05$) different medians for different concentrations according to the Kruskal-Wallis test and post hoc Mann-Whitney pairwise comparisons. Table S3: Summary statistics of the assessed chlorophyll fluorescence induction parameters after 72 h-long Cr(VI)-treatments of the S. polyrhiza UD0401 clone. The table summarizes minimums (Min), maximums (Max), arithmetic means (Mean), standard deviations (SD) and coefficients of variation (CV) of pooled data, expressed as percentage of their respective control means, from 3 independent experiments with 4-4 parallel treatments at each applied Cr(VI) concentraiton (n = 12). Different upper cases indicate significantly ($p < 0.05$) different medians for different concentrations according to the Kruskal-Wallis test and post hoc Mann-Whitney pairwise comparisons. Table S4: Summary statistics of the assessed chlorophyll fluorescence induction parameters after 72 h-long NaCl-treatments of the S. polyrhiza UD0401 clone. The table summarizes minimums (Min), maximums (Max), arithmetic means (Mean), standard deviations (SD) and coefficients of variation (CV) of pooled data, expressed as percentage of their respective control means, from 3 independent experiments with 4-4 parallel treatments at each applied NaCl concentraiton (n = 12). Different upper indicate significantly ($p < 0.05$) different medians for different concentrations according to the Kruskal-Wallis test and post hoc Mann-Whitney pairwise comparisons. Table S5: Summary statistics of the assessed basic chlorophyll fluorescence induction parameters after 72 h-long treatments of the S. polyrhiza UD0401 clone to Ni, Cr(VI) and NaCl, respectively. The table summarizes minimums (Min), maximums (Max), arithmetic means (Mean), standard deviations (SD) and coefficients of variation (CV) of pooled data expressed as percentage of their respective control means, from 3 independent experiments with 4-4 parallel treatments at each applied concentration (n = 12). Different upper cases indicate significantly ($p < 0.05$) different medians for different concentrations according to the Kruskal-Wallis test and post hoc Mann-Whitney pairwise comparisons. Figure S1: Kinetics in the chlorophyll fluorescence yield of the S. polyrhiza UD0401 clone during the first saturation pulse after 20 min of dark adaption, and the consecutive 10 min-long actinic irradiation routine with 77 E m^{-2} s^{-1} (left charts). The test plants were cultured in either pure Steinberg medium (control), or Steinberg medium containing 10 g l^{-1} NaCl for 3 days, respectively. The photochemical [Y(II)], regulated non-photochemical [Y(NPQ)] and non-regulated non-photochemical [Y(NO)] quenching coefficients (right charts) were calculated using the basic, mutually independent chlorophyll fluorescence yields indicated in the top left subfigure. Figure S2: Measured and modelled responses of the assessed chlorophyll fluorescence induction parameters to 72 h-long Ni-treatments of the S. polyrhiza UD0401 clone, as compared to their respective control data. Cyrcles denote means (n = 4) of the repeated experiments (n = 3) at the applied concentrations. Thin black lines denote the best-fitting non-linear regression model for each ChlF parameter with 95% confidence intervals (gray shaded areas). Figure S3: Measured and modelled responses of the assessed chlorophyll fluorescence induction parameters to 72 h-long Cr(VI)-treatments of the S. polyrhiza UD0401 clone, as compared to their respective control data. Cyrcles denote means (n = 4) of the repeated experiments (n = 3) at the applied concentrations. Thin black lines denote the best-fitting non-linear regression model for each ChlF parameter with 95% confidence intervals (gray shaded areas). Figure S4: Measured and modelled responses of the assessed chlorophyll fluorescence induction parameters to 72 h-long NaCl-treatments of the S. polyrhiza UD0401 clone, as compared to their respective control data. Cyrcles denote means (n = 4) of the repeated experiments (n = 3) at the applied concentrations. Thin black lines denote the best-fitting non-linear regression model for each ChlF parameter with 95% confidence intervals (gray shaded areas).

Author Contributions: Conceptualization, V.O. and I.M.; methodology, V.O.; formal analysis, V.O.; investigation, V.O., A.H. and M.I.; resources, V.O., A.H. and M.I.; data curation, V.O.; writing—original draft preparation, V.O. and I.M.; writing—review and editing, V.O. and I.M.; visualization, V.O. All authors have read and agreed to the published version of the manuscript.

Funding: This research was funded by the National Research, Development and Innovation Office—NKFIH—of the Hungarian Ministry for Innovation and Technology, grant number OTKA FK 134296.

Data Availability Statement: The datasets used in the present study are available from the corresponding author on reasonable request.

Acknowledgments: The authors are grateful to the Agora Science Centre (Debrecen, Hungary) for providing the opportunity to use the Imaging-PAM instrument. Muhammad Irfan is thankful for the support of the Tempus Public Foundation (Hungary) within the framework of the Stipendium Hungaricum Scholarship Programme.

Conflicts of Interest: The authors declare no conflict of interest.

References

1. *Environment Canada Biological Test Method-Test for Measuring the Inhibition of Growth Using the Freshwater Macrophyte* Lemna minor, 2nd ed.; Environmental Protection Service Report EPS 1/RM/37; Method Development and Applications Section, Environmental Technology Centre, Environment Canada: Ottawa, ON, Canada, 2007; ISBN 978-0-662-43340-8.
2. Acosta, K.; Appenroth, K.J.; Borisjuk, L.; Edelman, M.; Heinig, U.; Jansen, M.A.K.; Oyama, T.; Pasaribu, B.; Schubert, I.; Sorrels, S.; et al. Return of the Lemnaceae: Duckweed as a Model Plant System in the Genomics and Postgenomics Era. *Plant Cell* **2021**, *33*, 3207–3234. [CrossRef] [PubMed]
3. Ziegler, P.; Adelmann, K.; Zimmer, S.; Schmidt, C.; Appenroth, K.-J. Relative in Vitro Growth Rates of Duckweeds (Lemnaceae)-the Most Rapidly Growing Higher Plants. *Plant Biol. J.* **2015**, *17*, 33–41. [CrossRef] [PubMed]
4. ISO 20079:2005. *Water Quality-Determination of Toxic Effect of Water Constituents and Waste Water to Duckweed (*Lemna minor*)—Duckweed Growth Inhibition Test*, 1st ed.; International Organization for Standardization: Geneva, Switzerland, 2005.
5. *OECD Guidelines for the Testing of Chemicals, Revised Proposal for a New Guideline 221, Lemna Sp. Growth Inhibition Test*; OECD: Paris, France, 2006.
6. Mkandawire, M.; Teixeira da Silva, J.A.; Dudel, E.G. The Lemna Bioassay: Contemporary Issues as the Most Standardized Plant Bioassay for Aquatic Ecotoxicology. *Crit. Rev. Environ. Sci. Technol.* **2014**, *44*, 154–197. [CrossRef]
7. Henke, R.; Eberius, M.; Appenroth, K.-J. Induction of Frond Abscission by Metals and Other Toxic Compounds in *Lemna minor*. *Aquat. Toxicol.* **2011**, *101*, 261–265. [CrossRef]
8. Park, A.; Kim, Y.-J.; Choi, E.-M.; Brown, M.T.; Han, T. A Novel Bioassay Using Root Re-Growth in Lemna. *Aquat. Toxicol.* **2013**, *140–141*, 415–424. [CrossRef]
9. Brain, R.A.; Cedergreen, N. Biomarkers in Aquatic Plants: Selection and Utility. In *Reviews of Environmental Contamination and Toxicology*; Springer: New York, NY, USA, 2009; Volume 198, pp. 1–61. ISBN 978-0-387-09646-9.
10. Ralph, P.J.; Smith, R.A.; Macinnis-Ng, C.M.O.; Seery, C.R. Use of Fluorescence-Based Ecotoxicological Bioassays in Monitoring Toxicants and Pollution in Aquatic Systems: Review. *Toxicol. Environ. Chem.* **2007**, *89*, 589–607. [CrossRef]
11. Ziegler, P.; Sree, K.S.; Appenroth, K.-J. Duckweed Biomarkers for Identifying Toxic Water Contaminants? *Environ. Sci Pollut. Res.* **2019**, *26*, 14797–14822. [CrossRef]
12. Murchie, E.H.; Lawson, T. Chlorophyll Fluorescence Analysis: A Guide to Good Practice and Understanding Some New Applications. *J. Exp. Bot.* **2013**, *64*, 3983–3998. [CrossRef]
13. Schreiber, U.; Quayle, P.; Schmidt, S.; Escher, B.I.; Mueller, J.F. Methodology and Evaluation of a Highly Sensitive Algae Toxicity Test Based on Multiwell Chlorophyll Fluorescence Imaging. *Biosens. Bioelectron.* **2007**, *22*, 2554–2563. [CrossRef]
14. Muller, R.; Schreiber, U.; Escher, B.I.; Quayle, P.; Bengtson Nash, S.M.; Mueller, J.F. Rapid Exposure Assessment of PSII Herbicides in Surface Water Using a Novel Chlorophyll a Fluorescence Imaging Assay. *Sci. Total Environ.* **2008**, *401*, 51–59. [CrossRef]
15. Ralph, P.J.; Macinnis-Ng, C.M.O.; Frankart, C. Fluorescence Imaging Application: Effect of Leaf Age on Seagrass Photokinetics. *Aquat. Bot.* **2005**, *81*, 69–84. [CrossRef]
16. Wilkinson, A.D.; Collier, C.J.; Flores, F.; Mercurio, P.; O'Brien, J.; Ralph, P.J.; Negri, A.P. A Miniature Bioassay for Testing the Acute Phytotoxicity of Photosystem II Herbicides on Seagrass. *PLoS ONE* **2015**, *10*, e0117541. [CrossRef]

17. Chen, Y.-E.; Wu, N.; Zhang, Z.-W.; Yuan, M.; Yuan, S. Perspective of Monitoring Heavy Metals by Moss Visible Chlorophyll Fluorescence Parameters. *Front. Plant Sci.* **2019**, *10*, 35. [CrossRef]
18. DalCorso, G. Heavy Metal Toxicity in Plants. In *Plants and Heavy Metals*; Furini, A., Ed.; Springer Briefs in Molecular Science; Springer: Dordrecht, The Netherlands, 2012; pp. 1–25. ISBN 978-94-007-4440-0.
19. Appenroth, K.-J.; Krech, K.; Keresztes, Á.; Fischer, W.; Koloczek, H. Effects of Nickel on the Chloroplasts of the Duckweeds Spirodela Polyrhiza and *Lemna minor* and Their Possible Use in Biomonitoring and Phytoremediation. *Chemosphere* **2010**, *78*, 216–223. [CrossRef]
20. Yusuf, M.; Fariduddin, Q.; Hayat, S.; Ahmad, A. Nickel: An Overview of Uptake, Essentiality and Toxicity in Plants. *Bull. Environ. Contam. Toxicol.* **2011**, *86*, 1–17. [CrossRef]
21. Andresen, E.; Opitz, J.; Thomas, G.; Stärk, H.-J.; Dienemann, H.; Jenemann, K.; Dickinson, B.C.; Küpper, H. Effects of Cd & Ni Toxicity to Ceratophyllum Demersum under Environmentally Relevant Conditions in Soft & Hard Water Including a German Lake. *Aquat. Toxicol.* **2013**, *142–143*, 387–402. [CrossRef]
22. Naumann, B.; Eberius, M.; Appenroth, K.-J. Growth Rate Based Dose–Response Relationships and EC-Values of Ten Heavy Metals Using the Duckweed Growth Inhibition Test (ISO 20079) with *Lemna minor* L. *Clone St. J. Plant Physiol.* **2007**, *164*, 1656–1664. [CrossRef]
23. Oláh, V.; Hepp, A.; Mészáros, I. Comparative Study on the Sensitivity of Turions and Active Fronds of Giant Duckweed (*Spirodela Polyrhiza* (L.) Schleiden) to Heavy Metal Treatments. *Chemosphere* **2015**, *132*, 40–46. [CrossRef]
24. Cervantes, C.; Loza-Tavera, H.; Torres-Guzman, J.C.; Moreno-Sanchez, R. Interactions of Chromium with Microorganisms and Plants. *FEMS Microbiol. Rev.* **2001**, *13*, 335–347. [CrossRef]
25. Shanker, A.; Cervantes, C.; Lozatavera, H.; Avudainayagam, S. Chromium Toxicity in Plants. *Environ. Int.* **2005**, *31*, 739–753. [CrossRef]
26. Hasanuzzaman, M.; Nahar, K.; Fujita, M. Plant Response to Salt Stress and Role of Exogenous Protectants to Mitigate Salt-Induced Damages. In *Ecophysiology and Responses of Plants under Salt Stress*; Ahmad, P., Azooz, M.M., Prasad, M.N.V., Eds.; Springer: New York, NY, USA, 2013; pp. 25–87. ISBN 978-1-4614-4746-7.
27. Jajoo, A. Changes in Photosystem II in Response to Salt Stress. In *Ecophysiology and Responses of Plants under Salt Stress*; Ahmad, P., Azooz, M.M., Prasad, M.N.V., Eds.; Springer: New York, NY, USA, 2013; pp. 149–168. ISBN 978-1-4614-4746-7.
28. Sree, K.S.; Adelmann, K.; Garcia, C.; Lam, E.; Appenroth, K.-J. Natural Variance in Salt Tolerance and Induction of Starch Accumulation in Duckweeds. *Planta* **2015**, *241*, 1395–1404. [CrossRef] [PubMed]
29. Oláh, V.; Hepp, A.; Gaibor Vaca, N.Y.; Tamás, M.; Mészáros, I. Retrospective Analyses of Archive Phytotoxicity Test Data Can Help in Assessing Internal Dynamics and Stability of Growth in Laboratory Duckweed Cultures. *Aquat. Toxicol.* **2018**, *201*, 40–46. [CrossRef] [PubMed]
30. *Walz GmbH IMAGING-PAM M-Series Chlorophyll Fluorometer: Instrument Description and Information for Users*, 5th ed.; Heinz Walz GmbH: Effeltrich, Germany, 2019.
31. Roháček, K. Chlorophyll Fluorescence Parameters: The Definitions, Photosynthetic Meaning, and Mutual Relationships. *Photosynthetica* **2002**, *40*, 13–29. [CrossRef]
32. Klughammer, C.; Schreiber, U. Complementary PS II Quantum Yields Calculated from Simple Fluorescence Parameters Measured by PAM Fluorometry and the Saturation Pulse Method. *PAM Appl. Notes* **2008**, *1*, 27–35.
33. Lichtenthaler, H.K.; Buschmann, C.; Knapp, M. How to Correctly Determine the Different Chlorophyll Fluorescence Parameters and the Chlorophyll Fluorescence Decrease Ratio R_{Fd} of Leaves with the PAM Fluorometer. *Photosynthetica* **2005**, *43*, 379–393. [CrossRef]
34. Park, J.; Brown, M.T.; Depuydt, S.; Kim, J.K.; Won, D.-S.; Han, T. Comparing the Acute Sensitivity of Growth and Photosynthetic Endpoints in Three Lemna Species Exposed to Four Herbicides. *Environ. Pollut.* **2017**, *220*, 818–827. [CrossRef]
35. Kalaji, H.M.; Schansker, G.; Brestic, M.; Bussotti, F.; Calatayud, A.; Ferroni, L.; Goltsev, V.; Guidi, L.; Jajoo, A.; Li, P.; et al. Frequently Asked Questions about Chlorophyll Fluorescence, the Sequel. *Photosynth. Res.* **2017**, *132*, 13–66. [CrossRef]
36. Ritz, C.; Streibig, J.C. Bioassay Analysis Using R. *J. Stat. Softw.* **2005**, *12*, 1–22. [CrossRef]
37. R Core Team. *A Language and Environment for Statistical Computing*; R Foundation for Statistical Computing: Vienna, Austria, 2015.
38. Brain, P.; Cousens, R. An Equation to Describe Dose Responses Where There Is Stimulation of Growth at Low Doses. *Weed Res.* **1989**, *29*, 93–96. [CrossRef]
39. Cedergreen, N.; Ritz, C.; Streibig, J.C. Improved Empirical Models Describing Hormesis. *Environ. Toxicol Chem* **2005**, *24*, 3166. [CrossRef]
40. Babani, F.; Lichtenthaler, H.K. Light-Induced and Age-Dependent Development of Chloroplasts in Etiolated Barley Leaves as Visualized by Determination of Photosynthetic Pigments, C02 Assimilation Rates and Different Kinds of Chlorophyll Fluorescence Ratios. *J. Plant Physiol.* **1996**, *148*, 555–566. [CrossRef]
41. Láposi, R.; Veres, S.; Lakatos, G.; Oláh, V.; Fieldsend, A.; Mészáros, I. Responses of Leaf Traits of European Beech (*Fagus sylvatica* L.) Saplings to Supplemental UV-B Radiation and UV-B Exclusion. *Agric. For. Meteorol.* **2009**, *149*, 745–755. [CrossRef]
42. Oláh, V.; Lakatos, G.; Bertók, C.; Kanalas, P.; Szőllősi, E.; Kis, J.; Mészáros, I. Short-Term Chromium(VI) Stress Induces Different Photosynthetic Responses in Two Duckweed Species, *Lemna gibba* L. and *Lemna minor* L. *Photosynthetica* **2010**, *48*, 513–520. [CrossRef]

43. Pietrini, F.; Bianconi, D.; Massacci, A.; Iannelli, M.A. Combined Effects of Elevated CO2 and Cd-Contaminated Water on Growth, Photosynthetic Response, Cd Accumulation and Thiolic Components Status in *Lemna minor* L. *J. Hazard. Mater.* **2016**, *309*, 77–86. [CrossRef]
44. Kumar, K.S.; Han, T. Physiological Response of Lemna Species to Herbicides and Its Probable Use in Toxicity Testing. *Toxicol. Environ. Health Sci.* **2010**, *2*, 39–49. [CrossRef]
45. Perreault, F.; Oukarroum, A.; Pirastru, L.; Sirois, L.; Gerson Matias, W.; Popovic, R. Evaluation of Copper Oxide Nanoparticles Toxicity Using Chlorophyll a Fluorescence Imaging in *Lemna gibba*. *J. Bot.* **2010**, *2010*, 1–9. [CrossRef]
46. Lahive, E. Frond Development Gradients Are a Determinant of the Impact of Zinc on Photosynthesis in Three Species of Lemnaceae. *Aquat. Bot.* **2012**, *101*, 55–63. [CrossRef]
47. Flores, F.; Collier, C.J.; Mercurio, P.; Negri, A.P. Phytotoxicity of Four Photosystem II Herbicides to Tropical Seagrasses. *PLoS ONE* **2013**, *8*, e75798. [CrossRef]
48. Dewez, D.; Goltsev, V.; Kalaji, H.M.; Oukarroum, A. Inhibitory Effects of Silver Nanoparticles on Photosystem II Performance in *Lemna gibba* Probed by Chlorophyll Fluorescence. *Curr. Plant Biol.* **2018**, *16*, 15–21. [CrossRef]
49. Pietrini, F.; Passatore, L.; Fischetti, E.; Carloni, S.; Ferrario, C.; Polesello, S.; Zacchini, M. Evaluation of Morpho-Physiological Traits and Contaminant Accumulation Ability in *Lemna minor* L. Treated with Increasing Perfluorooctanoic Acid (PFOA) Concentrations under Laboratory Conditions. *Sci. Total Environ.* **2019**, *695*, 133828. [CrossRef]
50. Park, J.; Lee, H.; Han, T. Comparative Paraquat Sensitivity of Newly Germinated and Mature Fronds of the Aquatic Macrophyte Spirodela Polyrhiza. *AJPS* **2020**, *11*, 1008–1024. [CrossRef]
51. Lee, H.; Depuydt, S.; Shin, K.; Choi, S.; Kim, G.; Lee, Y.H.; Park, J.T.; Han, T.; Park, J. Assessment of Various Toxicity Endpoints in Duckweed (*Lemna minor*) at the Physiological, Biochemical, and Molecular Levels as a Measure of Diuron Stress. *Biology* **2021**, *10*, 684. [CrossRef]
52. Park, J.-S.; Brown, M.T.; Han, T. Phenol Toxicity to the Aquatic Macrophyte Lemna Paucicostata. *Aquat. Toxicol.* **2012**, *106–107*, 182–188. [CrossRef]
53. Kummerová, M.; Zezulka, Š.; Babula, P.; Tříska, J. Possible Ecological Risk of Two Pharmaceuticals Diclofenac and Paracetamol Demonstrated on a Model Plant *Lemna minor*. *J. Hazard. Mat.* **2016**, *302*, 351–361. [CrossRef]
54. Hepp, A.; Oláh, V.; Sipos, O.; Adorján, B.; Mészáros, I. Békalencse fitotoxikológiai tesztek eredményét befolyásoló tényezők vizsgálata (Assessment of factors affecting the results of duckweed phytotoxicity tests). In Proceedings of the XIV Environmental Scientific Conference of the Carpathian Basin, Gödöllő, Hungary, 5–7 April 2018; pp. 127–131.
55. Oláh, V.; Hepp, A.; Lakatos, G.; Mészáros, I. Cadmium-Induced Turion Formation of *Spirodela Polyrhiza* (L.) Schleiden. *Acta Biol. Szeged.* 2014; 58, 103–108.
56. Alkimin, G.D.; Daniel, D.; Frankenbach, S.; Serôdio, J.; Soares, A.M.V.M.; Barata, C.; Nunes, B. Evaluation of Pharmaceutical Toxic Effects of Non-Standard Endpoints on the Macrophyte Species *Lemna minor* and *Lemna gibba*. *Sci. Total Environ.* **2019**, *657*, 926–937. [CrossRef]
57. Christensen, M.G.; Teicher, H.B.; Streibig, J.C. Linking Fluorescence Induction Curve and Biomass in Herbicide Screening. *Pest. Manag. Sci.* **2003**, *59*, 1303–1310. [CrossRef]
58. Calabrese, E.J.; Blain, R.B. Hormesis and Plant Biology. *Environ. Pollut.* **2009**, *157*, 42–48. [CrossRef]
59. Agathokleous, E. Environmental Hormesis, a Fundamental Non-Monotonic Biological Phenomenon with Implications in Ecotoxicology and Environmental Safety. *Ecotoxicol. Environ. Saf.* **2018**, *148*, 1042–1053. [CrossRef]
60. Cedergreen, N.; Streibig, J.C.; Kudsk, P.; Mathiassen, S.K.; Duke, S.O. The Occurrence of Hormesis in Plants and Algae. *Dose-Response* **2007**, *5*, 150–162. [CrossRef]
61. Belz, R.G.; Cedergreen, N.; Sørensen, H. Hormesis in Mixtures—Can It Be Predicted? *Sci. Total Environ.* **2008**, *404*, 77–87. [CrossRef]
62. Agathokleous, E.; Mouzaki-Paxinou, A.-C.; Saitanis, C.J.; Paoletti, E.; Manning, W.J. The First Toxicological Study of the Antiozonant and Research Tool Ethylene Diurea (EDU) Using a *Lemna minor* L. Bioassay: Hints to Its Mode of Action. *Environ. Pollut.* **2016**, *213*, 996–1006. [CrossRef] [PubMed]
63. Zhong, Y.; Li, Y.; Cheng, J.J. Effects of Selenite on Chlorophyll Fluorescence, Starch Content and Fatty Acid in the Duckweed Landoltia Punctata. *J. Plant Res.* **2016**, *129*, 997–1004. [CrossRef] [PubMed]
64. Agathokleous, E. The Rise and Fall of Photosynthesis: Hormetic Dose Response in Plants. *J. For. Res.* **2021**, *32*, 889–898. [CrossRef]
65. Hulsen, K.; Minne, V.; Lootens, P.; Vandecasteele, P.; Hofte, M. A Chlorophyll a Fluorescence-Based *Lemna minor* Bioassay to Monitor Microbial Degradation of Nanomolar to Micromolar Concentrations of Linuron. *Environ. Microbiol.* **2002**, *4*, 327–337. [CrossRef]
66. Küster, A.; Pohl, K.; Altenburger, R. A Fluorescence-Based Bioassay for Aquatic Macrophytes and Its Suitability for Effect Analysis of Non-Photosystem II Inhibitors. *Environ. Sci. Pollut. Res. Int.* **2007**, *14*, 377–383. [CrossRef]
67. Juhel, G.; Batisse, E.; Hugues, Q.; Daly, D.; van Pelt, F.N.A.M.; O'Halloran, J.; Jansen, M.A.K. Alumina Nanoparticles Enhance Growth of *Lemna minor*. *Aquat. Toxicol.* **2011**, *105*, 328–336. [CrossRef]
68. Lahive, E.; O' Halloran, J.; Jansen, M.A.K. Differential Sensitivity of Four Lemnaceae Species to Zinc Sulphate. *Environ. Exp. Bot.* **2011**, *71*, 25–33. [CrossRef]
69. Zezulka, Š.; Kummerová, M.; Babula, P.; Váňová, L. *Lemna minor* Exposed to Fluoranthene: Growth, Biochemical, Physiological and Histochemical Changes. *Aquat. Toxicol.* **2013**, *140–141*, 37–47. [CrossRef]

70. Senavirathna, M.D.H.J.; Takashi, A.; Kimura, Y. Short-Duration Exposure to Radiofrequency Electromagnetic Radiation Alters the Chlorophyll Fluorescence of Duckweeds (*Lemna minor*). *Electromagn. Biol. Med.* **2014**, *33*, 327–334. [CrossRef]
71. Gomes, M.P.; Juneau, P. Oxidative Stress in Duckweed (*Lemna minor* L.) Induced by Glyphosate: Is the Mitochondrial Electron Transport Chain a Target of This Herbicide? *Environ. Pollut.* **2016**, *218*, 402–409. [CrossRef]
72. Di Baccio, D.; Pietrini, F.; Bertolotto, P.; Pérez, S.; Barcelò, D.; Zacchini, M.; Donati, E. Response of *Lemna gibba* L. to High and Environmentally Relevant Concentrations of Ibuprofen: Removal, Metabolism and Morpho-Physiological Traits for Biomonitoring of Emerging Contaminants. *Sci. Total Environ.* **2017**, *584–585*, 363–373. [CrossRef]
73. Paolacci, S. The Invasive Duckweed Lemna Minuta Kunth Displays a Different Light Utilisation Strategy than Native *Lemna minor* Linnaeus. *Aquat. Bot.* **2018**, *7*, 8–14. [CrossRef]
74. Chen, X.; O'Halloran, J.; Jansen, M.A.K. Orthophosphate Modulates the Phytotoxicity of Nano-ZnO to *Lemna minor* (L.). *Environ. Technol.* **2019**, *40*, 2446–2454. [CrossRef]
75. Grenni, P.; Patrolecco, L.; Rauseo, J.; Spataro, F.; Di Lenola, M.; Aimola, G.; Zacchini, M.; Pietrini, F.; Di Baccio, D.; Stanton, I.C.; et al. Sulfamethoxazole Persistence in a River Water Ecosystem and Its Effects on the Natural Microbial Community and *Lemna minor* Plant. *Microchem. J.* **2019**, *149*, 103999. [CrossRef]
76. Mateos-Cárdenas, A.; Scott, D.T.; Seitmaganbetova, G.; Frank, N.A.M.v.P.; John, O.; Marcel, A.K.J. Polyethylene Microplastics Adhere to *Lemna minor* (L.), yet Have No Effects on Plant Growth or Feeding by *Gammarus duebeni* (Lillj.). *Sci. Total Environ.* **2019**, *689*, 413–421. [CrossRef]
77. Nunes, B.; Veiga, V.; Frankenbach, S.; Serôdio, J.; Pinto, G. Evaluation of Physiological Changes Induced by the Fluoroquinolone Antibiotic Ciprofloxacin in the Freshwater Macrophyte Species *Lemna minor* and *Lemna gibba*. *Environ. Toxicol. Pharmacol.* **2019**, *72*, 103242. [CrossRef]
78. Liu, Y.; Xu, H.; Wang, Y.; Tang, X.; He, G.; Wang, S.; Ma, Y.; Kong, Y.; Yu, C.; Zhou, G. A Submerged Duckweed Mutant with Abundant Starch Accumulation for Bioethanol Production. *GCB Bioenergy* **2020**, *12*, 1078–1091. [CrossRef]
79. Shi, H.; Duan, M.; Li, C.; Zhang, Q.; Liu, C.; Liang, S.; Guan, Y.; Kang, X.; Zhao, Z.; Xiao, G. The Change of Accumulation of Heavy Metal Drive Interspecific Facilitation under Copper and Cold Stress. *Aquat. Toxicol.* **2020**, *225*, 105550. [CrossRef]
80. Walsh, É.; Kuehnhold, H.; O'Brien, S.; Coughlan, N.E.; Jansen, M.A.K. Light Intensity Alters the Phytoremediation Potential of *Lemna minor*. *Environ. Sci. Pollut. Res.* **2021**, *28*, 16394–16407. [CrossRef]
81. Walsh, É.; Coughlan, N.E.; O'Brien, S.; Jansen, M.A.K.; Kuehnhold, H. Density Dependence Influences the Efficacy of Wastewater Remediation by *Lemna minor*. *Plants* **2021**, *10*, 1366. [CrossRef]

plants

MDPI

Article

A Machine-Learning Method to Assess Growth Patterns in Plants of the Family Lemnaceae

Leone Ermes Romano *, Maurizio Iovane, Luigi Gennaro Izzo and Giovanna Aronne

Department of Agricultural Sciences, University of Naples Federico II, 80055 Portici, Italy; maurizio.iovane@unina.it (M.I.); luigigennaro.izzo@unina.it (L.G.I.); aronne@unina.it (G.A.)
* Correspondence: leoneermes.romano@unina.it

Abstract: Numerous new technologies have been implemented in image analysis methods that help researchers draw scientific conclusions from biological phenomena. Plants of the family Lemnaceae (duckweeds) are the smallest flowering plants in the world, and biometric measurements of single plants and their growth rate are highly challenging. Although the use of software for digital image analysis has changed the way scientists extract phenomenological data (also for studies on duckweeds), the procedure is often not wholly automated and sometimes relies on the intervention of a human operator. Such a constraint can limit the objectivity of the measurements and generally slows down the time required to produce scientific data. Herein lies the need to implement image analysis software with artificial intelligence that can substitute the human operator. In this paper, we present a new method to study the growth rates of the plants of the Lemnaceae family based on the application of machine-learning procedures to digital image analysis. The method is compared to existing analogical and computer-operated procedures. The results showed that our method drastically reduces the time consumption of the human operator while retaining a high correlation in the growth rates measured with other procedures. As expected, machine-learning methods applied to digital image analysis can overcome the constraints of measuring growth rates of very small plants and might help duckweeds gain worldwide attention thanks to their strong nutritional qualities and biological plasticity.

Keywords: duckweed; machine learning; image analysis; machine training; aquatic plants; Lemnaceae; Lemna

check for updates

Citation: Romano, L.E.; Iovane, M.; Izzo, L.G.; Aronne, G. A Machine-Learning Method to Assess Growth Patterns in Plants of the Family Lemnaceae. *Plants* **2022**, *11*, 1910. https://doi.org/10.3390/plants11151910

Academic Editors: Viktor Oláh, Klaus-Jürgen Appenroth and K. Sowjanya Sree

Received: 28 June 2022
Accepted: 20 July 2022
Published: 23 July 2022

1. Introduction

Image analysis has changed the way scientists experiment in numerous fields [1]. The image analysis approach allows scientists to frame time-specific data that can be analysed later. This methodology has been adopted in multiple plant science research fields [2]. Image analysis software is the go-to technology to correctly satisfy the needs of modern research data. In several areas of study (including genetics), the requirement of image analysis software that could quantify tiny differences among plants' phenotypes has been mandatory and has led us to enter the so-called "big data era" in plant science [3,4]. Thanks to the breach made by the genetic field, this software analysis method soon became mandatory in numerous other areas, such as botany, agronomy, and forestry [5–10].

Within the "big data era", scientists are now facing the new challenge of extrapolating scientific-sounding data from the monstrous amount produced by image analysis [11,12]. This represents the birth of the "artificial intelligence era" [13], in which computer intelligence substitutes humans to extrapolate scientific-quality data among large quantities. The advent of artificial intelligence in plant science is already paying off [14]. In numerous fields of plant science, this technology is speeding up the process and excluding numerous errors made by human operators [11,15–18]. Although artificial-intelligence interfaces are still too complicated for most plant biologists, some software leads to the better use of this technology in numerous research fields [14,19]. Among others, ilastik® is a supervised machine-learning software (learning from training data) that brings machine-learning-based image analysis to end-users without extensive computational expertise [20].

Ilastik® provides end-users with a supervised machine-learning experience without requiring extensive training data. This is achieved thanks to the accurate machine-training feature of the software that can fine-tune training via a "paint"-like interface [21]. Ilastik® contains predefined workflows that can be used for image segmentation, object classification, counting, and tracking [20]. Moreover, a specific setup of the machine-training ilastik® process can be reutilised numerous times, and applying a particular feature of the program, "batch analysis", can be performed theoretically with an infinite number of images [21].

This paper proposes the use of ilastik® in a low-cost setup aimed at getting a new standardised method to perform image analysis of the aquatic plant family Lemnaceae. These hydrophytes have been often mentioned referring to their small size and fast growth [22–24]. However, their current appreciation is moving toward these plants' exceptional nutritional qualities [23]. Additionally, Lemnaceae are gaining worldwide attention in numerous other fields, such as phytoremediation, plant science, biomonitoring, and closed bioregenerative systems [25–27]. Due to the simplicity of these biological systems, numerous scientists are evaluating the possibility of using these plants as a model [28]. Research in all these fields is constrained by the extremely small size of individuals that prevents applying the methods that are commonly used for biometric measurements and plant growth rates in all other flowering species. We suggest a new image analysis method via machine learning to boost knowledge and standardise the scientific analysis of the Lemnaceae plant's growth. This approach can increase confidence in experimental results and speed up image analysis techniques by offloading the image analysis process to a machine [20]. Due to scientists' strong interest in this family of plants, we shared the view that it is mandatory to standardise analysis methods [29] and decided to contribute to achieving such a goal.

In particular, we focused on methods able to identify fine changes in growth phenological processes more effectively than those based on the number of fronds used in the past [30]. The new method had to be applied in any growth-related tests, such as bioassay and laboratory tests, for Lemnacae and other floating aquatic plants.

More specifically, our work aimed to validate the utilisation of Ilastik® software in monitoring the growth rate of Lemnaceae. Our approach was to highlight the possible effects of two light treatments on plant growth by applying the previously used analyses and the newly proposed method.

2. Results

To evaluate the use of the machine-learning system, we cultivated Lemna minor (9440) under different light-quality treatments. Moreover, we studied the growth rates with different methods. More specifically, the standard gold method has been defined by the ISO 20079 protocol. This method requires counting the number of fronds over a period of time at constant time intervals. Furthermore, two digital methods were also investigated, one previously described by Haffner et al. [29] and the newly described ilastik® method. The three methods used the Naumann et al. [30] equation to calculate growth rates. The three methods' results were compared to appreciate any existing differences. Additionally, the Ilastik® method's results were compared with those produced in Fiji.

Plants cultivated under the two different combinations of light conditions (white and white + red) grew healthy, with no visual sign of overall differences. Images of plants at the beginning and end of the experiment were used to calculate the relative growth rates (RGR) of Lemna minor by applying the three different methods (the ISO 20079 frond number evaluation, the Fiji image analysis software, and the ilastik® machine-learning method). The ANOVA results showed no significant difference (p = 0.985) in the mean growth rates calculated with the three methods (Figure 1). Therefore, they are equally valid in calculating the growth rate of duckweeds.

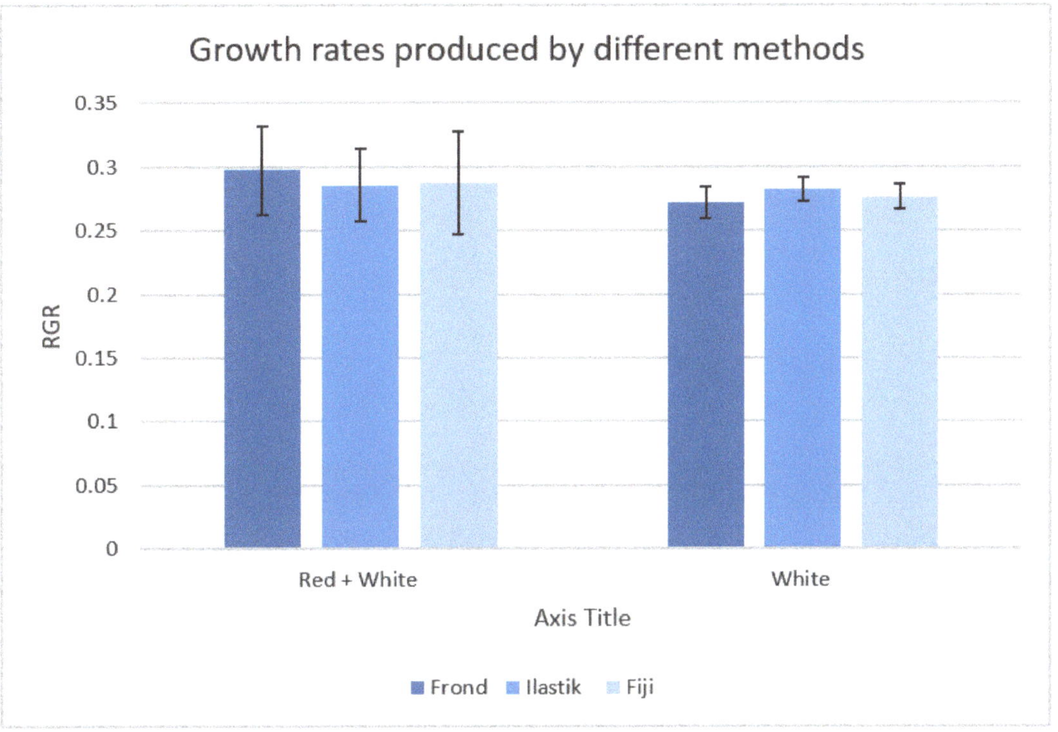

Figure 1. Graph compares RGR calculated with three methods (frond number, ilastik® and Fiji) for the two experiment setups (red-light treatment and control).

Unlike the ISO 20079 method, the other two allowed us to calculate plant growth throughout the experiment by analysing a series of photos taken at regular intervals. We used these data to compare the two computerised methods further. Data showed a high correlation between the measurements by ilastik® and Fiji methods, as represented in the scatter diagram (Figure 2). The strong correlation is supported by calculated coefficients of 0.99 for the Pearson coefficient and 0.99 for the Spearman coefficient. More specifically, the Pearson coefficient showed an almost perfect strength of agreement among data compared, while Spearman's rho coefficient of rank correlation is 0.995. The 95% confidence interval ranges from 0.993 to 0.998. The conclusion is that there is a significant relationship between the two variables.

Due to the presence of outliers to the median line, we compared results by means of difference. The Bland–Altman (B&A) analysis is reported in Figure 3.

The results from the Heteroskedascity test with the White method have a p-value of 0.06; we fail to reject the null hypothesis and conclude that residuals show a homoscedastic distribution.

The three statistical analyses performed show that the two analysis methods can be used interchangeably.

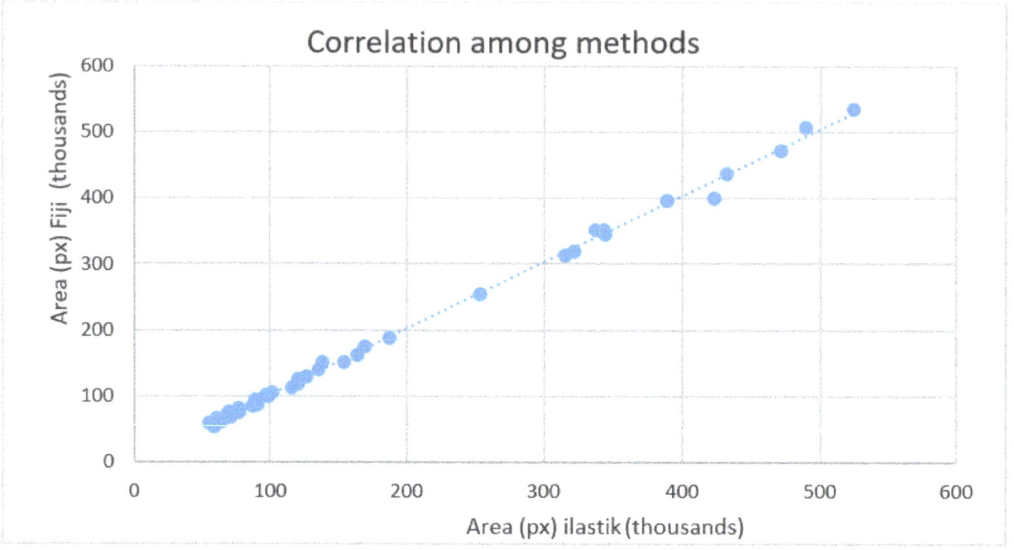

Figure 2. Graph shows plotted output data (pixel) of the same images analysed with the two methods (Fiji and ilastik®). As demonstrated by the data visualisation, a strong correlation between the two methods is present.

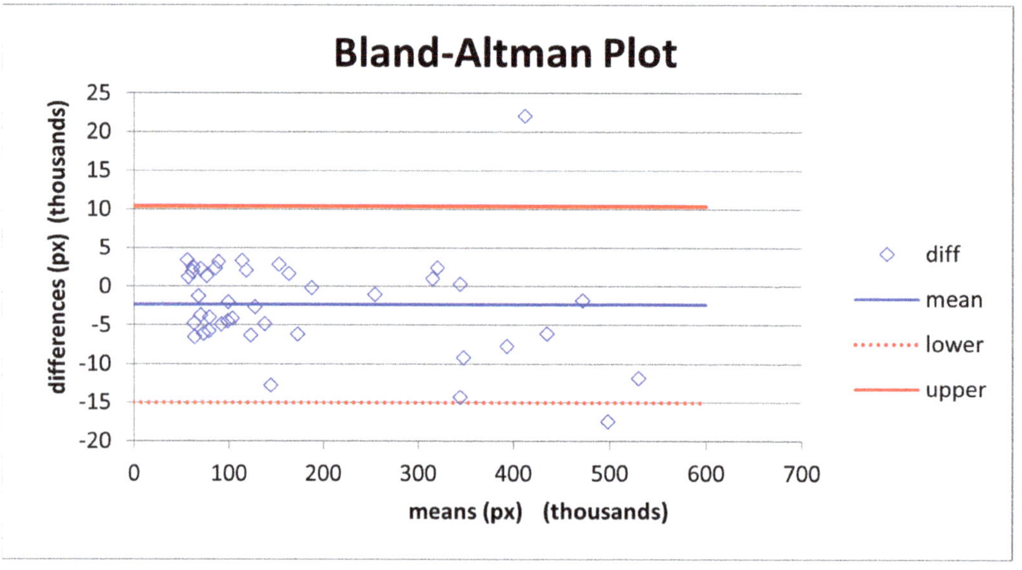

Figure 3. The B&A plot can be evaluated as in good agreement according to the scatter dispersion. The scattering of points is reduced, and points lie relatively close to the line representing mean bias. It is essential to consider the big numerical difference existing among data; this difference in the two outlier cases might (outside the limits of agreement) be due to human error during the Fiji analysis [3].

Time to Data

The previous paragraph shows that the newly described method is perfectly coherent with the results produced with the previously described method (Fiji) regarding data outcomes. We now consider the possible benefits and advantages in terms of time to produce data.

The time required to run the analysis with both software by the same operator was 95.4 s for Fiji software and 300 s for ilastik® per picture. The main difference between the two methods was that the operator who wanted to run the additional analysis with Fiji needed to start over again. This required the same amount of time per analysis (Figure 4). Differently, the operator that trained the machine by using ilastik® to analyse the first image needed a time longer than that for one image with Fiji, but the operator could immediately run any batch analysis with no additional time required because the machine performed the same task for any number of pictures selected (Figure 4).

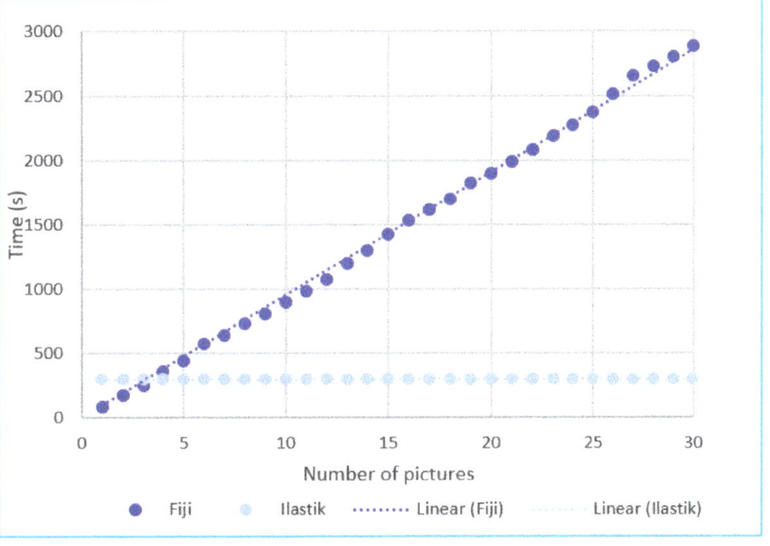

Figure 4. The time required by the operator to analyse the area occupied by fronds of Lemnaceae. The dark blue line represents the time required by the operator by employing the Fiji software; the light blue line represents the time requirements with the ilastik® software. The x-axis represents the number of pictures; the y-axis represents the time requirement per picture.

Furthermore, in the case of ilastik®, it was possible to save the machine-training parameters to be applied to possible future pictures taken under identical conditions (light, distance from the camera, camera setup, etc.).

According to the data recorded during the tests, the predictive analysis of time necessary to measure plant growth as a function of the number of images is in perfect accordance with the time model described by Formulas (1) and (2). Equation (1) describes plotted data from the Fiji software:

$$y_1 = 95.4x \quad (1)$$

where y refers to the time (in seconds) required by the operator to perform the analysis and x is the number of images to be analysed. Equation (2) describes the plotted data from the analysis conducted with the ilastik® software:

$$y_2 = 300 \quad (2)$$

3. Discussion

In this paper, we have demonstrated how a newly described method can be effective in calculating the biological effect of utilising machine learning during image analysis. Plants grown with different light recipes have not shown different growing patterns by means of relative growth rates differently from other studies conducted on crop species [31,32]. Furthermore, the data showed a strong correlation between our newly designed method with the older one (Fiji). Such a strong correlation maximises confidence in the new adoption of the method. Moreover, we have demonstrated how the presence of outliers has been warded off by testing for heteroskedasticity.

It is important to remark on the importance of RGR calculation in the field of study of the Lemancaeae plants as one of the few growth-monitoring tools. Upon having compared the RGR outcomes obtained with the different methods, we can conclude that the three methods are equally valuable for studying the growth rate of duckweed plants.

Nevertheless, the methods relying on image analysis do not require destructive measurements and can facilitate other in vivo analyses such as genetics [2].

The newly proposed approach of using a machine-learning system has numerous advantages and fewer disadvantages than previously proposed methods based on image analysis [29]. In fact, with the initial setup of a photo booth box, researchers can rely on coherent methods that discard human input in the analysis process. However, the newly described method has been applied in laboratory conditions, performing image analysis of plants from one and not multiple species. Further studies could provide helpful insight into its applications in open-field scenarios to evaluate the growth parameters of different species of Lemanaceae in the same photo. In this framework, researchers might utilise technologies that can picture entire ponds (drone photography) and define a single area to be analysed through ilastik® software. This would be helpful in monitoring growth rate over time in raceway ponds for the massive production of Lemnaceae species [33].

It is important to remark that the value of the so-far-used methods remain not lowered; however, thanks to the more substantial presence of open-source software and more available technologies, tweaking these systems to researchers' needs has become more accessible.

Our method is faster and as reliable as the other methods previously used to measure the growth rate of Lemnaceae [34,35]; however, our experiment did not compare results with a fresh or dry weight of plants because this was not the main aim of our work, and we considered reliable the correlation between weight and frond area [35]. The main advantage of applying our method is that it drastically reduces the time required by the operator to analyse the growth rate in Lemnaceae to only the time needed to train the machine. Noticeably, the latter corresponds to the time usually required to analyse a few images with the so-far-used image analysis methods. Moreover, by applying the ilastik® procedure to different experiments designed to use the same photographic conditions and identical clones, the saved machine training can be stored by the researchers and used, theoretically, an infinite number of times. In these cases, our machine-learning approach simplifies the analysis methods to the "click of a button".

Overall, both our result and what has been previously reported in the literature [29] confirmed the urge to use computer image processing to speed up the innovation process in Lemnaceae. As previously mentioned by other authors [29], the usage of this technology with low-cost hardware can define new qualitative standards in determining the growth rate of Lemnaceae.

4. Materials and Methods

Plants of *Lemna minor* were grown under identical environmental conditions through temperature, nutrient media, and background light conditions. The advent of light-emitting-diode (LED) technology allows scientists to provide plants with the exact amount and quality of light needed to maximise growth and efficiency. We experimented with different light recipes to validate the machine-learning method. More specifically, half of the plant samples were treated with a background light (white) and the other half with the integration of red light (white + red). Details of the cultivation and experiment setup, as well as of the data analyses, are reported below.

4.1. Plant Cultivation

Lemna minor (9440) was cultivated for 168h in a controlled-temperature chamber FOC 200IL Velp scientifica® (Monza, at a constant temperature of 25 ± 0.5 °C. Five plants with two or more fronds were cultivated in a 150 mL glass beaker with 100 mL of Murashige and shook growth medium (Sigma-Aldrich—Murashige and Skoog basal medium, St. Louis, MO, USA) (pH adjusted to 5.8). The beakers were covered with a Petri dish to avoid water evaporation. The growth chamber was illuminated with a background white LED light.

Pictures of the growing plants were taken every 24 h with a Sony® alpha 7 II camera equipped with a Sigma® 50 mm Art F1.4, mounted on a fixed stand. Photos were shot under an illuminated photo booth with a white background to guarantee optimal sample illumination and contrast. Furthermore, camera photo parameters were kept constant throughout the experiment.

4.2. Photo Booth Setup

To maintain a constant photo-shooting environment, we set up a photo booth in a dark room of our laboratory. This approach guarantees stable light conditions and centring the samples to the camera frame. We achieved this by buying a photo booth online and a camera tripod. Both components were fixed to a table to keep the camera distance and centring constant throughout the experiment.

4.3. Light Quality and Quantity

We opted for a different light-quality setup to stimulate differences in growth; we decided to use the following light treatment setup. Plants were exposed to the same white background light. The existing difference among samples was due to providing extra monochromatic lighting to the red treatment. More specifically, single 3w red-coloured LEDs (not branded) were installed to achieve light treatment. Light quality and quantity are described in the following table. They are expressed as averages among the three replicas per treatment. Light quality and quantity were measured with a spectroradiometer (SS-110, Apogee Instruments Inc., Logan, UT, USA) to control the emission spectrum of each light treatment Table 1.

Table 1. Total photon flux density (PFD) ($\mu mol{\cdot}s^{-1}$), photosynthetic photon flux (PPF) ($\mu mol{\cdot}m^{-2}{\cdot}s^{-1}$), yield photon flux (YPF) ($\mu mol{\cdot}s^{-1}$), and photosynthetic photon efficacy (PPE) (PPF umol/watts); R/FR is the red (R) light relative to the amount of far-red (FR) light.

	Total PFD	Stdev.	PPF	Stdev.	YPF	Stdev.	PPE	Stdev.	R/FR	Stdev.
Red treatment	128.34	1.1	126.55	1.07	110.72	0.94	0.88	0	10.62	0.03
Control	129.69	0.98	126.53	0.61	107.06	0.66	0.83	0.02	7.08	0.01

4.4. Measuring Systems

In this study, we adopted three different methods to evaluate the relative growth rates of *Lemna minor* during the experiment. As described by the ISO 20079, we used frond number as an evaluation method for growth during the investigation [30]. The other two approaches were achieved via computer software (Fiji and ilastik® (Figure 5)) installed on Lenovo E480, intel CORE i7 8th gen, 16 GB of memory (8 GB minimum required by the software). Both methods produced quantitative information on the area occupied by the plant (in pixel).

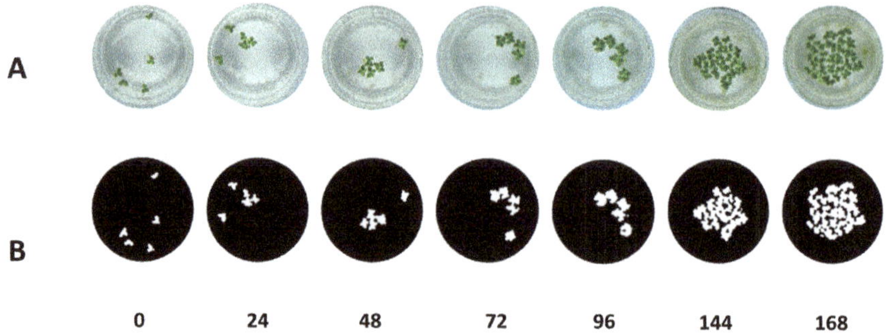

Figure 5. (**A**) pictures of *Lemna minor* at different time intervals (**B**) pictures processed by the ilastik® software (version 1.3.3).

A plant's growth rate can be calculated from the area or number of fronds described by Haffner et al. [29]. We employed Fiji software as defined by Haffner et al. [29]. The ilastik® software was used following the protocol described in Supplementary Materials. When training the machine, we started from the pictures where the frond number was the highest to better train the machine in understanding the picture composition. The output file from the software and the feature selected for the first picture was saved and could be used for other analyses. The three measuring systems were used to calculate the relative growth rate (RGR) described by Naumann et al. [30].

4.5. Statistical Analysis

Statistical analysis was conducted following four-step phases; first, we compared the three methods' relative growth rates with the formula described by Naumann et al. [30]. In this phase, the outcome for the six growth rates compared to performing a one-way ANOVA analysis was performed with the SPSS Statistics 27 software (IBM Inc., Armonk, NY, USA). The ANOVA was fundamental in confirming that the three methods' growth rates were in accordance. More specifically, the two computed methods were in accordance with the gold standard defined by the ISO 20079 protocol. Subsequently, we compared the proposed method (ilastik® software) with the previously described one (Fiji). Agreement among the two computed methods (Fiji and ilastik®) was shown by calculating correlation coefficients with Pearson and Spearman. As described by Mcbride, the correlation can be defined as almost perfect because the value of ρc is 0.999 in the range > 0.99 [36].

Furthermore, we compared utilising differences between the two measurement techniques with the Bland–Altman plot to underline the presence of bias between the two methods. As described by Dogan [29], Bland and Altman's limits of agreement (LoA) have conventionally been used in medical research to evaluate the agreement between two methods of measurement for quantitative variables [37]. Nevertheless, Bland and Altman's LoA method may be misleading in the presence of heteroskedastic distributions [38]. Due to the fact of outliers in the Bland and Altman graphic representation [39], we opted to test heteroskedasticity with the White test because it can better perform in the presence of nonlinear forms of heteroskedasticity (presence of outliers).

4.6. Time Analysis of the Software Utilisation

The following formula was utilised to compare the time usage of the two software:

$$y = a + bx \tag{3}$$

where y is the time required by the operator to perform analysis with the software under evaluation, a is the time to set up the analysis with the given software. The letter b indicates the time for the operator to analyse a single image, and x is the number of images analysed.

Consequently, the equation can be solved, respectively, for Fiji (4) and ilastik® (5).

$$y_1 = bx \tag{4}$$

$$y_2 = a \tag{5}$$

To validate what was modelled by the mathematical equations, we provided quantitative data about the time to analyse the two computed methods.

4.7. Relative Growth Rate

Growth was calculated following the growth rate calculation, Ziegler et al., 2014 [40].

$$RGR = (\ln X_{tn} - \ln X_{t0}) / (t_n - t_0) \tag{6}$$

where X is the pixel in the area as described by Haffner et al. [29] and t_0 and t_n represent the start and the end of the test, respectively.

5. Conclusions

We presented a novel method designed to rapidly and inexpensively quantify the Lemnaceae growth rate by tracking frond surface variance at different time intervals. We showed that machine-learning technology can substitute traditional methods and document biological phenomena if well applied to the observed system. The proposed approach helps the researcher train the machine ad hoc to the specific requirements. This customisable method can be used across different Lemnaceae applications and other surface-floating aquatic plants.

It is important to remark on the importance of RGR calculation in the field of study of the Lemancaeae plants as one of the few monitors for growth. Upon having compared the RGR outcomes obtained with the different methods, we can conclude that the three methods are equally valuable for studying the growth rate of duckweed plants.

Nevertheless, the methods relying on image analysis do not require destructive measurements and can facilitate other in vivo analyses such as genetics [2].

The newly proposed approach of using a machine-learning system has numerous advantages and fewer disadvantages than previously proposed methods based on image analysis [29]. In fact, with the initial setup of a photo booth box, researchers can rely on coherent methods that discard human input in the analysis process. It is important to remark that the value of the old methods remains not lowered; however, thanks to the more substantial presence of open-source software and more available technologies, tweaking these systems to researchers' needs has become more accessible.

Supplementary Materials: The following supporting information can be downloaded at: https://www.mdpi.com/article/10.3390/plants11151910/s1; here, you will find a step-by-step guide to perform the RGR analysis with ilastik software. Futher information can be found at: https://www.ilastik.org/documentation/objects/objects (accessed on 22 July 2022).

Author Contributions: L.E.R. conceived and designed the work and performed the acquisition and analysis; G.A. interpreted the data; L.E.R. drafted the work; G.A. substantively revised it; M.I. and L.G.I. revised and formatted the work. All authors have read and agreed to the published version of the manuscript.

Funding: This research work was funded by the European Space Agency with a grant to the University of Naples Federico II—Department of Agricultural Sciences, ESA Contract No. 4000133778/21/NL/CBi.

Institutional Review Board Statement: Not applicable.

Informed Consent Statement: Not applicable.

Data Availability Statement: Not applicable.

Conflicts of Interest: The authors declare no conflict of interest.

References

1. William, W.; Ware, A.; Basaza-Ejiri, A.H.; Obungoloch, J. A Review of Image Analysis and Machine Learning Techniques for Automated Cervical Cancer Screening from Pap-Smear Images. *Comput. Methods Programs Biomed.* **2018**, *164*, 15–22. [CrossRef] [PubMed]
2. Lobet, G. Image Analysis in Plant Sciences: Publish Then Perish. *Trends Plant Sci.* **2017**, *22*, 559–566. [CrossRef] [PubMed]
3. Popescu, G.v.; Noutsos, C.; Popescu, S.C. Big Data in Plant Science: Resources and Data Mining Tools for Plant Genomics and Proteomics. *Methods Mol. Biol.* **2016**, *1415*, 533–547. [CrossRef] [PubMed]
4. Berman, J.J. *Principles of Big Data*; Elsevier: Amsterdam, The Netherlands, 2013; ISBN 9780124045767.
5. Oliveira, M.C.; Osipitan, O.A.; Begcy, K.; Werle, R. Cover Crops, Hormones and Herbicides: Priming an Integrated Weed Management Strategy. *Plant. Sci.* **2020**, *301*, 110550. [CrossRef]
6. Chlingaryan, A.; Sukkarieh, S.; Whelan, B. Machine Learning Approaches for Crop Yield Prediction and Nitrogen Status Estimation in Precision Agriculture: A Review. *Comput. Electron. Agric.* **2018**, *151*, 61–69. [CrossRef]
7. McCabe, M.F.; Tester, M. Digital Insights: Bridging the Phenotype-to-Genotype Divide. *J. Exp. Bot.* **2021**, *72*, 2807–2810. [CrossRef]
8. Zhang, D.; Tsai, J. *Advances in Machine Learning Applications in Software Engineering*; IDEA Group Publishing: London, UK, 2007. [CrossRef]
9. Cortés, A.J.; Restrepo-Montoya, M.; Bedoya-Canas, L.E. Modern Strategies to Assess and Breed Forest Tree Adaptation to Changing Climate. *Front. Plant. Sci.* **2020**, *11*, 583323. [CrossRef]
10. Ma, C.; Xin, M.; Feldmann, K.A.; Wang, X. Machine Learning–Based Differential Network Analysis: A Study of Stress-Responsive Transcriptomes in Arabidopsis. *Plant Cell* **2014**, *26*, 520–537. [CrossRef]
11. Ma, C.; Zhang, H.H.; Wang, X. Machine Learning for Big Data Analytics in Plants. *Trends Plant Sci.* **2014**, *19*, 798–808. [CrossRef]
12. Marx, V. The Big Challenges of Big Data. *Nature* **2013**, *498*, 255–260. [CrossRef]
13. Schatz, M.C. Computational Thinking in the Era of Big Data Biology. *Genome Biol.* **2012**, *13*, 177. [CrossRef]
14. Ratner, B. *Statistical and Machine-Learning Data Mining*; Routledge, Taylor & Francis Group: Abingdon, UK, 2017; ISBN 9780367573607.
15. Hesami, M.; Jones, A.M.P. Application of Artificial Intelligence Models and Optimization Algorithms in Plant Cell and Tissue Culture. *Appl. Microbiol. Biotechnol.* **2020**, *104*, 9449–9485. [CrossRef] [PubMed]
16. Van Dijk, A.D.J.; Kootstra, G.; Kruijer, W.; de Ridder, D. Machine Learning in Plant Science and Plant Breeding. *iScience* **2021**, *24*, 101890. [CrossRef] [PubMed]
17. Hesami, M.; Alizadeh, M.; Jones, A.M.P.; Torkamaneh, D. Machine Learning: Its Challenges and Opportunities in Plant System Biology. *Appl. Microbiol. Biotechnol.* **2022**, *106*, 3507–3530. [CrossRef] [PubMed]
18. Singh, A.; Ganapathysubramanian, B.; Singh, A.K.; Sarkar, S. Machine Learning for High-Throughput Stress Phenotyping in Plants. *Trends Plant Sci.* **2016**, *21*, 110–124. [CrossRef] [PubMed]
19. Bassel, G.W.; Glaab, E.; Marquez, J.; Holdsworth, M.J.; Bacardit, J. Functional Network Construction in Arabidopsis Using Rule-Based Machine Learning on Large-Scale Data Sets. *Plant Cell* **2011**, *23*, 3101–3116. [CrossRef] [PubMed]
20. Berg, S.; Kutra, D.; Kroeger, T.; Straehle, C.N.; Kausler, B.X.; Haubold, C.; Schiegg, M.; Ales, J.; Beier, T.; Rudy, M.; et al. Ilastik: Interactive Machine Learning for (Bio)Image Analysis. *Nat. Methods* **2019**, *16*, 1226–1232. [CrossRef]
21. Sommer, C.; Straehle, C.; Kothe, U.; Hamprecht, F.A. Ilastik: Interactive Learning and Segmentation Toolkit. In Proceedings of the 2011 IEEE International Symposium on Biomedical Imaging: From Nano to Macro, Chicago, IL, USA, 30 march–2 April 2011; pp. 230–233. [CrossRef]
22. Sree, K.S.; Sudakaran, S.; Appenroth, K.J. How Fast Can Angiosperms Grow? Species and Clonal Diversity of Growth Rates in the Genus Wolffia (Lemnaceae). *Acta Physiol. Plant.* **2015**, *37*, 204. [CrossRef]
23. Appenroth, K.J.; Sree, K.S.; Böhm, V.; Hammann, S.; Vetter, W.; Leiterer, M.; Jahreis, G. Nutritional Value of Duckweeds (Lemnaceae) as Human Food. *Food Chem.* **2017**, *217*, 266–273. [CrossRef]
24. Appenroth, K.-J.; Sree, K.S.; Bog, M.; Ecker, J.; Seeliger, C.; Böhm, V.; Lorkowski, S.; Sommer, K.; Vetter, W.; Tolzin-Banasch, K.; et al. Nutritional Value of the Duckweed Species of the Genus Wolffia (Lemnaceae) as Human Food. *Front. Chem.* **2018**, *6*, 483. [CrossRef] [PubMed]
25. Fujita, M.; Mori, K.; Kodera, T. Nutrient Removal and Starch Production through Cultivation of Wolffia Arrhiza. *J. Biosci. Bioeng.* **1999**, *87*, 194–198. [CrossRef]
26. Gupta, C.; Prakash, D. Duckweed: An Effective Tool for Phyto-Remediation. *Toxicol. Environ. Chem.* **2014**, *95*, 1256–1266. [CrossRef]
27. Romano, L.E.; Aronne, G. The World Smallest Plants (*Wolffia* Sp.) as Potential Species for Bioregenerative Life Support Systems in Space. *Plants* **2021**, *10*, 1896. [CrossRef]
28. Acosta, K.; Appenroth, K.J.; Borisjuk, L.; Edelman, M.; Heinig, U.; Jansen, M.A.K.; Oyama, T.; Pasaribu, B.; Schubert, I.; Sorrels, S.; et al. Return of the Lemnaceae: Duckweed as a Model Plant System in the Genomics and Postgenomics Era. *Plant Cell* **2021**, *33*, 3207–3234. [CrossRef] [PubMed]
29. Haffner, O.; Kučera, E.; Drahoš, P.; Cigánek, J.; Kozáková, A.; Urminská, B. Lemna Minor Bioassay Evaluation Using Computer Image Analysis. *Water* **2020**, *12*, 2207. [CrossRef]

30. Naumann, B.; Eberius, M.; Appenroth, K. Growth Rate Based Dose-Response Relationships and EC-Values of Ten Heavy Metals Using the Duckweed Growth Inhibition Test (ISO 20079) with *Lemna minor* L. Clone St. *J. Plant Physiol.* **2007**, *164*, 1656–1664. [CrossRef]
31. Lin, K.H.; Huang, M.Y.; Huang, W.D.; Hsu, M.H.; Yang, Z.W.; Yang, C.M. The Effects of Red, Blue, and White Light-Emitting Diodes on the Growth, Development, and Edible Quality of Hydroponically Grown Lettuce (*Lactuca sativa* L. Var. Capitata). *Sci. Hortic.* **2013**, *150*, 86–91. [CrossRef]
32. Metallo, R.M.; Kopsell, D.A.; Sams, C.E.; Bumgarner, N.R. Influence of Blue/Red vs. White LED Light Treatments on Biomass, Shoot Morphology, and Quality Parameters of Hydroponically Grown Kale. *Sci. Hortic.* **2018**, *235*, 189–197. [CrossRef]
33. Stejskal, V.; Paolacci, S.; Toner, D.; Jansen, M.A.K. A Novel Multitrophic Concept for the Cultivation of Fish and Duckweed: A Technical Note. *J. Clean. Prod.* **2022**, *366*, 132881. [CrossRef]
34. Hayes, A.F.; Cai, L. Using Heteroskedasticity-Consistent Standard Error Estimators in OLS Regression: An Introduction and Software Implementation. *Behav. Res. Methods* **2007**, *39*, 709–722. [CrossRef]
35. Mazur, R.; Szoszkiewicz, K.; Lewicki, P.; Bedla, D. The Use of Computer Image Analysis in *a Lemna minor* L. Bioassay. *Hydrobiologia* **2018**, *812*, 193–201. [CrossRef]
36. McBride, G.B. A Proposal for Strength-of-Agreement Criteria for Lin's Concordance Correlation Coefficient. 2005. Available online: https://www.medcalc.org/download/pdf/McBride2005.pdf (accessed on 4 July 2022).
37. Doğan, N.Ö. Bland-Altman Analysis: A Paradigm to Understand Correlation and Agreement. *Turk. J. Emerg. Med.* **2018**, *18*, 139. [CrossRef] [PubMed]
38. Sedgwick, P. Limits of Agreement (Bland-Altman Method). *BMJ* **2013**, *346*, f1630. [CrossRef] [PubMed]
39. Ludbrook, J. Confidence in Altman-Bland Plots: A Critical Review of the Method of Differences. *Clin. Exp. Pharm. Physiol.* **2010**, *37*, 143–149. [CrossRef] [PubMed]
40. Ziegler, P.; Adelmann, K.; Zimmer, S.; Schmidt, C.; Appenroth, K. Relative In Vitro Growth Rates of Duckweeds (Lemnaceae)—The Most Rapidly Growing Higher Plants. *Plant Biol.* **2015**, *17* (Suppl. S1), 33–41. [CrossRef]

MDPI
St. Alban-Anlage 66
4052 Basel
Switzerland
www.mdpi.com

Plants Editorial Office
E-mail: plants@mdpi.com
www.mdpi.com/journal/plants

www.ingramcontent.com/pod-product-compliance
Lightning Source LLC
Chambersburg PA
CBHW061451220926
51850CB00004B/62
* 9 7 8 3 0 3 6 5 9 0 6 9 1 *